# Contents

# Preface

The 2023 Rapid Excavation and Tunneling Conference (RETC) will be held in Boston, MA and will include the presentation of 133 papers divided into 20 technical sessions during the 3 day conference that were selected from over 200+ papers that were submitted to the committee. The selected papers are included within this book to provide a comprehensive recap of the conference while also providing a source of reference for contractors and engineers in the future. The selected papers will include contributions from all divisions of the tunneling industry plus risk management and best contract practices.

Boston is known to the tunneling industry for its complex and difficult tunneling history, specifically of recent memory the Central Artery/Tunnel Project also know as the "Big Dig" of the 1980s and early 1990s. The selection of Boston as a host city was fitting from a historical perspective and as avenue for the city to present its upcoming tunnel projects.

RETC began in 1972 with the idea of sharing ideas, design and project construction experience, and new technology, between contractors, designers, owners and suppliers. The ability to provide a single location every other year for all facets of the tunneling industry to gather and discuss the aforementioned was and still is viewed as a way to help the tunnel industry grow. The lessons learned from completed projects, the advancement of technology which could be implemented into design and the presentation of new projects by owners/designers and contractors allows for the limits and complexity of upcoming projects to be expanded beyond our current capabilities.

Our industry has never been as busy as it currently is, and we are seeing more future projects coming out for solicitation across all business sectors. With the good fortune that all of us have been blessed with one of the biggest challenges that we have been faced with is the shortage of future management on all levels for our projects including the labor force. We all want our industry to continue to grow and succeed, our most important assets are the people that we employ. Investing in the education and continued training of our work forces to be up to date with the latest technology and methods is the key to our success. This also includes develop, recruiting and mentoring of young students graduating college to join our industry so that we can all continue to grow.

We hope that all those attending the conference will have the ability to attend as many sessions as possible and to expand their network within our industry. As in past years, scholarship students will be in attendance and we encourage all to offer these students an opportunity to visit jobs, attend learning sessions or future internships to expand their interest in our industry that needs an injection of young engineers to contribute and lead our future Rapid Excavation and Tunneling Conferences.

The program chairs wish to thank the session chairs, co-chairs, authors, and other members of the RETC Executive Committee for their dedication to this conference. Last but certainly not least, we give special thanks to the SME staff for their hard work and enthusiastic support in making this conference possible.

Frank Pepe
Nestor Garavelli

# Executive Committee

**Chair:**

**Shemek Oginski**
Vice President, J.F. Shea Construction
Vice-Chair

**Anthony Del Vescovo**
Vice President of National Tunnel Group, Walsh Group

**Colin A. Lawrence**
Global Practice Leader—Tunnels, Executive Vice President, Mott MacDonald

**Chris D. Hebert**
Division Manager, Traylor Bros Inc

**Scott Hoffman**
Project Manager, Skanska USA Civil Northeast Inc

**Gregg Davidson**
Principal Engineer/Chief Operating Officer, McMillen Jacobs Associates

**Jarrett Carlson**
Mott MacDonald

**Program Co-Chair:**
**Nestor Garavelli**
Project Manager, Frontier Kemper Constructors Inc

**Program Co-Chair:**
**Frank Pepe Jr.**
Senior Vice President, National Director Tunneling and Geotechnical Engineering, STV

**Samer Sadek**
Principal Tunnel Engineer, Jacobs Engineering Group Inc

**Brain Hagan**
Vice President, Jay Dee Contractors Inc

**Christian Heinz**
Project Manager, J.F. Shea Construction

**Matt L. Swinton**
Senior Vice President, District Manager, Kiewit Infrastructure Co.

# Session Chairs

**Ehsan Alavi**
Jay Dee Contractors

**Youyou Cao**
STV Inc

**Tony Cicinelli**
Kiewit

**Dave Dorfman**
Walsh Group

**Dan Dreyfus**
McMillen Jacobs Associates

**Vinay Duddempudi**
Traylor Bros Inc

**Manan Garg**
Austin Transit Partnership

**Da Ha**
STV Inc

**William Hodder**
North Tunnel Constructors ULC

**Steve Kramer**
Cowi North America

**David Lacher**
Traylor Bros Inc

**Micheal Lang**
Frontier-Kemper Constructors Inc

**Glenn Larose**
Jacobs

**Chris Lynagh**
McNally

**Mo Magheri**
Kiewit

**Julio Martinez**
Skanska

**Leah McGovern**
STV Inc

**Dan McMahon**
Traylor Bros Inc

**Ben McQueen**
Frontier-Kemper Constructors Inc

**Veronica Monaco**
Jacobs

**Sergio Moya**
Frontier-Kemper Constructors Inc

**Alston Noronha**
Black and Veatch

**Luis Piek**
Arup

**Seth Pollak**
Arup

**Steve Price**
Walsh Group

**Rebecca Reeve**
Traylor Brothers Inc

**Joe Rigney**
Parsons

**Jean-Luc des Rivieres**
JF Shea

**Jay Sankar**
Amtrak

**Colin Sessions**
Jacobs

**Lisa Smiley**
Jay Dee Obayashi JV

**David Smith**
WSP

**Kevin Smyth**
Frontier-Kemper Constructors Inc

**Bade Sozer**
McMillen Jacobs Associates

**Kaveh Talebi**
Jay Dee Contractors Inc

**Zeph Varley**
WSP USA

**Boris Veleusic**
Michels

**John Yen**
Skanska

# International Committee

| | |
|---|---|
| Argentina: | Nestor Garavelli |
| Australia: | Harry Asche |
| Austria: | Norbert Fuegenschuh |
| Brazil: | Tarcisio Barreto Celestino |
| Canada: | Tim Cleary |
| Chile: | Alexandre Gomes |
| Czech Republic: | Karel Rossler |
| France: | Francois Renault |
| Italy: | Remo Grandori |
| Germany: | Klaus Rieker |
| India: | R. Anbalagan |
| Japan: | Hirokazu Onozaki |
| Mexico: | Roberto Gonzalez Izquierdo |
| New Zealand: | Bill Newns |
| Norway: | Tone Nakstad |
| Singapore: | Leslie Pakianathan |
| Spain: | Juan Luis Magro |
| Sweden: | Per Vedin |
| Switzerland: | Frederic Chavan |
| United Kingdom | Ross Dimmock |

PART 

# Contract Practices

*Chairs*

**Seth Pollak**
Arup

**David Smith**
WSP

# Atlanta—Plain Train Tunnel West Extension Project—Progressive Design-Build Risk Management Approach

**Walter Klary** ▪ Gall Zeidler Consultants
**Dominic Reda** ▪ Gall Zeidler Consultants
**Vojtech Gall** ▪ Gall Zeidler Consultants
**Elmira Riahi** ▪ Gall Zeidler Consultants
**Jake Coibion** ▪ Guy F. Atkinson Construction

## ABSTRACT

Progressive Design-Build is becoming a more common delivery method in the tunneling industry. This approach allows the Owner to select a design-builder primarily based on qualification instead of lowest price. The delivery method promotes flexibility and collaboration at all levels from the initial design stage through construction. Using the example of the Atlanta Airport Plane Train Tunnel West Extension, value-added risk management was added during the initial design phase with an independent reviewer. This paper presents the independent reviewer transitioning from constructability review in the very early stages of design to independent design verification to on-site supervision during construction.

## INTRODUCTION

The Plane Train Tunnel West Extension is located within the Hartfield-Jackson Atlanta International Airport in Georgia, USA. The purpose of the extension is to increase the capacity of the Automated People Mover (APM) system which is connecting domestic and international terminals to significantly decrease the turnaround time of the trains. The project consists of a approx. 900-ft. long, 22-ft. high and 20-ft. wide tunnel, providing the space to add a turnback switch to the APM system. As shown in Figure 1, the tunnel starts as a single tube tunnel from a secant piled wall shaft with a diameter of 36-ft and a depth of approximately 55-ft.. After approximately 400-ft, the tunnel transitions through a widened area (bifurcation) into two single tube tunnels up to the connection with the existing terminal station.

## GROUND CONDITIONS

### Fill

Encountered in the majority of project borings, Fill consists of very loose to dense silty sands (SM) with variable rock fragments, clayey sands (SC and SC-SM), and some sandy silt (ML). The Fill typically underlies asphalt or concrete at the ground surface. Fill along the single tube tunnel section consists primarily of silty sands (~60% SM), with remaining material consisting of clayey sands (SC), silts (ML) and clays (CL). The N-Values along the sections mentioned above, average 28 bpf (normalize for a penetration depth of 12-in) indicating medium dense material.

### Residuum/Saprolite

Residuum/Saprolite consists predominantly of coarse-grained soils of silty sand (SM) with gravel and some fine-grained soils of sandy silt (ML). Layers and lenses of rock and Partially Weathered Rock (PWR) can occur locally within the Residuum/Saprolite

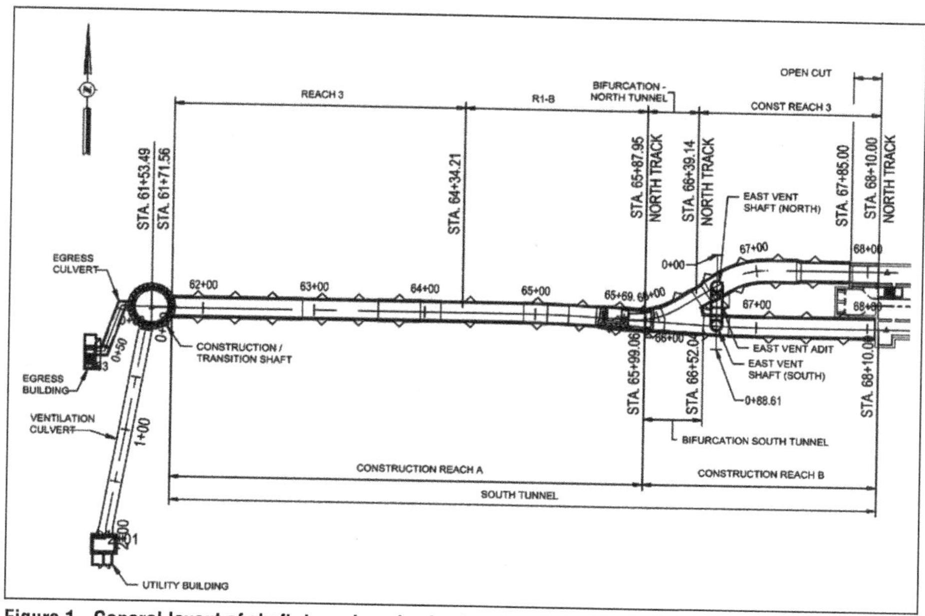

Figure 1. General layout of shaft, tunnel, and culverts

and have STP N-values of >100 bpf. The Residuum/Saprolite may retain the geologic structure of the underlying parent rock including foliation and jointing, acting as pathways for groundwater and along which failure may occur. Average SPT N-Value for Residuum/Saprolite in the single tube section is 32 bpf (medium density).

## Partially Weathered Rock (PWR)

The PWR consists of rock-like remnants of decomposed gneiss surrounded and separated by Residuum/Saprolite. The PWR consists primarily of coarse-grained soils of silty sand (~87–100% SM) with some sandy silts (ML) and occasional gravel (GP). PWR thickness is variable along the tunnel and lenses of PWR were encountered within the Residuum/Saprolite. SPT N-values are consistently between 100 and 200 bpf; the upper values likely attributable to the presence of gravel to boulder-size rock fragments within a soil matrix.

## Bedrock

Local bedrock consists of the Stonewall Gneiss, which is a medium- to high-grade metamorphic crystalline rock with well-developed foliation. Secondary quartz veins up to 3-ft in thickness are common through the rock. A weathered bedrock zone ranging in thickness from ~1.5-ft to 10-ft below the top of bedrock is indicated by borings. Depth to bedrock along the alignment varies as shown in Figure 2.

## Groundwater and Hydrology

According to the GBR-C [1], the baseline groundwater table is at El. 1,001 ft with typical variation in the groundwater table depth ranging from 21.1 ft below ground surface (bgs) (El. 989.7 ft) and 29.0 ft bgs (El. 998.9 ft). Dewatering from ground surface will not be performed to lower the groundwater table during the construction of the

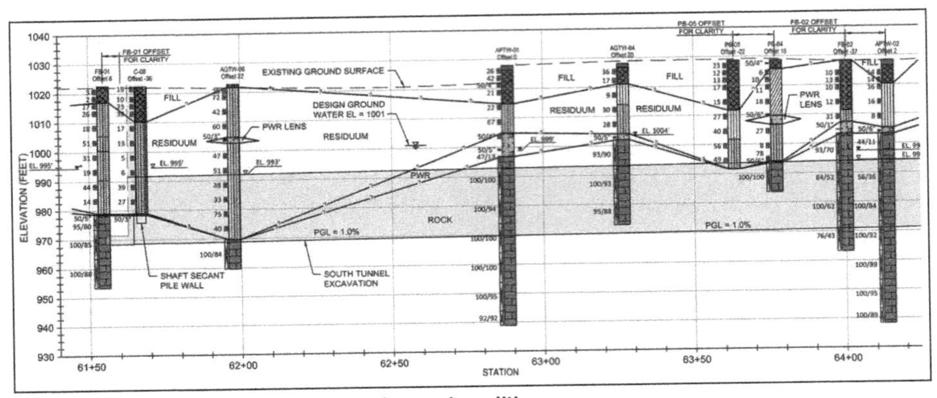


**Figure 2. Geological long section—mixed ground condition**

bifurcation portion but are required for single tube tunnel section. In-tunnel gravity or vacuum dewatering will be performed, if required, to minimize seepage forces in the around tunnel openings prior to installation of the initial shotcrete lining. In addition, weep holes in the initial shotcrete lining will be required to prevent buildup of groundwater pressure behind the lining.

## Mixed Ground Conditions

In the first stretch of the tunnel starting from the shaft mixed ground conditions were present. The upper part of the tunnel was in PWR (Partially weathered rock) and was excavated utilizing an excavator with a hammer attachment. The lower part comprised of bedrock (Gneiss) and was excavated with blasting. In the area with soft ground pocket excavation was required and fiberglass face bolts were installed. Also, in the last stretch of the mixed ground conditions were encountered. The soft ground close to the end wall of the tunnel included backfill material which consisted of trash and steel from the building of the terminal.

# TUNNELING CHALLENGES

## Existing Structures

The tunnel was excavated in proximity of existing structures servicing the airport. The MARTA Station and the Skytrain foundations were selected for further evaluation of the impact of tunneling on existing structures. MARTA station is composed of a heavy rail system that connects the airport to the center of Atlanta. The Skytrain is connecting the terminal of the airport to the rental car center. Both have a station right outside of the domestic terminal (see Figure 3).

## Blasting and Vibration Control

The owner restricted blasting to windows from midnight to 0400 where Skytrain and MARTA were in reduced operation and the activities in the terminal were at a minimum. Seismographs were installed on the surface and on the buildings to monitor the impact of blasting. To minimize impacts of excavation adjacent to existing structures, blasting was prohibited to protect foundations of buildings, the elevator shaft and the micropiles beneath the terminal building.

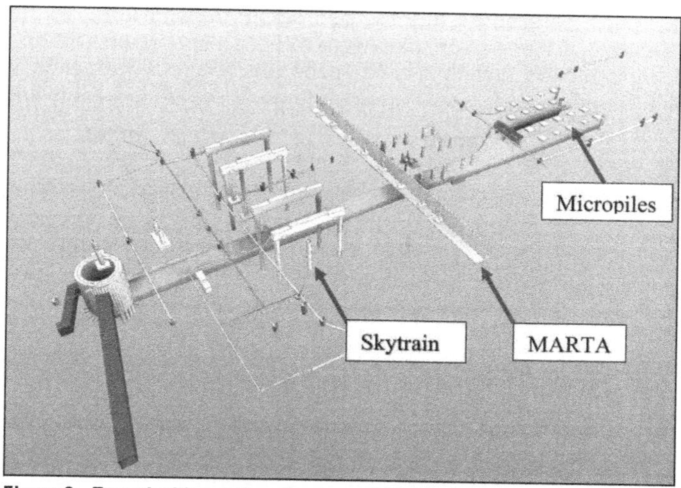


Figure 3. Tunnel with existing structures (MARTA, Skytrain) provided by McMillen Jacobs

## Bifurcation

Due to low rock cover and proximity to the surface, the bifurcation was the most challenging area of the tunnel. During the design phase the geometry was adjusted, and a flat roof implemented to maximize the rock cover above the bifurcation. For this area Gall Zeidler Consultants (GZ) performed an independent design review (IDR) of this change by creating a fully independent three-dimensional (3D) model based on the available geotechnical data for the project. The IDR team used a different software package (Midas GTS NX) from the one that EOR has used to confirm the adequacy of the geometry and support. The results of the 3D analyses confirmed that the change in the design was beneficial. Figure 4 shows the 3D model set up in Midas for the Bifurcation; Figure 5 shows bifurcation during construction.

## Excavation Adjacent to Micropiles

The last approximately 120 ft. of the tunnel was excavated beneath the domestic airport. Due to structural concerns and the nature of the existing foundation of the

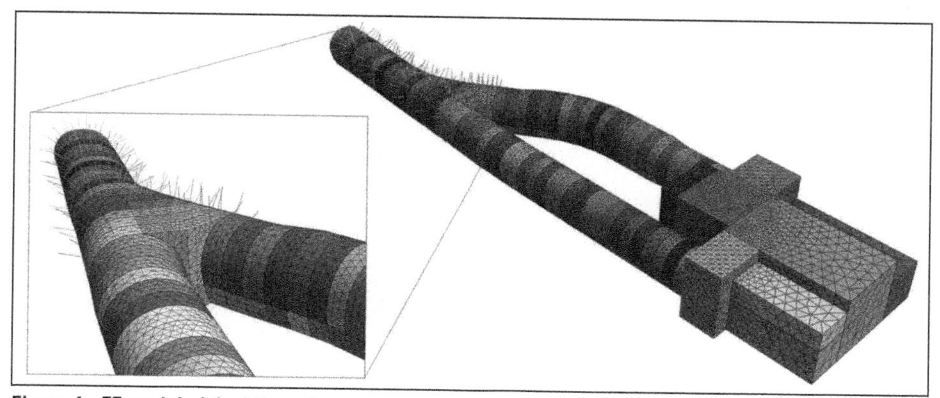

Figure 4. FE model of the bifurcation

building, micropiles were drilled into the bedrock to transfer the load of the building to bedrock. Several of these micropiles were exposed during excavation. No blasting zones were implemented adjacent to the piles and the rock was excavated with an excavator with hammer attachment; exposed micropiles during excavation is shown in Figure 6.

**Figure 5. Excavation of the bifurcation**

**Figure 6. Exposed micropiles**

## PROGRESSIVE DESIGN-BUILD

Progressive Design-Build allows the Owner to select a design-build team during the procurement process based on qualification rather than costs. Typically, a project is divided into two phases; preliminary design (Phase 1) and detailed design, construction and commissioning (Phase 2). In Phase 1, the bridging documents, Geotechnical Baseline Report (GBR) and Geotechnical Data Report (GDR) are reviewed and subsequently, a lump sum for the project is negotiated. Phase 2 then proceeds with the final design, construction and commissioning, with milestones to review construction costs and schedule as the design is further developed. These Phases can be broken down into more granular components depending on how complicated the Project may be.

The City of Atlanta and the department of Aviation as the project Owner selected the Clark/Atkinson/Technique Joint Venture (JV) as the design-builder with McMillen Jacobs Associates (MJA) as the lead designer. The JV assigned Gall Zeidler Consultants (GZ) for peer review services, independent design reviews and on-site support services. In Progressive Design-Build, the design/builder is not obliged to hire an independent reviewer as part of the design process. During the negotiations between the Owner and the JV, it was agreed that due to the complexity and risk of the project an independent reviewer would help to mitigate the risk. This allowed GZ to be involved in the project starting from early in the design until the completion of the tunnel works.

Close collaboration with the JV and the lead designer MJA in the early phase allowed for refinements, and in some instances, optimization of the design. During construction, GZ provided on-site support which included SEM Superintendents for each shift and the SEM Engineer.

The constructability review included a review of the initial lining, the final CIP lining and waterproofing systems, recommendations for the distribution of support types along the tunnel alignment, the excavation and support sequence, groundwater control, pre-support measures, toolbox measures and specific support elements such as Lattice Girders, Roof Ribs and bolts.

As mentioned previously, GZ was chosen as independent reviewer in the early phases of the design. The main tasks performed by GZ was to review the design of the shaft at the breakout for the tunnel, the design for the tunnel excavation with temporary support, final lining and waterproofing. In addition, GZ performed independent design analyses/modeling for the tunnel including a 3D model for the bifurcation.

The independent design analyses/modeling included the verification of the adequacy of the structural stability of the excavation and support and the final lining. A three-dimensional finite element (FE) analysis has been performed, utilizing Midas GTS NX, to check temporary and permanent support. The bifurcation structure, the North and South Tunnels were constructed following the principles of the Sequential Excavation Method (SEM). The objectives for the FE analyses were to validate the adequacy of the SEM initial lining thicknesses for all support types for the portion of the APM Tunnel—Single tube tunnel and bifurcation—and to investigate the structural capacity of the initial lining and the final CIP lining.

The main changes during the design phase were condensing the support types into fewer support types, optimization of the support, change in rock bolt pattern and rock bolt types, eliminating roof ribs, changes in pre-support measures and optimization of the waterproofing system. The efficiency of the changes were verified in the field by the on-site SEM team during construction.

## EXPERIENCE DURING CONSTRUCTION

### Excavation Overview

The tunnel was excavated with a Top Heading and Bench excavation sequence. The full length of the Top Heading for both tunnels were excavated up to the terminal station prior to bench excavation. A combination of Drill and Blast and soft ground excavation techniques were utilized based on the ground conditions encountered. Four ground classes were developed during the design phase for the tunnel and one ground class for the bifurcation. The main challenges during excavation were mixed ground conditions, vibration control, shallow cover, and existing structures. GZ provided an on-site team during excavation, waterproofing and final lining. This included a SEM Engineer and SEM Superintendents for each shift. In the daily Required Excavation Support (RES) meetings the ground conditions encountered, probe drilling, instrumentation and monitoring, blast vibration monitoring, shotcrete test results were discussed by the SEM Engineer/Superintendent, the JV, the Engineer of Record's (EOR) onsite representative and the Owner's Engineer and the ground support type for the next excavation round was determined. This highly experienced onsite team was able to adjust support type etc. if required and verified that the design and adjustments/optimizations made during the design were adequate. This meeting also allowed for the review of quality items such as shotcrete performance and blast results from the previous shift. The main benefit of the Progressive Design-Build is that the design/builder in conjunction with the independent designer work together in the design phase as well in the construction phase.

### Shotcrete

Fiber reinforced shotcrete was used for the initial support of the excavation. Extensive preconstruction testing was performed by the JV to validate and prepare the mix, the nozzlemen and the equipment for the demands of the Project. A batching plant was installed on the construction site, this guaranteed the permanent availability of shotcrete. During construction an extensive testing program was executed, this included early strength testing, compressive strength testing and round panel tests. After spraying shotcrete an exclusion zone was introduced where personnel were not allowed to enter. This was managed by the SEM Superintendents and eliminated the risk of personnel being exposed to fresh shotcrete. As in most projects the round panel testing was challenging, and special care had to be taken in handling the panels. The shotcrete was applied utilizing a Normet Minimec concrete spraying manipulator.

### Instrumentation and Monitoring

A thorough monitoring program was implemented on the surface and in the tunnel. Surface monitoring included ground monitoring points, structural monitoring points on buildings, tiltmeters on building columns, inclinometers, extensometers, strain gauges on the foundation underpinning, hydrostatic level cells, and piezometers. In-tunnel monitoring utilized convergence monitoring arrays on 50-ft intervals. All data were compiled in real time on a web portal and were available online. Due to ongoing construction on the surface as well as in the terminal building, special care had to be taken in protecting the instruments and keep line of sight between targets and total station. All available date were reviewed in the daily RES meeting and potential issues discussed in a weekly meeting. The in-tunnel monitoring did not show any significant movement of the tunnel lining, therefore negligible settlements on the surface were observed. An important part of the monitoring included vibration monitoring during blasting. Based on the seismograph recorded data, the drill and blast design were adjusted and optimized in order to stay within the allowable limits.

## Mixed Ground Conditions

In the first stretch of the tunnel adjacent from the shaft and along the last stretch beneath the Terminal building, mixed ground conditions were encountered. In particular, beneath the Terminal building soft ground was encountered in the upper half of the Top heading excavation (Figure 7). The lower part consisted of fresh, hard Gneiss. In the soft areas, pocket excavation was utilized and supported with shotcrete, the lower part was excavated by drill and blast. Figure 8 shows the highly weathered soft ground where pocket excavation was utilized. The number and sizes of the pockets was driven by the stability of the soil during mechanical excavation and was determined by the GZ Superintendent and SEM Engineer. Due to the highly experienced team, geological overbreak or instabilities of the face were not observed, geological overbreak was also prevented by the installation of grouted tube spiles. As shown in Figure 7 the lower part of the face comprised of fresh, hard Gneiss with only few discontinuities. In areas where blasting was restricted, a trial was performed to utilize rock splitters. It turned out that due to few discontinuities, this method was less effective than using a excavator with a chipping hammer.

## Toolbox Items (Spiles, Pocket Excavation)

Toolbox items in the design included gravity dewatering, vacuum dewatering, pocket excavation, face wedge, spiling, grouting and face bolting. Due to favorable ground conditions only gravity dewatering, and pocket excavation were utilized. By design, the heavier support classes (3-A and 3-B) included 21 grouted pipe spiles (L = 12 ft.) for pre-support. However, in daily RES meetings, the ground conditions and required pre-support was discussed and due to favorable ground conditions in certain locations, the grouted pipe spiles were determined to be unnecessary as part of the excavation cycle. When unfavorable ground conditions were encountered, the RES meeting determined grouted pipe spiles in conjunction with pocket excavation was required. In total, a significant number (692) of spiles were eliminated. This led to cost savings and schedule benefits. Spiling impacted the cycle time significantly and to meet the narrow

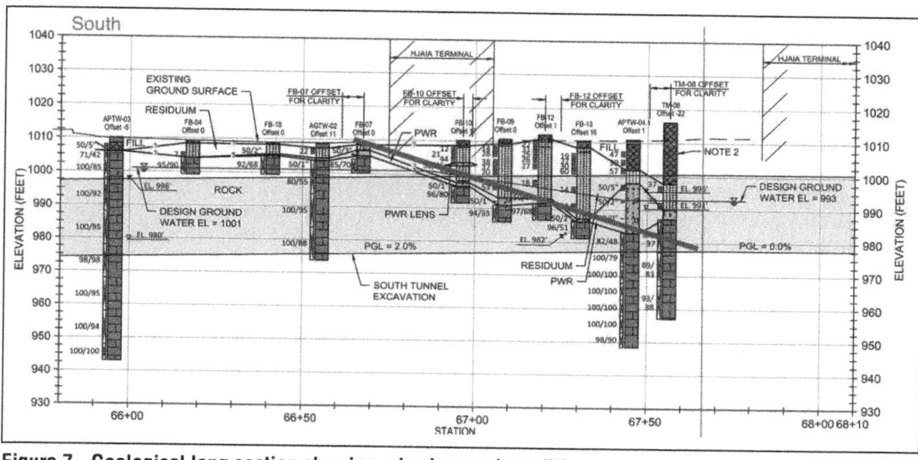


Figure 7. Geological long section showing mixed ground conditions

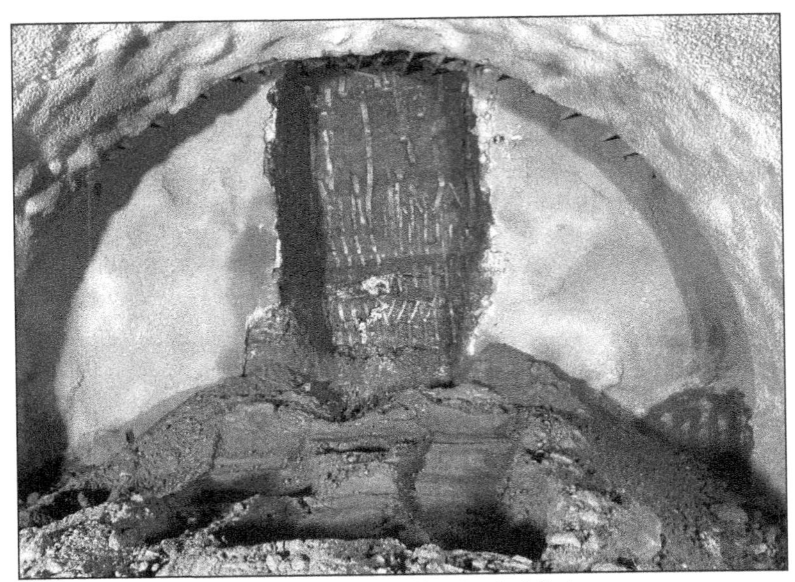

Figure 8. Pocket excavation in soft ground with spiles installed

blasting window each night adjusting the number or eliminating the spiles was one key for the success of the project. Spiling as toolbox items in ground support classes without spiles were not utilized. Toolbox measures included also pocket excavation. This was utilized in soft ground conditions and mixed face conditions. The number and sizes of pockets driven by the stability of the ground during excavation. Other toolbox items shown in the design drawings such as vacuum dewatering, face wedge, grouting and face bolting were not utilized due to the stable ground conditions.

### Unexpected Utilities

During the excavation of the last 15 feet of the south tunnel backfill material from the original terminal foundation was encountered during tunnel excavation. During bench excavation in this location, an undocumented 4-inch steel pipe filled with water was encountered. The initial water inflow was measured at approx. 40 gpm. The Designer and Owner were immediately informed, and an investigation started. This unknown utility was not documented in any existing drawings, and it was decided to let the water drain and observe the change in quantity. Pouring the protection slab stopped the water inflow only temporarily and a grouting program was implemented. The grouting included chemical grouting of the invert slab and lower bench walls. Since the water pressure was building up water inflow returned, and it was decided to install a permanent drainpipe at one side of the tunnel and divert the water to a sump in the construction shaft.

## CONCLUSION

Progressive Design-Build allows the independent reviewer of the contractor being involved in the very early stage of the project. The constructability review allowed to make impactful changes at very early stages of the project. In addition, the collaboration with the design team and the contractor established a relationship between the different parties which was beneficial for the project. The independent design review, especially of the bifurcation, built confidence that the changes in geometry and support was adequate and beneficial for the project. The on-site team with SEM Engineer and SEM Superintendent collaborated closely with the Designer Representative on site and the JV. This trust-based environment developed over years of collaboration led to open discussions when challenges emerged, or changes were required. This collaborative environment between the Owner, the JV, the EOR, and the IDR resulted zero major issues during construction for the Plane Train Tunnel West Extension Project.

## REFERENCES

Murray, J. 2021. Concept for Contingency Measures of Bifurcation Excavation. Memorandum from J. Murray to R. Gould and L. Gilbert, CAT JV. April 9, 2021.

Sun Y, Murray J, Elbin D, Gould R, Gilbert L, 2022. Design and Construction of a Shallow Twin Tube Bifurcation—Atlanta Plane Train West Extension, NAT 2022.

Gould R, Murray J, Elbin D. 2022. Benefits and Challenges of Progressive Design-Build Procurement-Atlanta Plane Train Project, NAT 2022

Buckley J, Cidlik R, Sibley R, Pallua T, Schrank J, Giltner S, 2022. Analyses of Near Field Vibration Monitoring in Construction Blasting, NAT 2022

Gall Zeidler Consultants. 2020 Summary Memorandum 1711 MEM005 R00, Independent 3D Finite Element Analysis of APM Tunnel (Design Package 04 Bifurcation to the East End and Design Package 06 Final Lining), November 17, 2020. Prepared for: CAT JV.

Loulakis, M.C., Kinsley, P., Neumayr, G.W., Unger, C.H., and The Progressive Design-Build Task Force. 2017. Progressive Design-Build, Design- Build Procured with a Progressive Design & Price, A Design-Build Done RightTM Primer. A Design-Build Institute of America Publication. https://dbia.org/wp-content/uploads/2018/05/Primer-Progressive-Design-Build.pdf.

# CMGC Delivery of the I35W SSF Project—Fostering Collaboration to Meet Stormwater System Resiliency Challenges

**M.B. Haggerty** ▪ Barr Engineering Co.
**Joe Welna** ▪ Barr Engineering Co.

## ABSTRACT

The Minnesota Department of Transportation is constructing a stormwater storage facility (SSF) along Interstate 35W (I35W) south of downtown Minneapolis, MN. The project will provide storm surge capacity reducing the frequency with which major precipitation events result in flooding of I35W. The design consists of six, 42-foot diameter, 90-foot deep, adjacent underground shafts providing 14.7-acre feet of storage. Construction challenges include high groundwater with variable glacial outwash deposits. Constructability and ground variability were managed using construction manager general contractor (CMGC) project delivery. CMGC provides an environment for collaboration and innovation starting in the design phase and carrying through construction. This paper provides an overview of key efficiencies and collaboration examples implemented on the I35W SSF project which were aided by CMGC delivery.

## BACKGROUND

### Description of the Project

The Minnesota Department of Transportation (MnDOT) is currently constructing a stormwater storage facility (SSF) along Trunk Highway Interstate 35W (I35W) south of downtown Minneapolis. The purpose of the SSF is to provide storage volume for stormwater surge capacity thereby reducing the flooding frequency of the traffic lanes during significant precipitation events. During the conceptual design phase, MnDOT reviewed project delivery method options for the project. Both design-build (DB) and construction manager-general contractor (CMGC) were evaluated and ultimately CMGC was chosen as the delivery method of the project. The Federal Highway Administration (FHWA) describes CMGC as a project delivery method which allows an owner to engage a construction manager (CM) during the design process to provide constructability input. The CM input can inform areas such as scheduling, pricing, and constructability. A CMGC delivery method can be advantageous for complex projects where the owner desires to maintain control during the design phase of work. The I35W SSF conceptual design resulted in a variety of options to further evaluate in the final design. Additionally, the site poses geotechnical risks due to the ground conditions and limited work area. These collective reasons were the basis for which CMGC was selected to deliver this project.

The concept selected for final design and eventual construction has a volume of 17,269 cubic meters (14 acre-foot) and consists of a subsurface storage structure located in the east shoulder of I35W in an area roughly bordered by northbound I35W, 42nd Street, 39th Street, and 2nd Avenue. The SSF consists of six, 12.8-meter (42-foot) diameter cells constructed using a diaphragm wall (D-wall) construction technique (Figure 1). The D-wall was constructed from a surface grade of approximate elevation

**Figure 1. Design layout of SSF cells and connection to mainline I35W stormwater infrastructure**

**Figure 2. Two-sided weir located in cell 1 of the SSF and engaged during large precipitation events**

251 meters (824 feet) above mean sea level (AMSL) and extends at depth to a D-wall tip elevation of 217 meters (711 feet) AMSL. The bottom of the cell excavation will be approximately elevation 223 meters (731 feet).

The SSF operates by diverting water from the existing 2.0-meter (78inch) diameter stormwater pipe located in the median of I35W. The diverted water flows through a two-sided weir structure (Figure 2) located within the southernmost cell of the SSF (cell 1). Base stormwater flows do not activate overtopping of the two-sided weir and are routed back to the existing 2.0-meter (78inch) diameter stormwater drainage system which drains to the north and discharges to the Mississippi River. However, during larger precipitation events, the stormwater overtops the weir and is captured within the

Figure 3. Diaphragm wall panel construction in 2020

SSF. A lift station located in the northernmost cell (cell 6) of the SSF pumps the water back into the existing stormwater drainage system after the storm has passed and the mainline system can handle the additional water.

In general, underground construction contains a level of variability due to ground conditions that can be challenging to predict even with an extensive geotechnical investigation. A summary of the geotechnical data for this project was presented in a geotechnical data report (GDR). In addition to the GDR, a geotechnical interpretive report (GIR) was developed for the project which was referenced by the contractor during bidding. The GIR serves as a baseline of ground conditions to be anticipated during construction. Geotechnical conditions at the site heavily influenced the consideration and eventual selection of construction techniques to achieve the design goal and provide risk mitigation given the ground conditions.

### Current Construction Progress

Construction of the SSF began in Fall 2019. D-wall for all six cells were installed in 2020 utilizing a combination of clamshell and hydro-mill excavation methods (Figure 3). Mass excavation of the cell interiors began in 2021 and was completed in 2022. Construction of the project is anticipated to be completed in 2023. Use of the GDR and GIR have aided owner, contractor, and engineer in resolution of unforeseen conditions due to variable ground during construction.

## GEOLOGIC SETTING

### Stratigraphy

The immediate vicinity of the project site consists of Pleistocene glacial outwash over glacial till and earlier outwash deposits (Meyer 1989; Mossler 1979). The outwash and till deposits present in the area are primarily sourced from two main glacial advances and retreats, first the Superior Lobe, followed by the Des Moines Lobe (Meyer 1989). Although the deposits of both lobes have distinguishing features, their engineering properties are similar (Mossler 1979). Glacial outwash deposits in the area are typically sand, sometimes with significant zones or gravel or fines. The glacial till tends to be sandy clay, silty sand, or clayey sand. (Mossler 1979) indicate that undifferentiated glacial till, outwash, and alluvium typically found in buried bedrock valleys may underlie the outwash. A sequence of till and outwash deposits may be present in these bedrock valleys, where alluvium and colluvium initially present in the valley is buried by till deposits of the Superior lobe. Later the retreat of the Superior lobe deposited outwash before the advance of the Des Moines lobe would have deposited more till. Together, the outwash and till are expected to be 15 to 55 meters (50 to 180 feet) thick within the vicinity of the project area (Bloomgren 1989; Bloomgren 1979).

Sedimentary rock of Ordovician age underlies the site, consisting of Platteville Limestone, Glenwood Shale, and St. Peter Sandstone Formations (Olsen 1979). The St. Peter Formation consists of sand that is nearly pure quartz with fine- to medium-sized rounded grains and its thickness varies due to the erosional nature of its lower interface but is typically about 46 meters (150 feet) feet in the Twin Cities (Dittes 2015). The uppermost region of the St. Peter Sandstone grades to conform to the Glenwood Formation, a thin shale layer (ranging from less than 0.1 to 2 meters (0.5 feet to 6 feet) thick) separating the St. Peter and overlying Platteville Formations. The Platteville Formation is partially dolomitized limestone.

After the deposition and lithification of the sedimentary strata and before the advancement of glaciers across the project site, stream erosion cut deep valleys in the site's sedimentary rock. These valleys are partially the result of the relative softness of the St. Peter Formation compared to the overlying Platteville Limestone. The contrasting erodibility of these rock formations makes the Platteville Limestone a "cliff-forming" unit. Due to the steep preexisting valleys cut through the St. Peter Formation, the bedrock topography in the vicinity of the project site can be quite steep (Bloomgren 1989). Through successive glacial advances and retreats, enough sediment was deposited as till or outwash to fill most topographic lows that existed before the arrival of glaciers. As a result, the surficial expression of the bedrock topography is almost nonexistent near the project site. Review of existing data provided contradicting bedrock maps at the initial time of the investigation. Figure 4 presents the mapped extent of the top of Platteville formation from two separate Minnesota Geological Survey (MGS) maps. The 1989 extents indicate the entire project area to be underlain by the Platteville

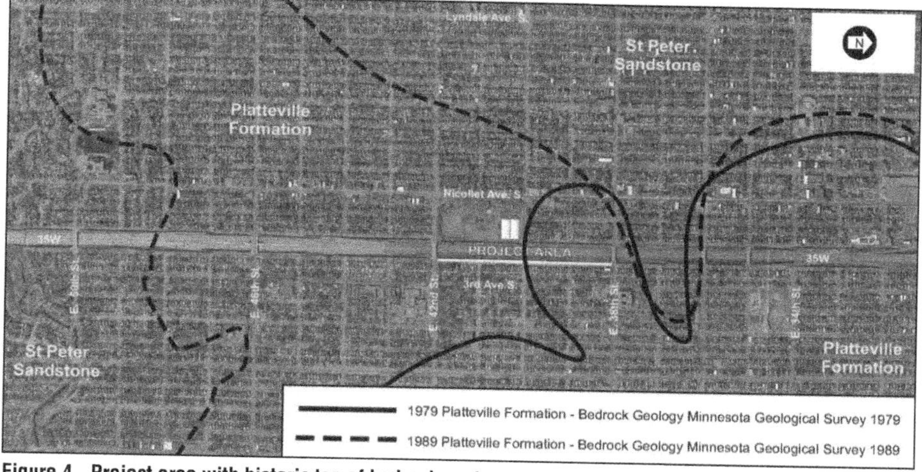


**Figure 4. Project area with historic top of bedrock contours**

formation, but the 1979 extents indicate only part of the project to be underlain by the Platteville formation. One goal of the geotechnical investigation was to determine the actual bedrock extent location.

### Groundwater

Based on review of data from installed vibrating wire piezometers, the groundwater at the project site exhibits a downward gradient and a northward gradient. Due to the bedrock valley, the St. Peter sandstone is in direct hydraulic connection with the overlying glacial outwash. Piezometer data collected between August 2018 and January 2019 indicate baseline total head elevations fluctuate between elevation 248 and 250 meters above mean sea level (815 and 820 feet) near the northbound, east-side shoulder of I35W. The high groundwater table and limited space in the project area influenced the decision to construct the SSF cells utilizing D-wall construction techniques. Figure 5 presents a geologic cross section developed during the design phase of the work and illustrates the steep glacial erosion valley, high groundwater water, and variable geology.

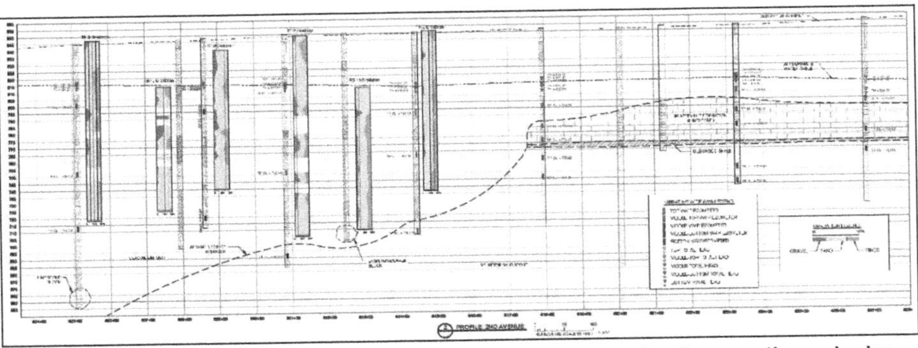

**Figure 5. South (left) to north (right) geologic section developed across the site presenting grain size variability**

## CMGC Delivery

### *Process and Benefits*

CMGC (also referred to as construction manager-at-risk or CMAR) has been defined as an integrated team approach to planning, design, and construction; to controlling schedule and budget; and to ensuring quality for the project owner (Gransberg, 2015). This type of delivery requires an intentional approach to collaboration between all parties: owner, designer, and contractor and it provides a mechanism to determine and allocate risk between parties. Similar to design-build, CMGC brings a contractor to the design phase for construction, schedule and cost inputs however different that design-build, CMGC allows the owner to retain control over the design. Given the subsurface risks associated with ground conditions and the potential associated implications on cost and schedule, CMGC provided a beneficial project delivery method for I35W SSF.

In 2012, Minnesota legislature enacted legislation authorizing MnDOT to use CMGC as a delivery method. The law authorized an award of up to four CMGC projects per year with a total project maximum of ten. In 2021, as MnDOT was approaching ten projects in the program life, the state legislature updated the statute to increase the CMGC project maximum to twenty (MnDOT CMGC program website). Utilization of

CMGC produces an expectation of benefits to the overall project as the design phase work can result in more iterative design and option considerations given the number of parties involved. Benefits expected include the following: constructability, reduction/management of project risk, and construction schedule efficiency. Various benefits were realized and continued to be realized during the CMGC process for the I35W project, but a few specific items are highlighted in the following sections.

## Early Bid Package for Site Preparation

The SSF project site location was driven by proximity to the natural low area along I35W and where the most hydrologic benefit could be obtained relative to historical flooding problems. Adjacent to the SSF project and occurring concurrently during early phases of the SSF construction was a major design-build interstate project (Crosstown to Downtown project) which impacted all lanes of traffic along I35W. MnDOT desired to prevent either of the adjacent projects to negatively impact the other during construction. As a result, the site for the SSF project was aligned to be on the east should of I35W on MnDOT's right-of-way. This narrow project area required site preparation to occur ahead of primary diaphragm wall construction. The site preparation included a soil nail wall to be constructed which would serve as earth retention during construction and also be integrated as a long-term earth retention option at the end of the project. The site preparation activities were identified as a schedule risk during the CMGC design phase, and as a result, steps were taken to mitigate the risk including:

- Separate the site preparation work to be its own work package—work package 1 (WP1). This allowed for contractor pricing to be conducted and bid ahead of the major SSF structure design package. WP1 design was completed in spring of 2018 and bid during the summer with construction starting in September 2018. Work package 2 (WP2) design continued through August of 2018.
- Coordination with City of Minneapolis to have existing utilities upgraded ahead work. Minneapolis initiated structural lining of an existing water utility ahead of WP1. The water utility was in the potential zone of influence of the proposed soil nail wall.
- Soil nail wall design was selected for the WP1 design because it allowed for the noise wall to remain in place during construction to minimize disturbance to residential neighborhoods just east of the site. This also allowed 2nd avenue (residential street just east of the project site) to remain open during the majority of construction (Figure 6).

## Management of Groundwater

The design of the SSF includes extending the D-walls below the design floor elevation of 223 meters (731 feet) by approximately 6 meters (20 feet) to increase the cut-off effect and reduce the uplift pressure gradient during construction. However, supplemental depressurization is required to maintain stability during the construction phases until the floor slab is poured and cured. During the collaboration of entities in the design phase of the CMGC process, various issues were identified and addressed which allowed for management of risk during construction. Some of these issues included:

- **Contractual progressive depressurization targets:** Performance of two pump tests to inform design and baseline expectations of a depressurization system. Results of the pump test and hydraulic modeling of the depressurization system led to the development of specific staged depressurization targets that are contractually required to be achieved and maintained during excavation

Figure 6. Installation of soil nail wall during WP1 site preparation. Note the existing noise wall remaining in place

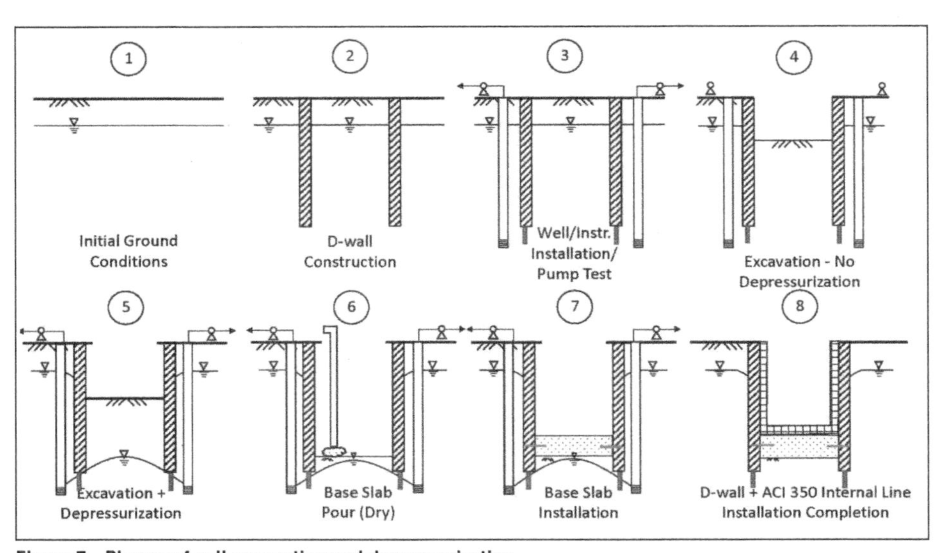


Figure 7. Phases of cell excavation and depressurization

of each cell. The depressurization is achieved through a series of screened wells installed around the perimeter of each SSF cell. Remote monitoring of piezometer total heads allows for confirmation that the pressure requirements are achieved prior to advancing excavation below specific elevations. Figure 7 provides a summary of sequencing of the depressurization work.

Multiple vibrating wire piezometers were installed prior to and during construction for monitoring purposes and are utilized to monitor groundwater pressures at each cell. Specific piezometers near the diaphragm wall tips are utilized for the contractual depressurization measuring point. Each SSF cell has redundancy of instruments at these read point elevations to allow for review of conditions around perimeter of each SSF cell and prevent impacts if a single sensor is disabled or destroyed. The owner, contractor, and engineer can view the total head data through a remote website and confirm that depressurization targets are sufficient to allow excavation to proceed.

- **Incorporation of depressurization standby equipment:** The pump test data confirmed the expectation that in the event of loss of the depressurization system due to pump system loss of power would be unacceptable given the risk to the cell excavations at depth. As a result, contractual requirements and commensurate pricing were incorporated into the project such that standby equipment be onsite, installed, and available for immediate operations if any of the depressurization system became inoperable. Alarm notifications were implemented for the remote monitoring system transmitting vibrating wire piezometers total head data. If action trigger levels were exceeded, automated notifications were sent to key stakeholders on all teams if either of the following conditions were met: increase of four feet of total head over a two-hour period or increase of five feet of total head compared to recent total head values.

## Construction of the D-Wall

Construction using a diaphragm wall approach was selected based on high groundwater table and limited work area, and, for the target depths, a diaphragm wall provides a higher degree of certainty for overlap between panels versus a secant pile wall construction for the tank diameters. Figure 8 shows the site at the beginning of D-wall operations.

The D-wall panels were excavated in upper elevations with clamshell and lower elevations with hydro-mill techniques. The anticipated elevation for change in density and

**Figure 8. Initial construction of the D-Wall equipment on the work pad after completion of WP1**

excavation ease/production were based on the geotechnical investigation. The decision was made to not have shared walls between the cells because of constructability concerns around cell closure and panel overlaps with a shared segment (Figure 9). As a result, jet grouting and hand excavation was performed to tunnel between the cells and create final structural connections.

Once the D-Wall panels were constructed, and the depressurization system was operating, excavation of soil from inside the cells commenced. Upon completion of the mass excavation, the cell floors were doweled into the D-walls and an interior liner constructed in each cell.

## CONCLUSION

The I35W SSF is currently under construction and will provide stormwater resiliency to a major Twin Cities traffic corridor. The SSF presents a significant subsurface project

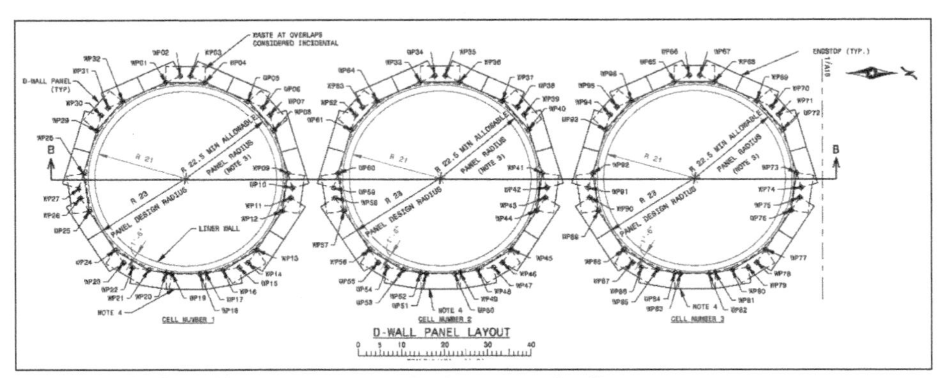


**Figure 9. D-Wall panel layout—note separation between each cell**

**Figure 10. Construction of the cells—cell 1 through 5 have been excavated and cell 6 is just beginning excavation (May 2022)**

which, by nature of the work, creates an inherent increase in risk due to the probability that ground variability may have an impact on construction which, if negative, can result in significant schedule and cost impacts. The utilization of a CMGC alternative delivery format provided a means to collaborate between owner, engineer, and contractor on issues such as constructability, staging, risk, management, and innovative design to better manage risk associated with the project and while also baselining some conditions for reference during the bidding process. While ground variability challenges have occurred during the construction phase, the CMGC delivery method helped establish a collaborative working team during the design phase which carried over into construction. Currently, the I35W SSF is anticipated to be completed in the second half of 2023.

## ACKNOWLEDGMENTS

A project of this scale is only successful due to the efforts of a broad, dedicated team. A large thank you to the entire Minnesota Department of Transportation staff who endeavored to think creatively and whose implementation of the CMGC program set a tone for collaborative effort on this project, the contractor team of Kraemer North America and Nicholson Construction joint venture, and everyone on the design team including Barr Engineering, Brierley Associates, TKDA, HZ United, Braun Intertec, Olson Engineering, and Traut Companies.

## REFERENCES

Bloomgren, Bruce A. and Poppe, James R. Plate 4 Thickness of Drift. Geologic and hydrologic aspects of tunneling in the Twin Cities area, Minnesota (Miscellaneous Investigations Series Map I-1157). s.l.: Minnesota Geological Survey, 1979.

Bloomgren, Bruce A., Cleland, Jane M. and Olsen, Bruce M. Plate 4 Depth to Bedrock and Bedrock Topography. C-04 Geologic atlas of Hennepin County, Minnesota. s.l. : Minnesota Geological Survey, 1989.

Dittes, Michael E. Mechanical Properties of St. Peter Sandstone a Comparison of Field and Laboratory Results (Master's Thesis). s.l.: University of Minnesota, December 2015.

Gransberg, Douglas D. and Jennifer S. Shane. *Defining Best Value for Construction Manager/General Contractor Projects: The CMGC Learning Curve*. J. Manage. Eng., 2015, 31(4).

Meyer, Gary N. and Hobbs, Howard C. Plate 3 Surficial Geology. C-04 Geologic atlas of Hennepin County, Minnesota. s.l.: Minnesota Geological Survey, 1989.

MnDOT. "Innovative contracting—Construction Manager/General Contractor." (https://www.dot.state.mn.us/const/innovative-contracting/construction-manager-general-contractor.html).

Mossler, John H. and Olsen, Bruce M. Plate 5 Geologc Sections and Engineering Characteristics of Quarternay Units. Geologic and hydrologic aspects of tunneling in the Twin Cities area, Minnesota (Miscellaneous Investigations Series Map I-1157). Minnesota Geological Survey: s.n., 1979.

Olsen, Bruce M. and Bloomgren, Bruce A. Plate 2 Bedrock Geology. C-04 Geologic atlas of Hennepin County, Minnesota. County Atlas Series. s.l.: Minnesota Geological Survey, 1989.

# Packaging and Contract Delivery Methods for the Horizon Lateral

**Adriana Ventimiglia** ▪ Southern Nevada Water Authority
**Ray Brainard** ▪ Black & Veatch
**Amanda Kerr** ▪ Black & Veatch

## ABSTRACT

The $2 billion Horizon Lateral includes over 30 miles of trenched and tunneled pipeline with diameters up to 120 inches. Related infrastructure includes large and small pumping stations, reservoirs, and flow control facilities. This paper will discuss the strategy and processes used to determine contract packaging and contracting delivery methods for the various project elements with an emphasis on the tunnels and trenchless crossings. Prior Authority experience and the rigorous risk-based approach to develop the preferred alignments played a large role in determining the outcomes along with schedule concerns and permitting and easement acquisition.

## INTRODUCTION

The Southern Nevada Water Authority (SNWA) was established in 1991 to address the water needs of the Las Vegas Valley (Valley) on a regional basis. SNWA currently operates and maintains the South Valley Lateral (SVL), which has the capacity to provide 306 million gallons per day (MGD) to the City of Henderson (COH) and the Las Vegas Valley Water District (LVVWD). SNWA is now proceeding with the Horizon Lateral Program to improve the reliability and flexibility of water delivery in the Valley's southern portion for existing customers and to ensure capacity to accommodate projected water needs.

The Horizon Lateral would be a new conveyance transmission system capable of delivering up to 375 MGD of water from the existing River Mountains Water Treatment Facility (RMWTF) site to the southern portion of the Valley. As an initial step to execute the Program, SNWA retained a consultant team led by Black & Veatch Corporation (the BV Team) to conduct a Feasibility Study. The purpose of the study was to identify preferred pipeline routes and related facility configurations, to develop costs, schedules, and to identify key activities critical for successful Program implementation.

As part of the feasibility investigation, the BV Team provided engineering and planning services including data collection; rights of way (ROW); corridor analysis; public outreach planning; hydraulics and hydraulic optimization; environmental assessment; hard and soft ground tunnel evaluations; facility engineering; construction methods; construction cost estimating, and a risk-based alternatives evaluation. The BV Team also participated in or led extensive workshops with SNWA and a wide variety of stakeholders. The Feasibility Study was finalized in 2021. It concluded that further alignment evaluations should be considered during the preliminary design phase as new or updated Program information become available, allowing for either a refinement of prior alignment choices or selection of newly preferred pipeline and tunnel segments. The Project Stakeholders selected two alignment alternatives to advance into preliminary design: the North Alignment Alternative and the South Alignment Alternative. Both alternatives are shown on Figure 1.

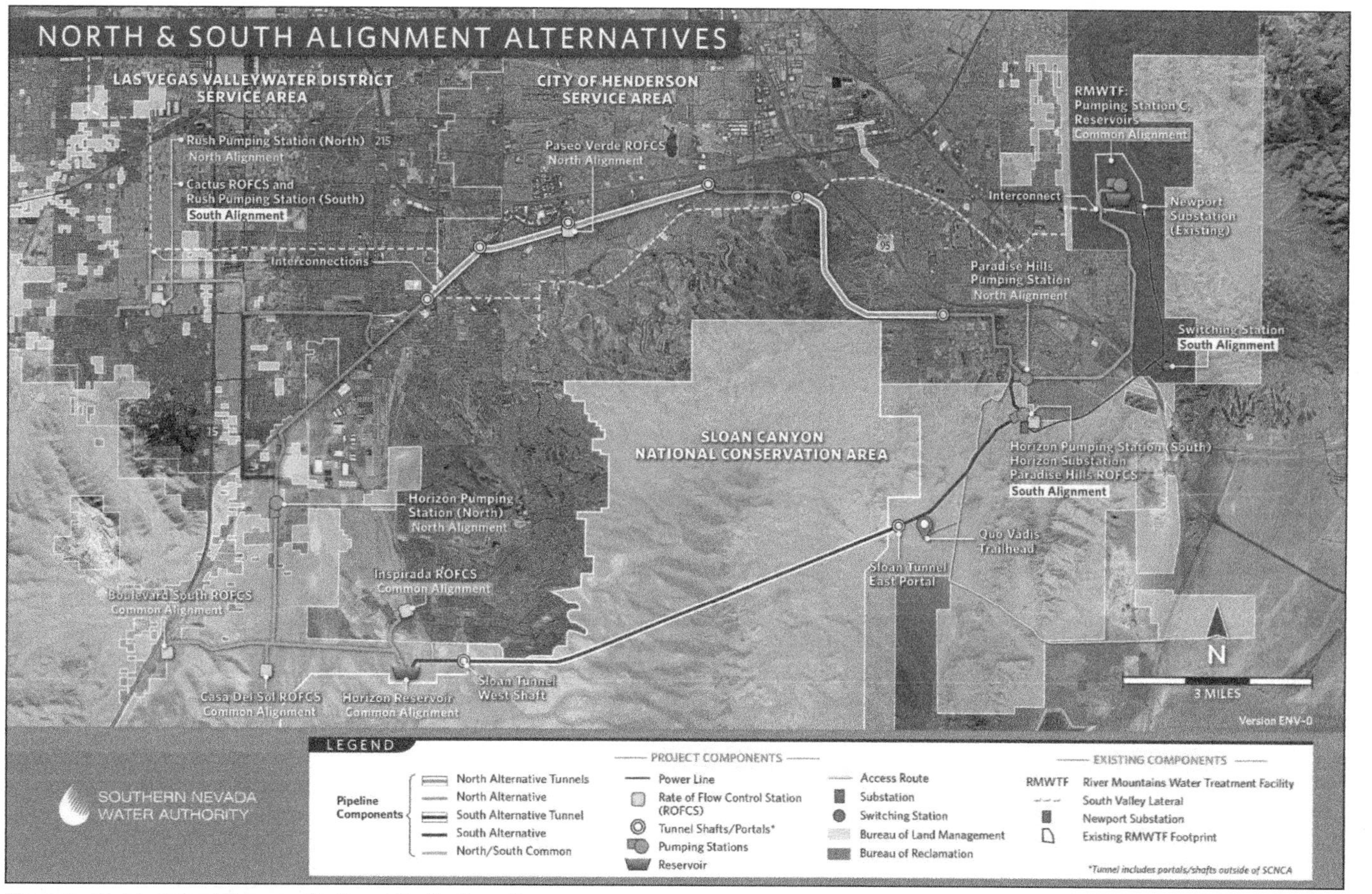


**Figure 1. Horizon Lateral Program, North and South alternatives**

## CURRENT DESIGN PHASE

Further analysis and evaluation shaped the North and South alignments alternatives. Geotechnical investigations were performed for the Northern Alignment Alternative, while the permissions process to physically investigate the Southern Alignment Alternative is still underway.

## CONTRACT PACKAGE EVALUATION

The approach to evaluating contract packaging options involved creating individual pipeline and facilities packages for both the North and South Alignment Alternatives. The contract packages are valued between approximately $25 million to $450 million and consist of either traditional design/bid/build and Construction Management at Risk (CMAR) construction delivery methods. The individual pipeline segments and facilities associated with both the North Alignment Alternative and the South Alignment Alternative were separated into three contract packaging alternatives for each alignment and are described below. Contract Packaging Plan 1 was a carryover from the Feasibility Study and was the baseline case. The other two alternatives were created with the goal of optimizing the number of contract packages; i.e., to reduce the number of packages and lessen the administrative load while not creating so few packages that they become excessively large in scope and/or contract value.

The three contract packaging plans evaluated further for the North Alignment Alternative were:

- Plan 1: 12 contract packages with values ranging from $35 million to $375 million.
- Plan 2: 7 contract packages with values ranging from $110 million to $325 million.
- Plan 3: 8 contract packages with values ranging from $117 million to $360 million.

The three contract packaging plans evaluated further for the South Alignment Alternative were:

- Plan 1: 11 contract packages with values ranging from $25 million to $375 million.
- Plan 2: 8t contract packages with values ranging from $100 million to $375 million.
- Plan 3: 6 contract packages with values ranging from $200 million to $625 million.

Breaking the work down into a relatively large number of comparatively small contract packages for a program of this size would have the advantage of spreading the work across a larger number of bidders and lessens the overall risk associated with any individual package. However, with the comparatively large number of packages comes the disadvantages of more administrative burden (and cost) to execute the work, additional interfaces between packages which introduce some risk and complexity, and the general risk that comes along with anything that has "more moving parts."

Ultimately, Contract Packaging Plan 2 was selected because it was expected to create the best contract package size (dollar volume) for the number of packages to be administered, while not excessively concentrating the risk into very few, very large packages.

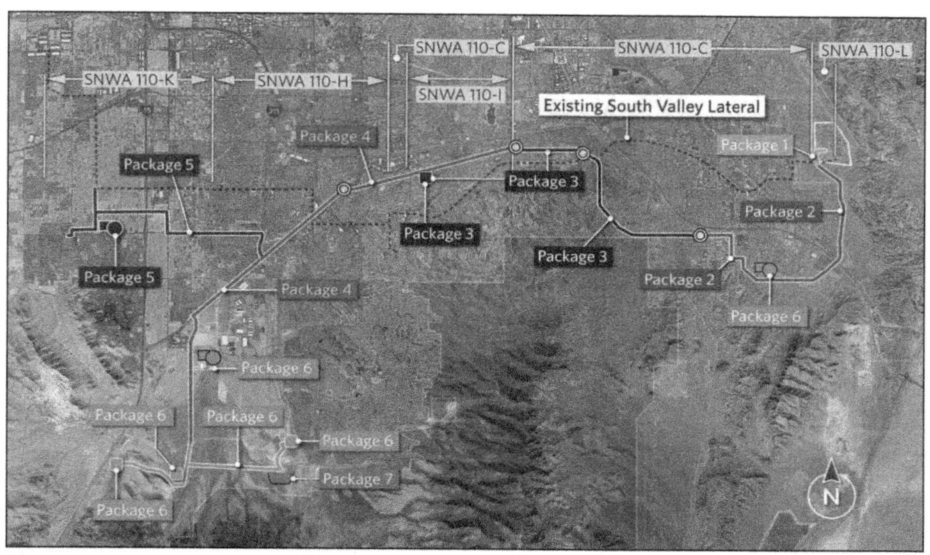


**Figure 2. North alternative contract packages**

With respect to delivery methods, a mix of design-bid-build and CMAR, was selected, depending on the risks specific to each contract package.

## DELIVERY METHODS

In the Feasibility Study, multiple program delivery methods were evaluated and discussed with SNWA. The three delivery methods primarily used by SNWA and allowed in Nevada are: traditional design/bid/build, construction manager at risk (CMAR), and design/build.

For the Horizon Lateral Program, the following construction methods were recommended:

- Pipelines and Interconnects—Design/Bid/Build
- Tunnels—Design/Bid/Build or CMAR*
  **BLM requires near complete designs before issuing right of way permits, therefore Design-Build would not be suitable for tunneling contracts on BLM lands.*
- Large Pumping Stations—CMAR
- Reservoir—Design/Bid/Build
- Small Pumping Stations—Design/Bid/Build
- Rate of Flow Control Stations—Design/Bid/Build
- Power Transmission Line—Design/Bid/Build (with a procurement phase) or CMAR

### North Alignment Alternative

The North Alignment Alternative, if selected, would route most of the program through fairly urban areas of the City of Henderson. See Figure 2 for reference. The packages include tunnels, trenchless crossings, and open cut installation methods, as well as a variety of pumping stations and rate of flow control structures (ROFCS). The packages including underground infrastructure are emphasized below.

### *North Alignment—Package Two*

North Package Two will consist of approximately 22,000 linear feet (LF) of 120-in. diameter pipe commencing with the connection to the South Valley Lateral, and approximately 12,000 LF of 108-in. diameter pipe following the interface with a proposed pumping station. This package will also include approximately 200 LF of 60-in. diameter pipe. Construction will be open cut except where the pipeline crosses Interstate 11 and the Union Pacific Railroad (UPRR) Tracks, totaling about 700 LF of proposed trenchless crossings.

### *North Alignment—Package Three*

North Package Three will consist of approximately 19,000 LF of 108-in. diameter pipe installed via tunnel in rock conditions north-westerly under McCullough Hills. This package will also include about 7,200 LF of open cut and 1,000 LF of trenchless. This package will also include the construction of a 60 MGD ROFCS.

### *North Alignment—Package Four*

North Package Four contains the VEA/St. Rose tunnel, expected to be tunneled in soft ground conditions for approximately 26,000 LF. Pipe installed in the tunnel will be either 108-in. or 96-in. This package will also contain about 7,200 LF of 96-in. diameter pipe and 16,500 LF of 84-in. diameter pipe installed via open cut, and approximately 400 LF of 72-in. diameter pipe to create another connection to the South Valley Lateral. About 2,000 LF of trenchless crossings are anticipated at various locations throughout the package.

### *North Alignment—Package Five*

This package will consist of approximately 25,000 LF of 66-in. diameter pipe and a pumping station. Additionally, about 3,600 LF of trenchless crossings are proposed for this package.

## Estimated Construction Cost

A thirty percent (30%) design OPCC was developed based on the design drawings, technical memos and other design information developed in the Preliminary design phase. These OPCC values include anticipated construction costs, construction contingency and permitting fees.

Construction costs were estimated for budgeting purposes using historical cost data that includes costs from past projects; cost estimating team experience in the industry; and numerous other sources. This level of definition allows for complete transparency throughout the lifecycle of design and will aid the industry in forming potential partnerships due to bonding capacities (Table 1)

**Table 1. Northern alternative contract packages, opinion of probably construction cost**

| Contract Package No. | OPCC, 2022 Dollars |
|---|---|
| 2 | 250M–275M |
| 3 | 250M–275M |
| 4 | 350M–400M |
| 5 | 125M–150M |

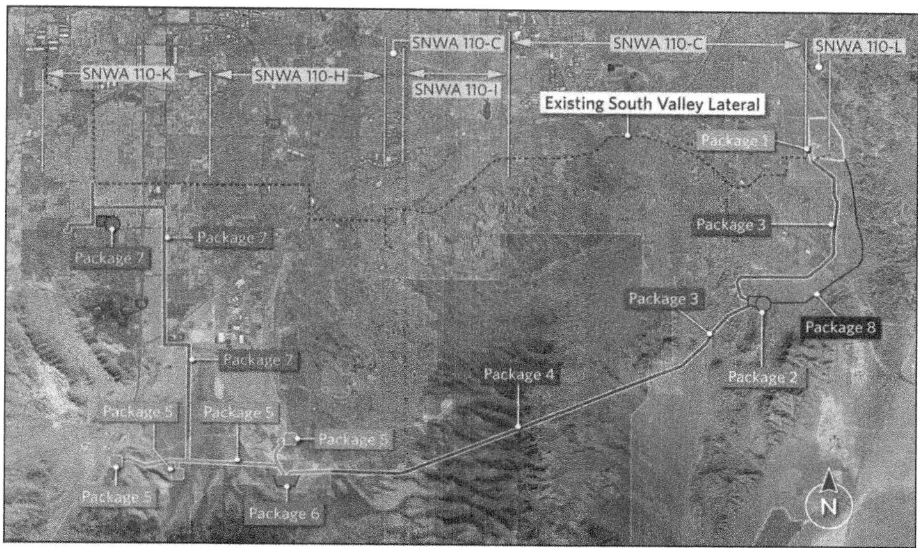


Figure 3. South alternative contract packages

## South Alignment Alternative

The South Alignment Alternative, if selected, would route most of the program through less developed areas, including a tunnel under the Sloan Canyon National Conservation Area. See Figure 3 for reference. The packages include tunnels, trenchless crossings, and open cut installation methods, as well as a variety of pumping stations and rate of flow control stations (ROFCS). The packages including underground infrastructure are emphasized below.

### *South Alignment Package 3*

South Package Three will consist of approximately 22,000 LF of 120-in. diameter pipe commencing with the connection to the South Valley Lateral, and approximately 26,000 LF of 108-in. diameter pipe. This package will also include approximately 200 LF of 60-in. diameter pipe. Construction will be open cut except where the pipeline crosses Interstate 11 and the Union Pacific Railroad (UPRR) Tracks, totaling about 700 LF of proposed trenchless crossings.

### *South Alignment Package 4*

This tunnel would have a finished inside diameter of 108-inches and an approximate length of 7.1 miles. The tunnel would traverse the northern McCullough Mountains. The tunnel would be constructed in a single drive, utilizing a two-pass system whereby the tunnel excavation and initial support system is constructed, followed by the installation of the 108-inch finished inside diameter carrier pipe.

### *South Alignment Package 7*

This package would include approximately 44,000 LF of 66-in. diameter pipe and 3,400 LF of 60-in. diameter pipe installed by open cut methods, a pumping station, and about 1,100 LF of trenchless crossings.

### *Estimated Construction Cost*

A thirty percent (30%) design OPCC was developed based on the design drawings, technical memos and other design information developed in the Preliminary design phase. These OPCC values include anticipated construction costs, construction contingency and permitting fees.

Construction costs were estimated for budgeting purposes using historical cost data that includes costs from past projects; cost estimating team experience in the industry; and numerous other sources. This level of definition allows for complete transparency throughout the lifecycle of design and will aid the industry in forming potential partnerships due to bonding capacities (Table 2).

**Table 2. South alternative contract packages, opinion of probably construction cost**

| Contract Package No. | Approximate Value, 2022 Dollars |
|---|---|
| 3 | 250 M–300 M |
| 4 | 275 M–325 M |
| 7 | 200 M–250 M |

## ESTIMATED PROGRAM SCHEDULE

The contract phasing was evaluated using a preliminary schedule, at the time of this document preparation, and work durations based upon the cost estimating effort for the pipeline segments, tunnels, and facilities. Key items evaluated were the following:

- First contract advertising—Midyear 2025
- Program construction completion Midyear 2030
- Program startup Midyear 2031
- Advertising, Bidding, Award—60 days
- Longest duration contract checked against respective program completion date.
- Should the program start be delayed by NEPA Permitting, ability to bid multiple contracts simultaneously should be considered.

The development of the program schedules for all of the alternatives resulted in an understanding that all of the packaging alternatives work, and all allow the work to be completed within the desired time horizon. It is also of note that the key driver controlling the alternative schedules is the NEPA permitting process. Knowing that the NEPA permitting process is the driver, managing that process and outcome is the program's critical path. One could say that the North Alignment Alternative is slightly advantageous from this standpoint, but the effect is minimal, providing approximately 6 months of additional float that the South Alignment Alternative does not have. The NEPA process will continue to be monitored from an overall program standpoint and the schedule will be updated as the process advances into the final design phase.

## CONCLUSION

The $2 billion Horizon Lateral program includes over 30 miles of trenched and tunneled pipeline with diameters from 60 inches up to 120 inches, as well as various other infrastructure needed to pump the water to the desired locations. Contract packaging and delivery methods were evaluated to best suit the design and construction of a program of this size, under the required timeline. Prior Authority experience and the rigorous risk-based approach to develop the preferred alignments played a large role in determining the outcomes along with schedule concerns and permitting and easement acquisition. As the design process advances, further updates will be provided to the tunneling community on this groundbreaking program.

# Progressive Design-Build in the Tunneling/Underground Construction Industry—Perspective from the Private Sector

**Carlos Tarazaga** ▪ AZTEC Engineering Group, Inc (TYPSA Group)

## ABSTRACT

Progressive Design-Build (PDB) is an emerging variation of alternative delivery programs in the underground construction industry. PDB refers to the way a construction project design is developed by the Owner and the Design-Builder in a step-by-step process. According to the Design-Build Institute of America (DBIA), Progressive Design-Build allows the design and construction team to collaborate during the earliest stages of project development. This enables engagement between the three key players in a construction contract: the owner, the designer, and the contractor.

Is the PDB the best alternative program delivery for large contracts? What is the perspective of the private sector?

This paper will provide an overview of the different alternative delivery programs used in the underground industry, pros and cons of each of programs, and the perspective of the designers and contractors. The paper will focus on how the PDB works, reasons why owners should consider PDB for their projects, best collaboration strategies among owners, designers, and contractors to deliver a successful project, and how to manage the project risks using a PDB method.

## INTRODUCTION

Project delivery and procurement methods have generally evolved from the traditional Design-Bid-Build (DBB) approach as the "baseline" most used by public entities. In recent decades, the various collaborative delivery methodologies have emerged as viable alternatives to traditional DBB delivery. These alternatives to DBB seek to better allocate risk and responsibility, save time, and support a selection methodology beyond low-bid price.

In today's market, a collaborative approach is necessary to be able to obtain fair pricing, maintain the budget, and finish the project on schedule. PDB is growing in popularity in select infrastructure markets amongst owners. PDB currently represents much of the Design-Build market for water and wastewater projects, and its use is becoming ever more common in the aviation industry.

Public authorities have expressed interest in using PDBs over the coming years, with few of them advancing procurement of one or more PDB Projects. Several of these projects are part of large capital programs, such as North Civil-Pape Tunnel and Underground Stations for the Ontario Line in Toronto in November 2022, Metrolinx and Infrastructure Ontario approved a competitive solicitation of a PDB contract to achieve the proposed design approach, specific project futures and functions, and other project criteria in addition to price. Also, the Santa Clara Valley Transportation Authority's (VTA) BART Silicon Valley Extension Program (BSV Program) is the largest public infrastructure Program ever constructed in Santa Clara County (California), using a PDB solicitation for the Contract Package 2 (Tunnel and Trackwork Project).

## What Is Progressive Design-Build?

According to the Government of the State of California, Progressive Design-Build means "a project delivery process in which both the design and construction of a project are procured from a single entity that is selected through a qualifications-based selection at the earliest feasible stage of the project." (1).

Essentially, PDB is a two-phase process where the project owner engages the contractors and design team members early in the life of the project and before the design has been developed past a conceptual level. The owner and the design-build team work collaboratively together to first develop the project's overall design and clarify the programming and priorities. The design-build team is generally selected on qualifications and pricing of components such as development of Phase One Services and Overheads for construction, but the final project cost/price and schedule commitments are not part of the selection process. Once that is agreed upon, the parties mutually develop the project design to an adequate point where a realistic schedule and cost can be developed.

## Procurement and Delivery Model Process

**RFQ/RFP:** The PDB procurement model begins with a Request for Qualifications (RFQ) and/or Request for Proposals (RFP). The owner often selects the design-build team based on the qualifications and pricing of components such as development of Phase One Services and Overheads for construction. The qualifications-based process allows owners to reduce the cost and length of the procurement process, compared with other alternative delivery models.

**Phase One Services:** the design-build team collaborates with the owner and with its own consultants to validate existing design concepts, propose design alternatives, evaluate construction phasing alternatives, and develop the project's overall design and define the programming requirements. Design, schedule, quality, operability, maintainability, and other project decisions such as owner preferences are established and mutually agreed upon with the owner's participation. The design-build team provides ongoing cost estimates to ensure the owner's budgetary requirements are achieved, enhancing the project's transparency with an open-book approach. When the design has been advanced to an appropriate level of definition that aligns with the owner's requirements, the design-build team provides a commercial proposal to undertake the Phase Two Services, including the Guaranteed Maximum Price (GMP) for the project. In recent projects when PDB has been used, the commercial proposal was established when the design was at 80%-85% complete. The level of completion of the design can depend on the amount of control the owner desires to maintain over the design definition.

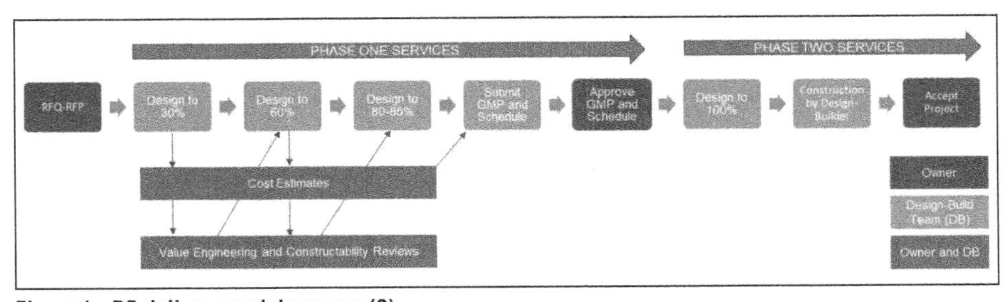


Figure 1. DB delivery model process (2)

**Phase Two Services:** Phase 2 consists of the final design, construction, start-up/commissioning, and handover to the owner, in accordance with those commercial terms. Phase 2 will be initiated only after completion of Phase 1 and successful negotiation of the GMP and schedule. If, for any reason, the parties cannot reach agreement on the Phase Two commercial terms, the owner typically has an "off-ramp" and can opt to proceed in a traditional manner and have the design-builder complete the design. The owner can then solicit construction bids from contractors and complete the project in the traditional design-bid-build manner. On some occasions, the design-builder may remain in place and provide the owner traditional construction management services.

## ADVANTAGES OF PROGRESSIVE DESIGN-BUILD

PDB brings together the best of the Design-Build (DB) and Construction Management at Risk (CMAR) models. PDB is an excellent opportunity for complex underground projects to reduce project contingencies and make more realistic pricing assumptions. The main advantages of PDB are:

### Collaboration Throughout the Process

The PDB model fosters the maximum collaborative relationship between owner, designer, and builder. This team works together throughout the entire project to review the progress and direction of the project from their respective viewpoints. The decisions are based on overall project critical success factors versus a single factor, such as cost. Phase 1 allows the owner to provide substantial inputs on the design and allows the builder to perform constructability reviews of the design, optimization of means and methods, cost and schedule estimating and use of Value Engineering to reduce time and cost, and increase efficiencies.

### Owner Control of the Budget

The biggest concern for owners is unexpected change orders during the design and construction stages. In a Progressive DB, preliminary GMP cost-models are provided at design completion milestones throughout the Phase 1, using an open-book basis. This open-book approach provides the owner transparent access to project costs and the ability to factor cost and schedule considerations, allowing the owner to use non-self-performed work by the builder, thus maximizing the work that may competitively procured.

### Single Point of Responsibility

One entity assumes the responsibility and risk for the project from beginning to end once concept, program and budget are aligned. Providing owners with a single point of contact, the responsible entity manages the project team and maintains project knowledge continuity throughout the design and construction of the project.

### Flexibility for Phasing Construction Activities

The Owner has substantial flexibility in authorizing design-builder to perform early construction and procurement work packages as the design develops during Phase One. These early work packages may be proposed by the owners based on availability of ROW and third-party agreements or proposed by the contractor to expedite the construction in Phase Two. Recent examples of early work packages in tunneling projects are related to utility relocation works and other enabling works with long leads, as procurement of the Tunnel Bored Machine.

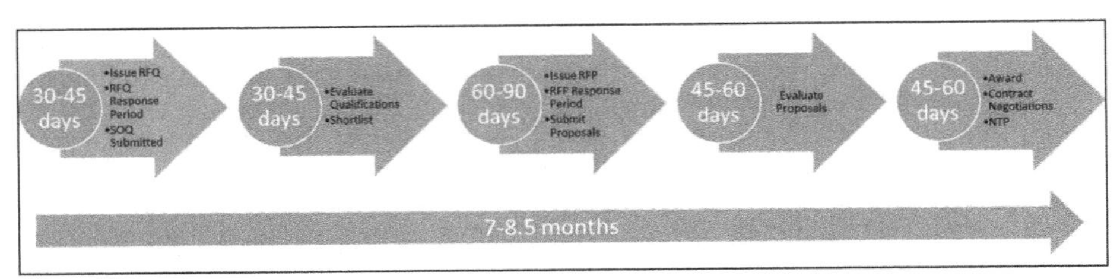


Figure 2. Typical PDB procurement duration

### Faster Decision Making in Procurement, Design and Construction

One of the benefits of progressive design-build is that it streamlines and shortens the overall project schedule with a quicker procurement process which encourages competition. As shown in Figure 2, a typical PDB procurement may last among 7 and 8.5 months

As example, Metrolinx and Infrastructure Ontario (the Sponsors) considered different project delivery methods and selected PDB method for the North Civil-Pape Tunnel and Underground Stations for the Ontario Line (3) to enhance bidder interest and allow for a collaborative advancement of design, joint risk identification and mitigation, early constructability input and identification of innovations. The Sponsors characterized the main features of a PDB method as:

- Streamlined procurement phase;
- Development Phase to advance design, begin early works, understand project risks, and establish a Target Price(s);
- Concludes with the execution of a Target Price (Design Build) Agreement (TPA), which marks the beginning of the Implementation Phase to finalize the design and construction of the Works;
- Target Price with a gainshare/painshare mechanism; and
- Open-book approach for the Development Phase and Implementation Phase.

## DISADVANTAGES OF PROGRESSIVE DESIGN-BUILD

### Construction Costs Are Not Known at the Time of Initial Contract Signing

When the contractor or design-builder is selected using a PDB process, the price is still unknown at the time of the initial contract award. Thus, when the initial budget is established, it can not only include the value of the technical scope of work, but also the value of the risk as mutually agreed.

### Cost Determined Through Negotiated and Competitive Processes

PDB calls for the owner to select the design-build team largely based on qualifications, without the benefit of price competition on the overall design-build contract price. Some owners may not be in favor in negotiating the commercial terms of the arrangement without a GMP in place.

### New Delivery Method in the Tunneling Arena

PDB models have been used mainly in water and wastewater projects. An effective education program for public owners may be needed to overcome concerns with the PDB process and final construction cost determination.

## MANAGEMENT OF RISKS

During the design development in a PDB (see Figure 3), the construction cost is progressively developed by the design-build team often in conjunction with the 30- and 60-percent levels of design detail. The design-build team schedules and conducts meetings with the owner and any other necessary individuals or entities to discuss and review the scope of work to establish a design-build preliminary evaluation of the Project and to provide any and all preliminary engineering required to design the Project, as detailed in the Scope of Work or determined by the design-builder, in consultation with the owner, to be necessary to complete preliminary engineering for the Project.

The iterative, “design to budget” approach to cost estimates from the design-builder helps ensure the project budget is not exceeded. As compared to the often single, value engineering step within a DBB or CMAR-CM/GC models, this continuous price feedback allows the owner to constantly evaluate decisions and adjust as needed to deliver the best overall project value.

After enough design definition is achieved, often between the 60- to 85-percent range, a proposed GMP or fixed price is prepared by the design-builder for owner’s approval. Owners typically retain an independent cost estimating (ICE) consultant to furnish a second opinion of probable project costs during the GMP negotiation process. The preparation of the GMP at that level of design reduces risks and contingencies by all the parties, and schedules are generally more realistic.

Project design risk is shared by the design-builder and the owner because they work together during the design phase, with the owner providing continuous input to the design and at specific milestones. This collaboration reduces risk of the owner’s design decisions or preferences impacting constructability and reduces cost-growth risk to the builder because the owner’s input is included before the GMP is finalized. Because of the single-point responsibility for the design-builder, design coordination risks are shifted to the design-builder, which distinguishes PDB from DBB and CMAR-CM/GC delivery methods. These design coordination risks often result in project changes that the owner bears in the DBB and CMAR-CM/GC delivery methods

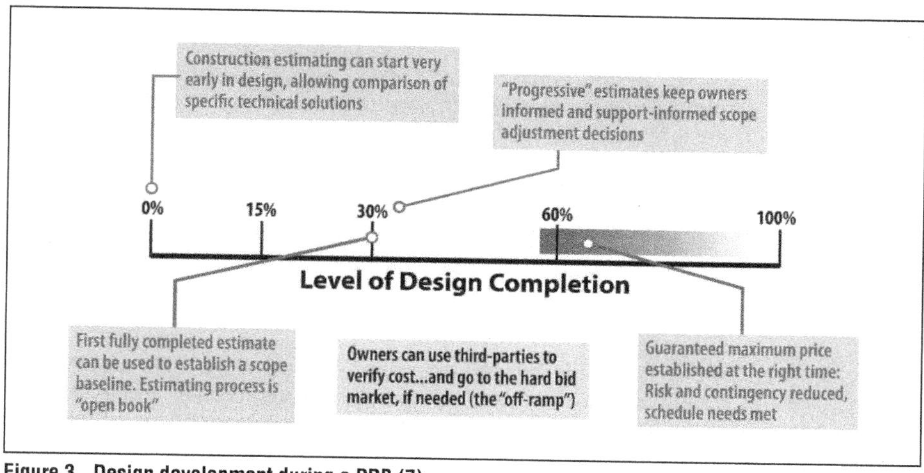


Figure 3. Design development during a PDB (7)

**Table 1. Risk allocation in PDB (Adapted from the Water Design Build Council's Water and Wastewater Design-Build Handbook, Fourth Edition)**

| Risk | Design-Build Team | Owner |
|---|---|---|
| Project Design | Shared | Shared |
| Coordination with existing infrastructures | Shared | Shared |
| Project Performance | X | |
| Schedule | Shared | Shared |
| Cost of constructed project beyond GMP | X | |
| Site conditions | | X |
| Construction Warranty | X | |
| Materials Cost Escalation | Shared | Shared |

because the designer's liability is limited to the "standard of care" and the contractor's liability is restricted to the as-bid construction documents.

The design-builder retains construction schedule risks, assuming the design reviews and other owner's responsibilities are met. The cost of the constructed project up to the GMP is the responsibility of the owner including such items as price escalation not explicitly addressed by the contract, owner-requested changes in scope, and changed conditions. Beyond the GMP limit, the design-builder is responsible for budget overages. The design-builder is responsible for project performance/acceptance and the testing and commissioning of the infrastructure.

The design-builder is responsible for demonstrating the project's performance through an acceptance-testing procedure that is agreed upon with the owner during the contract negotiation or GMP/fixed price negotiation stages. Performance risk is borne by the design-builder until the owner accepts the project, which represents the transfer of operation, maintenance, and performance risk to the owner.

## CONCLUSIONS

PDB delivery has been used mainly in water and wastewater projects. PDB procurements in tunneling and underground construction projects may be preferred when a project lacks definition, or still requires final permitting, or when an owner prefers to remain involved in the design-decisions process, all while leveraging the schedule, collaboration, and contractual advantages provided by a DB approach. This model may be also valuable in underground projects when regulatory permitting requires well-developed design solutions, or when an owner believes that it can lower cost by participating in design decisions and in managing risk progressively through the project definition phase.

PDB models can appropriately and reasonably allocate the risk for each party involved in a underground project. It achieves this goal by providing an opportunity for all parties to have more input earlier in the process which leads to more certainty when it comes to delivering a project on-time and on-budget with everyone earning a reasonable rate of return.

Additionally, the GMP in a PDB is based upon open-book pricing. This means the owner is provided a detailed breakdown of all costs, markups, profits, and overhead costs. All cost components are transparent. This provides the owner an opportunity to negotiate items that influence price. For example, material costs may have changed significantly due to inflation. Having the opportunity to view open book pricing and subcontractor bids will educate owner's staff on reasonable bid pricing. This approach will likely prove to be important to explain and justify any supply-chain issues and resulting increased costs associated with the COVID pandemic and the current market conditions.

## REFERENCES


1. California Legislative Information https://leginfo.legislature.ca.gov/faces/codes_displaySection.xhtml?lawCode=GOV§ionNum=13332.19.#:~:text=(2)%20%E2%80%9CProgressive%20design%2D,feasible%20stage%20of%20the%20project.

2. The Alternative of Alternative Delivery: Progressive Design Build. PNCWA 2012.

3. Request for Qualification. Applicant's meeting for Ontario Line-North Segment. November 3rd 2022.

4. Progressive Design-Build Done Right. Design Build Institute of America.

5. Progressive Design-Build. Design Build Procured with a Progressive Design&Price. Design Build Institute of America.

6. Five Reasons to Use Progressive Design-Build for Corrections and Municipal Projects. https://www.performanceservices.com/resources/5-reasons-to-use-progressive-design-build-for-corrections-and-municipal-projects.

7. What's so Progressive about Progressive Design-Build? Water Design-Build Council https://waterdesignbuild.com/wp-content/uploads/Whats-Progressive-about-DB-Alpert-Tunnicliffe.pdf.

8. Critical Comparison of Progressive Design-Build and Construction Manager/General Contractor Project Delivery Methods. https://journals.sagepub.com/doi/full/10.1177/0361198118822315.

PART 

# Design

*Chairs*

**John Yen**
Skanska

**Manan Garg**
Austin Transit Partnership

# Design Aspects of the Minneapolis Central City Parallel Tunnel

**Bruce D. Wagener** ▪ CNA Consulting Engineers
**Randall C. Divito** ▪ Hatch Associates Consultants, Inc.
**Craig R. Eckdahl** ▪ CNA Consulting Engineers
**Joseph Klejwa** ▪ City of Minneapolis Public Works Surface Water & Sewers

## ABSTRACT

The existing Central City Tunnel was built primarily in the 1930s with a small portion as old as the 1880s. It conveys stormwater in downtown Minneapolis. It is a 4,400-ft long tunnel located 80 feet underground. The Central City Tunnel has experienced surcharging and surface overflows during heavy rainfall events. This paper presents the details of tunnel and shaft design for the new Central City Parallel Tunnel to mitigate the overflows. The design considered hydraulics and geotechnical engineering for the structural geologic conditions including the weak Saint Peter Sandstone and overlying strong Platteville Limestone. Ground conditions resulted in tunnel designs including large flat-back box tunnels and unique cathedral-shaped tunnel sections.

## BACKGROUND

### History of Tunneling in Saint Peter Sandstone

The Minneapolis-St. Paul, Minnesota metro area has a long history of underground space utilization in the St. Peter sandstone. For as long as the area has been inhabited, both natural and human-made underground spaces have been utilized. The combination of easily excavated sandstone and unsaturated conditions near the Mississippi River led to numerous underground uses such as:

- Combined sewers, storm sewers, sanitary sewer tunnels
- Water supply, telecommunication, and power utility tunnels
- Headrace and tailrace tunnels
- Caverns and cellars for beer, cheese, mushrooms, archival materials, and other storage
- Silica mining for glass production
- Shelter

Early on, excavation was done with picks and shovels. As time progressed, other excavation methods such as high-pressure water lances, hydraulic hammers, boom mounted cutting heads, and tunnel boring machines have been used. Figure 1 shows a typical tunnel under construction.

### Existing Central City Tunnel System

The existing Central City Tunnel System consists of approximately 4.1 miles of St. Peter Sandstone tunnels and associated drop shafts shown by black lines in Figure 2. The Hennepin Avenue, Nicollet Avenue, Marquette Avenue, and 2nd Avenue tunnels all flow into the Washington Avenue tunnel. At Portland Avenue, the Washington Avenue tunnel turns north one block, then east two blocks where it turns north on Chicago Avenue. From there, it outlets to a historic mill district channel that flows into the Mississippi River.

Figure 1. Typical St. Peter Sandstone utility tunnel under construction (1938)

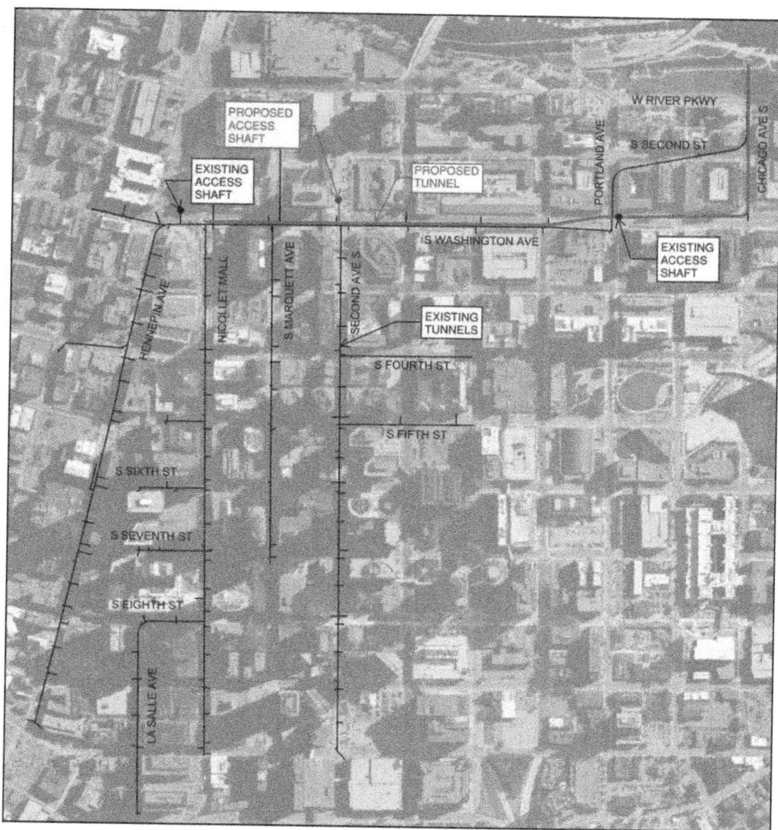


Figure 2. Minneapolis central city storm sewer tunnel system

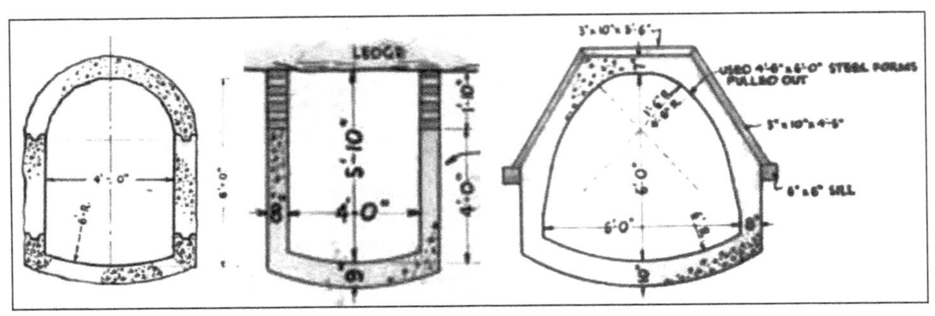


Figure 3. Cross sections of typical existing tunnels

Figure 4. Outlet structure (1890)

The existing tunnels are either horseshoe-shaped or flat-backed with sizes ranging from 6-foot tall by 4-foot wide to 9-foot tall by 7.5-foot wide. Typical tunnel cross sections are shown in Figure 3.

At the mill district channel near the Mississippi River, the Washington Avenue tunnel terminates at a large chamber that has an outlet to the channel. Figure 4 is an enhanced photo of the chamber during its construction. The structure is showing signs of distress, as evidenced by masonry separation in the crown.

## Central City Parallel Tunnel Project

"Parallel" is in the Project name because the proposed tunnel runs parallel to the existing tunnel in Washington Avenue as shown in Figure 5. Existing parallel tunnels on Chicago Avenue will be replaced with one larger tunnel. The total proposed tunnel length from upstream end to outlet is 4,200 ft.

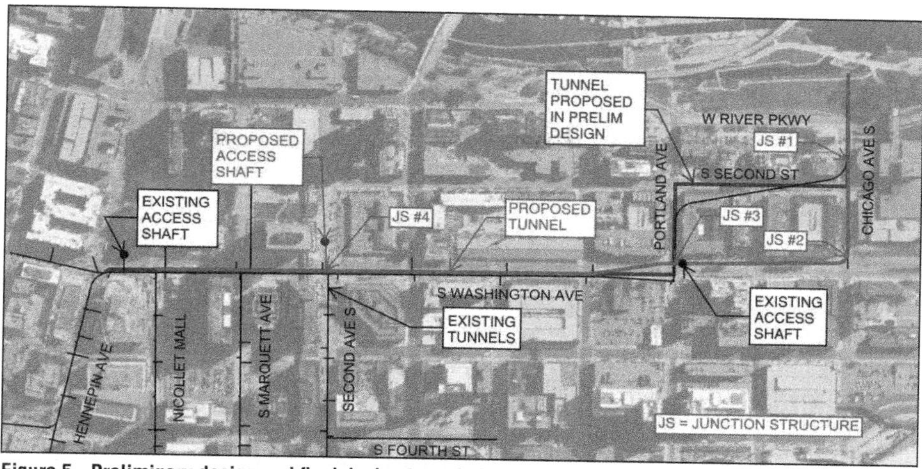


**Figure 5. Preliminary design and final design tunnel alignments**

Previous hydraulic studies by other firms were used by Hatch, who performed additional hydraulic investigations and identified tunnel sizes required to carry a 100-year design storm without pressurization. Minimum equivalent tunnel diameters were determined as follows:

- 9-foot diameter from Hennepin Avenue to Nicollet Mall
- 9.75-foot diameter from Nicollet Mall to 2nd Avenue
- 11-foot diameter from 2nd Avenue to Portland Avenue
- 12-foot diameter from Portland Avenue to Chicago Avenue

Actual proposed tunnel shapes are either rectangular- or cathedral-shaped and were driven by the geology, discussed later in this paper. Four junction structures connecting the existing Washington Avenue and Chicago Avenue tunnels were required to connect the existing mainline tunnels to the proposed tunnel. Four cross-connections, located at Nicollet Ave, Marquette Avenue, and 2nd Avenue connect existing perpendicular tunnel systems to the proposed Washington Avenue parallel tunnel. Also required were seven drill hole connections to existing drop shaft drifts that collect stormwater from localized shallow storm sewer systems and a new outlet structure.

## Regional Geology and Ground Conditions

The Twin Cities metropolitan area, including the area of direct interest to the Project, is underlain by nearly 1,000 feet of sedimentary rocks from the Ordovician Age. These gently dipping to near horizontally bedded rocks form the Twin Cities structural and hydrologic basin. A mantle of glacial and post-glacial deposits covers the area.

The present topography and surficial deposits of the Twin Cities area are a result of geologic processes active during Pleistocene time (11,000 to 2 million years ago) with minor modifications during Holocene (most recent 11,000 years) time. Several glacial advances and retreats typically left up to 100 feet of gravel, sand, silt, and clay deposits over the area. Glacial deposits, up to 400-feet thick, fill river valleys to the west of downtown Minneapolis and east of downtown St. Paul (Norvitch & Walton, 1979). The Mississippi River and its associated alluvial deposits reflect Holocene post-glacial drainage modifications. Bedrock underlying the glacial deposits is exposed along the steep Mississippi River bluffs.

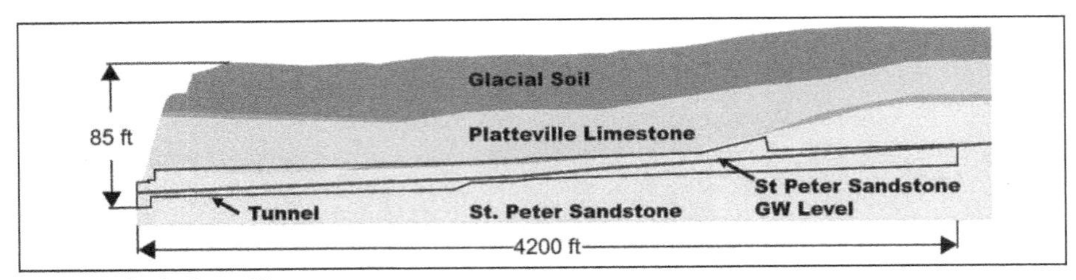


**Figure 6. Geologic profile**

The main structure in the region is a broad basin of marine sedimentary rocks deposited in an embayment of early Paleozoic seas. The embayment extended northward through the metropolitan area from southern Minnesota over part of a preexisting Precambrian basin. These formations consist of four rock types that provide a record of at least five advances of Paleozoic seas into the embayment area. These rocks generally have maintained their original, nearly horizontal bedding, except for a monoclinal flexure which trends roughly coincidental with the buried ancestral valley east of downtown Minneapolis (Norvitch & Walton, 1979). In general, there is a high degree of predictability to the general bedrock geology except where cut by buried valleys.

The stratigraphic section underlying the Project area consists of glacial deposits underlain by approximately 1,000 feet of sandstone, shale, and carbonate rocks of Cambrian and Ordovician age. The first bedrock unit is the Platteville Formation, an approximately 30-foot-thick formation which is divided into several dolomitic limestone and dolomite members. At the base of the Platteville Limestone is the thin (2-feet to 5-feet thick) Glenwood Shale which overlies the St. Peter Sandstone. The tunnels will be constructed in the Glenwood Shale and the St. Peter Sandstone, which is a 150-foot-thick massive sandstone unit composed of fine-grained to medium-grained, well-rounded, and uniformly graded quartz sand.

Groundwater in the soil is perched by the Platteville Limestone. Groundwater in the Platteville Limestone is perched by the Glenwood Shale. These conditions create unsaturated zones in the Platteville limestone and St. Peter Sandstone. The unsaturated zone in the St. Peter Sandstone has been utilized for low-cost tunneling and underground space construction in the Twin Cities area for over 100 years. The St. Peter Sandstone groundwater level at the Project location is located within or just above proposed tunnel profile, as shown in Figure 6, the geologic profile of the tunnel.

## SELECTION OF PREFERRED OPTION

This section describes the investigations and evaluations that informed the selection of the preferred tunnel size, location, and cross-sectional shapes.

### Additional Geotechnical Investigations

Geotechnical investigations for the Project included both review of existing information and new geotechnical exploration. Previous projects constructed in the St. Peter Sandstone that were reviewed include underground space, storm sewer tunnels, and sanitary sewer tunnels. All of which CNA Consulting Engineers had involvement with the design and/or construction. Of particular interest was the Minnesota Department of Transportation (MnDOT) 2nd Street Storm Sewer Tunnel, which was constructed in 1977 and is located one block north of the proposed tunnel. Several aspects of that project were similar to this Project, including similar-sized cathedral-shaped tunnel and required deep well dewatering. The information was used in the ground support, liner design, and dewatering design.

Building owners in the area were contacted to obtain available geotechnical reports. Several owners responded. Most borings taken did not extend into the bedrock, but they provided useful information on the overburden soil and associated groundwater conditions and the top of bedrock elevation.

The City of Minneapolis maintains excellent plan and profile drawings of its tunnel systems. Besides tunnel and shaft construction information, the profiles show geologic information.

New geotechnical investigation for the Project included borings and geophysics. Five borings were taken for the Project. Drilling included SPT sampling in the soil and HQ wireline coring of the limestone, shale, and sandstone. Four of these included installation of monitoring wells for groundwater level and contaminant testing. Finding locations for borings that avoided existing utilities and that did not unacceptably disrupt traffic proved to difficult and required a significant amount of planning and relocating.

The geophysical investigation was comprised of Multichannel Analysis of Surface Wave (MASW) seismic surveys. They were conducted on the slope upstream of the tunnel outlet to identify the location of the buried erosional bedrock edge at the Mississippi River gorge. The survey included five lines and were conducted by 3Dgeophysics.

## Additional Hydraulic Investigations

Preliminary hydraulics investigations were performed by other consultants. During final design, Hatch evaluated the new parallel tunnel system hydraulic operation including the outlet structure, using a one-dimensional model, under normal flow conditions for several horizontal and vertical tunnel alignment options to refine and optimize the design. This included incorporation of the existing Central City Tunnel into the design. These evaluations revealed the potential for tunnel surcharging leading to additional hydraulic investigations by Hatch in association with Hydra-Tek of Vaughn, Ontario Canada, to assess hydraulic surge risk during tunnel filling and peak flow periods using a surge response analysis assessment approach.

The results of the additional hydraulic investigations demonstrated the most suitable tunnel alignment and optimal tunnel cross-section for efficient hydraulic operation of the tunnel, including surge risk control. The surge risk hydraulic investigation analyses indicated low surge risk and low risk of surface overflows (i.e., geysering) in the new parallel tunnel system during peak flow periods. The surge risk analyses demonstrated that surge risk was eliminated in the Hennepin and South Marquette tunnels located upstream of the existing Central City Tunnel. For the 2nd Avenue and Nicollet Mall tunnels, located upstream of the existing Central City Tunnel, the surge risk is remains unchanged from the existing conditions and is to be addressed during subsequent planned phases of the project.

The outlet structure hydraulics were investigated to optimize the hydraulic grade line at the outlet and avoid tunnel surcharging due to outlet structure hydraulic constriction. Hydraulic constriction is mitigated through the use of tunnel widening upstream of and at the outlet structure. A related design issue for the outlet structure included flow energy dissipation to protect the receiving flow channel against erosion due to tunnel discharge. Flow energy dissipation involved partial submergence of the outlet structure. The receiving flow channel is also protected against erosion through the use of rock rip-rap or armor stone.

## Environmental Investigations

Environmental investigations were conducted to identify contaminated soil, bedrock, and groundwater. American Engineering Testing, Inc. provided environmental services, including:

- Review of the Minnesota Pollution Control Agency's (MPCA) "What's in My Neighborhood: online database.
- Environmental sampling of soil and groundwater.
- Applying for National Pollutant Discharge Elimination System (NPDES)/State Disposal System (SDS) Contaminated Groundwater General Permit.

The environmental investigation revealed that groundwater with petroleum impacts and volatile organic compounds (VOCs) was present at two of the four monitoring well locations. Because of this, groundwater treatment was specified at these two areas. In addition, the NPDES/SDS permit requires monthly testing of all dewatering well discharges.

## Shaft Siting Evaluation

No new shafts are required for conveying stormwater, but access shafts are required for construction. A shaft siting study was conducted to determine the best locations for access. Considerations included:

- Tunnel constructability
- Neighborhood concerns, avoiding residential areas as much as possible
- Traffic disruption, including vehicles, pedestrians, and bicycles
- Existing utilities
- Permitting or easement requirements
- Surface workspace availability
- Proximity to the proposed tunnel

In collaboration with the City, one new access shaft, a drive-in access at the river portal, and utilization of two existing shafts were chosen as access points. The new access shaft is located on 2nd Avenue, approximately 100 ft north of the new tunnel as shown in Figure 5. The shaft is 12-ft inside diameter and lined with a steel casing grouted in place. At completion of the Project, the shaft will have a 48-inch diameter casting at the street surface and a full diameter precast concrete slab 3 ft below the street surface, which can be removed in the future for access for maintenance activities.

The two existing shafts are reinforced concrete pipe-lined and 8 ft in diameter, located at Hennepin Avenue and Portland Ave. as shown in Figure 5. At Hennepin Avenue, access is through an existing drift tunnel and the existing storm tunnel. At Portland Avenue, the new tunnel passes directly under the shaft.

Access is also possible through seven existing smaller shafts. These are typically 14-inch diameter and can be used for utility or concrete slick line access.

To minimize disruption to the neighborhood, work hours at all shafts are limited to 7:00 AM to 7:00 PM Monday through Friday and 9:00 AM to 5:00 PM on Saturday. Work is allowed 24 hours per day at the drive-in portal access because that location is more remote.

## Selected Tunnel Alignments and Enlarging of Existing Tunnel

The City and another consultant developed a preliminary tunnel plan and profile as part of their preliminary design contract. CNA and Hatch refined the design during final design. The final alignment modifications are shown in Figure 5.

Modifications included the following:

- Ending the upstream end of the Project at Nicollet Avenue instead of Hennepin Avenue. Additional hydraulic modeling showed no benefit in enlarging the tunnel in this area. Also, additional tunnel construction and a complicated junction structure was avoided near Hennepin Avenue.
- From Washington Avenue, follow Chicago Avenue instead of Portland Avenue. The revised route eliminated one horizontal radius in the tunnel, which are more time consuming to construct and hydraulically inefficient. This change required removal of the existing Chicago Avenue tunnel between Washington Avenue and 2nd Street in addition to sections further downstream as shown in Figure 5.
- Removing and replacing existing parallel tunnels and portal chamber in Chicago Avenue between 2nd Street and the river. Removal and replacement were chosen because the portal chamber was showing signs of distress. In addition, right-of-way limitations did not allow adequate space between the existing and proposed tunnels to allow construction of a parallel tunnel, shown in Figure 7. This change also required removal and replacement of the existing outlet structure.

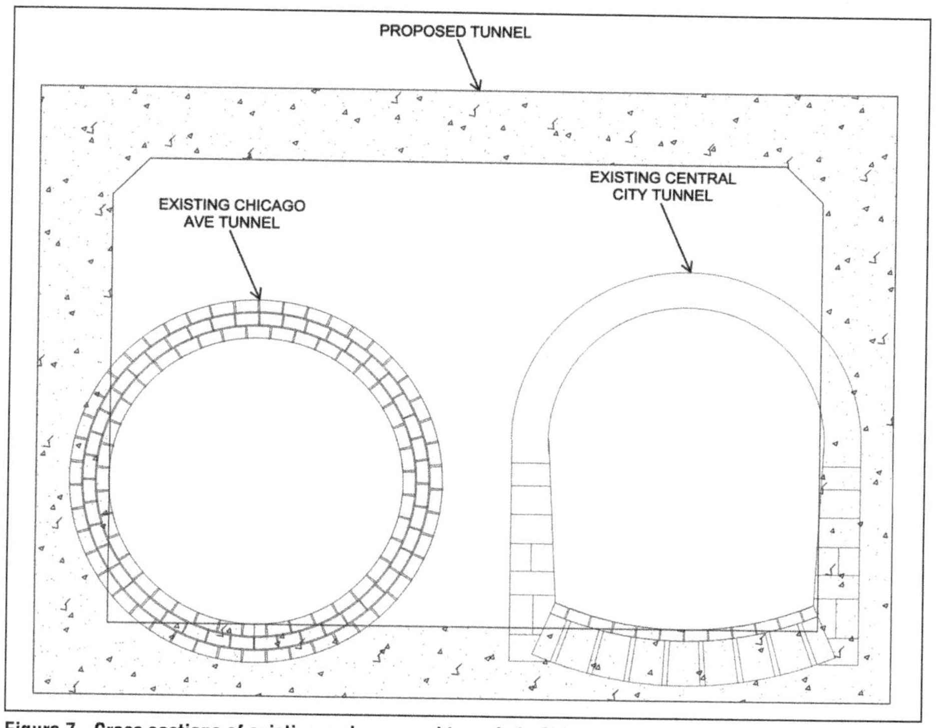


**Figure 7. Cross sections of existing and proposed tunnels in Chicago Avenue**

## SEM Versus TBM Tunneling Methodology Evaluation

The selection of tunneling methodology between the Sequential Excavation Method (SEM) and Tunnel Boring Machine (TBM) tunneling was considered for this project given the unique and interesting requirements and constraints presented by the required structures, tunneling equipment and technology, downtown urban environment, hydrological (tunnel flooding) conditions, and the ground conditions.

### *Tunneling Equipment and Procurement*

TBM tunneling equipment and procurement requires sourcing/purchase, refurbishment, acceptance, transportation and delivery to the site, assembly, commissioning, and launching. TBM tunneling requires special machine selection with unique features to suit the ground conditions, tunnel excavation, and tunnel support requirements. This process typically requires up to nine (9) months to complete resulting in a significant delay to the start of tunnel excavation and construction. TBM tunneling equipment also requires specialized personnel and workforce that are trained to efficiently operate and maintain the TBM.

SEM tunneling equipment and procurement is much more economical and simpler than TBM tunneling equipment. SEM tunneling equipment (e.g., hydraulic rock hammers, transverse and axial cutting units, Brokk machines, and roadheader machines) is usually readily-available, uncomplicated, and construction-ready. It can also be efficiently utilized for tunneling by a broader range of experienced construction personnel. This equipment and its procurement and start-up processes are far shorter than a TBM and allows tunnel excavation work to begin without delay and nearly immediately upon notice to proceed with construction.

### *Flooding Impact Risk*

Since the required construction will be near the existing Central City Tunnel and downstream of several stormwater drainage tunnels, flooding risk in the new parallel tunnel excavation is a significant concern for security and productivity. As a result, tunneling equipment that is mobile and robust is better suited to protect against this risk. SEM tunneling equipment is easily demobilized and secured in flooding events and is less subject to flooding damage impacts. On the other hand, TBM flooding is unavoidable and will result in flooding water exposure impacts to electrical and mechanical systems that are highly-susceptible to water damage and will result in significant post-flooding delays in excess of typically delays for SEM tunneling equipment and operations.

### *Shaft Site Surface Requirements*

TBM tunneling operations require a tunnel work shaft site for TBM assembly, TBM launching, tunneling logistical support, and tunnel mucking. TBM tunneling also requires a reception shaft site to receive the TBM and remove it from underground. A TBM tunnel work shaft site requires at least twice the shaft diameter and twice the available surface space of an SEM tunnel work shaft site, although more SEM tunnel work shafts are required for multiple tunnel headings, if needed.

The tunnel alignment and project sites are located in the central business district (CBD) of downtown Minneapolis. Downtown Minneapolis has limited available surface space for larger tunnel work shafts and reception shafts. Larger TBM tunnel work shaft sites will have greater impact to surface and underground utilities, businesses, and residents than SEM tunnel work shaft sites.

Regarding utility impacts, logistics and associated delays, SEM tunneling presents less risk of underground and surface utility relocation-related project delays. Smaller SEM tunnel work shafts and sites have a lower probability of utility conflicts due to lesser site footprint. SEM tunneling requires lower electrical power demand and provides more flexible usage of available electrical power compared to TBM tunneling since a new electrical power supply line is not required.

### *Tunnel Constructability*

Due to the required in-line replacement of the existing Chicago Ave. Sewer tunnel, TBM tunneling is only usable for 76% (3,200 LF out of 4,200 LF) of the project tunnel alignment. TBM tunneling must be performed in a single drive to be economical for the relatively short tunnel drive length and due to the limited available surface space for TBM tunnel work shafts. SEM tunneling can advance multiple tunnel drives concurrently from small SEM tunnel work shaft sites. SEM tunneling further allows concurrent excavation for drill holes and cross-connects during excavation of the new parallel tunnel, whereas TBM tunneling operations will conflict resulting in work delays.

Geological ground conditions present higher risk to TBM tunneling with a circular bore due to ground relaxation-induced loosening and associated overbreak of the weak sandstone rock above the tunnel bore before it can be supported. This type of overbreak can result in chimney void development above the TBM shield and is difficult to support during TBM tunneling without using full-circle, ring beam steel sets which would still require significant secondary grouting to backfill voids above the overbreak failed rock wedge or chimney void. The use of steel sets or the use of conventional rock reinforcement support will result in delays and slow production yielding an estimated average TBM tunneling advance rate of 40 feet per day in the authors' opinions. The use of SEM tunneling can mitigate these issues by accommodating the rock overbreak into a stabilized excavation shape resembling a harmonic hole consisting of a bifurcated ellipse or egg-shaped opening (referred to as cathedral-shaped) thereby providing a unique optimized section for the finished tunnel. The estimated average SEM tunneling advance rate is 10 feet per day. Based on the project tunnel alignment and access configuration, it is possible to advance four tunnel headings concurrently.

Another TBM tunneling constructability issue is the rock contact monocline resulting in a flat-back tunnel near the contact between the Platteville Limestone (average intact rock UCS = 17,000 psi; overlying the sandstone) and the St. Peter Sandstone (average intact rock UCS = 125 psi; underlying the limestone). The limestone hard rock above the tunnel bore could locally dip down and intersect the tunnel bore, which could jam the TBM cutterhead and stop the TBM advance since it would be unable to cut the harder rock in the crown due to tooling and other mechanical limitations. SEM tunneling is inherently designed with the flexibility to mitigate this issue as it is encountered.

### *Tunneling Methodology Evaluation Summary*

Based on the technical reasoning and justification presented in the foregoing paragraphs, it is apparent that SEM tunneling is the more suitable tunneling methodology over TBM tunneling. Further, SEM tunneling is the historically proven tunneling method for small tunnels in the local geological and ground conditions in downtown Minneapolis and in the St. Peter Sandstone.

Given the previously stated TBM procurement delay and estimated advance rates for TBM and SEM tunneling with a single TBM tunnel drive and four concurrent SEM tunnel drives, the overall construction schedule duration to the completion of tunnel excavation is expected to be 20 months. Therefore, the main difference between TBM and SEM tunneling is the higher initial equipment investment cost of the TBM which makes TBM tunneling more expensive than SEM tunneling by approximately 10% of the project construction cost without considering the inherent flexibility and economical benefits of SEM tunneling in the project ground conditions.

## SEM Tunneling with Unique Geologically-Controlled Geometries

The tunnel vertical alignment is notably controlled by the tunnel-junction and tunnel-cross-connect intersection elevations with the existing Central City Tunnel which is integral with the new parallel tunnel. The tunnel section shape is controlled by the vertical alignment position relative to the monocline comprised of the rock contact between the Platteville Limestone and the Saint Peter Sandstone. In general, the tunnel section for the upstream 800 linear feet (LF) of tunnel alignment is a cathedral-shaped section, as shown in Figure 8, including the drill holes and cross-connects. The remaining 3,400 LF of the tunnel is a box-shaped section with a flat back due to the close proximity to the monocline limestone rock contact, as shown in Figure 8, including the tunnel junction structures.

The cathedral tunnel section is a direct result of the harmonic hole (self-stabilized) opening created by excavation and subsequent immediate ground relaxation of weak cemented sandstone of the St. Peter Sandstone which results in progressive crown arch failure and high overbreak leading to a bifurcated elliptical or cathedral-shaped excavation. Overbreak leads to a stable harmonic or neutral hole tunnel excavation cross-section referred to as a cathedral-shaped tunnel section. This geologically-controlled excavation shape is taken advantage of in design and construction through optimization by maximizing the tunnel storage and conveyance volume (relative to the excavated width) and minimizing the required structural strength requirement of the tunnel initial support and the final concrete lining.

The box tunnel section is a direct result of the cut-and-cover tunnel for the outlet structure and the horizontally-bedded limestone rock contact closely overlying the tunnel in St. Peter Sandstone from the outlet structure to 2nd Avenue. St. Peter Sandstone overbreak propagation upward to the shale and limestone rock contact leads to near vertical excavation sidewalls and a flat back for a box or rectangular tunnel excavation cross-section, as shown in Figure 8.

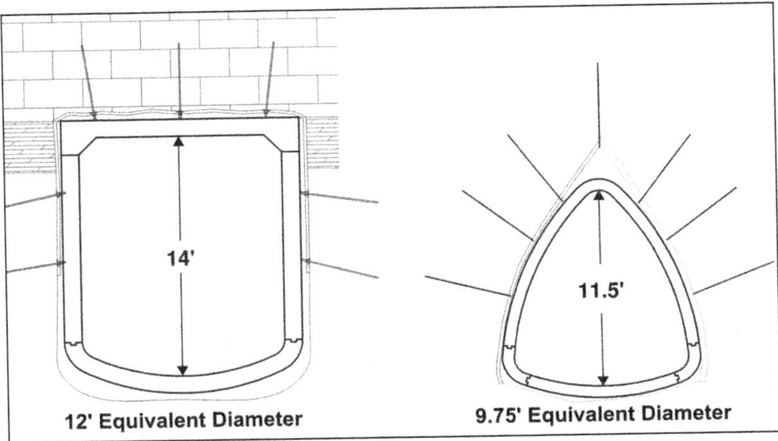


**Figure 8. Cross sections of proposed box (left) and cathedral (right) tunnel sections**

## TUNNEL DESIGN ELEMENTS

### Permitting

The Project required coordination and permitting with numerous agencies, some of which are listed below.

#### *Permits and Easements Obtained by Owner*

Minnesota Department of Natural Resources (MnDNR)Water Appropriation Permit. This permit is required for the deep wells for dewatering the St. Peter Sandstone for the tunnel on Washington Avenue. On many projects, dewatering design and implementation is left to the contractor. For this Project, CNA chose well locations and prepared preliminary dewatering well design before construction so that construction would not be delayed.

Minnesota Pollution Control Agency (MPCA) Contaminated Groundwater General Permit. This permit contained requirements for testing and treatment of dewatering well discharges.

US Army Corps of Engineers (USCOE) Nationwide Permit 7—Outfall Structures and Associated Intake Structures. This permit covers reconstruction of the outfall structure. The pre-construction notification triggered a review of the effect on the Mill District historic property. The permit required an archeological construction monitoring and unanticipated discoveries plan.

Minneapolis Parks and Recreation Board (MPRB) Temporary Construction Easement. The outlet structure and downstream end of the tunnel is located on MPRB property. A Temporary Construction Easement was obtained.

Metropolitan Council Environmental Services (MCES) Sanitary Sewer Modifications. MCES and City of Minneapolis share access to their sanitary sewer (MCES) and existing Central City storm sewer tunnel (City). A joint use agreement was executed to cover modifications to the sanitary sewer required for the proposed tunnel construction.

#### *Permits Obtained by Contractor*

Minnesota Department of Natural Resources (MnDNR)Water Appropriation Permit. Design of any dewatering in the tunnel outlet area was left to the contractor and the contractor was required to obtain this permit.

Minnesota Department of Health (MDH)—Well Construction and Sealing. This permit is for the actual construction and sealing of all dewatering wells and monitoring wells.

City of Minneapolis Permits. These include:

- Obstruction/Right-of-Way
- Lane use
- Hooding parking meters
- Sidewalk use
- Utility connections for power, water, etc.
- After Hours Work Permits

## Excavation and Support

The Contract Documents specified minimum permanent ground control requirements with the ability to add ground support based on observed conditions. The base ground support included sodium silicate spraygrout, rockbolts, steel straps at some locations, and welded wire fabric—identified as "Minimum Ground Control." "Heavy Ground Control" included Minimum Ground Control with the addition of 3.5 inches of shotcrete. To allow for flexibility in ground support, the Contract included unit prices for additional ground support elements, including additional rockbolts, steel straps, and shotcrete.

In addition to the specified permanent ground support, the Contract Documents specified the maximum allowable distance to excavation face without all temporary ground control installed. The distance varied depending on the excavated shape, proximity to junction structures, and proximity to the tunnel portal.

The Specifications required the contractor's professional engineer to prepare and certify a temporary ground control plan, which include details of the temporary ground control plan for all underground work, including observations by a competent person, personnel access control, trimming, and scaling.

## Groundwater Control

Groundwater control was required for three general conditions:

Lowering the St. Peter Sandstone groundwater level to below the tunnel invert. This was accomplished by installing deep dewatering wells. The Contract included five deep wells as the base dewatering system and five additional deep wells if required to supplement the base system. Well locations and design were chosen during the design phase to allow for the initiation of required MnDNR and MPCA permits before the award of the construction contract. Well discharges are routed back into the existing storm sewer tunnel through existing shafts.

Portal open cut groundwater control: The Contract Documents requires the contractor to design and implement groundwater control at the portal worksite to allow them to choose the methods that best suit their preferences and capabilities. They are required to obtain the MnDNR Water Appropriations permit for this area.

Handling any Platteville limestone groundwater leaking from the back of the flat-backed tunnel. Previous experience has shown that groundwater can leak through vertical Platteville limestone joints. The Contract Documents contained provisions installing panning below the joints to divert nuisance water or urethane chemical grout injection in fractures to stop flow.

## Tunnel Flow Control

The proposed tunnel will connect to the existing tunnels at numerous locations. Flow from the existing tunnel needs to be controlled or completely isolated from the proposed tunnel construction. The Contract Documents contain the following flow control requirements to ensure the construction can progress:

No work is allowed in the Chicago Avenue tunnel during the summer months. Tunnel construction in the Chicago Avenue tunnel involves removing the existing tunnels, which are still active. Work is allowed in the winter when high rainfall is unlikely. The Contract Documents also baseline a number of days each month when the tunnel will need to be evacuated due to excessive flow from snowmelt or rainfall.

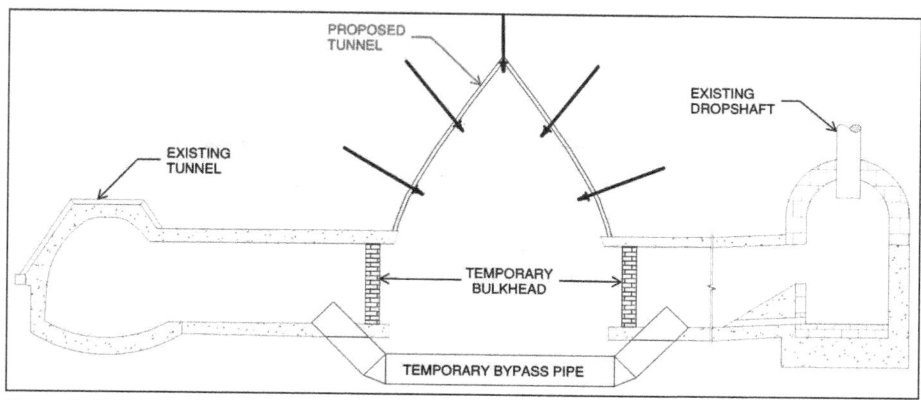


**Figure 9. Temporary flow bypass**

Partial bulkheads and bypass piping is required in the existing Chicago Avenue tunnel. The tunnel carries dry weather flow, which consists of groundwater leakage into the existing system and discharge from the proposed St. Peter Sandstone dewatering wells. Construction will progress upstream from the portal, removing the existing Chicago Avenue tunnels as the work progresses. Flow from upstream will need to be passed though the work to the river.

The proposed Washington Avenue tunnel is to remain isolated from the existing tunnel flow, allowing work to be done year-round. Isolation is accomplished by bypassing the flow from intersecting drifts under the proposed tunnel as shown in Figure 9.

## Permanent Structures

The permanent structures associated with the new Central City Parallel Tunnel include: (1) parallel tunnel, (2) tunnel junction structures, (3) cross-connects, (4) drill holes, and (5) outlet structure. All permanent structures are lined with a conventional, steel bar-reinforced, cast-in-place, concrete lining in direct, uniform contact with the surrounding rock.

### *Parallel Tunnel*

The parallel tunnel is the mainline tunnel between the structures. The tunnel is lined with a box tunnel section from the outlet to 2nd Avenue, including the tunnel junction structures, and a cathedral tunnel section from 2nd Avenue to Nicollet Mall, including the drill holes and cross-connects. Technical details that determined the tunnel sections have been presented previously.

### *Tunnel Junction Structures*

Tunnel junction structures serve to provide structural connection of the new parallel tunnel with the existing Central City Tunnel, Chicago Ave. tunnel, 2nd Ave. access adit, cross-connects, and drill holes. There are four tunnel junction structures. These structures are lined with a thickened concrete and heavily reinforced box tunnel shape. An example is Junction Structure #4 shown in Figure 10.

### *Cross-Connects*

Cross-connects serve to provide a cross passage or interconnect between the new parallel tunnel and the existing Central City Tunnel. There are four cross-connect

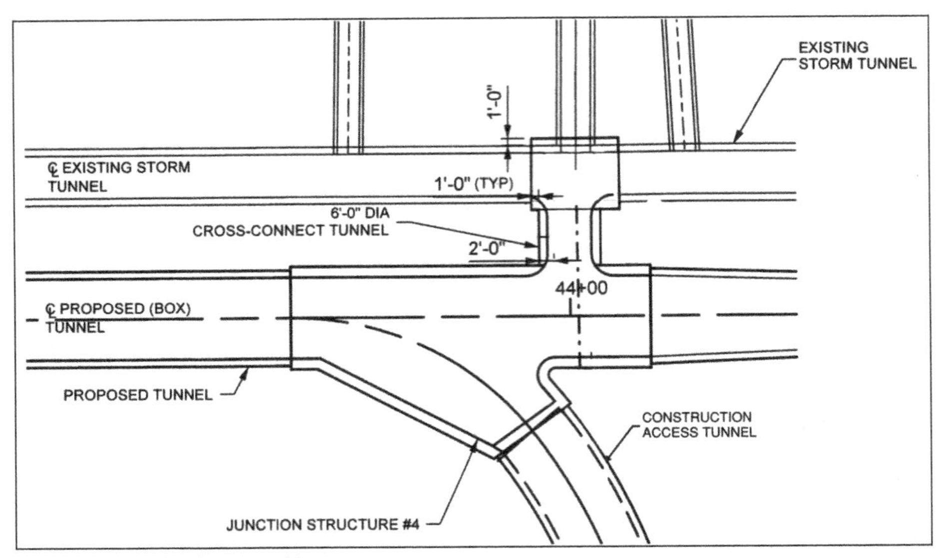


**Figure 10. Junction Structure #4**

structures. These structures are concrete lined with a cathedral or horseshoe tunnel shape.

### *Drill Holes*

Drill holes serve to provide hydraulic plunge flow drop vertical and horizontal intakes to the new parallel tunnel and the existing Central City Tunnel. There are seven drill hole structures connected to the new parallel tunnel. These structures are concrete lined with a cathedral or horseshoe tunnel shape depending on their location relative to the geologic conditions.

### *Outlet Structure*

The outlet structure is located at the downstream 128 LF of the tunnel and it serves to provide a smooth discharge flow transition from the tunnel to the Mississippi River flow channel and to protect against receiving channel erosion. It is partially submerged to provide energy dissipation and optimize the hydraulic grade line to prevent flow backup and tunnel surcharging. It is designed to eliminate the existing sump to provide maintenance-free operation and reduce unwanted access points. The outlet structure will be constructed in a cut-and-cover tunnel excavation and concrete lined with a box tunnel shape.

## CONTRACT RISK MANAGEMENT PROVISIONS (GBR, RR, EBD, DRB)

Contract risk management provisions are important for project risk management through identification, quantification, delegation, avoidance and mitigation strategies. The City of Minneapolis has taken proactive steps and implemented several provisions, including the Geotechnical Baseline Report (GBR), Risk Register (RR), Escrow Bid Documents (EBD), and Dispute Resolution Board (DRB).

### Geotechnical Baseline Report (GBR) Highlights

The Contract Documents include a Geotechnical Baseline Report (GBR) that provides a Project description, sources of geotechnical information, Project geologic setting, previous construction experience. Baselined elements include:

- Tunnel flow conditions, including tunnel evacuation days by month for high flows due to rainfall or snowmelt for areas exposed to existing tunnel flow
- Soil classification in soil excavation areas
- Bedrock unit elevations and locations along the alignment
- River level fluctuations
- Rock properties for each rock unit including compressive strength, density, jointing, bedding, sandstone gradation, and sandstone abrasivity
- Rock behavior, including standup time for unsupported rock
- Site hydrogeology, including groundwater levels in soil, Platteville Limestone, and St. Peter Sandstone; sandstone hydraulic conductivity; aquitard elevations

The GBR was developed and published in accordance with the guidelines presented in Essex (2007) and Wilson (2019).

### Risk Register (RR)

A Risk Register was developed to identify and mitigate major and minor risks to the project. Risks include technical issues, security issues, commercial issues, schedule delay issues and project procurement issues. The risk register was developed in accordance with the guidelines presented in Wilson (2019) and O'Carroll & Goodfellow (2015).

### Escrow Bid Documents (EBD)

Escrow Bid Documents requirements in the Contract were developed and implemented in accordance with the guidelines presented in Wilson (2019).

### Disputes Review Board (DRB)

Dispute Review Board requirements in the Contract were developed and implemented in accordance with the guidelines presented in Wilson (2019). Due to the relatively small size of the project, the use of a three-member or fewer member panel is provided as an option to potentially improve the effectiveness of a DRB for the project.

## CONCLUDING STATEMENTS

The Central City Parallel Tunnel will provide much needed capacity to the downtown Minneapolis storm drainage system. The unique geology of easily excavated sandstone with manageable groundwater conditions lends itself to effective tunnel construction. Considerations such as required connections to existing storm tunnels, equipment procurement, potential flooding from existing active storm sewer tunnels, stable tunnel shapes and available staging areas makes SEM tunneling the most viable construction option.

## ACKNOWLEDGMENTS

The Project was envisioned and funded by the City of Minneapolis Department of Surface Waters and Sewers, including site investigation, design studies, and project design were. The authors greatly appreciate the City managers and Legal Department for their diligent partnering in developing a successful design.

## REFERENCES

Barr Engineering (2015a). *Central City Tunnel System Feasibility Study: Central City Tunnel System Pressure-Mitigation Options*. Prepared for the City of Minneapolis.

Barr Engineering (2015b). *Central City Tunnel System Hydrologic and Hydraulic Modeling Using XP-SWMM: Central City, Eleventh Avenue, and Chicago Avenue Tunnel Systems*. Prepared for the City of Minneapolis.

CDM Smith (2018). *Preliminary Design Report—Central City Tunnel System*. Prepared for the City of Minneapolis, February.

Dittes, M. (2015). *Mechanical Properties of St. Peter Sandstone—A Comparison of Field and Laboratory Results,* M.S. Thesis, University of Minnesota, Minneapolis, MN.

Essex, R.J., ed. (2007). *Geotechnical Baseline Reports for Construction: Suggested Guidelines*, Reston, VA: American Society of Civil Engineers.

Norvitch, R. F., & Walton, M. S. (1979). *Geologic and Hydrologic Aspects of Tunneling in the Twin Cities Area, Minnesota.* Reston, Virginia: U.S. Geological Survey.

O'Carroll, J and R. Goodfellow (2015). *Guidelines for Improved Risk Management on Tunnel and Underground.Construction Projects in the United States of America (GIRM)*, UCA of SME, Englewood, CO.

Sanders, G., M. Gilbert, M. Khwaja and B. Lueck (2019). "Rehabilitation and Expansion of the Central City Tunnel System in Minneapolis, MN," *Tunneling & Underground Construction*, Vol. 13, No. 2, UCA of SME, pp. 22-28, June.

Wilson, S.H. ed. (2019). *Recommended Contract Practices for Underground Construction, Second Edition.* SME, Englewood, CO.

# Gateway Program—Cut and Cover Tunnels in Manhattan

**David Smith** ▪ WSP
**Matteo Ferrucci** ▪ WSP
**Drew Bazil** ▪ WSP
**David Pittman** ▪ Amtrak

## ABSTRACT

The Gateway program will provide additional rail capacity between New Jersey and Manhattan. This paper describes how a 70-foot-deep section of cut and cover tunnel was designed to directly support future high-rise towers. The tunnel will link previously completed tunnel sections under Hudson Yards to future tunnels under the Hudson River. The design aims to minimize constraints on future overbuild and future tunnels, while maximizing constructability within a small worksite. The tunnel will pass under the High Line elevated walkway, next to an active rail yard, and close to a subway tunnel.

The design was developed by a tri-venture of WSP, STV and AECOM. WSP led the design of major underground works; STV designed utility relocations and the underpinning concept for the High Line; and AECOM designed temporary in-tunnel systems. Construction is expected to start in 2023.

## INTRODUCTION

The Hudson Yards Concrete Casing—Section 3 Tunnel (HYCC-3), when constructed, will continue to preserve an underground right-of-way on Manhattan's west side for a future rail connection into Penn Station, New York (PSNY). This section of tunnel is bounded by 11th Avenue and West 30th Street within the western railyard area of the Hudson Yards (Figure 1). The cut and cover tunnel will create an approximately 400-ft long reinforced concrete structure to accommodate two railroad tracks (Figure 2). The final design for this project is complete. Construction is expected to begin in 2023, with an anticipated construction duration of about three years.

## GATEWAY PROGRAM

The Gateway Program is a comprehensive program of phased strategic rail infrastructure improvements designed to preserve and improve current rail services and create new capacity that will allow the doubling of passenger trains on Amtrak's Northeast Corridor (NEC) between Newark, New Jersey, and PSNY. The objective of the Gateway Program is two-fold: (1) to modernize existing infrastructure, including repairing infrastructure elements that are damaged due to age or events such as Superstorm Sandy; and (2) to increase track, tunnel, bridge, and station capacity, eventually creating four mainline tracks between Newark and PSNY to allow the doubling of passenger trains in this section of the NEC.

## PREVIOUS GATEWAY PROGRAM CONSTRUCTION

In cooperation with the MTA Long Island Rail Road (LIRR) and Related Companies (a developer that holds a 99-year lease to develop an overbuild on the two city blocks that comprise the rail yards), Amtrak delineated an underground easement for the Hudson

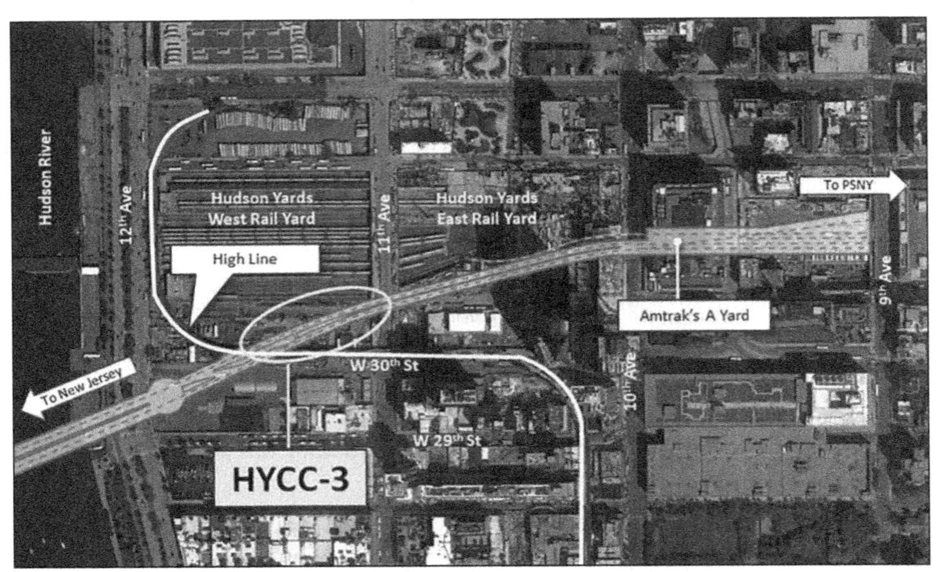


Figure 1. Location of Hudson Yards concrete casing (Section 3) Project

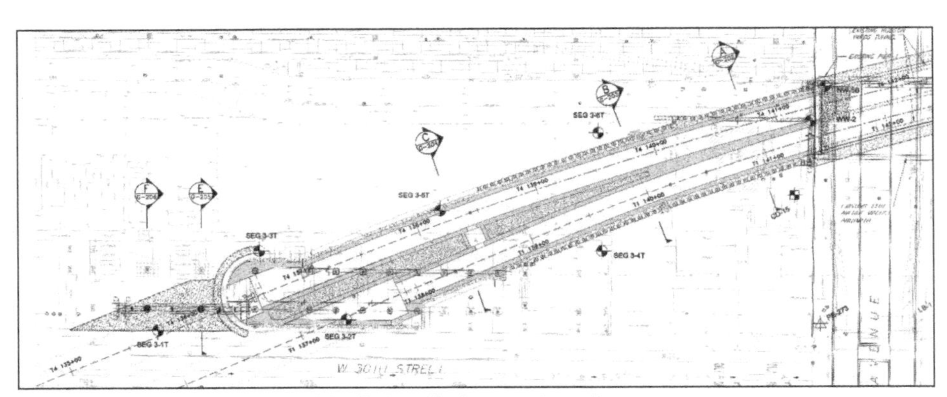

Figure 2. Plan of final structure of the Hudson Yards concrete casing

Yards Concrete Casing tunnel structure within the Hudson Yards. Construction of the first and second reinforced concrete box tunnel sections began in October 2013 and was completed in May 2018 (including punch list and closeout). This infrastructure is bound by 10th and 11th Avenues, between West 33rd and West 30th Streets. The two-cell tunnel is approximately 925 ft long by 52 ft wide by 35 ft tall and was constructed by the cut and cover method. The first section, between 10th and 11th Avenues, was constructed utilizing secant piles as the support of excavation system providing a strut free area within the excavation (Figure 3). The tunnel was designed to support massive loads imposed by overbuild foundations within the tunnel footprint, as coordinated with the overbuild developer. The second section, below the footprint of the overhead 11th Avenue viaduct, was constructed utilizing a support of excavation system of jet grout and mini piles due to the low headroom of the overhead superstructure. The viaduct superstructure was temporarily underpinned, a pier was removed, and the tunnel box was constructed before rebuilding the pier integral with the tunnel roof (Figure 4).

Figure 3. 2015 Aerial view of Hudson Yards looking west from 10th Avenue

Figure 4. 11th Avenue viaduct underpinning box girder and overburden excavation activities during construction of the second Gateway Tunnel section

## PROCUREMENT PROCESS

HYCC-3 is following a design-bid-build project delivery method. Selection of a contractor is anticipated in 2023. Contractor procurement will be performed collaboratively between Amtrak and the site Developer, in coordination with LIRR. Construction of HYCC-3 will maintain full activity of the adjacent LIRR rail yard and maintenance of equipment facilities, and may be concurrent with construction of the overbuild development and other onsite capital improvements, requiring detailed planning and coordination for laydown space, contractual interfaces, and the responsibilities of each party.

## COORDINATION WITH EXTERNAL PARTNERS

The HYCC-3 tunnel will be excavated entirely within an existing active LIRR railyard, in the vicinity of various existing structures:

- 11th Avenue is carried on an elevated structure over the LIRR West Side Storage Yards. This viaduct is owned and maintained by the New York City Department of Transportation (NYCDOT) and carries six lanes of southbound vehicular traffic with two outside sidewalks. The superstructure consists of longitudinal steel girders, transverse bracing, and a reinforced concrete bridge deck.
- The NYCT No.7 Line tunnels are aligned under 11th Avenue. They were excavated in rock using a tunnel boring machine (TBM), have a diameter of 21 feet-2 inches, and are lined with 10-inch thick precast concrete segmental liners. The crowns of these tunnels are approximately 90 feet below the grade of the LIRR Yards. The closest vertical distance between the extrados of the westernmost #7 Line Tunnel and the proposed excavation for the HYCC-3 tunnel is approximately 35 feet.
- The High Line was built in the 1930s as part of a public-private infrastructure project to carry freight trains along the west side of Manhattan. It accommodated freight traffic on an elevated steel structure 30 feet above the ground, removing trains from the streets of Manhattan's largest industrial district. Now converted into a park and walkway, it is much-loved by locals and tourists and must generally remain open throughout construction. This steel structure spans over 11th Avenue just north of West 30th Street within the project limits. The foundations of the High Line in this area are either large diameter concrete piers founded on rock or groups of timber piles with concrete caps.

Multi-milestone reviews of the contract documents have been conducted by the owners of the above-mentioned structures during the design phase. Similarly, during the construction support phase a joint submittal review process between Amtrak's Owner Representative, the EoR and the external partners will be conducted, as was the case for the previously constructed Gateway Tunnel sections.

The LIRR owns and operates the rail yard at the surface. The existing South Access Road is being utilized by LIRR personnel to support active rail operations at the West Rail Yard and needs to be maintained active at all times. This existing access road overlaps with a portion of the excavation at the excavation's northeast corner. To guarantee a 24/7 uptime of the South Access Road, the HYCC-3 contractor will be required to design and construct a temporary access bridge over the corner of the open excavation. A temporary detour road to the south side of the LIRR Access Road will be necessary for the installation of the temporary access bridge, and subsequently again during the removal of the temporary access bridge after the tunnel construction is completed. In addition, a second temporary vehicular bridge will be located midway along the excavation spanning from north to south.

## UTILITIES AND ANCILLARY STRUCTURES

The existing LIRR Emergency Service Building (ESB) falls within the footprint of the HYCC-3 excavation and must be relocated to allow for cut and cover tunnel construction. As part of an ongoing construction contract, a new interim ESB compound is being constructed south of the proposed HYCC-3 tunnel footprint. Demolition of the existing LIRR ESB and associated utilities is included in the HYCC-3 project scope.

All utility services are to be maintained throughout construction. Numerous underground utilities require maintenance and protection, and some will need to be supported across the open excavation. Others will require relocation and/or replacement during construction. In the final stages of construction, most utilities will be restored to their original configuration and alignment.

The contractor is required to provide and maintain a safe and continuous five-foot wide minimum pedestrian path between the South Access Road and the interim ESB compound for LIRR personnel to utilize and maintain their relocated equipment.

### Storm Water

A 43-inch by 68-inch elliptical reinforced concrete pipe (ERCP) sewer will be intersected by the open-cut excavation. During construction, the sewer will be diverted into a temporary 48-inch steel pipe sewer that will be suspended from the South Access Road temporary bridge and support of excavation (SOE) walls. After completion of the tunnel, the original sewer will be reinstated.

### Sanitary and Domestic Water

An existing 8-inch sanitary force main running east-west along the site at the northeast corner will intersect the SOE wall alignment. If there is direct interference with the mini piles, the sanitary force main will be locally offset to facilitate the installation of the mini piles. A 6-inch domestic water line will be diverted away from the SOE walls with bends and fittings by the LIRR ESB Relocation Contractor.

### AC Power and Commination Lines

There are medium voltage AC power and communication duct banks within the open excavation limits of the HYCC-3 tunnel construction. New duct banks are being constructed by the LIRR ESB Relocation Contractor to generally bypass the HYCC-3 area. The existing duct banks will be deactivated and demolished during the open excavation work by the HYCC-3 Tunnel Contractor. Some of the AC and communication conduits of the relocated LIRR ESB are routed and suspended over the tunnel excavation on a utility bridge and need to be maintained and protected during HYCC-3 construction. A jet grout curtain is proposed in the secant pile wall (or slurry wall) at the location of the utility bridge to create closure below the utilities.

## SUBSURFACE CONDITIONS

Various subsurface investigation programs have been conducted over the years within or in the proximity of the HYCC-3 project limits, including investigations for HYCC-3 (performed in August 2015), 11th Avenue Extension, NYCT No. 7 tunnel construction, and the Trans-Hudson Express Tunnel (THE Tunnel).

The thickness of overburden soils along the project Section 3 alignment ranges from 30 feet to over 80 feet. Review of existing subsurface investigation data indicates the overburden stratigraphy generally consists of the four strata described below, listed in descending order:

- *Stratum F (Fill):* urban fill consisting of a generally loose heterogeneous mixture of mostly sand, with silt, gravel, and miscellaneous debris, such as rock fragments, concrete, brick, cinders and roots.

- *Stratum S (silt, silty sand, sand):* brown to red brown medium dense to dense, inorganic silt, silty sand to sand. This stratum is discontinuously found below the fill and above the estuarine deposits (where present) and above the bedrock.
- *Stratum C (silty clay/clay):* 17 feet to over 40 feet thick estuarine deposit consisting of generally soft normally consolidated, predominantly inorganic gray to dark gray, silty clay to clay with trace amounts of fine-grained sand and shell fragments.
- *Stratum T (glacial till/decomposed rock):* relatively dense materials from either glacial till or decomposed rock. Recovered soil samples in this stratum consist of gray fine to coarse sand, little silt or clayey silt with trace to some gravel with SPT N-values ranging from 20 bpf to refusal (>50 bpf).
- *Bedrock:* excavations along the tunnel alignment within the HYCC-3 are expected to encounter several distinct rock types, principally mica schist, granitic rocks, and pegmatite. Mica schist occurs as medium gray or dark gray to black, fine to medium-grained or fine to coarse-grained. Granitic rocks occur as pink to gray granitic rocks consist primarily of granodiorite but also including true granite, quartz monzonite, and quartz diorite Pegmatites typically occur as 5- to 10-foot thick lenses or veins within the mica schist and within the granitic rocks.

Groundwater levels observed within the overburden soils in borings drilled inside the Hudson Yards during Hudson Yards Tunnel design phase are typically 5 to 7 feet below existing ground level. Groundwater levels can be expected to approach ground surface during periods of heavy sustained rainfall along the entire alignment.

The contractor will be required to perform sub-bottom cores spaced at 10 feet max along the SOE to verify top of rock elevation, rock quality and permeability to inform the final design of the SOE.

## INSTRUMENTATION AND MONITORING

Underground construction has the potential to impact adjacent structures and facilities by inducing movement of the ground supporting existing foundations. The construction of the HYCC-3 tunnel will require installation of SOE elements, soil excavation, mechanical excavation and controlled blasting of rock, and jet grouting operations, all contributing to potential settlements/movements or vibrations. A comprehensive geotechnical and structural instrumentation and monitoring program will detect movement and vibrations.

The HYCC-3 Contract Documents include minimum requirements to maintain, protect, instrument, secure and, if necessary, safely support or underpin existing facilities and structures including the High Line structure, the 11th Avenue Viaduct, the Eleventh Avenue Hudson Yards Tunnel Section, the NYCT No. 7 Line Subway Tunnel, substations, switch gear, LIRR operating track and associated signal/control systems, and underground utilities.

Table 1 includes the different type of instrument required by the Contract Documents, together with their purpose and location.

The instrumentation and monitoring program also requires pre-and post-construction condition surveys for all structures within or adjacent to the West Rail Yard (HYCC-3). Crack gauges will be installed at existing cracks identified during the pre-construction condition survey.

**Table 1. Type, purpose, and location of geotechnical and structural instruments**

| Instrument | Purpose | Structure or Facility |
|---|---|---|
| Deep Benchmark (BM) | Establish and maintain a stable survey point reference for other monitoring instruments | N/A |
| Deformation monitoring points | Monitor vertical and horizontal movement | ▪ Pavement<br>▪ 11th Avenue Viaduct deck<br>▪ High Line Viaduct deck<br>▪ Tracks |
| Tiltmeters (TM) | Measure the change in inclination of structural members such as walls and support columns | ▪ High Line Viaduct columns,<br>▪ 11th Avenue Viaduct pier and abutment. |
| Inclinometers | Monitoring deformations and lateral movement of support walls towards excavation | ▪ Soil and rock surface<br>▪ Excavation support wall |
| Geophones/Seismographs | Measure the vibration waves generated by blasting propagating through the ground and structures | ▪ Hudson Yards Tunnel<br>▪ 11th Avenue Extension<br>▪ High Line Viaduct<br>▪ No. 7 Line Subway Tunnels<br>▪ 11th Ave. Viaduct |
| Dynamic Strain Gauges (DSG) | Monitor the actual strain (or relative displacement) of structural elements during blasting | ▪ 11th Ave. Viaduct<br>▪ High Line Viaduct |
| Crack Gauges (CG) | Monitoring cracks in concrete or plaster or brickwork | ▪ 11th Ave. Viaduct<br>▪ High Line Viaduct<br>▪ No. 7 Line Subway Tunnels |
| Top of Rail Monitoring Point (TOR) | Monitor rail movement adjacent to excavation | ▪ LIRR railroad tracks |
| Observation Wells (OW) | Monitor groundwater levels outside excavation | ▪ Ground adjacent to excavation |
| Vibrating wire Strain Gauges | Monitoring strut loads during excavation | ▪ Minimum of three excavation support wall strut arrays within the High Line area at every strut level |

The Contract Documents require the Contractor to interpret the instrumentation data collected by making correlations between instrumentation data and specific construction activities and careful evaluation to determine whether the response to construction activities is as expected. The Contract Specifications include predetermined Review and Alert Levels for the various instruments and/or structures being monitored. Prior to the start of demolition and construction, the Contractor is required to submit a detailed plan of action to include specific actions, responsible parties, and timeframe to be followed in the event of an exceedance.

## SUPPORT OF EXCAVATION SYSTEM

The HYCC-3 Contract Documents provide an indicative design for the support of excavation (SOE) system. The design of the SOE will be finalized by the contractor, per criteria defined in the contract. The indicative design envisions three different support-wall types: slurry walls, secant pile walls and jet grout walls.

### Slurry Walls

Slurry walls are proposed at the west end of the structure. The selection was informed by:

- Depth to top of rock (up to 75-feet)
- Low-headroom under the High Line viaduct (approximately 18-feet at beams)

Reinforcement in slurry walls (also referred to as soldier-pile-tremie-concrete (SPTC) walls) could be wide-flange steel sections or rebar cages. Wide-flanged steel sections can have significant bending capacity and allow for relatively easy connection of walers and temporary cross-lot bracing.

The support wall at the west end of the structure is arched in plan which minimizes the bracing requirement in the deepest part of the excavation. Steel reinforcement is not permitted in the part of the arch where a future tunnel will be mined into the HYCC-3 box. It is anticipated that glass-fiber reinforcement or plain concrete will be used.

### Secant Pile Walls

Secant pile walls are proposed at the east end of the structure. This selection was informed by:

- Depth to rock of between 30-feet and 50-feet
- Sequential stages of wall construction to maintain vehicular access for LIRR
- Gaps in the secant-pile wall at a utility crossing
- Multiple changes in wall direction
- Limited space for slurry separation plant

Reinforcement could be steel sections or rebar cages.

### Jet Grout Walls

Jet grout walls are proposed above the existing box tunnel (with minipile reinforcement), at utility crossings, and at other discontinuities in the perimeter walls. This selection was informed by:

- Ability to create jet grout closure panels below utilities
- Limited headroom under 11th Avenue viaduct
- Avoiding damage to the existing tunnel box under 11th Ave

### Support of Excavation Requirements

Slurry walls and secant piles are required to penetrate at least 2-feet into sound rock. This will establish toe fixity for the support of excavation system and promote an effective groundwater cut-off. The support walls are required to restrict water ingress, thereby inhibiting drawdown-induced settlements in any adjacent compressible soils. If permeability of the rock mass, based on packer tests at each panel location, is below the contractual threshold, then rock mass grouting is required to limit potential seepage below the wall. This could be performed through steel pipes tied to the core beams installed in the secondary piles.

In areas where rock excavation will be required, secant pile walls and SPTC walls will be offset a minimum of 2'-6" from the permanent box structure. This will provide a rock ledge that will help maintain the stability of toe of the wall.

The support of excavation walls will be combined with appropriate bracing and/or tieback systems to provide the necessary stiff lateral support to limit settlements caused by wall deflections. Internal bracing systems are likely to be the main method of supporting the support of excavation walls, especially at the deeper, western, end of the structure. Tiebacks can provide better internal construction clearance for construction of the tunnel structure, but are subject to various constraints, including that they must remain within railroad property, and avoid the existing footings of the High Line. Tiebacks are temporary elements but could impact construction of future deep foundations. To minimize their footprint, they are prohibited above the elevation of the tunnel roof slab and must be installed at inclination of no less than 45-degrees.

As excavation proceeds, support walls will span vertically between the previously installed level of struts or tiebacks, and a passive-resistance wedge in the unexcavated ground. The low strength of the soft clay could result in high displacements of the passive wedge, and high bending moments in the walls. To reduce these moments, the soil between the north and south support walls may need to be strengthened. Jet grouting could be used to create either a continuous block or struts. The jet grout would be excavated as excavation proceeds.

The designers generated a 4-dimensional schedule-loaded Virtual Design and Construction (VDC) model of the temporary and permanent works using Navisworks by Autodesk (Figures 5 and 6). This model allowed the designer to better visualize and resolve constructability challenges and optimize suggested construction sequences. The VDC model shows an indicative design for the temporary support of excavation (SOE) elements and other temporary works which is one possible solution both in terms of the geometry/sizes as well as sequence of construction.

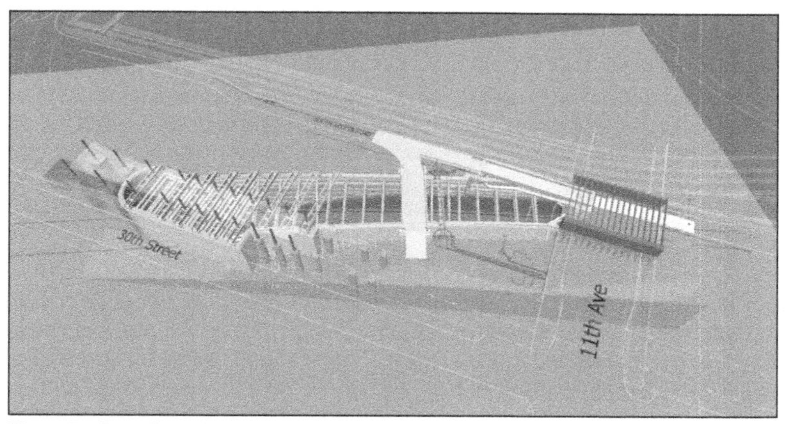


**Figure 5. Overall spatial domain of VDC model**

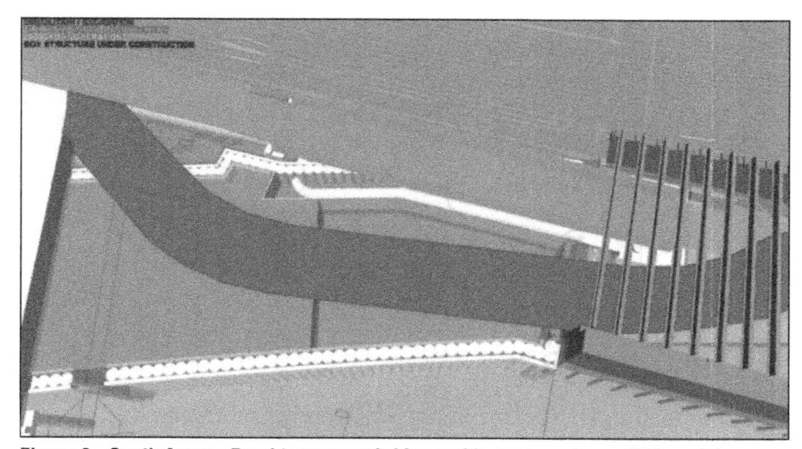

**Figure 6. South Access Road temporary bridge and bypass road, per VDC model**

### Overburden Excavation

Construction of the support of excavation walls, and subsequent bulk excavation, may encounter large obstructions, including manmade materials, previously deposited or abandoned construction debris, existing support of excavation wall system and caissons. At the west end of the structure, in-situ soft ground below invert level will need to be over-excavated to top of sound rock and replaced with lean concrete to provide an evenly stiff bearing surface for the HYCC.

### Rock Excavation and Support

The anticipated top of rock elevation is higher than the tunnel invert slab along most of tunnel box. Rock excavation will use mechanical and/or controlled blasting techniques. Channel drilling or reduced charges could be used to restrict blasting vibrations and strains in adjacent structures. Rock excavation vibrations will be closely monitored by a comprehensive array of instruments installed on adjacent structures.

The stability of the rock faces will be maintained by installing pattern rock bolting with shotcrete. Rock mass grouting may be required to reduce any significant groundwater flows emerging from discontinuities within the rock mass. Pre-excavation rock mass grouting was successfully used in the previous Hudson Yards excavations.

### Ground Improvement Outside the HYCC-3 Excavation Limits

All surface construction work required for the Gateway Program, within the West Rail Yard site, needs to be performed as part of HYCC-3 Contract. The HYCC-3 contract therefore includes some preparatory works for future Hudson Tunnel construction. Jet grouting (from current grade) will be required within a portion of soft soil outside of the cut and cover tunnel western limit, such that future tunnels can be mined through.

## REINFORCED CONCRETE FINAL STRUCTURE

The tunnel structure will accommodate two railroad tracks alignments, T1 and T4, with lateral spacing that increases from east to west. These will be achieved with a permanent reinforced concrete box structure. The east end will be a two-cell structure that connects with the existing tunnel structure under the 11th Avenue Viaduct. Further west, it widens and transitions into a three-cell box structure. The middle cell will be backfilled with Controlled Low Strength Material (CLSM). The west end of each outer cell steps down and widens to accommodate future sequential excavation tunnels from the west.

### Design Provisions for Future Overbuild

The final reinforced concrete structure is independent of the support of excavation system. It is designed to resist the following primary loads:

- Soil, rock, and groundwater
- Buoyancy
- Future overbuild loads
- High Line viaduct loads

Other loads, including seismic loads, are also considered.

The Hudson Yards developer (Related Companies) plans to construct a building west of the 11th Avenue viaduct and north of High Line that will be supported on the HYCC-3 tunnel structure. This development, while not yet programmed and designed, is anticipated to include a high rise(s) comparable to the completed overbuilds located on the eastern block of Hudson Yards between 11th Avenue and 10th Avenue. The overbuild loads are unusual, both in their magnitude (column loads up to 16,000 kips service load) and distribution (the future high-rise structures are not yet designed, and the column layout is not known). The future developer worked collaboratively with Amtrak to develop the design criteria for the tunnel and future overbuild. While seeking to minimize the limitations on future overbuild design, the parties recognize that design criteria are necessary. These include:

- On the tunnel roof
  - Delineated loading zones
  - Maximum primary-column load
  - Maximum average load
  - Maximum lateral loads (limited to 1% of gravity loads)
- Below the tunnel box
  - Maximum downward bearing pressure on rock
  - No uplift loads to act on the tunnel box
- Adjacent to the tunnel box
  - Delineated loading zones
  - Limits on future excavation
  - Influence lines for deep foundations

The railroad alignments slope down to the west at approximately a 2% grade. If the tunnel roof and invert slab were sloped, vertical loads from overbuild structures could result in lateral/sliding component forces. To avoid this, the roof and invert are stepped. Almost all lateral loads (wind, seismic) acting on the future overbuild structures will be transferred to adjacent structures founded on rock. However, the associated vertical tension in certain building columns may need to be transferred to bedrock below the tunnel. The tunnel box design makes provision for tie-down anchors by including steel tubes at regular intervals along the exterior and interior walls. In order to protect the tunnel roof from the impacts of SOE for the excavation work for the future overbuild, Amtrak directed the tri-venture to design temporary retaining walls supported on the HYCC-3 roof at specific areas according to the preliminary footprint of the planned building cellar footprint.

The west end of the tunnel box will support the High Line viaduct (Figure 7). Although the High Line is now a walkway, the tunnel is designed to accommodate railroad loads on the High Line, per legal requirements.

3D finite-element analyses of the cut and cover tunnel were performed using SAP 2000 (software by Computers and Structures, Inc). Given the unknown configuration of the future overbuild columns, multiple load cases were considered to identify the worst-cases.

Final design of the reinforced concrete structure was performed in accordance with ACI 318. Earlier design stages had used ACI 350, which limits flexural stresses and controls the width of cracks. While this can enhance structural durability, it resulted in undesirable rebar congestion. It was determined that the equivalent protection could be provided by using an external waterproofing membrane in combination with internal waterstops and grout tubes. Cover concrete of 3-inches to primary bars will enhance durability and promote concrete flow around areas with multiple layers of rebar.

The tunnel invert slab is generally 7'6" thick, and the tunnel roof is generally 10'-0" thick with local deviations. To avoid potentially damaging heat of hydration temperature developing, the contractor will be required to develop and comply with an approved

**Figure 7. View of the High Line structure looking northeast from West 30th Street**

thermal control plan. Concrete will have a compressive strength of 6,000 psi and primary reinforcement will have a minimum yield strength of 80 ksi. Up to three layers of flexural reinforcement is required. The higher-than-typical reinforcement grade reduces rebar congestion and results in a significant reduction in the in-place cost.

The design anticipates that the permanent structure will include temporary windows (holes) in the perimeter and interior walls to accommodate temporary bracing. After the permanent structure is completed, and any gap between the exterior walls and SOE walls is filled, the temporary struts will be removed and the windows will be filled.

### Temporary In-Tunnel Systems

The design requires the installation of temporary lighting, ventilation, and a sump pump. These systems will allow for safe maintenance of the structure until permanent systems are installed as part of the larger Gateway Program.

## UNDERPINNING OF THE HIGH LINE

The cut and cover box will be partially constructed below the High Line viaduct. Fifteen foundations for nine pier-bents of the High Line fall within the excavation limits of the HYCC-3 structure. These will be underpinned to allow cut and cover construction of the tunnels to proceed directly below. An indicative underpinning design has been developed and presented in the Contract Documents. This design proposes that braced pairs of underpinning beams will span between the north and south excavation support walls as shown in Figure 8. Each pair of beams will be centered on a High Line bent. After the columns are disconnected from their foundations, the load from each column will be jacked into the underpinning system during short night-time closures of the walkway. This will allow the existing footings to be removed, and for excavation and tunnel box construction to proceed.

After completion of the tunnel box, support plinths will be constructed on the roof at each column location. Column loads will then be transferred onto the permanent structure, and the underpinning system will be removed, as seen in Figure 9.

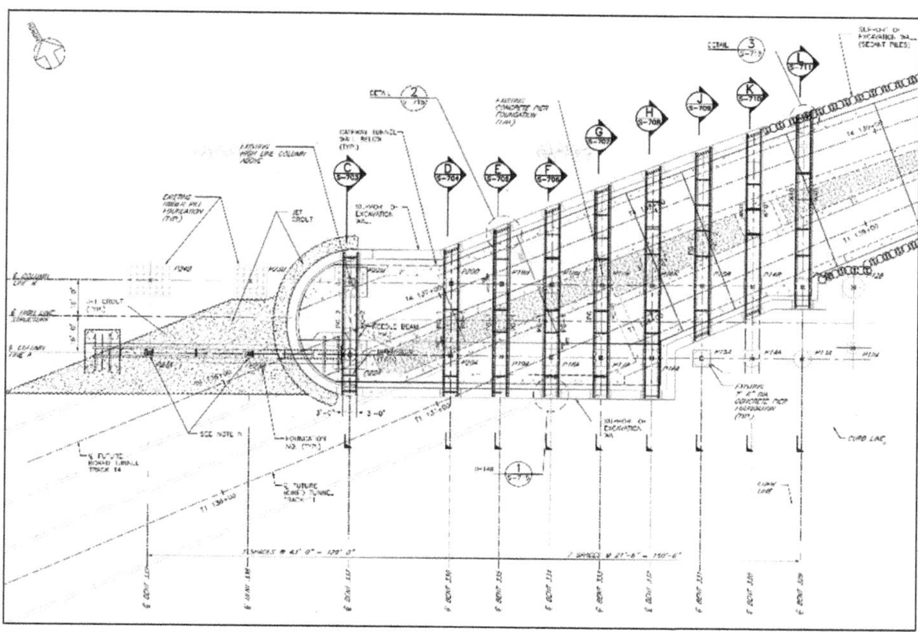

Figure 8. Proposed underpinning framing plan

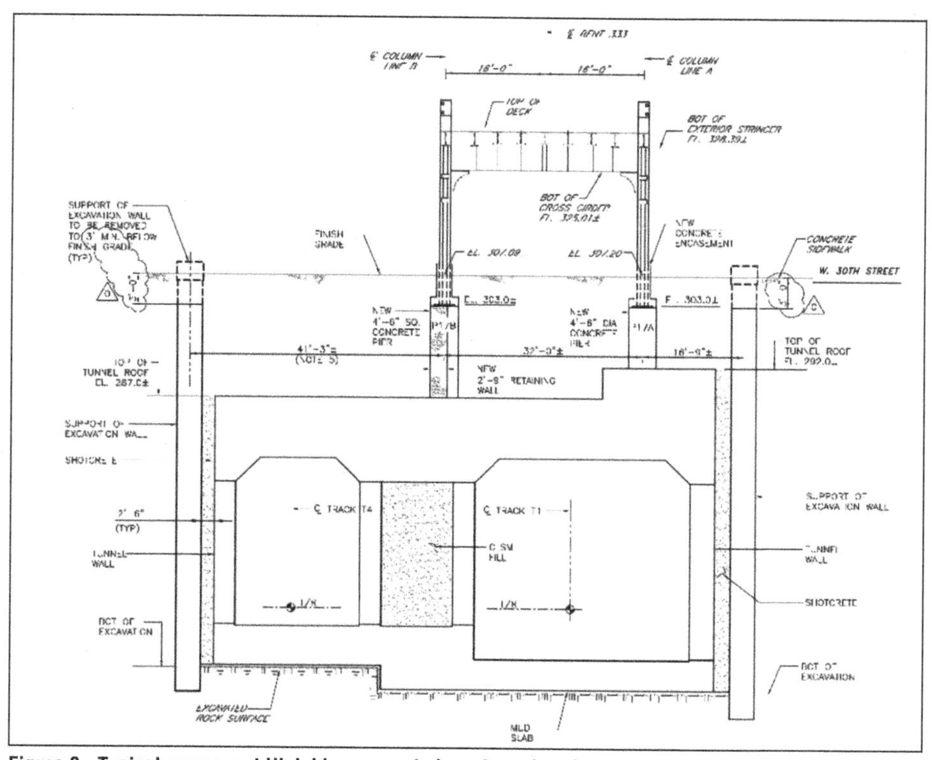


Figure 9. Typical permanent High Line support above tunnel roof

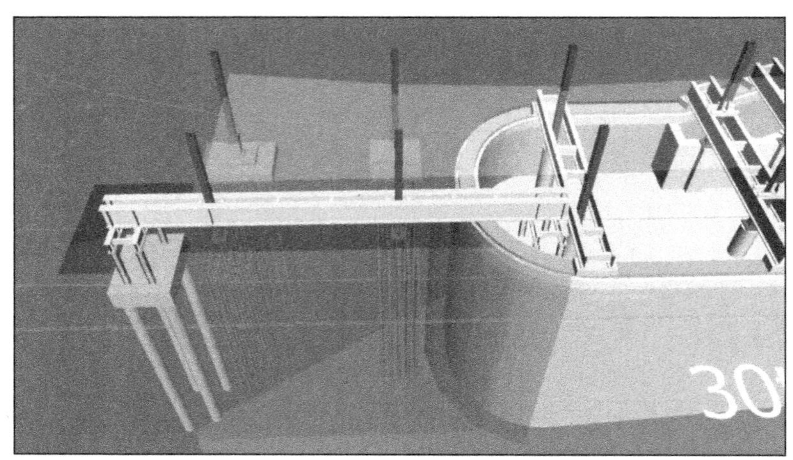


**Figure 10. VDC-model excerpt of temporary load transfer beam outside HYCC-3 excavation**

Immediately west of the cut and cover box, two existing High Line column footings are located within the alignment of future Sequential Excavation Method (SEM) tunnels. To enable these footings to be removed, the HYCC-3 box contractor will install a new piled foundation outside the footprint of the future tunnels. After completion of the main HYCC-3 box, a pair of temporary load transfer beams will be installed spanning from the new pile-cap to the HYCC-3 box roof (Figure 10). Load in the existing footings will be jacked into the transfer beams. A reinforced concrete below-grade beam will permanently support the two columns.

## NEXT STEPS

Award of the HYCC-3 contract is anticipated in early 2023. The contractor will commence design of the SOE system, with construction starting as soon as possible following NTP. Construction is expected to last approximately three years.

Other sections of the Gateway Program are also expected to be procured shortly after HYCC-3, including the Tonnelle Avenue Bridge, the Palisades Tunnels, the Hudson Tunnels and the Manhattan Shaft and Tunnels (referenced herein for connection to HYCC-3).

## ACKNOWLEDGMENTS

The authors would like to thank Amtrak for the permission to publish this paper. In addition, the contributions to the project from the design team, the Long Island Rail Road, Related Companies and other stakeholders are gratefully acknowledged.

# Gravity Sewer Tunnel Liner Corrosion Protection—Part Two

**Jon Y. Kaneshiro** ▪ Parsons Corporation
**Pooyan Asadollahi** ▪ Parsons Corporation
**Eric Dawson** ▪ Parsons Corporation
**Steven Hunt** ▪ Black and Veatch

## ABSTRACT

This paper provides updates to the paper published by the lead author at RETC 2011 with a summary of developments in technologies, products, and approaches to corrosion analyses of gravity sewer tunnels and microtunnels. This paper presents approaches in evaluating degradation of reinforced concrete liners. Also, the carbon footprint of conventional concrete liners and alternative "green" liners are compared and their life cycle costs are considered. A few case histories from the past decade are provided indicating the latest and emerging trends in liner protection.

## INTRODUCTION

Kaneshiro et al. (2011) (also referred here as the 2011 RETC paper) presented external corrosion considerations, an overview of the pros and cons one-pass versus two pass liner alternatives in corrosion protection and the options for thin membranes embedded in one-pass segmented liners, properties of polymer concrete nascent emergence as a one-pass liner, a review of various corrosion resistant carrier pipes, and case histories of corrosion resistant liners at the time of its writing. Joye et al. (2016) also presents durability design requirements of underground structures.

This paper attempts to update external corrosion considerations, discuss internal corrosion considerations and the effects of hydrogen sulfide concentrations on corrosion potential, make corrections, further discuss developing technologies, and present state-of-the-practice (precedence) since 2011 for providing corrosion protection for gravity wastewater tunnels and micro tunnels.

## EXTERNAL AND INTERNAL CORROSION REVISITED

The 2011 paper discussed evaluating external corrosion and the migration of chlorides (e.g., in saltwater environs) through multiple layers: ground, grout, and concrete. Internal corrosion was discussed in terms of alternatives for internal embedded plastic liners (fully tanked situation) and other state-of-the-art materials to resist sulfuric acid attack created by sewer gases. Some additional discussions are provided in this paper in terms of external migration of corrosive chloride and sulfide ions, methodology for evaluating internal corrosion, and updates on the state-of-the industry.

### Influent Molecule Size and Time/Diffusion Rate

When evaluating the groundwater quality, in terms of the concentration of sulfides and chlorides that would penetrate the concrete over the design life of the structure, it is important to understand the permeability of the concrete and the pore space diameter. All conventional Portland cement concretes are permeable/porous, although infinitesimally so (e.g., as low as $k < 10^{-18}$ m/second and voids in the $10^{-7}$ to $10^{-11}$ m in diameter). Consideration of available concrete materials and their pore space is required;

such as Type I-II cement, pozzolan fly ash (PFA), ground granulated blast furnace slag (GGBFS), and silica fume. The culprits for external corrosion are the migration of chlorides and sulphates in the groundwater (and the ground) through the concrete and access to oxygen. As shown in Figure 1, the corrosive ion sizes are more than four orders of magnitudes smaller than the pore sizes of typical Type I-II cement and PFA. GGBFS and silica fume have pore space that are three and two orders of magnitude larger than the typical corrosive ions, respectively. Figure 1 also shows ultrafine and microfine cements typically used for permeable rock fracture or permeable sandy material grouting (also, with three to 3.5 orders of magnitude). However, flow of ions through the mezopores depend on more than just pore diameter, it is a complex process with many variables, including, to name a few, the viscosity of the medium carrying the ions, diffusivity of ions, temperature, ion binding properties, and surface ion concentrations.

The penetration of ions may be taken as a series of flow paths through various media following Darcy's law as shown, for example, in Figure 1 of the 2011 RETC paper. Knowing the permeability and the corrosive ion concentration (e.g., in parts per million), it is possible to calculate theoretically the time to full saturation of corrosive ions as well as groundwater leakage (Doran et al., 1987).

The key would be to stop the ion flow as well as the groundwater flow. This is possible by providing an external coating (bitumen, polyurethane or epoxy), an internal impermeable membrane (embedded plastic or steel liners). High density polyethylene (HDPE) embedded plastic liners available on the market (standard anchor spacing) have shown that the external hydrostatic pressure can only withstand about 1.5 bars. Casting a steel liner within a large diameter opening would be very expensive. A

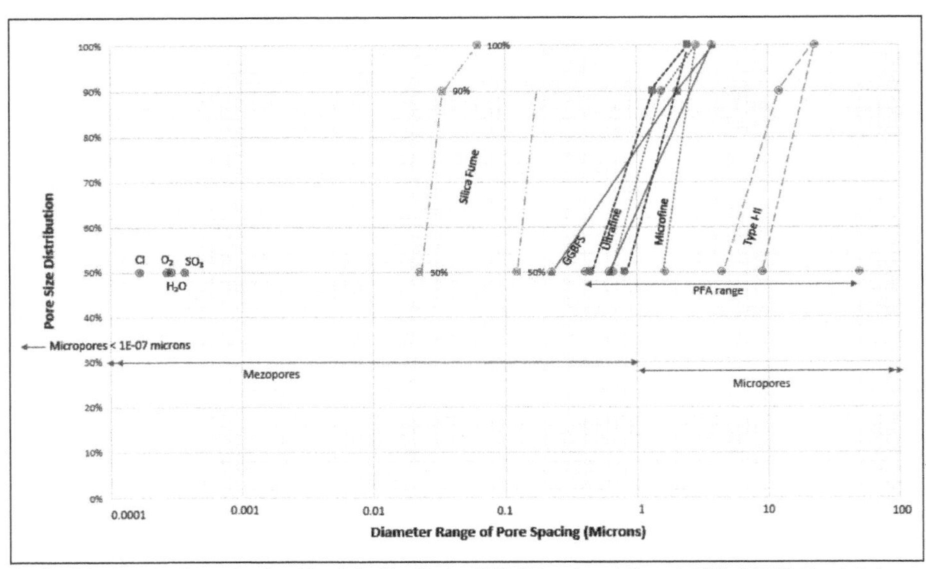


1. Particle size: ultrafine, microfine, Type I-II, after Clarke, 1982; Shimoda and Ohmori, 1982; GGBFS, Yusuf et al. 2014; Silica fume: Panesar (2019); PFA: USACOE (1961)
2. Pore diameter dependent on particle packing arrangements: between 0.225/0.285 and 0.414 of the particle size (Gupta and Larson, 1979; Wu, 1987)

**Figure 1. Cementitious material particle size, pore size, versus corrosive ion sizes**

practical alternative that might be considered to fill the tiny pores (and microcracks) is the use of an admixture that reacts with water and crystallizes to fill voids (e.g., manufactured by Xypex, Kryton, Penetron, Sika, Master Builders, Mapei).

## Spalling

The presence of corrosive ions (e.g., Cl) in the groundwater and their rates of diffusion impact the service life of reinforced concrete tunnel liners and selection of the type of reinforcing. Without a protective layer on the extrados of the liner, Fick's 2nd law of diffusion (ca. 1855) determines the rate of transfer of free molecules or ions (e.g., Chlorides) in a fully saturated medium from regions of higher to lower concentration. Fick's law simplifies in tidal/splash zone marine conditions, which is analogous to extrados of the tunnel liner in a marine environment, to:

$$C_R = 0.46\,(e^{1.84\,C_x}) \quad \textbf{(EQ 1)}$$

where, $C_R$ = corrosion rate (microns/year)
$C_x$ = ion concentration (% by mass of binder) at depth of steel reinforcement

Using the approach by Doran et al. (1987), the time to saturate the concrete tunnel liner can be calculated for a particular chloride concentration where degradation of steel reinforcement is of concern. However, this approach ignores the complexity of many variables discussed above regarding the flow of corrosive ions. The British Standards of 0.4% Cl concentration by weight may be overly conservative (De Witt, 2017). At an Institute of Concrete Technology Seminar in 2014 on "Chloride Transport in Concrete" as reported by Goodier and Dunne (2015), a relationship between the probability of corrosion versus chloride was proposed as shown in Figure 2.

Once the Chlorides begin to attack uncoated steel, spalling stresses are initiated, which is given by:

$$\sigma_b' = C_R\, Y\, E_c/x \quad \textbf{(EQ 2)}$$

where $Y$ = years of Cl accumulation (ignoring the time for Cl accumulation to 0.4%)
$E_c$ = Youngs' modulus of concrete
$x$ = thickness of cover over steel

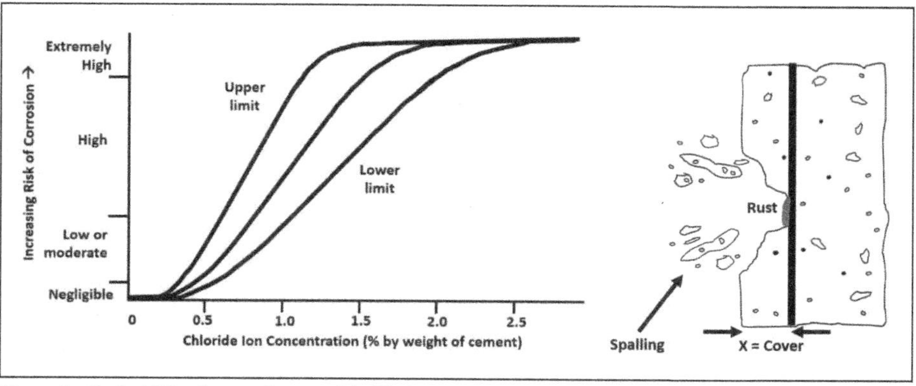


**Figure 2. Probability of corrosion vs. chloride ion concentration (modified after Goodier and Dunne, 2015) and Spalling**

The spalling stress may be compared to the bearing capacity of plain concrete $\sigma_b$ allowable = 0.3 $f'_c$ (ACI-318-19 (22), Ch. 14, Plain Concrete). Utilizing the above relationship, the required theoretical concrete cover over rebar can be calculated. For steel fiber reinforcing, concrete cover is not relevant. Wimpenny et al. (2009) note that because of discrete nature and small diameter of steel fiber, galvanic corrosion, carbonation, and associated spalling appear to be eliminated. Even if steel fibers corrode, the small volume and distributed nature of the fibers does not cause the type of bursting forces as rebar.

## Overview of Hydrogen Sulfide Generation and Wastewater Tunnel Ventilation

The potential for $H_2S$ corrosion of gravity sewer tunnel liners is affected by several factors. To have $H_2S$ corrosion, sulfide must be produced in the wastewater. $H_2S$ can only be produced in wastewater when dissolved oxygen (DO) levels are very low. When DO is greater than 1 mg/L, sulfate reduction is eliminated and $H_2S$ will not be produced. DO is gained through reaeration at the surface and through turbulence induced by junctions, drops, etc., and is a function of the oxygen deficit and hydraulic conditions (depth and area of flow, velocity, and slope). DO is lost through consumption by microorganisms present in the wastewater and the slime layer during oxidation of organic matter. The oxygen consumption rate can be estimated from an empirical relationship that is a function of DO concentration, pipe slope, flow velocity, and hydraulic radius. Evaluation of the internal corrosion potential for tunnel reaches can be estimate based on oxygen consumption and the reaeration rates (USEPA, 1985).

For this analysis, the oxygen consumption rate is compared with the reaeration rate at a DO level of 1 mg/L, since this is the threshold where $H_2S$ production is inhibited. When reaeration is greater than oxygen consumption, the DO will be greater than 1 mg/L. Sewer tunnels are often oversized with respect to average flows to accommodate large peak flows. Since corrosion is a long-term process, the evaluation of corrosion potential should be based on average flows instead of peak flows which typically have short durations. Average flows are often characterized by shallow flow depths, where the air/water interface is large compared to the depth, which is a condition favorable for reaeration. Conversely, tunnel slopes can be relatively flat, limited by geology or elevations of connections to existing wastewater infrastructure. Flat slopes result in lower velocities and greater depths, which tend to favor oxygen consumption over reaeration.

In cases where the hydraulics result in the potential for $H_2S$ production, corrosion potential is evaluated by assessing the predicted corrosion rate that would result from low DO concentrations. The corrosion rate can be calculated based on the flux of $H_2S$ to the pipe wall and the alkalinity of the cement bonded material (USEPA, 1985). The flux of $H_2S$ to the wall is a function of hydraulic conditions, pH, average dissolved sulfide in wastewater, organic content of the wastewater, and the ratio of width of the wastewater stream to exposed perimeter of the pipe wall above the water surface. This calculation can be done using conservative values for coefficients and assuming the sulfide concentration in the wastewater is equal to the theoretical limiting value, which will be greater than the actual concentration. To account for uneven corrosion, where maximum rates may be higher than average, the average corrosion rate may be multiplied by a factor (e.g., 1.5). Without liner protection, the expected corrosion rate is compared with a thickness of sacrificial concrete to determine the expected life of the tunnel liner.

For many projects, the tunnel will have a low potential for corrosion, however the tunnel areas near the drop structures will require protection against corrosion (e.g., several hundreds of meters either side of the drops) through the use of corrosion resistant coatings, materials, or additives. In borderline cases or cases where conditions may change in the future, it may be prudent to protect the entire length of the tunnel since inspection and repairs are very difficult once the tunnel is put in service.

Wastewater tunnels commonly end in wet wells where wastewater typically has residence time in the wet well to prevent fast cycling of the pumps. If the pump station is operated with long-residence time, significant generation of $H_2S$ is possible.

Additionally, the airflow in wastewater tunnels should be considered. Care must be taken to avoid stagnant headspace, particularly at drop structures, where $H_2S$ can be stripped from wastewater that drops to the tunnel from near-surface sewers. Similarly, lift station wet wells can provide a dead-end for headspace air flow, leading to corrosion and odor issues. Management of headspace air at the downstream end of the tunnel is an important consideration for gravity sewer tunnel design.

## UPDATES ON CORROSION RESISTANT LINERS

The following is an update of the 2011 paper on case histories and developments of PVC, HDPE and fiberglass thin membranes linings both integrated/embedded with segments and cast-in-place (CIP), polymer concrete liners, two-pass liners, and other measures such as additives to protect the intrados of tunnel liners from corrosive effluent and sewer gases. There are no updates on nascent (embryonic) integrated PVC and HDPE liners within precast concrete liners discussed in the 2011 paper.

### Thin Membrane Liners Updates

For HDPE CIP liners, the Singapore Deep Tunnel Sewerage System (DTSS) was previously cited and recommended for future projects for ease of handling by Marshall et al. (2007). Since that time Abu Dhabi STEPs program has incorporated HDPE as a second pass liner and is proposed as an alternative for Dubai's DTSS. The Doha Sewage Infrastructure Program (IDRIS) originally incorporated a CIP HDPE liner as part of the design, but the JV changed to an integrated HDPE liner like that which occurred on the PVC integrated liners in Sacramento, CA and Panama City. The 16 km of tunnel required over 250 km of HDPE extrusion welding of joints located above the bottom 30-degree angle (Najder Olliver and Lockhart, 2017). It should be noted that the authors are not aware of any major tunnel projects incorporating PVC (Ameron T-lock®) as integrated into precast concrete segmented liner (or as integrated as a second-pass CIP tunnel liner) since the project in Sacramento, CA (Samuelson-Klein et al., 2010) and Panama City, Panama (Wilshusen et al., 2012). On the Singapore DTSS, it is our understanding that PVC liner was listed as an alternate; however, PVC was effectively written out of the project specifications by the requirement of higher tensile strengths and other material requirements that PVC liners could not meet. Whether the higher strength of HDPE is actually required is debatable at a higher cost than PVC, although Wilshusen et al. noted that on the Panama City tunnel for curves the trailing gear edges scraped the PVC near the invert and had to be spot repaired. Also, it was noted in the 2011 paper, that HDPE has a "higher temperature resistant for fires that may occur during construction" is misleading. HDPE, while having a higher ignition temperature, is still combustible and will burn completely unless fire retardants are put into the material at manufacture. Even with fire retardants HDPE is still flammable. By comparison PVC is flame resistant and self-extinguishing when the flame or fire is removed. PVC will not burn without a continuous source of excessive heat or flame. Those interested in learning more about HDPE flammability should read the Executive Summary, Metro Red Line Fire, the City of Los Angeles dated July 13, 1990.

Herrenknecht's Combisegment®, originally introduced as an integrated fiber glass (GRP) one-pass liner in 2011, is now incorporating HDPE as integrated liner, primarily because handling of the segments makes the corners subject to damage and chipping. The Combi-segment® system eliminates the need for welding of the plastic liner by wrapping the plastic liner with the segment sealing gaskets. Combisegment® projects with HDPE include: a 3 m ID × 3.4 m OD × 1.2 m width × 110 m length with a 4+2 ring configuration for the West Trunk Sewer Phase 2, Region of Peel, Mississauga, Canada completed in 2018; a 3 m ID × 3.5 m OD × 1.2 m width × 9.1 km length with 5+1 ring configuration for the West Tehran Sanitation Project, Iran, completed in 2021. Herrenknecht has successfully tested the bonding capacity of off-the-shelf HDPE anchored liner systems on their segments up to 1.5 bars; however, they have a one-way valve drainage plug that is welded to the back side of the liner to relieve pressure if needed.

## Polymer Concrete in Segmented Liners

### *Brief History in Tunneling Industry*

The use of polymer concrete in the tunnelling industry was first considered in the early 1970s by the U.S. Bureau of Reclamation (Cowan et al., 1972). Research at the time was geared toward polymer impregnated concrete, which is no longer used in the polymer concrete industry in general because of poor cost/performance balance. Maguire and Iskander (2008) and Lang (2008) discuss the first polymer concrete segmented liner construction completed tunnels in 2002 in Offenbach, Germany (Figure 3). These were constructed by Tauber Construction using Meyer Polycrete® with internal diameters of 1.3 and 2.0 m, with glued joints. Tauber's brochures indicated capabilities to 3.0 m diameter.

The Sacramento Regional County Sanitation District approved construction based on rigorous testing provided by iNTERpipe and SolidCast Polymer Technology for an alternate design to conventional concrete with a PVC mechanically anchored plastic liner though the PVC integrated liner won (Maguire and Iskander, 2008, and Samuelson-Klein et al. 2010). Demonstration polymer concrete segments were cast at the time as shown in Figure 4 and 5.

## Polymer Concrete Developments

Asadollahi (2021) discusses the Oakland-Malcomb North Interceptor East Arm Tunnel where an existing 1.4 km long by 5.3-m ID tunnel with an average depth of 15 m to invert constructed in the early 1970's within the Detroit Metropolitan Sewer Services District will be partially relined for 85 m with 4.88-m ID by 76-mm thick by 2.13-m long (with tongue and groove joints) that are glued together (Figure 6).

**Figure 3. Tabuer Construction (2002) 3-piece 1.3 m dia. polymer concrete segments**

Figure 4. Demonstration polymer concrete segments with bolts and no gasket grooves (Osborn and Espeland, 2007; PPT, 2006) using Hollywood Bypass Tunnel, CA and Benbrook Tunnel, TX molds

Figure 5. Demonstration polymer concrete segments with dowel connectors and gaskets (SCP, 2008) using the TRex New Mississippi Outfall, CO/Lower Sacramento River Crossing, CA molds

Figure 6. Precast polymer concrete tunnel reline segments. Washington, Indiana Plant

The Dubai DTSS is proposed to allow for two alternatives: two-pass CIP HDPE liner or a polymer concrete liner for the entire tunnel or for portions near the splash zone at drop shafts (Monks, 2021). The project in Dubai includes 75 km of deep sewer tunnel, ranging from 3.5 to 6 m diameter, and 22 to 100 m deep. Approximately 3.36 km of the 75 km would be polymer precast concrete segmented liner if only used in the splash zone.

The Silicon Valley Clean Water in 2018 considered a polymer concrete liner versus a two-pass system with 3 m and 3.4 m ID corrosion resistant pipe; for example, centrifugally cast fiber reinforced polymer mortar pipe (CCFRPM) or filament wound fiber reinforced plastic pipe (FWFRP) in the progressive design build 5.3 km by 4.1 m

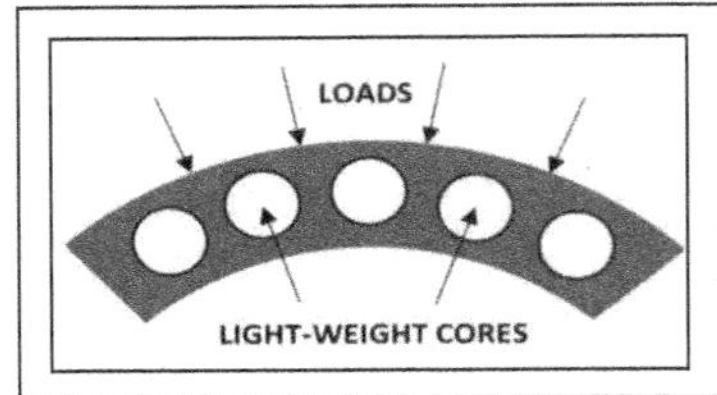

**Figure 7. Precast concrete polymer segments using patented Solid Core Technology (RockHardSCP®) and samples tested in 4-point bending**

ID/4.9 m OD tunnel. Polymer concrete design innovations as discussed below could have provided a cost savings; however, it is understood that the role of precedence in civil engineering practice led to SCVW staying with the two-pass CCFRPM system. Project details of the two-pass system installed can be found in Sucilsky et al, 2021).

Besides the savings in schedule and cost from a smaller diameter bore and elimination of a second pass, a design innovation that has made polymer concrete more cost competitive includes the introduction of lightweight Styrofoam cores (Figure 7) (Smith, 2018). Some manufacturers of polymer concrete that release no volatile organic compounds during production and have a significantly reduced carbon footprint by as much as about four to five times less than conventional concrete (Cubeta and Kaneshiro, 2021) and 2.5 times less than environmentally "green" concretes (Kaneshiro, 2022). Also, polymer concrete does not consume water. While polymer concrete may have higher capital costs, life-cycle costs of polymer concrete can be more beneficial.

## Corrosion Resistant Pipe

The 2011 paper provided an extensive review of corrosion resistant pipes to be considered in one- and two-pass installations. While curved microtunneling has been used in the USA for PVC lined RCP or clay pipe as well as in other countries for quite some time, Table 1 may be useful when evaluating precedence of CCFRPM and polymer pipes jacked on curves.

**Table 1. Curved jacking of CCFRPM and polymer pipes**

| Project | Location | Year | Provider* | Diameter, m | Radius, m | Length, m |
|---|---|---|---|---|---|---|
| Leverkusen–Schloss MorsBroich | Germany | 1992 | Meyer | 0.8 to 1.2 | 115 | 407 |
| Hamburg Graumannsweg | Germany | 2000 | Meyer | 2.6 | 600 | 437 |
| Zieolona Gora, Sewer | Poland | 2001 | Hobas | 1.22 | 350 | 1149 |
| Attendorn-Sondern | Germany | 2003 | Meyer | 1.6 | 600 | — |
| Hamburg–Eichenstr. | Germany | 2006 | Meyer | 1.6 | 300 | 327 |
| Hamburg–Wendenstr. | Germany | 2009 | Meyer | 1.2 | 500 | 455 |
| Freiberg–ZKA collector | Germany | 2010 | Meyer | 1.4 | 345 | 1368 |
| Honolulu-Beach Walk Force Main | Honolulu, Hawaii | 2012 | Meyer | 1.8 | 274 | 390 |
| De 2160 in Paris, Storm Water | France | 2013 | Hobas | 2.11 | 500 | 205 |
| Chelford City Diversion, West Harris Co. Service Area | Houston, TX | 2022 | Hobas | 1.52 | 1566; 1224 | 225; 179 |

* List of Curved jacking projects provided by Hobas and Meyer pipe except the last row, which was provided by BRH-Garver Construction, Houston, TX (Al Hajj et al., 2022; Griffin, 2022)

## Other Internal Corrosion Resistant Measures

### *Linings*

Redner et al, (2004) of the Los Angeles County Sanitation Districts (LACSD) performed extensive testing from 1983 to 2004 on 96 coating and liner systems including bonded systems (Coal Tar, Coal Tar Epoxy, Coal Tar Epoxy Mortar, Coal Tar Urethane, Concrete Sealers, Epoxy, Epoxy Mortars, Phenolic, Polyester, Polyester Mortars, Polyurea, Silicone, Specialty Concrete, Urethane, Vinyl Ester, and Vinyl Ester Mortars) and liner systems (including mechanically anchored PVC, polyethylene, HDPE), one poly coated Corrugated Metal Pipe, Glass Reinforced Plastic (GRP), Sulphur concrete, and two on polymer concrete pipes. The evaluation program found that 39 of the 96 coatings and liner systems successfully completed rigorous corrosive testing known as the "pickle jar test." It is noted that most coating manufacturers, and in many cases with justification, cite application problems as being the cause of coating failure, particularly in less-than-ideal conditions.

Many of the coating applied lining evaluated by LACSD are often used for manholes or repair of manholes. Spray-on or trowel on linings such as calcium aluminate cements (e.g., CALUCEM-SewerCem) have applications for tunnels as well. Singapore DTSS phase 2 is incorporating calcium aluminate microbial induced corrosion (MIC) resistant admixture in their concrete mixes for the tunnel's secondary CIP liner embedded with HDPE (Hin et al., 2016).

On the Region of York's South East Collector Trunk Sewer, the baffle drop structure treated with an anti-microbial experienced excessive corrosion early in its life. To protect the shaft, a spray on self-priming polyurethane lining (SprayRoq, Spraywall®) is being considered. For large scale tunnel lining operations consideration, though costs are in the same range as one-pass and two pass solutions, safety during application will required diligence.

Also, a brief word should be said about geopolymer concrete (Davidovits, 1999; Luhar and Luhar, 2021). Geopolymer concrete is reported to have properties in terms of strength, cure time, corrosion resistance, savings on water use, and a reduced $CO_2$ footprint rivaling polymer concrete. It should be noted, however, that the reported savings in $CO_2$ emissions is by use of fillers, such as flyash or other coal bi-products wastes. Moreover, the mixtures have a potential for voids (vugs) and requires addition of sodium silicate and caustic NaOH. Because of the mix' collapse potential, it is best suited for retrofit and spray on applications.

It is our understanding that the Oakland Macomb rehabilitation project now in progress in Detroit, as discussed for polymer concrete, is relining with segmented 4.9 m ID GRP (Channelline™) for about 366 m and with carbon fiber reinforced polymer (QuakeWrap™) for about 55 m.

### *Anti-Microbial Additives*

Concrete additives employing antibacterial admixtures have had success and have been specified for cast-in-place linings in the vicinity of deep drop shafts where the turbulence from the drop emits corrosive sewer gases. It should be noted that antibacterial admixtures are banned in some locations (e.g., in California) because the additive is considered an insecticide. While anti-microbial additives like CONSHIELD® or ConBlock MIC inhibit growth on treated concrete, it does not prevent them from growing on things like the underside of manhole covers or ladders, etc., which can result in acid running down or dripping down onto unprotected concrete. Also, such additives do not offer the same kind of protection/efficacy from acids (compared to membranes and polymer concrete) once they are generated or from industrial or laboratory wastes (though may be prohibited from dumping down the drain).

An anti-microbial additive was specified for 600 ft upstream and downstream of drop shafts for the 19 ft ID CIP liner for Metropolitan St. Louis Sewer District's Deer Creek Sanitary Tunnel project (2023 estimated completion) with due consideration for wastewater tunnel ventilation as discussed above. An anti-microbial additive combined with a crystalline water proofing is being considered for the upcoming Toronto's Regions of York's West Vaughn Tunnel. It should be noted that anti-microbial additives, will not prevent bacteria grown on other non-concrete surfaces. For example, on Ontario's Region of York Queensville-Holland Landing project in East Gwillumbury, evidence of bacteria growth on other surfaces may have led to acid generation and attack of the anti-microbial protected concrete lining.

## COST BENEFIT—LIFE EXPECTANCY

Often initial capital costs for corrosion resistant linings are the primary driving force when selecting the corrosion resistance lining. The efficacy of corrosion protection measures should be evaluated more closely and the whole-life cycle costs, including repair and rehabilitation of systems should also be evaluated when choosing the final solution.

In July 2020, the City of Houston convened a Tiger Team, consisting of Jack Canfield, Mohammad Haider, Markos Mangesha, and Hasnian Jaffri (City of Houston); Roger Weber (Atkins); Anil Dean (Stantec); Bryan Gettinger (Freese & Nichols); Eric Dawson and Pooyan Asadollahi (Parsons); and Steve Hunt (Black & Veatch), to evaluate corrosion protection alternatives for strategic tunnel enhancement and disaster mitigation for wastewater facilities impacted by Hurricane Harvey. The study provides illuminating information useful during the planning process when considering alternative corrosion protection measures for tunnel liners. Ten alternative linings were considered by the Tiger Team as shown in Table 2. The approach used by the Tiger Team was to update Carroll and Ivy (2010) review of various corrosion protection systems 2009 costs to 2020 dollars for a 14 ft ID tunnel using Engineering News Record's 20 Cities Construction Cost Index. This paper updates the square footage costs from the 2020 Tiger Team study to 2022 US$ costs. For a 14 ft ID tunnel for a one-pass solution the added unit costs ranged from $23 to $53/sf and for two pass solutions from $18 to $68/sf.

The Tiger Team concluded the following for durability of liners greater than 25 years:

- Higher added corrosion protection costs:
  - One-pass segments with plastic liner
  - One-pass polymer concrete segments (solid)
  - Two-pass segments with corrosion resistant carrier pipe
- Lower added corrosion protection costs:
  - One-pass integrated plastic liner
  - Two-pass 330o plastic liner wrap
  - Two-pass sacrificial concrete secondary liner

It should be noted, however, that polymer concrete segments were based on solid polymer segments without Styrofoam cores, and, for example, on Doha IDRIS, the

**Table 2. Alternative corrosion protection for tunnel liners and added cost (2022 $US) per square foot for 14 ft ID tunnel**

| (1) Option | (2) Examples (Product type, Project, &/or Comments) | (3) Range, $US | (4) Development Stage | (5) Durability§, yrs | (6) Efficacy |
|---|---|---|---|---|---|
| 1-pass corrosion resistant concrete design mix* | Antimicrobial (internal), crystalline water proofing (external) | $23–38 | Mature | 10–30 | 50% |
| 1-pass jacked corrosion resistant pipe† | VCP; RCP w/ PVC T-Lock®; Polycrete®; CCFRPM Hobas® | Not evaluated | Mature | Not evaluated | 95% |
| 1-pass precast polymer segments | Polycrete®; RockHardscp®; Polymer Pipe Technology; iNTERPipe® | $42–53 | Nascent | 60–80 | 100% |
| 1-pass PVC & HDPE precast segments | Sacramento (PVC), Panama City (PVC), Doha IDRIS (HDPE) (welded in tunnel after embedded in precast) | $38–53 | Emerged | 40–60 | 90% |
| 1-pass precast GRP/HDPE Combisegment® | Moscow (GRP); Ontario Region of Peel (HDPE); Tehran (HDPE) (no welding required) | $30–45 | w/HDPE Emerged | 20–40 | 95% |
| 2-pass CCFRPM or FWFRP backfill grouted carrier pipe‡ | Kaneohe-Kailua, HI; North Sewer Relief, Houston, TX; Silicon Valley Clean Water, CA; (CCFRPM Hobas®, FWFRP = Infrastructure Solutions International Fiberstrong™, or Future Pipe Industries Flowtite®) | $45–68 | Mature | 25–50 | 100% |
| 2-pass sprayed or troweled coating w/PVC sheets | BWARI, Columbus, OH (see 2011 RETC paper for application problems) | $18–38 | Nascent-emerged | 10–15 | 80% |
| 2-pass 330° CIP HDPE inside precast segments | Singapore DTSS, Abu Dhabi STEP | $38–45 | Emerged- Mature | 60–80 | 90% |
| 2-pass sacrificial CIP concrete inside precast segments | Lee Tunnel on Thames Tideway, London: 7.8 m ID precast segments, 300 mm thick sacrificial liner | $18–36 | Emerged-Mature | 30–40 | 90% |
| 2-pass Acid Resistant Sprayed on Lining | Calcium aluminate; shotcrete with antimicrobial and crystalline water-proofing additives; geopolymers | $27–33 | Emerged; Mature repair | 15–30 | 90% |

* Silica fume additive was also evaluated at cost of $5–8/sf but durability judged to be 10–20 years.

† Maximum common carrier pipe diameters may be found in the RETC 2011 paper. Not evaluated, but can be very economical, particularly for microtunneling. Jacked pipes have limitations for curves, length and diameter, record set is 2.5km × 3 m ID for Europipe (1994) in Germany.

‡ Maximum slip lined or backfill carrier pipes to date is 3.35 m, but 3.66 m is possible according to Hobas®.

§ Durability is revised years and subjective. Most products and projects will claim durability life between 50–100 years, and some products like polymer concrete claim on the order of 300 years.

Column Explanation:
(1) Alternative linings considered by the Tiger Team.
(2) Indicates the respective product type, or major projects and comments, some of the pros and cons.
(3) Estimated added costs for corrosion per square foot in 2022 US dollars.
(4) Authors' judgement of the stage of development in terms of engineering practice for tunnels.
(5) Estimated durability by the Tiger Team.
(6) Efficacy of the alternative in terms of reliability for corrosion resistance and potential for low life-cycle costs and life expectancy.

economics on that job were that it was cost effective to weld the HDPE liner rather than use the Combisegment® system, which integrates the gasket with the HDPE liner and no welding is required.

## SUMMARY AND CONCLUSIONS

Gravity Sewer Tunnel Liner Corrosion Protection—Part Two provides an update of the 2011 paper with developments in the tunneling industry regarding improvements, lessons learned, as well as innovation in corrosion protection for tunnel liners, evolving in a way that makes each option more competitive and more competitive with the other. While detailed cost-benefit and life-cycle costs (repair and maintenance) are required when evaluating specific tunnel projects and specific local marketplaces, in the authors' professional opinion, considering past project experiences, CCFRPM and FWFRP and polymer concrete are the most reliable and durable solution for corrosion protection of sewer underground infrastructures, especially when considering life-cycle costs.

In extremely severe environment of sewer tunnels, sacrificial concrete cover alone does not always provide reliable corrosion protection in splash zone as indicated by Austin's Govalle Tunnel and also was observed for the Region of York's Southeast Collector Tunnel/Shafts. On the other hand, although anti-microbials additives are proven to be able to delay concrete deterioration, they can hardly meet the anticipated design life of the infrastructures.

Past experiences in Toronto and Houston as well as other areas demonstrated that epoxy coating is not able to extend the design life beyond few decades. HDPE lined segments such as combi-segments have limitations too as discussed in the earlier sections of this paper. Coatings, if applied properly, can increase design life by 50 years but there are concerns regarding proper application and utilizing it as corrosion protection measure in large diameter tunnels and shafts since minor leakage through the concrete liner will results in damage/failure of the coating.

## ACKNOWLEDGMENTS

Thanks to City of Houston's Tiger Team on corrosion protection evaluations of tunnel liners and to Praveen Krishna (formerly of Parsons, presently with Arcadis) for his diligence in preparing corrosion resistance calculations on past projects.

## REFERENCES

Al Hajj, R., J. London, J. Farmer, and D. Ellett. 2022. Embracing innovative technology to push construction limitations associated with large diameter deep gravity sewer tunnels. PPT presentation. Water Environment Federation of Texas Conference.

Asadollahi, P. 2021. Abstract and Presentation: Detroit's Oakland-Macomb Interceptor Rehabilitation, Building Materials and Construction Technologies Conference. April 7.

Carroll, J.B. and H.M. Ivory. 2010. Corrosion protected systems for tunnels and underground structures, North American Tunneling Conference.

Clarke, W.J. 1982. Performance based characteristics of microfine cement. ASCE preprint 84-023. Atlanta, GA. May 14–18.

Cowan, W., L. Carpenter, and R. Spencer. 1972. Polymer-impregnated concrete tunnel support and lining systems, Rapid Excavation and Tunneling Conference Proceedings. Chapter 37.

Cubeta, R. and J.Y. Kaneshiro, 2021, Abstract and Presentation: Polymer concrete in wastewater, transportation, mining and chemical processing industries and its carbon footprint, Building Materials and Construction Technologies Conference, April 7.

Davidovits, J. 1999. Chemistry of geopolymeric systems, terminology. Second International Conference, Geopolymer '99. Geopolymer Institute. Saint-Quentin: France, pp. 9–39.

De Wit, Inken. 2017. Understanding corrosion processes in concrete. Phys.org, ETH Zurich. August 3. https://phys.org/news/2017-08-corrosion-concrete.html.

EPA [Environmental Protection Agency]. 1985. Design Manual Odor and Corrosion Control in Sanitary Sewerage Systems and Treatment Plants.

Goodier, C. and D. Dunne. 2015. Chloride transport in concrete: specification, testing and modeling, presented at Institute of Concrete Technology, report on half-day seminar in September 2014 on 'Chloride Transport in Concrete'. article provided by Loughborough University Institutional Repository. *Concrete magazine.* March 2015. Evolving Concrete. 6–7 May 2015. www.evolving-concrete.org.

Griffin, J. 2022. Curved microtunnel project saves time, effort. *Underground Construction Magazine*. Sept. Vol. 77. No. 9.

Gupta, S. C. and W. E. Larson. 1979. A model for predicting packing density of soils using particle-size distribution. Soil Sci. Soc. Am. J. 43:758–764.

Hin, Y.W., W.L. Lynn, J. Boev, and G. Piggot. 2016. Overview of Singapore's Deep Tunnel Sewerage System. Director of DTSS 2 Department, PUB, Singapore's National Water Agency. July 13.

Joye, M., P. Asadollahi, and P. Krishna. 2016. Durability design requirements for reinforced concrete structures. World Tunnelling Congress Proceedings. June.

Kaneshiro, J.Y., 2022, Abstract and Presentation: Carbon Footprint of Tunnel Projects (with emphasis on Tunnel Liner), Lunchtime lecture series—#21, International Tunnelling Association, Committee on Education and Training, December 13.

Kaneshiro, J.Y., K. Kuhr, and D. Yankovich. 2011. Gravity sewer tunnel liner corrosion protection. Rapid Excavation and Tunneling Conference. San Francisco, CA. June.

Lang, G. 2008. Long distance (micro-) tunneling with polymer concrete. May 23. E Journal. UNITRACC.com https://www.unitracc.com/e-journal/news-and-articles/long-distance-micro-tunneling-with-polymer-concrete-en.

Luhar. I. and S. Luhar, 2021. A comprehensive review on fly ash-based geopolymer. Journal of Composite Sciences. July.

Maguire, C. and E. Iskander. 2008. The use of polymer concrete in sewage tunnels for one-pass segmental lings. North American Tunneling Conference. San Francisco. p. 146 to 151.

Monks, G. 2021. Abstract and presentation: Dubai Strategic Sewerage Tunnel—Tunnel Design Consideration. Building Materials and Construction Technologies Conference. April 7.

Najder Olliver, A.M. and T. Lockhart. 2017. Innovative One-pass Lining Solution for Doha's Deep Tunnel Sewer System. World Tunnel Congress. Bergen, Norway.

Osborn, L. and Espeland, J. 2007. Polymer concrete products for new construction. No-Dig. April.

Panesar, D.K. 2019. Supplementary cementing materials. In S. Mindess, ed. Developments in the formulation and reinforcement in concrete. 2nd edition. Elsevier.

PPT [Polymer Pipe Technology, LLC], iNTERpipe, 2006, Trade Show Presentation, January.

Redner, J.A., R.P. Hsi, E.J. Esfandi, R. Sydney, R.M. Jones, and D. Won. 2004. Evaluation of protective coatings for concrete. August 2002, update. Paper was also presented at Water Environment Federations National Conference in Los Angles (1986) and Philadelphia (1987), California Water Pollution Association Annual Conference in Sacramento (1987), National Conference and Exposition of the Steel Structures Painting Council in Baltimore (1988) and in Long Beach (1991).

Samuelson-Klein, G., M. Monaghan, S. Gambino, D. Deutscher, and R. Guizar. 2010. Lined concrete segments: an alternative construction method for large diameter sewer tunnel. North American Tunneling Conference Proceedings. Portland, OR.

Shimoda, M. and H. Ohmori. 1982. Ultrafine grouting material. in W.H. Baker, ed. Grouting in geotechnical engineering practice. ASCE Proceedings. New Orleans, LA. February 10–12.

Smith, K., 2018, Have segments turned the corner, Tunnelling Journal, April/May.

Sucilsky, J., M. Jaeger, and N. Sokol. 2021. Progressive design build tunnel procurement in Silicon Valley. North American Society for Trenchless Technology. No-Dig Proceedings. Orlando, FL.

Tauber (accessed July 14, 2021) https://www.minitunnel.de/minitunnel.html, www.meyer-polycrete.com and Tauber brochures entitled: Mini Tunnel and Tauber Tunneling with Segments Made of Polymer Concrete.

USACOE [United States Army Corps of Engineers]. 1961. Technical Report No. 6-583. Nature and distribution of various fly ash.

Wilshusen, T.P., R.J. Guardia, J. Gil. 2012. Panama's first wastewater tunnel, North American Tunneling Conference. Indianapolis, IN. June.

Wimpenny, D., W. Angerer, T. Cooper, and S. Bernard. 2009. The use of Steel and Synthetic Fibres in Concrete under Extreme Conditions. 24th Biennial Conference of the Concrete Institute of Australia.

Wu, L. 1987. Relationship between pore size, particle size, aggregate size and water characteristics. Masters' thesis. Oregon State University. June 17.

Yusuf, M.O., M.A.M. Johari, Z.A. Ahmad, and M. Maslehuddin. 2014. Evolution of alkaline activated ground blast furnace slag-ultrafine palm oil ash-based concrete. *Materials and Design,* Elsevier. Vol. 55.

# Milestone Reservoir Quarry Project: Collaborative Design Development Using CMAR

**Steve Miller** ▪ Schnabel Engineering, LLC
**James O'Shaughnessy** ▪ Arcadis
**Mike Hanna** ▪ Black & Veatch
**Nathan Scalla** ▪ Clark Construction
**Savita Schlesinger** ▪ Loudoun Water

## ABSTRACT

Loudoun Water's Quarry A (Milestone Reservoir) project involves conversion of a rock quarry to a billion-gallon raw water reservoir. The project includes a new 40 MGD pump station, deep intake shaft, and three drill and blast intake tunnels for accessing the reservoir. Due to concerns regarding constructability and construction risks, the project design was paused while a Construction Manager At-Risk (CMAR) contractor was procured. The CMAR contractor was engaged collaboratively with the design team to provide risk and constructability input for refining the design. Aspects of the original design, design modifications, and benefits of the CMAR collaboration will be presented.

## INTRODUCTION

The Milestone Reservoir and Raw Water Pumping Station project is the conversion of a retired rock quarry to a 1.24-billion-gallon raw water reservoir and connection of the reservoir to Loudoun Water's existing raw water transmission main between the Potomac River Raw Water Intake and the Trap Rock Water Treatment Facility (Trap Rock WTF). The project is part of the Potomac Water Supply Program (Figure 1) to meet the demands of the rapid population growth of Loudoun County. Upon completion, the quarry will provide several modes of operation to meet the water supply demands, water quality, and provide a resilient water supply to residents.

The Milestone Reservoir project will be delivered using the Construction Manager-at-Risk (CMAR) delivery method with Arcadis as the Prime Design Engineer, Schnabel Engineering as the Geotechnical and Tunnel Design Engineer, and Urban LTD as the site civil designer. Black and Veatch is the Owner's advisor. Clark Construction and their subsidiary, Guy F. Atkinson Construction is serving as the CMAR. Additional third parties involved in the project development include Northern Virginia Regional Parks Authority (NVRPA), Dominion Energy, and Luck Stone.

## PROJECT HSTORY AND EARLY DESIGN

The concept for this quarry conversion began in the early 2000s, with initial discussions between Loudoun Water and Luck Stone of the potential use of one of their existing quarries as a new raw water storage reservoir once mining had ceased. There are four quarries in the area (Figure 2), the smallest being Quarry A. Mining operations for Quarry A were completed in 2019 and is planned to become the Milestone Reservoir. An agreement was negotiated for Loudoun Water to obtain a perpetual lease for Quarry A, while Quarries B, C, and D will remain in operations for the next 50+ years.

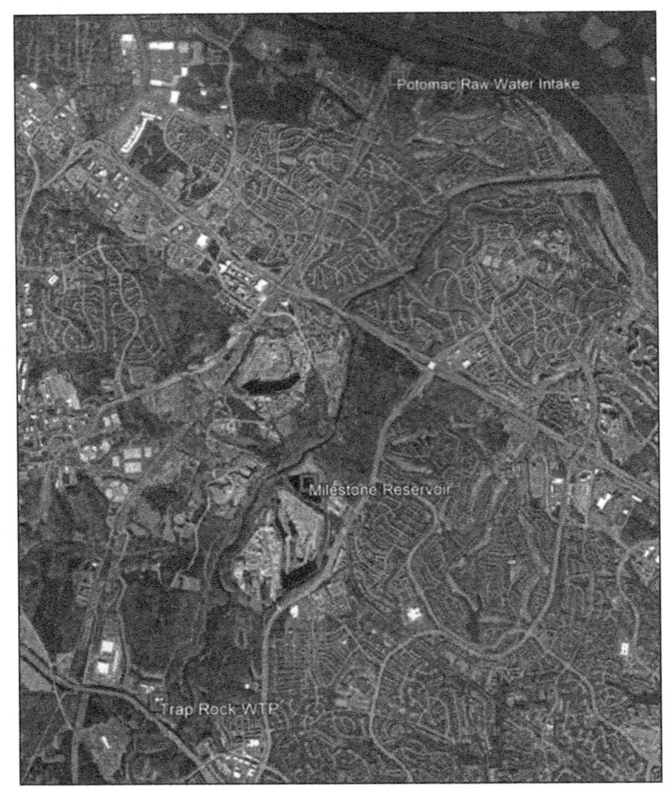


**Figure 1. The Potomac Water Supply Program Overview**

In 2011, preliminary evaluations were performed on Quarry A for the potential use as a raw water storage reservoir. In 2015 design began with an initial geotechnical investigation conducted to confirm the overall feasibility of the project and to aid in the selection process for location of the key features included in the project. The design progressed to 60% in 2018 and consisted of a deep shaft with three intake tunnels to the reservoir along the East Rim of Quarry A (Figure 3).

The project was paused after the 60% design milestone was reached to evaluate options for procurement. With the cost of the project increasing and several construction risks associated with the construction of the pump station and routing of the RWTM along the quarry rim, the project had grown to be one of Loudoun Water's largest projects. Based on the size and risk associated with the project, the decision was made to use CMAR to complete the project. Under CMAR, a General Contractor (GC) is engaged based on qualifications during the design process, provides input during the design, develops a guaranteed maximum price (GMP) for performance of the work, managing or self-performing the construction work, and is responsible as the Construction Manager (CM) for the cost and schedule control in construction (i.e., "at risk"). The pricing for the work, either self-performed or subcontracted, is typically done in an open-book approach so the owner has confidence the work is priced competitively. The decision for using the CMAR method for procurement stemmed from multiple factors. One factor was the size of the project, which would require significant oversight during construction allowing Loudoun Water to have the oversite they desired while reducing the burden in managing and coordinating the construction phase. The CMAR method provided a higher level of cost control and aid in development of contingencies and allowances for the GMP. It also provided the ability to provide input to design decisions from a constructability standpoint early in the design process and provide feedback on construction options to help control budget and schedule.

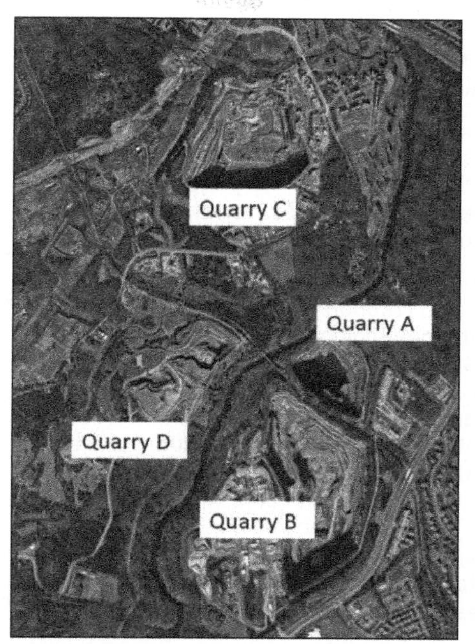


Figure 2. Location of quarries (Image: Landsat/ Copernicus, Google Earth)

Once the decision to use CMAR was made, Black & Veatch was added to facilitate the procurement process and overall project team coordination. Criteria and procurement documentation for the selection of the CMAR were developed, with procurement beginning in 2020 and included a shortlist of three firms and interviews, with selection made in November 2021. Once the CMAR contractor was under contract, the Quarry A process design was restarted to develop the foundation for the team approach, with the first project action being review of the previous design, constructability, risk, and new alternatives were developed.

## ONE TEAM APPROACH TO CMAR

Several partnering meetings were held to establish a "One Team" mentality, which included the addition of overall project goals as set by Loudoun Water. The main goals for the project included:

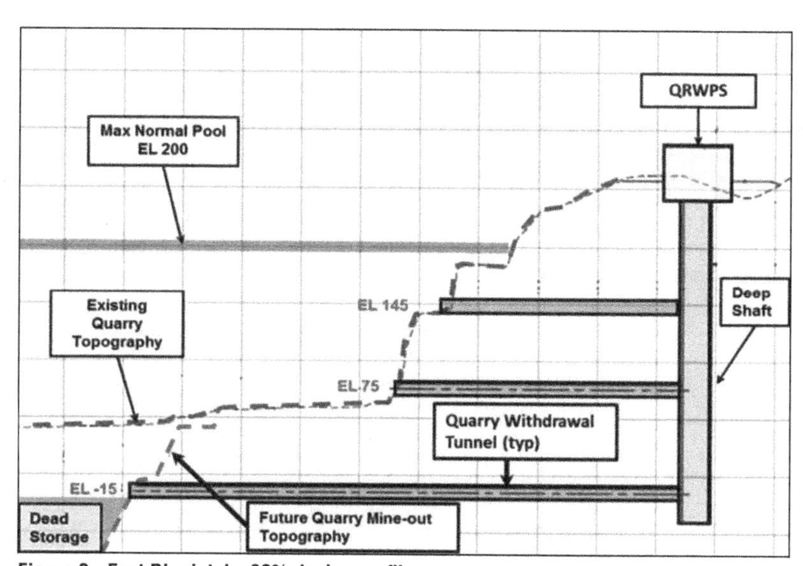


Figure 3. East Rim intake 60% design profile

1. Create momentum for One Team project success and build critical relationships
2. Reinforce the purpose, goals, and impact of the project
3. Align the One Team around common goals and define expectations for how the team will work together
4. Identify lessons learned from previous projects and experiences and discuss how to implement them
5. Ensure commitment to partnering and the One Team concept moving forward
6. Ensure all team members feel heard in the decision-making process

Additional discussions at the partnering meetings lead to the development of goals developed by the project team. Each of these goals set the results to which the project team is striving for in addition to meeting the main goals. The project will serve the community, and these are the biggest impacts on the community once the project is completed. The project development team project specific goals for the design and development stages of the project include:

- Safety through the Facility Lifecycle: Through a One Team safety culture with a clearly defined approach and communication focused on everyone's safety (project team, public, and staff), incorporate safety into design, planning, construction and operations.
- Make Good Decisions as One Team: Drive timely decisions in the best interest of the project through a mentality of compromise, including "the right people in the room" at the appropriate levels. Maintaining credibility and strong partnerships, ensuring it is clear to stakeholders why decisions have been made.
- Achieve High Standards of Quality, Performance, and Environmental Impact: Uphold Loudoun Water's standards of service, striving to provide water to customers under all circumstances with a positive impact on the community and environment. Deliver a streamlined start-up process that enables efficient operations and maintenance and minimizing changes after project turnover.
- Lifecycle Return on Investment for Customers: Leverage the CMAR process to provide strong ROI for stakeholders by making informed, value-based decisions, weighing short-term benefits and long-term planning.
- Keep Schedule Commitments without Sacrificing Quality: Engage in deliberate project execution such that were is a consistent story to report on progress towards the team's goals.

These goals were developed using inputs and recommendations from all team members. The purpose was to define the priorities for the design and development of the project moving forward. In recognition of how to best achieve these goals, the team made commitments to working on this project with other team members. The team commitments decided upon are:

- Communicate as One Team: Engage in open information sharing and inclusion such that all team members are comfortable speaking up. Share bad news in a transparent, solutions-oriented manner.
- Assume Positive Intent: Trust that goals are aligned and value each other's perspectives.
- Help each other Succeed: Be open to feedback and take ownership of mistakes. Forgive mistakes but expect corrections.
- Maintain "Opportunity: Mindset: Stay positive, focused on the success of the project, and celebrate wins and positive accomplishments.

The partnering sessions were a way of engaging all members of the project team including senior leadership, project managers, the design team, the operations team, and the CMAR contractor team. This unified the project team early in the process and served as guidance moving forward during the design process, with collaboration and communication serving as key components critical for success.

## CMAR DESIGN PROCESS

### Alternative Sites

The project restart as a CMAR project required a review of the previous design considerations and decisions. A series of Alternative Design Workshops were held to review the site selection criteria for the location of the pump station, as well as the review of previous concepts for the design under consideration. The Owner, the Owner's Advisor, the CMAR team, and the Design Team attended the workshops where a total of seven alternative sites were considered as part of the evaluation efforts (Figure 4).

There were five alternative site locations under consideration that were from the original selection process. Three of the alternative sites were located around the rim of the quarry (Alts 1a, 1b, and 1c), which were originally considered as part of the previous site selection process. These three sites all share the 60% design concept, involving a deep shaft with three intake tunnels as presented previously in Figure 3. Of those three locations around the quarry rim, the East Rim site (Alt 1b) was selected due to the length of tunnels and overall rim stability over the North and Southeast Rim sites.

The Two Creeks site (Alt 2) was added as a potential location as part of the reevaluation. When the original site selection process was performed back in 2017, the Two Creeks property was not available to the project, and therefore was not considered feasible at the time. It wasn't until recently in 2021, just prior to the restart of the project, that it was determined that it would now be a viable option for consideration. The Two Creeks site would consist of a concept similar to the original 60% design concept, involving a deep shaft and three intake tunnels (Figure 5).

The Bypass Vault Deep Shaft site (Alt 3) was originally considered as an option for the project and was included in the evaluation for the site. The site could possibly support three intake tunnels, or a single transmission tunnel from an intake structure or shaft and 3 tunnels located near the western rim of the quarry, similar to Alt 2.

1. Quarry Rim Deep Shaft
   1a. North Rim
   1b. East Rim
   1c. Southeast Rim
2. Two Creeks Deep Shaft
3. Bypass Vault Deep Shaft
4. Quarry Rim Intake Tower
5. Southwest Rim Site

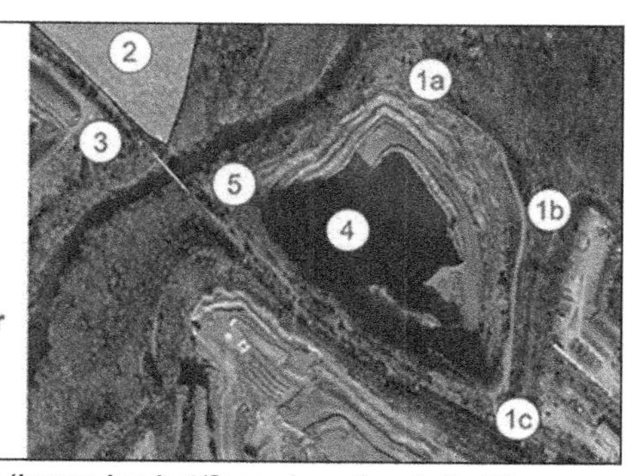


Figure 4. Alternative site locations (Image: Landsat/Copernicus, Google Earth)

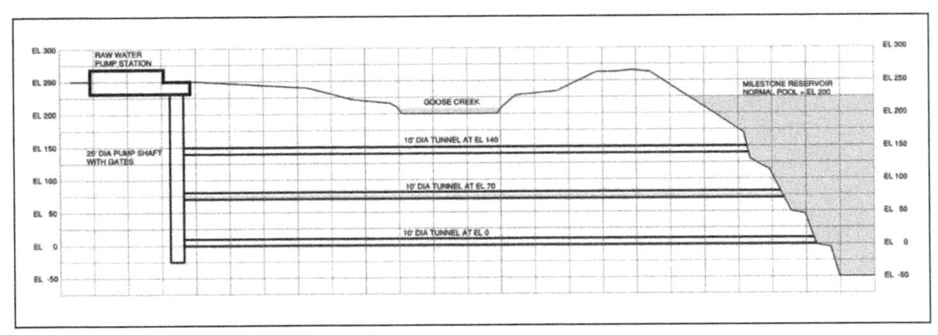


**Figure 5. Two Creeks deep shaft concept**

The Quarry Rim Intake Tower site (Alt 4) was added to the evaluation process, which would be located in the western corner of the quarry itself. This site and the concept were considered due to the flexibility offered for the exact placement in the quarry, and proximity to the Bypass Vault. This alternative would require a tower intake structure and pump station in the quarry itself, with a bridge connecting to the rim of the quarry for access to the pump station.

The Southwest Rim Site (Alt 5) was a location that was evaluated in the original site selection process. The site was not selected in the original evaluation due to issues with the available space for pump station placement, issues with construction access, and overall stability concerns that would result in significant risk and cost to the project and was not considered a viable location. The CMAR contractor agreed with this assessment, and Alt 5 was removed from consideration without further evaluation.

## Alternatives Scoring and Down Selection

Out of the seven alternatives, Alt 1b, Alt 2, Alt 3, and Alt 4 were considered viable and were included as part of the evaluation. Each of these sites were evaluated by the project team for the location of the pump station. General items for consideration for the selection of the pump station site included: cost, available space at the site, and proximity the Bypass Vault where the project would connect into the transmission mains to the water, and overall constructability. The sites were then scored by the project team based using a raking system of Poor, Average, Good, or Fatal Flaw on several criteria, with the results of the workshop scoring presented in Figure 6.

Based on the results of the scoring, the project team determined that both Alt 3, and Alt 4 had several fatal flaws associated with their design and were eliminated from further consideration. Alt 1b selected previously during the initial design was still considered as a viable, however the new Alt 2 was also considered viable and potentially had some advantages when directly compared to the Alt 1b. Based on the results of the workshops, both locations were selected for additional analysis by the project team for consideration.

### *Alternatives Criteria*

In the next phase of site alternative evaluation, the criteria were updated, and additional criteria weights were added to the scoring system (Table 1). The scoring criteria was developed to capture the main factors contributing to the project form the Owner's point of view. During the initial evaluation and discussion of the two remaining sites, it

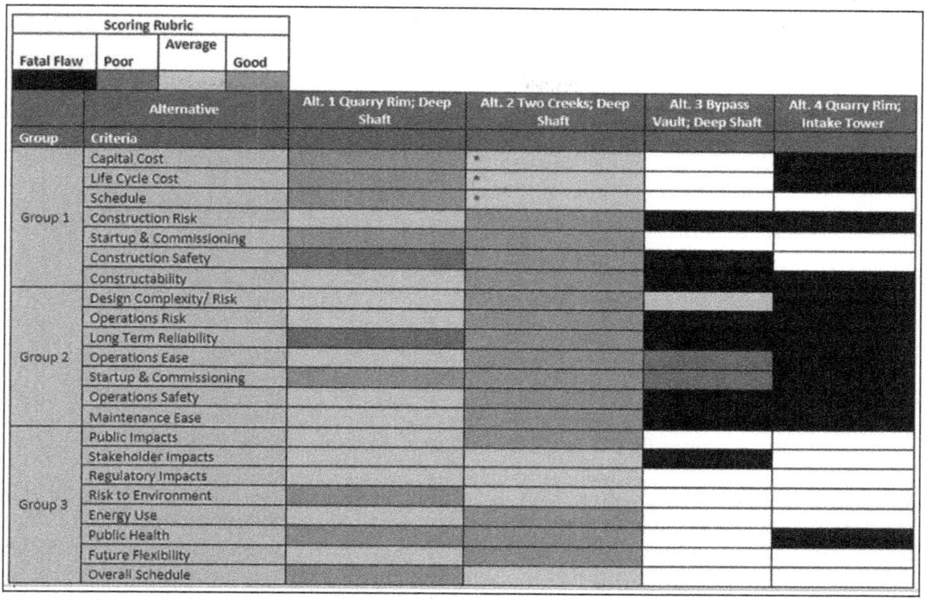

| Scoring Rubric | | | |
|---|---|---|---|
| Fatal Flaw | Poor | Average | Good |

| Group | Alternative / Criteria | Alt. 1 Quarry Rim; Deep Shaft | Alt. 2 Two Creeks; Deep Shaft | Alt. 3 Bypass Vault; Deep Shaft | Alt. 4 Quarry Rim; Intake Tower |
|---|---|---|---|---|---|
| Group 1 | Capital Cost | | * | | |
| | Life Cycle Cost | | * | | |
| | Schedule | | * | | |
| | Construction Risk | | | | |
| | Startup & Commissioning | | | | |
| | Construction Safety | | | | |
| | Constructability | | | | |
| Group 2 | Design Complexity/ Risk | | | | |
| | Operations Risk | | | | |
| | Long Term Reliability | | | | |
| | Operations Ease | | | | |
| | Startup & Commissioning | | | | |
| | Operations Safety | | | | |
| | Maintenance Ease | | | | |
| Group 3 | Public Impacts | | | | |
| | Stakeholder Impacts | | | | |
| | Regulatory Impacts | | | | |
| | Risk to Environment | | | | |
| | Energy Use | | | | |
| | Public Health | | | | |
| | Future Flexibility | | | | |
| | Overall Schedule | | | | |

**Figure 6. Scoring for alternative sites**

**Table 1. Alternative evaluation criteria**

| Criteria Weight | Topic | Criteria | Sub Criteria Weight |
|---|---|---|---|
| 20% | Cost | Capital | 40% |
| | | Life Cycle | 60% |
| 20% | Construction | Schedule | 25% |
| | | Constructability | 25% |
| | | Construction Safety | 25% |
| | | Construction Risk | 25% |
| 10% | Design | Design Complexity | 30% |
| | | Design Risk | 30% |
| | | Design Schedule | 40% |
| 25% | O&M | Operations Risk | 20% |
| | | Operations Safety | 20% |
| | | Long Term Reliability | 15% |
| | | Operations Ease | 15% |
| | | Maintenance Ease | 15% |
| | | Startup & Commissioning | 15% |
| 10% | Stakeholders | Public Impacts | 30% |
| | | Stakeholder Impacts | 40% |
| | | Regulatory Impacts | 30% |
| 10% | Triple Bottom Line | Risk to Environment | 30% |
| | | Public Health | 40% |
| | | Energy Use | 30% |
| 5% | Future | Future Flexibility | 100% |

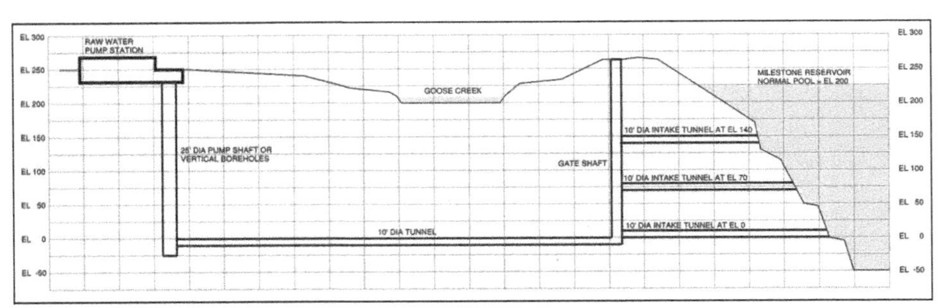


**Figure 7. Two Creeks Alt2b concept**

was determined there was a second design option for the Two Creeks Deep Shaft site, which would involve a second gate shaft near the quarry, with three tunnels connecting from the quarry and a single tunnel to the pump station site (Figure 7). This design option would be called Alt 2b, while the original Alt 2 would be called Alt 2a.

Alt 2b was considered as an option to potentially reduce the total underground work required for the project. The alternative would shorten the intake tunnels and connect to a gate structure shaft, which would then flow to a single transmission tunnel. The concept was feasible from a design and technical standpoint and would need to be evaluated for constructability and cost estimation along with the other two alternatives in consideration to determine the best alternative.

### *Cost and Schedule Evaluations and Site Selection*

The CMAR developed conceptual cost estimates and schedules based on the alternatives in consideration. A previous cost estimate was developed in 2018 for Alt 1b, and the CMAR evaluated and updated the estimate to the current market conditions in 2021. The new 2021 cost update resulted in an increase of nearly 32% over the 2018 cost estimate, with no increases due to changes in design. This increase was due to the current heavy civil construction market conditions. This cost information was important in understanding the current market conditions in developing cost estimates for the two additional alternatives in consideration, as well as understanding the significant increase to the Alt 1b cost estimate. The cost for the three alternatives in consideration were developed, and the direct comparisons are presented in Table 2.

The results of the direct cost comparison of the three alternatives showed a lower cost for both Alt 2a and Alt 2b when compared to Alt 1b. The project component area with the most significant savings in cost was the raw water transmission main (RWTM) and site utilities. The main difference was due to the difference in the required length of raw water transmission pipe necessary for each alternative. Both Alt 2a and Alt 2b had

**Table 2. Alternatives cost comparison**

| Cost Component | Alt 1b | Alt 2a | Alt 2b |
|---|---|---|---|
| **Project Components** | | | |
| Pump Station | $28.4M | $27.4M | $27.4M |
| RWTM/Site Utilities | $14.0M | $5.8M | $5.9M |
| Site Improvements | $12.4M | $13.3M | $13.7M |
| Geotechnical | $45.2M | $42.9M | $46.2M |
| DVP Allowance | $2.0M | $0.8M | $0.8M |
| **Subtotal** | **$102.0M** | **$90.2M** | **$94.0M** |

a significantly shorter requirements for water transmission pipe, which, based on current market conditions for the cost of steel, had the largest impact on the overall cost. The conceptual schedule was also developed for each alternative in consideration (Figure 8).

The East Rim Site (Alt 1b) alternative has a slightly longer schedule compared to the other two alternatives due to the sequencing of construction between the tunnel and shaft work, and the pump station work. Both alternatives for the Two Creeks site allow for construction of the pump station while the underground work is in progress, which reduces the overall schedule even though the tunnel and shaft work is expected to take an additional 9 months to complete. Factoring in the cost and schedule information available, the three sites were then scored by 19 different groups/individuals. The results of the individual scoring are presented in Table 3.

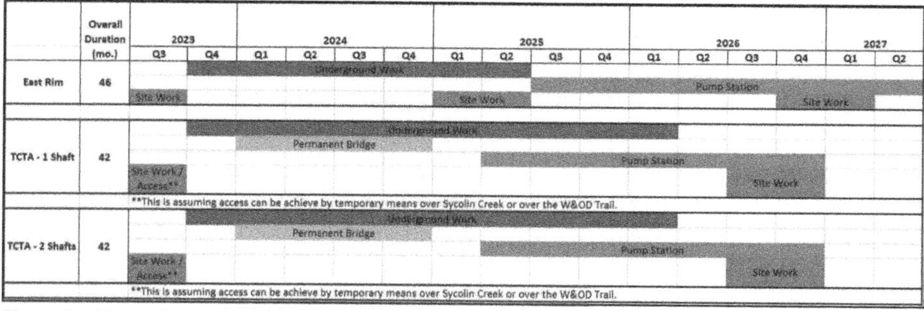


**Figure 8. Alternatives schedule comparison**

**Table 3. Alternatives scoring**

| Scorer | Alt. 1: Quarry Rim | Alt. 2a: TCTA- 3 Full Length Tunnels | Alt. 2b: TCTA- Gated Shaft at Quarry Site |
|---|---|---|---|
| Scorer 1 | 2.8 | 3.6 | 3.4 |
| Scorer 2 | 3 | 3.6 | 3.1 |
| Scorer 3 | 2.3 | 3.2 | 3.1 |
| Scorer 4 | 3 | 3.8 | 3.3 |
| Scorer 5 | 2.8 | 3.5 | 3 |
| Scorer 6 | 2.5 | 4.4 | 4.2 |
| Scorer 7 | 2.8 | 3.7 | 3.1 |
| Scorer 8 | 3.1 | 3.6 | 3.4 |
| Scorer 9 | 3 | 3.9 | 3.1 |
| Scorer 10 | 2.7 | 4.4 | 3.8 |
| Scorer 11 | 2.7 | 4.3 | 3.8 |
| Scorer 12 | 2.5 | 2.8 | 2.5 |
| Scorer 13 | 2.5 | 2.6 | 2.5 |
| Scorer 14 | 2.7 | 3.7 | 3.4 |
| Scorer 15 | 2.5 | 3.5 | 3 |
| Scorer 16 | 2.7 | 3.4 | 2.5 |
| Scorer 17 | 2 | 3.7 | 3.3 |
| Scorer 18 | 2.9 | 3.4 | 3.4 |
| Scorer 19 | 2.7 | 3.8 | 3 |

Based on the results of the scoring, the Two Creeks Deep Shaft alternative with Three Tunnels (Alt 2a) site was consistently the highest-ranking alternative by all scorers. The overall difference in the scores between alternates per scorer did vary, but the overall order of the rankings of each alternative were consistent between scorers. As part of the final alternatives selection process, the alternatives were reevaluated during a workshop to determine if there were any cost reduction measures or other design modifications for consideration that would significantly change the scoring of each alternative. It was determined that there were no significant modifications that could be considered to any of the alternatives to reduce the cost, or provide any additional benefits to the alternative. The recommendation by the project team was to select the Two Creeks Deep Shaft with Three Tunnels (Alt 2a) as the project site location moving forward. The largest drivers behind the selection were the overall O&M benefits to the facility location and design, the constructability of the project on the site, and possibility for future expansion of additional maintenance buildings available at the site for Loudoun Water.

## Pump Station Design Alternatives Evaluation

Once the alternative site was selected, the pump station building was reevaluated using a similar process that was used to select the location for the pump station. The new Two Creeks site had fewer constraints for the pump station design compared to the previously selected site location. It was determined that several options were available for consideration in the pump station design. A total of 5 pump station alternatives were considered (Figure 9).

Alternative 1 was the original 60% concept from the East Rim Site. This design was feasible at the new site but was designed to meet the site constraints at the East Rim location. Alternative 2a was a longer building that was decoupled from the pump station shaft. Alternative 2b was similar to Alternative 2a, but the building was wider and not as long. Alternative 2c had the same building as Alternative 2b but considered the use of individual smaller bored shafts for the intakes instead of a larger single shaft. Alternative 3 placed the pumps and valves in an inground vault away from the pump station building. Each of the pump station alternatives was scored and the overall results are presented in Table 4.

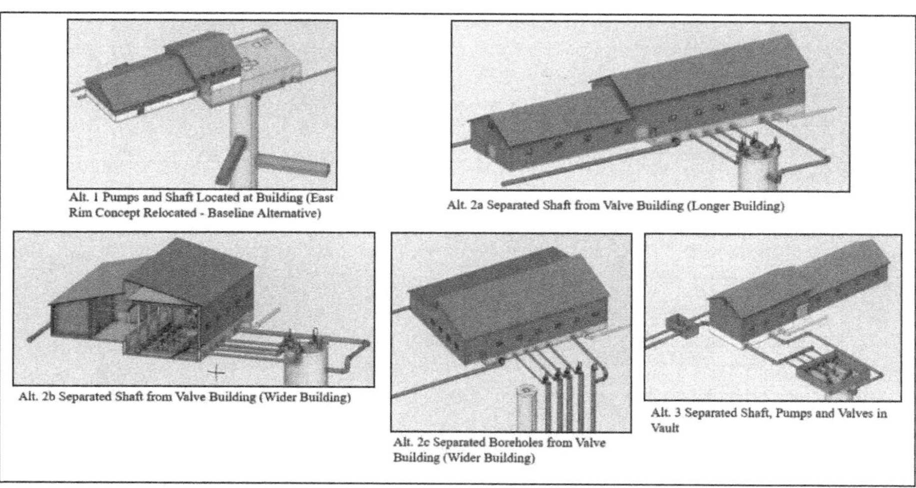


**Figure 9. Pump station design alternatives**

Table 4. Pump station alternative scoring

| Scoring Summary | | Alternatives | | | | |
|---|---|---|---|---|---|---|
| Criteria Topic | Weight | Alt. 1 | Alt. 2a | Alt. 2b | Alt. 2c | Alt. 3 |
| Construction | 33% | 2.6 | 3.4 | 3.6 | 2.9 | 2.9 |
| Design | 17% | 3 | 3.3 | 3.4 | 3 | 3 |
| O&M | 42% | 2.8 | 3.3 | 3.7 | 3.3 | 2.9 |
| Future | 8% | 2.4 | 3.3 | 3.8 | 2.9 | 2.7 |
| Weighted Sum | | **2.7** | **3.3** | **3.6** | **3.1** | **2.9** |

Table 5. Pump station alternative costs

| Alternative | Alt. 1: Pumps and Shaft Located at Building (Baseline) | Alt. 2a: Separated Shaft from Valve Building (Longer Building) | Alt. 2b: Separated Shaft from Valve Building (Wider Building) | Alt. 2c: Separated Boreholes from Valve Building (Wider Building) | Alt. 3: Pump and Valve Vault Over Shaft, Support Building, Flow Meter Vault |
|---|---|---|---|---|---|
| Total Capital Cost | $114,293,000 | $115,648,000 | $115,198,000 | $117,301,000 | $113,611,000 |
| % Cost Compared to Baseline | — | 101% | 101% | 103% | 99% |
| Cost Differential from Baseline | — | $1,355,000 | $906,000 | $3,008,000 | ($681,000) |

The costs associated with the construction of each pump station design alternative were developed by the CMAR (Table 5).

The pump station designs had a slight impact to the overall pump station costs. The construction schedule for each of these designs also was a factor in the selection process. Alternative 1 was projected to take 54 months to construct, while the other alternatives were projected to take 42 months to construct. Alternative 1 had a longer construction schedule due to the placement of the pump station building over the pump station shaft. The other alternatives separated out the pump station building from the pump station shaft, which resulted in a shorter construction schedule. Using this information, each alternative was scored by each member of the project design team (Table 6).

The Operations and Maintenance staff at Loudoun Water heavily favored Alternative 2b, but the alternative was also favored by the other internal organizations at Loudoun Water. Alternative 2b was selected by Loudoun Water as the design of choice, as it provided the most benefit to their operational needs.

## SUMMARY AND CONCLUSIONS

The selection of a CMAR process for the conversion of a mined-out quarry to a water storage reservoir has shown to be beneficial to the design and current success of the project. This could not have been accomplished without the "One Team" approach that was formed early in the updated design process, and the design workshops and meetings to evaluate and review alternatives for the project. The early partnering meetings brought together the owner, owners' advisor, designers, and the CMAR contractor to discuss the goals for the project, as well as the concerns or risks for this project, and set the standards for the design process for all team members to adhere to.

At the time this paper was written, the project has reached 60% design. Together, the CMAR, the owner, and the design team have worked through the initial project design

Table 6. Scoring results by organization

| Organization | Alt. 1: Pumps and Shaft Located at Building (Baseline) | Alt. 2a: Separated Shaft from Valve Building (Longer Building) | Alt. 2b: Separated Shaft from Valve Building (Wider Building) | Alt. 2c: Separated Boreholes from Valve Building (Wider Building) | Alt. 3: Pump and Valve Vault Over Shaft, Support Building, Flow Meter Vault |
|---|---|---|---|---|---|
| Arcadis/Schnabel | 2.2 | 3.2 | 3.3 | 2.9 | 2.6 |
| Black & Veatch | 2.9 | 3.7 | 3.7 | 3.8 | 3.3 |
| Clark/Atkinson | 3 | 2.9 | 2.9 | 3 | 2.9 |
| Loudoun Water | 2.3 | 3 | 3.4 | 2.6 | 2.4 |
| LW-Engineering–Cap Const | 2.4 | 2.9 | 2.8 | 2.6 | 2.6 |
| LW–Engineering–Cap Design | 2 | 2.8 | 3.4 | 2.3 | 2.1 |
| LW–Engineering Management | 2.1 | 2.9 | 3.5 | 3 | 2.2 |
| LW–GMO | 2.7 | 3 | 3.7 | 1.8 | 2.3 |
| LW–Water Operations | 2.3 | 3 | 3.3 | 2.9 | 2.3 |
| LW–O&M | 2.8 | 3.1 | 4.3 | 2.4 | 3.4 |

and goals to resolve any potential issues or setbacks, and they will continue the process developed to complete the design. The pump station location and pump station design were two aspects of this project that were successfully updated by the project team. Given market conditions that are currently affecting the heavy civil construction field, having the ability to adapt to changing market conditions has and will play a significant role in the success of this project, which would not have been as feasible without the integration of a CMAR contractor.

The CMAR procurement method has been key to the development of this project, to deliver a solution that is beneficial to the owner on many levels. The early involvement of the CMAR contractor created a collaborative work environment for all parties and allowed development of a design focused on constructability while maintaining technical requirements for the system as a whole that meets the long-term goals for Loudoun Water. The reevaluation of the site location with the CMAR collaboration was key to validating and providing information for optimization of the design as well as development of the construction logistics. These aspects of the process will aid in the development of the GMP and project schedule and therefore reduce overall risk to the project regarding technical, cost, and schedule considerations.

## ACKNOWLEDGMENTS

The authors would like to acknowledge the contributions of Loudoun Water's staff and all members of the project team for the collaboration and contributions thus far in the development of the Milestone Reservoir Project.

# Tunneling for High-Energy Physics in Menlo Park, CA

**Justin Lianides** ▪ Mott MacDonald
**Derek Penrice** ▪ Mott MacDonald
**Irene Bendanillo** ▪ SLAC National Accelerator Laboratory
**Canon Cheung** ▪ SLAC National Accelerator Laboratory

## ABSTRACT

The SLAC National Accelerator Laboratory's two-mile-long particle accelerator generates the world's brightest X-rays from its Linac Coherent Light Source (LCLS) free-electron laser. This facility has driven groundbreaking discoveries in medicine and industry. The LCLS-II-HE project will provide a significant increase in laser energy, allowing cutting-edge research in fields including biology and environmental science.

A key project component is the Low Emittance Injector Tunnel (LEIT), a 240-foot-long tunnel to be built alongside the existing accelerator and connected to it via a 30-foot-long Transfer Tunnel.

This paper discusses site and scientific constraints that led to the selection of the LEIT configuration: tunnels to be constructed using the sequential excavation method in mixed-face conditions comprising native rock and placed backfill, and critical design considerations including protection of overlying and adjacent historic structures and structural accommodation of near-fault seismic loading.

## INTRODUCTION

SLAC National Accelerator Laboratory (SLAC) is a facility operated by Stanford University in an unincorporated area of San Mateo County in the San Francisco Peninsula. SLAC is under contract to the United States Department of Energy (DOE) to construct and operate particle-colliding facilities and conduct research in areas of physics, accelerator science, particle physics, photon science, and astrophysics. SLAC users have made cutting-edge discoveries in material sciences, catalytic sciences, structural molecular biology, molecular environmental sciences, and energy generation.

SLAC has operated since the 1960s, beginning with the construction of a two-mile-long underground particle linear accelerator (Linac). In 2009 SLAC completed a major facility upgrade, the Linac Coherent Light Source (LCLS). LCLS achieved the world's first hard X-ray free electron laser, allowing sub-nanometer ($1 \times 10^{-9}$ meter) and femtosecond ($1 \times 10^{-15}$ second)-scale research.

In 2011 SLAC began plans for a Linac upgrade: the Linac Coherent Light Source–II (LCLS-II), quickly followed by plans for the LCLS-II High Energy (LCLS-II-HE) in 2014. LCLS-II-HE will provide additional research stations, higher X-ray energy, increased time resolution, and improved coherence and control. An overall figure of the SLAC facility is shown on Figure 1, with the location of the LCLS-II upgrade indicated at its western limit.

A relatively recent addition to the LCLS-II-HE concept has been to add a Low-Emittance Injector (LEI) parallel to the existing LCLS-II injector, which will improve beam

Source: Linac Coherent Light Sourcea—LCLS-II: A World-Class Discovery Machine (2022).
**Figure 1. SLAC facilities layout**

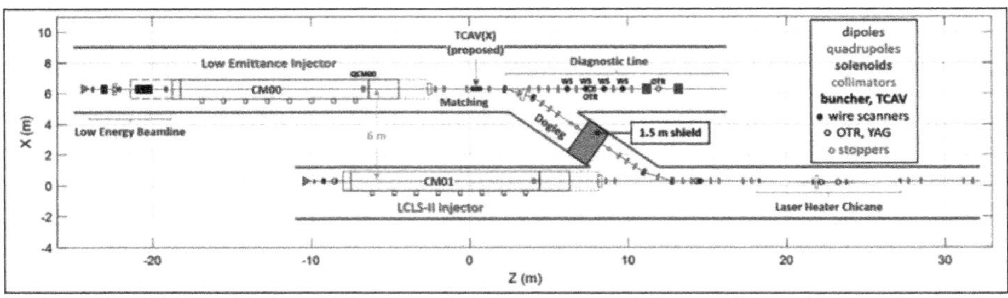


Source: Nosochkov et al. (2022)
**Figure 2. Schematic plan view of LCLS-II-HE LEIT depicting physics installations**

brightness and increase the research availability of the LCLS-II accelerator. The LEI will be installed within a new 240-foot-long tunnel beginning at the west end of the Linac and parallel to it. The LEI tunnel (LEIT) will connect to the Linac with a beamline underground Transfer Tunnel. A schematic of the physics installations within the LCLS-II-HE LEIT is shown in Figure 2.

## LEIT PROJECT BACKGROUND

### Linac Construction

The Linac was designed and constructed between 1962 and 1966, and upon its completion was the largest of its kind in the world (Neal et al. 1968).

The principal structures of the Linac are the Accelerator Housing and the Klystron Gallery. The Accelerator Housing is a cut-and-cover concrete structure, buried approximately 22 feet below grade with a portal entry at its western end. The Accelerator Housing was constructed at the base of a sloped and benched rock excavation (see Figure 3) and backfilled with imported drain rock and embankment material obtained from the site (see Figure 4 and Figure 5). Subdrains at the base of the Accelerator Housing were installed to relieve the structure of groundwater pressure and limit infiltration into the housing. The water captured from the subdrains require radiation monitoring and treatment.

**Figure 3. Completed Linac excavation**

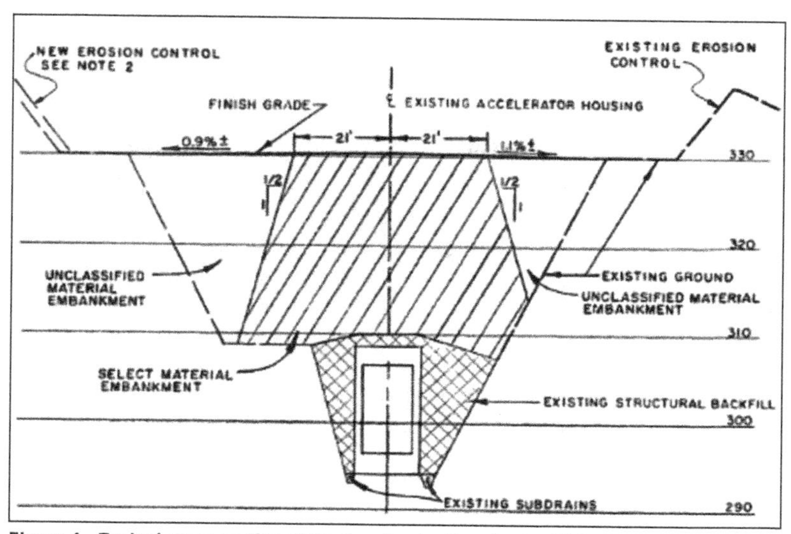

Figure 4. Typical cross section of the Accelerator Housing backfill

At the surface directly over the Accelerator Housing is the equally long Klystron Gallery, which provides an enclosed working area for housing technical systems and for servicing and operating the Linac. Access shafts and vertical utility penetrations connect the two structures. The two-mile-long Klystron Gallery above the underground beam line is the longest building in the United States and is included in the National Register of Historic Buildings.

Figure 5. Placement of drain rock around the Accelerator Housing

## Geologic Conditions

The Linac is located within the California Coast Ranges Geologic Province, an area dominated by tectonic deformation and crustal uplift. The site is at the western base of the Santa Cruz Mountains with bedrock consisting of Miocene to Eocene-age (11- to 58-million-year-old) sedimentary rock and Miocene-age (11-million-year-old) basalt (SLAC 2006). The site is approximately one mile from the San Andreas Fault zone, a major strike strip fault that trends along the western side of California and is capable of generating earthquakes greater than moment magnitude 7.0.

The LEIT will be constructed within the Eocene-age Whiskey Hill Formation (36 to 58 million years old). The formation is mostly composed of very weak to weak weathered sandstone, with abundant shears and fractures. Gravel and colluvium beds are along the north side of the Accelerator Housing excavation. Backfill of the Accelerator

**Figure 6. Excavations for the damping structure along the Accelerator Housing (left); breakthrough into the Accelerator Housing wall (right)**

housing excavation consists of pea gravel drain rock, which directly surrounds the Accelerator Housing, then is overlain by compacted cohesive backfill.

In the 1980's excavations were made for a damping ring structure a few hundred feet east from the LEIT site. The area demonstrated good standup behavior of the rock and soil, and little if any groundwater production (see Figure 6). The bulk of that excavation was performed in an open-cut fashion; however, some localized mining was required under the existing Klystron Gallery. To facilitate mining, a series of needle beams was installed between the excavation support and the top of the Accelerator Housing. These beams were supported on drilled shaft foundations. The space below the Klystron Gallery was hand-mined. Construction photos show the drain rock standing vertically and overhead.

## LEIT PROJECT DESIGN

### Principal Elements and Configuration of the LEIT

The LEIT contains the following principal elements, which are depicted on Figure 7:

- LEIT west portal: The primary entrance to the LEIT will be through the west portal structure. It will contain control racks to operate the LEIT injector gun. The structure will need to accommodate delivery of a 50-foot-long cryomodule into the LEIT.
- West portal structure: The structure provides storage for control racks and other technical equipment in a non-radiation zone.
- West maze shielding: A maze-style concrete wall system prevents radiation from the injector gun area from being released to the west portal. The west maze will contain a sliding concrete door.
- Laser pipe penetration: An approximately 12-inch-diameter casing that will connect the LEIT to the Accelerator Housing for installation of a 7-inch-diameter laser transport pipe.
- Transfer Tunnel: This tunnel transfers the beamline from the Low Emittance Injector to the Linac.
- East Stairwell Shaft: The East Stairwell Shaft will provide a personnel exit route and fire protection emergency access to the LEIT and deliver physics utilities to the LEIT. Utilities that will pass through the shaft include helium, low-conductivity water, cable trays, control racks, and cryogenic distribution

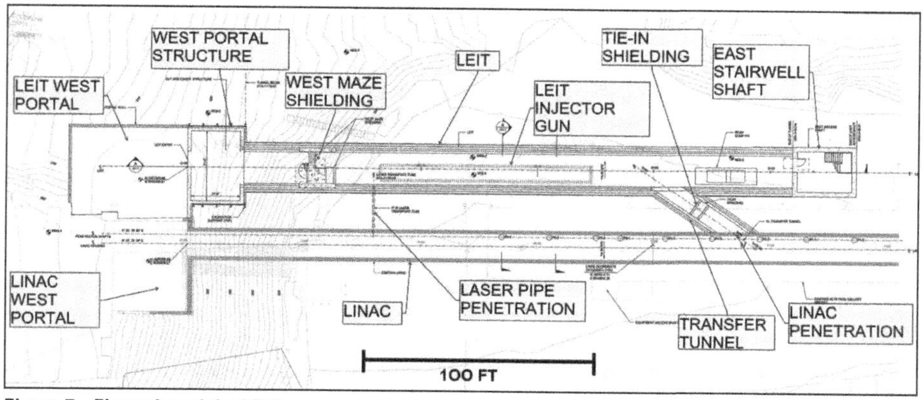


Figure 7. Floor plan of the LEIT

piping, as well as mechanical, electrical, plumbing, and communication lines (MEPC) for facility operations. Entry into the shaft from the ground surface will be through a steel-framed headhouse.

- Radiation protection: The injector area of the LEIT will need to be enclosed by radiation protection elements as described below:
  - Tunnel structure:
    - Initial support and final lining with a minimum combined thickness of two feet and a minimum density of 150 lb/ft$^3$.
    - Earth materials surrounding the LEIT and Transfer Tunnel to serve as radiation protection.
  - Internal structures:
    - West shielding maze, which will also contain a two- to three-foot-thick sliding concrete door.
    - Tie-in shielding inside the Transfer Tunnel, essentially a tunnel plug consisting of one-meter-thick mass concrete and nine inches of steel plates on each side of the shield.
    - Beam dump pit and concrete wall within the pit.
    - Concrete wall separating the LEIT from the East Stairwell Shaft.
    - Concrete wall separating shaft utilities from personnel access through the East Stairwell Shaft.
    - Concrete stairs and a concrete platform at the top of the East Stairwell Shaft.
- Tunnel finishing installations: Once the shielding has been constructed, the tunnels will be fitted with MEPC, technology, and fire protection. The work will be done after the conclusion of the LEIT heavy civil construction window and during operation of the Linac beamline. Vibrations from these installations are anticipated to be inconsequential to Linac operations within the Accelerator Housing .
- Physics installations: SLAC engineering and physics staff have been developing details for the cryogenic distribution system (CDS) piping, laser gun, laser pipe, waveguides, and controls. Final arrangement within the LEIT cross section is shown in Figure 8.

A cross-section of the LEIT in relation to the Linac and Klystron Gallery is shown in Figure 9.

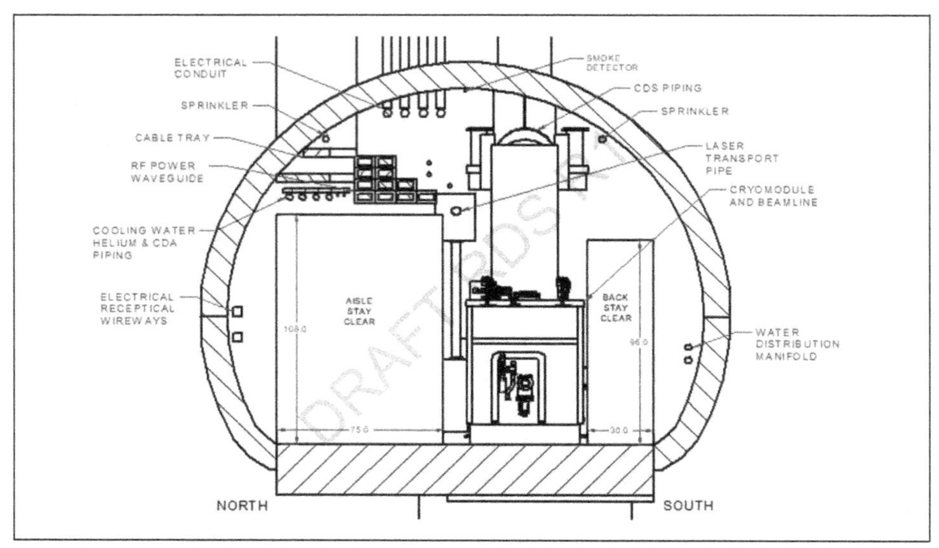


**Figure 8. LEIT cross section with SLAC physics installations**

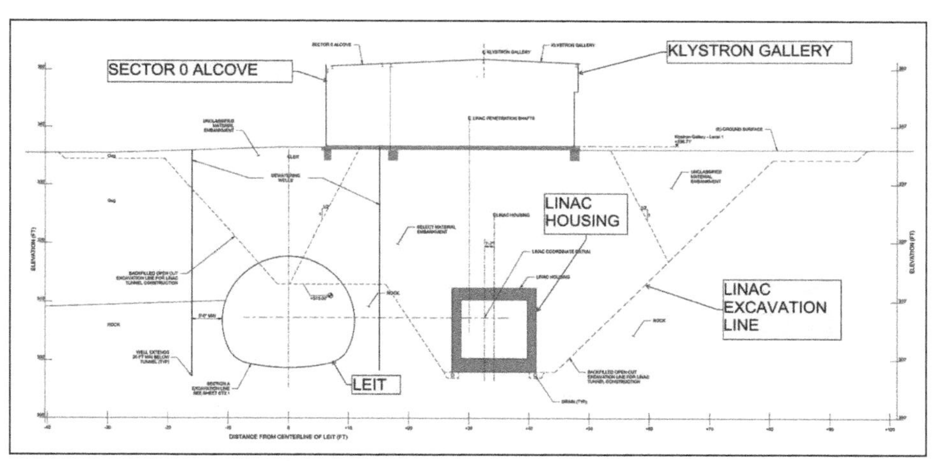


**Figure 9. Subsurface cross section of the LEIT**

## Geotechnical Investigations and Characterization

Two phases of geotechnical investigations have been recently performed at the LEIT project site. The first phase was conducted by another engineering contractor in 2021 and the second phase by Mott MacDonald in 2022. The investigations included the following exploration techniques: rotary wash and hollow-stem auger drilling through soils, rock coring, downhole suspension logging, installation of four vibrating wire piezometers, laboratory testing, and two seismic refraction geophysical surveys. The locations of the explorations are shown on Figure 10.

The soil materials encountered in the explorations are summarized in Figure 11, organized by Unified Soil Classification System (USCS) soil group (ASTM 2017). Soil was found to have Standard Penetration Test (SPT) (ASTM 2018) blow counts generally ranging from 20 to 50+, with a vast majority above 35 blows per foot, indicating hard consistency (Terzaghi and Peck 1967).

The blow counts observed support the assumption that backfill was compacted well during the original Linac construction, likely to the targeted 90% to 95% relative compaction. No attempt was made to sample drain rock due to the need to protect the functionality of the drain system. Records of the drain rock particle size distribution are not readily available at this time.

Rock has been characterized in terms of Rock Quality Designation (RQD) (ASTM D6032) and Geological Strength Index (GSI) (Marinos and Hoek 2000). The distribution of RQD and intact rock unconfined compression strength (UCS) (ASTM D2166 and ASTM D7012D) encountered in the borings is shown in Figure 12 and Figure 13, respectively.

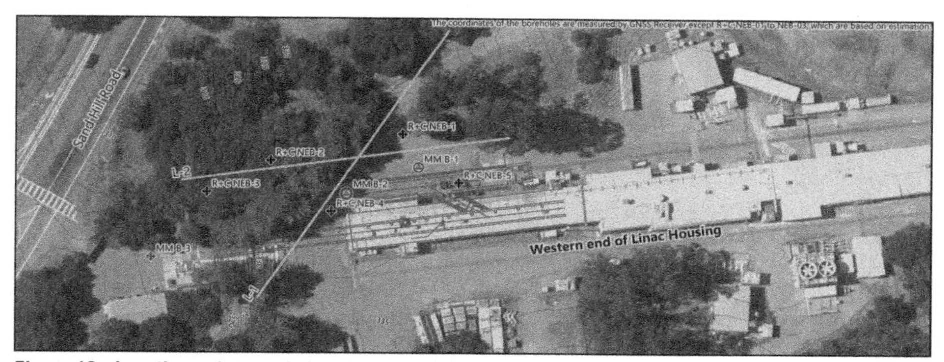


Figure 10. Locations of geotechnical investigations

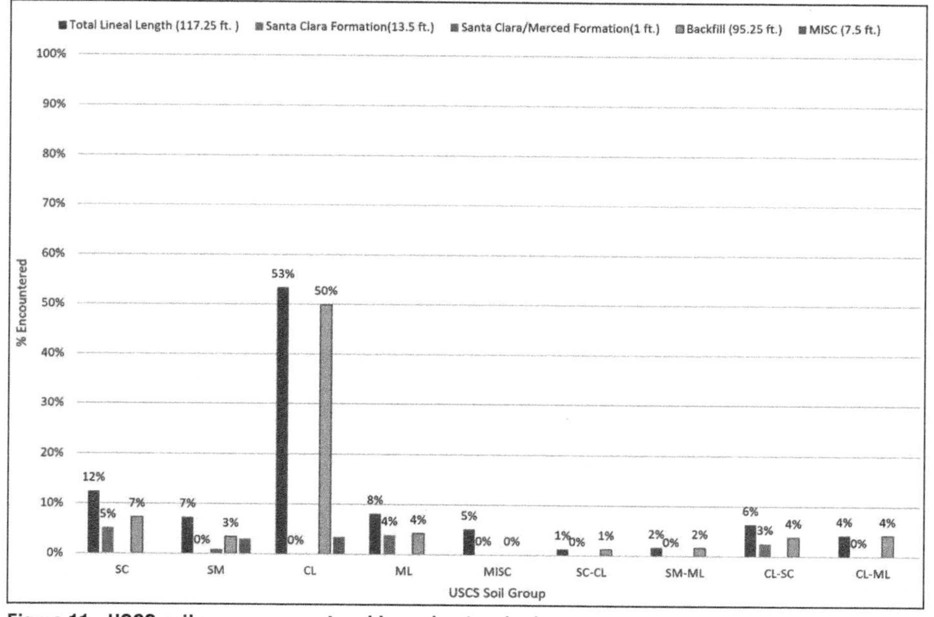


Figure 11. USCS soil groups encountered in exploratory borings

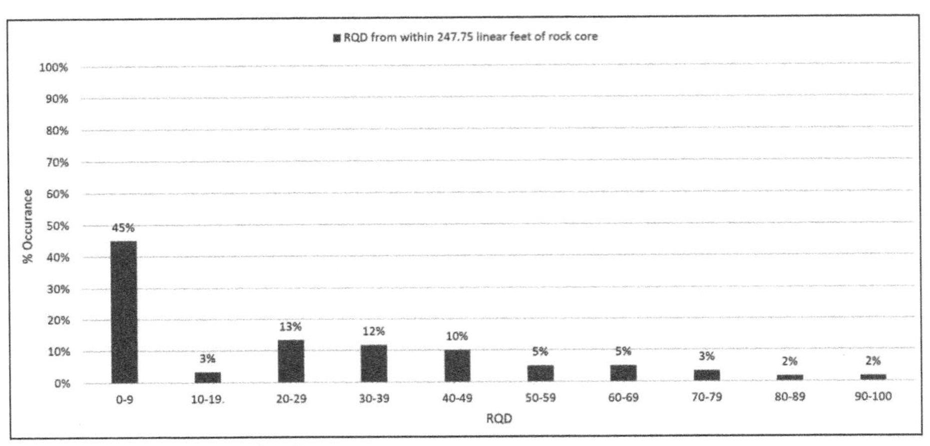


**Figure 12. RQD of recovered rock core**

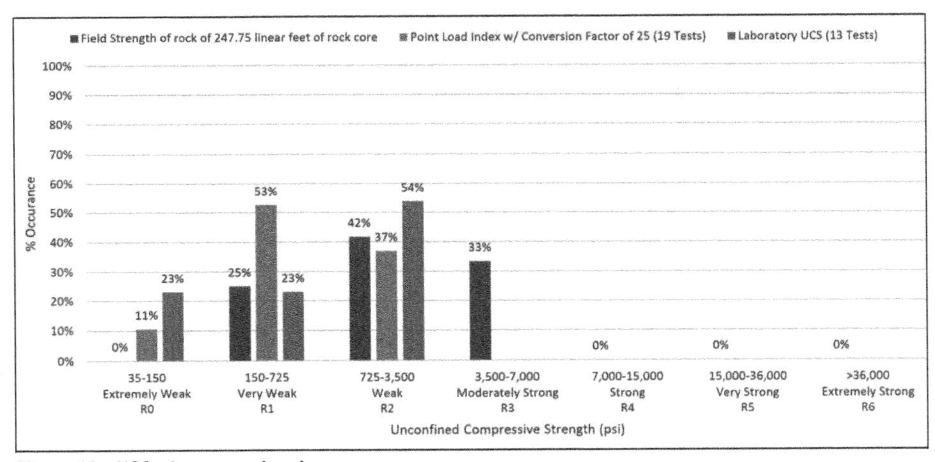


**Figure 13. UCS of recovered rock core**

Groundwater readings from the VWP have been recorded daily since March 2022, suggesting a generally stable depth to the groundwater table at between 34 and 40 feet below the ground surface. These depths are within rock and are coincident with the planned invert depth of the LEIT. The VWP will be monitored through spring 2023 to document changes in groundwater levels during the California rainy season.

## Alignment Selection

SLAC commissioned a feasibility study in 2021 to assess several alignments and construction methods for the LEIT and Transfer Tunnel that could be accommodated during a future 12-month-long scheduled outage of the Linac beamline. From a scientific standpoint, the preference was to construct a 14-foot-wide LEIT relatively close to the Accelerator Housing. The LEIT and Accelerator Housing would be connected to each other by a Transfer Tunnel angled at 35 degrees. This configuration is indicated in Figure 2. Minimizing the offset between the two tunnels would maximize space for racks and controls west of the LEIT injector gun, a piece of equipment that requires a fixed amount of floor space.

The feasibility study concluded that both sequential excavation mining (SEM) and cut-and-cover construction methods were feasible for the LEIT alignment separated from the Accelerator Housing by eight feet. Mott MacDonald is evaluating the preferred configuration against various tunneling and open cut construction methods, providing an analysis of several optimized alignments, considering management of cost and schedule risk, and completing the design of the selected alternative, along with considerations applicable to the 12-month scheduled Linac outage.

Conceptual engineering (30%) commenced on February 4, 2022. To accommodate SLAC requirements for project implementation, alternatives evaluation and recommendation for a preferred alternative were due to be complete on May 20, 2022, and 60% design of the preferred alternative to be complete on September 9, 2022. This required a collaborative design approach involving frequent workshops and coordination meetings.

During 30% design, SLAC and Mott MacDonald worked collaboratively to bring tunneling and physics expertise together, weigh alternatives for the LEIT alignment, and assess constructability options to reduce project risk. At targeted risk workshops, the project team identified and discussed the major project risks associated with each alignment and construction alternative and determined appropriate mitigations. The following major conclusions from the risk workshop and subsequent construction methods analysis guided the selection of the preferred alternative:

- The desired minimum eight feet of separation between the two tunnels required mining directly below the Section 0 alcove and the northern spread footing of the Klystron Gallery and posed a high constructability risk. A continuous pipe canopy would be required over the LEIT to mitigate risk of ground movement and building damage. However, the mitigation would come at a high cost. The East Stairwell Shaft also would require underpinning or structural modifications to the Kystron Gallery to allow space for construction.
- Shifting the LEIT northwards to provide 22 feet of separation would result in only the northern foundation of the Sector 0 Alcove overlying the LEIT, reducing settlement risks to a more manageable level. SEM techniques for both the LEIT and the Transfer Tunnel would be used to construct the new facility.
- Further increasing the separation between the Accelerator Housing and LEIT to 28 feet would allow either open cut construction methods or SEM for the LEIT. Compared to SEM, open cut construction would result in greater disruption to SLAC facility operations. The Transfer Tunnel would still need to be mined or the Klystron Gallery underpinned, as previously done for the damping ring project.
- Multiple contractors for the LEIT would increase risk of schedule slip and require more staging and coordination. Thus, a single construction method was preferred as mitigation.
- Increasing the separation between the tunnels would result in lost space west of the west maze shielding would need to be mitigated by construction of an enlarged west portal structure.

Thirty-percent design-level alignments were developed for 8-foot, 22-foot, and 28-foot separations. Following a detailed alternatives ranking, again involving collaborative effort to identify evaluation criteria and the relative importance of each, the project team reached consensus to proceed to detailed design with the 22-foot separation concept. The proposed alignment was approved to move forward with preliminary design.

## SEM Tunneling Methods for the LEIT and Transfer Tunnel

The SEM excavation method offers flexibility to adapt to the varying ground conditions anticipated along the project alignment. However, a number of specific items related to the LEIT tunnel construction needed to be addressed as part of the project design:

- The LEIT will be approximately 20 feet below the ground surface (one tunnel diameter) and for approximately 50 linear feet it will be below the Sector 0 alcove. Development of a tunnel excavation and support sequence that controls ground movement is a major aspect of the LEIT construction.
- The tunnel face will consist of backfill over rock and the contact location and composition may fluctuate along the alignment. The mixed-face ground conditions also necessitate careful excavation and support techniques.
- A zone of "chaotic" rock was mapped during the original Linac construction at the eastern side of the tunnel, including the East Stairwell Shaft. The chaotic rock will have a greater number of lithologies, increased shearing, and lower intact rock strength.
- At least one non-active fault lineament has been mapped obliquely crossing the LEIT.
- Excavation and support is currently planned to start at the west portal, but a second heading might be operated from the east shaft if needed.

Structural modeling of the LEIT SEM excavation and initial support was performed using Itasca's FLAC 8.0 finite difference software. Soil properties were idealized to vary with depth and are summarized as follows: undrained shear strength between 750 and 3,500 pounds per square foot (psf), 0-degree friction angle, undrained Young's moduli (E) between 250,000 and 700,000 psf, and unit densities between 128 and 135 pounds per cubic foot (pcf). Rock mass parameters were developed using the Hoek-Brown failure criterion (Hoek 2007) for three design types of rock mass conditions, later converted to equivalent Mohr-Coulomb parameters for FLAC model implementation:

- Te-1 (average): GSI=55, UCS of 1,800 pounds per square inch (psi), and E of 680,000 psi.
- Te-2 (average): GSI=30, UCS of 700 psi, and E of 100,700 psi.
- Te-2 (lower bound): GSI=20, UCS of 180 psi, and E of 68,000 psi.

The three-dimensional effects of tunnel excavation ground relaxation were based upon ground reaction curves generated in FLAC and empirical methods for longitudinal displacement profiles described by Vlachopoulos and Diederichs (2008). Settlement mitigation measures were evaluated in FLAC and have been incorporated into the detailed design of the LEIT.

Mining using top heading and bench for Te-2 (average) rock conditions, with a maximum of a 40 ft bench length, has been confirmed using FLAC to satisfy allowable criteria of 0.5 inches of vertical displacement and $1.3 \times 10^{-3}$ in/in of angular distortion of overlying structures. Mining within Te-1 (average) could be performed using full-face excavation and still meet settlement requirements as confirmed by structural modeling. Te-2 (lower bound) rock conditions will require additional pre-excavation support measures during tunneling. For detailed design, however a single excavation sequence will be shown with additional SEM toolbox methods.

The LEIT initial support system will consist of 12 inches of fiber reinforced shotcrete and steel lattice girders installed at three-foot spacing. Both ends of the LEIT will be pre-supported by a pipe canopy system: a 70-foot-long pipe canopy at the west portal

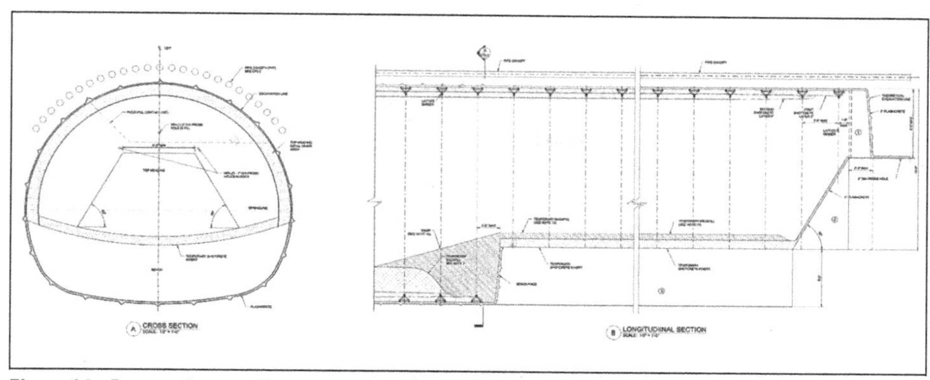


**Figure 14. Proposed excavation sequence of the LEIT under a pipe canopy**

where shallow cover exists, in order to reach an overburden depth of one tunnel diameter and a 30-foot-long pipe canopy at the East Stairwell Shaft.

Other SEM ground control requirements include continuous probe hole drilling, installation of weep holes, maintaining a face wedge, and prescriptive spiling when outside the pipe canopy areas. For poorer ground conditions, such as Te-2 (lower bound), additional SEM toolbox measures include installation of face dowels, pocket excavation to rapidly install the initial support, and tunnel face drain holes. The proposed excavation sequence is shown in Figure 14.

The Transfer Tunnel has both a smaller cross section and a shorter length than the LEIT. This tunnel will be mined to the north wall of the Accelerator Housing once the LEIT is fully excavated and supported. Mining will be performed under protection of a 34-foot-long pipe canopy and is expected to consist of a full-face excavation using a face wedge, probe holes, and SEM toolbox items as indicated for the LEIT.

Spiling for the LEIT will need to be removed in the Transfer Tunnel breakout area for installation of the pipe canopy and subsequent mining sequence. The Transfer Tunnel will be mined directly below the Klystron Gallery for several feet through the Accelerator Housing drain rock. Figure 15 shows a proposed excavation sequence of the Transfer Tunnel towards the Accelerator Housing. Prior to breakthrough of the

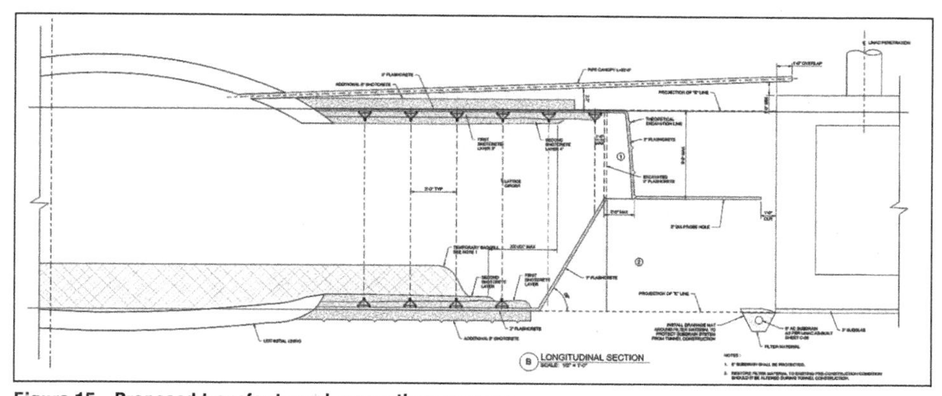


**Figure 15. Proposed transfer tunnel excavation sequence**

Transfer Tunnel into the Accelerator Housing, sensitive scientific equipment in the Accelerator Housing will need to removed or otherwise be protected from dust, vibrations, and damage.

## Tunnel Waterproofing

Groundwater must be prevented from entering the underground structures due to stringent requirements for handling, treatment, and disposal of radiation contaminated water. To meet these objectives, a waterproofing membrane will be installed between the initial support and final lining to completely seal all underground structures, including the LEIT, Transfer Tunnel, and East Stairwell Shaft.

The proposed membrane must be capable of withstanding deterioration from radiation exposure, even once protected from direct exposure by the tunnel final lining. SLAC has expressed a preference for materials that can better withstand radiation exposure, thus from a technical standpoint very low-density polyethylene (VLDPE) was initially preferred. Polyvinyl chloride (PVC) waterproofing membrane has also been evaluated by the SLAC radiation protection team and has demonstrated satisfactory long-term performance when placed behind 12 inches of normal density concrete.

Approximate radiation tolerances of individual polymeric materials provided by G. Von White et al. (2013) were reviewed in conjunction with SLAC. Based upon this reference and upon conversations with waterproofing suppliers with experience of installations at CERN, the European Organization for Nuclear Research and the Hinckley Point nuclear facility, a 100-mil PVC membrane is expected to meet the project durability requirements, while providing constructability and cost benefits when compared to similar-performance VLDPE membranes.

## Tunnel Final Lining

The LEIT and Transfer Tunnel will be finished with a 12-inch-thick cast-in-place conventionally reinforced concrete lining designed in accordance with the 2019 California Building Code. A shotcrete final lining is also being considered as an alternative to cast-in-place concrete. SLAC radiation physics have used computational models to verify that the combined thickness and geometry of the final lining and initial support are sufficient to contain radiation emissions produced during laser operation.

Calculations for the tunnel final lining were performed using FLAC software, built using the same model developed for design of the initial support. For long-term static conditions the final lining was modeled using two-dimensional beam elements. The initial support was removed later, for load transfer from the initial support to the final lining. The waterproofing membrane was modeled using an interface element with a 10-degree friction angle to allow slippage between the ground elements and the final lining. The groundwater was raised from below the tunnel to 15 feet below the ground surface, soil properties were adjusted for long-term behavior as shown on Figure 16, and permanent dead and live loads were added to the lining.

Seismic analyses of the tunnel final lining have been performed in accordance with ASCE-07. A probabilistic seismic hazard assessment for the site was conducted, then seven spectrally matched acceleration time histories were developed according to the assessment results. The time histories were applied to one-dimensional (1D) ground profiles and modeled using nonlinear soil deformation properties, and mean maximum ground deformations versus depth were obtained.

The resulting surface mean peak ground accelerations were 0.53 g (rock only) and 0.65 g (fill over rock) and surface displacements were between four inches (rock only) and six inches (fill over rock). The 1D ground deformation results were then applied to the FLAC models for the LEIT and the Transfer Tunnel using a pseudo-static seismic loading approach (see Figure 17). The resulting ovaling deformation of the tunnel

| Long-term | | | | | | | | | | | | |
|---|---|---|---|---|---|---|---|---|---|---|---|---|
| Unit | Sublayer | Elevation Range | Total Unit Weight, $\gamma_t$ | E | K | G | Cohesion Intercept, c' | Friction Angle, $\phi$ | Dilatancy, $\psi$ | Poisson's Ratio, $\nu$ | Tension Cutoff | At-Rest Earth Pressure Coefficient, $K_0$ |
| | | (ft) | (pcf) | (psf) | (psf) | (psf) | (psf) | (°) | (°) | | (psf) | |
| Unclassified Material Embankment | UME 01 | Above +332 | 128.0 | 250,000 | 416,667 | 89,286 | 1,500 | 25 | 0 | 0.4 | 0 | 1.00 |
| | UME 02 | +332 to +327 | 130.1 | 250,000 | 416,667 | 89,286 | 1,500 | 25 | 0 | 0.4 | 0 | 1.00 |
| | UME 03 | +327 to +322 | 135.0 | 250,000 | 416,667 | 89,286 | 1,500 | 25 | 0 | 0.4 | 0 | 1.00 |
| | UME 04 | +322 to +317 | 135.0 | 250,000 | 416,667 | 89,286 | 1,500 | 25 | 0 | 0.4 | 0 | 1.00 |
| | UME 05 | Below +317 | 135.0 | 250,000 | 416,667 | 89,286 | 1,500 | 25 | 0 | 0.4 | 0 | 0.87 |
| Select Material Embankment | SME 01 | Above +332 | 135.0 | 380,000 | 633,333 | 135,714 | 20 | 42 | 0 | 0.4 | 0 | 0.84 |
| | SME 02 | +332 to +327 | 135.0 | 380,000 | 633,333 | 135,714 | 20 | 42 | 0 | 0.4 | 0 | 0.55 |
| | SME 03 | +327 to +322 | 135.0 | 380,000 | 633,333 | 135,714 | 20 | 42 | 0 | 0.4 | 0 | 0.45 |
| | SME 04 | +322 to +317 | 135.0 | 380,000 | 633,333 | 135,714 | 20 | 42 | 0 | 0.4 | 0 | 0.40 |
| | SME 05 | Below +317 | 135.0 | 380,000 | 633,333 | 135,714 | 20 | 42 | 0 | 0.4 | 0 | 0.37 |
| Drain Rock | - | +317 to +299.5 | 135.0 | 750,000 | 500,000 | 300,000 | 500 | 33 | 0 | 0.25 | 0 | 0.33 |
| Rock | Te-2 (AVG) | - | 135.0 | 14,500,800 | 9,667,200 | 5,800,320 | 432 | 49 | 0 | 0.25 | 49 | 0.67 |

**Figure 16. Soil and rock parameters modeled in FLAC to capture long-term conditions**

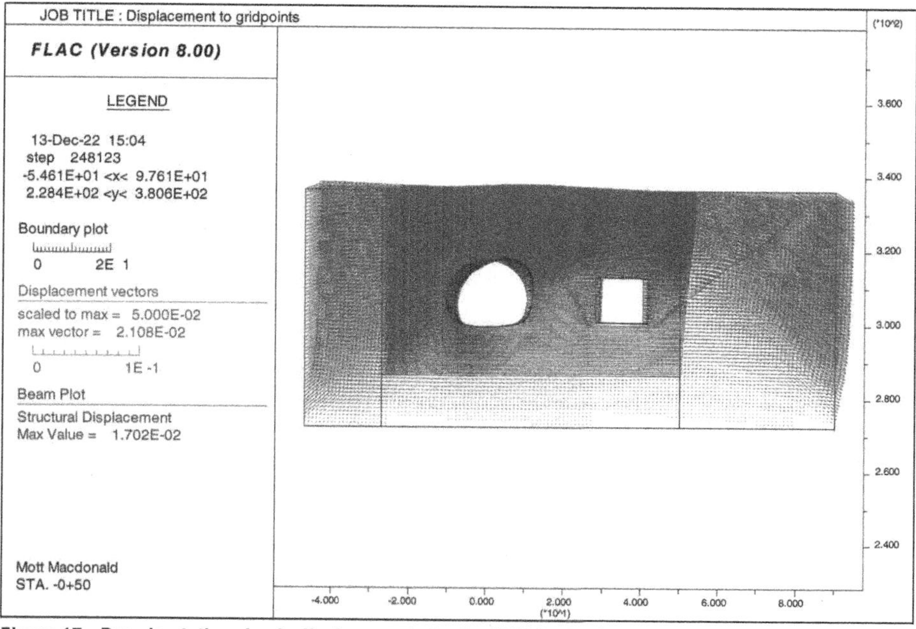


**Figure 17. Pseudo-static seismic displacements for the LEIT (in feet)**

required additional shear reinforcement of the final lining at the 5 o'clock and 7 o'clock positions along the tunnel perimeter.

### Tunnel Junction Between the Accelerator Housing and Transfer Tunnel

The junction of the Transfer Tunnel and Accelerator Housing will be constructed in the following sequence:

- SLAC will remove existing physics equipment in the Accelerator Housing from the opening location.
- Dust control and structural monitoring instruments will be installed within the Accelerator Housing around the opening location.
- Steel bracing will be installed around a 15-foot-wide by 7-foot-tall rough opening of the Accelerator Housing north wall. The bracing will consist of three columns and one header beam connected to the ceiling of the Accelerator Housing between the left-in-place CDS piping and the north wall of the Accelerator Housing. Bracing materials and installation equipment will need to be delivered through an existing equipment access shaft near the junction.
- The Transfer Tunnel will be mined from the LEIT to the wall of the Accelerator Housing.
- The needed area of the Accelerator Housing wall will be removed by concrete coring and saw cutting.
- The Transfer Tunnel final lining will be constructed, including the waterproofing membrane and a headwall.
- A permanent concrete collar will be constructed around the opening of the Accelerator Housing wall.
- A full-perimeter expansion joint and waterproofing seal will be installed around the opening.

### 3D BIM Modeling

Design decisions have been continually verified using Building Information Modeling (BIM) methods. Software used to develop the BIM framework include Autodesk Revit for structures, MEPC, and architecture; and Autodesk Civil 3D for topographic surfaces, site utilities, and incorporation of a survey base map. The existing Accelerator Housing and Klystron Gallery were modeled in Revit based on original construction as-builts. Changes to the 3D Revit model are periodically uploaded to the BIM 360 web domain and used as a collaboration tool by the Mott MacDonald design team and SLAC subject matter experts to inform space-proofing design in the shaft and tunnel, clash detection, science equipment installation, and coordination for discussion in meetings and workshops. A snapshot of the Revit model is shown in Figure 18.

## CONSTRUCTION SCHEDULE

The Linac beam outage is scheduled to last 12 months. During this time the following project activities are planned:

- Site clearing and tunnel portal development.
- Clearing of Linac installations at the location of the junction with the Transfer Tunnel. The CDS piping within the Accelerator Housing will need to be protected in place.
- Installation of temporary support for the Accelerator Housing tunnel breakout.
- Construction of the LEIT, Transfer Tunnel, and Accelerator Housing permanent collar.
- Construction of the west portal.

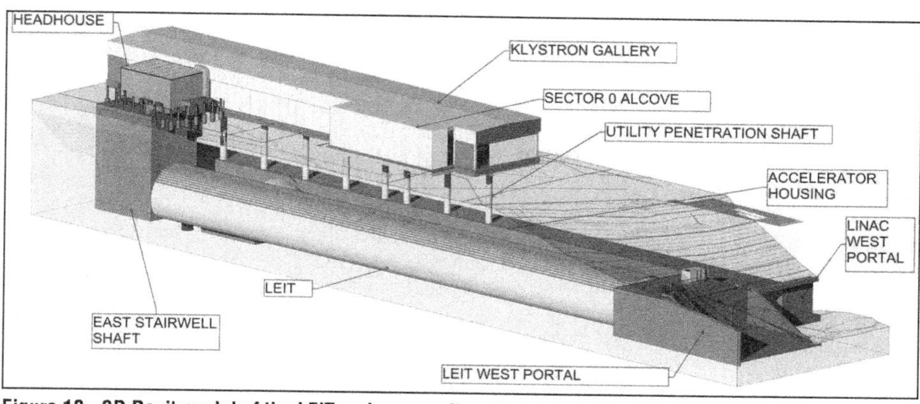


**Figure 18. 3D Revit model of the LEIT and surrounding structures**

- Construction of the East Stairwell Shaft.
- Installation of tunnel radiation shielding.

Once the radiation shielding is installed, the Linac beamline will be reconstructed at the Transfer Tunnel penetration area and can resume operation. The following activities can occur beyond the radiation shielding after the beam is operational:

- MEP, communications, technology, and fire protection installations.
- Installation of LEIT physics equipment.
- Commissioning of the LEIT.

## CONCLUSION

SLAC is implementing a major facility upgrade via the LCLS-II-HE Project. This will provide a significant increase in laser energy, allowing cutting-edge research in fields including biology and environmental science. A collaborative design approach has been implemented for the project, which has enabled decisions on preferred alignment and construction methods to be made in an expedited manner to maintain the overall schedule. A tunneled solution comprising SEM construction methods is proposed for the LEIT and Transfer Tunnel to achieve the construction schedule goals. Once the LEIT tunnel is commissioned, the SLAC LCLS-II-HE will offer increased research availability to the LCLS-II accelerator and improved beam brightness.

## REFERENCES

AACE International (AACE). "Cost Estimate Classification System—As Applied in Engineering, Procurement, and Construction for the Mining and Mineral Process Industries." AACE International Recommended Practice No. 47R-11. Revised August 7, 2020.

American Society of Civil Engineers (ASCE). Minimum Design Loads for Buildings and Other Structures, ASCE/SEI 7-16. Reston, VA: ASCE. 2016.

ASTM International (ASTM). D1586-18: Standard Test Method for Standard Penetration Test (SPT) and Split-Barrel Sampling of Soils. West Conshohocken, PA: ASTM International. ASTM International (ASTM). 2018.

ASTM International (ASTM). D2166-16: Standard Test Method for Unconfined Compressive Strength of Cohesive Soil. West Conshohocken, PA: ASTM International. 2016.

ASTM International (ASTM). D2487-17: Standard Practice for Classification of Soils for Engineering Purposes (Unified Soil Classification System). West Conshohocken, PA: ASTM International. 2017.

ASTM International (ASTM). D6032-17: Standard Test Method for Determining Rock Quality Designation (RQD) of Rock Core. West Conshohocken, PA: ASTM International. 2017.

ASTM International (ASTM). D7012-14: Standard Test Methods for Compressive Strength and Elastic Moduli of Intact Rock Core Specimens under Varying States of Stress and Temperatures. West Conshohocken, PA: ASTM International. 2014.

Hoek, Evert. "Practical Rock Engineering." Lecture Notes. University of Toronto. 2007.

Linac Coherent Light Source—LCLS-II: A World-Class Discovery Machine. 2022. https://lcls.slac.stanford.edu/lcls-ii. Accessed November 2022.

Marinos, P. and Hoek, E. 2000. "GSI: a geologically friendly tool for rock mass strength estimation." Proceedings from the GeoEng 2000 Conference, Melbourne.

Neal, R. B. , D. W. Dupen, H. A. Hogg, and G. A. Loew. 1968. *The Stanford Two-Mile Accelerator.* New York, New York: W.A. Benjamin, Inc. Digital copy of print.

Nosochkov Y., C. Adolphsen, R. Coy, C. Mayes, T. Raubenheimer, M. Woodley. 2022. Electron Transport for the LCLS-II-HE Low Emittance Injector. In *13*th *Int. Particle Acc. Conf,* Bangkok, Thailand. June 12–17. JACoW Publishing. doi:10.18429/JACoW-IPAC2022-TUPOPT046.

SLAC. 2006. *The Geology of the Stanford Linear Accelerator.* ES&H Division. SLAC Report SLAC-I-750-3A33X-002. Digital copy.

Terzaghi, K. and R. Peck. 1967. *Soil Mechanics in Engineering Practice.* 2nd Edition. New York, New York: John Wiley & Sons, Inc. 1967. Print.

Vlachopoulos, N. and M.S. Diederichs. "Improved Longitudinal Displacement Profiles for Convergence Confinement Analysis of Deep Tunnels." Rock Mechanics and Rock Engineering (2009) 42: 131–146. 2009.

White, II, Gregory Von, Tandon, Rajan, Serna, Lysle M., Celina, Mathias C, and Bernstein, Robert. 2013 "An Overview of Basic Radiation Effects on Polymers and Glasses." United States: N. p. Web.

PART 

# Design–Build Projects

*Chairs*

**Jay Sankar**
Amtrak

**Julio Martinez**
Skanska

# A Case Study in Successful Progressive Design Build Tunneling

**Leo C. Weiman-Benitez** ▪ Barnard Bessac Joint Venture
**Nik Sokol** ▪ Arup US, Inc.
**Mike Jaeger** ▪ Tanner Pacific, Inc. representing Silicon Valley Clean Water

## ABSTRACT

Silicon Valley Clean Water signed a progressive-design-build (PDB) contract with Barnard Bessac Joint Venture (BBJV) to design and construct the Gravity Pipeline tunnel in Redwood City, CA. Owner and Design-Builder teamed with designer Arup to provide a final product: 3.3 miles of 182 in. O.D. precast concrete segmentally lined tunnel with 10 ft. and 11 ft. I.D. Fiberglass Reinforced Polymer Mortar (FRPM) carrier pipe, complete with two FRPM Drop Structures. BBJV and Arup collaborated with SVCW to evaluate alternative design concepts, negotiate an agreeable price and reduce contingencies through a transparent estimating process, and deliver the project ahead of schedule.

## INTRODUCTION

### Silicon Valley Clean Water

Silicon Valley Clean Water (SVCW) is dedicated to conveying and treating wastewater from more than 220,000 people and businesses in the service area in San Francisco Bay Region of California. They are a Joint Powers Authority serving the communities of Belmont, Redwood City, San Carlos, and the West Bay Sanitary District. They value environmental stewardship and innovation and, by effectively treating wastewater at their advanced treatment facility, helping to keep San Francisco Bay clean and environmentally healthy.

### Development of RESCU program

SVCW is undertaking improvements to the reliability of its wastewater conveyance system, which is deteriorating and needs to be replaced. An exhaustive conveyance system alternatives analysis identified over 140 alignments, pumping arrangements and pipeline installation methods. Based on a feasibility assessment, these 140 alternatives were reduced to 15 feasible alternatives. Figure 1 provides an overview of the selected pipeline alignment.

Due to the complexity of the RESCU program, the scope was divided into the Gravity Pipeline Tunnel (GP); Pump Stations Improvements (PSI); and Front of Plant (FoP) Projects. This division of scope streamlined the design and construction process. This manuscript details only the GP scope of the RESCU program.

### Solicitation of the Gravity Pipeline Tunnel Project

SVCW determined the most efficient manner of procuring the GP was through a progressive design-build delivery. The project was solicited to the industry via a Request for Qualifications (RFQ), followed by a Request for Proposal (RFP) issued to three teams pre-qualified based on their RFQ responses. The Barnard Bessac Joint Venture with Arup as their designer was selected as the Design-Builder based

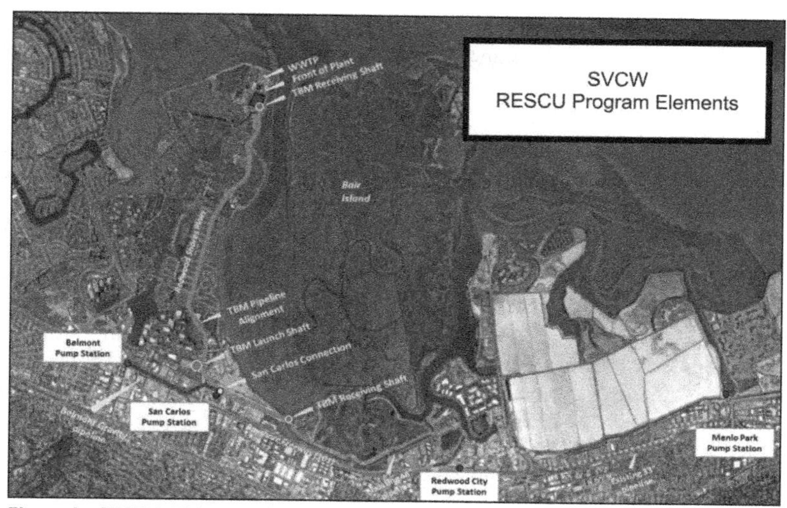


**Figure 1. SVCW RESCU program elements. The Gravity Pipeline extends between the two TBM receiving shafts**

on their qualifications and their proposal. The design-build contract was signed on October 2017.

## BBJV & ARUP INVOLVEMENT

### Progressive Design Build Contract Structure

The Design-Builder: Barnard Bessac Joint Venture is comprised of 70% Barnard Construction Co. Inc., a major North American heavy civil construction contractor, and 30% Bessac, Inc. a world-renowned tunneling contractor headquartered in France. Barnard's familiarity with construction projects of this magnitude, complexity and location and Bessac's specialization in tunnel construction with similar dimensions and geology facilitated the ultimate success of the project.

BBJV contracted with Arup North America Ltd. as the Designer Consultant. Arup drew from local and global resources to provide expertise in tunnel engineering and alternatives analysis, utilizing a wide breadth of engineering disciplines to deliver time and cost efficient designs for the GP project.

SVCW generally structured the PDB contract with BBJV into Stage 1 and Stage 2 services. The goal of Stage 1 was to advance the design far enough to allow a construction contract to be agreed upon, while the goal of Stage 2 was to finalize the design alongside the complete construction of the project. Certain design elements were accelerated during Stage 1 to allow purchase of long-lead project elements, such as the Earth Pressure Balance Tunnel Boring Machine (TBM) and precast concrete tunnel lining segment moulds. Stage 1 was signed in October 2017. Stage 2A was signed in August 2018, and Stage 2B was signed in November 2018.

### Scope of Work & Schedule

Figure 1 is a summary of the features of work which comprise the GP project. Major milestone dates are included to emphasize the "quick turnaround" between design completion (Issued for Construction: IFC) and the start of construction. Minimal lag time between design completion and start of construction is where the PDB contract structure gains an advantage over other contract structures.

### *Airport Access Shaft (AAS)*

This 58-ft diameter, 52-ft deep structure utilized reinforced concrete slurry wall support of excavation (SOE). This structure was used as the means to access both branches of the GP system: Tunnel Drive 1 (TD1) and Tunnel Drive 2 (TD2). The (TBM) launched from this shaft for both tunnel drives. The Carrier Pipe was also installed into both Tunnel Drives from this shaft. See Figure 2 for a photo of the Airport Access Shaft.

To facilitate mucking, a 9-ft ID, 246-ft long inclined conveyor tunnel was constructed by pipe-jacking a steel casing pipe at a 10-degree incline from a surface trench until it intersected the AAS near the base of the shaft. This inclined tunnel also served as an additional means of access to the base of the shaft.

To allow additional space and reduce the amount of umbilical work required for TBM launch, two TBM Starter Galleries were constructed using Sequential Excavation Methods. Twin 18-ft tall × 17-ft wide mined excavations extended 42-ft horizontally out from the bottom of the AAS, in the direction of the future tunnel drives. Ground support for the Starter Galleries was provided by a pipe canopy, steel lattice girders and steel fiber reinforced shotcrete.

Major milestone dates in the design and construction of this feature of work are as follows:

- 60% Design Completed: June 2018
- IFC Design Completed: December 2018
- Construction Start Date: December 2018
- Construction Complete: May 2019

**Figure 2. AAS after excavation, during construction of the base slab, prior to TBM Starter Gallery construction**

Subcontractors involved with this feature of work and their scopes:

- Tri-Valley Excavating Co., Inc.: worksite development including: clearing & grubbing, grading, stormwater management, & paving
- Malcolm Drilling Company, Inc.: slurry wall SOE
- Drill Tech Drilling & Shoring, LLC: Inclined Conveyor Tunnel pipe-jacking, SEM mined TBM Starter Galleries

### *Tunnels*

An approximately 16-ft OD Earth Pressure Balance (EPB) TBM was used to excavate the tunnels and install a 15-ft, 2-in OD, 10-in thick steel fiber reinforced precast concrete segmental liner. The TD1 is approximately 5,100-ft long, and TD2 is approximately 12,300-ft long. Depth of the crown of the tunnel structure varied from approximately 30-ft below surface grade at the beginning of the Gravity Pipeline on Bair Island to about 50-ft below surface grade at the end of the Gravity Pipeline, where it interfaces with the Front of Plant project at SVCW's new headworks facility.

See Figure 3 for a photo of a section of completed tunnel.

Major milestone dates in the design and construction of this feature of work are as follows:

- Precast Concrete Tunnel Liner 60% Design Completed: June 2018
- TBM Order Date: August 2018
- TBM Arrival Date: July 2019
- Precast Tunnel Liner Supply Agreement Signature: January 2019
- First Precast Concrete Tunnel Liner Segments Arrive: July 2019
- Tunnel Drive 1 Start Date: September 2019
- Tunnel Drive 1 Hole-through: March 2020
- Tunnel Drive 2 Start Date: June 2020
- Tunnel Drive 2 Hole-through: June 2021

**Figure 3. The "S-Curve" in TD2, after mining completion, prior to temporary utility removal and carrier pipe installation**

Suppliers & Subcontractors involved with this feature of work and their scopes:

- *Herrenknecht Tunneling Systems USA, Inc.*: TBM
- *Traylor Shea Precast, AJV*: precast concrete segmental tunnel liner

***Bair Island (BI) Shaft***

This 44-ft deep, 60-ft long × 32-ft wide structure utilized steel sheet piles and bracing as SOE. This excavation was used as the retrieval shaft for the TBM after TD1. The TBM break-in was accomplished using an innovative approach whereby a grout block was cast inside of the shaft, the shaft was flooded, the TBM mined into the grout block, landed onto a steel cradle pre-installed on the shaft invert, and was retrieved after the shaft was de-watered.

After the Carrier Pipe was installed, the BI FRPM Vortex Drop Inlet was installed in the shaft, as detailed below. The SOE was removed upon backfilling the drop structure. The Vortex Drop Inlet serves as one of the two connections where the existing sewer network flows into the GP system.

See Figure 4 for a photo of the BI Shaft.

**Figure 4. BI Shaft after excavation, during casting of the grout block, prior to receiving the TBM**

Major milestone dates in the design and construction of this feature of work are as follows:

- 60% Design Completed: November 2018
- IFC Design Completed: August 2019
- Construction Begin: August 2019
- Construction Complete: January 2020

Subcontractors involved with this feature of work:

- *Blue Iron Foundations and Shoring, LLC*: steel sheet pile & bracing SOE

### *Carrier Pipe*

Installed inside the tunnels is a pipeline of approximately 900 pieces of 11-ft and 10-ft Inside Diameter (ID) × 20 ft long, 15,000-lb Fiberglass Reinforced Polymer Mortar (FRPM) pipe. A custom pipe placing machine was designed and fabricated to handle the pipe with care, while crews installed each pipe section, one piece at a time. At the inaccessible far end of a pipe section, the gasketed joint was designed to support the weight of the pipe on one end, while the accessible near end of the pipe was propped into place using small jacks designed to be left in place during grouting. After completion of the entire pipe installation, the annulus between the tunnel liner intrados and the carrier pipe extrados was grouted with a Low Density Cellular Concrete.

Figure 5 shows a photo of the pipe carrier and pipe installation activity.

Major milestone dates in the design and construction of this feature of work are as follows:

- 60% Design Completed: June 2018
- Carrier Pipe Supplier first solicited: October 2018
- Supply Agreement Signature: June 2019
- First pipe installed: November 2020
- Last pipe installed: April 2022
- Final backfill of carrier pipe (Project Conditional Substantial Completion): May 2022

**Figure 5. A 20-ft long by 11-ft ID FRPM pipe staged for installation into TD2**

Suppliers & subcontractors involved with this feature of work:

- *Future Pipe Industries, Inc.*: FRPM pipe
- *Kelly Engineered Equipment, LLC*: design & fabrication of the custom pipe placing machine
- *Pacific International Grout Co.*: annulus grouting with low density cellular concrete

***San Carlos Shaft, Adit, and Basement Connection***

This complex feature of work is comprised of a vertical shaft, one horizontal connection from the bottom of the shaft to the nearby TD1, and one horizontal connection mid-shaft into the adjacent basement of an existing pump station. The vertical shaft is 14-ft ID (oval) × 48-ft deep utilizing secant piles as SOE. The shaft was required to be excavated in an "un-crewed" fashion via a telescoping clam shell bucket due to the presence of hazardous levels of Ammonia Nitrogen Gas released by the soil. The safety precautions proved fruitful as construction was completed incident-free.

The Adit is a 7-ft ID steel casing pipe which was horizontally jacked from the bottom of the shaft, intersecting the tunnel at a 60-degree angle approximately 25-ft away. A 6-ft ID FRPM carrier pipe was installed into the Adit for the final condition. A special Y-section of FRPM pipe was installed inside the tunnel to form the inlet into the Gravity Pipeline.

See Figure 6 for a photo during SC Adit construction.

**Figure 6. SC Adit pipe jacking frame installation**

Major milestone dates in the design and construction of these features of work are as follows:

- 60% Shaft & Adit Design Completed: November 2018
- IFC Shaft & Adit Design Completed: September 2020
- Shaft Construction Start Date: September 2020
- Shaft Construction Complete: March 2021
- Adit Construction Start Date: July 2021
- Adit Construction Complete: December 2021

The Basement Connection is a horizontal excavation mid-way up the San Carlos shaft wall, which penetrates the Secant Pile SOE of the shaft and the reinforced concrete basement wall of the Pump station, a few feet away. The soil above is supported via ground support consisting of steel spiling and shotcrete. A rectangular horizontal galvanized steel sleeve was installed inside the excavation to form the permanent Basement Connection. Two parallel branches of FRPM carrier pipe—33-in ID and 29-in ID—a were installed inside of the steel sleeve and grouted into place.

See Figure 7 for a photo taken during Basement Connection construction.

Major milestone dates in the design and construction of this feature of work are as follows:

- IFC Basement Connection Design Completed: March 2023
- Basement Connection Construction Start Date: February 2023
  - (Due to our close working relationship, SVCW, Arup & BBJV agreed to start construction on certain items, even though the IFC design package was not complete)
- Basement Connection Construction Complete: April 2023

After the San Carlos Shaft, Adit, and Basement Connection were fully constructed, the San Carlos FRPM Drop Structure, detailed below, was installed in the shaft.

**Figure 7. Spiling installed for ground presupport prior to SC basement connection excavation**

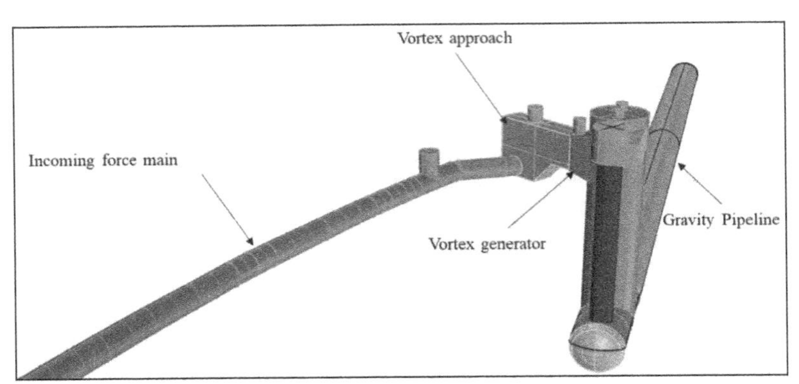


Figure 8. 3-D visualization of the BI drop structure

Subcontractors involved with these features of work:

- *Drill Tech Drilling & Shoring, LLC*: secant pile SOE & clamshell bucket excavation
- *Kelly Engineered Equipment, LLC*: design & fabrication of the adit casing pipe jacking equipment

### *Drop Structures*

Flows enter the GP system at two locations: San Carlos and Bair Island. The main function of the drop structures is to accept flows from the existing network of near the surface, dissipate the energy of the flow and direct it into the deeper GP Carrier Pipe at depth. Directing this flow at both locations is accomplished through custom FRPM structures. The Drop Structure at Bair Island is a vortex generator and vortex shaft consisting of six pieces of FRPM with a total depth of approximately 46-ft deep, and a combined weight of 82,000-lb. The Drop Structure at San Carlos is baffled to allow flow from two inlet pipes to cascade down. It consists of two pieces of FRPM with a total depth of approximately 53-ft, and a combined weight of 68,000-lb.

Figure 9. Lowering the bottom-half of the SC Drop Structure into the SC Shaft

See Figure 8 for a visualization of the BI Drop Structure, Figure 9 for a photo of the SC Drop Structure during installation, and Figure 10 for a visualization of flows entering the gravity pipeline through the Drop Structures.

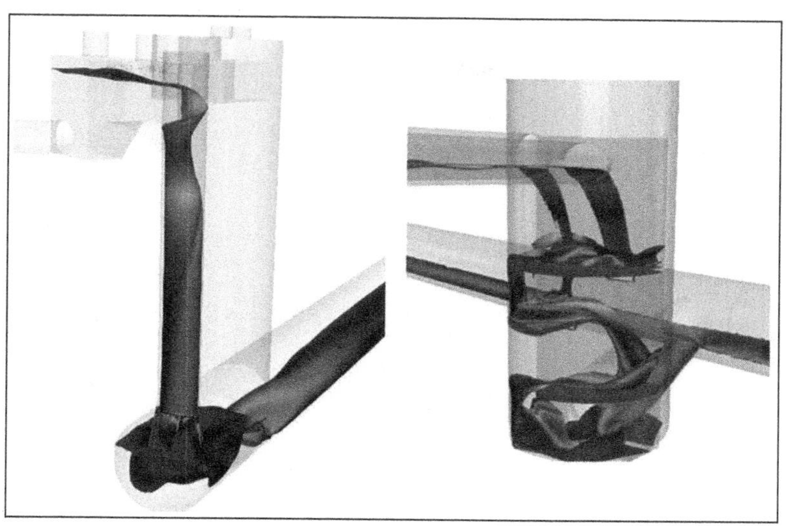

**Figure 10. Computational Fluid Dynamics models of the San Carlos baffle drop structure (right) and the Bair Island vortex drop structure (right)**

Major milestone dates in the design and construction of these features of work are as follows:

- IFC design completed: November 2020
- Supply agreement signature: January 2021
- Drop Structures arrive on-site: January 2022
- Drop Structures final backfill: April 2022

The supplier of the Drop Structures was the same as the carrier pipe: *Future Pipe Industries, Inc.*

### Lookahead

As of the fourth quarter of 2022, commissioning activities are all that remains to complete the Gravity Pipeline project. Commissioning includes the collection of data from flows introduced into the GP system, and evaluation of system performance for design compliance. These commissioning activities will commence once the other RESCU Program FoP and PSI projects are completed, anticipated to be during the third quarter of 2023.

## UNIQUE PROGRESSIVE DESIGN BUILD CHARACTERISTICS

### Importance of Owner/Designer/Design-Builder Collaboration

Progressing the design from the 10% level through IFC required input and agreement from SVCW, Arup and BBJV at every step of the design development. The parties collaborated through brainstorming sessions, workshops, submittal review, and general correspondence. Side-by-side, all parties were able to voice expectations, propose alternatives and respond to each other's concerns. This ultimately led to a design that satisfied SVCW RESCU Program objectives while being optimized for BBJV's means and methods. SVCW was satisfied with the functionality of the end product, Arup was satisfied as Engineer of Record, and BBJV was satisfied with the constructability.

**Figure 11. SVCW, Arup & BBJV project leaders and members of SVCW's advisory team from Tanner Pacific, Kennedy Jenks, Mott MacDonald and JCK in front of the TBM cutting wheel during the TBM naming ceremony**

The PDB environment limited contractual conflict common to the construction industry replaced it with a common interest in getting the job done right. "Finger pointing" was replaced with a team ownership of all challenges encountered. Co-location of SVCW, BBJV and Arup achieved a "level playing field," as everyone's office was only a few steps away. See Figure 11 for a photo of representatives from SVCW, BBJV and Arup standing in front of the TBM after it arrived on site.

## Alternative Design Evaluation

The Progressive Design Build model allowed for the design-build team to work collaboratively with SVCW and their advisory team to evaluate project alternatives through co-located working and workshops. Through Big Room discussions and open-book cost estimating, a number of impactful decisions were made by the collective project team. As described in Sokol, et al., 2020, some of the critical decisions open to evaluation at the beginning of the project include:

- TBM launch shaft location, which determines whether one or two tunnel drives are required to construct the tunnel. The two tunnel drives option was selected in part because it would not require modification to existing permits and EIR.
- Approach to microbial induced corrosion (MIC) protection of the tunnel lining interior, which determines whether the tunnel will be constructed using a onepass or twopass lining method.
- Inlet/hydraulic connection type between the existing system and the Gravity Pipeline; the style and geometry of the inlets has a marked influence on the turbulence and associated hydrogen sulfide released into system.
- Shaft support of excavation (SOE) method.
- Interface between the Gravity Pipeline and Front of Plant projects and the associated shaft(s) configuration at the Receiving Lift Station (RLS).

## Schedule Acceleration

Breaking the contract up into two stages allowed for the critical design elements to be expedited thus allowing construction of some temporary structures necessary for tunnel construction and installation of the pipeline to commence sooner than originally planned. The following critical elements for the project were expedited during this phase.

### *TBM Design & Procurement*

- During the early months of the project, the team worked towards achieving a 60% design level by Oct 2018. Concurrently, TBM manufacturers were contacted and it was determined that at least 12 months was required for manufacture of the EPB TBM required for this project. Through collaborative pull planning workshops and feedback from the several TBM manufacturers who were provided tender documents, the team determined that TBM procurement was a critical path item.
- The project team evaluated MIC protection options and selected the base case conventional pipe-in-pipe option to advance to final design. While more expensive, this option provided the most proven track record, which was an important consideration for SVCW.
- Through a contractual amendment, the tunnel lining design was advanced during Stage 1 to a level of design adequate for determining the lining thickness and segment configuration, the two most critical lining parameters needed for purchase of the TBM. The parties selected Herrenknecht (HK) and began the early procurement process. This proved critical to the success of the contract, as it allowed the machine to be manufactured and shipped to the site by July 2019, facilitating the start of tunneling in September 2019, approximately 6 months earlier than originally planned.

### *Concrete Segment Fabrication*

- Similarly, while the Stage 2 contract was still being negotiated, Arup and BBJV worked together with the SVCW Owner Advisor (OA) team to confirm final design of the concrete segments. During this time various approaches to reinforcing of the segments were reviewed and evaluated. All parties agreed to steel fiber reinforcement for ease of manufacturing and to provide the required strength for the segments.
- Acceleration of segment production was beneficial to the project. BBJV was able to enter into an agreement with Traylor Shea Precast JV in December of 2018 to commence manufacture of the segments. This contract included several milestones, including owner and designer representatives visiting the manufacturing site in Stockton, CA to evaluate the manufacturing process, steel fiber mixing in the concrete and QA/QC processes in place to confirm proper segment manufacture. All of these elements proved acceptable and segment manufacture proceeded allowing segments to be delivered on a "Just-In-Time" basis to allow tunneling to proceed unhindered.

### *Tunnel Launch Shaft Design*

- In parallel with the segment design work, the project team also utilized a contractual amendment to accelerate the design of the temporary TBM launch shaft. The launch shaft site, known as the Airport Access Shaft (AAS), also served as the main site for construction of the Gravity Pipeline (GP). While this was a temporary structure, it was the deepest and largest in diameter for the entire project. This was required to reach the proper depth for the tunnel alignment and to allow the TBM and all the gantries to be installed in one location.

- The ground conditions in this area include a significant depth of a soil identified as "Young Bay Mud" (YBM). The YBM material has very low strength and can reach depths of over 100 ft below finish grade along the San Francisco Bay shoreline. At the AAS site the YBM was approximately 15 ft thick. Additionally, groundwater levels were near surface, within 5 ft below finish grade and can fluctuate with the tides. These conditions created a difficult design for a deep shaft that must remain in place for almost 3 years. The design team and OA team worked together to successfully consider and mitigate these known challenging conditions through selection of a slurry panel SOE approach. Construction of the AAS commenced in January 2019 and was complete by late June 2019 just prior to the arrival of the TBM and gantries.

The alternatives analysis and schedule acceleration could not have been accomplished in a traditional Design-Bid-Build environment. These key elements of the project allowed for the on time and on budget completion of the tunnel, allowing the pipe installation to proceed as planned.

### Transparent Estimation Process

BBJV prepared an estimate and schedule for the work was developed concurrently with design progression during Stage 1. The schedule and estimate achieved higher levels of accuracy as facets of the design reached greater stages of progression. All schedule and estimate assumptions were discussed openly between SVCW and BBJV throughout this process. A complete estimate of all labor, materials, equipment, subcontractor, overhead, and profit costs were delivered along with the 60% design at the end of Stage 1. This included all price quote documentation. From this submitted estimate and schedule, SVCW and BBJV were able to negotiate and amicably reach a final fair lump sum price for Stage 2 services.

## PDB LESSONS LEARNED

The Progressive Design Build procurement of tunnel projects presents many opportunities for innovation and alternatives-analysis through co-location and collaborative working between the Owner, Contractor and Designer. On the SVCW Gravity Pipeline project, the potential for project success was facilitated by:

- Identifying a suitable representative from the Design Builder's team to moderate technical workshops. The moderator must be a subject matter expert, an excellent communicator and unbiased in their treatment of design alternatives.
- Identifying a suitable representative from the Owner's team that will commit to multiple criteria decision-making, consensus building and maintaining a position once a decision is made.
- Acknowledgment that bias can suppress innovation. The fewer constraints imposed on the concept design, the more innovation the PDB can introduce to a project. To realize innovation in the alternatives-analysis process, avoid bringing a pre-conceived notion of an optimal solution.

The up-front and lifecycle costs, schedule, operations and maintenance impacts should be considered fairly for each alternative. Risk aversion can suppress innovation and increase costs. An equitable approach to risk sharing creates an environment in which the owner feels confident and the Design-Build team feels respected. Use of moderated risk workshops and sophisticated risk registers can facilitate discussions around risk sharing and tolerance.

Design alternatives should be selected during Stage 1 to avoid design change orders in Stage 2. The Design-Builder's Stage 1 scope and fee should allow for adequate alternatives analysis, but once in Stage 2, design direction should be defined and no further alternatives considered, unless new information is introduced. All decisions on design direction should be completed within Stage 1; additional alternatives analysis requested during Stage 2 is basis for a change order and may cause project delays.

During Stage 2, design work was taking place concurrently with construction. Proper planning must reconcile the completion of IFC design with the start of associated critical path elements of the schedule. Adequate schedule time must be added between IFC completion and construction start to account for activities such as procurement and planning. When executed correctly, this is where PDB outshines other contractual models due to minimizing the time any given work package spends in the period between IFC and construction. Conversely, when improper planning may result in schedule delays if IFC completion pushes critical path items. This is the "Achilles Heel" of PDB which can be avoided with proper project planning and resource allocation.

# Design and Construction of the NBAQ4 Project—First Urban EPB TBM Tunnel in Metro Manila

**Andrew Raine** ▪ ARUP
**Abegail Endraca** ▪ ARUP
**Allan Patdu** ▪ ARUP
**Allan Sebastian** ▪ ARUP

## ABSTRACT

The project involves the construction of a new aqueduct tunnel by an Earth Pressure Balance TBM tunnel for the Novaliches-Balara Aqueduct No 4 (NBAQ4) project in Metro Manila. It comprises a 7.3 km long and 3.1 m internal diameter tunnel that can deliver up to 1,000 MLD of raw water and the tunnel passes under urban residential and urban populated areas adjacent to and along Commonwealth Avenue. A new reservoir intake structure was constructed in the La Mesa reservoir by a cofferdam and the TBM breakthrough connected the reservoir to the outlet surge shaft structure and the downstream pipe network system connecting to the Balara Water Treatment Plant. The TBM incorporated a double articulated head configuration to allow tight 80 m radius curves to navigate some of the difficult alignment considerations. The alignment passed very close to some existing foundations for MRT7 and needed to clear several tight curves to remain within government land ownership. The project excavated through the local Guadalupe weak rock formations and there was minimal ground movements observed as a result of the work.

## PROJECT BACKGROUND AND OBJECTIVES

The NBAQ4 project aims to improve the reliability and security of the raw water transmission system from La Mesa Dam Reservoir to Balara Treatment Plant 1 and 2 and to allow the staged shutdown of the existing three (3) raw water aqueducts. Built in 1929, 1956 and 1968 respectively, these aqueducts are already nearing, if not already exceeding, the service life of 50 years for concrete structures and therefore are now in need of assessment and rehabilitation.

The existing aqueduct system is composed of Aqueducts 1, 2 and 3 and it is required to continuously deliver 1,600 MLD of raw water to Balara Treatment Plant 1 and 2; therefore, the supply aqueducts need to be operational at all times and cannot be shut down to facilitate the required assessment and rehabilitation works. The Project shall provide a new aqueduct to enable the sequential temporary suspension of service for inspection, assessment and subsequent rehabilitation of the existing aqueducts. Upon its completion, the Project will also provide emergency redundancy in the event of a failure in any of the existing aqueducts.

The Project is composed of a new intake facility at the La Mesa Dam Reservoir, a water tunnel from La Mesa Dam Reservoir, passing under Commonwealth Avenue to Balara Treatment Compound, and an outlet facility at the Balara Water Treatment Plant. Once completed and commissioned, it will be capable of delivering 1,000 MLD to both Balara Treatment Plant 1 and 2, allowing assessment and rehabilitation of the existing aqueducts.

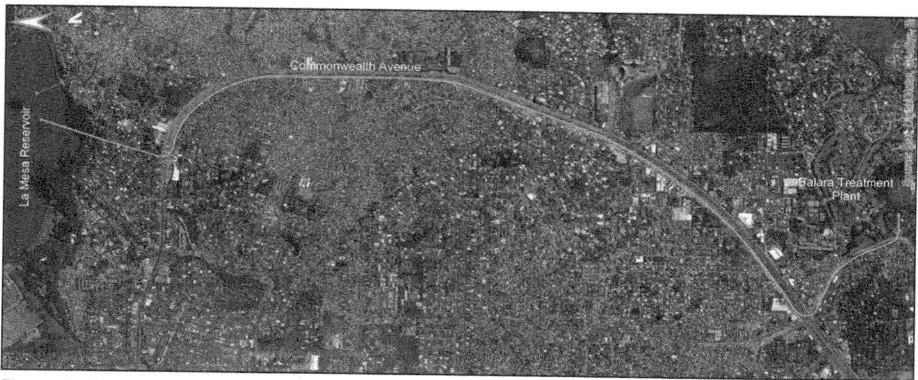


**Figure 1. Project location**

Manila Water Company Inc. (MWCI), the concessionaire assigned by the Metropolitan Manila Water and Sewerage System (MWSS) to provide water supply to the East Zone Service Area of Metro Manila, is currently undertaking various service improvement and expansion projects in the entire East Concession Area and this project forms part of those improvements.

MWCI had already engagaed the services of a local consultant in association with an international consultant to prepare a reference engineering design of the Project before Arup were awarded the final design review, tender procurement services and construction supervision contract.

## CONTRACT ARRANGEMENT

Ove Arup & Partners Hong Kong Ltd. (Philippines Branch and Hong Kong Office) (Arup) was appointed by MWCI in 2016 to undertake an engineering study and the procurement management to develop the reference design, specifications, and form of contract for the project.
During the engineering stage, Arup was tasked to perform the following activities:

- Review of the existing project documents issued by MWCI and notify MWCI of any deficiency in the documents, including providing MWCI the necessary recommendations of actions and measures that need to be addressed to bridge the gaps.
- Verify if the assumptions, constraints, and conditions considered in the preparation of the original project documents are still valid and notify MWCI of any changes and provide recommendations to address the change.
- Provide recommendations to MWCI on the most appropriate methodologies to be adopted for the project and provide updated construction technologies that may be used and applied to optimize the overall project cost.

Following the review of initial documents, Arup carried out the development of reference documents comprising the tunnel alignment and sections, geotechnical data report summarizing the available site and subsurface information obtained, civil and structural components, and electrical and mechanical equipment.

The procurement strategy and approach were also developed for the project through a number of discussions with MWCI. It was recommended that due to the inherently complex nature of the project, a well-renowned, technically competent contractor should be procured. It was agreed to procure the project through a Design and Build approach.

In 2017, Arup was engaged to undertake the contractors detailed design review, the overall construction supervision and the contract administration services for the project.

## BASIS OF DESIGN REQUIREMENT AND THE EMPLOYER'S REQUIREMENTS

After the review of the initial reference design, Arup optimized the design based on the current site conditions, latest code and standards, appropriate construction methodologies, and operational requirements of the project. The following subsections outline the high-level overview of the basic design requirements and Employer's Requirements for the main project elements.

### Hydraulic Design

The hydraulic features of the project shall be designed to convey the required flows of 1,000 MLD from the Intake Structure to the Outlet basin in a safe and reliable manner. Hydraulic design and performance shall consider the impact of sedimentation in the tunnel.

The Intake Structure was proposed to have three intake levels composed of six nos. of 1 $m^2$ sluice gates, each provided with a trash screen. The inlet of the Intake Structure was set to prevent unacceptable vortex formations at maximum discharge and the connection to the Tunnel shall be designed to minimize head loss to reduce potential for flow separation at the intake. Similarly, the geometry, size, orientation, and connection of the Outlet Structure to the Tunnel was designed to minimize head losses and any flow separation. In addition, the diameter and height of the Outlet Structure was determined by considering transient conditions and the surge of water due to any instant closure of the flow from the system.

The friction losses through the tunnel are the major contributor to total head losses through the system. An indicative roughness value of 0.5 mm was assumed for reference design.

### Ground Conditions and Seismic Conditions

The geology of the project site is largely located in an area characterized by a complex, heterogeneous assemblage of flat lying, medium to thinly bedded pyroclastic and tuffaceous deposits. The materials are relatively weak and have quite high porosity. The available ground investigation data was considered adequate and acceptable for the production of a reference design for procurement. However, additional geological/geotechnical studies were to be conducted by the Contractor as necessary to complete his detailed design for construction of the project.

The project is located in a highly seismic zone and in close proximity to the Marikina Valley Fault System and therefore seismic design is critical to the success of the project. Therefore, all work shall be designed for seismic loading conditions. The earthquake loads shall be obtained in accordance with the National Structural Code of the Philippines (NSCP). The required performance criteria for the project is operational after design earthquake. The design earthquake is defined as an earthquake with 10 percent chance of being exceeded in 50 years (return period of 475 years), which shall be used in the design of all structures. The project is assigned an Occupancy

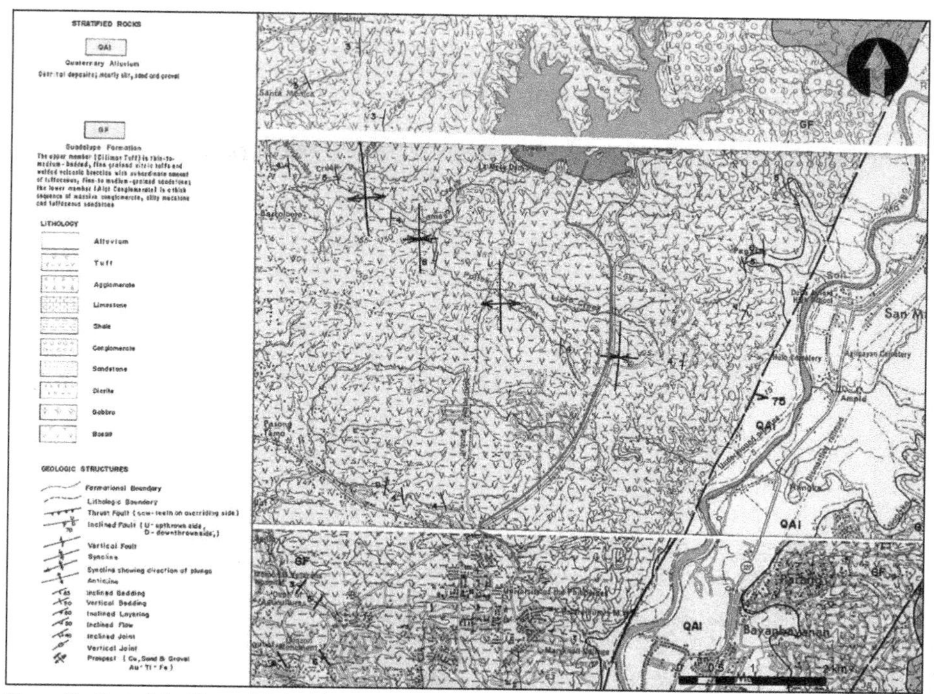


Figure 2. Part pint of the published geological maps (map sheet 3263 IV—geological map of Manila and Quezon City quadrangle (scale 1:50,000), 1st Edition, 1983 and map sheet 3264 III—geological map of Montalban quadrangle (scale 1:50,000), 1st Edition, 1983)

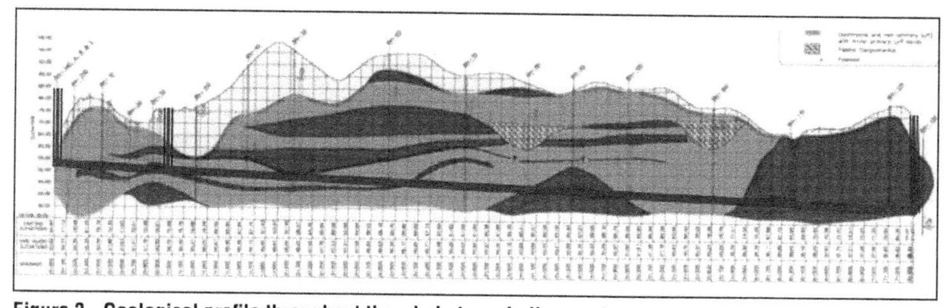

Figure 3. Geological profile throughout the whole tunnel alignment

Category of "Category One—Essential Facilities" with Seismic Importance Factor of 1.5.

## Intake Access Bridge

The access bridge is to be provided from the shore of the lake to the Intake Structure to allow all operation and maintenance activities to be carried out at the Intake Structure such as the removal of intake screens, trash screens, and actuators without the need for a boat or barge mounted crane. The minimum carriageway width is 2.50 m and typical span lengths as per the Department of Public Works and Highways (DPWH) requirements.

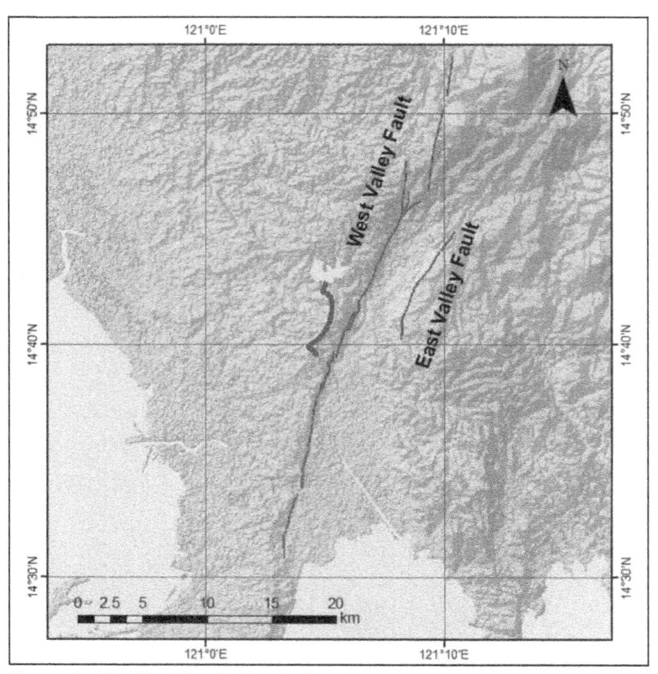


Figure 4. The Marikina Valley fault system

## Intake Structure

The Intake Structure was configured to draw water horizontally at three levels each opening is provided with a trash rack to prevent ingress of divers, debris, or fishes by considering future maintenance requirements. The Intake Structure to Tunnel connection shall be capable of withstanding the full range of all loads up to and including when the lake level is at 82.5 mRL with the tunnel fully drained.

General marine works requirements were provided in detail for the technical specifications for the construction of Intake Structure and its connection to the tunnel within the lake. Compliance to all environmental protection measures and relevant regulations shall be strictly followed to maintain and protect the water quality of the reservoir.

The requirements for following major activities were also discussed in the Employer's Requirements.

1. Onshore staging area to be used in support of offshore operations
2. Selection of suitable offshore construction equipment
3. Excavation and disposal of excavated lakebed materials
4. Connection of the Aqueduct Tunnel to the Intake Structure
5. Cast-in place concrete construction
6. Grouting/Backfilling
7. Steelworks
8. Fill material for marine works
9. Underwater survey works and inspection works
10. Dredging
11. Flooding the tunnel

Figure 5. The intake structure and the bridge

## Tunnel

The tunnel alignment was revised due to road impact of road widening works along Commonwealth Avenue and of a new major interface, the construction of MRT7 metro line. The tunnel is approximately 7.3 km long with an internal diameter of 3.1 m.

The tunnel was intended to have a segmental concrete lining and constructed using a Tunnel Boring Machine (TBM). The TBM was required to be a fully shielded TBM articulated with the ability to operate in a pressurized mode and excavate through all ground conditions anticipated based on geological interpretation along the tunnel alignment. The selection of a suitable type and components of TBM and construction of the tunnel shall take into account the limited ground investigation and potential variations of ground condition along the tunnel alignment. The minimum horizontal alignment radius curve was specified to be 80 m. Other critical TBM components and supporting equipment were required to provide a complete system and safe working space during the tunneling works and TBM's breakthrough to the Intake Structure.

## Outlet Structure and Downstream Network System

The indicative internal diameter of Outlet Structure is 8.0 m. The Outlet Structure shall be designed to consider transient conditions due to instant closure of valve installed between the Outlet Structure and outlet basin. The structure shall be capable of withstanding the full range of all loads up to and including when the lake level is at 82.5 mRL with the tunnel in operation or Tunnel and Outlet Structure fully drained.

The Outlet Structure shall be reinforced concrete lined and constructed to minimize groundwater inflows during excavation and lining of the Outlet Structure.

The Outlet Structure shall be connected to the existing treatment plants through steel pipes and connection basins. The connection to the existing basins was flagged as a

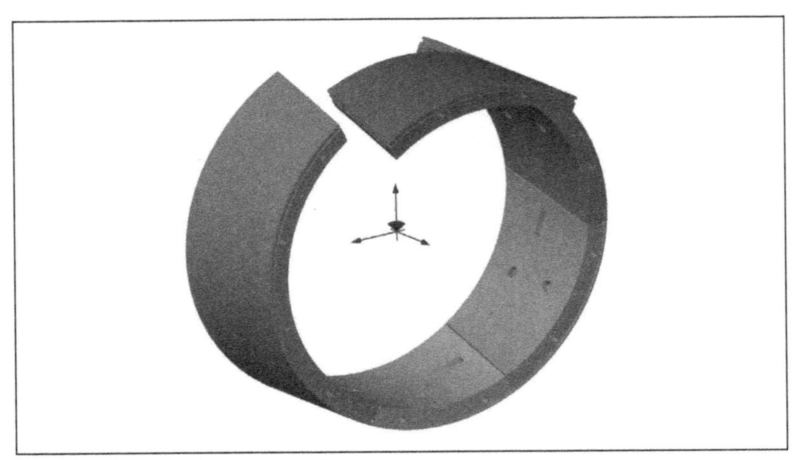

Figure 6. 3D view of segmental lining

high-risk construction issue, with the possible implications to the successful operation of the existing aqueducts and supply of water.

### Electrical and Mechanical (E&M)

Electrical and mechanical works shall include piping, valves, gates, flowmeters, controls, and all necessary power supply, connection, and communication system. The overall operation and monitoring system shall be available at both Intake Structure and Outlet Structure and downstream network system through a communication system acceptable to the operations.

All equipment must be designed for safe and efficient operation and maintenance. Power supply shall be provided with a back-up power supply to ensure continuous plant operation.

## TENDER, PROCUREMENT AND AWARD

The project was procured as a Design and Build contract, based on a lumpsum basis where the contractor is responsible for the design and construction of all elements of the project. MWCI General Conditions of Contract, a contract form based on FIDIC Yellow Book, was used and supplemented by Special Conditions of Contract. It was recommended that the contract be procured as a single package in order to limit risks associated with the potential design and construction interfaces. In terms of risk allocation strategy, it was recommended to adopt the risk allocation principles of Manila Water's GCC, and the use of a GBR is not recommended.

A two-stage selective tendering process was proposed with a prequalification exercise being carried out prior to inviting a limited number of tenders. The prequalification process was proposed to ensure that only applicants with that are suitably experienced in the type of work and construction technology involved, that are financially and managerially sound, and that can provide all the key personnel required were invited to submit a tender.

In November 2016, the prequalification process was completed and three Tenderers were selected to proceed with the tender submission. Finally, two tenders were received by April 2017.

The technical assessment was undertaken by various discipline expert reviewers in accordance with the marking scheme issued to the Tenderers in the tender documentation. Eventually, MWCI awarded the Design and Build contract to the most compliant Tenderer, NovaBala JV Corporation (NBJVC)—a joint venture between CMC Ravenna, Chun Wo, and First Balfour. The Contractor's JV subsequently awarded their design contract to GHD.

## CONSTRUCTION WORKS

The following sections discuss the construction work methodologies, challenges and corresponding solutions implemented to resolve the issues encountered for the main structures such as the intake tower, tunnel, outlet shaft, and downstream network system.

### Intake Structure

#### *Features*

The submerged structure including the inlet was installed in the excavated floor of the La Mesa Dam Reservoir that connects intake inlet with tunnel aqueduct. It consists of 16 m diameter, 7.3 m high cylindrical base with a 2 m high octagonal pedestal on top of it supporting 21.5 m high intake shaft having a circular cross section measuring 6.0 m internal diameter. The base slab thickness measures 1.3 m. The shaft and TBM receiving pit wall (Breakthrough Area) thicknesses are 0.50 m and 1.2 m, respectively.

The structure is situated at 150 m from the lake shoreline wherein the inlet invert level at bottom shaft is 26 m depth (RL 53 m) below lake surface water level (average RL 79 m). The top of the shaft is at RL 82.5 m which is also the dam crest level. The spillway level is at RL 80.15 m.

The intakes are configured to draw water horizontally at three levels. These levels are RL 76 m, RL 72.5 m and RL 68.5 m. Each inlet is consisting of a 1×1 m square clear opening which are formed by cast in stainless steel bell mouths with a trash racks and sluice gates mounted outside the tower.

#### *Cofferdam and Intake Construction*

The Intake Tower was constructed inside circular cofferdam with a diameter of 19.51 m and consisting of 56 nos. of steel piles interconnected as shown in Figure 7 with a diameter of 914 mm. It was built by the used of modular barge having dimensions of 36 m × 16 m × 2.5 m carrying 165T crawler crane. The steel piles were driven in sequence to the lakebed hard strata and an additional boring for the bored pile up to the required design strata (RL 48 m). The depth of embedment from lake surface water is roughly 30 m depending on the climate condition. The cofferdam was also built with 13 well pumps and 4 piezometers to provide stable working condition and to monitor ground water pressures.

Furthermore, the cofferdam was constructed in parallel with the access bridge having a total length of 153 m. The span between piers is measured 21.15 m and a width of 4 m. This bridge was built as part of Employer's Requirement to carry out all operations and maintenance without the need of boat or barge crane. The bridge was also designed to allow heavy operation at cofferdam and intake tower construction.

The dewatering and excavation inside the cofferdam were executed with respect to waler beam elevations and installed simultaneously. These are welded and bolted

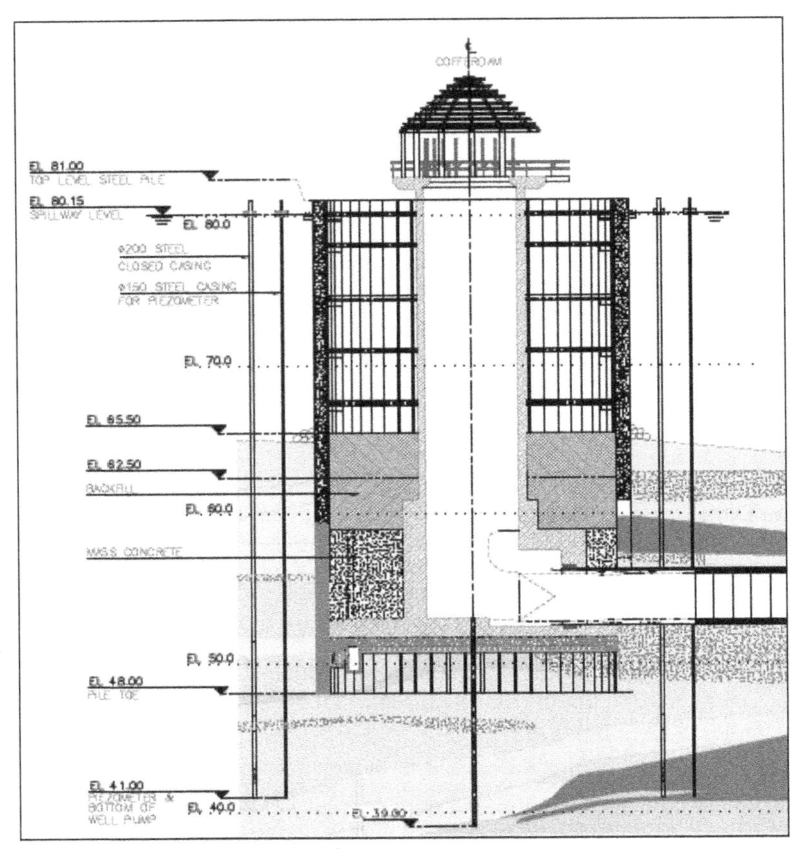


Figure 7. Intake and cofferdam section

connections to ensure stability of the cofferdam against the horizontal load of the water and ground. Excavation below lakebed walers was done alternating until reaches the bottom slab using excavators lowered inside cofferdam. Then, the excavated material is transported through muck pit lifted by 165T crane and loaded to dump truck prior to disposal to nearby stockpile area.

However, at excavation level 54.7 m during preparation and installation of waler beam, the cofferdam encountered high ingress of water between pile 1 and 2 equalizing the dam level in a matter of hours. The leakage spot created depression of 5 m depth below lakebed with an area size of 7 m × 1 m as shown in Figure 7. The remedial works were done through filling depression area with similar lakebed material, additional sand filling around cofferdam and concrete plugging the leak spot by tremie pipe and sandbags.

For the remaining excavation, shotcrete and additional grouting were introduced between the pile gaps. After established, pre-excavation of certain area and grouting at various depths were performed prior excavation advancement.

After completion of excavation, construction of temporary base slab and TBM receiving pit commenced in preparation for the breakthrough.

Figure 8. Cofferdam construction

Figure 9. TBM breakthrough at the intake

The succeeding concrete lifts were completed after TBM breakthrough and retrieval.

## Outlet Structure

### *Features*

The tower shaft's diameter is 8 meters, and its overall length is 56 meters. The secant pile's diameter is 8 meters, and its length is 27.34 meters. The building of the outlet shaft was divided into three stages: secant piling work, excavation and wall lining work, and tower construction.

### *Outlet Shaft Construction*

The secant piling works are the first stage of the outlet shaft construction. The secant pile wall was built using 40 interlocking piles, divided into 20 primary piles (each measuring 1.5 meters in diameter) and 20 subsidiary piles (each measuring 0.8 meters in diameter), for a total of 80 interlocking piles. The "Kelly-drilling method with slurry support," the drilling operations were carried out by self-erecting hydraulic drill rigs of type BG40 or BG30.

A capping beam was casted on the pile cap to secure the secant pile's opening. Afterwards the bulk excavation was carried out utilizing excavators and specially constructed muck skips.

Slab and wall lining concreting was done after the design level was reached. For this activity, climbing scaffold and formworks were employed.

Once the ground level is reached, double-sided formworks were used all the way up to the tower's top level. Concreting begun with the creation of a special hydraulic shape that enabled the connection of the outlet shaft to the downstream distribution system.

Following lifts were implemented with the same mechanism as described above, with the exception that the formworks were two-sided; outside and inside the shaft. In this phase, most lifts were done every 3.5 m up to the top elevation of the outlet shaft.

## Tunnel

### *Access Shaft*

The tunnel's designed elevation is 40 meters below the working site, therefore a secant piling solution was used to construct the access shaft for this project. There are 44 interconnected piles, 22 primary and 22 secondary piles. The bulk excavation took only two weeks due to secant piling, which eliminated the need for beams and shotcrete work.

**Figure 10. Outlet shaft construction**

### *Mined Tunnel*

A mined tunnel was required to provide space for TBM assembly and daily operations. Before any development in the mined tunnel, which was built using manual excavation and a lattice girder as a circumferential support and shotcrete as the tunnel lining, horizontal canopy tubes were drilled and grouted to the crown as presupport to increase the strength of the ground.

### *Segmental Lining*

Parallel to the construction of access shaft and mined tunnel, the manufacturing of Tunnel permanent concrete lining was done outside of the project site. Production and delivery of the rings were correlated with the tunnel progress. There are two types of ring design used: S-type for straight advancement and T-type for curve advancement.

### *Tunnel Construction*

The Earth Pressure Balance (EPB) TBM was chosen due to the ground conditions throughout the tunnel layout. EPB TBM technology is based on the use of the excavated soils as the supporting medium in the excavation chamber.

**Figure 11. Tunnel access shaft**

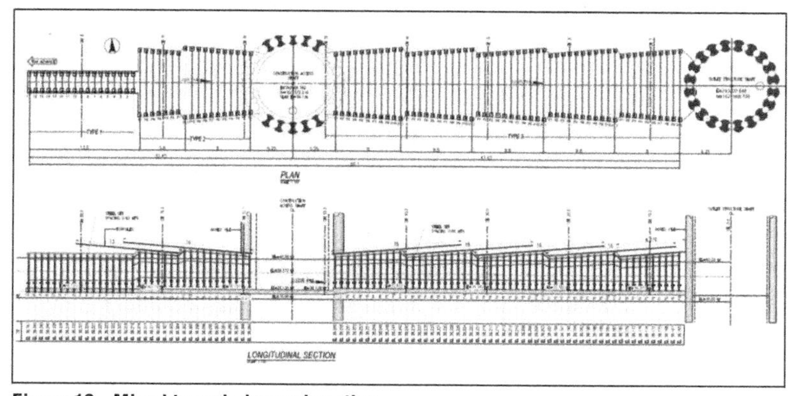


**Figure 12. Mined tunnel plan and section**

**Figure 13. Precast segments at the fabrication yard**

The excavated soil is mixed with ground conditioning foam and polymer in the excavation face is extruded through the openings of the cutterhead to fill the excavation chamber and screw conveyor during TBM advance.

By synchronizing of the advance speed of the shield and spoil, extraction rate of the screw conveyor, it is possible to establish a controlled face support pressure.

The tunnel works progressed sequentially, with the cutter head excavating in front of the TBM and sending the spoil into the waiting locomotive via screw conveyor and over-head conveyor. When an adequate space is excavated a man-controlled erector arm will assemble the ring, locking the ring by inserting the key segment. The locomotive entering the TBM delivers segments and carries the excavated soil going out.

The annular grouting unit injects grout following each advancement. The grout component A is stationed at the surface with a batching plant and supplies the TBM via a series of utility lines, while the component B is stored in the TBM.

The tunnel advancement was consistently monitored by both manual survey and survey system installed in the TBM. Whilst the installed segments were inspected in the post-installation inspections, defects on the rings were identified and repaired.

The following key geotechnical features have been identified through ground investigation

1. Fault
2. Weak rock strata with "soil like" strength properties
3. Shallow ground water conditions, high ground water inflow rates in sub-vertical fractures
4. Presence of swelling clays

**Figure 14. TBM assembly inside the mined tunnel**

Tunnel excavation was completed in August of 2021, with the TBM breaking through at the La Mesa Intake tower, where it was disassembled and transferred to the storage facility. The total length excavated is 7,268 m, with 6,194 rings. The average number of rings installed per day was 19 rings or 23 m.

## Downstream Network System

### *Features*

The Distribution Network System (DNS) consists of Civil/Structural works such as Basin, Electrical and Instrumentation building and Pipelaying of DN2500 and DN2000 including Plunger Valve and Sluice Gates. Due to constraints related to the site proximity which the alignment is being close to the existing Water Treatment Plant, the Construction Methodology including the temporary works were precisely established to reduce risk and meet completion dates. In addition, comprehensive work program to analyze the right flow of the construction sequence was developed using Primavera.

### *Pipelaying*

Site investigations included manual trial pits and Ground Penetration Radar (GPR) scanning to confirm the locations and identify underground structures such as concrete and steel pipelines that crossed the alignment. The Contractor provided design and analysis for temporary supports, mined tunnelling, and conventional steel supports for live connections.

Earthworks were completed using a combination of manual and machine excavation. The steel support system, shotcrete, and soil nailing were in place during structural and pipeline deep open cut excavation to mitigate the risk of collapse and damage to the existing utilities.

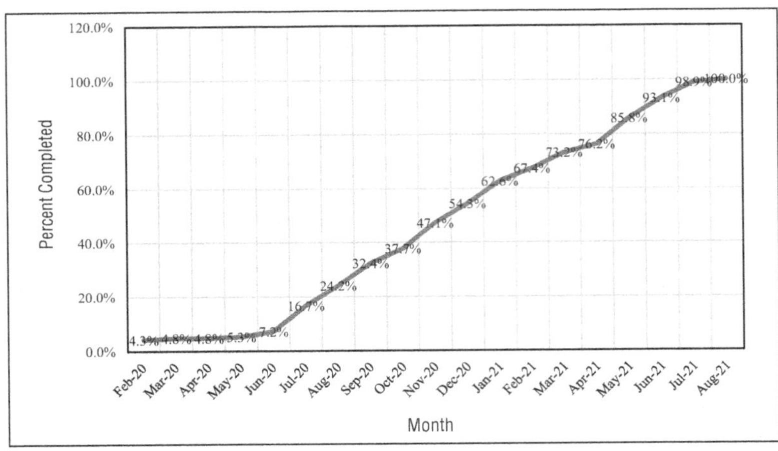


**Figure 15. Tunnel progress**

**Figure 16. Tunnel construction**

Backfilling was done with sand for the pipeline and fill materials for the structures. A section of mined tunnelling was also used to replace the traditional construction for around 20% of the DNS extent.

Pipeline construction with a larger diameter such as DN2500, DN2400 and DN2000 within the proximity of the existing water treatment plant is critical including the old existing concrete aqueduct.

All welding works (SMAW) including tacking and welding support in accordance with ASME B31.1 was applied. Non-destructive test such as ultrasonic test (UT) and Joint Pneumatic Test were conducted on the completed welds.

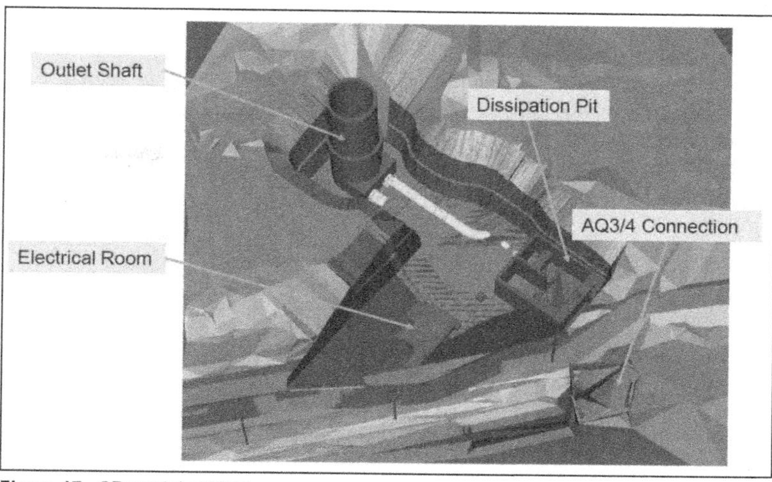


**Figure 17. 3D model of DNS**

**Figure 18. Pipelaying at DNS**

### *Downstream Structure Construction*

The following structures such as Dissipation, Flowmeter 1 & 2, AQ1 and AQ2 basins were constructed in a conventional RC method. The reinforced concrete wall was casted in stages. Puddle pipes were used for all openings where pipelines connect to structures.

Live connections to existing basins were completed using a cofferdam. Prefabricated steel supports were installed, and dewatering system was in place prior to demolition. Tie-ins were completed successfully without disrupting the operation of the live water treatment plant.

## Electrical and Mechanical

### *Features*

The Electro-Mech system composes of hydromechanical components like sluice gates, steel pipelines, valves and actuators, flowmeters, power supply systems, instrumentation, control systems, and communication systems to provide overall control and monitoring at both upstream and downstream operation.

### *Intake Structure*

There are three levels in intake structure, each level has six (6) sluice gates operated by motorized electric actuators. Level sensors were also installed to monitor the level of water at the reservoir.

### *Downstream Network System*

The Outlet structure at the Balara site is linked to the dissipation/diversion basin by two pipelines, the major line, DN2000, and the secondary line, DN2400. A Plunger Valve is a component of DN2000 line that regulates the water flow. There are also Stop logs for the pipelines at the dissipation. DN2500 twin pipeline was connected between dissipation basin and AQ2. A concrete culvert connected the dissipation basin to the current AQ3 system.

Flow meters were installed on the lines. All three pipelines were made of carbon steel pipe. Chambers were built for the Ultrasonic Flow Meters.

### *Power Distribution System*

A 34.5 kV line was constructed to link the new electrical rooms to the current distribution line. To reduce power outages, all MV distribution line were insulated. The design of the system was in accordance with standard requirements.

The electrical rooms were equipped with transformers of 3-Single Phase Pole Mounted Type, 460 Volts, 3-Phase Secondary Load Side, and 34.5 kV Primary Line Side and Type Tested Form 3B, Type 2 free standing MCC panels.

**Figure 19. Plunger valve installed downstream of outlet shaft**

A generator set with an Automatic Transfer Switch was installed in each electrical room as a backup. The generator is rated for prime or standby operation. Fuel storage tank holds 1,500 litters of fuel sufficient to for 2 days continuous operation.

Each electrical room has fire alarm system, lightning and grounding protection, and an air conditioning system at the PLC /UPS room. The NBAQ 4 Project was constructed with two primary control systems that will use Programmable Logic Controllers (PLC) at three sites. The SCADA's PLC was interfaced with via fibre optic cable. Radio antennas was used for communication between two sites.

## TESTING AND COMMISSIONING

There has been three phases in the testing and commissioning process for the project, each has requirements that outline the prerequisites for the system's commissioning, and all must be based on the Employer's Requirements. All Electro-Mech and Instrumentation equipment needs to be commissioned.

### Phase 1 Pre-commissioning

- Pre-commissioning is where all equipment and components are installed, pre-commissioned and tested for operation.

### Phase 2 Commissioning

- The Commissioning Test will see the entire system subjected to a simulation of real operational conditions.

### Phase 3 Testing after completion

- The system must pass a series of tests to ensure that it meets both operational and performance objectives.

### Commissioning Procedure

Wet testing ensures that leakage stays within the range specified in the design seating heads. Water must fill the intake above the third level of the sluice gates. During wet testing, DN2000 and DN2400 BFV's must be evaluated for open/close ability to hold a flow of water. Sluice gates must be fully field operated, and leakage must be verified. Water level in the dissipation pit must be maintained at the design head. The AQ2 and AQ1 basin will be filled with water from the dissipation pit. Flowmeters are used to monitor water flow.

### Commissioning for Electrical

La Mesa, Balara AQ2 and Balara AQ3/4 Electrical Rooms

Testing and commissioning the electrical system:

1. 34.5 kV MV cable, transformer energization
2. Energizing of MCC and PFC panel
3. UPS and generator set
4. Sluice gates, BFV, Plunger Valves, Flowmeters and Level Transmitters
5. Pumps
6. Communication, PLC, and SCADA system

## CONCLUSION

The outstanding achievement of the completion of the Design and Construction of NBAQ4 Project is a benchmark project not just in the water industry in the Philippines but in the entire civil infrastructure industry, the practical demonstration of how to adequately manage stakeholder risks in a complex tunnelling environment such as Metro Manila demonstrates not only the foresight and vision of the owner, it also clearly demonstrates the engineering and management capabilities of the PMC, and the resilience and tenacity of the Contractor, in what has been an extremely challenging project to execute during the peak of the global Covid 19 pandemic emergency.

It's the first use of a double articulated head shield tunnel boring machine with a tight 80 m radius in the Philippines. It's the first EPB machine to be used in the Philippines and it has set excellent records of production and progress as a benchmark for other projects to emulate.

The project has had a number of other physical and construction challenges and in many cases those have been encountered on other underground projects of similar nature elsewhere in the world, there have been commercial as well as construction challenges, it is worth to mention that one of the most important aspects contributing to the success of the Project is the collaborative and partnering approach encouraged by MWCI between all the parties involved.

NBAQ4 is a historical bench mark which now paves the way for larger and more complex projects such as the Manila Metro Subway Project and other tunnelling related projects for the city—the future is looking bright for urban tunnelling in Metro Manila.

# Designing and Constructing the Advance Tunnel for the Scarborough Subway Extension

**Michael Dutton** ▪ Arup
**Andreas Raedle** ▪ Arup
**Uhland Konrad** ▪ Strabag Inc.
**Jorge Ferrero** ▪ Strabag Inc.
**Keivan Pak Iman** ▪ GHD
**Walter Trisi** ▪ Metrolinx

## ABSTRACT

Strabag, in partnership with Arup, is constructing the Advance Tunnel for the Scarborough Subway Extension (ATSSE); a project by Metrolinx and Infrastructure Ontario (Sponsors). The 10.7 m internal diameter single bore tunnel represents a significant departure from the traditional arrangement of twin subway tunnels in North America. ATSSE is the first-time steel fibers are used to solely reinforce a tunnel of this size in Eastern Canada.

This paper discusses the challenges with design and acceptance of SFRC segments and the state-of-the-art numerical models employed to check the joint performance. The advance procurement approach for TBM and PCTL's procurement to accelerate the project schedule is also presented.

## INTRODUCTION

The Scarborough Subway Extension, a project by Metrolinx and Infrastructure Ontario (the Sponsors), is a 7.8 km long extension to the existing Toronto Transit Commission (TTC) Bloor-Danforth Line 2. A much-needed subway service to Scarborough, that will make it faster and easier for riders to get where they need to be each day. The project consists of a large diameter single bore tunnel from Sheppard Avenue East and McCowan Road to Midland Avenue and Eglinton Avenue East.

The tunnel will be constructed utilizing an earth pressure balance (EPB) tunnel boring machine (TBM) of 11.93 m bored diameter, that will install a 10.7 m internal diameter precast concrete tunnel liner (PCTL). The machine will be driven from the north and will pass through layers of till and glaciolacustrine and glaciofluvial deposits resulting from intermittent advancing and retreating glacial activities.

Along with the bored tunnel, launch, and retrieval shafts, headwalls for future stations and emergency exit buildings are also to be constructed. Once completed, the tunnel will house two trains within this single tube and will be the largest diameter metro tunnel in Canada.

## ADVANCE PROCUREMENT AND WORKS

In the interest of accelerating project delivery, the Sponsors elected to deliver the project under two main scope packages: the delivery of the Design Build Finance (P3) package for the Advance Tunnel or ATSSE (the subject of this paper), and a second package for the delivery of the Stations, Rail and Systems, or SRS.

The Request for Proposal (RFP) period included a series of interactive meetings between the Contracting Authority (CA) and the three shortlisted proponents where a Reference Concept Design was utilized as a proof of concept and conceptual design. The contracting process was in collaboration with the Technical Advisor, OneT+, a joint venture partnership between Gannett Fleming Canada, IBI Group and GHD. OneT+ provided advisory services to support the overall planning, preparation, procurement and construction of the project prior to and after Financial Close (FC).

A goal was set by the Sponsors to have the launch shaft fully constructed approximately one year after FC, and to have the TBM on-site, assembled, and ready to mine shortly thereafter. To achieve the desired schedule, the in-market period saw the proponents advancing their designs to a minimum 30% stage with some critical elements, such as PCTL design, even further. In addition, the CA decided to continue to fund the advancement of critical design during the RFP evaluation process and then employ the use of an Early Works Agreement (EWA) to mobilize the First Negotiation Proponent in immediate design and site works. The objective was to have design progression equivalent to the following design stages by FC:

- 100% PCTL design
- 100% launch shaft design including utilities
- 90% TBM design by TBM manufacturer (ready to start manufacture)

Project Co, Strabag and their prime designer Arup, was selected as the First Negotiations Proponent (FNP) to commence early works in March 2021 while working towards FC in May 2021. During this period, critical design packages, construction work submittals and procurement activities were advanced. If for whatever reason, commercial and financial close were not to proceed, Project Co would be reimbursed for the direct costs incurred for these particular activities.

## Tunnel Boring Machine

TBM engineering began during the EWA period in order to allow for a shorter lead time for the eventual delivery. Strabag elected to purchase the large diameter EPB TBM from Herrenknecht Germany in March 2021 right after FNP. The TBM selected was an 11.9 m diameter, mixed face EPB TBM and designed to handle the geological conditions which were to be encountered along the ATSSE tunnel alignment. With this advance work, orders for long lead items such as the main bearing, drive motors, and electrical components could be placed which would allow Herrenknecht to manufacture the TBM in record time.

The TBM arrived and began its assembly in the launch shaft in April 2022. Table 1 outlines some of the TBM specifications.

## Molds for the Segmental Tunnel Lining

The internal diameter of the tunnel was prescribed by the CA to be 10,700 mm. This encapsulates the train clearance envelope of two adjacent tracks separated by a dividing wall with outer safety walkways. Project Co selected a 400 mm thick tunnel lining and was validated for structural capacity against all load cases. The rings are 2,000 mm in length to optimize navigating the alignment and efficiency of ring builds. As shown in Figure 1, the ring geometry consists of six 45° rectangular segments, two 45° trapezoidal segments.

Molds were procured from CLECO and began manufacturing in June 2021.

**Table 1. Overview of key TBM specifications**

| | | | |
|---|---|---|---|
| TBM Cutting Diameter: | 11,910 mm | Max unlocking Torque: | 43,711 kNm |
| TBM Weight: | 1,500 t | Propulsion Cylinders: | 48 No. × 2,155 kN |
| Overall Length: | 87 m | Total Nominal Thrust:<br>Maximum Thrust Force: | 101,787 kN<br>142,502 KN |
| Cutterhead Type: | Mixed Face | Screw Conveyor Diameter: | 1,400 mm |
| Disc Cutters/Scrapers | 58 No. × 483 mm dia./156 | Screw Capacity: | 1,800 m$^3$/h |
| Main Drive Type: | Electric VFD | Segment Erector: | Vacuum pick up |
| Main Drive Power: | 16No. × 350kW = 5600kW | Primary Voltage: | 13.8 kV |
| Main Drive Speed: | 0–2.6 rpm | Transformers: | 2 × 3,500 + 2,000 kVA |
| TBM Airlock | 2 chamber | Total Installed Power: | 6,750 kW |
| TBM Material lock | 1 | Annular Void Grouting: | 2 component (A + B) |
| Tunnel Guidance | VMT system | Ground Conditioning | yes |
| Probe Drill and Ports | Yes | Automatic Bentonite | Plenum injection |

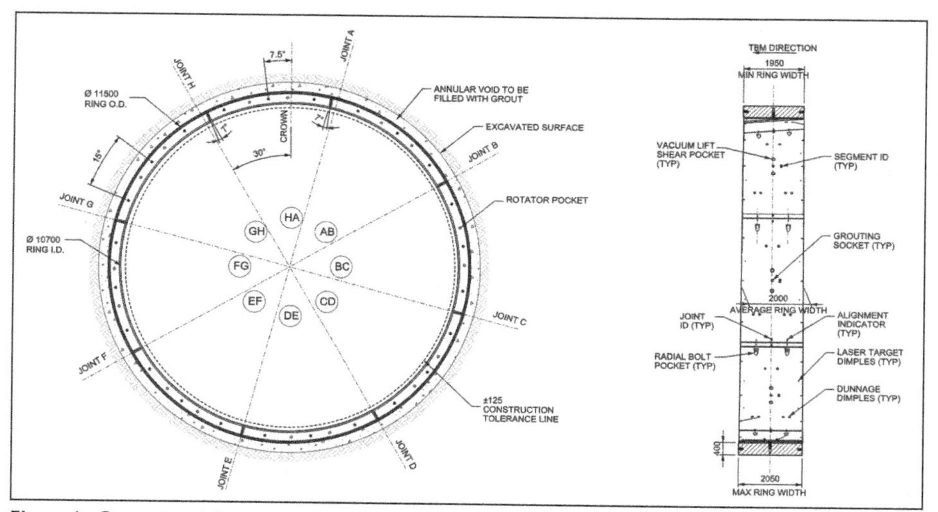


**Figure 1. Geometry of the segmental ring of PCTL**

## Launch Shaft

The project agreement (PA) specified that the TBM launch shaft was to be constructed as "water-tight"; as defined in Haack's water tightness criteria 1991—Tightness class 3. The design also called for requirements with respect to maintaining stability of the shaft faces for TBM launch and breakthrough, minimum size requirements to host future SRS structures, as well as limits to the use and type of internal bracing to accommodate future SRS contract integrations. The launch shaft was constructed using watertight secant piles and tie-backs. Specifications with respect to the future SRS contract integration were implemented in accordance with the PA and through coordination with CA and TA. Hybrid structural elements were used for the king piles located in the location of the tunnel eye. They consisted of a steel-fiber-steel pile, where the fiber was located in the area of the future TBM bore, creating a "soft-eye" to allow the machine to mine through the wall. A similar type of pile was used at the location of the headwalls that were to be constructed along the alignment for the future

**Figure 2. Launch shaft and construction laydown area for ATSSE**

stations and emergency exit buildings. The launch shaft dimensions are approximately 28 m in width, 80 m in length, and 26 m in depth as shown in Figure 2.

## PCTL MIX DESIGN

Considering the successful precedents use of SFRC for PCTL across the globe (ITA, 2016), the CA and OneT+ made an effort prior to the RFP release to allow for innovation by the proponents and efficiency in tunnel liner design. As a result, the PA permitted SFRC to be used for the first time in the Toronto subway system.

According to the PA, proponents could elect the type of reinforcement in the PCTL based on their design and chosen means & methods. This included traditional reinforcing bars, steel fibre reinforcement or a hybrid. To implement such a change from the historically established bar reinforcement, mandatory codes and standards (e.g., ACI 533.5R, ACI 544.7R and FIB Model Code 2010) were included in the PA for the proponent to comply with.

Any specific conditions, in addition to codes and standards, were included in the PA to set the baseline for design compliance. This includes any specific design loads, load factors, water tightness requirements, and stakeholders and SRS contract interface requirements. One particular Owner requirement was that the peak load from the SFRC beam test shall be 120% of the cracking load to ensure a flexural hardening response. This provided an enhanced sense of safety and alleviated safety concerns around SFRC ductility.

The PCTL design life is specified as 100-years per the PA. Although the design should achieve the specified requirements in accordance with CSA S478, further analysis was required to demonstrate the 100-years design life is achieved as CSA A23.1-19 does not extend beyond 50-years. Project Co, submitted a PCTL Durability Report as part of their submittals including further analysis on the mix design to satisfy this important requirement by the CA.

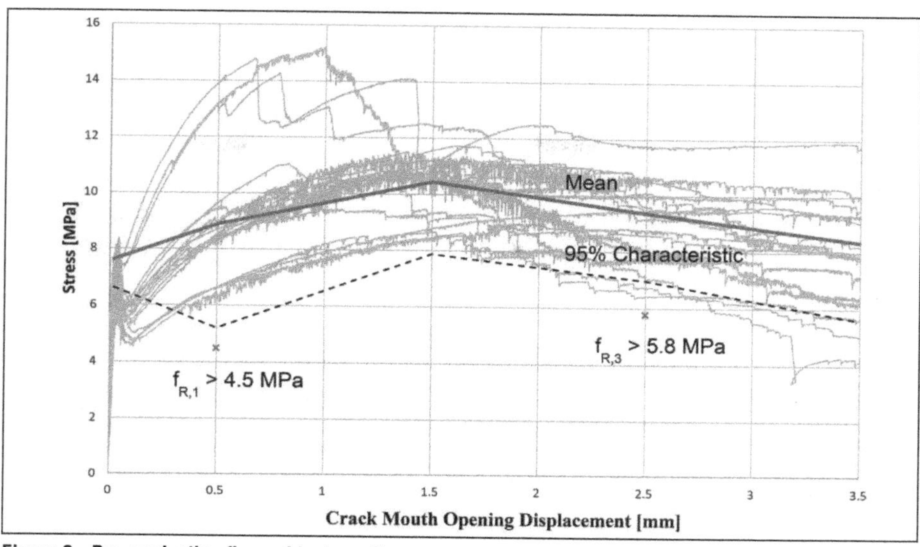


**Figure 3. Pre-production flexural test results**

Project Co elected to use SFRC due to its superior crack control behavior and cost-effective segment construction (Susetyo, et al. 2018). The minimum compressive strength of the concrete to satisfy the structural demand under service and at demolding is 60 MPa and 12 MPa, respectively. The design determined a minimum tensile strength of the concrete is 4.5 MPa per ASTM WK73348 Standard Double Punch Test. The concrete mix contains 40 kg/m$^3$ of Dramix 4D 80/60 BGP steel fibres to achieve a characteristic residual flexural strength of $f_{R,1}$ = 4.5 MPa and $f_{R,3}$ = 5.8 MPa according to EN 14651. Figure 3 plots the mean and characteristic flexural response obtained during pre-production mix trials.

## PCTL DESIGN HIGHLIGHTS

To meet the project's aggressive schedule and milestone dates, all aspects of the PCTL design were accelerated. During the RFP period, a mixture of analytical and numerical design methods was employed. All load cases were examined with suitable levels of complexity. For instance, the flexural design utilized numerical models and established workflows to quickly gain confidence and support in the chosen steel fiber reinforced concrete (SFRC) scheme. While on the other hand, TBM loading on the joint was demonstrated by analytical methods to be acceptable and deferred a more rigorous check to after FNP.

Work continued after the tender submission and before FNP was announced. This focused on improving Project Co's confidence in the radial joint's performance with steel fiber reinforcement and the selection of the ring thickness. The following gives highlights of key aspects of the PCTL design.

### Flexure

The numerical modelling of the PCTL was performed using FLAC3D developed by Itasca Consulting Group. The software is an explicit finite volume program to study the

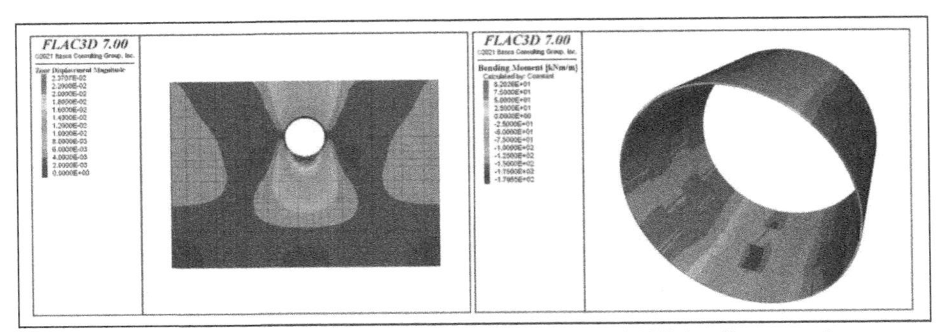


**Figure 4. Typical FLAC3D results left) ground displacement and right) PCTL bending moment**

mechanical behavior of continuous three-dimensional mediums as it reaches equilibrium. The FLAC3D model explicitly simulate the PCTL segments, circumferential and radial joints and their skew angles. Interface elements are located at all joints with prescribed compression-only frictional behavior. This allows the elements between the segments to separate freely but not slide against each other unless the interfacial joint shear force overcomes the frictional resistance.

The PCTL ring model contains 4 consecutive rings and does not explicitly simulate the sequential advance of the TBM. Nevertheless, the effects of ground stress redistribution are critical to the PCTL ring analysis and thus are implicitly simulated in the plane-strain analysis. The magnitude of ground convergence was estimated by the Convergence-Confinement Method (CCM) (Panet, 1995), which has been widely used in the industry to evaluate the effect of the ground relaxation in lining design. The method was developed for open-face mined tunnels primarily, but a study by Almog et al. (2015) further honed its capabilities to be applicable to full-face bored tunnels (termed CCM-TBM). This approach represents an analytical elasto-plastic axis-symmetric solution specially tailored for closed-face TBMs and accounts for the applied face pressure through a confinement ratio to the effective ground stresses.

The FLAC3D model is capable of incorporating a wide range of load cases and staging them together where applicable. For ATSSE, this included primary and secondary grouting, ground and water loads, construction and operational loads, thermal gradients, and seismic. Typical results are illustrated in Figure 4.

## Circumferential Joint

After assembly of a complete PCTL ring, the TBM advances forward by thrusting against the most recent assembled ring. When the TBM's hydraulic jacks bear against the jacking pads, they in turn bear against the circumferential joint. High compressive stresses can be expected to develop under the jacking pad which result in the formation of tensile bursting stresses within the segment. Additionally, tensile forces may develop on the surface of the leading edge between jacking pads.

The tensile forces were first estimated using the commonly accepted elastic approach by Leonhardt (1964) and compared against the concrete's tensile strength. During detailed design, additional complexities and post-cracking behavior was added using non-linear numerical models in the software package LS-DYNA developed by Livermore Software Technology Corp.

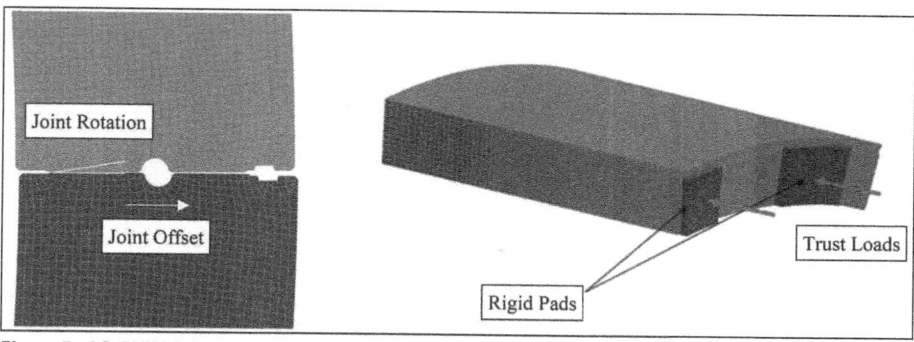


**Figure 5. LS-DYNA joint models left) radial and right) circumferential**

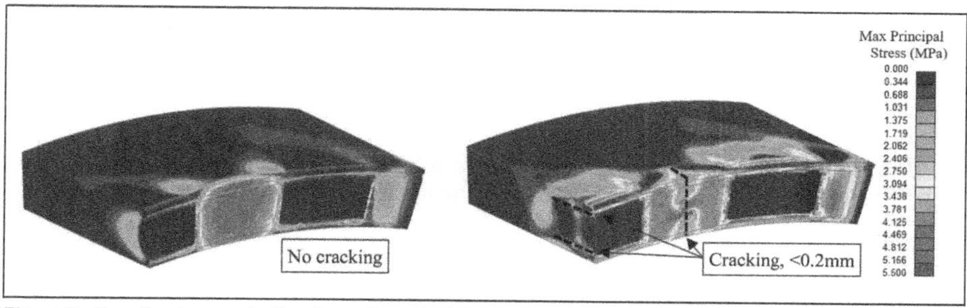


**Figure 6. Tensile stresses in the circumferential joint model at left) design load and right) 2× design load**

The PCTL was represented using the Winfrith concrete model. Originally developed by Broadhouse (1995), it has been extensively validated against experimental data. As a smeared crack model, Winfrith captures concrete cracking in tension, crushing in compression, tension stiffening, compression softening and overall failure. It is suitable for SFRC design and was calibrated against simulated material tests (i.e., compression cube, split cylinder, and flexural beam).

Construction errors may occur during ring-building which have to be considered within the design and which lead to unintended misalignment of the joints such as offsets and rotations, or the relative placement of the TBM's thrust rams such as on curves. These were considered in the models as illustrated in Figure 5.

The TBM of ATSSE has a maximum installed thrust of 142 MN, which when spread over 24 pads, may be exerted as an average 22.9 MPa on the bearing surface. Under the design load, tensile stresses are seen in Figure 6 between the pads and beneath the bearing area which mirrors the elastic analysis. As the load increases and cracks form, the design width limit of 0.2 mm is reached only at twice the design load level.

## Radial Joint

The erection of the segments into rings within the TBM is anticipated to have minimal radial distortions. Mitigation measures include dowels on the circumferential joints, guide rods along the full radial joint length, two-component annulus grout and a 5 mm maximum limit in the lipping between adjacent segments. Nonetheless, a 25 mm

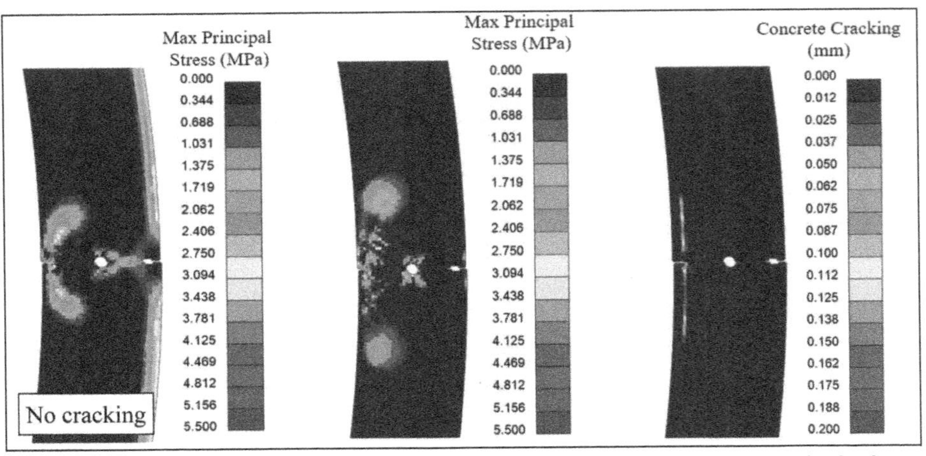


**Figure 7. Tensile stresses in the radial joint model at left) design load, and middle) 2× design load. Right) crack pattern and widths at 2× design load**

radial distortion design limit was set which is equivalent to 9.4 mrad in total joint rotation. These were considered in the models as illustrated in Figure 5.

An onerous design scenario was envisioned whereby the TBM exerts the highest possible face pressure at the deepest alignment location. Thus, the ground is left with little in-situ stress relaxation and a factored design hoop load of 4,400 kN/m is established. Joint opening is estimated from FLAC3D and found to be negligible in this case.

The non-linear plastic analysis of the radial joints indicated that their ultimate capacity exceeds the factored design loads (Figure 7). Additionally, only internal microcracks (<0.03 mm), which are self healing cracks and won't be visible, were predicted at the design load level.

## SUMMARY

The paper describes key procurement activities and design highlights of the precast concrete tunnel lining for the Advance Tunnel for the Scarborough Subway Extension. Upon completion, it will be the first single bore transit tunnel in Toronto and the largest SFRC tunnel in Canada. The design considers the sequential stages of the tunnel segment's loading from the erection inside the TBM tunnel to long-term effects such as fire. A combination of analytical and advance numerical models was used to assess and validate the ground-structure interaction behavior.

## REFERENCES

Almog, E., Mangione, M., and Cachia, G. 2015. Ground relaxation in segmental lining design using the Convergence-Confinement Method, Underground Design and Construction Conference, IOM3 Hong Kong Branch, Hong Kong, 335–345.

Broadhouse, B.J. 1995. The Winfrith Concrete Model in LS-DYNA3D, AEA Technology. United Kingdom.

ITA Report, 2016. Twenty years of FRC tunnel segments practice: Lessons learnt and proposed design principles ITA Working Group 2. International Tunnelling and Underground Space Association, Avignon. France.

Leonhardt, F. 1964. Prestressed Concrete: Design and Construction, 2nd ed., Wilhelm Ernst & Sohn, Berlin, Germany.

Panet, M. 1995. Calcul des Tunnels par la Méthode Convergence-Confinement, Presses de l'Ecole Nationale des Ponts et Chaussées, Paris, France.

Susetyo, J., Dutton, M., Gregor, T. and Mongeau, M. Designing Reinforcement for Precast Concrete Tunnel Lining Segments, North American Tunnel Conference, Society for Mining, Metallurgy, and Exploration, Washington, DC, USA, 524–531.

# Qualifications and Design of Waterproofing Systems for Underground Structures

**Stefan Lemke** ▪ Renesco Inc.
**Tim Kearney** ▪ Renesco Inc.
**Enrico Pavese** ▪ Renesco Inc.

## INTRODUCTION

Watertightness is of central importance in hydraulic and underground structures but has always been associated with a risk, as demonstrated by the very high impact that water inflows have on the damage of linings and infrastructure (Howard, 1991; Sandrone and Labiouse, 2011). Concrete is vulnerable to groundwater and chemical contaminants in the soil. These can leach into the pores and cracks of the concrete and degrade its integrity, causing spalling and corroding the reinforcing steel. Moisture migrating into the building can also degrade structural elements over time and lead to mold and other hazards to occupants' health. In this connection, the demands placed on the setup and detailed design of the seal depending on the negative influence by water on the one hand and on the other hand on the function of a construction in view to its watertightness classification. Thus, the reliable functioning of a seal is of particular significance in the case of traffic tunnels, which are not easily accessible for all subsequent repairs after the construction in seepage water and especially when located in a pressure water zone (risk–cost assessment).

Requirements for the long-term efficiency and quality of underground structures have increased in recent decades. Today, the standard lifetime requirement for an underground infrastructure project is 100 to 120 years and, in some cases, even higher. This calls for high increased attention to the durability of the whole structure and of each individual element.

Standards, which started as application guidelines given by the manufacturers, are available in the market since the 1980s and are constantly adapted to the current developments in various technical committees. Differentiations are made between:

- Manufactures application guidelines, case studies or published articles
- Test standards for the laboratory to guideline the testing procedures (air temperature, type of test equipment, dimensions, test speed, test duration, concentration, type of liquids, monitoring, tolerances, interpretation of results, documentation, etc.), developed by committees based on material and laboratory experts
- Application standards, which include test standards but also the related minimum performance requirements (values) for material, processing, and quality assurance (practical tolerances), developed by committees based on a wide range of experts (e.g., designers, construction companies, applicators, manufactures, owners, test institutes, quality engineers) and therefore also based on a wide field of experience, different views and up-dated case studies.

In the following there are some examples of guidelines and published information's, ranked according to the year of issue, which are relevant for the development of waterproofing specifications in the tunnel industry:

- manufactures application guidelines, case studies or published articles
  - 1957: Lufsky, K., Sealing of buildings with plastic, construction planning—construction technology, H. 5, page 205–211; VEB Verlag Technik, Berlin, Germany
  - 1969: Girnau, G.; Haack, A., tunnel sealing; STUVA Forschung + Praxis, Band 6, Germany
  - Nov. 1978: Heller, G., Experiences with tunnel and shaft sealing during the construction of large projects in Austria, conference, sealing of rock cavities, Dynamite Nobel AG, Troisdorf, Germany
  - 1979: STUVA, Investigation of the possibility of seam testing for single-layer plastic sheet waterproofing in tunnel construction, in: STUVA Research Report, 12/79, Germany.
- test standards for the laboratory
  - 1977: DIN 1910-3: Welding; Welding of Plastics, Processes,
  - 1986: DIN 16726: Plastic roofing felt and waterproofing sheet—Testing
  - 2004: EN 13491: Geosynthetic barriers - Characteristics required for use in the construction of tunnels and associated underground structures
- application standards/ guidelines
  - 1960: Guideline German Railway, Sealing of buildings, instructions for engineering structures (AIB)
  - 1977: SIA 280: plastic geomembranes, requirement values and material test, Swiss Society of Engineers, and Architects
  - 1983: DIN 18195: Waterproofing of buildings
  - 1986: DIN 16938: Plasticized polyvinyl chloride (PVC-P) waterproofing sheet incompatible with bitumen
  - 1988: Heft 365: Basic documentation on the construction and testing of tunnel sealing Systems, Federal Ministry for Economic Affairs—Austria, investigation of highways
  - 1992: Fascicule n°67 Titre III, Waterproofing of CCTG underground works, French Ministerium of Transportation
  - 1993: DS 853, Guideline German Railway, planning, building, and maintaining railway tunnels
  - 1995: ZTV-ING, Additional technical contract terms and guidelines for engineering structures, Federal Highway Research Institute, Germany
  - 1997: Recommendation Double Layer Sealing, Tunnel—EDT, Ernst & Sohn, German Society of geosynthetics (DGGT)
  - 2012: OEBV Guideline, tunnel waterproofing, Austrian Construction Society (English translation 2015, www.bautechnik.pro)

Some key findings from these published information, guidelines, and standards, which are important for the further perspective on sealing systems for underground structures.

- Independent of the construction and installation technology, the waterproofing sealing system acts generally as a permanent elastoplastic flexible skin. It is characterized by the fact that it adapts to the subsequent deformation of the structure or its components and that it bridges over system-specific cracks of the final structure without any damages or leakages. Furthermore, the rheological behavior is of great importance. Substances with pronounced flow behavior, a time-dependent constant load deformation, offer in this case additional benefits. They decline, over some time, the imposed tension due to building deformation by corresponding material flow operations.
- Material thickness increases the dynamic perforation/ puncturing resistance against mechanical damages, bridging capability and acts also as

a buffer layer against chemical attacks, which boosters also the material's ageing behavior.

- In case of a double shell tunnel lining, incorporating a geomembrane, the membrane system (loose-laid, sport-wise fixed) acts as a sliding surface between the two shells, avoiding shrinkage cracks inside the final concrete lining and gives a further contribution to the watertightness, beside a repair possibility for defined areas/ compartments in case of remedial grouting, even in inaccessible areas (e.g., under the track).
- Stiffer, non-reinforced materials tend to have a greater linear thermal expansion/ contraction amplitude in case of temperature changes. This phenomenon is used—for example—for the installation procedure of loose-laid HDPE geomembranes ($250 \times 10^{-6}$ $K^{-1}$) for landfills incorporated in soil, wherein the change in temperature between day and night caused material tension (shorten of sheet length) and finally a smooth membrane surface. In contrast to the use in landfills—these temperature-related stresses may cause in the application field of underground structures debonding and failure of the spot-wise fixation.
- A possible laying close/ tight to the entire surface of the substrate (as cavities-free as possible) should be aimed, so that no folds within the sealing sheet are created during concreting. Stresses above the material yield point produce a significant reduction in service lifetime (chemically and physically). This can be avoided in different ways: (a) placing the fixing points in the low points, (b) increasing the number of fixing points, (c) using elastoplastic flexible materials and (d) reducing the waviness of the substrate.
- In order to achieve a close fit to the entire surface of the (shotcrete) substrate, the materials have to be very flexible (low stiffness), ideally with a secant module of $E_{1-2} \leq 20$ N/mm$^2$. Further on, in order to achieve a continuing sealing system, their welding properties are essential, and the plastic sealing systems may be not too sensitive to mechanical stresses in the construction and operating phases.
- A proper joint-sealing between two sheets is characterized by a) behavior in shear test (e.g., according to EN 12317-2), which should generate a fracture outside the seam (joint is stronger than the sheet) and b) the peel strength should be ≥6.0 N/mm (e.g., according to EN 12316-2) for flexible elastoplastic materials like PVC-P or FPO and even higher for stiffer (PE based, $E_{1-2} >$ 80 N/mm$^2$) products.
- Another key principle in view to a proper seal is not to change the waterproofing system and material in the same cross-section of a structure caused by the problematic of transition details and different system/ material behavior.

Looking back to the history of waterproofing, a wide range of different materials were used in the field of tunnels with more or less success, e.g., bitumen, ethylene- copolymerized bitumen (ECB), ethylene- vinyl-acetate-copolymer (EVA), rubber, modified high density polyethylene (mod. HDPE), polyvinylchloride (PVC-P). In the early time, new products came up based on polyethylene e.g., low density, linear low density or extremely (very) low density polyethylene (PE-LD, PE-LLD or PE-VLD), flexible polypropylenes (PP- flex) and a lot of mixed combinations, so-called compounds. These wide ranges of products are called TPO or FPO (thermoplastic-polyolefin or flexible polyolefin).

Considering the application of membranes in tunnels, the long-time experience of sealing, the practical welding behavior, the economics, and the technical characteristics of all above mentioned materials, in general two of them have convinced: PVC-P and FPO/TPO.

Between both materials exists naturally differences in characteristics and behavior, mechanical properties, chemical resistance, and welding behavior, which are essential for the material specification and test procedures.

On top of all, the material must fulfill the environmental requirements, e.g., according to REACH (Registration, Evaluation, Authorization and Restriction of Chemicals), wherein the use of substances in relation to its concentrations are restricted.

## MATERIAL DURABILITY

During the last decades, different waterproofing materials have been studied and tested in tunnel applications, mainly in Switzerland and Germany.

Material's long-term ageing behavior is not given per se and must be explored extensively by suitable tests, which simulate—under accelerated processes—real exposures and loads. Hereby the understanding of the individual ageing processes is curial beside the long-term experience (references) with these materials to ensure the transfer of the laboratory results and performance to site practice (test calibration) and vice versa.

The long-term resistance of the waterproofing membrane defines its ability to maintain the required characteristics under exposure to the predictable environmental influences during its service life. Hereby the plastic material durability depends on various mechanisms that cause a kind of degradation, which obviously can act in synergy. However, specific material groups are affected by these various degradation mechanisms in different ways, which has to find its way in the relevant guidelines and standards. Further on, entirely different chemical processes take place at higher temperatures than at lower temperatures, which means that temperature-accelerated ageing processes and natural ageing processes are not parallel. Therefore, it is recommended to limit the test temperatures in view to the in-situ environment along with an extension of the test period and an introduction of identification tests. Otherwise test procedures tend more to a material Olympiad with a destructive character and without simulating the natural ageing.

There are many accelerating tests available, but none of them can guarantee an exact lifetime prediction of a given material which strictly depend on the real condition of use. Those tests are useful for comparative analysis between different products or between a given product and a naturally aged sample of the same or similar material, which evidence is shown in the NEAT (new railway tunnels through the Swiss Alps) or BASt (German Federal Highway Research Institute) evaluation programs and which can exclude unsuitable materials.

Especially the NEAT has investigated in a 2-, 5- and 10-years research program for the waterproofing systems of the new tunnels, Loetschberg (35 km) and Gotthard (57 km), which must remain watertight and drained over the design lifetime of 100 years. Permanent temperatures up to 45–50°C were detected especially in the deep-seated Gotthard tunnel. These service temperatures are high compared to tunnels with less rock mountain above the tunnel construction and are challenging for regular waterproofing and drainage materials. Ten different polymeric waterproofing membranes (PVC-P, PE and PE-copolymers, e.g., EVA-Ethylene Vinyl Acetate) and eight drainage materials on the base of PE, PP, and PA were included in the evaluation. All materials studied were commercially available products.

Immersing and leaching of materials in various media under different conditions allowed judgment of the thermal, chemical, and biological long-term properties of the investigated materials. Leaching tests in water at 23°C, 45°C and 70°C resulted in information on the influence of elevated temperatures and data for life-time predictions. Ageing in oxygen-enriched water at 70°C under low pressure (3 bars) allowed a study of the influence of oxygen, in particular on polyolefin materials. In order to simulate the acidic water that may locally be present in the mountain, the polymeric waterproofing membranes and geotextiles were immersed in a 0.5% solution of sulfuric acid at 50°C. Simulating the effects of alkaline water from the concrete environment was accomplished by immersing the products in a saturated calcium-hydroxide solution at 50°C. Within the evaluation program further aspects were addressed regarding system behavior, in particular of their drainage capacity by studying long-term properties in compression and compression/shear behavior of systems, and installation applicability of the waterproofing system (1:1 in-situ field test incl. re-construction, shotcrete evenness in comparison to the membrane flexibility and fixation technology).

The findings of these studies and programs are integrated in the up-dated relevant standards, like SIA 272 (Switzerland), OEBV (Austria), ZTV-ING (Germany) and in the project specifications of NEAT (Switzerland), Brenner-Base-Tunnel (Austria-Italy), and TELT-Turin-Lyon (France- Italy).

## FULLY BONDED SYSTEMS

Full surface bonded (adhered) sealing systems are mostly used in the building industry for concrete foundations, basements, walls (structural waterproofing), floors, and roofs or for bridge-deck waterproofing, wherein a wide range of different systems and materials exist, like bituminous membranes, liquid applied coatings (PU, polyurea, epoxy, etc.), spray-applied membranes (acrylic, EVA, bituminous, cement, etc.), asphalt or sealing paintings. All these bonded systems require:

- a defined substrate quality (smoothness, roughness, porosity, grain-size, moisture content, free of loose particles, limited gaps/ voids, minimum pull-off strength, etc.), on which the material is applied (bond) and which is part of the system and cost evaluation,
- a defined surrounding environment (air-temperature, humidity, rain, wind, dust, UV, etc.) which will influence the material, curing, bonding, durability and finally the system performance,
- system relevant details for construction and expansion joints.

In view to the above, some of the systems require a substrate pre-treatment (primer, levelling layer, etc.), a multilayer application, water management, pre-injection, and stringent environmental application conditions.

The technical advantage of the fully bonded systems is the non-water-migration (lateral water underflow) between the adhered layers and the intake of shearing loads. On the other side, these systems have a limited activated multi-axial elongation behavior based on the bonding character in comparison to a loose-laid system and therefore a limited crack-bridging ability (Figure 1).

Investigation programs in UK (Diez, 2018) shows for some sprayed systems that the bonding strength to the concrete substrate is reduced of >80% in permanent contact with water after 3 weeks, plus a significant material creeping under permanent loads,

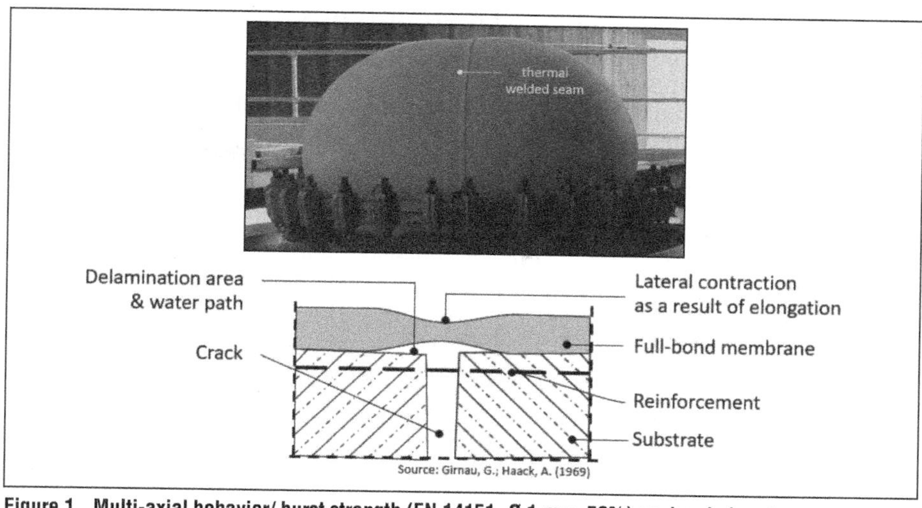


**Figure 1. Multi-axial behavior/ burst strength (EN 14151, Ø 1 m: > 50%) vs. bonded system**

which must be considered in case of statical aspects. Also, system related laboratory tests to measure the hydrostatic pressure resistance of waterproofing membranes in view to lateral water underflow or migration, for example according to ASTM D 5385 (mod.), should be critical reviewed. Hereby the surface structures of the different layers inside the test mold (membrane vs. concrete, membrane vs. steel flange) plays an essential criterium for the system performance. Laboratory simulated smooth surface structures are not jobsite related conditions (concrete roughness, honeycombs, cracks, membrane waves, winkles, etc.), especially when the adhesive layer acts a) as a bonding component, for example between the film fabric and the concrete structure, b) as a levelling layer and c) as a water sealing/ coating component itself.

Systems, where the bonding depends in full or partly on adhesives and/or adhesive (joint) tapes, it is essential that the adhesive behavior of the composite is also maintained along the service of the structure. This needs additional proof for the material compatibility and durability, also in view to the surrounding environment (concrete aggressive water, microbiological attack, hydro carbonates/ methane-gas, temperature, etc.). The understanding of the adhesive ageing processes, especially in view to the chemical ageing in combination with accelerated temperatures and the presents of water, beside the long-term experience (references) with these relative new materials are rare at this stage, especially in view to long terms of >120 years.

Therefore, some manufactures of fully bonded systems have limited the application range to 5–10 m water pressure head and recommend its use only in combination with watertight concrete structures, wherein the cracks are considered as self-sealing (<0.2 mm), whereas manufactures of liquid applied membranes for the roof industry limited the free-maintenance period to 20–25 years.

Since the aspect of sustainability is more commonly discussed in building structures, especially combined products (e.g., sheet membranes with adhesive layers) are critical reviewed in relation to the separation of raw materials.

## CUT AND COVER STRUCTURES

Open constructions are often underestimated as an engineering structure, especially with regard to the waterproofing system, which shows parallels to the waterproofing of buildings and roofs. But which is indeed rather complex, and the error rate is high. Attached are some samples of planning principles.

A distinction is made between:

- Horseshoe profile, box profile,
- Portals, free-standing construction, galleries, and open cuts using timber/ sheet walling, shotcrete, bored pile walls or diaphragm walls,
- Top-down construction,
- Pressurized water or permanent drained.

The waterproofing system can be divided in:

- External or structural (internal), watertight concrete structure with internal joint profiles,
- Post-applied or pre-applied,
- Loose laid (spot-wise fixed) or fully bonded over the entire surface,
- Industrial pre-manufactured (e.g., Sheet membranes) or in-situ manufactured (liquid-/fluid-/spray/ brushed-applied, cold/ hot applied),
- Seamless (day joint) or in-situ jointed.

In addition, the sealing system of a cut & cover structure is maybe subject to certain loads, for example:

- Re-filling (material compaction, shearing),
- Traffic loads (at overburden <1 m), e.g., shear loads caused from braking process,
- UV (ozone), wind, rain, moisture, snow and ice loads, dirt, temperature, mechanical impact, etc.,
- Sliding, friction, shear loads (e.G., Slopes)
- Concrete aggressive soil/ water (alkaline, sulfurous acid, hydrogen sulfide, seawater, etc.), microorganism/ roots, hydrocarbons (diesel, fuels, methane) depending on exposure time and concentration rate, fire loads

Other key elements for the design:

- 120 years design lifetime, which has an influence on the choice of material, on the system and in view to the risk-cost assessment on the repair possibility
- In case of post applied sealing systems
  - Environment, also in view to the application period
  - Substrate of the sealing system is the concrete structure itself with all the advantage/ disadvantages (smooth surface, sharp edges in the construction joint area, movement joints, honeycombs, formwork penetration, etc.)
- Low overburden
  - Limitation of bonding activation/ pressure-sensitive adhesive, which maybe requires a change in the sealing system on the upper walls/ roof part in comparison to the lower walls and foundations
  - Limitation of possible (repair) injection pressure, especially in the roof area to avoid up-lifts and material consumption
- Possible continuation of the sealing system from the mined tunnel to the portal area, otherwise transition details are required
- The permanent drainage must be designed against clogging from soil particles instead clogging from sintering (mined tunnel)

- In case of using external water barriers (profiles)
  - Integrity of the anchor ribs inside the concrete matrix is essential for its performance
  - In view to the roof area, air-capsulation between/under the anchor ribs must be avoided, otherwise this will cause a non-integrity of the ribs inside the concrete matrix, which requires a change to an adhesive strip (post applied).
  - The maximum water head which the profile will be able to withstand will be influenced from the geometry (labyrinth principle). In case of pressurized water, min. 6 Anchor ribs with a height of min. 30 Mm.
  - In view to a correct thermal-plastic-welding (handmade seam, min. width of 30 mm) with a sheet membrane, the lateral flaps/ edge-area of the profile (external water barrier) should have a width of minimum 40 mm and a similar thickness like the sheet membrane.

Some examples from various international standards:

- According to OEBV-tunnel sealing (Austria), SIA 272 (Switzerland) and ZTV-ING (Germany), the thickness of a loose-laid sheet membrane for cut & cover structures is set to 3 mm, even for permanent drained structures. This requirement is based on the risk assessment for a possible mechanical damage during the installation, but also during the re-filling process.
- According to ZTV-ING (Germany), the deviation of the longitudinal axis of anchor ribs of the external water barrier are limited to ±2 mm and the inclined angle is limited to ±5° (Figure 2, right hand-side). The reason behind this requirement is to guarantee the integrity of the ribs inside the concrete matrix during the concrete process. The same approach was developed some decades ago in the US industry (Figure 3). To avoid "curved" anchor ribs, the manufacture established a water barrier with more dimensional stable (thicker) ribs.
- Several new built cut & cover structures in Switzerland (e.g., Eppenberg, Belchen, Cholfirst) were designed in the wall and roof part with a combined system, consisting of a 2-K-PU based adhesive and a 2 mm thick standard sheet waterproofing membrane on the existing concrete structure. The system combines the advantage of a flexible sheet waterproofing with the "fully

**Figure 2. Examples of external water barriers (profiles)**

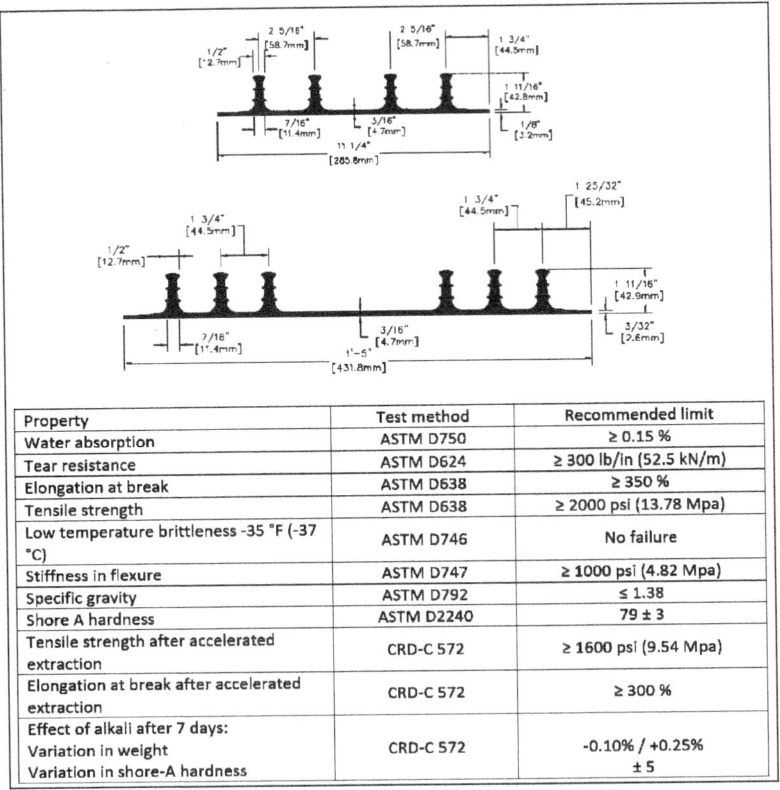

| Property | Test method | Recommended limit |
|---|---|---|
| Water absorption | ASTM D750 | ≥ 0.15 % |
| Tear resistance | ASTM D624 | ≥ 300 lb/in (52.5 kN/m) |
| Elongation at break | ASTM D638 | ≥ 350 % |
| Tensile strength | ASTM D638 | ≥ 2000 psi (13.78 Mpa) |
| Low temperature brittleness -35 °F (-37 °C) | ASTM D746 | No failure |
| Stiffness in flexure | ASTM D747 | ≥ 1000 psi (4.82 Mpa) |
| Specific gravity | ASTM D792 | ≤ 1.38 |
| Shore A hardness | ASTM D2240 | 79 ± 3 |
| Tensile strength after accelerated extraction | CRD-C 572 | ≥ 1600 psi (9.54 Mpa) |
| Elongation at break after accelerated extraction | CRD-C 572 | ≥ 300 % |
| Effect of alkali after 7 days:<br>Variation in weight<br>Variation in shore-A hardness | CRD-C 572 | -0.10% / +0.25%<br>± 5 |

**Figure 3. US external water barrier section detail with physical characteristics**

adhered" approach, with a documented product lifetime expectancy of >120 years, permanent seaming by thermo-welds, substrate inspection/ testing, surface levelling with the PU resin, prevention of lateral underflow between the concrete structure and the membrane in case of leakage.

- According to OEBV -tunnel sealing (Austria) and SIA 272 (Switzerland), the use of stiff membranes with an $E_{1-2}$ > 65N/mm$^2$ is restricted based on the environmental stress cracking, the welding behavior, especially in view to hand welding and its possibility to form details and the multi-axial elongation/ elastic behavior.

## CONCLUSION

One of the key differences of a waterproofing system between the tunnel and the building industry is the design lifetime of 120 years, which has an influence on the choice of material, on the system and in view to the risk-cost assessment on the repair possibility. Thus, the reliable functioning of a waterproofing system is of particular significance in the case of traffic tunnels, which are not easily accessible for all subsequent repairs after the construction, especially when located in a pressure water zone.

The effectiveness of a sealing system depends on a potpourri of factors, which includes design, material (product), application, substrate, and environmental conditions, etc. The challenge for the industry and owners is to develop qualification tests, requirements, acceptance criteria for products and quality assurance processes to ensure the service lifetime performance, which should be integrated as a routine in the relevant standards. The specifications should also consider the needs of a successful application like welding processes, qualification/ certification of the welding operators and environmental conditions.

Open constructions are often underestimated as an engineering structure, especially with regard to the waterproofing system, which shows parallels to the sealing of buildings and roofs. But which is indeed rather complex, and the error rate is high. Modern standards and guidelines for those waterproofing systems are available in the international industry, which gives a good overview about design aspects (toolbox approach).

## REFERENCES

Batty E., Bond N., Kentish E., Skarda A.& Webber S. Comparison Between Sprayed and Cast in Situ Concrete Secondary Linings at Bond Street and Farringdon Stations, WTC, 2016.

Diez, R. 2018. Testing of Sprayed Waterproof membranes and implications for composite action, BTS Conference 2018, UK.

Howard, A. J. 1991. Report on the damaging effect of water on tunnels during their working life. Tunnelling and Underground Space Technology incorporating Trenchless, 6 (1), pp. 11–76.

Lemke, S.; Schaelike, H. and Gerstewitz, T. 2015. Bonded strip termination using plastic waterproofing sheet membranes on segmental tunnel linings as an effective and economic alternative to clamping constructions. RETC, New Orleans, USA.

Lemke, S. and Moran, P. 2015. A controversial discussion regarding the use of spray-applied waterproofing for tunnel applications, RETC, New Orleans, USA.

Lemke, S., Eckl, M. and Londschien, M. 2016. 120-Year Design Lifetime of Plastics, WTC 2016, San Francisco, California, USA.

Sandrone, F. and Labiouse, V. 2011. Identification and analysis of Swiss National Road tunnels pathologies, Tunnelling and Underground Space Technology. Elsevier Ltd, 26 (2), pp. 374–390. Doi: 10.1016/j.tust.2010.11.008.

# Successful Completion of LA Metro Regional Connector Transit Project

**Mike Harrington** ▪ LA Metro
**Tung Vu** ▪ VN Tunnel and Underground, Inc

## ABSTRACT

The Regional Connector Transit Project is a 1.9-mile-long underground light rail system that will connect LA Metro's A (Blue), E (Expo), and L (Gold) Lines in downtown Los Angeles. This $1.75-billion design-build project is scheduled for revenue service in Spring 2023. The project consists of 21-foot diameter twin-bored tunnels, a 300-foot-long crossover SEM cavern, three new underground stations (at 1st Street/Central Avenue, 2nd Street/ Broadway Avenue, and 2nd/Hope Streets), and cut-and-cover tunnels along South Flower, Alameda, and 1st Streets. This paper, which is the continuation of our previous paper presented in RETC 2019—LA Metro Regional Connector Transit Project: Successful Halfway-Through Completion, will provide insight regarding the construction of major components, and the system integration and testing work undertaken for this complex transit project.

## INTRODUCTION

The Regional Connector Project (Project) is a complex subway light rail project designed and constructed by Regional Connector Constructors (RCC), a joint venture between Skanska USA Civil West California District, Inc. and Traylor Brothers Inc., with Hatch Mott McDonald (HMM) as the Engineer of Record (EOR). Please refer to the RETC 2019 paper—LA Metro Regional Connector Transit Project: Successful Halfway-Through Completion, for details regarding the initial design and construction phases of the project. This paper mainly focuses on the remaining construction, system integration, and testing parts of this project.

### Tunnels

The twin bored tunnels were completed in January 2018. After completion of the tunnel invert slab, RCC started work on the emergency walkways inside bored tunnels in August 2018. Prefabricated steel forms were used to accelerate the construction. Each segment of prefab form was 10 feet long to be able to fit the tunnel design curves. The form segments utilized wheels running on tracks installed along the tunnel invert. Once positioned, additional dowels were installed to resist lateral load from concrete pour. The top of frame was anchored to the tunnel lining to prevent lateral movement. The prefabricated forms proved to be efficient in expediting the emergency walkway construction and being reusable for other projects. Figure 1 and Figure 2 show the emergency walkway under construction and at completion.

The Regional Connector tunnels were designed as a watertight system with double gaskets installed at the segment joints as well as at the ring joints. At the interfaces between the tunnels and cast-in-place (CIP) structures—such as the cut-and-cover (C&C) tunnels, stations, and SEM cavern—an Omega seal system was installed to seal the gap between the separate structures. The seal assembly utilized a continuous reinforced rubber sheet tightly clamped into galvanized steel plates. These

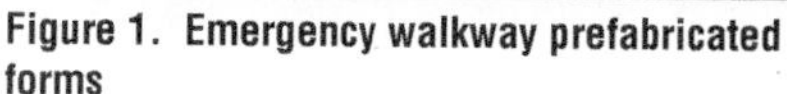

**Figure 1. Emergency walkway prefabricated forms**

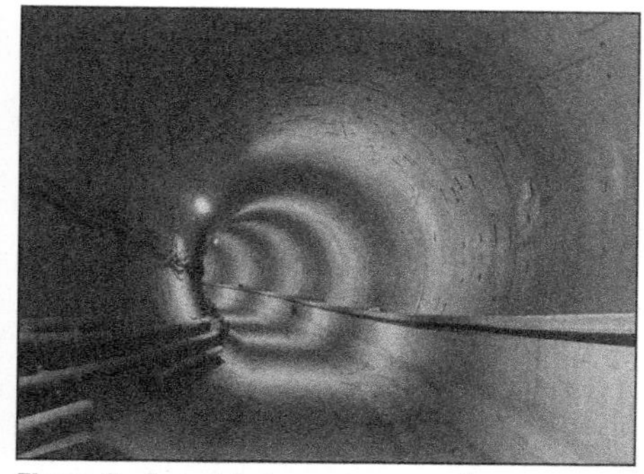

**Figure 2. Completed emergency walkway**

galvanized plates were attached to the structures by embedding them into the CIP concrete along one side of a joint, and by bolting them to the tunnel rings on the other side. The seal assembly was designed to withstand the anticipated groundwater pressures while also allowing for the Omega seal sheets to be shaped to accommodate seismic ground movements and differential settlement. Figure 3 shows a mockup of the Omega seal assembly. Figure 4 shows the installation of an Omega seal assembly in the field.

**Figure 3. Omega seal mockup assembly**

**Figure 4. Omega seal installation**

## Tunnel Cross Passages

The typical construction sequence for the cast-in-place concrete cross passages consisted of installing 60-mil thick High Density Polyethylene (HDPE) waterproofing membranes and geotextiles atop temporary shotcrete linings, and then placing reinforcing steel and formwork to accommodate CIP concrete placements. The waterproofing membranes were compartmentalized utilizing closed rings of HDPE water barriers (with re-injectable grout hoses, or Fuko hoses, installed at the mid-length), remedial grout hoses within each compartment, and mechanical connections to the precast concrete tunnel lining via anchor bolts, steel straps, and hydrophilic strips.

Waterproofing system installations were especially challenging due to the complicated shapes of the cross passage excavations and the stiffness of the HDPE membranes. Installation crews were unable to bend the membranes at tight corners and instead had to cut and weld the sheets using multiple smaller pieces. An example of a

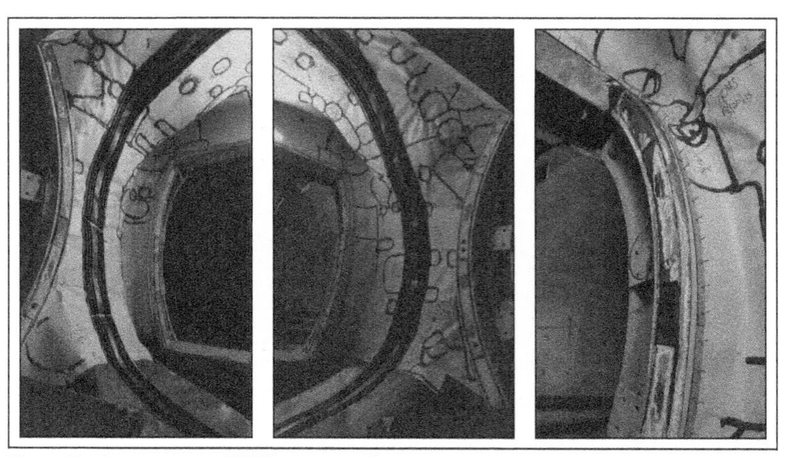

Figure 5. Crosspassage HDPE waterproofing system

completed membrane consisting of numerous small pieces and welds is presented in Figure 5. Not surprisingly, water intrusion at some of these completed cross passages has since been observed, possibly due to HDPE weld imperfections or the tearing of seams during concrete placement. Further discussion of the cross passage water intrusion challenges is presented below.

Placement of the CIP concrete for the final cross passage linings also proved to be difficult. Concrete placement for the first cross passage (CP3) was unsuccessful due to the incorrect use of a low slump concrete mix and the crew's ongoing learning curve. After curing and form removal at CP3, voids were found at the crown and along the surfaces of the CIP lining. To correct the defects, the Contractor had to pump large quantities of cement grout into the voids, and perform significant concrete surface repairs along final lining.

Construction of the two other cross passages (CP1 and CP2) did improve however, as lessons learned from the first cross passage were incorporated. The use of a higher-slump and smaller aggregate concrete mix for CP1 and CP2 resulted in fewer voids and surface defects.

## SEM Cavern

The SEM cavern is located at the eastern end of the Historic Broadway Station and is an important structure of the project. It consists of 1'-6" thick CIP reinforced concrete final lining with a thickened invert slab of up to 5'-6" at the middle. Inside the cavern, a plenum slab was designed and constructed to separate the trainway from the ventilation plenum in the upper part of the cavern, which is connected to the emergency fan room located at the eastern end of the Broadway station.

The construction sequence of the SEM permanent structure consisted of constructing invert slab first, followed by the side walls, plenum slab, and cavern crown. The work started with the invert slab sections at the eastern end of the cavern and progressed toward the Broadway station. As soon as the first sections of invert slab were completed, the side wall work started. The same work pattern was repeated for the plenum slab and cavern crown.

The final lining was designed with a compartmentalized water proofing system consisting of a 60-mil thick HDPE membrane on the exterior face. This HDPE water barrier was installed with re-injectable Fuko hoses at each of the construction joints, and with remedial grout hoses inside each compartment.

Nonwoven geotextile was first installed on the face of initial shotcrete lining to smoothen out the substrate and provide a cushion for the HDPE membrane. Hanging the wall and crown HDPE membrane prior to concrete placement was a challenge. The contractor proposed to use Velcro strips adhering on the back of HDPE membrane to hang it to the nonwoven geotextile. It was accepted by Metro after the mockup sample showed it could be a good solution. It worked well in the field except some sagging of HDPE at localized locations. A bigger challenged came when it was recognized that the concrete pours at the lower portions of the invert slab and walls pulled down the HDPE membrane that has been installed on the upper parts. As the membrane does not have enough slack to accommodate the stretch, the crew had to cut the wall HDPE and add a new piece along the side to provide sufficient slack and avoid tearing during concrete pours for the walls and cavern crown. Figures 6 shows the HDPE installation in progress.

The side walls and SEM crown were constructed using prefabricated steel forms with rolling wheel on tracks installed at invert or plenum slab. The form has both wall mounted vibrators and ports for needle vibrators. Figures 7 and 8 show the prefabricated forms for the walls and cavern crown being installed and ready for concrete pours.

The final lining was designed using 4,000-psi concrete; however, the contractor elected to use a 5,000-psi concrete mix design to allow early form stripping time. The concrete mix also used superplasticizer and small aggregate to allow for better workability. The concrete work came out successfully without any major defects. The SEM permanent lining construction took 13 months to complete, started in April 2019 and completed in May 2020. Figure 9 shows the completed SEM cavern at track level.

Upon completion of the final lining, the contractor performed contact grouting through the grout tubes that were installed along the cavern crown to allow grouting both front and back of the HDPE simultaneously to avoid damage to the waterproofing membrane due to unbalanced grouting pressure. Thicker grout of 4:1 and 6:1 ratio (i.e., a ratio of 4 gallons of water to 1 bag of 94 lbs of Portland cement) was first used at the ends of the SEM cavern to create a barrier to prevent grout from filling the Omega seal or travel to the station structure. For the remaining middle section of the SEM cavern, a thinner grout of 8:1 ratio was used to allow grout to travel farther and fill all voids.

**Figure 6. SEM cavern HDPE installation**

**Figure 7. SEM cavern prefabricated wall form**

Figure 8. SEM cavern crown form

Figure 9. Completed SEM cavern at track level

A total of 36.16 cubic yards of grout was injected for a 283 feet long cavern, of which 23.9 cy was injected inside the HDPE and 12.26 cy was injected outside the HDPE.

## Cut and Cover Tunnel and Underground Stations

There are three C&C tunnel segments along 1st Street, Alameda Street, and Flower Street, and three underground stations which are the Little Tokyo/Arts District Station, Historic Broadway Station, and Grand Ave./Bunker Hill Station. The C&C tunnels are typical CIP reinforced concrete structures with a compartmentalized HDPE waterproofing membrane on the exterior face. The construction of C&C tunnels went smoothly without any major issues. The main challenge was to construct the roof of tunnel under extremely low headroom. Figure 10 show the C&C structure underneath the 1st and Alameda St. intersection where the top of concrete is only a couple inches below the support of excavation waler or existing utilities.

The construction of each underground station has unique challenges by its own. All three stations were designed with a fluted wall finish at the track level. The fluted finish was created by using a thermoformed plastic liner as shown in Figure 11 to attach to the interior face of a regular formwork system. The first couple of concrete wall pours at the Grand Ave./Bunker Hill Station were not very successful as there were a lot of honeycombs and voids on the fluted surface. This was caused due to a combination of factors, including the use of a low slump mix design by the contractor, congested wall rebar, and poor workmanship. The concrete pours were later improved with the use of a concrete mix having higher slump using superplasticizer and small aggregate to allow for better workability.

Figure 10. Construction of C&C tunnel roof

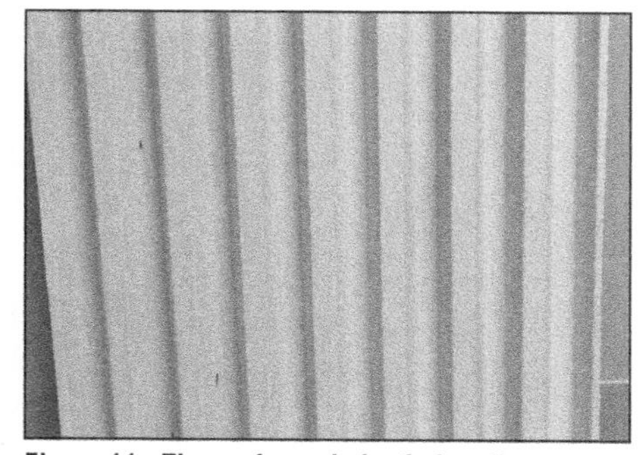

Figure 11. Thermoformed plastic formliner

Figure 12. Example of station wall rebar

Figure 13. Example of embedded conduits in roof slab

Figure 14. Threaded rods for future overbuild

Figure 15. Protection of threaded rods prior slurry fill

At the Historic Broadway Station site, the challenges came from the extremely congested rebar of the entrance structure as it was designed to support a high-rise overbuild structure in the future. Figures 12 shows an example of wall rebar of the entrance structure. Another challenge is that there a great number of conduits that are embedded in the roof and invert slab for the MEP and communication system as shown in Figure 13. Despite the extreme congested rebar and embedded conduit conditions, most of the concrete pours came out satisfactorily. Some wall pours were less successful with a presence of voids and honeycombs on the surface. Coring of the full wall thickness was then required for quality verification purposes. The testing and visual inspection of the concrete cores determined that the structural walls were satisfactory and only surficial repairs were necessary.

The Historic Broadway entrance structure also accommodated a load transfer system (LTS) to allow for structural connection of the future overbuild structure. After multiple design alteration and evaluation, it was decided to extend the vertical threaded rods above the top of LTS that were then be protected with plastic wrap and slurry fill to avoid corrosion. Figures 14 and 15 show the extended threaded rods on top of LTS for future overbuild.

The architectural finish and artwork started at all three stations as soon as the major structure works were completed. Figures 16 show the 17 by 60 feet mosaic artwork piece and high-speed elevator head house installed at the Grand Ave./Bunker Hill Station.

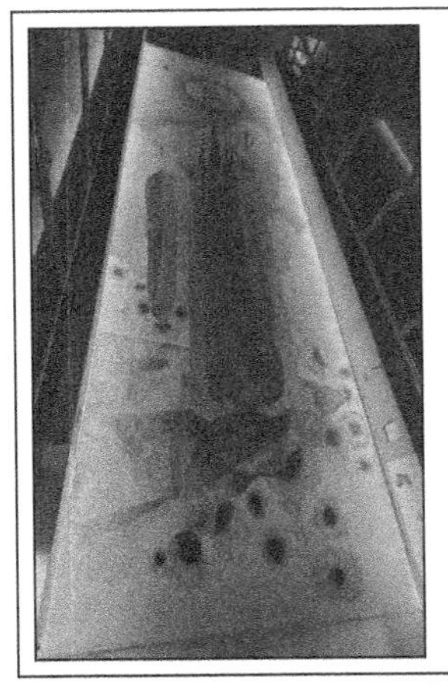

**Figure 16. Mosaic artwork and high-speed elevator headhouse at Grand Ave./ Bunker Hill Station**

## Trackwork

The Regional Connector project required special trackwork along some segments to mitigate the potential for ground borne noises and vibrations to impact certain sensitive adjacent properties. A floating slab track section was designed and installed at the Grand Ave./Bunker Hill Station and along 1,100 feet of each tunnel where the alignment crosses beneath The Broad Museum, Colburn School of Music, and the Walt Disney Concert Hall. Stared in November 2019, this floating slab system utilizes precast concrete segments supported by vertical and lateral elastomeric pads. The precast slab segments were delivered to the site where vertical support pads were then adhered to (i.e., glued) into 1-foot diameter recesses cast into the bottoms of the slabs. The precast segments were then lowered to the track level and installed atop a finished invert slab. Once the precast floating slab segments were installed, rails and direct fixation anchors were placed and grouted into place. Figures 17 and 18 show the floating slab and rail installation work.

In addition, a Low Vibration Track (LVT) system was also required for the tunnel section at the eastern end that goes under the Japanese Village Plaza and adjacent to the residential buildings in the area. For the rest of the alignment, traditional concrete

**Figure 17. Floating slab track installation**

**Figure 18. Completed floating slab track**

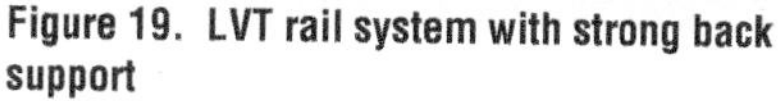

**Figure 19. LVT rail system with strong back support**

**Figure 20. Infill concrete placement**

plinth was required per the contract. However, for construction convenience and schedule benefits, the contractor elected to use the LVT system for the rest of the alignment except at the special trackwork locations. The rails and LVT blocks were first delivered and installed on the tunnel invert. The rail system was then set into the position and supported with the pour-in-place (PIP) strong back systems at every 4th resilient tie. "Cow pie" concrete was poured to lock the resilient ties next to strong back systems in place. Infill concrete was delivered by on-track buckets to the pour location to complete the LVT embedment. Figures 19 and 20 show the LVT ties and rail supported with the PIP strong back systems and the infill concrete placement in operation.

## Emergency Ventilation, Traction Power, and Other Systems

A Fire Dynamics Simulation (FDS) and computational fluid dynamics (CFD) package was used to design the station and tunnel emergency ventilation systems. This package modeled the three-dimensional aspects of the smoke and heat movement from a medium-growth fire to simulate the impact that Emergency Ventilation Fans (EVF's) would have on air movement throughout the project corridor. The CFD analysis allowed the engineering team to determine the emergency ventilation system capacities (as per the LA Metro Design Criteria and NFPH 130) necessary to provide tenable environments for passengers during a fire evacuation event.

A key finding of the CFD analysis was that two fan plants—each plant having a nominal capacity of 500 kcfm—were needed at each of the three new stations. The analysis also showed that an additional fan plant was necessary at the Alameda Wye where the two track alignments diverged. To achieve these flow rates, 600-HP and 400-HP vane axial reversable fans were installed, in pairs, at each end of the new stations and at the Wye. Figure 21 shows a 600-hp EVF assembly being installed.

The systems were also designed to accommodates 60 separate Emergency Ventilation Evacuation Operations scenarios (EVOP's). In the unlikely event of a fire emergency, a Metro Controller or first responder can activate any one of the 60 EVOP's using a single-button command. These commands can be executed either locally at the station Emergency Management Panels (EMP's), or remotely from Metro's Rail Operations Center (ROC) in central Los Angles.

In addition to the emergency ventilation equipment, water suppression systems were installed to assist with the control of heat and smoke from a fire. The water suppression systems include a standpipe distribution network connected to two separate city mains, for redundancy, along with under-train deluge systems at each of the station platforms. Provisions for removing traction power are provided during a water discharge event to help protect passengers, first responders, and other personnel. Figure 22 shows the undercar deluge systems adjacent to the Historic Broadway station platform.

The Regional Connector Traction Power (TP) Supply System was designed and constructed to be compatible with the existing equipment utilized on the Metro A-Line and L-Line. Primary power to each of the substations, which are located in back-of-house areas of the new stations, is provided by the Los Angeles Department of Water and Power (LADWP) at 34.5kV. These substations then covert the 34.5kV AC power to 750-volt DC, which is then distributed to trains along an Overhead Contact System (OCS). A unique feature of the project is that instead of utilizing traditional wire-based centenary assemblies for the OCS, an Overhead Catenary Rail (OCR) system was used. The OCR technology is relatively new for Metro but was chosen due to its durability and reduced maintenance requirements compared to wire-based systems. Figures 23 and 24 shows the OCR at the Alameda-leg crossover, and a typical traction power distribution equipment installation, respectively.

A Communications System consisting of audio, video, and data circuits was constructed to connect the new stations and tunnel areas with Metro's existing ROC in central Los Angles. Construction of this new system was especially challenging as it had to remain compatible with the existing A-Line, L-Line, and L-Line Eastside Extensions systems, each of which were constructed at different times. Some of the existing communication equipment on the A-Line, for example, is analog-based and more than 30-years old. The existing Supervisory Control and Data Acquisition (SCADA) cable runs, conduits, and communications network equipment had to be extensively evaluated to determine design requirements that would provide reliable

**Figure 21. Emergency ventilation fan installation**

**Figure 22. Under-car deluge systems**

**Figure 23. Overhead catenary rails at crossover**

**Figure 24. Traction power substations room**

SCADA communications for the new train control, seismic detection, radio, CCTV, and other systems interfaces.

An Automatic Train Control (ATC) System and related components were installed to provide for fully integrated operations of the new system with the existing Metro system, so that train operations would be seamless. This new system provides continuous ATC operation from the existing lines into the Regional Connector corridor without train operators having to intervene. Also included is an Automatic Train Protection (ATP) system that provides for vital train detection, automatic protection, and interlocking control. All train control and related system elements interface with the existing SCADA system to allow for control requests from and to the ROC.

## LEAK SEALING

One of the major challenges that the project team had to face was the water leakage that mainly occurred at the tunnel interfaces, cross passages, SEM cavern, and underground stations.

At the tunnel interfaces with SEM cavern or C&C structures, water leaks were observed soon after completion of the Omega seal installation as the groundwater pressure started building up behind the tunnel lining. Water intrusion was observed at the gaps between the steel ring plate and precast segmental lining. Leakage was also observed at the lining segment and ring joints near the interfaces where relaxation of ground confinement pressure occurred due to the station excavation. Water was also observed dripping from the bolt holes at the tunnel interfaces as well as along the tunnel reaches. To seal these leaks, the contractor had to drilled through the lining and inject two-component polyurethane grout to limit the water infiltration. Additional packers were also used to target the leaked segment joints and ring joints. The leaked bolt holes were cleaned and filled with leak master sealant before the bolts were reinstalled and tightened. After a significant sealing leak grouting operation, the contractor successfully sealed off all the leaks around the tunnel interfaces. Figure 25 shows water leakage at the Omega seal at the western end of the Grand Ave./Bunker Hill Station before and after the leak repairs.

All CIP structures of the project used the same waterproofing system that consisted of compartmentalized HDPE waterproofing membrane with water barrier and re-injectable grout hose installed at the construction joints and remedial grout hoses within the compartments. This compartmentalized system was design to allow for isolating the leak areas and injecting grout into the back of structures to seal the leaks.

**Figure 25. Water leaks at 2nd/Hope tunnel interfaces before and after leak repair**

**Figure 26. Cross passage CP3 water leaks before and after leak repairs**

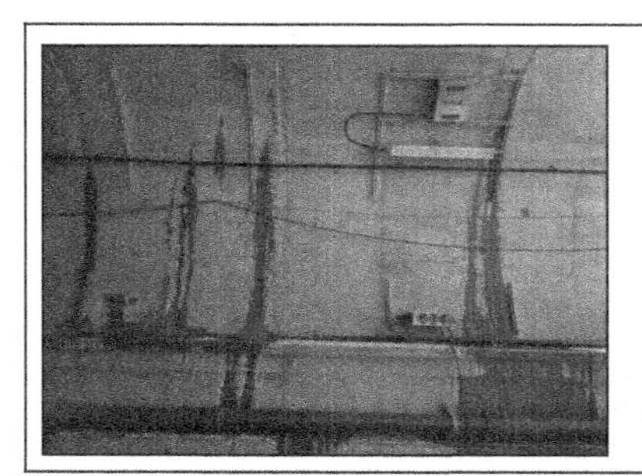
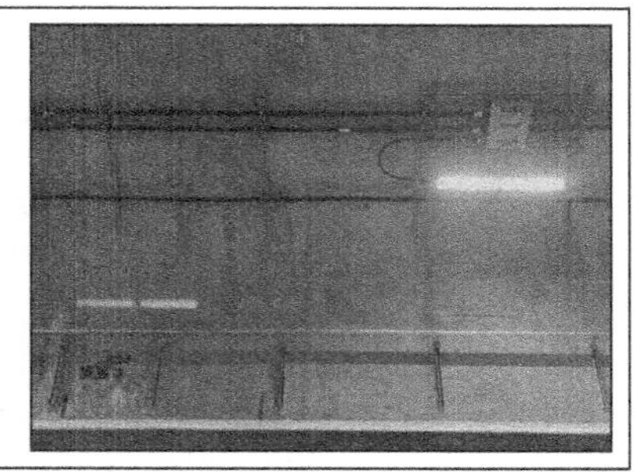

**Figure 27. SEM cavern water leaks before and after repairs**

Although the waterproofing system was installed with care and strict QA/QC procedure, some significant water leaks were observed at the cross passages, SEM cavern, and station structures after some rain falls in 2020 to 2022. The contractor had to inject two-component polyurethane grout into the Fuko hoses and remedial grout hoses to seal the leaks. Additional holes were also drilled to target the individual cracks. After multiple grouting passes, the water leaks in all cross passages, SEM cavern, and stations were successfully sealed. Figures 26 and 27 show the cross passage CP3 and the SEM cavern, prior to and after completion of the water intrusion repairs.

## Field and Systems Integration Testing

Testing of all systems-related project elements was performed utilizing Local Field Acceptance Testing (LFAT) procedures and a 2-phase Systems Integration Testing (SIT) approach. Local testing, or LFAT's, were performed directly by the Contractor and their subcontractors, and the results documented and submitted to Metro for approval. Upon successful completion of an LFAT, the first phase of systems integration testing (i.e., SIT-1) of that component could commence. The purpose of this SIT-1 work was for the Contractor—with Metro witnessing—to verify that specific components were performing as designed while operating in a local-mode configuration. Upon a successful

SIT-1 test, Metro staff would then perform the second phase of integration testing (i.e., SIT-2), where proper remote communication and control through the SCADA system was confirmed. During the 14-month testing period more than 600 systems data points and commands (i.e., SCADA points) were tested between the project corridor and the ROC.

Figure 28. SIT-2 testing of power systems

In addition to the LFAT and integration testing, extensive train testing through the Regional Connector corridor was performed. These series of tests included Minimum Headway Verifications, Signal Sightings Confirmations, and Operational Stress Tests, among others. The purpose of this train testing was to confirm that Metro could successfully operate its trains in compliance with the project operational specifications and per the predictive models. Figures 28 and 29 show systems integration testing at an electrical substation, and a Light Rail Vehicle (LRV) traveling though the SEM cavern crossover during headway verifications, respectively.

## CONCLUSION AND LESSONS LEARNED

The overall construction of Metro's Regional Connector Project was highly successfully; however, valuable lessons were still learned regarding the underground construction techniques and design details employed. These lessons learned are being

Figure 29. SIT-2 headway verification testing

shared throughout the transit agency and will be considered as Metro's advances its multi-decade construction program. Examples of lessons learned include:

1. The tunnel cross passage geometry was too complex. As discussed previously, field crews had much difficulty getting the simi-ridged HDPE liner to conform to and fit into the tight corners and bends, resulting in voids and weld tears when the CIP final liner was placed. Utilization of a simplified cross passage geometry, with fewer and sweeping bends, would allow the HDPE installation to be performed more smoothly and improve the overall quality of completed work.
2. The design of the interfaces between the bored tunnels and the cut-and-cover structures must be improved. Compression of the tunnel lining joints near the interfaces with the cut-and-cover structures tend to become relaxed, primarily due to a reduction of ground confining pressures from the adjacent excavations. Additional compression in the circumferential and longitudinal directions of the bored tunnel lining, in the form of pre-tensioned cables or rods, should be considered to ensure proper compression and the effective control the water leakage through the joints.
3. Do not wait until the end of construction to address water infiltration issues. Water leak sealing work should start as soon as possible, as the outcome of leak repairs is not readily predictable. Leak repairing is a time-consuming process and the work may require several passes before the leaks can be sealed successfully. It also requires great care to protect the other completed works in the vicinity of the leak sealing, as grout may expand into or spill over adjacent systems elements or architectural finish works.

PART 

# Difficult Ground

*Chairs*

**Chris Lynagh**
McNally

**Dave Dorfman**
Walsh Group

# Eisenhower Memorial Tunnel—50 Years Later

**Adam Bedell** ▪ Stantec Consulting Services
**Neal Retzer** ▪ Colorado Department of Transportation
**Nick Cioffredi** ▪ Stantec Consulting Services

## ABSTRACT

The first wagon trail across Loveland Pass was built in 1869 thus starting the struggle between the Colorado Rockies and interstate travel. The 1st Pioneer Bore underneath Loveland Pass at the Continental Divide in Colorado was attempted between 1941 and 1943 and was abandoned due to ground conditions at the Loveland Fault. Construction on the nearby Eisenhower Tunnel began in 1967 and was completed in 1973. Numerous challenges associated with tunneling, including at 11,000 ft were overcome. The purpose of this paper is to summarize the history of investigation and construction challenges associated with the Eisenhower Tunnel to celebrate its 50th anniversary.

## HISTORICAL BACKGROUND

The first wagon trail was made in 1869 across Loveland Pass and the Continental Divide at almost 12,000 ft elevation. Challenges associated with negotiating high mountain passes such as steep grades, hairpin turns, snow drifts, avalanches, and icy conditions resulted in difficult travel during the winter months. The road was abandoned in 1906 until the U.S. Forest Service restored the road for vehicular use in 1920. This road was converted to an automobile highway in the 1930s.

Discussion and planning for a vehicular tunnel under the Continental Divide began in the late 1930s and an exploratory tunnel known as the Pioneer Bore was excavated below Loveland Pass and the Continental Divide between 1941 and 1943. This exploratory tunnel was a 7-foot by 7-foot modified horseshoe and driven approximately 5,500 feet parallel to the Loveland Fault and in extremely poor ground conditions before it was abandoned.

In 1943, the Colorado Highway Department advertised for a tunnel under the Loveland Pass, but all bids came in over the $10M Engineer's Estimate ($172,261,271 in 2022) thus shelving the project. In 1950, the Highway Department paved Loveland Pass. In 1952, the Highway Department hired two New York City engineering firms to assess the costs and benefit of tunneling under the Continental Divide. Also, in 1953, the Colorado state legislature reorganized and was renamed to the Colorado Department of Highways (CDOH). During 1955, a geotechnical investigation was performed to drill along both Straight Creek and Vasquez Pass alignments. Following solicitation in 1956 for tunnel contractors, a low bid was accepted and awarded to the Colorado Tunnel Constructors of Denver. Following project award, nothing happened, and the bid expired subsequently placing the project on hold (not much is known about the project).

## STRAIGHT CREEK PILOT TUNNEL

In 1960, a location study was performed focusing on Berthoud Pass, Vasquez Pass, Devil's Thumb, Jones Pass, Loveland Pass, and Straight Creek. The Straight Creek option was judged the preferred option for continuation of I-70 west into Utah. This was chosen as it provided an 11-mile savings over the Loveland Pass option and it was considered the least difficult and lowest cost option.

Following tunnel alignment selection, construction began on approach roadways from the east and the west to Eisenhower. During construction of the open-cut section of the eastern tunnel approach, a large landslide developed north and west of the portal area.

Construction of the Pilot Tunnel started in fall of 1963 and the 10-foot by 10-foot 8,400 ft long pioneer bore was completed in 1964. The purpose of the Pilot Tunnel was to provide a geologic blueprint of what was to be encountered during subsequent excavation of the Eisenhower Tunnel. The distances between the Pilot and the future Eisenhower tunnels is between 120 feet at the portals to 250 feet near the center. During construction of the Pilot Tunnel, mapping was performed by Colorado Division of Highways and the U.S. Geologic Survey. Features noted included the following:

- Rock type
- Foliation, faults and shear zones, fracture orientation
- Groundwater infiltration
- Resistivity and seismic velocity measurements
- Projections to the North Tunnel (Eisenhower).
- Load cell measurements in steel ribs

During construction of the Pilot Tunnel, additional investigations continued into groundwater conditions and the landslide near the eastern portal. The landslide investigation was carried out during 1963 through 1965. This investigation revealed the estimated landslide extent through mapping and installation of eight inclinometers at 60 to 240 feet deep. Following data analysis, the selected solution was to buttress the slide with 365,000 CY of compacted fill on top of the cut-and-cover section of tunnel.

The groundwater investigation was started during construction in 1962. This early investigation was in response to challenging groundwater conditions in the Moffat Tunnel (1929) and the Roberts Tunnel (1960) which yielded 1,800 gpm and 200 gpm, respectively.

Two hydraulically different zones of groundwater infiltration were observed in the Pilot Tunnel—an active zone below the topographic surface (defined below) and a passive zone within the rockmass below the active zone. The differentiation was based on changes in groundwater discharge with the season and tunnel station and on groundwater chemistry. The active zone was defined as the thickness of porous and permeable rock material capable of transmitting seasonal recharge. The passive zone was defined as the rockmass below the active zone that is relatively impermeable but can store water within fractures.

The investigation results suggested that the Eisenhower Tunnel should be portaled in late fall or early winter, to allow flush flows from the passive zone to drain, and joints spacings between 0.1- to 1-foot should be grouted in both the active and passive zones.

An observation from the Pilot Tunnel was that 26% of the Pilot Bore was in self-supporting rock and 74% required support.

## CONSTRUCTION OF THE EISENHOWER TUNNEL

The original tunnel design was performed by Tibbets, Abbets, McCarthy and Stratton of New York. During tunnel design one factor received a lot of scrutiny. Because of the location at elevation 11,000 feet, ventilation was considered a huge consideration. At the time in the late 1960's, the effects of elevation on behavior of carbon monoxide from mechanized underground excavation equipment was not what it is today.

### Geology

Both the Eisenhower and Johnson Memorial Tunnels pass through 75% granite of the Silver Plume Granite and 25% schists and gneisses of the Idaho Springs Formation. The schists and gneisses occur as migmatitic inclusions within the granite. At the Loveland Fault, the rock has been crushed and altered to almost a soil-like condition. The terrain was mapped in 1962 by the U.S. Geological Survey.

During detailed design, the tunnel was divided into five zones based on characteristics developed during design. The following zones are broken down as follows, from west to east:

- Station 38+06 to 81+57 (West Portal to Loveland Fault—Zone 1)
- Station 81+57 to 100+30 (Zone 2-Loveland Fault)
- Station 100+30 to 109+07 (Zone 3)
- Station 109+07 to 118+05 (Zone 4)
- Station 118+05 to 122+30 (Ends at Eastern Portal--Zone 5)

### Procurement

Pre-qualification requirements were let prior to bid. The major requirements included:

- At least 5 years of experience in major construction work similar to Straight Creek Tunnel
- Satisfactory completion of $10 million project
- Completed 10,000 ft of tunnel excavation
- Establish a net quick asset condition of at least $1,500,000. All bids must be accompanied by a certified or cashiers check for $1,000,000.

Colorado DOT received six bids in October 1967 from joint ventures comprised of 18 of the country's major contractors. The Eisenhower Tunnel contract was awarded at $50 million, and construction started in 1968. The winning bid was submitted by Straight Creek Constructors, a joint venture of Al Johnson Construction Company (Minneapolis, MN), Sponsor--Gibbons and Reed (Salt Lake City, UT), Western Paving (Denver, CO), and Kemper Construction (Los Angeles, CA). The bidding phase lasted 6 weeks and Straight Creek Constructors spent $50,000 on consultants and pre-bid work (over $420,000 in 2022). Straight Creek Constructors understood that the complexity of geology was a key component and hired key consultants such as a Construction Geologist and a Rock Mechanic Consultant. The maximum cover above the tunnel was approximately 1,450 ft and the tunnel was going to be approximately 48 feet wide and 50 feet tall.

In 1967, this represented the largest contract awarded for the Interstate Freeway System as Federal funding accounted for 92% of the project cost.

### Rock Tunnel Initial Support

The western heading began in 1968 using a top heading and bench excavation sequence and advanced approximately 4,300 feet without issue. The top heading had steel ribs in the arch, set on I50 wall plates. Concrete was placed between steel ribs 4 feet above the wall plate. Rockbolts (10-foot) were installed between the ribs and through the concrete. It was estimated that 30 feet of bench could be safely removed prior to placing posts under the wall plates. Using this sequence, the bench was removed without issue for 1,600 feet from the portal. Because of good ground conditions, the bench length was increased from 30 to 60 feet. This resulted in a collapse of approximately 60 feet of arch. The collapse was judged to be due to joints with talc, clay, and calcite dipping into the tunnel causing lateral loads on the ribs and wall plates. Original rock reinforcement was supplemented with 20-foot grouted rock bolts and in some cases the ribs were pinned with 10-, 15-, and 20-foot long rockbolts. This continued up until the heading reached the Loveland Fault.

The eastern heading started in late 1968 and proceeded through the landslide area. Spiling was required once blocky ground was encountered past the landslide area. As the western edge of Zone 3 was advanced, it was noted that large loads were unsymmetrically loading the ribs and deforming and twisting the ribs in Zones 5, 4, and 3. Wall plates were deforming in Zone 3 and allowing the arch ribs to settle.

In September 1969, a shield was to be used to advance the heading from the west into the Loveland Fault. The shield concept was also to be used in the eastern heading. As excavation progressed without issue on the western heading (up to 1,600 feet in from the portal), the contractor abandoned the shield concept for the eastern heading as they felt that experience gained from the western heading would allow for a top heading and bench excavation. As it turned out, great difficulties were encountered with use of the shield approaching the Loveland Fault, and it was abandoned.

On December 6th, 1969, the east heading stopped due to unexpected convergence and deformation of the steel support sets. All underground construction stopped for just over a year, resuming on January 7th, 1971. The difficulties from both the east and west headings caused a re-evaluation of tunnel initial support and procedures to complete tunnel excavation. Figure 1 shows a partial schedule of excavation sequences up to the re-evaluation of tunnel initial support and excavation sequences.

## Re-Evaluation

The immediate need was to stabilize the heading as most severe conditions existed in Zone 3 from the eastern heading. The bearing failure and collapse for the wall plates and distress in the arch ribs was first observed. The first remedial action was to install 5-foot wide by 6-foot high concrete buttresses. In addition to buttresses, jacket sets and invert struts were installed but did not arrest ground movement.

During design re-evaluation, it was determined that the amount of overbreak and cribbing between the steel sets and the rockmass created a "soft spot" and rock in close contact with the steel sets created a "hard spot." The substantial cribbing used acted as a cushion and allowed for deflection of the steel sets. This condition was corrected through encasement of the ribs with either concrete or shotcrete and filling all remaining voids in the cribbing and blocking with grout. No 8 bars, 20 feet long were installed on a 5- by 5-foot pattern over the entire arch. The invert was also lowered throughout Zone 3.

Removal of the eastern heading bench had 2 issues—first undermining of the arch support that had already experienced movement and distress. The original planned support did not account for horizontal ground movements (the design intended to use straight side posts). This was complicated as curved side-posts and struts could not

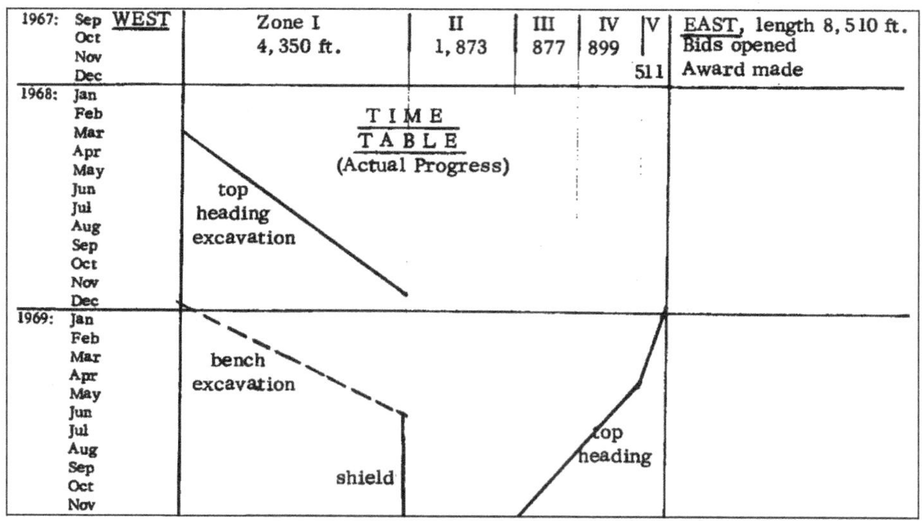


**Figure 1. Reproduced from 1974 Rapid Excavation and Tunneling. Project progress versus calendar time. West portal is on the left and the east portal is on the right, with the Loveland Fault in the middle. The calendar is stopped towards the end of 1969**

easily be adapted. Rock dowels 20 feet long were installed through the wall plates and concrete was placed between the invert struts. Concrete footings from previous drifts added safety by reducing the time supporting the arch and bench face. It was determined that concrete footage drifts would be driven for Zone 4 in addition to the concrete buttress. Additional rock support stabilizing the ground was used.

Excavation in Zone 2 would encounter conditions worse than Zone 3 due to squeezing ground in the Loveland Fault. A full shield driven on concrete footings from the footing drifts (also called foundation drifts—8- by 8-foot excavations) with arches and invert struts on 3-foot centers would support the excavation. Best ground conditions would use 12WF106 arch ribs and the worse conditions would receive 14WF287 invert struts. When excavated, the footing drifts confirmed the ground conditions indicated by the pilot tunnel.

Given ground conditions in Zone 3, and observed extensometer readings, ground loads in Zone 2 could be considerably higher than both primary tunnel liner or the shield had been designed for. The project team also looked at the Carley V. Porter Tunnel in California for potential risks. Due to the low probability of successfully driving a shield through Zone 2, a multiple drift attack was chosen for tunnel excavation through Zone 2. Dr. R.L. Taylor and Dr. T.L. Brekke in Berkely, California utilized quadrilateral "linear strain elements" to determine high strain gradients.

For most of Zone 2, five drifts were used and were composed of a crown drift, two wall plate drifts, and two footing drifts. In portions where bearing capacity of concrete wall plate was estimated to be too low, an additional sidewall drift was driven on top of the existing footing drift and backfilled with concrete. Within the Loveland Fault, 13 stacked drifts were used continuing across the crown. Each drift was backfilled with concrete before the next drift was excavated. In essence, a concrete arch was constructed before the core of the tunnel was excavated. The crown drift experienced the greatest distortion in initial support members than any of the other drifts excavated in the Loveland Fault.

The zones were heavily instrumented using multiple position borehole extensometers, miniature weldable strain gauges, rock pressure cells, load cells, rockbolt deformeters, and tape extensometers. Locations of instrumentation were based on the geology and construction history for the various zones within the tunnel. As one would expect, the poorest ground conditions were the most heavily instrumented. Figure 2 shows a model that depicts the tunnel and generalized support styles through the excavation.

Excavation progressed from west to east utilizing "Big Mac," the breast board jumbo. No real difficulty was noted following the re-design and utilization of stacked drifts. Hole through occurred in July 1972. The tunnel was opened for traffic 8 months later in March 1973.

## Permanent Tunnel Lining

Concrete lining started in early 1970. Permanent tunnel lining is behind superficial architectural elements within the Eisenhower Tunnel. The permanent reinforcing steel varied by tunnel zone and is presented on Table 1.

The permanent tunnel lining thickness varied by tunnel section type and are presented on Table 2.

Figure 3 shows the plenum arrangement during construction.

**Figure 2. CDOT had a model built following construction of Eisenhower. Model shows different support styles from the east end of the excavation. The 13 stacked drifts are shown in Zone 2**

**Table 1. From the as-builts on permanent lining reinforcing steel**

| Zone | Section Type | Segment Nos. | Long Bars | Radial Bars |
|---|---|---|---|---|
| 1A | | | #5 @ 12" | #8 @ 10" |
| 2 | | | #6 @ 12" | #6 @ 12" |
| 3 | | | #6 @ 12" | #10 @ 12" |
| 4 | | 141–145, 151–156 | #5 @ 12" | #8 @ 10" |
| 4 | | 146–150, 157–158 | #5 @ 12" | #9 @ 10" |
| 5 | Floor to 60°<br>60° to crown | | Not shown | #6 @ 12"<br>#6 @ 18" |
| | III a | | #5 @ 12" | #8 @ 12" |
| | III b | | #5 @ 12" | #9 @ 12" |

**Table 2. From the as-builts on permanent tunnel concrete**

| Zone | Section Type | Compressive Strength (psi) | Full Concrete Lining |
|---|---|---|---|
| | III a | 5,000 | 2'-8" thick |
| | III b | 5,000 | 4'-2" |
| | IV a and IV b | 5,000 | |
| 5 | | | 5'-6" |

**Figure 3. Historical photo from the Denver Post. Installation of the plenum dividing wall and floor slabs during construction progressing from west to east. View is looking west. The left plenum is the exhaust air and the right side of the plenum is fresh supply air**

## Schedule

Once the decision was made to utilize multiple drifts, substantial overruns in cost and schedule were apparent. In 1970 while the structural redesign was occurring, the Contractor evaluated three different schedules, all of which assumed various lengths of challenging ground conditions. Even under the most favorable scenario, opening before spring 1974 seemed extremely unlikely. At the time, Colorado Governor John Love expressed hope that "the controversial tunnel" could be opened in 1972. The project team accepted this as a mandate to think outside the box and find other options. Due to the elevation and health and safety limits, additional manpower or equipment was not possible.

In September of 1971, a revised schedule was issued and approved. Changes to the original design were submitted and accepted. Changes include:

- DIP and PVC pipes instead of vitrified clay and cast iron—Due to construction scheduling and sequencing, the contractor needed to find materials that could withstand heavy loads under less than 2 feet of backfill and without protection by the pavement. DIP was selected for water and sewer lines rather than cast iron and vitrified clay. Thick-walled PVC pipe was selected for drainage mains and all collectors and crossovers instead of vitrified clay.
- Outside pre-assembly of rebar mats using tension-splice with cable clamps instead of continuous rebar. Due to construction scheduling, sequencing, and designed construction joint locations, a traditional lap-splice at the centerline of the tunnel was problematic. A splice with only 16 inches of lap with two cable clamps torqued to 250 foot-pounds was chosen, tested by United States Bureau of Reclamation, and accepted for use.
- Modified arch concrete placement lining 4,322 feet in 10 weeks
  - Lower ribs were reinforced to withstand accelerated pour rates
  - Belt conveyors used to place concrete up to 10-to-2 o'clock, then pumped to fill crown
  - Homogenized concrete mix to prevent bleed in 30 minutes in the event of pump stoppage.
  - Embedded items were placed in block-outs keeping continuity of placements
- Pre-cast and epoxy coated safety barriers rather than cast-in-place and tile faced. The barriers were pre-cast in Cheyenne, Wyoming and hauled to site to facilitate tunnel construction concurrent with tunnel excavation.
- Replaced in-place glazed tile and cast-in-place concrete ceiling with prestressed hollow ceiling slabs with porcelain enamel coated steel sheets assembled off-site. Panels were cast in Salt Lake City months ahead of time incorporating porcelain enameled sheets fabricated in Los Angeles.
- Glazed tile wall panels assembled offsite. Design option utilized by subcontractor to replace hand-laid tile with pre-assembled panels made in Denver then trucked to the site. This activity was completed the night before the tunnel opened.
- Full depth asphalt paving concurrent with other work instead of reinforced concrete and asphalt overlay. This was originally scheduled to occur 1 month before opening the tunnel and would have caused an interruption to numerous finishing activities. Asphalt paving took less than 4 weeks to complete and allowed continuation of all other activities.

A key component to the schedule savings involved the overlapping of activities. This included the overlap of finishing work and tunnel excavation from February 1971 through 1972 and the overlap of finishing work and concrete lining for arch and invert from August 1972 through November 1972.

Figure 4 presents the complete construction schedule as presented in the 1974 RETC paper.

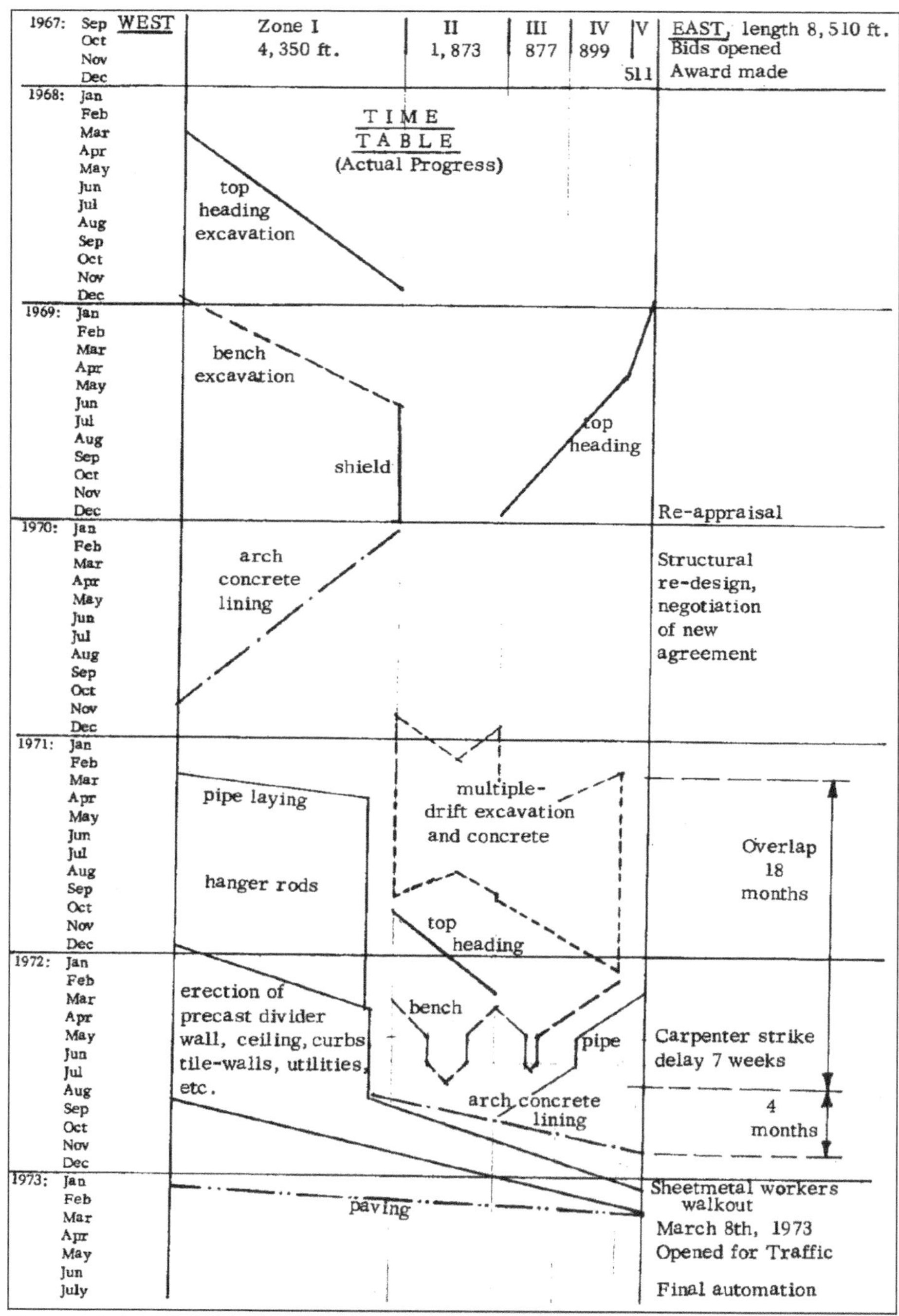


Figure 4. Reproduced from 1974 Rapid Excavation and Tunneling. Project progress versus calendar time. West portal is on the left and the east portal is on the right, with the Loveland Fault in the middle

## CONCLUSION

Completion was delayed by 2 years and construction costs escalated from $52 M to $108 M ($327 M and $678 M in 2022, respectively). The original completion date was 1971 (3 years) and it took 5. The Eisenhower Tunnel represents one of the most challenging tunnel projects constructed in the United States. Its scope and breadth are remarkable by today's standards, but are made even more so considering the time period in which it was constructed—the late 1960s and early 1970s. Solutions to challenges met and overcome during the construction have filtered into all aspects of our industry, including the creation and use of Dispute Review Boards and Escrow documents (implemented for construction of the Johnson Memorial Tunnel). While length restrictions of the manuscript prohibit the inclusion of all the details associated with the construction of Eisenhower, the authors invite all readers interested in learning more to read papers presented in references.

## ACKNOWLEDGMENT

The authors thank Mike Salamon of Stantec Consulting Services for his contribution to this paper.

## REFERENCES

Hopper, R., Lang, T., and Mathews, A, Construction of the Straight Creek Tunnel, Colorado, Rapid Excavation and Tunneling, 1972.

Hurr, R., and Richards, D., Ground-Water Engineering of the Straight Creek Tunnel (Pilot Bore), Colorado, Bulletin of the Association of Engineering Geologists, Vol. 3, Nos 1 and 2, 1966.

Jordan, Kathy, Notion of a Tunnel under Divide Predates Reality by 100-Plus Years, The Daily Sentinel, December 2011.

Kellog, J., The Straight Creek Tunnel-An Engineering and Constructor's Challenge, American Society of Civil Engineers, National Structural Engineering Meeting, 1969.

Lee, Fitzhugh, et al., Geological, Geophysical, and Engineering Investigations of the Loveland Basin Landslide, Clear Creek County, Colorado, 1963–1965, Geological Survey Professional Paper 673-A, B, C, D, E F, G, United States Government Printing Office, Washington, 1972.

Lee, Fitzhugh, et al., Engineering Geologic, Geophysical, Hydrologic and Rock Mechanics Investigations of the Straight Creek Tunnel Site and Pilot Bore, Colorado, Geologic Survey Professional Paper 815, United States Government Printing Office, Washington, 1974.

Lee, Fitzhugh, Mystkowski, Walter., Chapter 13-Loveland Slide Basin, Colorado, USA, Developments in Geotechnical Engineering, Volume 14, Part B, 1979, pages 473–514.

McOllough, Phillip, Contractual Requirements and Design Philosophies Employed to Minimize Adversary Relationships : Eisenhower Memorial Tunnel, Second Bore, Transportation Research Record, Issue 792, 1981.

Merten, F., Straight Creek Tunnel Construction, Route I-70, Colorado, Highway Geology Symposium, 1975.

Robinson, C., and Lee, F., Geologic Research at the Straight Creek Tunnel Site, Colorado, U.S. Geologic Survey.

Robinson, C., and Lee, F., The Validity of Geologic Projection; a Successful Example : The Straight Creek Tunnel Pilot Bore, Colorado, U.S. Geologic Survey, 1966.

Ruemmele, Walter, Lining and Finishing Eisenhower Memorial Tunnel, Rapid Excavation and Tunneling, 1974.

Trapani, R., Kaneshiro, J., Jurich, D., Salamon, M., and Quick., S., Structural Inspections of Colorado's Eisenhower Johnson Memorial Tunnel, Hanging Lake Tunnel, and Reverse Curve Tunnel, North American Tunneling, 2010.

Water Inflows and Drainage at the Eisenhower Memorial Tunnel, Technical Memorandum 237.120-6, 1974.

# Gas Extraction and In-Situ Oxidation for TBM Tunneling of the Purple Line Extension, Section 1, Los Angeles

**Richard McLane** ▪ Traylor Bros., Inc.
**Matt Neuner** ▪ Golder Associates, now with Ecometrix Incorporated
**Hugh Davies** ▪ Golder Associates, now with Newmont Corporation
**James Corcoran** ▪ Traylor Bros., Inc.
**Joseph DeMello** ▪ Los Angeles County Metropolitan Transportation Authority

## ABSTRACT

The Los Angeles County Metropolitan Transportation Authority (LA Metro) Purple Line Extension Section 1 (PLE1) is a $3.12 billion USD design-build, underground heavy rail project, connecting an existing station at Wilshire Blvd. and Western Ave. and extending 3.92 miles under Wilshire Blvd, terminating approximately 550 feet west of Wilshire and La Cienega Blvds. The project alignment runs immediately adjacent to the La Brea Tar Pits, world renowned for Pleistocene fossils and continuous seepage of asphaltum (locally known as tar) and gas in an urban setting.

This paper presents a successful case history of the planning, design, and construction efforts to safely construct the bored tunnels through Reach 3 of the alignment (between Wilshire/Fairfax Station, and Wilshire/La Cienega Station), in a designated Methane Zone in the City of Los Angeles. Historical context on decades of challenges and studies is summarized. Further, this paper outlines the pro-active partnership between the Owner: LA Metro, the Design Builder: Skanska-Traylor-Shea Joint Venture, and geotechnical design services provided by Golder Associates to safely manage pressurized face tunneling through a confined gas zone near the La Brea Tar Pits. Specifically, a program of in-situ oxidation and soil gas extraction was designed to test and analyze risks associated with pressurized face TBM tunneling through the complex subsurface geological conditions, then implemented during construction. In advance of tunneling, a network of wells was used to draw air into the gas zone to oxidize the 6,500 ppm hydrogen sulfide gas, as well as determine methane recharge rate. During TBM tunneling, the wells were converted to a gas extraction system to evacuate, filter ($H_2S$) and burn ($CH_4$) the gas as the Tunnel Boring Machine passed, providing a preferential pathway for the gas to migrate toward extraction wells, and preventing the migration to adjacent buildings. The team extracted and treated 3,857,000 cubic feet of mostly methane gas over 40 days as the TBMs passed through the zone. Gas behavior, pressures, and concentrations were monitored not only from the extraction system, but also in basements and utility vaults along this section of the alignment.

## INTRODUCTION

The highly anticipated Purple Line Extension (formerly Westside Extension, renamed D Line Extension in 2020) is an underground heavy rail segment of the Los Angeles County Metropolitan Transportation Authority (LA Metro) rail transit system being constructed to extend a high-speed link from downtown to the busy Westside of Los Angeles. The first of three sections, Section 1, is a $3.12 billion (US) design-build project, connecting an existing station at Wilshire Blvd. and Western Ave. in the city of Los Angeles, extending 3.92 miles to the west under Wilshire Blvd, and

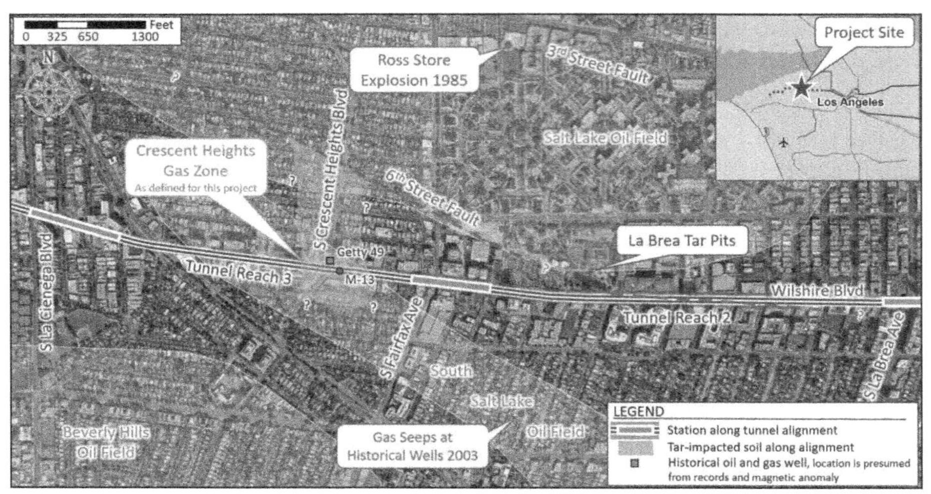


Note: Oil and gas wells not shown, except for Getty-49 (API 03715144); their locations are available at CalGEMs Well Finder

**Figure 1. Location of the tunnel alignment and the Crescent Heights Gas Zone**

terminating approximately 550 feet west of Wilshire and La Cienega Blvds in the city of Beverly Hills. It was considered the backbone route for Metro's rapid transit network as early as 1961 (Kaiser 1961), but construction was delayed for decades due to challenges with difficult ground conditions and other factors (City of Los Angeles 1985; Cobarrubias 1992).

Los Angeles was built over oilfields that comprise what was one of the world's most productive basins in the early twentieth century, the Los Angeles Basin. Long before development of the city, heavy oil seeping from the ground at what is now called the La Brea Tar Pits was used and traded by the Chumash indigenous people (Bilodeau et al. 2007). Drilling in the area began around the turn of the twentieth century, and by the 1930s hundreds of wells had been completed in the Salt Lake Oil Field and the South Salt Lake Oil Field (Figure 1; Crowder 1961, Crowder and Johnson 1963). During this period, urban areas expanded west from downtown and developers such as Gaylord Wilshire, J. Harvey McCarthy and Moses Sherman had started developing the Wilshire corridor; McCarthy had envisioned a subway under Wilshire Blvd. as early as the 1920s. After initial planning in the 1960s of a subway line along the Wilshire corridor, an extensive geotechnical program was completed in the late 1970s and early 1980s (CWDD 1981; Kaiser and Gage-Babcock 1983; Proctor 1985).

On March 24, 1985, a methane ($CH_4$) seep led to an explosion and fire at the Ross Dress-for-Less department store several blocks north of the La Brea Tar Pits (Figure 1). While there was some controversy as to the source of the gas initially, the City of Los Angeles designated a methane high potential risk zone (Figure 2; City of Los Angeles 1985; Cobarrubias 1992) that corresponded generally with mapped extents of the Salt Lake Oil Field, the 6th Street Fault, and the 3rd Street Fault (Figure 1). In 1985, a moratorium was placed on federal funding for tunneling through the methane potential risk zones in Los Angeles by congressional order (Section 321 of US Congress Public Law 99-190, 1985) and by resolution of Metro's predecessor transit agency. Endres et al. (1991), Hamilton and Meehan (1992), and Chilingar and Endres (2005) concluded that the source of the gas that caused the explosion was the re-injection of gas and saltwater brine back into the underlying oil field along with the faulting in the area.

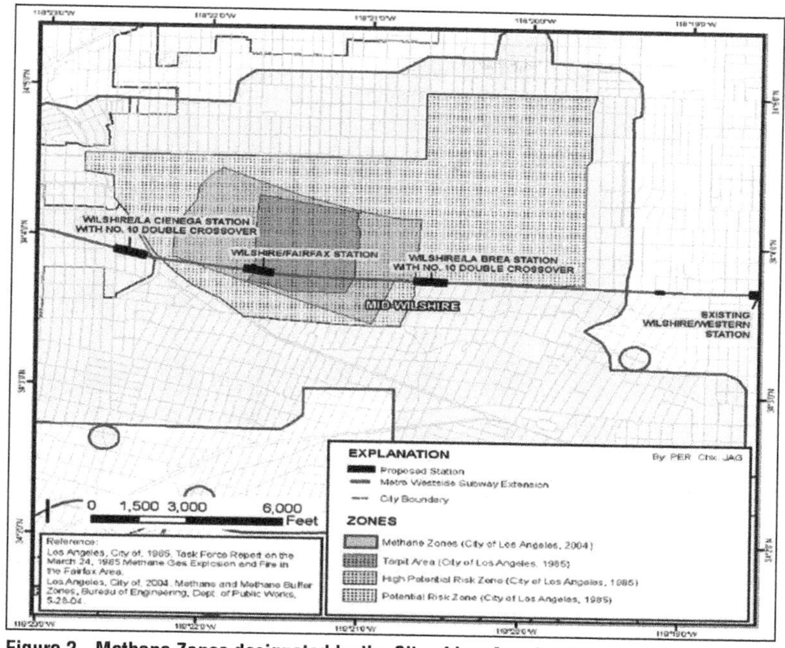


**Figure 2. Methane Zones designated by the City of Los Angeles (source: LA Metro 2014)**

After the explosion, the Red Line (renamed B Line in 2020) was re-routed to the east, and north, of the methane potential risk zone in the Fairfax District (Figure 2; SCRTD 1987). LA Metro began operating the Red Line, the city's first rapid transit line, in 1993.

In the 1990's, geotechnical investigations were completed for a re-route of the Western extension subway line south of the methane potential risk zone associated with the Salt Lake Oil Field (Elioff et al. 1995; Enviro-Rail 1996, Metro 1996). Following identification of gas in an unsaturated zone several blocks south of Wilshire Blvd. (Pico and San Vicente Blvds) with concentrations of hydrogen sulfide ($H_2S$) gas as high as 20,000 parts per million (ppm), 200 times greater than the amount immediately hazardous to life or health (IDLH), construction was again delayed, due to funding constraints.

In 2005, a panel of experts engaged by the American Public Transportation Association (APTA 2005) concluded, after cursory review, that tunneling along the Wilshire Blvd. corridor could be done safely due to advances in technologies for gas detection and pressure face tunnel boring and experience tunneling in Los Angeles. Successful completion of the Gold Line (renamed L Line) Extension in East Los Angeles, constructed by Traylor Frontier Kemper JV, was cited as a basis of current tunneling technology. The congressional moratorium was repealed in 2007 (H.R.2764 Section 169 of US Congress Public Law 110-161, US Congress 2007). LA Metro carried out additional geotechnical studies from 2009 to 2014, and a second APTA peer review report (APTA 2012) concluded that tunneling using pressure face technology (EPB or Slurry TBM) past the La Brea Tar Pits could be done safely. During these investigations, pockets of pressurized gas were identified (LA Metro 2011) and gas in an unsaturated, sandy confined layer with approximately 95% by volume (% vol.) $CH_4$ and 6,500 ppm $H_2S$ was identified at a single well (M-13; Figure 1) along the alignment west of the La

Brea Tar Pits. When the Geotechnical Baseline Report (GBR; Metro 2014) was issued, the extent of the gas in this sandy confined layer, which would become known as the Crescent Heights Gas Zone (Figure 1) was not well defined. In 2014, Metro awarded the design-build contract to the Skanska-Traylor-Shea Joint Venture (STS). STS hired the geotechnical consultants Golder Associates to further investigate risks associated with $CH_4$ and $H_2S$ gases at this location as well as in the Fairfax Station area to the east, adjacent to the tar pits.

While tunneling through gassy ground is not new, particularly in California where detailed tunneling safety standards have been in place for decades, relatively few case studies of tunneling through gassy ground have been published. Proctor (2002) described the fatal San Fernando Tunnel explosion in June 1971, which occurred during tunneling in gassy ground associated with oil fields in northern Los Angeles (Sylmar Tunnel Explosion). This incident led to strengthening of the tunnel safety standards in California, which now form the state's Tunnel Safety Orders (Cal/OSHA Title 8 subchapter 20, Tunnel Safety Orders). This incident and 10 others involving various tunnel types in 8 countries are summarized by Copur et al. (2011). Most of these incidents involved tunneling through or near formations with oil and natural gas and inadequate ventilation, gas detection, automation, and/or spark prevention.

For the Westside Purple Line Extension Section 1 (WPLE1), LA Metro and its consultants formed a proactive partnership with STS and Golder that facilitated building on the knowledge gained from decades of investigations while working collaboratively to develop a safe approach to tunneling past the La Brea Tar Pits and through the Crescent Heights Gas Zone. The main objectives of these efforts were to assess the risks and develop mitigations around two potential scenarios:

- Ingress of explosive and acutely toxic gases ($CH_4$ and $H_2S$) into areas of the tunnel boring machines (TBMs) where workers were present and where ventilation exited the TBMs and would be emitted to the ambient air, and
- Migration of explosive and acutely toxic gases ($CH_4$ and $H_2S$) due to normal operation of EPB TBMs, from the pressure face to adjacent building basements or utility vaults—effectively inducing another Ross-Dress-for-Less explosion by the TBM.

This paper summarizes activities completed by the WPLE1 team over five years leading up to and including successfully and safely tunneling through exceptionally gassy ground. The following sections describe results of investigations, design and testing of an in-situ oxidation and gas extraction system, operation of the TBMs, and measured conditions during tunneling.

## INVESTIGATIONS

The WPLE1 team carried out investigations from 2015 to 2019 to characterize the Crescent Heights Gas Zone, study potential gas sources and connections, develop the special mitigations of in-situ oxidation and gas extraction, and test these mitigations in advance of tunneling. Prior to these investigations, several key uncertainties remained:

- The extents of the gas zone around the M-13 well were not well-defined, such that possibilities ranged from it potentially being a small gas pocket around the well (Getty 49) to it possibly being a large gas zone connected to the underlying oil fields.
- It wasn't clear whether there might be other gas zones like the one at the M-13 well that were possibly between the previous drill holes.

- If the gas zone at the M-13 well was extensive, it would not be feasible to treat $H_2S$ at 6,500 ppm entering the ventilation system at the planned rates of excavation.
- The implications of the potential for flowing gas from an extensive gas zone on both gas release inside the TBM and gas migration from the TBM were not well understood and presented a greater potential challenge than smaller gas bubbles that would release proportionally to the TBM advance rate.
- While it was expected that the pressure face at the EPB TBM cutterhead would limit gas ingress to the working area inside each TBM (due to over-pressure), it was unclear whether the face pressure or injected foam required to maintain the pressure and condition the muck could cause gas to migrate to a building basement or utility vault.

The WPLE1 team proactively investigated each of these uncertainties prior to tunneling.

## Potential Gas Source and Connections

The extents of the gas zone around the M-13 observation well were likely related to the source of the $CH_4$ and $H_2S$ gases. The WPLE1 team investigated the potential for connection of the gas zone to the underlying oil fields, with potential routes of connection including historical oil wells, exsolution from the tar, and gas flow up an adjacent fault.

### *Oil Wells*

Historical records (CalGEM Well Finder, formerly DOGGR, API 03715144) indicate that an oil well was drilled in 1907 near the intersection of Wilshire Blvd. and S. Crescent Heights Blvd. by a predecessor of the Getty Oil Company (Arcturus Oil Company, Figure 1). Referred to as the Getty-49 (Chevron 03715144) well, it was drilled into the South Salt Lake Oil Field and was immediately abandoned ("Stopped Work") that same year with final plugging in 1913; the well record indicates abandonment involved explosives at 10 depths, pulling the three casings, and "filling [the] hole." The Getty-49 well was drilled into the upper limb of the main oil-bearing unit of the South Salt Lake Oil Field, such that it could potentially form a direct connection for gas migration from the oil field to the Crescent Heights Gas Zone (Figure 3). The WPLE1 team attempted to locate the Getty-49 well using various methods from the ground surface and with magnetometer surveys from both vertical and horizontal borings drilled at the depth of the tunnel alignments. A magnetic anomaly was found near the intersection, which may have been the Getty-49 well. The alignments of the tunnels were modified so that the TBMs would tunnel to the south of the anomaly (Figure 1) and avoid it by approximately 10 feet.

Several oil wells were drilled directionally from 1970 to 1974 from the Packard Drill Site into the upper limb of the South Salt Lake Oil Field (Figure 3; Samuelian 1990; CalGEM Well Finder). These wells were produced mainly in the 1970s to 1990s, and some of them remain active. From 1999 to early 2003, 1.2 billion cubic feet of gas was *injected* at an average pressure of 575 pounds per square inch (40 bar) into one of these wells (P-70) to enhance oil production. It is possible that activities at these wells might have contributed to the formation of the Crescent Heights Gas Zone, for example potentially by pressurizing the production zone in the oil field combined with migration up the Getty-49 well or a fault. Gas injection to P-70 and P-65 was suspended in early 2003 because gas seepage was discovered at surface at locations of old wells 0.5 mile southeast of the intersection of Wilshire Blvd. and South Crescent Heights Blvd. (Figure 1; Chilingar and Endres 2005) and tracer gas (perfluorodimethylcyclobutane and perfluoromethylcyclcohexane) that was injected was

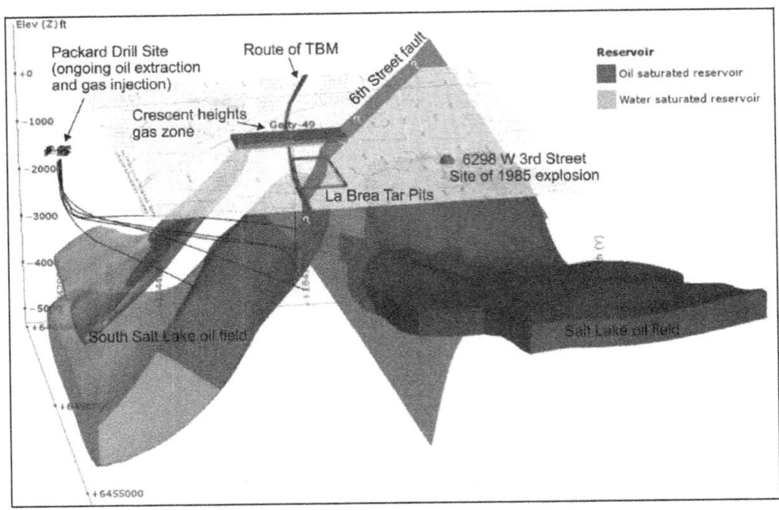


Figure 3. Geologic setting of the Crescent Heights Gas Zone

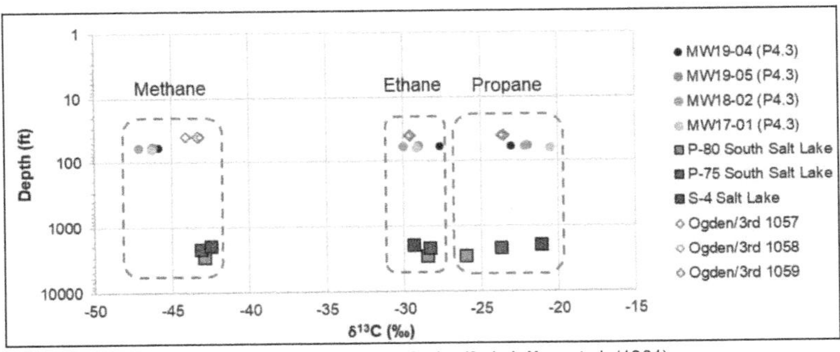


Note: Source for samples from the oil fields and Ogden/3rd: Jeffrey et al. (1991)

Figure 4. Select isotopic results from gases sampled from the Crescent Heights Gas Zone, the gas zone at the Ross Store, and underlying oil fields

not detected—essentially the injected gas migrated out of the oilfield (CalGEMs Well Finder, API 03721161 pg. 81).

The WPLE1 team analyzed gas samples collected from shallow gas wells in the Crescent Heights Gas Zone (i.e., tunnel alignment depth) for isotopes of carbon ($\delta^{13}C$ in $CH_4$, ethane [$C_2H_6$], propane [$C_3H_8$], and $CO_2$) to aid with interpretation of gas sources. These analyses clearly identified the $CH_4$ gas in the Crescent Heights Gas Zone as thermogenic gas, that is, the type of natural gas present in an oil field. The isotopic signature of the light hydrocarbon gases in the Crescent Heights Gas Zone was similar to gases reported by Jeffrey et al. (1991) that were sampled from the underlying oil fields (from oil wells shown in Figure 3) and from the gas zone that caused the 1985 Ross-Dress-for-Less explosion (Figure 4). Further, the ratios of straight-chain n-butane (n-$C_4$) to branched iso-butane (iso-$C_4$) in gas samples from the Crescent Heights Gas Zone were indicative of non-biodegraded thermogenic gases, similar to the n-$C_4$ to iso-$C_4$ ratios in samples from the South Salt Lake Oil Field (Jeffrey et al. 1991) and from the most active gas vent at the La Brea Tar Pits (Etiope et al. 2017).

Note the WPLE team did test for the tracer gases from P-65 and P-70 and did not detect either.

The WPLE1 team investigated the source of $H_2S$ in the Crescent Heights Gas Zone using isotopes of sulfur ($\delta^{34}S$) in $H_2S$ gas, in sulfate in groundwater, in tar, and in bulk soil samples. The team was able to rule out bacterial reduction of sulfate from anhydrite and thermochemical reduction of sulfur from anhydrite or organosulfur compounds in the tar. But the exact source of the $H_2S$ wasn't determined, in part because information on $H_2S$ in the oil fields was not available. The source was interpreted to likely be due to biodegradation of the tar with associated release and reduction of sulfur. But other potential sources that could not be ruled out were similar biodegradation of oil in the oil field and mantle gases (Jung et al. 2015).

### *6th Street Fault*

The oil, tar, and gas seeping at the ground surface at the La Brea Tar Pits reportedly migrate from the Salt Lake Oil Field along the 6th Street Fault (Hamilton and Meehan 1992; Khilyuk et al. 2000; Chilingar and Endres 2005; Bilodeau et al. 2007; Etiope et al. 2017). The 6th Street Fault (Figure 3) is a seismically inactive fault that dips steeply to the northeast and forms a structural trap for oil and gas in the Salt Lake Oil Field (Lang and Dreessen 1975; DOGGR 1992). Studies of regional tectonics (Shaw and Quinn 1986; Wright 1987) indicate that the 6th Street Fault was likely seismically active until the late Pleistocene, after which time it became inactive and was concealed by deposition of younger alluvium. Continuous seepage of oil, tar, and gas at the site now known as the La Brea Tar Pits was reported as early as 1792 (Bilodeau et al. 2007) and in a 1906 USGS report (Arnold 1906), and the rate of seepage of $CH_4$ from the site makes it one of the most active hydrocarbon seeps in North America (Etiope et al. 2017).

The WPLE1 team drilled 43 wells into the Crescent Heights Gas Zone, and consistently measured positive gas pressures (i.e., greater than atmospheric) that decreased from north to south. These measurements, together with development of a calibrated multiphase flow model (discussed below), indicate that gas naturally flows from north to south through the gas zone. Since this natural flow of gas within the Crescent Heights Gas Zone appears to originate to the north of and beyond the likely area of the Getty-49 well, the WPLE1 team interpreted the likely pathway of gas migration to be from the Salt Lake Oil Field along the 6th Street Fault to the gas zone (Figures 1 and 3). It is noted that the location of the 6th Street fault as shown and projected to surface is the subject of interpretation and open to debate.

## Gas Extraction Trials for the Mid-City Project, Early 1990s

During investigations in the early 1990s for the re-routed tunnel alignment south of the methane potential high risk zone associated with the Salt Lake Oil Field, LA Metro encountered a gas zone beneath the intersection of Pico Blvd. and San Vicente Blvd. This gas zone had $CH_4$ concentrations greater than 99 % by volume and $H_2S$ concentrations as high as 20,000 ppm in unsaturated sandy sediments of the San Pedro Formation confined by clayey sediments of the relatively impermeable Lakewood Formation (Metro 1996). Metro tested a method of in-situ oxidation of $H_2S$ by extracting gas continuously while introducing atmospheric air into the gas zone (Elioff et al. 1995). Within two days, $H_2S$ concentrations within a radius of 100 ft were lowered to approximately 100 ppm. More than a month after the gas extraction and air introduction was terminated, $H_2S$ concentrations were 300 ppm at the extraction well and $<1$ ppm at the air introduction wells. However, $CH_4$ concentrations recovered to greater than 99 % vol. within days of turning off the system. Due to the similarity of this gas zone to the Crescent Heights Gas Zone, this experience was valuable in understanding an effective mitigation of $H_2S$ concentrations.

## Gas Extraction Trials, 2018 to 2019

The WPLE1 team carried out gas extraction trials in 2017, 2018 and 2019 to better characterize the Crescent Heights Gas Zone and to trial in-situ oxidation and gas extraction systems prior to tunneling. A summary of the gas extraction trials is provided in Table 1.

Drilling and testing during the trials provided detailed characterization of the geologic conditions of the Crescent Heights Gas Zone under Wilshire Blvd, with 31 wells drilled by the end of the third phase of testing. Predominantly fine-grained sediments of the Lakewood Formation, typically dominated by silt and clay, were encountered from surface to a depth of approximately 60 feet (Figure 5). The contact between the Lakewood Formation and the top of the San Pedro Formation typically consisted of clayey ground underlain by sand or silty sand. The upper 5 to 20 feet of the San Pedro Formation was coarse grained, unsaturated, forming the gas zone. Water-saturated sands formed the bottom of the central and western portions of the gas zone, and tar-saturated sands formed the bottom of the eastern portion of the gas zone (Figure 5).

These drilling programs revealed that the Crescent Heights Gas Zone extended approximately 600 feet from east to west and coincided with the depth of the tunnel alignments (Figure 5). The north to south extents of the gas zone were confirmed to be at least 420 feet but were thought to potentially extend approximately 1,500 feet or more (Figure 1).

Gas extraction Trial 1 (Phase 4.1) was carried out with two days of gas extraction using wells that were installed during baseline geotechnical programs and the initial gas investigation. Approximately 117 thousand cubic feet (Mcf) of gas were extracted from the M-13 observation well while 8 Mcf of air was passively introduced by opening two nearby gas probes once suction was confirmed. The $CH_4$ concentrations in the extracted gas remained high (average of 91 % vol.) during the testing, but the $H_2S$ concentrations decreased from 6,000 ppm to less than 1 ppm after extraction. *This trial demonstrated that the gas zone was more extensive and that $H_2S$ could effectively be treated with in-situ oxidation by passively introduced air.*

**Table 1. Gas extraction from the Crescent Heights Gas Zone during trials and tunneling**

| Project Phase | Duration Extracting, days | Gas Extracted, Mcf | Average Methane, % Vol. | Air Passively Introduced, Mcf | Summary |
|---|---|---|---|---|---|
| Trial 1 | 2 | 117 | 91% | 8 | Gas zone recognized as extensive and transmissive; $H_2S$ treated in-situ near wells |
| Trial 2 | 6 | 731 | 87% | 119 | $H_2S$ treated across the alignment, remaining low after a month, but $CH_4$ returned after extraction |
| Trial 3 | 28 | 3,435 | 36% | 1,489 | Suction maintained and $H_2S$ treated across gas zone, but $CH_4$ cannot be treated before tunneling |
| Tunneling | 40 | 3,857 | 62% | 155 | TBMs safely tunneled through the gas zone with only minor $CH_4$ ingressions to buildings and TBM |

Mcf = thousand cubic feet; % vol. = percent by volume; $H_2S$ = hydrogen sulfide; $CH_4$ = methane

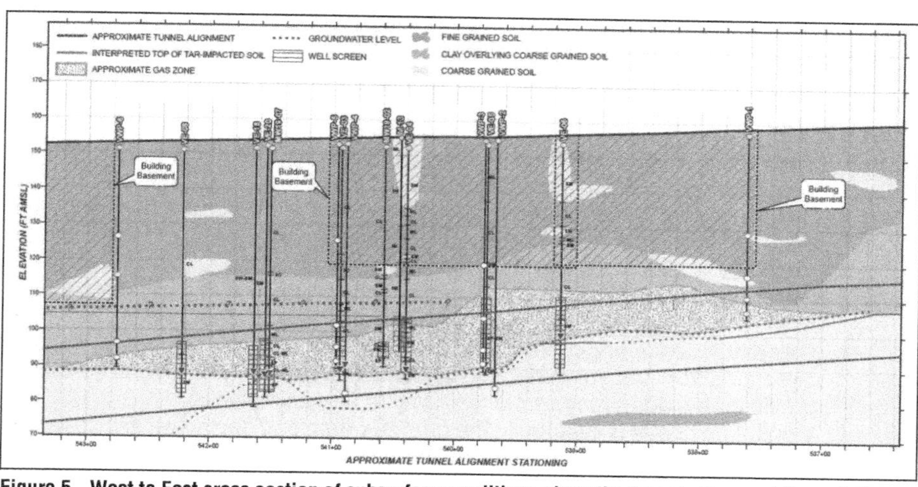


**Figure 5. West to East cross section of subsurface conditions along the tunnel alignment through the Crescent Heights Gas Zone**

Gas extraction Trial 2 (Phase 4.2) included installation of a network of six wells that were used for gas extraction and passive air introduction, and installation of three monitoring wells. Additional borings encountered water-saturated San Pedro sands, which identified the eastern and western boundaries of the gas zone (borings not shown on Figure 5). Gas was extracted for six days (731 Mcf) by extracting from wells on one side of Wilshire Blvd. while allowing air to be passively introduced through wells on the other side of Wilshire Blvd. The configuration of gas extraction and passive air introduction was changed twice during the trial to maximize the footprint of in-situ oxidation. Concentrations of $H_2S$ ranged from 88 to 8,000 ppm before the trial and were lowered to a maximum of 50 ppm in wells on Wilshire Blvd. 30 days after extraction was stopped, indicating successful in-situ oxidation over a period relevant for tunneling. Concentrations of $CH_4$, however, returned to levels near 88 to 95 % vol. within 10 days after the end of extraction.

Gas extraction Trial 3 (Phase 4.3) was carried out over 28 days of extraction, with the goal of investigating whether a longer gas extraction and passive air introduction duration could result in complete depletion of $CH_4$ and replacement with air. An additional 10 extraction/introduction wells and 7 monitoring wells were installed with another drilling program. A total of 3,435 Mcf of gas was extracted over 28 days, while 1,489 Mcf of air was passively introduced. During gas extraction, suction was maintained throughout the Crescent Heights Gas Zone under the Wilshire Blvd/South Crescent Heights Blvd. intersection. Concentrations of $H_2S$ declined further and were approximately 2 ppm across much of the gas zone two weeks after the end of extraction. Concentrations of $CH_4$, however, remained approximately 20 to 40% vol. during extraction. Within 10 days of the end of extraction, $CH_4$ concentrations and gas pressures returned to baseline levels of approximately 95% vol. and positive pressure of up to 20 inches of equivalent water column. *This trial confirmed that gas extraction and in-situ oxidation could adequately mitigate the $H_2S$ hazard but not the $CH_4$ hazard in advance of tunneling and that the source was not a small pocket of gas, but a much larger, near infinite source.*

## PLANNED MITIGATION, THEORY, AND EXPECTED OUTCOME

Following the gas extraction trials, the WPLE1 team decided to mitigate the risks associated with tunneling through the Crescent Heights Gas Zone with the following measures:

- Ventilation within each TBM at a nominal rate of 100,000 cfm, which passed through specified activated carbon scrubbers to treat $H_2S$
- Gas detection systems, automated alarms and shut-off switches, and electrical equipment in accordance with Cal/OSHA requirements for gassy and extra-hazardous tunnels within each TBM
- Gas extraction and passive air introduction in advance of tunneling using a network of wells to further remove $H_2S$ by in-situ oxidation
- Gas extraction during tunneling using a network of wells to apply suction and direct gas from the subsurface to systems to treat $CH_4$ and $H_2S$ at surface—providing a guided preferential pathway to extraction wells
- Monitoring of gas pressures in the ground within and above the gas zone with a network of vibrating wire piezometers (VWPs) and monitoring wells
- Monitoring of $CH_4$ and $H_2S$ concentrations in building basements and utility vaults adjacent to the gas zone using portable open-path infrared (Heath RMLD) and traditional gas detection equipment
- Contingency plan prepared to respond to gas detections if needed, including triggers and an action plan for notification of responders, evacuation of affected area, supplemental ventilation systems, and other measures as might have been needed

The WPLE1 team developed a conceptual model for how tunneling through the Crescent Heights Gas Zone could potentially result in gas migration and how gas extraction could mitigate this risk. As an EPB TBM advances, it applies mechanical stress (pressure) to the ground as thrust jacks push it forward. The mechanical stress is borne by grain-to-grain contacts, and soil deformation limits the influence of these forces to a radius on the order of tens of feet or less around the cutterhead, dependent on soil or rock strengths. Hydraulic pressure is applied to the ground around the cutterhead by the pressure bulkhead and the injection of fluids (water, surfactant, and air) to the excavation chamber and directly to the ground in front of the cutterhead.

Injection of water, surfactant, and compressed air to the excavation chamber results in a plastic muck with fluid pressure maintained at the desired levels, which are typically the hydrostatic pressure (i.e., equivalent to a water column the height of the depth below ground surface) plus any load to support the face and a factor of safety. Planned depths of the bottom of the tunnels through the gas zone ranged from 65 to 78 feet, such that planned pressure to be maintained in the excavation chamber of each TBM was approximately 2 to 3 bar (800 to 1,200 in. $H_2O$). Openings in the cutterhead allow the hydraulic pressure applied to the muck in the excavation chamber to also be applied to the ground at the cutting face. The equivalent pressure on both sides of the cutterhead define the EPB pressure, a highly successful mitigation to minimize ground settlement and groundwater inflows to a TBM. The EPB pressure also provides mitigation to ingress of gases to the TBM and tunnel (due to pressure gradient). But since the pressure maintained at the cutterhead (800 to 1,200 in. $H_2O$) is much greater than the gas pressures in the Crescent Heights Gas Zone (up to 20 in. $H_2O$), this pressure gradient causes increases in groundwater levels around the TBM that could potentially drive gas migration away from the TBM and potentially into receptors such as building basements or utility vaults. Additionally, compressed air injected for soil conditioning in the excavation chamber can release into the ground and potentially pressurize or displace ground gases, which can cause gas migration. Previously, the WPLE1 team had measured a zone of influence of 600 feet when injecting compressed air in the chamber during a long stoppage.

Injection of foam directly to the ground in front of the cutterhead posed a greater risk of operation of the TBM potentially causing gas migration—thus creating another Ross-Dress-for-Less explosion. Foam is injected to the ground through nozzles on the cutterhead to reduce permeability of the ground, which inhibits dissipation of the pressure in the chamber, and for conditioning of the muck so that it forms a viscous paste. Foam is produced in the TBM by mixing a surfactant (i.e., soap) into water and injecting compressed air; amounts of each of these can be varied to optimize performance of the foam. The amount of compressed air required, expressed as the foam expansion ratio (FER), is important for foam performance but also had potential to influence gas behavior in the ground. A necessary function of operating an EPB TBM, foam injection into the Crescent Heights Gas Zone posed a risk for gas migration.

A final potential mechanism for gas migration that could possibly be caused by tunneling was the mining of the clay directly overlying and capping the gas zone and/or potential fracturing of clay overlying the gas zone. The WPLE1 team ranked the potential for this mechanism lower than other potential gas migration mechanisms due to the approximately 50 feet of predominantly fine-grained sediments that would remain overlying the tunnels. However, the vertical thickness of fine-grained sediments between the tunnels and multi-level building basements ranged from approximately 3 to 16 feet (Figure 5).

The primary mitigation for $H_2S$ in the Crescent Heights Gas Zone was in-situ oxidation treatment in advance of tunneling. The concept for in-situ oxidation of $H_2S$ was that oxygen in air passively introduced to the ground during gas extraction would displace and rapidly oxidize $H_2S$ in the gas phase and then partition into the soil moisture and oxidize the aqueous sulfide dissolved in the soil moisture. The effect was long-lasting because as $H_2S$ flowed back into the gas zone with the $CH_4$ and other natural gases that rapidly returned after gas extraction stopped, the dissolved oxygen in the soil moisture provided a reservoir of treatment capacity to continue oxidizing $H_2S$.

The primary mitigation for $CH_4$ in the Crescent Heights Gas Zone was gas extraction during tunneling to maintain suction in the gas zone and collect and treat/oxidize gases at surface. Based on the estimated gas permeabilities and transmissivities and measured radius of influence of gas extraction during the trials in 2018 and 2019, the WPLE1 team anticipated that suction could be applied across the gas zone with on the order of 10 gas extraction wells. To understand the potential interactions between the EPB TBMs and the gas zone, the WPLE1 team developed a three-dimensional (3-D) multi-phase flow model using the code PFLOTRAN (developed by the US national laboratories; www.pflotran.org).

Modeling the Crescent Heights Gas Zone consisted of four steps: constructing a 3-D geologic model using Leapfrog (www.seequent.com), constructing a 3-D multiphase flow model in PFLOTRAN, calibrating the flow model to results from the gas extraction trials, and simulating EPB TBM tunneling to forecast gas flows around the TBMs. Key observations for setting up the flow model were that positive pressures existed in the gas zone, there is a north to south gas pressure gradient, and gas pressures returned to a steady state, positive levels soon after gas extraction stopped. To simulate these key observations, a constant gas source north of Wilshire Blvd. was required in the model, as shown in Figure 6. Gas was simulated in PFLOTRAN to be flowing from the gas source, which represents natural gas flowing from the underlying oil fields north to south through the unsaturated San Pedro sands (i.e., the gas zone), and then

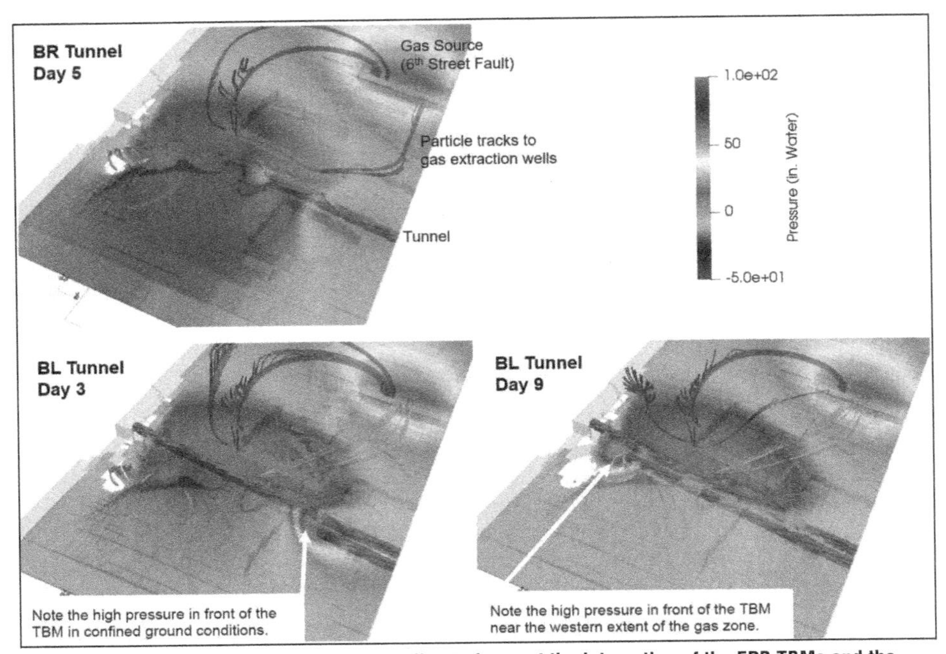


**Figure 6. Modeling completed prior to tunneling to forecast the interaction of the EPB TBMs and the planned gas extraction system**

upward as a broad dispersed flow through the overlying fine-grained sediments of the Lakewood Formation to emit into the deep building basements and at the ground surface. Hence, gas migration to receptors was already occurring naturally at rates that were being managed with sufficient barrier walls and ventilation systems in many of the basements. Gas permeabilities were adopted from well test analyses done using the reservoir engineering software SAPHIR (Kappa Engineering), and the PFLOTRAN model was calibrated to the gas extraction rates, gas pressure responses, and groundwater level changes measured during the gas extraction trials in 2019 and baseline measurements of $CH_4$ concentrations in basements. It should also be noted that no building owners reported issues with gas intrusion prior to tunneling.

To simulate interaction of the TBMs and tunnels with gas extraction from the gas zone, the WPLE1 team modeled the advancing TBM with injection rates of air (up to 5 cfm) and water (up to 2 cfm) to represent foam injection at the cutterhead and a no-flow condition at the walls of the tunnels. Foam injection rates were based on laboratory testing that the WPLE1 team completed on samples of San Pedro sands from boreholes. Operation of the gas extraction system was simulated in the model with the planned gas extraction rates from the extraction wells (discussed in the next section). Findings from the modeling completed prior to tunneling are illustrated in Figure 6 and summarized as follows:

- Gas flows naturally through the gas zone, and baseline $CH_4$ flows into four building basements adjacent to the tunnel alignments in the gas zone were on the order of 20 kg/d
- Suction could be maintained across much of the gas zone and particularly across the more permeable central portion of the gas zone, even close to each TBM as it advanced through these areas in the model

- In more confined areas in the eastern and western portions of the gas zone after the first tunnel is constructed, gas pressures were predicted to be high (on the order of 100 in. $H_2O$)
- Gas migration into one basement was predicted to be increased on the order of 40 kg/d for a period of several days due to operation of the TBM, likely a gas ingress rate that could easily be managed to safe levels by the existing ventilation systems in the building basements
- After construction, the tunnels could partially impede the natural north-to-south gas flow, potentially resulting in higher gas flow to basements and utilities on the north side of Wilshire Blvd

The expected outcome of tunneling through the Crescent Heights Gas Zone, based on over three years of investigations and considerable prior experience, was that gas extraction would effectively mitigate the risk of gas migration. The residual risk was characterized as a moderate-to-high probability that the TBMs would cause minor, acceptable, temporary, additional gas migration to building basements and there was a very low probability that tunneling would result in catastrophic gas migration.

## GAS EXTRACTION SYSTEM

The gas extraction system consisted of 11 gas extraction wells, lateral piping in backfilled trenches, and 3 modular gas extraction and treatment systems. Monitoring points consisted of nested VWPs at six locations, six monitoring wells, and pressure and flow gauges on each gas extraction well. The layout of the system is shown in Figure 7.

The wells and VWPs were installed by Golder and Gregg Drilling, lateral piping was installed by Golder and Lonestar West, and the gas extraction systems were assembled and operated by Envent Corporation with input on well hydraulics from Golder. Extraction wells were 4-inch and 6-inch diameter PVC. Nine of the eleven extraction wells were fitted to 2-inch diameter PVC pipes placed in backfilled tranches that terminated in well vaults in the middle lane of Wilshire Blvd. or in the median of McCarthy Vista Blvd. (Figure 7). Gas extraction equipment was set up in the central three lanes of Wilshire Blvd. and in the median and central two lanes of McCarthy Vista Blvd.

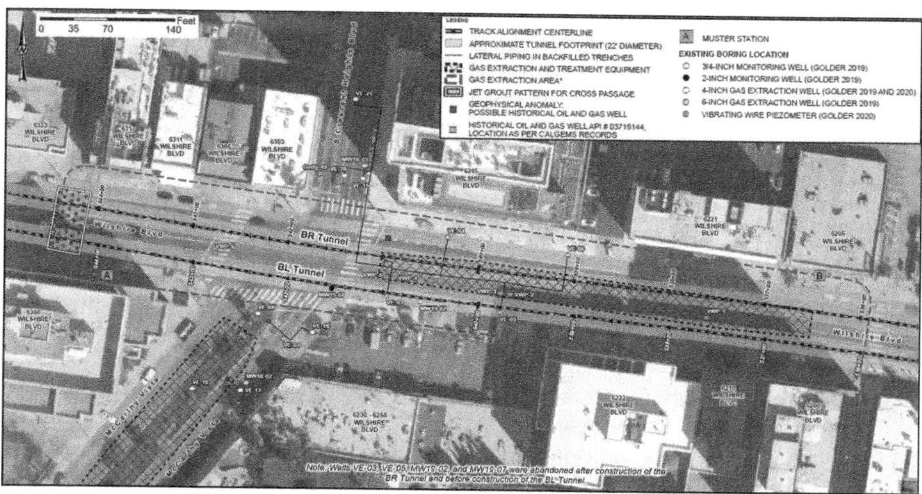

Figure 7. Plan view of the gas extraction system

The gas extraction equipment consisted of three systems, each having a positive displacement blower and a variable frequency drive, a moisture knockout tank, a dual liquid caustic scrubber to treat $H_2S$, and a thermal oxidizer to combust $CH_4$ and volatile organic compounds located in the middle of Wilshire Blvd. Each system was operated continuously, day and night, from the start of extraction on May 24, 2020, until an unplanned shutdown on May 30, 2020. Extraction resumed on June 15, 2020, and continued for 15 days during construction of the north (BR) tunnel. Equipment was demobilized and then re-mobilized, and extraction resumed on July 21, 2020, for 18 days during construction of the south (BL) tunnel. Between construction of the two tunnels, four wells in the tunnel alignment were abandoned.

Prior to tunneling, 155 Mcf of air was passively introduced to the gas zone over 10 days. Concentrations of $H_2S$ under Wilshire Blvd. were less than 7 ppm after the air introduction. But to the north and south of the footprint of the in-situ oxidation, $H_2S$ concentrations typically ranged from 1,000 to 5,000 ppm. During construction of the BR tunnel (Figure 7), the total gas extraction rate averaged 104 cfm, and a total of 2,653 Mcf of gas was extracted. Gas pressures were maintained at between approximately –40 and –120 in. $H_2O$ in extraction wells, which maintained gas pressures at monitoring locations in the gas zone between approximately –20 and –90 in. $H_2O$ with exceptions discussed in the next section. These suction ranges targeted optimization of gas flow to extraction wells while intending to avoid induced groundwater level changes that could interfere with gas extraction. During construction of the BL tunnel (Figure 7), the total gas extraction rate averaged 50 cfm, and a total of 1,204 Mcf of gas was extracted. Gas pressures were maintained at between approximately –25 and –120 in. $H_2O$ in extraction wells, which maintained gas pressures below atmospheric levels in the gas zone except close to the TBM and in confined portions of the gas zone (described in the next section).

## TBM OPERATION AND MEASURED CONDITIONS

The TBM referred to as the Purple TBM (BR—Right side looking west) mined through the 840 ft-long gas zone (included buffer zones within water- or tar-saturated sand) from June 19 to 29, 2020, to construct the BR Tunnel (Figure 7) at an average rate of approximately 72 feet per day. STS operated the TBM continuously on three shifts per day. Pressure responses to tunneling measured at two of the monitoring locations (VWP-4 and VWP-5) in the highly-permeable portion of the gas zone adjacent to 6245 Wilshire Blvd. (Figure 7) are presented in Figure 8. Operation of the gas extraction system resulted in gas pressures of approximately –70 in. $H_2O$ and a 1.5 foot increase in groundwater levels in the days prior to the TBM passing 6245 Wilshire Blvd. As the TBM passed 6245 Wilshire Blvd, gas pressures increased by 30 in. $H_2O$, but suction was maintained in the gas zone.

Gas pressures were approximately 10 in. $H_2O$ greater when the Purple TBM mined through the more permeable central and western portions of the gas zone, than when it mined through the less permeable eastern portion (Figure 8). Groundwater levels rose as much as two feet, but this response was more localized to the TBM with a radius of approximately 100 feet in front and behind the cutterhead based on responses at VWP-4. Gas pressures responded to the Purple TBM as much as 300 feet away, because gas pressures increased in response to TBM operation across the entire central and western portions of the gas zone. This widespread but low magnitude pressure response correlated with an increase in the FER, or amount of air in the soil conditioners that was required to maintain target face pressures in the permeable sands (Figure 8).

Methane concentrations varied from 0 to 90 % vol. and $H_2S$ concentrations were typically less than 1 ppm under Wilshire Blvd. as the Purple TBM mined through the gas zone. The WPLE1 team monitored gas concentrations three times daily in basements of 13 buildings and 18 utility vaults during tunneling through the gas zone. Concentrations of $CH_4$ and $H_2S$ remained low in the basements during construction of the BR Tunnel, and there were no measured increases in the basements or utility vaults correlated with TBM operation.

The TBM referred to as the Red TBM mined through the gas zone from July 27 to August 8, 2020, to construct the BL Tunnel (Figure 7) at an average rate of 79 feet per day, again mining continuously with three shifts per day. Gas pressure responses to TBM operation in the east and west portions of the gas zone, which were confined by the newly-constructed BR Tunnel and the boundaries of the gas zone, are presented in Figure 9. While the TBM passed 6222 Wilshire Blvd, gas pressures increased by up to 675 in. $H_2O$, with substantial pressure increase in both the partially-tar saturated sands in that area of the San Pedro Formation and the overlying clayey silts of the Lakewood Formation. When the TBM mined through the more permeable central and western portions of the gas zone, gas pressures increased by up to 170 in. $H_2O$. Gas pressures remained high as the TBM mined past 6330 Wilshire Blvd. Similarly, as for the BR Tunnel, gas pressure increases correlated with the higher FER required for the BL Tunnel in the clean sands of the central and western portions of the gas zone (Figure 9).

Concentrations of $CH_4$ under Wilshire Blvd. were lower for the BL Tunnel, apparently due to compartmentalization/barrier of the gas zone by the newly-constructed BR Tunnel and gas extraction. But as the Red TBM passed 6222 Wilshire Blvd. and 6300 Wilshire Blvd, minor ingressions of $CH_4$ were measured at the lowest level of each of these building basements. Each of these $CH_4$ ingressions lasted for less than one day, correlating with the passing of the TBM, and were limited to small areas at basement sumps (i.e., detectable in covered sumps but not in the basement atmosphere). There was also one instance of ingression of $CH_4$ into the TBM during tunneling through the gas zone, but it caused only minor delay (i.e., less than an hour) until concentrations lowered by ventilation and authorization was given to resume mining.

## CONCLUSIONS

Potential risks associated with tunneling through gassy ground adjacent to the La Brea Tar Pits in Los Angeles were key factors leading to decades of delays during planning stages for a critical portion of the city's rapid transit system. During baseline characterization and further investigations for Section 1 of the Purple Line Extension, an extensive gas zone with 95 % vol. $CH_4$ and up to 6,500 ppm $H_2S$, referred to as the Crescent Heights Gas Zone, was identified west of the La Brea Tar Pits. This gas zone at the depth of the tunnel alignments may be connected to the underlying oil fields by a fault or a historical oil and gas well.

To mitigate the risk of explosive and acutely toxic gases entering the TBMs or migrating away from the pressure-face TBMs into adjacent buildings or utility vaults, the WPLE1 team adopted several mitigations. A ventilation system and gas detection system were operated inside each TBM, and a gas extraction system was operated outside the TBMs and paired with monitoring of in-situ ground conditions and building basements and utility vaults adjacent to the gas zone. Air in the soil conditioning system was turned off.

The performance of the gas extraction system and tunneling through the exceptionally gassy ground west of the La Brea Tar Pits are summarized as follows:

- In-situ oxidation by passive introduction of air during gas extraction successfully lowered $H_2S$ concentrations by a factor of 1,000 such that the TBMs mined through ground with safe levels of $H_2S$, and $H_2S$ was not detected inside the TBMs or adjacent building basements.

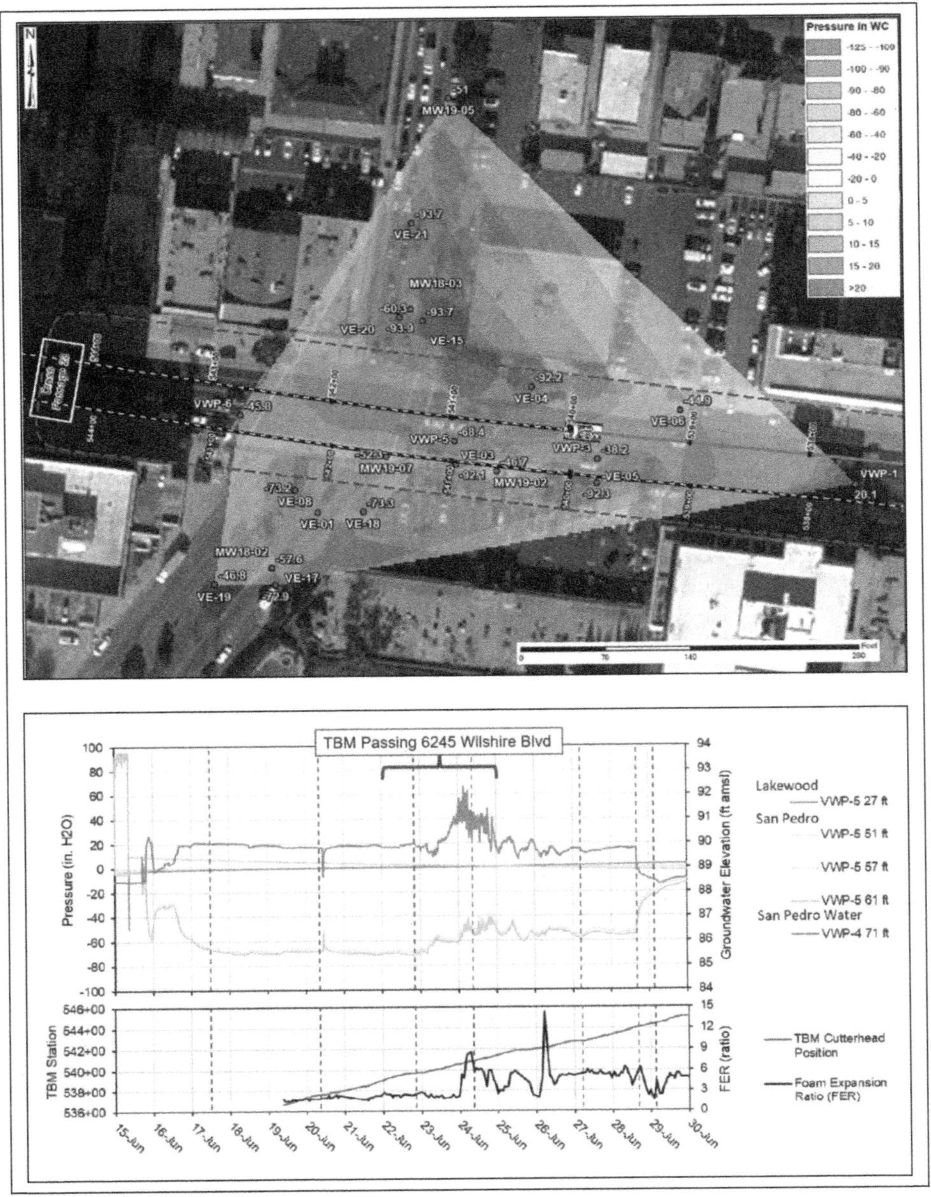


**Figure 8. Gas pressure responses as the TBM passed 6245 Wilshire Blvd**

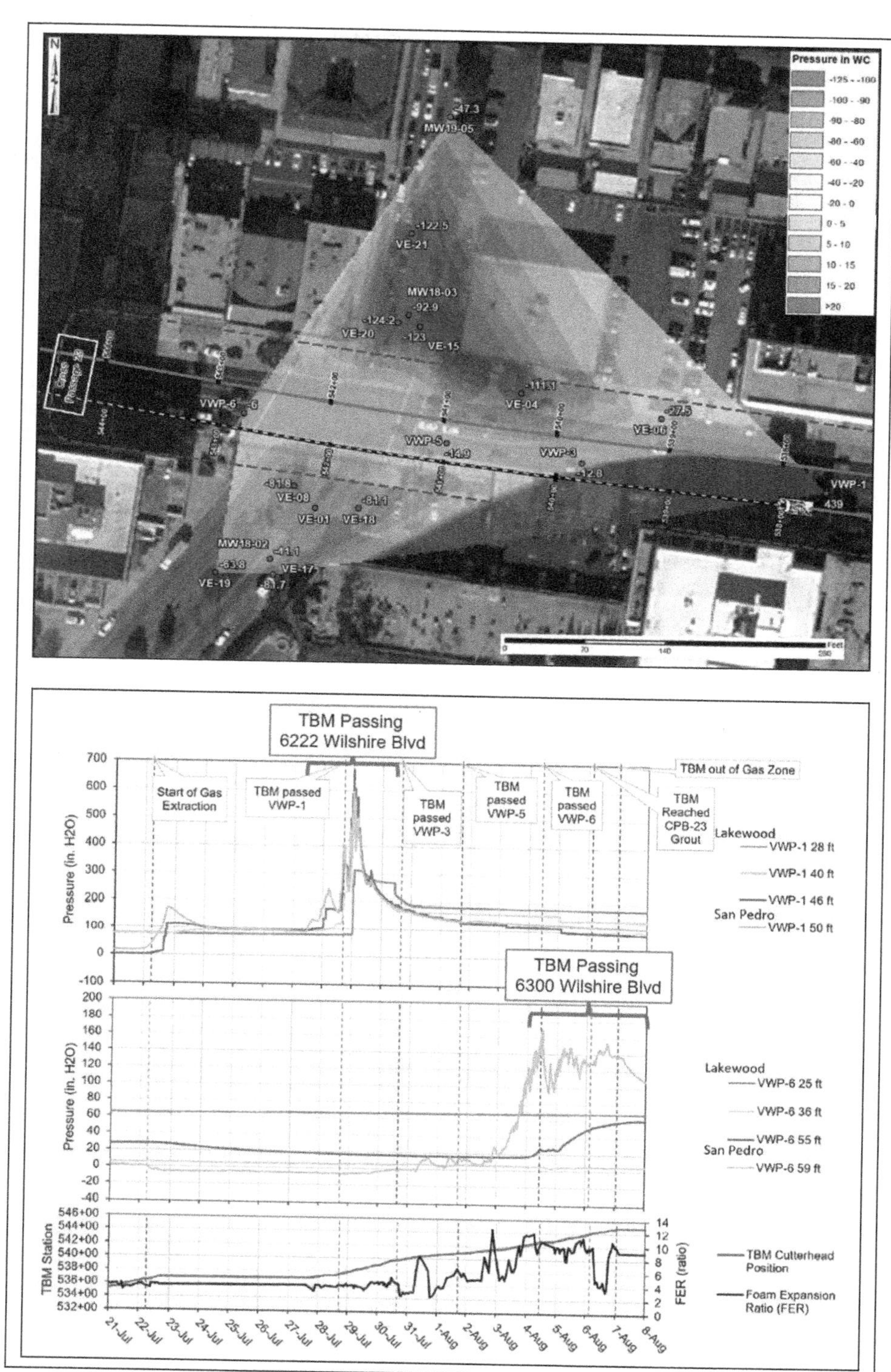


**Figure 9. Gas pressure responses as the TBM passed 6222 and 6300 Wilshire Blvd**

- Operation of a gas extraction system successfully maintained suction as the TBMs mined through the more permeable portions of the gas zone. While the TBMs mined through less permeable and more confined portions of the gas zone, particularly when further bounded by the newly-constructed tunnel, TBM operation caused gas pressures to increase substantially and there were minor, short-lived ingressions of $CH_4$ into the sumps at the deepest level of two building basements.
- Safely tunneling through the gassy ground was made possible by effective planning, investigations, operation of a gas extraction system, and coordination between the members of the WPLE1 team.

## ACKNOWLEDGMENTS

The authors are grateful for contributions to the project by the entire PLES1 team, including Jim Cohen and others at LA Metro; Pat Nicholson and others at the Skanska-Traylor-Shea Join Venture; Cameron Beul, Mark Naugle, and others at Golder Associates (now WSP); Amanda Elioff, Roy Cook, and others at WSP and their subconsultants; Kroner Environmental Services; Gregg Drilling; Lonestar West; Envent Corporation.

## REFERENCES

American Public Transportation Association (APTA). 2005. Peer Review Panel Report on the Wilshire Corridor Tunneling Project. Report prepared for Los Angeles County MTA, November 2005. 28p. http://libraryarchives.metro.net/DPGTL/peerreview/2005_apta_wilshire_corridor_tunneling_final_report.pdf.

American Public Transportation Association (APTA). 2012. Peer Review Report on the Wilshire Corridor Westside Extension Tunneling Project—Part-II. Report prepared for Los Angeles County MTA, November 1, 2012. 19p.

Arnold R. 1906. The Salt Lake Oil Field Near Los Angeles, Cal. Chapter in Eds. Arnold R, Clapp FG. Contributions to Economic Geology, 1905: Petroleum and Natural Gas. USGS Bulletin 285-G. https://pubs.er.usgs.gov/publication/b285G.

Bilodeau WL, Bilodeau SW, Gath EM, Oborne M, Proctor RJ. 2007. Geology of Los Angeles, California, United States of America. Environmental and Engineering Geoscience, XIII(2):99–160. https://www.aegweb.org/assets/docs/la.pdf.

California Geologic Energy Management Division (CalGEM). Well Finder. Online geographic information system. https://maps.conservation.ca.gov/doggr/wellfinder/#/. API 03715144: https://filerequest.conservation.ca.gov/WellRecord?api=03715144. API 03721161 2018 data file: https://filerequest.conservation.ca.gov/WellRecord?api=03721161.

Chilingar GV, Endres B. 2005. Environmental hazards posed by the Los Angeles Basin urban oilfields: an historical perspective of lessons learned. Environmental Geology, 47:302–317. https://link.springer.com/article/10.1007/s00254-004-1159-0.

City of Los Angeles. 1985. Task Force Report on the March 24, 1985 Methane Gas Explosion and Fire in the Fairfax Area, City of Los Angeles. Report prepared by special task force for the mayor and city council. http://libraryarchives.metro.net/DPGTL/losangelescity/1985_methane_gas_explosion_task_force_report_fairfax.pdf.

Cobarrubias JW. 1992. Mathane gas hazard within the Fairfax District, Los Angeles. Engineering Geology Practice in Southern California, Association of Engineering Geologists, Special Publication No. 4, 131–143.

Converse Ward Davis Dixon Earth Science Associates Geo/Resource Consultants (CWDD). 1981. "Geotechnical Investigation Report," Volume 1, Southern California Rapid Transportation District Metro Rail Project. Available at UCLA library collection.

Copur H, Cinar M, Okten G, Bilgin N. 2011. A case study on the methane explosion in the excavation chamber of an EPB-TBM and lessons learnt including some recent accidents. Tunnelling and Underground Space Technology, 27:159–167. doi:10.1016/j.tust.2011.06.009.

Crowder RE and Johnson RA. 1961. Los Angeles City Oil Field. Chapter in Summary of Operations California Oil Fields, Annual Report No. 47, Department of Natural Resources, Division of Oil and Gas. pp. 67–77.

Crowder RE. 1963. Recent Developments in Jade-Buttram Area of Salt Lake Oil Field. Chapter in Summary of Operations California Oil Fields, Annual Report No. 49, Department of Natural Resources, Division of Oil and Gas. pp. 53–58.

Division of Oil, Gas, and Geothermal Resources (DOGGR). 1992. California Oil and Gas Fields: Vol. II—Southern, Central Coastal, and Offshore California Oil and Gas Fields. DOGGR Publication TR12. 645p.

Elioff MA, Smirnoff TP, Ryan PF, Putnam JB, Ghadiali BM. 1995. Geotechnical investigrations and design alternatives for tunneling in the presence of hydrogen sulfide gas—Los Angeles Metro. Proceedings of the 1995 Rapid Excavation and Tunneling Conference, Ch. 19, 299–318.

Endres B, Chilingarian GV, Yen TF. 1991. Environmental hazards of urban oilfield operations. Journal of Petroleum Science and Engineering, 6(2):95–106. https://doi.org/10.1016/0920-4105(91)90030-Q.

Enviro-Rail, 1996. Phase II Western Extension Reassessment Study. Report prepared for Los Angeles County Metropolitan Transportation Authority. March 1996. 574p. Figure 3-2.

Etiope G, Doezema LA, Pacheco C. 2017. Emission of methane and heavier alkanes from the La Brea Tar Pits seepage area, Los Angeles. Journal of Geophysical Letters: Atmospheres, 122:12,008-12,019. https://doi.org/10.1002/2017JD027675.

Hamilton DH, Meehan RL. 1992. Cause of the 1985 Ross Store explosion and other gas ventings, Fairfax District, Los Angeles. Engineering Geology Practice in Southern California, Association of Engineering Geologists, Special Publication No. 4.

Jeffrey AWA, Alimi HM, Jenden PD. 1991. Geochemistry of Los Angeles Basin Oil and Gas Systems. Chapter 6 in Ed. Biddle KT, Active Margin Basins. American Association of Petroleum Geologists. https://doi.org/10.1306/M52531C6.

Jung B, Garven G, Boles JR. 2015. The geodynamics of faults and petroleum migration in the Los Angeles Basin, California. American Journal of Science, 315:412–459. DOI 10.2475/05.2015.02.

Kaiser Industries Corporation (Kaiser). 1961. General Description of Rapid Transit System Backbone Route for Los Angeles Metropolitan Transit Authority. July 1961. http://libraryarchives.metro.net/DPGTL/lamta/1961_kaiser_general_description_rapid_transit_system_backbone_route.pdf

Kaiser Engineers California (Kaiser) and Gage-Babcock & Associates, Inc. (Gage-Babcock). 1983. Study of Methane and Other Combustible Gases Effect on Underground Operation of the Metro Rail Project. March 1983. http://libraryarchives.metro.net/DPGTL/scrtd/1983-study-of-methane-and-other-combustible-gases-effect-on-underground-operation-of-the-metro-rail-project.pdf

Khilyuk LF, Chilingar GV, Endres B, Robertson JO. 2000. Gas Migration: Events Preceding Earthquakes. Gulf Professional Publishing, Houston, TX. ISBN 0884154300, 9780884154303, 389p.

Lang HR and Dreessen RS. 1975. Subsurface structure of the northwestern Los Angeles Basin. California Department of Conservation Division of Oil and Gas Publication No. TP01. https://www.conservation.ca.gov/calgem/pubs_stats/Pages/technical_reports.aspx

Los Angeles County Metropolitan Transportation Authority (LA Metro). 2011. Westside Subway Extension Project, Wilshire/Fairfax Station Construction. Paleontological Resources Extraction. Attachment 3 to Appendix G Memorandum of Understanding for Paleontological Resources, Final environmental Impact Statement/Environmental Impact Report, Vol. 4.

Los Angeles County Metropolitan Transportation Authority (LA Metro). 2014. Westside Subway Extension Project, Section 1: Contract 1045 Geotechnical Baseline Report. Conformed November 3, 2014. Prepared by Parsons Brinkerhoff.

Proctor RJ and Monsees JE. 1985. Los Angeles Metro Rail Project: Design issues related to gassy ground. Proceedings of the 1985 Rapid Excavation and Tunneling Conference, vol 1, ch.30, 488–505.

Proctor RJ. 2002. The San Fernando Tunnel explosion, California. Engineering Geology, 67(1-2):1–3. https://doi.org/10.1016/S0013-7952(02)00042-X

Samuelian RH. 1990. South Salt Lake Oil Field. Chapter in California Department of Conservation, Division of Oil and Gas Publication No. TR32. Originally submitted 1984. https://www.conservation.ca.gov/calgem/pubs_stats/Pages/technical_reports.aspx

Shaw AC, Quinn JP. 1986. Rancho La Brea: A look at coastal southern California's past. California Geology, 39:123–133.

US Congress. 1985. Public Law Statute 99-190. December 19, 1985. https://www.govinfo.gov/content/pkg/STATUTE-99/pdf/STATUTE-99-Pg1185.pdf

US Congress. 2007. Public Law Statute 110-161. December 26, 2007. https://www.congress.gov/110/plaws/publ161/PLAW-110publ161.pdf

Wright T. 1987. Geological setting of the Rancho La Brea Tar Pits. In AAPG Pacific Section 2009—Petroleum Geology of Coastal Southern California (1987), 87–91.

# Los Angeles, California JWPCP Effluent Outfall Tunnel Project—Tunneling Under Extremely Challenging Conditions

**R. Schuerch** ▪ Pini Group Ltd.
**P. Perazzelli** ▪ Pini Group Ltd.
**M. Piemontese** ▪ Pini Group Ltd.
**P. Halton** ▪ Pini Group Ltd.
**N. Karlin** ▪ Dragados USA
**C. Cimiotti** ▪ Dragados USA

## ABSTRACT

The new Los Angeles, California JWPCP effluent outfall tunnel will transport secondary-treated effluent from the Joint Water Pollution Control Plant in Carson to White Point Manifold. The tunnel will be approximately 7 miles (11 km) long with a finished internal diameter of 18 ft (5.5 m). The TBM is designed to cope with high water pressure up to 10 bar. The geotechnical challenges are due to the heterogeneous subsurface conditions: in the first half of the alignment, the tunnel runs through soils with a low overburden under urban areas, and, in the second half, the tunnel runs through extremely weak, intensely faulted rock masses with high overburden. Following an overview of the project, this paper focuses on the northern portion of the alignment and presents a decision-making procedure developed during pre-excavation phase in order to cope with the expected heterogeneous conditions during mining. The risk assessment and the TBM operational design of the southern portion will be dealt with in a future paper.

## PROJECT OVERVIEW

The new Los Angeles, California JWPCP effluent outfall tunnel (shown in red in Figure 1) will transport secondary treated effluent from the Joint Water Pollution Control Plant in Carson to White Point Manifold at the coast. The aim of the project is to provide relief and redundancy to the existing tunnels (shown in green in Figure 1) as well as providing additional overall conveyance capacity (GBR 2018). The tunnel will be approximately 7 miles (11 km) long with a finished internal diameter of 18 ft (5.5 m).

The tunnel will be excavated using a Mixshield TBM with precast segmental concrete lining designed to cope with the baseline maximum water pressure of 9.1 bar (according to the Geotechnical Baseline Report—GBR 2018).

The northern portion of the tunnel alignment is beneath relatively flat terrain and extends from the JWPCP shaft site to approximately the intersection of North Amelia Avenue and West Capitol Drive. It passes parallel to and under crosses Machado Lake within the Harbor Regional Park at approximately 40 ft depth. To the south of this intersection, the terrain steepens through the Palos Verdes (PV) Hills and falls again to the White Point Shaft. The overburden varies between 30 ft on approach to the White Point Shaft to 450 ft in the PV Hills. The tunnel slope is approximately 0.1%. The alignment runs 68% underneath urban development including Harbor Freeway and 32% underneath greenfield.

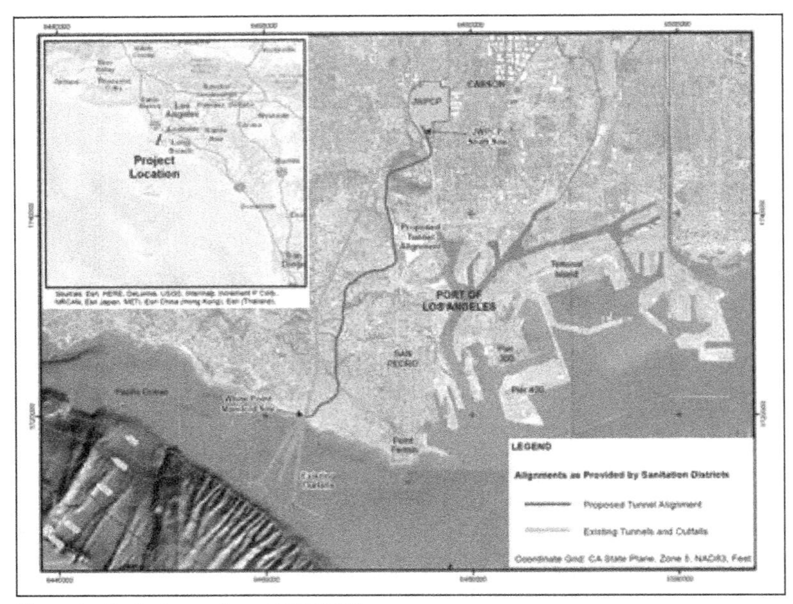


Figure 1. Project area (from GDR 2017)

## GEOLOGICAL CONDITIONS EXPECTED ALONG THE TUNNEL ALIGNMENT

### Geology

For approximately 53% of the alignment, the TBM is expected to pass through heterogeneous soils of the Alluvium, Lakewood Formation and San Pedro Sand; 5% of the alignment through very weak Transitional Rocks of Lomita Marl, Malaga Mudstone and Valmonte Diatomite; and 42% of the alignment through weak and intensely fractured Altamira Shales and Point Fermin Sandstone. Table 1 describes the geologies that the TBM is expected to encounter along the tunnel alignment.

The overburden in the northern portion varies approximately from 45 feet to 115 feet and it mainly consists of heterogeneous soils. In the southern portion, the overburden varies approximately from 115 feet to 450 feet and it mainly consists of transitional rocks and highly fractured, weathered rock mass. The tunnel will run through two major fault zones (Palos Verde Fault Zone and Cabrillo Fault Zone) and it may cross a number of other unnamed faults and fault splays associated with the Palos Verde Anticlinorium and the Gaffey Anticline.

### Hydrogeology

The tunnel runs through different soil and rock aquifers (GDR 2018). Low permeability ground zones along the alignment act as a partial barrier to ground-water flow between the different aquifers resulting in significant changes in groundwater levels along the tunnel. The northern portion of the alignment is located within the soil aquifers of the West Coast Basin (WCB). The southern portion of the alignment is located within the rock aquifers of PV Hills. They are not considered as a significant source for regional water supply purposes (GBR 2018) and are not subject to significant variation from water supply pumping or injection wells (differently to the groundwater basins).

**Table 1. Geologies expected along the TBM tunnel alignment (adapted from GBR 2018)**

| Geology | Age | Thickness, ft | Comment |
|---|---|---|---|
| Alluvium | Recent | ~4 to ~40 | Associated with rivers and streams. Clays, silts and sands with organic debris, gravels, cobbles and boulders of subangular to subrounded weathered rock fragments. |
| Lakewood Formation | Middle to Late Pleistocene | ~180 to ~280 | Bedded and cross-bedded flood plain and alluvial fan deposits. Predominantly of sand, sand with varying amounts of silt and clay and lenses of gravels and shells. Unconformity at upper and lower boundaries and within strata. |
| Sand Pedro Sand | Middle Pleistocene | ~300 | Bedded and cross-bedded flood plain and alluvial fan deposits. Predominantly of sand, sand with varying amounts of silt and clay and lenses of gravels and shells. Unconformity at lower boundary and within strata. |
| Timms Point Silt | Middle Pleistocene | ~120 | Predominantly of silt, siltstone, claystone and mudstone with seams of sand, organic fragments, shells and gravel. Locally interbedded with San Pedro Sand. |
| Lomita Marl | Middle Pleistocene | ~275 | Unfractured silty sandstones, and sandstones with dolomite layers. On the northern side of the PV Hills, deposits consist of unfractured calcareous siltstone and sandstone with shell fragments and cobble-size calcareous concretions. Boulder-size mudstone clasts and gravel- and cobble-size cemented siltstone clasts are found in calcareous sand layers through the deposits and marking the basal boundary with the Valmonte Diatomite. Unconformity at lower boundary and within strata. |
| Malaga Mudstone | Miocene | ~200 to ~600 | Associated with deep sea. Mudstone, siltstone and claystone with layers of diatomite, diatomaceous shale, limestone and chert. Locally contains pockets of organic material, sand and dolomite bands. Unconformity at upper boundary and within strata. |
| Valmonte Diatomite | Middle to Late Miocene | ~300 to ~500 | Diatomite and laminated diatomaceous mudstone, claystone and siltstone with occasional layers of limestone, shale, black chert and volcanic ash. Bands of dolomitic- and calcitic-cemented layers, seams of gravel-size sandstone, black chert and schist clasts. |
| Altamira Shale | Middle Miocene | ~100 to ~900 | Siltstone, claystone, dolomite with locally chert, quartz-filled vugs, diatomite, limestone, mudstone and shale. Bands of gravel-size clasts of Point Fermin Sandstone, bentonite, blue schist and siltstone, and brecciated siltstone. Miocene volcanic rocks of basalt irregularly intrude as sills and dikes. Brecciated bands indicating extensive weathering planes and zones. Several unconformities within strata. |
| Point Fermin Sandstone | Miocene | ~120 | Coarse-grained sandstones with gravels and cobbles of blueschist, quartzite and basalt. Brecciated bands containing rip-up clasts of the Altamira Shale, volcanic tuff and intruded basalt. Several unconformities within strata. |

Within the PV Hills, the groundwater table generally mimics the topography of the hills, but at a lower elevation (GBR 2018).

The hydraulic head along the alignment was determined from the installation and monitoring of 28 observation wells and 12 Vibrating Wire Piezometers (VWP) in selected boreholes and is shown in Figure 2. The maximum undisturbed water pressure expected at the tunnel elevation varies from 0.7 to 8.4 bar.

The hydraulic conductivity of the soil aquifers was evaluated by 14 slug tests. The hydraulic conductivity of the rock masses in the PV Hills were evaluated from 126 Packer tests. The permeability of the soils is highly variable ranging from very low to very high ($10^{-9}$ to 1 cm/s). The permeability of the Lomita Marl, Malaga Mudstone and the Valmonte Diatomite varies between $1\times10^{-5}$ and $1\times10^{-3}$ cm/s, while the permeability of the Altamira Shale, Point Fermin Sandstone and Basalt varies between $1\times10^{-7}$ and $1\times10^{-1}$ cm/s. There are major changes in ground permeability in fault zones from intensively fractured rock (high permeability) to major clayey fault gauge

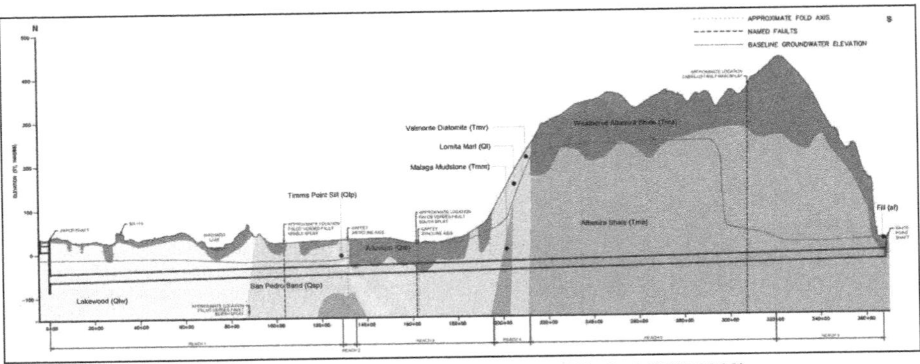


**Figure 2. Geological profile of the JWPCP Effluent Outfall Tunnel Project (GBR 2018)**

(low permeability). In particular, the low permeability rocks of the Gaffey Anticline and the CFZ act as barriers inside these aquifers (GBR 2018).

## MAIN GEOTECHNICAL HAZARDS OF THE TBM EXCAVATION AND MITIGATION MEASURES

The expected geotechnical hazards, without considering support pressure, within the soils of the northern section (Station 0+00 to Station 195+50) are shown in Figure 3.

### Tunnel Face Instability

The instability of the tunnel face is the main hazard scenario to be expected when tunneling through soft soils. Figure 3 shows that the risk is expected along the entire alignment in the northern portion. In water-bearing ground, the face stability depends on ground behavior (drained, undrained or partially drained). The response (drained, undrained or partially drained) of water-bearing ground to excavation depends on the material permeability and to the advance rate. For typical TBM advance rates (about 5–15 m/day) and permeability values higher than $10^{-4}$ cm/s, drained conditions have to be expected during excavation (Anagnostou et al. 2017, Schürch 2016). Partially drained conditions are expected around the tunnel face for permeability values lower than $10^{-4}$ cm/sec. On the basis of these considerations and for the permeability values provided be the GBR, it is plausible to expect ground response of the soils of San Pedro Sand and Lakewood will be from partially drained to drained.

### Blow-Up

Figure 3 shows the location of the sections potentially critical with respect to the risk of blow-up failure. Excess face pressure may lead to ground uplift failure during advance mode and blow out of applied compressed air during hyperbaric interventions. Consequently, the sudden loss of support pressure takes place and large settlements may occur at the surface. Due to the low overburden, the risk is expected along the entire northern portion of the tunnel alignment.

### Large Surface Settlements

Tunnels in an urban environment are often located at small depths underneath densely populated areas and their construction may affect existing structures at the surface and subsurface (e.g., utilities). Tunneling operations in soft soils at shallow depth may cause surface settlements resulting in damage to existing subsurface and/or surface

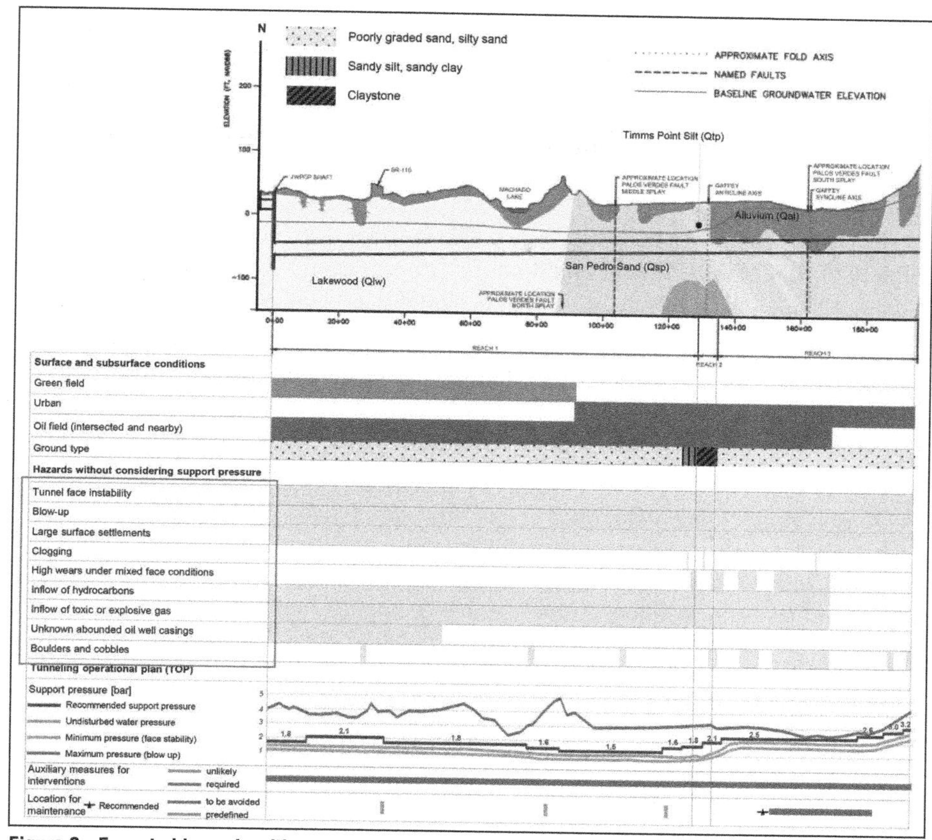

**Figure 3. Expected hazards without considering support pressure for the northern section (Stations 0+00 to 195+50) and extract from the Tunneling Operational Plan (TOP)**

structures. Large settlements may be caused by large convergences of the tunnel (i.e., big volume loss) and/or by the reduction of the pore pressure in case of TBM advance with support pressure lower than the in-situ water pressure.

Figure 3 shows that the risk of large surface settlements is present along the entire northern portion of the tunnel alignment.

## Clogging

Figure 3 shows the location of the sections potentially critical with respect to the risk of clogging. These sections correspond to the portions of the tunnel in high plasticity clayey and silty soils of San Pedro Sand, Lakewood deposits and the Timms Point Silt.

The excavated material, at its natural moisture content, may stick to the steel of the cutter face and obstruct the machine. This phenomenon, commonly referred to as "clogging," leads to the wear of the cutterhead and muck transportation system, causing downtimes and economic loss.

The clogging potential of a material increases with increasing clay content. The adhesion strength between clay and steel is the leading factor of clogging problems in tunneling. It depends on the characteristics of the clay (grain size, mineralogy and water content) and the roughness of the steel surfaces. The risk of clogging tends to increase with higher face support pressure. This behavior is assumed to be attributed to the pressure in the excavation chamber acting on the material stacked on the steel surfaces (cutterhead, cutting tools) and thus increasing the adhesion force. The risk of clogging can be mitigated by the design of the TBM—including type and arrangement of cutting tools, and cutterhead structure opening ratio—and by the use of ground conditioners.

### High Wear Under Mixed Face Conditions

During advance in mixed face conditions, excessive wear of the cutting tools and, in the worst case, damages to the cutterhead may be caused by the sudden loading changes (and/or asymmetrical load) on the cutting tools induced by the wide contrast of mechanical rock mass properties (e.g., strength and stiffness).

Figure 3 shows the location of the sections potentially critical with respect to the risk of high wear under mixed face conditions. These sections correspond to the portions of the tunnel over which the face is expected to include at the same time Alluvium, San Pedro Sand and Timms Point Silt.

### Boulders and Cobbles

Figure 3 shows the location of the sections potentially critical with respect to the risk of boulder and cobbles. They are mainly located in Alluviums, less frequently in Lakewood deposits and San Pedro Sand.

Boulders may jam against the crusher of the Slurry TBM making difficult the mucking-out of the material from the excavation chamber. Boulders and cobbles may lead to severe wear of the cutterhead and also local instability in case of over-excavated face, frequent stops for maintenance works, downtimes and economic loss.

### Inflow of Hydrocarbons

The tunnel alignment is near, or crosses known active and dormant oil fields. The inflow of hydrocarbons into the tunnel may cause the clogging of the spoil transport system and separation plant. The risk of clogging may lead to long stops for maintenance works.

Figure 3 shows the location of the sections potentially critical with respect to the risk of inflow of hydrocarbons. In the northern portion of the alignment, they correspond to the sections within Torrance Field and near the Wilmington Oil Field.

### Inflow of Toxic or Explosive Gas

The entire length of the tunnel is classified as gassy as defined by OSHA Standard 29 CFR 1926.800. During TBM maintenance works under atmospheric conditions, gas may flow into the tunnel exposing the workers to unsafe gas levels. Long stoppages may be required to allow sufficient ventilation to restore the air inside the tunnel to safe conditions.

Figure 3 shows the location of the sections potentially critical with respect to the risk of inflow of gas and they correspond to the northern alignment with the sections within Torrance Field and near the Wilmington Oil Field.

### Unknown Abounded Oil Well Casings

Oil well casings within the excavation area are physical obstructions to the TBM requiring removal and may cause sudden loss of face pressure causing face collapse. Inspection and removal activities are very time consuming and may cause downtimes. Although no known well casings have been identified in the pre-excavation phase, undocumented wells may exist along the tunnel alignment. Figure 3 shows the location of the sections potentially critical with respect to the risk of unknown abounded wells and they correspond to the portion within the Torrance Oil Field.

## TUNNELLING OPERATIONAL PLAN AND DECISION-MAKING DURING ADVANCE

Based upon the identified hazards, a tunneling operational plan (TOP) was produced during the pre-construction phase of the project. The TOP defines how the TBM shall be operated along the entire alignment and which additional measures are required during the advance and for the interventions. Figure 3 shows an extract from the TOP of the project.

It is expected that the hydrogeological conditions encountered during the TBM advance will be highly heterogeneous. Therefore, the TBM operational parameters collected during excavation may differ from the values assumed in the design phase (as shown in the TOP). In order to manage the impact of this variation, a Decision-Making Tree was created (Tamburri et al. 2019). This tool (Figure 4) allows the early assessment of the ground conditions encountered during the advance and the adjustment of the most appropriate operational parameters (for TBM and Slurry Separation & Treatment Plant—STP). Additionally, it can be used to identify suitable locations to carry-out interventions (i.e., safe havens) for maintenance works in advance.

The data collected during advance for the real-time ground condition assessment include:

- Real-time check of the TBM operational data
- Monitoring of the geotechnical instrumentation data
- Check of the separation treatment plant data
- Muck inspections
- Face mapping and inspections in front of the TBM
- Large-scale permeameter tests carried out using the TBM

### Definition of TBM Operation

#### *Support Pressure*

In order to prevent the risk of face instability, pressurization of the face is required over the entire northern portion of the alignment. In particular in soft grounds, face stability is ensured by applying a support pressure sufficiently high in order to compensate the in-situ pore pressure and support the soil skeleton (so-called effective support pressure)—the latter depending on the strength and on the effective weight of the soil.

In urban excavation, the support pressure at the face plays a significant role in the control of the surface settlements. Hence, the recommended pressure is the maximum between the pressure required to encsure face stability and the one required to avoid excessive surface settlements, sufficiently low in order to avoid blow-up failure. The recommended values take into account tolerances due to pressure deviations.

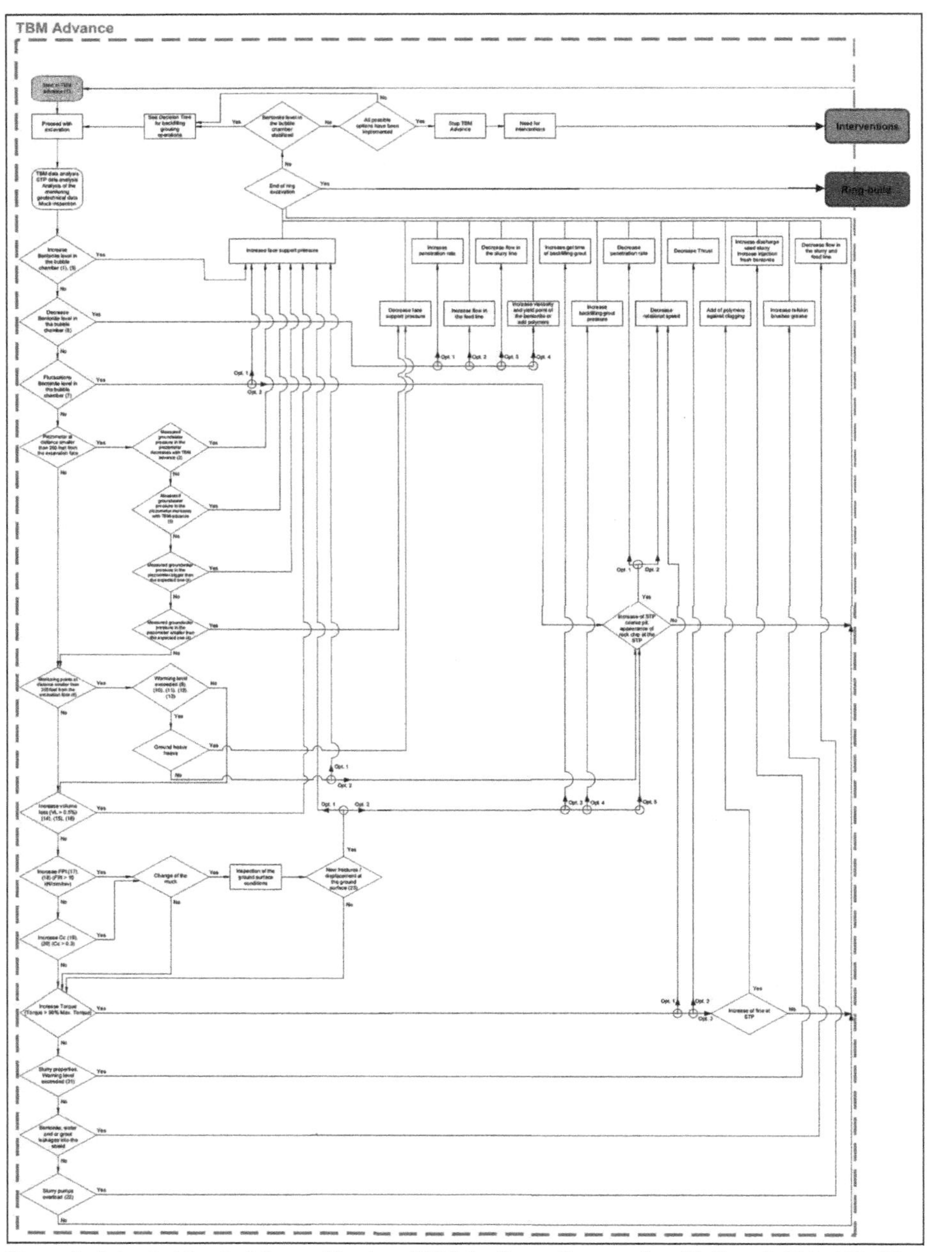


Figure 4. Extract of the decision-making tree (DMT) for the northern section (Stations 0+00 to 195+50)

For the expected ground conditions over the northern portion of the tunnel alignment, the recommended support pressure during advance varies between 1.5 and 3.2 bar as indicated in the TOP.

Depending on the encountered geotechnical conditions, the support pressure may be increased or decreased with respect to the design support pressure (i.e., the recommended pressure provided in the TOP) during tunneling operations.

Support pressure higher than the recommended support pressure will be required in order to avoid instabilities and inadmissible settlements in case of:

- In situ undisturbed groundwater pressure higher than the one assumed in the design;
- Groundwater excess pore pressure induced by the excavation process during TBM advance (Broere 2001);
- Presence of zones at worse conditions than the ones assumed in the design.

The piezometer measurements and bentonite levels in the working chamber will provide an indication of the groundwater pressure. The monitoring of the ground surface, structures and utilities, TBM parameters (e.g., high torque, low penetration rate, and high over-excavation) and STP data will indicate the presence of conditions differing from the ones assumed in the design.

### *Quality of the Bentonite Slurry*

In a slurry shield the bentonite is used to ensure the micro-stability of the soil grains and to seal the tunnel face in order to transfer the support pressure onto the soil skeleton. To seal the tunnel face the bentonite slurry has to infiltrate the soil and form a filter-cake. The bentonite slurry will be tested at each excavation ring. Large penetration of the bentonite into the soil may cause the instability of tunnel face. This hazard will be indicated by decreasing bentonite in the babble chamber and/or estimates of loss of bentonite at the STP. This risk is mitigated by modifying the slurry mix design. The used bentonite will be adjusted by discharging part of it and by adding fresh bentonite or water in order to fulfill the target values of the bentonite slurry parameters.

In case of tunneling through unexpected very coarse soils, it may not be feasible to guaranty the minimum required stagnation gradients for stability. In this case, the risk of instability will be mitigated by decreasing the duration of standstills and by increasing the TBM advance rate.

### *Measures in Case of Boulders*

The presence of boulders is usually indicated by the presence of rock chips or the increase of coarse material discharged at the STP, the sudden decrease of the penetration rate or the increase of the torque. Smaller boulders are usually not critical. They pass through the cutterhead openings and are evacuated by the slurry line. However, smaller boulders may block the crusher or the bubble gate.

Larger boulders are usually not critical if the ground is firm enough to keep their position in front of the cutterhead and progressively be broken by the disc cutters. Larger boulders are critical if the soil matrix is weak as the boulders may be dislodged and moved: they may either be pushed radially outward by the rotary action of the cutterhead and beyond the tunnel periphery or may remain in the face area and eventually block further progress of the machine. This scenario is critical as the boulders may cause:

- An increase of the volume loss and so of the surface settlements. It occurs because the boulders tend to displace the soil creating voids above and in front of the tunnel axis;
- An increase of wear of the cutting tools and/or the cutterhead structure (strength contrast/asymmetric loading of the cutters and dynamic loads during cutterhead rotation).

The usual recommended action when boulders are encountered is to reduce the thrust applied on the cutterhead (and front shield) and, eventually, increase the cutterhead rotation speed, in order to avoid damage to the cutterhead and cutters.

In the event of blocked crusher and/or bubble gate or in case of large boulders trapped in front of the cutterhead, compressed air interventions for maintenance are required. Entering the cutterhead chamber is subject to the same conditions as for other hyperbaric interventions. The boulders shalle be broken up working outside the cutterhead. This is usually achieved by use od hydraulically-powered rock splitters (electrical, hydraulic and pneumatic power plugs are available on the front shield bulkhead); alternately, expanding mortar can be used as a means of hydrostatic fracturing.

***Auxiliary measures for interventions***

Interventions under atmospheric conditions are only possible within safe havens. Interventions under hyperbaric conditions are required outside the safe havens or within the safe havens if the water inflows and the risk of local instabilities is beyond acceptable under atmospheric conditions. In highly permeable zones, pre-excavation grouting may be required in order to reduce the loss of pressurized air during cutterhead inspections and interventions under hyperbaric conditions. The TOP indicates along the tunnel alignment where the need for auxiliary measures is unlikely, possible or probable.

In case of interventions within the safe havens, the following approach is recommended:

- Excavation until the target section (such that at least two rings are located inside the safe heaven and that the grouted portion ahead of the tunnel face is larger than the minimum required thickness).
- Execution of large scale piezometer tests (to be performed prior to any planned inspection or intervention activity);
- When the extrapolated water inflow under atmospheric conditions is lower than 50 m3/h, the excavation and working chamber will be totally depressurized and inspection of the tunnel face and of the cutterhead will be carried-out;
- When the water inflow under atmospheric conditions extrapolated from the water test is higher than the assumed limit value of 50 m3/h: Option 1—re-pressurization of the excavation chamber, execution of the secondary backfill grouting of the first two rings installed inside the grouted body (to ensure the sealing of the annular void between the segmental lining and the grouted body), and execution of a new large scale piezometer test; Option 2—execution of the cutterhead inspection and interventions under hyperbaric conditions;

- When interventions will be executed under hyperbaric conditions, fresh bentonite shall be injected into the excavation chamber to ensure formation of an appropriate filter cake before entering the excavation chamber; afterwards slurry level will be gradually reduced to expose cutterhead and tunnel face;
- When regular interventions are required, the interventions will be executed without auxiliary measures and with a systematic monitoring of the water inflows, air pressure, and face conditions. In case of raveling of the face or unexpected phenomena, the interventions shall be stopped and the chamber pressurized again. In case of water inflows bigger than the assumed limit value of 50 m3/h: Option 1—re-pressurization of the excavation chamber, execution of the secondary backfill grouting of the first two rings installed inside the grouted body and execution of a new large scale piezometer test; Option 2—execution of the cutterhead inspection and interventions under hyperbaric conditions.

In case of interventions outside the safe havens, the following approach is recommended:

- The inspection of the cutterhead and interventions will be executed under hyperbaric conditions; fresh bentonite will be injected into the excavation chamber to ensure formation of an appropriate filter cake before entering the excavation chamber; afterwards slurry level will be gradually reduced to expose cutterhead and tunnel face;
- When regular interventions are required, the interventions will be executed without auxiliary measures and with a systematic monitoring of the water inflows, air pressure, and face conditions;
- When regular interventions are required in a high permeability zone and the intervention cannot be delayed, pre-excavation grouting in advance remains the only option for reducing loss of air and performing interventions;
- When special interventions are required (i.e., involving activities in front of the cutterhead), and the intervention cannot be delayed, pre-excavation grouting may be required in order to reduce the risk of face instabilities.

## CONCLUSION

The main challenges of the JWPCP Effluent Outfall Tunnel Project are related to the extremely variable conditions expected along the alignment. The ground conditions are expected to be highly challenging below groundwater level with approximately 53% of the alignment through heterogeneous soils at low overburden (northern portion), 5% through very weak rocks and 42% through highly fractured rock mass at high overburden (southern portion). Along the northern section, the main risks are related to the occurrence of face instability, blow up failure and large surface settlements. A slurry Mixshield TBM has been selected for the excavation and it is expected to be operated over the northern portion with a face support pressure varying between 1.5 and 3.2 bar, as indicated in the Tunneling Operational Plan (TOP) of the project.

A Decision-Making Tree has been elaborated to cope with the highly heterogenous conditions expected during advance. Through the early assessment of the geological conditions, this tool allows a timely reaction to any deviation of the encountered conditions from the ones assumed in the design. Therefore, the operational parameters of the TBM and of the Slurry Separation & Treatment Plant can be adjusted accordingly during advance.

## REFERENCES

Anagnostou, G., Schuerch, R., Perazzelli, P., Vrakas, A., Maspoli, P. & Poggiati, R. 2017. Tunnel face stability and tunneling induced settlements under transient conditions. Eidgenössisches Departement für Umwelt, Verkehr, Energie und Kommunikation UVEK, Bundesamt für Strassen.

Broere, W. 2001. Tunnel face stability & new CPT applications. PhD Thesis Technische Universiteit Delft.

JWPCP Effluent Outfall Tunnel Project, Geotechnical Data report (GDR), May 2017. Prepared by Fugro USA Land for the Sanitation Districts of Los Angeles County.

JWPCP Effluent Outfall Tunnel Project, Geotechnical Base Line report (GBR), May 2018. County Sanitation District No. 2 Los Angeles County California.

Schuerch, R. 2016. On the delayed failure of geotechnical structures in low permeability ground. PhD Thesis ETH No. 23681.

Tamburri, E., Pizzarotti, L., Perazzelli, P., Schuerch, R. and Moranda, G. 2019. The Three Rivers Protection and Overflow Reduction Tunnel (3RPORT)—Decision-making during construction. Rapid Excavation and Tunneling Conference, Chicago (IL), USA, pp 703–714.

# Second Narrows Water Supply Tunnel—Conventional Deep Shaft Excavation in Variable Weak Rock

**J. Andrew McGlenn** ▪ Delve Underground
**Bruce Downing** ▪ WSP Golder
**Murray Gant** ▪ Metro Vancouver
**Brian McInnes** ▪ Traylor AECON General Partnership

## ABSTRACT

The Second Narrows Water Supply Tunnel crossing beneath Burrard Inlet is an important component of the Greater Vancouver Water District's plans to increase seismic resilience and meet increasing water demands. The project included the 110-meter-deep (350 ft) South TBM Receival Shaft, one of the deepest shafts to date in the sedimentary rock unit underlying much of Metro Vancouver. The shaft encountered varied sedimentary rock types, including fractured rock, with accompanying high groundwater inflows. This paper discusses the design and construction challenges associated with this shaft and the necessary means to complete the shaft in time for TBM arrival.

## INTRODUCTION

Metro Vancouver, through the Greater Vancouver Water District (GVWD), provides a reliable supply of safe, high-quality drinking water to approximately 2.6 million people in its eighteen member municipalities, one electoral area, and one treaty First Nation. GVWD's Second Narrows Water Supply project will improve seismic resiliency of the district's water supply system, meet the future demand of Metro Vancouver's continued population growth, and provide long-term scour protection.

The Second Narrows Water Supply Tunnel will replace three of Metro Vancouver's existing water mains between North Vancouver and Burnaby ("the Second Narrows Crossing"). These are Capilano Main No. 7 (1,220 mm ID [48 in.]), Seymour Main No. 2 (1,370 mm ID [54 in.]), and Seymour Main No. 5 (1,980 mm ID [78 in.]). These water mains are linked by a system of underground valve chambers and large-diameter pipe connections in Beach Yard to three existing submarine pipelines that cross the Burrard Inlet (Second Narrows Crossing Nos. 1, 2, and 3). The three submarine crossings are buried in a shallow trench, passing beneath Burrard Inlet within a relatively narrow corridor, and come up under the Canadian Pacific Railway (CP Rail) tracks along the south shore of Burrard Inlet in the City of Burnaby. Previous seismic vulnerability studies found that the existing crossings may fail during a moderate earthquake. Once the Second Narrows project is complete, drinking water will be conveyed from the Seymour and Capilano watersheds on the north shore, under Burrard Inlet, to Burnaby and other municipalities south of the inlet.

Delve Underground is Engineer of Record (EoR) for all underground structures and prime consultant for the design team, along with our two main subconsultants, WSP-Golder (geotechnical engineering) and AECOM (surface works, mechanical and electrical engineering). Procurement of the tunnel contractor commenced in late 2017, and the contract was awarded to Traylor Aecon General Partnership (TAGP) in late

2018. Site mobilization commenced at the North Shaft site in January 2019. Since then, the EoR team has been providing on-site engineering and inspection services.

The approximately 1,100-meter-long (3,610 ft) tunnel was constructed with a pressurized tunnel boring machine (TBM). The project includes two deep shafts: the 72-meter-deep (236 ft), 16-meter-diameter (52.5 ft) North Shaft, which served as a launching shaft for the TBM and pipe installation, and the 110-meter-deep (350 ft) 10-meter-diameter (33 ft) South Shaft, which served as the receiving shaft for the TBM. The latter is one of the deepest shafts to date in the sedimentary rock unit underlying much of Metro Vancouver and is the focus of this paper. Its excavation encountered varied sedimentary rock types including fractured rock, which is likely associated with a fault underlying Burrard Inlet. Design and construction challenges associated with this shaft and the means necessary to complete the shaft in time for TBM arrival are discussed herein. A broader discussion of the project is provided in the 2021 RETC paper "Second Narrows Water Supply Tunnel—Final design, procurement and construction."

## PROJECT OVERVIEW

The project configuration consists of a 1,100-meter-long (3,610 ft), 6.5-meter (21.3 ft) outside diameter (OD) bored and segmentally lined tunnel mined between two deep vertical shafts. Two new valve chambers will be constructed over or adjacent to each shaft collar to control water flows through the three pipes in the new tunnel. Two of the steel pipes will be 2.4 meters (m) (8 ft) in diameter and one will be 1.5 m (5 ft) in diameter. Figure 1 shows the cross section of the tunnel with the pipe configuration for the three steel water mains. Figure 2 provides an illustration of the project vertical alignment.

The key features of the project are shown in Figure 3.

### South Shaft Design

The 110 m deep (360 ft), 10 m diameter (33 ft) South Shaft served as the receiving shaft for the TBM. It is located in Second Narrows Park in the City of Burnaby and was excavated through a series of interbedded sandstones, claystones, and siltstones, with minor conglomerate, coal,

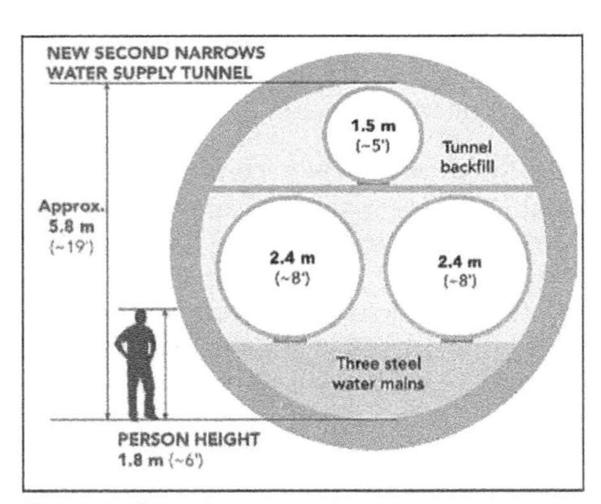


Figure 1. Tunnel cross section

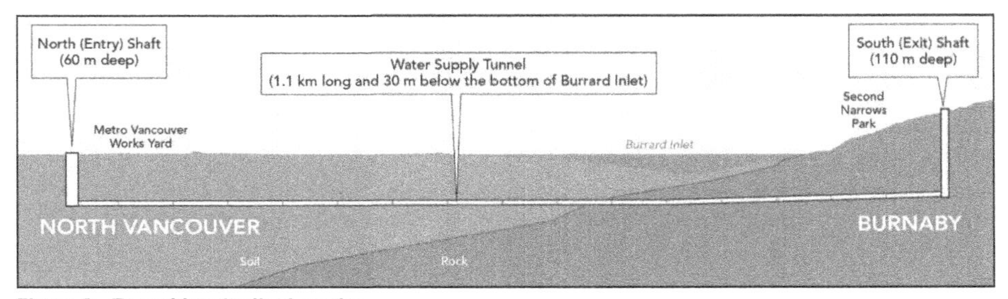


Figure 2. Tunnel longitudinal section

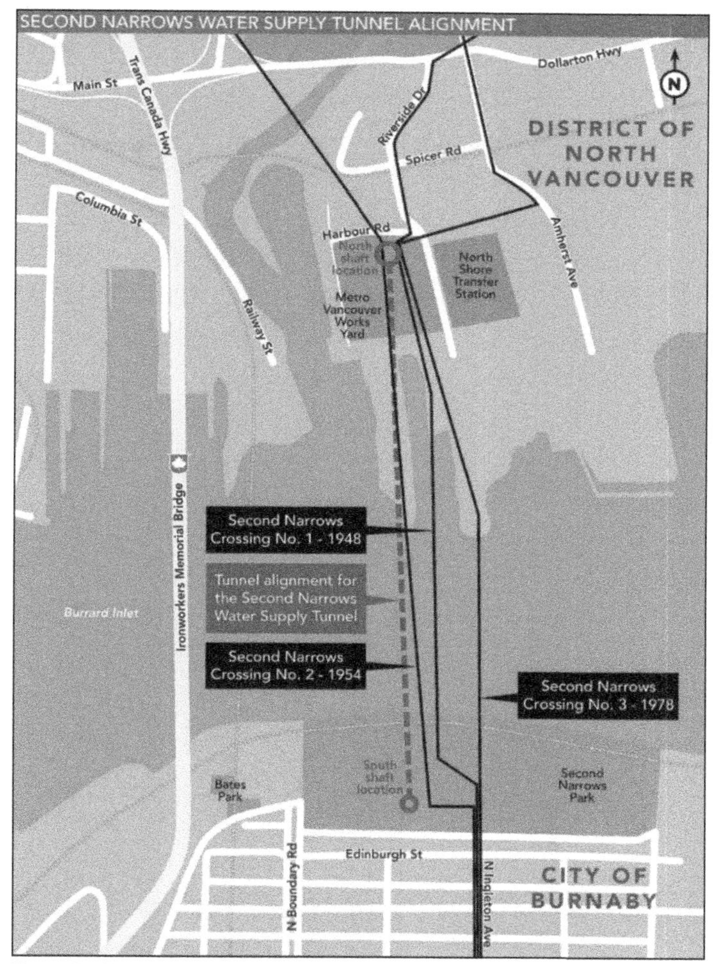


Figure 3. General plan of project alignment

and calcareous sandstone. Conventional mechanized top-down excavation methods were used with support provided by steel lattice girders and wide flange beams followed by shotcrete. The three pipes are due to be installed in the shaft, and the remaining space around them will then be backfilled with lean mix concrete.

The South Valve Chamber is an independent structure located near the South Shaft, which houses the necessary mechanical and electrical equipment for the system operations. It was constructed through a series of overburden soil and conglomerate, similar to what was found at the top of the South Shaft. Temporary excavation support is necessary to construct the South Valve Chamber.

Work on the South Shaft began in spring 2019 with the construction of a soil nail wall along the southern site boundary to create a level area of sufficient size to construct the shaft support of excavation (SOE). Excavation and lining installation progressed steadily through the upper conglomerate unit and then into the mudstone and sandstone sequences. The SOE consists of shotcrete encased steel lattice girder

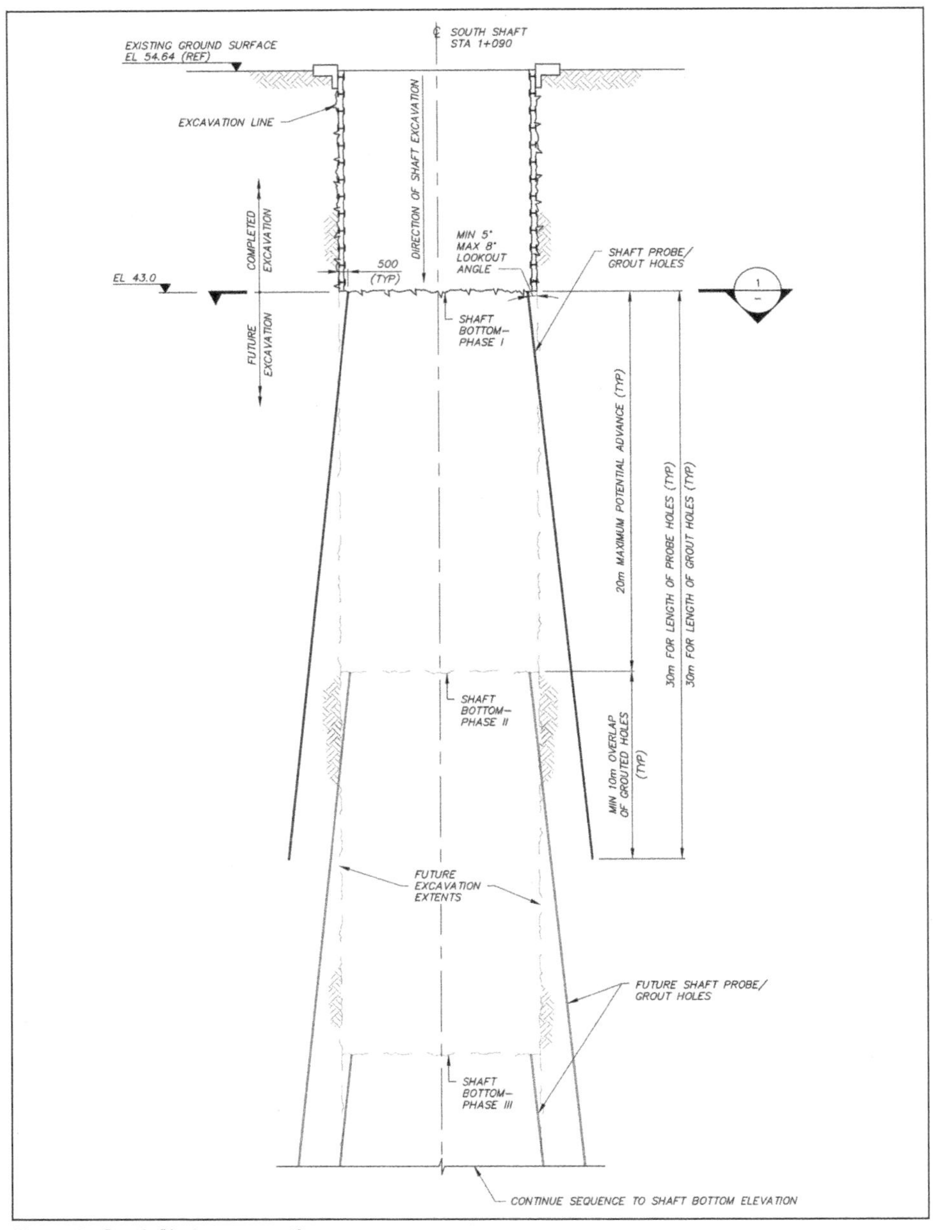


**Figure 4. South Shaft cross section**

ring beams spaced at 1 m centers vertically. The lattice girders transition to steel wide flange beams approximately 70 m (230 ft) down due to higher ground loads. Pre-excavation grouting was performed every 20 m (66 ft) of excavation to help control groundwater inflows. The fifth and last ring of pre-excavation grouting was performed at a depth of approximately 86 m (282 ft) in February 2021. Shaft excavation and lining installation was completed in spring 2021.

## SUBSURFACE CONDITIONS

### Geology

The project is located in an area of geological complexity with the site underlain by bedrock of varying composition (intrusive crystalline bedrock to the north of the site and sedimentary bedrock to the south of the site) and surficial soil deposits of glacial, marine, and fluvial origin. The bedrock at the site has been affected by faulting, which has been identified in several of the boreholes drilled.

During the Tertiary age (starting about 80 million years ago), the Coast Crystalline bedrock that forms the present North Shore Mountains was eroded, resulting in the deposition to the south of a sequence of sedimentary and organic materials into a depression known as the Georgia Basin. These sediments became lithified into a sedimentary sequence of sandstone, mudstone, shale, and coal, referred to as the Huntingdon Formation, of which the Kitsilano Member is a part. The Kitsilano Member consists of interbedded sandstone, mudstone, and conglomerate with occasional thin coal layers, and calcareous cemented beds or lenses. The rocks are typical of the estuarine/floodplain depositional environment described above. The variability in the bedrock lithology and in situ properties is indicative of the complex depositional environment. An east–west trending fault in Burrard Inlet has been inferred by geologists since the early 1900s. This was based on the spatial relations of the sedimentary rocks making up the Huntingdon Formation encountered south of Burrard Inlet and the granitic rocks that underlie the southernmost Coast Mountains north of Burrard Inlet, and the inferred relative displacements between the two formations. The presence of faulted bedrock was confirmed in the site investigations where several zones of sheared and broken rock were encountered. The fault is not active.

### Ground Conditions at the South Shaft

Four rock units within the Middle and Lower Units of the Kitsilano Member of the Huntingdon Formation were encountered in the South Shaft (Figure 5).

- Unit 6A (0–22 m depth): Conglomerate. The conglomerate is poorly graded and mainly clast supported in an extremely weak to weak lightly cemented sandstone matrix. The gravel to cobble-sized clasts are subrounded to rounded and primarily igneous in origin. The clasts are typically strong to very strong (International Society of Rock Mechanics [ISRM] R4 to R5) while the rock as a whole is very weak to weak (ISRM R1 to R2).
- Unit 6B (22–47 m depth): Mudstone (with minor Sandstone Layers). The mudstone is typically brown-gray to green-gray, laminated to medium bedded, very fine grained, very weak (R1) to weak (R2). Rock quality was typically good, with rock quality designation (RQD) typically between 90 and 100%. Minor Sandstone Layers were encountered in this unit. There was a notable layer of coal and highly weathered mudstone from 38.2 m to 39.3 m depth; a loss of circulation occurred at this location during drilling of the exploratory borehole at the shaft.
- Unit 6C (47–80 m depth): Sandstone/Mudstone sequence. This rock is typically brown-gray to green-gray, thickly bedded, fine to very coarse grained, very weak (R1) to weak (R2) argillaceous sandstone intermixed or grading to very fine grained, weak (R2) Mudstone. The rock strength is more uniform than Unit 6B. Rock quality was typically good, with RQD typically between 90 to 100%.

- Unit 6D: (80 m—bottom of shaft): Sandstone (with minor Mudstone Layers). This rock is typically brown-gray, very fine to very coarse grained, weak (R2) sandstone. Generally, rock strength increases with depth. A fault was encountered between 88.1 and 90.3 m. Brecciated and slickensided features were noted between 89.2 m and 90.1 m. Circulation loss occurred during drilling between 88.1 and 90.3 m depth.

On the south side of Burrard Inlet, groundwater generally flows in a south to north direction. The water table in the bedrock in which the South Shaft is located is above the water level in Burrard Inlet. Groundwater flow is generally downward through the bedrock with discharge to Burrard Inlet. Tidal effects are negligible in the bedrock; small variations in seasonal water levels have been observed. Figure 5 shows the groundwater conditions at the South Shaft. Perched conditions were encountered in the upper 38 m of the shaft, above the permeable coal layer. Pre-excavation packer testing, through a single vertical borehole at the center of the shaft, indicated that high inflows could also be expected in association with the fault at 88 m depth. The vertical borehole was found to have not encountered a small number of steeply inclined and northerly dipping open subvertical joints, which produced larger and more distributed flows, as discussed below in "Groundwater Control and Grouting."

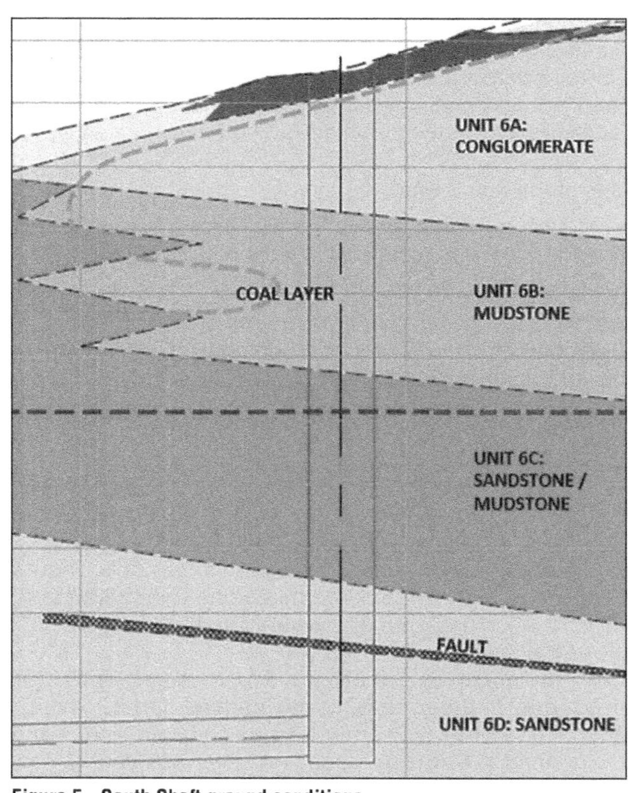


Figure 5. South Shaft ground conditions

## DESIGN CONSIDERATIONS

### South Shaft

Four construction methods for the South Shaft were evaluated during detailed design: (1) conventional top-down with an excavator, (2) modified conventional (with the addition of a muck hole so that muck can be hauled away from the bottom of the shaft), (3) raise bore, and (4) blind bore. In the end, the conventional shaft excavation method was selected because it offers the most flexibility in terms of the excavation methods, support options, and construction sequencing. While this method is relatively inefficient, in terms of excavation rates and volume of muck requiring removal, it can be used to excavate through all the anticipated units, including overburden, conglomerate, and other sedimentary bedrock.

A significant consideration in the construction of the South Shaft and South Valve Chamber is their location within and next to the GVWD right-of-way in Second Narrows Park, directly adjacent to a residential neighborhood. A site plan of the area is provided in Figure 5. GVWD and the design team worked closely during detailed design to interface with the City of Burnaby, knowing that construction could create a significant disruption to the adjacent residents and public who frequently use the park.

## CONSTRUCTION

### Overview and Excavation Methods

#### *Rationale for Means and Methods Selection*

The South Shaft has essentially three purposes: (1) facilitate removal of the TBM, (2) provide secondary means of access to the tunnel after boring is completed, and (3) provide a conduit to install and backfill the shaft pipes. The same methods considered during the design phase were evaluated for actual construction of the shaft. Note that blasting was not permitted per the contract documents. Essential variables that were considered when choosing a method included impact to the critical path, suitability of the rock, water management, size of site, spoil removal rate, and practicality of pipe installation.

Both raise boring and blind boring methods would prevent removal of the TBM, require a larger on-site footprint, trigger additional community impacts pertaining to noise and high-volume spoil removal, and impact the constructability of the shaft pipe installation. In addition, raise boring impacted the critical path of the project.

Ultimately, top-down excavation with an excavator was chosen as the preferred method of excavation because of the following:

- There was no impact to the critical path schedule (excavation could occur during tunneling).
- Baselined rock strength was suitable for conventional excavation.
- Water inflows could be managed using conventional methods.
- The TBM could be retrieved through the shaft.
- It facilitated shaft pipe installation.
- It reduced noise impacts to surrounding residents.
- Spoil removal rate was more easily managed.

Figure 6. South Shaft site located within Second Narrows Park

### *Location and Site Setup*

The site is located in Second Narrows Park (Figure 6), which is a forested urban area in Burnaby, BC. There is a walking path (Trans Canada Trail) and a shortcut through the site, both of which are regularly used by local residents. The area slopes toward the Burrard Inlet and is within close proximity to many homes that overlook the park. The closest homes were approximately 15 m to the edge of site and 50 m to the shaft.

The site is narrow but long from east to west. As shown in Figure 7, a 10-m high soil nail and shotcrete retaining wall was installed on the side of the hill to provide a level working area. Some of the material excavated was stockpiled on the east end of the site to be used for backfill during site restoration. The actual working area is around 40 m wide by 100 m long. Key components of the site, as shown in Figure 7, include a muck storage area, water treatment plant with ponds, wheel wash for exiting trucks, a shotcrete plant setup area, and ventilation setup on the west side of the shaft. The crane was able to reach the middle of the shaft and muck pit area to hoist and dump a 4.5 $m^3$ muck box. There was no room for site trailers, lunchrooms, and washrooms so they were positioned on the access road, which is the only way in and out of site. As such, the road only facilitates one-way traffic for deliveries, muck haulage, and concrete deliveries.

## Equipment and Sequence

This section describes steps required for excavating and shoring the soil overburden and rock shaft, as well as the pre-excavation grouting methodology.

### *Excavation and Support of Overburden Soil*

The 4 meters of overburden was supported with four flanged, seven-gauge liner plates with full-diameter W150x36 steel ribs. The liner plate system was designed to handle the loading imposed by the ground conditions, as well as a surcharge load anticipated for TBM removal. The entire 4 m excavation was completed from the surface with a 36-ton excavator, and the liner plates were preassembled on the surface, installed as one unit, and backfilled with a low-strength concrete.

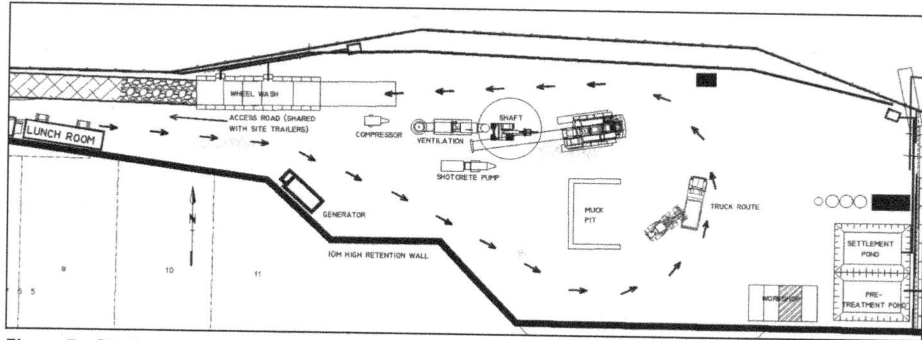


Figure 7. Shaft excavation site setup

### *Excavation and Support of Rock*

Upon completion of the liner plate support, a 14-ton excavator was used to excavate the shaft (Figure 8). The excavator was equipped with a digging bucket, hydraulic hammer, or rotating cutterhead attachment, depending on the strength of rock. Once enough material was broken, the excavator loaded a 4.5 $m^3$ skip bucket that was hoisted out by the crane into the muck bin for truck haulage. The sequence of excavation generally involved working in a rotating pattern because of the tight confinement within the shaft, as shown in Figure 9, with a spotter outside the equipment managing pump hoses and cables, and rigging the muck bucket. Typically, excavation and support occurred in 1 m lifts.

Figure 8. Excavation of the South Shaft

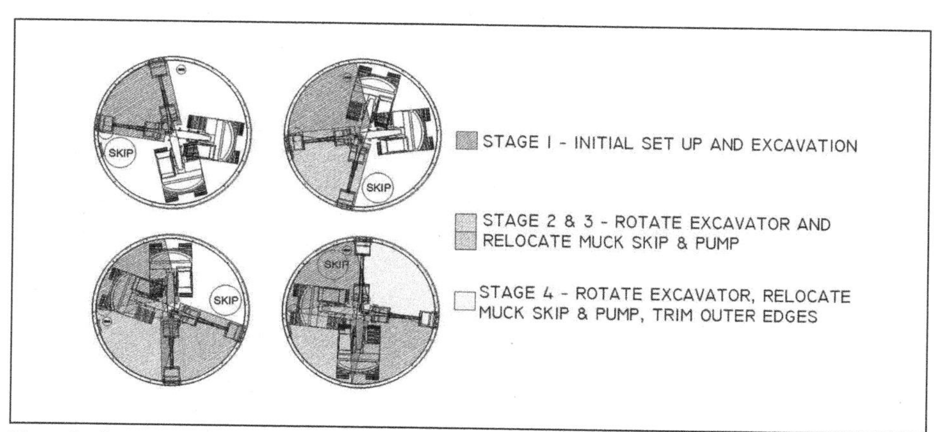


Figure 9. Excavation sequencing with 14-ton excavator

The rock support consisted of a two-stage shotcrete system. Once excavation was completed, a 25 mm thick flash coat of high early strength shotcrete provided preliminary support. The next day, a six-piece ring beam or lattice girder was bolted together in the middle of the lift to form a full-in beam (see Figure 10 for ring-beam installation). The final thickness of the shotcrete was 150 mm, with a minimum 25 mm cover over the ring beam. The ring beam was originally specified as a W200x52 beam; however, the contractor proposed using CP230 lattice girders to prevent shadowing and ensure full shotcrete encapsulation of the outside face of the beam. Lattice girders were used for the first 74 m and W-beams were used for the bottom 30 m to do the anticipated higher load expected on the ring beam.

Wet shotcrete was delivered by trucks and lowered in by concrete buckets to the shotcrete pump. In areas of weak rock, fault zones, and mudstone, shotcrete was applied to the rock no later than 3 hours after exposure.

The typical excavation and shoring sequence was as follows:

1. Excavate 1 vertical meter with the excavator and drilling attachment.
2. Install dimple board/mira drain vertically against the rock wall.
3. Apply a flash coat of shotcrete (approximately 25 mm) to the newly excavated wall face.
4. Apply a final coat of shotcrete (approximately 125 mm) to the previous day's lift, fully encapsulating the ring beam by 25 mm.
5. Install drain holes and polyvinyl chloride (PVC) pipes every other lift at 2 m × 2 m spacing.
6. Install a ring beam around the newly excavated lift.

Wet shotcrete was delivered by trucks and delivered to the shotcrete machine either on surface or at the base of the shaft. While the shaft was still relatively shallow, shotcrete could be pumped down a 50 mm slick line to the nozzle held by the operator. However, as the depth of excavation exceeded approximately 60 m, this process became infeasible because of excessive clogging. The shotcrete pump was then located at the base of the shaft and was loaded by crane-hoisted concrete buckets. This resulted in slower shotcrete durations. In areas of weak rock, fault zones, and mudstone, it was believed that shotcrete would need to be applied within several hours of excavation. A dry mix shotcrete machine was on standby if the excavation face degraded. It would allow quick application of shotcrete if the face stability began to deteriorate. However, the ground was relatively stable in all ground conditions encountered and this face saving method was never required.

**Figure 10. Ring beam installation (left) and shotcrete application (right)**

The contract also specified the installation of 4 m long #8 rock dowels in areas of sub-horizontal bedding planes and fault zones or extremely weak rock layers. However, the ground proved to be sufficiently self-supporting and never required rock dowels. If layers of weak rock were observed, the general approach was to avoid leaving a vertical face exposed for extended periods of time, such as over a weekend. In these cases, a 1 vertical × 1 horizontal berm was left unexcavated over an extended period and only excavated prior to a flash coat.

Water infiltration was controlled with high head submersible pumps at the base of the shaft, which pumps shaft water into the surface ponds for treatment and discharge. A primary pump and secondary pump were set up in a steel sump and equipped with floats to activate the pump when the water level reached a certain height. Small sump pumps were set up around the shaft to transfer water into the main steel sump. Pump maintenance became a critical activity due to the harsh environment and laden material being pumped. Often, the synthetic fibers would pass through the filters and damage the pumps, causing delays and requiring regular maintenance. A backup generator was also required to ensure that the pumps would remain in operation following a power outage.

## Construction Issues and Production Rates (Performance During Construction)

### *Construction Issues (Challenges During Excavation)*

- Excavation rates were sensitive to ground type and took longer than anticipated.
- There were higher groundwater flows than expected.
- Equipment maintenance: Due to the harsh excavation conditions, a backup excavator and several backup pumps were required.
- Delivery of shotcrete to the bottom of the shaft was slow.
- The relatively small size of the shaft made it very difficult to switch between excavation, ring beam installation, and shotcreting as equipment needed to be removed and reinstalled regularly.

### *Production Rates*

Daily production was largely based on the ground conditions. There were generally three general classifications of rock encountered during excavation (Table 1).

The conglomerate was encountered directly below the overburden and was much slower to excavate than expected. Most of the excavation required a hydrsaulic hammer on the boom of the excavator to break it free. The duration to excavate took anywhere from four to six shifts to excavate a 1 m lift.

**Table 1. Rock type versus excavation rates**

| Rock type | Description | Average Duration to Excavate a 1 m Lift |
|---|---|---|
| Conglomerate | This material was hard, stable, and competent ground with little risk of ground loss and face instability. | 5 shifts |
| Sandstone | This material was hard, stable and competent ground with little risk of ground loss and face instability. | 2 to 3 shifts |
| Mudstone | This material was weak but stable and competent. | 1 shift |

The sandstone varied in rock strength and also required the use of a hydraulic hammer to excavate. In areas of weaker sandstone, the excavation took two shifts, whereas areas of harder sandstone took up to five shifts to complete excavation.

The mudstone was the weakest of the material encountered and it could typically be excavated with the bucket.

After the excavation was completed, the application of the shotcrete flash coat and installation of the ring beam would take about one-half to one shift. The final layer of shotcrete would take approximately one-half to one shift to complete, depending on the depth of the excavation.

## Groundwater Control and Grouting

While most of the bedrock in Unit 6 has low permeability, zones of higher groundwater inflows were anticipated, and pre-excavation grouting was specified. Pre-excavation cover grouting, carried out in five stages over the length of the South Shaft, was required under the contract documents as a means to manage groundwater inflows to the shaft and mitigate potential stability issues caused by groundwater inflows in areas of poor rock (Figure 11).

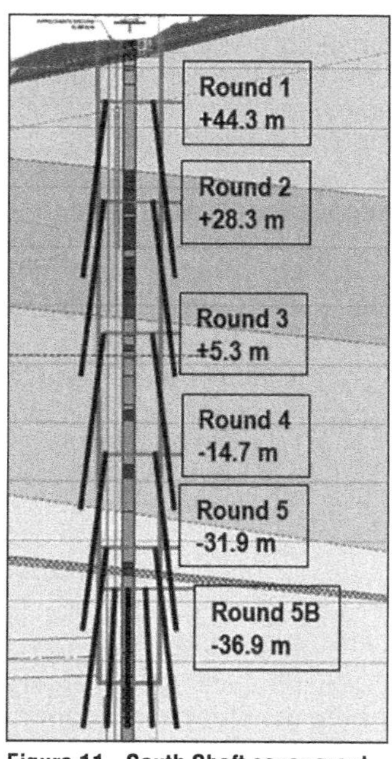


Figure 11. South Shaft cover grouting layout

Initially, the South Shaft encountered low groundwater inflows similar to expectations; however, starting at Lift 77 (El –25.7 m to –26.7 m), groundwater inflows increased markedly. Geotechnical mapping has shown that the increases in inflow are primarily related to inflows through open subvertical joints dipping to the north. TAGP began to measure the shaft inflows with the installation of a dedicated flowmeter for the shaft on November 19, 2020 (during excavation of Lift 82). The increased groundwater inflows were observed to be above the baselined inflows in the Geotechnical Baseline Report (GBR) and led to the issuance of a notice of a potential Differing Site Condition (DSC) from TAGP on November 20, 2020.

The unanticipated high inflows resulted from inadequate grouting in Ring 4, as observed with the open flowing joints visible in the walls of the excavation below El –27 m. As a result, pre-excavation grouting effort was increased for Round 5, compared to Round 4. The main changes that were incorporated into the Ring 5 pre-excavation grouting are as follows (see also Figure 12 for hole layout of Rings 5 and 5B):

- All drilled holes were pressure grouted. For Ring 4 and above, only those holes that met inflow criteria were pressure grouted. All other holes were backfill grouted (lower pressure).
- Additional secondary and tertiary holes were required depending on performance during primary and secondary grouting.
- Enhanced thickening sequence and Celbex 653 admixture were used in the final stage of grout thickening.
- A maximum grout volume cap was placed on each grout hole because of the persistence and aperture of the open joints.

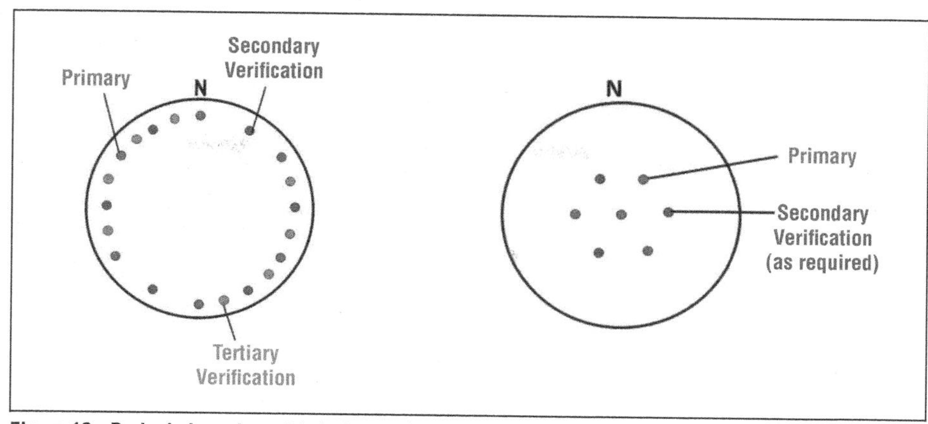


Figure 12. Probe hole and grout hole layout: Ring 5 (left) and Ring 5B (right)

The fifth stage of cover grouting (Round 5) was initiated on November 25, 2020. The Round 5 shaft invert elevation (El −31.9) was within the zone where numerous locations of groundwater inflow were observed. In particular, groundwater was flowing into the shaft floor through two subvertical joints that crossed the shaft floor in an east–west orientation. These shaft inflows created challenges to grouting, namely the potential for thinning and washout of injected grout (depending on the location of the grout hole with respect to the flowing features).

Figure 13. Grout infilled joints visible on shaft walls

As anticipated, substantial amounts of grout were injected for Round 5. Some issues were encountered with communication between grout holes and washout of grout to the shaft floor; however, the conditions improved as the grouting moved through the closure sequence from primary to tertiary holes. A total of 20 holes were grouted, with about 170 $m^3$ of grout injected, compared to 16 holes and 60 $m^3$ of grout injected in Ring 4. The grouting was highly effective for Ring 5, as there was very little additional groundwater inflow below El −34 m; in other words, the entire zone covered by Ring 5. Figure 13 shows the steeply dipping joints that were fully grouted, and Figure 14 shows the high inflows from the open joints that were not adequately grouted.

## SUMMARY

The design and construction of the South Shaft, while conventional and relatively straightforward in terms of approach, encountered several challenges as excavation progressed. Mitigating impacts on the public and the environment was a particular challenge in the setting of this site within a well-used city park. This was achieved through environmental and public impact mitigation measures and public outreach.

Figure 14. High water flows through open joint (intersected by drain pipe) at El –31 m

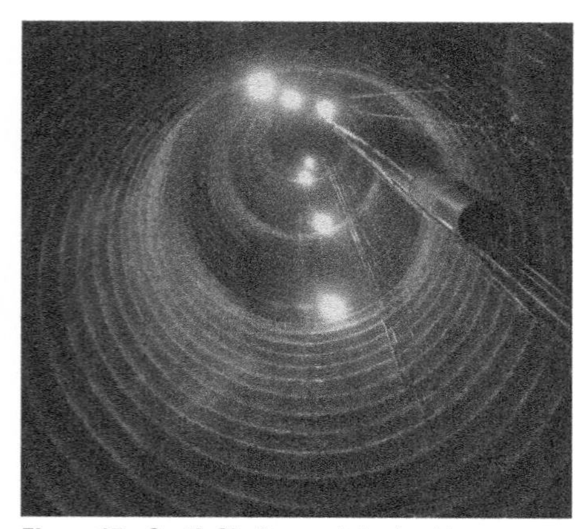

Figure 15. South Shaft complete: Looking up shaft with lattice girder and shotcrete support

Shaft excavation and support were challenging because of the high degree of variability in the ground conditions and the changes in methods needed to deal with the changing conditions. Control of groundwater inflow near the bottom of the excavation proved to be a key challenge in terms of impacts to methods and schedule while maintaining stability and safety of the excavation. The successful implementation of pre-excavation grouting was key to completing the work. The pre-excavation grouting program was adapted during construction to meet the particular challenges associated with grouting of the large, open continuous fractures that were present at the shaft location. Collaboration between Metro Vancouver, the design team and TAGP, and the ability to adapt and adjust methods as the work progressed were essential, and the shaft was completed well in advance of the arrival of the TBM. Installation of the shaft pipe is scheduled for May 2023. Backfilling will follow concurrently and sequentially.

## ACKNOWLEDGMENTS

The authors wish to acknowledge the many contributions of Metro Vancouver staff; the detailed design team of Delve Underground, Golder Associates, and AECOM; the construction management team of Mott MacDonald; and the project contractor, Traylor Aecon General Partnership.

## REFERENCE

Huber, Frank, Davidson, Gregg W,. and McGlenn J. Andrew. 2021. Second Narrows Water Supply Tunnel—Final design, procurement and construction. In *Proceedings of the 2021 Rapid Excavation and Tunneling Conference*. Littleton, CO: SME, 358–366.

PART

# Future Projects

*Chairs*

**Mo Magheri**
Kiewit

**Rebecca Reeve**
Traylor Brothers Inc

# Allegheny County Sanitary Authority—Ohio River Tunnel Project Updates

**Mike Lichte** ▪ ALCOSAN
**Shawn McWilliams** ▪ ALCOSAN
**Greg Colzani** ▪ Jacobs
**Aini Sun** ▪ Jacobs

## ABSTRACT

The Allegheny County Sanitary Authority (ALCOSAN) adopted its Clean Water Plan (CWP) in May 2020, as part of a federal consent decree to comply with the USEPA's Combined Sewer Overflow (CSO) Control Policy. Planned improvements include constructing a new regional storage/conveyance tunnel system, promoting green infrastructure/source control, and expanding the treatment plant. Preliminary planning for the tunnel system was completed in October 2020. The Ohio River Tunnel (ORT) project is the first of three tunnel projects that make up the regional tunnel system. The ORT project began final design in 2021. This paper discusses the advancements made to the ORT preliminary planning concepts considering property availability, alignment optimization, tunnel boring machine evaluation, contract packaging and delivery schedules.

## INTRODUCTION

The Allegheny County Sanitary Authority (ALCOSAN) is the regional wastewater authority serving 83 communities throughout the Greater Pittsburgh Area. The ALCOSAN system illustrated in Figure 1 includes over 90 miles of interceptors, up to 10.5 feet in diameter and 100 feet deep, built in the 1950s running mainly along the three main rivers of Pittsburgh: the Ohio, Allegheny, and Monongahela Rivers. These interceptors are serviced by numerous sewer regulator structures that send flow to the existing interceptors and overflow through existing sewer overflow outfalls for storm events that exceed the regulator capacities. The interceptor system sends flows to the Woods Run wastewater treatment plant (WWTP) via an existing main pump station located at the treatment plant. The existing WWTP has a treatment capacity of up to 250 million gallons per day (MGD).

### Clean Water Plan

ALCOSAN has adopted a Clean Water Plan (CWP) (ALCOSAN, 2019) to improve and protect the water quality of the region's streams and rivers by controlling combined sewer overflows (CSOs) and eliminating sanitary sewer overflows (SSOs). The CWP is part of a federal consent decree to comply with the United States Environmental Protection Agency's CSO control policy. The consent decree was entered in May 2020 (United States District Court for the Western District of Pennsylvania, 2020).

The first stage of the CWP involves the implementation of an Interim Measures Wet Weather Plan (IWWP). The IWWP is made up of four components: plant expansion, regionalization of multi-municipal sewers, source control and regional conveyance upgrades. The regional conveyance upgrades involve constructing a new regional

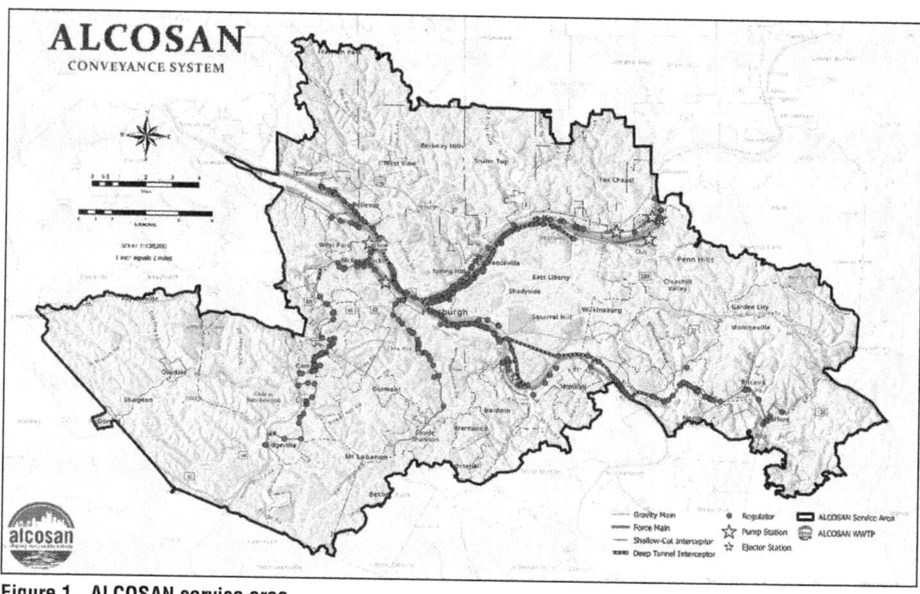


Figure 1. ALCOSAN service area

tunnel system to convey flow to the expanded WWTP via a new wet weather pump station with a capacity of 120 MGD and to provide temporary storage.

The IWWP must be implemented by December 31, 2036. It is anticipated to involve approximately two billion dollars in investment (2010 dollars). The implementation of the IWWP is anticipated to reduce the volume of wet weather overflows by approximately seven billion gallons per year by the end of 2036. Upon completion of the IWWP, post construction monitoring and modeling will be conducted to assess the need for additional controls to meet the full requirements of the consent decree.

## Regional Tunnel System

The regional tunnel system includes three main tunnel projects with supporting consolidation sewer and drop shaft structures as shown in Figure 2. The system must be constructed and operational by December 31, 2036.

The proposed facilities include the following major components:

- Ohio River Tunnel (ORT), adits, consolidation sewers, regulators, drop shafts and river crossings
- Allegheny River Tunnel (ART), adits, consolidation sewers, regulators and drop shafts
- Monongahela River Tunnel (MRT), adits, consolidation sewers, regulators and drop shafts

Combined, these improvements include 16.4 miles of main tunnel and river crossings with a storage capacity of approximately 160 MG, 30 shafts, over 40 regulators and approximately 23,000 feet of consolidation sewers ranging from 24 to 144 inches in diameter.

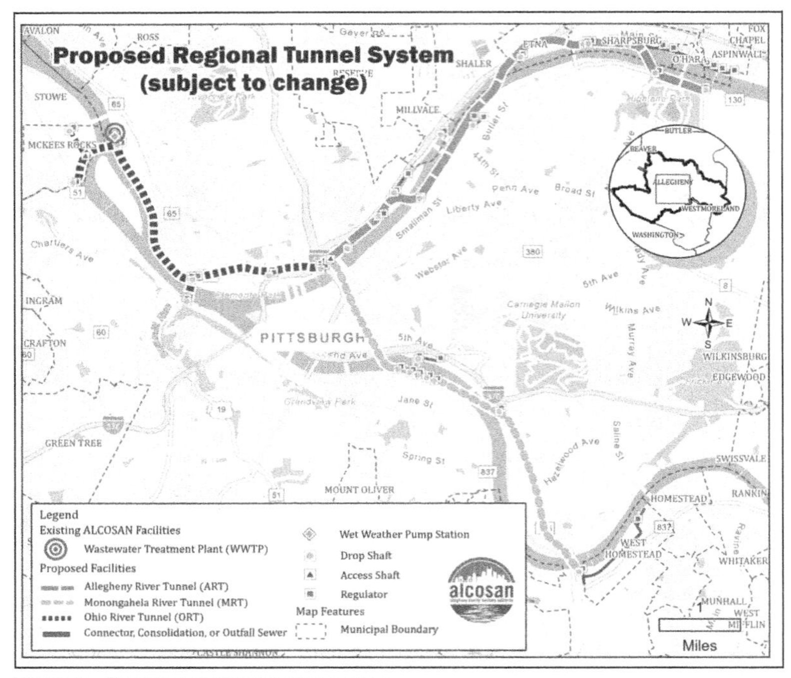


Figure 2. Proposed regional tunnel system

## ORT PROJECT COMPONENTS

The ORT project is currently at 30% design. It is anticipated to be delivered with up to six construction contracts as described below. Contract packaging is still being evaluated to minimize contractor interference and cost.

### ORT Tunnel Contract

- Tunnel boring machine (TBM) launch and retrieval shafts and access shafts (see Table 1)
- ORT main tunnel—18-foot finished diameter, 19,990 feet long, 120 to 150 feet below ground surface
- Chartiers Creek Tunnel (CCT) river crossing—14-foot finished diameter, 4,420 feet long
- Saw Mill Run Tunnel (SMRT) river crossing—14-foot finished diameter, 1,600 feet long
- Dewatering tunnel (i.e., connector tunnel to the new WWPS)
- Drop shafts and internal structures (see Table 1)
- Near surface facilities (NSF) at O27 and O07 sites, including regulators, conveyance sewers and tunnel emergency outfalls

### ORT Near Surface Facilities (NSF) Contracts

- NSF Package 1—O-06 & O-06A
- NSF Package 2—O-40 & O-41
- NSF Package 3—O-14
- NSF Package 4—A-48
- NSF Package 5—A-58

Table 1. ORT project shaft summary

| Shaft Reference | Shaft Type | Internal Shaft Radius | Minimum Concrete Lining Thickness | Support of Excavation Type |
|---|---|---|---|---|
| ORT-A48 | Drop shaft | 13'-3" | 2'-6" | Secant pile |
| ORT-A58 | Drop shaft | 11'-9" | 2'-6" | Secant pile |
| SMRT-014 | Drop shaft | 17'-6" | 3'-0" | Secant pile |
| ORT-027 | Drop shaft; TBM launch and retrieval | 25'-0" | 3'-6" | Slurry wall |
| ORT-041 | Drop shaft | 17'-6" | 3'-0" | Secant pile |
| CCT-007 | Drop shaft; TBM retrieval | 23'-0" | 3'-6" | Slurry wall |
| CCT-006A | Drop shaft | 12'-6" | 2'-6" | Secant pile |
| CCT-006 | Access shaft | 12'-6" | 2'-6" | Secant pile |
| ORT-AS-1 | Access shaft; TBM launch shaft | 25'-0" | 3'-6" | Slurry wall |
| ORT-AS-2 | TBM launch shaft (for the ART project) | 25'-0" | 3'-6" | Slurry wall |

# SUBSURFACE CONDITIONS

## Geologic Setting

The project area lies at the western edge of the Appalachian Plateau Physiographic Province. In general, the terrain is hilly and rugged except for floodplains along the Ohio, Allegheny and Monongahela Rivers which are nearly flat. The valley bottoms are relatively narrow and constrained by uplands rising abruptly on each side. Upland areas are dissected by numerous narrow, steep-sided secondary drainages. Dissected terraces occur above the floodplain of the lower Ohio River. Total relief between the valley bottoms and upland surfaces is usually less than 200 feet but is as much as 600 feet in some locations.

In general, the project area is underlain by unconsolidated deposits overlying sedimentary bedrock deposits of mid to late Pennsylvanian (approximately 285 to 300 million years before present) age. The unconsolidated deposits consist of alluvium and terrace deposits representing older alluvial (fluvioglacial) deposits related to various glacial episodes. They are overlain at many locations by recent artificial fill materials. Regionally these unconsolidated materials have been observed to range up to 150 feet in thickness.

The bedrock underlying the project area includes the lower portion of the Glenshaw Formation of the Conemaugh Group. The Conemaugh Group is reported to be 600 to 650 feet in total thickness, with approximately the lower half representing the Glenshaw Formation. Rock types within the Glenshaw Formation include complex, alternating, repetitive sequences of claystone, shale, sandstone with beds of coal and limestone known as cyclothems. These sequences were deposited in deltaic and swamp environments by meandering, low energy streams entering a shallow sea—the sequences, as related individual lenses, can be tracked for large distances.

However, individual beds are often discontinuous, interlayered and grade into one another. For example, the thick-bedded channel sandstones which may be prominent at one location may be thin or even absent at another.

The bedrock units generally dip gently to the south, from virtually horizontal to approximately 5% on the flanks of broad folds with axes that trend north to northeast. Few faults are noted in prior studies and those that were have minor displacement offsets.

Jointing within the bedrock units has been described as generally being near vertical with greater variability within the red beds. Fractures within red and gray shale units have been characterized to be frequently tight to very tight in un-weathered rock. Joints in the sandstone and limestone have been described as open and sometimes enlarged by solution activity in the limestone.

## Geotechnical Investigations

A preliminary geotechnical investigation program was undertaken to support planning and preliminary engineering for ALCOSAN's CWP. ALCOSAN had performed an initial geotechnical investigation prior to bringing the preliminary planning (PP) team on board. This initial program, referred to as Phase I, was conducted in 2017 and included drilling 12 widely spaced geotechnical borings, water pressure packer testing in rock, installation of ground water monitoring wells and piezometers and laboratory soil and rock testing. A phased geotechnical investigation program was developed and implemented by the PP team. This initial phase of the PP team's program, referred to as Phase II, was developed based on the results of the Phase I program implemented previously and preliminary layouts of the proposed tunnels.

The Phase II geotechnical investigation program began in June 2018 and field work concluded in May 2019. Advancing testing was performed through 32 test borings along the preliminary tunnel alignments, in situ permeability testing, geophysical logging, installation of observation wells and piezometers, and laboratory testing of soil and rock properties.

Phase III, the final PP team phase of geotechnical investigations, started in October 2019 and was completed in March of 2020. The Phase III investigation plan supplemented and built upon the subsurface information gathered from the Phase I and II investigations. The Phase III investigation sought to "fill in the gaps" in data and overall stratigraphic correlation where necessary. The Phase III borings were located along the preferred tunnel alignments. Relatively shallow test borings were also included along preliminary alignments near surface consolidation sewers and locations of various near surface structures. Like the previous phases, Phase III consisted of advancing 33 test borings, in situ permeability testing, geophysical logging, installation of observation wells and piezometers and laboratory testing of soil and rock properties.

The data collected by the PP team was used to further refine the layouts of the proposed near surface structures and horizontal and vertical alignments of the proposed regional tunnel. In addition, these data will enable future designers to scope additional geotechnical investigations needed to support final design of the proposed regional tunnel system.

A summary discussion of the geologic investigations was provided in the previously published manuscript (Kennedy K. et al., 2019).

## Preliminary Ground Condition Related Construction Considerations

Perhaps the most significant construction consideration in response to the preliminary planning geotechnical findings is the method of tunneling for the proposed tunnels.

For planning purposes, it is prudent to assume that the tunnels will be excavated with a fully shielded TBM using a one-pass lining system consisting of bolted and gasketed precast concrete segments with probing and pre-excavation grouting.

This tunneling method will result in significant reduction in risk for:

- Excessive ground water infiltration and potential ground water lowering along the tunnel alignments
- Work stoppages due to the presence of hazardous gases
- Instability of Claystone strata where encountered, relative to other feasible tunneling methods

Other significant construction methods assumptions with the objective of reducing risk of ground water drawdown include: pre-excavation grouting ahead of the TBM during tunnel construction to reduce risk of excessive ground water inflows during tunneling, use of relatively impermeable temporary support of excavation (SOE) systems for deep shafts to reduce risk of unacceptable ground water drawdown, and use of micro-tunneling (i.e., pipe jacking using a pressurized face microtunnel boring machine) for trenchless reaches of consolidation sewers to reduce risk of unacceptable ground water drawdown.

## GEOLOGIC CONDITIONS FOR ORT

### Overburden Soils

The soil deposits overlying bedrock in the project area, as disclosed by the Phase I and Phase II geotechnical investigations, were generally as anticipated. These deposits include alluvium and terrace deposits representing older alluvial (fluvio-glacial) deposits related to various glacial episodes. They are overlain at many locations by recent artificial fill materials. Archive records for ALCOSAN's existing interceptors indicate that thicknesses of between 20 and 80 feet were encountered during construction.

Fill materials are anticipated at the surface at thicknesses that could reach 50 feet. They will be highly variable in nature and extent. Many of the fills are byproducts from the mining and/or steel-making operations and contain materials such as ash, slag, cinders and/or lime. Boulders, cobbles and debris, including bricks or metal, are commonly encountered within the fill along with variety of natural soil materials.

The alluvial deposits consist of interbedded clay, silt, sand and gravel with the sand and gravel becoming more prevalent with depth. They range in density/consistency from loose/soft to dense/hard and exhibit a significant amount of variation. Cobbles and/or boulders are also commonly encountered. Significant accumulations of alluvial materials are present along each of the main river drainages and occur as terraces.

The key engineering properties of the overburden soils from a preliminary planning perspective include the high permeability of the alluvial sand and gravel deposits; the potential for consolidation-induced settlement of the alluvial clays and silts if ground water levels are lowered for a significant period of time; and the presence of cobbles and boulders.

### Bedrock Stratigraphy

Based on PP and the final design 30% stage evaluations, all tunnels on the ORT project will be constructed deep in bedrock, at depths reaching approximately 120 feet to 150 feet below the ground surface.

The bedrock is known to consist of sandstone, claystone, siltstone, and shale and is typically described as weak to very strong, highly weathered to fresh, laminated to thinly bedded but can be indistinctly bedded, particularly in claystone. Joints are found to be predominantly parallel to sub-parallel with the bedding dipping 0 to 10 degrees, and near vertical joints observed at irregular spacing. The bedrock generally exhibits a higher degree of weathering near the top of rock surface. Typical pertinent rock engineering properties are:

- Core recovery: Approximately 25 to 100 percent
- Rock quality designation: Approximately 25 to 100 percent
- Rock core strength: Uniaxial compressive strength averages approximately 10,000 psi, with claystone approximately 3,000 psi.
- Cerchar Abrasivity Index for the sandstone ranging from 1.2 to 2.6, indicating medium to high abrasivity
- MOH's hardness ranging from 1.0 to 7.5, indicating moderately abrasive to highly abrasive rock

Boreholes and construction records from the existing Deep Tunnel Interceptor (DTI) indicate the presence of explosive gases such as methane. The geology indicates that hydrogen sulfide would also be present. Shaft and tunnel construction will, therefore, be considered "potentially gassy" according to Occupational Safety and Health Administration definitions.

### *Ground Water*

The ground water table ranges from elevation 705 feet to 715 feet at approximately 10 to 20 feet below ground surface (bgs). Geographically, the alignment of ORT will be either fully under the Ohio River or in near proximity of the Ohio/Allegheny Rivers. Both the CCT and SMRT are nearly 100 percent under the Ohio River.

All the ORT project component tunnels will be constructed fully under the ground water table and are expected to experience hydrostatic pressure as high as four to five bars. Without the proper means of ground water control, significant ground water inflows are expected during tunnel excavations.

This is substantiated by the construction records of the existing DTI and the Preliminary Basis of Design Report (PBODR) (ALCOSAN, 2020) findings summarized as follows:

### *DTI Construction Observations*

Existing DTI Contract 46 that was built along roughly the ORT alignment between Shaft O27-DS and beyond Shaft O41- DS:

- Required continuous probing and grouting over approximately 40 percent of the Contract 46 alignment
- Experienced an average pumping rate 330 gallons per minute (gpm) per 1,000 feet of tunnel. This would equal about 6,600 gpm of flow in ORT based on a length of roughly 20,000 feet. It is assumed the reported pumping rate was "post grouting."

Existing DTI Contract 47 which extended the DTI beyond the planned ORT TBM launch site (ORT-AS1):

- Required probing and grouting along the entire alignment.
- Significant inflow at single feature estimated at over 400 gpm which flooded the tunnel when encountered.

The final designer's evaluation using the same data resulted in a higher estimated flow of 6,700 to 9,800 gpm. Heading inflows/discreet features are expected to experience 130 to 400 gpm. These flow values assume excavated bare rock without grouting.

Shorter CCT and SMRT inflows are expected to be under 1,000 to 2,000 gpm. Adits would experience proportionally less inflow, although they could experience flush flows of 130 to 400 gpm if discrete jointing features are encountered.

## ADVANCEMENTS TO ORT SINCE PRELIMINARY PLANNING

### Property Availability

After issuance of the PBODR, a high level of uncertainty arose regarding the availability of the preferred site for use as the ORT TBM launch and retrieval sites. A study was performed to evaluate ORT launch and retrieval site alternatives to assist ALCOSAN in property acquisition efforts. Ten sites were studied and presented to ALCOSAN for consideration on May 17, 2021, accounting for site requirements such as space needs and ease of access.

Each of the ten sites were evaluated based on their rating in the categories shown in Table 2, with different weights assigned to each.

**Table 2. Evaluation criteria, weights, and rating definition criteria**

| Criteria | Attribute | Attribute Weight | Definition of Rating |
|---|---|---|---|
| Site Area | N/A | 20% | 5 = 3+ Acres—Optimal<br>3 = 2.5–3 acres—Acceptable<br>1 = <2.5 acres—Not Recommended |
| Economic | Cost | 15% | 5 = No anticipated impacts on environment<br>3 = Anticipated environmental impacts can be easily mitigated<br>1 = Anticipated environmental impacts difficult to mitigate |
| Environmental | N/A | 10% | 5 = Limited sensitive receptors could be impacted by noise and vibrations, Limited traffic impacts<br>3 = Some sensitive receptors could be impacted by noise and vibrations; Moderate traffic impacts<br>1 = Many sensitive receptors could be impacted by noise and vibrations; Significant traffic impacts |
| Noise and Vibration | Noise, vibration, traffic impacts | 10% | 5 = Limited sensitive receptors could be impacted by noise and vibrations; Limited traffic impacts<br>3 = Some sensitive receptors could be impacted by noise and vibrations; Moderate traffic impacts<br>1 = Many sensitive receptors could be impacted by noise and vibrations; Significant traffic impacts |
| Green Leave Behind | Green leave behinds/ co-benefits | 10% | 5 = Excellent candidate for green leave behind/co-benefit<br>3 = Limited green leave behind potential/co-benefit<br>1 = No green leave behind potential/co-benefit |
| Operation and Maintenance | N/A | 10% | 5 = Limited operational impacts, routine maintenance anticipated, good access provided<br>3 = Some operational impacts, moderate maintenance anticipated, limited access provided<br>1 = Significant operational impacts, frequent maintenance anticipated, difficult access provided |
| Implementation and Construction | N/A | 10% | 5 = Routine construction with relatively low risk of issues<br>3 = Moderately complex construction requiring some risk mitigation<br>1 = Complex construction requiring special risk mitigation measures |
| Property/ Easement Acquisition | N/A | 15% | 5 = All or majority of required property/easements are in public property/or already ALCOSAN owned—no atypical property costs<br>3 = Requires some property/easements acquisition for private property<br>1 = Requires significant property/easements for private property |

Source: ALCOSAN, 2022b

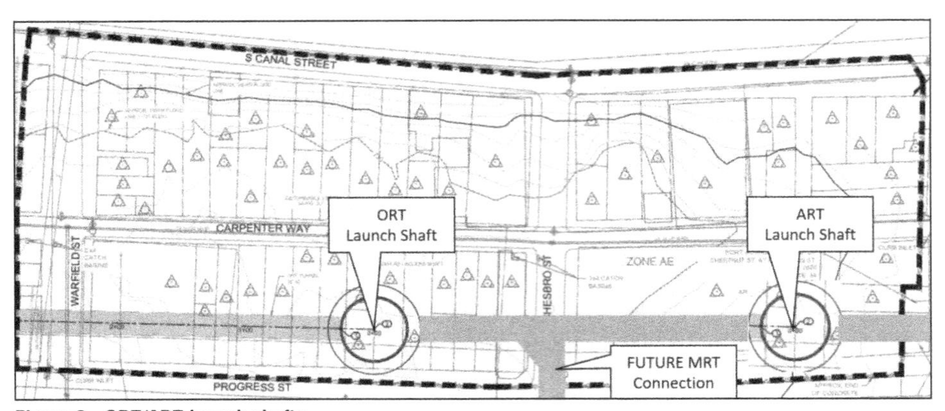


**Figure 3. ORT/ART launch shafts**

Full consideration was given to each potential site with regards to constructability, risk, cost, and other key criteria. After discussion and coordination, a privately owned 4.2-acre site located east of the Veterans Bridge property was chosen as the TBM launch site for ORT and ART (Figure 3).

This selection provides the following advantages:

- Common site for TBM power supply
- Reduces the schedule potential contract interface conflicts at the originally proposed ORT retrieval/ART launch shaft
- Provides ample space, easy access
- Potential for a Green Leave Behind project

This selection had the following disadvantages:

- The ORT tunnel length was decreased by approximately 4100 feet which will be transferred to the ART tunnel alignment.
- Due to property and schedule constraints, the ORT will have to be mined down grade from the launch site.

## Alignment Optimization

Analyses have been completed to optimize the ORT alignment with considerations for:

- Alignment length
- Hydraulic point of connections
- Order of magnitude cost estimates
- Number of easements and/or perceived difficulty in obtaining public or private easements and permits

The final designer studied three alternative alignments (Figure 4):

- Base Case Alignment—Modified PBODR Alignment: The Base Case Alignment incorporated certain modifications to the PBODR alignment as initiated by or agreed to with ALCOSAN. The section from the WWTP to the West End Bridge was moved from the shore slightly into the river. The portion of the alignment east of West End Bridge running longitudinally under the I-279 right-of-way (ROW) was kept unchanged from the PBODR alignment except for the eastern portion, which was modified to accommodate a launch

Source: ALCOSAN, 2022c

**Figure 4. ORT alternative alignments**

from near the Veterans Bridge. The CCT alignment was also optimized to facilitate adit mining.

- Alternative 1—Northern Alignment: This alternative follows the same alignment as the Base Case Alignment from the west terminus to the vicinity of West End Bridge, but then turns northeast across State Route 65 to clear the I-279 horizontal ROW, and then turns back across I-279 to end at the same terminus as the PBODR alignment.
- Alternative 2—Southern Alignment: This alternative follows the same alignment as the Base Case Alignment from the west terminus to West End Bridge and continues along the Allegheny River bank to the south of the Acrisure Stadium; from thereon, the alignment turns north inland in parallel to but south of I-279 and runs east to end at the same terminus as the PBODR alignment.

The alignment preference was driven by two factors. First, the relocation of A-48 connection site from the parking lot of Residence Inn by Marriott to the north of I-279/PBODR alignment, a different parking lot area owned by the same property owner. Second, Pennsylvania Department of Transportation (PennDOT) and Federal Highway Administration (FHWA) requirements discussed with the two authorities. These requirements include a Highway Occupancy Permit permitting application document, no commercial enterprise within a limited access highway, justification for PennDOT to allow the facility within limited access, and possibly FHWA approval and National Environmental Policy Act documentation.

After considering the above, ALCOSAN decided to proceed with the Alternative 1—Northern Alignment as the preferred alignment. The final designer then further refined that portion of the main ORT alignment, as well as the two branch tunnels, the CCT and the SMRT, to streamline the alignment and flow adit connections and minimize property encroachments. The updated tunnel alignment is shown in Figure 5.

## Tunnel Boring Machine Evaluation

### *Change in Tunneling Technology*

Three types of tunneling methods were possible: drill and blast (D&B), mechanical excavation, or tunnel boring machine (TBM). Three types of TBMs were considered: open face, single mode pressurized face, and multi-mode pressurized face.

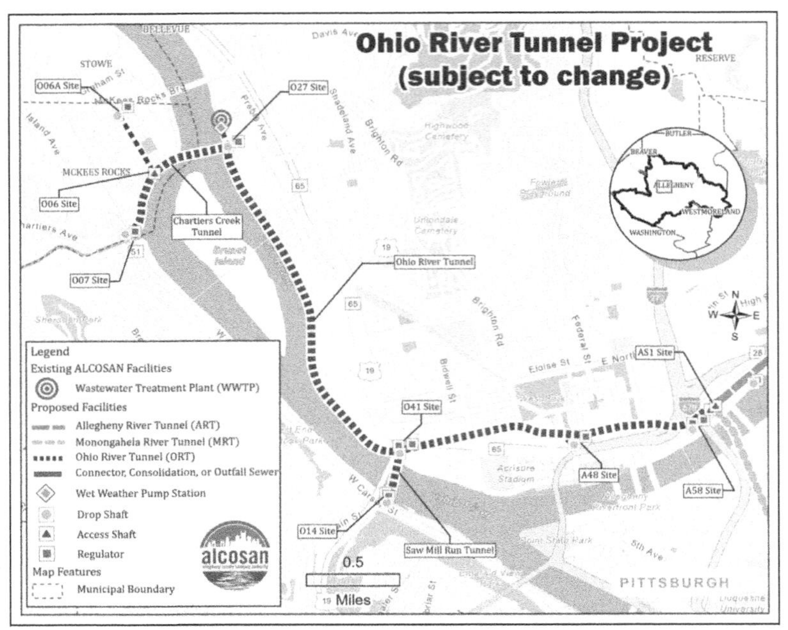


**Figure 5. Updated ORT alignment**

Table 3 summarizes the advantages and disadvantages of each method for ORT (ALCOSAN, 2022a).

Several critical factors controlled the decision for the tunneling method:

- Ability to efficiently control expected ground water inflow, made more important by the downgrade direction of mining
- Ability to deal with variable rock strata, i.e., sandstone, shale, claystone, coal seams
- Ability to deal with "potentially gassy" conditions during tunneling
- Ability to help mitigate schedule risk due to the above factors. The critical path for the entire project goes through the main ORT tunnel.
- Successfully install a precast concrete one-pass segmental lining and grout the annular space in rock

The key aspect will be the ability to control ground water in front of the TBM without having to rely on probing and grouting, while providing a reliable means for segment grouting. The final designer recommended the use of a pressurized face TBM, including a slurry shield TBM, or a multi-mode TBM, similar to a crossover EPBM, or the modified non-continuous pressurized EPBM.

As design progresses and more is known with respect to overall project schedule and project ground and ground water conditions, the suitability of these types of TBMs will be further evaluated. The ultimate choice of the type of pressurized face TBM(s) will be left to the bidding contractors, in collaboration with the TBM manufacturers, subject to the TBM specification requirements that will capture all the critical considerations outlined herein.

**Table 3. Comparison of advantages and disadvantages of feasible tunneling methods for ORT**

| Tunnel Methodology | Advantages | Disadvantages |
|---|---|---|
| D&B | ▪ Flexibility in alignment, variable cross sections<br>▪ Lower up-front investment than a roadheader or TBM, making it potentially more economical | ▪ Larger overbreak than roadheader and TBM an issue, unless mitigations such as efficient blast pattern design or line drilling used<br>▪ Invert might not be as even (compared to a tunnel excavated by a roadheader), temporary concreting may be necessary<br>▪ Efficient use of equipment (e.g., drill rig, mucker) is necessary<br>▪ Advance rates vary widely based on multiple factors that also may apply to mechanical excavation |
| Roadheader (mechanical excavation) | ▪ Less overbreak than D&B<br>▪ Better profile control with the use of improved/advanced guidance system<br>▪ Flexibility in alignment (such as staying within ROW)<br>▪ Less noise and vibration compared to D&B<br>▪ Typically, can be combined with D&B (allowing the contractor to have the flexibility to manage work, which may reduce the cost) | ▪ Largest roadheader (with a size close to the 15 feet inner diameter of the tunnel) required for more powerful/efficient excavation in higher strength rock<br>▪ Increased downtime for roadheader maintenance and pick changes in abrasive rock (may be prohibitive)<br>▪ Roadheader size may require expanded tunnel niches for roadheader to back out into in order to move in the drill jumbo in for each cycle of the pre-excavation probing/grouting<br>▪ Requires initial support then installation of permanent lining, two pass system<br>▪ Pumping required<br>▪ Abrasive rock increases downtime for roadheader cutter pick changes and maintenance |
| Open face TBM | ▪ Less overbreak than D&B mining and other mechanized excavation methods<br>▪ Generally faster advance rates than D&B mining and roadheader/other mechanized excavation methods | ▪ Higher up-front investment than roadheader/D&B<br>▪ Longer set-up and dismantling time than other methods<br>▪ No provision for ground water control, relies on probing and grouting<br>▪ Pumping required<br>▪ Abrasive rock increases downtime for TBM maintenance and cutter disc changes<br>▪ Requires two-pass support system with main beam or "finger shield" type<br>▪ In rock with high water flow potential, lining installation can be very problematic |
| Earth pressure balance TBM (EPBM) | ▪ Less overbreak than D&B and other mechanized excavation methods<br>▪ Faster advance rates than hand mining and other mechanized excavation methods<br>▪ One-pass support system<br>▪ Face ground pressure control through balanced excavation<br>▪ Disk cutter change in sound rock zones | ▪ Higher up-front investment than roadheader/D&B<br>▪ Longer set-up and dismantling time than other methods<br>▪ Abrasive rock increases downtime for TBM maintenance and cutter disc changes<br>▪ Difficulty in dealing with high ground water inflow from rock tunnel face<br>▪ Difficult to grout segmental linings in areas of high water flow<br>▪ May require hyperbaric intervention for disc cutter change |
| Slurry shield TBM | ▪ Less overbreak than D&B mining and other mechanized excavation methods<br>▪ Generally faster advance rates than D&B mining and roadheader/other mechanized excavation methods<br>▪ No forward probing/grouting required<br>▪ One-pass support system<br>▪ Face earth pressure and ground water control through slurry system | ▪ Higher up-front investment than roadheader/D&B<br>▪ Longer set-up and dismantling time than other methods<br>▪ Abrasive rock increases downtime for TBM maintenance and cutter disc changes<br>▪ May require hyperbaric intervention for disc cutter change and/or extensive pumping of the face |
| Multi-mode TBMs | ▪ Tunnelling mode adaptable to changing ground conditions<br>▪ Can offer (as one example) conversion to allow earth pressure balance mining in rock with high water flows with the ability to fully grout segmental lining<br>▪ Optimum safety and flexibility during the entire tunnelling process | ▪ Higher up-front investment than roadheader/D&B, and single-mode TBMs<br>▪ Longer set-up and dismantling time than other methods.<br>▪ Abrasive rock increases downtime for TBM maintenance and cutter disc changes<br>▪ Downtime required to switch between varying modes<br>▪ May require hyperbaric intervention for disc cutter change |

Source: Adapted from ALCOSAN, 2022a.

### Updated Mining Rate and Supply Chain Uncertainties Impacting TBM Availability

The mining rate assumption made in the PBODR was 55 ft/WD. Since deviations in mining rate could significantly impact the construction schedule and program schedule, a rate study was completed to assess whether revisions needed to be made to the mining rate assumption.

For the study, historical mining rate data was examined, and discussions were held with related contractors to gain an understanding of the actual mining rates achieved in similar tunnel projects. Through this research, it was determined that the study would consider TBM excavation rates from 30 ft/WD to 60 ft/WD.

The impact of varying TBM mining rates on schedule was investigated (Figure 6). Key assumptions included:

- Developed mining durations for TBM excavation rates at 5 feet intervals
- Assumed 24 months for TBM procurement, delivery and installation, to the point where the TBM is ready to excavate
- Assumed a 3-month learning curve, about the first 500 feet of TBM tunnel drive
- Assumed 6 months post-tunneling duration for ORT and ART. Assumed 8 months for post-tunneling duration for MRT to complete adit connections for drilled drop shafts.

As a result of this study, it was decided to assume a mining rate of 45 ft/WD after learning curve.

These rate assumptions were applied to the baseline program schedule. The schedules with mining rates lower than 55 ft/WD were indicated to experience various degree of delays with tunnel completion dates for either of the three tunnels. Some of these delays could impact critical program milestones. To relieve the schedule pressure, ALCOSAN is studying alternative project schedules for ART and MRT.

### Contract Packaging

Several contract packaging options are being considered to optimize schedule and reduce risk. These options will look at the potential incorporation of NSF contract(s) with the ORT tunnel contract. The goal is to optimize construction, reduce contract interface on small drop shaft sites and reduce or eliminate multiple mobilizations by multiple prime contractors to selected sites.

## NEXT STEPS

Design for the ORT project was initiated in November 2021. Design is anticipated to be completed in June 2024 for start of construction in January 2025. End of construction and commissioning are anticipated in 2029 with Place in Operation by December 31, 2029.

## REFERENCES

ALCOSAN. 2019. *ALCOSAN Clean Water Plan.*

ALCOSAN. 2020. *Contract No. S-432. Regional Conveyance Facilities of the Interim Wet Weather Plan—Preliminary Basis of Design Report. Prepared by Wade Trim.*

ALCOSAN. 2022a. *Contract No. S-485. Ohio River Tunnel. Technical Memorandum: Assessment of Tunneling Methodology. Prepared by Mott MacDonald.*

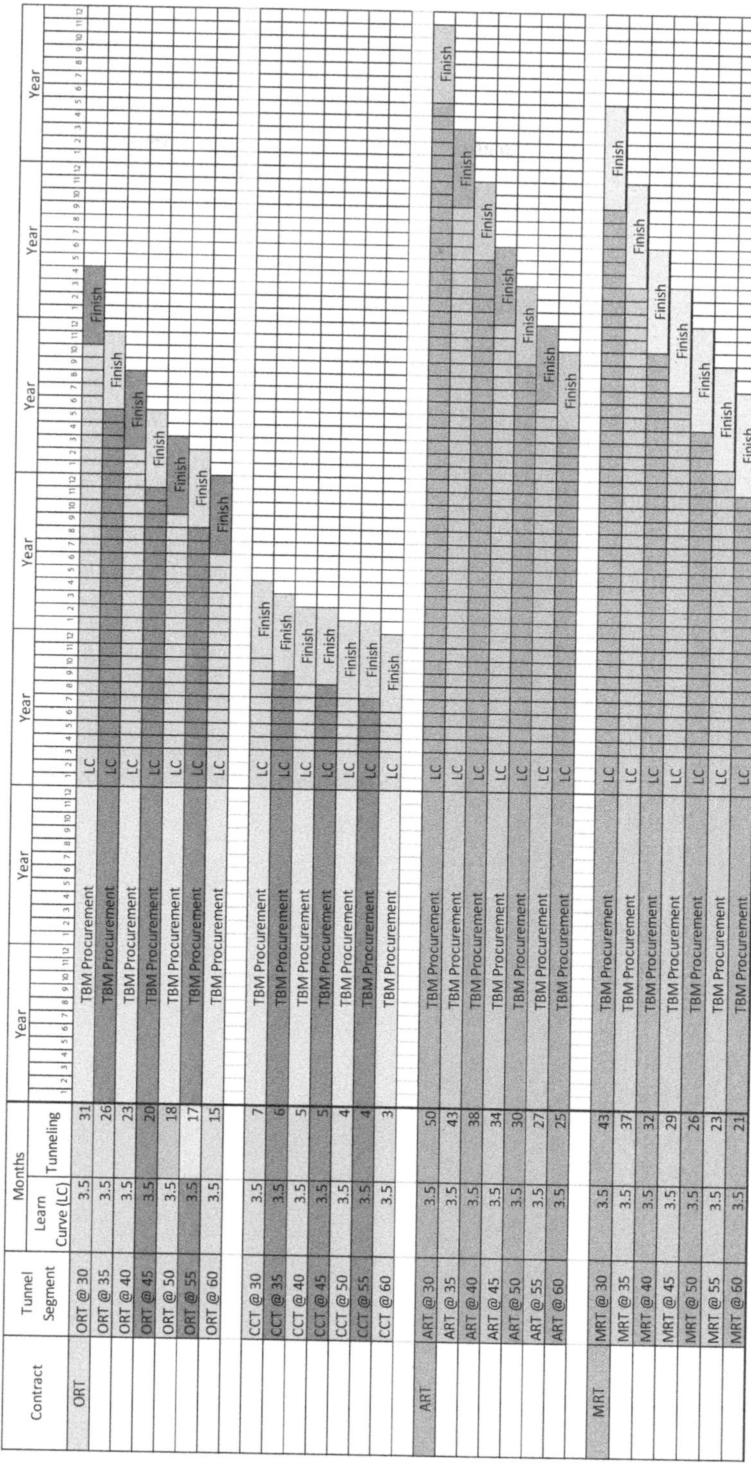

| Contract | Tunnel Segment | Months: Learn Curve (LC) | Months: Tunneling |
|---|---|---|---|
| ORT | ORT @ 30 | 3.5 | 31 |
| | ORT @ 35 | 3.5 | 26 |
| | ORT @ 40 | 3.5 | 23 |
| | ORT @ 45 | 3.5 | 20 |
| | ORT @ 50 | 3.5 | 18 |
| | ORT @ 55 | 3.5 | 17 |
| | ORT @ 60 | 3.5 | 15 |
| | CCT @ 30 | 3.5 | 7 |
| | CCT @ 35 | 3.5 | 6 |
| | CCT @ 40 | 3.5 | 5 |
| | CCT @ 45 | 3.5 | 5 |
| | CCT @ 50 | 3.5 | 4 |
| | CCT @ 55 | 3.5 | 4 |
| | CCT @ 60 | 3.5 | 3 |
| ART | ART @ 30 | 3.5 | 50 |
| | ART @ 35 | 3.5 | 43 |
| | ART @ 40 | 3.5 | 38 |
| | ART @ 45 | 3.5 | 34 |
| | ART @ 50 | 3.5 | 30 |
| | ART @ 55 | 3.5 | 27 |
| | ART @ 60 | 3.5 | 25 |
| MRT | MRT @ 30 | 3.5 | 43 |
| | MRT @ 35 | 3.5 | 37 |
| | MRT @ 40 | 3.5 | 32 |
| | MRT @ 45 | 3.5 | 29 |
| | MRT @ 50 | 3.5 | 26 |
| | MRT @ 55 | 3.5 | 23 |
| | MRT @ 60 | 3.5 | 21 |

**Figure 6. Mining rate analysis**

ALCOSAN. 2022b. *Contract No. S-485. Ohio River Tunnel. 30% Drawings. Prepared by Mott MacDonald.*

ALCOSAN. 2022c. *Contract No. S-485. Ohio River Tunnel. Technical Memorandum: Ohio River Tunnel Preferred Alignment. Prepared by Mott MacDonald.*

Kennedy, K., Lichte, M., O'Rourke, T., Dobbels, D. 2019. *ALCOSAN Tunnel System Preliminary Planning*. North American Tunneling Conference 2020 Proceedings. Track 2, Session 5, p. 387.

United States District Court for the Western District of Pennsylvania. 2020. Modified Consent Decree Case 2:07-cv-00737-NR, Document 33-1, Filed 05/14/2020.

# Design of the Akron Northside Interceptor CSO Tunnel

**David Mast** ▪ AECOM
**Heather Ullinger** ▪ City of Akron, Ohio
**Amanda Foote** ▪ AECOM
**Juan Granja** ▪ GPD Group
**Dominick Mandalari** ▪ AECOM

## ABSTRACT

The City of Akron will construct the Northside Interceptor Tunnel project, a U.S. EPA/DOJ CSO Long Term Control Plan Project. The system will provide at least 10.3 million gallons of storage in a 16.5-foot finished ID and 6,660-foot-long rock tunnel. The system will reduce both overflow volume and number of overflow structures on the Cuyahoga River. The project includes deep drop shafts and more than 3,000 linear feet of 48 to 96-inch I.D. sewers installed by trenchless construction methods. This paper will discuss the project background, ground conditions, and other challenges overcome during the design phases.

## PROJECT BACKGROUND

The City of Akron is located in the northeast region of Ohio and is the fifth-largest city in the state. The City's population is approximately 198,000 and is approximately 62 square miles in size. The City was founded in 1825 and, because of the Ohio and Erie Canal, was the fastest growing city in America during the 1910s and 1920s. Once known as the "Rubber Capital of the World," the city is now a world-renowned center for polymer research and development. Akron is also the hometown of NBA superstar Lebron James.

Portions of Akron's sewer system were constructed as combined sewers in the early 1900's. Some of the sewers in the city are over 90 years old, and over 85% of them are more than 50 years old. In a combined sewer system, stormwater and sanitary sewage from domestic, commercial, and industrial sources flow together in a single pipe to the Water Reclamation Facility to be treated. However, during rainfall events, the pipes become full of stormwater and sewage. CSO outfalls allow excess flow to discharge to nearby water bodies, preventing backups at storm drains and in basements. These untreated wastewaters are discharged to the receiving streams, including the Cuyahoga River, along with large quantities of stormwater. The City of Akron currently has 34 CSO outfalls in the sewer collection system, shown as block dots on Figure 1. Akron is one of over 770 cities in the United States which have had combined sewer overflows.

In 2014, the City entered into a Long-Term Control Plan Consent Decree agreement with the US Environmental Protection agency, U.S. Department of Justice, and the Ohio Environmental Protection Agency. The agreement requires the City to make significant improvements to its sewer system and wastewater treatment plant to reduce combined sewer overflows and bypasses. The plan included 5 sewer separation projects to eliminate overflow points, multiple storage basins, two tunnels to store overflows during wet weather events, and major upgrades to the City's Water Reclamation Facility to increase treatment capacity. The estimated cost for the 26 major projects under the Consent Decree was estimated in 2022 to be nearly $1.2 billion (in 2010 dollars).

As of October 2019, over 92% of all Consent Decree projects were either completed or under construction. (Akron Waterways Renewed! 2019) The City has invested over $700 million on the projects to date. The projects completed have resulted in keeping an estimated 2.2 billion gallons of untreated overflow and partially treated sewage bypasses out of the local waterways in a typical year. (Horrigan, 2022)

## LOCATION

The Northside Interceptor Tunnel (NSIT) is one of two large diameter storage tunnels required by the Consent Decree. The first large diameter storage tunnel built under this program was the 27-foot ID Ohio Canal Interceptor Tunnel (OCIT) and associated collection system stores combined sewage flowing into downtown Akron and the Ohio Canal (the CSO receiving waters). The tunnel alignment is shown on Figure 1. The OCIT project was awarded to a joint venture of Kenny Construction and Obayashi in 2015 and was placed in service in June 2020 (Tunnel Business Magazine, 2019). The total project cost is listed on the City's website as $303.2 million and, by itself, the OCIT system is estimated to remove 467 million gallons of combined sewage overflows in a typical year.

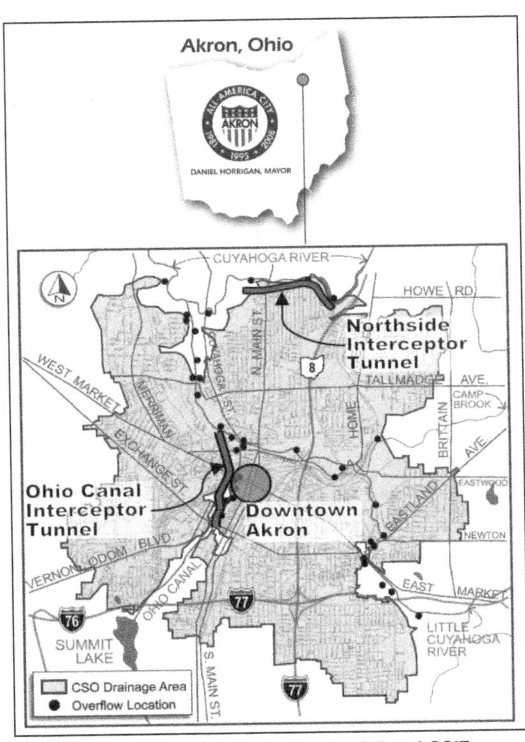


Figure 1. City of Akron CSO Outfalls, NSIT and OCIT tunnel locations

The NSIT will be located in north central Akron, just south of the neighboring City of Cuyahoga Falls and the Cuyahoga River, as shown generally on Figure 1. The project will address combined sewer overflows on the Cuyahoga River and at the north end of the North Hill neighborhood, shown in detail on Figure 3. Residential properties in the North Hill range from circa 1920s to 1970s (see Figure 2). The project area is entirely residential with the tunnel alignment mostly following a residential street named Riverside Drive. A majority of the project staging and construction activities

Figure 2. Typical Residential Home in North Hill Neighborhood

will occur at the westernmost and downstream end of the tunnel, in the Cascade Valley Metro Park The upstream end of the tunnel is located just east of Front Street, near its intersection with Howe Avenue.

## PROJECT COMPONENTS

The North Hill neighborhood and immediately surrounding areas are currently served by the Northside Combined Sewer Interceptor (NSI), which is generally in fair to poor condition and extends along the Highbridge Trail and many unstable portions of the Cuyahoga River gorge hillside. The NSI experiences flow volumes in excess of the system capacity, and consequently, overflows frequently at Rack Structures. A Rack Structure is a regulator structure which diverts dry weather flow through a bar screen ("rack") in the base of the structure, and out of the structure via an underflow pipe. The underflow pipes convey flow to the NSI for conveyance to the City's Water Reclamation Facility (WRF). However, during large storm events, the NSI becomes surcharged, cannot accept flow from the Rack Structures, and so excess flow is directed past the bar rack, out the side of the Rack Structure, and into an overflow pipe leading to the Cuyahoga River to the north and west of North Hill. The NSIT will address sewer outfalls currently discharging from the racks named in Table 1, which currently include about 755 acres of combined sewer service area.

**Table 1. Service Areas Tributary to NSIT**

| Tributary Area Name | Area, acres |
|---|---|
| Rack 32 | 212.0 |
| Rack 33 | 33.1 |
| Rack 34* | 76.5 |
| Rack 35 | 434.0 |
| **Total** | 755.6 |

* New storm sewers being installed to separate Rack 34.

In order to achieve the Consent Decree requirement of no combined sewage overflows to the Cuyahoga River in a Typical Year of storms, the large diameter tunnel must be able to hold a minimum of 10.3 million gallons. However, the City plans to take advantage of this project to further improve the project with the following additional project goals.

- The NSIT is being designed to accept both dry and wet weather flows from the Racks. This will allow the City to abandon over 6,000 feet of the failing Northside Interceptor and eliminate the potential risk of catastrophic slope failures. The NSIT tunnel slope will induce minimum flow velocities of 2 to 3 feet per second under average dry weather flows. The portions of the NSI which will remain in service to receive flows from the NSIT are to be relined and are indicated on Figure 3 with a magenta highlight.
- The NSIT will be designed to allow abandonment of all other combined sewage relief points on the Cuyahoga River between Rack 35 Rack 32. To achieve this, the NSIT has been designed with a conveyance capacity greater than the sum of theoretical maximum flow rates for sewers flowing into Racks 32, 33, 34, and 35. During all Typical Year storms, the tunnel will receive and store the combined sewage from the specified Racks. During storm events larger than Typical Year storms, the tunnel will again receive all of the flow which can be conveyed to the Rack structure. Excess combined sewage for the entire service area will be conveyed through the NSIT tunnel to a single, newly constructed overflow structure. This will allow the existing Cuyahoga River combined sewage outfalls for Racks 32, 33, and 35 to be completely abandoned. The current Rack 34 combined sewage outfall will be converted to a stormwater-only outlet.

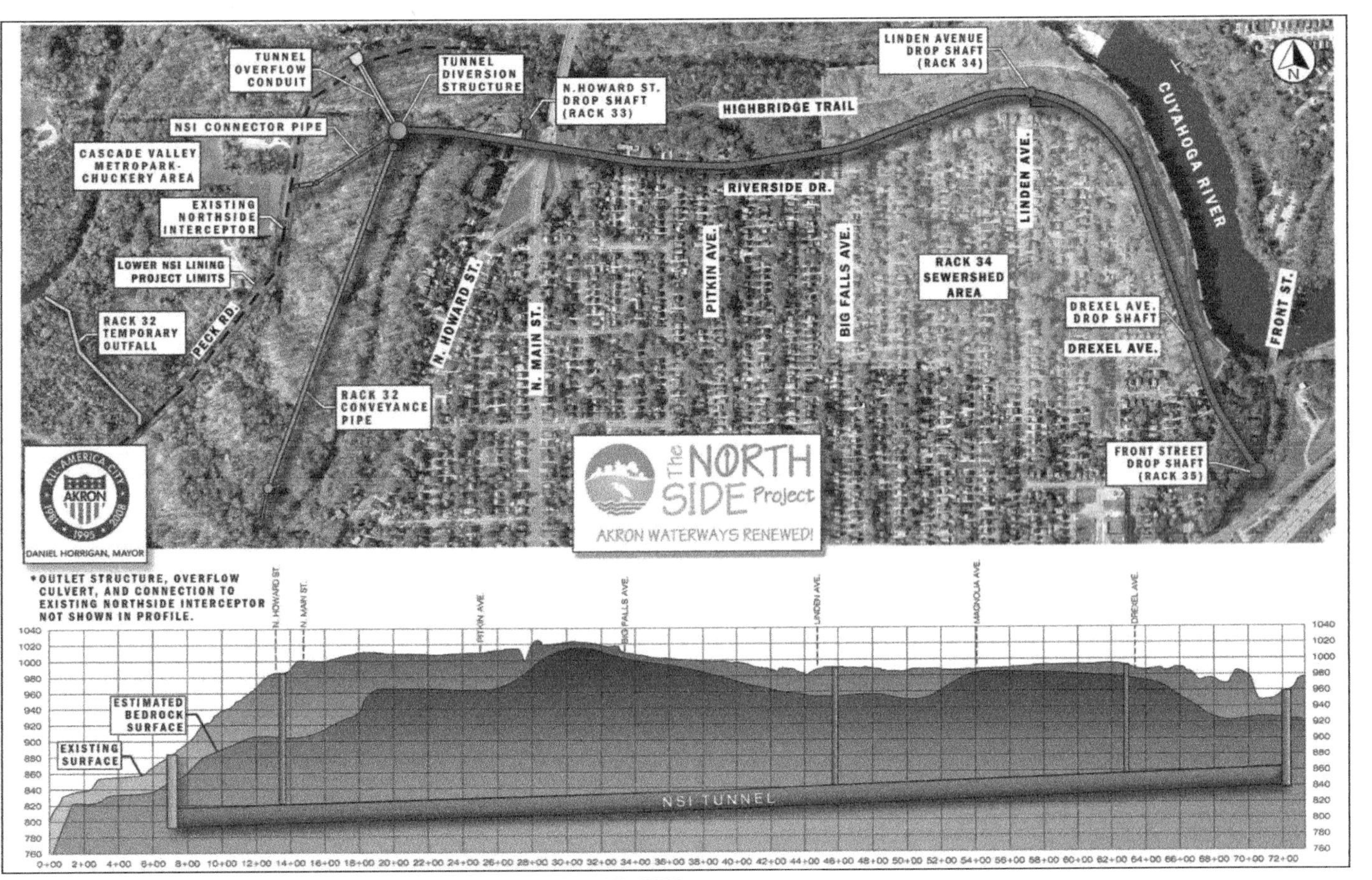


Figure 3. Schematic of Northside Interceptor Tunnel plan and profile, 10/14/2022

Figure 3 illustrates the storage tunnel alignment as well as the collection sewers and drop shafts which will bring combined sewage into the tunnel.

- Rack 32 flows will be collected in the southwest part of the project, dropped about 80 feet below grade in a 28-foot ID shaft, and conveyed to the west end of the NST by a 96-inch ID conveyance pipe ("Rack 32 Conveyance Pipe").
- Rack 33 flows will drop about 170 feet in the North Howard St. Drop Shaft and be conveyed to the NSIT through a 6-foot ID adit tunnel.
- Rack 34 dry weather flows will drop about 150 feet to tunnel elevation in the Linden Avenue drop Shaft (an 18-inch ID plunge drop pipe) and will be conveyed to the NSIT through a 6-foot ID adit tunnel.
- Flows from several homes near Drexel Avenue will be collected, dropped about 150 feet to tunnel elevation in the Drexel Avenue Drop Shaft, and conveyed to the NSIT through a 6-foot ID adit tunnel.
- At the east end of the tunnel, Rack 35 flows will drop about 100 feet to tunnel elevation in the Front Street Drop Shaft.

The flows from all Racks will be conveyed to a single control structure, the Tunnel Diversion Structure (see Figure 4). During periods of dry weather and low volume storm events, all flows will be directed out the southwest quadrant of the shaft into the NSI Connector Pipe, and then into the existing NSI. The NSI will convey those flows to Akron's Water Reclamation Facility for treatment.

Once the NSI capacity is reached, the City will close valves on the NSI collector. Combined sewage will begin to collect in the NSIT system. The TDS is designed as a donut-shaped vertical shaft with a slightly offset center core. Sewage levels will rise in the TDS outer ring until either the storm event ends, or the sewage levels top the internal weir walls. If the storm event ends before reaching the internal weir, the City

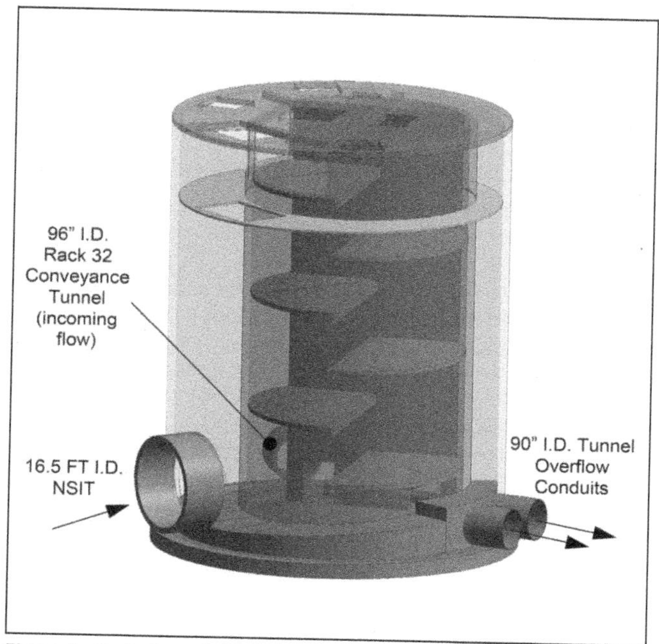


**Figure 4. Preliminary 3-D model of tunnel diversion structure (looking Southwest)**

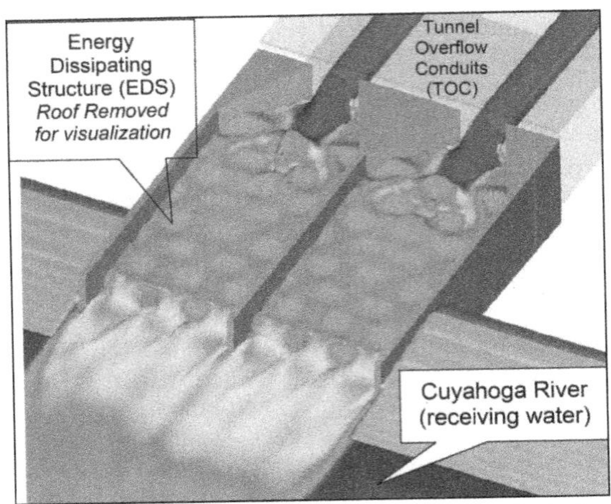


**Figure 5. Preliminary FEM model for NSIT energy dissipating structure to Cuyahoga River**

waits until the NSI has conveyance capacity, then dewaters the NSIT at a controlled rate into the NSI for treatment.

If the sewage levels in the NSIT Diversion Structure rise to the internal weir elevation, the combined sewage will begin falling into the inner shaft of the TDS, where they will cascade over baffle drop shelves to the bottom of the TDS, and then flow to the northwest via the Tunnel Overflow Conduits (see Figure 3). The drop shaft helps to reduce air entrainment and control flow velocities as the sewage leaves the TDS. The TOC consists of two (2) 90-inch ID pipes which control velocities of the overflow (see Figure 5). Finally, the overflow will transition through a box culvert-like Energy Dissipating Structure and into the Cuyahoga River. The TDS, TOC and EDS are designed to convey up to 1.08 billion gallons per day (BGD) of flow to the Cuyahoga River if necessary.

## STORAGE TUNNEL ALIGNMENT, PROFILE, AND GROUND CONDITIONS

The NSIT Advanced Facility Plan performed in 2013 identified four (4) potential storage tunnel alternatives which met the 23-million-gallon storage volume believed to meet the Consent Decree performance requirement. (Haywood, 2013. The tunnel diameters ranged from 20 to 24 feet. The lengths of bored tunnel ranged from 6,850 to 9,700 feet. Alternatives 1 and 2 required a 20 to 22-foot ID bored tunnel built through both soft ground and rock conditions. Alternative 3 required a 24-foot ID and 6,550 feet long tunnel but would be bored through rock materials only. Alternative 4 was a 22-foot ID rock tunnel with an additional 1,650 feet of shallow box culvert to add more storage volume. Alternative 3 was the preferred AFP alternative. The currently chosen NSIT alignment and length is similar to AFP Alternative 3 and will allow for tunneling entirely in rock materials.

After the AFT was completed, the City and their Consultants performed extensive field investigations, flow metering, and additional modeling. Based on the updated results, it was confirmed that in the hydraulic models, a storage tunnel with 10.3 million gallons will achieve the performance requirement of zero (0) overflows in the 1994 adjusted typical year. As a result, the NSIT will be a concrete-lined conveyance and storage tunnel with a minimum internal diameter of 16.5 feet, a length of 6,660 fee, and will hold at least 10.3 million gallons of combined sewage behind the downstream weir. The horizontal alignment, shown on Figure 3 was chosen from about three alternates. The horizontal alignment met the following City preferences:

- The tunnel alignment generally extends along the north edge of North Hill where flows from Racks 33, 34, and 35 are generally combined into a single large-diameter pipe. This allows to drop flows almost directly into the storage tunnel (with very short adits at depth). This is not true for Rack 32 flows, which requires over 2000 feet of conveyance tunnel in order to connect to the storage tunnel (discussed in detail later in this paper.)
- Over 75% of the chosen tunnel alignment is below City right of way, City-owned property, or property of the Summit County Metroparks, a partner agency to the City also interested in improving the Cuyahoga River and associated parklands.
- From the outside face of the launch shaft, the first curve is more than 350 feet away. The final 320 feet of mining into the retrieval shaft is on a straight alignment
- Each of the 6 curves along the horizontal alignment have a radius of at least 1000 feet and at least 100 feet of straight alignment into and out of the curves.
- The west end of the alignment was generally controlled by the position of the launch shaft, which will also be the location of the permanent Tunnel Diversion Structure. In general, the launch shaft location was position away from existing high tension electrical transmission lines, at a location to facilitate the Tunnel Overflow conduit alignment to the Cuyahoga River, and in a location visually screened from residences and park areas by existing forested parkland.

The vertical profile of the tunnel was generally controlled by the three major items listed below.

- The rock cover at the storage tunnel's west end (the downstream end) is the shallowest present along the entire alignment. The chosen profile allows the TBM to start mining from the launch shaft in a full face of rock, with more than one tunnel diameter of rock cover.
- The first 3,210 feet of tunneling will be at a 0.63% slope, which is necessary to create sewage flow velocities in the 2 to 3 feet per second range during dry weather conditions. At about 3,210 feet from launch, the TBM will pass the sewage connection adit for Rack 34, and there will be less sewage flow in the tunnel to the upstream end. Therefore, the final 3,430 feet of the tunnel will be mined at a slightly higher slope of 0.78% to achieve the desired flow velocities.
- Except immediately at the launch shaft, the rest of the alignment is deep enough into competent rock that the TBM diameter could be increased to about 20 feet and there would still be more than a tunnel diameter of rock cover. Although some treatment might be needed at the launch face, this item gives the City the option to allow Contractors to propose a reconditioned TBM's which is slightly larger than the minimum 16.5-foot ID tunnel.

Once the vertical and horizontal alignments were established, the tunnel diameter was back calculated to achieve the 10.3-million-gallon storage capacity, assuming the downstream weir was set at the tunnel crown elevation at the upstream end.

The NSIT will be tunneled through the Cuyahoga Formation, a Mississippian-Age Shale and Siltstone formation with sandstone (Siebert, 2015). Specifically, the TBM is expected to encounter the Meadville Shales, the Sharpsville Sandstone, and the

Orangeville shale units. Natural gas is known to occur in the bedrock formations in this part of Ohio, although no significant gas has been encountered at test boring locations to date. The dominant structural feature of the bedrock is expected to be horizontal bedding due to the depositional history of the area. Although the Rock Quality Designation RQD for rock in the tunnel zone is high (typically 95 to 100), the shale portion of the geologic formation has a low unconfined compressive strength (expected to range from about 1,500 to 15,000 psi). However siltstone partings and layers several feet thick will be present within the shale matrix. Siltstone samples tested for the nearby OCIT project were found to have unconfined compressive strengths between 6,400 and 32,500 psi (Siebert, 2015). TBM cutterhead design will have to consider this variation of strengths, and typically a combination of cone picks and rock disc cutters are incorporated into the face to address the range in rock conditions.

## CONSENT DECREE MILESTONE DATES AND ROCK TBM AVAILABILITY

The NSIT project has several challenges to overcome as part of the design process. The first challenge is the design and construction schedule. Design for the tunnel started in January 2022 with the requirement to meet the first project Consent Decree milestone.

- Bidding of Control Measure (Advertisement)—April 30, 2023.

The combined pre-design stage and detailed design schedule has been compressed to 16 months. After the pre-design stage, design submittals to the City were modified to include only 30%, 75%, 100%. The City and AECOM have adopted a "No Surprises During Review Periods" and increased communication frequency. Progress meetings are held with the City every other week to provide updates on various project components.

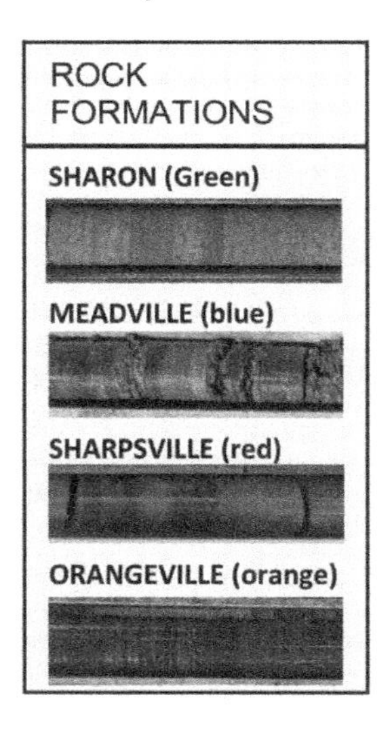


In addition to a tight design schedule, the construction schedule is also compressed. In addition to the Consent Decree Advertisement date requirement, the project also has a Consent Decree construction date requirement.

- Achievement of Full Operation (AFO)—December 31, 2026.

If the City, their consultants, or Contractors do not meet a milestone before the required date, the Consent Decree agreement permits the U.S. Federal Courts to assess financial penalties on the City. The penalty amounts range

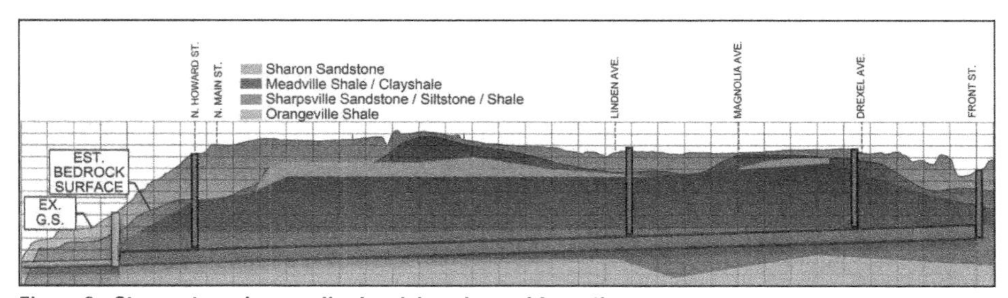


**Figure 6. Storage tunnel generalized rock layering and formations**

from $3,000 to $5,000 dollars per day, depending on the number of days after the required date that the milestone is achieved.

Given the current market conditions, where materials take longer to produce due to reduced workforce and rising costs, as well as longer shipping times, available float in the construction schedule is limited. The aggressive construction schedule will force the construction contractor to be creative with solving problems quickly as they arise during construction in order to keep the project on track. Based on the 75% design estimated construction schedule, critical path construction activities are expected to be related to the procurement of the Storage Tunnel rock TBM, preparation of the rock TBM launch shaft (the TDS shaft), rock tunnel mining and construction, and construction of the TOC outlet pipes and EDS. Assuming a 14-month procurement schedule for the rock TBM, current schedule estimates show less than a month of float for the AFO milestone.

In order to increase the amount of float in the construction schedule, the City and AECOM will incorporate several items into the construction contract. One item was introduced to the city by the independent peer review panel.

### Construction Schedule Opportunity 1—Refurbished Tunnel Boring Machines

The City of Akron is considering allowing Contractors to utilize refurbished rock TBMs installing a precast concrete segment tunnel up to 19 feet ID. Over the past 20 years, numerous tunnels have been built in similar rock conditions in Ohio (including the cities of Cleveland, Akron, Columbus, and Lorain) and in the province of Ontario, Canada. In addition to several sewage tunnels close to the NSIT tunnel diameter, the majority of single-track subway tunnels in the region require TBMs in the 6.5-meter (21 feet) diameter. As discussed above, the lowest rock cover for the storage tunnel occurs at the launch shaft and at the upstream retrieval shaft (Rack 35). The cover at the upstream retrieval shaft is more than 25 feet above the current tunnel crown, and the rock conditions at the launch shaft improve less than 100 feet out of the shaft because of the steep rise in the top of rock elevation. By allowing the contractors to build the storage tunnel with a finished inside diameter between 16.5 and 19 feet, the City hopes Contractors will be able to reduce the TBM procurement time.

### Construction Schedule Opportunity 2—Construction of TOC Pipes in Mined Tunnel

The storage tunnel launch shaft/TDS structure was located by the City in an area which will be screened by old growth forests from existing apartment buildings to the east and the Chuckery Park activity and parking areas to the west. To maintain this visual buffer, and to be able to launch the storage tunnel TBM into a full face of rock, the NSIT contractor will have to launch from the TDS shaft, which is mid-way up a 140-foot high hill between the NSIT construction staging area in Chuckery Park and the closest North Hill private property up the hill. On the other hand, as shown in Figure 7, in order to convey sewage out of the TDS, twin 90-inch ID Tunnel Overflow Conduits must be installed through the hillside in mostly rock materials using trenchless construction methods to reach the EDS structure on the Cuyahoga River. The peer review panel suggested that the Contractors may be able to use the Tunnel Overflow Conduit excavation as a tail tunnel to the TDS, and in fact it could be daylighted to the north of the TDS, which would permit drive-in access. One challenge for this item is the TOC proximity to a set of stone stairs which were built during the Works Project Administration (WPA) in the early 1900s. Provided these can be protected, the construction documents will likely suggest this option.

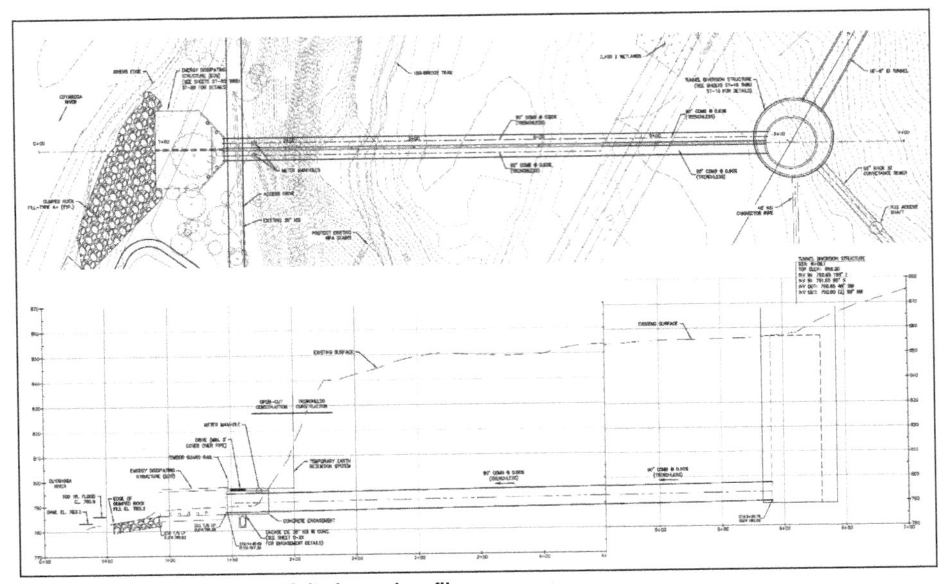

**Figure 7. Tunnel overflow conduit plan and profile**

## Construction Schedule Opportunity 3—Separate Launching Shaft for Rack 32 Tunnel

The TDS structure is required to collect and distribute flows to all the tunnels at the west end of the project. In addition, the largest available staging area for the Contractors is at the Chuckery Park. Finally, the upstream Rack 32 shaft is located in a heavily forested and sensitive park area, and construction staging areas are limited. For all these reasons, it is very likely that the Contractor selected for this job will have to launch the Rack 32 conveyance tunnel TBM from the same area as the TDS (see Figure 8). Depending upon how fast the Contractor can build access to the TDS site and construct the Storage Tunnel Launch shaft, there may not be space and time to build the Rack 32 tunnel form that same shaft. In order to gain additional float, the city is requiring a smaller Rack 32 launch shaft, mining from that shaft, and then it will be finished to provide Operations and Maintenance access.

# RACK 32 SOFT GROUND TUNNEL CHALLENGES

The calculated design flows from the Rack 35 and Rack 32 sewersheds account for 94% of the total 1.08 BGD overflow rate used to design the TDS, TOC, and EDS. Fortunately, Rack 35 flows will drop directly into the NSIT storage tunnel at the upstream baffle drop. However, the Rack 32 combined sewer pipes (which account for 31% of 1.08 BGD flowrate) currently converge at the southwest quadrant of the project area. Figure 9 is a portion of the Ohio Department of Natural Resources Division of Geologic Survey top of bedrock topography for the west end of the NSIT project. Its shows the relative position of the Rack 32 drop shaft and the proposed Tunnel Diversion Structure at the end of the NSIT. The map suggests a buried bedrock valley, located just south of the Rack 32 upstream shaft, extends west to east from the Cuyahoga River into the North Hill neighborhood. In order to be able to overflow to the Cuyahoga River, the TDS structure has to be founded around EL 792. This was

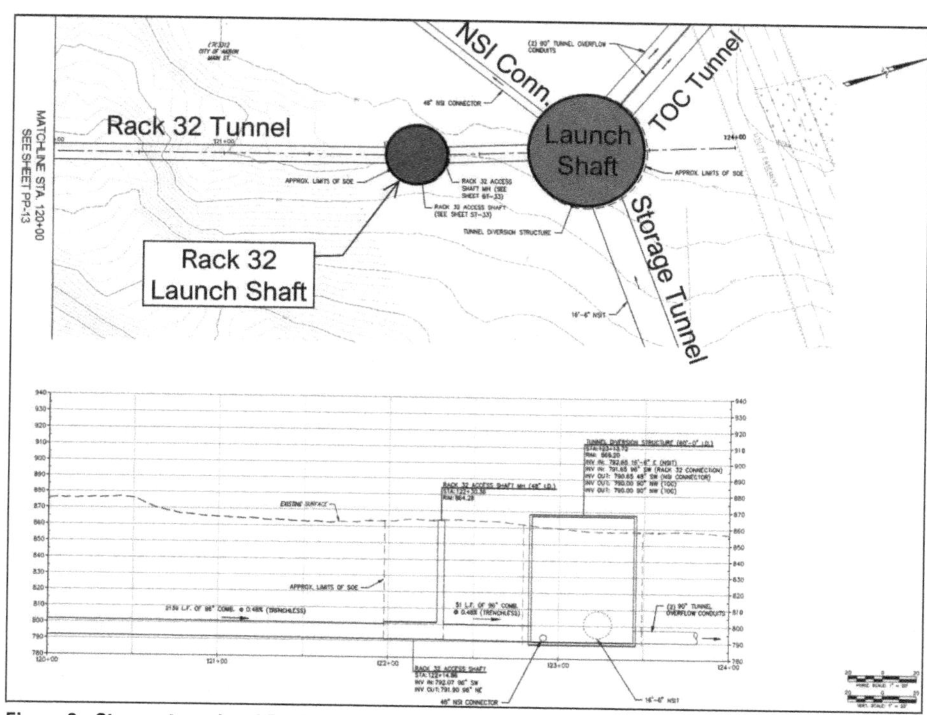


Figure 8. Storage tunnel and Rack 32 tunnel launch shaft locations

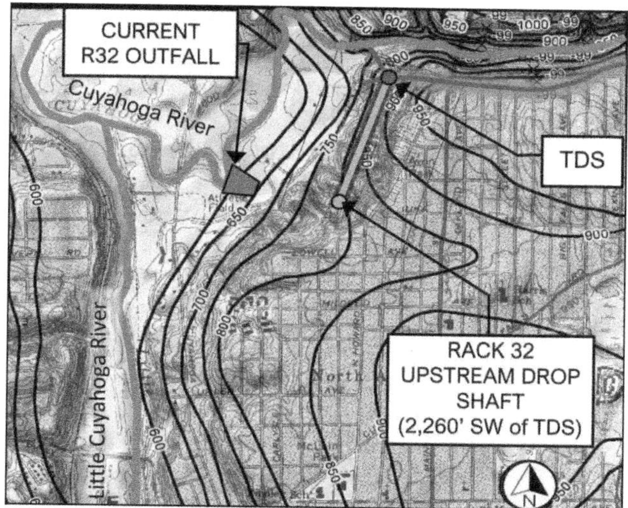


Figure 9. Annotated bedrock surface topography map (Volkmeiker, 1996)

sufficiently deep to keep the NSIT entirely in rock. However, to achieve minimum velocities in the northward flowing Rack 32 conveyance sewer, the upstream end must be at EL 800 or higher, meaning the Rack 32 conveyance tunnel would have to be constructed in mixed face or entirely in soft ground conditions.

To further complicate the Rack 32 conveyance sewer planning, the straight-line alignment shown on Figure 9 extends below an existing heavily forested park and crossed below several steep ravines which extend from east to west (draining towards the Cuyahoga River).

Following completion of the Basis of Design Report (BODR), the Rack 32 alignment was expected to be constructed using shallow profile and installed using microtunneling construction methods (MTBM) along a curved alignment which bowed outwards to the east. The curved alignment kept the tunnel profile entirely in soft ground conditions while also maintaining minimum cover over the top of the pipe in the deep ravines along the route. A flow drop structure was to be added at both the upstream and downstream ends of the alignment. The R32 conveyance pipe was to be 2,480 LF long, constructed using 96-inch ID precast concrete jacking pipe, along an alignment with a single 1,000-foot radius curve. Cover varied from about 30 to 120 feet (under a ridgeline), except at one valley where cover was expected to be 10 feet. Ground conditions included shale, a transition zone to stiff silty clays, and compact to very compact silts and silty fine sands. The groundwater table was anticipated to be at the ground surface, which was as

In August 2022, the City of Akron convened an independent group of experienced geotechnical and tunnel construction professionals to provide a peer review of the 30% design plans. The peer review panel member expressed concern with the length and diameter of the curved microtunnel option and suggested opening up the bid to other construction methods. The options suggested included the following:

- The same shallow alignment from the 30% Design but allow use of an MTBM on straight line alignments with a turning shaft at the mid-point.
- A deeper straight-line alignment between the Rack 32 upstream shaft and the TDS. This would likely require use of either a MTBM or a pressurized face Tunnel Boring Machine. The peer review panel noted that the straight-line options may allow the city to utilize jacking pipes made from materials which are more corrosion resistant.

AECOM proceeded with additional geotechnical investigations along the straight-line Rack 32 alignment to assess ground and groundwater conditions. After some key borings were completed, in early October 2022 a workshop was held discuss the advantages, disadvantages and risks of the alignment options and construction methods (open-face TBM vs. closed pressurized face TBM vs MTBM). The generalized subsurface profile and alignments under consideration are shown in Figure 10. The city decided to include only the straight-line, deep tunnel alignment in the design document and to require the Contractors to utilize a TBM with pressurized face capabilities and installing precast concrete tunnel liner segment. The biggest driver of this decision was accessibility of the tunnel alignment for any type of emergency rescue shaft.

## UNIQUE PROJECT CHALLENGES AND CONSTRAINTS

The NSIT project is being implemented at the same time as several other significant City and State projects. The most significant of these projects is the removal of the Gorge Dam, a 60-foot high by 420-foot-wide concrete hydropower dam located on the Cuyahoga River adjacent to and the NSIT tunnel alignment. In addition to the

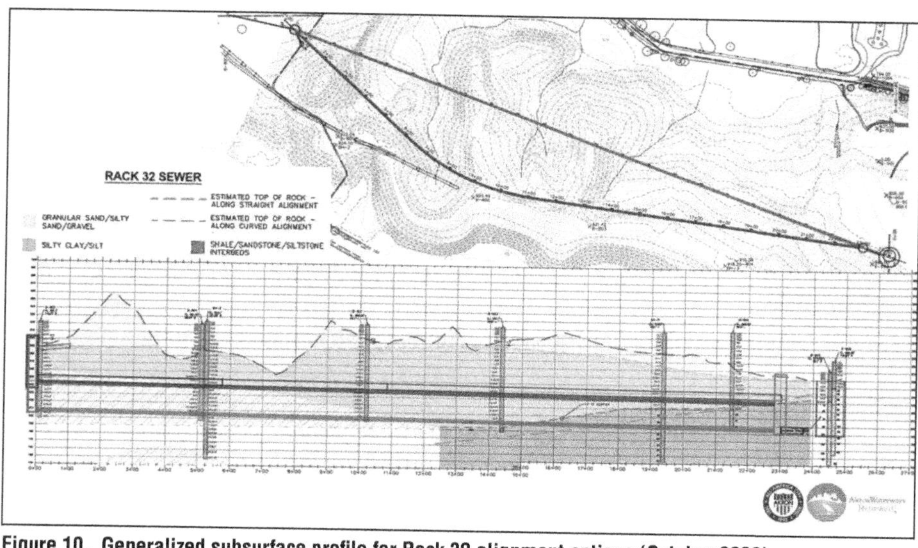


**Figure 10. Generalized subsurface profile for Rack 32 alignment options (October 2022)**

residents of the North Hill neighborhood, Summit County Metroparks operates the park and trail areas in which the NSIT will be built, and the City of Cuyahoga Falls is directly across the Cuyahoga River from the project area as well as contributes sewage to the NSI system. The following sections describe some of the coordination efforts performed by the City of Akron and the AECOM design team.

## Adjacent Projects

The Chuckery Area is a critical location that will require extensive coordination between many contractors. Peck Road is the only access to this 200-acre park area which is bound on two sides by the Cuyahoga River. The park is almost entirely wooded except for two sports fields and two parking lots.

Several major construction projects share the same project area and schedule. The Gorge Dam is located on the Cuyahoga River directly north of the tunnel alignment. The dam and approximately 1,000,000 cy of sediment behind the dam will be removed with construction starting in Spring 2024. The sediment will be pumped west in three 8" HDPE pipes along the southside of the Cuyahoga River to the Chuckery Area. Once the pumped sediment reaches this area, the pipeline will traverse through the main NSIT construction area and down a steep grade before crossing Peck Road. Here, it will be stabilized using a Pneumatic Flow Tube Mixing (PFTM) process. The PFTM process, which involves the addition of Portland cement, will be located just south of the NSIT staging and storage yard. Once stabilized, the sediment will be placed over a 30-acre site between Peck Road and the Cuyahoga River.

A new bridge is planned to be built across the Cuyahoga River on a north-south line which will run between the proposed Rack 33 drop shaft and the existing Main Street Bridge (see Figure 3). The NSIT Rack 33 drop shaft area is being designed to place new NSIT structures or tunnels near proposed bridge abutment locations. A traffic roundabout is also planned to be constructed connecting the existing Main Street Bridge, Main Steet, North Howard Street, and Riverside Drive. The new Main Street Bridge will also be connected once completed. This new roundabout will be in close proximity to the NSIT Rack 33 drop structure. Although the tunnel and shaft are likely to be constructed first, AECOM is coordinating with the other project owners in an attempt to avoid future interferences.

## Parkland

A portion of the Northside Interceptor Tunnel (NSIT) was originally planned to be located within the Chuckery Area of Cascade Valley Metroparks. The land is owned by the City of Akron (City) but is leased/managed by Summit Metroparks (SMP). This park site is a viable location for the tunnel since the City of Akron owns the land and because the Northside Interceptor (NSI) sewer (which the tunnel dewaters into) routes through the Chuckery park. Although this area is owned by the City, the design team was met with additional challenges that altered the design and alignment of the NSIT.

## Stakeholders

- For coordination with the Gorge Dam Removal project, the team interacts with USEPA, their design consultant, and stakeholder group.
- Also heavily involved in the Gorge project, Summit Metro Parks manages the Chuckery Area and owns and manages Gorge Metro Park.
- The City of Cuyahoga Falls is located just north of the NSIT project area and relies on the City of Akron to receive their wastewater at two locations in the NSIT project area. Babb Run Park Bird and Wildlife Sanctuary, which is located across the Cuyahoga River, is managed by the City of Cuyahoga Falls.
- Several agencies are involved for permitting the project including USACE, USFWS, ODNR, Ohio EPA, OHPO.
- The NSIT will be located in a residential neighborhood. Throughout the design, the team has taken steps to inform the public about the future project. Door hangers, mailed letters, social media posts, public open houses, yard signs, and Ward newsletters have been used to communicate with the residents here.
- The ultimate owner and operator of the NSIT will be the City of Akron Water Reclamation Services. With a fast-tracked design schedule, bi-weekly meetings were held to expedite input and address issues ahead of plan delivery.

## Historical Places on and Near the NSIT Project Site

One of the first challenges the design team encountered was the existence of the Works Progress Administration (WPA) stairs. Two sets of stone steps (North and South) were built in 1937 as a WPA project and have been since incorporated into the SMP Chuckery Trail.

A preliminary assessment was conducted to find potential historical and/or cultural resources that may be within or near our project area and require additional investigation. The WPA stairs were determined eligible for the National Register of Historic Places (NRHP) by the Ohio Historic Preservation Office (OHPO). Due to the NRHP eligibility, the WPA stairs could not be impacted by construction of our tunnel without consultation with OHPO, SMP, and other parties.

To mitigate impacting both sets of WPA stairs the tunnel was designed to maintain a minimum 50-foot distance from the stairs. This distance was determined using a one-to-one slope to calculate the zone of influence from the depth of the tunnel. Settlement and vibration monitoring will also be utilized during construction to confirm that the stairs are not being impacted.

Figure 11. South WPA stairs (view from top)

Figure 12. North WPA stairs (view from bottom)

The challenge of not impacting the WPA stairs was greater than expected once the actual rock depth was encountered during our geological exploration. At one time, the City and design team considered an open cut launch portal. However, the rock face for mining would have to be 300 feet farther east than expected. The location of the South WPA stairs was too close to the preferred portal alignment. To protect the WPA stairs and reduce long term visual impact, a launch shaft (TDS) was designed at the point where the TBM could launch into a full face of rock.

The North WPA stairs also posed a challenge during design because they limited the Tunnel Overflow Conduits' (TOC) route to the Cuyahoga River. The TOC had to be routed East or West of the WPA stairs to avoid impacting them, and that in turn dictated the launch shaft (TDS) location. The TOC has been routed east of the WPA North stairs, the launch shaft was shifted an additional 100-feeteast, which in turn shortened the storage tunnel. The ripple effect was that originally planned 16-foot ID tunnel now needed to be 16.5 feet ID to achieve the minimum required storage capacity of 10.3 million gallons.

## Wetlands, Rare and Endangered Species

Approximately 20 acres of tree and forest clearing is anticipated for the NSIT project. Ohio Department of Natural Resource (ODNR), United States Fish and Wildlife Service (USFWS), and Summit Metroparks have reviewed and commented on the project's impacts. Mussel reconnaissance surveys were performed in the Cuyahoga River at the proposed outfall locations to determine the presence of endangered freshwater mussels. Fortunately, no mussels were found during the surveys. Summit Metroparks provided a recent map of rare and endangered plants and flowers and the design team performed surveys throughout the growing season. Fortunately, no populations or individuals of identified potential threatened or endangered species were identified during the surveys. Due to the large number of trees required to be removed, USFWS requested that a bat survey be conducted to determine if any threatened or endangered Indiana or Long-Eared bats were in the project area. A mist net survey was performed, and no threatened or endangered bats were found during the survey. The project has been requested by ODNR to limit tree clearing to within the normal off-season dates. Clearing that may need to occur outside the off season will have to be approved by ODNR before occurring, which could impact the construction schedule. Multiple category 1, 2, and 3 wetlands exist within the Chuckery Park area. The Rack 33 conveyance sewer design was directly impacted by the presence of a Category 3 wetland. However, the design was revised after the Peer Review to avoid the wetlands to avoid potential permitting delays.

## REFERENCES

AWR!. (2022, October 30). *City of Akron celebrates progress in consent decree projects*. Akron Waterways Renewed! Retrieved December 18, 2022, from https://www.akronwaterwaysrenewed.com/news/city-akron-celebrates-progress-consent-decree-projects.

Haywood, D. (2013). (rep). Akron CSO Program 2012 Northside Interceptor Tunnel Advanced Facility Plan. (pg 3-4 to 3-8). Akron, Ohio: City of Akron.

Horrigan, Mayor. D. (2022, April 19). Letter to U.S. EPA Administrator Michael Regan. Akron, Ohio.

Siebert, A. J. (2015). (rep.). *City of Akron Ohio Canal Interceptor Tunnel Geotechnical Baseline Report* (Final Submittal, pp. 10–118). Akron, Ohio: City of Akron.

Tunnel Achievement Award, Ohio Canal Interceptor Tunnel. (2019, August). Tunnel Business Magazine. Retrieved from www.tunnelingonline.com.

Vormelker, J.D., and Swinford, E.M., 1996, Bedrock topography of the Akron West, Ohio, quadrangle: Ohio Division of Geological Survey, Digital Map Series BT-3B Akron West (supersedes Open-File Map version), scale 1:24,000.

# Evaluating the Tradeoffs Between Microtunnelling and Tunnel Boring Machines for Small Diameter Tunnels—A Case Study of the Ferry Road and Riverbend Combined Sewer Relief Project

**Dani Delaloye** ▪ Mott MacDonald
**Kas Zurek** ▪ City of Winnipeg

## ABSTRACT

Evaluating and selecting the appropriate excavation method is a key to the success of any tunnelling project. The benefits and risks of different methods should be well understood by the designer and communicated to the project owner to enable selection of a preferred method from a technical, social and environmental impact, cost and constructability perspective.

For tunnels that have an internal diameter around 3 m, depending on the tunnel purpose and ground conditions, both microtunnelling with pipe jacking and tunnel boring machine excavation with a one- or two-pass lining may be feasible construction methods. Using the Ferry Road and Riverbend Combined Sewer Relief Project located in Winnipeg, MB as a case study, this paper will discuss the benefits and drawbacks of each method, including technical design, social, environmental, cost, schedule, and risk considerations.

## INTRODUCTION

The City of Winnipeg is in the preliminary design phase for a 1.6-km-long 3-m-diameter tunnel within an urban environment as part of the Ferry Road and Riverbend Combined Sewer Relief Project (the Project). The alignment follows the narrow, north-south residential Rutland Street. To the north, the alignment extends beyond the northern terminus of Rutland Street beneath the St. James Rods football fields to the western projection of St. Matthews Avenue. To the south, the alignment extends beyond the southern terminus of Rutland Street beneath the St. James Library parking lot, a track field, and Bourkevale Park to the northern shore of the Assiniboine River. The last approximately 60 m of the alignment is planned to be constructed using open cut methods. The sewer alignment has a gradual downslope from north to south, approximately following the existing ground surface.

The alignment traverses beneath four intersections with east-west roads. From north to south, these are: Silver Avenue, Ness Avenue, Bruce Avenue, and Portage Avenue. It is our understanding that traffic at all these intersections, apart from Bruce Avenue, cannot be interrupted by construction of the sewer.

Due to the length and diameter of the Project, both microtunnelling with pipe jacking and tunnel boring machine excavation with a precast or one- or two-pass lining may be feasible construction methods. This paper describes the evaluation and comparison of these two construction methods, using the Project as a case study.

## TUNNEL SUPPORT OPTIONS

Before the excavation method can be evaluated, it is important to understand what methods of tunnel support are acceptable. Tunnel support for small diameter tunnels excavated in soil may consist of pipe jacked behind the tunnelling machinery, segmental lining erected behind the machine, or steel ribs and lagging typically erected in traditionally mined tunnels.

### Pipe Jacking

Pipe jacking is a trenchless construction technique in which prefabricated pipe segments are installed by pushing (jacking) them longitudinally through the ground behind a tunnelling shield at the same time as excavation takes place within the shield. A hydraulic jacking frame in the jacking shaft advances the pipe string (with the tunnelling machine at the front) and when retracted, a new pipe section is inserted. The pipe and shield are jacked from the jacking shaft to the receiving shaft.

Intermediate jacking stations (IJS) are typically installed within the pipe string to advance shorter sections of pipe. An IJS consists of a series of hydraulic rams installed between two jacking pipes within a telescoping steel shell. When needed, the IJS can be used to advance just the pipes in front of it, thereby splitting the full jacking thrust among multiple sections of pipe.

Tunnel lengths for pipe jacking are typically limited by jacking forces approaching the capacity limits of the jacked pipe or jacking frame. Because the pipe string is jacked from the frame within the jacking shaft, jacking forces generally increase with increasing frictional forces between the ground and the lengthening pipe string. Techniques such as using IJSs and injecting lubrication in the annular space between the ground and casing pipe can help to maintain low jacking forces for longer lengths of tunnel. Typical tunnel lengths using pipe jacking are 600 m or less, however longer tunnels have been constructed.

Shafts are required at each end of each drive, therefore, because the length of drive is more limited with pipe jacking than with other tunnel support methods, more shafts are required when using pipe jacking. The jacking shaft will also include a thrust block, typically a reinforced concrete mass at the backside of the shaft to support the jacking forces imposed on the jacking frame. The thrust block is cast directly against the shoring, therefore the shoring and the soil immediately behind the shoring must be able to resist the anticipated thrust forces from the jacking operations.

Pipe jacking as a method of tunnel support is also typically limited to a maximum diameter of approximately 3 m due to the weight of the jacking pipe segments. However, advances are being made to expand capabilities with recent pipe jacking projects being completed with diameters in the 3.4 m range.

### Segmental Lining

Segmental lining is a tunnel construction technique in which prefabricated lining segments are installed in place immediately behind the tunnelling shield. Several segments are installed to form a completed ring. Unlike with pipe jacking, the rings are installed in their final position and the tunnelling machine pushes off the completed rings. Segments are typically precast reinforced concrete using reinforcing bars or fibers. Rubber gaskets are placed in the joints between segments to reduce water inflow into and outflow out of the tunnel. The shield is advanced from the launch shaft to the reception shaft.

Tunnel lengths for segmental lining are generally only limited by the muck disposal cycle. Muck may be transported from the face to the launch shaft using conveyor belts or muck cars (on rubber tires or rail). A single tunnel length of 2000 m is well within the capability of a segmental lining tunnel.

Due to the capabilities for longer tunnel alignments, a segmentally lined tunnel in the Project's range would only require a shaft at each end. Nevertheless, intermediate shafts between launch and reception shafts can be used for cutterhead inspection and maintenance. Segmental lining tunnels do not require a thrust block or shoring designed to carry thrust forces.

Segmental lining can be used for a wide range of tunnel diameters—at the smaller end, diameters less than approximately 3 m in diameter are uncommon due to challenges with segment handling and operations within the tunnel. Hence, the Project diameter is near the limit of the applicability of using segmental lining.

### One-pass Versus Two-pass System

With both pipe jacking and segmental lining, the tunnel support may be part of a one-pass system or a two-pass system. With a one-pass system, the tunnel support installed also acts as the completed pipeline. In a two-pass system, an oversized tunnel support system is initially installed followed by the installation of the carrier pipe or product pipe. The annular space between the tunnel support and the carrier pipe is then typically grouted. The benefit of a two-pass system is that the carrier pipe can be installed to exact line and grade within the tunnel which may have deviated from line and grade during excavation. A larger tunnel diameter is required for a two-pass system to allow for the installation of the carrier pipe on spacers or skids, and therefore increases tunnelling costs as well as pipe material costs as the carrier pipe is a separate entity from the tunnel support. For pipe jacking in particular, the two-pass system is not considered feasible for the Project because the larger tunnel diameter required for the two-pass system is at the limit of available microtunnel boring machine (MTBM) diameters on the market.

## EXCAVATION METHOD SELECTION

Once the tunnel support options are understood, the excavation method can be evaluated. For small diameter tunnels in soil, various types of shield tunnelling techniques could be used for excavation with pipe jacking and segmental lining. Although the contractor is typically responsible for means and methods, it is important to evaluate the potential tunnelling techniques during the design process for several reasons. First, it is necessary to anticipate what methods contractors might use in order to establish an accurate basis for the engineer's cost estimate. Second, it may be in the owner's interest to rule out certain construction methods based on their associated risks and constructability issues. Finally, the bid documents should baseline the risks that contractors must address with their means and methods to help ensure that the bids will accurately reflect the constructability challenges associated with a given project.

### Types of Tunnelling Machines

There are three main groups of available tunnel boring machines (TBMs): open shield, earth pressure balance (EPB), and slurry. An open shield machine does not provide active support to the tunnel face, so this approach is generally best suited for stable ground such as stiff clay or granular soils above the groundwater table. Both EPB and slurry machines have the capability of continuously balancing the external earth and water pressure at the tunnel face by applying a positive face stabilization pressure through different methods. All three main groups—open shield, EPB, and slurry—are compatible with pipe jacking and segmental lining tunnel support systems.

### Open Shield Machine

The three main types of open shield machines are open face TBM, cutter boom shield, and backacter (backhoe) shield. Due to the potential for unstable face conditions, the cutter boom and backacter shield options were ruled out during the conceptual stage. Open face TBMs can be equipped with hydraulically operated flood doors which can be used to help control the amount of material excavated through the TBM cutterhead and to close off the tunnel face, preventing unstable ground and groundwater from entering the machine. Even with flood doors, open face TBMs are typically not effective in unstable ground below the groundwater table.

### Earth Pressure Balance Machine

EPB TBMs have a sealable excavation chamber behind the cutterhead where the excavated material can be mixed with additives and pressurized. EPB TBMs balance external pressures by adjusting the machine advancement against the rate of material extraction from the excavation chamber using a screw conveyor. EPB TBMs are best suited to soils with appreciable amounts of fines (silt and clay; particles passing 60 microns or a No. 200 sieve) because these particles help create the low-permeability, plastic mixture needed to control the face pressures. Excavated material in the excavation chamber is conditioned using additives such as foam, bentonite and polymers to provide a consistency capable of exerting an even balancing pressure against the tunnel face.

### Slurry Tunnelling Machine

Slurry TBMs balance external pressures by mixing the excavated material in the excavation chamber with a bentonite suspension (or slurry) and pressurizing the contents of the chamber to create a support medium for the tunnel face. The slurry mixed with excavated soil is then pumped through a pipe system to a separation plant, where the excavated material is separated out of the slurry, and the slurry is recirculated through the slurry circuit. The contents of the excavation chamber are pressurized by regulating the pressure in the slurry circuit and, in some cases, by means of an air bubble in a special chamber behind the excavation chamber. Slurry TBMs perform best in clean sands and gravels because separating fines from slurry is a complex, costly and time-consuming method. A fines content exceeding 20% to 30% can create difficulty for the slurry separation system.

### Microtunnelling Machines

Both EPB and slurry machine types can be used for microtunnelling, which for the purposes of this paper is defined per ASCE/CI 36-15 "Standard Design and Construction Guidelines for Microtunneling" (ASCE, 2015), as a specific trenchless method with the following four characteristics:

- Remotely controlled: The operator of the microtunnel boring machine (MTBM) is located in a control cabin on the surface, rather than inside the tunnel like with a conventional TBM.
- Guidance: The MTBM utilizes a guidance system, typically a laser system for shorter drives and total stations for longer drives, and an articulated cutterhead to steer the tunnel to design line and grade.

- Pipe jacking: Prefabricated pipe segments sequentially jacked into place behind the MTBM.
- Continuous support: A positive pressure is applied, either through EPB or slurry, to control ground and groundwater inflow at the face.

Although EPB MTBMs do exist, slurry MTBMs are much more available and more widely used than EPB MTBMs.

### Machine Selection

TBM selection is typically completed by evaluating the range of geotechnical parameters anticipated to be encountered, including the following:

- Fines content
- Permeability
- Consistency
- Relative density
- Confinement pressure
- Swelling potential
- Abrasivity

A range in geotechnical parameters can be assessed for each geologic unit and then compared to the capabilities of each TBM type, placing it within one or multiple of the following categories:

1. Anticipated parameters are a limited application of the TBM type where precautionary methods must be adopted.
2. Anticipated parameters are an extended application of the TBM type where the application depends on particular details of the ground conditions.
3. Anticipated parameters are within the main field of application of the TBM type, or within the common procedures and methods of the TBM type.

A quantitative evaluation system can be applied to evaluate the relative capabilities of each machine type. For example, each geotechnical parameters can be weighted on a scale of 1 to 3 based on importance. A value can then be assigned ranging from 1 to 3 considering whether the machine type capabilities (i.e., limited application = 1, extended application = 2, and main field of application = 3). An overall score for the application of each TBM type in each ground type is calculated from the following:

Overall score = Weight of parameter 1 × rating of parameter 1 + weight of parameter 2 × rating of parameter 2 + …

## DESIGN AND CONSTRUCTION CONSIDERATIONS

In addition to the technical considerations of tunnel support and excavation method, constructability of a tunnel is essential to consider during the design. This section

**Table 1. General capabilities of machine types**

| Machine Type | Suitable Ground Types | Capability Below the Groundwater Table |
|---|---|---|
| Slurry tunnelling machine | Soft ground, rock, mixed ground | Yes |
| Earth pressure balance machine | Soft ground with low permeability preferred, conditioning can increase ground type range | Yes |
| Open shield machine | Stable soft ground and rock | Requires dewatering or ground improvement in permeable ground |

discusses the major factors affecting the design and constructability of pipe jacking and segmental lining tunnels and provides insight on how these will influence tunnelling on this project.

Tunnel construction minimizes most surface related construction impacts, but it does not eliminate them all. The following are several of the key requirements for tunnel construction sites, which have constructability, as well as social and environmental impacts:

## Shaft Requirements

For both pipe jacking and segmental lining options, the shield/tunnel will be advanced from the jacking or launch shaft to the receiving or reception shaft.

### *Shaft Sizes*

For pipe jacking, the footprint of the jacking shaft must be sized to accommodate the thrust block, jacking frame, tunnel eye, slurry pumps, the length of one pipe segment, and the length of the longest MTBM "can" (the MTBM typically consists of a few separate segments called cans). The footprint of the receiving shaft is used to remove the MTBM and must be sized to accommodate the tunnel eye and the length of the longest MTBM can.

As discussed previously, the pipe jacking method requires a thrust block cast against the shoring. The shoring should be designed to carry the thrust forces considering the strength of the soils immediately behind the shoring and thrust block.

For 3-m-diameter tunnels, the shafts sizes for segmental lining tunnels are typically similar to MTBM tunnels.

### *Shaft Locations*

General shaft locations should be selected considering typical tunnel lengths for pipe jacking and segmental lining, available laydown area, traffic impacts, and community impacts. When selecting the exact locations of shafts, consideration should be taken for existing underground utilities, existing overhead utilities, and crane placement and swing radius.

For multiple tunnel drives in a series, mobilization time and costs are reduced by having two drives launch from the same jacking shaft, and conversely two drives received from the same receiving shaft. Once one drive has been completed, the equipment can be turned around, thrust blocks coordinated, and the MTBM relaunched from the same location. This also saves on shaft construction costs as the number of jacking shafts, which have larger footprints than receiving shafts, is reduced.

For pipe jacking, the shafts are located assuming a typical maximum drive length of 600 m, and by locating shafts near intersections where more space is available for laydown area. Longer pipe jacking lengths than 600 m are feasible depending on ground conditions. To confirm the maximum drive length pipe jacking force calculations for the given ground conditions are required (these should be completed at the next stage of design if pipe jacking is carried forward). Longer drive lengths also require a highly skilled tunnelling contractor utilizing a great amount of jacking force control.

### *Shaft Access*

When shafts are open, they should be isolated from accidental entry using barriers, covered with traffic plates, or otherwise fenced. A crane is required at each shaft location and to have, as a minimum, an unobstructed swing for shaft construction, pipe transfers into the shaft, and all equipment set-up and tear-down.

## Staging Areas

The American Society for Civil Engineers (ASCE) and Construction Institute (CI) recommends a minimum 650 $m^2$ at the jacking shaft to accommodate cranes, loaders, bentonite mixers, pallets of bentonite, the slurry separation plant, the control room, the power plant, utility pipes, slurry pumps, trucks, and wastewater treatment (ASCE/CI 2015). This does not include area for casing pipe laydown which can be located at the jacking shaft or at an off-site location. For the receiving shaft, ASCE/CI recommends a minimum of 375 $m^2$. ASCE/CI does not indicate the range of tunnel sizes that these minimums apply to, and larger diameter tunnels require larger surface equipment (cranes, separation equipment, etc.) which in turn require more laydown area. Typically, less surface laydown area is required for EPB than slurry as EPB does not require a slurry separation plant which has a fairly large footprint.

Laydown areas with similar width and length (e.g., square shape) are preferred over long, narrow laydown areas. As with small laydown areas, narrow laydown areas reduce efficiency and can lengthen cycle times and overall schedule. Both jacking/launch and receiving/reception staging areas should be sized and shaped to accommodate construction of the shafts.

## Noise, Vibrations and Odors

The major sources of noise and vibrations assuming unrestricted equipment are the crane, shaft construction equipment, generators, trucks, and slurry separation equipment. If needed, sound walls can be erected in compliance with local building codes. A contractor may elect to replace a crane with an electric gantry crane to reduce operational noise during the tunnelling operation. Slurry separation plants are also particularly noisy. The city noise ordinance should be consulted, but sound walls will likely be required for shaft sites located in residential neighborhoods.

Most generated odors will be from diesel equipment, which may be used on pipe jacking or segmental lining tunnels. The slurry separation plant for a slurry TBM may cause odors to be released.

## Truck Access and Parking

18-wheeled truck access is required for the import and export of equipment, materials, workers, and spoils during construction. The trucks typically will need to either drive through the site or turn around to reenter existing traffic. Trucks may also back-in to the site, if feasible.

The number of trucks required to access the site will depend on the excavation rate, tunnel size, working hours, amount of storage available at staging area, etc. For example, for the Project, the number of daily trucks is expected to range between 20–40.

In addition, the project workers will need road access and parking as well as other appropriate facilities near each work location.

### Waste Disposal and Discharge of Water

The spoils may include additives. All additives, if available, can be required to align with regulation, (i.e., for the Project to be NSF/ANSI Standard 60 clean water approved or meet local regulatory requirements). The approval is to be based upon the additive and not upon removal of the material. Other waste materials should be contained and removed daily.

Construction related wastewater will be generated from tunnelling activities such as changing water in slurry tanks, therefore discharge locations and treatment requirements should be identified in the contract documents, and any required permits obtained. If there is no suitable disposal location at the site, the water will need to be trucked off site.

### Power Requirements

Tunnelling equipment requires a large amount of power and may require access to the power grid. Alternatively, contractors may use portable generators. The need for power grid access should be assessed early on and power grid availability described in the Contract Documents. Power requirements for TBM tunnelling are typically higher than what is required for microtunnelling operations.

## COST, SCHEDULE, AND RISK

A general comparison between the cost and schedule for different machine types is provided in Table 2.

Other considerations for comparison of project cost, schedule and risk include the following:

- The number of shafts can significantly increase project cost and construction schedule, especially due to their impact on the surrounding areas and the permitting timeline and cost.
- The biddability of a project is a key consideration when selecting the preferred construction method for the design. The designer should evaluate the local contractor familiarity as well as the appetite for contractors from other regions to bid on the work.
- Geotechnical baseline reports should be used for any tunnel project to manage and clearly allocate geotechnical risk
- Two-pass lining can provide better hydraulics and more certainty on gradient than one-pass lining systems.
- The tunnel lining type selection can have significant cost implications—for example, the cost of a ribs and lagging system is about 60% of the cost of a segmental lining.
- Intervention locations can be selected along the alignment in advance as a planned risk mitigation strategy.
- The owner/designer can stipulate requirements for use of IJSs for long microtunnel drives—although the IJSs may not all be used, including them in the pipe string reduces the risk of the pipe freezing.

**Table 2. General cost and schedule comparison of machine options**

| Machine Type | Cost and Schedule |
|---|---|
| Slurry tunnelling machine (STM) | Higher cost, slower production |
| Earth pressure balance machine (EPB) | Higher cost, slower production |
| Open shield machine | Lower cost, faster production |

Table 3. Parameters per geologic unit

| Geotechnical Parameter | Alluvium | Lacustrine Soil | Silt Till |
|---|---|---|---|
| 1. Fines content (%) | 50 to 99% | 99 to 99.9% | 92 to 97% |
| 2. Permeability (m/sec) | $1\times10^{-10}$ to $4\times10^{-5}$ | $1\times10^{-10}$ to $1\times10^{-8}$ | $2\times10^{-6}$ to $3\times10^{-5}$ |
| 3. Consistency | 0 to 0.67 | 0.04 to 0.68 | N/A |
| 4. Relative density (%) | N/A | N/A | Loose to very dense |
| 5. Confinement pressure (bar) | 0.2 to 0.68 | 0.39 to 0.93 | 0.38 to 0.90 |
| 6. Swelling potential | Little | Little | Little |
| 7. Abrasivity | Very low to medium | Very low to low | Medium to v. high |

It is important to note that the selected contracting method (i.e., Design-Bid-Build, Design-Built and Early Contractor Involvement/CMGC) also has impacts on project cost, schedule, and risk, but evaluation and comparison of these methods is beyond the scope of this paper.

## FERRY ROAD AND RIVERBEND COMBINED SEWER RELIEF PROGRAM CASE STUDY

The geotechnical conditions for the Project are summarized in Table 3.

### Tunnel Support

Due to the soil and groundwater conditions, steel ribs and lagging are not considered an acceptable solution for the lining of the Ferry Road tunnel. Either pipe jacking or segmental lining are expected to be applied.

### Excavation Method

Based on geotechnical setting alone, EPB TBM is considered the preferred tunnelling method as it is well-suited to the cohesive soils that are most prevalent along the alignment, and it also typically works well in the cohesionless glacial till. The anticipated high abrasion potential of the cobbles and boulders in the glacial till may be problematic, and precautionary measures should be adopted.

Slurry TBM is less capable in alluvium and lacustrine clay than the EPB TBM but on par within the glacial till. In general, slurry TBM is not as well suited to cohesive soils as EPB TBM and requires anti-clogging additives and more expensive slurry separation techniques.

Open shield TBM is less suited to all ground types present on the Project, in particular the loose glacial till which is below the groundwater table and anticipated to exhibit flowing ground conditions. The risk of flowing ground could be reduced through extensive dewatering of the ground or ground improvement. Without extensive dewatering or ground improvement performed from surface along the alignment, open shield TBM is not considered a feasible option. Hence, as surface impacts are to be limited open shield excavation is not considered feasible.

Based on case histories, contractor experience, and machine availability, both EPB and slurry machines are viable and available, though for pipe jacking, slurry machines are much more widely used than EPB machines due to availability of the machines and contractor familiarity.

## Shaft Sizes and Locations

For preliminary engineering of the Project, the footprints of the jacking and receiving shafts have been evaluated based on using an AVND3000AB model MTBM, 3,000 mm length pipe segments, and 1 m thick shoring:

- Jacking shaft: 13.5 m long × 8 m wide rectangular shaft or 14 m diameter circular shaft
- Receiving shaft: 10 m long × 7 m wide rectangular shaft or 10.5 m diameter circular shaft

Given the requirements of the Project, the above recommended dimensions are anticipated to be sufficient for the launch and reception shafts for the segmental lining option as well.

Due to the length of the alignment and area restrictions, three to five shaft locations are expected for a microtunnelling options whereas only two shafts would be required for a TBM tunnel.

## Other Considerations and Comparable Projects

Local underground contractors in the Winnipeg area have limited experience with pipe jacking operations in the diameter range considered in this project; however, there are enough larger Canadian tunnel contractors with the experience and equipment required for the Project to create a competitive bidding environment. For example, Ward & Burke constructed the City of Winnipeg's Cockburn Sewer Relief Project. Where the option exists and compatible equipment is available, contractors typically refurbish and use equipment they already have from a past project. This presents the opportunity for cost savings for the contractor and the owner. For the Wacker Drive Tunnel project in Chicago, Illinois, a Caterpillar EPBM previously used as a conventional precast segmental lining TBM was refurbished to a reinforced concrete pipe jacking machine in 2011 (Trisi et al. 2012). On the South Interceptor Parallel Phase III Project in King County, Washington, the same machine was used (with modifications) for precast segmental lining installation and reinforced concrete pipe jacking on different parts of the same tunnel alignment (Breeds et al. 2003).

Similar to pipe jacking, local underground contractors in the Winnipeg area do not have experience with segmental lining tunnels in the diameter range considered for this project; however, there are enough larger Canadian tunnel contractors with the experience and equipment required for the Project to create a competitive bidding environment. For example, Michels Corporation, a contractor with experience in small diameter segmental lining tunnels, has locations in British Columbia, Alberta, and Ontario, and McNally International, headquartered in Ontario, constructed the Port Mann Water Supply Tunnel in British Columbia.

Whereas 3 m inside diameter is considered on the large side for pipe jacking projects, 3 m inside diameter is considered on the small side for segmental lining projects. One very similar and well-documented project is the Blacklick Creek Sanitary Interceptor Sewer Tunnel constructed by Michels Corporation. The tunnel 3 m ID tunnel and was constructed in Columbus, Ohio using a 3.7 m diameter EPB TBM and precast concrete segmental lining. The tunnel 6.88 km alignment included two intermediate shafts for future works which also served as access points for TBM maintenance, and 13 curves with radii of 335 m or less. The ground conditions consisted of alluvium, glaciolacustrine, glacial till, and more, with clusters of boulders anticipated. The owner of the project required the use of an EPB TBM equipped with a manlock for hyperbaric tunnel face and cutterhead access. Due to the small inside diameter, the contractor could not use a conveyor belt system for muck handling and the TBM was 16 m long to accommodate the necessary equipment and hyperbaric manlock. The length of the TBM, the line-of-sight for an optical based TBM guidance system such as a theodolite or laser was diminished. This, compounded with the number of curves in the alignment, made a theodolite or laser system unreliable, and the contractor opted to use a gyroscope and hydraulic level for guidance. The use of a gyroscope resulted in a 23 mm deviation from the design tunnel alignment at reception (Kerr et al, 2019).

## CONCLUSIONS

Evaluating and selecting the appropriate construction method is a key to the success of any tunnelling project. For tunnels that have an internal diameter around 3 m, depending on the tunnel purpose and ground conditions, both microtunnelling with pipe jacking and tunnel boring machine excavation with a precast one- or two-pass lining may be feasible construction methods. Each project needs to be evaluated on an individual basis to determine which method is best suited, considering the geotechnical conditions, social and environmental constraints, and local construction experience. The cost, schedule and risk considerations are also key to selecting the appropriate method. In some cases, such as the Ferry Road and Riverbend Combined Sewer Project, both methods are technically feasible and have similar cost and schedule estimates, therefore, to improve biddability of the project, the choice of construction method can be left to the contractor. The key is for the owner to be well informed by their designer of the tradeoffs for each method to allow for appropriate selection.

## REFERENCES

American Society of Civil Engineers (ASCE)/Construction Institute (CI) 2015. *Standard Design and Construction Guidelines for Microtunneling*. ASCE/CI 36–15.

Breeds, C.D., Molvik, D., Gonzales, D. and Fulton, O. 2003. South Interceptor Tunnel Construction Using a Lovat EPBM. *Proc., Rapid Excavation and Tunneling Conference (RETC), Society for Mining, Metallurgy & Exploration (SME)*, Englewood, CO, 1080–1085.

Kerr, A., Ross, M., Whitman, E., Alavi, E. 2019. Challenging Geology: Blacklick Creek Sanitary Interceptor Sewer Project in Columbus, Ohio. *Proc., Rapid Excavation and Tunneling Conference (RETC), Society for Mining, Metallurgy & Exploration (SME).*

Trisi, W.R., Sawicki, W. and Barkowski, J. 2012. Large diameter pipe jacking for the MWRD By-Pass tunnel project in Chicago, USA. *Proc., TAC 2012 Conference, Tunnelling Association of Canada (TAC)*, Richmond, BC.

# Metropolitan Water Tunnel Program in Massachusetts

**Rafael C. Castro** ▪ JCK Underground, Inc.
**Kathleen M. Murtagh** ▪ Massachusetts Water Resources Authority
**Paul V. Savard** ▪ Massachusetts Water Resources Authority

## ABSTRACT

Preliminary design is underway for the Massachusetts Water Resources Authority's (MWRA) next major tunnel program, the Metropolitan Water Tunnel Program (MWTP). The MWTP is the latest part of MWRA's ongoing efforts to achieve redundancy for its metropolitan water transmission system including its historic deep rock City Tunnel, City Tunnel Extension and Dorchester Tunnels. The MWTP comprises several deep rock tunnels that will be approximately 10-foot to 12-foot finished diameter totaling just over fourteen miles long and ten connections to existing large diameter surface piping and pumping stations. This paper will present an update on current engineering and permitting efforts, packaging schemes, and the schedule to complete the MWTP.

## INTRODUCTION

MWRA is a public authority that provides wholesale water and sewer services to 3.1 million people and more than 5,500 large industrial users in 61 communities in eastern and central Massachusetts. For its wholesale water system, MWRA currently operates over 200 facilities including the John J. Carroll Water Treatment Plant, with a capacity of 405 million gallons per day (mgd), 12 pump stations, and 14 below- or above-ground storage tanks. Interconnecting these facilities with the communities they service is a water transmission system which includes 105 miles of active tunnels and aqueducts (mostly 10 to 14 feet in diameter) and 39 miles of standby aqueducts. [Murtagh, 2019]. Figure 1 shows the general layout of the major components of MWRA's water transmission system.

### Genesis of the Metropolitan Water Tunnel Program

The existing Metropolitan Tunnel System consists of 10-foot and 12-foot diameter concrete-lined pressurized tunnels ranging in depth between 250 feet and 450 feet below ground surface (bgs). They were constructed as deep rock tunnels using primarily drill and blast methods from the 1950s to 1970s. These tunnels connect to MWRA surface pipe network through a series of steel- and concrete-lined shafts and underground valve chambers. The Metropolitan Tunnel System carries approximately 60% of the total system daily demand to the communities within and around the metropolitan Boston area.

MWRA and its predecessor organizations have been planning for and constructing a fully redundant system east of the Wachusett Reservoir since at least the 1930s. Construction of the Cosgrove Tunnel and MetroWest Water Supply Tunnel (MWWST) provided redundancy for approximately 25 miles from the Wachusett Reservoir to the beginning of the existing Metropolitan Tunnel System. The Metropolitan Water Tunnel Program (MWTP), which consists of two deep-rock tunnels, will provide full redundancy for the remainder of the system, allowing the Metropolitan Tunnel System to be taken offline for inspection and repair as necessary.

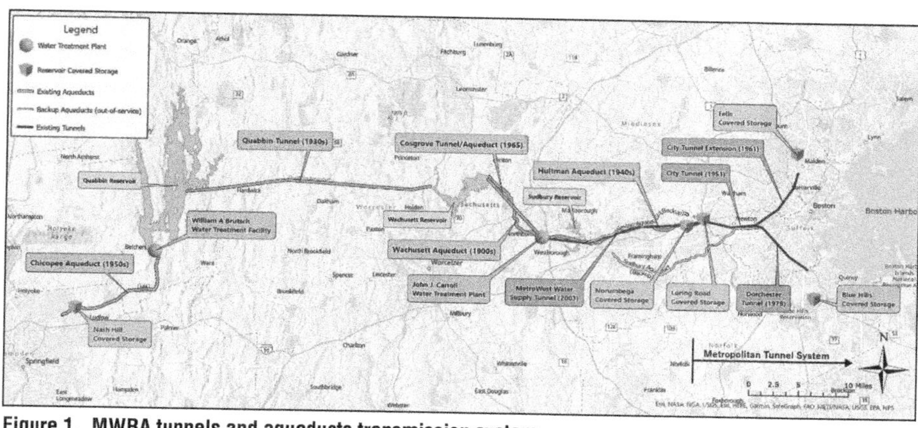


Figure 1. MWRA tunnels and aqueducts transmission system

## PROJECT SETTING

Two independent tunnel systems will comprise the Metropolitan Water Tunnel Program, one to the north and one to the south. The two tunnels will connect the Hultman Aqueduct and MWWST to surface transmission mains and pumping stations currently being serviced by the Metropolitan Tunnel System.

### North Tunnel

The North Tunnel will provide a connection from the Hultman Aqueduct near the Interstate 90 and Interstate 95 interchange (I90/I95 Interchange) in the Town of Weston to MWRA's existing Weston Aqueduct Supply Main No. 3 (WASM 3) transmission main near the border of the City of Waltham and the Town of Belmont. The North Tunnel will be approximately 4.5 miles long, roughly parallel the Charles River and pass nearby the Stony Brook Reservoir, which is a water supply for the City of Cambridge.

### South Tunnel

The South Tunnel will provide a tunnel connection from the Hultman Aqueduct (separate and isolated connection from the North Tunnel and the existing City Tunnel) in Weston to the southern terminus near the Dorchester Tunnel Shaft 7C in Mattapan, a neighborhood in the city of Boston. The South Tunnel alignment will be approximately 10 miles long and in addition to the two communities at each end, will pass beneath the communities of Wellesley, Needham, Newton, and Brookline. Approximately 4 miles of tunnel will parallel the Charles River.

### Geologic Setting

As a result of the city of Boston's long history of coastal settlement and as a locus of general scientific study, knowledge of the Boston-area and eastern Massachusetts bedrock geology extends back into the mid-19th century, yet only recently has the examination of new deep rock borings been able to provide a better understanding of the stratigraphic rock relationships and location of faults within the poorly-exposed bedrock found in metropolitan Boston. The wide variety of rock types present are considered part of the regional Avalon Terrane platform that extends from Rhode Island to Atlantic Canada and Newfoundland, reflecting a long geologic span of igneous activity and sedimentary rock accumulations.

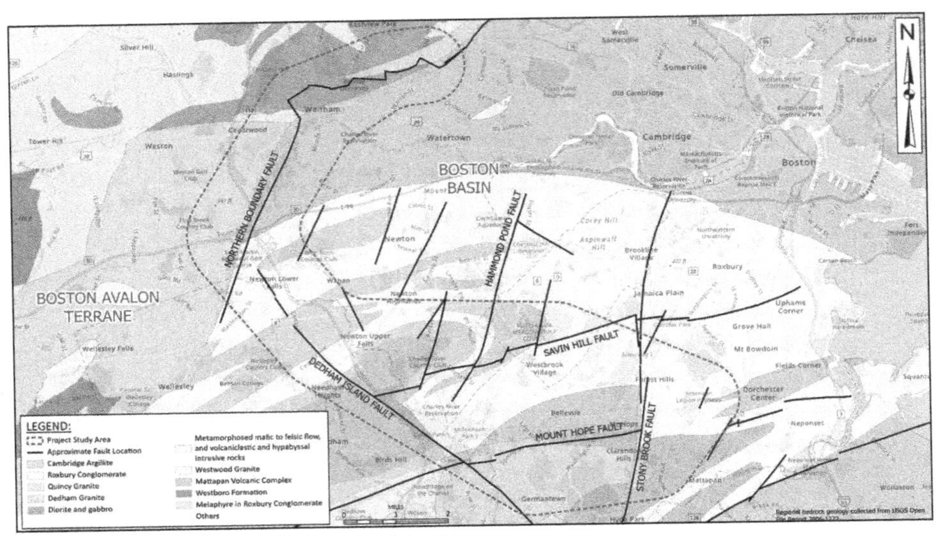


**Figure 2. Tunnel program preliminary geology map**

Locally, the bedrock setting for the MWTP area consists principally of uplifted Late Precambrian-age igneous rocks that form an upland border ring that surrounds the topographically-low Boston Basin in a northerly, westerly and southerly direction. Figure 2 indicates the geology in the area of the tunnels. Separated from the Boston Basin by several major fault systems (including the Northern Boundary Fault to the north and west, and the Savin Hill Fault to the south), the igneous upland rocks consist of granites, a series of undifferentiated gabbro/diorite/tonalite intrusives, complex volcanic rocks, a meta-sedimentary quartz-rich rock unit enclosed within the igneous blocks and an older, highly-weathered sedimentary unit that predates the igneous rocks. These rocks all exhibit a wide range of strengths, hardnesses, abrasivities and variable degree of weathering and alteration. The North Tunnel will likely be excavated principally in these rock types.

Within the fault-bounded and topographically low Boston Basin are a complex arrangement of slightly-folded and faulted sedimentary rocks (conglomerates, sandstones, argillites) and interbedded intrusive and extrusive mafic volcanic rocks. The South Tunnel is anticipated to cross these Boston Basin rocks, along with uplifted blocks of the silica-rich Mattapan Volcanic Complex. Similarly, the South Tunnel bedrock is expected to exhibit a wide range of strengths, hardnesses and abrasivities, and variable degrees of weathering and alteration.

ºSubsequent glaciation has deposited a wide variety of unconsolidated soil deposits on the irregular bedrock surface. Along with the aforementioned fault systems (Northern Boundary Fault and Savin Hill Fault) are other recognized fault zones, such as the multi-splay Stony Brook Fault (crossed by the Dorchester Tunnel) in Forest Hills, the Dedham Island Fault in the western limit of the Boston Basin and other unnamed faults that cross rocks within and beyond the Boston Basin.

## ALTERNATIVES EVALUATION, PERMITTING AND PRELIMINARY DESIGN

The Program has progressed to the preliminary design phase that comprises Alternatives Evaluation, Environmental Permitting, and the Preliminary Design. This work includes joint efforts by MWRA's Tunnel Redundancy Department, and its Program Support Services consultant and Preliminary Design Engineer consultant. The following describes the efforts to date for each of the three key aspects of the preliminary design effort.

## Alternatives Identification and Evaluation

An alternatives evaluation was completed to identify a preferred alternative that would be carried through to environmental permitting and preliminary design. Prior to selecting alternatives to evaluate, connection points to the existing system and available land for both temporary construction support and permanent facilities were identified. Figure 3 shows the results of the alternatives evaluation, the preferred alternative (Alternative 4) including shaft site intended use. Note that the South Tunnel comprises two tunnel segments. MWRA must construct and operate both segments to provide a South Tunnel that meets the redundancy requirements.

### *Connection Points*

The connection points to the existing water system to provide the desired redundancy were selected using MWRA's hydraulic model under future high day water demands of years 2040 and 2060.

The North Tunnel and the South Tunnel both connect to the Hultman Aqueduct, which is the source of water for the tunnels, in the vicinity of I90/I95 Interchange in Weston. The downstream end of the North Tunnel connects to an existing large diameter transmission water main (Weston Aqueduct Supply Main No. 3, WASM 3) in the vicinity of the Waltham and Belmont border. There are two other connection points planned along this tunnel: the first to the Lexington Street Pumping Station (through the School Street Shaft Site); and the second to the Cedarwood Pumping Station in Waltham.

The downstream end of the South Tunnel connects to the near surface distribution pipelines at the Dorchester Tunnel Shaft 7C. There are four other connection points planned along this tunnel: the first to the Hegarty Street Pumping Station in Wellesley;

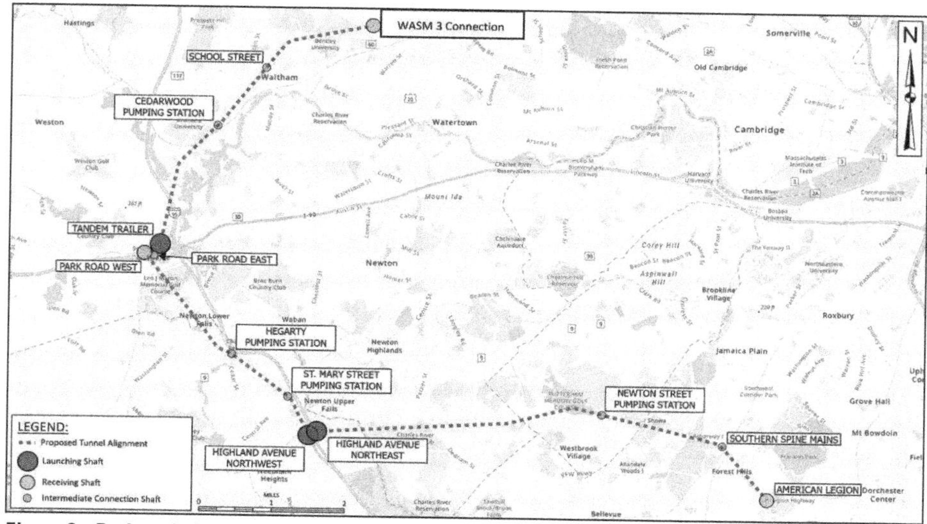


**Figure 3. Preferred alternative tunnel alignment**

the second to the St. Mary Street Pumping Station in Needham; the third to the Newton Street Pumping Station in Brookline; and the fourth to the Section 39 Southern Spine Mains in Boston.

#### *Land Use and Availability*

MWRA initially identified land suitable for possible shaft locations in the vicinity of the connection points. Initial screening to evaluate suitability of land for connection, launch, and receiving shafts considered the following site characteristics:

1. Approximately 5 acres for a launch shaft, 2 acres for a receiving shaft, and 0.2 or more acres for shafts not used as a launch or receiving shaft
2. Undeveloped land suitable for construction staging, shaft excavation, excavated material removal, and concrete operations
3. Near major transportation roadway facilities for transport of excavated material for off-site re-use or disposal
4. Launching shaft sites suitable for construction dewatering and with a nearby receiving water suitable for discharge of anticipated volume of construction water
5. Preferably isolated away from residential areas or other noise sensitive receptors

In addition to the connection points discussed above, to facilitate tunnel construction of the longer South Tunnel, an additional construction shaft site was identified. This construction site would serve as a launch or receiving shaft site located midway along the South Tunnel at the Highland Avenue Interchange on I-95 in Needham.

#### *Evaluation of the Alternatives*

Alternatives were developed using the land identified near the connection points and the I95/Highland Avenue Interchange. Each tunnel alternative considered various combinations of TBM launching and receiving shafts that when taken together would constitute a complete North and South Tunnel. After an initial screening, ten alternatives were furthered in the evaluation.

A multicriteria decision tool was developed to consistently apply the evaluation criteria and subcriteria to compare the 10 alternatives. Figure 4 shows a summary of the multicriteria results comparing the final set of 10 alternatives that lead to shortlisting to the three most favorable.

All 3 shortlisted alternatives are included in the Draft Environmental Impact Report, and the selected preferred alternative is being carried forward into preliminary design.

### Environmental Permit

In Massachusetts, it is necessary to demonstrate that construction of a project of this magnitude has considered the environmental impacts in conformance with the Massachusetts Environmental Policy Act (MEPA). The MEPA Office conducts reviews of environmental impacts of development projects and other activities that require one or more state agency actions and provides opportunities for public review of these potential environmental impacts. The MEPA review process requires the Proponent (the MWRA) to study the environmental impacts and to use all feasible measures to avoid, minimize, and mitigate damage to the environment or, to the extent damage to the environment cannot be avoided, to minimize and mitigate damage to the environment to the maximum extent practicable. MWRA was required to assess the

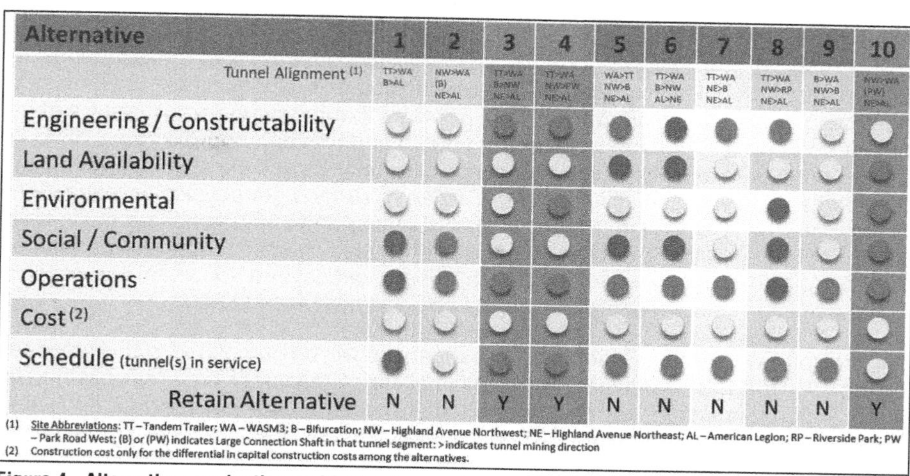


**Figure 4. Alternatives evaluation multi-criteria overview results**

environmental impacts of the project and publish a detailed report on these impacts within an Environmental Impact Report (EIR) to the MEPA Office.

MWRA completed and submitted the Draft EIR (DEIR) in October 2022 for public comment that ended in December 2022. The DEIR demonstrates that MWRA consulted with the MEPA Office and other permit agencies and held appropriate outreach to impacted stakeholders. As part of the process, MWRA developed a set of clear commitments of mitigation measures to implement, estimated the individual costs of each proposed measure, identified the parties responsible for implementation, and provided a schedule for implementation. These measures must first look to avoid and minimize impacts and then to mitigate impacts when needed. Recent changes to MEPA regulations require projects to assess environmental justice considerations and include outreach to the affected communities as well as an assessment of impacts on climate change. Other important impacts and mitigation steps looked at in the DEIR include: rare species and wildlife habitat, wetland and waterway resources, water supply, cultural and historical resources, hazardous materials, land use, open space and community resources, transportation, air quality and greenhouse gas (GHG) emissions, and noise and vibration.

A Final EIR (FEIR) will address comments received on the DEIR and finalize the mitigation commitments that will be further developed during final design and incorporated into the construction contracts as needed. It is estimated that the FEIR will be completed in November 2023.

## Preliminary Design

Preliminary Design will take the DEIR preferred alternative to approximately a 30-percent level design, culminating in a Preliminary Design Report (PDR) and Drawings, along with a Geotechnical Data Report (GDR) and a Preliminary Geotechnical Design Report (PGDR). These documents, among others, will then be used to inform the final design. The status and approach to key elements of the Preliminary Design are discussed herein, including geotechnical investigations, hydraulics and transient analysis, horizontal alignments, liner design and vertical alignments, construction and professional services contract packaging and contract practices.

### *Geotechnical Investigations*

At the onset of the Program, it was clear that the geotechnical investigations for approximately 14 miles of deep rock tunnel would be on the critical path to completing the design and starting construction. We estimate that the program will require about a hundred deep test borings (greater than 400 feet deep) and approximately 35 to 70 shallow borings or test pits to support design of structures and pipelines as well as site developments. It will also include over 10 miles of surface geophysical investigations to complete the final design. MWRA began planning and implementing a multiphase program in early 2020. The preliminary design phase investigations are focused on supporting selection of shaft sites, the preliminary vertical and horizontal tunnel alignments and inform engineering, constructability and cost evaluations.

The preliminary design geotechnical investigations will be completed in three phases (phase 1A, 1B and 1C) and include review of historical geotechnical data collected by others, outcrop mapping, approximately 20 deep borings, approximately 62,000 linear feet surface geophysical surveys, downhole geophysical logging, in-situ water pressure testing and laboratory testing. In general, the phase 1A borings were primarily focused on selected potential shaft and connection sites identified during the alternatives evaluation, while phase 1B and 1C focus on collecting information along the potential tunnel alignments between the shafts sites. There are a number of faults and key formation features that are important in selecting the preliminary horizontal and vertical alignment, which are also the focus of the investigations.

Since the geotechnical investigations have such significant influence on the Program schedule, MWRA has elected to procure a Geotechnical Support Services (GSS) consultant that will focus on collecting geotechnical subsurface data and fill even a short time gap between preliminary design and final design to keep the investigation program moving forward.

### *Horizontal and Vertical Alignments*

The horizontal alignment at the preliminary design phase is based in large part on proximity to the necessary connections and availability of land (for temporary construction use and permanent facilities), as well as considerations of the geologic conditions anticipated, TBM steering capabilities, and other factors. For example, the North Tunnel crosses and runs in close proximity to the published location of the Northern Boundary Fault. This fault distinguishes between the meta-sedimentary rock of the Boston Basin (e.g., argillite, shale, sandstone, conglomerate) which lie to the east of the fault and crystalline rocks (e.g., granite, diorite, gabbro, etc.) which are being thrust upward on the west side of the fault. The preliminary design will include one approach to the horizontal alignment through this area, considering the angle at which the tunnel will cross the fault and which side of the fault presents rock conditions with the least risk to the project. As another example, the horizontal alignment looked to reduce the number of crossings below surface waters as these can influence groundwater infiltration rates especially when they exist in the vicinity of the many faults known and unknown along the tunnel alignment. To mitigate some of this risk, MWRA has identified local water supply wells, irrigation wells, and geothermal wells based on current records as these may be impacted by groundwater infiltration during construction.

The vertical alignment is based on several factors including geotechnical conditions affecting the soil/rock cover, the maximum slope for constructability, proximity to the existing tunnel systems, operational considerations, and anticipated direction of TBM excavation. The soil/rock cover criterion is based in part on the tunnel internal operation pressure, hydraulic grade line (HGL), and whether the pressure can be contained by the vertical in-situ stress of the ground with an appropriate safety factor. The objective at the preliminary design stage is to determine the tunnel vertical alignment for which an unreinforced (plain) concrete liner is viable while minimizing tunnel depth and the length of tunnel that would require reinforcement and/or steel pipe.

For construction purposes, the maximum practical grade for the tunnel will be set based on consideration of a maximum slope for which rail can be supported, which is about 4%. To ensure the most efficient use of rail, if selected by contractors, the maximum grade used for the preliminary design vertical alignment has been based on 1%.

Further adjustments to the tunnel horizontal and vertical alignments will continue as the design progresses and more is learned about the geologic profile along the tunnel alignment (e.g., top of rock, rock quality, fault locations) from subsequent investigations. As the design progresses and the final location of the horizontal alignment is set, it is conceivable that adits may be needed to make connections to the shafts. The vertical profile will also continue to be refined during the final design phase as the design is refined and more geotechnical information is collected.

### *Tunnel and Shaft Design Approach*

The tunnel is planned to be 10 to 12-foot finished diameter. The diameter will be finalized considering the hydraulic analyses, water quality as measured by water age, constructability considerations and cost. The main consideration related to the finished diameter for hydraulics and water quality is water age. The diameter of the tunnel is a key factor on how fast water is delivered to its destination and directly related to water age. Based on preliminary design water age analysis, both 10-foot and 12-foot finished diameter tunnels may deliver water within MWRA standard for water age. Constructability and costs associated with delivery of concrete to construct the tunnel liner, capability of TBMs, and tunnel ventilation during construction of the longest tunnel segments, among other factors were considered in evaluating the various finished tunnel diameter range determined by the hydraulics analyses. The preliminary design is planned to conclude with a single finished diameter for the tunnels.

The tunnel liner design approach anticipates a plain (unreinforced) cast-in-place concrete final lining along the majority of the tunnel alignment. The lining will not be designed to contain the internal tunnel water pressure as mentioned when discussing the approach for setting the tunnel vertical profile. Rather, the concrete lining will provide a smooth internal wall for hydraulic efficiency and long-term durability. This approach is well documented in the literature [Chen, 2022]. It is a proven approach for the geologic conditions anticipated based on the previous tunnels already completed by MWRA.

Reinforcement may be added to the cast-in-place concrete lining at locations where the rock quality, strength or permeability are insufficient for a plain concrete lining, however, the tunnel alignment will be selected to limit the number and length of such locations in order for the economy of the overall tunneling approach to remain favorable. The reinforced concrete lining may require further upgrades to a steel lining at locations where rock quality and strength are particularly low, where rock mass permeability is particularly high, where the operational pressure of the tunnel exceeds the minimum in situ confining stress, or where required for hydraulic separation from adjacent infrastructure.

During tunnel excavation, the primary rock support would include support structures such as rock dowels, shotcrete, and steel sets and lagging. The tail and starter tunnels and connection adits, if needed from intermediate connection shaft locations to the main tunnel or between tunnel segments, may be constructed by drill and blast method or other mechanical means. The type of TBM will be further evaluated as additional geotechnical information is collected, however, based on the limited information collected to date, a main-beam machine with a robust probing and grouting capabilities would be a reasonable conclusion.

Shafts are planned to be excavated down through overburden and rock to a depth of approximately 250 to 400 feet below the surface. Launching and receiving shafts are large enough that they are expected to be constructed through rock by drill and blast method (top-down), while some connection shafts would be constructed through rock using raisebore (bottom-up), blind drilling, or other techniques.

For intermediate connection shafts, the finished shaft diameters would range from 2 to 10 feet (to be optimized during final design phase). The raisebore construction method is attractive for intermediate connection shafts because it requires a relatively small construction footprint and limits the amount of material that needs to be removed from the surface at the intermediate connection shaft site. Notwithstanding potential surface impact restrictions, the means and methods of shaft construction are expected to be left to the contractor to select.

***Contract Packaging***

The three tunnel segments of the preferred alternative are each less than 7 miles long, which is advantageous from a construction mobilization and phasing perspective. The preferred alternative (Figure 3) would require six construction shaft sites, three for launching and three for receiving. MWRA is considering whether to package construction in 2 or 3 main tunnel construction contracts. Considerations for the tunnel construction contract packaging, to be finalized during preliminary design, include contract size to attract appropriately qualified contracting teams, mitigation of contract interface risks, efficient use of available labor resources, and other factors including ensuring adequate flexibility in the overall Program schedule. The tunnel construction contracts would follow Massachusetts General Law (MGL) Chapter 30 regulations that govern linear construction in Massachusetts.

Other potential early works packages to relocate existing facilities may be beneficial to complete prior to starting the tunnel construction packages. These could include relocation of major utilities and reconfiguring the Tandem Trailer parking facility at the Tandem Trailer launch shaft site. Early works contract packaging could include installation of temporary works that would support shaft and tunnel construction at launch shaft sites, such as installation of temporary (possibly permanent) pipelines for shaft and tunnel construction groundwater discharge lines such as at the Highland Avenue shaft sites where trenchless methods are envisioned to install these lines, and TBM power installations.

Professional services packaging schemes are currently being evaluated. MWRA is considering one or two Final Design Engineers contracts and one or two Construction Management contracts. There are many factors being considered in evaluating the number of professional services contracts including, size and duration of the contracts, availability of qualified consultant resources, MWRA resources to manage the work, consistency among the design and construction admiration approach and other factors including ensuring adequate flexibility in the overall program schedule.

### *Contract Practices*

MWRA has previously constructed a number of deep rock tunnels to convey water and wastewater throughout its system and was typically in the forefront of adopting industry practices. However, the most recent construction of a potable pressurized water tunnel was the MetroWest Water Supply Tunnel completed in 2003. Construction contract practices and risk management have evolved considerably since that time, and MWRA intends to update its practices to follow many of the most recent practices. Figure 5 overlays how various risk management construction practices have evolved over time along with past MWRA tunnel projects.

MWRA is intending to follow many of the industry practices established in *Recommended Contract Practices for Underground Construction* [Wilson 2019] and *Guidelines for Improved Risk Management on Tunnel and Underground Construction Projects in the United States of America* (GIRM) [O'Carroll 2007] along with conducting outreach to other owners, engineers, and contractors to incorporate the latest lessons learned on such practices.

To be in-line with current practice, MWRA developed and is executing a program Risk Management Plan (RMP). The RMP and the associated risk registers identify risks and assigns the risks to the party in the best position to mitigate the risks. The RMP was written with consideration of the GIRM and identifies participant roles and responsibilities, establishes the risk management process and the roles and responsibilities (MWRA, and its design consultants, the Construction Manager, and the contractors.)

It is expected that the latest guidelines and practices will be followed with respect to allocation of geotechnical risks, including the inclusion of Geotechnical Data Report (GDR) and Geotechnical Baseline Reports (GBR) as contract documents. The GDR is expected to systematically document the data from the subsurface investigations. The GBR would provide the contractual understanding (interpretation) of the subsurface site conditions and baseline geotechnical conditions that a construction contractor can rely on in preparation of its bid.

MWRA has historically included a differing site conditions (DSC) clause in its construction contracts to equitably address physical conditions that differ from those

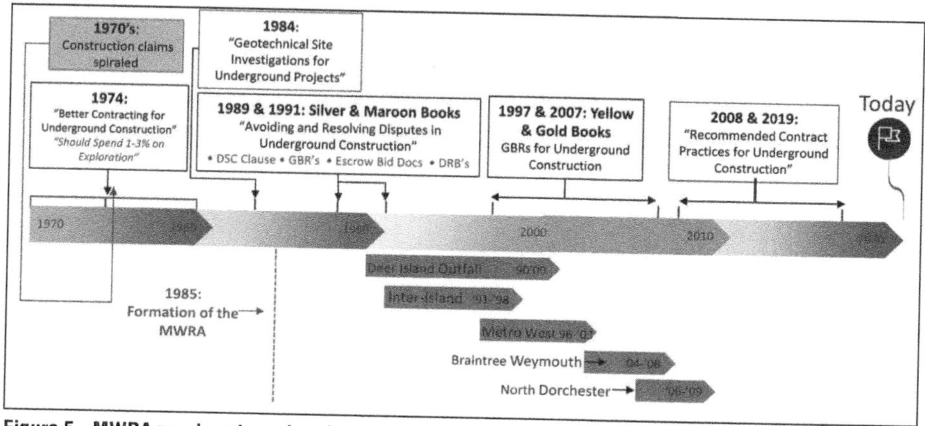


Figure 5. MWRA previous tunnel projects and industry practice changes timeline

anticipated at the time a contract is bid and that impact project cost and/or schedule (e.g., subsurface conditions, utilities, environmental conditions, existing structures). For the Program, MWRA intends to supplement the standard DSC clause to indicate how the GBR and GDR should be employed in sync with the DSC clause and to coordinate with other contract provisions.

MWRA is also considering whether to incorporate other contract practices into the construction contract documents, such as a Dispute Resolution Board (DRB), Escrow Bid Documents (EBD), Partnering, Value Engineering Change Proposal, fair escalation provisions and several other practices. Decisions regarding incorporating these practices will be made and detailed language incorporated into the contracts during the final design phase.

## FUTURE PHASES LEADING TO CONSTRUCTION

It is expected that the preliminary design phase would be completed early in 2024, with final design to start in middle of 2024. MWRA would then advance final design to prepare procurement documents, including bid-ready plans and specifications, and updates to the detailed construction cost estimates. Based on completion of these documents, MWRA would initiate the procurement process. Procurement for the first construction contract is anticipated to begin in 2027. Subsequent procurements may be staggered depending on the construction packaging considerations.

Program construction is estimated to take approximately 8 to 12 years. MWRA expects that the proposed new deep-rock tunnel system will be placed into service by 2040.

## REFERENCES

Murtagh, K., and Brandon F. 2019. MWRA Metropolitan Boston Tunnel Redundancy Program, Program Update. *Rapid Excavation and Tunneling Conference Proceedings*. Society for Mining, Metallurgy, and Exploration, Inc.

Chen, W. N. 2022. Hard Rock Pressure Tunnel Final Lining Design Guide Summary. *North American Tunneling Proceedings*. Society for Mining, Metallurgy, and Exploration, Inc.

Wilson, S. 2019. *Recommended Contract Practices for Underground Construction*. Society for Mining, Metallurgy, and Exploration, Inc.

O'Carroll, J. and Goodfellow, R. *Guidelines for Improved Risk Management on Tunnel and Underground Construction Projects in the United States of America*. Underground Construction Association, Society for Mining, Metallurgy, and Exploration, Inc.

# San Francisco Downtown Rail Extension Project: Evolution of Mined Tunnel Construction Approach and Design

**Kush Chohan** ▪ Delve Underground
**Yiming Sun** ▪ Delve Underground
**Meghan Murphy** ▪ AECOM Program Management Project Controls/Transbay Joint Powers Authority

## ABSTRACT

DTX is a 1.3-mile-long rail extension being constructed by Transbay Joint Powers Authority to extend Caltrain rail service and future California High Speed Rail service to downtown San Francisco. It includes a 3,352-foot-long mined tunnel with a cross section varying from 50 to 60 feet wide and 43 feet high to accommodate two- and three-rail tracks. DTX will be excavated in mixed-face and Franciscan Formation with ground cover ranging from 40 to 85 feet. This paper discusses how mined tunnel design has evolved since the initial preliminary design phase over 10 years ago. The current design concept for tunnel construction is also discussed.

## INTRODUCTION

The Downtown Rail Extension (DTX) will connect Caltrain's regional rail system and the California High-Speed Rail Authority's statewide system to the Salesforce Transit Center in downtown San Francisco. The rail alignment will be constructed principally below grade to provide a critical link for Peninsula commuters and travelers on the state's future high-speed rail system.

The project is being developed by the Transbay Joint Powers Authority (TJPA) in collaboration with the Program's major stakeholders: the Metropolitan Transportation Commission (MTC), San Francisco County Transportation Authority (SFCTA), Peninsula Corridor Joint Powers Board–Caltrain, California High-Speed Rail Authority (CHSRA), the City and County of San Francisco, and the California Department of Transportation.

The TJPA Board adopted a rebranding strategy for the Downtown Rail Extension at their board meeting in December 2022. At that meeting, the project was rebranded as "The Portal," with the tagline "Uniting the Bay. Connecting California." More information about the rebranding of DX will be released in 2023.

### DTX Alignment and Major Elements

The DTX alignment begins in the below-grade Salesforce Transit Center rail station at First and Mission Streets. At the west end of the station, the station's six tracks transition to two tracks through a cut-and-cover throat structure and continue in a mined tunnel southward under Second Street and westward under Townsend Street to a new underground station at Fourth and Townsend Streets. West of the station, near Seventh and Townsend Streets, the tracks ascend to grade via a U-shaped retained cut (referred to as the "U-wall"), and the alignment continues southward at grade to 16th Street, just south of the existing Caltrain terminal station and railyard. A tunnel stub box extends side-by-side with the U-wall to allow for a connection to the future

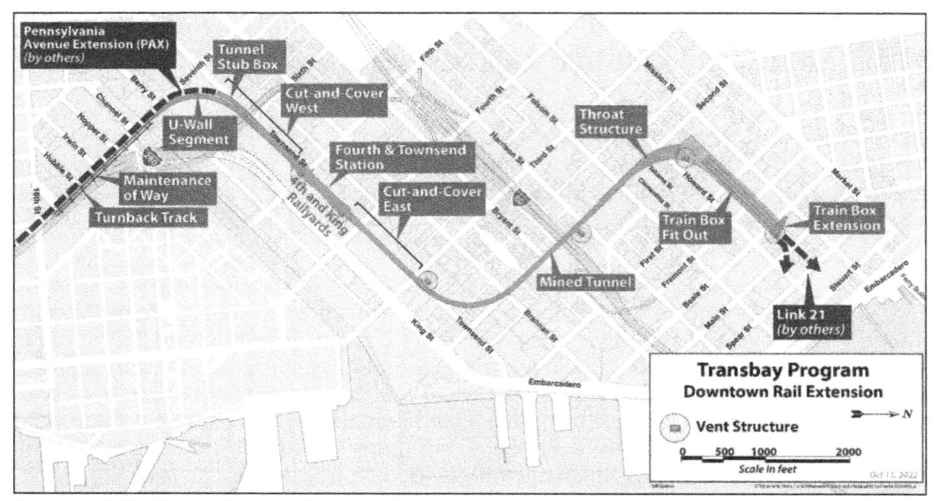


**Figure 1. DTX alignment and major elements**

Pennsylvania Avenue Extension—a tunnel being developed by the SFCTA that will grade-separate the rail alignment from surface streets. Including revenue and non-revenue at-grade trackwork and stations, the total construction length of the DTX is 2.2 miles. Figure 1 provides an overview of the DTX alignment and the major elements of the project.

The main elements of the DTX are as follows:

Salesforce Transit Center train box extension: The train box extension will extend the existing below-grade structural box of the Salesforce Transit Center eastward from the east side of Beale Street to the TJPA's property line to extend the platform lengths and provide ventilation and exiting from the east end of the train box. The train box extension will be constructed under TJPA property with an open-cut method.

Salesforce Transit Center fit-out: The fit-out of the two-level below-grade rail station at the Salesforce Transit Center will include facilities for rail operations, customer service, and ancillary support. The lower concourse, one level below the grand hall, will house ticketing, passenger waiting, and support spaces for Caltrain and the CHSRA—the primary tenants—as well as leasable retail space. On the level below, six tracks and three center platforms will serve commuter and high-speed trains. Back-of-house support spaces will also be built on this level to support rail service.

Cut-and-cover structures: Cut-and-cover construction will be used along Second Street, Townsend Street, and in portions of the 4th and King Railyards for the following structures:

- Throat structure located at Second and Howard streets at the northern end of the DTX alignment where the two-track alignment widens to six tracks at the west end of the Salesforce Transit Center
- Tunnel east of the Fourth and Townsend Street Station along Townsend Street
- Fourth and Townsend Street Station
- Tunnel west of the Fourth and Townsend Street Station along Townsend Street

- U-wall and tunnel stub box along Townsend Street west of Sixth Street to bring the tracks to grade and allow for a connection to the Pennsylvania Avenue Extension, a planned grade separation tunnel project being led by the SFCTA

Mined tunnel: Sequential Excavation Method (SEM) mining is being considered for the tunnel along portions of Townsend Street and Second Street. The mined tunnel extends from the west side of Third and Townsend Streets to Clementina and Second Streets. The tunnel is primarily two tracks but expands to three tracks as it approaches the throat structure. The length of the mined portion of the tunnel is approximately 0.65 mile.

Fourth and Townsend Street Station: The Fourth and Townsend Street Station will serve Caltrain and high-speed rail passengers with destinations in the South of Market area or transferring to the Muni Central Subway. The street-level station entrances and exits along Townsend Street will lead to two levels below grade: a concourse and a train platform level. The concourse level will accommodate passenger amenities such as ticketing machines, maps, and schedule information. This level will also house mechanical and electrical rooms and staff areas. The platform level will have two tracks: an 875-foot center platform for Caltrain passengers and two 800-foot side platforms for high-speed rail passengers. The underground station will be constructed using cut-and-cover techniques.

Ventilation and emergency egress: Ventilation and emergency egress structures will house equipment for the ventilation of the tunnels and include emergency egress to allow passengers to evacuate safely from the tunnels to grade in the event of an incident. Ventilation shafts will be located at either end of the Fourth and Townsend Street Station and the underground station at the Salesforce Transit Center. Two standalone ventilation and emergency egress structures will be located along the tunnel alignment. These will be constructed on parcels next to the DTX tunnel outside the street right-of-way, one at Third and Townsend Streets and the other at Second and Harrison Streets.

Trackwork: Trackwork includes the mainline tracks through the tunnel and stations as well as 0.4 mile of at-grade maintenance-of-way and turnback tracks within the existing Caltrain right-of-way.

Systems: Systems include rail systems such as traction power, overhead contact, train control, signaling, radio, and network systems; mechanical, electrical, plumbing, fire-life safety, and security systems for the tunnel, stations, and ventilation and emergency egress structures. Also included are other support systems, such as closed-circuit television, fare collection, and passenger display information systems.

## BENEFITS OF THE DTX PROJECT

The DTX project is recognized as a project of statewide, regional, and local importance in the California State Rail Plan, MTC's Regional Transportation Plan, CHSRA'S Business Plan, and the San Francisco Transportation Plan. The project will close the gap to downtown San Francisco by extending regional and future statewide high-speed rail systems into the new multimodal Salesforce Transit Center (STC). The connection to the Transit Center will allow transit connections to 11 regional and local transit systems and provide access to and from the East Bay and the Peninsula/South Bay. The project also will create an essential link in the state's rail strategy to improve mobility for an expanding workforce that is vital to the Bay Area's continued economic growth. The project also acts as a lynchpin for realizing the new Link 21 Transbay Rail Crossing that increase capacity and connect rail through the Northern California megaregion.

The DTX project will save San Francisco Caltrain commuters one hour per day on average and will allow passengers on California High Speed Rail a one-seat ride between San Francisco and Los Angeles. An estimated 90,000 passengers are predicted to use the DTX portion of project each day. Passengers will be offered a seamless transfer at and near the transit center, including BART, Muni, and buses serving the East Bay, North Bay, and other regional and long-distance destinations. The neighborhood where the Transit Center is located is projected to have the greatest growth in San Francisco over the next 30 years.

From an environmental and sustainability perspective, the project will improve air quality and reduces greenhouse gases (CO2e) by an estimated 3 million metric tons in the first year of Caltrain service and by over 5 million metric tons in the first year of high-speed rail service. In the opening year of DTX service, the project will remove vehicles from US Highway 101 (US 101) in San Francisco with an estimated reduction of 9.9 billion vehicle miles traveled. The transit-oriented development from the DTX project will provide an alternative to the predominant pattern of low-density sprawl that results in dependency on automobile travel.

Economically, the project builds two rail stations in areas targeted for investment, new affordable homes, and job growth. The rail extension enhances access to retail, entertainment, and employment, which allows improvement of opportunities for disadvantaged communities. For the construction of the Transit Center, the TJPA has made $145 million in payments to disadvantaged business enterprises and $137 million to small business enterprises. The project to date has had 7,900 construction workers employed, with nearly one-third identified as minorities and 300 women. With the completion of DTX, an estimated 125,000 jobs will be directly and indirectly created and open up a significant number of employment opportunities for people who live in San Francisco and along the Peninsula.

## EVOLUTION OF MINED TUNNEL DEVELOPMENT

The mined tunnel development has evolved through three major phases:

- Initial Preliminary Engineering (PE) Design, completed in 2010
- Tunnel Options Study, completed in 2017
- Updated 30% PE Reference Design, completed in 2022

### 2010 Preliminary Engineering Design

The 2010 PE Design called for approximately 3,200 feet of the mined tunnel using the Sequential Excavation Method (SEM) in the reach underneath Second Street and onto Townsend Street. Cut-and-cover structures were recommended for the northern approximately 700 feet of tunnel from the Salesforce Transit Center to the intersection of Clementina Street and Second Street as well as for the approximately 1,400 feet of tunnel east of the Fourth and Townsend Street Station. The 2010 PE Design was based on recommendations of the Final Tunnel Evaluations Report (Delve Underground 2006) and the project Environmental Impact Report (TJPA 2004). The SEM tunnel configuration was ultimately selected because it had the lowest construction costs and durations, and because the tunnel boring machine (TBM) method was not considered feasible, primarily because the operators required three tracks, right of way to incorporate the three tracks, and lack of ground cover. The three-track configuration

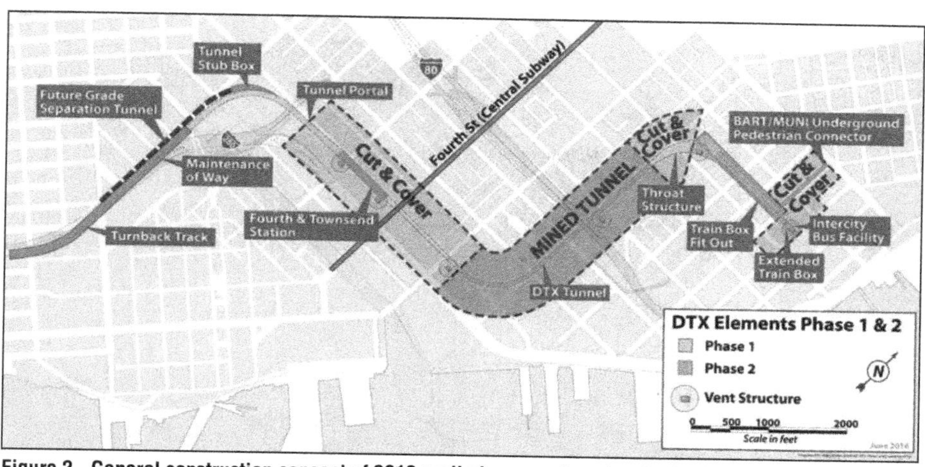


Figure 2. General construction concept of 2010 preliminary engineering design (Parsons et al. 2010)

was required to provide hold-out capacity in Second Street for trains awaiting berthing at transit center platforms and a passing track in Townsend Street for high-speed rail trains to bypass commuter trains stopped at the Fourth and Townsend Street Station. Based on anticipated ground conditions, structure width, and rail profile, the station must be constructed using cut-and-cover methods as the top of the structure is essentially at grade. West of the station, cut-and-cover construction is used until the alignment becomes too shallow for this method, at which point it transitions to retained cut U-wall construction and then to an at-grade segment. Both the U-wall and the at-grade works are in the Caltrain right-of-way. Figure 2 illustrates general arrangement and concept of the 2010 PE Design.

In the 2010 PE Design, the mined tunnel had a three-track configuration for the entire length (see Figure 3 for a typical cross section). The tunnel would be about 56 to 61 feet wide and 42 feet high (Figure 3). Ground cover above the tunnel crown ranged from 30 to 70 feet. An SEM excavation sequence was developed for the mined tunnel section, including a top heading excavation followed by bench and invert excavations. Ground support measures include fiber-reinforced shotcrete, lattice girders, rock reinforcement, face support, and a pipe canopy as well as spiling presupport.

To facilitate design of the mined tunnel, a detailed ground characterization was developed that identified rock mass types (RMTs), soil types (STs), and ground classes (GCs) based on available geotechnical data and interpretations from the project-specific geotechnical investigation. RMTs and STs were then grouped into four ground classes based on the similarity of anticipated ground behavior upon tunnel excavation. Each ground class was composed of RMTs expected to respond similarly to tunnel excavations and to require similar support systems.

The soil deposits present along the DTX mined tunnel alignment include artificial fill, Colma Formation and residual soil. The underlying bedrock is composed of Franciscan Formation (also referred to as Franciscan Complex in some published studies). The distribution of these units along the alignment (within the tunnel horizon) is indicated in Figure 4. The majority of the mined tunnel will encounter the Franciscan Formation, as shown in Figure 4. The Franciscan Formation along the DTX mined tunnel alignment is a highly variable unit in terms of its lithology and rock mass structure. It is composed of fractured, folded, and extensively sheared sandstone; siltstone; shale; and

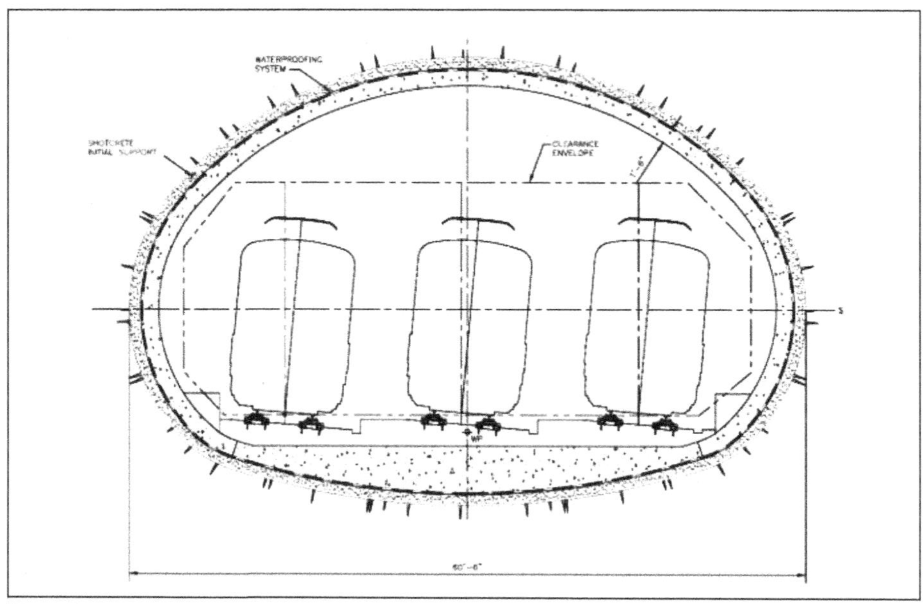


**Figure 3. Typical mined tunnel cross section of preliminary engineering design (Parsons et al. 2010)**

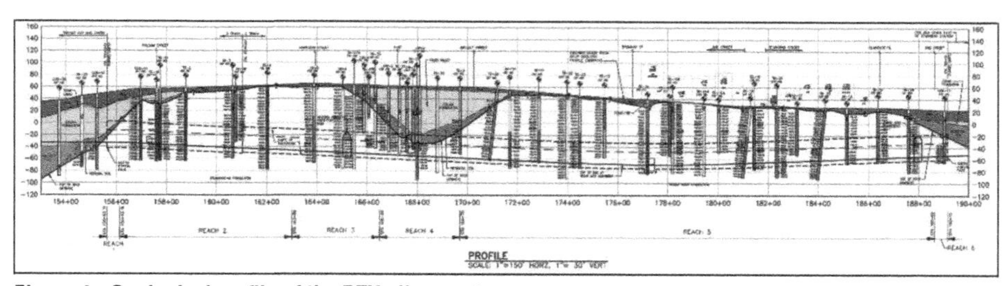


**Figure 4. Geological profile of the DTX alignment**

sandstone, shale, and siltstone breccia. The northern portion of the tunnel alignment is mainly in weak sheared shale, and the southern portion includes a mix of rock types but is mainly in stronger sandstone and interbedded sandstone/siltstone. The north end of the tunnel and a paleovalley in the central portion of the tunnel alignment will encounter soil deposits over Franciscan Formation bedrock. A section of mixed face conditions may be encountered at the location of the paleo valley. The paleo valley dips steeply across the alignment to the southwest overlies the Franciscan Formation and contains residual soils and the Colma Formation, which consists of medium dense to very dense gravelly, clayey sand with random layers of stiff to very hard clay.

Four support types and associated excavation sequences were developed in the 2010 PE Design for the mined tunnel segment corresponding to four ground classes defined (see Figure 5). These support types were developed based on the ground classes, past experience with similar excavation, and empirical support design methods, specifically the Q and Rock Mass Rating (RMR) classification systems.

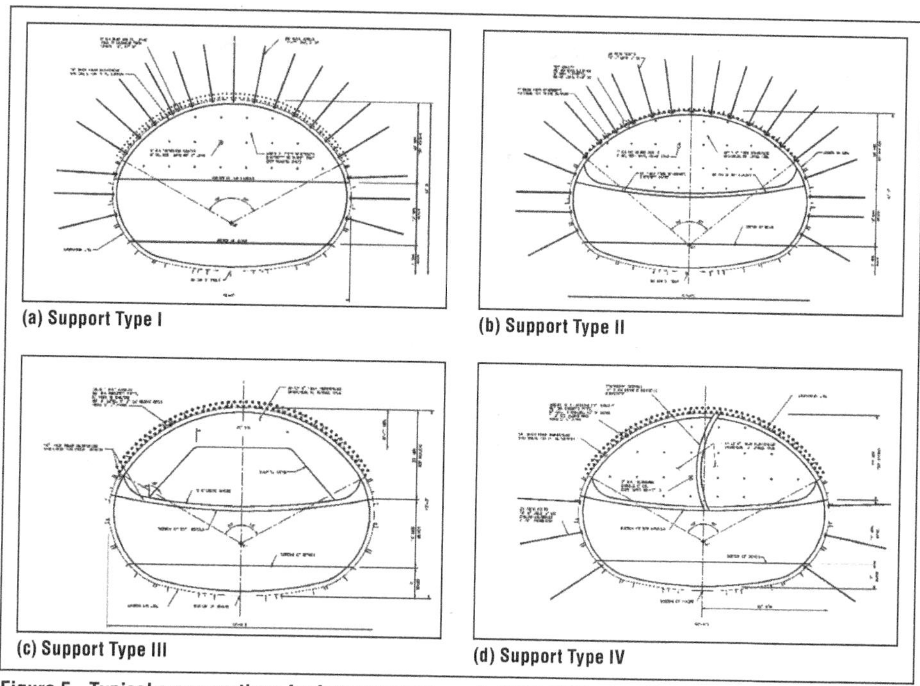


**Figure 5. Typical cross sections for four support types in 2010 PE Design (Parsons et al. 2010)**

## 2017 Tunnel Options Study

Following the 2010 PE Design concerns were raised by some City officials regarding the potential impacts to the City from the DTX cut-and-cover construction after recent local experience during the construction of San Francisco Central Subway, which garnered complaints from adjacent business and residents about the noise and inconveniences caused by the subway construction. The TJPA initiated review of alternative construction methods that could have the opportunity to minimize surface disruptions and socio-economic impacts. In this study, various options were evaluated for cost, construction risks, and schedule impacts, and the 2010 PE Design was defined as the Baseline Design for purposes of comparison with the studied options. The Tunnel Options Study (Parsons et al. 2017) divided the tunnel alignment into five segments, A to E, as shown in Figure 6.

- Segment A was a wide throat section from the Salesforce Transit Center to Clementina Street with a length of 658 feet. The width of this segment varied from 75 feet at the intersection with Clementina Street to 144 feet just north of Howard Street. In the 2010 PE Design, this segment was designed as a cut-and-cover box structure. To accommodate the requirements of the California High-Speed Rail Authority (CHSRA), the cross section of this segment was widened subsequent to the 2010 PE Design completion. Underpinning schemes were developed for three structures with interfaces with the cut-and-cover construction. Because of the complexity of this segment, it was further divided into northern and southern sections and two alternative approaches were explored for each of these two sections.

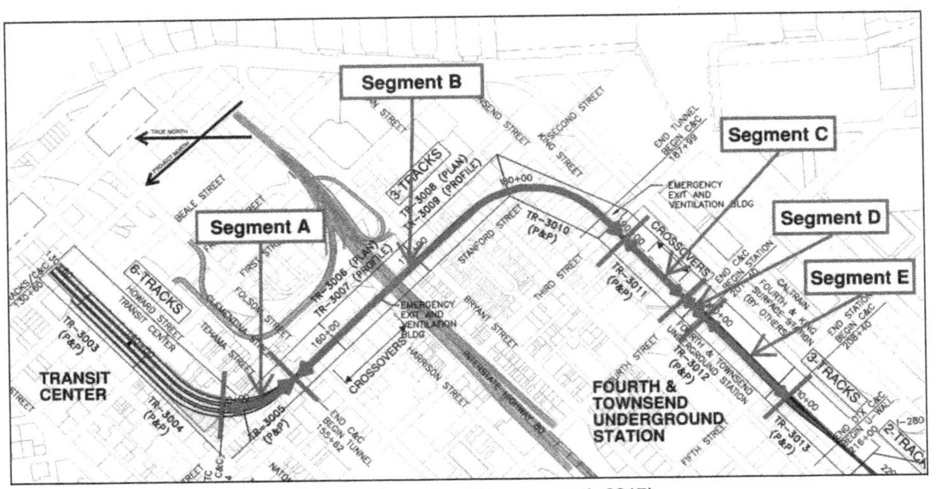


**Figure 6. Tunnel Options Study alignment overview (Parsons et al. 2017)**

– North Throat Section (Howard Street to Tehama Street with a length of 350 feet) was the wider portion of the throat section crossing Howard Street and included the construction parcels on the southeast corner of Howard and Second Streets. Two alternative construction methods for this section were evaluated:
  - Approach A—Pilot tunnels with mining: This method would involve dividing the wide throat structure into four smaller caverns by using a series of pressured face MTBM pilot tunnels to create underground bearing walls. The four caverns would then be mined between these bearing walls using a pipe arch presupport for the excavation. This approach would limit the maximum cavern arch width to less than 45 feet and reduce surface settlements as well as the risk of damaging existing street utilities. Figure 7 shows a typical Approach A section for Segment A North Throat Section.
  - Approach B—Pipe roof with mining: This method would use the same four-cavern layout as Approach A to keep cavern widths less than 45 feet. In this approach, horizontal interlocking steel pipes would be pushed below Howard Street to form a continuous presupport roof above the top of the final tunnel structure. Once the pipe roof was installed, pilot drifts similar to the pilot tunnels in Approach A would be hand mined just below the pipe roof. The pipe roof would provide initial support so hand mining of the drifts could be done at atmospheric pressure. The main caverns would be mined to the underside of the pipe roof presupport without additional excavation support. This would allow the mining of the main caverns to advance more quickly than Approach A. Figure 8 illustrates a typical Approach B section for Segment A North Throat Section.

– South Throat Section (Tehama Street to Clementina Street with a length of 212 feet) would extend primarily on Second Street between Tehama and Clementina Streets. The eastern side of the tunnel also would extend under an existing property on the east side of Second Street. The width of the tunnel would vary from 115 feet at the intersection of Second and Tehama Streets to 75 feet at the southern interface with Segment B. This

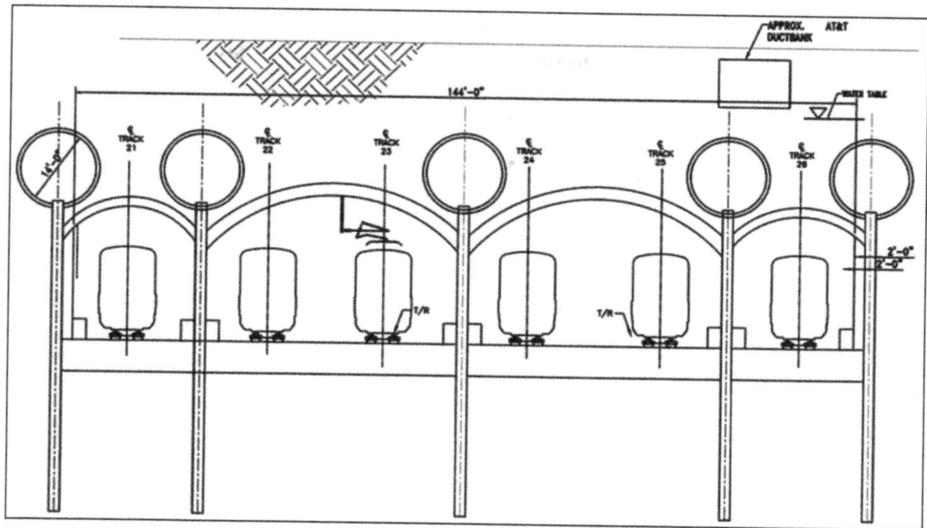


Figure 7. Typical Approach A Section for Segment A North Throat Section (Parsons et al. 2017)

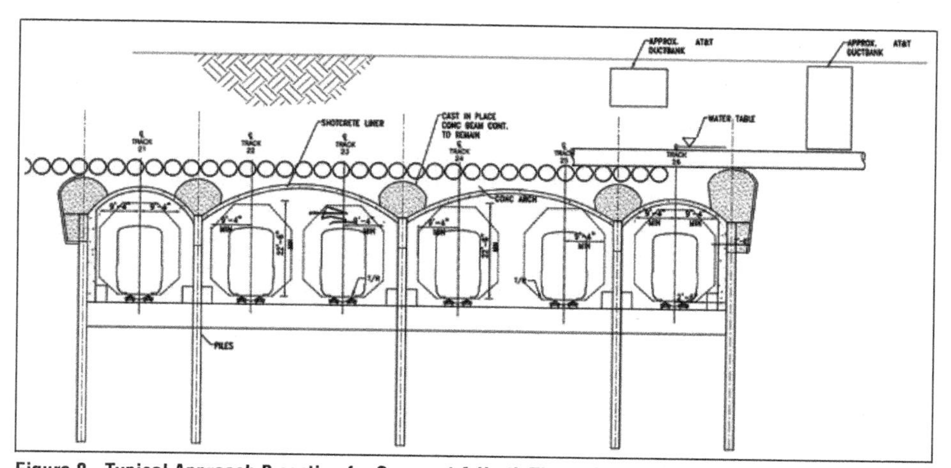


Figure 8. Typical Approach B section for Segment A North Throat Section (Parsons et al. 2017)

portion of the throat segment interfaces with the TBM running tunnels outlined in the Segment B tunneling option. These TBMs are integrated into both of the South Throat Section tunneling approaches. Two approaches were proposed for this South Throat Section for Segment A:

- Approach A—TBMs used for interior walls: The running tunnel TBMs from Segment B would be extended to Tehama Street. The TBMs would maintain an approximately 45-foot center-to-center spacing through the South Throat Section. As the Throat Structure widened, side walls would be installed outside the limits of the TBMs to create the additional width required at the Throat Section tunnel. On the west side of Second Street, the side wall would be formed by installing secant piles from the sidewalk of Second Street. On the east side of the throat, an additional pilot tunnel would be driven under the property.

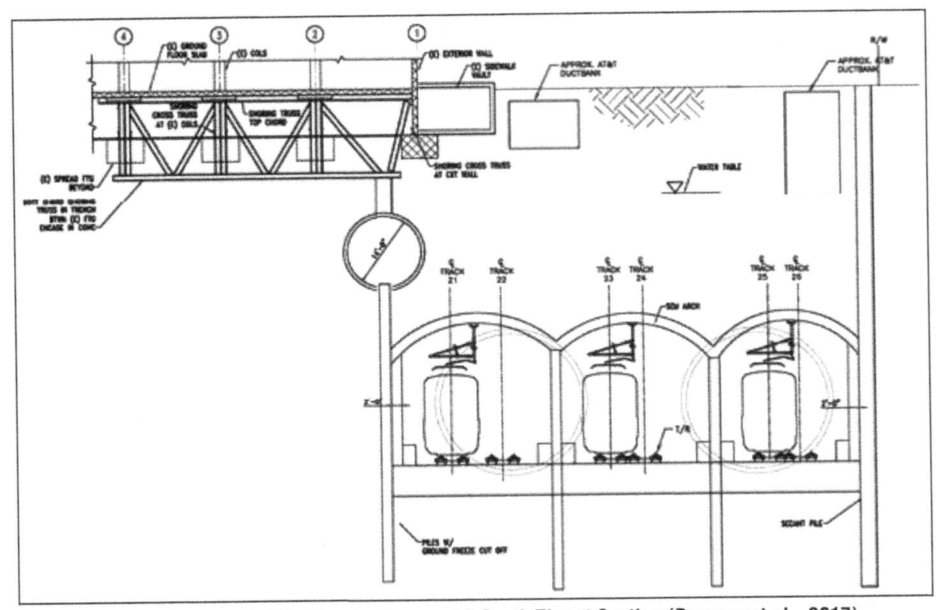

**Figure 9. Typical approach A Section for Segment A South Throat Section (Parsons et al., 2017)**

Vertical supports would be placed inside the TBM tunnels, and three caverns would be mined between vertical supports using pipe arch canopies. Each of these caverns would have a span of less than 50 feet. Figure 9 illustrates a typical Approach A section for Segment A South Throat Section.

- Approach B—TBMs used for outside walls: This approach would also require that the TBMs from Segment B be extended to Tehama Street. In this method, the TBM alignments would follow the outside limits of the Throat Section as the tunnel widens. The center-to-center spacing of the tunnels would increase to 80 feet at Tehama Street. Since the TBM tunnels would follow the outside limits of the Throat Section, Throat side wall support would not be required. To control surface settlements at the wide center cavern between TBM bores, the main cavern arch would be supported by internal shoring supports and the TBM liner on each end of the arch. A final concrete structure would be installed to create three permanent caverns, each with a span of less than 50 feet. Figure 10 illustrates a typical Approach B section for Segment A South Throat Section.

- Segment B would extend between Segment A and Segment C, and run primarily along Second Street starting at Clementina Street, turning onto Townsend Street, and ending at Third Street for a length of 3,388 feet. Two approaches were considered:
  - TBM with SEM: The proposed method for the segment was to create a three-track tunnel using running twin TBMs in combination with an SEM center bore between the TBM tunnels. This method would have the potential to reduce surface settlement and improve emergency egress relative to the 2010 PE Design. Figure 11 illustrates typical cross sections of twin bore TBM with SEM.

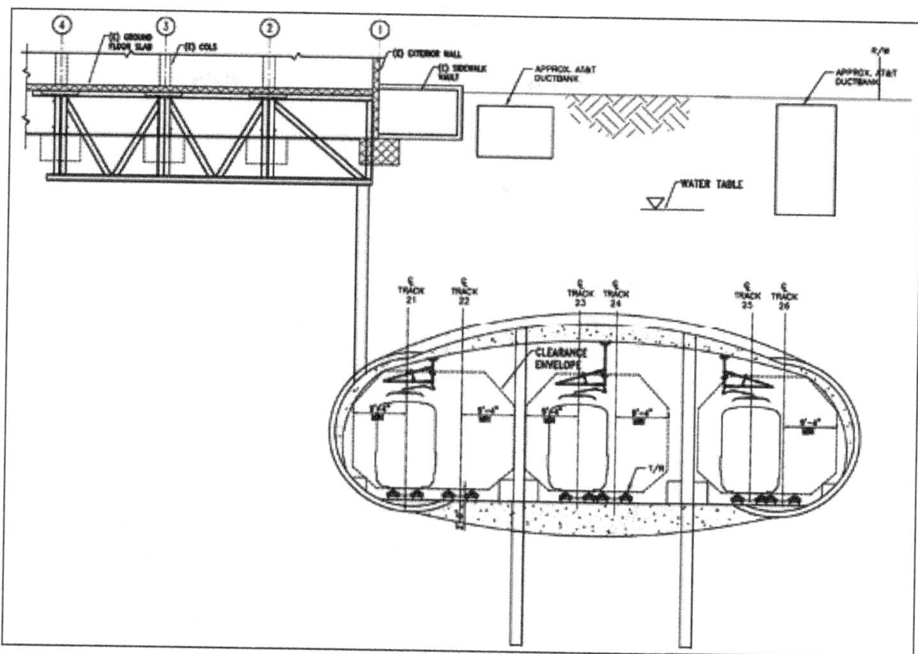


Figure 10. Typical Approach B section for Segment A South Throat Section (Parsons et al. 2017)

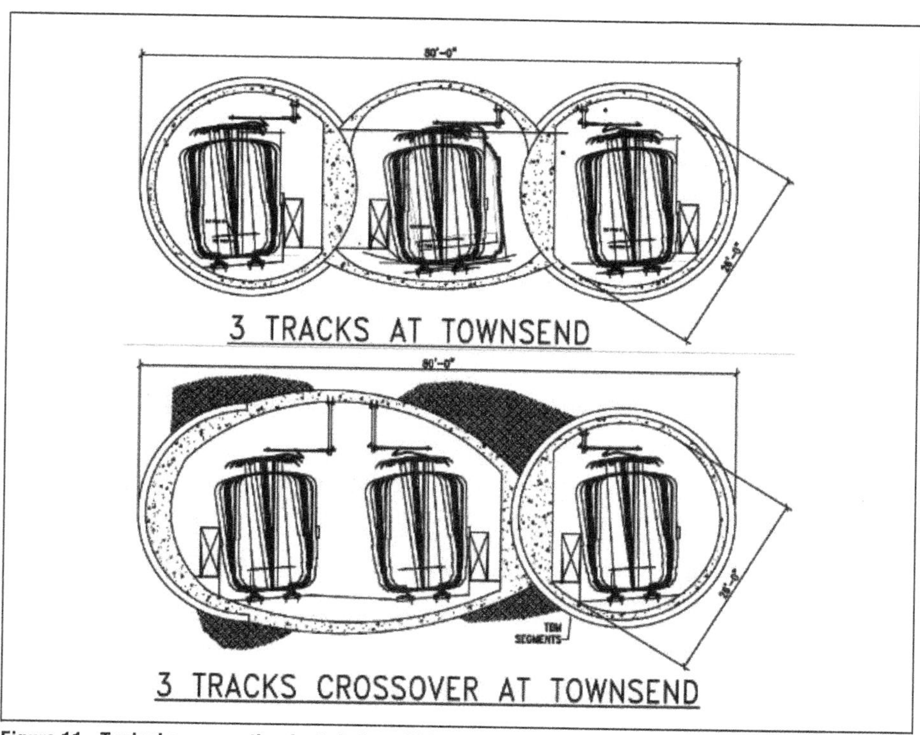


Figure 11. Typical cross section for twin bore TBM with SEM for Segment B (Parsons et al. 2017)

   - SEM tunneling in rock: This method was proposed in the 2010 PE Design and retained as the other construction method considered for Segment B. Refer to Figure 3 for a typical cross section with SEM in rock.
- Segment C would extend along Townsend Street from the west side of Third Street to the east side of Fourth Street for a length of 700 feet. The 2010 PE Design consisted of a cut-and-cover tunnel under Townsend Street. Three alternatives were considered in the Options Study:
- TBM with SEM
- Soft Ground SEM: This method was determined to be feasible with ground improvement. The minimum width required for this option would be 65 feet. The approach would start from the headwall of the Fourth and Townsend Station, just west of Fourth Street and consist of a single-cell cavern constructed using soft ground SEM. Under Fourth Street, a pipe arch canopy will be used to improve ground and reduce settlements at utilities and the Muni crossing. Ground improvement east of Fourth Street will also be required to control water and settlements during construction. The tunnel excavation sequence was expected to require nine drifts or headings to advance the tunnel face. Excavation support was expected to include heavy spiling, steel sets, and shotcrete.
- Pipe Roof Tunnel: Like the soft ground SEM, this method was determined to be feasible with ground improvement. A MTBM would be used from impervious headwall to headwall to install a series of interlocking steel pipes which form a watertight temporary ground support system, below which the final section of the box tunnel would then be mined and constructed.
- Segment D would represent the Fourth Street crossing on Townsend Street which passes under the MUNI Central Subway and a large, combined sewer and the approach to the Fourth and Townsend Street Station with a length of 490 feet. For the Fourth Street crossing, two construction methods were considered:
   - Pipe Arch Canopy: This concept used a pipe canopy umbrella in combination with fiberglass face reinforcement and horizontal permeation grouting techniques along the 210-foot-long segment. Specifically, due to the presence of saturated marine sand and soft Bay mud, a pipe canopy alone will not be able to stabilize the tunnel excavation. Therefore, it was anticipated that ground improvement along with pipe arch canopy would be used from the headwall of the cut-and-cover structure to form a minimum 9-foot-thick grout curtain around the entire excavation envelope to cutoff water infiltration and provide preconfinement for the planned excavation. Then two rows of 10-inch diameter micropiles will be installed to form a pipe canopy around the tunnel that would allow for tunnel excavation.
   - Jacked Box: This construction method consisted of a jacked box using an anti-drag system to keep the drag forces low and eliminate disturbance of adjacent soil. The approach would include casting four monolithic box structures in the launch zone. These box structures would consist of rear end jacking and intermediate jacking systems installed in a special joint between sections. After ground improvement, the entire structure would be jacked into place and cast in place concrete would be used to complete the tunnel structure.
- Segment E was the Fourth and Townsend Street Station with a length of 700 feet and would be cut-and-cover construction. With the option of utilizing TBM with SEM for Segments B and C, this cut-and-cover site could serve as the launch pit for the TBM. The cut-and-cover construction could occur while waiting for delivery of the TBM to the project site.

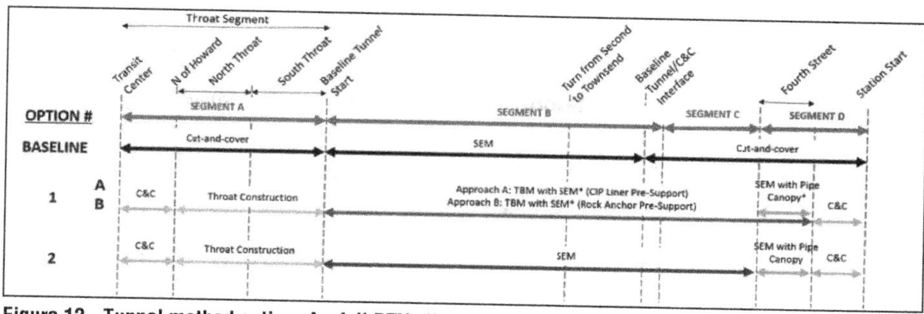

**Figure 12. Tunnel method options for full DTX alignment (Parsons et al., 2017)**

Compared to the PE Design, the Tunnel Options Study significantly increased the length of mined tunnel. Figure 12 shows a comparison of the construction methods between the PE Design (Baseline) and the Tunnel Options Study.

## 2022 Updated 30% PE Reference Design

Following the Tunnel Options Study, the TJPA held an Industry Sounding with 13 contractors in 2020 to inform strategy and future decision-making efforts for Phase 2 of the Transbay Program. Among the participating contractors, a majority indicated a preference for the SEM tunneling method and some preferred the TBM-SEM hybrid option. However, none of the contractors were enthusiastic about the mined Howard Street crossing (the Throat Structure). Contractors indicated a strong preference for cut-and-cover construction at this location.

Therefore, prior to commencing the Phase 2 Reference Design, the TJPA held a Howard Street Crossing Charrette with the DTX designers and stakeholders to further evaluate alternative tunneling options. This charrette focused on the risks, costs, and impacts of the mined options compared to the cut-and-cover option for constructing the Throat Structure. The conclusion was that the cut-and-cover option with an accelerated bridge construction approach would be able to limit the time for street closure, save overall construction costs, and most importantly, remove the risks associated with the mined tunneling method. Therefore, it was recommended that the cut-and-cover method be carried forward to the Phase 2 30% Reference Design.

In addition, rail operations analyses, performed as part of a phasing study, determined that a reconfiguration of the trackwork through the tunnel and the Fourth and Townsend Street Station would allow the tunnel to be reduced from three running tracks to two running tracks. As a result, the DTX design team conducted a tunnel excavation method study under the direction of TJPA to evaluate several previously considered tunneling methods for the two-track mined tunnel:

- Single bore TBM with SEM
- Twin bore TBM with SEM
- Single bore SEM
- Twin bore SEM

This study evaluated some key factors that affect the mined tunnel construction, impacts to adjacent community, vertical and horizontal alignment, right-of-way, operations and operational flexibility (crossovers, alcoves, and cross passages; egress and ventilation; etc.), costs and construction schedule, and risks. The study recommended that the Phase 2 Reference Design be developed based on single bore SEM and SEM mining with ground improvement under the Third Street crossing.

With these two studies, the cut-and-cover method for the Throat Structure and the SEM tunneling for the mined tunnel section as opposed to the TBM with SEM option are retained. Compared to the 2010 PE Design, the following are the two major changes to the mined tunnel section:

- The vertical alignment was lowered by approximately 20 feet at the buried valley so the anticipated exposure of soft ground in the tunnel face is significantly reduced.
- The south portal of the mined tunnel was extended to the west edge of the Third Street crossing. This change will avoid potential closure of the intersection of Townsend Street and Third Street during construction but introduced a mixed face condition at this location.

For the 2022 Reference Design, which was completed in October 2022, the mined tunnel extends from the west side of Third and Townsend Streets to Clementina and Second Streets. The tunnel is primarily two tracks but expands to three tracks as it approaches the Throat Structure. The length of the mined tunnel is approximately 3,350 feet (0.65 mile). Two standalone ventilation and emergency egress structures will be located along the mined tunnel alignment, one at Third and Townsend Streets and the other at Second and Harrison Streets.

Similar to the 2010 PE Design, the ground along the mined tunnel alignment is grouped into four ground classes based on the similarity of anticipated ground behavior. Three support types with associated excavation sequences have been designed to address the initial support requirements for the anticipated ground conditions associated with four ground classes. Support Type I is applicable to Ground Classes 1 and 2, and Support Type II is applicable to Ground Classes 3 and 4 (at the buried valley). Finally, Support Type III is only applicable to Ground Class 4 at the south and north portals. Table 1 presents a comparison of the support types and their anticipated distribution over the mined tunnel alignment between the 2010 PE Design and the 2022 Reference Design.

Figure 13 shows typical cross sections for three support types developed in the Reference Design. Compared to the 2010 PE Design, the initial support requirements have been streamlined to account for the potential beneficial effect from lowering the vertical alignment and improved understanding of anticipated ground behavior based on additional analyses and experiences gained from relevant recent SEM projects either completed or still ongoing. One of the philosophies for the Reference Design

**Table 1. Comparison of 2010 PE Design and 2022 Reference Design support types**

| | 2010 PE Design | | 2022 Reference Design | |
|---|---|---|---|---|
| Ground Class | Support Type | Tunnel Length, feet | Support Type | Tunnel Length, feet |
| 1 | I | 700 (22% of tunnel) | I | 2,330 (70% of tunnel) |
| 2 | II | 1,400 (44% of tunnel) | | |
| 3 | III | 460 (14% of tunnel) | II | 890 (27% of tunnel) |
| 4 | IV | 635 (20% of tunnel) | III | 130 (3% of tunnel) |

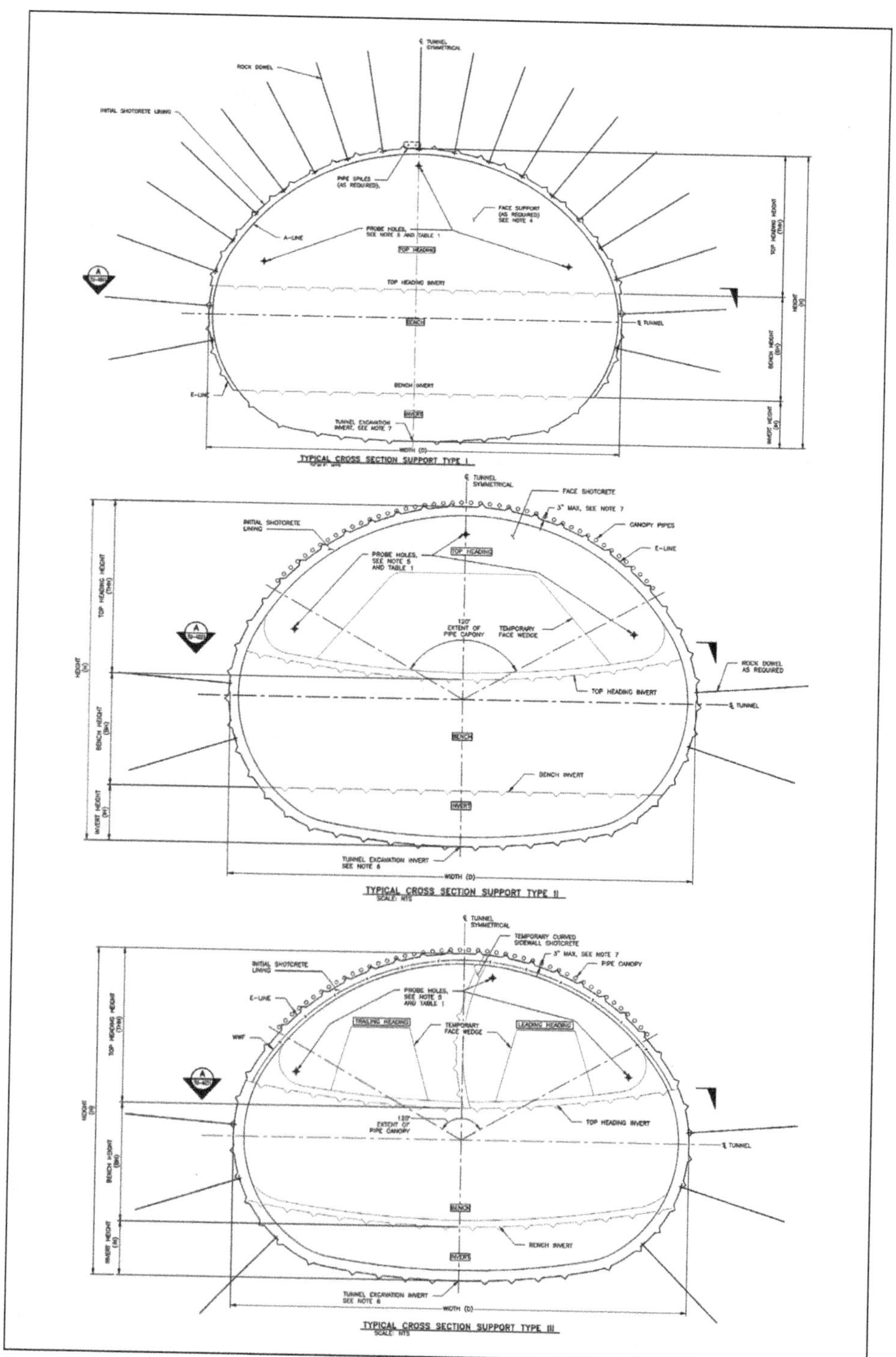


Figure 13. Typical cross sections for three support types in the 2022 Reference Design

in terms of the initial support requirements is to rely on the contractor's means and methods and toolbox items for addressing actual encountered ground conditions as the tunnel delivery method will be a progressive design-build. Over-prescription in the requirements is considered unnecessary.

## MILESTONES, PROJECT DELIVERY METHODS, AND PROPOSED SCHEDULE

The project is currently finalizing the project development of the project including packaging for different delivery methods, finalizing cost estimates, and contractor outreach. The cost estimate and project schedule are currently being updated and will be used to solicit funding for the project. The 30% design level cost estimate and updated schedule for the project will be made public in early 2023. Figure 14 presents the past and future milestones for the project, including design, procurement, construction, and anticipated ready for service. The timeline is subject to approval of funding.

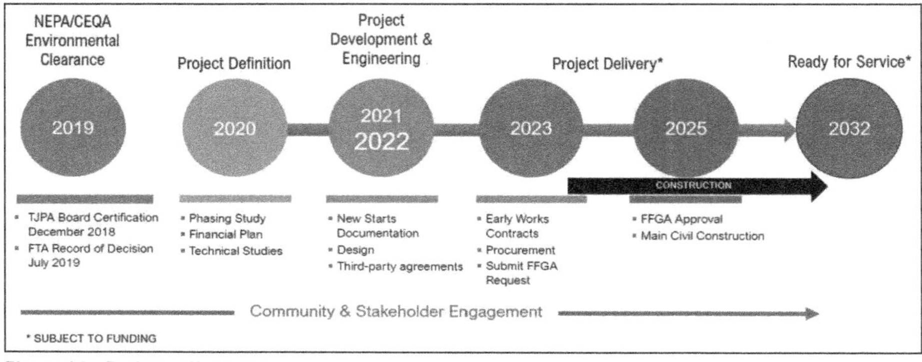


**Figure 14. Project milestones**

In 2022, TJPA undertook a comprehensive analysis of potential delivery methods through the conduct of a Project Delivery Alternatives Study. The full range of alternatives from Design Bid Build to Public Private Partnerships were evaluated. Agency goals were defined, partner agency requirements were catalogued, and industry outreach was conducted. The recommended approaches, described below, were reviewed with, and confirmed by, the TJPA Board of Directors.

The proposed delivery methods for the various anticipated contract packages of the project are summarized below:

- Utility Relocation: Design-Bid-Build (DBB) or Construction Manager/General Contractor (CMGC)
- Advance Site Works: DBB or CMGC
- Building Demolition: DBB or CMGC
- Main Civil Package: Progressive Design-Build (PDB)
- Track and Rail Systems: CMGC
- Station Fit-out (including building systems and above-grade station structures): CMGC

In 2022, the project was focused on engaging the market and industry and refining the delivery options for the various options on the project. The design team along with the TJPA and Program Management Project Controls team has been refining the budget and schedule, reviewing risk, and completing the 30% design (Reference

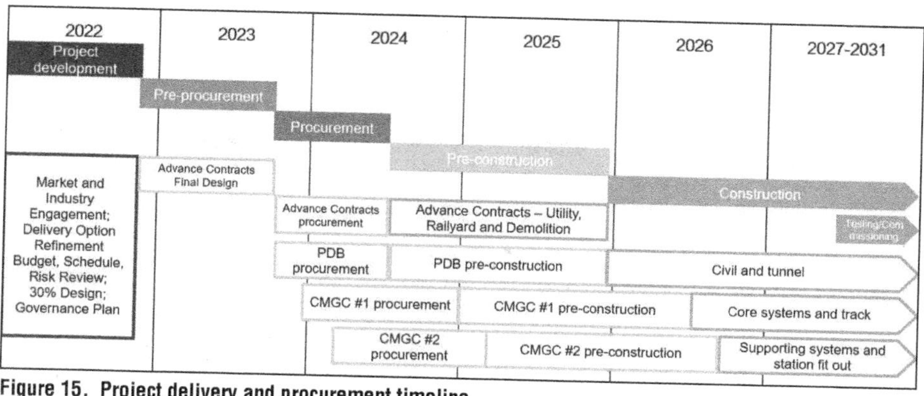


**Figure 15. Project delivery and procurement timeline**

Design). Figure 15 presents the procurement timeline, subject to funding, of the project as it enters 2023 and prepares for solicitations from the industry. Figure 15 shows the phases relative to time with overlap between pre-procurement and procurement, which allows for issuances of Request for Qualifications while finalizing the Request for Proposals. This also allows for a staggered procurement process between the different contracts to manage delivery agency resources and acknowledges the different timelines of the package construction. This allows for a complex problem to be simplified in order to highlight priorities and manage major project interfaces.

## ACKNOWLEDGMENTS

The authors acknowledge the contributions of Transbay Joint Powers Authority, San Francisco County Transportation Authority, AECOM, Mott MacDonald, and Parsons Transportation Group and their subconsultants, including Delve Underground, in the design and delivery of this project. This paper's contents reflect the views of the authors, who are responsible for the facts and accuracy of the data presented herein, and do not necessarily reflect the official views or policies of the Transbay Joint Power Authority and/or the San Francisco County Transportation Authority.

## REFERENCES

Delve Underground (formerly Jacobs Associates). 2006. Final Tunnel Alternatives Evaluation Report. Vol. I–III. Report prepared for the Transbay Joint Power Authority as a subconsultant to Parsons Transportation Group.

Parsons Transportation Group, ARUP, and Delve Underground (formerly Jacobs Associates). 2010. Preliminary Engineering Report. Report prepared for the Transbay Joint Power Authority.

Parsons Transportation Group and Delve Underground (formerly McMillen Jacobs Associates). 2017. Tunnel Options Study for the Downtown Rail Extension Project. Report prepared for the Transbay Joint Power Authority.

U.S. Department of Transportation Federal Transit Administration. City and County of San Francisco, Transbay Joint Powers Authority, and San Francisco Redevelopment Agency. 2004. Final Environmental Impact Statement/Environmental Impact Report and Section 4(f) Evaluation (FEIS/EIR).

# Understanding the Subsurface Conditions for the Cemetery Brook Drain Tunnel Project

**Mahmood Khwaja** ▪ CDM Smith
**David Polcari** ▪ CDM Smith
**Michael S. Schultz** ▪ CDM Smith
**Frederick McNeill** ▪ City of Manchester, New Hampshire
**Timothy Clougherty** ▪ City of Manchester, New Hampshire
**Benjamin Lundsted** ▪ City of Manchester, New Hampshire

## ABSTRACT

The City of Manchester, New Hampshire, is embarking on one of its largest public works projects in history to construct a 12,000 foot long, large-diameter, conveyance tunnel through downtown Manchester. This area has challenging geology with undulating bedrock (granite, schist, and gneiss) profile overlain by fluvial deposits. Understanding the subsurface conditions is key to mitigating project risks, establishing plan and profile for the tunnel, and selecting the most cost-sensitive and technically appropriate tunneling method. This paper presents the project team's approach for conducting the ground investigation, developing preliminary subsurface profile, and the considerations for selecting the tunneling method.

## PROJECT BACKGROUND

The Cemetery Brook Drain (CBD) project is the cornerstone of the Phase II Combined Sewer Overflow (CSO) program, which will provide the City of Manchester, NH, (City) with the stormwater infrastructure to implement brook removal, sewer separation, and address the capacity related backups and street flooding issues in the Cemetery Brook CSO basin. Cemetery Brook is the largest of the five brooks that currently enter the combined sewer system. Removal of the brook is one the primary 2010 Long Term Control Plan (LTCP) objectives. The base flow from the brook during dry weather minimizes capacity in the combined sewer system which results in excessive flows and CSOs during wet weather. The Cemetery Brook flows into the City's sewer system, reduces the system capacity, and increases the operation and maintenance (O&M) costs at the Manchester Wastewater Treatment Plant (MWWTP).

The City's combined sewer system was constructed to collect both sanitary sewage and stormwater flow as one pipe network. During dry weather, the sewage flows are conveyed to the MWWTP. During wet weather, stormwater flows enter the collection system and combine with sewage flows. The combined flows can exceed pipe capacity, which limits the amount of flow conveyed to the MWWTP, and results in CSOs that discharge to the Merrimack River via the Cemetery Brook CSO outfall. The goal of sewer separation is to have two separate pipe systems with one dedicated for sanitary sewage and the other for stormwater to mitigate overflows discharging into the Merrimack River.

In July 2020, the City, the State of New Hampshire, and the United States Environmental Protection Agency (USEPA) finalized a Consent Decree that requires the City to implement CSO control measures based on the projects identified in the LTCP as components of the Phase II CSO abatement program. The CBD is one of the key components of the LTCP and the Consent Decree mandates the design and construction of the CBD project. The CBD will serve as the new large-diameter drain for the Cemetery Brook basin to redirect the basin flows approximately 2.3-miles from the inlet area to the Merrimack River, where it will discharge via a new stormwater outfall south of the Queen City Bridge. The CBD will remove Cemetery Brook from the existing Cemetery Brook Conduit (CBC), which experiences significant surcharge during rain events, and will eliminate the Cemetery Brook flow to the WWTP. In addition, the CBD project will capture stormwater flows from, and reduce flooding events within, the abutting separation areas. The project will maximize benefits to both the Merrimack River and the City's collection system.

## PROJECT AREA

The project area is part of the Cemetery Brook CSO drainage basin in central Manchester on the east side of the Merrimack River. The 4,500-acre drainage basin is the largest in the city and includes nearly 3,000 acres served by the combined sewer system which represents more than 50 percent of the entire combined sewer system. The remaining acreage is served by separate sanitary and stormwater collection systems. One third of the basin is a combined system without an existing drainage system; another third has separate drainage systems that re-connect to the combined sewer system at various locations. A significant portion of collection system in the Cemetery Brook CSO basin was constructed in the early 1900s as the city developed and most of the older combined pipes remain in operation today. The Cemetery Brook CSO drainage basin, the project area including the proposed CBD alignment, the future sewer separation areas, and other major features in the overall CSO mitigation program are shown in Figure 1.

The CBD project will capture and convey flows from Cemetery Brook and from future drains in the sewer separation areas. The future separation projects will ultimately separate over 1,700 acres in the basin. This includes the Stevens Pond and the upstream open channel Cemetery Brook area (approximately 900 acres), and the sewer separation areas, which represent over 800 acres that are mostly combined stormwater and sewer system.

The topography in the basin consists of land sloping westerly across a valley that was formerly a natural drainage channel for Cemetery Brook to convey surface water from Stevens Pond to the Merrimack River. In addition to Cemetery Brook, there were numerous additional streams and brooks that used to flow to this valley and the Merrimack River prior to the growth of the city. As the city developed, Cemetery Brook was redirected into culverts under city streets which were eventually joined with a series of pipeline projects. The CBC generally follows the former route of the brook and is a closed conduit for much of its run through the city from Porter Street to the East Interceptor North along Merrimack River.

## GEOTECHNICAL INVESTIGATION PROGRAM

A project specific geotechnical investigation program (GIP) approach was developed to support concept through final engineering phases. The subsurface explorations included test borings through soft ground and rock, seismic refraction survey, televiewer logging, groundwater observation wells, soil sampling for environmental and engineering purposes, and geotechnical and environmental laboratory testing.

The GIP was executed in three phases. Phase 1 and Phase 2 investigations were conducted during conceptual engineering phase between April 2000 and October

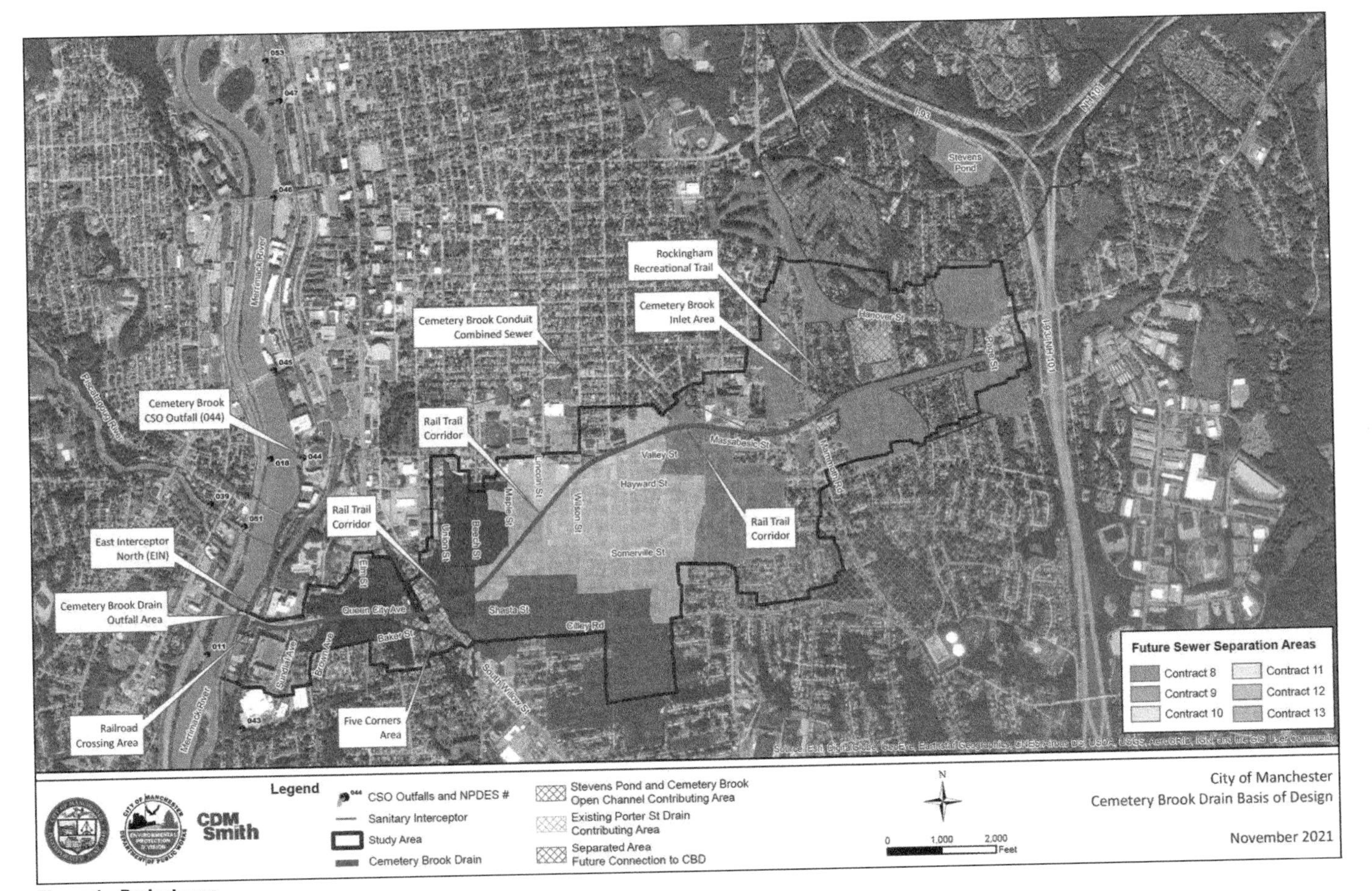


**Figure 1. Project area**

2020. Phase 3 investigation was started in July 2022 and is expected to be complete by early spring 2023. The project was originally conceived using open-cut construction method and the CBD was to be completed in four segments over a period of 14 years [1]. Phase 1 of the GIP was planned to support design and construction of the first segment limits were defined between the outfall on the Merrimack River and Union Street, approximately 3,800 feet from the Merrimack River's western bank. While the alignment for majority of the drain was well established, the outfall location had two potential discharge locations. This required evaluation of at least two alignments within the first segment of CBD (Figure 2). For that reason, Phase 1 of the GIP included A-series and B-series borings (Figure 3). Phase 2 of the GIP covered the project limits extending, approximately, 8,100 feet from Union Street to the CBD inlet point, approximately 250 feet west of Mammoth Road.

The design team was tasked with evaluating an alternative method of construction for the first segment to mitigate construction impacts related to the open-cut construction method. The Phase 1 GIP consisted of deeper borings, up to 80 feet, to allow evaluation of trenchless/tunneling options. The Phase 2 GIP borings were primarily shallower borings to support using open-cut approach.

For Phase 1, ten (10) A-series borings, and seven (7) B-series borings were drilled, with four (4) of the borings converted to observation wells; Phase 2 consisted of drilling fourteen (14) borings, with four (4) of the borings converted to observation wells. In addition to the borings, a geophysical survey using the seismic refraction method was conducted during Phases 1 and 2 to help estimate top of bedrock and rock quality along most of the preferred alignment. Phase 1 included 3,560 linear feet of seismic refraction survey from Sundial Avenue to Union Street; Phase 2 included 8,100 linear feet of survey along alignment corridor between Union Street and Mammoth Road—see Figure 3 for the seismic refraction limits. In-situ testing included packer test to measure the hydraulic conductivity of the rock. Laboratory testing included:

- Water Content in accordance with D2216
- Grain-size analyses including hydrometer in accordance with D6913/D7928
- Atterberg Limits in accordance with D4318
- Cerchar Abrasivity Index (CAI) rock analyses in accordance with D7625
- Uniaxial Compressive Strength and Elastic Moduli in accordance with D7012
- Drillability Test Suites for rock including Drilling Rate Index, Bit Wear Index, and Cutter Life Index in accordance with NTNU/SINTEF's 13A-98 Drillability Test Methods
- Splitting (Indirect or Brazilian) Tensile Strength in accordance with D3967
- Point Load Index tests in accordance with D5731
- Petrographic analyses

The Phase 1 and 2 GIP information was used to evaluate concept feasibility for the open-cut and tunnel alternatives and to select a preferred option. The test borings and seismic refraction data has been used to better define the bedrock profile and better understand the subsurface strata along the full CBD alignment for both alternatives.

The interpreted data from the GIP indicates soil and groundwater conditions that generally consists of granular fill, overlying glacial deposits consisting of boulders, gravel, sand, silt, and clay (glaciofluvial deposits, glaciolacustrine deposits, glacial till), weathered bedrock, and bedrock along the alignment. Bedrock was encountered at test borings with depths ranging from 7.5 to 87.5 feet below ground surface (bgs) and the bedrock surface undulates considerably based on the seismic refraction survey data. The undulations are more prominent in the downstream portion of the alignment,

Figure 2. Alignment alternative overview

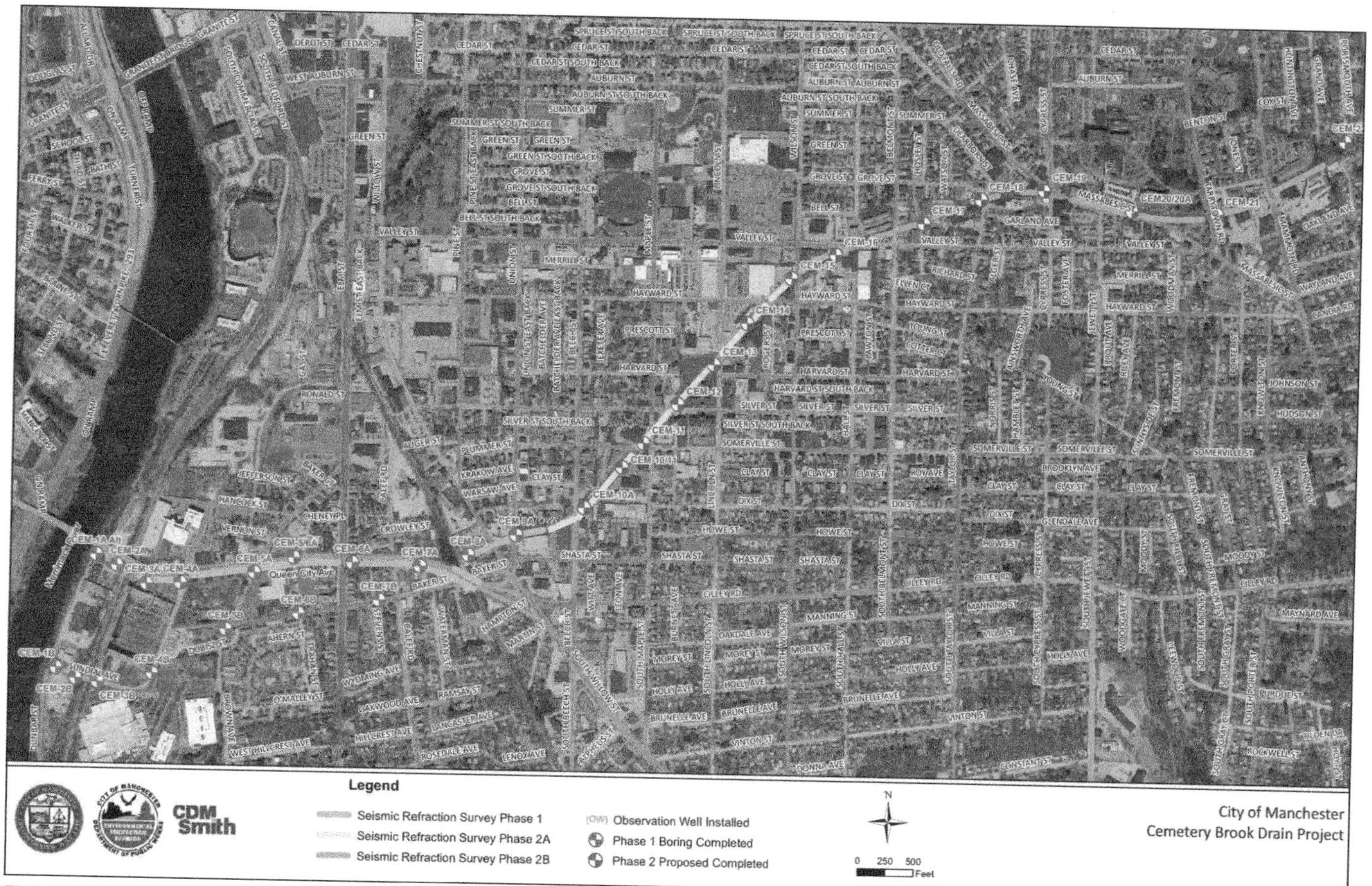


Figure 3. Phase 1 and 2 boring and seismic refraction location plan

with exposed rock outcroppings on the east side of Sundial Avenue near Queen City Avenue and bedrock depths approaching 100 feet bgs to the east on Queen City Avenue only 2,000 feet away. The bedrock surface is shallower along upstream portion of the alignment from the rail trail corridor at Valley Street to the Cemetery Brook inlet area with depths ranging from 7.5 to 41.0 feet bgs.

The preferred alignment for the first segment of the CBD was selected along the A-series borings. The GIP Phases 1 and 2, in combination with constructability and construction impacts to the environment and community, eventually led to the decision to select tunneling as the most appropriate and technical sound method of construction for the entire CBD alignment [1]. To support this decision on selection of alignment and construction approach, a targeted Phase 3 of the GIP was developed and implemented. The GIP will be completed in spring 2023. The planned Phase 3 of the GIP consists of fifty-seven (57) boring through soft ground and rock, televiewer logging, in-situ packer tests, and laboratory testing. Three (3) of the borings are within the Merrimack River to support design/construction of the outfall structures; six (6) borings are located around the proposed tunnel boring machine (TBM) launching trench and the proposed CBD transition structure to the outfall structure east of the launching trench. The remainder of the Phase 3 borings are located along the alignment to support design/construction of the CBD tunnel, the seven (7) drop shafts, and three (3) deaeration chambers and adits.

A total of seventy-four (74) borings were drilled along the preferred route; fifty-two (52) borings are within the 11,600 feet (approx.) of the tunnel alignment and penetrate through the tunnel horizon, providing an average spacing of approximately 225 feet.

## REGIONAL GEOLOGY

New England is thought to consist of several distinct terranes or composite terranes that have different ages and metamorphic histories including the Rowe-Hawley, Connecticut Valley, Bronson Hill, Central Maine, Merrimack, Putnam-Nashoba, and Avalon. As viewed from west to east, the terranes were affected increasingly by the Taconic, Acadian, and Alleghenian orogeny's respectively. Rocks of the Avalon composite terrane are exposed throughout southeastern New England and are thought to represent the result of a fragment of North Africa colliding with North America approximately 625 million years ago during the Acadian Orogeny. The Massabesic Gneiss Complex in Manchester, New Hampshire is one of the western most exposures of the Avalon terrane and is thought to have been the leading edge of the North African landmass that collided with North America due to evidence that ocean basin material is present within the rock formation.

The Massabesic Gneiss Complex has been found to be from Late Devonian (380 million years ago) to Pre-Cambrian (625 million years ago) in age depending on the location and mineral assemblages that are analyzed as part of the dating process. The Complex is typically made up of rock that is coarse to medium grained, foliated, occasionally banded, with pink gneiss composed of microcline quartz, biotite, oligoclase feldspar, muscovite, and magnetite. The rock in the Complex can also be composed of white gneiss that contains oligoclase, quartz, biotite, garnet, sillimanite, and muscovite.

The project site is located within the Eastern New England Upland physiographic region, specifically the Merrimack Valley. The main region extends from northern Maine to eastern Connecticut. The physiographic province is generally composed of glacial till deposits overlain by varying thicknesses of unconsolidated glaciolacustrine sediments. The project site is located within the boundaries of prehistoric glacial Lake Merrimack which was present in the area from approximately 12,000 to 10,000 years ago. The sediments from Glacial Lake Merrimack dominate the landscape as we see it today, except for Quaternary age alluvium that has been post-glacially deposited by rivers and streams.

## Glacial Lake Merrimack

Glacial lake and glacial stream deposits were laid down during deglaciation of present-day Manchester, New Hampshire by meltwater from the margin of the continental ice sheet as it retreated northward. Material for these sediments was derived mostly from within the ice sheet with small contributions from meltwater erosion of the land near the margin. Glacial Lake Merrimack (Figure 4) was formed when the Laurentide Ice Sheet in the Merrimack Valley began to retreat around 14,000 years ago during a period of rapid warming in the northern hemisphere. The ice sheet passed by the location of modern-day Manchester, New Hampshire around 13,200 years ago and created an ice blockage that dammed the lake at the terminal moraine of the glacier. The lake continued to expand in size from Boylston, Massachusetts to as far north as Plymouth, New Hampshire. As the ice sheet melted northward, the depth eventually reached several hundred feet, and at some point drained to create the present-day Merrimack Valley and associated Merrimack River.

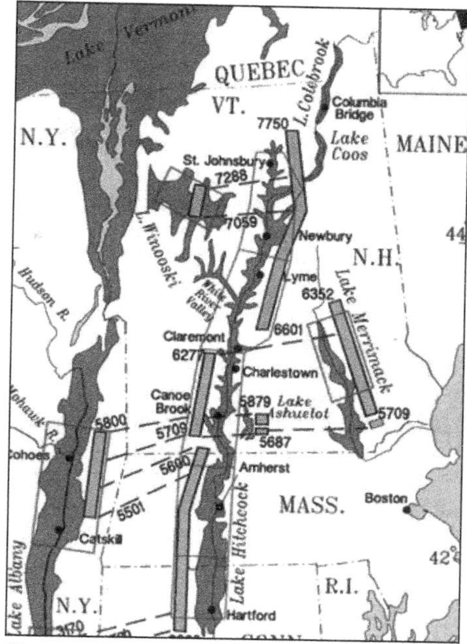

Figure 4. Glacial Lake Merrimack map

As the glacier continued to retreat, the flow velocity into the lake slowed dramatically and the transport of larger particles was replaced by fine sediments. The sediments deposited into the lake were mostly silts and clays from nearby mountain runoff with poorly vegetated slopes due to extensive glaciation. The streams and rivers that flowed into the lake contained similar fine sediments along with glacial streams that were located at the base of the glaciers. Silt particles settled to the bottom of the lake during summer months. During winter months the lake would freeze over and calm to allow clay-sized particles to be deposited. The finely laminated silt-clay sediments deposited in the lake by this yearly cyclical depositional environment are known as varve deposits.

## Site Specific Geology

The geology at the project site from the Merrimack River on the west to Portsmouth Avenue on the east in Manchester, New Hampshire mimics the regional geologic setting for overburden and bedrock alike. The overburden along the proposed alignment is dominated by the presence of fine-grained Glacial Lake Merrimack sediments on the west and glacial till and glaciolacustrine sediments on the eastern end.

The stratigraphic sequence begins in bedrock with the Massabesic Gneiss Complex which consists of metamorphic gneiss and schist as well as intrusive granites. The main body of the complex has been dated to be pre-Cambrian age, approximately 625 million years ago. The intrusive granites are significantly younger age, thought to be around 390 million years ago. The samples tested for petrographic analysis were found to be Altered Schist consisting of altered biotite, quartz, potassium feldspar, plagioclase schist, and an altered rock consisting of biotite, quartz, potassium feldspar, and plagioclase feldspar from a protolith or parent rock of granodiorite or arkose (variety of sandstone).

The bedrock has a surface layer of weathered rock in localized areas and is not present throughout the entire alignment. The weathered bedrock was found to be soft enough to be drilled easily and split spoon sampled where found during the program. The bedrock was overlain by glacial till deposits which are generally less than 15-foot in thickness. The glacial till was not present between Sundial Avenue and Beech Street except for the railroad crossing at Willow Street. The till material was prevalent throughout the Phase 2 borings from Beech Street to Tarrytown Road except near Somerville Street. The glacial till was not present at the far eastern end of the alignment which is most likely due to the presence of shallow bedrock. The glacial till at the project site is a traditional New England style till, very dense, moderate to well bonded silty sand, with small amounts of fine gravel. The project site was nearly devoid of oversized material with very few cobbles and boulders drilled through or split spoon sampled.

A deep layer of glaciofluvial deposits were seen sporadically throughout the alignment and is thought to have been deposited after the glacial till. However, the two layers were not seen in the same exploration together. The glaciofluvial deposits were only seen in 5 of the 34 borings drilled and were present at the base of the glaciolacustrine deposits on the western end of Alignment A. The deeper glaciofluvial deposits are like the upper glaciofluvial material, medium dense well-graded sand with gravel, with minimal cobbles present.

The glacial till and deeper glaciofluvial deposits were overlain by glaciolacustrine deposits placed during the existence of Lake Merrimack during the end of the last ice age, approximately 11,000 years ago. The glaciolacustrine deposits were overlain by the shallow layer of glaciofluvial deposits once the material in Glacial Lake Merrimack completed deposition. The final layer of glaciofluvial deposits was likely deposited by the Merrimack River as it was cutting into the landscape over the last 10,000 years. A layer of manmade fill ranging up to 10-ft thick, with exceptions, is present on top of the glaciofluvial deposits.

## Generalized Subsurface Conditions

The generalized subsurface conditions have been interpreted from the geophysical survey and the GIP test borings with an average spacing of 225-feet. This spacing is adequate to support final engineering and construction of a tunnel project. If additional information is needed during final design, the design team will evaluate adding to Phase 3 of the GIP. A sample of the interpreted subsurface conditions and preliminary tunnel profile along the full alignment is depicted in Figure 5. The undulating bedrock profile, initially interpreted after Phase 1 and 2 of the GIP, is reconfirmed based on the finding of the GIP Phase 3. The geophysical interpretation performed during Phase 1 and 2 was re-interpreted based on the findings of GIP Phase 3, which provided an updated top of bedrock estimate. Approximately 70 percent of the 11,600-feet of the tunnel alignment is in rock, of which nearly 60 percent occurs in a contiguous

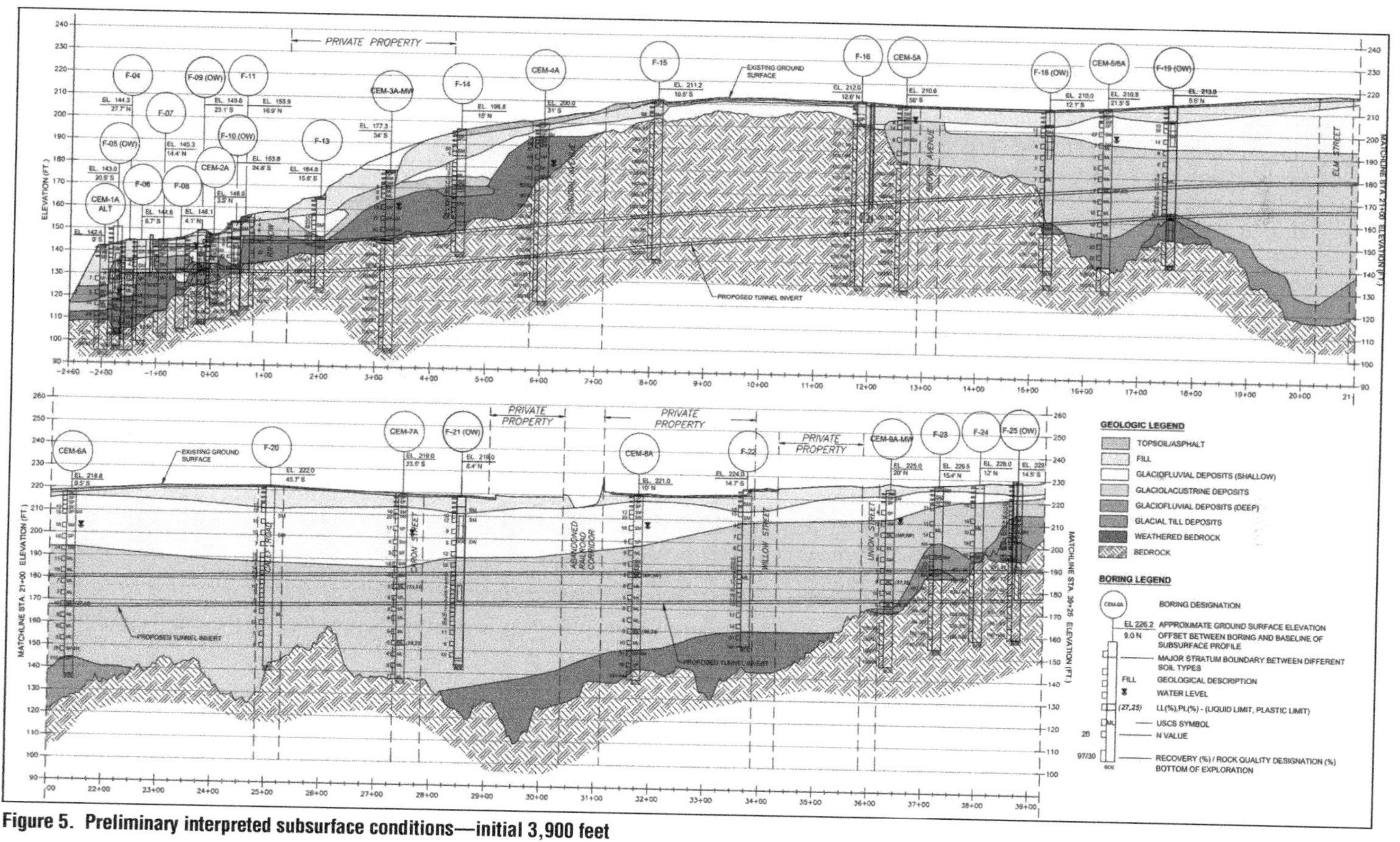


Figure 5. Preliminary interpreted subsurface conditions—initial 3,900 feet

run in the final 5,500-feet. The remainder of the alignment is in soft ground and/or mixed face.

The strata that are anticipated to be encountered within the proposed CBD tunnel horizon and alignment include Glaciofluvial Deposits, Glaciolacustrine Deposits, Glacial Till, Weathered Bedrock, and Bedrock. In addition, localized zones of cobbles and boulders are anticipated to be encountered within or near the proposed tunnel alignment within the glacial soil strata. Table 1 summarizes the strata that are anticipated to be encountered within or near the proposed tunnel alignment (within approximately ± one tunnel diameter of the tunnel horizon) and the approximate length of the tunnel alignment where those strata are to be encountered. The predominant strata are briefly discussed below:

**Table 1. Strata summary for the tunnel alignment**

| Anticipated Strata | Estimated length of Strata Along Tunnel Alignment, ft* |
|---|---|
| Bedrock | 8,100 to 9,800 |
| Glaciolacustrine Deposits | 3,050–3,350 |
| Glacial Till | 400–850 |
| Glaciofluvial Deposits (Shallow) | 600 to 1,900 |
| Glaciofluvial Deposits (Deep) | 100 to 450 |
| Weathered Rock | 50 to 100 |

* Sum of the estimated lengths is greater than the length of the tunnel alignment due to locations where strata overlap, transition, or underlies the alignment

**Fill**—this stratum was encountered in most of the test boring locations at depths ranging between 0 and 18 feet bgs. The fill typically consisted of very loose to loose, poorly graded sand and silty sand, with varying amounts of gravel, sand, and silt. The fill contained boulders, cobbles, woodchips, and roots.

**Glaciolacustrine Deposits**—this stratum was encountered in several of the test borings at depths ranging from 1.2 to 82 feet bgs. The Glaciolacustrine Deposits typically consisted of loose to very dense and medium stiff to very stiff, poorly graded sand, sandy silt, silt with sand, and silty sand. With varying amounts of gravel, sand, silt, and clay. Most of the Glaciolacustrine Deposits encountered along the proposed CBD tunnel alignment were loose to medium dense silt, and medium stiff to stiff silt with varying amount of clay.

**Glaciofluvial Deposits**—this stratum was encountered in almost all test borings at depths ranging from 1.4 to 47.5 feet bgs. A lower layer of Glaciofluvial Deposits was encountered at several test borings at depths ranging from 46 to 82 feet bgs with generally higher blow counts. The Glaciofluvial Deposits typically consisted of loose to very dense, poorly graded sand, sandy silt, or silty sand, with varying amounts of gravel, sand, and silt. The Glaciofluvial Deposits contained very few cobbles.

For the purpose of the preliminary profile interpretation and parameter development, the Glaciofluvial Deposits were divided into two sub-strata based on its geologic events occurred: Glaciofluvial Deposits (Shallow) with thickness ranging between zero and 45 feet; and Glaciofluvial Deposits (Deep) with thickness varying from zero to 15 feet. The Glaciofluvial Deposits (Deep), if encountered, was typically below the Glaciolacustrine Deposits and had higher blow counts than the Glaciofluvial Deposits (Shallow).

**Glacial Till**—this stratum was encountered in many of test borings at depths ranging from 3 to 78 feet bgs. The Glacial Till typically consisted of medium dense to very dense, poorly graded sand, well graded gravel, poorly graded gravel, or silty sand with gravel, with varying amounts of gravel, sand, and silt. Boulders and nested cobbles may be present in this stratum.

**Weathered Bedrock**—Weathered Bedrock was designated for the stratus that was typically encountered immediate on top of the Bedrock stratus and consisted of very dense multicolored (black, gray, pink) highly weathered bedrock of both gneiss and granite, with frequent medium to coarse sand and occasional sandy silt. The Weathered Bedrock was encountered in several of the test borings at depths ranging from 7.5 to 81.5 feet bgs.

**Bedrock**—While not observed in all borings due to boring depth, it is anticipated that the Bedrock predominantly consisting of decomposed to fresh, hard to very hard, gneiss, schist, or granite will be present. The recovery ranged between 60 and 100 percent. The RQD ranged between 15 and 100 percent. The bedrock encountered is part of the Massabesic Gneiss Complex. The Complex typically consists of a multicolored (white, black, gray, green, pink) locally rusty weathering, quartzose-feldspathic gneiss, biotite schist, and calc-silicate rocks intruded by a pink gneissic granite of pre-Cambrian age, approximately 625 million years of age. See Figure 6 for a typical rock core recovered. These different types of rock have similar engineering properties based on laboratory testing results. For discussions within this paper, the bedrock is classified as a single stratum.

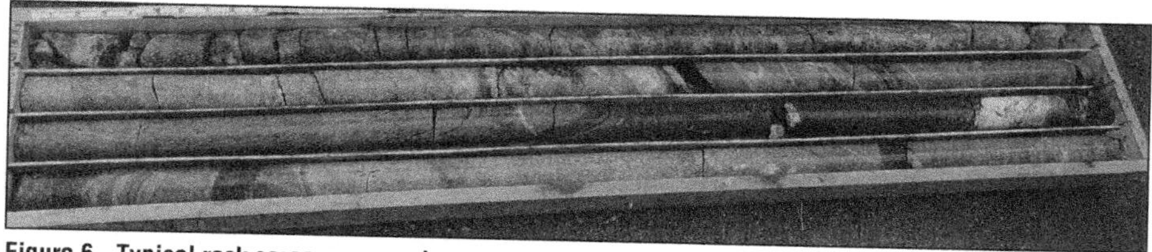

**Figure 6. Typical rock cores recovered**

Based on the currently available information, the tunnel alignment is proposed to be preliminarily delineated into eleven (11) reaches, as noted in Table 2. Each reach limit is based on the interpreted subsurface conditions, and when the tunnel horizon transitions from one type of ground to another. At present, the reaches are only identified as either soft ground, mixed face, or competent rock. As the design progresses, the reach limits and the description will be refined and better described.

Groundwater level varies along the tunnel alignment based on the readings from observation wells and test borings. The groundwater level ranges in depth between 5 and 25 feet bgs and generally follows the surface elevation change along the tunnel alignment where it is at El.145 near the launch trench and rises to approximate El. 290 at the reception shaft (Drop Shaft 7).

It is noteworthy that at the current locations all drop shafts, deaeration chambers, and adits are within competent bedrock.

**Table 2. Preliminary tunnel reaches**

| Reach | | |
|---|---|---|
| **No.** | **Limits** | **Ground Type** |
| 1 | 0+00 to Sta. 0+60 | Soft ground |
| 2 | Sta. 0+60 to Sta. 4+75 | Mixed face |
| 3 | Sta. 4+75 to Sta. 15+55 | Competent rock |
| 4 | Sta. 15+55 to Sta. 15+75 | Mixed face |
| 5 | Sta. 15+75 to Sta. 37+75 | Soft ground |
| 6 | Sta. 37+75 to Sta. 38+25 | Mixed face |
| 7 | Sta. 38+25 to Sta. 50+30 | Competent rock |
| 8 | Sta. 50+30 to Sta. 50+50 | Mixed face |
| 9 | Sta. 50+50 to Sta. 61+00 | Soft ground |
| 10 | Sta. 61+00 to Sta. 62+50 | Mixed face |
| 11 | Sta. 62+50 to Sta. 119+00 | Competent rock |

## Tunnel Profile

The geotechnical information obtained for this project provided the basis for developing the vertical tunnel profile and identifying the appropriate tunneling methodology. The tunnel will be bored through hard rock, mixed face, and soft ground conditions below groundwater water level. Stormwater will be captured and channeled to the tunnel using seven drop shafts. The tunnel invert elevation is set at the outfall end since the CBD profile has vertical constraints including the existing railroad tracks and the 72-inch (internal diameter) East Interceptor North (EIN) conveyance pipe. The railroad tracks are located 400 feet east of the Merrimack River with an approximate top of rail elevation of 156.00 and the EIN is located 100 feet east of the Merrimack River with an approximate top of pipe elevation of 130.00.

The proposed CBD must navigate between these constraints and maintain adequate cover under the railroad tracks. Based on these constraints, the top of invert for the tunnel at the launch trench is elevation 135.00 and will be set at a slight gradient of 0.5 percent. The tunnel profile depth to crown varies between 35 and 65 ft and typically set to maintain at least 1 tunnel diameter rock cover when in rock. The team also worked to minimize mixed face and/or transitions between soft ground and rock.

The preliminary tunnel profile has been established to minimize the number of mixed face conditions, as well as shorten the length of the transition between soft ground/mixed face/rock. This design team varied the slopes along the alignment; this is currently being investigated against hydraulic flow requirements, particularly at drop shaft locations where the flows in the tunnel are impeded as the additional flows are introduced via the drop shafts. The tunnel reaches are summarily defined as:

### Reach 1—Sta. 0+00 to Sta. 0+60

The CBD tunnel starts at Reach 1 at the launch trench. Reach 1 is anticipated to encounter primarily soft ground consisting of loose to medium dense silty sand or poorly graded sand with gravel, with some risk of weathered bedrock near the invert; weathered bedrock consisting of very dense weathered rock of mostly medium to coarse sand-size particles with occasional intermixed sandy silt pockets.

### Reach 2—Sta. 0+60 to Sta. 4+75

Reach 2 consists of mixed face conditions of soils and bedrock before the tunnel transitions into full face of bedrock. The soils that are anticipated to be encountered include silty sand, sand with gravel, silt with sand, silty sand with gravel, sand with silt and gravel, and weathered bedrock. The bedrock anticipated will be Gneiss with thin bedding dipping low to moderate angle, and joints.

### Reach 3—Sta. 4+75 to Sta. 15+55

Reach 3 is anticipated to be in full face of bedrock. The bedrock at this reach consists of Granite and Schist with interbedded Granite. The bedrock joints are typically at low angle, close to moderately spaced.

### Reach 4—Sta. 15+55 to Sta. 15+75

Reach 4 is anticipated to be short in mixed face conditions with a sharp transition from the bedrock into soft ground consisting of silts with sands. Bedrock in this section is anticipated to be Granite with typically at low angle, moderately to widely spaced joints.

### Reach 5—Sta. 15+75 to Sta. 37+75

Reach 5 is anticipated to be in full face of silts and gravel.

### Reach 6—Sta. 37+75 to Sta. 38+25

Reach 6 is anticipated to be short in mixed face conditions with a sharp transition from soft ground to Gneiss bedrock.

### Reach 7—Sta. 38+25 to Sta. 50+30

Reach 7 is anticipated to be in full face of bedrock of fresh Gneiss/Granite. The bedrock joints are typically at low angle, close to widely spaced.

### Reach 8—Sta. 50+30 to Sta. 50+50

Reach 8 is anticipated to be short in mixed face conditions with a sharp transition from bedrock into sands with silts and gravel (Glacial Till) and silts with clay.

### Reach 9—Sta. 50+50 to Sta. 61+00

Reach 9 is anticipated to be in full face of soft ground (silts with sands and poorly graded sands with silt and gravel).

### Reach 10—Sta. 61+00 to Sta. 62+50

Reach 11 is anticipated to be mixed face conditions with a transition from soft ground consisting of silt with gravel to bedrock that consists of very hard, fresh, fine-grained Gneiss.

### Reach 11—Sta. 62+50 to Sta. 117+00

Reach 11 is anticipated to be in full face of bedrock.

## Tunneling Method

The proposed CBD has a 12-foot internal diameter requirement to convey the stormwater flows; an excavated tunnel diameter of between 14 and 14.5-feet will be required. The subsurface strata are challenging ground conditions for tunneling, particularly through the soft ground and mixed face. The sections of the soft ground portion of the tunnel alignment are through ground that is considered “flowing,” where the ground does not have the inherent strength to maintain stable excavation without some type of pre-support. The preliminary interpretation of the subsurface condition suggests highly undulating bedrock profile along the tunnel alignment, particularly within the first 6,500 feet. As noted in Table 1, the tunnel horizon transitions between soft ground, mixed face, and bedrock multiple times. For large segments of the tunnel alignment, open face tunnel cannot be achieved. In fact, active face pressure needs to be applied when tunneling through the soft ground and mixed face portions.

A pressurized face TBM will be required when tunneling through the soft ground and mixed face conditions. In addition, the excavation will be completed as single pass, segmentally lined tunnel. The TBM will likely be operated in open mode for the tunnel constructed within the competent rock section. This means the face is not pressurized, and the cutterhead can be modified to have larger openings dressed with cutting tools that are specifically designed to excavate through rock. A “cross-over” or a “mixshield” machine capable of excavating soft and hard ground will likely be required for the project.

The tunneling will commence in closed/pressurized face mode as the machine breaches the launching trench headwall and starts to excavate the ground. As it excavates, the machine will erect segmental lining behind the cutting face within a protective shield and will use the lining for reaction to generate forward thrust. Pressurized face mode will be maintained in soft ground and in mixed face conditions. When excavating in rock with a minimum of one-half diameter competent bedrock and low probability of groundwater inflow, the machine will likely be switched to open mode. Open mode or open face tunneling is generally executed at a faster rate without the complexities of maintain face pressures. However, it can only be done in ground that exhibits good stand-up time (medium to stiff clay) or is self-supporting (rock) with manageable groundwater inflows. The proposed tunnel vertical profile is below the groundwater table for majority of the alignment. The tunnel alternative assumes that segmental lining will be used, and the finished tunnel size of 12-foot internal diameter will be maintained for the entire alignment. These assumptions will be further evaluated during the final design phase to identify potential cost reductions that could be incorporated into the project.

## SUMMARY

The Cemetery Brook Drain Tunnel Project will be one of the most ambitious and challenging projects undertaken by the City of Manchester's Department of Public Works. The preliminary information presented within the body of the paper will need to be further evaluated and ground conditions will need to be thoroughly understood to develop the most technically sound and cost-effective approach to tunneling. Although majority of the alignment appears to be within good quality bedrock, the profile of the tunnel horizon along the alignment experiences several mixed face zones, as well as significant contiguous segments of soft ground below groundwater table. For that reason, a mix-shield or cross-over type tunnel boring machine is under consideration, and one that can handle soft ground and hard rock excavation. At present, the design will require a 12-foot inside diameter tunnel that is 11,700 feet long with seven drop shafts.

## ACKNOWLEDGMENT

The authors would like to acknowledge the City of Manchester, Department of Public Work—Environmental Protection Division, and the prime consultant CDM Smith for the opportunity to present this paper. CDM Smith looks forward to supporting the City to successfully deliver the Cemetery Brook Drain Tunnel Project and help meet project goals and objectives.

## REFERENCES

1. Khwaja, M., Polcari, D., Harper, J., Lavoie, S., McNeill, F., Clougherty, T., A Tunnelled Solution for the Cemetery Brook Drain Tunnel Project, New Hampshire, North American Tunneling Proceedings, 2022.

PART 

# Geotechnical Considerations

*Chairs*

**Dan McMahon**
Traylor Bros Inc

**Da Ha**
STV Inc

# Adapting to Project Needs: Frozen Cross Passages and Adits

**Aaron K. McCain** ▪ SoilFreeze, Inc.
**Kyle Amoroso** ▪ SoilFreeze, Inc.
**Larry Applegate** ▪ SoilFreeze, Inc.

## ABSTRACT

The use of frozen soil as ground improvement and temporary ground support for the excavation and construction of cross passages and adits has gained in popularity over the last decade. Frozen soil provides a number of advantages for this application such as being waterproof, structurally stable, and working well with any soil type. One of the greatest advantages is the ability for the system to be adaptable to the project geometry and site limitations. Multiple case histories illustrate that freezing can occur from the ground surface, from within the tunnel, or as a combination.

## INTRODUCTION

Ground freezing techniques have been used in civil construction projects since first developed in the late 19th century to provide temporary ground improvement and groundwater control. Improvements in technology in the last 25 years have allowed for more efficient and modular systems to be established for a wider variety of projects ranging from a wine cave in a private residence to large diameter shafts for mining or tunneling operations extending more than 100 feet in depth. The system is ideal to support excavations extending below the groundwater table by providing both engineered support of excavation and groundwater control.

Freeze systems typically consists of a series of close-ended, small diameter steel pipes that are arranged around the perimeter of the proposed excavation. The pipes are plumbed with an interior pipe to allow for downward and upward flow within the freeze pipe and everything is connected to a manifold system. The manifold system also includes pumps, electrically powered chillers, and expansion/contraction tanks that accommodate the volume change of the system at different temperatures. Typically, a calcium chloride brine is circulated through the manifold and is used as the primary medium to remove heat from the ground. The temperature of the brine is reduced by the mechanical chiller and pumped through the pipes at temperatures well below the freezing point of water. The brine absorbs the heat from the surrounding soils and returns to the chiller where the heat is removed into the surrounding atmosphere. The brine is rechilled and circulated back into the ground in a loop that runs continuously for as long as the support of excavation is required.

In the ground, the soil freezes around the individual pipes due to the brine circulation. The extent of the frozen soil gradually increases around each pipe until the frozen zones around the individual pipe overlaps with that of the adjacent pipe and a continuous barrier is formed. As the in-situ groundwater transitions from liquid to solid ice, it creates a bond with the existing soil particles that is not significantly different from cement in Portland cement concrete. The bond isn't as strong as Portland cement, but the mechanism is similar. Once fully formed, the frozen bond is impermeable and increases both the stiffness and strength of the frozen material as the temperature of the ground decreases.

## ADVANTAGES OF FROZEN SOIL

With the unique ability to create a completely temporary barrier, frozen soils present many advantages for heavy civil construction applications. The primary benefit is the ability to provide a waterproof, structural mass with predictable engineering properties that can be used as temporary shoring or initial support of excavation. In addition, a frozen soil barrier is completely temporary and has minimal long-term impacts to the in-situ ground and the groundwater regime. The ability to perform both structural support and groundwater cutoff can make frozen soil shoring a cost competitive solution.

Frozen soil shoring is suitable for most any soil condition as the process only requires changing the temperature of the existing material. Nothing is required to be injected, permeated, or displaced into the ground to create the barrier. This is especially helpful for excavations that extend through a transition of sand overlying cohesive soils where dewatering is not feasible. In fact, where frozen soil is used, the need for significant dewatering can often be eliminated along with the risk of induced settlement, movement of a contamination plume, or bulky treatment systems.

Temperature monitoring devices are installed at the onset of the freezing process to continuously track the temperature of the circulating brine solution, freeze pipes, and the ground at strategic locations and depths. Groundwater levels are also monitored during the freezing process. As a result, verification of groundwater cutoff and the formation of the frozen soil mass can be achieved prior to the mobilization of excavation equipment to the job site.

Perhaps one of the most overlooked advantages of the frozen soil shoring system, and the focus of this paper, is the geometric flexibility of the system. Without the need to physically connect system elements during installation, there are almost limitless potential orientations of a frozen soil system. The system can create curved walls as easily as straight ones, be routed to avoid obstructions, and straddle existing and active utilities. This geometric flexibility can also minimize the footprint of the required ground improvement as the freeze pipes can be located exactly where they are needed. The only limitation to the geometry is installing the small diameter freeze pipes at a relatively uniform spacing from top to tip. However, freeze pipes have been successfully installed vertically, horizontally, and battered at every angle in between.

## ADAPTING GEOMETRY FOR CROSS PASSAGES AND ADITS

The excavation and shoring requirements for cross passages and adits are sufficiently unique to require a specialized approach for each one. These short, horizontal excavations, typically located well below the ground surface require creativity to meet project constraints. These constraints generally include depth, groundwater intrusion, limited access, location of the TBM, and the need to connect two man-made structures that are often neither rectangular nor planar. For this reason, ground freezing, with all of its advantages, can be the ideal solution for these challenging additions to bigger tunneling projects.

This paper presents examples of three projects where ground freezing was used to provide groundwater cutoff and initial support of excavation for cross passage or adit excavations. Each one presented unique challenges that were met and overcome by a different geometric freeze pipe array. These projects are as follows:

- Cross Passage No. 5, Third Street Light Rail Project—San Francisco, CA
- Cross Passage No. 31, SoundTransit Northlink Project—Seattle, WA
- 4th Street Adit, DC Clean Rivers, Northeast Boundary Tunnel—Washington DC

## Cross Passage No. 5, Third Street Light Rail Project—San Francisco, CA

The successful excavation of Cross Passage No. 5 was completed as part of the Third Street Light Rail Project for the San Francisco Municipal Transportation Agency in 2014. The Barnard Impregilo Healy Joint Venture acted as the general contractor for the project. The cross passage was located in downtown San Francisco, 80 to 100 feet directly below the busy 4th Street. The cross passage had an oval cross-section that was approximately 13 to 15 feet in diameter with about 15 feet of lateral distance between the spring line of the two adjoining tunnels. In addition, the excavation included a sump that extended about 15 feet below the horizontal cross passage.

Ground freezing was selected as the initial support of excavation for this project primarily for two reasons. The first is that the soil conditions included a transition from dense silty sand to very stiff clayey silt near the spring line of the cross passage. All the other cross passages for the project were able to stay above the clayey silt and were conventionally dewatered. However, at Cross Passage No. 5, the soil and groundwater conditions necessitated a waterproof shoring system. The second reason was the busy right-of-way at the ground surface which significantly limited the work that could be completed from the surface.

The adaptability of the ground freezing system allowed for the installation and operation of a waterproof shoring system completely from within the confines of the two tunnels. The resultant design consisted of six arrays of up to 12 freeze pipes each, staged at a 4-foot horizontal spacing along the alignment of the tunnels. The freeze pipes in each array varied in length from 16 to 30 feet and were installed at angles ranging from 70 degrees below horizontal to 55 degrees above horizontal as shown in Figure 1. The outer two arrays were located beyond the lateral extents of the cross passage excavations, while the inner four arrays were within the limits of excavation. This placed some of the freeze pipes within the path of the excavation and when encountered, the pipes needed to be cut and rerouted or worked around. The intent of the design was to freeze all of the soils within the cross passage excavation. The freeze pipe arrays all originated from the southbound tunnel and extended through the concrete liner.

To maintain a frozen bond with the back of the concrete liner in the northbound tunnel, small diameter lines were attached directly to the interior face of the liner. The layout of the small diameter surficial freeze lines in the northbound tunnel is shown in Figure 2. The gap between the curvature of the small diameter lines and the surface of the tunnel was filled with packable grout to increase the efficiency of the heat transfer. The flow of chilled brine through these "surficial" freeze lines prevented the warmth from the air circulation within the tunnel from thawing soils at the interface assuring no groundwater seepage at this location. In addition, the lines were eventually covered with a layer of insulation to further protect the bond between the liner and the frozen soil.

The design challenges of ground freezing generally revolve around the installation process of the freeze pipes. The tunnels were 18 feet in diameter and actively being mined during the freeze pipe installation. Therefore, the southbound tunnel could not be blocked for longer than a shift and all installation equipment needed to be removed from the tunnel during inactive shifts. In addition, a conveyor belt was located at the spring line opposite to the cross passage portal and needed to be protected. The northbound tunnel could not be blocked by equipment at any time. To comply with these constraints, the drill rig was mounted on a trailer bed for a rail locomotive that operated on wheels instead of tracks. This allowed for some consistency in the height of the drill machine and relative distance from the face of the tunnel liner, however

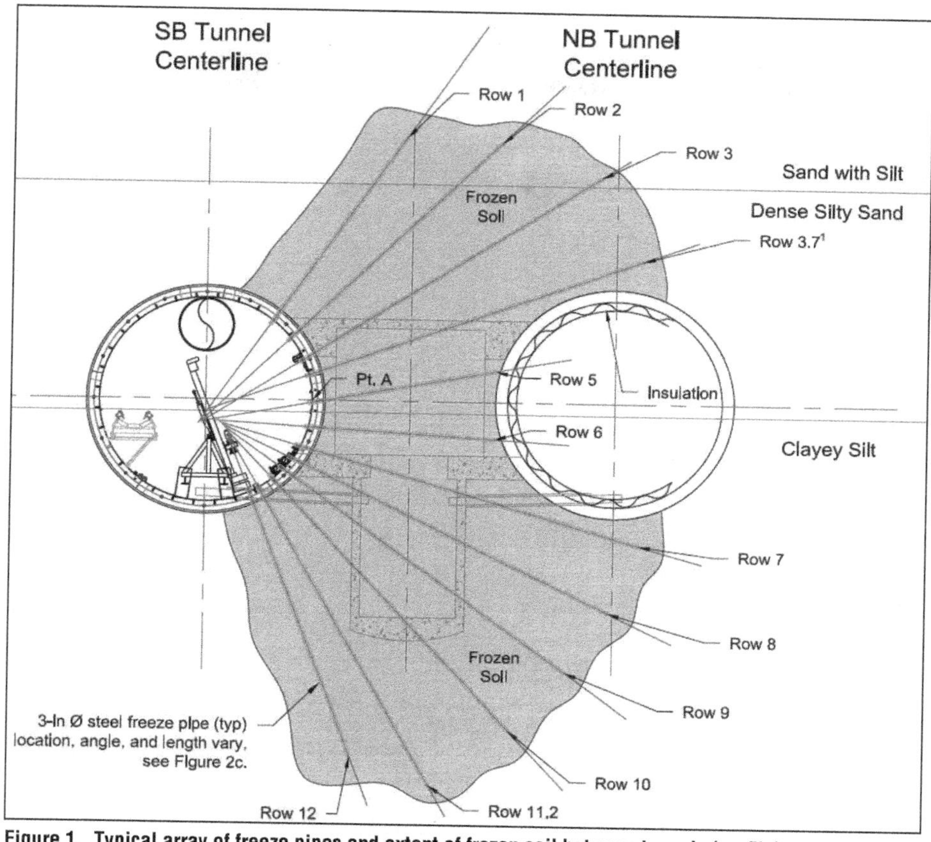


**Figure 1. Typical array of freeze pipes and extent of frozen soil between tunnels (profile)**

adjustments were required along the alignment of the tunnel when the drill rig was mobilized to the project area each morning.

In addition to the increased mobility of the drilling system, the presence of the conveyor and diameter of the tunnel limited the length of installed freeze pipe sections to 5 feet. Drilling holes through the side of a perfectly good tunnel under at least 50 feet of groundwater head also required packers and guillotines to limit groundwater inflow and manage drilling spoils. A system was developed using epoxy on the threaded joints of the drill casing and a tapered plug at the end, so that the drill casing could be abandoned in place and serve as the external portion of the freeze pipes.

Once installed, the freeze system was plumbed and connected to the chillers that were located within each of the tunnels. Special chillers were obtained with a narrow footprint so that they could sit on shelves up at the spring line of the tunnel to allow room for the passage of equipment during the freezedown process. A photo of the freeze pipe array covered in ice is provided as Figure 3. The chiller set-up can also be seen in Figure 3 tucked up tight to the left side of the tunnel. After seven weeks of freezing, excavation began by cutting into the face of the tunnel liners, and within three weeks the cross passage and deep sump were successfully excavated.

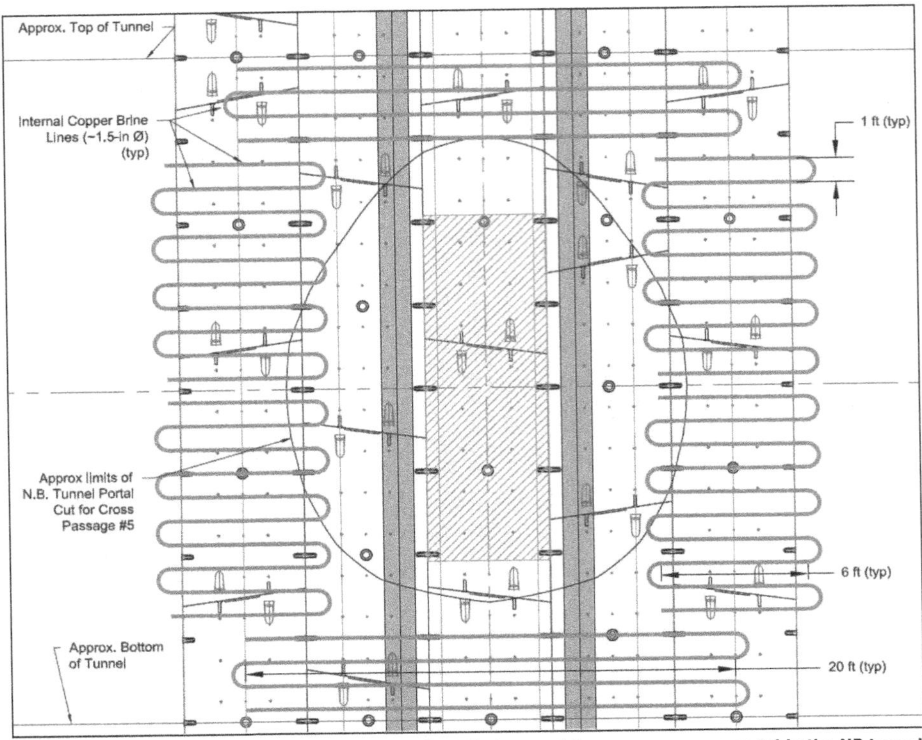


Figure 2. Layout of small diameter freeze lines installed around the cross passage portal in the NB tunnel

Figure 3. The array of freeze pipes (covered in ice) with the chiller set up in the distance

### Cross Passage No. 31, SoundTransit Northlink Project—Seattle, WA

The successful excavation of Cross Passage No. 31 was one of ten cross passages completed using frozen soils as a method of ground improvement on the Northgate Link Lightrail Extension project for SoundTransit in 2016. The general contractor for the project was JCM Northlink LLC (JCM) which was a joint venture between Jay Dee Contractors, Frank Coluccio Construction Company, and Michels Corporation.

Cross Passage No. 31 was located approximately 115 feet beneath the residential University District neighborhood located north of downtown Seattle, Washington. The cross passage had an oval cross-section approximately 18.5 feet high and 17 feet wide, spanning a distance of about 20 feet between the spring line of the two tunnels. This excavation also included a sump that extended about 16 feet below the center-line of cross passage. The other cross passages for the project were similar in size to CP#31, however none of the others that utilized frozen soil shoring also included a sump excavation.

JCM opted to use ground freezing for all the cross passages requiring "Category 3" ground support systems in the project specifications. Category 3 cross passages were typically located in the most difficult soil conditions and at Cross Passage No. 31, the soil materials within the limits of excavation included a transition between very permeable sand and gravel over the less permeable clay and silt layers. Similar to Cross Passage No. 5 in San Francisco, dewatering was not feasible at Category 3 locations. JCM decided to utilize ground freezing instead of jet grouting at these locations to minimize the community and environmental impacts within the relatively dense residential neighborhood. Ground freezing also presented a more flexible scheduling solution because it could be decoupled from the tunneling process and be completed before or after the TBM had passed through the project area.

In order to be decoupled from the TBM progress, the freeze pipes had to be installed from the ground surface and at CP31, installation began before the TBM arrived. The bulk of the freezing was achieved by installing a grid of freeze pipes arranged into six equal rows extending along the alignment of the tunnels at a spacing of about 4.5 feet. Each row contained 5 freeze pipes spaced at approximate 4-foot centers as shown in Figure 4. The outer row of freeze pipes were outside the limits of excavation while the interior freeze pipes passed through the limits of the excavation and were deemed temporarily sacrificial. As shown in Figure 4, some adjustments to the grid were required due to the presence of near surface utilities which included a 42-inch diameter cast iron water line and an 8-inch diameter pressurized sewer line. The freeze pipes were moved to provide sufficient clearance from the active utilities and then angled so that the pipe more closely matched the grid location at the depth of excavation.

It was not necessary to freeze the entire soil column. Only the soils in the vicinity of the cross passage excavation needed to be improved and waterproofed. This was accomplished by using a patented zone freeze pipe that isolates and insulates the upper portion of the freeze pipe, concentrating the freeze within the target zone at depth. This specialized freeze pipe has two advantages: 1) it significantly reduces the chilling demands on the system which in turn reduces the electrical consumption and 2) it minimizes the potential of freezing existing infrastructure near the surface.

While the vertical zone freeze pipes were able to freeze the bulk of the soils, the extent of the freeze extending beneath and above the curvature of the tunnel was limited. Therefore, two smaller secondary freeze systems were installed in both tunnels: haunch pipes and a surficial freeze system. The "haunch" freeze pipes were installed in both tunnels and arranged into two approximately horizontal rows of 5 freeze pipes as shown on Figure 5. The haunch pipes were located between the columns of freeze pipes extending down from the ground surface. The first row was located approximately 7–8 feet above the spring line of the tunnel and were angled upwards 45 degrees from horizontal. The second row was located approximately 8.5 feet below the spring line of the tunnel and were angled downwards approximately 60 degrees from horizontal. Due to the depth of the excavation for the sump, two additional haunch pipes were included at CP31 to provide additional freezing at key

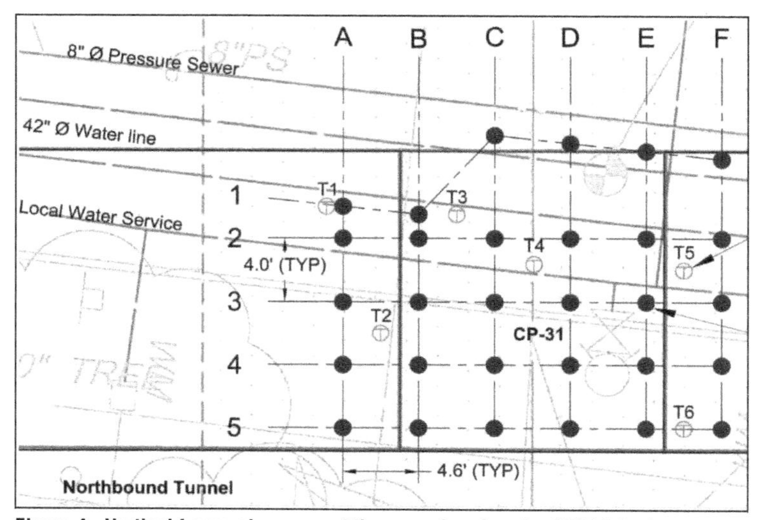

Figure 4. Vertical freeze pipe array at the ground surface for CP31 (plan)

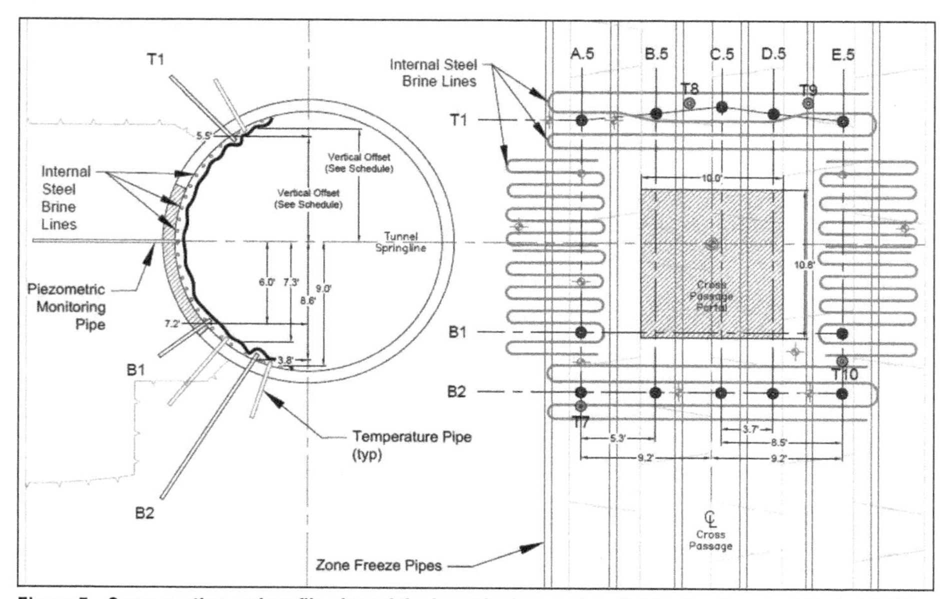

Figure 5. Cross-section and profile view of the haunch pipes and surficial brine lines installed at CP31

locations. The two additional freeze pipes were located about 6 feet below the spring line of the tunnel and angled 36 degrees below horizontal. These haunch freeze pipes were installed through packers mounted directly to the interior face of the concrete tunnel liners which successfully controlled groundwater flow and drilling spoils during the drilling and installation process.

A surficial freeze system was also attached to the interior face of both tunnels in a manner similar to that done at the cross passage in San Francisco previously discussed. Steel pipes (lines) were attached directly to the concrete tunnel liner, however grout was not used around the lines at this project. The array of freeze lines around the portal entry for the excavation is shown in Figure 5 and a photo of the lines covered in frost after the cross passage was completed is shown as Figure 6. This surficial freeze ensured that the frozen bond between the exterior face of the concrete tunnel liner and the frozen soil remained robust and intact. This bond prevented groundwater from traveling along the circumference of the existing tunnels into the excavation of the cross passage. Additionally, insulation blankets were placed over both the haunch pipes and the surficial freeze manifold to reduce the exposure to the circulating warm air in the tunnel.

After installation, an electric chiller located at the ground surface was used to supply the chilled brine to the vertical zone freeze pipes. The chiller was fitted with low-noise fan blades and housed in a wooden enclosure that was lined with sound dampening blankets to ensure that the white noise produced by the equipment remained at or below acceptable levels for the surrounding neighbors. Inside the tunnels, the chilled brine was circulated through small chillers located on shelves placed near the tunnel spring line. In both locations the freeze pipes and chiller units were positioned to allow for the passage of traffic and equipment during the freezedown process.

Once sufficient freezing was achieved and the TBM had passed through the area, excavation began. As work progressed the vertical freeze pipes immediately in front

**Figure 6. The array of freeze lines and haunch freeze pipes covered in ice, near the end of the cross passage construction**

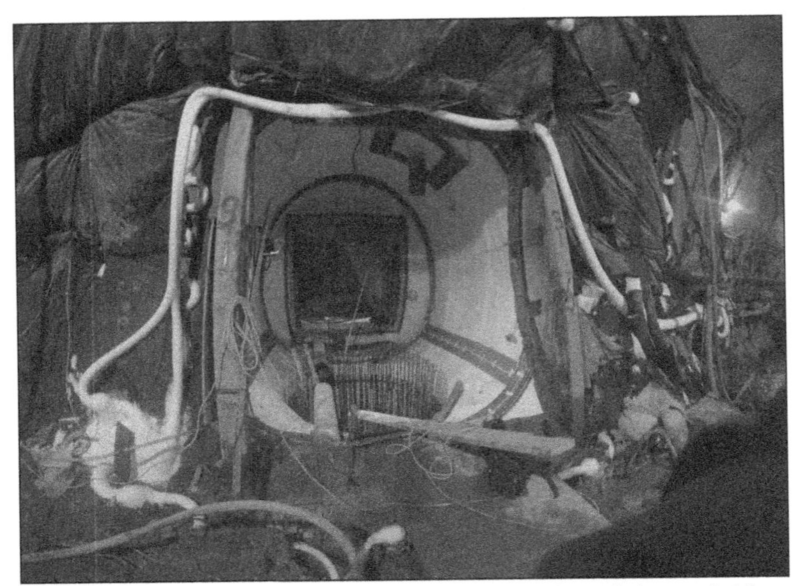

**Figure 7. Cross passage 31 nearing completion after excavation and placement of the waterproofing membrane**

of the excavation were deactivated and the brine removed using compressed air at the ground surface. This allowed the removal and rerouting of the the freeze pipes within the limits of the excavation while working in the cross passage. The freeze pipes were then brought back online by capping and attaching hoses that extended around the perimeter of the cross passage and allowed for the supply of chilled brine into the lower portion of the freeze pipe. The contractor used shotcrete and rebar ribs to provide the primary support after the excavation was complete. The rerouted hoses from the cut vertical freeze pipes were able to be buried in the shotcrete to help keep the perimeter cold for the remainder of the successful excavation and construction process. A photo of the cross passage after the excavation is complete and waterproofing has been installed is provided as Figure 7.

## 4th Street Adit, DC Clean Rivers, Northeast Boundary Tunnel—Washington DC

Unlike the previous two examples, the 4th Street Adit excavation did not consist of a cross passage between two concrete lined tunnels. Instead, this freeze project allowed the successful excavation of a 17-foot diameter adit that extended approximately 73 feet between a vertical shaft and the Northeast Boundary Tunnel in Washington DC. This project was completed as part of the DC Clean Rivers Project, Northeast Boundary Tunnel for the District of Columbia Water and Sewer Authority in 2019 and 2020. This project was designed to be completed ahead of the TBM passage and located at a depth of about 95 feet below Rhode Island Avenue. The excavation was designed to extend out into the path of the TBM. The construction of the permanent adit liner would initially be advanced just short of completion and the path of the TBM. A larger chamber was excavated at the end of the initial adit advancement. This chamber was within the path of the TBM, and back filled with flowable fill concrete. The TBM then mined through the large mass of flowable fill on its own schedule and the connection to the nearly completed adit was made. The adit and the chamber are visible in the profile view of the ground freezing plan as shown in Figure 8.

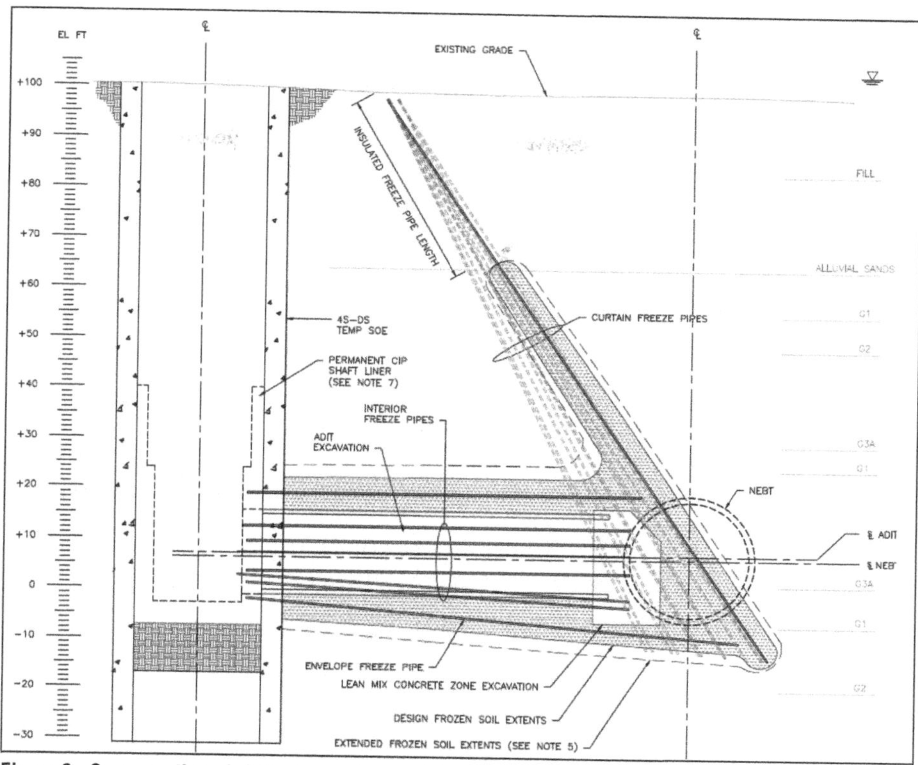


**Figure 8. Cross-section of shaft, adit and ground freezing approach at 4th Street**

Ground freezing was selected as the initial support of excavation and groundwater control primarily due to the ability of the system to be installed with any geometry. The system's ability to be installed at any orientation allowed for the temporary shoring to be installed exactly where it was needed for the complex geometry of a deep horizontal excavation with a dead end. As with the previous two examples, the soil materials consisted of layers of sand and gravel mixed with layers of silt and clay where dewatering would have been ineffective.

The primary "envelope" freezing was installed horizontally and in a pattern that encircled the circumference of the adit as shown in profile view down the alignment of the adit, Figure 9. The pipes were installed from within the confines of the 24-foot diameter slurry panel shaft which wasn't wide or deep enough to allow for the installation of freeze pipes that ran parallel with the alignment of the adit. Nearly all of the 24 envelope freeze pipes were installed with either a slight horizontal and/or vertical angle which increased the center to center spacing of the freeze pipes as the distance from the shaft also increased. This is visible in the plan view provided as Figure 10. The freeze pipe lengths ranged from 67 to 96 feet and were spaced approximately 4 feet apart out near the path of the TBM. One of the challenges of the freeze pipe installation was the preparation of the rough and uneven surface of the slurry panel for the mounting of a rubber gasket and backflow preventer. This work required chipping out cavities, that were not perpendicular to the face of the shaft. Survey and a device called an axis aligner, were used to ensure that the drilling operations began on the proper alignment. A photo of the drilling set up with the shaft is provided as Figure 11.

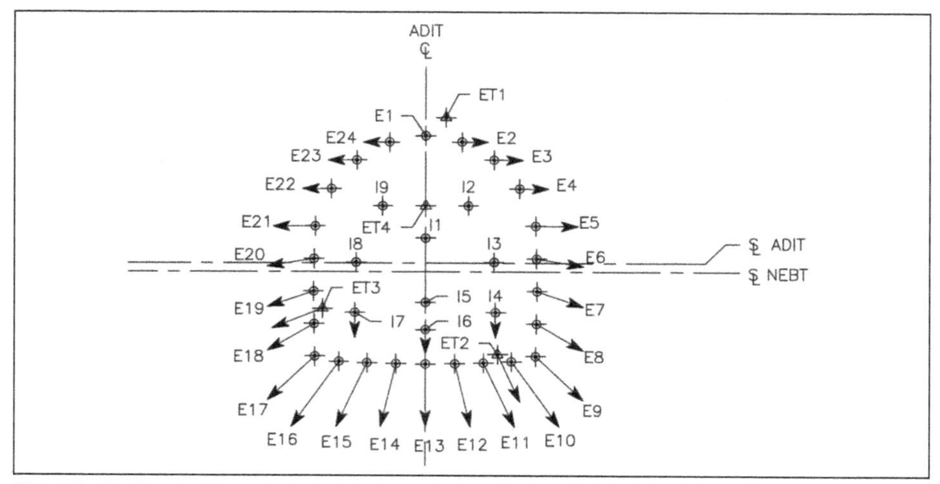


**Figure 9. Profile view of the freeze pipe layout as view from the bottom of the shaft, down the adit alignment**

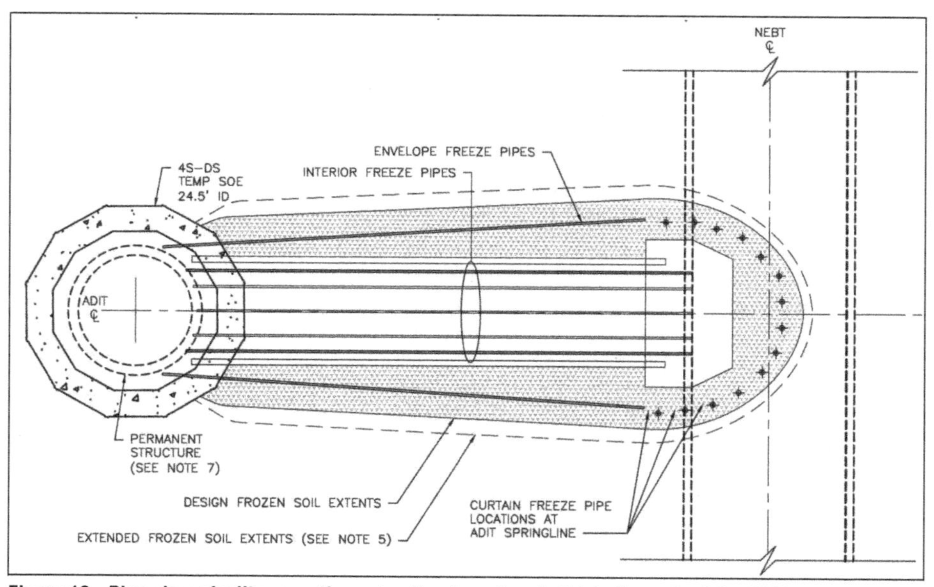


**Figure 10. Plan view of adit excavation extending from the shaft into the path of the TBM**

Due to the challenges of drilling a horizontal hole beneath the groundwater table in permeable soils, the nearly horizontal envelope freeze pipes were grouted in place after installation and could not be removed once installed. Therefore, none of the steel envelope freeze pipes extended into the path of the TBM.

In addition to the envelope freeze pipes, the client requested that the interior of the adit be completely frozen to ensure face stability during excavation. An additional nine pipes were added within the limits of the adit excavation. These interior freeze pipes were later removed during the excavation process.

**Figure 11. Drilling set up for the installation of the horizontal envelope freeze pipes**

Without the presence of the concrete lined tunnel at the end of the excavation, there was nothing in the ground to "close" the end of the frozen envelope cylinder and prevent groundwater from running parallel with the axis of the adit. This was solved via the installation of "curtain" pipes extending down from the ground surface to intersect the envelope pipes as shown in Figure 8. These curtain pipes could not be installed within the busy Rhode Island Avenue right-of-way per the project specifications. The pipes were therefore installed from within the project limits on 4th Street and inclined at angles ranging from 56 to 65 degrees below horizontal. At the ground surface the pipes were arranged in nearly a straight line, however at depth the pipes fanned out to create a sloped arch that encompassed the envelope freeze pipes as shown in Figure 10. These pipes were removed at the end of the project and therefore could be installed into the path of the TBM. As a final requirement, the upper 40–45 feet of the curtain freeze pipes were insulated to minimize the freezing impacts in the upper aquifer and near surface utilities.

The system required approximately 10 weeks to achieve freezedown. The perimeter freeze was completed after about 6 weeks, however additional time was necessary to freeze the interior of the adit excavation solid. Once excavation began, the interior

Figure 12. Adit excavation underway with frozen soil exposed at the end of the adit

freeze pipes were permanently deactivated and removed in pieces as the excavation progressed. A photo of the adit excavation underway is provided as Figure 12. The primary support of excavation consisted of shotcrete and steel girders, installed at approximately 5-foot spacing along the alignment of the adit. Once the chamber at the end was reached, no steel was used and only fiber reinforced shotcrete was placed against the frozen soil. The freeze system continued to operate during the entire excavation process and was deactivated after the chamber was backfilled with flowable fill. No significant deformation of the primary lining was observed during the freezing process. The removal of the curtain pipes went smoothly and took about a week to complete.

## CONCLUSION

Ground freezing technology has many advantages as a method for providing initial support of excavation and groundwater control in heavy civil construction projects. These advantages include being waterproof, suitable in most soil conditions, verifiable before excavation begins, and cost competitive with other traditional shoring methods. However, as demonstrated by the three examples provided in this paper, perhaps the most overlooked advantage is the flexibility in the geometric layout of the system. This is especially useful in excavations for cross passages or adits where the geometry of the curved surfaces can be challenging to accommodate without massive grouting efforts that more than encompass the entire work zone. Whether installation is done from within a tunnel or shaft, the ground surface, or any combination in between, ground freezing can provide specialized temporary shoring and groundwater control exactly where it is required.

# Design-Build Project Delivery Method Selection and Implementation of a GBR-B and GBR-C for the Pawtucket Tunnel

**Kathryn Kelly** ▪ Narragansett Bay Commission
**Julian Prada** ▪ Stantec
**Todd Moline** ▪ MWH Constructors
**Robin Dill** ▪ AECOM
**Brian Hann** ▪ Barletta Engineering Co.

## ABSTRACT

Narragansett Bay Commission's (NBC) Pawtucket Tunnel Project includes construction of the 11,600-ft long, 30-ft ID, deep rock tunnel. This paper presents insight into the features and benefits of lump-sum, design-build that led NBC to select it as the project delivery method. The use of a design-build contracting process that included proprietary meetings to discuss and obtain feedback on Alternative Technical Concepts as well as to facilitate reaching consensus on modifications to the GBR-B is also described. The discussion includes features and benefits of design-build process that not only addressed NBC's desired outcomes, but also allowed the Design-Builder to propose several innovative changes to the Base Technical Concepts included in the bid documents, and to develop a GBR-C that was compatible with their proposed means and methods of construction and addressed what they believed to be the more significant subsurface risks on the project.

## PROJECT BACKGROUND AND DESCRIPTION

The Narragansett Bay Commission (NBC) owns and operates Rhode Island's two largest wastewater treatment facilities and provides wastewater collection and treatment services for approximately 360,000 customers in the greater Providence metropolitan area. There are ten-member communities in the NBC district. Three of these communities, Providence, Pawtucket and Central Falls, have combined sewer systems which can overflow during intense wet weather events when the volume of flow exceeds the capacity of the pipe and relieves through outfall overflows. In 1992, NBC entered into a Consent Agreement with the Rhode Island Department of Environmental Management to significantly reduce occurrences of CSOs by constructing CSO control facilities in a three-phase CSO Control Program. NBC is currently in the third phase of its CSO Control Program.

Phases 1 and 2 of the Program included construction of CSO control facilities primarily in the City of Providence. Phase 1 which included a deep rock storage tunnel known as the Main Spine Tunnel (refer to Figure 1) was completed in 2008 at a cost of $360 Million and Phase 2 was completed in 2015 at a cost of $197 Million. All facilities constructed during Phases 1 and 2 of the Program were completed using the traditional design-bid-build delivery method. Construction documents for the storage tunnel and the 1,800-ft. rock adit constructed during Phase 2 included geotechnical baseline reports prepared by NBC.

Phase 3 CSO control facilities will primarily address CSOs in the Cities of Pawtucket and Central Falls and will be constructed in four sub-phases over a period of 22 years. Facilities to be constructed during Phase 3A of the CSO Program include the Pawtucket

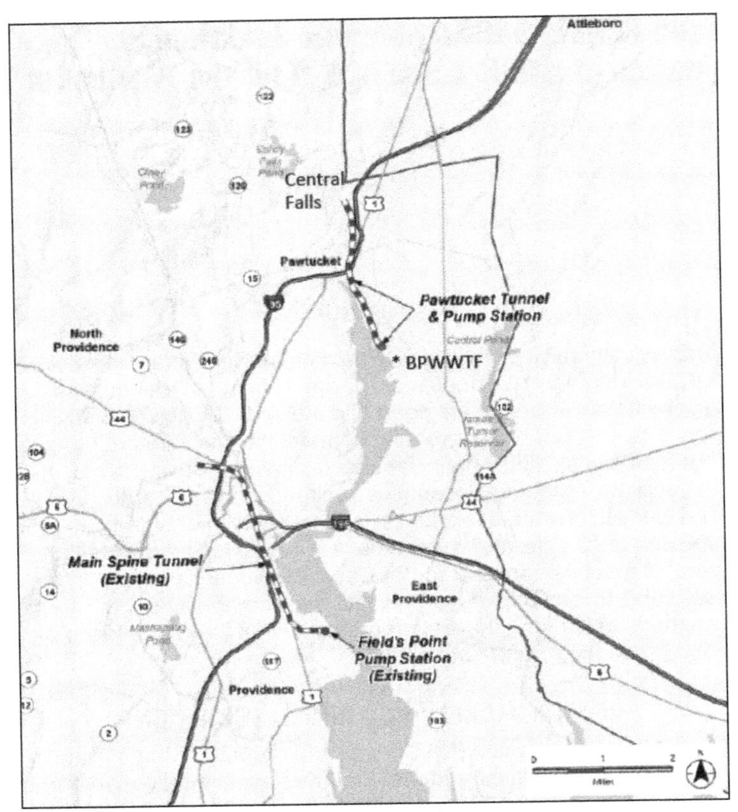


**Figure 1. Pawtucket Tunnel location map**

CSO Storage Tunnel and accompanying tunnel pump station, near-surface facilities, and green stormwater infrastructure. Figure 1 shows the location of the Phase 3A Storage Tunnel and Pump Station.

The Pawtucket Tunnel Project of Phase 3 consists of a new tunnel and ancillary underground components including a launching shaft, receiving shaft, drop shafts, tunnel pump station shaft, and corresponding adit tunnels to support the functionality of a tunnel system to serve as a CSO storage facility. The purpose of the tunnel system is to provide volume to store all contributing overflows during a storm event up to the three-month storm for subsequent pump out and treatment at the Bucklin Point Wastewater Treatment Facility (BPWWTF). The required minimum storage volume to achieve the defined hydraulic criteria is 58.5 million gallons (MG).

The tunnel is a rock tunnel, 115-ft to 155-ft below the ground surface, located north of the BPWWTF in Pawtucket, RI, adjacent to the Seekonk and Blackstone Rivers as shown in Figure 2. The tunnel is approximately 11,600 feet in length with a 30-foot inside finished diameter. The tunnel was procured and designed with a single-pass, gasketed, precast-concrete, segmental tunnel liner. This lining system was selected as best suited to control groundwater inflows, maintain rock stability, control quality, and reduce time of installation.

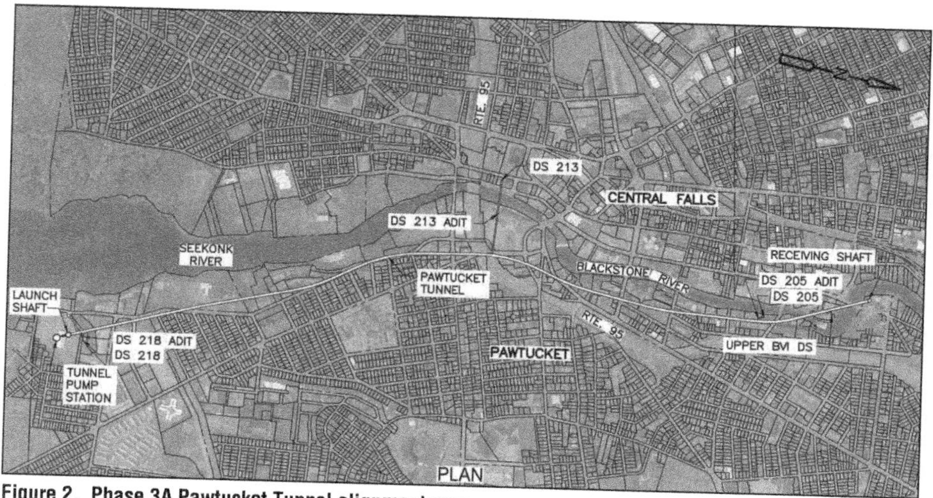

Figure 2. Phase 3A Pawtucket Tunnel alignment map

NBC's Program Management (PM) team of Stantec/Pare developed the preliminary design of the Pawtucket Tunnel. The preliminary design included a plan and profile for the Pawtucket Tunnel and adits which showed the tunnel depth with a minimum of two tunnel diameters of bedrock cover. The preliminary design also displayed locations and dimensions for all shafts. As part of their preliminary design efforts, the team undertook an extensive geotechnical exploration program that included vertical and angled drilling of borings and utilized this data to develop a geotechnical data report and a Geotechnical Baseline Report for bidders (GBR-B). The winning Design-Build Team consisted of a joint venture of CBNA (Bouygues Construction) and Barletta Corporation with AECOM as lead designer. The Design-Build team was responsible for completing the final design including preparing the Geotechnical Baseline Report for Contracting (GBR-C).

## GEOLOGY

The Narragansett Basin within which the Pawtucket area is located, is a structural and topographically controlled depression trending northeast-southwest. The basin contains Pennsylvanian sedimentary rock units that have been variably metamorphosed and deformed prior to periods of glaciation. Faulting in the region is believed to be in the western and eastern flanks of the Narragansett Basin which developed as unconformities. The basin is bound by basement rocks with variably metamorphosed sedimentary rock comprising the central portion of the Narragansett Bay Group.

The Narragansett Basin is approximately 55 miles long, 15 to 25 miles wide, and is made up of several thousand feet of non-marine sedimentary rocks that have been folded, faulted, and slightly to moderately metamorphosed. Sedimentary rocks of the Narragansett Basin have been divided into five formations, of which the Wamsutta Formation and the Rhode Island Formation occur within the depths of the Pawtucket Tunnel project. Bedrock is overlain by glacial deposits consisting of till and glaciofluvial deposits as well as man-made fill deposits near surface.

The Rhode Island Formation is widespread in the Narragansett Basin and consists mainly of gray sandstone and siltstone, with lesser amounts of gray to black shale, conglomerate, and coaly rock. These sediments were deposited in medial to distal alluvial fan environments, and are comprised of meandering stream deposits, bank and flood plain deposits, and swamp deposits. Sediments of the Rhode Island Formation generally are finer grained than those of the Wamsutta Formation. Rock types are laterally and vertically discontinuous, a characteristic of the environment in which they formed. Along the tunnel alignment, bedrock consists predominantly of the Rhode Island Formation from the south end of the project to the second river crossing on the north end.

The Wamsutta Formation occurs in the northern part of the Narragansett Basin, where it underlies and partially interfingers with the Rhode Island Formation (Skehan, Rast, and Mosher 1986). The Wamsutta Formation consists of a sequence of red conglomerates, sandstones, and shales up to 3,000 feet thick. Volcanic fragments are common. Conglomerate in the Pawtucket quadrangle contains boulders as large as 4 feet in diameter (Quinn 1971). These sediments were deposited in an alluvial fan environment, and are composed of braided stream deposits, with related crevasse splay deposits and flood plain deposits. Bedrock of the Wamsutta Formation is present along the northern 200 to 400 feet of the tunnel alignment and at the receiving shaft.

## PROCUREMENT PROCESS

### Use of Alternative Project Delivery Method

As indicated above, Phases 1 and 2 were delivered using the design-bid-build method of project delivery. For the Pawtucket Tunnel Project, NBC explored using an alternative delivery method to mitigate risks associated with design, schedule, and cost. NBC procurement regulations limited the method of project delivery of the Pawtucket Tunnel to conventional design-bid-build or design-build. State procurement regulations do not allow the use of the construction management at risk (CMAR) method of project delivery for heavy civil construction. When exploring alternative methods of project delivery, NBC was limited to two variants of design-build: fixed price design-build or progressive design-build.

#### *Fixed-Price Design-Build*

In fixed-price design-build, the design-build entity (typically a team of a contractor, its subcontractors, a design firm, and subconsultants) agrees to perform the final design and construction of the project for a lump-sum price under a single contract. The scope of work for the lump-sum price is described in tender documents developed by the owner. The scope of work is delineated in performance or prescriptive requirements, or a combination of both. Performance requirements, in their purest form, describe the owner's needs with respect to the performance of the constructed facility with minimal limits or direction on the design and use of equipment or materials. Prescriptive requirements set forth the owner's mandates on how the facility will function, the use of materials and equipment, as well as other necessary constraints. Prescriptive requirements are usually set forth in preliminary drawings and specifications for mandated equipment and materials.

Design-build based strictly on performance criteria is optimal when the owner can turn over the entire design development effort to the design-builder after providing performance criteria. For the Pawtucket Tunnel Project, NBC maximized the use of performance requirements whenever possible, but needed to provide prescriptive criteria as well. Prescriptive criteria are necessary to limit project layouts to within rights-of-ways and easements, and to mandate the use of standard materials and equipment that the owner is already prepared to maintain and service.

Unless the special provisions are included in the design-build agreement, the owner has little to no input during design development in fixed-price design-build. The owner may review design documents at stages of design development for adherence to the project criteria and performance and prescriptive design requirements. However, if the design is compliant with the original requirements, the owner cannot change the design-builder's design without a change to the design-build agreement.

### *Progressive Design-Build*

Progressive design-build is ideal for owners who want to be engaged in design development. Progressive design-build is similar to the construction manager at-risk (CMAR) method of project delivery in that the owner is actively engaged with the design-builder during design development. What differentiates progressive design-build from CMAR is that the design is performed by the design-builder and not by a designer contracted directly by the owner. Estimates performed by the design-builder at each stage of design development allow the design to be modified to meet the owner's budget. Owners do not know the final price of the project when selecting a design-builder for progressive design-build. The design-build agreement is amended to include a lump-sum price or guaranteed maximum price (GMP) is agreed upon by the owner and design-builder when an advanced or final stage of design is achieved.

### *Selection of Fixed-Price Design-Build*

NBC explored using an alternative delivery method for the Pawtucket Tunnel Project to reallocate the inherent project risks typically associated with design-bid-build projects. Although design-build cannot eliminate project risks, it can facilitate the allocation of risk to the party best suited to manage the risk Note that assumption of additional risks by the design-builder will likely be incorporated in the design-builder's lump-sum price in the form of a contingency.

NBC selected fixed-price design-build to reallocate inherent risks resulting from the design process and also to leverage its other benefits to mitigate additional risks:

1. Speed of Delivery—NBC must complete the Pawtucket Tunnel and facilities of Phase 3A of the CSO Program in accordance with a consent decree with a regulatory agency. The consent decree includes a milestone date by when the facilities of Phase 3A must be operational. Design-build allows the procurement of a design-builder prior to final design of the project. It also allows construction to be "fast-tracked" by prioritizing the design of the initial elements of the project such as construction of shafts and starting construction of those elements prior to finishing the design of all elements of the project.
2. Single Point of Accountability—NBC wished to minimize the amount of contractor's change requests resulting from omissions, unclarity, and gaps in interdisciplinary coordination of the design documents. Contractor's requests for change based on omissions, unclarity or gaps in coordination in the owner-provided design results in the expenditure of considerable amounts of owner's time and resources to manage and resolve such requests. Use of design-build relieves the owner of a great deal of involvement in the design change management process.
3. Contractor Involvement in Design Process—Preferences for means and methods of underground construction can vary between contractors. The owner can reduce risk of duplicated design efforts by the contractor related to preferred construction methods by allowing the design-builder to determine the best design to accommodate its means and methods.

4. Price Certainty—As with all capital projects, NBC strongly desired to limit costs for the Pawtucket Tunnel Project to within the contract amount and its contingency. With the exception of differing site conditions, the use of design-build mitigates the risk of cost increases resulting from changes to an owner-provided design.

## GBR-B DEVELOPMENT

During the preliminary engineering and in the development of the procurement documents, a Geotechnical Baseline Report for Bidding (GBR-B) was prepared to describe subsurface physical and geotechnical conditions expected to influence construction. The GBR-B was prepared by Stantec in general conformance with "Geotechnical Baseline Reports for Underground Construction; Suggested Guidelines," (Essex 2007). As part of their preliminary design efforts, Stantec undertook an extensive geotechnical exploration program that included vertical and angled drilling of borings and utilized this data to develop a geotechnical data report and the GBR-B. The process for modification of the GBR-B to the GBR-C was provided through the Tunnel Design-Build Agreement.

In addition to the GBR, the procurement process also established procedures for Alternative Technical Concepts (ATCs) which are design concepts and/or construction methods that offer alternative approaches to completing the project according to the criteria provided in the RFP. This process was enabled through Proprietary Meetings between the Owner, the Program Manager, and the Design-Builder in an effort to allow the Design-Build team to provide innovation and propose alternatives that suit their preferred design and means and methods.

The GBR-B was provided to all parties bidding on the project and was a critical document which contained statements describing the geotechnical conditions anticipated or assumed to be encountered during construction. The baseline generally comprises: 1) the geotechnical data, 2) interpretations of existing geotechnical conditions, and 3) expected ground response to construction. Risks associated with conditions consistent with (or less adverse than) the baselines are allocated to the winning Bidder and those that are more adverse than the baselines are allocated to NBC.

The GBR-B produced for the procurement provided all Design-Build teams with the same set of physical baseline conditions to be used in their design and construction planning during their bid. The process was set up to allow the selected Design-Build team to provide their interpretations of the various baselines expressed in the GBR-B in developing their design and construction approaches, identifying any gaps, and providing any supplemental information gained through additional geotechnical investigations performed by the Design-Build team. The Owner and the Design-Build team then negotiated revisions to the GBR-B to formulate the Geotechnical Baseline Report for Contracting (GBR-C). The GBR-C reflects the joint understanding of NBC and the selected Design-Build Team regarding the interpreted baselines that are consistent with their design approach, equipment and construction means and methods.

Benefits of this approach included:

- Provided opportunity for NBC to allocate risks based upon known subsurface conditions
- Allows the Design-Builder the opportunity to supplement the GBR-B with information collected during design and adapted to their preferred means and methods of construction

- Allows NBC and the Design-Builder the opportunity to mutually agree to baseline conditions memorialized in the GBR-C, thus reducing the potential for claims and project delays.

Following the ASCE guidelines, the GBR included sections for Ground Characterization and Construction Considerations. Various topics regarding ground and groundwater conditions were discussed within each section and were the subjects where revisions were addressed between the GBR-B and GBR-C. The text below describes some of the more significant risks in the GBR-B development from the Owner's perspective regarding ground and groundwater conditions.

## Suspected Fault Presence

During the preliminary engineering site investigation, the presence of a fault was inferred from the difference in lithologies in nearby borings. The gray, Rhode Island Formation was encountered in one area and the dark red Wamsutta Formation was encountered in the adjacent area near the end of the tunnel drive upon crossing the Blackstone River. The possibility of a fault under the river was suggested by a mapped north-south trending fault that lies about 3 miles to the north and directly aligns with the river course. In an attempt to locate this fault, three angle borings were drilled from the riverbanks, but none were successful in penetrating a fault or a contact during the preliminary engineering phase. However, the angled borings did reduce the likely area of the fault crossing.

A baseline of conditions expected in the potential fault zone was established and a suspected interval was presented in the GBR-B figures (see Figure 3). Conditions in the fault zone were stated as unknown since the borings did not intersect the fault. As a baseline, this fault was described as a near-vertical fault, composed of a major slip zone one-foot wide, surrounded on each side by a damage zone comprising a network of low-throw faults trending subparallel to the major fault. Each damage zone was estimated to be up to 20 feet wide, for a total width of the fault zone of 40 feet. Baseline conditions for groundwater inflows from the fault zone were defined within the report and to mitigate the risk of excessive inflows while tunneling through the fault zone, probing and grouting ahead of the tunnel excavation was specified within the "area of potential fault."

## Groundwater Inflows

Groundwater control is paramount on all tunnel projects. Detailed analyses to predict groundwater inflows were conducted through packer testing performed during the preliminary engineering geotechnical investigation program. The results of those estimates of hydraulic conductivity were 10 to 20 times lower than those derived from packer testing data from the Phase 1 tunnel in Providence. Despite the moderate to low values of hydraulic conductivity for the Phase 3A tunnel, construction for the Phase 1 tunnel and shafts had encountered excessive amounts of groundwater that had negatively impacted treatment operations at the NBC treatment facility. The NBC therefore dictated that strict groundwater control measures be required for the Pawtucket tunnel and shafts construction.

The GBR-B included baseline groundwater inflow criteria and established that pre-grouting in advance of excavation to reduce inflow to the shafts during construction be the Design-Builder's decision. Groundwater control at the shafts required that cutoff methods such as secant piles be implemented to control groundwater and for the efficiency of this method at the depth to rock under consideration, and the ability to drill

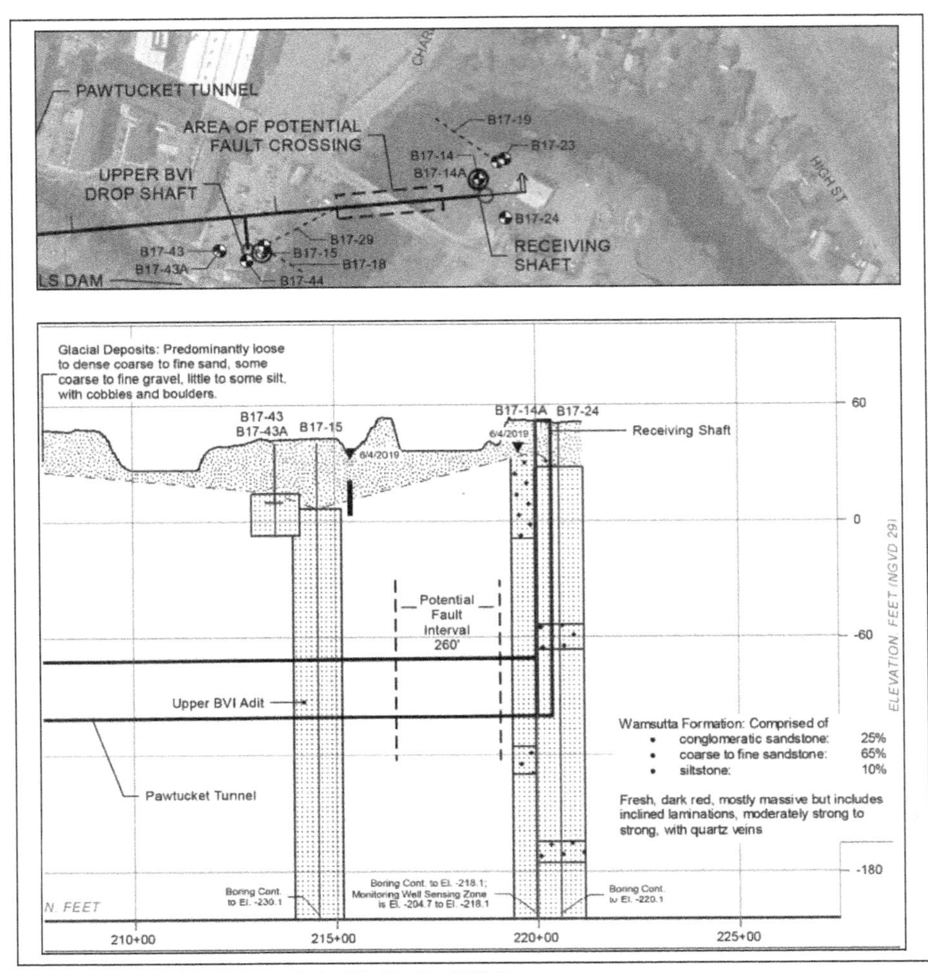


**Figure 3. Potential fault interval identified in the GBR-B**

through boulders in the till. Secant piles were required to extend far enough into sound rock to secure the cutoff. A slurry diaphragm wall was also allowed.

For the tunnel, a single-pass lining approach was required to not only minimize groundwater discharge volumes to the Bucklin Point Wastewater Treatment Facility but also to maintain rock stability, control quality, and reduce time of installation.

## PROPRIETARY MEETINGS AND GBR-C DEVELOPMENT

The Design-Build Team welcomed the opportunity to hold proprietary meetings prior to submitting their final technical and price proposals. Not only was it an opportunity to discuss and obtain feedback on proposed ATCs, but it was also an opportunity to discuss proposed modifications to the GBR-B to be incorporated into the GBR-C with the Owner Team's approval. It was felt that this second purpose would benefit all stakeholders in reducing bid contingencies while adequately managing subsurface

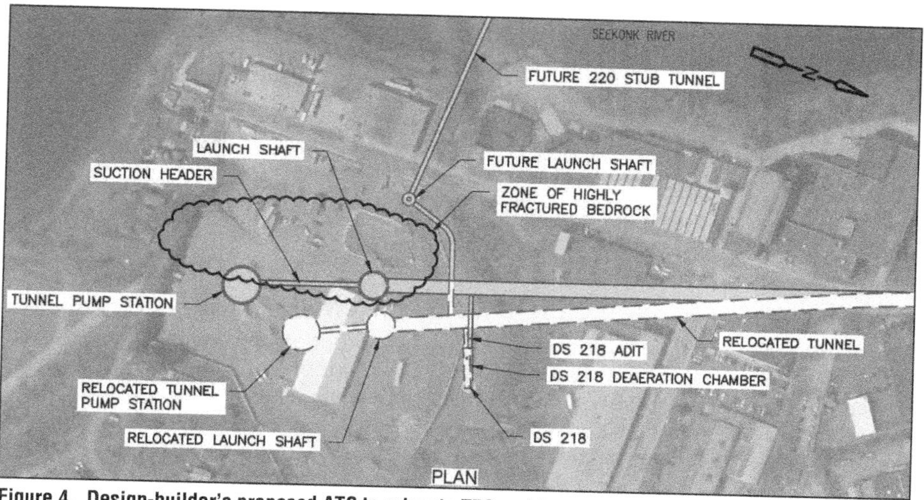


**Figure 4. Design-builder's proposed ATC to relocate TPS and launch shaft**

risks using baselines that were realistic and applicable to the means and methods being contemplated by the Contractor. In total, five proprietary meetings were held.

The Design-Build Team developed a list of ATCs that were discussed at the initial proprietary meetings. The ATCs were proposed to reduce costs or risks to the project. In total, six ATCs were submitted for Owner review and approval. One of the ATCs proposed was related to adverse ground conditions near the tunnel pump station (TPS) and launch shaft (LS) locations and was designed to reduce both project risk and costs. As shown on Figure 4, the proposed ATC was to relocate the launch shaft 100 feet to the north along the tunnel alignment and shifting it at least 80 feet to the east of the existing shaft. In addition, the TPS was to be moved closer to the launch shaft to be in line with the re-aligned tunnel and to shorten the suction header tunnel. This alternative would move these elements away from the highly fractured rock (potential fault zone) as observed in the Owner's geotechnical explorations. Other benefits included:

- Shortening the overall length of the mainline tunnel by 100 feet
- Reducing the depth of SOE in overburden for both the TPS and LS due to higher bedrock surface at the relocated positions.
- Shortening the length of the OF-218 adit.

To validate the assumptions regarding ground conditions and provide necessary information for design and construction of the various downstream project elements, the Design-Builder also proposed a substantial supplemental geotechnical investigation program (GIP) to be performed following NTP.

Two proprietary meetings were also devoted to discussing subsurface risks and development of modified and/or new baselines to incorporate into the GBR-C. One of the major concerns of the Design-Builder expressed during the first meeting was that the supplemental GI could potentially disclose different physical conditions than indicated in the GBR-C which was required to be finalized prior to contract execution. However, NBC acknowledged that if different conditions were discovered after notice to proceed, the DSC claim(s) would be addressed through the normal contract procedures and administered through a change order if found to be valid.

The Design-Builder proposed the following key issues be addressed with modifications and/or new baselines to the existing GBR-B:

1. Fault presence and groundwater inflows
2. Distribution of rock types in tunnel zone
3. Baseline key bedrock properties by rock type
4. Deep erosional channels in bedrock

Further discussion on finalization of the GBR-C is included in the following section of this paper.

## GBR-C FINALIZATION

Based on collaborative discussions during two proprietary meetings between the NBC, their Consultants and the Design-Build team, general consensus was reached regarding the establishment of modified and/or new baselines to be incorporated into the GBR-C. The following text describes the discussions held and consensus reached regarding the issues listed above.

### Fault Presence and Groundwater Inflows

As previously discussed, the GBR-B had baselined a potential for a fault to be present under the Blackstone River at the north end of the tunnel alignment, despite the fact that the Owner's geotechnical investigation had not confirmed its presence. The ground conditions to be encountered in the fault were described in the GBR-B and the baseline for groundwater inflow through the fault zone was established to be up to 200 gpm.

The Contractor had concerns regarding the potential for significant inflows under the river and mostly due to this reason, decided to use a hybrid TBM for the tunnel excavation capable of converting to a pressurized closed-face EPB if very poor ground conditions and significant groundwater inflows were encountered.

The Contractor was also concerned that other potential faults if encountered, could substantially impact the TBM progress, and therefore suggested a new baseline be created to address this risk. After some discussions it was agreed to add the following baseline: "*If other faults or shear zones are encountered by the TBM (other than the baselined one) it is assumed that groundwater inflows will not exceed the baselined values and that no soil-like fault gouge will be encountered by the TBM within the other faults or shear zones.*"

The Design-Builder also proposed to further investigate the presence of the baselined fault under the river during their supplemental GIP.

#### *Distribution of Rock Types in Tunnel Zone*

To provide an objective basis for the Contractor to develop TBM productivity estimates for the price proposal, it was felt that a quantitative baseline for the various lithologies to be encountered in the tunnel face was warranted. This was especially important to the Design-Builder since the rock properties were significantly different for the various lithologies (e.g., conglomeritic sandstone was much stronger than shale). After considerable discussions, it was decided to only state the percentages of different lithologies for all rock cored rather than baseline what would be encountered within the tunnel zone. This approach was used separately for both the Rhode Island and Wamsutta formations.

### *Baseline Key Bedrock Properties by Rock Type*

The GBR-B had grouped both the Rhode Island and Wamsutta formation rock types together (i.e., shale, siltstone, sandstone and conglomeritic sandstone) in baselining rock properties including UCS, BTS and Cerchar Abrasivity Index (CAI). The UCS results were summarized in a data plot and histogram and covered a wide range of values (60 to 23,000 psi). Mean and standard deviation values were also provided. The Design-Builder was concerned about this generalized approach because of the significance of these properties on all excavation work (TBM advance rate, cutter disk wear, shaft drilling, and drilling/blasting work). Based on their independent analyses of testing results, the Design-Builder proposed to baseline individual values of the respective properties by rock type. This approach would enable a more accurate basis of estimating TBM advance rates based on the distribution of rock types anticipated along the tunnel zone.

The Owner and their Consultant did not approve of this approach. After considerable discussion it was decided that specific baseline statements would be created for grouped lithologies to characterize both UCS and CAI that could be used by the Design-Builder to base their bid on. The following baseline statements were established

*For baseline purposes related to the TBM penetration and progress rate, the following overall rock strength values shall apply:*

- *Average uniaxial compressive strength (UCS) of 8,000 psi.*
- *No more than 10% of the rock samples will have a UCS above 17,000 psi.*
- *The maximum UCS shall be equal to 24,000 psi.*

*For baseline purposes, the following values of Cerchar Abrasivity Index shall apply:*

- *For sandstone and conglomerate—No more than 10% of the rock samples will have CAI above 5.0*
- *For siltstone and mudstone - No more than 10% of the rock samples will have CAI above 2.0.*

### Deep Erosional Channels in Bedrock

Due to the relatively wide spacing of borings and large diameter of the proposed tunnel, the Design-Builder felt there was a significant risk associated with the potential presence of buried valleys within the bedrock which could have significant impacts on TBM progress if mixed face conditions were to be encountered. Consequently, the NBC agreed to establish a baseline statement stating that no such erosional channels would be encountered.

## CONTRACTOR'S SUPPLEMENTAL GEOTECHNICAL INVESTIGATION PROGRAM

As indicated above, The Design-Builder decided to perform a substantial supplemental geotechnical investigation program (GIP) to not only characterize ground conditions in the area of the relocated TPS and LS, but also to further investigate areas of perceived risks or where boring coverage was not deemed adequate. These areas included:

- Fault zone under Blackstone River
- DS-213 adit crossing under Seekonk River
- Gap in tunnel borings near station 110+00 (potential shear zone)
- Footprint of receiving shaft

The supplemental GIP included twelve (12) test borings, totaling approximately 457 feet of soil and approximately 1,379 feet of rock. Despite difficult access, two water borings were performed, one in the Blackstone River and one in the Seekonk River. The program also included performing geophysical surveys, packer testing, installation of two fully grouted vibrating wire piezometers and one observation well, and laboratory rock testing.

The information provided by the supplemental GIP was used by the Design-Builder to execute the final design as well as to finalize the construction means and methods to address the areas of risk listed above. Of particular interest was that the boring performed in the Blackstone River (in the baselined fault zone area) did not find any evidence of a fault. Three more borings were drilled to locate the fault, including one angled boring. None of these additional borings encountered the fault zone, although the contact between the two rock formations was successfully located. Due to the Contractor's decision to use a hybrid TBM and based on the likely absence of a fault zone, the NBC decided to allow the Contractor to forego the probing and grouting requirement under the river provided they operate the TBM in the closed-face EPB mode.

## CONCLUSIONS

The use of fixed price Design-Build contracting approach offered several benefits to the NBC including speed of delivery, single point accountability, Contractor involvement in the design process, and more price certainty as compared to the design-bid-build model used on their previous large CSO control projects. With Design-Build delivery, the use of a GBR-B and GBR-C was also adopted and implemented in accordance with industry recommended guidelines.

The use of proprietary meetings prior to bid submission to discuss ATCs and GBR-C development added significant value for all project participants, including the Owner, Program Manager, Contractor and Contractor's Designer. This approach allowed the various stakeholders to collaboratively discuss the perceived risks on the project and develop an understanding of each other's concerns. These meetings promoted the development of several cost-saving ATCs as well as a consensus on GBR-C baselines by the NBC, Stantec, and the Design-Builder.

## REFERENCES

Essex, R., ed. 2007. *Geotechnical baseline reports for construction: suggested guidelines*. Technical Committee on Geotechnical Reports of the Underground Technology Research Council. Reston, Virginia: American Society of Engineers.

Quinn, A. W. 1959. *Bedrock geology of the Providence quadrangle, Rhode Island* [map and text]. Washington, D.C.: U.S. Geological Survey.

Skehan, J. W., N. Rast, and S. Mosher. 1986. Paleoenvironmental and tectonic controls of sedimentation in coal-forming basins of southeastern New England. In *Paleoenvironmental and tectonic controls in coal-forming basins of the United States*. GSA Special Paper 210. Ed. P. C. Lyons and C. L. Rice; p. 9–30. Boulder, Colorado: Geological Society of America.

# Ground Improvement in Glacial Soils for the Lower Olentangy Tunnel—Columbus, OH

**Jeremy Cawley** ▪ City of Columbus
**Horry Parker Jr.** ▪ Black & Veatch
**Jeff Murphy** ▪ DLZ
**Jake Keegan** ▪ McMillen Jacobs Associates
**Brock Gaspar** ▪ Schnabel Geostructural Design & Construction

## ABSTRACT

The Lower Olentangy Tunnel project includes 47,000 cubic yards of planned ground improvement for earth pressure balance machine (EPBM) tunneling (5.2 km) and microtunneling (335 m) in glacial soils. Jet grouting started in late 2021 and will continue through 2023 to complete operations at 5 shafts, 14 safe havens, 5 water-main ground improvement zones, and a river crossing. Typical jet grout columns are installed at depths between 15–24 meters. Discrete soil-cement blocks range from 267–8,900 cubic yards in volume. This paper discusses the methodology, testing, and technical considerations associated with jet grouting operations in glacial overburden across an urban environment.

## INTRODUCTION

### Project Overview

The Lower Olentangy Tunnel (LOT) is a key component of the City of Columbus' Integrated Plan, which addresses deficiencies within the collection system by providing hydraulic relief to several large diameter trunk sewers within the Olentangy River watershed. The 5.2 km (17,066 ft), 3.7 m (12 ft) finished inside diameter tunnel will create additional capacity necessary to intercept overflows and mitigate wastewater discharge into local waterways. LOT will link to the upstream shaft of the OSIS Augmentation and Relief Sewer (OARS) tunnel completed in 2017.

The LOT project consists of the following work:

- 5 Shafts
  - 1 slurry panel shaft—Gowdy Field
  - 4 secant pile shafts—Vine Street, Tuttle Park, FMN, 2nd Ave
- Shaft pre-excavation grouting (karstic limestone)
- 5.2 km (17,066 ft) earth pressure balance machine (EPBM) excavation
  - South drive (1,518 m; 4,981 ft)
  - North drive (3,684 m; 12,085 ft)
- 104 m (341 ft) Sequential Excavation Method (SEM) Connector Tunnel in karstic limestone (link to OARS shaft 6)
- 335 m (1,100 ft) microtunnel excavation (3 m excavated diameter; 90 inch inside diameter)—Gowdy Field shaft to 2nd Avenue shaft
- 47,000 cubic yards of ground improvement
- 290 m (950 ft) Kinnear Subtrunk (KST) Relief Sewer (36 inch inside diameter)—open-cut and jack-and-bore

- Diversion Structures
- Odor Control (Biofilter upgrades)

The project team includes:

- Owner—City of Columbus
- Design Engineer—DLZ/McMillen Jacobs Associates
- Construction Manager—Black & Veatch
- Contractor—Granite Construction
  - Ground Improvement subcontractor—Schnabel Geostructural Design and Construction
  - Secant pile wall and pre-excavation grouting subcontractor: Goettle Inc.
  - Diaphragm slurry wall subcontractor: Keller NA

This paper focuses on the ground improvement program, which has four (4) main objectives related to TBM tunneling and microtunneling:

- Minimize ground settlement at shaft break-in and break-out zones
- Provide impermeable safe havens for TBM maintenance as needed
- Minimize risk of ground settlement in the vicinity of a 42-inch water main along portions of the tunnel alignment
- Minimize mixed-face conditions at the transition from glacial overburden to shale (river crossing zone) at the northern end of the tunnel.

The LOT project consists of two EPBM tunnel drives to meet the Ohio Environmental Protection Agency (OEPA) Consent Order milestone for flow diversion of July 1, 2025. As of December 2022, shaft excavation and base slab concrete placement at Gowdy Field are complete and the Vine Street shaft excavation is in progress. The Gowdy Field shaft consists of a double-lobed slurry panel wall (two intersecting 17.3 m/56 ft diameter lobes) designed to accommodate TBM and microtunneling operations. The south tunnel drive is scheduled to begin in January 2023 from the Gowdy Field south shaft lobe. After completion of the south drive, the TBM will be extracted from the Vine Street shaft and relaunched at the Gowdy Field north lobe for the north tunnel drive. The microtunneling operation will then commence from the south lobe once the north tunnel drive is underway.

The jet grout locations are shown in Figure 1. As of October 31, 2022, the contractor has completed approximately 51% of the ground improvement program, including jet grouting at the Gowdy Field shaft, seven (7) safe havens, and three (3) water main ground improvement zones. The three (3) safe havens on the south tunnel drive are complete. Jet grouting at the river crossing zone on the north tunnel drive is currently underway. The contractor will then mobilize to Vine Street to complete the shaft break-in zone outside of the 16.8 m (55 ft) diameter secant pile shaft. Ground improvement productivity, technical challenges, and implications for tunneling on the south drive from Gowdy Field to Vine Street are discussed in detail below.

## Project Geology

The geology along the tunnel alignment is characterized by 10–30 m (33–99 ft) of glacial soil (till/outwash) overlying sedimentary rock (limestone or shale). The jet grout blocks are located between 15–25 m (50–83 feet) below ground surface (the tunnel invert ranges from EL 677 at Tuttle Park to EL 660 at Vine Street). Typical jet grouting will be performed entirely within the glacial overburden. However, at the northern end of the tunnel alignment, some of the grout columns terminate in shale at the river

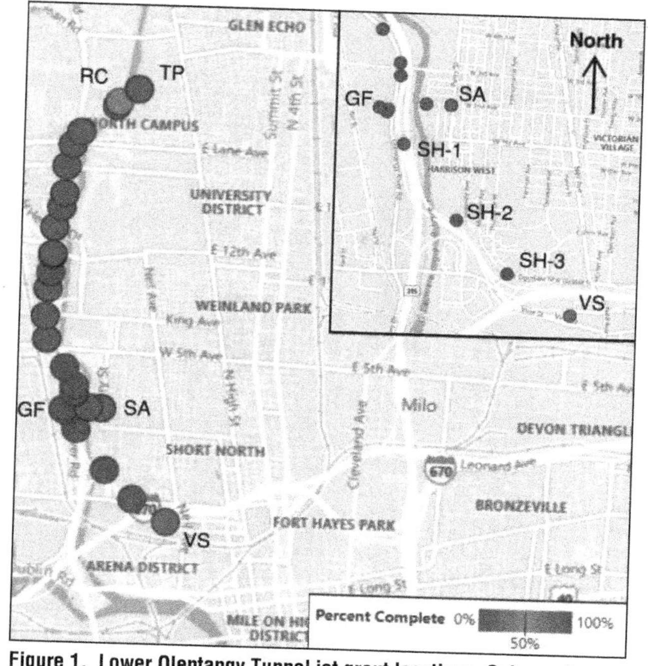


Figure 1. Lower Olentangy Tunnel jet grout locations. Color scale indicates percent completion at each site as of October 31, 2022. The inset map shows safe havens on the south tunnel drive between the Gowdy Field (GF) shaft and Vine Street (VS) shaft. The test program at Vine Street is complete, and production column installation is pending. The river crossing zone (RC) on the north tunnel drive is partially complete. TP: Tuttle Park; SA: Second Avenue

crossing zone and Tuttle Park shaft break-in zone. Groundwater levels across the project area vary from 0–15 m (0–50 ft), and jet grouting is consequently performed below the water table.

Based on the LOT geotechnical investigation, the soils were grouped into six (6) units.

- Unit 1—Fill
- Unit 2—Landfill
- Unit 3—Sand and Gravel
- Unit 4—Silty Sand and Gravel
- Unit 5—Cohesionless Silt
- Unit 6—Cohesive Silt and Clay

All glacial soil units contain cobbles and boulders. The baseline soil characterization was derived from regional glacial geology, previous experience on tunnel projects in Columbus, and the LOT geotechnical investigation. At the Gowdy Field and Vine Street shafts, observations during shaft excavation confirmed the ubiquitous presence of boulders and provided insight on impacts to jet grouting operations. Figure 2 shows the range of hydraulic conductivity values for each soil unit. The required values of hydraulic conductivity for ground improvement overlap with values for in-situ cohesive silt and clay.

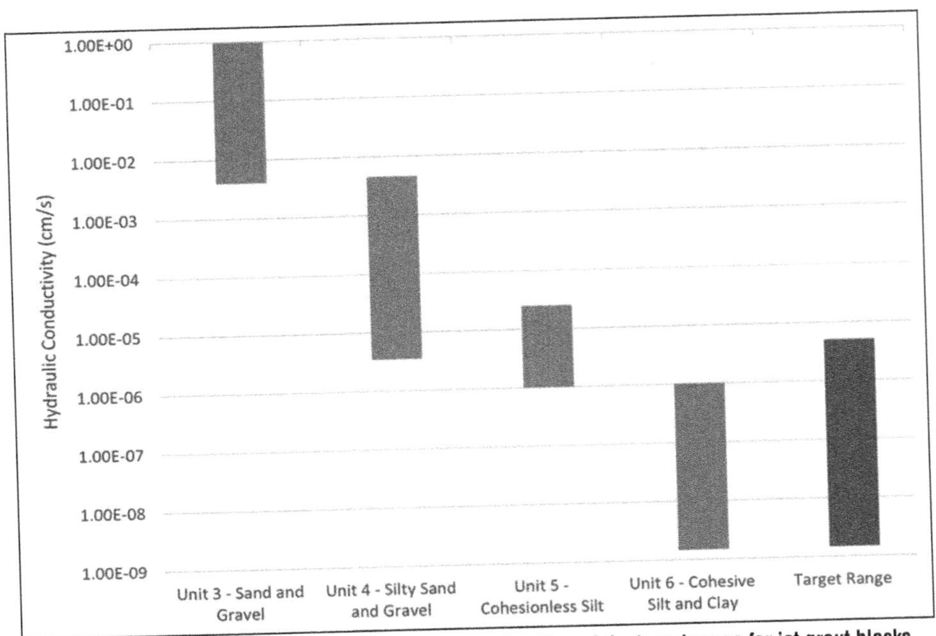


**Figure 2. Hydraulic conductivity values for baseline soil units and the target range for jet grout blocks. The required minimum hydraulic conductivity of 5×10$^{6}$ cm/s correlates with cohesionless silt and cohesive silt/clay identified in the geotechnical investigation**

## GROUND IMPROVEMENT—JET GROUTING

### Methodology

The ground improvement subcontractor performed jet grouting using a Comacchio MC-28-HD jet grout drill and Tecniwell grout pumps/mixers. Drilling in glacial soils was performed with a 20.3 cm (8 in) roller cone drill bit, and jet grouting was performed using a double-fluid jet grout approach (Shen et al., 2013). Drilling and grouting parameters, including drill hole inclination, were monitored using a Jean-Lutz data acquisition system. In some cases, additional columns were added to account for excessive drilling deviations and ensure closure at depth.

The typical jet grout work cycle consisted of bentonite fluid mixing, drilling, grout mixing, and jet grouting. Potholing at each column location was also typically required prior to drilling at most sites to ensure there were no utility conflicts. During jet grouting, spoils returned to the surface were pumped to a holding container for temporary storage, and vacuum trucks were used to haul the material to settlement basins located at the Vine Street shaft site. The Gowdy Field site was located within a former landfill, and jet grout spoils were pumped to settlement bays for contaminant testing prior to removal from site.

### Test Program

Initial field tests at all shaft locations and three (3) safe havens were required to establish jet grouting parameters, and subsequent testing of the production columns was required (2 columns per 1000 CY of improved ground). The jet grout program was designed to meet the following criteria:

- Hydraulic conductivity: less than $5 \times 10^6$ cm/s, averaged over a 3-day period
- Minimum compressive strength: 150 psi for break-in/out zones and safe havens; 75 psi for water main ground improvement zones and the river crossing

In-situ hydraulic conductivity testing was required at safe havens to verify that selected jet grout parameters produced a nearly impermeable block of improved ground. Coring of the grout columns was also required for compressive strength testing and any ungrouted zones within the core were to be tested for Atterberg limits, particle size, and cohesion/plasticity.

### *Initial Testing*

Prior to production work, the Contractor installed test columns to determine the best parameters for jet grouting. Given the considerable effort required for site mobilization, the testing program was an ongoing effort as equipment mobilization to different sites progressed. Although there were many variables to consider, the lift rate and water/cement ratio (W/C) were systematically varied for the test columns. Table 1 shows typical jet grouting parameters used to obtain the proposed 3 m (10 ft) column diameter and 0.6 m (2 ft) overlap. The verification testing procedures also evolved as the project team learned more about the ground conditions, coring limitations, and hydraulic conductivity testing uncertainties.

After the test columns were completed, the drilling subcontractor used hollow-stem augers to drill 1.5 m (5 ft) into the top of the jet grout zone (typically 15 m (50 ft) below ground surface). A PVC casing extending from the jet grout zone to the surface was then grouted in place to facilitate subsequent coring and hydraulic conductivity testing. Hydraulic conductivity testing was initially successful at Safe Haven 3 (SH-03) and Safe Haven 5 (SH-05), but retrieval of intact cores at jet grout columns using HQ triple-barrel coring proved to be difficult. In general, disintegration of the soil-cement cores during sampling was attributed to gravel content, though one core sample from SH-05 was retrieved and yielded a compressive strength of 756 psi. In lieu of compressive strength testing for intact cores (the specification required 3 samples from each core for testing), the Contractor also collected and tested wet-grab samples from spoils returned to the surface during jet grouting. The average compressive strength of the wet-grab samples at SH-05 was 403 psi, exceeding the requirement of 150 psi for safe havens. Given the lack of cores, the Contractor also performed sonic coring to retrieve samples in the jet grout zone at SH-05. Sonic coring was performed in the column center as well as the column overlap zone to confirm the presence of cement. Although the samples were disturbed, phenolphthalein spray indicated that cement

**Table 1. Typical jet grout parameters**

| Parameter | Value |
|---|---|
| Design Diameter (m) | 3.05 |
| Withdrawal rate (cm/min) | 15.24 |
| Air pressure (MPa/psi) | 0.7 / 120 |
| Water-cement ratio (W/C) | 1.5 |
| Number of nozzles | 1 |
| Grout flow rate (gpm) | 130 |
| Nozzle diameter (10^-3 m) | 7 |
| Rotation speed of rod (rpm) | 1.9 |
| Jetting pressure (psi) | 6100 |

was typically present throughout the jet grout interval. The project team accepted the alternative approach to maintain schedule, and the Contractor selected a conservative 10 cm/min (4 in/min) lift rate and W/C of 1.5 based on initial tests at SH-03 and SH-05.

In subsequent testing at Water Main Ground Improvement Zone 2 (WMGIZ-02) in February 2022, the Contractor demonstrated acceptable column diameter and permeability using a 15.2 cm/min (6 in/min) lift rate. Therefore, the lift rate was modified to 15.2 cm/min for the remaining safe havens and water main ground improvement zones. At the Gowdy Field shaft site, the 10 cm/min (4 in/min) lift rate was maintained based on initial testing at the break-out zones.

#### *Production Testing*

After establishing parameters through initial field tests, subsequent testing of production columns was required, including coring and hydraulic conductivity testing (2 columns per 1000 CY of improved ground; overlap verification was not required). This phase of testing was typically performed after the entire zone was completed due to limitations on access for the testing subcontractor while jet grouting was in progress. As indicated above, retrieval of intact core samples was variable, and hydraulic conductivity testing was primarily used to assess the effectiveness of the grout program. Production testing results are discussed in more detail below.

## PRODUCTION AND TESTING RESULTS

### Production Summary

Jet grouting began in August 2021 with one drill rig, and a second rig was mobilized in December 2021 (Figure 3). The typical work schedule was five 10-hour days. The overall combined production rate for both rigs through October 2022 was 1.4 columns per calendar day. The 30-day moving averages (based on calendar days) show the variability in production throughout the project (Figure 3). Although consistent production rates of 2–3 columns per workday were common with favorable ground conditions, the combined effects of site demobilization and mobilization efforts, traffic control set-up, utility conflicts, poor weather, winterization efforts, night work, and difficult ground conditions resulted in lower overall average productivity. Issues with ground conditions included difficult drilling in granular soils (gravels, cobbles, and boulders), groundwater, drilling fluid loss, and lost drill tooling. In three instances, the drill bit and inclinometer assembly were lost at depth, and a separate drilling subcontractor retrieved the assembly using a customized grabber. The loss of drill tooling was attributed to excessive deflection of the drill steel at depth caused by cobbles and boulders.

#### *Drilling Rates*

The observed drilling rates highlighted the variable nature of glacial soils across the tunnel alignment (Figure 4). The average drilling rate to date for all sites is 24.8 m/hr (81.2 ft/hr), though significant variation was attributed to heterogeneous ground. Relatively high drilling rates were associated with sandy or silty soils. Gravel, cobbles, isolated boulders, and nested boulders reduced drilling productivity. Pressure spikes and loss of fluid return during drilling were attributed to borehole collapse, and numerous boreholes in poor ground were redrilled. Bentonite drilling fluid was also modified as needed to maintain progress.

A subset of data from WMGIZ-01 and WMGIZ-02 illustrates the effect of ground conditions on drilling (Figure 4). The average drilling rate at WMGIZ-01 was 38.4 m/hr

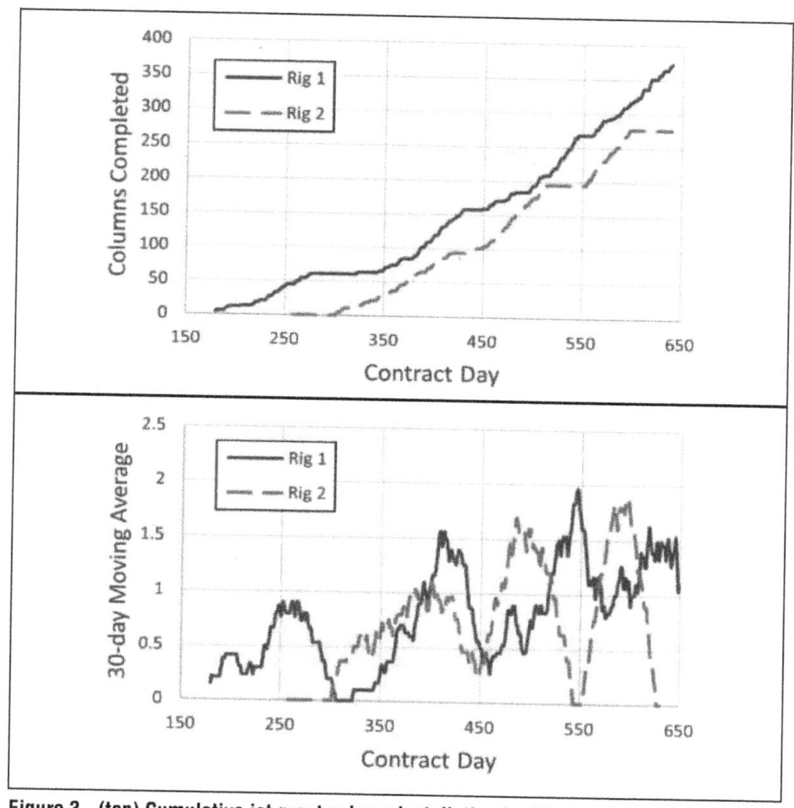


**Figure 3. (top) Cumulative jet grout column installation for Rig 1 and 2. Production rates for Rig 1 were impacted by mobilization delays and winter weather (contract days 275–350) and the addition of a jet grout curtain wall at WMGIZ-01 (contract days 425–450). The curtain wall was installed to mitigate risk associated with the 42-inch water main in this location. Rig 2 did not operate in August 2022 during double-shift operations at Gowdy Field (contract days 515–553).(bottom) The 30-day moving averages (by calendar day) highlight the inconsistencies in production associated with work in an urban setting with highly variable ground conditions**

(126.0 ft/hr) compared with 21.2 m/hr (69.6 ft/hr) at WMGIZ-02. High drilling rates at WMGIZ-01 were attributed to predominantly sandy soils, though extremely low rates for several columns likely reflect the presence of boulders at depth. Results from WMGIZ-02 (218 columns) show that rates can vary significantly from each end of the site because of geologic variation. On the south end of the jet grout zone, the entire drill steel became stuck at two different columns. The second drill rig was mobilized to the site to assist with freeing the assembly. The drilling rates were also influenced by driller experience and equipment operation.

Redrilling of boreholes also affected productivity. At Gowdy Field, thirteen boreholes were redrilled because of unstable borehole conditions (Figure 5). In some cases, jet grouting was terminated because of borehole collapse or clogging accompanied by pressure spikes and/or loss of jet grout spoils return. Redrilling was also required after excessive bentonite slurry loss during drilling and grout loss during jet grouting at the

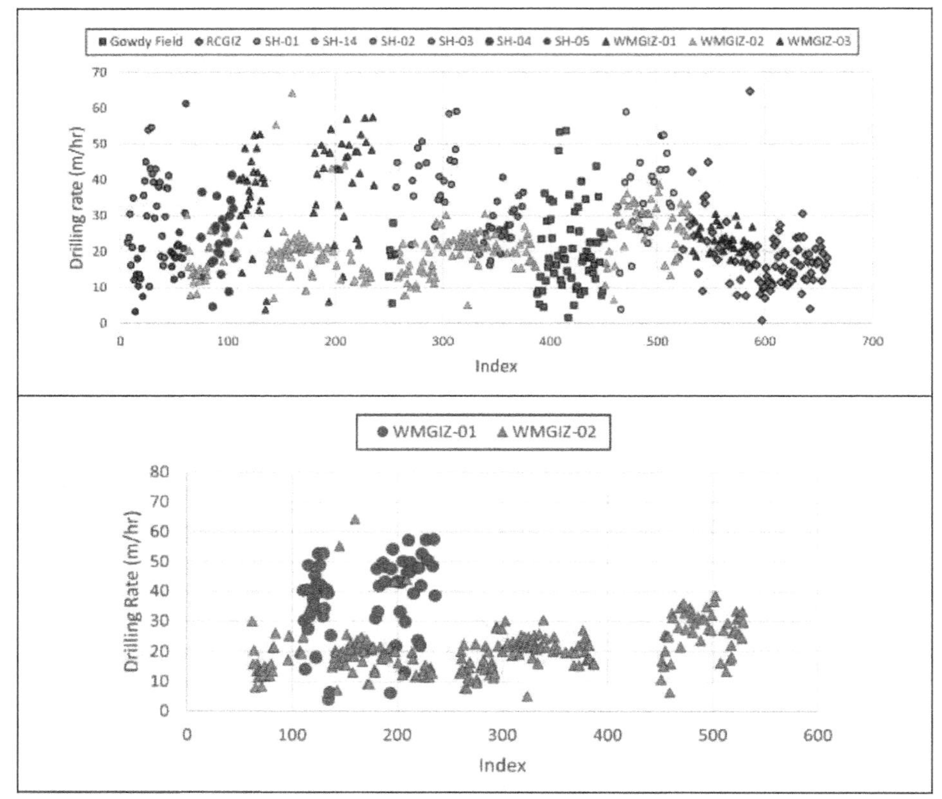


**Figure 4. (top) Drilling rates for jet grout zones completed to date. The overall average rate is 24.2 meters per hour. (bottom) Drill rates for WMGIZ-01 and WMGIZ-02 highlighting the variability between sites**

Gowdy Field south break-out zone. Fluid and grout loss at Gowdy Field was consistent with previously observed slurry loss in highly permeable ground during construction of the slurry panel shaft wall. At SH-01, excessive groundwater returns to the surface during jet grouting forced the Contractor to terminate jet grouting and redrill several columns. The groundwater issues were attributed to hydraulic connectivity of permeable soil layers (glacial outwash) with the Olentangy River. Two drill head assemblies were lost at depth at SH-01 because of deflection of the drill steel in gravelly and boulder-laden ground.

### *Grout Volumes*

Jet grouting volumes are dependent on the pre-set lift rate, column height, and grout flow rate. Figure 6 shows the grout volumes recorded for each column. For a 10 cm/min (4 in/min) lift rate, the average grout volume for a typical column is approximately 10,000 gallons. The switch to a 15.2 cm/min (6 in/min) lift rate at WMGIZ-02 after additional testing resulted in a reduction in grout volume to approximately 6,000 gallons. The variation in volumes for the river crossing columns are related to the decrease in column height as the bedrock elevation changed. However, the scatter within the Gowdy Field sites is attributed to difficulty maintaining grout flow (pressure) or

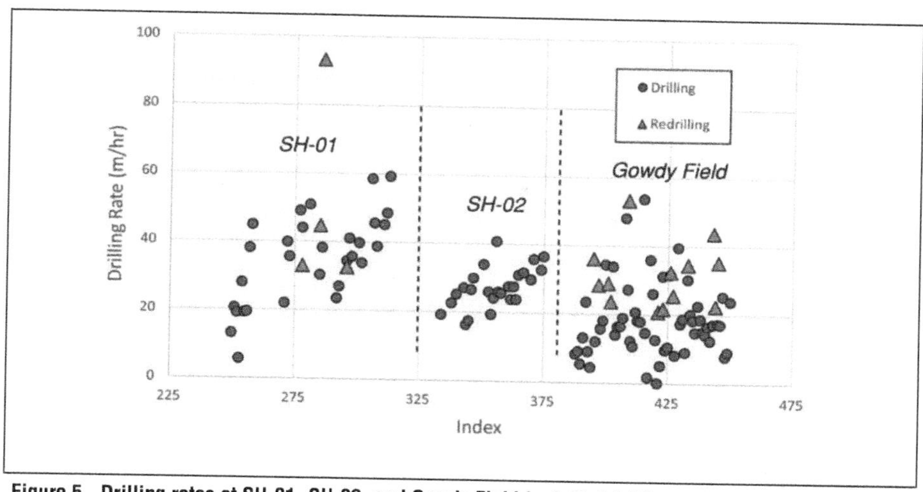


**Figure 5. Drilling rates at SH-01, SH-02, and Gowdy Field for initial drilling and redrilling. The lack of redrilling at SH-02 indicates comparatively better ground conditions, whereas redrilling at SH-01 and Gowdy Field illustrates the effects of difficult ground conditions. At SH-01, excessive groundwater return with spoils during jet grouting required the contractor to stop and redrill several columns. At Gowdy Field, unstable borehole conditions and pressure spikes during jet grouting also forced the contractor to terminate jet grouting and redrill thirteen (13) boreholes**

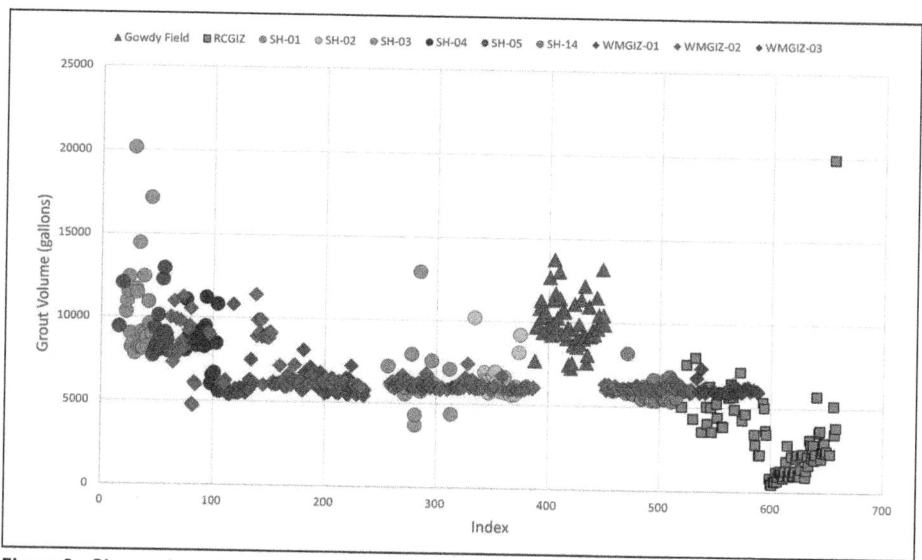


**Figure 6. Observed grout volume per column for LOT jet grout zones. The variation primarily reflects modifications to the jet grout lift rate and column height. The relatively small grout volumes for the river crossing (RCGIZ) are indicative of shorter column heights as bedrock was encountered across the zone. The scatter observed at the Gowdy Field breakout zones is partially attributed to multiple sweeps during jet grouting associated with redrilling in difficult ground**

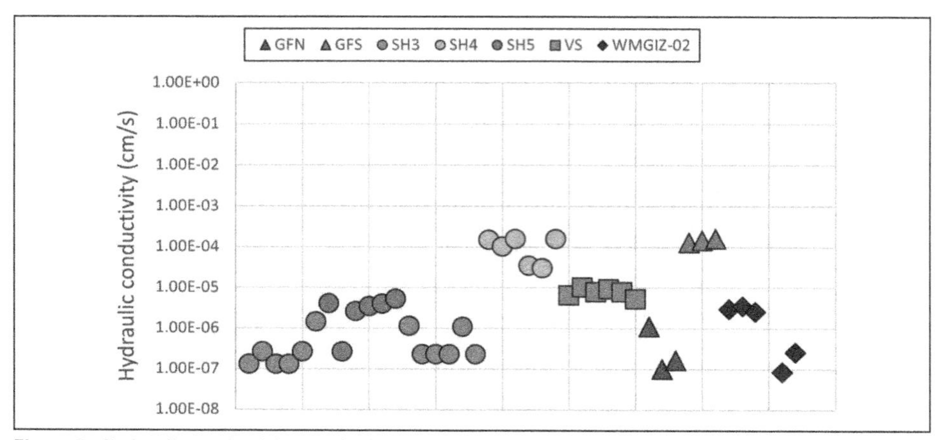


**Figure 7. Hydraulic conductivity results for the Gowdy Field break-out zones (GFS and GFN), Vine Street (VS) break-in zone, water main ground improvement zone 2 (WMGIZ-02), and selected safe havens (SH). Additional testing is tentatively scheduled for the Gowdy Field site to verify initial results. The data set includes pre-production and post-production test results**

clogging at the nozzle in difficult ground. In many cases, grouting was temporarily halted because of ground conditions and equipment problems. The holes were then redrilled to complete grouting. The additional 'sweep' of the hole to complete jet grouting resulted in higher grout consumption for some columns.

## Production Coring and Hydraulic Conductivity Testing

For post-production testing, core recovery over the 3 m (10 ft) interval was variable. At SH-03, two small sections of core were suitable for compressive strength testing (1298 psi and 780 psi). At SH-02, core recovery from column 202 was approximately 65%. The variation in core retrieval may reflect differences in driller experience.

Hydraulic conductivity testing also yielded varied results (Figure 7), and there was uncertainty if testing failures were attributed to the jet grout procedures and parameters, testing procedures, or a combination of both. The contractor attributed some failures to leakage of the grout seal at the base of the PVC casing during testing. Alternatively, shadowing effects in zones with high boulder density may have resulted in ungrouted seams in the jet grout block (Bellato et al., 2018). Some zones that exhibited high hydraulic conductivity are tentatively scheduled for retesting based on concerns with testing equipment. In addition, the contractor performed hydraulic conductivity testing of intact core samples and wet-grab samples retrieved from Safe Haven 2 (SH-02) for comparison with field results to gain additional insight.

# DISCUSSION

## Gowdy Field Break-out and the South Tunnel Drive

The break-out zones at the Gowdy Field shaft were designed to minimize settlement and groundwater inflow during the TBM launch. The first launch from the south shaft lobe is scheduled for early 2023, and the hydraulic conductivity results from the south jet grout zone are still under review. The project team continues to weigh factors including jet grouting parameters, ground conditions, and testing procedures to evaluate the effectiveness of grouting and the conditions outside the slurry panel wall for the TBM launch.

Slurry panel shaft construction, jet grouting, and shaft excavation provided insight on the challenges associated with underground construction in glacial soils. The slurry wall construction provided an initial indication of ground conditions at the depth of the tunnel envelope and jet grout zone (EL 660–685). In addition to cobbles and boulders encountered during slurry panel excavation (inferred from hydromill excavation), slurry loss at depths of 16.5–21 m/54–70 ft (EL 668–684) indicated relatively permeable ground in the break-out zone. In subsequent drilling for jet grout columns, heavy chatter indicated the presence of cobbles and boulders. On the north break-out zone at the Gowdy Field shaft (GFN), slow drilling in columns adjacent to the slurry panel wall was also attributed to concrete over-pour exterior to the shaft wall. On the east break-out zone (GFE) for the microtunnel, there was drilling fluid communication at the surface indicative of poor ground conditions. Difficult drilling, bentonite fluid loss, and grout loss on the south break-out zone (GFS) also indicated more permeable ground (low fines content in a matrix of gravel, cobbles, and boulders). After completion of jet grouting, observations of shaft excavation at Gowdy Field were consistent with an increase in boulder content at approximate elevation 680–660, coinciding with slurry loss and difficult drilling.

Downstream of the Gowdy Field shaft, conditions at SH-01 suggest that significant groundwater may be transmitted through permeable layers connected with the Olentangy River. The river crossing between the Gowdy Field shaft and Vine Street shaft is located at approximate STA 43+55. At face value, drilling observations, borehole data, and hydraulic conductivity results at SH-02 and SH-03 suggest less permeable ground conditions. Vine Street shaft excavation observations also indicated that nested boulders and layers of cobbles were not as prevalent as observed at Gowdy Field.

## River Crossing

The river crossing site downstream of the Tuttle Park shaft required the construction of a 76 m × 12 m (250 ft × 40 ft) causeway in the Olentangy River consisting of approximately 1,800 CY of granular fill. The Contractor installed 30.5 cm (12 in) diameter, 2.4 m (8 ft) steel sleeves through the causeway fill at each column location to facilitate drilling and minimize the inadvertent discharge of drilling fluid or grout to the river during operations. The geotechnical investigation indicated that the bedrock elevation increased from southwest to northeast, and drilling observations (e.g., shale cuttings in return fluid) on the causeway were consistent with the borehole data. The bedrock topography resulted in a wedge-shaped jet grout block at the river crossing site. Seven planned columns were eliminated on the northeast end of the causeway due to the actual rock depth.

The contractor designed and fabricated a temporary holding tank to connect with the vertical steel casing at each column to contain the jet grout spoils, and the river was continuously monitored for inadvertent releases. Typical jet grouting was accompanied by air release through the subsurface to the river, and the resulting silt disturbance was contained by a turbidity curtain installed downstream of the causeway. For two columns, grouting operations were terminated because of communication with the river.

Sinkhole formation at column locations was also prevalent on the causeway. Subsidence around the steel casings (during or after jet grouting) was attributed to the collapse of non-cohesive granular material into the top of the 3 m (10 ft) diameter jet grout column. In multiple instances, the steel casing dropped into the borehole as a sinkhole formed (estimated 1–2 CY volume). At the river crossing, there was less cover (<6 m; 20 ft) compared with other locations, possibly affecting the ability of the overburden to bridge the 3 m (10 ft) diameter columns. This effect was not observed at other locations. The casings will be retrieved during removal of the causeway.

The average production rate at the river crossing site was approximately 1.3 columns per calendar day. Production at the onset of causeway operations was affected by efforts to control inadvertent fluid discharge through the causeway fill and in-situ, permeable glacial material. During the initial work, 20 additional 6.1 m (20 ft) vertical sleeves were installed at columns on the southwest end of the causeway to evaluate the effect on communication. In terms of the ground conditions, non-cohesive soils with cobbles appeared to contribute to sinkhole formation. On the northwest end of the causeway, the contractor completed 4–5 columns per workday with the increase in bedrock elevation and decrease in column height. Productivity was also impacted by high rainfall and dam releases from Delaware Lake, forcing the contractor to temporarily demobilize from the causeway until the river level dropped.

Tunnel boring machine performance in the wedge-shaped jet grout block above shale bedrock will be evaluated further. This section on the north tunnel drive is expected to be completed in July 2024.

## CONCLUSION

The LOT jet grouting program has yielded a comprehensive data set of drilling rates in glacial soils, highlighting the variability of drilling in glacial till and outwash with boulders. Production rates averaged over the course of the program offer an indication of realistic production rates for a long-duration grouting program involving complex glacial geology, numerous mobilizations in an urban environment, and multiple drilling rigs. Although favorable drilling conditions can result in elevated short-term productivity, long-term progress is ultimately tempered by numerous factors including winter weather, site mobilizations, equipment maintenance, and decreased productivity associated with challenging ground conditions. To achieve a balance between productivity and quality control, testing programs should be carefully sequenced to account for the lag time between test column installation and hydraulic conductivity testing.

The variability in hydrologic/geologic conditions along the alignment highlights the importance of ground improvement for TBM tunneling in glacial soils. The potential for high groundwater inflows in permeable ground during tunneling was confirmed by observations of slurry wall construction, jet grouting, and shaft excavation. The effectiveness of the jet grout program will be evaluated as tunnelling commences from the south break-out zone at the Gowdy Field shaft.

## REFERENCES

Bellato, D., Schorr, J., and Spagnoli, G. 2018. Mathematical Analysis of Shadow Effect in Jet Grouting. Journal of Geotechnical and Geoenvironmental Engineering. 144(12):04018088.

Shen, S., Wang, Z., Yang, J., and Ho., C. 2013. Generalized Approach for Prediction of Jet Grout Column Diameter. Journal of Geotechnical and Geoenvironmental Engineering. 139:2060–2069.

# Submerged Penetrations Through a Frozen Soil Shoring System

**Kyle Amoroso** ▪ SoilFreeze, Inc
**Aaron K. McCain** ▪ SoilFreeze, Inc

## ABSTRACT

Frozen soil shoring has been used for decades in the tunneling industry to provide temporary support of excavation for shafts and cross-passage excavations. However, horizontal penetrations through the frozen soil can prove to be challenging. The key to creating a stable portal through frozen ground is managing and mitigating groundwater flow into the excavation through the annular space surrounding the drill stem or casing. This paper will review different approaches that have been successfully used to penetrate frozen soil shoring on multiple projects.

## INTRODUCTION TO ARTIFICIAL GROUND FREEZING

The earliest commercial applications of artificial ground freezing date back to the middle of the 19th century. The practice was further developed in the later portion of the century with the first patent on ground freezing techniques submitted by German engineer F.H. Poetsch in 1883 (Chang & Lacy, 2008). While the technology has changed and been updated, the basic principles of the freezing process laid out in Poetsch's patent application are still used in ground freezing today (Poetsch 1883).

The main components of an artificial ground freezing system typically include a chiller unit, a pump unit, a series of 3 to 5-inch diameter steel pipes, calcium-chloride brine, a manifold system to carry the brine, insulation, and temperature monitoring devices. The artificial ground freezing process consists of continual circulation of chilled brine (generally –15°F to –25°F) through a series of close-ended steel pipes installed around the perimeter of an excavation. This process gradually draws heat out from the ground, cooling a targeted volume of soil. As time passes, the ground eventually cools below the point of freezing (~32°F) the moisture within the soil experiences a phase change to ice. The ice acts similarly to cement within concrete and binds the soil particles together to create a solid mass of frozen soil. Ice formation begins at the interface between the steel pipe and the surrounding soil and extends radially outward. As ice accumulation from neighboring pipes intersect, a unified and interconnected volume of frozen soil forms. The interconnected frozen volume of soil is impermeable and experiences a significant increase in strength and stiffness. The ground freezing process generally takes 3 to 7 weeks to complete. The exact duration depends on the soil's moisture content and the spacing of the freeze pipes. Frozen soil systems can be active and maintained for as long as a project requires.

The primary advantage of artificial ground freezing, or ground freezing, is that the method provides both structural shoring and groundwater control. The frozen soil volume creates an impermeable barrier or plug within the ground. This can be especially useful when excavations extend through a soil transition of high permeability soils (sands and gravels) overlying very low permeability soils (silts and clays), where conventional dewatering is not feasible. Ground freezing systems are very adaptable and effective in nearly all soil types (gravels, sands, silts, clay), including organic-rich materials such as peat. Ground freezing can be used to create various geometries of frozen soil, from simple circular shafts to complex shapes and areas. Ground freezing systems are also verifiable, using an array of sensors constantly monitoring ground temperature, brine temperature, chiller unit performance, groundwater level (if necessary), and more. These readings are entirely automated and do not require personnel on-site.

Another strong advantage of ground freezing is that the systems are environmentally friendly. Furthermore, freeze systems are entirely self-contained, nothing is injected into the ground, and all components of the system can be completely removed, recycled, or reused. Additionally, artificial freeze systems are fully electric and require only 480-volt three-phase power to operate. In most ground freezing systems, a 30% solution calcium chloride brine (i.e., very salty water) is used as the cooling agent, which is non-toxic and environmentally benign. The impermeable barrier generated with artificial ground freezing can prevent off-site migration of contaminated groundwater or the introduction of regional contaminants onto the site.

## SUBMERGED PENETRATIONS THROUGH FROZEN SOIL SHORING SYSTEMS

While frozen soil shoring has been used historically to support excavations that extend below the groundwater table, it is typically overlooked when the support of excavation needs to be penetrated for horizontal installations. Normally, shafts initially shored with frozen soil during excavation are subsequently lined with a secondary shoring system, and the frozen soil system is deactivated before mining efforts begin. However, this is not always necessary because frozen soil can be used as the primary temporary shoring during horizontal penetrations, providing both structural stability and groundwater control.

The primary concern while drilling or tunneling through frozen soil shoring is limiting and controlling the flow of groundwater through the frozen soils. Water flow introduces warmth to the system and, left unchecked, can cause thermal and mechanical erosion and expand the size of the penetration. Therefore, specific attention to critical details is necessary to perform horizontal installations successfully. The approaches laid out within this paper were developed through observations made during the penetration of concrete tunnel liners and shafts for horizontal freezing applications.

When freezing around a concrete or steel shaft/tunnel, key design considerations include, first, ensuring that the initial penetrations for the installation of freeze pipes extend through a packer or backflow preventer that is attached and sealed to the concrete/steel face. Once frozen, the next consideration is to maintain the frozen bond between the soil and the back of the concrete/steel surface. The wall or lining of the shaft is exposed to heat from the circulation of interior ambient air. This heat can cause the interface between the soil and tunnel walls to thaw, potentially creating seepage pathways for water to enter the excavation. This is best mitigated by installing freeze pipes to match the geometry of the shaft/tunnel to freeze this interface effectively. Insulation can also be installed along the interior walls of the shaft/tunnel to keep away heat from circulating ambient air.

When conducting submerged horizontal penetrations through a frozen soil shoring system, the critical focus remains similar. A mechanical seal around the drill steel or TBM that is also sealed to a flat concrete surface is necessary. Maintaining the frozen bond between the flat concrete, sealable surface, and frozen soil shoring is the process's most critical element. Multiple techniques are available to maintain this bond, and more than one approach is often used on a specific project.

Successful submerged penetrations have been achieved within the following projects highlighted below:

1. Cross-Passage No. 32, Sound Transit Northlink—Northgate Expansion
2. Portland International Airport (PDX) Terminal Core Redevelopment
3. Carlls River Watershed Sewer Expansion

The first project outlines the details of horizontal penetrations through a concrete tunnel liner to support the excavation of frozen cross-passages. The following two projects illustrate how similar techniques are used to facilitate the penetration of a frozen soil shoring system for smaller diameter tunneling work.

## CROSS-PASSAGE NO. 32, SOUND TRANSIT NORTHLINK—NORTHGATE EXPANSION

### Project Background

In Seattle, Washington, as a part of the Northgate Link Extension project, Sound Transit extended the existing light rail system north by about 4 miles from the University of Washington to the neighborhood of Northgate. This extension consisted predominantly of two parallel 18-foot 10-inch diameter tunnels with 23 cross-passages connecting the two tunnels. Each cross-passage was elliptical in shape (approximately 18.5 feet in height and 17 feet in width) and approximately 20 feet in length, measured from the spring line. The project's general contractor was JCM Northlink LLC (JCM), a joint venture between Jay Dee Contractors, Frank Coluccio Construction Company, and Michels Corporation.

The project documents identified three categories of ground support requirements for each of the cross-passage excavations. These categories were based on the soil materials anticipated at each cross-passage. The soils considered most difficult for cross-passage excavation were listed as Category 3 soils. These soils were identified to be at transitions between dissimilar soils, generally between high-permeability coarse sands and gravel and low-permeability silts and clays. These transitions rendered dewatering efforts ineffective because the necessary groundwater drawdown could not be achieved. Engineers determined that cross-passages within Category 3 materials would require some form of ground improvement prior to excavation. To meet these demands, a shoring system was needed that could accomplish three goals: (1) provide soil stabilization during excavation, (2) provide groundwater cut-off, and (3) improve the soils so that transition zones encountered during excavation would not generate issues for the TBM. Ground freezing was selected over other solutions, such as jet grouting because it could accomplish all three goals, provide superior design adaptability, and generate the lowest environmental impact. While the ground freezing process was similar in each of the cross-passages, this paper will highlight cross-passage #32. The base of cross-passage #32 was located at a depth of approximately 85 feet below the existing ground surface (bgs) and subject to approximately 55 feet of hydrostatic pressure.

### Design and Installation

Due to spatial constraints within the existing northbound and southbound tunnels, the decision was made to divide the artificial ground freezing system into two portions. The primary/larger portion was installed outside the existing tunnels, from the ground surface. This portion consisted of a grid (5x6) of 30 vertical steel freeze pipes located

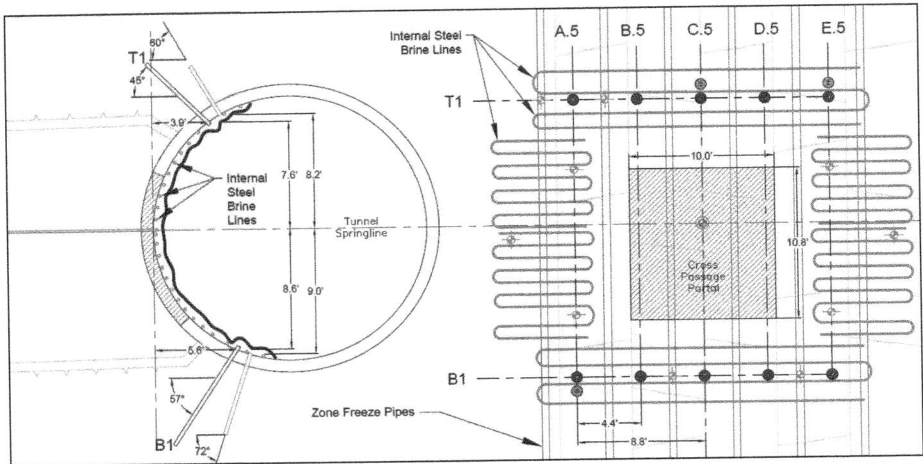


Figure 1. Cross-section and profile view of the haunch pipes and accompanying manifold lines at CP32

between the two existing tunnels. The purpose of this system was to provide the bulk of the freezing capacity and generate a frozen soil volume spanning the two tunnels.

A secondary and smaller freeze system was installed from within the tunnels to extend the freeze into the areas around the haunches of the tunnels, above and below the spring line, as shown in Figure 1. These areas could not be effectively reached by the primary freeze system but were necessary to freeze to ensure soil stability and groundwater cut-off for the excavation. Within each tunnel, at the face of the cross-passage portals, the "haunch" freeze pipes were arranged into two horizontal rows of seven. The first row was located approximately 7 feet above the cross-passage portal and were angled upwards 45 to 60 degrees from horizontal, as shown in Figure 1. The second row was located approximately 7 feet below the cross-passage portal and were angled downwards approximately 57 to 72 degrees from horizontal, as shown in Figure 1. These freeze pipes were installed using an articulating drill rig capable of meeting the various design inclinations. Once the liner was penetrated, the interior of the tunnel would be exposed to the full hydrostatic pressure of the groundwater column.

Therefore, drilling was completed through a packer assembly with a ball valve that allowed regulation for the removal of spoils and the inflow of water, as shown in Figure 2. Multiple layers of rubber wipers were installed at the entrance to the packer to allow the drill steel to move in and out of the packer. A rubber gasket was also placed between the packer and the concrete surface of the tunnel liner. Groundwater inflow was anticipated during drilling operations; however, this set up allowed the drillers to shut off the flow once drilling was complete. Once drilling activities were completed, the annular space on the exterior of the drill steel was grouted closed.

Heat is applied to the tunnel liner as ambient warm air circulates within the enclosed space, potentially developing a thawed seepage path at the extrados of the tunnel liner. To combat this the tunnel liner was cooled attaching a manifold system, carrying the chilled brine directly to the interior face of the liner around the cross-passage portal, as shown in Figure 3. Insulation blankets were placed over the manifold system to further reduce the exposure to the circulating warm air.

Figure 2. Haunch freeze pipes installed through the packer assembly

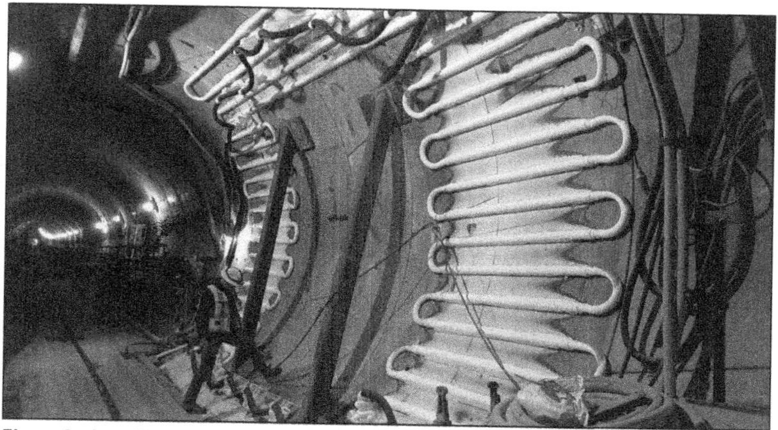

Figure 3. Installed haunch freeze pipes; and manifold brine lines attached to the interior tunnel liner

Along with the primary freeze system, these horizontal penetrations through the concrete liner and the manifold system applied directly to the face of the liner enabled the successful completion of the cross-passage excavation. This innovative approach allowed freezing above and below the excavated cross-passage to be maintained. Following the completion of tunneling operations, the freeze system maintained operation for an additional eight weeks to facilitate the construction of the cross-passage, which included installing waterproofing, placement of reinforcing steel, and the concrete pour for the final cross-passage liner.

## PORTLAND INTERNATIONAL AIRPORT (PDX) TERMINAL CORE REDEVELOPMENT

### Project Background

The PDX Terminal Core Redevelopment project, also known as PDX NEXT, is a series of upgrades and additions to increase the capacity of Portland International Airport (PDX) in preparation for an estimated doubling or tripling of demand by 2045. The project is a joint venture between general contractors Hoffman and Skanska, with Coffman Excavation as the primary earthwork and excavation contractor. Included as a part of these upgrades was the installation of a new aviation fuel line. This fuel line was installed with an 18-inch diameter steel casing and extended below two existing and separate concourses (Concourses C and D). These concourses needed to remain fully operational for the length of the project; thus any impact on these structures needed to be negligible. The width of each concourse was approximately 130 feet, and the aviation fuel line was to be installed approximately 15 feet below the existing ground surface (bgs). Coffman proposed digging four pits, two per concourse, approximately 22 feet in depth, to act as jacking and receiving pits for the new fuel line. Groundwater across the airport was determined to be at a depth of 7 to 10 feet bgs. The geotechnical reports also highlighted the presence of loose flowing sands within the upper 20 feet bgs. Beneath these sands, however, was a thick layer (approximately 40 feet) of low-permeability silts. This soil transition between high-permeability and low-permeability soil materials prevented the use of conventional dewatering methods. These constraints led Coffman to choose ground freezing to supply structural shoring and groundwater control for the excavation of the jacking and receiving pits. Ground freezing was also selected to provide temporary ground improvement, a uniform soil condition for tunneling, and groundwater control for installing the fuel line casing beneath both concourses.

## Design and Installation

The depth of the freeze pipes at each of the four pits extended the freeze through the saturated loose flowing sands and into saturated silt layers. While these depths were needed to shore the pits structurally, they were also vital in providing groundwater cut-off. The potential for horizontal groundwater flow into the pits was successfully "cut-off" by generating a continuous frozen soil wall around the perimeter of the excavation. However, the potential for vertical groundwater flow into the pits was successfully "cut-off" by designing the freeze pipes to extend approximately 15 feet into the low-permeability silt strata. The permeability of this silt layer was low enough to act as a "plug," preventing groundwater from being pushed beneath the frozen soil walls and up into the excavation by differential head pressures between the interior and exterior of the drained excavation.

The horizontal ground freezing and temporary soil improvement for the installation of the aviation fuel line casing was achieved by installing four to five freeze pipes between the jacking and receiving pits using pilot tube drilling methods. The pipes were arranged in a circular pattern, equally spaced apart, approximately 2.5 feet from the center of the proposed 18-inch casing, as shown in Figure 4. Once activated, these freeze pipes would gradually form a uniform cylinder of frozen soil materials that connected the frozen soil shoring walls at the jacking and receiving pits to create a unified water-tight zone for tunneling operations.

While the corridor would be frozen during the casing installation, the key concern was the installation of the freeze pipes and managing groundwater flow after penetrating through the frozen soil shoring. The pilot tube drill head was under approximately 8 feet of hydrostatic head pressure, and the drillers (Armadillo Boring) were concerned with water blowing out around the drill stem as they progressed.

Before the horizontal freeze pipes were installed, a concrete bulkhead was erected at the headwall for both the jacking and receiving pits, as shown in Figures 5 and 6. These concrete walls were installed against the frozen soil shoring so that a rubber seal around the pilot tube could be pressed against a smooth concrete face. However,

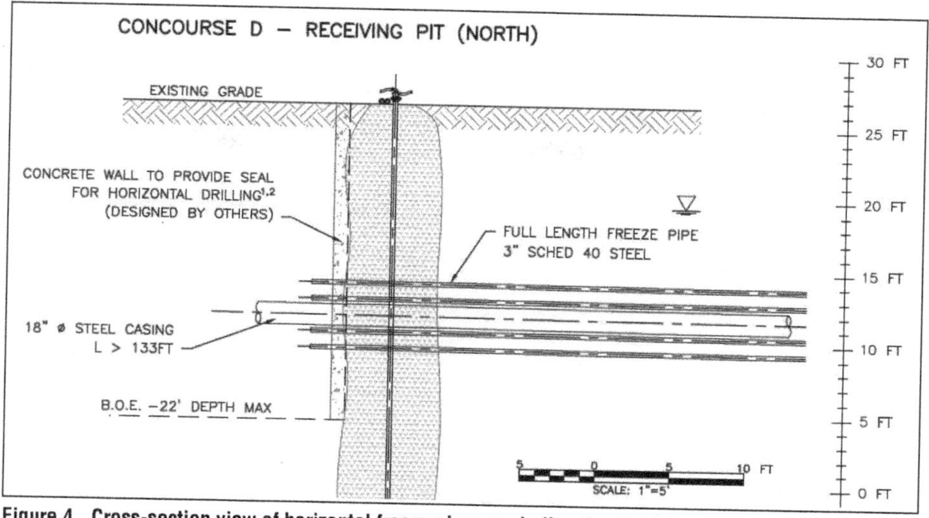


**Figure 4. Cross-section view of horizontal freeze pipes encircling the 18-inch steel casing**

**Figure 5. Installation of horizontal freeze pipes through the concrete headwall**

maintaining a frozen bond/seal at the interface between the concrete and the frozen soil shoring interface would be complicated by the heat generated from the concrete hydration and exposure to the summer sunshine. Similar to the extrados of the tunnel liner for the cross-passage excavation, an insufficient bond could create pathways for groundwater to enter the excavation from behind the wall or render the concrete wall free-standing and unstable. To combat this a ¾-inch diameter copper pipe was placed at the interface between the frozen soil wall and the proposed cast-in-place concrete bulkhead/headwall. The copper pipe wholly encompassed the perimeter of the horizontal freeze pipe installation. Once the concrete walls were formed, liquid nitrogen could be continuously dripped into the copper line until the installation of the horizontal freeze pipes was complete. The extreme cold of the liquid nitrogen ensured a frozen, water-tight bond remained between the concrete bulkhead/headwall and the frozen soil shoring.

**Figure 6. Completed installation of the horizontal freeze pipes installed through the concrete bulkhead**

Combining the rubber seals at the concrete face and the copper loop at the back of the concrete wall allowed for the successful installation of the freeze pipes beneath the existing concourses without significant groundwater inflow. Once freezedown was complete, installation of the 135-foot aviation fuel casing began and was completed at both concourses within one week. The frozen soil system was successful in controlling groundwater throughout each of the drilling operations.

## CARLLS RIVER WATERSHED SEWER EXPANSION

### Project Background

The Carlls River Watershed Sewer Expansion project was located within a residential neighborhood near the town of West Babylon on Long Island, New York. The owner, Suffolk County Department of Public Works, wanted to expand public sewer access and capacity within the area by installing new manholes, pipelines, and other sewage related infrastructure. The critical task, however, was connecting two sewer systems separated by a major highway (Southern State Parkway). Posillico, the general contractor, proposed using a micro-tunneling boring machine (MTBM) to tunnel beneath the highway and install a 48-inch sewer pipeline connecting the two systems. Posillico selected temporary ground freezing as a singular solution to provide structural shoring and groundwater control for the construction of the jacking and receiving shafts and during MTBM operations.

### Design and Installation

The geotechnical data for the project site indicated that the groundwater table was approximately 7 feet below the existing ground surface (bgs), and the soil conditions were generally homogenous, consisting of sands with gravel, to a depth of 50 feet.

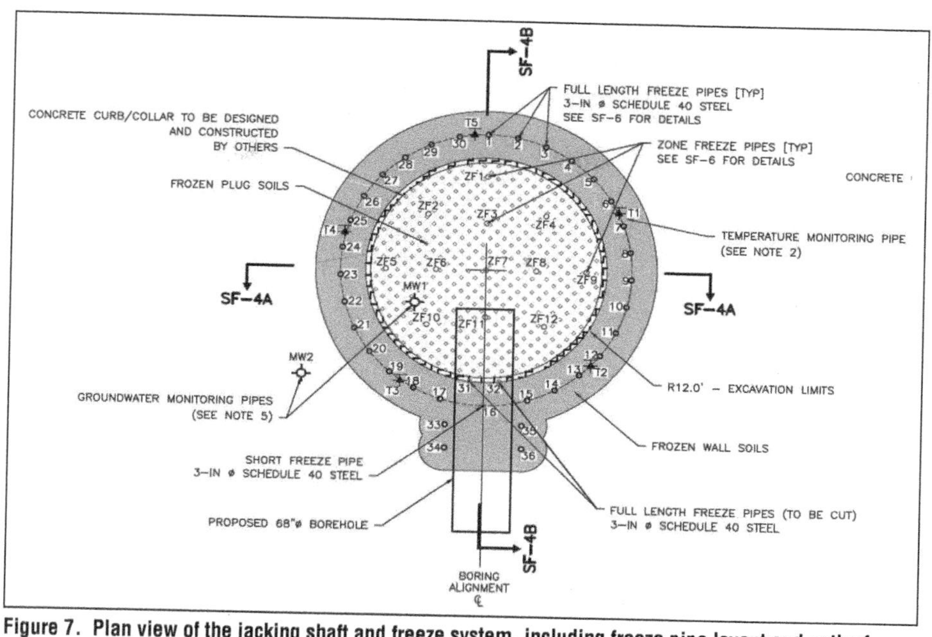


**Figure 7. Plan view of the jacking shaft and freeze system, including freeze pipe layout and path of MTBM**

Since the bases of excavation at the two shafts were approximately 18 feet to 22 feet below the water table, the artificial ground freezing system was designed to freeze a singular uniform wall of soil around the circumference of the shafts and a plug of soil immediately beneath and equal to the area of the base of the excavation, as shown in Figure 7. The installation and operation of the freeze pipes allowed for the successful excavation of both shafts in preparation for micro-tunneling operations.

As with the previous projects, it is critical during MTBM penetrations through a frozen soil shaft, that groundwater does not pass unchecked around the shaft liner and into the excavation causing thermal and mechanical erosion to shaft walls. Similarly to the systems previously described, this was avoided using a combination of two separate seals. It is nearly impossible to place a mechanical seal against a frozen wall; therefore a concrete bulkhead or headwall was constructed at both shafts. The concrete walls did not line the entire shaft, as they extended only slightly wider and taller than the mechanical rubber seal and top hat used with the MTBM. This seal, typically used with MTBM operations, was the primary defense against the inflow of groundwater through the frozen soil shoring system. However, a second seal was provided by looping a hose (as shown in Figure 8) with chilled brine around the limits of the tunneling operations prior to the placement of the concrete for the walls. This ensured that active freezing energy was located exactly where it was needed, and the bond at the interface between the concrete walls and the frozen soil shoring walls remained robust and intact.

In addition to the seals installed around the entrance and exit of the MTBM, the freeze pipe layout for the shoring system also had to incorporate the passage of the 48-inch diameter tunneling machine. During the design phase, two of the vertical steel freeze pipes were located within the path of the large MTBM at both the jacking and receiving shafts, as shown in Figure 7. They blocked the path of the MTBM; however, they

**Figure 8. Heavy-duty rubber hose used to reconnect the cut pipes and fastened to the frozen soil shoring**

were required to obtain the design thickness of frozen soil shoring. These pipes were "adjusted" and located at the face of the excavation, instead of offset 2.5 feet from the face and in the middle of the frozen soil shoring wall, as shown in Figure 7. Additionally, one "short" freeze pipe was added in the middle of the frozen shoring wall, as shown in Figures 7 and 9. This short pipe was centered along the MTBM path and installed to a depth of 1 foot above the proposed crown. While still in the path of the MTBM, the "adjusted" freeze pipes allowed excavators to easily expose them as part of the shaft excavation.

Once freezedown was complete, these two pipes were temporarily deactivated and pumped dry. The general contractor then cut away the exposed upper portion of these two pipes to provide space for the MTBM and to form the concrete headwall/bulkhead. The bottom portions of these cut pipes, still embedded in the ground, were reconnected using flexible-heavy duty rubber hoses. As previously mentioned, this was the rubber hose that looped around the proposed MTBM path and was fastened to the frozen soil face, as shown in Figure 8. This allowed the hose, carrying chilled brine to the cut pipes below, to sit between the frozen soils and the concrete bulkhead/headwall and help maintain the frozen bond between them. The reconnected pipe bottoms and the "short" freeze pipe allowed for freezing to continue both above and below the MTBM and bulkhead/headwall with no steel in the machine's path, as shown in Figure 8. These steps, in combination with the MTBM's mechanical rubber seal of the MTBM, were designed to prevent potential seepage pathways from forming at the concrete-soil interface.

To install the concrete bulkhead/headwall (as shown in Figure 10), workers excavated approximately 1-foot into the existing frozen soil wall to better anchor the two systems

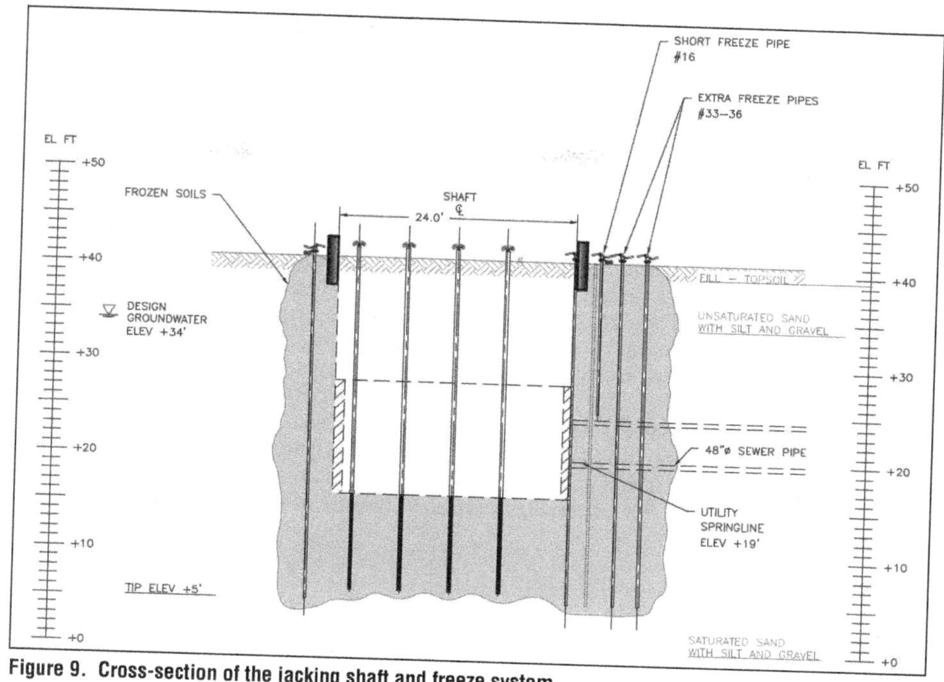


**Figure 9. Cross-section of the jacking shaft and freeze system**

**Figure 10. Installed concrete bulkhead**

together. To compensate for the thinner wall, four steel freeze pipes were installed at the perimeter of the frozen soil shoring at the intersection with the MTBM path. These four freeze pipes were arranged in a 2×2 pattern, in-line with, and 12 inches off the MTBM path. These freeze pipes increased the freezing capacity over this small area, causing a small frozen "bulge" or "bump out," adding a couple feet of additional frozen soil. This addition reduced the potential for groundwater infiltration and maintained the minimum design thickness of the freeze system. After installing the concrete bulkhead and headwall, the MTBM was fitted with a mechanical rubber seal, and drilling operations commenced. Operations continued unimpeded until successful completion approximately six days later.

## CONCLUSION

For more than a century, artificial ground freezing has built a reputation of success, often providing temporary structural shoring and groundwater control for excavations extending below the water table. The examples noted within this paper illustrate the additional capabilities of freeze systems regarding submerged horizontal penetrations. As noted within the Carlls River and PDX examples, freeze systems can also be installed as the primary temporary shoring on projects where submerged horizontal penetrations through shoring are required. The Sound Transit example highlights the potential for freeze systems installed around existing concrete structures and used as the penetrated medium. Not only can ground freezing systems be used in these circumstances, but when selected provide several advantages, including being cost competitive, easy to install, environmentally friendly, and effective in almost all soil conditions.

## REFERENCES

Chang, Dong K. and Lacy, Hugh S., "Artificial Ground Freezing in Geotechnical Engineering" (2008). *International Conference on Case Histories in Geotechnical Engineering*.

Poetsch, F.H. (1883). Method of and Apparatus for Sinking Shafts Through Quicksand (U.S. Patent No. 300,891). *U.S. Patent and Trademark Office*. https://patents.google.com/patent/US300891A/en.

PART 

# Ground Control Approaches and Methods

*Chairs*

**Luis Piek**
Arup

**Vinay Duddempudi**
Traylor Bros Inc

# Bay Park Conveyance Project, New York—Construction Update

**David I. Smith** ▪ WSP
**Brian Lakin** ▪ Delve Underground
**Bryan Aweh** ▪ Western Bays Constructors/John P. Picone
**Vincent Falkowsk** ▪ Nassau County Department of Public Works
**Kenneth Arnold** ▪ Nassau County Department of Public Works
**Andy Fer** ▪ New York State Department of Environmental Conservation

## ABSTRACT

Nassau County, New York, is undertaking several projects to improve the water quality in the Western Bays of Long Island. A major step forward will be the completion of the Bay Park Conveyance Project, which will divert most treated effluent away from the Western Bays to an existing ocean outfall at the Cedar Creek Water Pollution Control Plant. The ongoing works include 3.6 miles of microtunneling, 7 miles of sliplining, 14 shafts, and a 75 MGD pump station. This paper provides an overview of the design and an update on construction progress through December 2022. The project is a collaborative effort between Nassau County, New York State Department of Environmental Conservation, the design-builder, and various consultants.

## PROJECT PURPOSE

The Nassau County Department of Public Works (County) is responsible for the operation and maintenance of two sewage treatment facilities on the south shore of Long Island: the South Shore Water Reclamation Facility (SSWRF) in Bay Park, and the Cedar Creek Water Pollution Control Plant (CCWPCP).

The SSWRF currently treats approximately 50 MGD and discharges its treated effluent into Reynolds Channel, which is part of an estuarine complex called the Western Bays. These bodies of water are significantly cut off from the Atlantic Ocean by barrier islands that limit the mixing of bay water and ocean water.

High levels of nitrogen in the Western Bays violate state and federal water quality standards. Among other things, excess nitrogen contributes to two notable problems in these waters:

- The nitrogen rich environment leads to a proliferation of macro-algae (specifically *Ulva lactuca*, or “sea lettuce”). Decomposing *Ulva* depletes oxygen from the water, and it washes onshore where it emits noxious odors, leaving beaches unsuitable for recreation.
- Excess nitrogen is taken up by the near-surface roots of marsh grasses, reducing the growth of deeper roots. The weakened root systems render marshes more susceptible to erosion, leading to diminished protection of shorelines from coastal storm surge and waves.

Nassau County has been upgrading the SSWRF to reduce the amount of nitrogen discharged. It is also part-way through the Bay Park Conveyance Project, which will divert most of the treated effluent through new tunnels and piping to an existing ocean outfall 2.5 miles offshore.

## PROJECT OVERVIEW

New York State Department of Environmental Conservation (NYS DEC) and Nassau County have entered into a partnership to divert up to 75 million gallons per day (MGD) of effluent from the SSWRF to the CCWPCP, from where the effluent will be discharged via an existing ocean outfall (Figure 1). The Bay Park Conveyance Project is currently under construction and comprises the following principal elements, which are shown in Figure 2:

- Effluent Diversion Pump Station at the SSWRF
- 2.0 miles of force main between SSWRF and Sunrise Highway (Bay Park Force Main)
- 7.2 miles of repurposed aqueduct pipe beneath Sunrise Highway (Sunrise Highway Aqueduct)
- 1.6 miles of force main between Sunrise Highway and CCWPCP (Cedar Creek Force Main)
- A receiving connection at the CCWPCP, including a receiving tank and standpipe, with connection to the existing CCWPCP outfall
- Replacement of pumps and controls at the CCWPCP

Major portions of the project will utilize trenchless techniques (specifically microtunneling and sliplining) to minimize disruption to the public.

Nassau County retained WSP to perform preliminary engineering services, including preparation of an Environmental Assessment, and bridging documents for design-build procurement. Program Management on behalf of Nassau County is provided by a joint venture of Arcadis and Hazen & Sawyer.

Nassau County and NYS DEC partnered with a common interest in improving the water quality in the Western Bays. This partnership allowed the NYS DEC to lead the project delivery and implement its design-build authority.

In February of 2021, a design-build contract was awarded to Western Bays Constructors (WBC), a joint venture of John P. Picone, Inc. and Northeast Remsco Construction, Inc. For final design, WBC retained Delve Underground to lead the design team. Working for Delve Underground were Greeley and Hansen and others.

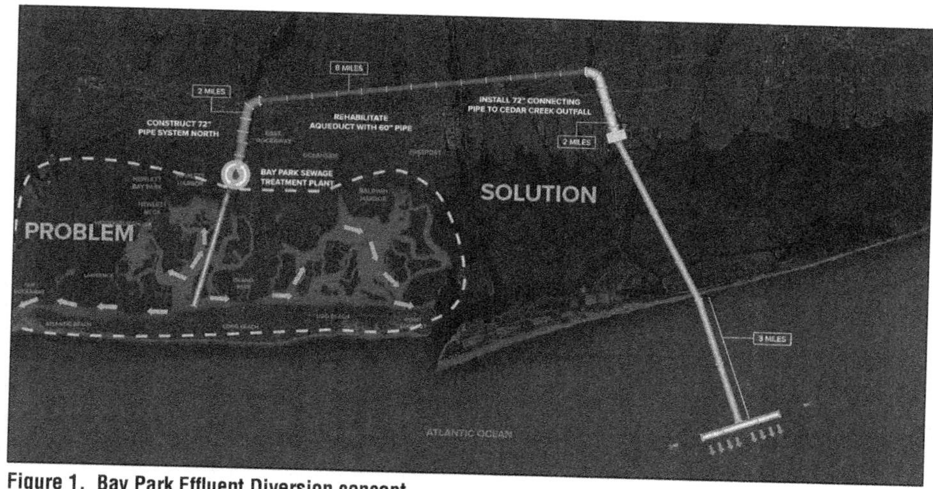


Figure 1. Bay Park Effluent Diversion concept

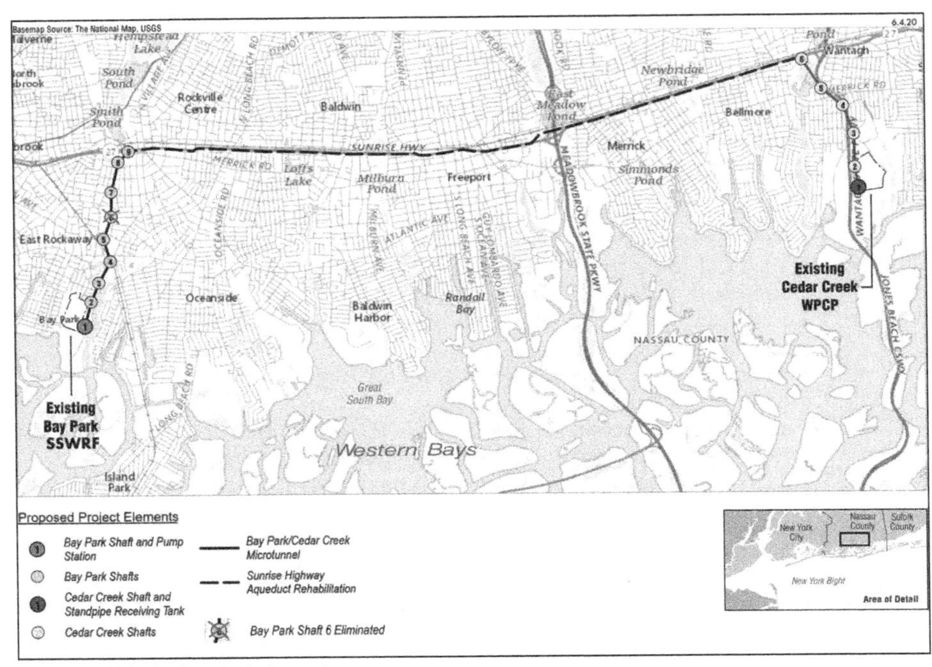


**Figure 2. Principal project elements**

## SURFACE AND SUBSURFACE CONDITIONS

The south shore of Long Island is, for the most part, densely developed with primarily low-rise residential and commercial structures. The area between the SSWRF and the CCWPCP, has minimal topographical variation in elevation.

The project lies approximately 7 miles south of the terminal moraines from the last ice age. The near-surface geology was formed by outwash from the melting glaciers, which is overlain by marsh deposits. The findings from a subsurface investigation program are presented in a Geotechnical Data Report (GDR) and a Geotechnical Baseline Report (GBR). The soils in the face of the microtunnels are anticipated to be composed predominantly of sand. The sand along the Bay Park alignment has a variable amount of gravel and silt, and is typically very dense. The Cedar Creek microtunnels are anticipated to also generally encounter sand, with progressively more clay expected when closer to the CCWPCP. The sand along the Cedar Creek alignment generally has a low fines content, and is loose to very dense. The sands and clays are overlain, in most areas, by organic soils and fill. Groundwater is generally 2 to 13 feet below ground surface and is considered brackish in some areas.

## BAY PARK EFFLUENT DIVERSION PUMP STATION

At the SSWRF, treated effluent will be diverted into the Bay Park Effluent Diversion Pump Station (BPEDPS). This facility will have a 75 MGD firm capacity, achieved using four 900 hp variable frequency drive (VFD) pumps. A circular wet well will be divided into two parts, each accommodating two pumps. The split wet well allows for maintenance of up to two pumps while dry weather flows are accommodated by the remaining two pumps. A rendering of the pump station design is shown in Figure 3.

Figure 3. Revit model of Bay Park Effluent Diversion Pump Station (from RFP Reference Design)

One of the main challenges of completing the BPEDPS design was the tight footprint available for the facility. There were also many active and abandoned buried utilities that needed to be identified during design, some of which were actively under construction as part of adjacent plant electrical and SCADA upgrade contracts. Developing a complete picture of utility status around the BPEDPS site required extensive coordination with plant personnel, County representatives, and designers and constructors of adjacent contracts. The final designers of the BPEDPS began with a utility survey completed during preliminary design, then incorporated new information from adjacent contracts. Any remaining utility information gaps were addressed with an extensive test pitting program undertaken with input from plant and County personnel. In some instances, new utility conflicts surfaced late in the design process through test pitting and coordination efforts, which required the team to advance several design options until an acceptable solution could be found.

WBC and its affiliates successfully installed an extensive support of excavation system for the effluent diversion structure and wet well, consisting of secant pile walls and jet grout bottom plugs for groundwater cutoff (Figure 4). Considerations for this system during the planning phase included sealing around an active existing pile-supported effluent conduit and managing grouted pile installation in a thick layer of organic soil. In total, 239 secant piles are required at this site. Currently, all permanent piles and foundations have been installed, subgrade building elements are waterproofed and in place, and the above-grade structure is in progress (Figure 5). Long lead equipment including vertical turbine pumps and controls are in production and nearly complete. At the time of writing this paper, WBC is preparing for tie-in of the new structure to the existing effluent conduit in the upcoming months.

## BAY PARK FORCE MAIN

Treated effluent from the Effluent Diversion Pump Station will be piped via a 2-mile-long, 71-inch-inside-diameter (ID), microtunneled force main to convey it toward Sunrise Highway (Figure 6). The force main alignment generally parallels the Mill River. The depth of cover ranges from 54 feet (at the southern terminus) to approximately 25 feet

Figure 4. Bay Park effluent diversion structure (foreground) and pump station (background, cylinder)

Figure 5. Bay Park effluent pump station, superstructure construction

(near the northern terminus), which is largely controlled by overlying pile-supported structures, the sloping alignment, surface topography, and bathymetry as the alignment crosses the Mill River at several locations. Key surface features along the Bay Park Force Main include (south to north): SSWRF, SSWRF floodwall, a navigable channel, Mill River, Long Island Rail Road, Woodcrest Village Park apartment buildings, East Rockaway High School, and Merrick Road.

Microtunneling of the Bay Park and Cedar Creek force mains will be performed using two Herrenknecht AVN 1800TB microtunnel boring machines (MTBMs). Each MBTM is equipped with a skin kit to increase its diameter to accommodate the 87.5-inch-outside-diameter (OD) jacking pipe. The cutterhead face is pressurized with bentonite slurry for settlement control and spoil removal. For the MTBMs, local schoolchildren selected the names MARSH MELLOW and P.O.S.E.I.D.O.N (Project of Science to Expel an Incredibly Damaging Overabundance of Nitrogen).

The Flowcrete jacking pipe consists of fiberglass reinforced polymer mortar (FRPM) carrier pipe encased in reinforced concrete. Jacking lengths vary from approximately 830-ft to 2,300-ft. Intermediate jacking stations have been installed within selected drives, up to two per drive, where jacking loads are anticipated to approach the capacity of the jacks at the launch shaft.

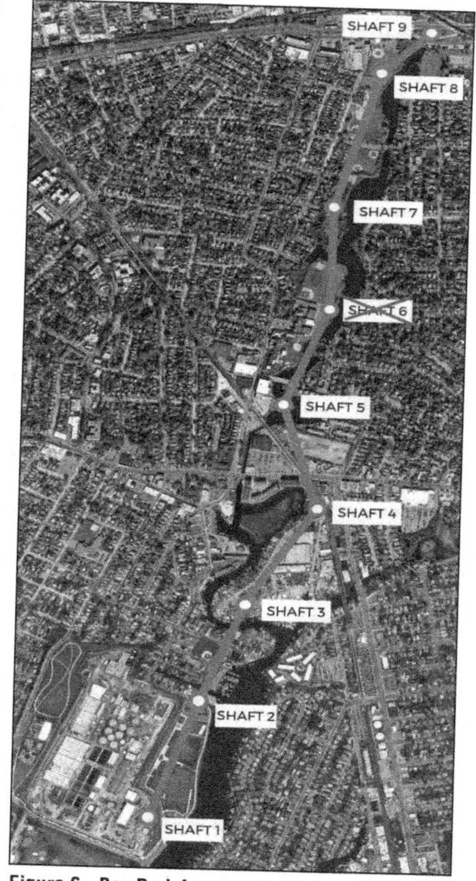


**Figure 6. Bay Park force main alignment (Shaft 6 eliminated)**

## Bay Park Shaft Design

Eight construction shafts, between 25 feet and 30 feet in diameter, are required to accommodate pipe jacking and reception. The shafts are generally temporary and need to be sufficiently watertight for construction operations and to prevent drawdown of the adjacent groundwater table. Two of these shafts will be converted to permanent access shafts, using stacked manhole rings, to enable dewatering of the force main and maintenance.

Design considerations for the type of support of excavation to be used at each location included the soil types, depth to groundwater, amount of brackish water within the soil, and the shaft size necessary to fit tunneling equipment. For most shafts, secant piles (Figure 7) and cutter soil mixing (CSM) (Figure 8), are viable alternatives. Selection of the preferred method at each site was based on a combination of geotechnical conditions, productivity and availability of equipment.

Each shaft included the construction of a jet grout base plug. This plug was constructed prior to the shaft being excavated in order to both stabilize the support of excavation walls and to provide a nearly watertight seal for the shaft floor. During excavation, the shaft walls and floor prevented groundwater from flowing into the workspace.

## Bay Park Microtunnel Design

Seven reaches of microtunneling are being constructed along the Bay Park portion of the project. The preliminary design for the microtunnel included nine construction shafts. Shaft 6 was originally intended to be located on the East Rockaway High School property, such that two straight microtunnel drives could be aligned around the school buildings. This shaft was eliminated by replacing two straight microtunnel drives with one longer, curved, microtunnel drive having a length of approximately 2,300 feet. The microtunnel includes two straight sections, one at each end of a 4,800-foot radius curved section. This radius was selected because it is compatible with the selected

Figure 7. Bay Park Shaft 1—MTBM arrival at secant pile shaft

Figure 8. Bay Park Shaft 3—MTBM launch from CSM shaft

microtunnel pipe being used on the project, and it will provide enough clearance from the surrounding structures so as not to impact foundations or subsurface utilities while still remaining within the previously approved project easement. The elimination of an access shaft on school property avoided the impacts of many work-hour restrictions that had been included in the original contract so that students would not be disturbed by construction work.

Geotechnical and structural instrumentation is being installed throughout the project in the ground, on adjacent structures, and above utilities. These are being monitored by manual and automated methods, with the data being stored on a web-accessible app. The data provides assurance that any movement of surface and subsurface facilities is within allowable limits. If readings exceed predetermined alert levels, construction methods are adjusted in accordance with response plans.

### Construction Progress

Microtunneling has generally proceeded as planned, but on the early drives, WBC encountered difficulty maintaining the watertightness of the rubber seal when breaking into shafts. On this project, jet grout blocks were not used at break-in/out locations, which increased the reliance on effective seals. To resolve the watertightness issue, WBC procured a revised exit seal (Figure 9), which includes an inflatable bladder to provide positive sealing pressure against the outer diameter of the MTBM. Additionally, this system uses a prefabricated steel "cartridge," which is engaged with the seal prior to break through. The combination of these two elements has been successful at dealing with the high hydrostatic groundwater pressure and cohesionless soils.

## CEDAR CREEK FORCE MAIN

The Cedar Creek Force Main, which runs from Sunrise Highway to the CCWPCP, consists of approximately 1.6 miles of 71-inch ID microtunneled pipe (Figure 10). It is being constructed using a Herrenknecht AVN 1800TB MTBM, similar to the one used on the Bay Park Force Main. Flowcrete jacking pipe is being used.

The force main alignment generally parallels the Wantagh State Parkway. The depth of cover ranges from 46 feet (at the southern terminus) to approximately 23 feet (at the northern terminus), largely controlled by the sloping alignment and surface topography. Key surface features along the Cedar Creek Force Main generally include (south to north): CCWPCP, Cedar Creek Park, a 96-inch concrete interceptor sewer, Wantagh State Parkway, and Millpond Park.

**Figure 9. Revised exit seal and cartridge system, shown here at Cedar Creek Shaft 1**

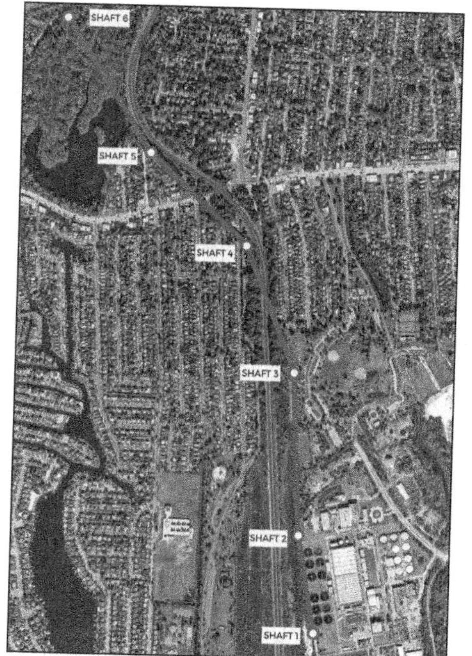


**Figure 10. Cedar Creek Force Main alignment**

## Cedar Creek Shaft Design

Six construction shafts are required to allow for pipe jacking or reception. Two of these shafts will be converted to permanent access shafts. Similar to the Bay Park side of the project, options for the type of support of excavation were either secant pile or CSM walls. The geology on the Cedar Creek side provided some flexibility to the project team regarding temporary shaft inverts. At Cedar Creek Shaft 1, geotechnical investigation results showed that a clay layer below that shaft was thick and strong enough to provide the seal necessary for a dry working shaft, without the need for a jet grout plug. The support of excavation walls were extended slightly deeper than usual to engage with the clay layer. This approach saved time and money, and reduced trucking though residential areas. Figure 11 shows Shaft CC-1, a receiving shaft that incorporates a clay base plug.

## Cedar Creek Microtunnel Design

Five reaches of microtunneling are being constructed along the Cedar Creek portion of the project. One of these reaches will traverse beneath an existing, active 96-inch diameter sewer main that must remain in service during the tunnel drive. The microtunnel will cross under the sewer with approximately 5.5 feet of clearance between the two. To minimize the risk of settlement below the existing sewer, other drives are being performed and completed to ensure that crews and equipment are through any learning curve prior to tunneling under the existing sewer. Depending on the success achieved during preceding tunnel drives, the team may determine that no other preemptive actions are necessary to protect the sewer. Alternatively, the team may implement one or more ground improvement methods at this location to prevent settlement during tunneling.

## Construction Progress

Tunneling production, alignment control, and settlement control (except at break-in/out) have generally been good. On one of the Bay Park drives, the MTBM was briefly unable to advance and organic material was found in the spoil arising from the MTBM. At the time of writing this paper, an investigation is ongoing to determine the source of the organic material. Jacking pressures have remained low on both sides of the project and intermediate jacking stations, while installed, have not been needed thus far. To date, approximately 4,500 linear feet (LF) of tunnel (3 drives) has been completed of the 12 total drives on the project (Figures 12 and 13).

Figure 11. Cedar Creek Force Main, Shaft CC-1

Figure 12. Cedar Creek Force Main, Shaft CC-2, MTBM launch

Figure 13. Cedar Creek Force Main—Jacking pipe from Shaft CC-2

Figure 14. Removed Sunrise Highway aqueduct pipe section showing riveted joints

## SUNRISE HIGHWAY AQUEDUCT

The economic viability of the Conveyance Project was achieved by repurposing 7.2 miles of abandoned 72-inch steel pipe that exists below Sunrise Highway. The pipe, or aqueduct, was constructed in 1909 to convey drinking water from abstraction wells on Long Island to Brooklyn. Based on an evaluation completed in 2017, the hydraulic integrity of the steel pipe was known to be compromised, but the overall stability is good. Figure 14 shows a section of the existing pipe that was removed during construction of a jacking pit.

### Aqueduct Survey

The project includes sliplining the existing aqueduct with structurally independent, segmented fiberglass reinforced polymer mortar (FRPM) pipe. WBC conducted a comprehensive survey of the interior of the existing pipe, including sections that could not be accessed prior to award of contract. This survey identified more ovalization and bends than previously known. This resulted in selecting a 57.6-inch ID pipe, rather than the anticipated 60-inch pipe. To offset the resulting hydraulic system head loss, the BPEDPS pump horsepower was increased from 4×800 hp to 4×900 hp.

### Design of Slipline and Permanent Works

Air valves and maintenance access holes are required at various locations throughout the project. The air valves will allow trapped air or vacuum to release, maintaining hydraulic pressures within system limits. The maintenance access holes will enable access to the force main and air valves.

Preparatory work prior to sliplining included removal of twelve 48-inch cast iron gate valves, as well as removal of standing water, silt and debris, and rust/tuberculation. Twenty-four jacking and reception pits are being constructed along Sunrise Highway. Once valves, large tubercles, and other obstructions have been removed from the existing pipe, and a pipe segment is open at either end, the sliplining effort can begin.

Low-friction rails are installed in the pipe to help guide the new pipes, reduce jacking forces, and keep the slipline pipe sections clear of rivet heads. Each FRPM pipe segment is approximately 19 feet long and incorporates fiberglass bumpers cast around the perimeter that align the pipe and which prevent any abrasion from affecting the overall structural capacity. Once an initial pipe segment is lowered into the excavation and placed inside a jacking frame, it is pushed into place within the aqueduct pipe. The jacking frame then receives the second pipe segment, which is connected to the first with a push-on joint setup, sealing the connecting gasket. The second segment is then pushed into the carrier pipe, displacing the first segment. This process is repeated until the entire length is installed for the portion of work being completed. In areas where the existing aqueduct has tight curves (either vertical or horizontal) it is necessary to carry in individual pipe sections using a purpose-built cart.

After installation of a complete length of pipe between jacking pits, the annulus between the carrier pipe and host pipe is filled with cellular grout. The pipe section is then filled with water and pressure-tested for leaks.

## Construction Progress

### *Slipline Work Pits*

Sunrise Highway is a critical east-west transportation link on Long Island, so much of the work is performed at night (Figure 15). Extensive traffic control measures are used, in coordination with New York State Department of Transportation (NYSDOT) and local authorities. Work pits in the middle of the highway are covered over in peak hours using precast concrete planks.

Crews continue to make good progress installing steel sheeted pits along Sunrise Highway. Sixteen of 24 pits have been installed. Seven have been backfilled.

## Slipline Pipe Installation

Good progress continues to be made on the sliplined FRPM pipe installation. Over 22,000 LF of pipe has been installed thus far (Figures 16 and 17).

**Figure 15. Nighttime construction of slipline work pits on Sunrise Highway**

Figure 16. Sliplining with FRPM pipes within Sunrise Highway

Figure 17. Sliplining with FRPM pipes adjacent to Sunrise Highway

## PROGRESS ON OTHER PROJECT COMPONENTS

At the time of writing this paper, the following work is ongoing.

### CCWPCP Upgrades

Work at CCWCPS has primarily been procurement of the new mechanical and electrical systems. Delivery of the new effluent pumps is anticipated for spring 2023, kicking off the start of major work in the pump station.

### Wet Tap at CCWPCP and Outfall Work

Crews are currently installing support of excavation for the wet tap work scheduled for spring 2023. Once the existing 84" ID outfall pipe is excavated and exposed, the final survey of the pipe can be conducted by a wet tap specialist.

## UPCOMING WORK THROUGH PROJECT COMPLETION

Major upcoming work on the project includes installation of the receiving tank at the CCWPCP, yard piping from the microtunnel to the receiving tank, and the above grade structure at the Bay Park Effluent Diversion Pump Station, as well as restoration of completed microtunnel shafts and pits along Sunrise Highway.

Preliminary hydrostatic testing of installed slipline pipe has begun and will continue in the upcoming months. Microtunnel pipe pressure-testing is scheduled to begin in spring 2023. Commissioning of the new mechanical equipment is anticipated to start in late spring/summer 2024.

For the latest project information, please visit: https://www.bayparkconveyance.org/

## ACKNOWLEDGMENTS

The authors wish to thank all the team members at Nassau County DPW, New York State Department of Environmental Conservation, Western Bays Constructors, the Delve Underground–led design team, WSP, Arcadis, Hazen & Sawyer, AECOM, and many other individuals and firms who are collaboratively working towards the successful completion of this amazing project.

# Evaluation and Construction Effects of a 22-Story Tower on Adjacent Metro Tunnels in Los Angeles

**S. Amir Reza Beyabanaki** ▪ Delve Underground
**Yiming Sun** ▪ Delve Underground
**Stan Tang** ▪ Geotechnologies Inc.
**N. Sathi Sathialingam** ▪ LA Metro

## ABSTRACT

The Kurve mixed-use development project including a 22-story tower at 2801 Sunset Place was constructed on a site consisting of a roughly rectangular-shaped lot at the intersection of Wilshire Boulevard and Hoover Street in Los Angeles. Two Metro Red Line tunnels, with depths of approximately 30 feet and 58 feet, respectively, cross below the building at distances of approximately 1 foot to over 63 feet from the site. This paper presents numerical modeling performed to evaluate the ground and tunnel behavior in response to building construction. Results of geotechnical instrumentation and monitoring undertaken during the building construction are also discussed.

## INTRODUCTION

The Kurve mixed-use development project at 2801 Sunset Place in downtown Los Angeles (LA) includes construction of a 22-story tower on a site consisting of a roughly rectangular-shaped lot at the southwest corner of Wilshire Boulevard and Hoover Street (Figure 1). The site excavation for construction of the new tower foundation varied in depth from about 5 to 15 feet, with the deeper portion being on the west side of the site. Within a portion of the site (north and northeast corner), the excavation was limited to about 5 feet in depth. Two Red Line tunnels—AL (upper) tunnel and AR (lower) tunnel—operated by Los Angeles Metropolitan Transportation Authority (Metro), cross below the north edge and northeast corner of the building (Figure 1). The nearest sidewalls of the AL and AR tunnels range at distances from approximately 1.0 foot near the northeast corner of the site to over 63 feet near the northwest corner. The vertical separation between the two tunnels is approximately 8.5 feet. The crowns of the AL and AR tunnels are in depths of approximately 30 feet and 58 feet, respectively, from the ground surface.

LA Metro requirements dictate that any new construction adjacent to Metro tunnels should not adversely affect either performance of tunnel structures or the safe and efficient operation of trains. Jamison Properties, LP (JP) contracted Delve Underground to evaluate the potential impacts of construction of the Kurve project on the adjacent Metro Red Line tunnels. The evaluation was carried out using two-dimensional (2D) and three-dimensional (3D) numerical continuum models for an evaluation of the performance of the tunnel structures and the rail cars operating through the tunnels. Validation of the numerical continuum model was performed using historical observations of performance of the tunnels adjacent to similar development as well as utilizing data from the geotechnical instrumentation and monitoring implemented during the building construction. This paper discusses the numerical modeling and comparison of the predicted and observed responses of ground movements around the tunnels to the excavation and construction of the new tower.

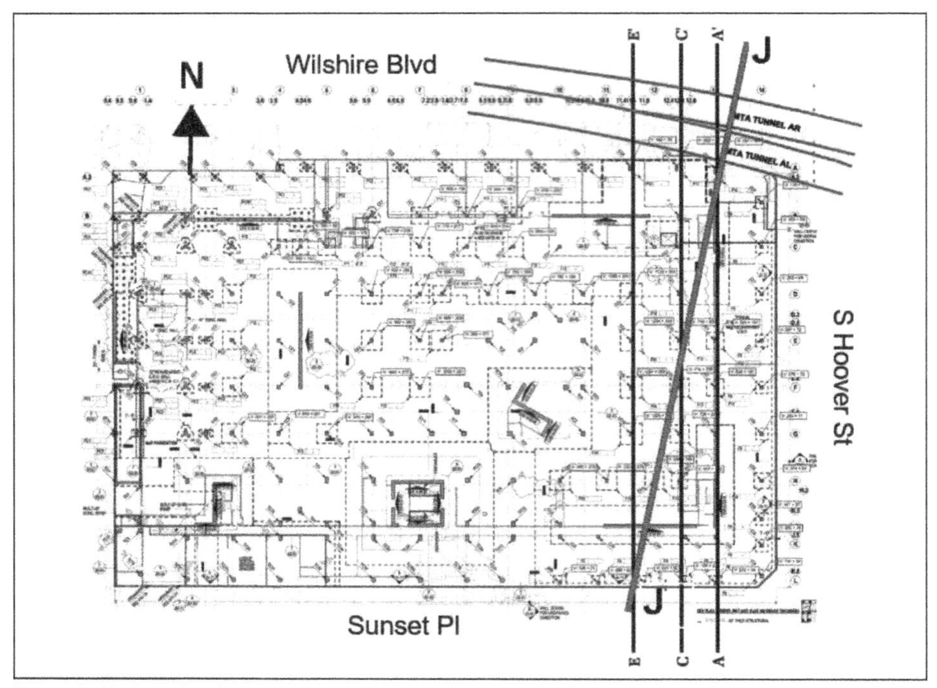


Figure 1. Project site plan and adjacent Metro Red Line tunnels

## LA METRO RED LINE TUNNELS AND PERFORMANCE CRITERIA

### Red Line Tunnels

According to the LA Metro's as-built civil drawings (Metro 1990), the reach of tunnels (Stations AL/AR 294+00 to 299+00) near the project site were constructed as part of LA Metro's construction contract B201, which extended from Wilshire/Alvarado Station to Wilshire/Vermont Station. This reach of tunnels is along a curved section with a radius of approximately 1,500 feet (Metro 1990). The tunnels had an excavated diameter of approximately 22 feet. The initial ground support for the tunnels consisted of 9-inch-thick, 4-foot-wide expanded precast concrete segments (Mahar 1994).

Following completion of tunnel excavation and initial support, the tunnels were lined with a high-density polyethylene (HDPE) gas and water barrier. Then a 12-inch-thick final cast-in-place reinforced concrete lining with an inside diameter of about 18 feet was constructed. Therefore, a design thickness of 12 inches for the tunnel final linings was assumed in this evaluation.

### Performance Criteria

The performance of the Red Line tunnels was evaluated using the following four criteria:

- Structural capacity of the final linings
- Allowable distortion of the tunnel cross section
- Allowable deviation of the rail line from the design line and grade
- Allowable changes of stresses in the soil/rock around the tunnels (stress criteria)

**Table 1. Limits of tunnel lining radius change in soil (Metro 2017)**

| Soil Type | ΔR/R-Range |
|---|---|
| Stiff to Hard Clays | 0.15–0.40% |
| Soft Clays or Silts | 0.25–0.75% |
| Dense or Cohesive Soils, Most Residual Soils | 0.05–0.25% |
| Loose Sands | 0.10–0.35% |

1. Add 0.1 to 0.3% for tunnels in compressed air, depending on air pressure.
2. Add appropriate distortion for effects such as passing neighbor tunnel.
3. Values assume reasonable care in construction, and standard excavation and lining methods.

**Table 2. Limits of rail track displacements (Metro 2019b)**

| Type | Action Level of Deviation for Existing Track | | Maximum Level of Deviation for Existing Track | |
|---|---|---|---|---|
| Horizontal Track Alignment | Tangent track: | ±9⁄16 inch on 62-foot-long line | Tangent track: | ±3⁄4 inch on 62-foot-long line |
| | Curve track: | ±2⁄7 inch on 31-foot chord<br>±3⁄8 inch on 62-foot chord | Curve track: | ±3⁄8 inch on 31-foot chord<br>±1⁄2 inch on 62-foot chord |
| Vertical Track Alignment | ±3⁄4 inch from uniformed profile at middle ordinate in a 62-foot chord | | ±1 inch from uniformed profile at middle ordinate in a 62-foot chord | |
| Cross Level | Zero cross level on tangent track: ±3⁄4 inch | | Zero cross level on tangent track: ±1 inch | |
| | Reverse cross level elevation on superelevated curves: 3⁄4 inch | | Reverse cross level elevation on superelevated curves: 1 inch | |

The structural performance of the tunnel final linings was evaluated by comparing the total factored forces (thrusts and bending moments) developed in the linings, including the loads induced by the project construction, with the design capacity. The design capacity envelope for the 12-inch-thick reinforced concrete final lining with a 28-day compressive strength of 4,000 psi was generated based on ACI 318 (2019). A load factor of 1.35 was used in accordance with the Metro Rail Design Criteria (Metro 2017).

The Metro Rail Design Criteria also provides guidelines for the limits of tunnel distortions in terms of change of lining radius (expressed as a percentage, see Table 1). Based on the ground conditions discussed in the soil report (GI 2016), and in the tunnel construction documentation from Metro (2019a), the bedrock (Puente Formation) is present around the tunnels and is considered a soft rock, like stiff to hard clay, in its behavior to tunneling. For stiff to hard clays, the range of allowable tunnel distortion is 0.15 to 0.40%, which was adopted as the limit of tunnel distortion for this study. Using these criteria as a guide, the percentage of change in tunnel radius (or diameter) should be kept below 0.40% for the Kurve development construction.

The action and maximum levels (limits) of tunnel rail track deviations from design alignment were provided by Metro (2019b). These limits were specified to ensure the safe and smooth operations of rail trains and are listed in Table 2.

The stress criteria for limiting the changes of stresses in the soil/rock around the affected tunnels were provided by Metro as follows (2019b):

> *Estimate the stress increase (imposed surcharge loads) on the tunnel due to the proposed development by calculating the increase/decrease in the vertical, horizontal and shear stresses on the soil around the tunnel. Based on the estimated percent stress increase/decrease, the suggested actions are as follows:*

a. *Category 1—Less than 5%—Proceed with no further action by the developer*
b. *Category 2—Between 5% and 10%—Proceed with pre-and post-construction surveys and periodic monitoring during construction by the developer*
c. *Category 3—Greater than 10%—Perform additional analyses to estimate the anticipated incremental deformation/movement of the tunnels due to the proposed development. If the anticipated incremental deformation/movement is within the Metro Rail Track Displacement limits (action level and maximum level) as attached, then proceed with pre-and post-construction surveys and periodic monitoring during construction; if not proceed with mitigation to reduce the stresses.*

## SITE CONDITIONS AND GEOLOGIC PROFILE

Information on the site geologic conditions and geotechnical parameters was provided by Geotechnologies, Inc. (GI), the project's geotechnical consultant. Typical geologic conditions at the project site are shown in Figure 2 and consist of artificial fill, native soils, and bedrock (presumably weathered Puente Formation). The groundwater table was at the interface between native soils and bedrock (Figure 2) and is assumed to be at a depth of 19 to 20 feet from the surface (GI 2016). The top of bedrock is at a depth of about 10 feet from the ground surface. At the project site, both tunnels are located entirely within the bedrock with a rock cover of more than 25 feet. This information is also consistent with the construction records obtained from Metro (1990, 2019a). The upper half of the AL tunnel face and crown is in weathered Puente and firm to stiff clayey alluvium, while the lower half of the tunnel face is in unweathered Puente (Metro 2019a).

The geotechnical properties including the in situ stress conditions used in the numerical analyses were provided by GI and are presented in Table 3.

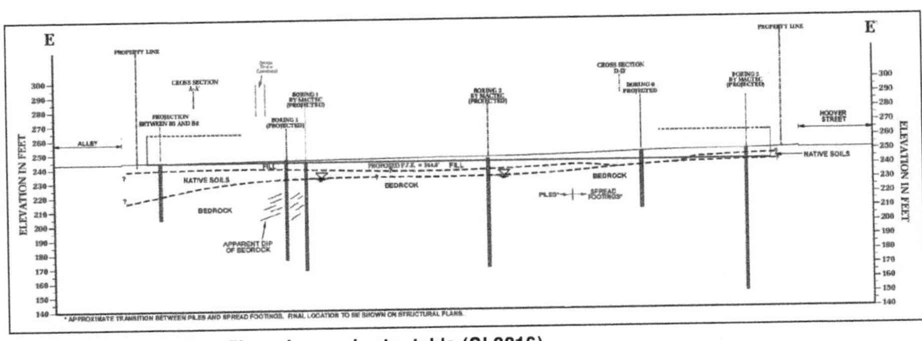


**Figure 2. Geological profile and groundwater table (GI 2016)**

**Table 3. Soil and rock mass properties (GI 2019)**

| Geologic Unit | Total Unit Weight, pcf | Young's Modulus, ksf | Poisson's Ratio | Cohesion, psf | Friction Angle, degrees | Coefficient of Earth Pressure, K0 |
|---|---|---|---|---|---|---|
| Fill | 121.1 | 500 | 0.30 | 370 | 27 | 0.43 |
| Native Soils | 121.6 | 500 | 0.30 | 300 | 27 | 0.43 |
| Bedrock | 120.0 | 2,500 | 0.25 | 810 | 29 | 0.60 |

## NUMERICAL MODELING AND RESULTS

The numerical analysis was performed using the 2D finite-difference program Fast Lagrangian Analysis of Continua (FLAC) Version 8 (Itasca, 2016). In addition, a 3D model using Fast Lagrangian Analysis of Continua in Three Dimensions (FLAC3D) Version 5 (Itasca 2012) was also carried out to estimate the longitudinal displacement profiles of the tunnels caused by the project construction. Both the FLAC and FLAC3D analyses were based on a ground-structure interaction approach to simulate the sequence of excavation and support installation. The analyses were performed to examine the following aspects of the project construction:

- Potential changes in the stresses in the soil/rock around the tunnels that occur in response to the excavation and application of the building surcharge loads
- Potential movements and deformations of the existing tunnels induced by the excavation and building surcharge loads
- Potential changes in the forces and moments in the tunnel final linings that occur in response to the excavation and application of the building surcharge loads
- An assessment of the effects of excavation and building surcharge loads on the overall performance and stability of the tunnels

In the 2D and 3D analyses, three different conditions were analyzed that simulated the baseline, excavation stage, and building construction or permanent condition. These three different conditions are defined as follows:

- **Baseline** represents the existing conditions prior to any project construction activities in terms of stresses in soil and deformations and forces in tunnel final linings. The baseline conditions were calculated based on the assumed soil profile, soil/rock properties, in situ stress conditions, and tunnel excavation and initial lining installation sequence, condition of initial lining, and interaction of final lining with the ground. For the purpose of evaluating the effects of new development, the baseline deformations or displacements of tunnel final linings were set to zero prior to any excavations associated with the new development.
- **Excavation Stage** represents the conditions when all excavation activities associated with the building foundation construction are completed prior to applying any building surcharge loads for the new development.
- **Permanent Condition** represents the final conditions when all surcharge loads of the new development are applied.

### 2D Analysis

A critical cross-section J-J′ was selected for the 2D numerical analysis. This cross section is approximately perpendicular to the tunnel axes (Figure 2). As shown in Figure 3, the AL (upper) tunnel has a ground cover of about 34 feet, measured from the tunnel crown to the existing ground surface. After the excavation was completed, the minimum clearance between the invert of the excavation and the AL tunnel crown was approximately 29 feet. As shown in Figure 3, the tunnels are located below the pad foundation. The excavation of the site would result predominantly in an unloading condition above the tunnels.

The geological profile shown in Figure 2 was used as a basis the analysis. Figure 4 shows the 2D FLAC model at the final stage of the analysis for Section J-J′.

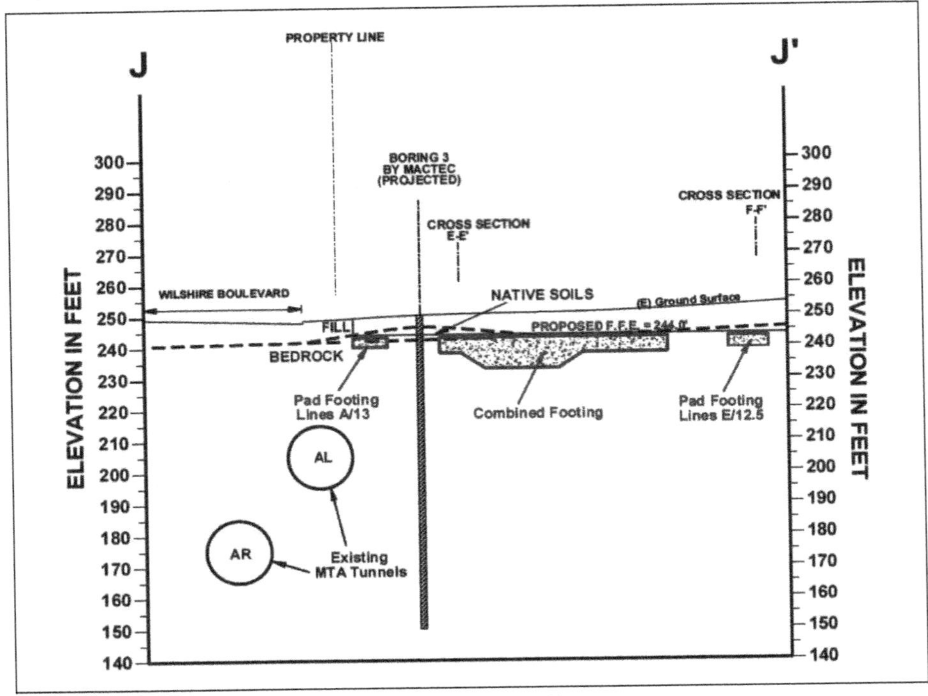


Figure 3. Cross Section J-J′ selected for 2D analysis (GI 2016)

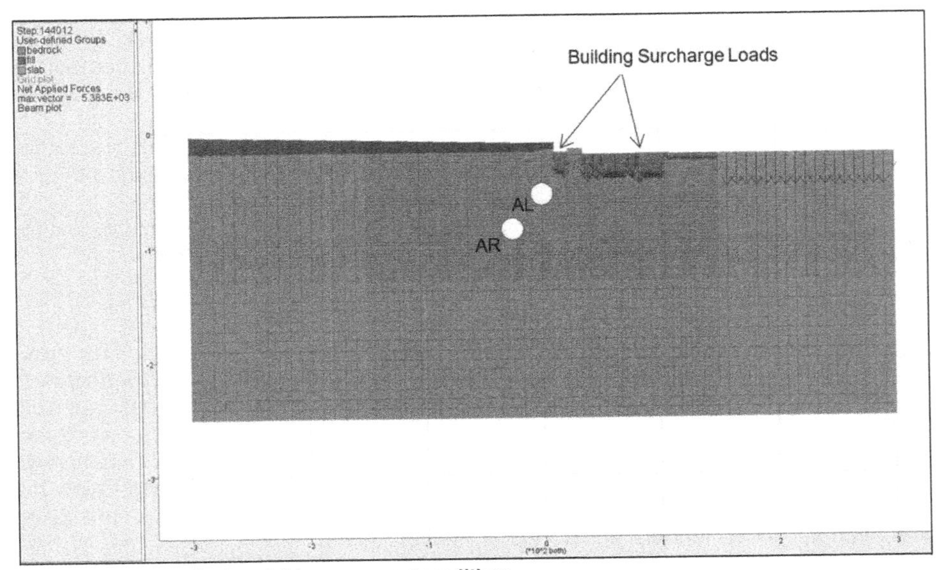


Figure 4. 2D Numerical model for permanent conditions

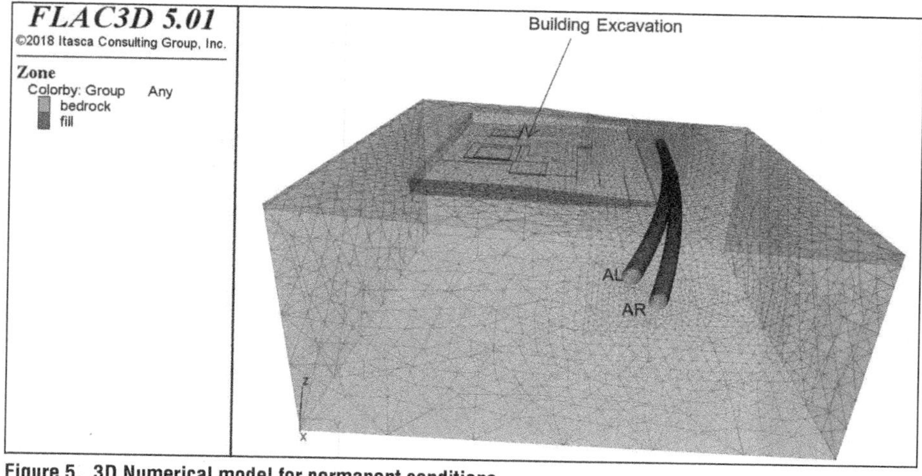


Figure 5. 3D Numerical model for permanent conditions

## 3D Analysis

In the 3D numerical analysis, the entire building foundation plan and associated excavation and surcharge loads were considered. The geological profile shown in Figure 2 was assumed applicable to the entire project site and used in the 3D model. The locations of the AL and AR tunnels relative to the building foundation and curvatures of these tunnels were modeled based on the information provided by JP. Figure 5 shows the 3D model configuration and associated geological conditions.

## Construction Sequence

The general sequence assumed in the numerical analyses was as follows:

1. Set up the initial ground conditions.
2. Excavate tunnels AL (upper) and AR (lower) simultaneously.
3. Install the final linings in both tunnels (Baseline).
4. Excavate to the bottom of foundation levels (Excavation Stage).
5. Apply the building surcharge loads to simulate the construction of buildings (Permanent Condition).

## Modeling Assumptions

To facilitate the analyses, the following key modeling assumptions were made:

- The response of soil and rock to static loading is modeled to be elasto-plastic. The plastic response is governed by the Mohr-Coulomb yield criterion.
- The groundwater table is assumed to be at a depth of 19 feet from the ground surface (GI 2016).
- The initial linings (precast concrete segments) have completely degraded in both tunnels, meaning that there is no structural contribution from the initial lining to the stiffness of the tunnels. This is a conservative assumption in terms of the evaluation of the deformation of the tunnel section in response to the project development. It should be noted that this assumption could result in an overestimation of the thrusts developed in the final linings for the baseline condition (prior to the project construction) but would not

significantly affect the changes of the forces in the final linings caused by the project construction.
- The unconfined compressive strength of the final lining was assumed to be 4,000 psi (Metro 1985).

## Results of Analyses

The results obtained from the 2D and 3D modeling are presented below.

### *Changes of Stresses in Rock around Tunnels*

The horizontal, vertical, and shear stresses—as well as changes of these stresses as a result of the project construction at eight points in rock around the AL and AR tunnels—were calculated from the 2D analysis for three different conditions (Baseline, Excavation Stage, and Permanent Condition) and are summarized in Table 4.

The observations from the calculated changes in stresses are provided below:

- The most critical condition for the project development was expected to be associated with the excavation stage, when the excavation for the foundation

**Table 4. Summary of changes in stresses**

| | | Excavation Stage*,† | | | | Permanent Condition | | | |
|---|---|---|---|---|---|---|---|---|---|
| | | AR Lower Tunnel | | AL Upper Tunnel | | AR Lower Tunnel | | AL Upper Tunnel | |
| Model | Location‡ | Sxx Change, % | Syy Change, % | Sxx Change, % | Syy Change, % | Sxx Change, % | Syy Change, % | Sxx Change, % | Syy Change, % |
| 2D Section J | (1) Crown | −6.1 | −4.4 | −7.3 | −13.1 | −1.6 | 1.9 | −0.1 | 9.8 |
| | (2) | −2.2 | −13.5 | −4.0 | −9.2 | 1.7 | −8.0 | 3.3 | 4.8 |
| | (3) Springline | −5.9 | −2.1 | −10.7 | −1.8 | −0.7 | 1.7 | 4.3 | 2.0 |
| | (4) | −1.8 | −11.6 | −12.3 | −7.9 | 6.3 | −5.5 | 6.6 | 6.9 |
| | (5) Invert | −3.9 | −9.2 | −7.5 | −12.3 | −2.5 | 1.5 | −3.8 | 5.2 |
| | (6) | −4.1 | −11.1 | −11.9 | −9.4 | −1.6 | −8.3 | −3.4 | -9.8 |
| | (7) | −4.8 | −0.7 | −9.6 | −7.8 | 3.4 | 0.4 | 4.2 | 7.4 |
| | (8) | −3.7 | −4.0 | −20.9 | −9.7 | 4.7 | 1.9 | 12.1 | 2.3 |

* Sign convention: Normal stresses (horizontal: Sxx; vertical: Syy) are positive in tension and negative in compression.
† Change in normal stresses is positive for increase and negative for decrease.
‡ Locations of measured changes in stresses are shown below:

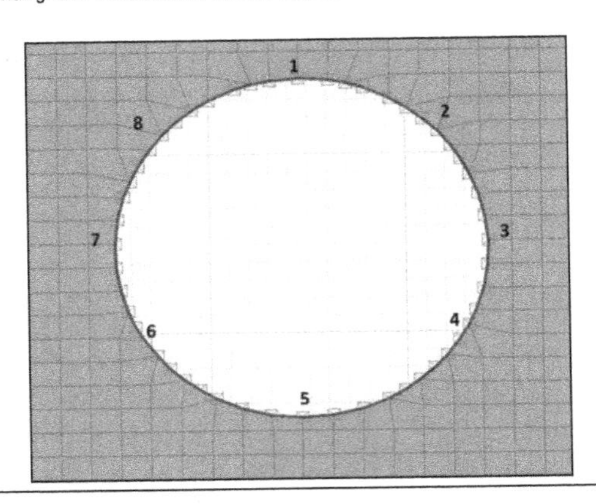

construction would result in a stress relief (unloading) from the soil around the existing adjacent Metro Red Line tunnels. This unloading condition would cause a decrease in the (confining) stresses around the tunnels. The changes were predicted to vary from –2% to –21% (decrease) for the AL (upper) tunnel and from –1% to –14% (decrease) for the AR (lower) tunnel.

- Upon completion of new building(s), the surcharge loads from the new building(s) would create a reloading condition. This reloading would reverse to a certain degree what had occurred during the unloading caused by the excavations, offsetting some of the changes in stresses occurring during the excavation stage. From this perspective, the building surcharge loads were considered a "beneficial" factor, which would result in a reduction of changes in the stresses. The changes were predicted to vary from –4% (decrease) to +12% (increase) for the AL (upper) tunnel and from –8% (decrease) to +6% (increase) for the AR (lower) tunnel.
- Changes in the normal (both horizontal and vertical) stresses calculated at the excavation stage would be higher, generally over 10%, than those associated with the permanent condition, less than or nearly equal to 10%.
- Therefore, percentage changes in stresses caused by the excavation for building foundation construction and permanent building surcharge loads could also be affected by those assumed inputs.
- In the permanent condition for this project, the changes in stresses in soil around the tunnels were calculated to be generally below 10%, except for a few isolated points and the shear stresses, which would not jeopardize the structural capacity of the tunnel final lining.

### *Forces in Tunnel Final Lining*

Forces (bending moments and thrusts) in the circumferential direction were extracted from both 2D and 3D analyses. These forces result in ovaling of a tunnel final lining. Figure 6 shows the moment-thrust interaction diagrams for the final linings in the AL (upper) and AR (lower) tunnels for three different conditions based on the 2D analysis. Figure 7 shows the moment-thrust interaction diagrams for the final linings in the AL (upper) and AR (lower) tunnels, respectively, for three different conditions based on the 3D analysis. Since the steel reinforcement in the final linings is not continuous on both extrados and intrados, two capacity envelopes are plotted in these interaction diagrams. The capacity envelope marked by the solid line represents the section capacity with reinforcement in compression, while the envelope marked by the dashed line represents that with reinforcement in tension.

As shown in Figures 6 and 7, the calculated total forces (including those from full ground loads and groundwater hydrostatic pressure based on the assumed groundwater table) fall within the design capacity envelopes. This suggests that the tunnel final linings were expected to have adequate structural capacity during and following the project building construction.

### *Tunnel Displacements*

The predicted maximum tunnel convergence and transverse horizontal and vertical displacements of the AL and AR tunnels are summarized in Table 5. As indicated, maximum tunnel horizontal and vertical displacements of about 0.45 and 0.50 inch, respectively, were predicted. Note, however, that these magnitudes of displacements were predicted from the 2D analysis. The maximum tunnel horizontal and vertical displacements predicted from the 3D analysis were about 0.12 and 0.35 inch, respectively (see Table 5), much smaller than those predicted from the 2D analysis. The

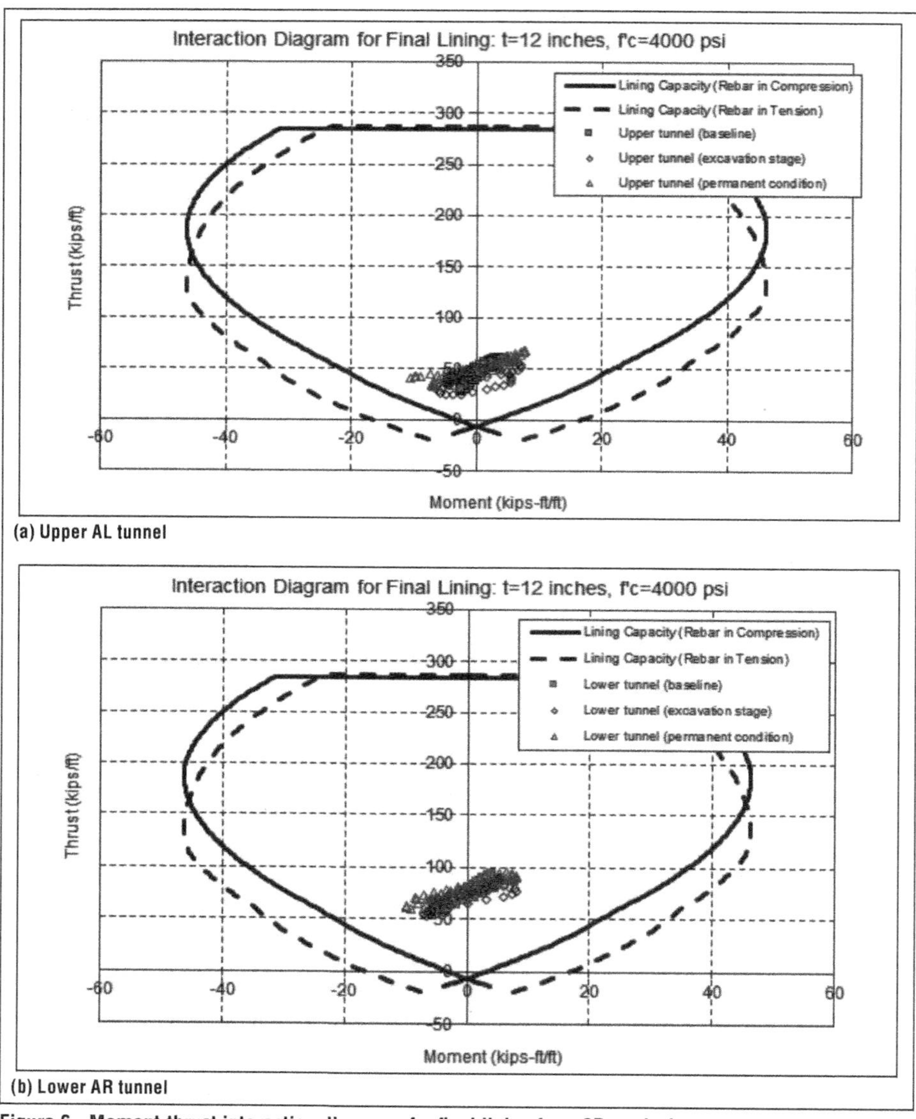


**Figure 6. Moment-thrust interaction diagrams for final lining from 2D analysis**

3D analysis usually gives more realistic prediction of the effect of building construction to adjacent tunnels as it considers the effects of size of construction, variations in the distance between the tunnels and the construction, and boundary conditions. The 2D analysis assumed a plane strain condition and ignored the changes in the distance between the tunnels and the construction along the site length, so it was more conservative.

The profiles of longitudinal track deflections along the tunnels were estimated based on the results from the 3D analysis and are presented in Figures 8 and 9 for the AL (upper) and AR (lower) tunnels, respectively.

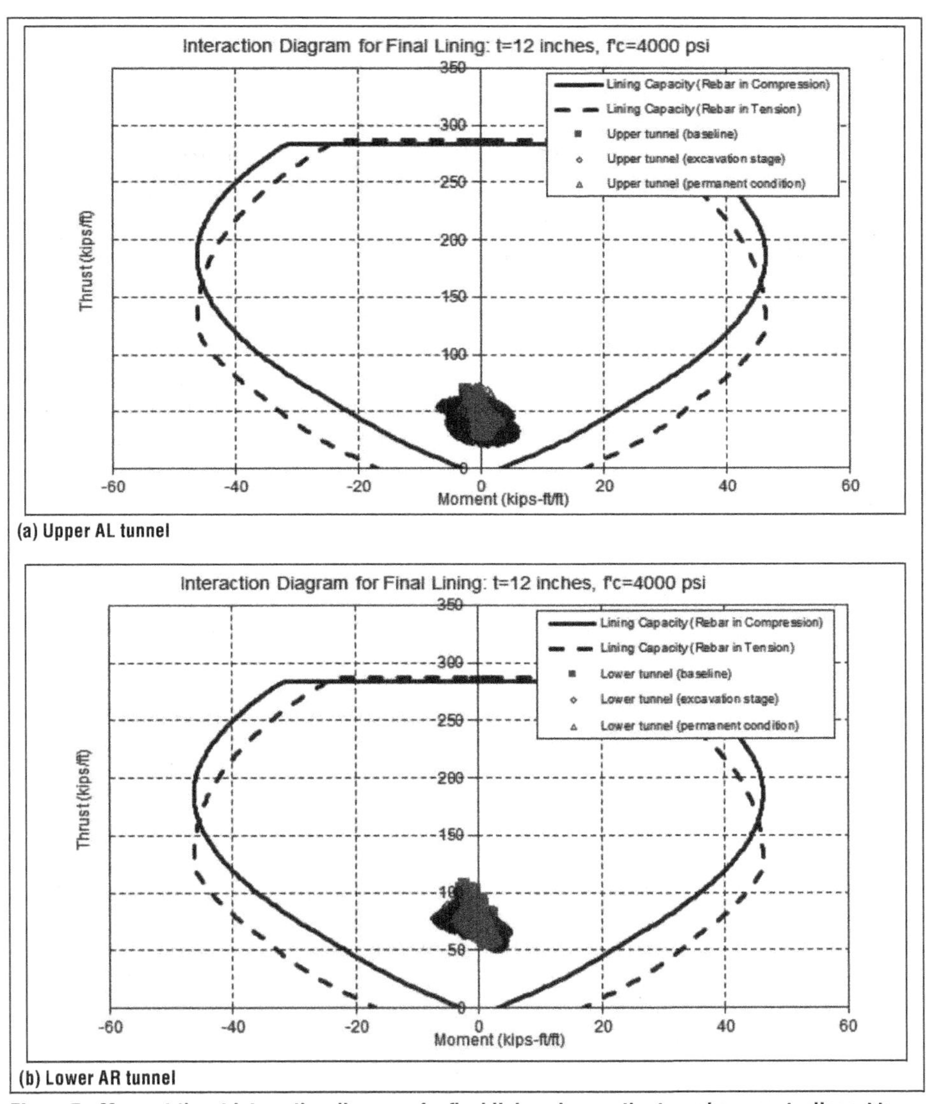


**(a) Upper AL tunnel**

**(b) Lower AR tunnel**

**Figure 7. Moment-thrust interaction diagrams for final lining along entire tunnel segment adjacent to building from 3D analysis**

The results obtained from the numerical analyses indicate that:

- The primary displacements were in the vertical direction, resulting in a vertical ovaling as a result of the unloading associated with the excavation for building foundation construction.
- The maximum tunnel lining distortions caused by the construction were predicted to range from approximately less than 0.01% to 0.03% (see Table 5), which are below the limit of 0.40%, as listed in Table 1.
- The calculated maximum track or rail horizontal displacements were expected to vary from 0.02 to 0.12 inch from the 3D analysis. These are below the

**Table 5. Maximum tunnel and track displacements**

| Tunnel | Condition* | Convergence, in. | Maximum Diameter Change, % | Track or Rail Horizontal Displacement, in.† | Track or Rail Vertical Displacement, in.†,§ | Cross Level, in. |
|---|---|---|---|---|---|---|
| AL (Upper) | Excavation Stage | 0.00 | 0.00 | 0.32 (0.06) | 0.50 (0.35) | 0.07 |
| | Permanent | 0.04 | 0.02 | 0.16 (0.03) | −0.09 (0.06) | 0.01 |
| AR (Lower) | Excavation Stage | 0.03 | 0.01 | 0.45 (0.12) | 0.29 (0.26) | 0.04 |
| | Permanent | 0.06 | 0.03 | 0.23 (0.02) | −0.06 (0.04) | 0.01 |

* Refer to Section *Numerical Modeling and Results* for definition of these conditions modeled.
† Values in parentheses are the displacements calculated from 3D analysis.
§ Negative values indicate downward vertical displacements.

action level of ±0.375 inch and maximum level of ±0.5 inch for tracks along a curved alignment over a 62-foot-long chord, as listed in Table 2. As indicated in Figures 8 and 9, the upper end of this estimated deflection would occur in response to the excavation and the movements would reduce following the building construction.

- The calculated maximum track or rail vertical displacements were calculated to vary from about 0.04 to 0.35 inch, which are below the action level of ±0.75 inch and well below the maximum level of ±1.0 inch, as listed in Table 2. As indicated in Figures 8 and 9, the upper end of this estimated deflection would occur in response to the excavation and the movements would reduce following the building construction.
- The calculated maximum track cross level changes were predicted to vary from 0.01 to 0.07 inch. These are within the action level of ±0.75 inch and well below the maximum level of ±1.0 inch for tracks along a curved alignment, as listed in Table 2. As indicated in Table 5, the upper end of these estimated track cross level changes would occur in response to the excavation and the movements would reduce following the building construction.

## COMPARISON OF MODELING RESULTS AND OBSERVED RESPONSES

Because of the adjacent construction over the duration of the project construction, a geotechnical instrumentation and monitoring program was developed and implemented to monitor the ground movements and tunnel displacements using inclinometers and deep settlement points (single point with Borrow-type anchor). The observed ground and tunnel displacements are compared to Metro's allowable tunnel displacement criteria specified for the project (see Table 2) to ensure that the responses of the tunnel structures and rail tracks to the construction activities can satisfy Metro's criteria. A comparison was also made to those predicted by the numerical analyses and served as a model validation.

Figure 10 shows two locations of geotechnical instrumentation installed adjacent to the Red Line Tunnels. Figure 11 shows the cross sections of inclinometers and deep settlement points at these two monitoring locations with respect to the adjacent AL tunnel. Three levels of monitoring trigger levels and construction responses that would be implemented are listed in Table 6.

The measured horizontal and vertical displacements from two monitoring locations are presented in Figures 12 and 13, respectively. A summary of actual tunnel performance compared to the predicted is given as follows:

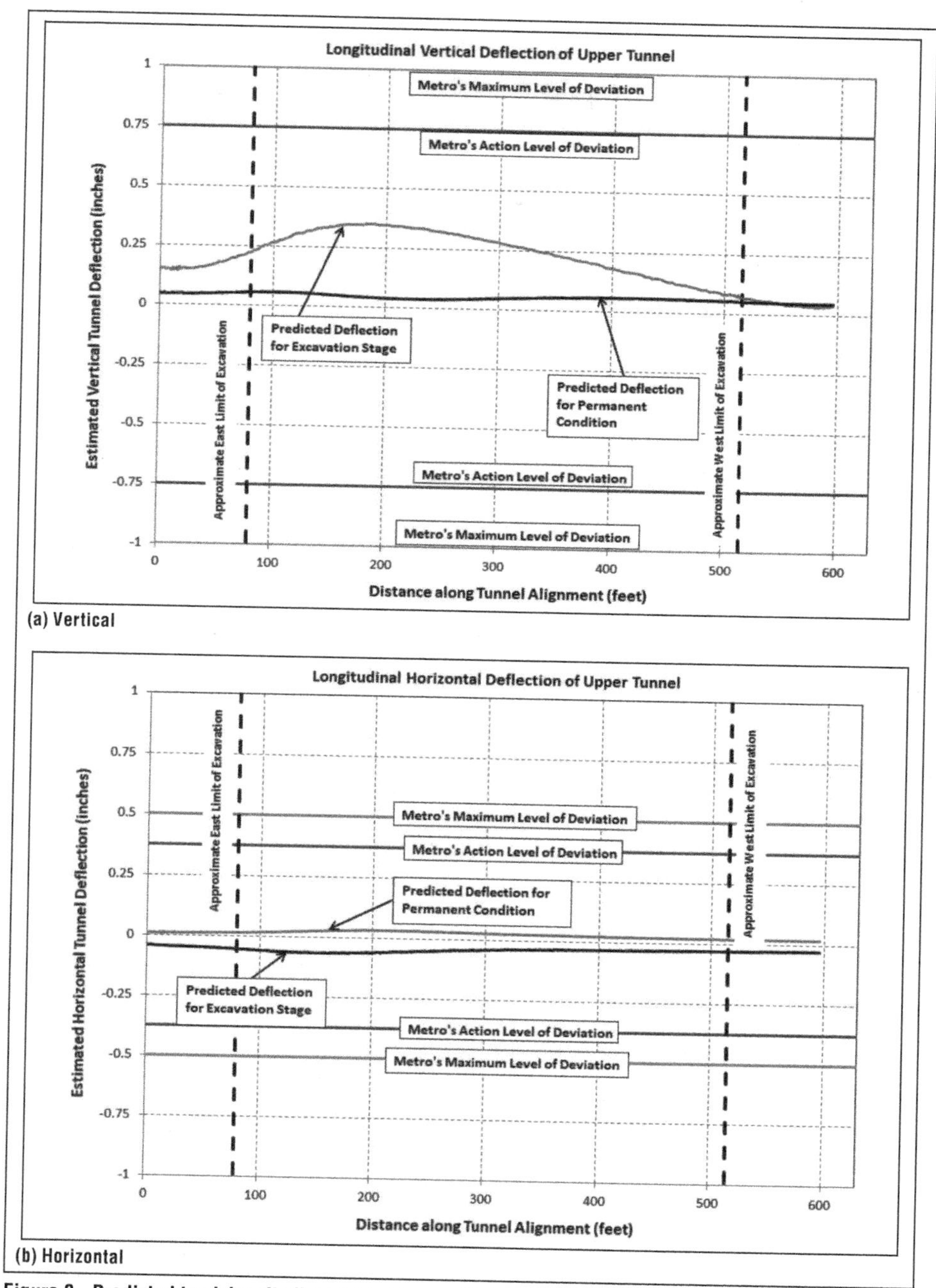


Figure 8. Predicted track longitudinal deflections of AL (upper) tunnel

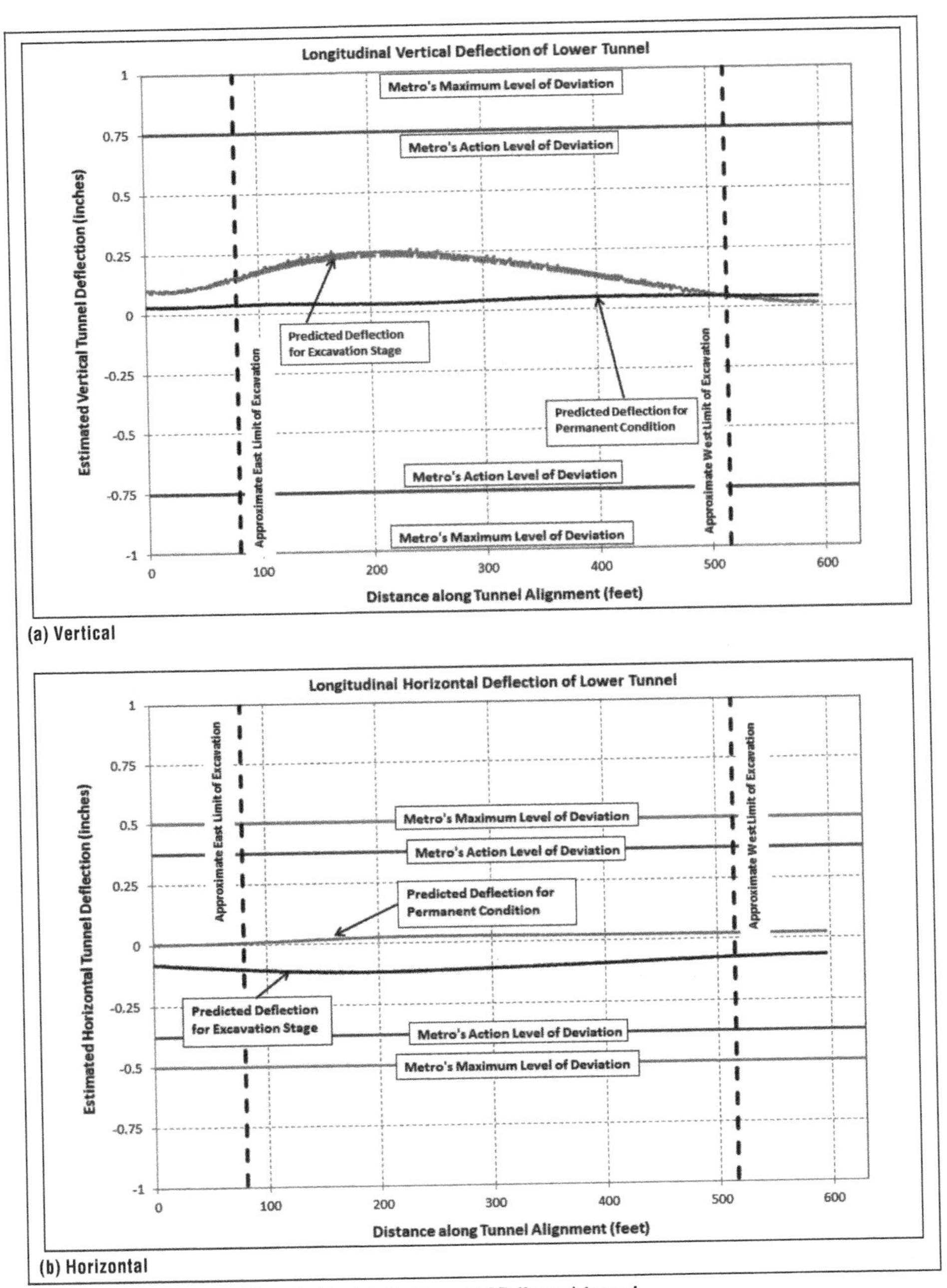


**Figure 9. Predicted track longitudinal deflections of AR (lower) tunnel**

**Table 6. Construction response trigger levels**

| Instrument Type | Movement Direction | Level 1: Construction Threshold Value | Level 2: Construction Response Value | Level 3: Construction Shutdown Value |
|---|---|---|---|---|
| Inclinometers | Horizontal | 0.375 inch horizontal change | 0.4375 inch horizontal change | 0.5 inch horizontal change |
| Settlement Points | Vertical | 0.75 inch vertical change | 0.875 inch vertical change | 1.0 inch vertical change |

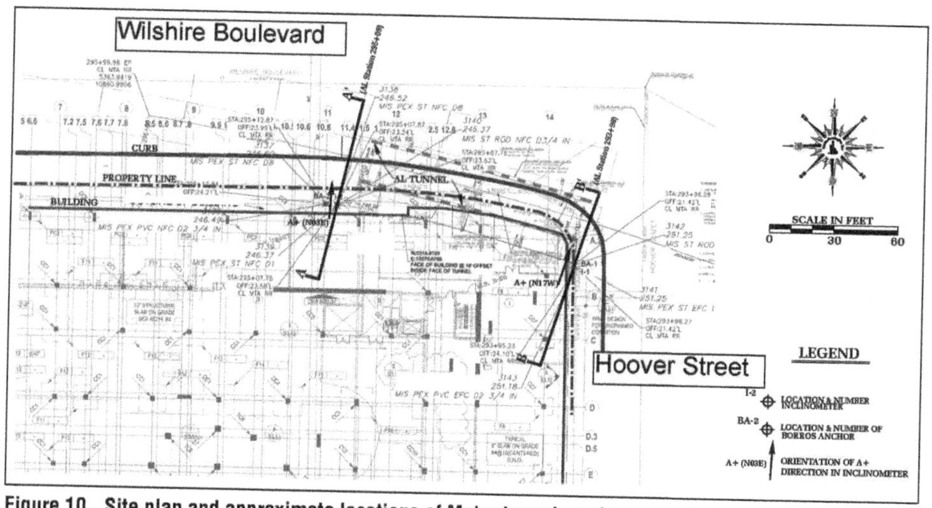


Figure 10. Site plan and approximate locations of Metro tunnels and two monitoring points

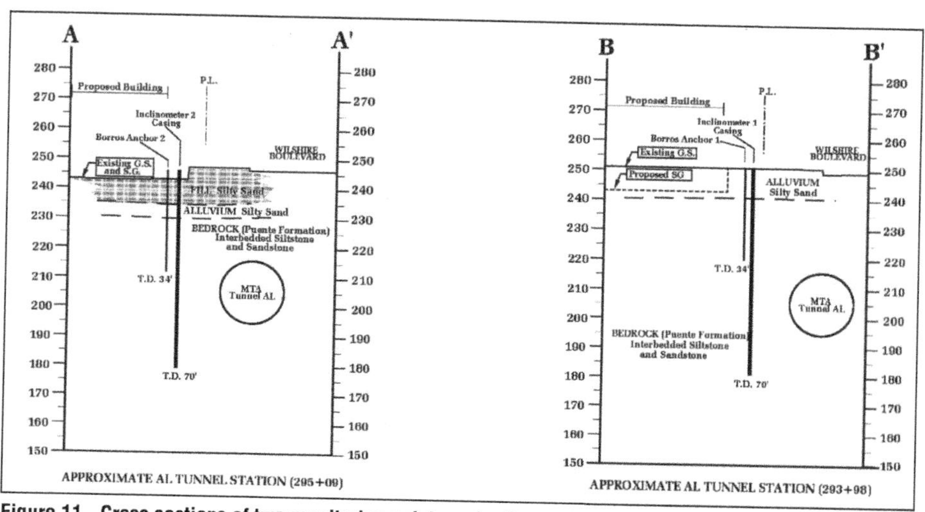


Figure 11. Cross sections of two monitoring points and adjacent Metro tunnel AL

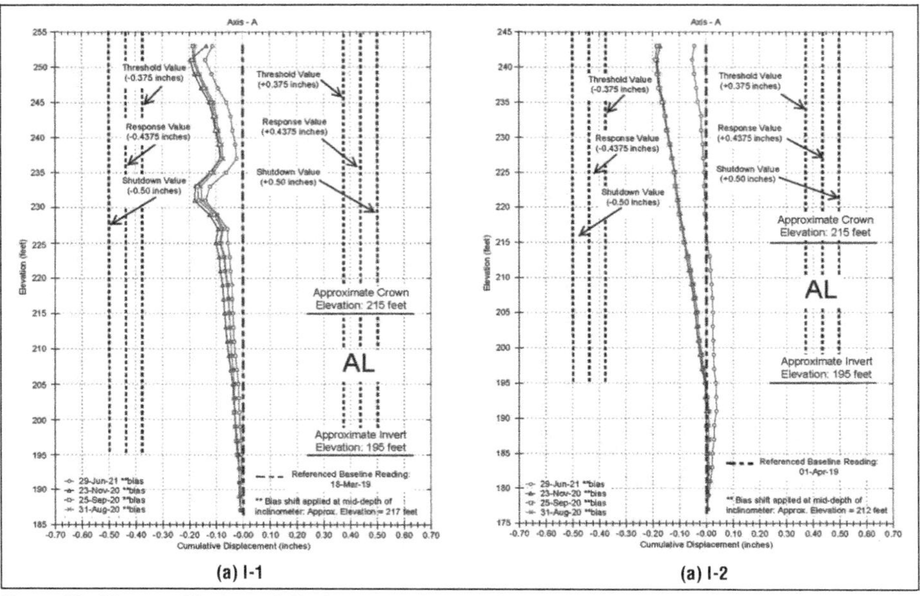


(a) I-1 (a) I-2

**Figure 12. Measured horizontal displacements from inclinometers**

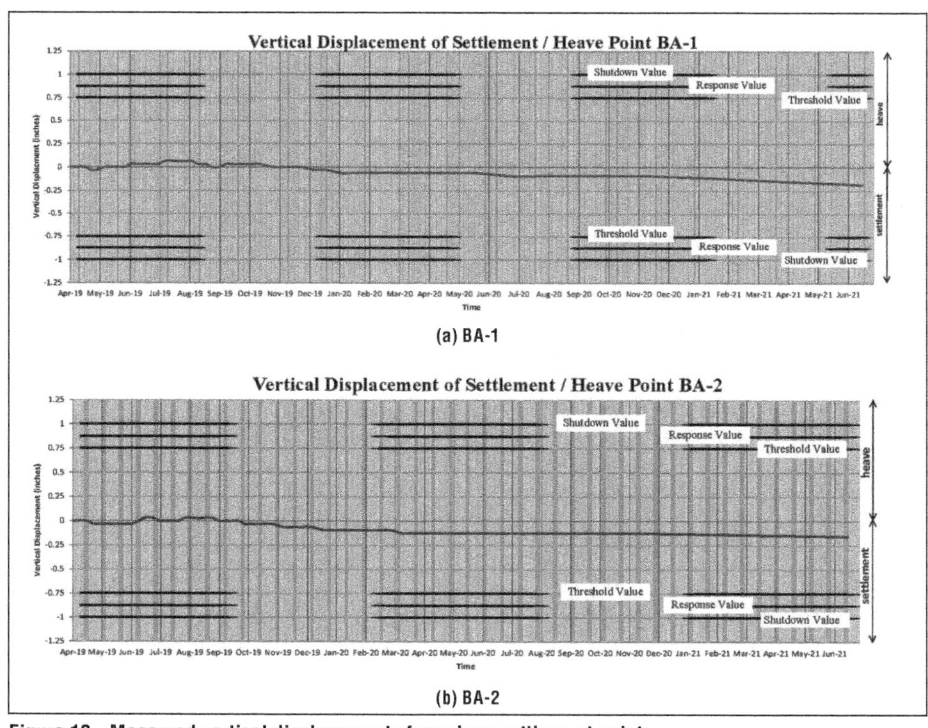


(a) BA-1

(b) BA-2

**Figure 13. Measured vertical displacements from deep settlement points**

- The measured horizontal displacements from both inclinometers I-1 and I-2 were well below the Level 1 construction threshold value of 0.375 inch (Table 6). They were also approximately at the predicted maximum horizontal displacements at the AL tunnel invert elevation of 0.06 inch and below 0.32 inch from 3D and 2D analyses, respectively.
- The measured vertical displacements from both deep settlement points BA-1 and BA-2 were well below the Level 1 construction threshold value of 0.75 inch (Table 6). They were also well below the predicted maximum values of 0.35 inch and 0.50 inch from 3D and 2D analyses, respectively.
- Judging by the minimal ground movements observed in close proximity (see above for approximate distances) to the Metro tunnels, the impact of construction to the tunnels was considered to be minimal.

## CONCLUSIONS

Construction of the Kurve development was started in March 2019 and completed in June 2021. The effects of the building construction on the existing adjacent Metro Red Line tunnels were evaluated by numerical analyses and monitored by geotechnical instrumentation during the course of construction. The key findings from these analyses and monitoring are:

- In the permanent condition for this project, the changes in stresses in soil around the tunnels were calculated to be generally below 10%, except for a few isolated points and for the shear stresses (which would not jeopardize the structural capacity of the tunnel final linings).
- The construction was predicted to induce displacements in the existing tunnel structures in addition to what they had already experienced. The maximum displacement of the tunnels was predicted to be approximately 0.12 and 0.35 inch in the horizontal and vertical directions, respectively, based on the 3D analysis using the site plan, excavation geometries, and building surcharge loads. The maximum track or rail displacements in both tunnels were predicted to be below the limit specified by the Metro criteria.
- The construction was predicted to induce additional forces in the existing tunnel final linings. The total combined ovaling forces developed in the final linings of the tunnels were predicted to be well within the design structural capacity of the linings.
- The distortion of the tunnel lining in response to excavation and building construction was well below the limits adopted by Metro. These displacements and deformation of the tunnel linings might cause minor cracking or widening of existing cracks in the final linings and were not expected to affect the functionality of the tunnel linings or the HDPE membrane behind the final concrete lining.
- Overall, the total permanent horizontal and vertical displacements at the tunnel crown elevation caused by the building construction are small, well below the specified threshold of 0.375 and 0.75 inch for the horizontal and vertical displacements, respectively. These observed displacements are also generally consistent with those predicted by the numerical analyses (MJ 2019). Based on these observed levels of ground movements, it can be confirmed that the effects of the building construction on the adjacent Metro tunnels were expected to be insignificant and the tunnel final linings were expected to be structurally sound, well within their design capacity when subjected to normal operational conditions.

## ACKNOWLEDGMENTS

The authors would like to acknowledge the many contributions and support of Jamison Properties, LP; Geotechnologies Inc.; Delve Underground; and LA Metro for this project. The contents of this paper reflect the views of the authors, who are responsible for the facts and accuracy of the data presented herein, and do not necessarily reflect the official views or policies of either Jamison Properties, LP or LA Metro. This paper does not constitute a standard, specification, or regulation.

## REFERENCES

American Concrete Institute (ACI). 2019. ACI 218—Building Code Requirements and Commentary. Farmington Hills, MI: ACI.

Geotechnologies, Inc. (GI). 2016. *Geotechnical Engineering Investigation*. Proposed Mixed-use Development 2900 Wilshire Boulevard, Los Angeles, CA, October 13, 2016.

Geotechnologies, Inc. (GI). 2019. *Geotechnical Parameters,* Proposed Mixed-use Development 2900 Wilshire Boulevard, Los Angeles, CA, January 25, 2019.

Itasca. 2016. Fast Lagrangian Analysis of Continua (FLAC), Version 8.0. Minneapolis, MN:. Itasca Consulting Group.

Itasca. 2012. Fast Lagrangian Analysis of Continua in 3 Dimensions (FLAC3D), Version 5.0. Minneapolis, MN: Itasca Consulting Group.

Los Angeles Metropolitan Transportation Authority (Metro).1985. Metro Red Line: Structural Standard Tunnel Subway—Cast in Place Concrete Lining Typical Section and Details, Sheet 184. As-built Records Received from Metro, January 2019.

Los Angeles Metropolitan Transportation Authority (Metro). 1990. Metro Red Line: Right of Way Map, Contract B201, Sheets 20 and 21, As-built Records Received from Metro, January 2019.

Los Angeles Metropolitan Transportation Authority (Metro). 2017. Rail Design Criteria, Section 5 Structural/Geotechnical, Revision 12.

Los Angeles Metropolitan Transportation Authority (Metro). 2019a. Geotechnical Design Summary Report for Construction Contract B201 Tunnels, February 14, 2019.

Los Angeles Metropolitan Transportation Authority (Metro). 2019b. Metro Performance Criteria and Tolerances for Existing Track Works, February 06, 2019.

Mahar, J. 1994. Field Observations and Evaluation of Final Lining Conditions in L.A. Metro Red Line Segment 1 Tunnels for Tunnel Review Panel appointed by Los Angeles County Metropolitan Transportation Authority, March 1994.

McMillen Jacobs Associates (MJ). 2019. Evaluation of Potential Effects of 2900 Wilshire Blvd Development Construction on Adjacent Metro Tunnels (Final). Technical Memorandum from A. Beyabanaki et al. to G. Lee and A. Park, April 3, 2019.

# SVCW Gravity Pipeline—Ground Movements During Construction

**Jon Hurt** ▪ Arup
**Eric Sekulski** ▪ Arup
**Nik Sokol** ▪ Arup
**Shrinidhi Vijayakumar** ▪ Arup
**Phaidra Campbell** ▪ JCK Underground
**Leo Weiman-Benitez** ▪ Barnard Bessac Joint Venture

## ABSTRACT

The Silicon Valley Clean Water (SVCW) Gravity Pipeline (GP) is the first Progressive Design Build (PDB) tunnel completed in North America. As part of the Regional Environmental Sewer Conveyance Upgrade (RESCU) Program, the tunnel is a key component to replacing and rehabilitating components of the existing conveyance system in San Mateo County, California. The GP consists of a large diameter Fiberglass Reinforced Polymer Mortar (FRPM) pipe installed inside a 4.1 m (13.5 ft) inside diameter, 5.3 km (3.3 mile) long precast concrete segmentally lined tunnel. The tunnel was constructed using a 4.6 m (15.2 ft) diameter Earth Pressure Balance (EPB) Tunnel Boring Machine (TBM) through heterogenous silts, clays and sands. Construction included two separate TBM launches from a centrally located access shaft, and a tunnel alignment below but in close proximity to very soft Young Bay Mud. Settlement criteria and mitigation strategies focused on protection of an existing 1.42 m (56 inch) diameter force main and adjoining structures along the alignment. This paper discusses the results of the theoretical settlement analysis versus the actual observed ground movements, while also presenting mitigation measures and adjustments made during tunneling to reduce the risks of adverse settlement.

## PROJECT BACKGROUND AND SETTING

### Project Purpose

SVCW is completing the RESCU Program through the delivery of a series of projects that will improve the reliability of SVCW's wastewater conveyance and treatment system. The $218M GP is a key project in the program. It is being delivered using PDB, among the first for the tunnel industry.

The RESCU Program objectives are:

- Replace the existing wastewater infrastructure and construct other improvements to ensure reliable operation of the overall wastewater conveyance system in accordance with San Francisco Bay Regional Water Quality Control Board (RWQCB) permit conditions.
- Reduce the likelihood of spills and discharges of untreated sewage to the surrounding environment.
- Implement a Program that minimizes adverse environmental effects, adverse impacts to public health and private property owners, utility interference and disruption during construction, and short-term and long-term costs.

- Improve SVCW's Wastewater Treatment Plant process reliability and increase operational readiness.
- Meet future regulatory requirements imposed by the RWQCB.

RESCU consists of three key components: the GP project, the Front-of-Plant (FoP) project, and the Pump Stations Improvements (PSI) project. The GP replaces the 1.2 m (48 inch) and 1.4 m (54 inch) diameter force main segments between Bair Island and the SVCW treatment plant. The FoP project includes a Surge and Flow Splitter (SFS) shaft, a Receiving Lift Station, a Headworks, and other support facilities, all located adjacent to the current SVCW treatment plant. The PSI project includes the rehabilitation of the existing Menlo Park and Belmont Pump Stations and the replacement of the Redwood City Pump Station. With the completion of the GP and associated drop structures, the function of the existing San Carlos Pump Station is shifted from housing the force main pumps to housing the connecting pipework between the existing San Carlos gravity sewer and the GP.

The project has been designed to provide long-term wastewater resilience for Silicon Valley, with a 100-year design life for the GP, which is located just six kilometers (four miles) from the San Andreas fault.

The GP project was contracted to the Barnard-Bessac Joint Venture (BBJV), with Arup as Lead Designer. SVCW, BBJV, and Arup co-located at site offices to collaboratively develop seismically resilient designs for TBM and Sequential Excavation Method (SEM) tunnels, temporary sheet pile, secant pile and slurry diaphragm wall Support-of-Excavation (SOE) construction shafts, vortex and baffle sewage drop shafts, temporary diversions, and connecting pipework.

Most of the construction is being performed adjacent to the San Carlos Airport and along the environmentally sensitive San Francisco Bay margin.

## Project Configuration

Key features of the GP project indicated in include:

- Airport Access Shaft (AAS): 17.7 m (58 ft) diameter, 15.8 m (52 ft) deep slurry diaphragm wall SOE temporary TBM launch shaft;
- Twin, 5.5 m (18 ft) tall × 5.2 m (17 ft) wide excavations mined via SEM protruding 12.8 m (42 ft) horizontally out from the bottom of the AAS, in the direction of the future tunnel drives;
- Gravity Pipeline: A 5.3 km (3.3-mile) long, 4.6 m (15.2 ft) outside diameter (OD), 4.1 m (13.5 ft) inside diameter (ID), precast concrete segment lined tunnel with 3.7 km (2.3 miles) of 3.4 m (11 ft) ID and 1.6 km (1.0 mile) of 3.0 m (10 ft) ID FRPM pipe installed;
- Bair Island Shaft and Inlet: 18.3 m (60 ft) long, 9.8 m (32 ft) wide, 13.4 m (44 ft) deep rectangular sheet pile SOE retrieval shaft at the end of Tunnel Drive 1 with a permanent 1.2 m (4 ft) diameter FRPM hydraulic vortex drop, 3 m (10 ft) diameter FRPM access structure and associated connecting pipework to connect with the existing force main;
- San Carlos Shaft and Inlet: 4.3 m (14 ft) diameter ID (oval), 14.6 m (48 ft) deep shaft at the end of Tunnel Drive 1 with a permanent 3.4 m (11 ft) diameter FRPM access / hydraulic baffle drop structure and associated connecting pipework to connect with the existing gravity collection system; and
- 1.8 m (6 ft) ID connecting pipe between San Carlos drop shaft and GP.

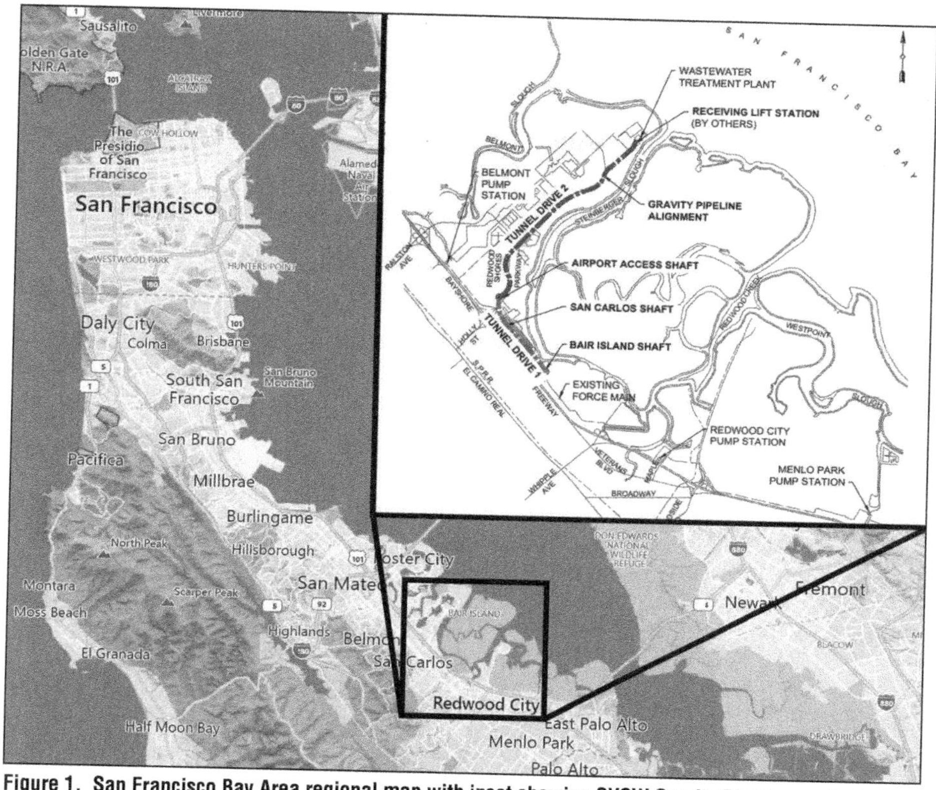


**Figure 1. San Francisco Bay Area regional map with inset showing SVCW Gravity Pipeline project alignment and key features**

The TBM was retrieved at the end of Tunnel Drive 2 via the SFS/Reception shaft, which was built as part of the FoP project, concurrently with the GP project.

The vertical alignment of the tunnel was based on several considerations:

- Placement of tunnel within favorable ground for tunneling, maintaining minimum 1 m (3 ft) clearance below low points in the very soft Young Bay Mud.
- At the Pulgas Creek water crossing, maintaining minimum one tunnel diameter (4.8 m) clearance between the tunnel crown and the creek bed determined via bathymetry survey.
- Minimum grade required to achieve systemwide gravity flow hydraulic requirements for sediment transport during full range of dry weather and wet weather flow conditions.
- Maximum grades allowable using proposed tunneling methods as advised by BBJV.

## Construction Approach

The construction methods and sequence of the GP are fully described in detail in Weiman-Benitez, et al (2023). A bulleted list of major features of work and completion dates, along with select photos have been included below.

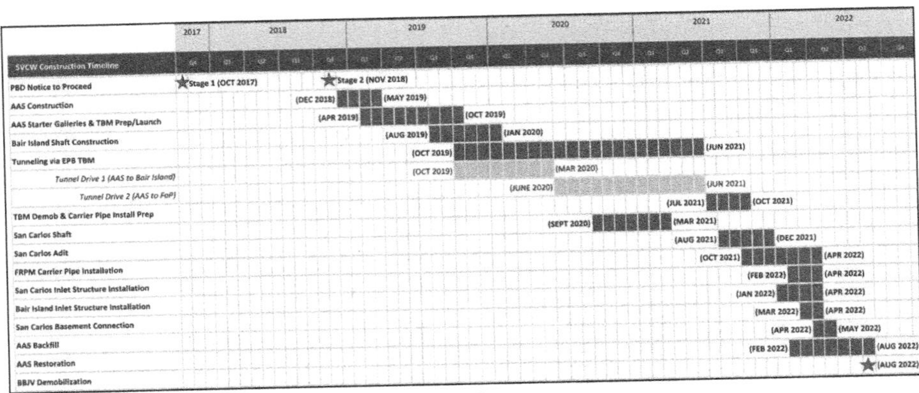


**Figure 2. SVCW Gravity Pipeline construction schedule**

- PBD Notice To Proceed: October 2017 (Stage 1) and November 2018 (Stage 2)
- AAS completed: May 2019
- Bair Island Shaft completed: January 2020
- Tunneling via EPB TBM:
  - Tunnel Drive 1 (AAS to Bair Island) Hole-through: March 2020;
  - Tunnel Drive 2 (AAS to Receiving Lift Station at FoP) Hole-through: June 2021
- San Carlos Shaft completed: March 2021
- San Carlos Adit completed: December 2021
- FRPM Carrier Pipe installation completed: April 2022
- San Carlos and Bair Island Inlet Structures installation completed: April 2022
- San Carlos Basement Connection completed: April 2022
- AAS Backfill completed: May 2022
- AAS restoration & BBJV demobilization completed: August 2022

Figure 3 and Figure 4 show the TBM during assembly in the AAS and the installed PCTL after completion of Tunnel Drive 1.

## GROUND CONDITIONS AND INVESTIGATIONS

The conditions of San Francisco Bay margin soils are notoriously challenging for underground construction. Arup planned and executed ground investigations to supplement existing Owner-supplied geotechnical data. These additional investigations targeted key data gaps, including shear wave velocity testing for seismic design dilatometer testing for in-situ soil stiffness, new piezometers for groundwater levels and hydraulic conductivity testing, confirming bottom elevation of the Young Bay Mud in critical locations, and collecting soil samples for abrasion and soil conditioning testing for understanding TBM performance. Existing and supplemental data was then compiled to form the project Final Geotechnical Data Report and Geotechnical Baseline Report.

### Geologic Setting

The project site is located within the Coast Ranges geomorphic province on the San Francisco Peninsula. The major structural element of the San Francisco Bay Region is the San Andreas fault system, which stretches over an area approximately 150 km (90 miles) wide from the San Gregorio fault along the coastline to the Coast Ranges Central Valley blind thrust at the western edge of the Central Valley. The San Andreas

Figure 3. Tunnel boring machine during assembly prior to launch for Tunnel Drive 2

Figure 4. Precast concrete tunnel lining (PCTL) after completion of Tunnel Drive 1

fault is the dominant fault in the system, spanning nearly the full length of the state of California. The site is situated on the San Francisco Bay structural block located to the east of the San Andreas fault (Nilsen and Brabb, 1979).

The tunnel alignment underlies historically reclaimed land from Inner Bair Island through Redwood Shores peninsula and is underlain by Quaternary Alluvium and marine deposits. The primary geologic units along the project alignment include Artificial Fill (Fill), Young Bay Mud (YBM), Upper Layered Sediments (ULS), and Old Bay Deposits (OBD). The ULS and OBD were not categorically differentiated in the geologic profiling developed for the tunnel (and presented later in Figure 9).

A geologic profile of the project alignment is presented.

- Artificial Fill (Fill) was encountered in thicknesses ranging from 0.5 m (1.6 ft) to over 2.5 m (8 ft) thick. Fill was observed to be predominantly silt and clay with varying amounts of sand and gravel. Fill can be highly variable and may include organic material, cobbles, and other man-made debris.
- Young Bay Mud (YBM) was encountered below the Fill in thicknesses typically ranging from 0.5 m (1.6 ft) (along San Carlos Airport) up to 14 m (46 ft) at the east end of the alignment near the receiving shaft. YBM comprises very soft to soft, highly compressible fat clays, containing variable amounts of shell fragments and organics. The YBM is often described as having strong organic or sulfuric odor.
- Underlying the YBM are a sequence of alluvial and marine sediments called the Upper Layered Sediments (ULS). The vertical alignment of the tunnel was set within the ULS and below the YBM. The ULS comprise alternating layers of lean to fat clays, sandy clay, silts, clayey sand, and silty sand. Within the tunnel zones, the ULS strata are predominantly clays that vary from medium stiff to stiff. The sand layers present within the ULS are often inferred to be of limited vertical and horizontal extent; however, significant and potentially interconnected sand deposits are observed in the geotechnical investigations at some locations.
- Old Bay Deposits (OBD) underlie the ULS and the tunnel zone and are often encountered as a more uniform marine deposit comprised of fat clay or lean clay with (typically) fewer sand layers and occasional shells fragments.

Since the tunnel alignment underlies developed land historically reclaimed by placing fill above mud flats and tidal marshes, groundwater condition can vary significantly, between ground surface and the bottom of the Fill. Piezometric pressures within the deeper sand layers have been observed to fluctuate with the San Francisco Bay tides.

The area demonstrates signs of ongoing consolidation settlement, which is believed to be primarily occurring in the YBM. The YBM is generally expected to be normally consolidated; however, recent fill placement may have instigated consolidation that is still ongoing.

## SETTLEMENT IMPACT ANALYSIS AND PREDICTION

Numerous factors influence ground movements associated with deep excavations and tunnels. These factors include but are not limited to subsurface conditions, loads and surcharges, support of excavation, excavation sequencing, construction means and methods, and quality of construction.

The software program Xdisp (Oasys, 2019) was used to evaluate building and utility deformations adjacent to the shafts and tunnel. The program follows procedures outlined in CIRIA C580 (2003) for ground movements beside embedded retaining walls and follows Gaussian distribution for ground movements due to tunnel excavation.

The Gaussian distribution developed using the O'Reilly and New (1982) method was used as input in Xdisp for ground movements above the tunnels. The tunnels are modeled as cylindrical excavations in soil. The program defines the settlement profile at the surface or sub-surface based on a user defined volume loss (VL) during tunnel excavation.

The O'Reilly & New (1982) method assumes that the ground settlements follow a Gaussian Curve shaped trough centered above the tunnel alignment, as shown in Figure 5. The trough width is defined as a function of the parameter "$i$," where $i = kz_0$, and $k$ and $z_0$ are the soil type factor and the depth to tunnel springline, respectively. The maximum settlement ($S_{max}$) is calculated from the percentage volume loss (VL), the trough parameter ($i$), and the tunnel geometry, TBM outer diameter.

The impacts of settlement on buildings and utilities was assessed following the methodology described by ITA-AITES WG2 (2007), which is based on the work of Burland (1995) and Mair et al. (1996). This uses a three-stage approach—(i) preliminary assessment to determine the Zone of Influence (ZOI), (ii) second stage assessment to identify and buildings and utilities within the ZOI that exceed the project damage criteria based on a conservation analysis using "green field" ground movements, and (iii) a detailed evaluation for any buildings and utilities identified in the second stage.

For the preliminary and second stage assessments, an upper bound VL of 1% was assumed - a standard industry benchmark value considered readily achievable with

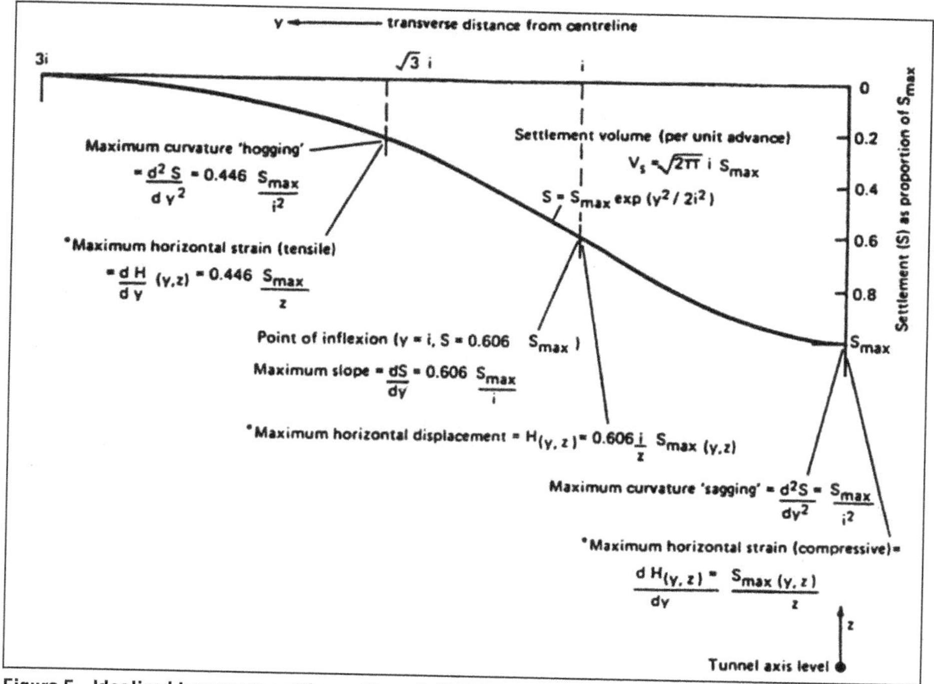


Figure 5. Idealized transverse settlement trough (ITA/AITES WG2 (2007))

modern EPB TBM equipment and methods. The predominant soil type overlying the tunnel alignment is clay, typically stiff clay except near the SVCW wastewater treatment plant where the majority of overlying soil is soft clay. The soil type factor ($k$) varies from 0.5 for stiff clay to 0.7 for soft clay. Lower $k$ values are typically attributed to granular soil. For the same volume loss, the higher soil type factor for soft clay produces in a wider but flatter settlement trough. Since building and utility damage occurs as the result of maximum curvature from differential settlement, a constant soil type factor of $k = 0.5$ (deeper trough) was adopted for analysis purposes.

The first stage in the settlement analysis was to define the ZOI. Transverse settlement profiles were calculated by Xdisp along the GP alignment. A 6 mm (0.25 inch) settlement criteria was selected for defining the ZOI, with the distance from the tunnel centerline where the predicted settlement reaches 6 mm settlement criteria determining the width of the ZOI. An offset of 7.6 m (25 ft) from each side of the tunnel centerline was adopted for the ZOI to encapsulate all settlement profiles along the alignment.

The next stage of the analysis was to perform a screening assessment for building and utility damage. With the alignment primarily running under roads, only two residential buildings and the pump station at San Carlos Shaft were within the ZOI. The building damage assessment methods built into XDisp were used, based on Burland et al (1977) and Burland (1995) and using the concept of limiting tensile strain to study the onset of cracking in simple weightless elastic beams undergoing sagging and hogging modes of deformation. The two residential buildings identified within the ZOI were predicted to be within the Category Damage 0, "Negligible." The existing pump station at San Carlos Shaft was identified as Category Damage 2, "Slight"; however, that included construction impacts from both tunnel and shaft excavations—further analysis of the secant pile shaft SOE concluded the pump station would experience less settlement.

There are a significant number of utilities above the tunnel alignment. Utility damage assessments were performed by assessing the extent of rotation of joints, pullout of joints, and axial and flexural strains. Utilities were modeled at their specific locations, with appropriate dimensions and acceptance criteria for rotation, pullout and strain. Xdisp uses the analysis data at joints and reports whether the rotations, pullouts and strains calculated at joints satisfy their respective limiting criteria established by CIRIA Project Report 30 (1996). The results indicated that utilities subject to tunnel-induced settlement were not expected to exceed the damage criteria. The condition of the existing SVCW force main was known to be fragile (the reason for the project) and there was a history of leakage and repairs. The existing SSFM was not quantitatively assessed as it was not possible to establish specific damage criteria to reflect the current condition of the pipe and joints, which could not be inspected.

## CONSTRUCTION OBSERVATIONS

### Instrumentation

The aim of the construction phase instrumentation program was to generally follow industry standards for urban tunneling projects and provide a cost-effective program that would serve three key purposes:

- Verify ground behavior is consistent with design assumptions and allowable damage criteria limits.
- Document pre-existing conditions and construction-induced ground movement to help protect the Owner and Contractor from spurious third-party claims.

- Provide feedback to calibrate the TBM operation to the actual ground response observed, as soon as possible following launch of the TBM. This calibration would allow the contractor the opportunity to modify mining operations and lining installation to minimize ground settlement, maximizing the protection of overlying utilities and structures.

The GP project used the following instrumentation types:

- Surface Settlement Points (SSP): A PK-Nail survey point installed on pavements and intended to monitor changes to surface elevation in response to tunnel and shaft construction.
- Subsurface Monitoring Points (SMP): similar monitoring principle to SSPs, though consists of a rigid rod embedded in native soil in unpaved areas.
- Subsurface Settlement Points (SSMP): identical to SMPs, though installed within native soil below paved areas, accessible via flush-mounted traffic-rated well-box.
- Building Monitoring Points (BMP): A survey point installed on buildings and exposed utilities to monitor changes to building elevation in response to tunnel and shaft construction.
- Utility Monitoring Points (UMP): survey point installed on specific utilities/ at key locations to monitor utility response to tunnel and shaft construction.
- Multi-Point Borehole Extensometers (MPBX): rod extensometers installed in boreholes to monitor ground displacements at various depths above tunnel, primarily for use in close proximity to TBM launch areas and to confirm/calibrate TBM effect on site-specific ground behavior. Located with SSP arrays along tunnel alignment for ongoing verification of TBM zone of influence.
- Inclinometers: grooved casing is installed in boreholes to accommodate a wheeled instrument designed to accurately measure borehole shape; for use a verification that lateral ground movements in the vicinity of shafts are consistent with design assumptions.
- Vibrating Wire Piezometers: sensors installed at distinct levels in fully-grouted boreholes to provide measurement of pore water pressure response during shaft and tunnel excavation.
- Observation Wells: open standpipes installed with a slotted screen at specific depths within boreholes; allows for measurement of groundwater levels manually or by automated data logger.
- Various structural monitoring, including strain gauges, crackmeters, tiltmeters and structural monitoring points.

The instrumentation was arranged along the tunnel alignment as follows:

- A centerline surface point (SSP, SSMP or SMP as appropriate) every 15 m (50 ft) along the alignment
- Transverse arrays of surface settlement points orthogonal to the tunnel alignment, typically with an MPBX installed above the tunnel centerline. There were 20 arrays in total along the alignment typically located at cross-streets, with a higher concentration at the start of the tunnel drives (to allow the trough width behavior to be confirmed as early as possible).
- Various BMPs on buildings and exposed utilities along the alignment. Utilities were only exposed at the Bair Island Shaft and San Carlos Shaft.
- Additional MPBXs in the vicinity of AAS to capture the effect of TBM launch.
- Additional instrumentation within and around shaft excavations.

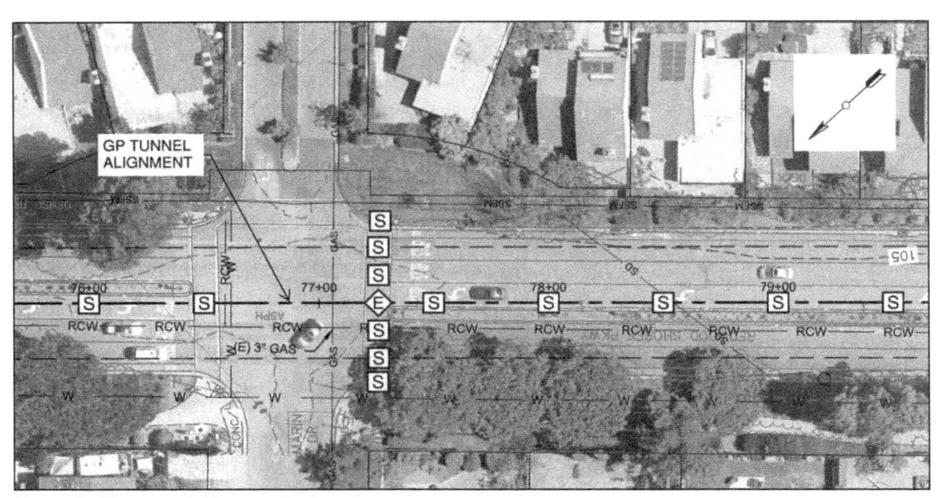


**Figure 6. Representative monitoring plan**

## Settlement Behaviors

### *Shape of Settlement Trough*

The data from the settlement arrays was collected and assessed to determine the shape of the settlement trough. The assessment method started with re-baselining the recorded movements to a new zero datum. This took place generally within 60 m (200 ft) prior to the TBM passing underneath the settlement monitoring instrument. The re-baselining of instruments prior to TBM arrival was to eliminate confusion of settlement related to tunnel construction from the historical ongoing settlement which has been resulting from consolidation of YBM versus what was induced by tunnel excavation. Typically, the re-baselining resulted in corrections ranging from 0 to 5 mm (0.2 inches). Instruments were typically monitored on a daily basis and data collected, plotted, and flagged if issues were observed. Once the TBM had passed and ground movements stopped, the data recorded at transverse array locations were plotted against their offset from the tunnel centerline. A 'best-fit' gaussian settlement curve was then obtained by manually varying the $k$ and volume loss parameters. The results for a typical array are shown in Figure 7.

The $k$ value that was derived from using the best-fit results for each of the settlement arrays generally fell within the range of 0.26 to 0.55. To better understand this range, the ground conditions at each array were used to review the results. All results generally fell between the typical $k$ values for cohesive ground and granular soils, but with variation along the alignment, as shown in Figure 8. To investigate this in more detail, the method for layered soil proposed by Selby 1998 was used, but with different $k$ values:

$i_0 = k_1 z_1 + k_2 z_2 - 0.1$ meters ($i$, $z_1$ and $z_2$ in meters).

where

$i_0$ = distance from tunnel centerline to point of inflexion of settlement trough. at surface
$k = i_0/z_0$, in meters
$z_1$ = thickness of upper layer, in meters
$z_2$ = thickness of tunnel stratum. in meters
$z_0$ = distance from tunnel springline to surface = $z_1 + z_2$, in meters

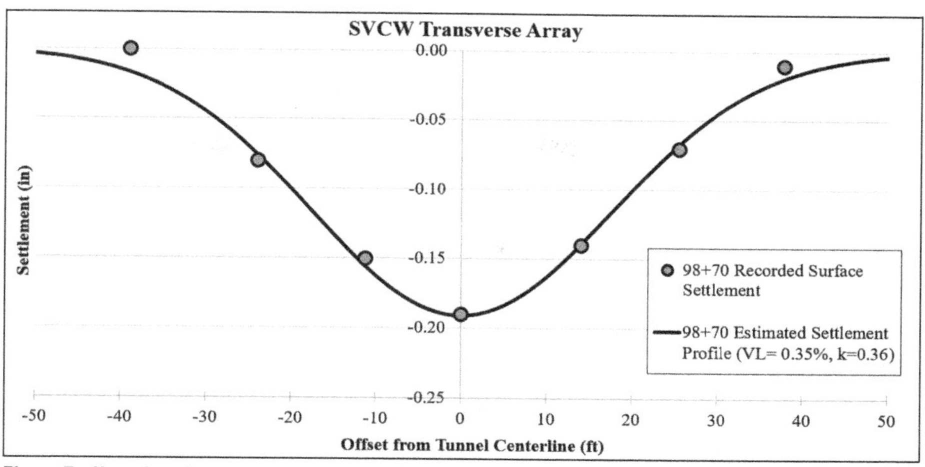


Figure 7. Use of settlement array data to obtain best-fit gaussian settlement curve

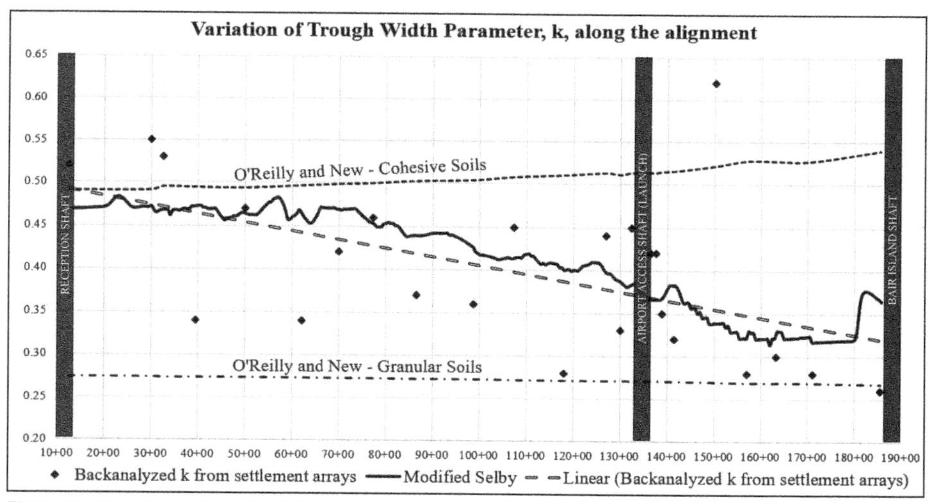


Figure 8. Variation of trough width parameter, *k*, along the alignment and correlation with *k* derivation methods

This allowed a very close correlation with the observed trough shapes when used with a $k_1$ of 0.55 in the Young Bay Mud and a $k_2$ of 0.26 in the Upper Layered Settlements/ Old Bay Mud as shown in Figure 8. This implies that the Upper Layered Settlements/ Old Bay Mud, which are generally cohesive materials with granular lenses, behaved similarly to granular soils. The response of the YBM—the thickness of which increased in the direction of the Reception Shaft—was more typical of a stiff cohesive stratum.

### *Volume Loss*

The centerline settlements were generally measured every 15 m (50 ft) along the tunnel alignment. Using the k value calculated in accordance with the Selby (1998) method described above at each of the more than 300 locations, the volume loss was estimated for each point using the Gaussian settlement curve equation:

From $V_s = \sqrt{2\pi} \cdot i_0 \cdot S_{max}$ (O'Reilly and New (1982); $S_{max} = \dfrac{VL \cdot \pi R^2}{\sqrt{2\pi} \cdot k \cdot z}$

where $S_{max}$ = maximum settlement at the center of the settlement trough
$V_s$ = settlement volume (ie volume of the settlement trough)
$VL$ = volume loss, ie settlement volume as a percentage of excavated tunnel volume
$R$ = Excavated tunnel radius

As can be seen in Figure 9 there were a number of locations along the alignment where volume losses were over 1%, and each are discussed below:

1. Start of drive Tunnel Drive-1, stations 136+00 to 136+50: This was within the area where the TBM was launched in a series of steps, with slow progress and extended stops for addition of trailing gantries. It is also close to the launch shaft, where ongoing consolidation settlement was resulting from shaft and adit construction.
2. 125+00 to 122+50: The monitoring point indicating the peak settlement in this zone corresponds to the location of the TBM during the reconfiguration of the muck transportation system from a pumped slurry to conveyor belt. This consisted of an extended stop as well as a period of inconsistent advance rate when mining resumed. Production of slurry-pumped muck requires different TBM operating parameters than that of belt-conveyed muck, which must be determined experientially.
3. 91+50 to 88+00: In this area, as explained in the longitudinal settlement section shown in , some ongoing movement was occurring before and continued after the tunnel excavation passed by, likely due to ongoing consolidation. Additionally, this zone corresponded with documented mixed-soil conditions of clays and sands as indicated in the observed muck. Further, the monitoring point indicating the peak settlement in this zone is the closest point to an abandoned cone penetrometer test borehole (CPT) that was installed by others, thus which was intersected by the TBM, indicating that it was improperly abandoned (by others). Therefore, it was determined when the TBM passed by the remaining abandoned CPTs, continuous mining would be required to avoid the risk of induced settlement.
4. 84+00 to 79+00: This zone corresponded to potentially entering and exiting a zone of mixed-soil conditions of clay and sand. There were multiple recommendations on how to limit settlement within this reach, which consisted of the following: increase settlement readings, adjust face pressures, increase grouting pressures, and use bentonite injection within the steering gap. Ultimately adjusting face pressures and using bentonite injection was adopted as a means and method to help reduce the risk of settlement. TBM data (face pressures, grouting pressures, backfill grouting, etc.), was analyzed although there was no conclusive evidence for reason for the settlement except for the mixed-soil conditions and sensitivity of the sand lenses.
5. 74+00 to 63+00: As above, this zone also corresponded to a documented mixed-soil condition of clay and sand. Again, since this was a known mixed-soil zone, the contractor utilized adjusted face pressures and the bentonite injection system.
6. 53+00 to 41+50: Increased downtime caused by conveyor belt issues was a major contributing factor to inconsistent advance rates, known to increase the chances of surface settlement.
7. 39+00 to 37+50: This section exhibited similar behavior to 91+50 to 88+00. Further, this zone corresponds with the location where the TBM's tail shield

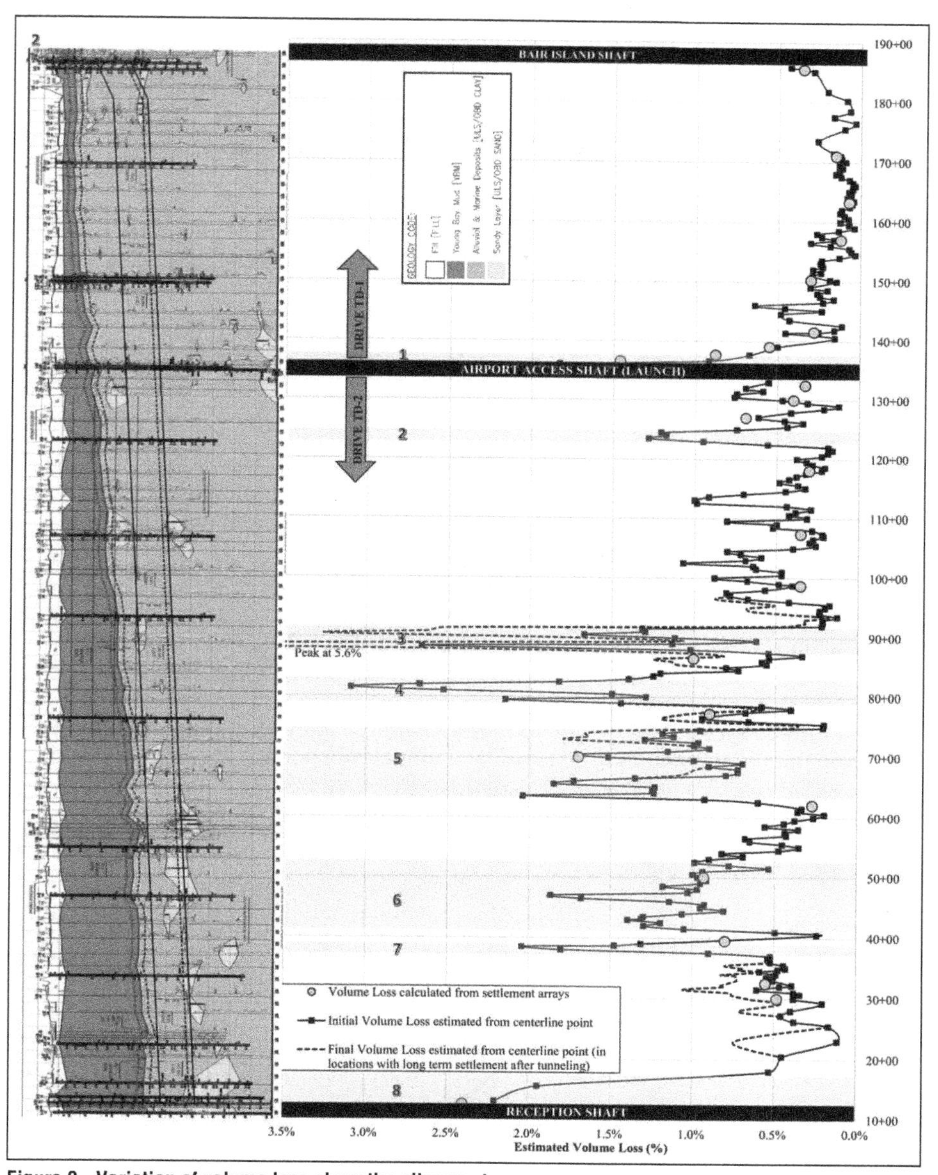


**Figure 9. Variation of volume loss along the alignment**

seal brushes were replaced. General wear and tear diminished the effectiveness of this seal, which led to nuisance groundwater inflow. The team took the proactive measure to replace the brushes at this location, which was prior to passing between and near the two residential buildings.

8. End of the drive—this area exhibited similar behavior to the launch shaft.

Figure 10 shows the distribution of volume loss over the whole project.

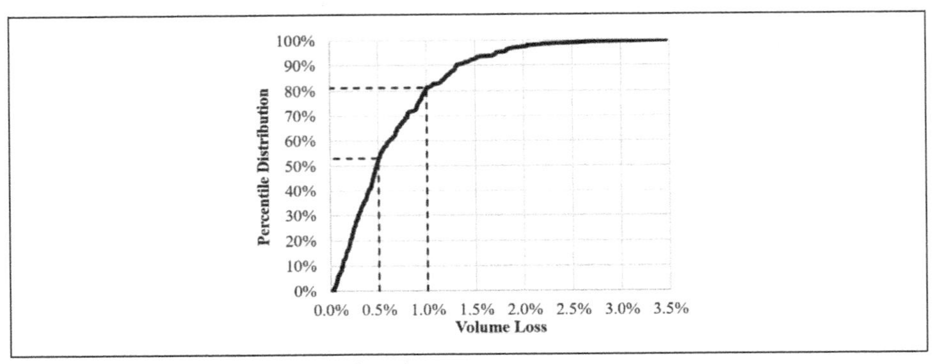


**Figure 10. Percentile distribution of volume loss**

### *Longitudinal Settlement Profile*

The longitudinal settlement behavior was assessed by plotting the normalized settlement ($S/S_{max}$) for each centerline point against the distance from the location of the TBM cutterhead. While there was considerable scatter in the data, the following general observations can be made:

- In many, but not all locations, some minor heave could be observed. This was typically in the order of a few millimeters (0.1 inches) and occurred about eight meters (25 ft) ahead of the tunnel face.
- In most locations, the longitudinal settlement profile followed a normal distribution curve, as shown in Figure 11. This figure also shows, as is typical in pressurized face tunneling, the location of start of the curve is delayed compared to a theoretical curve for a tunnel with no face support. The length of the longitudinal profile was also extended compared with the theoretical profile, occurring over a total length of around $15i_0$ compared with the theoretical $6i_0$ ($i_0$ is distance from tunnel centerline to point of inflexion of settlement trough at surface).

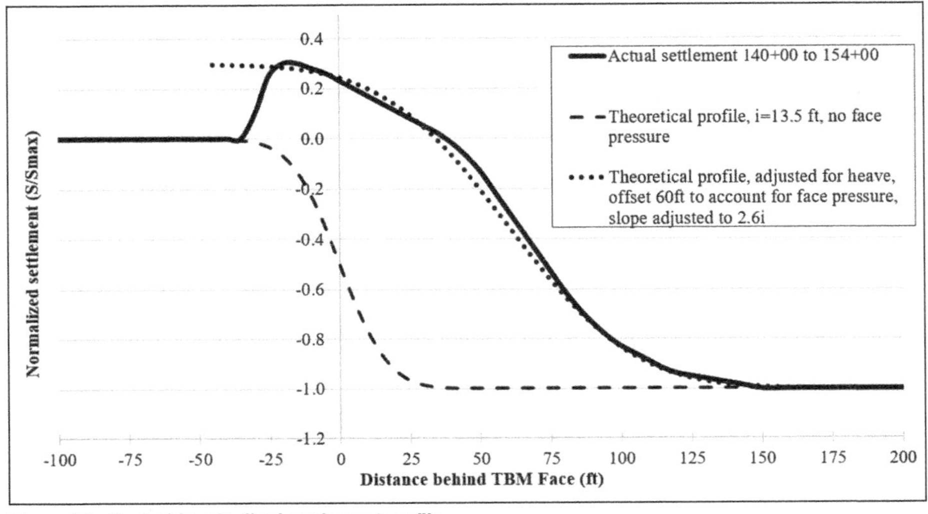


**Figure 11. Typical longitudinal settlement profile**

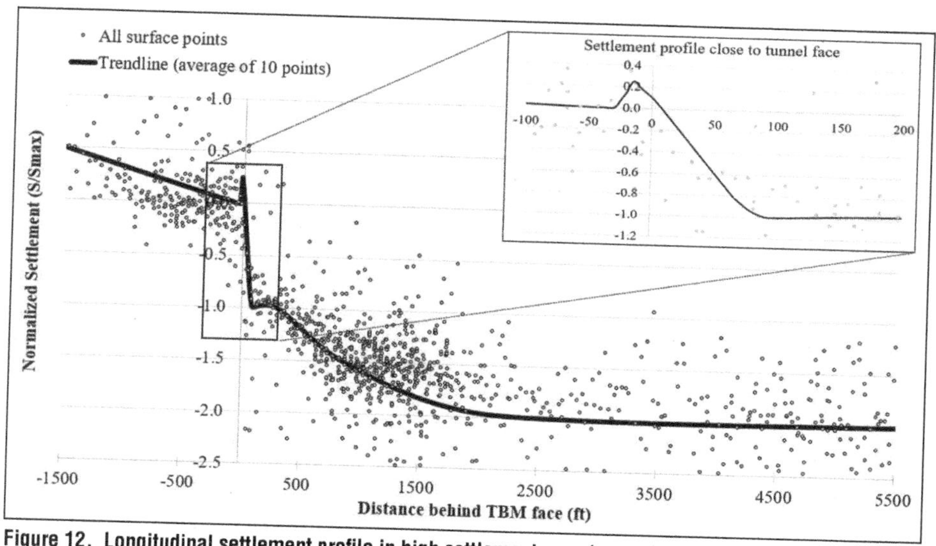


**Figure 12. Longitudinal settlement profile in high settlement area (normalized against settlement $S_{max}$ from approximately 200 ft behind the TBM, when immediate settlement due to tunneling is complete)**

- For about one third of the TD-2 drive, an atypical longitudinal settlement profile was observed. All these locations were where the distance from the tunnel crown to the anticipated interface between YBM and the OLS was less than 3 m (10 ft) (the converse was not necessarily true—i.e., not all areas where the distance to the interface was less than 3 m demonstrated this behavior). This was most pronounced around station 90+00.
- As shown in Figure 12 it appears that there was ongoing movement, attributed to the likely consolidation, before the TBM arrived. It is possible that the recorded settlements as the TBM passed is a combination of normal ground movements combined with some accelerated consolidation, which then paused before restarting later. This would explain the higher volume losses measured after tunnel construction.

### MPBX Results

The MPBX data was analyzed in a similar manner to the surface settlement points, and compared with the theoretical settlement that would be caused by the tunnel depth, and the VL and k factors already calculated at the relevant location. A typical MPBX displacement is shown in Figure 13.

The theoretical settlement of an MPBX point was calculated as

$S_{MPBX} = VL \cdot \pi R^2 / \sqrt{2\pi} \cdot i_{MPBX}$ (Attributed to Harris and Alverado, see XDisp Manual)

where

$$i_{MPBX} = i_0 \left( \frac{z_{inv} - z_0}{z_{inv}} \right)^m$$

m = user defined, 0.5 for stiff clay

$z_{inv}$ = depth to tunnel invert from MPBX point.

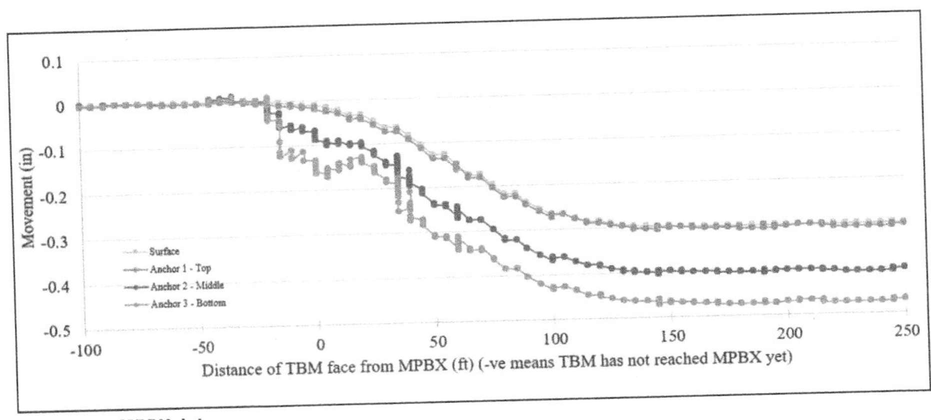


**Figure 13. MPBX data**

However, when used to back-calculate *m*, this method gave a wide scatter of values for *m* from 0.15 to 0.9, with an average of 0.44, not providing very good correlation with the theoretical approach.

While the movement of the anchor points in the Upper Layered Settlements/Old Bay Mud typically stabilized a few diameters back from the tunnel face, as would be expected from typical tunnel settlement, the MPBX anchors in the Young Bay Mud demonstrated long term movement between the upper two anchors. This is another indication of the ongoing consolidation settlement in the area.

## MITIGATION ACTIONS

As the TBM tunnel progressed, the TBM data was monitored continuously and reviewed against the daily measured settlement data and the anticipated subsurface conditions presented in the Final GDR and GBR. The TBM crew, engineers, and superintendents interpreted dozens of data points, and reacted in real-time. Additionally, each morning, the Tunnel Task Force, comprised of representatives from the Owner, Designer and Design-Builder, met to discuss multiple different parameters such as machine face pressures; backfill grouting pressures, quantities & quality; bentonite injection pressures and quantity; and totalized muck removed per advance. These key parameters were determined by the Tunnel Task Force to be of most relevance to surface settlement.

From discussions above, it was noted that there could potentially be a correlation between the locations of previously abandoned CPTs and increased settlement. Therefore, continuous mining was required when the TBM passed underneath an abandoned CPT, which was observed to solve the issue of increased settlement at those locations.

TBM face confinement pressure has a direct effect on surface settlement. The target operating face confinement pressure of the TBM was back-calculated based on anticipated groundwater and subsurface ground conditions. This generally ranged between 0.6 to 2.1 bar depending on TBM location. In tunnel zones in which higher settlement was anticipated, TBM face pressures were set close to the maximum pressure of 150% of the designed target range. Mining with increased face pressures is known to cause a variety of subsequent issues, so face pressures were reduced back to the target value when a given settlement-prone zone was passed. The decision to maintain, increase or decrease face confinement pressure was made daily during Tunnel Task Force Meetings.

One settlement mitigation tool primarily useful during mixed-face conditions, was bentonite injection into the Steering Gap of the TBM. YBM consolidation due to the draining sand lenses within the steering gap was a risk discussed early in the planning process of the project. The TBM was procured with a bentonite injection system as a proactive solution to this consolidation risk. The bentonite injection system, consisting of a batching system, tanks, pumps, lines, ports, and pressure sensors which work together to inject a thin film of bentonite between the outside of the TBM shield and the surrounding strata at a specific pressure. The intent of this pressurized layer of bentonite slurry is to ensure the surrounding strata is held in place as the TBM passes by. Maintaining this pressure also provides the added security of reducing ground movement during periods of downtime and during shift changes. The introduction of bentonite injection did not appear to adversely affect the rate of production of tunneling and also seemed to reduce the risk of increased settlement along the tunnel alignment. The downside to bentonite injection is suffering downtime due to possible failure and maintenance of each component of the system.

Tail skin backfill grouting parameters appeared to have less contribution to settlement control along the alignment.

The tail-skin brush replacement was another settlement mitigation strategy which was executed during TBM mining. The tail-skin brushes are a hoop of wire brushes saturated with thick tailskin grease, which presses against the extrados of the precast concrete ring as the TBM advances the ring past the tail-skin of the TBM. This forms a seal to the strata, groundwater, and backfill grout which is constantly exerting pressure on the outside of the TBM. Inflows of strata, groundwater and backfill grout into the TBM tailskin are possible if the seal is worn. These inflows are known to induce surface settlement. Prior to reaching the residential buildings at the far end of the alignment, tail skin brushes exhibited symptoms of wear. As the TBM approached these high-risk buildings, the Tunnel Task Force discussed the integrity of the seal decided to stop mining and replace the brushes. This allowed the TBM to mine past the residential buildings with a seal in ideal condition. The TBM passed underneath the residential buildings without impact.

Another settlement mitigation strategy was continuous mining. As the TBM is advancing, it is constantly exerting pressure on the strata beyond. When the TBM stops advancing, even though face pressures are maintained, the strata is given an opportunity to "relax" around the TBM which is a risk to causing settlement. Extended stoppages were undesired near high-risk locations of the alignment. Each time a critical location of the alignment was approached, the team implemented continuous mining plans. This included stopping a certain distance before the critical location to perform all routine maintenance. Having all TBM systems in ideal operational condition reduces the risk of downtime caused by equipment failure. Further, the working hours of the crew were modified to ensure the TBM production did not cease until the TBM passed the critical location. Typical working hours induced a 4 hour period of no mining between the end of night shift and the beginning of day shift. The modified working hours ensured no gaps between night shift and day shift, and also implemented a "hot-change" where crew members handed their duties off directly mid-advance. The modified working hours were reduced back to the typical working hours after the critical location of the alignment was passed.

Responding to settlement is crucial in reducing the risk of settlement at that specific location and to understand how to control settlement at future locations along the tunnel alignment. The methods that were observed to work in reducing settlement included: adjusting pressures, introducing bentonite in the steering gap, replacing tail skin brushes to reduce leakage, and continuous mining It shall be observed that the largest mitigation success was collaboration with the contractor, designer, construction managers, and owner to achieve an approach that reduced the risk of further and future settlement.

## CONCLUDING REMARKS

Protecting existing facilities is a priority to Owners and Contractors alike. Developing an effective design for instrumentation and monitoring, and continuing to optimize EPB tunneling parameters during construction, is vital to preventing damage to structures by taking proactive measures to mitigate settlement. The open communication developed between stakeholders during the Progressive Design Build process helped establish the trust and relationships necessary to implement flexible solutions during construction to mitigate settlements. The open communication led to a set of tools implemented to proactively limit settlement. The project achieved the goal of no damage to utilities and no spurious third-party claims using the collaboratively developed settlement mitigation tools.

## REFERENCES

CIRIA. 2003. Embedded retaining walls—guidance for economic design. Report C580.

ITA/AITES WG2. 2007. Settlements induced by tunneling in Soft Ground. Report 2006. In *Tunnelling and Underground Space Technology* 22 (2007) 119–149.

Nilsen, T.H. and Brabb, E.E., 1979. Geology of the Santa Cruz Mountains, California. Geological Society of America, Cordilleran Section.

Oasys. 2019. XDisp Help guide.

O'Reilly, M.P., and New, B.M. 1982. Settlements above tunnels in the United Kingdom—their magnitude and prediction. In *Tunnelling'82*, London. IMM, pp. 173–181.

Selby A.R. 1988. Surface movements caused by tunnelling in two-layered soil. In *Engineering geology of underground movements*. Edited by Bell et al. Geol. Soc. Engineering Geology Special Publication No. 5, pp. 71–77.

Weiman-Benitez, L., Jaeger, M. and Sokol, N. 2023. A Case Study in Successful Progressive Design Build Tunneling. Rapid Excavation Tunneling Conference.

PART 

# Ground Support and Final Lining


*Chairs*

**Jean-Luc des Rivieres**
JF Shea

**William Hodder**
North Tunnel Constructors ULC

# Acclimatization: Adapting Conventional Tunnel Lining Techniques to Overstress Rock Conditions

**Jean-Luc des Rivières** ▪ J.F. Shea Co. Inc.
**Ross Goodman** ▪ J.F. Shea Co. Inc.

## ABSTRACT

The Ohio River Tunnel Project encountered multiple unexpected geotechnical challenges which, continuing to develop over time, hampered excavation completion and impacted the cast-in-place final lining technique and work execution sequence for the entire project. The continued degradation of the excavated hard rock tunnel conditions led to a unique, dynamic, geologic setting that forced the project to innovate new and adaptable measures allowing for a conventional cast-in-place liner to be completed in a safe and timely manner and minimize delay.

## INTRODUCTION

### Project Purpose & Outline

The Ohio River Tunnel Project (ORT) is a combined sewer/wastewater outfall storage and conveyance tunnel project that was constructed in the vicinity of downtown Louisville, KY. The goal of the tunnel project was to provide a combined sewer storage and conveyance method to the aging combined sewer system of Louisville. Prior to the project's completion, Louisville's sewer system would regularly exceed holding and conveyance capacity during even minor precipitation events and would overflow to the Ohio River. This condition resulted in an EPA Consent Decree directing the city to correct the condition which Louisville Metro Sewer District (MSD), the city sewage controller, chose to solve partially through the construction of a combined sewer storage & conveyance tunnel. At project completion, sewer flow that exceeds the system capacity is captured by the ORT as well as other consent decree related projects and pumped to the Morris-Foreman municipal treatment facility when plant capacity is available.

The ORT is a finished 20 ft. diameter, hard rock, cast-in-place concrete lined CSO tunnel owned by Louisville MSD. The project was constructed by a Shea-Traylor Joint Venture (ST-JV), and the Engineer of Record is Black and Veatch. A hard rock, main beam TBM was utilized to excavate the 22 ft.-4in. diameter tunnel, and telescoping concrete forms were used to place a 14 in. thick cast-in-place final liner.

The ORT contract consisted of two large diameter shafts approximately 220 ft. deep, five deaeration chambers with corresponding surface intake structures, a retrieval shaft, and 13,000 ft. of TBM mined tunnel with concrete cast-in-place final liner. After construction started, an additional 7,200 L.F. of tunnel were added by change order, along with two drop shafts with intake structures (Figure 1). Notice to proceed was awarded in Winter 2017 with final completion achieved in the Summer of 2022.

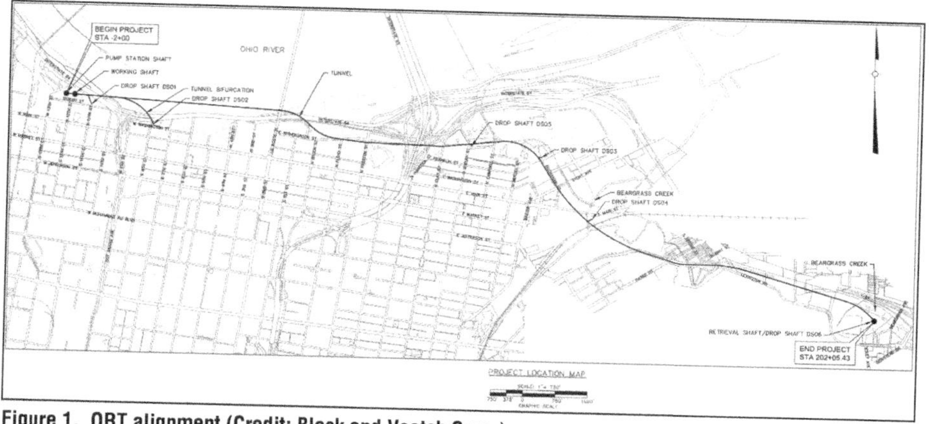


Figure 1. ORT alignment (Credit: Black and Veatch Corp.)

## Project Ground Difficulties

The ORT faced time and orientation-dependent dynamic hard rock ground conditions in the TBM excavated tunnel that impacted initial excavation and follow-on work. This paper serves to outline the changing geologic conditions as the project progressed, how those geologic changes impacted the Design-Bid-Build project scope, and what changes were made to the project construction methods to accommodate the adverse rock conditions encountered. The tunnelling and lining activities on ORT were most hampered by in-situ rock stress and rock unit interface interactions, although ORT also experienced rock strength and durability conditions of a lesser value than base-lined in the GDR/GBR

Beginning shortly after TBM tunnel excavation and continuing through the completion of tunnel cast-in-place lining, overstress rock conditions were experienced on ORT and manifested themselves through crown and invert rock compressive failure modes as well as rock integrity degradation between rock layers that defined excavation and lining operations. ORT tunnel construction scope began as a crown supported (rock dowels in crown array) hard-rock bored tunnel in primarily competent limestone and dolomite conditions, but throughout unforeseen and changing rock conditions experienced on the project, the scope had to be expanded to include multiple passes of ground stabilization improvement to facilitate safe excavation, tunnel travel, and lining activities. During TBM excavation, additional crown support elements including mine straps, welded wire fabric, rolled channel, hardwood blocking, additional crown bolting arrays at reduced longitudinal spacing, as well as necessary relief of failed rock masses to facilitate continued TBM excavation were all undertaken.

## Project Innovations Outline

The difficulties outlined above meant that the telescoping forms used for the cast-in-place final lining were unable to function as designed. To allow for the lining operations to continue, significant modifications to the lining operations were necessitated including reversal of sequential lining direction to meet owner schedule, removal of fractured and broken invert rock, use of customized screw jacks to hold form work in place where contiguous rock conditions were not available, safety shutdowns to undertake secondary rock support operations, reduction in concrete placement rate due to rock stability, float pin supports to distribute loads, and more.

# GEOLOGICAL OVERVIEW

## Contractual Geotechnical Baseline Condition and Setting

The geologic setting of the ORT project was in the sedimentary hard rock conditions in the Ohio River Valley approximately 200 ft. beneath ground elevation. The project alignment traversed the downtown region of Louisville, KY passing under the Ohio River for portions of the alignment as seen in Figure 1. Louisville lies near the southern edge of the greater Great Lakes glaciated zone and evidence of such geologic history abounds on the surface in the soil as well as the rock units found therein. The rock at or near the tunneling elevation included layers of limestones, dolomitic rocks, and shales. The sedimentary rock units were found to be striking in a general N–S orientation and dipping to the west throughout the alignment. (Black and Veatch,2017)

Due to the number of rock layers encountered, the project engineer of record elected to consolidate the numerous distinct rock units and associated measured analytical qualities into three rock engineering units composed of rock units of roughly similar theoretical quantitative qualities for the purposes of establishing a geologic baseline report for the project contract documents. The rock engineering units established were as follows: Rock Engineering Unit (REU) 1 composed of Limestone and Laurel Dolomite Formations, REU 2 composed of Waldron Shale, and REU 3 Composed of a variety of rock types including limestone and dolomite units interspersed with shale, chert, and calcite. The rock units were deposited in the following order shallow to deep: REU 2, REU1, REU 3.(Black and Veatch,2017)

Despite the wide variety of material characteristics housed within the units, rock conditions were baselined using average and percentile rock characteristics. Additionally, the tunnel alignment, both original and final extended, were expected to stay within REU 1 & 3 thereby avoiding the more difficult excavation conditions associated with shale found in REU 2. The baselined conditions throughout the alignment were expected to be competent rock with high average unconfined compression testing results (in excess of 15 ksi), high slake durability of rock units, low occurrence of faulting, and moderate rock support requirements (mostly 4 to 7 crown rock dowel support rows at 5 feet on center). (Black and Veatch, 2017)

## Changing Geological Conditions Through Excavation

The tunnel alignment was planned to be excavated from West to East beginning in REU 1 and transitioning through the sloping sedimentary rock beds to REU 3 near the halfway point of the alignment shown in Figure 1. As excavation started in REU 1 in an eastward direction, the crown rock compression strength was not adequate to support the stress found around the bored rock opening and the crown rock fractured upwards to REU 2 which, despite its shale characteristics, had high compressive strength and massive, contiguous characteristics and did not fail further upwards in the initial stress condition of the excavation. As tunnel excavation moved forward through REU 1 into the initial bifurcation curve, crown and invert conditions degraded severely as the alignment transitioned from a primarily eastward heading to a South-East heading. Initially as the curve began, the rock pieces falling from the crown were thin, intact shields of rock with sharp edges comprising multiple bedding planes of limestone/dolomite shearing off from the crown with a slight preference in appearance location being the more southern edge (right side of the bore looking up station) of the tunnel crown. As the boring of the curve progressed, larger pieces of rock began coming loose from the tunnel crown in addition to those coming loose from tunnel invert with significant pieces of rock falling from the crown onto the TBM shield before initial support could be installed. Unlike the first stretch of tunnel immediately after beginning the excavation in which the overstress fractures stopped at the interface with REU 2, these overstress conditions found in the tunnel bifurcation broke upwards in the typical cathedral effect which allowed for much greater size of rock masses to break free and fall into the tunnel or onto the machine. These conditions significantly slowed TBM excavation as additional support was erected and fallen rock was removed as necessary to advance excavation.

Figure 2. Overstress invert rock failure (Credit: Ross G.)

At this point in the tunnel drive, the contractor in consultation with a third-party geologic consultant investigation (R. Heuer) found that overstress conditions were likely to blame for the seemingly boring-bearing dependent failure of the rock and the types of rock failure that was occurring. Evidence of high horizontal stresses and material failure at the crown and invert were telling evidence. These rock failure modes, similar to that found in the bifurcation, can be clearly seen in attached Figures 2 and 3. Pre-excavation in-situ stress testing was not undertaken by the owner or their representatives prior to project bid, so these conditions were not a part of the as-bid baseline conditions and were thus not anticipated.

Figure 3. Overstress crown and quarter arch failure (Credit: Ross G.)

Figure 4. Invert voids arising from overstress rock failure (Credit: Ross G.)

After completing the tunnel drive of the tunnel bifurcation, the TBM had to be backed up and during this process the invert continued to degrade from overstress fractures in the rock which vertically crossed horizontal bedding planes. The addition of loading the fractured invert rock with rubber tire and railroad equipment during the TBM reversing operation led to the loosening of these invert rock pieces and rock fallout (Figure 4), that left voids in the invert and up the rib of the TBM bore to approximately the 5 o'clock and 7 o'clock positions that had to be physically improved prior to tunnel lining operations and is discussed later in this paper. As TBM excavation continued to the East throughout REUs 1 and 3 a pattern emerged where when tunnel bearing took a more southeastern bearing, crown and invert conditions degraded in a fashion similar to the initial bifurcation overstress conditions experienced.

The excavation difficulties due to overstress conditions incurred during the TBM excavation were significant (totaling over 6,500 L.F. throughout the alignment) and during this time the rock conditions continued to degrade in some previously unaffected areas that had previously been excavated and supported in accordance with contract documents and contractor best practice. These failures in the rock were very similar or identical to the overstress conditions encountered during TBM excavation but simply had an extended standup time due to a more favorable tunnel excavation orientation given the overstress conditions or stiffer and/or stronger rock layer interface than others. Rock strain was visible in the tunnel as some rock units within the same rock engineering units would be seen to have moved laterally when compared with neighboring bedding units as seen in Figure 5. In rock engineering units comprised of dissimilar rock types, such as REU 3, the differences in behavior of the units under stress became apparent as some units would fracture earlier or deeper than others under the overstress loading. This time delayed rock failure occurred during all phases of construction and lead to repeated resupport efforts due to rock movement and the ensuing safety concerns.

Figure 5. Strain between rock unit bedding planes

## Rock Failure Modes

Failure modes of overstress conditions on the ORT project were as follows: as previously discussed, in the initial area where overstress conditions were found as the TBM began its drive, overstress conditions at the crown position were greater in magnitude than the stress resistance capacity of the rock found between the crown and the overlying REU 2 material which were able to withstand the in-situ stress. During the excavation of the tunnel bifurcation and frequently throughout the tunnel, the rock was unable to withstand the in-situ stress loads in both the invert and crown tunnel positions and as such the cathedral condition occurred wherein the rock continued fracturing and thereby relieving the in-situ stress until the stress occurring around the broken tunnel opening was less in magnitude than the strength of the rock. In the crown position this meant that rock would fall unless adequately supported from falling while in the invert position this meant that the broken rock mass would not necessarily be visible or identifiable until load was placed upon it through equipment traverse or tunnel lining formwork needed to be placed upon it.

Additionally, in areas of tunnel excavation in REU 3, the rock types layered within each other behaved significantly different under the overstress condition than the baselined conditions would suggest. The rock layers housed within REU 3 had greatly different relative stiffness in relation to the encountered horizontal stress. It was observed that the near-horizontally bedded rock layers moved different amounts in the presence of horizontal stress thereby causing vertical cracks to develop through the horizontally bedded rock units and allowing the so-called quarter arch and invert areas of the tunnel to break off wedges of material that were difficult to secure with rock support due to the relatively small size of the rock pieces that would further degrade under rock drill loading and the location below typical roof bolted rock support typical of hard rock tunnel and main beam TBMs. The quarter arch and invert failures were a particular difficulty to handle because they prevented safe and effective tunnel lining form use as well as impeded support equipment travel within the bored tunnel.

## TUNNEL LINING

ORT was lined using a conventional cast-in-place lining technique. The cast-in-place lining operation used a 30 ft. long, 20 ft. diameter circular concrete form, with seven crown sections and eight invert sections for a total possible single pour distance of 210 ft. A telescoping form carrier stripped, transported, and set the forms (Figure 6). A 30 ft. long bulkhead car was used as a staging area for other equipment necessary to the operation and allowed crews access to construct bulkheads. The final lining operation experienced numerous delays due to the overstress conditions throughout the alignment. This paper aims to identify how the changing in-situ rock conditions prevented conventional hard rock excavation and lining techniques from being implemented and forced the project team to develop adaptable measures that allowed for a conventional cast-in-place liner to be completed in a safe and timely manner while minimizing delays in adverse geological conditions.

### Cathedral Failure in Crown

As discussed in the Geological Overview, the overstress conditions in the tunnel manifested themselves in classic cathedral failures in the crown and invert. The cathedral failure observed in the arch dramatically impacted the cast-in-place lining operation. Welded wire mesh installed during the mining operation proved insufficient in confining the ground resulting in large areas of the tunnel where loaded mesh had sagged beyond the confines of the cast-in-place final liner. This loaded mesh had to

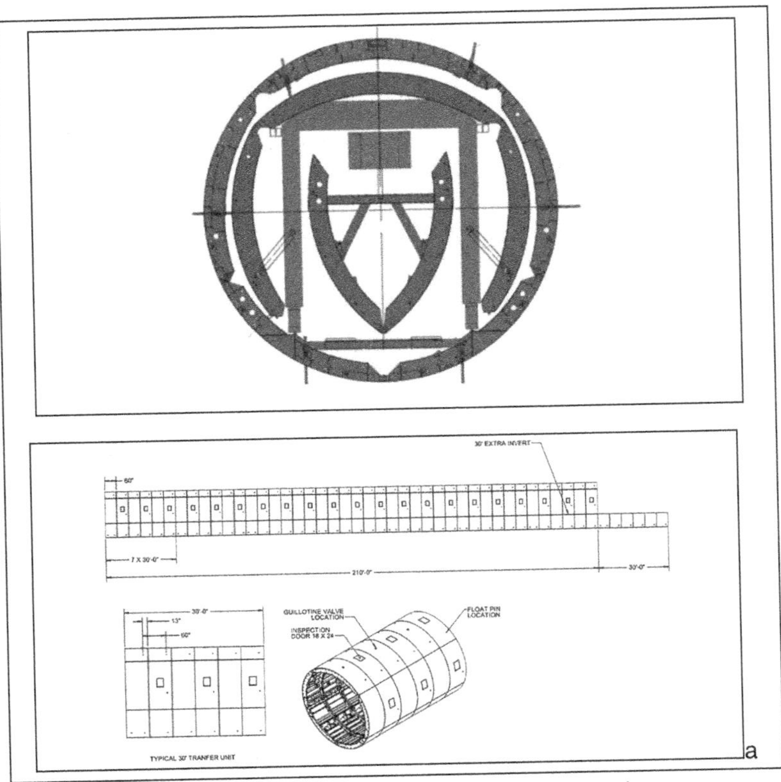


Figure 6. Cast-in-place tunnel concrete forms (Credit: MSP Structures)

be addressed for the tunnel to have the requisite physical clearance for tunnel formwork to advance through the tunnel and carry out concrete placement. Additionally, the telescoping concrete forms experienced a shifting event (hydraulic displacement) that was directly attributable to the observed cathedral failure in the crown. The ST-JV team developed several abatement techniques to address the two major operational issues listed above.

### *Loaded Mesh*

The only course of action in confronting the loaded mesh was for crews to cut the mesh, release the ground, install new mesh, and install further temporary ground support. A rebar jumbo was retrofitted with a large steel chute (Figure 7). This chute allowed crews to safely cut the mesh and release the loaded ground into large rail-mounted muck boxes. New mesh was then pinned with a mixture of 4 ft. long "split-set" rock bolts and 8 ft. fully epoxy-encapsulated rock dowels. Once a section had been resupported, the jumbo would advance, and the cycle would repeat.

**Figure 7. Jumbo mounted shoot (Credit: Jean-Luc d.)**

Initially, ST-JV believed that this resupport operation would be able to be completed off the critical path and would have no effect on the concrete lining operation. However, due to the extensive amount of ground requiring resupport, the tunnel lining operation was paused for 23 working days as part of a safety shutdown following the rapid and sudden degradation of some areas of the bored tunnel. The safety shutdown allowed additional shifts to work on resupporting the ground and limited unnecessary traffic traveling through the regions of concern ensuring employee safety. Once a significant portion of the resupport had sufficiently been completed, the tunnel lining operation continued. While the tunnel lining crew continued, a limited set of crews continued working on resupporting the tunnel upstream. ST-JV ultimately resupported approximately 6,500 L.F., a majority of which took place along a South-East heading bearing.

### *Form Shifting*

To accommodate for the poor rock conditions in the arch, ST-JV fabricated a 5 in. × 5 in. float pin support pad to help distribute the pressure the float pins placed on the rock during the final liner placements and prevent localized rock failure under the float pins during pour events. These pads were installed between the float pins and the rock through pre-existing doors on the arch forms. However, early in the tunnel lining operation, while pouring tunnel liner, an overstress-caused failure in the arch resulted in the forms shifting due to insufficient static support against hydraulic loading. After the shift, ST-JV teams observed that small pieces of tunnel crown had been displaced at the location of the crown float pins allowing displacement of that form section.

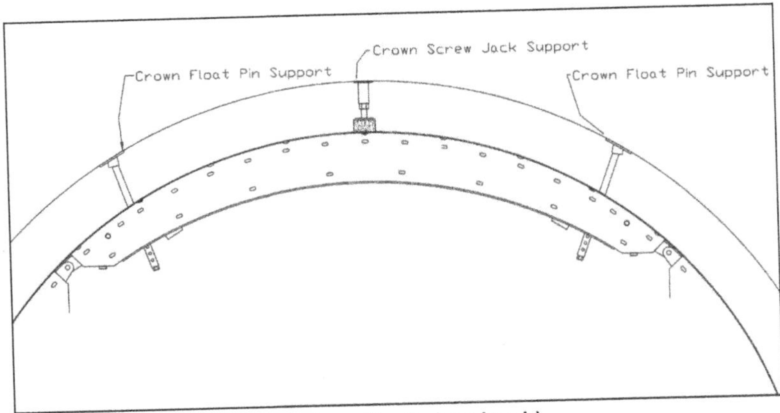


Figure 8. Tunnel form support systems (Credit: Jean-Luc d.)

Following the initial shifting event, ST-JV made immediate changes to the final lining operation, as crews were already aware that ground conditions in the arch would continue to deteriorate as they continued with the concrete placement operation Eastward. The minimum number of float pin supports that were installed was increased, additionally, a custom screw jack was fabricated that would be installed at the 12 o'clock position between the concrete forms and the tunnel arch (Figure 8). Through the worst ground these jacks were installed on 5 ft. centers and as ground improved were spaced to 10 ft. centers. These jacks were designed to act as a third float pin to further stabilize and distribute hydraulic loading and were cast-in-place. Crews were also directed to install additional jacks on as-needed bases depending on the ground conditions they encountered. The additional work of adding float pin support pads and screw jacks had a measurable effect on production rates. Crews that were tasked with setting forms had to be expanded to accommodate the additional work required.

### *Additional Abatement Techniques*

Additional abatement techniques were proposed to further mitigate any future shifting event, but due to the project's EPA imposed time constraints any solution that would have had extended the project's operational schedule was ruled out. Other techniques proposed to continue pouring concrete in the poor rock conditions included slowing the placement rate of concrete along with placing concrete in multiple lifts.

## Cathedral Failure in Invert

The over-stress conditions also manifested themselves in the tunnel invert. The cathedral effect experienced in the invert posed extensive challenges to the tunnel concrete lining operation. These challenges can be broken down into four categories: poor invert regions identified prior to tunnel lining operation, poor invert regions created by tramming the bulkhead car over effected ground, poor invert regions created by the form setting operation, and fabrication of custom bulkheads.

### *Poor Invert Regions Identified Prior to Tunnel Lining Operation*

Along the tunnel alignment, there were regions of invert that experienced failing before the tunnel lining operation approached. These regions could be easily identified, and crews could work ahead of the tunnel lining, removing the broken material and preparing the area for the cast-in-place concrete (Figure 9). While trying to advance the bulkhead car through the regions, the bulkhead car would often become wedged on the ribs of the tunnel. Crews would then spend significant amounts of time working to dislodge the car.

### *Poor Invert Regions Created by Tramming Bulkhead Car Over Effected Ground*

While some regions of invert failure could be identified before the arrival of the bulkhead car, various regions were only discovered while tramming the bulkhead car over them (Figure 10). Due to the way the invert broke, it often became impossible to tram the bulkhead car back over that region. This resulted in extensive time delays as crews were forced to hand muck the material that had been broken by the weight of the bulkhead car. Ultimately, after consulting with the bulkhead car manufacturer, the supplied

**Figure 9. Pre-identified regions of invert degradation (Credit: Jean-Luc d.)**

**Figure 10. Bulkhead car damaging tunnel invert (Credit: Jean-Luc d.)**

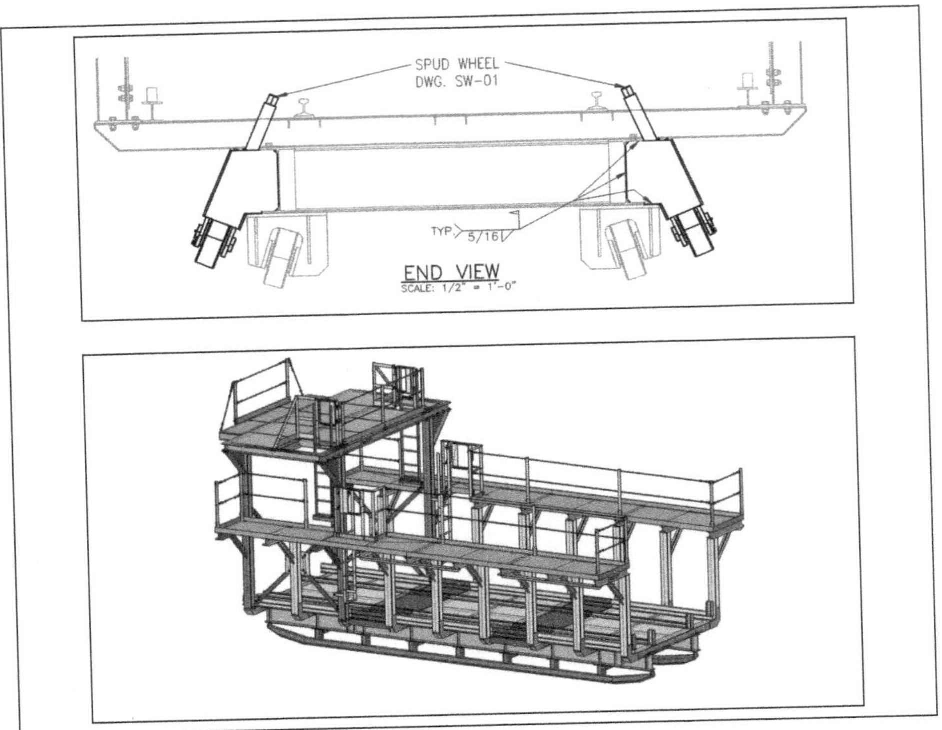


**Figure 11. Bulkhead car modifications (Credit: J.F. Shea and MSP Structures)**

wheels were removed in favor of a skid to limit the ground failing under the weight of the bulkhead car and mitigate the risk of wedging the bulkhead car (Figure 11).

Additionally, ST-JV chose to install a set of auxiliary telescoping wheels before installing the bulkhead car skid (Figure 11). If the bulkhead car became wedged, these wheels could be extended, and the bulkhead car could be freed. Over repeated use, these auxiliary wheels could not withstand the harsh conditions of the uneven ground and were ultimately removed in favor of the skid runners.

***Poor Invert Regions Created by the Form Setting Operation:***

In much the same way the invert fractured under the weight of the bulkhead car, the invert would also fracture under the weight of the tunnel forms. Float pin support pads were installed on invert float pins, along with repurposed screw jack supports to prevent the forms from shifting during the form setting operation. Float pin baring pads were required on both leading-edge float pins and alternating every other invert float pin. Invert float pin supports were installed through custom invert doors that were designed by the manufacture and installed by ST-JV following the delivery of the forms. The repurposed screw jacks were installed on the leading edge of the form on an as needed basses at the 5, 6, and 7 o'clock positions.

If a tunnel form required adjustment due to ground failures after being initially set, the form would be picked up and reset. Often the form would have to be removed entirely, and crews would have to hand muck the invert in the same procedure performed in the previous section. To help facilitate the mucking, a walk-behind skid-steer was

**Figure 12. Hand muck tunnel invert to competent ground (Credit: Jean-Luc d.)**

implemented, along with a small portable conveyer system that could transport material through the bulkhead car. Once sufficient material was mucked to allow the float pins to have a competent bearing surface, the form would be reset (Figure 12). Approximately 7,000 float pin bearing pads and 1,800 screw jacks were used throughout the tunnel alignment. An additional unforeseen consequence of removing large sections of tunnel invert is that the initially purchased invert float pins were not long enough to reach across the broken-out areas of tunnel invert and new longer float pins had to be purchased to place the forms correctly and maintain the specified tunnel alignment and grade.

### *Custom Bulkheads*

Along with significant delays during the form-setting operation, the cathedral failures caused by the overstress conditions caused additional delays due to the time required to build custom reinforced bulkheads. A prefabricated aluminum bulkhead was designed to bolt directly to the tunnel forms and create a seal with the excavated perimeter; however, the fractured invert and arch meant a good seal could not be made without additional work. Crews were required to construct custom wooden bulkhead extensions in the regions with broken ground. Building wooden bulkheads on an as-needed basis was time consuming and exposed the tunnel lining operation to additional risk. On several occasions, the wooden bulkhead extensions failed due to the surrounding rock giving way under the new hydraulic pressures introduced during the concrete placement.

### *Additional Abatement Techniques*

Along with the abatement techniques implemented by the contractor, several other options were considered. It was proposed to pre-chip large sections of tunnel invert and pour mud slabs to prevent any unforeseen delays in tunnel lining. These slabs would allow the bulkhead car to easily travel through the tunnel and prevent any forms shifting during the setting operation. However, due to schedule constraints, an extended shutdown to perform this work was deemed an unsuitable solution.

## SCHEDULE IMPACTS

This paper references the schedule impacts the TBM excavation and cast-in-place lining operations experienced due to the changing and unforeseen ground conditions and the abatement techniques that were implemented to overcome them. When ground conditions matched as-bid conditions, ST-JV was able to match or exceed the anticipated production rates. However, when ground conditions varied from the baselined conditions, tunnel lining production rates were on average 50% of the anticipated rate. Furthermore, this average included times when tunnel conditions prevented lining work from progressing.

## CONCLUSION

The ST-JV team encountered several challenging ground conditions during the excavation and lining of the ORT. By implementing numerous adaptable abatement techniques, ST-JV was able to complete the cast-in-place final liner and deliver the project to the owner within the timeframe stipulated by Louisville's commitment to the EPA. ORT began receiving flow in the Summer of 2022 and plans to prevent 439 million gallons of overflow from being discharged in a typical year.

For future projects in glaciated zones or those where the threat of overstress exists, ST-JV can recommend that preconstruction in-situ stress testing be undertaken, despite cost and time delays, to provide better context than historical precedent as to whether overstress conditions will manifest themselves during construction so all parties are more familiar with the possible geologic and construction risks that can accompany these conditions and work to mitigate them where possible. While ST-JV understands that results of in-situ overstress testing results have the potential for significant magnitudes of error, the relative results can still be an effective tool to help limit risk to all contract parties and to determine the bearing and magnitude of any overstress forces acting upon a proposed tunnel alignment.

ST-JV found that, consistent with professional opinion, supporting fractured/failed rock in-situ when possible, provides for a safer and more productive work environment acknowledging though that due to the constraints of excavation and construction techniques, it is sometimes recommended to remove failed rock to allow for rock support to be more effectively installed and for the safe operation of follow-on activities such as tunnel lining. Additionally, through the implementation of novel adaptations to traditional hard rock tunnel lining techniques, adverse conditions can be overcome, and successful lining work can be completed.

## REFERENCES

*Geotechnical Data Report for the Construction of Ohio River Tunnel Project* (Vol. 4A of 4, pp. 1–1828, Rep.). (2017). Louisville, KY: Black and Veatch.

# Bypass Tunnel Shafts—Shotcrete Lining

**Paul Madsen** ▪ Kiewit-Shea Constructors, AJV
**Bade Sozer** ▪ Delve Underground
**Thomas Hennings** ▪ Delve Underground
**Eileen Test** ▪ Delve Underground

## ABSTRACT

The Rondout Bypass Tunnel in New York has two access shafts. The upper sections of the shafts are lined with steel pipe to resist a substantial net internal water head . Initial design included a ¾-inch-thick cement mortar lining (CML) for all three components. Because of concerns with CML application on large-diameter pipes, the protective lining was redesigned for shotcrete application. This paper discusses the details of that design, mock-ups, and shotcrete application.

## PROJECT DESCRIPTION

The Rondout-West Branch Tunnel (RWBT), a segment of the Delaware Aqueduct (the Aqueduct), was built from 1937 to 1944 and provides about 50% of New York City's total water supply. The tunnel is concrete lined and has a finished inside diameter of 13.5 feet. It is about 45 miles in length and runs in the southeasterly direction from the Rondout to the West Branch Reservoirs. Monitoring during tunnel operations consistently demonstrated that the RWBT is leaking up to 35 million gallons per day (MGD), mainly through locations at Roseton and Wawarsing. Leaks at the Wawarsing area will be mitigated through an extensive grouting program of the surrounding rock. The leaks originating from the Roseton area are being mitigated by constructing the Rondout-West Branch Bypass Tunnel (Bypass Tunnel). This paper discusses the design, mock-ups, and construction of the protective lining for the two access shafts of the Bypass Tunnel.

### Alternate Lining Design

Bypass Tunnel Access Shafts 5B and 6B are located on the west and east sides of the Hudson River, respectively, and are each over 700 feet in depth. The portions of the shafts, are lined with steel pipe. The steel lining resists net internal hydrostatic head and external groundwater head during operational and unwatered conditions, respectively. The original design specified a minimum of ¾-inch-thick cement mortar lining (CML) for the steel access pipe for protection against corrosion. During early planning stages for the application of CML to the surfaces of the access pipe system, the contractor (Kiewit-Shea Constructors, AJV [KSC]) had concerns about CML maintaining adequate adherence during construction. The specific concern was that crews would be required to work hundreds of feet below the mortar in the later stages of the project.

To save time on the schedule, KSC had planned to apply the CML to the access pipe sections prior to shaft installation. To demonstrate the adequacy of CML application and performance through a mock-up, KSC used a section of steel pipe that was a spare from the Bypass Tunnel final lining operation. In April 2021, KSC pneumatically applied CML without any reinforcement to an area of the spare liner pipe in accordance with the design. On a different area of the same pipe, KSC tack welded 4-inch × 4-inch welded wire reinforcement (WWR) before applying CML. See Figure 1 for

Figure 1. Welded wire reinforcement tack welded to the steel pipe

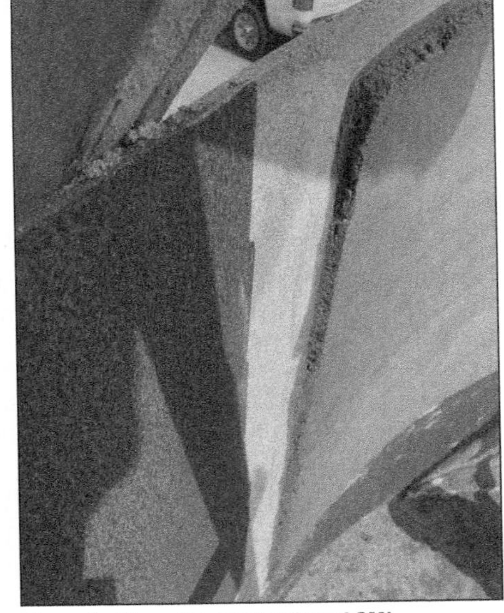

Figure 2. Pipe after application of CML

WWR arrangement. KSC planned to pick the mock-up pipe sections with a crane to evaluate the performance of the CML after it underwent any deformation changes as a function of the crane pick. However, this never became necessary as the unreinforced CML fell off the pipe before the crane arrived to do the test pick (see Figure 2). The mortar with the WWR stayed in place and did not show signs of distress following the test pick.

### Alternate Design Options

The New York City Department of Environmental Protection (NYC DEP) requested that Delve Underground (the department's tunnel consultant on the Bypass Tunnel Project) design an alternate shaft protective lining. Delve Underground considered reinforced CML, reinforced shotcrete, epoxy, and polyurethane based systems. In different lining evaluations, adequate surface preparation to install a suitable anchor profile, anticipated temperature/moisture conditions in the shaft, and the availability of skilled labor necessary for successful application were considered. Epoxy and polyurethane coatings became undesirable options since neither technologies would support an extended design life without maintenance. Of the reinforced CML and shotcrete options, reinforced shotcrete was selected because of Owner preference and the historically poor performance of CML when subjected to wet/dry cycles caused by tunnel unwaterings and restarts.

## ALTERNATE DESIGN APPROACH

The primary purpose of the reinforced shotcrete lining is to protect the steel access pipe from corrosion and to optimize the design life. The steel access pipe was designed solely to withstand the full net internal pressure head during operation and the external groundwater pressure head upon unwatering without any load sharing from the surrounding rock. Although the reinforced shotcrete lining serves as a protective barrier and is not required as a structural component, a portion of the loads will inherently be transferred to it during both operating and unwatered loading conditions based on its relative stiffness to the other components of the final system. In addition, gravitational loads need to be accounted for. To support self-weight of the shotcrete and its reinforcement, headed concrete connectors were required. The headed concrete connectors are welded to the steel components evenly throughout the interior of the steel access pipe to provide locations for the reinforcement cage to be secured during shotcrete application.

## Design and Serviceability Requirements

### *Design Requirements*

The design requirements of the reinforced shotcrete lining included meeting the minimum required shotcrete compressive strength per ACI 318 (2019). The shotcrete lining was designed conservatively following plain concrete design requirements, including applying a reduction factor of 0.6. Therefore, only nominal reinforcement was required for early age shrinkage and crack control. Minimum reinforcement requirements were set at 0.25% of the gross cross-sectional area of the shotcrete lining, consistent with project criteria. This minimum reinforcement requirement was set considering a combination of ACI 318 (2019) and ACI 350 (2020) requirements. The minimum clear cover requirement for the shotcrete reinforcement and headed concrete connectors was set at 2.5 inches, which is also consistent with project criteria.

### *Serviceability Requirements*

Reinforced shotcrete lining serviceability requirements included a smooth trowel finish primarily to achieve a visual similar to a formed finish and to control crack widths. Design for crack control included limiting the stresses in the steel reinforcement following ACI 350 (Eq. 10-4) requirements and considering normal environmental exposures. Following this criterion prevents both early age shrinkage cracking and cracking due to internal hydrostatic pressures upon Bypass Tunnel operation.

## Loading Conditions

Loading conditions include operational (when the Bypass Tunnel is in service) and unwatered (when the Bypass Tunnel hydraulic grade line [HGL]) is lowered below the access pipe elevation. Gravitational loads include the shotcrete lining and reinforcement self-weight. The operational and unwatered loading conditions are further defined in the following sections.

### *Operational Loading*

During Bypass Tunnel operation, the access shafts will be filled with Aqueduct water that is pressurized based on the HGL at the shaft locations and the elevation of the shaft pipe system. Because of external groundwater pressures, a net internal hydrostatic pressure (total internal hydrostatic pressures minus external groundwater pressures) will act on the access pipe. The steel access pipe was conservatively designed for the maximum net internal hydrostatic head without any load sharing based on the assumption that gaps could form between the pipe and the refill concrete, preventing any load transfer to the surrounding rock. The shotcrete lining, although not structurally required to support the net internal hydrostatic pressure, will experience load within the access pipe. Steel reinforcement is required to control the width of the cracks.

### *Unwatered Loading*

When the HGL is lowered below the bottom of the access pipe, the total net pressure will be acting externally on the access riser pipe. That is, the external groundwater pressure will be greater than the internal pressure based on the position of the HGL. Upon loading, the access pipe will deform inward against the shotcrete lining, which will absorb a portion of the load based on the relative radial stiffness between the two liner components. About 40% and 30% of the total external hydrostatic load is estimated to be transferred to the shotcrete lining at the Shaft 5B and Shaft 6B locations, respectively.

## Shotcrete Design

The tensile hoop stresses in the steel reinforcement during operation and the compressive hoop stresses in the shotcrete lining during unwatering were estimated using Roark's closed form solutions for hoop stresses due to uniform loading on a cylindrical shell (Eq. 1). Even though no load sharing was considered for the access pipe design to check reinforcement and shotcrete stresses for the protective liner, load sharing between the access pipe and shotcrete reinforcement and between the access pipe and shotcrete liner was considered during Aqueduct operation and unwatering, respectively. The load sharing distribution as a percentage was estimated based on the relative radial stiffness between the two load sharing components (Eq. 2).

Hoop stress of a circular pipe due to uniform pressure is as follows:

$$S_h = \frac{P \times R}{t} \qquad \textbf{(EQ 1)}$$

where,

$S_h$ = hoop stress (ksi)
$P$ = applied uniform pressure (ksi)
$R$ = pipe radius (inch)
$t$ = thickness of pipe (inch).

Load sharing between two shaft lining components resisting hoop stresses is as follows:

$$LS_{1,2} = \frac{\dfrac{E_{1,2} \times A_{1,2}}{R_{1,2}^2}}{\dfrac{E_1 \times A_1}{R_1^2} + \dfrac{E_2 \times A_2}{R_2^2}} \times 100\% \qquad \textbf{(EQ 2)}$$

where,

$LS_{1,2}$ = load share of components 1 or 2 (%)
$E_{1,2}$ = elastic modulus of components 1 or 2 (ksi)
$A_{1,2}$ = cross-sectional area of components 1 or 2 (square inch)
$R_{1,2}$ = radius of components 1 or 2 (inch).

The compressive stress in the shotcrete lining was checked against requirements per ACI 318 (2019). A final minimum comprehensive strength and thickness of the shotcrete lining was determined to be 4,000 psi and 4.5 inches, respectively. A WWR of 4x4 -D4xD4 was selected based on the Contractor's means and methods. During Aqueduct operation, the tensile stress in the shotcrete reinforcement was confirmed to be less than that required by ACI 350 Eq. 10-4 and meet crack control requirements.

### Headed Concrete Connector Design

The headed concrete connectors were designed to withstand the self-weight of the reinforced shotcrete. The vertical load induced on each shear connector is dependent on the circumferential and vertical spacing of the connectors along the access pipe steel surface. The nominal shear resistance of the connector embedded in the shotcrete liner was determined following AASHTO LRFD Chapter 6.10.10 (2020).

The final required spacing of the headed concrete connectors to support the reinforced shotcrete lining was determined to be 2 feet by 2 feet along the entire shaft height and circumference. Headed concrete connectors with ½ inch diameter and ultimate strength of 61,000 psi were selected.

## SHOTCRETE MOCK-UP

A mock-up was necessary because, although shotcrete is a viable alternative, it requires a high degree of quality control. If not performed correctly, the results could be undesirable. Goals of the mock-up included demonstrating the following:

1. Surface preparation of the steel plate using high pressure water washing is adequate for proper shotcrete application.
2. Welded headed concrete connectors are installed and pass verification testing.
3. Circumferential bar reinforcement is adequately secured to the welded headed concrete connectors.
4. Welded wire reinforcement (WWR) is adequately secured to the welded headed concrete connectors/circumferential bar system prior to shotcrete application.
5. Approved shotcrete mix performs satisfactory during application, (i.e., bonds well to steel plate, no evidence of sagging through set-up, no excessive rebound).
6. Shotcrete minimum depth and cover are met and adequately encapsulate the reinforcing steel.
7. Shotcrete as installed over the WWF contains minimal voids, sand pockets, or debonded material.
8. Shotcrete surface finish by troweling is smooth.
9. Certified nozzlemen used to apply mock-up shotcrete are prequalified and subsequently assigned to install shotcrete during the actual work for consistency.
10. Qualified and skilled individual providing oversight of the mock-up is also present to provide oversight during the actual work.

The mock-up was performed in February 2022 using an available 10-foot section of the steel interliner initially procured for the tunnel. The mock-up included all components of the design (pressure washing of the steel pipe, shear connector and WWR installations, and shotcrete) but on a smaller scale. Enough WWR sheets to overlap and form at least three circumferential (vertical) laps in addition to a longitudinal lap at each circumferential lap location (i.e., three representative locations for each nozzleman of the worst-case lap of three sheets layered on top of each other should exist) were installed and shotcreted in the Engineer's presence. The mock-up quality control included core testing to confirm "very good" steel encasement per ACI 506.6T-17, panel testing to confirm shotcrete mix strength, and thickness/cover verification. Figures 3 and 4 are from the mock-up. See Figure 5 for core quality which was "very good."

This mock-up was successful, and the design was finalized and issued for construction.

Figure 3. Mock-up: shotcrete application

Figure 4. Mock-up: completed

## SHOTCRETE LINING INSTALLATION

### Access

Key to a productive operation is good access to the work. Considering that the work was to take place more than 600 feet above the bottom of the shafts posed some unique challenges. The project team had extensive knowledge of working throughout the shaft, from placing concrete shaft liner to placing and installing the steel access pipe itself. Two work decks were designed and fabricated to be used at each shaft, to allow for concurrent operations.

Figure 5. Core from mock-up

The access decks were designed to handle loading for all the steps of the operations listed and described in the following sections. To ensure the safety of the workers on the deck, a roof was incorporated into the lifting frame above the deck. The roof was designed to withstand the loading from dropped objects. The lifting frame located above the deck also had lighting installed, plus the open diamond grading of the work deck allowed for air movement up and down the shafts to reduce the concentration of airborne particles from the shotcreting operation.

The decks were suspended from a Favco crane on the 5B side and a Cobelco crane on the 6B side. As secondary access to the deck, a backup crane with a bullet cage was available.

### Work Sequence

There were four major components of work to install the shotcrete lining, all performed in June/July 2022: pressure washing of steel lining; headed concrete connector and wire mesh installation; applying finishing, and curing shotcrete; and final cleanup.

#### *Pressure Washing of Steel Lining*

Total quantity of work was 17,762 square feet (SF) between the two shafts, and the work was completed with a 3,000-psi pressure washer located on the work deck. Design required that any rust scales, grime, oil, and dirt be removed prior to shotcrete application.

### *Headed Concrete Connector (Nelson Studs) and WWR Installation*

Nelson studs were used as headed concrete connectors. KSC rented Nelson stud guns and received training in stud installation from the supplier. A total of 4,394 EA studs were installed in the two shafts: a row of 28 EA studs per 2 feet vertically. A #4 rebar was bent to the diameter of the inside of the studs every other lift of studs, at 4-foot centers. The purpose of this bar, which is nonstructural, was to provide stiffness to the WWR and prevent it from vibrating during the shotcrete application. Further, it expedited the WWR installation as the wire did not have to be aligned on the mesh exactly with the studs. Figure 6 is a view down the inside of the steel liner from the work deck, showing the #4 bar and stud detail, prior to WWR installation. WWR was inspected during installation to ensure stiffness and correct overlap. See Figure 7.

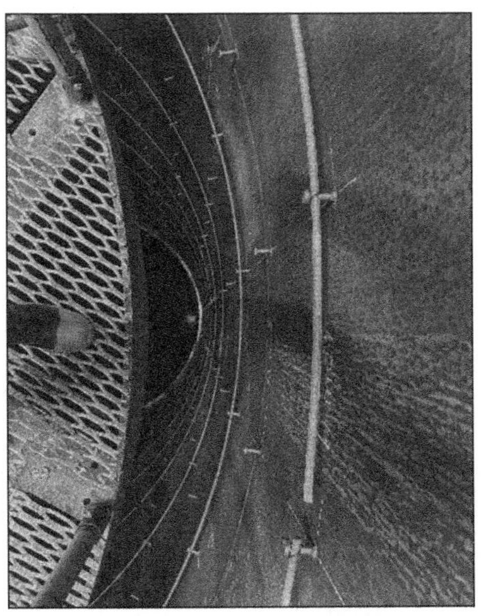

**Figure 6. Detail of #4 bar tied to the studs**

**Figure 7. Inspection to ensure the correct overlap of WWR**

### *Applying, Finishing, and Curing Shotcrete*

The shotcreting operation was performed by a specialty subcontractor on a single shift operation. The shotcrete crew consisted of six personnel on the deck: a bottom lander, a nozzleman, and four finishers. A second shift was used for applying curing compound to the finished shotcrete and to clean up and prepare the work deck for the following day's operation. The total quantity shotcrete for the work was 17,762 SF. At a thickness of 4.5 inches, this is 246 cubic yards (CY) neat.

Prior to starting the shotcrete operation, the specialty subcontractor tied a horizontal pencil rod to the wire mesh to serve as a screed for the finishers and to ensure the adequate cover of shotcrete over the WWR. The pencil rods were removed on the back shift.

The shotcrete equipment consisted of two high-pressure shotcrete pumps, one main and one spare; a 2-inch steel slick line to the top of the access pipe; and a 2-inch bull hose down to the work deck and the nozzle applicator. Air was delivered to the nozzle from KSC's on-site compressor though a ¾-inch hose. No accelerator was used for the shotcrete. See Figure 8 for a view down the finished 5B Shaft.

Figure 9. The completed Shaft 5B from above

### *Shaft Bottom Cleanup*

Despite having placed heavy plastic sheeting on the shaft bottom, the cleanup operation turned out to be more encompassing than anticipated. Lessons were learned from 6B, the first shaft to be shotcreted.

First, the quantity of waste shotcrete exceeded the volume anticipated. This was mainly due to the relatively thin lining where a few inches of overspray is a large percentage of the total. One inch of overspray on a 4.5-inch lining is over 20%. Typically, the subcontractor would overspray the pencil rods with an inch minimum and then trowel the surface back to the pencil rod to achieve the required smooth trowel finish. Over the depth of the access pipes this resulted in 50 CY of shotcrete ending up in the bottom of the shafts.

Second, the shotcrete would "splatter" up the sides of 6B when it hit the shaft bottom following a 600–800 foot drop. Therefore, at Shaft 5B, KSC installed sheets of heavy plastic 20 feet up the walls and covered the entrances to the Bypass Tunnel, which had received a good amount of "splatter" during the earlier shotcrete installation at Shaft 6B.

The third lesson learned was that the subcontractor would drop the discarded pencil rods down the shaft. These pencil rods would get embedded in the wet mix and create a shotcrete porcupine, which was a painful experience to remove.

## CONCLUSION

Mockups are valuable in identifying issues with an initial approach and later to verify a design and application method, including an evaluation of the personnel actually performing the work.

This small, but important, piece of work on a large project exemplified the importance of having a collaborative environment between project owner, Construction Manager, designer, contractor and specialty subcontractor, to tackle a technical and operational challenge in order not to delay the project.

Application and trowel finishing of shotcrete linings require highly skilled and trained personnel. Never underestimate the need for clean up after a shotcrete operation.

For Bypass Tunnel Shafts 5B and 6B, the quality of the shotcrete finish was exceptional and similar to a formed concrete surface.

## ACKNOWLEDGMENT

The authors wish to acknowledge the New York City Department of Environmental Protection for allowing to publish this work which was an example of collaboration on the Bypass Tunnel project and also Patriot Shotcrete, Jersey City, NJ for their contribution and hard work.

## REFERENCES

American Concrete Institute. 2019. *Building Code Requirements for Structural Concrete and Commentary (ACI 318-19).* Farmington Hills, MI: American Concrete Institute.

American Concrete Institute. 2021. *Code Requirements for Environmental Engineering Concrete Structures and Commentary (ACI 350-20).* Farmington Hills, MI: American Concrete Institute.

ACI Committee 506. 2017. *Visual Shotcrete Core Quality Evaluation Technote (ACI 506.6T-17).* Farmington Hills, MI: American Concrete Institute.

American Association of State and Highway Transportation Officials (AASHTO). 2020. *LRFD Bridge Design Specifications,* 9th edition. Chapter 6.10.10. Washington, DC: AASHTO. Budynas R.G. and Sadegh A.M. (2020). *Roark's Formulas for Stress and Strain, Ninth Edition.* McGraw-Hill.

# Direct Tensile Testing and CT-Scanning of Fiber Reinforced Shotcrete

**Mark Trim** ▪ Delve Underground
**Denis Tepavac** ▪ Delve Underground

## ABSTRACT

Structural performance of fiber reinforced concrete (FRC) depends on its ability to carry tensile forces after cracking. Australian Standards AS5100-2017 and AS3600-2018 are the only available design standards that establish a test procedure to determine strength and toughness of an FRC in direct tension. Direct tensile testing of fiber reinforced shotcrete samples was performed to determine if the model and guidelines in AS5100/3600 were applicable to shotcrete; all previous testing was from cast samples. In addition to direct testing, micro CT-scanning was also performed. The paper presents tests performed, results, and how the tests could be used on future projects. In addition, a brief summary is provided of the guiding structural mechanics used for estimating residual tensile strength and test methods utilized to verify those estimates.

## INTRODUCTION

This paper is based on a research project that compared residual tensile testing (direct tensile testing) of shotcrete with flexural beam testing (indirect tensile testing) and on the authors' experience using fiber reinforced shotcrete for permanent and temporary tunnel structural linings.

The flexural strength of fiber-reinforced concrete structural linings, specifically sprayed concrete (i.e., shotcrete), is determined after the concrete has cracked (i.e., exceeded the peak flexural tensile strength). This post-crack strength parameter is called "residual flexural tensile strength."

Verifying residual flexural tensile strength is challenging. A key reason for this is that concrete no longer behaves elastically post crack. The behavior of fiber reinforced concrete post crack will vary (strain hardening or softening) depending on the mix design and, particularly, the type of fiber used. To design and to verify performance assumptions for fiber-reinforced shotcrete linings, the designer must determine the performance needed from the lining and the test method—and it is critical to use a test method that matches the design method. That is, the test method must return a residual flexural tensile strength value that can verify the design, ideally with limited interpretation and/or post-processing of the test results.

Because of the approximations used in the current design and test methods (employing indirect test methods) and recent publication of the direct tensile test method in Australian Standards (AS) 3600 and 5100, Delve Underground partnered with a local contractor and the University of New South Wales (UNSW) to perform direct tensile tests on steel fiber reinforced shotcrete test specimens. The direct tensile test results were then compared to beam test results to better understand how flexural beam test approximations compare to direct tensile test results. This testing, its data, and interpretation were the scope of the aforementioned shotcrete research project.

Some of the material presented in this paper first appeared in an internal report: "Determination of Residual Tensile Strength and Fiber Orientation of Steel Fiber Reinforced Shotcrete" (Stephen Foster, Ahsan Parvez, Lakshminarayanan Mohana Kumar, Zhen-Tian Chang, and Mark Trim, 2020).

## STRESS BLOCK AND DETERMINATION OF TENSILE STRESS FOR SFRC

The design of steel fiber reinforced shotcrete (SFRS) at both the ultimate and serviceability limit states is based upon the post-cracking stress-strain relationship under imposed actions. The stress-strain behavior of SFRS sections is characterized by the tensile stress ($\sigma$) versus crack opening displacement (COD) relationship $\sigma$–*w*, where "*w*" is the width of the crack. The $\sigma$–COD relationship is used to develop the stress block, which is then used to determine equilibrium of forces and flexural capacity.

The flexural capacity of SFRS sections is determined based on the assumed geometry of the tensile stress block. Designers use the tensile stress block in combination with the compressive stress block to solve the equilibrium of forces equation and determine the flexural capacity. This is no different than design principles used to design conventionally reinforced concrete.

Prior to AS3600-2018/AS5100-2017, model codes produced by industry bodies defined the tensile stress blocks for use. The most commonly cited model codes and corresponding tensile stress blocks are the RILEM TC 162 (2003) and the *fib* Model Code 2010 (2013). The *fib* Model Code and RILEM stress blocks were developed assuming indirect prism beam testing and are based on residual flexural tensile stresses. Figure 1 shows three typical tensile (residual) stress blocks used by common model codes. This paper does not compare differences between the model codes, but in the authors' experience, there is little difference in outcomes in terms of flexural capacity between the varying stress blocks.

Determination of residual tensile strength is via testing. Various tests, including direct and indirect tensile tests, can be used to obtain the $\sigma$–COD relationship for SFRS.

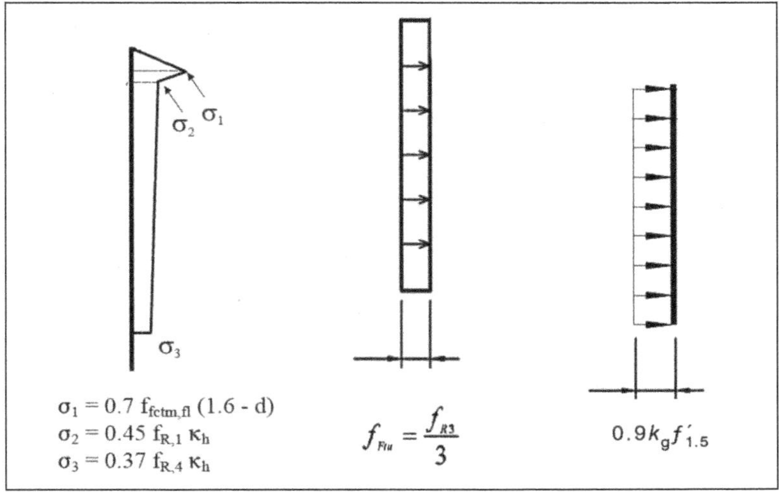


Source: Left to Right: RILEM 2003, fib Model Code 2010, and AS5100-2017/AS3600-2018.

**Figure 1. Common applied tensile stress blocks (from left to right, RILEM, *fib* Model Code, and Australian Standard AS5100/3600)**

## Determination of Residual Tensile Stress by Direct Tensile Testing

Direct tensile testing (e.g., uniaxial tensile test) is relatively uncommon in practice. Although there is discussion of direct tensile testing in literature such as RILEM TC 162 (2003), there is no consensus in the wider international community on the type and methodology for testing. The Australian Standards are the first internationally recognized design standards to codify direct tensile testing as part of AS5100-2017/ AS3600-2018.

The Australian Standards have adopted the "dog bone" test based on the research undertaken by van Vliet (2000). The research into tensile characteristics of unreinforced concrete was further developed by Htut (2010), Ng et al. (2012), and Amin et al. (2015), who similarly adopted the dog bone specimen and found results that agreed with van Vliet and van Mier (1999). The testing arrangement from the Australian Standards is shown in Figure 2. Using the AS5100/3600 uniaxial test geometry (dog bone), tensile strength prior to cracking and post-cracking is directly measured.

The results of direct tensile tests focus on the stress strain behavior post crack and the $\sigma$–COD relationship. If the designer is using the AS5100-2017/AS3600-2018 model as the basis of design, then COD values of most interest are defined at COD = 0.5 mm and 1.5 mm. The two stress values corresponding to the COD ($w$) at 0.5 and 1.5 mm are referred to as $f_{0.5}$ and $f_{1.5}$, and correlate to the COD at the serviceability limit state and ultimate limit state (as defined by AS5100-2017/AS3600-2018). Figure 3

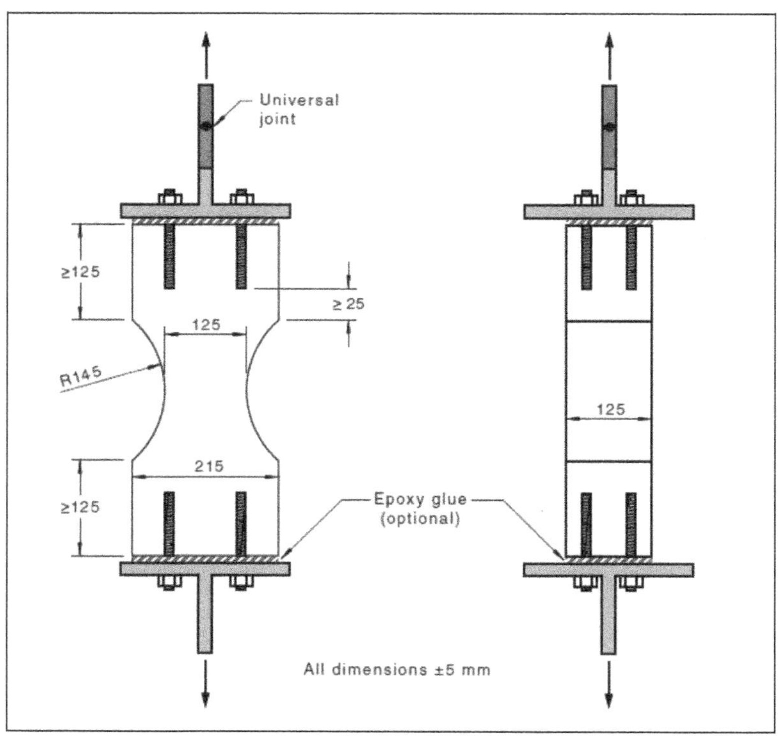


Source: AS5100-2017/AS3600-2018.

**Figure 2. Testing arrangement for direct tensile testing from Australian Standards**

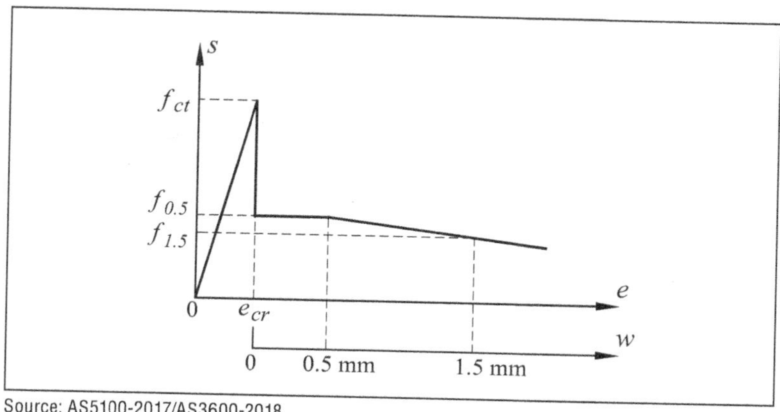


Source: AS5100-2017/AS3600-2018.

**Figure 3. σ–w relationship derived from direct tensile testing per Australian Standards**

presents the σ–*w* relationship from AS5100-2017/AS3600-2018, representing strain softening behavior.

Although the Australian Standards (AS5100/3600) have developed procedures and methods for direct tensile testing, direct tensile testing remains more complex and time-consuming than indirect tests. There is value in performing direct tests, but the owner, designer, and contractor need to validate if or how the test may be used in development of SFRS mix. At this time, the authors suggest using this test for reporting purposes only until more data and experience are gained from application of this test method to sprayed concrete test specimens.

## Determination of Residual Tensile Stress by Indirect Tensile Testing

Indirect tensile testing is common. On most major underground infrastructure projects in Australia and the United States, indirect tests are the only tests used. The common testing methodologies are the notched beam bending test EN14651 and the unnotched beam bending test according to ASTM C1609. The shotcrete research project used the EN14651 test and is discussed below. The ASTM C1609 follows similar principles in determining the residual strength of SFRS beams.

EN14651 derives the flexural tensile strength from the force (*F*) versus crack mouth opening displacement (CMOD). The CMOD is defined as the maximum crack width at the mouth of the crack. This differs from the COD (*w*), which is measured as the mean near-uniform width across the tension plane. The EN14651 test assembly is shown in Figure 4. Figure 5, per EN14651-2007, provides a definition of the key CMOD points used during testing and post-processing of the bending test results.

If the Australian Standard model (AS5100/3600) is used, the CMOD values associated with Serviceability Limit State (SLS) and Ultimate Limit State (ULS) are $CMOD_2$ and $CMOD_4$. The $CMOD_2$ and $CMOD_4$ values relate to maximum crack mouth opening displacements at 1.5 mm and 3.5 mm, respectively. The corresponding residual flexural tensile stress values are $f_{R.2}$ and $f_{R.4.}$ The authors note that $CMOD_2$ is used by the Australian Standard model for SLS. Other international models may use other CMOD values for SLS.

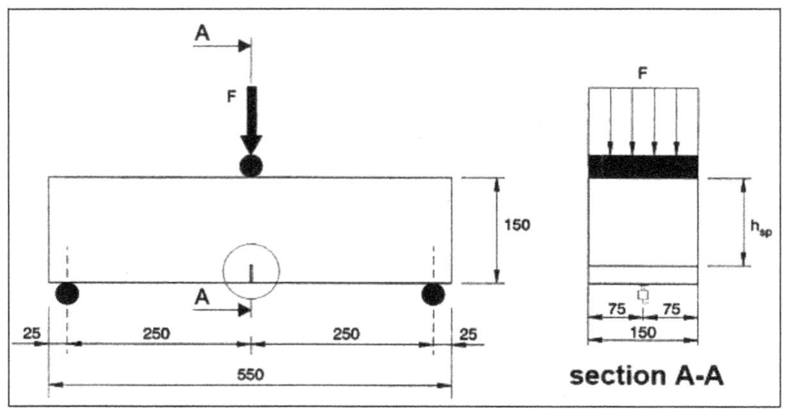


Source: European Standard, EN14651-2017.

**Figure 4. Three-point prism bending test arrangement**

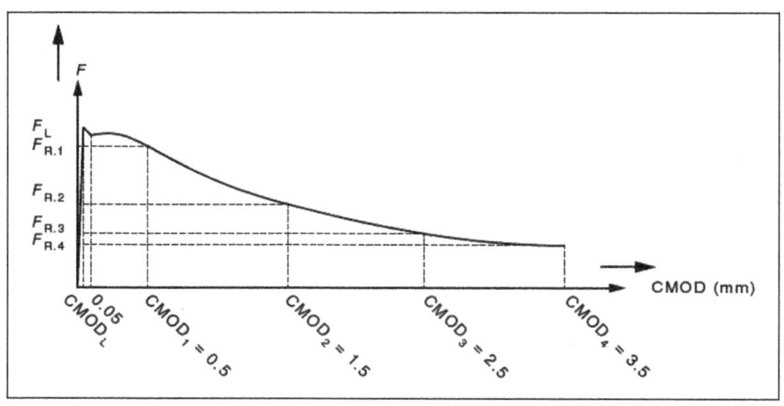


Source: European Standard, EN14651-2017.

**Figure 5. Definition of key points on the applied force versus CMOD curve for flexural testing**

The residual flexural tensile stress values at any CMODj, $f_{R \cdot j}$, can be determined from the relationship below, which is a function nominal test force $F_{Rj}$.

$$f_{Rj} = \frac{3F_{Rj}a}{bh_{sp}^2} \ldots\ldots j = 1-4 \qquad \textbf{(EQ 1)}$$

where $a$ = shear span, $b$ = width of the prism, and $h_{sp}$ = distance between tip of the notch and top of cross section.

## Inverse Analysis to Determine Residual Tensile Strength

As the name suggests, indirect testing does not directly determine residual tensile strength values. Instead, it estimates residual flexural tensile strength. In order to compare data (residual strength) from indirect test to direct test, inverse analysis is required. Inverse analysis converts the residual flexural tensile strength to direct tensile strength. Figure 6 is a simplified diagram illustrating the conversion (via inverse analysis) of common test methods to direct/uniaxial tensile ($\sigma$–$w$) relationship (post crack).

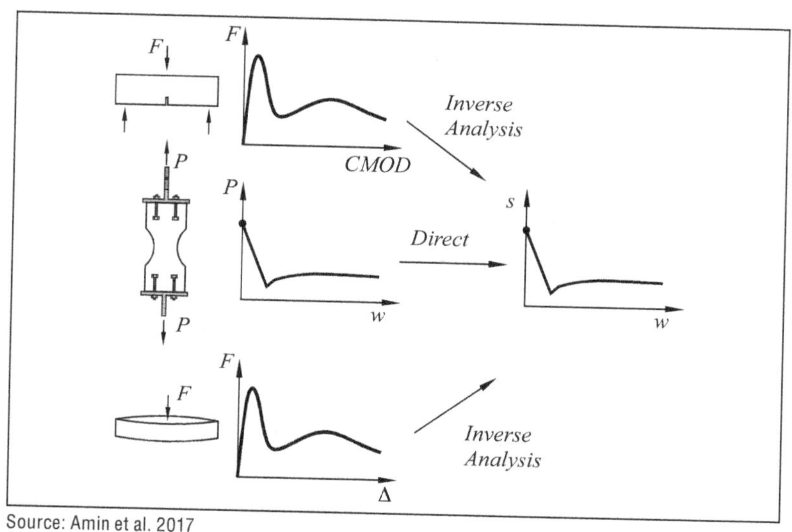


Source: Amin et al. 2017

**Figure 6. Indirect and direct methods for obtaining the tensile stress versus crack opening relationship of SFR concrete/shotcrete**

For this paper, the formulations used for the inverse analyses are from AS3600-2018. For details on the inverse analysis model used to inform this paper and AS3600-2018, refer to Amin et al. (2015), Foster et al. (2017), and Foster and Parvez (2019).

AS3600-2018 provides simplified formulations for conversion of residual flexural tensile stress to residual tensile stress, which are:

$$f'_{0.5}Ff = k_{3Db}(-0.04f'_{R,4} + 0.37f'_{R,2}) \leq k_{3Db}0.36\sqrt{f'_c} \quad \text{(EQ 2)}$$

$$f'_{1.5} = k_{3Db}(0.4f'_{R,4} - 0.07f'_{R,2}) \leq k_{3Db}0.36\sqrt{f'_c} \quad \text{(EQ 3)}$$

## EXPERIMENTAL TEST PROGRAM

The initial test program included testing of direct test specimens (dog-bone) and indirect test specimens (flexural prism/beam). UNSW incorporated micro-computed tomography (micro-CT) investigation into the test program to better understand fiber content and orientation in the dog-bone specimens. The test program for the direct and indirect testing divided test specimens into two series based on fiber dosage (refer to Table 1). Each series contained nine dog-bone test specimens, per AS3600-2018, and nine prism test specimens, per EN14651-2007. The following is the nomenclature for identification of test specimens: "SF" for steel fibers, "40" for the fiber dosage in kg/m$^3$, "D" for direct tensile test ("P" for prism bending test), and "1" for specimen's number.

**Table 1. Shotcrete properties**

| Test Series | Slump, mm | Density, kg/m$^3$ | Fiber Dosage, kg/m$^3$ | Steel Fiber Volume, %* | UCS Strength, $f_{cm}$ †, MPa |
|---|---|---|---|---|---|
| SF40 | 180 | 2,330 | 40 | 0.50 | 59.6 |
| SF45 | 160 | 2,330 | 45 | 0.57 | 48.0 |

* One kg/m$^3$ of polypropylene fibers was added, in addition to steel fibers.
† Average of three samples.

A total of 36 tests were divided into two equal groups of 18 specimens containing both Series 1 and Series 2 fiber dosages. One group of 18 was tested in accordance with the AS3600-2018 dog-bone test, the while the other followed the EN146512007 test methodology.

## Test Specimen Materials

The testing shotcrete specimens were sprayed at the construction site. The shotcrete was specified as grade of 40 MPa with maximum aggregate size of 10 mm. The compressive strength of the concrete was determined from 75 mm diameter × 150 mm high cylinder cores tested after 28 days. The results are summarized in Table 1.

The steel fibers used in this study were Dramix® end-hooked 4D-65/35-BG fibers. In addition to the steel fibers, 1 kg/m$^3$ of micro-synthetic monofilament polypropylene fibers was added to lower the risk of spalling at elevated temperatures. The micro-synthetic monofilament polypropylene fibers were not expected to influence the residual tensile strength. The properties of fibers are given in Table 2.

**Table 2. Properties of fibers**

| Fiber Type | Fiber Type | Length, mm | Diameter, mm | Aspect Ratio | Tensile Strength, MPa |
|---|---|---|---|---|---|
| Steel 4D | Hooked end | 35 | 0.55 | 65 | 1,850 |
| Polypropylene (pp) | Straight | 12 | 0.03 | 400 | – |

## Direct Tensile Testing

Eighteen dog-bone specimens in two series were prepared in accordance with Appendix C of AS3600-2018. Prior to casting, four 16 mm threaded rods were placed within 100 mm of each end of the formwork to act as anchors to bear tensile forces during testing (Figure 7a). The molds were then transported to the site for spraying of the shotcrete. The shotcrete spraying nozzle is shown in Figure 7b. The specimens were demolded after 24 hours and cured. After three weeks, the specimens were transported to the UNSW Heavy Structures Laboratory for testing. Authors note that trials were initially done with post-drilled anchors using the contractor manufactured dog-bone molds. Unfortunately, the level of accuracy needed to drill in straight dowels for the required embedment was difficult to achieve, and these molds and method of anchoring were not used. The molds used for the testing were provided by UNSW.

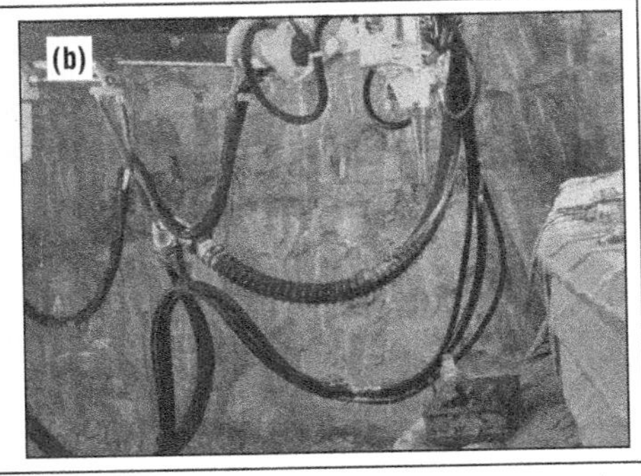


**Figure 7. Casting of dog-bone tensile specimens: (a) Dog-bone specimen molds; (b) Shotcrete spraying nozzle**

The dog-bone specimens were tested in a 250 kN Instron servo-hydraulic universal testing machine (UTM), as shown in Figure 10. The threaded rods protruding from the specimens were bolted to end plates and connected to the UTM. One end of the specimen was fixed in the grips and the other end was fitted with a universal joint to eliminate any residual tension that may develop during the gripping process. To measure the COD, four displacement transducers were attached, one to each face of the specimen in its mid height. The gauge length was 230 mm (Figure 10b). The COD ($w$) is calculated as the average of the four transducer measurements. Load was applied using displacement control. The progressive crack development during the test is shown in Figure 10.

## Prism Bending Test

Eighteen prisms were sprayed with SFRS, together with the dog-bone specimens, with nine prisms in each of SF40P and SF45P series. The nominal dimension of the prisms were 150 × 150 mm in cross section and 500 mm in length. The prism specimens were notched using a diamond blade saw. The prisms were tested at 28 days by a local certified testing laboratory following EN14651-2007.

## Micro-Computed Tomography (Micro-CT)

The potential for fiber orientation to be influenced by specimen preparation and geometry is well covered in literature and is equally represented in the various model codes and AS5100/AS3600. Using micro-computed tomography (micro-CT or µCT) technique described in Miletić et al. (2020), the failed dog-bone specimens were analyzed to ascertain the three-dimensional (3D) fiber orientation and distribution.

To conduct µCT analyses, two cores were taken, one from each batch. A core was taken from an undamaged region in one-half of the dog-bone specimens for specimen SF40D-6 and specimen SF45D-4, with their centers located approximately two fiber lengths away from the fracture plane (Figure 8). For both cores, the core nominal diameter was 62 mm and height 120 mm.

### *X-ray Micro-CT Imaging*

The specimens were scanned using the X-ray micro-CT facility available at Tyree Energy Technologies Building, UNSW. The original 3D images were scanned to a resolution of 363 per cubic millimeters (46,656 voxels/mm$^3$); that is, a resolution of 28 mm in any direction.

Figure 8. Micro-CT extraction of SFRS cores: (a) coring process, (b) one core from each batch

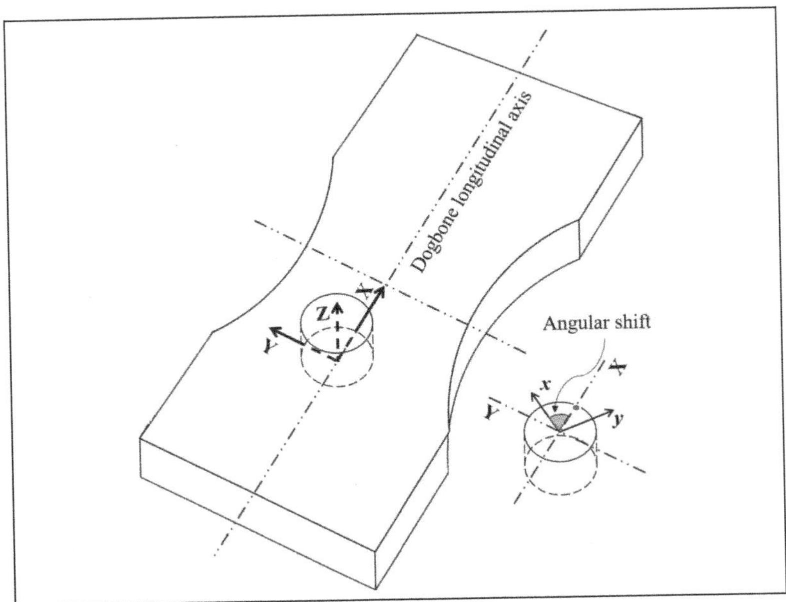


Figure 9. Coordinate systems of dog bone (global) and core specimen (local)

To manage the 120 mm length of the cores, the specimens were imaged in three parts: top, mid, and bottom, with the core aligned in the μCT field accordingly. The top, mid, and bottom images were then digitally stitched together using the software Avizo.

### Coordinate Systems

There are two coordinate axis systems used to define the fiber orientations observed in the specimen (image) to the coordinate system of the dog-bone specimen. The first is the global coordinate system, which is a right-handed system denoted by X, Y, and Z, with origin located at the bottom center of the dog-bone specimen (Figure 9). The Z axis is positive upwards, that is through the thickness of the dog bone. The second coordinate system, the local system, is denoted by *x*, *y*, and *z* axes, which are local to the images of core specimens defined by its orientation during the μCT scanner.

The core specimen is maintained in a vertical position during scanning, with the fiber orientations measured in the image coordinate system related to the dog-bone axes by observing a reference target placed on the core, and observable in the produced image. The local z-axis runs through the depth of the core and is aligned with the global Z-axis of the dog-bone specimen. Examples of core slices in each principal plane are provided in "Experimental Test Results."

## EXPERIMENTAL TEST RESULTS

As stated previously, a total of 36 test samples were produced and tested. Eighteen of the samples were dog-bone samples, and 18 were prism samples. The direct tensile dog-bone results are presented first, followed by the prism results, and finally the X-ray μCT scanning results.

Figure 10. Different stages of direct tensile test: (a) before cracking, (b) onset of cracks, (c) crack developing, and (d) end of test

## Direct Test (Dog-bone) Results

Seventeen of the 18 dog-bone specimens were deemed to be valid tests. Refer to Figure 11 for a photo of all 18 specimens tested. The one invalid test was from specimen SF45D-3. SF45D-3 failed near a potential defect/weakened plane near the anchor bolts; the results from this specimen were ignored during post-processing of the data. Figure 10 provides photographs taken at different stages in the test to illustrate how the crack develops during loading.

The post-cracking residual tensile strength for each specimen was determined and the $\sigma$–w curve found. The $\sigma$–$w$ curves are given in Figure 12a for series SF40D and Figure 12b for SF45D. In the processing of the data, the measured post-cracking

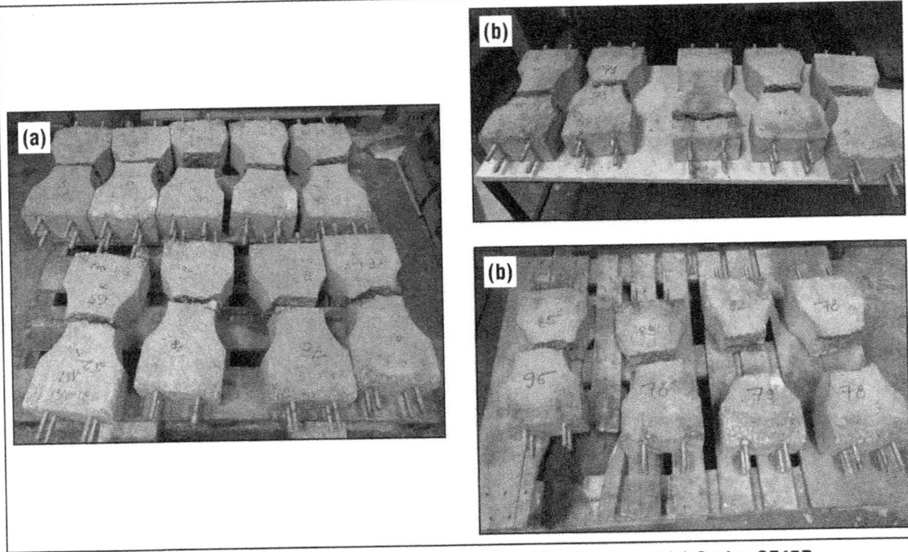


Figure 11. Views of tested dog-bone specimens: (a) Series SF40D, (b) and (c) Series SF45D

residual stresses were corrected for the wall (boundary) influence. The post-cracking residual tensile strength results are summarized in Table 3.

## Fiber Count at the Fracture Plane

At the conclusion of each test, a visual assessment of the crack plane for each dog-bone specimen was conducted and the numbers of fibers crossing the plane counted. The fiber count results are given in Table 4. Approximately 97% of the fibers failed by pull-out, and 3% ruptured. An example of the fracture surface at the conclusion of testing for Specimen SFD45-4 is presented in Figure 13.

For a three-dimensional randomly-orientate fiber distribution, the number of fibers (Nf) crossing a square fracture surface is given by (Aveston and Kelly 1973; Amin et al. 2017):

$$N_f = \frac{\rho_f A_c}{2k_{3Db} A_f} \quad \text{(EQ 4)}$$

where $A_c$ is the cross-sectional area of the failure surface, $\rho_f$ is the volumetric ratio of the fibers, $A_f$ is the cross-sectional area of the fiber, and $k_{3Db}$ is the fiber boundary effect factor (defined in AS3600-2018).

In Figure 14, the number of fibers crossing the crack planes are plotted against $f_{1.5}$ for each specimen; also plotted are the theoretical number of fibers for each mix calculated from Eq (4). It is seen that the residual stress is approximately linearly dependent on the number of fibers crossing the fracture plane. The ratio of fibers crossing the failure plane to the average number of fibers based on theory is 0.68 and 0.79 for SF40D and SF45D, respectively. This compares with the theoretical ratio for SFRC of 0.82, as demonstrated by Htut (2010) and Foster et al. (2014), and is the equivalent of the notch factor for notched prism bending tests. One potential explanation for the lower ratio found is that some fibers may be lost because of rebound during the initial phase of spraying the shotcrete into the model.

Table 3. Tensile strength and residual tensile strengths obtained from direct tensile test

| Specimen ID | Tensile Strength, $f_{ct}$, MPa | COD, w at Max Post-Cracking RTS, mm | Max Residual Tensile Strength, RTS, MPa | RTS at w = 1.5 mm, $f_{1.5}$, MPa |
|---|---|---|---|---|
| **Series SF40D** | | | | |
| SF40D-1 | 4.15 | 1.06 | 1.71 | 1.69 |
| SF40D-2 | 4.13 | 2.00 | 0.99 | 0.90 |
| SF40D-3 | 4.32 | 1.36 | 1.50 | 1.47 |
| SF40D-4 | 3.67 | 1.25 | 1.08 | 1.05 |
| SF40D-5 | 3.42 | 1.42 | 1.02 | 0.89 |
| SF40D-6 | 3.56 | 1.29 | 1.00 | 0.99 |
| SF40D-7 | 4.07 | 1.75 | 1.00 | 0.95 |
| SF40D-8 | 3.64 | 0.76 | 1.67 | 1.44 |
| SF40D-9 | 4.00 | 1.21 | 1.59 | 1.52 |
| Average | 3.88 | | 1.28 | 1.21 |
| CoV | 0.081 | | 0.251 | 0.259 |
| **Series SF45D** | | | | |
| SF45D-1 | 3.96 | 0.88 | 1.90 | 1.72 |
| SF45D-2 | 5.01 | 1.18 | 1.78 | 1.69 |
| SF45D-3* | — | — | — | — |
| SF45D-4 | 4.16 | 0.77 | 1.77 | 1.65 |
| SF45D-5 | 4.55 | 1.36 | 1.09 | 1.08 |
| SF45D-6 | 3.38 | 0.58 | 1.87 | 1.66 |
| SF45D-7 | 3.79 | 0.86 | 1.68 | 1.58 |
| SF45D-8 | 3.78 | 0.79 | 2.14 | 1.83 |
| SF45D-9 | 2.98 | 1.15 | 1.61 | 1.34 |
| Average | 3.95 | | 1.73 | 1.57 |
| CoV | 0.161 | | 0.176 | 0.155 |

* Specimen SF45D-3 failed at the anchor bolts.

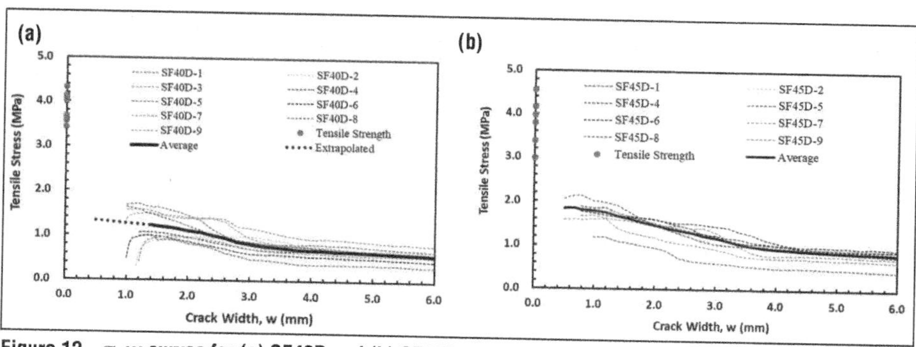


Figure 12. σ–*w* curves for (a) SF40D and (b) SF45D

## Prism Bending Test

A total of 18 prisms were sprayed with SFRS. To align with the dog-bone specimens for direct comparison, nine prisms from SF40P and nine from SF45P were prepared. The prisms were tested at 28 days by a certified local laboratory according to EN14651-2007. Load–CMOD graphs were obtained and the residual flexural tensile strengths (fR) calculated. The fR values from the laboratory results are shown in Table 5.

**Table 4. Number of fibers crossing the fracture plane in the uniaxial tension tests**

| Specimen ID | No. of Fibers on Each Face | | Ruptured Fibers | Total No. of Fibers | Residual Tensile Strength, MPa |
|---|---|---|---|---|---|
| | Face-1 | Face-2 | | | $f_{1.5}$ |
| **Series SF40D** | | | | | |
| SF40D-1 | 62 | 65 | 4 | 123 | 1.69 |
| SF40D-2 | 69 | 58 | 2 | 125 | 0.90 |
| SF40D-3 | 79 | 87 | 4 | 162 | 1.47 |
| SF40D-4 | 56 | 53 | 3 | 106 | 1.05 |
| SF40D-5 | 46 | 49 | 3 | 92 | 0.89 |
| SF40D-6 | 58 | 56 | 3 | 111 | 0.99 |
| SF40D-7 | 64 | 56 | 2 | 118 | 0.95 |
| SF40D-8 | 72 | 80 | 2 | 150 | 1.44 |
| SF40D-9 | 61 | 78 | 3 | 136 | 1.52 |
| Average | 63 | 65 | 3 | 125 | 1.21 |
| CoV | 0.153 | 0.211 | 0.271 | 0.175 | 0.259 |
| **Series SF45D** | | | | | |
| SF45D-1 | 80 | 95 | 4 | 171 | 1.72 |
| SF45D-2 | 71 | 73 | 3 | 141 | 1.69 |
| SF45D-4 | 91 | 89 | 3 | 177 | 1.65 |
| SF45D-5 | 76 | 53 | 4 | 126 | 1.08 |
| SF45D-6 | 85 | 95 | 3 | 177 | 1.66 |
| SF45D-7 | 83 | 76 | 2 | 157 | 1.58 |
| SF45D-8 | 92 | 79 | 4 | 167 | 1.83 |
| SF45D-9 | 78 | 78 | 2 | 154 | 1.34 |
| Average | 82 | 80 | 3 | 159 | 1.57 |
| CoV | 0.088 | 0.173 | 0.267 | 0.116 | 0.155 |

**Figure 13. Fracture surface of Specimen SF45D-4**

## Inverse Analysis from Prism Bending

Using inverse analysis and the data in Table 5, the $\sigma$–$w$ relationship has been developed (Figure 15). The results of analyses for each individual test are presented in Table 6 for $f_{0.5}$ and $f_{1.5}$. The averages and CoVs for each series are determined.

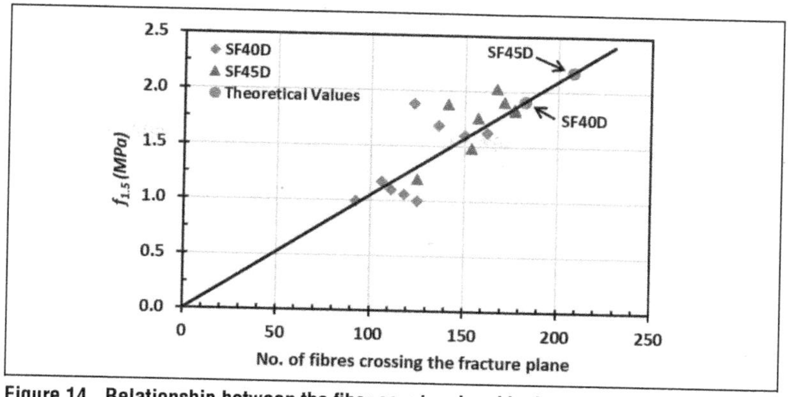


Figure 14. Relationship between the fiber count and residual tensile strength $f_{1.5}$

Table 5. Flexural residual tensile strength obtained from prism bending test

| Specimen | Residual Flexural Tensile Strength, MPa | | | |
|---|---|---|---|---|
| | $f_{R1}$, MPa | $f_{R2}$, MPa | $f_{R3}$, MPa | $f_{R4}$, MPa |
| **Series SF40P** | | | | |
| SF40P-1 | 4.8 | 5.3 | 4.5 | 4.0 |
| SF40P-2 | 4.4 | 4.7 | 4.3 | 3.8 |
| SF40P-3 | 4.4 | 4.1 | 3.4 | 3.0 |
| SF40P-4 | 4.7 | 5.6 | 5.3 | 4.9 |
| SF40P-5 | 4.5 | 5.2 | 4.6 | 4.1 |
| SF40P-6 | 4.1 | 4.7 | 4.1 | 3.2 |
| SF40P-7 | 3.6 | 3.5 | 3.2 | 2.7 |
| SF40P-8 | 4.8 | 4.5 | 4.2 | 3.7 |
| SF40P-9 | 3.9 | 4.2 | 3.9 | 3.4 |
| Average | 4.36 | 4.64 | 4.17 | 3.64 |
| COV | 0.095 | 0.142 | 0.152 | 0.182 |
| **Series SF45P** | | | | |
| SF45P-1 | 3.7 | 4.6 | 4.1 | 3.7 |
| SF45P-2 | 6.0 | 7.2 | 7.0 | 6.0 |
| SF45P-3 | 4.4 | 5.4 | 5.1 | 4.7 |
| SF45P-4 | 4.2 | 4.8 | 5.0 | 4.5 |
| SF45P-5 | 4.4 | 5.1 | 5.0 | 4.3 |
| SF45P-6 | 4.1 | 5.2 | 5.1 | 4.3 |
| SF45P-7 | 4.4 | 5.0 | 4.8 | 4.4 |
| SF45P-8 | 4.2 | 5.1 | 4.3 | 3.8 |
| SF45P-9 | 4.8 | 5.6 | 5.1 | 4.5 |
| Average | 4.47 | 5.33 | 5.06 | 4.47 |
| COV | 0.145 | 0.142 | 0.162 | 0.148 |

## Discussion on Residual Tensile Strength

As noted previously, the indirect tensile test method is simpler than direct tension for determining the residual tensile strength of fiber reinforced concrete/shotcrete. The data used to inform the Australian Standard model for the steel fiber reinforced section of AS5100-2017/AS3600-2018 are based on cast steel fiber reinforced specimens, and thus the accuracy of the direct tension test method has never been verified for SFRS.

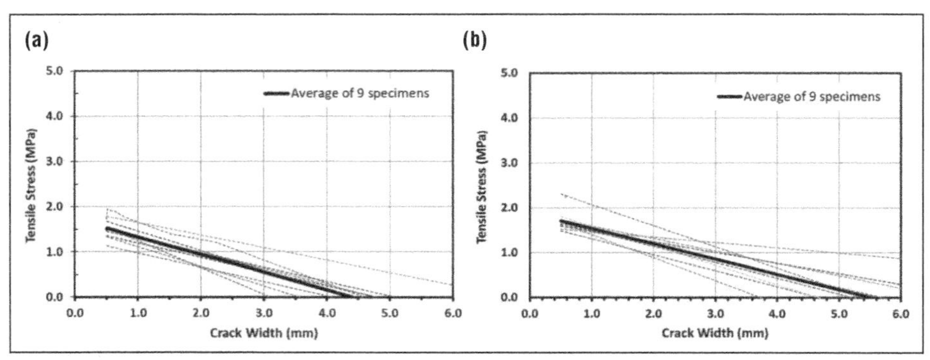


**Figure 15. Post-cracking residual tensile strength determined for each prism test**

**Table 6. Post-cracking residual tensile strength determined for each prism test**

| Specimen | $f_{0.5}$, MPa | $f_{1.5}$, MPa |
|---|---|---|
| **Series SF40P** | | |
| SF40P-1 | 1.72 | 1.21 |
| SF40P-2 | 1.51 | 1.16 |
| SF40P-3 | 1.33 | 0.90 |
| SF40P-4 | 1.79 | 1.51 |
| SF40P-5 | 1.68 | 1.24 |
| SF40P-6 | 1.54 | 0.95 |
| SF40P-7 | 1.13 | 0.82 |
| SF40P-8 | 1.45 | 1.13 |
| SF40P-9 | 1.35 | 1.04 |
| Average | 1.50 | 1.11 |
| CoV | 0.140 | 0.190 |
| **Series SF45P** | | |
| SF45P-1 | 1.48 | 1.13 |
| SF45P-2 | 2.31 | 1.84 |
| SF45P-3 | 1.73 | 1.45 |
| SF45P-4 | 1.52 | 1.40 |
| SF45P-5 | 1.64 | 1.32 |
| SF45P-6 | 1.67 | 1.32 |
| SF45P-7 | 1.60 | 1.36 |
| SF45P-8 | 1.66 | 1.14 |
| SF45P-9 | 1.81 | 1.37 |
| Average | 1.71 | 1.37 |
| CoV | 0.143 | 0.151 |

Figure 16 compares the results of the inverse analyses on prism bending tests conducted according to EN14651-2007 with the $\sigma$–$w$ diagrams determined by the uniaxial test for each series. The notch influence factor of kn = 0.82 (as used in SFRC) was applied to the prism results. It is seen that the inverse analysis procedure gives a good, generally lower bound, prediction for the key points $f_{0.5}$ and $f_{1.5}$ but underestimates the tensile strength at higher CODs. Also plotted in Figure 16 is the inverse analysis outcomes of AS3600-2018. It is seen that the Australian Standard's model provides for an appropriate level of conservatism using the inverse procedure but underestimates the contribution of the fiber to strength for design.

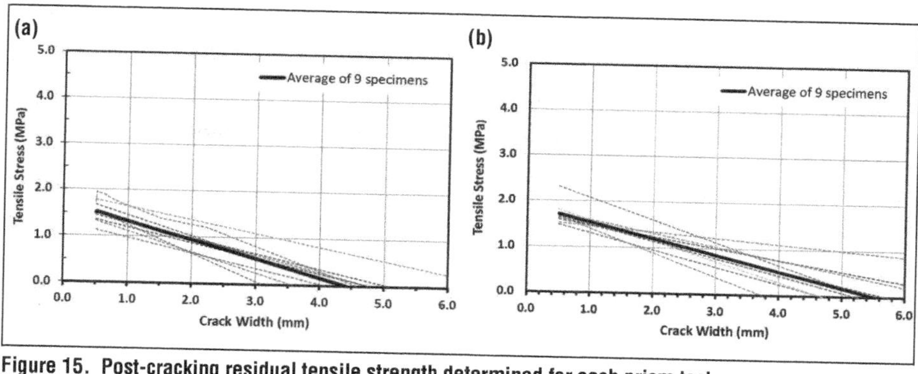


**Figure 15. Post-cracking residual tensile strength determined for each prism test**

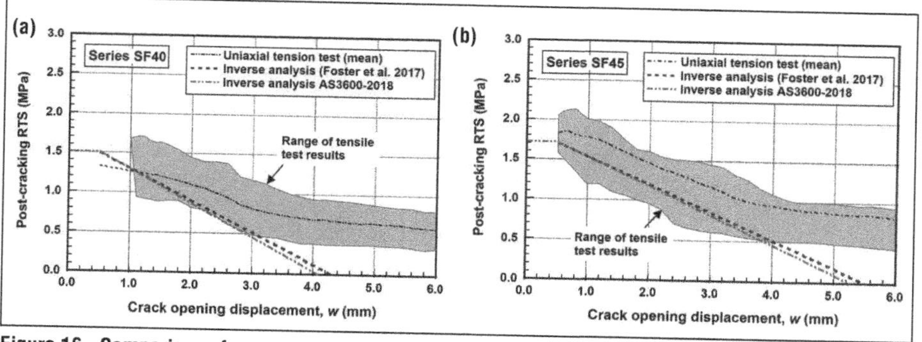


**Figure 16. Comparison of average σ–w relationship obtained from prism bending test with direct tensile test with correction factor (a) SF40 and (b) SF45**

Various factors influence the inverse model. In addition to manufacturing and consistency of fiber distribution, and relative dosages, through the mix we have the notch effect (which derives from the statistical variation of fibers when sampled at different sections), the ability for cracks to find paths of lower resistance, the influence of boundary walls on local orientation of fibers, and the effect of specimen preparation and casting on preferential alignment of fibers. In casting of the control specimens, preferential alignment of fibers in the direction of tension will increase strength, while preferential alignment normal to the direction of tension decreases strength. Casting of the direct tension and prism bending tests in this study followed similar processes; therefore, it is expected that any casting influence is similar in each and not causal to model differences.

The mean strengths are given in Table 7. Comparing the results shows that the uniaxial tests provide for a 10 to 20% higher strength at $w$ = 1.5 mm for the mean values.

**Table 7. Mean post-cracking residual tensile strengths determined from different procedures**

| Series | AS3600-2018 Direct Tension | | Prism Bending with Inverse Analysis | | AS3600-2018 with Inverse Analysis | |
|---|---|---|---|---|---|---|
| | $f_{0.5m}$, MPa | $f_{1.5m}$, MPa | $f_{0.5m}$, MPa | $f_{1.5m}$, MPa | $f_{0.5m}$, MPa | $f_{1.5m}$, MPa |
| SF40 | 1.33 | 1.21 | 1.50/1.56* | 1.11/1.15* | 1.51 | 1.09 |
| SF45 | 1.73 | 1.57 | 1.71/1.78* | 1.37/1.43* | 1.72 | 1.36 |

* First value obtained by averaging results analyzed from each test individually; second value obtained from averaged fR2 and fR4 test results.

AS3600-2018 provides a provides a mechanism for using inverse analyses calibrated to uniaxial tension for a given mix design (refer to AS3600-2018, in particular Section 16.3.3). Using equations 16.3.3.4 (1 & 2) and 16.3.3.6 (1 & 2 ) from AS3600-2018, the cross-calibration (AS3600-2018 reference factors, $k_{R,2}$ and $k_{R,4}$) was estimated for the given test series mix design. The results of the calibration for the SF40 and SF45 mix designs are presented in Table 8. This allows for a once-off calibration of the inverse analysis procedures, which may then be used for quality assurance testing (e.g., production testing using only indirect tests).

**Table 8. Reference factors $k_{R2}$ and $k_{R,4}$ for cross-calibration**

| Series | $f_{0.5m}$ | $f_{1.5m}$ | $f_{R2m}$ | $f_{R4m}$ | $k_{R,2}$ | $k_{R,4}$ |
|---|---|---|---|---|---|---|
| SF40 | 1.33 | 1.21 | 4.64 | 3.64 | 0.29 | 0.33 |
| SF45 | 1.73 | 1.57 | 5.33 | 4.47 | 0.33 | 0.35 |

## X-Ray Micro-CT Imaging Results (Observations from the Images)

Images of the core slices are shown in Figure 17 and Figure 18. Several key observations can be made by a visual inspection of these images. There is an observable gradation in size of the coarse aggregates used in the concrete. Interestingly, there is also a significant amount of void spaces between aggregates and surrounding mortar. These voids are concentrated primarily along interfacial transition zones (ITZ) of the large aggregate particles and less so in the paste between aggregates. Further, the shape of these gaps is nonspherical, unlike air bubbles commonly encountered in concrete. This is likely a result of the spraying process of the test specimens. The fibers are confined to pockets of mortar between aggregates, which results in clustering and influences their orientation.

Most fibers appear rather circular in the *xz*- and *yz*-planes while appearing elongated in the *xy*-plane. This implies that most of the fibers are oriented horizontally in the *xy*-plane (i.e., parallel to the face of the dog bones and normal to the direction of spray). In the case of core-specimen SF40D-6 in Figure 17, the number of fibers near the top of the specimen appears to be fewer than in mid and lower regions. This trend is less prominent in specimen SF45D-4. These aspects of fiber distribution are analyzed further below.

### *Image Analysis*

The procedure adopted for image processing is outlined in Miletić et al. (2020). The procedure allows identification of each individual fiber and its spatial position within the core, with minimal computational effort. The 3D images of the core specimens and the corresponding fiber-segmented images are presented in Figure 19. From the analyses, the fiber volume fraction in each core can be determined, together with their alignment and distribution statistics. Individual fibers are identified, and their location, diameter, and orientation are measured.

After processing of the images, SF40D-6 was found to have a fiber volume fraction of 0.48%, while specimen SF45D-4 yielded 0.57%. The void ratios in these specimens were determined as 4.2% and 3.4%, respectively. All the fiber location and orientation distribution is discussed in the following sections, with the results presented in global *XYZ* coordinate system, with the *X*-direction corresponding to the direction of the applied tension, the *Y*-direction normal to applied tension, and the *Z*-direction being through the thickness.

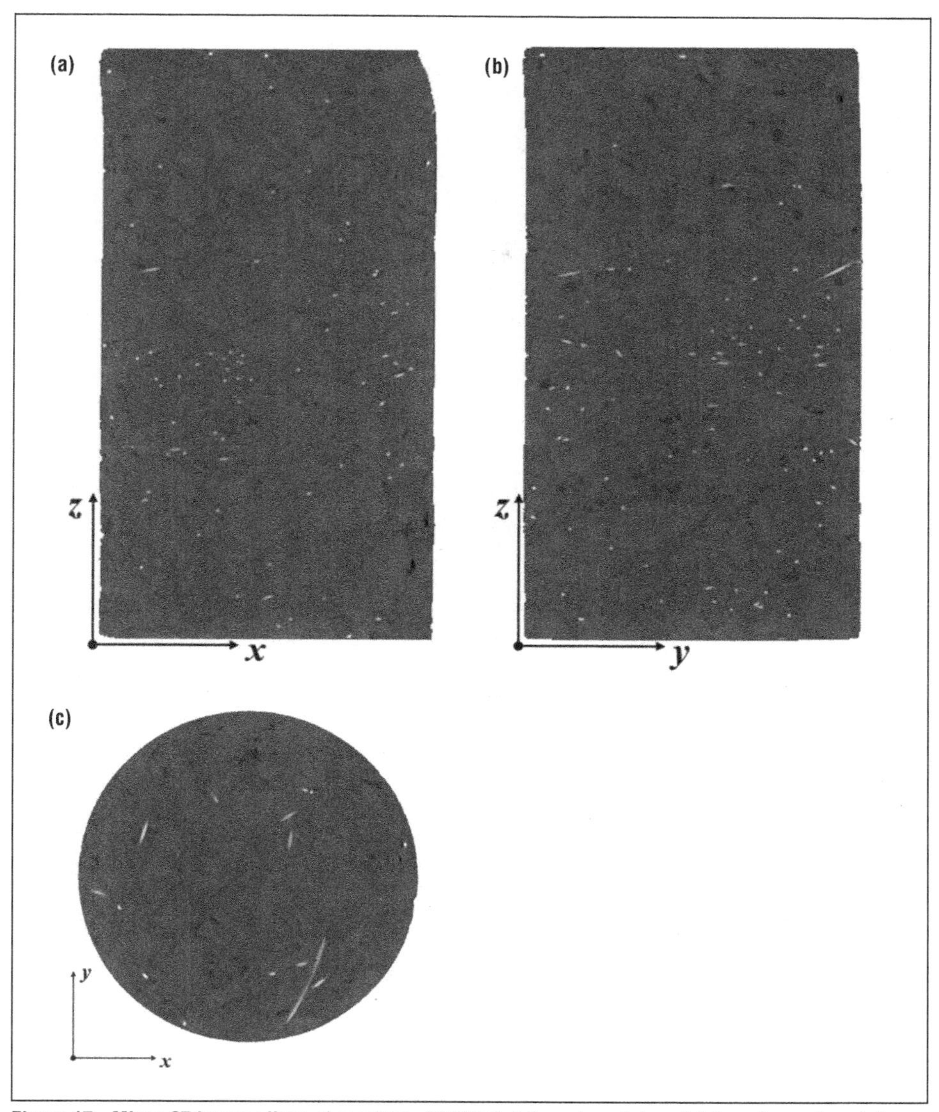


**Figure 17. Micro-CT image slices of specimen SF40D-6: (a) xz-plane (at y = 34.4 mm), *yz*-plane (at *x* = 36.7 mm), and (c) xy-plane (at z = 95.1 mm)**

### *Fiber Distribution*

The centroidal locations for each fiber are shown in Figure 20 and Figure 21 for specimen SF40D-6 and SF45D-4, respectively. The discrete probability of finding a fiber at a given *x*-, *y*-, or *z*-coordinate is presented as histograms. The probability plots are, essentially, histograms of relative frequency of occurrences of the different coordinate values in the fiber location data.

In considering measurements along any axis in the xy-plane, it is recognized that the width of a cut through the section is zero at the external surfaces and a maximum,

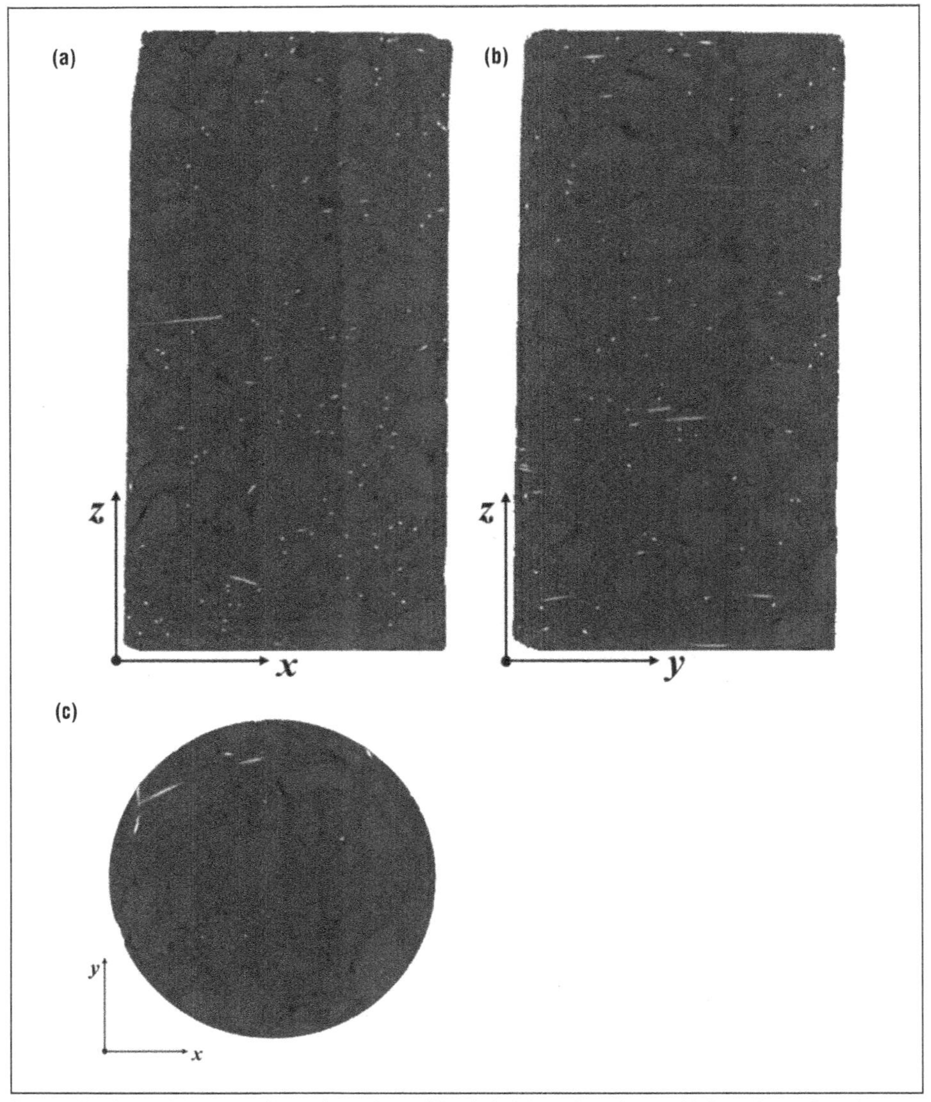


**Figure 18. Micro-CT image slices of specimen SF45D-4: (a) *xz*-plane (at $y = 39.3$ mm), yz-plane (at $x = 44.4$ mm), and (c) *xy*-plane (at $z = 78.7$ mm)**

equal to the diameter, at the core center. Thus, fewer fibers are encountered at the edges compared to a cut passing through middle of the specimen. To remove this bias, the probability value for each fiber centroid occurrence is normalized using the relative area of the specimen at the corresponding coordinate. The scatter plots indicate the presence of clusters of fibers (overlapping dots), as fibers are placed in mortar pockets between coarse aggregate particles.

In the *z*-direction, the mean probability of encountering a fiber at any point is $p(z)$ mean = 0.0086; this is 51% of that found in the *x*- and *y*-directions. The considerably lower probability of encountering a fiber in the *z*-direction shows a significant

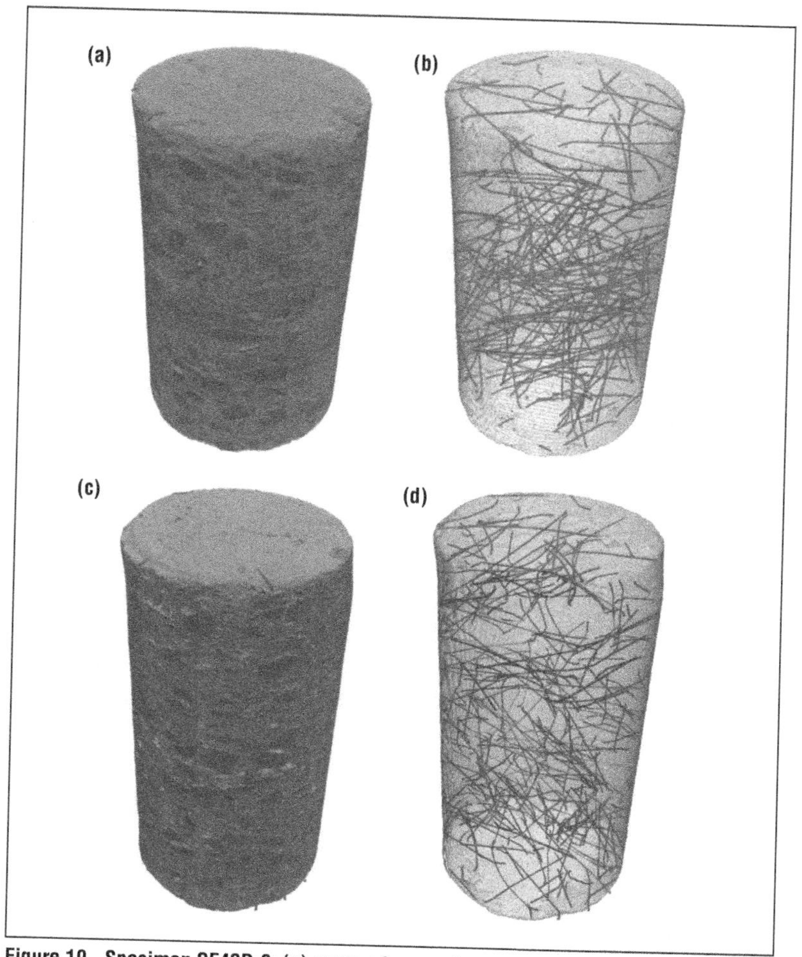


**Figure 19. Specimen SF40D-6: (a) opaque image of core and (b) fiber-segmented image; Specimen SF45D-4: (c) opaque image of core and (d) fiber-segmented image**

bias in fiber orientation toward the *xy*-plane, and is attributed to the layering of the spayed shotcrete and the velocity at which the new concrete impacts the concrete previously placed.

Bias in the fiber alignment may be further quantified from fiber angles; the vertical angle subtended by fibers with the *Z*-axis is denoted as $\theta$ and the in-plane angle subtended by fibers on the horizontal plane (*XY*-plane) with respect to the *X*-axis of the dog-bone is $\phi$ (Figure 22). The relative frequencies for the fiber orientations, $\phi$ and $\theta$, are presented as probability distributions in Figure 23.

Figures 23b and d show a high tendency for fibers to be parallel to the *XY*-plane; that is, the angle $\theta$ is close to 90 degrees. In the *XY*-plane, while some localized peaks and troughs are observed (Figures 23a and c), there is no obvious bias in the orientations.

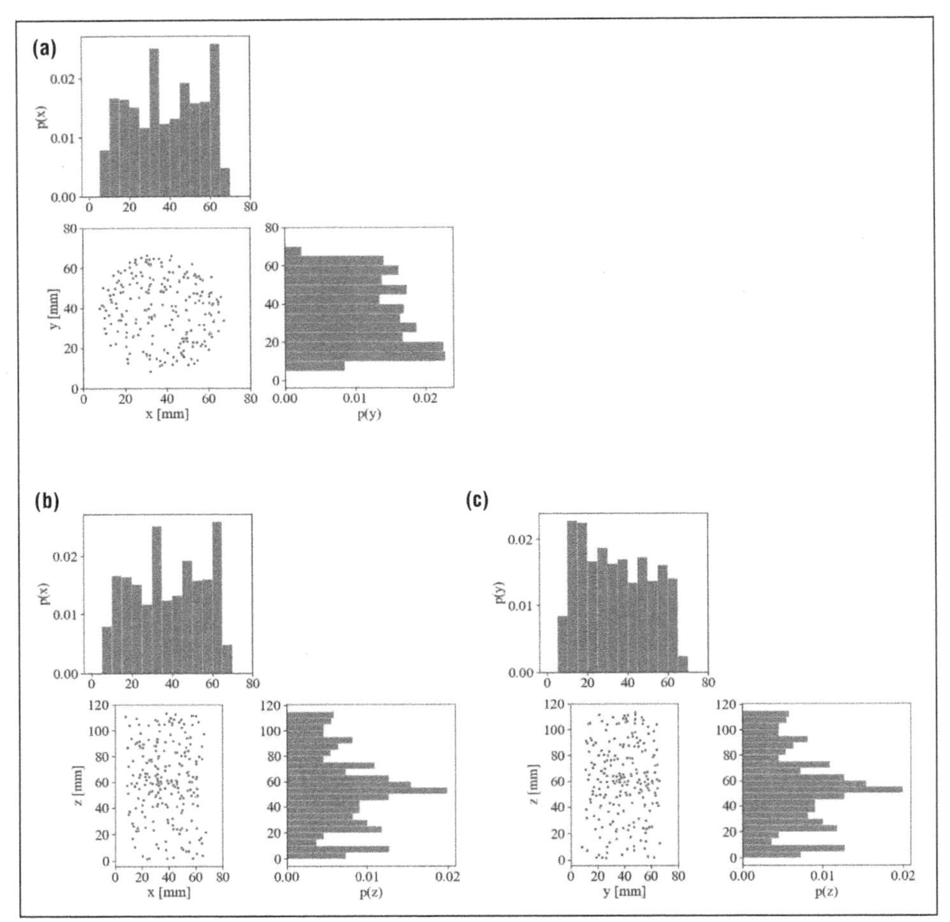


**Figure 20. Specimen SF40D-6: Locations of fiber centroids in (a) *xy*-planes, (b) *xz*-plane, and (c) *yz*-plane corresponding probability distributions**

### *Fiber Diameter*

The median value of fiber diameter determined from the images is 0.59 mm, as shown in Figure 24. This is close to the actual fiber diameter of 0.55 mm and consistent with slightly higher dimensions generally observed for steel fibers in their μCT images; the reasons for this are bleeding effects and down-sampling.

Some fiber diameter observations lie farther from the median value, seen as dots in Figure 24. These indicate fibers that are not fully separated from contact with each other and therefore a measure of efficiency of the image analysis scheme is adopted.

## CONCLUSION AND RECOMMENDATIONS

To design SFRS confidently, it is important to establish its post-cracking residual tensile strength. This can be obtained directly from uniaxial tensile tests or indirectly from prism bending tests combined with an inverse analysis procedure. The inverse analysis method requires calibration using sufficient comparison data between direct (uniaxial) tension tests and the indirect models.

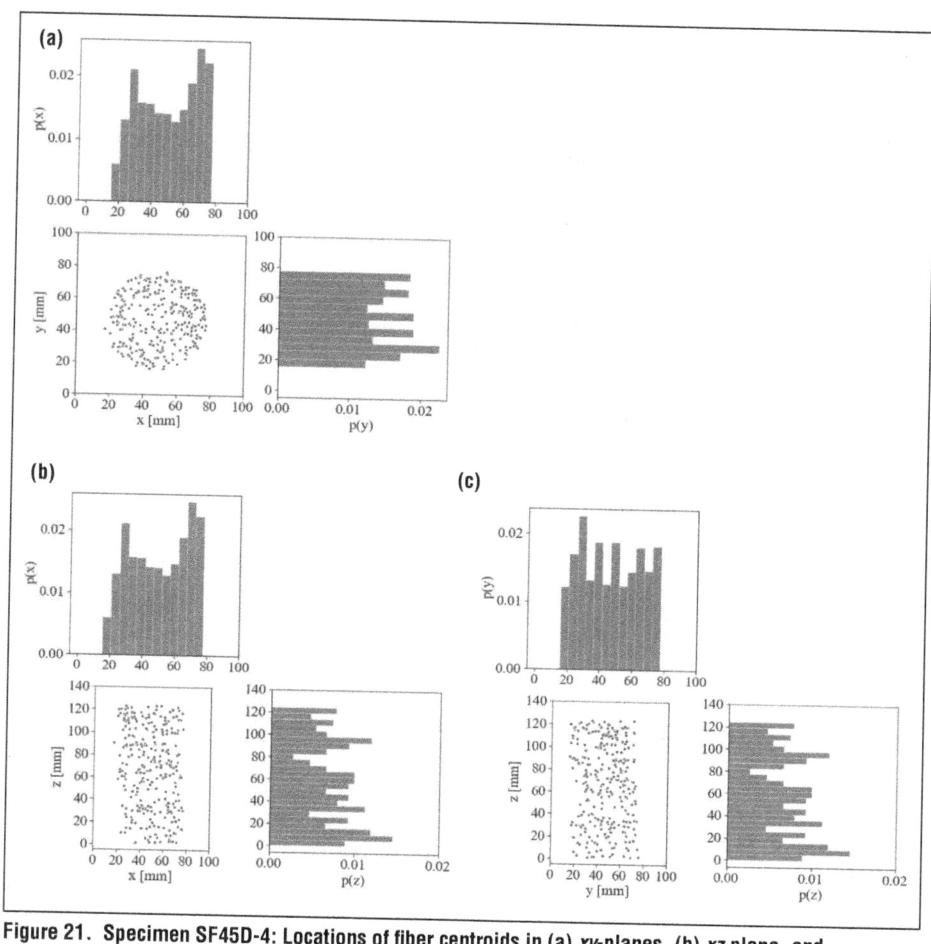


**Figure 21. Specimen SF45D-4: Locations of fiber centroids in (a) *xy*-planes, (b) *xz*-plane, and (c) *yz*-plane corresponding probability distributions**

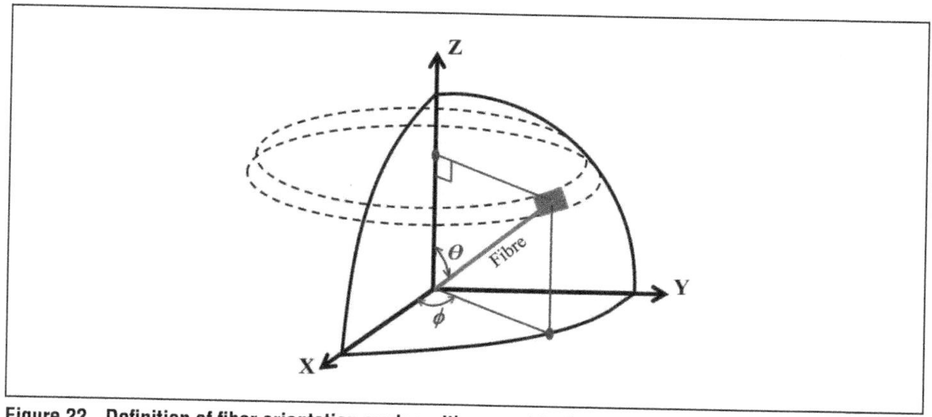


**Figure 22. Definition of fiber orientation angles with respect to dog-bone (global) coordinate system**

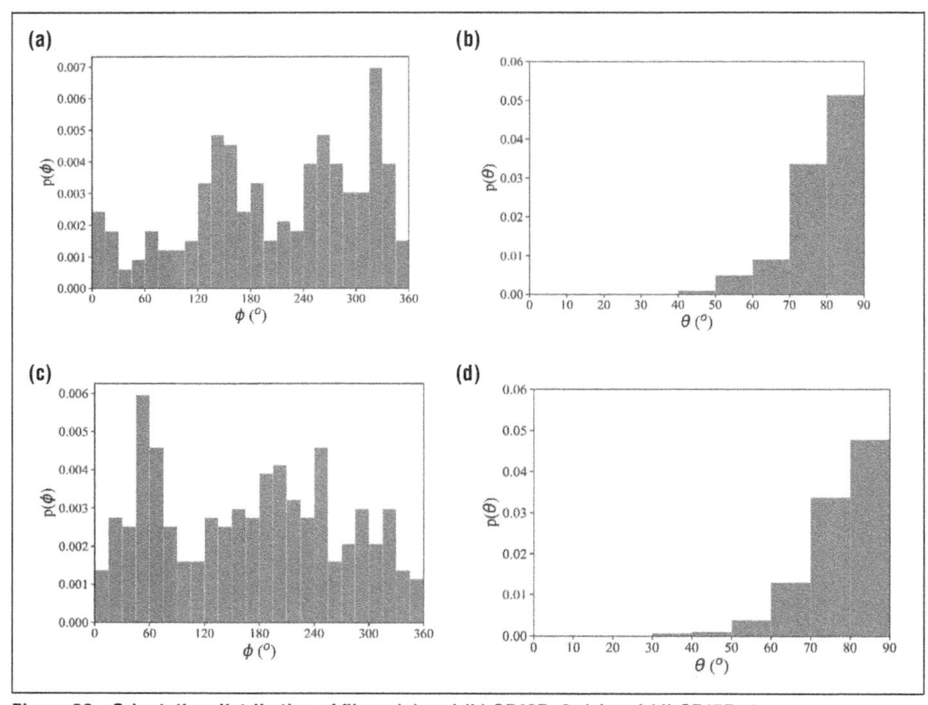


Figure 23. Orientation distribution of fibers (a) and (b) SF40D–6; (c) and (d) SF45D–4

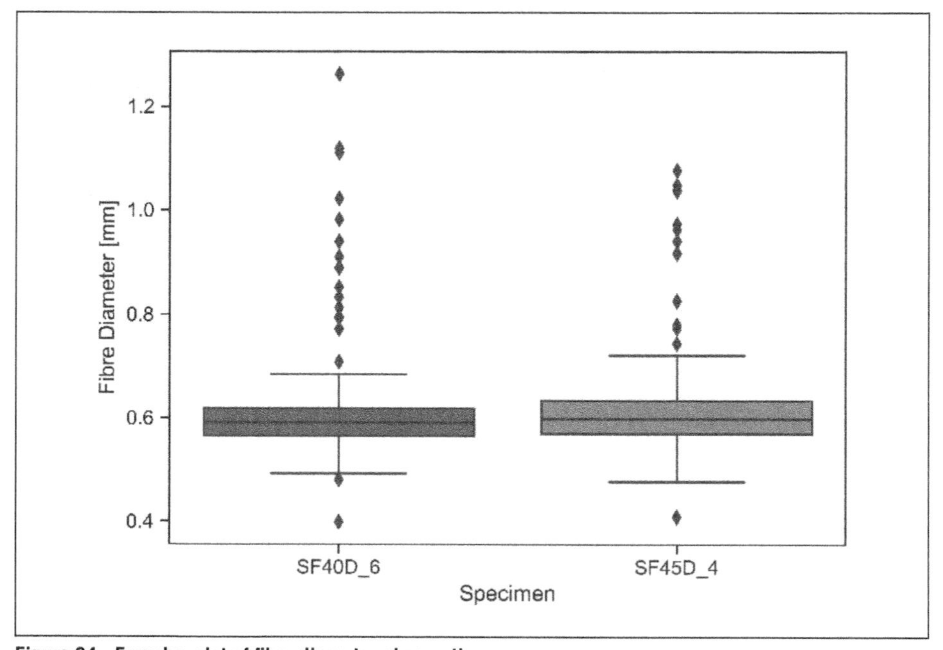


Figure 24. Error boxplot of fiber diameter observations

The direct tensile tests were conducted according to AS3600-2018. and the $\sigma$–$w$ relationships were obtained. This, however, was not a guarantee as sprayed specimens for the AS3600-2018 test geometry and molds used for cast samples were never done before this research project. Based on tests performed during this research project, the following conclusions are drawn:

- Both the direct tensile test method and the indirect tensile test method, by prism bending with inverse analysis, are suitable to determine the residual tensile strength of SFRS.
- The inverse analysis methods developed for SFRC in Foster et al. (2017) and AS3600-2018 can be applied for SFRS.
- The $\sigma$–$w$ relationships obtained from normal inverse analysis of prism bending test slightly underestimate the residual tensile strength of SFRS, as may be expected when using an indirect model.
- Micro-computed tomographic (µCT) analyses showed a strong bias in orientation of fibers toward a plane normal to the direction of spray (the *XY*-plane).
- No bias was found in fiber orientation in the *XY*-plane, the result being an anisotropic material, and although no tension tests were conducted in the through thickness direction, the strength in this direction (*Z*-direction) is likely to be substantively lower than for any axis in parallel with the *XY*-plane.
- The µCT images showed a significant number of voids occurring adjacent to paste-aggregate interfacial transition zones; the void ratio estimates were approximately 3% to 4%.
- The image analysis procedure demonstrated an efficiency of approximately 93 to 96% of identified fibers being perfectly segmented.

The research indicates that the AS5100/AS3600 direct test method can be used on spray shotcrete specimens. However, because this was the first application of the AS5100/AS3600 dog-bone direct test method for shotcrete, the authors recommend caution before specifying it on future projects until more experience and data are gained with tests on specimens prepared by spraying versus casting.

The µCT scanning may be valuable for asset management where limited as-built and design documentation exist for SFR concrete or shotcrete structure.

## REFERENCES

Amin, A., Foster, S.J., and Muttoni, A. 2015. Derivation of the $\sigma$-$w$ relationship for SFRC from prism bending tests. *Structural Concrete.* 16(1): 93–105.

Amin, A., Foster, S.J., and Kaufmann, W. 2017. Instantaneous deflection calculation for steel fibre reinforced concrete one-way members. *Eng. Struct.* 131: 438–445.

ASTM International. 2012. ASTM C1609—Standard Test Method for Flexural Performance of Fiber-Reinforced Concrete (Using Beam with Third-Point Loading). West Conshohocken, PA, USA: ASTM International, 9 p.

Australian Standards. 2017. AS 5100.5-2017—Australian Standard, Bridge Design Part 5: Concrete. Standards Association of Australia, Sydney.

Australian Standards. 2018. AS 3600-2018—Australian Standard for Concrete Structures. Standards Association of Australia, Sydney.

Aveston, J., and Kelly, A. 1973. Theory of multiple fracture of fibrous composites. *Journal of Materials Science* 8: 352–362.

European Committee for Standardization. EN14651. 2007. Test Method for Metallic Fiber Concrete—Measuring the Flexural Tensile Strength (Limit of Proportionality [LOP], residual).

Fédération Internationale du Béton (*fib*). 2010. *fib* Model Code for Concrete Structures. Ernst & Sohn, October.

Foster S.J., Agarwal A., and Amin A. 2017. Design of steel fiber reinforced concrete beams for shear using inverse analysis for determination of residual tensile strength. *Structural Concrete* 19: 129–140.

Foster S.J., and Parvez A. 2019. Assessment of model error for reinforced concrete beams with steel fibers in bending. *Structural Concrete* 1–12. https://doi.org/10.1002/suco.201800090.

Htut, T.N.S. 2010: Fracture processes in steel fiber reinforced concrete. PhD dissertation, School of Civil & Environmental Engineering, The University of New South Wales, Australia.

Miletić, M., Kumar, L.M., Arns, J.-Y., Agarwal, A., Foster, S.J., Arns, C., and Perić, D. 2020. Gradient-based fiber detection method on 3D micro-CT tomographic image for defining RILEM Final Recommendation TC-162-TDF, 2003. Test and Design Methods for Steel Fiber Reinforced Concrete; $\sigma$-$\varepsilon$ Design Method. *Materials and Structures*. 36:560–567.

Ng, et al. 2012.Fracture of steel fibre-reinforced concrete—the unified variable engagement model. UNCIV report R-460, School of Civil & Environmental engineering, The University of New South Wales, Australia.

RILEM TC 162-TDF. 2003. Test and Design Methods for Steel Fibre Reinforced Concrete - Design Method.

van Vliet, M.R.A., and van Mier, J.G.M. 1999. Effect of strain gradients on the size effect of concrete in uniaxial tension. *International Journal of Fracture*, 95: 195–219.

# Evaluation of Long Term Loads on Freight Tunnels in Chicago

**Alireza Ayoubian** ▪ Parsons
**Richard Finno** ▪ Northwestern University

## ABSTRACT

The network of freight tunnels in Chicago consists of about 62 miles of horseshoe-shaped tunnels constructed typically less than 50 feet below ground surface. These tunnels serve as repositories of much infrastructure and should remain serviceable. The tunnels are often impacted by adjacent excavations and the question arises as to the existing state of stress in the final liner. This paper discusses construction of these tunnels and presents the results of finite element analyses which were used to obtain ranges for axial forces and bending moments that have developed in the final liner of the freight tunnels since their construction.

## INTRODUCTION

The network of freight tunnels in Chicago were constructed in early 1900s to carry telephone and telegraph wires and cables, but were also used to transport merchandise and remove solid waste from building basements (Wren, 2007). Today, the freight tunnels are used to house highly sensitive utilities such as fiber-optics and electricity cables.

The Block 37 Site is located in the loop area of downtown Chicago and is bounded by State Street on east, Dearborn Street on west, Randolph Street on north and Washington Street on south. Major excavations and construction activities were undertaken by several contractors for development of the block from 2006 to 2008. A segment of the freight tunnel network in the vicinity of the Block 37 site, which is located below Randolph Street, was uncovered during excavation and information were collected regarding its construction and support system. This segment of the freight tunnel, which was a bypass tunnel, was constructed around 1940s. Details of the freight tunnel construction and the field records collected during uncovering of this segment of the tunnel during construction are discussed by Ayoubian (2015).

The results of the 2D finite element analyses presented in this paper were used to compute ranges for the axial forces, bending moments and internal normal stresses that were developed in the final liner of the freight tunnel over approximately 65 years (between 1940s and 2008) under long term loads due to consolidation of the ground and seepage forces under a steady state flow condition. The axial forces and bending moments were used to calculate internal normal stresses at the freight tunnel final liner and assess its condition.

Similar to the freight tunnels in downtown Chicago, the segment of the freight tunnel network below Randolph Street was excavated by hand and supported using steel ribs with wood laggings between the ribs. The unreinforced concrete final liner was constructed by hand packing concrete between forms and the wood lagging. A cross section of the freight tunnel below Randolph Street is shown in Figure 1. For the remainder of this paper, the freight tunnel below Randolph Street will be simply referred to as "the freight tunnel." The freight tunnel was approximately 7.7 feet wide at

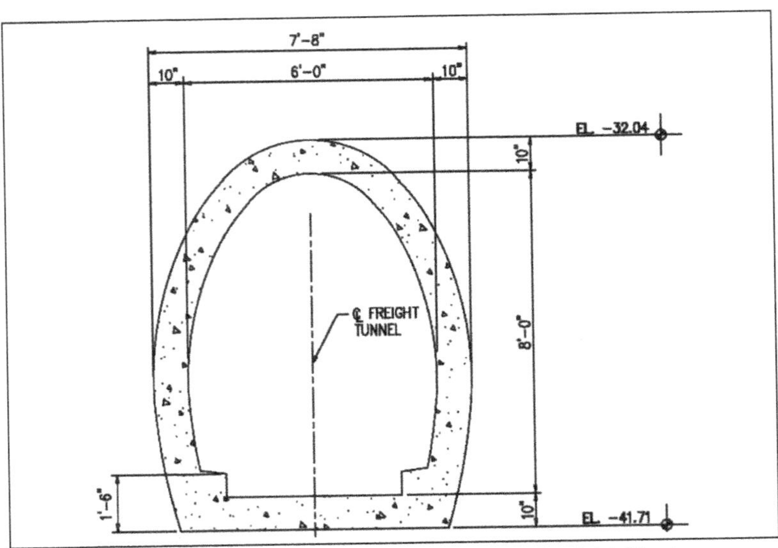


**Figure 1. A typical cross-section of the freight tunnel below Randolph Street**

the mid height and 9.7 feet tall with its crown and invert located at elevations –32.0 and –41.7 ft CCD (Chicago City Datum), respectively. The ground surface is at elevation of about 14.5 ft CCD in the vicinity of the freight tunnel.

The following details were deduced based on the field records collected during excavation and exposing of the freight tunnel (Ayoubian, 2015). These findings were used to develop the properties of the initial support system for the finite element modeling.

- The initial support for the freight consisted of steel channels installed at 5-foot spacing with wood lagging between the ribs for transfer of the earth pressure to the ribs.
- Steel ribs consisted of two pieces of C10x20 channels with flanges at the top and bottom.
- Excavation of tunnel was performed without air pressure, by hand and in "full-face" mode (no bench-and-heading excavation).
- The unsupported length of the freight tunnel during each round of excavation was about 5 feet.
- The initial support for the invert was a 10-inch-thick fresh concrete placed at the invert before erection of the steel ribs.

## CONCEPT OF STRESS RELEASE

Analysis of tunnel excavations inherently is a three-dimensional problem. To model the freight tunnel excavation in a two-dimensional finite element model and take into consideration the amount of stress distribution around and in front of the excavation face, some degree of ground stress release should be considered after tunnel excavation and before installation of the support system. An average stress release coefficient $\lambda$ (Panet and Guenot, 1982), associated with a softening factor n (achieved by dividing the modulus of the soil by "n" as discussed by Leca and Clough, 1992), is estimated using Equation 1.

$$1-\lambda=\frac{\overline{\sigma_r^n}}{\overline{\sigma_r^0}} \quad \text{(EQ 1)}$$

where, $\overline{\sigma_r}$ = an average value of the normal stress along the tunnel wall ($\overline{\sigma_r^0}$ = initial conditions; and $\overline{\sigma_r^n}$ = the value of $\overline{\sigma_r}$ after softening).

Ayoubian (2015), by performing analysis and considering the construction records collected at the Block 37 site after uncovering the freight tunnel, concluded that a stress release range of $\lambda$ =0.5 to 0.7 is reasonable for the freight tunnel.

## LONG TERM LOADS ACTING ON FREIGHT TUNNELS

The concrete liners of freight tunnels and Chicago subway tunnels contain joints and groundwater has been seeping through these joints over decades. Ground water table was encountered during several subsurface explorations at the Block 37 site at an average depth of about 14 feet below ground surface. However, limited piezometer readings from the Block 37 site indicated that the depth to groundwater was greater than 14 feet in the vicinity of the freight tunnel at the Block 37 site, suggesting a downward gradient of groundwater. Therefore, a main source of the long terms loads acting on freight tunnels in Chicago is believed to be the seepage forces due to leakage of water through the tunnel joints. Another source of long term loads on the freight tunnels in Chicago is consolidation of the cohesive soils after construction of the tunnels.

## SUBSURFACE CONDITIONS AT BLOCK 37

The geology of the Block 37 site is generally consistent with the regional geology of the downtown Chicago area. Several subsurface investigations were conducted at Block 37 prior to its development. The results of these investigations showed an upper layer of miscellaneous fill underlain by a series of glacial till layers. In descending order, the glacial till layers at the Block 37 site are Blodgett, Deerfield, Park Ridge, Tinley and Valparaiso strata. The fill stratum is a cohesionless stratum whereas the underlying glacial till strata are primarily cohesive. The cohesive strata tend to decrease in water content and increase in strength and stiffness with depth. The three cohesive strata of Blodgett, Deerfield and Park Ridge, which were deposited underwater, are called Lake Border clays (drifts) (Finno and Chung, 1992). The silty clayey tills of Tinley and Valparaiso are stiffer and contain more gravel than those of the Lake Border clays. These two tills are locally referred to as "Chicago hardpan" and constitute a hard bearing stratum overlying the bedrock. The bedrock in the Chicago region is of sedimentary origin and consists of stratified limestone and dolomite which were deposited from organic lime mud under the shallow Niagaran Sea (Bretz, 1939). The thickness of the soil deposits overlying the bedrock is generally 60 to 100 feet.

The existing grade at the Block 37 site is approximately 15 feet above the Lake Michigan surface water elevation. An idealized subsurface profile showing the subsurface strata and the elevations of top and bottom of each stratum were determined based on the subsurface investigation results and are presented in the idealized subsurface profile in Figure 2. The top and bottom of the freight tunnel is also shown in the subsurface profile. The ground surface was at elevation 14.5 feet CCD (CCD = Chicago City Datum) and initial hydrostatic ground water table was at elevation 0.5 feet CCD. The subsurface profile and ground water depth presented in Figure 2 represent the conditions during construction of the freight tunnel.

## FINITE ELEMENT ANALYSIS

The two-dimensional (2D) finite element PLAXIS 2D (2014) software was used for this study. This software was utilized to model the stages of the freight tunnel construction since 1940s until 2008 and included consolidation of soils and ground water seepage through the freight tunnels joints. The subsurface strata represented in the finite element model is shown in Figure 2. The phases that were used in the finite element analyses are summarized in Table 1.

Initialization of in-situ stresses in the Initial Phase involved calculating the initial vertical and horizontal stresses in the ground utilizing the soil properties and ground water table location defined for the model.

The distribution of ground stresses and deformations over a certain length behind a tunnel face is inherently a three-dimensional problem. In most tunnel projects, after excavation of ground and before installation of ground support, some degree of ground movements and stress release occur which results in reduction of ground stresses acting around the tunnel cavity. The amount of stress release during tunnel excavation depends largely on the details of construction and support delay length. In PLAXIS, the stress release is modeled by the factor λ that was introduced by Panet and Guenot (1982). Once the tunnel support is installed and comes in contact with the

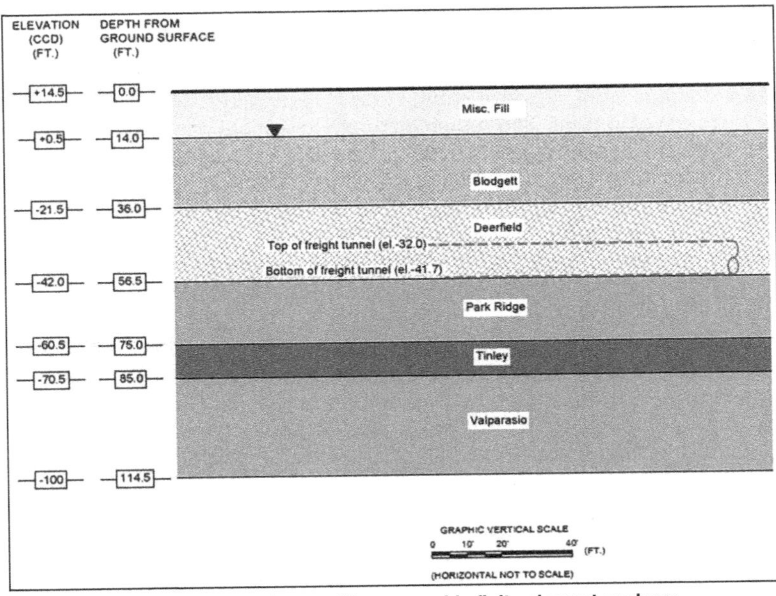


**Figure 2. Generalized subsurface profiles as used in finite element analyses**

**Table 1. Finite Element Phases for 2D freight tunnel analysis**

| Phase No. | Description of Phase |
|---|---|
| Initial | Initial Phase – calculation of the initial ground stresses |
| 1 | Excavation of freight tunnel with ground stress release of λ (λ<1) |
| 2 | Installation of initial support with release of the remaining ground stresses (1-λ) |
| 3 | Installation of final line, 65 year consolidation and steady state flow analysis (the drain element activated) |

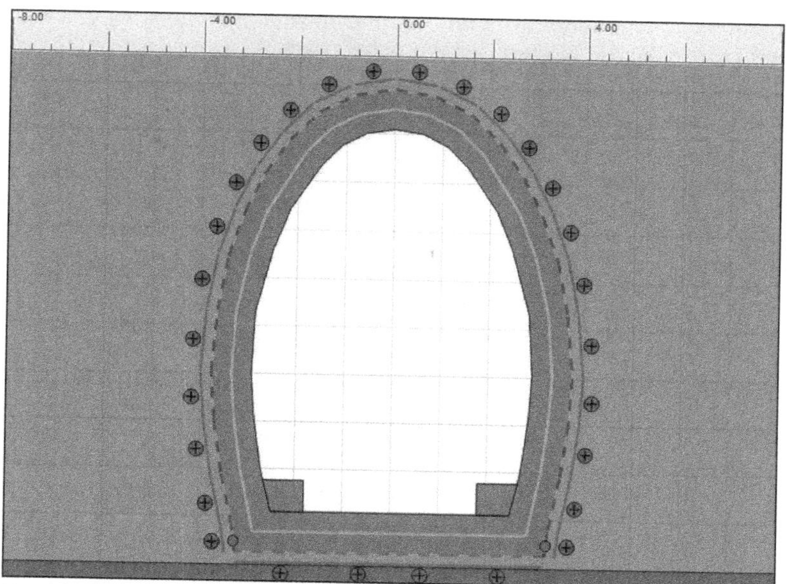


**Figure 3. A cross section of the freight tunnel at Phase 3 showing the initial support and final liner**

surrounding ground, the support only carries the remainder of the ground stresses. In Phase 1 of the finite element analysis, a stress release of λ was considered during excavation and prior to the installation of the initial support.

In Phase 2, the initial support was activated and the remaining 1–λ of the ground stresses was released. The initial support system consisted of steel ribs and wood lagging along tunnel walls and crown and freshly-poured concrete at the invert of the tunnel. The initial support was defined using plate elements. An interface element was introduced around the initial support to model the interaction between the support and the surrounding soil.

In Phase 3, the final liner was modeled using volume elements. A drain element was introduced along the perimeter of the freight tunnel and was activated in this phase to model the tunnel acting as a drain. Figure 3 presents a cross section of the freight tunnel in Phase 3 showing the initial support and final liner. The analysis in Phase 3 consisted of consolidation for 65 years for calculation of ground effective stresses and a steady state flow analysis for computation of final water pressures. It is thus assumed that the ground would reach a steady state flow condition 65 years after its construction. With activation of the drain element in Phase 3, excess pore water pressures was allowed to dissipate along the drain element during the consolidation analysis. The invert of the initial support system (the freshly poured concrete), which was activated in Phase 2, was deactivated in Phase 3. The invert of the freight tunnel in Phase 3 consisted of 10 inches of fully cured concrete.

## GEOTECHNICAL PARAMETERS

Mohr-Coulomb (MC) and Hardening Soil (HS) models were used for the soils encountered at the Block 37 project site. The parameters for MC and HS models for the subsurface soils at the Block 37 site are based on the results of various subsurface explorations and laboratory tests and calibration with field monitoring. Details of the MC and HS model parameters for the subsurface soils are provided by Ayoubian (2015).

Mohr-Coulomb (MC) model was used for the Fill, Tinley and Valparaiso strata. The Fill stratum was modeled as a drained material. The Tinley and Valparaiso strata were modeled as undrained soils with effective stiffness and strength properties for plastic analysis. HS model was employed to model the three compressible clays strata (Blodgett, Deerfield and Park Ridge). The Blodgett, Deerfield and Park Ridge strata were modeled as undrained soils with effective stiffness and strength properties for plastic analysis.

## STRUCTURAL PARAMETERS

For finite element modeling, the initial support (steel ribs and the concrete at the tunnel invert) was modeled using plate elements. The linear elastic material type was used for the initial support system. An interface element was introduced around the initial support to model the interaction between the initial support and the surrounding soil. Figure 3 shows the initial support and the interface element around the outside face of the support.

As discussed previously, the initial support for the freight tunnel walls and crown consisted of steel ribs (Channels C10x20) and wood lagging. To model this discontinuous system of support in a two-dimensional finite element model, the properties of the steel ribs (EA, EI) were divided by the spacing between them (i.e., five feet) and were used to represent a continuous support system in a two-dimensional model. The contribution of wood lagging was considered negligible and neglected.

The invert of the freight tunnel consisted of a 10-inch-thick concrete mat to act as the foundation for support of the steel ribs. The 10-inch-thick concrete mat later became part of the final liner when concrete was placed over and between the steel ribs.

The final liner of the freight tunnel consisted of a 10-inch-thick concrete liner which was constructed by placing concrete between the wood lagging (placed between the steel ribs) and wooden forms. The strength of freight tunnels' final liner was investigated by STS Consultants (2007) and the results indicated unconfined compressive strength values ranging from 8500 to 11700 psi. Therefore, an unconfined compressive strength of 9000 psi was assumed to develop the final liner properties for the finite element analysis. The linear elastic material type was used for the final liner.

## ANALYSIS RESULTS

As discussed before, Ayoubian (2015) showed that a stress release range of $\lambda$=0.5 to 0.7 is reasonable for the freight tunnels in Chicago. The results of the finite element analyses based on 50% stress release ($\lambda$=0.5) are presented in this section to illustrate the computed results for the initial support and final liner loads.

The axial forces, bending moments and deformations at different locations on the tunnel will be presented with respect to an angle ($\theta$) measured in counter clockwise direction from a horizontal axis at the spring line of the freight tunnel as shown in Figure 4. The spring line of the freight tunnel is assumed at elevation –36.9 feet (CCD). The freight tunnel crown and invert are at elevations –32.0 and –41.7 feet (CCD), respectively, as illustrated in Figure 4. Considering the symmetry of the freight tunnel with respect to its vertical axis, the angle $\theta$ varies from –90 at the center of the tunnel invert

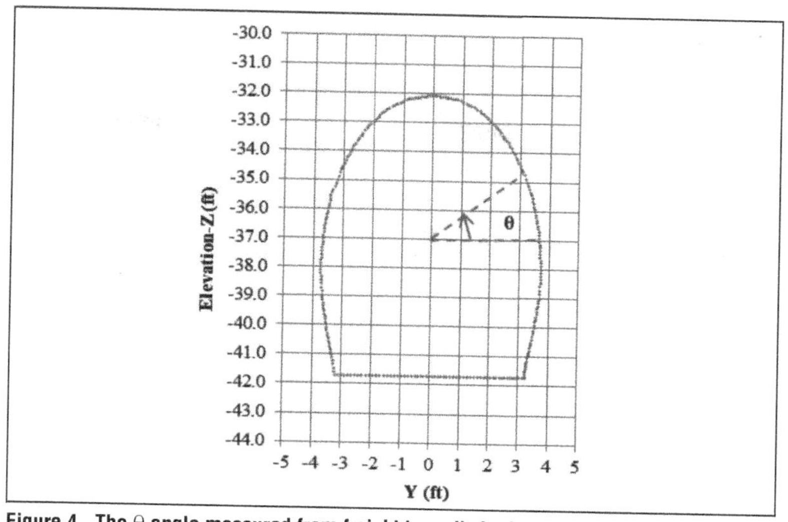


**Figure 4. The θ angle measured from freight tunnel's horizontal axis at spring line**

to +90 at the crown of the tunnel. The connection between tunnel wall and invert is located at $\theta = -56.3$ degrees for the unsupported tunnel and initial support.

In Phase 1 of the analysis, tunnel is excavated with 50% of the stresses released ($\lambda = 0.5$). During this phase, Convergence (inward displacement) of the unsupported ground along tunnel walls, crown and invert was resulted. In Phase 2 of the finite element analysis, the initial support consisting of steel ribs and wood lagging (along tunnel walls and crown) and freshly poured concrete (at the invert of the tunnel) were activated. The remaining 50% of the ground in situ stresses were released in this phase resulting in additional ground movements and deformation of the initial support.

In Phase 3 of the finite element analysis, the final concrete liner was activated, the invert of the initial support was deactivated and replaced by the final liner invert, the freight tunnel and the surrounding ground was subjected to 65 year consolidation, and the drain element around the tunnel was activated and seepage forces were generated assuming a steady state flow condition. The results showed that the initial support and final liner predominantly experienced downward movement in Phase 3 due to ground consolidation and seepage forces acting on the tunnel. Figure 5 presents the total displacement (relative to their original condition) distribution of the initial support and final liner (centerline of the final liner) at Phase 3 only. The initial support was only present along tunnel walls and crown in Phase 3. The invert of the initial support was deactivated and replaced by the final liner invert in Phase 3. It can be seen from this Figure 5 that both the initial support and final liner experienced total displacements of about 0.26 to 0.28 inches during Phase 3 due to ground consolidation and application of seepage forces. The deformations of the final liner in Phase 3 showed the tunnel "squatted" slightly, that is bulged horizontally, and mostly settled vertically under long term loads in Phase 3.

The total displacement distribution of the initial support in Phase 2 is also presented in Figure 5. The general modes of displacements shown in Figure 5 illustrate that in Phase 2, initial support experienced inward displacement where as in Phase 3 both the initial support and final liner underwent mostly vertical displacement. Inward displacements are positive in Figure 5 for the initial support. In Phase 2, the initial support and the surrounding ground are fully engaged and deform together under applied loads. The maximum inward displacement of the initial support in Phase 2 was about 0.18 inches at a θ angle of about –14 degrees according to Figure 5.

The long term axial force distribution in the final liner in Phase 3 is presented in Figure 6. Figure 6 also presents the long term axial force distribution for the initial support in Phase 3. It can be seen that the axial forces developed in final liner in Phase 3 were less than or equal to those developed in the initial support at the completion of Phase 3. The axial force in the final liner along tunnel walls and crown ranged from about 3 to 10 k/ft. Axial force in the final liner invert was about 11 k/ft. Figure 6 also shows that the final liner in Phase 3 is almost entirely in compression (positive forces) expect for very small areas at the edges of the tunnel invert. Compressive forces are positive in Figure 6 and the reminder of this paper. The long term axial force distribution for initial support in Phase 3 (presented in Figure 6) shows small changes in the walls and crown of initial support relative to those after support installation in Phase 2. As discussed previously, the invert of the initial support in Phase 2 was deactivated in Phase 3 since it became part of the final liner in that phase.

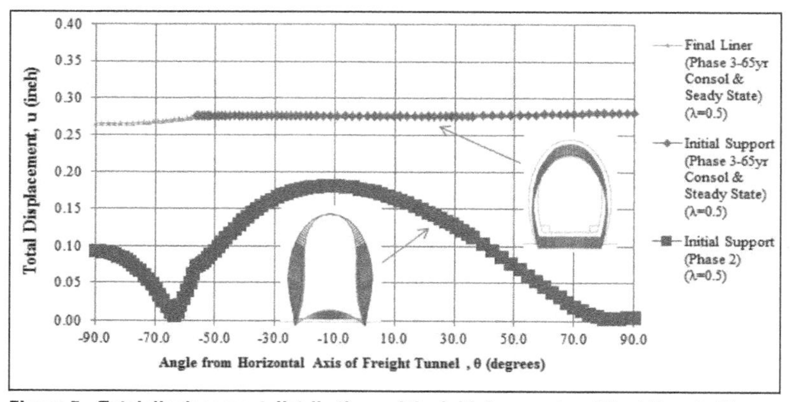


Figure 5. Total displacement distributions of the initial support and final liner in Phase 3 relative to displacement distribution of the initial support in Phase 2 only (λ = 0.5)

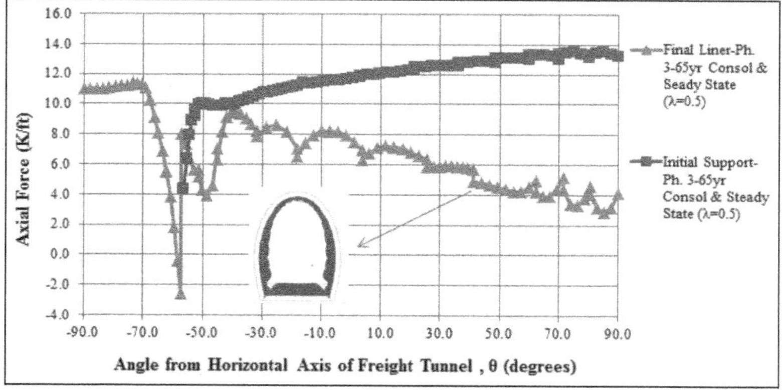


Figure 6. Axial force developed in initial support and final liner after 65 year consolidation and reaching steady state flow condition (Phase 3)

tensile strength (modulus of rupture) of plain concrete may be calculated based on its $f'_c$ according to Equation 2 recommended by The American Concrete Institute (ACI 318-05):

$$f_r = 7.5\sqrt{f'_c} \tag{EQ 2}$$

where,

$f_r$ = modulus of rupture of plain concrete
$f'_c$ = 28-day compressive strength of final liner concrete

According to Equation 3, the freight tunnel's final liner modulus of rupture is 691 to 811 psi assuming a compressive strength from 8,500 to 11,700 psi. Other research data suggest that the ratio of modulus of rupture to unconfined compressive strength ($f'_r/f'_c$) ranges from 0.11 to 0.23 (Wang et al., 2007). Therefore, a range of 800 to 1,200 psi was assumed for the modulus of rupture of the final liner for evaluation of the finite element analysis results.

The internal normal stresses due to the axial forces and bending moments obtained from the 2D finite element analysis were calculated at the completion of Phase 3 along the inner and outer faces of the final liner based on stress release values of $\lambda$ = 0.5 to 0.7 and are presented Figures 10 and 11, respectively. The figures illustrate that the impact of the stress release on the computed long term internal normal stresses is negligible. Also, shown in these figures is the estimated range for the modulus of rupture of the final liner. Compressive stresses are positive and tensile stresses are negative in Figures 10 and 11. The maximum tensile stresses were calculated at the inner face of the center of the invert ($\theta$ = −90 degrees) and the outer face of the bottom portion of the tunnel walls ($\theta$ = −47 degrees) and they were about 425 and 328 psi, respectively. The connection between tunnel walls and invert is at $\theta$ = −57.2 degrees. Figures 10 and 11 shows that after 65 year consolidation and application of seepage forces due to steady state flow condition, the calculated tensile stresses in the final liner are below the estimated range for the modulus of rupture. The calculated compressive stresses in Figures 10 and 11 are well below the $0.85f'_c$ compressive stress limit of the final liner. Therefore, no cracking or damages of the freight tunnels is expected under long term loads.

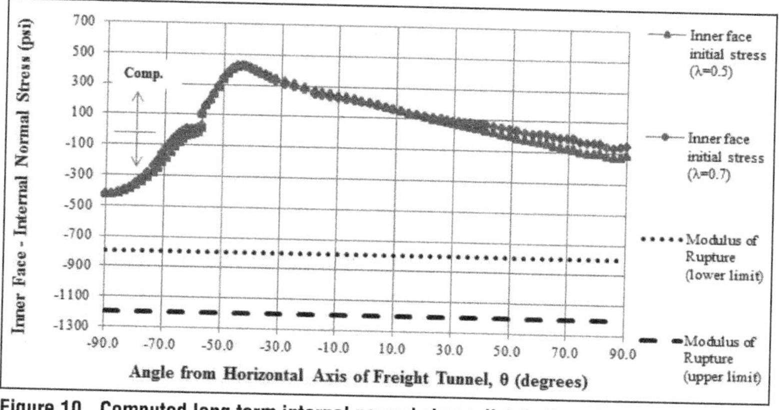


**Figure 10. Computed long term internal normal stress distribution along the Inner Face (extreme fiber) of the final liner**

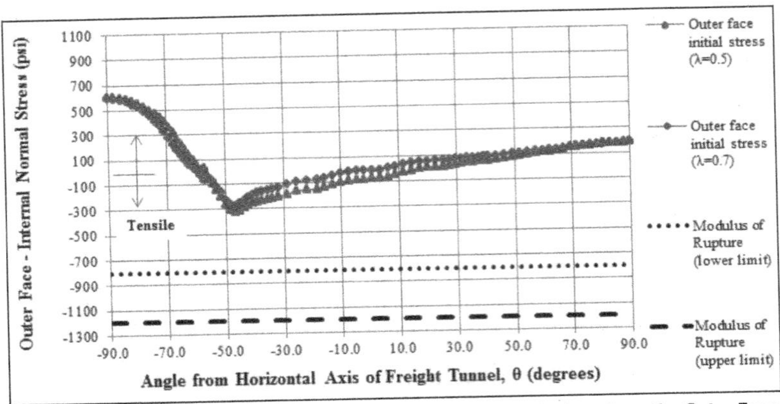


**Figure 11. Computed long term internal normal stress distribution along the Outer Face (extreme fiber) of the final liner**

It must be emphasized that consolidation of clays, reaching a steady state flow condition and curing of concrete are time dependent processes. The results presented in this section for Phase 3 only considers the loads acted on the freight tunnel at the completion of the 65 year consolidation when steady state flow condition is attained and the concrete at the final liner has reached the compressive strength of about 9000 psi.

During excavation and development of Block 37, the freight tunnel below Randolph Street underlying an excavation zone developed a 17-foot-long longitudinal crack at its crown and walls. Ayoubian (2015) developed three-dimensional finite element analyses to study the impact of the excavation on the freight tunnel that cracked. The pre-construction axial forces, bending moments and internal stresses presented in this paper were used to define the loading condition of the freight tunnel before construction in the 3D analyses. The results of the 3D finite element analyses, which are beyond the scope of this paper, were used successfully to explain the cracking of the freight tunnel. It was then concluded (Ayoubian, 2015) that the long term internal normal stresses and the corresponding axial forces and bending moments discussed in this paper were reasonably accurate.

## CONCLUDING REMARKS

The goal of this paper was to compute the axial forces, bending moments and normal stresses that have developed in the final liner of freight tunnels in Chicago. To achieve this goal, finite element analyses were conducted to model the construction of freight tunnels and the long term loads that acted on them since their construction.

The freight tunnel below Randolph Street was uncovered during Block 37 construction and data were collected on the properties of the initial support installed at the time of its construction. A two-dimensional finite element model was developed to represent the excavation of the freight tunnel and installation of its initial support system and final concrete liner. Sources for long term loads acting on the freight tunnel since its construction in 1940s consist of ground consolidation and seepage forces due to leakage of water through the joints of the tunnel. The consolidation of the ground and seepage forces under steady state flow condition were included in the finite element model. A range of stress release of $\lambda$=0.5 to 0.7 during construction of the freight tunnels was incorporated in the finite element analyses to take into consideration the impact of the three dimensional stress distribution during excavation. The results showed that the axial forces in the final liner were impacted by the amount of stress release whereas the bending moments showed relatively small variations with stress release. The analysis also showed that regardless of the amount of stress releases, the internal normal stresses computed based on axial forces and bending moments were within the tolerable limits and no cracking or damage of the freight tunnel was expected under long term loads.

It can be argued that the computed axial forces and bending moments for the final liner of the freight tunnel presented in Figures 8 and 9 (and the associated internal normal stresses presented in Figures 10 and 11) can be applied to other freight tunnels in downtown Chicago, if those tunnels were constructed at comparable depths and have similar shapes.

## REFERENCES

American Concrete Institute 2004. Building Code Requirements for Structural Concrete, ACI 318-05, ACI, Farmington Hill, Michigan.

Ayoubian, A. 2015. Effects of Excavation at Block 37 on Adjacent Freight Tunnels: Performance and Analysis. Ph.D. Thesis submitted to the Graduate School at Northwestern University in partial fulfillment of the requirements for the degree of Doctor of Philosophy.

Bretz, J. H. 1939. Geology of the Chicago Region: Part 1 – General. Illinois State Geologic Survey, Bulletin 65.

Popov, E. P. 1976. Engineering Mechanics of Solids. Prentice-Hall International Series.

Leca, E. and Clough, G. W. (1992). "Preliminary Design for NATM Tunnel Support in Soil," Journal of Geotechnical Engineering, ASCE, Vol. 118, No. 4, pp. 558–575.

Finno, R. J. and Chung 1992. "Stress-strain-strength responses of compressible Chicago glacial clays." Journal of Geotechnical Engineering, Vol. 118, No. 10, pp. 1,607–1,625.

Panet, M. and Guenot, A. 1982. "Analysis of convergence behind the face of a tunnel." Proceedings, International Symposium Tunneling '82, 197–204.

STS Consultants 2007. Personal Communications.

Wang, C., Salmon, C. G. and Pincheira, J. A. 2007. Reinforced Concrete Design. Hoboken, NJ, John Wiley & Sons.

Wren, J 2007. "The great Chicago flood, analysis of Chicago's 2nd great disaster." Structure magazine, August, pp. 35–40.

# Various Rock Tunneling Methods Utilized on the Doan Valley Storage Tunnel Project

**Collin Schroeder** ▪ McNally Tunneling Corp.
**Chris Lynagh** ▪ McNally Tunneling Corp.

## ABSTRACT

Determining the best excavation method is a critical step in successful tunnel construction. In rock tunneling, several factors can dictate the excavation method, such as: (1) rock characteristics, (2) tunnel size and length, (3) nearby utilities/infrastructure, and (4) contractor's preferred means and methods. This paper discusses the various rock tunnel excavation methods utilized on the Doan Valley Storage Tunnel Project (DVT), a project constructed as part of the broader Project Clean Lake in Cleveland, Ohio. The differences between each method's initial and final lining, geology, and cycle sequence will be analyzed and compared.

## INTRODUCTION

Rock tunnel excavation methods vary from project to project. Some of the main factors that determine the excavation methodology are:

- Geology (rock mechanical properties and conditions)
- Tunnel characteristics (size, length)
- Impacts to existing infrastructure (utilities, buildings, etc.)
- Means and methods (site/shaft constraints, initial support installation)

The Doan Valley Storage Tunnel project was a unique opportunity to utilize several different tunneling methods and pieces of equipment within the local Ohio Shale formations. On the project, 11 tunnels of varying length and size were excavated. The chosen excavation methods included a single shield tunnel boring machine (TBM) with a single-pass segmental liner, a main beam Robbins TBM with rib and timber lagging initial support, a roadheader, an auger bore machine (ABM), and various two-pass hand-mined tunnels with rock bolts and mesh or liner plate as initial support.

This paper will evaluate the criteria used to determine the method for each tunnel on the DVT project and how the tunneling cycles varied for each.

## PROJECT BACKGROUND

The DVT project is one of seven major tunnel contracts comprising the Northeast Ohio Regional Sewer District's (NEORSD) broader Project Clean Lake program. Project Clean Lake is a $3 billion, 25-year program implemented with the goal of capturing and treating 98% of wet weather flows entering the City of Cleveland's combined sewer system. This level of capture will significantly lower the raw sewage discharges that currently enter Lake Erie and associated waterways. DVT itself captures 365 MG/year and was successfully turned online in June of 2021.

## PROJECT GEOLOGY

The project's larger diameter tunnels (DVT, MLKCT, and WCT) all have a similar geologic setting of Chagrin Shale bedrock, with only a short portion of the WCT drive

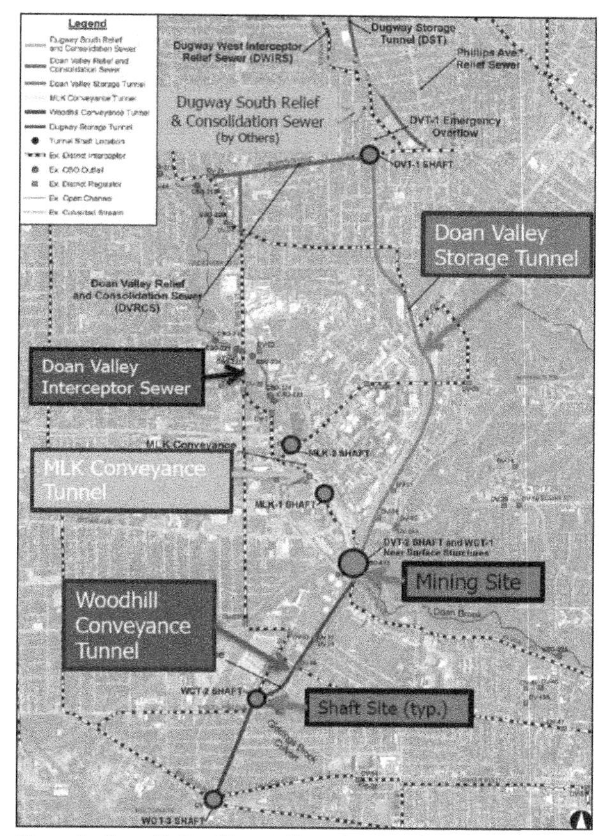


Figure 1. Doan Valley Storage Tunnel project layout

experiencing Cleveland Shale. Other tunnels on the project, namely the WCT-3 36" and 54" consolidation sewer auger-bored tunnels and the WCT-2 consolidation sewer tunnel, saw split faces of Chagrin Shale/Euclid Lentil Sandstone or Chagrin Shale/ Cleveland Shale.

In general, the Chagrin and Cleveland shale rock masses are characterized as having weakly bedded planes including clay-filled bedding joints and weathered interbeds, near vertical joint sets, and a tendency for slaking. The Euclid Lentil sandstone is characterized as a fine-grained sandstone with frequent shale seams. The average unconfined compressive strength (UCS) of shales and sandstones on the project was 2,800 psi and 11,500 psi, respectively. A summary of each tunnel's geological features, design requirements, and excavation methods are provided in Table 1.

## TUNNEL METHODS AND CHARACTERISTICS

As one can imagine, determining the tunneling method is greatly influenced by each tunnel's physical characteristics. What size is the tunnel? How long is the tunnel? What final lining is required for each tunnel? These are all questions that factor into the decision of which tunneling method will be utilized.

**Table 1. Summary of the geological features, design requirements, and excavation methods for each tunnel**

| Tunnel | Geology | | Design | | | Excavation Method |
|---|---|---|---|---|---|---|
| | Formation | UCS, psi (MPa) | Tunnel Length, LF | Depth, ft | Finished Diameter, ft | |
| DVT | Chagrin Shale | 2,800–11,500 (19–79) | 9,670 | 48–115 | 18 | One-pass TBM Excavation<br>▪ Segmental lining support |
| DVT Starter Tunnel | | | 335 | 68 | 18 | Two-pass Roadheader Excavation<br>▪ Rock Bolt/Wire/Mesh Shotcrete initial support<br>▪ Starter Tunnel—18' dia. CIP final lining<br>▪ Tail Tunnel—Backfilled and abandoned with cellular grout |
| DVT Tail Tunnel | | | 155 | 77 | N/A | |
| MLKCT | | | 2,970 | 52–97 | 8.5 | Two-pass TBM Excavation<br>▪ Ribs and lagging initial support<br>▪ 8.5' RCP final lining |
| WCT | Chagrin/ Cleveland Shale | | 6,200 | 27–77 | | |
| MLK-IAdit | Chagrin Shale | | 36 | 45 | 6 | Two-pass<br>▪ Rock bolts/mesh/mine strap initial support<br>▪ 72" CCFRPM(MLK-1) and 72" CIP lining (WCT-2) |
| WCT-2 Adit | | | 14 | 35 | | |
| DVT-1 EDS | | | 42 | 27 | 12'W × 17.5' H CIP Horseshoe Shaped | Two-pass<br>▪ Rock bolts/mesh/shotcrete initial support<br>▪ CIP horseshoe shaped final lining |
| WCT-1 FCS to DVT-2 Shaft | | | 13 | 18 | 6' W × 8' H Culvert | Two-pass<br>▪ 2-flange liner plates initial support<br>▪ 6'W × 8'H CIP Culvert (FCS to DVT-2 Shaft) and 72" RCP (WCT-2 CS) |
| WCT-2 CS | Chagrin/ Cleveland Shale | | 60 | 16 | 6 | |
| DVT-IDS | Chagrin Shale | | 72 | 41 | 4 | Two-pass ABM Excavation<br>▪ Steel casing used for initial support<br>▪ CCFRPM pipe used for final lining<br>▪ *See Table 2 for ABM linings broken down by tunnel |
| WCT-3 36" CS | Chagrin Shale/ Euclid Lentil Sandstone | 2,800–11,500 (19–79) 11,500–14,500 (79–100) | 70 | 16 | 3 | |
| WCT-3 54" CS | | | 42 | 9 | 4.5 | |

## [DVT] Single-Pass Segmentally Lined Tunnel Utilizing 20'-9" Herrenknecht Single Shield TBM

The Doan Valley Storage Tunnel (DVT) was the largest and longest tunnel on the project with a total length of 10,005 LF and an 18' finished diameter. DVT was designed as a one-pass tunnel with the tunnel's final precast concrete segmental (PCS) lining installed as the tunnel excavation progressed. While advancing the tunnel excavation, a two-component annular grout was placed between the extrados of the PCS and the excavation cut diameter. Locomotives were utilized to haul muck boxes that transported the excavated shale from the tunnel heading back to the mining shaft, the DVT-2 Shaft. In general, a mining cycle would consist of advancing 5 feet, building a PCS ring, and advancing another 5 feet while concurrently grouting the annulus of the previously built PCS ring. Utilities, power cable, ventilation, and rail were extended as needed during this mining cycle.

## [DVT] Two-Pass Tunnel Utilizing Antraquip Roadheader (DVT Starter and Tail Tunnel)

Project means and methods dictated that DVT be initially excavated as a 335' L × 24'-4" H × 23'-8" W horseshoe-shaped starter tunnel. This starter tunnel, along with a 155' L × 16' H × 15' W tail tunnel, would be long enough to fully assemble the 20'-9" diameter Herrenknecht (HK) TBM underground. The TBM would then excavate the remaining 9,670 LF of tunnel as discussed in the previous section. The initial support system for the tail and starter tunnel consisted of #8 rock dowels, welded wire mesh, and shotcrete. The starter and tail tunnel drives utilized an Antraquip roadheader machine along with an underground loader. A Commando remote hammer drill rig was used to install the initial support rock dowels. The cycles for both the starter and tail tunnel were generally the same, as they included an excavation advance with the roadheader, followed by an installation round of the tunnels initial lining. The one main difference between the two tunnels, is the starter tunnel utilized a split heading, excavating the entire top heading first (14'-4" H × 23'-8" W horseshoe), followed by excavating the entire bottom bench (10' H × 23'-8" W). In comparison, the entire tail tunnel profile was excavated in one pass. After the tunnel drive was completed, the tail tunnel was bulkheaded and backfilled with a cellular grout. The DVT starter tunnel was lined with a cast-in-place synthetic fiber reinforced concrete (Figure 3).

Figure 2. Doan Valley Storage Tunnel

## [MLKCT / WCT] Two-Pass Rib and Lagging Tunnel Using a Robbins Main Beam Gripper Machine

The Martin Luther King Jr. Conveyance Tunnel (MLKCT) and Woodhill Conveyance Tunnel (WCT) were both excavated similarly. At lengths of 2,970 LF and 6,200 LF respectively, both tunnels were excavated using a 145" Robbins Main Beam Gripper TBM. Utilizing a two-pass rock tunneling method, both tunnels were initially lined with steel ribs and lagging boards. By installing the steel ribs within the TBM's tail shield, the tunnel would always be supported once excavated. Similar to DVT, both the MLKCT and WCT used locomotives to transport muck boxes to remove tunnel spoils. In general, an advance cycle would consist of mining 5 feet, building a steel rib within the TBM's tail shield, installing lagging boards between the last two built steel ribs, and mining the next 5-foot advance. Utilities, power cable, ventilation, and rail were extended as needed during this mining cycle (Figure 4).

Following the excavation for both MLKCT and WCT, the linear plant was removed, and the initial lining was thoroughly cleaned. This was all done in preparation for installing the tunnels' final lining, an 8.5' diameter reinforced concrete pipe (RCP). A pipe carrier would be used to transport the 8.5' diameter sticks of RCP into the tunnel. Overall, 1,314 sticks of RCP were installed between the two tunnels.

Figure 3. DVT starter tunnel

Lastly, following the installation of the tunnels' final linings, a cellular grout was pumped into the annulus. This operation required a grout plant setup on surface to batch the cellular grout. This grout was pumped throughout the tunnel and injected into the annulus through grout ports that were cast into the RCP (Figure 5).

Figure 4. Woodhill conveyance tunnel

## Two-Pass Tunneling with Rock Bolts/Mesh (MLK-1/WCT-2 ADITS and DVT-1 EDS)

While the DVT project consisted of three TBM-excavated rock tunnels (DVT, MLKCT, and WCT), there were also several over non-TBM excavated tunnels. Two of which were adit tunnel connections that were excavated from a nearby shaft to an already-excavated tunnel. Both adit tunnel excavations on the project had similar characteristics, with both being 10' H × 10' W horseshoe shaped tunnel excavations. The 36'-4" long MLK-1 adit was excavated from the MLK-1 shaft to the previously excavated MLKCT, while the 14' long WCT-2 adit was excavated from the WCT-2 shaft to the previously excavated WCT. Both tunnels would be initially supported with a rock bolt/geogrid mesh/wire mesh/mine strap system and would be permanently lined with either a 72" centrifugally cast fiberglass-reinforced polymer mortar (CCFRPM) pipe (MLK-1 Adit) or a cast-in-place (CIP) concrete 72" pipe (WCT-2 Adit).

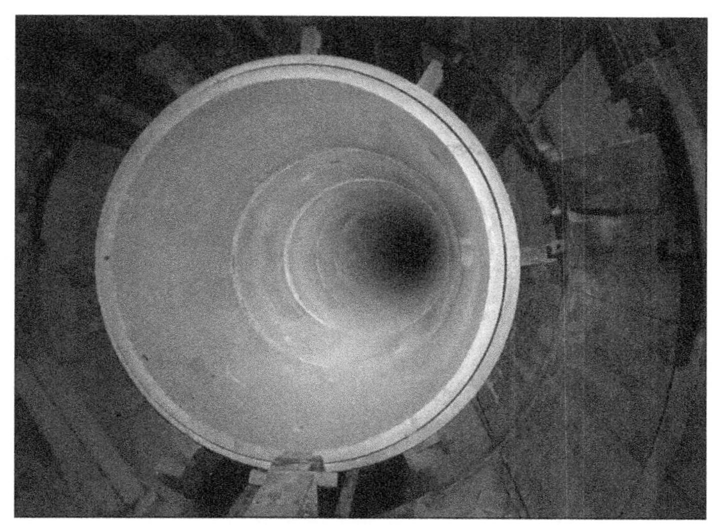

Figure 5. Martin Luther King Jr. conveyance tunnel RCP installation

Given the similar tunnel characteristics, both were excavated generally in the same manner. The entire adit tunnel excavation would be completed first, including the break-in to the MLKCT or WCT tunnel. Following the full excavation and breakthrough, the adit's final lining would be installed. In general, the excavation cycle consisted of hammering the tunnel face with an excavator, followed by pulling the muck out of the tunnel and shaft once enough muck had been broken up from the face. This process would be repeated until a clear 5-foot advance was made; at which time the next round of initial support would be installed.

While these tunnels were very similar in many ways, other factors contributed to which equipment would be utilized for excavation. The WCT-2 Adit was excavated primarily with a CAT 308 excavator, with a CAT 305 excavator utilized to retrieve the tunnel muck. In contrast, the MLK-1 Adit was excavated utilizing a Brokk robotic excavator and a Takeuchi mini-excavator to retrieve tunnel muck. Generally, spatial constraints were the limiting factor on which pieces of equipment were used (Figure 6).

Figure 6. MLK-1 adit tunnel excavation

The DVT-1 Energy Dissipation Structure (EDS) tunnel was a 21.75' H × 15.5' W × 42 LF horseshoe shaped tunnel that was excavated sequentially using a top heading and bottom bench. The top heading was 15' H × 15.5' W, while the bottom bench would consist of the remainder of the tunnel (6.75' H × 15.5' W). This tunnel was excavated using only a CAT 308 and was initially supported using rock bolts, wire mesh, and shotcrete. The cycle for this tunnel was generally the same as the adit tunnel excavations, with the major exception being the addition of a split heading. The final lining for the DVT-1 EDS tunnel was a cast-in-place concrete liner.

## Two-Pass Tunneling with Two-Flange Liner Plates (WCT-2 CS and WCT-1 FCS to DVT-2 Shaft Culvert)

Two other non-TBM excavated tunnels on the DVT project were the WCT-2 Consolidation Sewer (CS) and the WCT-1 Flow Control Structure (FCS) to DVT-2 Shaft Culvert. Both tunnels were initially supported using a 2-flange liner plate that was installed as the excavation advanced, followed by the final lining installation post-excavation.

The WCT-2 CS was a 10'-4" H × 11'W × 60' L horseshoe shaped tunnel that was primarily excavated using a Brokk robotic excavator and CAT 305. The 18" wide liner plates used for the initial lining support were installed using a cycle that would excavate 4.5' at a time, followed by the installation of three rows of plates. Once the three rows were installed, the annulus was backfilled with grout to ensure a solid contact area was seen throughout the initial lining coverage area. This tunnel was excavated from the WCT-2 Shaft to the WCT-2 Diversion Structure (WCT-2 DVS). Ultimately, this tunnel was utilized to convey the flows intercepted at the WCT-2 DVS to the WCT-2 Shaft. A 72" RCP was installed following tunneling operations to serve as the final lining.

Similar to the WCT-2 CS tunnel, the WCT-1 FCS to DVT-2 Shaft Culvert tunnel consisted of installing three rows of 18" wide 2-flange liner plates at a time. The cycles between the two tunnels were generally the same; however, the tunnel characteristics differed. The WCT-1 FCS to DVT-2 Shaft Culvert tunnel was larger than the WCT-2 CS tunnel at 13'-2" H × 15'-8" W, while the WCT-2 CS tunnel was longer, as the WCT-1 FCS to DVT-2 Shaft Culvert tunnel was just 13'-6" long. The final linings also differed between the two tunnels, as the WCT-1 FCS to DVT-2 Shaft Culvert tunnel required a final lining of 6' W × 8' H cast-in-place (CIP) concrete culvert.

## Two-Pass Tunneling with Auger Bore Machines (WCT-3 36"/54" CS, DVT-1 DS)

The final rock tunneling method used on the DVT project was an auger bore machine (ABM). There were three auger-bored tunnels on the DVT project: the WCT-3 36" Consolidation Sewer (CS), the WCT-3 54" Consolidation Sewer (CS), and the 48" DVT-1 Dewatering Sewer (DS). Each of these tunnels were excavated in a similar manner. This process would include using an ABM to excavate the tunnel and pushing a welded-on casing pipe to act as the tunnel's initial ground support. While the ABM cutterhead excavated, the spoils were transported back to the mining shaft using an auger. The spoils that were displaced into the mining shaft were then clam-bucketed out of the shaft and staged on surface to be loaded offsite. Each ABM excavated tunnel utilized 10' sticks of casing that were pushed into the bored tunnel (Table 2).

Following the excavation, the CCFRPM pipe was driven into the initial lining steel casing using the ABM. Once all sticks of pipe were installed through the tunnel, bulkheads were built on each end and grout was pumped to fill the annulus between the CCFRPM pipe and steel casing.

**Table 2. ABM-excavated tunnels summary**

| Tunnel | Initial Lining | Final Lining |
|---|---|---|
| DVT-1 DS | 60" Steel Casing | 48" CCFRPM (Hobas pipe) |
| WCT-3 36" CS | 48" Steel Casing | 36" CCFRPM (Hobas pipe) |
| WCT-3 54" CS | 66" Steel Casing | 54" CCFRPM (Hobas pipe) |

## EXISTING INFRASTRUCTURE IMPACTS/CONSIDERATIONS

Another factor that contributes to selecting the ideal tunneling method is to consider the impacts of existing infrastructure on the tunnel excavation, and vice versa. In fact, this was the main factor in several of the tunnels on the DVT project.

First and foremost, the project constraints altogether eliminated the option to drill and blast non-TBM tunnels, as a large portion of the project was in proximity to Case Western Reserve University and the University Circle area. Additionally, several high-profile utilities were in the vicinity of the tunnels and dictated the tunnel excavation methods. This was the case at both the WCT-2 and WCT-3 satellite sites.

The WCT-2 CS tunnel would require excavating underneath two 30" water mains that fed the City of Cleveland. Due to this, the 2-flange liner plate method was chosen as rows as small as 18" wide could be installed and backfilled at one time if any significant overbreak began to take place in the tunnel crown. These water lines also eliminated the possibility of using open-cut methods to install the consolidation sewer.

The WCT-3 CS tunnels both had several utilities located above the tunnel alignments. These utilities included an electric duct bank, an 8" gas main, a 12" gas line, a 10" water main, and two 30" water mains. Given the proximity to these utilities and the size required for the final pipe lining, an auger bore method was chosen as the preferred methodology. Similar to the WCT-2 CS, the overlying utilities eliminated the possibility of any open-cut excavation methods (Figures 7 and 8).

## MEANS AND METHODS

From the Contractor's perspective, understanding what the tunnel cycle will look like is of the utmost importance when determining the tunnel excavation strategy. Knowing how far an excavation will advance at one time, how long the initial lining will take to install, and what impacts these factors will have on the operation should be thoroughly evaluated. Two good examples of this took place with the MLK-1 Adit tunnel excavation and WCT-2 CS tunnel excavation.

The MLK-1 Adit tunnel excavation had several factors to consider as the mining shaft was only a 12' diameter drilled shaft. This meant that the tunnel operation would need to be optimized for a cycle and equipment that could efficiently and adequately work in such tight quarters. As a result of careful planning, a Brokk robotic excavator to hammer and a mini excavator to retrieve tunnel muck proved to be the most efficient equipment for the excavation cycle from a schedule and cost standpoint. Any other method would have resulted in slower advance rates and a longer operation overall.

There were other factors to consider when evaluating what cycle would be most efficient for the WCT-2 CS tunnel. As shown in Figure 9, the Contract Documents provided an initial support design that consisted of horseshoe shaped ribs, lagging boards, and 10'-long spile bolts. The project team evaluated the time it would take to install this initial lining system at each advance and determined that an alternative

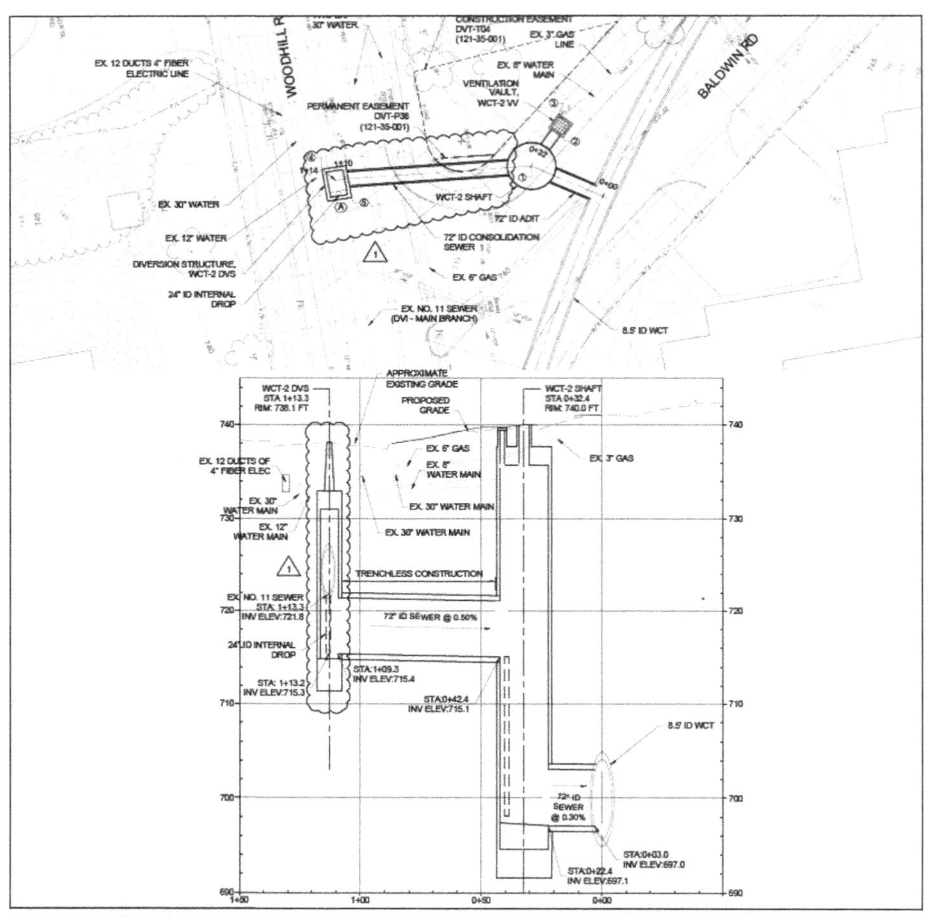

Figure 7. WCT-2 CS plan and profile

method should be pursued. Ultimately, an initial support system of horseshoe shaped two-flange liner plates was chosen, as this still provided the opportunity for shorter, 18" advances if needed while underneath sensitive overlying utilities (a major factor that led to the initial support design). Additionally, the two-flange liner plates would require much shorter installation time compared to the initial Contract design. This decision led to an overall shorter duration for the operation (Figure 10).

## CONCLUSION

Several different tunneling methodologies were used on the Doan Valley Storage Tunnel project. Various factors contributed to the method chosen for each tunnel excavation. Whether it was tunnel length, size, site/shaft spatial constraints, utility impact mitigation efforts, or a combination of them all, each tunnel was put through an evaluation process to determine which tunneling method would be the best fit. Overall, each tunnel was successfully completed and the entire DVT system came online in the summer of 2021.

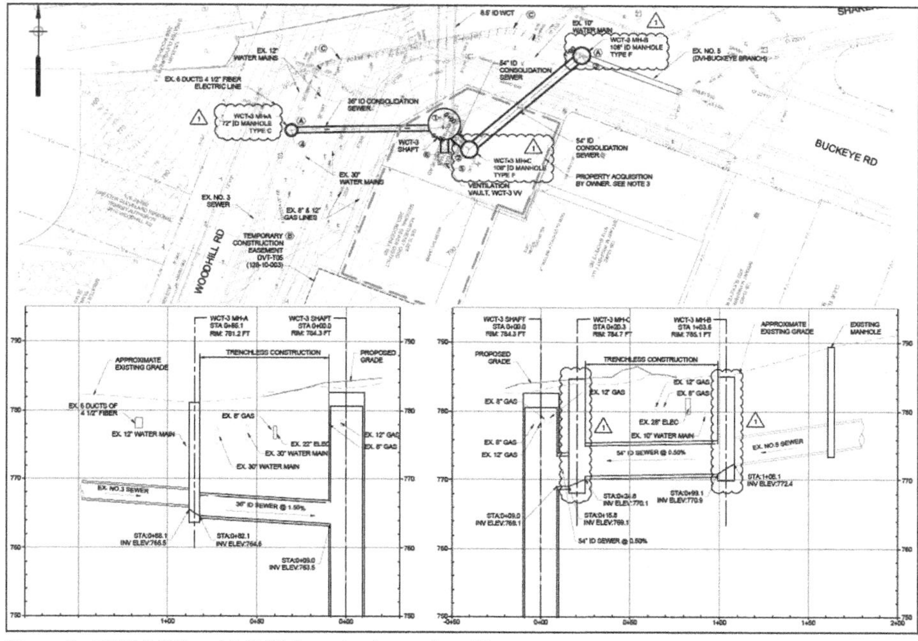


Figure 8. WCT-3 36" and 54" CS tunnels plan and profile

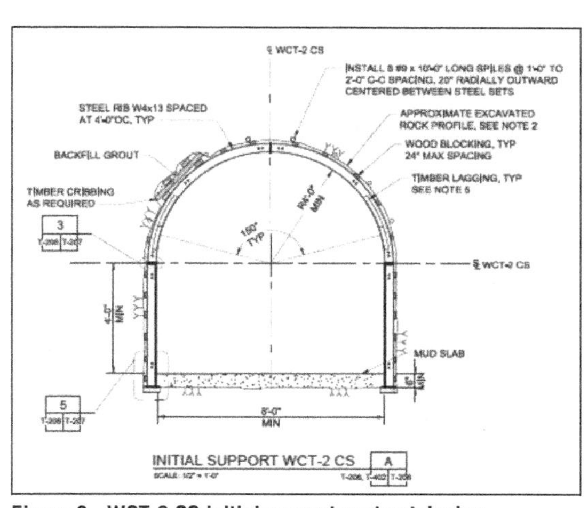


Figure 9. WCT-2 CS initial support contract design

Figure 10. WCT-2 CS 2-flange liner plates

PART 

# Hard Rock TBMs and Tunnels

*Chairs*

**David Lacher**
Traylor Bros Inc

**Ben McQueen**
Frontier-Kemper Constructors inc

# Construction of Coxwell Bypass Tunnel Project in Toronto, ON

**Ehsan Alavi** ▪ Jay Dee Contractors, Inc.
**Eren Kusdogan** ▪ Jay Dee Canada

## ABSTRACT

The City of Toronto's Coxwell Bypass Tunnel is currently being constructed by North Tunnel Constructors ULC, a joint venture of Jay Dee Canada, C&M McNally Tunnel Constructors and Michels Canada. This project includes construction of approximately 10.5 km of 6.3 meter finished diameter rock tunnel, five 20 meter diameter storage shafts and eleven tunnel connection drop shafts, along with associated deaeration and adit tunnels. This paper describes progress on the project to date and reviews the risk mitigation measures utilized on the project to move the project forward through tough geological conditions during shaft excavation and tunneling phase of the project.

## INTRODUCTION

The City Toronto has developed the largest and most significant storm water management program in the city's history. The Wet Weather Flow Master Plan (WWFMP) projects will eliminate the release of combined sewer overflows and polluted water to Lake Ontario, along with key infrastructure upgrades at Ashbridges Bay Treatment Plant to improve system capacity (Figure 1).

The Coxwell Bypass Tunnel is the first phase for 22 km Don River and Central Waterfront Wet Weather Flow System which will keep combined sewer overflows out of waterways by:

- Capturing the combined sewer overflows within the tunnel system
- Storing the combined sewer overflows during extreme rainstorms until the system capacity is restored and the water can be transported for treatment
- Transporting the flows to a new and dedicated high-rate treatment facility at the Ashbridges Bay Treatment Plant (ABTP)
- Treating the flows through a new ultra-violet disinfection system
- Discharging the flows through a new 7-meter diameter, 3.5-km long outfall into Lake Ontario.

The Coxwell Bypass Tunnel (CBT) consists of a 6.3 diameter 10.5 km long tunnel, 5 major work shafts with diameters varying from 20 to 22 m at average depths of 50 m, and 11 adit tunnels scattered throughout the main tunnel varying in diameter from 3.0 m to 8.45 m (Figure 2).

### Project Geology and TBM Selection

The Coxwell Bypass Tunnel (CBT) is constructed with a consistent slope of 0.15%. The tunnel invert is at depths ranging from 44 m to 69 m below ground surface. Tunnel Overburden consists of bedrock (Georgian Bay Formation), and the soil overburden. The TBM tunnel and all the adit tunnels are within the Georgian Bay Formation.

#### *Soil Over Burden*

Although the tunnel has been mined in bedrock, some sections of the tunnel have potential for buried valleys with less than a meter for rock cover (Figure 3).

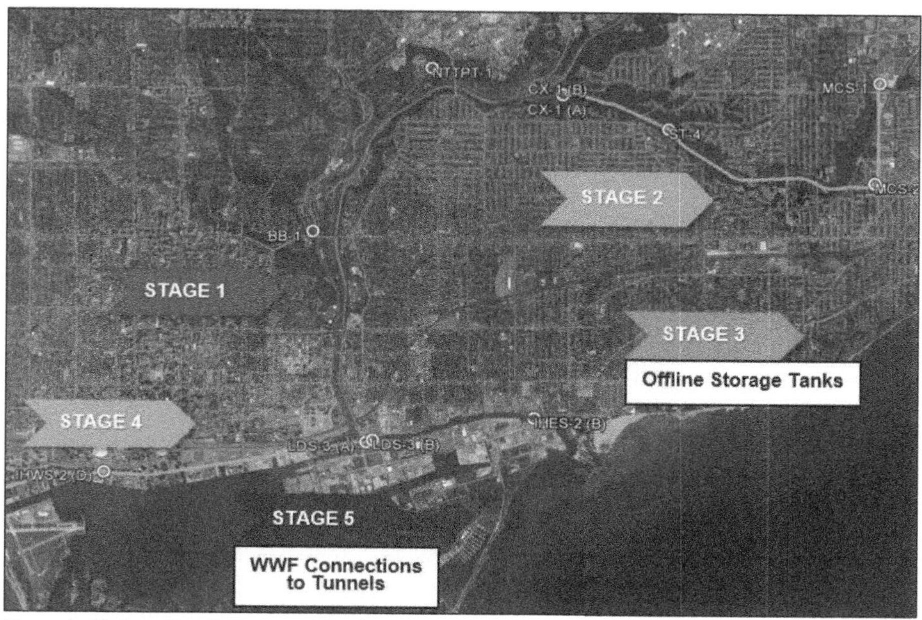


Figure 1. Wet weather flow master plan (WWFMP)

Figure 2. CBT project map

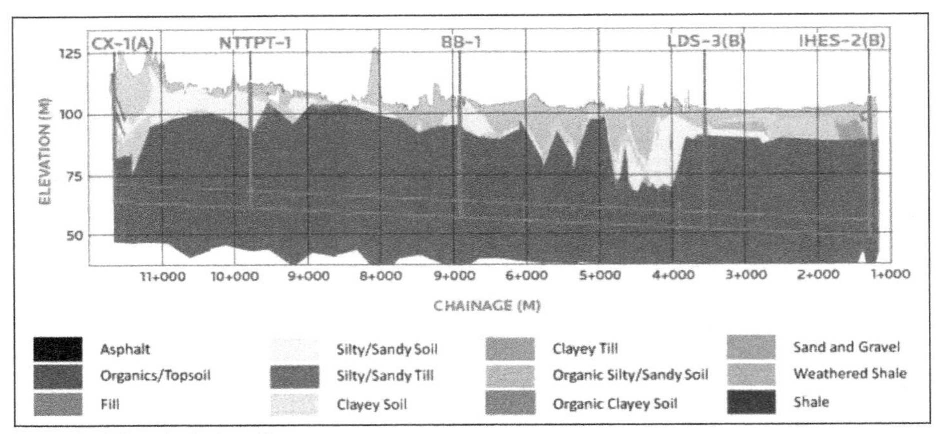


**Figure 3. BT geological profile**

### *Bed Rock (Georgian Bay Formation)*

The Georgian Bay Formation underlies the overburden soils mentioned above. The top of the bedrock was encountered at elevations ranging between approximately 42.6 m and 79.6 m across the CBT project. This formation consists of fresh to slightly weathered, thinly to medium bedded, bluish grey, very fine to medium grained, faintly porous, weak to moderately strong shale with occasional thin interbeds of harder locally fossiliferous siltstone and limestone. Typically, the top part of the bed rock (approximately 3 m to 6 m of bedrock) is more fractured and possibly more weathered then transitions to a fresh, more intact state at depth, with thicker zones of fracturing encountered as per the geotechnical baseline report of the project.

Some parts of the bedrock had Siltstone and Limestone interbedding Layers. In some instances, clayey fills have been seen in the bedrock which cause a formation of weak rock mass. Shale rock mass has horizontal bedding and vertical and subvertical joints. The average spacing of 2.5 m, the lengths between 2 m and 5 m is characterized in the GBR.

Ground stress condition varies based on the location and typically the in-situ minor horizontal stress is a minimum of 0.5 MPa, while the major horizontal stress is a maximum of 7.0 MPa. The uniaxial compressive strength of the bedrock varies between 6 to 40 MPa and while some areas encountered had strengths up to 70–80 MPa. Rock Quality Designation (RQD) at the tunnel depth is mainly 90–100% and at the few zones it has been seen to be between 60–70 and 80–90%.

Based on the geological conditions described in the GBR and the design considerations Hard Rock TBM which is also capable of operation in closed mode with 7.1 m cut diameter was chosen for the project (Figure 4).

The reason of having close mode capability is zones of increased weathering at depth are interpreted to occur between chainages 3+860 to 3+890, 3+950 to 4+250 and 4+575 to 4+660. Additional Seismic Investigation performed to model the zones where closed mode tunneling could be required. The birotational cutterhead is designed to have the best excavation performance in the geological conditions presented in the GBR and GDR. The focus of the cutterhead design is to choose the proper tools and have an appropriate opening area. The cutterhead is furnished with disk cutters and scrapers (Figure 5).

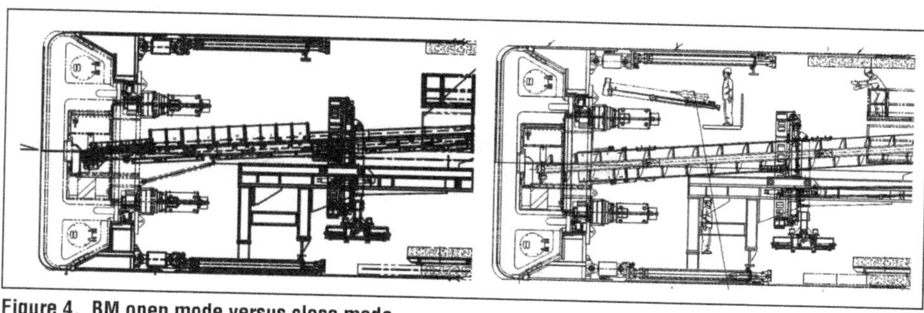

Figure 4. BM open mode versus close mode

Figure 5. BM cutter head

## TBM Tunneling Setup and Construction

### *Starter/Tail Tunnel*

The options are limited to start a TBM which is over 110 m in length from a 20 m diameter shaft. Rather than starting with umbilicals, NTC built starter and tail tunnels to have enough space to assemble the entire machine and conveyor belt system so there wouldn't be any interruption to mining for any changes of the set up.

120 m of Starter tunnel and 70 m of Tail tunnel were excavated. The temporary rock support consisted of ribs, bolts, welded-wire-mesh, and a 100 mm of shotcrete. The base was finished with a mud slab. Excavation was completed with a roadheader, top/bottom bench rock support was installed by utilizing an electric drilling unit, and the shotcrete was applied by utilizing an automated shotcrete sprayer (Figure 6).

### *Mucking System*

To transport muck generated during tunnelling operations, NTC procured a continuous conveyor which consists of multiple components including a horizontal discharge conveyor to provide flexibility for muck handling in the long and narrow Shaft 1 site,

Figure 6. BM walking in started tunnel

an inclined conveyor, a vertical shaft conveyor; and, a tunnel conveyor system. The entire generated muck along the length of the project was transported to the Shaft 1 site which makes the design and operation of the conveyor system quite challenging considering total length of the tunnel. The horizontal tunnel conveyor utilized on the project with overall length of 10.5 km is the longest single conveyor in North America.

### PCTL Design

The 300 mm thick Precast Tunnel Lining is a universal ring assembly comprised of four 67.5° rhomboidal segments and two 45° trapezoidal segments. The PCTL has been designed to facilitate ring build in a turning radius of 197 m, the concrete compressive strength is 50 MPa, and the steel fibers provided a minimum residual flexural strength of 3.4 MPa at L/150 of beam deflection. No rebar cages were used in the segments. The gaskets are designed to handle 12 bar of hydrostatic load. The segments are 1.8 m in length to facilitate longer excavation cycles and improve production. Time dependant deformation of shale in the Georgian Bay Formation was not found to increase the PCTL axial force and bending moment (Figure 7).

Rubber Tired Vehicles (RTVs) were procured for the project to be able to transport two sets of rings (twelve segments) at once so waiting time for the segments due to the length of the tunnel was minimized. Segment unloading system in the machine was also designed accordingly.

### TBM Tunneling and Challenges

The TBM started mining in March 2020 from IHES (Shaft 1). As shown in the project map (Figure 2) there are three intermediate shafts in the alignment. The TBM mined through the first intermediate shaft (LDS-3(B)—Shaft 2) and the first break through was at BB1 (Shaft 3). NTC made a strategic decision and moved the TBM logistics to BB-1 excluding the conveyor belt, so all the adit works between Shafts 1 and 3 did not have to wait for TBM tunneling to be completed.

**Blocky ground.** NTC faced unexpected ground conditions during TBM mining at two zones along the tunnel alignment. Tunnel operation experienced significant issues

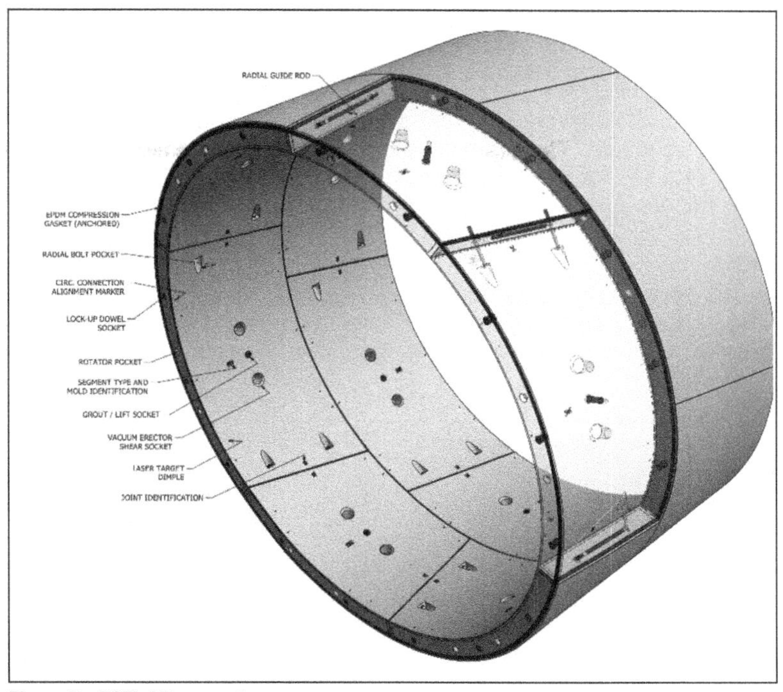


**Figure 7. PCTL 3D concept**

with the large and blocky ground encountered from STA 56+76.69 to 64+39.66 and STA 83+87.05 to 91+85.66 of the tunnel.

The subsurface conditions encountered negatively altered the breaking mechanism of the material, and as such, severely damaged the material handling mechanism. The designed material handling system was sized for material cut by the TBM, however the large blocks created by the subsurface condition encountered were outside of this design. The large size of the material caused irregular handling through the full material handling system and caused significant issues to the tunnel conveyor system functionality and delay to the TBM (Figure 8).

**Unknown geothermal system on the alignment.** the crew recovered a roughly 1.5" black pipe coming out of head and then started getting large inflows of water at approx. STA 76+20 and started getting large inflows of water. The TBM, the tunnel, the adits, the shafts flooded (Figure 9).

As per the contract document there were not supposed to be any utilities, however it was later confirmed that Brickworks had a geothermal HVAC system with 40 pipes drilled to 180 m, they reported shutting off the water, but NTC continues getting inflows.

**Soft ground.** NTC faced unexpected ground conditions during TBM mining at two zones, from 10+468 to 10+544 and from 10+869 to 10+930. The ground at these zones were significantly softer and contained loose interbedding which caused bigger rock pieces falling. Losing the interbedding prevents cutting the rock properly and cause the chunk of rock falls without breaking into smaller pieces and finally muck accumulating and travelling along the shield. Therefore the tail can was getting stuck and the pressure at passive articulation cylinders increased over 300 bar (vs. 100 bar during regular mining) resulting damages at the passive articulation cylinders, significant slower mining and ring building rates and several stoppages to unstuck the tail can.

The delays due to the unexpected ground conditions had a significant impact on the overall project schedule. The team developed and implemented several optimizations and logistics changes in adit constructions and shaft CIP works. These changes reduced the risk and additional cost to finish the project on time.

## Main Shafts

The project included five 20–22 m diameter shafts at depths of 50 m. Typically primary shafts have secant piles drilled from surface into the rock as the initial support of excavation. After removal of all overburden, a connection is made between the rock and

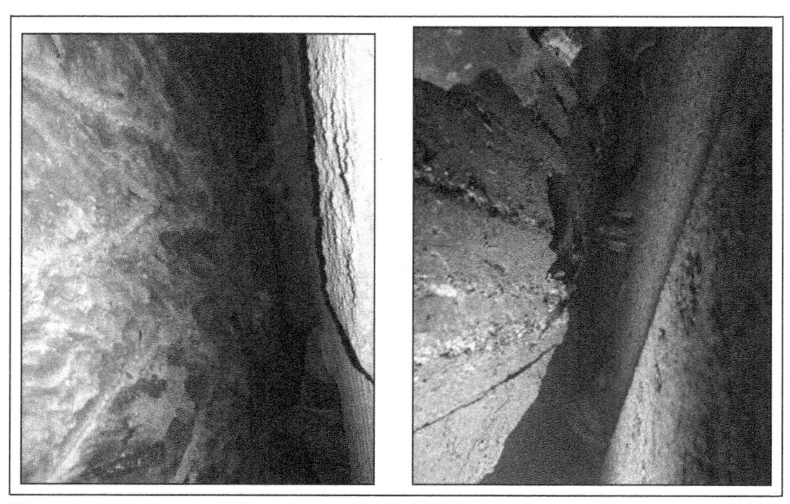

**Figure 8. Regular versus blocky ground excavation face**

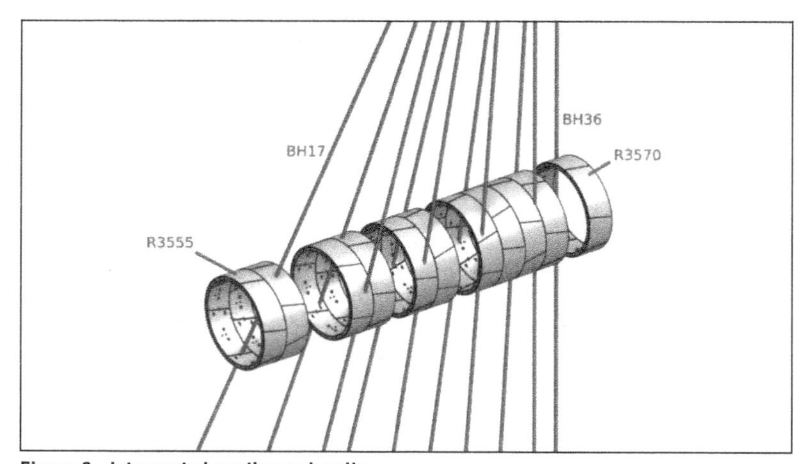


**Figure 9. Intercepted geothermal wells**

Figure 10. Opening PCTL tunnel as an access at shaft LDS-3(B)

secant piles using shotcrete. Rock support is performed during excavation by drilling, grouting in rock bolts, and applying welded wire mesh.

**Contaminated soil at LDS (Shaft 2) excavation.** During the test pit excavation at LDS NTC encountered heavily contaminated soil with free phase oil and hydrocarbon products. The "new" ground conditions caused delays and complications on the excavation of the shaft, therefore NTC decided to mine through the shaft with the TBM, until a new methodology for shaft excavation was developed.

After shaft excavation reached the tunnel elevation, NTC exposed and reinforced the rings and opened an access to allow the shaft access for supporting the adit works.

This delay effected the overall schedule of the shaft since it was always planned as adit support shaft. The final CIP works in this shaft could not be finalized until the adits are done (Figure 10).

### *Concrete Works*

The typical shaft features are shown in Figure 11. The base slabs of the shafts consisted of a 3 m thick steel reinforced concrete, with a hemispherical base profile. The base profile was used to optimize the design thickness to overcome hydrostatic pressures. On average, the base slabs consisted of approx. 100 MT of steel reinforcement and 900 $m^3$ of concrete. Concrete mix design optimization was used to offset the high temperatures encountered with mass concrete placement.

The shaft walls consisted of 700–800 mm of CIP concrete. The walls were reinforced with two layers of steel reinforcement, with some areas requiring horizontal layers of

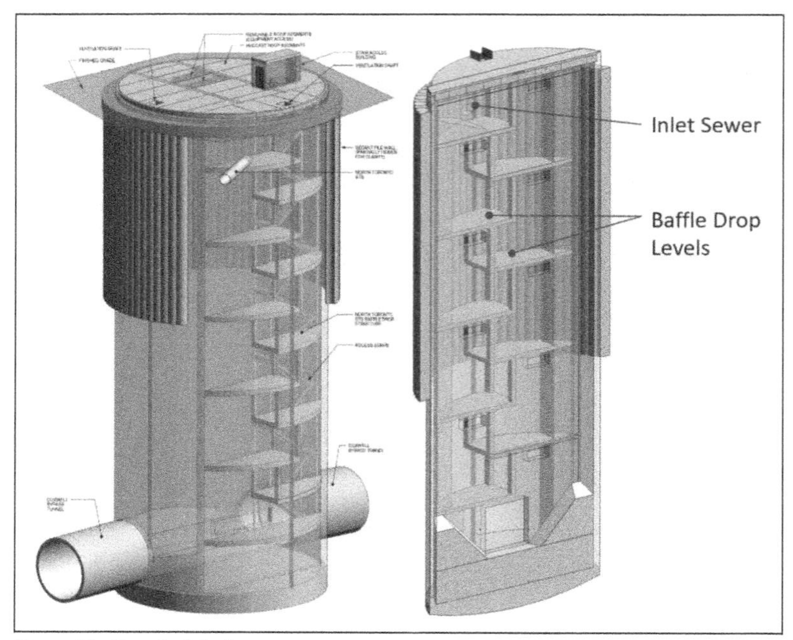


**Figure 11. Key shaft features**

35M @ 150 mm spacing. The forming system used was a steel jump form system. A low shrinkage concrete mix design was used to limit shrinkage cracking. Concrete pours ranged in volume from 200–240 $m^3$ per pour. To eliminate water leakage, a PVC waterstop was used in construction joints.

Within the large diameter storage shafts was a baffle structure with a vertical wall and alternating thick slabs. The baffles are meant to control the flow of sewer overflows to the shaft, by reducing the drop of the flows through the baffles (Figure 12). Air vent openings were installed under each slab to allow the air to escape. To increase the service life of the baffle structure, 50 MPa concrete was used, and the top horizontal steel reinforcement was eliminated to increase wear performance. The rebar in the baffle structures was tied into the shaft walls using embedded couplers. The baffle walls were 700 mm thick, while the slabs ranged from 400–550 mm thick. Various forming systems were used to construct the baffle structure. For the vertical baffle wall, a combination of jump forms and wall form panels were used. For the baffle slabs, various levels of shoring were used to withstand the weight of the thick slabs. The alternating design of the slabs, and the inaccessibility of previous slabs, increases the difficulty of the slabs construction.

The roofs of the storage shafts consisted of thick precast beams with a CIP topping above. The precast beams ranged in length from 16 to 18 m, and in thickness from 700–1,100 mm. Due to the weight of the precast beams, some are too heavy to ship and must be cast on site. The steel reinforcement in the precast beams consisted of 3–35M bundled bars in the bottom layer. The installation of the precast requires specialty equipment, as in some cases, site logistics does not allow for large cranes to be mobilized. After the precast is installed, a 200 mm CIP topping is poured over, and caps the large shafts. In one case, a future tunnel is proposed to connect to the shaft and requires the middle precast beams to be completely removable.

**Figure 12. Schematic of one of the large diameter storage shafts in the Coxwell Bypass Tunnel Project**

## Drop Shafts/Dearation and Adit Tunnels

The existing combined sewer overflows (CSOs) will be connected to the main tunnel via drop shafts, air shafts, dearation and adit tunnels.

As shown in the project map in Figure 2 (stars represents adits) the main tunnel will be connected to the existing CSOs at eleven locations.

The drop shafts and air shafts were drilled and lined from the surface prior to the adit excavations. The majority of the adits were located between Shafts 1 and 3. As mentioned before TBM logistics were moved to shaft 3 so adit works could get started concurrently with the remaining TBM tunneling. Shafts 1 and 2 effectively used for adits excavation, concrete lining, and pipe installation. After the TBM mining is completed, the operations expanded to four shafts.

The adits have been excavated with roadheaders which are capable of excavating various profiles and facilitate concurrent mucking. Rock bolts and welded-wire-mesh were used for the ground support.

The final product consisted of 35 MPa concrete with a dense rebar detail as per original design. NTC, in collaboration with the design engineers redesigned the final

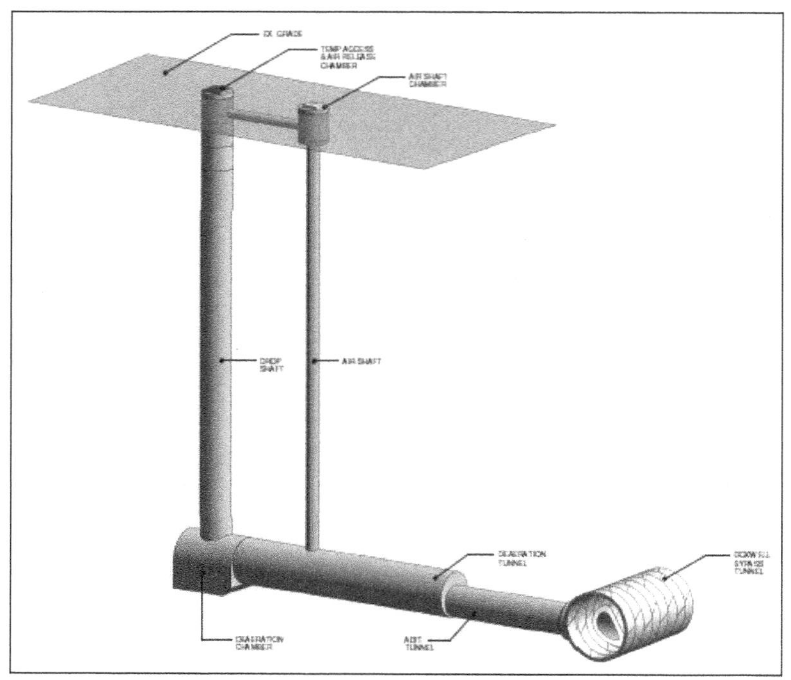


**Figure 13. Drop Shaft–Adit–Main Tunnel Connection 3D Concept**

concrete lining, and used 50 MPa concrete with steel fiber. The redesign significantly reduced the amount of rebar and increased the feasibility of the construction.

## CONCLUSION

Challenges are unavoidable in tunneling projects. The key to success is good planning, optimized design and most importantly, the collaboration of all the parties involved: the owner, project and construction management, designer, engineer and the field personnel who are building the job. At the time of writing this paper, the project is about 80% complete with majority of the remaining work consist of massive cast in place operation in access shafts and adit tunnels.

# Construction of TBM-Mined Segments of the Sister Grove Outfall Pipeline

**Greg Rogoff** ▪ Delve Underground
**Patrick Niemuth** ▪ Triad Engineering, Inc.
**Roshan Thapa** ▪ North Texas Municipal Water District
**Richard Yovichin** ▪ Delve Underground
**Ricky Chipka** ▪ Triad Engineering, Inc.

## ABSTRACT

A 4.5-mile-long, 96-inch-diameter outfall pipeline was constructed between North Texas Municipal Water District's proposed Sister Grove Regional Water Resource Recovery Facility and the Stiff Creek Discharge Structure. Two TBM-mined tunnel segments were excavated for the outfall pipeline through rock and mixed-face conditions within the Austin Chalk Formation. The TBM mining performance was tracked in detail and evaluated during mining. This paper discusses the TBM configuration, TBM utilization, advance rates, and challenges of both tunnel segments.

## INTRODUCTION

Tunnel boring machine (TBM) excavations in various geologic conditions have been studied in an effort to make predictions of penetration rates, cutter tool life, and overall TBM performance. The primary objective of these studies is to correlate TBM advance rates, defined as feet of advance per week, with ground conditions for future schedule predictions and cost estimates in similar geologic conditions. The two main rock mass parameters that affect the TBM penetration rate, defined as the ratio of advance distance to cutterhead revolution (mm/rev), are the intact rock strength and frequency and orientation of the discontinuities (Gong 2009). A more comprehensive indicator of TBM performance is the utilization rate (U), defined as the percentage of actual boring time to the total work time. The primary factors affecting U in addition to the geology are the methodology (type of TBM and mucking configuration, tools, and ground support) and process (TBM operation, maintenance, and project logistics) (Scialpi, Buxton, and Negrea 2020).

The Sister Grove Regional Water Resource Recovery Facility (SGRWRRF) Outfall Pipeline TBM mined segments were largely excavated in the favorable massive weak rock of the Austin Chalk Formation, thus limiting the geologic impact on penetration rates and TBM utilization. This project provided a unique scenario, where two tunnel drives were mined with the same TBM and cutterhead configuration, but with different mining shaft and backup gear configurations. This difference provided the opportunity to evaluate the impact of the different configuration on TBM advance rates and utilization.

## PROJECT BACKGROUND

The North Texas Municipal Water District (NTMWD) ("District") provides essential water, wastewater, and solid waste disposal services to member cities and customers in the North Texas region within the Dallas–Fort Worth metroplex. The District was created in 1951 with 10 original member cities and served approximately 32,000

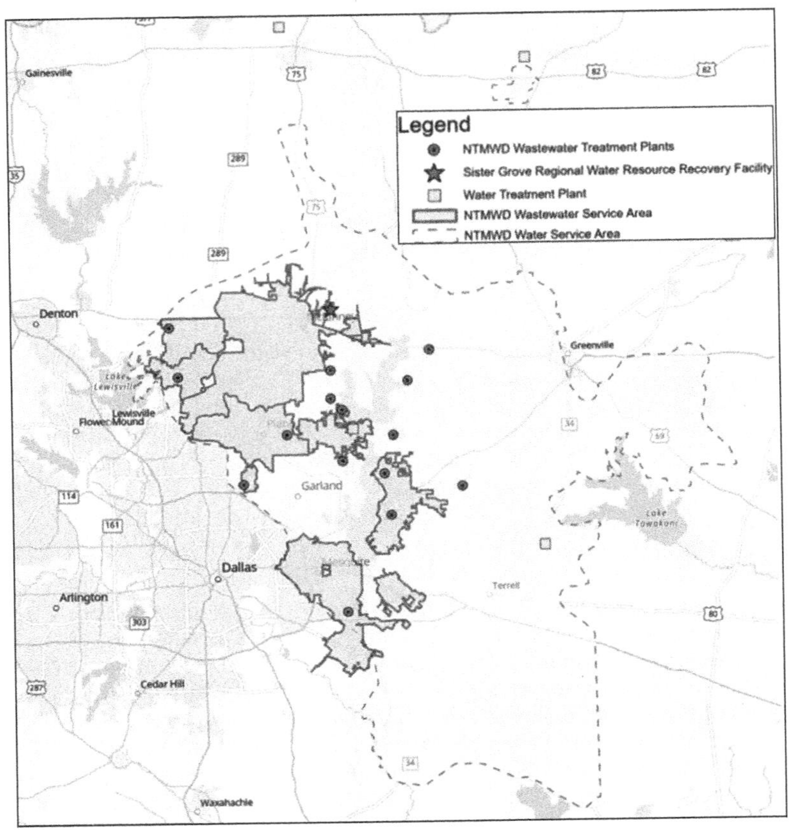


**Figure 1. NTMWD service area**

customers. Almost seven decades after its creation, the District serves more than 2 million people with safe, reliable drinking water and over 1.4 million with wastewater collection and treatment services. The communities within the 2,200-square-mile service area (as shown in Figure 1) are experiencing rapid population growth, especially in the northern part of the system, which has some of the fastest-growing communities in the nation. As the wastewater demand is growing and the two regional wastewater treatment plants currently serving this area are nearing available treatment capacity, the SGRWRRF project was necessitated. The planning phase of the project began in 2016 for facility siting considerations and selection, receiving water quality reviews and assessments, and facility planning concepts. The final design for the Phase I facility began in early 2019 by CDM Smith.

The biological-nutrient-removal SGRWRRF located on a 934-acre property in Collin County will provide an additional treatment capacity of 16 million gallons per day (MGD) annual average flow with a peak two-hour flow of 64 MGD for Phase I. Approximately 4 miles of 96-inch-diameter outfall pipeline will discharge the treated wastewater to Stiff Creek, which will ultimately make its way to the northernmost branch of Lavon Lake via Sister Grove Creek. Two reaches of the outfall pipeline were excavated by TBM: an 8,200-foot-long Rock Tunnel Segment and a 1,200-foot-long Mixed Face Tunnel Segment as shown in Figure 2. The project is being delivered through a Construction-Manager-at-Risk (CMAR) project delivery method by

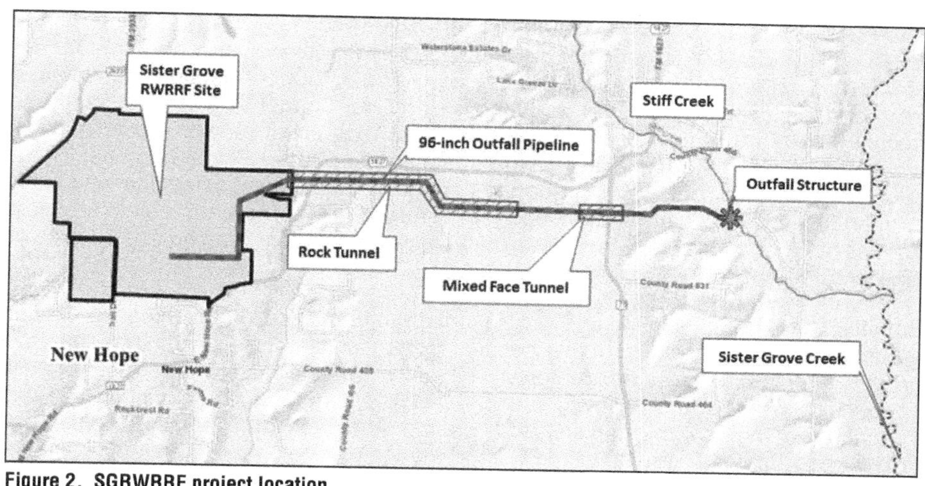


Figure 2. SGRWRRF project location

Garney Companies, which joined the project team in early 2020. The tunneling work package was selected by the CMAR and approved by the District Board of Directors in December 2020. The tunneling contractor, Triad McNally Midwest Joint Venture (TMMJV), mobilized to the site in mid-2021. Gresham Smith was hired to provide general inspection services and Delve Underground was hired by the District for specialized tunneling inspection and consulting during construction.

## SUMMARY OF SUBSURFACE CONDITIONS

The SGRWRRF outfall pipeline crosses three geologic formations: the Austin Chalk Formation, the Ozan Formation, and the alluvial deposits. The Austin Chalk Formation formed during the late Cretaceous Period and consists of chalk, marl, and limestone interbedded with interbeds of calcareous clays. Above the Austin Chalk Formation and also formed during the Cretaceous Period, the Ozan Formation is a medium gray calcareous shale and clay. The alluvial soils were deposited in floodplains and channels and mainly consist of clay, silt, sand, and gravels (McGowen et al. 1991).

The primary geologic setting for both tunnel segments was in limestone of the Austin Chalk Formation. However, fine-grained (typically high plasticity clay) alluvial deposits and weathered limestone with interbedded shale were encountered in the mixed face tunnel segment. The alluvial deposits primarily consisted of stiff to hard high plasticity clays. The weathered rock underlying the alluvial deposits ranged from poor to good quality weathered limestone to hard clay, and was defined in the Geotechnical Baseline Report (GBR) as transitional material consisting of soil with blow counts in excess of 50 to rock with rock quality designation (RQD) values less than 50. The fine-grained soil and completely decomposed weathered rock have a high stickiness potential according to criteria established by Thewes and Burger (2004). The limestone in the project area is typically slightly weathered and massive with RQD values typically above 90 in the tunnel envelope and widely spaced discontinuities. The limestone strength ranges from extremely weak to weak rock, with the average strength of weak rock being between 1,500 and 3,000 pounds per square inch (psi). According to test records provided in the Geotechnical Data Report (GDR), the limestone has medium to medium high slake durability and typically low abrasiveness potential (CDM Smith 2020).

The Rock Tunnel Segment consisted of two tunnel reaches (TRs) (as defined in the GBR): a 700-foot TR of poor to very good quality limestone, and a 7,500-foot TR of good to extremely good quality limestone with a 1,750-foot section of increased interbedded shale. Typical ground cover over the Rock Tunnel Segment was approximately 45 feet with the exception of a low 8-foot rock cover section under a creek crossing. The Mixed Face Tunnel Segment comprised three tunnel reaches: TR#1, consisting of a 325-foot reach of good to very good quality rock limestone with interbedded shale; TR#2, consisting of 630 feet of mixed face weathered rock to firm soil; and TR#3, consisting of 260 feet of mixed face weathered rock grading to a full face of alluvial clay deposits (CDM Smith 2021).

Both the Rock Tunnel and Mixed Face Tunnel Segments were largely mined under the groundwater table. Because of the massive rock mass structure, only minor sporadic groundwater inflows were encountered while mining within the in the rock. The most appreciable groundwater inflows, which were also minor, were encountered at the interface of alluvial overburden soils with weathered rock in the TR#2 and TR#3 of the Mixed Face Tunnel Segment.

## TBM CONFIGURATION

The TBM utilized to excavate both tunnel segments is a Lovat M123 Series 7100 (Figure 3). The TBM was fully refurbished for this project with many new parts for the main drive, stationary shell, and electrical and conveyor systems. The cutterhead, fabricated new by Lovsuns, included a new rotary fluid joint. The new bidirectional cutting head consisted of an eight-cylinder articulation system that was designed to provide two degrees of deviation in any direction. A triple row type main bearing with pressurized lubrication system was utilized. Power was provided by twin 350-horsepower (HP) electric motors coupled to a variable displacement pump.

The TBM propulsion system included eight 85-ton hydraulic cylinders that provided a maximum thrust of 680 tons at 5,000 psi hydraulic pressure. The cylinders provided a maximum stroke of 66 inches. Proportional pressure control to each cylinder was available to the TBM operator. This TBM utilized a push ring assembly that was

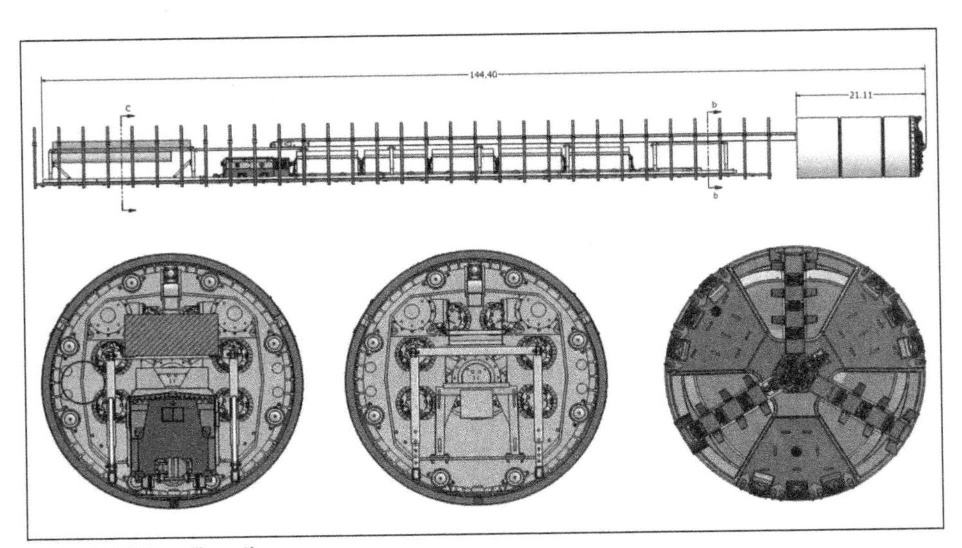


**Figure 3. TBM configuration**

designed and fabricated by Lovsuns, which transmitted all thrust force to the previously installed rib and lagging sets behind the machine. The tail shield was modified in-house to fit the new push ring and was outfitted with four gripping cylinders to mitigate roll and the total force being transferred to the rib and lagging initial shoring. TBM roll was also further mitigated with two stabilizer fins located in the stationary shell at the 7 and 4 o'clock positions. The TBM operator was able to hydraulicly extend these fins out into the TBM bore as needed.

The unitized conveyor setup was 24 inches wide by 100 feet long. Conveyor speed was variable, with a maximum speed of 300 feet per minute. The conveyor also had bidirectional capability. It was fully supported by the TBM gantry system and utilized hydraulic power for belt tensioning.

The electrical power for the TBM, unique to this project, was provided by a Caterpillar XQ1000 generator set capable of 910-kilowatt (kW) prime at 480-volt (V) three-phase voltage. The lack of adequate commercial power in the project area led TMMJV to run multiple pieces of equipment and other periphery equipment off genset supplied power. The 480 V power supplied by the genset was transformed to 4,160 V to mediate line loss over the length of the tunnel drives. It was then transformed back to 480 V via transformer to power the TBM in the excavation. Power was transmitted from the surface to the TBM with 1,000-foot total length 2/0 tunnel cables. Eaton soft starts were equipped for the 350-HP motors to prevent voltage spike. TBM lighting and all ancillary equipment were powered by 120 V single-phase power provided by the TBM transformer.

TBM safety measures included a keyed start to remove the chance of an unauthorized start, and emergency stop controls were in all major areas of work along the entire TBM configuration. An automatic tilt control was added so that all TBM power would be cut if the machine experienced 2.5 degrees of roll in either the clockwise or counterclockwise direction. Methane gas detection was provided by a Drager unit that shut down all power to the TBM if atmospheric conditions were at or above 20% lower explosive limit (LEL). Warnings would be given starting at 10% LEL.

The TBM was guided with a laser-mounted automated guidance system that utilized a robotic total station and electronic video target mounted in the TBM stationary shell to provide real-time TBM vertical and horizontal location. It also provided the operator with pitch and roll measurements.

The TBM cutterhead configuration for both tunnel segments consisted of an array of carbide-tipped bullet teeth for tunnel gauge and face penetration, and clay spades to clean the tunnel face moving muck to the cutterhead openings. The bore diameter measured 123 inches, with an initial overcut of 123.5 inches. The overcut diameter was eventually upsized to 124.5 inches to alleviate TBM bind, described in the next section.

## SUMMARY OF ROCK TUNNEL SEGMENT

The 8,200-foot Rock Tunnel Segment was excavated first and was mined upstream at a 0.15% grade to allow groundwater to flow out to the mining shaft. Although ground conditions would allow for spot dowels to support the ground, the initial tunnel support system consisted of W4x13 steel ribs and 5-foot-long 3×6-inch timber lagging to accommodate the TBM propulsion system. A 45-foot-long tail tunnel was excavated in the mining shaft to provide space for a switch rail and two sets of six muck cars in the shaft area. This mining shaft configuration allowed for muck to be removed via crane

Figure 4. Broken lagging observed during the Rock Tunnel TBM launch

while the TBM continued to mine with the additional set of muck cars. As the TBM advanced further from the shaft, train travel times increased, and an analysis was conducted to determine that it was advantageous to install a California switch approximately 5,500 feet into the tunnel drive. Once completed, three trains could service the TBM through the remaining tunnel excavation. During full production for the Rock Tunnel drive, the working weeks alternated between short (Monday to Thursday) and long (Monday to Saturday) with two 10-hour shifts. Mining started in December 2021 and was completed in 23 weeks.

Shortly after the machine launch, the TBM began to rise at a rate higher than the design grade, and thrust pressures simultaneously doubled. Experienced TBM operators struggled to correct the grade, and thrust pressures exceeded lagging capacity. This resulted in broken lagging boards in the crown (Figure 4). Discussions with the operators revealed that the TBM was not responding to steering or thrusting techniques that had worked on past excavations. Launch complications are expected in every operation, although the problems went beyond the usual learning curve inefficiencies. It was ultimately suspected that the TBM bore overcut was slightly undersized, causing the cutter shield to ride up on the rock in the invert and bind the TBM against the bore. TMMJV modified the overcut while the TBM was in the tunnel bore with longer bullet teeth to increase the overcut dimension by one inch to 124.5 inches. The increased bore size provided an immediate reduction in the thrust pressure required to advance the TBM, and the operators quickly regained control of the TBM grade. The remainder of the tunnel drive was largely uneventful, with the exception of a reduction in penetration rates for approximately 50 pushes that was attributed to several worn and damaged gauge cutters.

The advance rates for the Rock Tunnel Segment are shown in Figure 5. The average full production advance rate for the Rock Tunnel drive, excluding the weeks required to install the California switch, was 660 feet. The highest weekly advance rate was 890 feet during week #11, and the highest shift advance rate was 110 feet during week

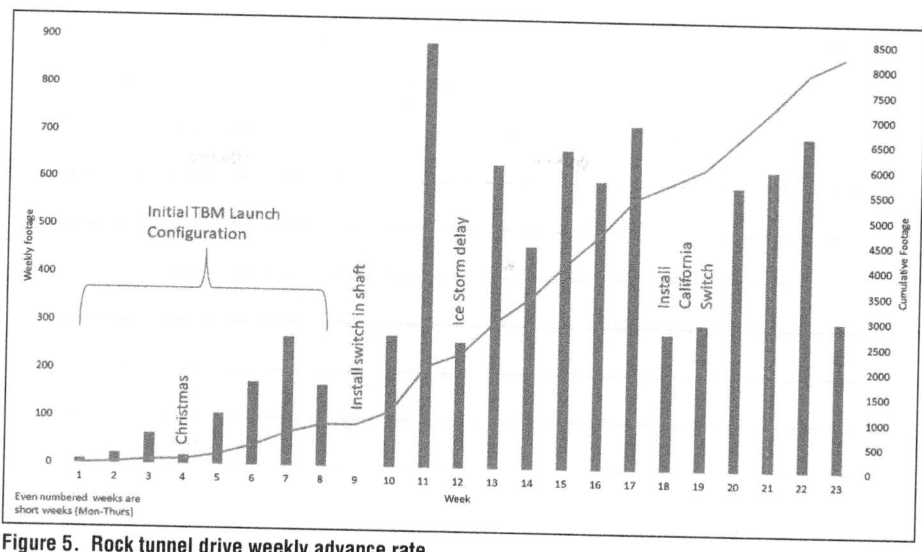


**Figure 5. Rock tunnel drive weekly advance rate**

#21. The relatively higher advance rates observed during week 11 are a result of several factors: two 10-hour shift workdays commenced, there was reduced time waiting for the muck train because the tunnel heading was located relatively close to the launch shaft, and there were minimal TBM and backup delays. During excavation of the rock tunnel, no discernable difference was observed in ground conditions between TRs, and the rock was observed to be generally massive and intact with widely spaced discontinuities. Therefore, the penetration rate for this tunnel was largely dependent on the intact rock strength. The penetration rate for the Rock Tunnel alignment typically ranged from 7 to 23 mm/rev, and averaged 15 mm/rev. This penetration rate is expected for these rock conditions (Gong and Zhao 2009). However, this TBM cutterhead was not equipped with roller cutters typically used on other TBM-mined rock tunnels. Cutterhead teeth were found to be in generally good condition at the end of the tunnel drive. Some bullet teeth were broken out of the cutting head from poor welds, and some gauge cutting teeth needed to be replaced. All cutting tools aside from that were left in place for the next 1,200-foot Mixed Face Tunnel Segment.

## SUMMARY OF MIXED FACE TUNNEL SEGMENT

Similar to the Rock Tunnel Segment, the shorter 1,200-foot-long Mixed Face Tunnel Segment was mined upstream. Excavating in this direction meant that the TBM would launch in a full-face clay condition in TR#3, which quickly graded to a full face of weathered rock in approximately 200 feet, and ultimately would encounter a full face of limestone in 890 feet. Shorter, 4-foot-long timber lagging boards were used for tunnel support in clay and weathered rock conditions, and 5-foot lagging was used in the limestone in TR#1. A single 10-hour day shift was conducted for the entirety of the Mixed Face Tunnel drive. Tunnel excavation started in June and was completed in 7 weeks.

Unlike with the long Rock Tunnel Segment, the contractor chose not to install a tail tunnel in the launch shaft. Several factors contributed to this decision: the tail tunnel excavation in a full face of clay would have been slow and costly; the benefits of using a tail tunnel would not be fully realized because of the relatively short Rock Tunnel length (1,200 feet); and the pipeline design grade would be increased to 3% in the tail tunnel area, adding complication to the proposed excavation. Without a tail tunnel, some equipment and operational changes were necessary to keep production at a reasonable pace. A shorter locomotive car with three muck cars was used in a two-step mucking process to allow the TBM to make a full advance during this drive. To start the advance, the locomotive, two muck cars, and a shortened flat car were transported to the tunnel heading. This train delivered the initial shoring and supporting materials to the tunnel heading and received the tunnel spoils produced by TBM excavation. After the two cars were filled, they were hauled to the shaft and lifted by crane to the surface. An additional three cars were then lowered onto the tunnel rail and taken back to the heading to complete the TBM advance. This two-step method for muck haulage decreased overall production rates but was necessary because of the lack of a tail tunnel.

During excavation of the launch shaft, the contractor observed the clay to be very stiff and competent when dry. Groundwater was not encountered in the launch shaft, and a nearby observation well showed the groundwater table to be below the invert of the tunnel. Considering the relatively short section of clay to be mined and the ground conditions observed in the mining shaft excavation, the Contractor decided to keep the same cutterhead configuration used in the rock tunnel. Some inefficiencies and complications were anticipated with this decision, although it was determined that the rock cutterhead configuration would overcome the initial setbacks once the TBM was in favorable ground. During the initial 200 feet of the excavation, the TBM production was slowed by the clay that would clog up the face of the machine. Perched groundwater was encountered at the transition seam between the clay overburden and weathered rock. On a daily basis, groundwater would soak the fractured clay left in the cutterhead chamber between shifts. The sticky clay had to be mucked out by hand every morning (see Figure 6). Once mucked out, the TBM could excavate the rest of the day without issue.

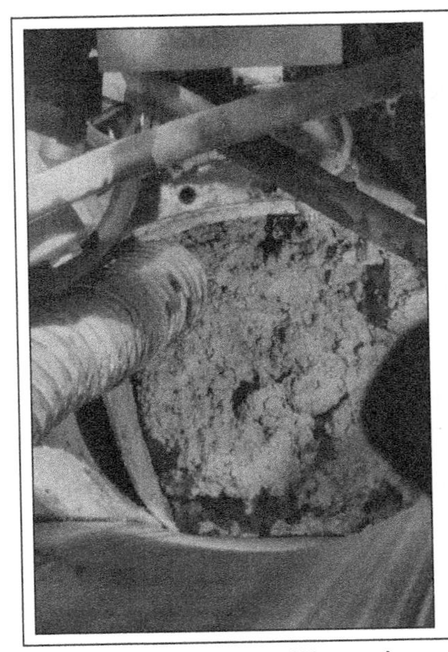

**Figure 6. Sticky ground conditions and groundwater inflows**

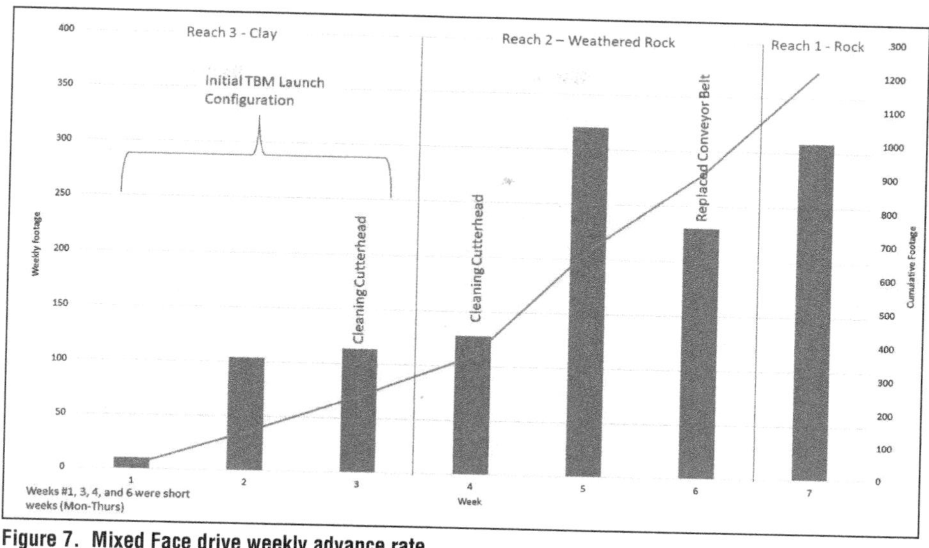


Figure 7. Mixed Face drive weekly advance rate

The weekly advance rates for the Mixed Face Tunnel are shown in Figure 7. The average weekly advance rates for the Mixed Face TR#3, TR#2, and TR#1 were 76, 226, and 309 feet, respectively. The lower advance rates for TR#3 are mainly a result of the TBM launch configuration and additional time to clean the cutterhead during week 3. The highest weekly advance rate was 320 feet, which occurred in TR#2 during week 5. The highest shift advance rate was 72 feet, which occurred in TR#2 during week 6. These maximum advance rates are relatively lower than the Rock Tunnel alignment because of single shift mining, the two-step muck hauling process, the use of shorter 4-foot rib and lagging sets, and additional time required to clean sticky ground from the cutterhead.

## TBM UTILIZATION

TBM performance data collection was conducted by the specialized tunnel inspectors and TBM data logger. The time study data collected by the inspection team included operation time (safety meetings, boring, rib building); cutterhead maintenance (scheduled maintenance, cutter changes, cleaning, other repairs); TBM delays (mechanical, electrical, TBM conveyor), TBM backup delays (gantries and gantry conveyor); and non-TBM delays (locomotive, crane, surveying, utility extension, inflows, site power). Scheduled maintenance was typically conducted during the day shift, while the night shift was dedicated to production.

The overall TBM operation time for the Rock Tunnel drive was 49%, and the average full production TBM utilization was 28% (Figure 8). The highest TBM utilization per shift was 41% and per week was 34%. These data lie within typical published data. TBM utilization can range between 5% and 55% in difficult to near-perfect conditions but is typically between 20% to 30% for rock tunnels (Rostami 2016). Nine out of ten shifts with the highest TBM utilization were night shifts. The majority of the non-TBM delays consisted of utility extension, waiting for the muck train, and crew transport. Delays resulting from TBM breakdowns and cutterhead maintenance accounted for a total of 14% of the time.

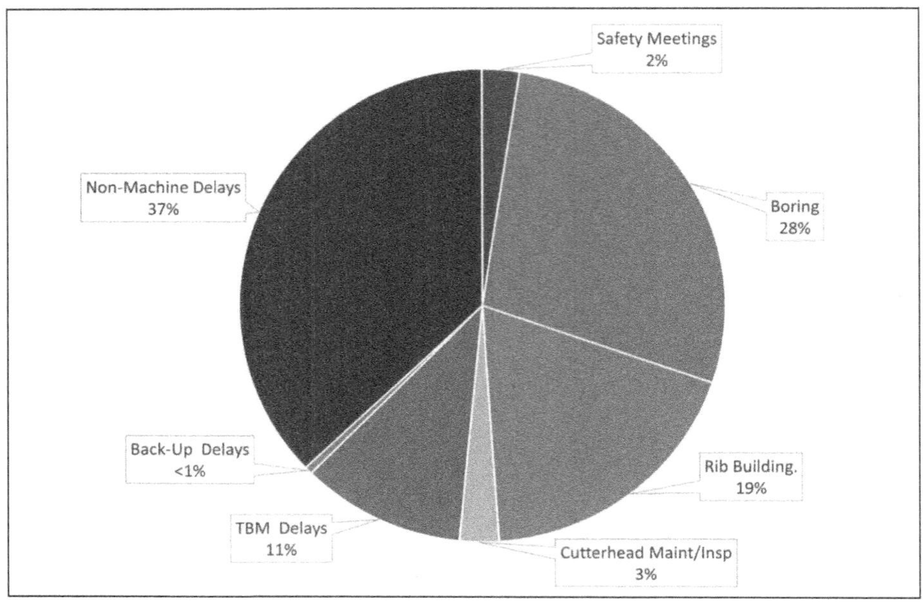


**Figure 8. Time study and utilization rate for the Rock Tunnel alignment**

**Table 1. Mixed Face Tunnel Segment TBM Time Study**

| Category | TR#3 | TR#2 | TR#1 |
|---|---|---|---|
| Safety meeting | 1% | 2% | 2% |
| Boring (U) | 17% | 23% | 21% |
| Rib Building | 15% | 16% | 13% |
| Cutterhead Maintenance and Inspection | 7% | 6% | <1% |
| TBM Delays | 2% | <1% | 1% |
| Backup Delays | 25% | 1% | 3% |
| Non-TBM Delays | 33% | 52% | 60% |

The time study and utilization rates for the Mixed Face Tunnel Reaches (TR) are listed in Table 1. The operation times for Mixed Face TR#3, TR#2, and TR#1 were 33%, 41%, and 36%, respectively. The highest utilization rates per shift and per week were 25% and 24%, respectively. As expected, the TR#3 operation time and utilization rates are relatively lower, which is accounted for in the TBM startup and backup assembly time. The relatively low operation time for TR#1 was generally due to the extra time waiting for the muck train and preparing the TBM receiving shaft. However, the operation time and utilization rates for the Mixed Face Tunnel are lower relative to the Rock Tunnel drive, which is expected because of the additional mucking time required for the two-step mucking process.

The cutterhead maintenance for TR#3 and TR#2 was exclusively due to the need to clean the cutterhead at the overburden and weathered rock contact. As expected, groundwater inflows were encountered at the contact, and sticky ground conditions resulted in increased time maintaining the cutterhead. Once the overburden and weathered rock contact were passed, no additional cutterhead cleaning was required for the remainder of the drive.

## CONCLUSION

The population of the Dallas metroplex area and Texas is expected to continue to grow for the foreseeable future. As cities grow and new infrastructure is constructed, the demand for future TBM excavations with similar tunnel diameters to the SGRWRRF outfall pipeline will simultaneously increase. It should be noted that such smaller diameter tunnels are designed and constructed largely based on tunneling experience in the local ground conditions. This paper was written with the intention of expanding on the TBM experience and performance in the Austin Chalk Formation. The following are notable take-aways specific to the SGRWRRF outfall pipeline TBM mined segments:

1. High penetration rates were achieved in the massive limestone with a cutterhead equipped with soft-ground tools.
2. Sticky ground conditions were encountered mainly at the contact of the alluvial soils and completely weathered rock. Outside of these areas, high TBM advance rates were achieved in the remainder of the tunnel drive without conditioning the ground.
3. Despite using the same TBM and cutterhead configuration for both tunnel segments, the TBM utilization and advance rates were lower in the Mixed Face Tunnel Segment because of the methodology and process of the mucking operation.

It should be recognized that the worldwide COVID-19 pandemic that impacted this project and many others in ways that are difficult to quantify. For the SGRWRRF outfall pipeline, the most significant impacts and challenges included the following:

1. Refurbishing the TBM cutterhead in China during a time with massive lockdown restrictions and when travel was difficult or banned.
2. Supply chain disruptions and companies going out of business that unprecedently impacted material cost and availability. Specifically to this project, tunnel timber lagging was not regionally available due to competition with rail companies buying out the supply.
3. Rapid inflation driving the significant rise of freight and trucking costs.

## REFERENCES

CDM Smith. 2021. Geotechnical Baseline Report, Sister Grove Regional Water Resource Facility Outfall Pipeline. North Texas Municipal Water District.

CDM Smith. 2020. Geotechnical Data Report, Sister Grove Regional Water Resource Facility Outfall Pipeline. North Texas Municipal Water District.

Gong, Q.M., and Zhao, J. 2009. Development of a rock mass characteristics model for TBM penetration rate prediction. *International Journal of Rock Mechanics and Mining Sciences*, 46:8–18.

McGowen, J.H., Hentz, T.F., Owen, D.E., Pieper, M.K., Shelby, C.A., and Barnes, V.E. 1991. Geologic Atlas of Texas, Sherman Sheet: The University of Texas at Austin, Bureau of Economic Geology, Geologic Atlas of Texas map scale 1:250,000.

Rostami, Jamal. 2016. Performance prediction of hard rock tunnel boring machines in difficult ground. *Tunneling and Underground Space Technology*, 57:173–182.

Scialpi, M., K. Buxton, B. Negrea. 2020. Mechanized excavation in shale formation—Performances comparison between the main beam TBM tunnels, the single shield

TBM tunnel and the roadheader tunnels at the Doan Valley Project. In *Proceedings of the North American Tunneling (NAT) Conference, Nashville*. Littleton, Colorado: Society for Mining, Metallurgy & Exploration, 659–669.

Thewes, M. and Burger, W. 2004. Clogging Risks for TBM Drives in Clay. *Tunnels & Tunnelling International*. June 2004: 28–31.

Ulusay, R., and Hudson, J., eds. 2015. ISRM Suggested Methods for Rock Characterization Testing and Monitoring: 1974–2006. ISRM Turkish National Group, Ankara, p. 628.

# Enhanced Probe Drilling and Pre-Grouting Design and Recommendations on Hard Rock TBMs

**Stryker Magnuson** ▪ Robbins

## ABSTRACT

While probe drills are not strictly necessary for all projects, the incorporation and use of probe drills and pre-grouting adds capability and insurance to boring operations. Water ingress and unstable ground can be resolved before becoming a problem and resulting in costly delays through the use of enhanced, 360-degree probe drilling set-ups. To do this, proper design of the array of drill ports in the shield, matched to the possible ground conditions, is critical. For ground with exceptional water and instabilities expected, additional probing locations are low-cost additions that can lower risk and increase efficiency. In this paper we will look at recent and ongoing projects including the Lower Meramec Tunnel and Jefferson Barracks Phase 2. We will detail the design of those probe drill arrangements, and our overall recommendations for probing/grouting systems that best suit challenging ground conditions and keep projects running smoothly.

## INTRODUCTION

Adverse geology and ground water can cause drastic problems for open gripper TBMs resulting in costly delays to the project and possible machine damage. Expected geology for a tunneling project is usually just a rough estimate based on a collection of core samples taken along the planned path and can very easily miss features such as voids, sinkholes, fissures and excessive ground water that can pose issues to the machine. Regular probe drilling in front of the machine can detect these features before they become a problem and when combined with pre-excavation grouting can consolidate unstable ground, fill voids and reduce material/water inrush to manageable levels, saving time and money over the course of the project.

## PROJECTS BACKGROUND

The two projects we'll be looking at, the Jefferson Barracks Phase 2 (JB2) are the Lower Meramec Tunnel (LMT) are both open gripper main beam TBMs with diameters Ø13'-6.5" (Ø4.13 m) and 11'-0" (Ø3.35 m) respectively.

The JB2 project is in St. Louis and will finish the last 10,000 ft (3,050 m) of tunnel from the original 17,770 ft (5,500 m) project. The tunnel is at a depth of 120 ft–220 ft (36 m-67 m) and will be lined with a Ø7' (Ø2.1 m) pipe to convey wastewater from Martigney Creek to the Lemay Wastewater Treatment plant. The project has been split into two phases as the first machine encountered an impassible karstic feature and was damaged by material inrush. The new machine (JB2) started boring from the other side of the feature to finish the remaining distance and complete the project. The geology expected for the remaining 10,000 ft is 90% good rock and 10% karstic limestone and dolomite formations (see Figure 1).

The LMT project is also in St. Louis Missouri, USA, and will be 6.8 miles (10.1 km) long at a depth of 78 ft–286 ft (24 m-87 m) and will be lined with an Ø8' (Ø2.4 m)

**Figure 1. Jefferson Barracks Phase 2 TBM**

pipe to convey sanitary flow from the Fenton Wastewater Treatment Facility to the Lower Meramec Wastewater Treatment facility at the junction of the Mississippi and Meramec Rivers. The geology expected is competent rock of shale and limestone.

## PROBE DRILL ARRANGEMENT, TBM FOR JEFFERSON BARRACKS 2 TUNNEL

The Jefferson Barracks Tunnel project was started with an 11'-0" Main Beam TBM (now called "JB1." This machine had good roof bolter drills, but no probe drill. The TBM encountered a large karst feature at about ,7000 feet in and became inundated with sand and mud. Efforts were made to stabilize the ground so the TBM could be recovered. The remaining length of the tunnel was let as a new contract. The new tunnel design was for a 14'-6" bore TBM, with facilities for comprehensive probe drilling and Pre-Excavation grouting. This new contract became known as "JB2."

On the JB2 machine, the probe drill assembly is fitted with two Timberock hydraulic feeds with Montabert HC95LQ drifter mounted on both lift and tilt frames on a carriage giving approximately 50 mm of lift to be used at –1.5° and the lift frame dropped to be used at 0° and +3.5° tilt. The carriage rides on 4 V-groove wheels and is driven by a pinion gear on an external ring gear. See Figure 2. The ports for drilling in front of the machine (0° and –1.5° drill tilt angles See Figure 3) have 4 ports each at the diagonals (±45° and ±135° from vertical). The two drilling angles, though slightly different, allow the drill holes to diverge when the holes reach 100–150 feet, to provide good coverage of the face area. The ports for drilling about the periphery (+3.5° drill tilt angle see Figure 4) have 19 positions with guide tubes at top dead center and spaced around with increasing gaps towards the bottom half to prioritize the umbrella. The addition of the dual 360° probe drill adds great performance benefits and insurance to this new JB2 machine.

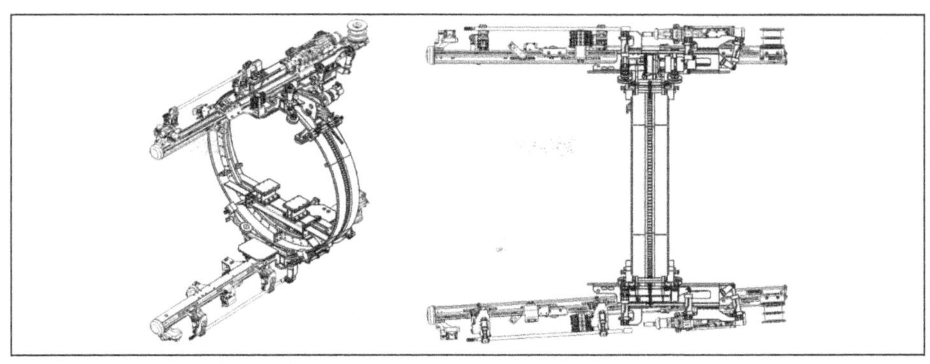

Figure 2. JB2 probe drill assembly

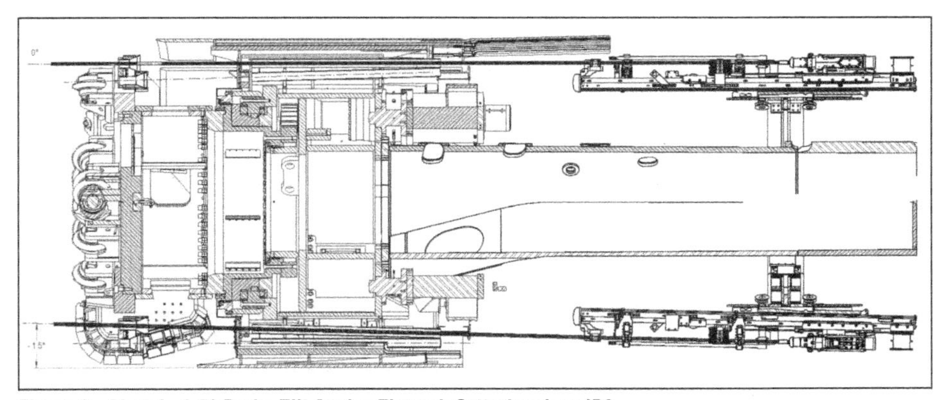

Figure 3. 0° and –1.5° Probe Tilt Angles Through Cutterhead on JB2

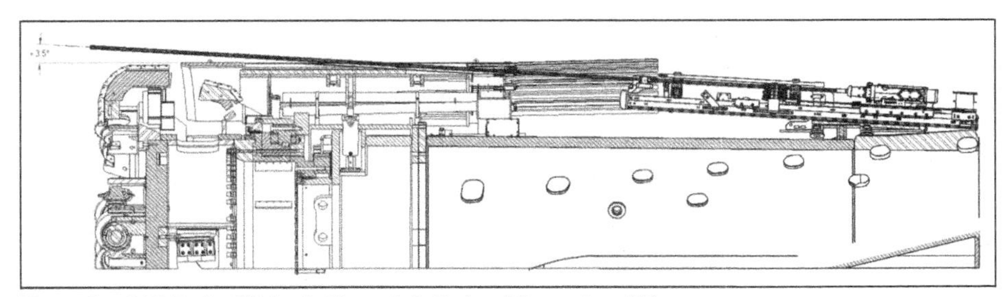

Figure 4. +3.5° Probe Tilt Angle Through Cutterhead Support on JB2

## PROBE DRILL ARRANGEMENT

TBM for LMT Tunnel The LMT machine is an 11'-0" bore High Powered TBM. There is very little room around the TBM for a 360 degree probe drill positioner. The TBM and positioner had special features to allow the 360 degree positioner to function in this confined space. Also, the LMT has many short radius curves. So the TBM length had to be kept as short as possible to allow the TBM to negotiate these curves. The probe drill ring for this TBM could be moved forward and aft for this reason. With the ring in the forward position, the probe drill could drill at 0 and 5 degree angles for drilling in the face and the perimeter, respectively. However, this forward position would conflict with movement of the roof drills during mining. . So the drill ring would normally be kept in the aft position. In the aft position, the drill could be used to drill peripheral holes at 3.5 degrees.

The LMT machine has a single hydraulic-powered chain feed with Montabert HC50 drifter mounted on a carriage with tilt angle 0° and +5°. The drill carriage rides on aluminized bronze bearings that double to locate the carriage on the ring and is driven with a single motor driving two pin gears on two pinwheel tracks. This gear design is desirable as it works with loose assembly tolerance, has high torque transmission, and allows for running with low lubrication and high contaminants. The ring is designed so the drill can be locked in position and the rest of the ring pulled back and out of the way. This configuration is for clearance of the roof drills as the machine thrusts. The ports for drilling straight in front of the machine (0° drill tilt angle) have four positions in the top half (guide tubes at ±20° and ±45° from vertical). The ports for drilling about the periphery (+5° drill tilt angle) have a total of 14 positions spaced roughly equally (guide tubes at ±10°, ±30°, ±60°, ±90°, ±112.5°, ±135°, and ±160° from vertical). For the drill ports at 0° tilt angle, the cutterhead has two guide holes that can be clocked to allow the drill to pass through. The major drawback to this design is the additional setup time for bringing the ring forward into position before the drill can be brought around to a drill port for investigative probing and pre-grouting. The reason for this design is due to the size constraints of LMT. This set up is still better and quicker set up than previous designs on small machines where the lower sections of the ring would have to be brought in from the tunnel for drilling. Being a small bore at 11', limited space is available and special features are needed to fit the equipment. These include the multipiece ring that can retract to an aft position and angled corners on the main beam (usually a flat top). The contractor for LMT is expecting to encounter a water bearing feature that will be beneath the tunnel at first, but then rise up across the face as the machine progresses. So, drilling in the invert positions will be as important as drilling in the crown positions. With this, the probe drill can be used in the invert to grout off water before encountering it. Major efforts were made to provide a functional 360° degree probe drill on this small TBM.

Both machines have 360° probe drilling with the ports orientation prioritized to the umbrella (above the machine) and boring ports angled horizontal or above (19/27 angled ≥0° on JB2 and 12/18 angled ≥0° on LMT). The focus on probe drilling above centerline/positive drill angle is because it's much easier to remove material drilling at an angle ≥0° (refer to Figures 4 and 5). On top of the ease of drilling, grout will propogate through the fissures in the rock and can successfully grout the path of the bore without the need to drill/grout in a full 360° pattern. Conversely, drilling down makes it harder to remove tailings from the drill hole as it is going against gravity. While probe drilling in the umbrella offers good coverage both for detecting problem geology infront of the machine and grouting unstable ground, it cannot compete with the capability full 360° probe drilling. Most previous machines are only capable of drilling in the umbrella (if they even have a probe drill) and can leave some potentially disastrous blind spots. Especially on JB2 where karst features are expected, the only way to accurately detect a void below centerline is by drilling in the area. A void under the machine could result in the machine diving or worse—becoming stuck. The addition of full 360° drill ports allows operators to detect problem features anywhere in front of the machine to prevent unwanted dives and provides coverage when going into a turn (via the periphery ports at stringline). Detecting unstable geology before encountering it is vital to smooth boring operations.

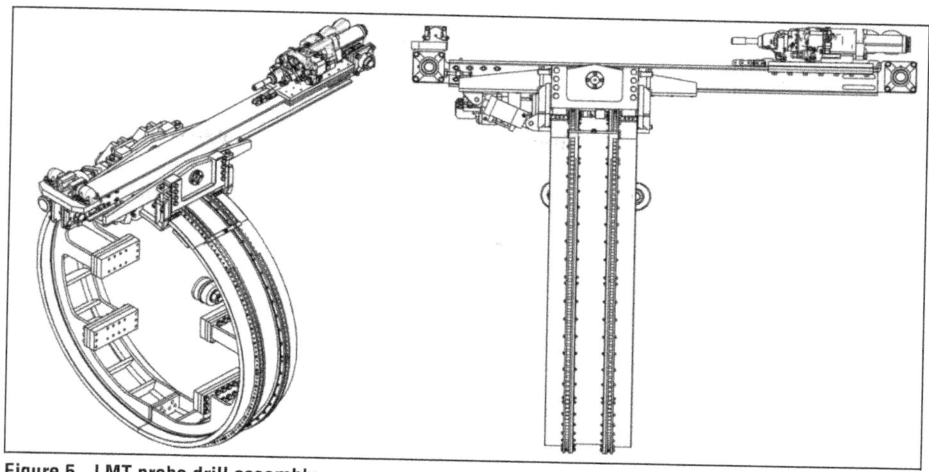

Figure 5. LMT probe drill assembly

## TRADEOFFS AND SACRIFICES

The primary roles of probe drills are investigating ground conditions in front of the machine and creating holes to inject ground treatment to consolidate unstable ground, fill voids and reduce water/material inrush as the machine bores forward. Identifying adverse conditions starts with surveyors conducting core drillings performed by a third party well before any boring begins. These core samples are a good starting place to get a general idea of what is to be expected but the tests can miss a lot and even close-together samples can vary wildly.

For example, on LMT from the tests conducted, ground is expected to be generally competent, but a probe drill is still necessary due to the possibility of karstic and other difficult features in this geology. Due to the small TBM size, only a single probe drill was fitted onto the LMT machine. This contrasts with JB2, where the first machine ran into significant obstacles (to the point a second machine was needed) and further features are expected. Therefore, dual probe drills with a fixed mount ring were specified so that probe drilling operations could be performed more quickly.

Probe drill assembly designs have a lot of constraints on them as well. First, the feed and drifter unit are large and, with the requirement to rotate 360° with clearance, presents a lot of potential interference. Furthermore, the assembly is vying for limited space at the front of the machine with ring erectors, roof drill, and working space for personnel. For JB2, the probe drill ring does not have fore/aft movement and is mounted on a longer main beam, aft of the roof drill travel zone. The drill can't be rotated around the main beam with the drill in the lowered and angled position. Because of this constraint on the +3.5° ports, the drill must be angled or lifted to clear the main beam. Given the geology expected on these two projects and probe assembly design on the machines, JB2 expects more adverse geology, and so the machine design prioritizes probe drilling compared to LMT with greater coverage, two drills, and a fixed mount ring.

Since probe drilling/pre-grouting can't be performed simultaneously with boring, and machine reset after a full push isn't enough time to drill a probe hole, probe drilling needs to be planned independent of the boring schedule. This downtime can be costly, but unstable ground and material inrush can also result in costly damage and delays, so the risk versus reward needs to be balanced to keep the machine on schedule and running.

## EFFECTIVENESS OF PRE-GROUTING

Evaluating the effectiveness of probe drilling and pre-grouting is not very straight forward as there is no way to have a baseline to reference. Whether that be boring 200 ft without probe drilling then 200 ft with probing, or two tunnels side by side, the conditions, formations and features encountered will be dramatically different between the two, and drawing conclusions based on data for the benefit gained from probe drilling and pre-grouting cannot be accurately determined. Although no direct comparison can be made for an overall project, it's clear to see that discovering and grouting adverse conditions before the machine reaches them can save immensely on costs and project schedule. By utilizing probe drills and pre-grouting, problems can be remedied before the machine encounters them. The alternative can lead to high material inrush damaging the machine or, in a worst case, a stuck machine.

## CONCLUSION AND RECOMMENDATIONS

With the aforementioned points in mind, probe drilling and pre-grouting should be looked at as an insurance policy to mitigate the potential issues that can arise from adverse geology. Specific recommendations for probe drilling should be made per individual projects as they need to factor in expected geology, schedule constraints, on site expertise, and available space on the machine. The available space pertains to the design of the probe drill assembly on the machine and will be dictated by expected geology and project requirements. Given that core samples for expected geology can miss a lot of potentially problematic features, and a damaged or stuck machine will result in costly delays, probe drills and pre-grouting are the best option for detecting adverse conditions in the bore path, consolidating ground and limiting material inrush for project insurance and risk mitigation. Enhanced 360° probe drilling with ports in front and around the periphery provide greater capability compared to existing machines that either only had a probe drill in the umbrella or none at all. These drill arrangements when put into use provide protection from unstable ground in any direction in front of the machine. Going forward, 360° drills should be used not just for inspection and pre-grouting but collecting measurements and data to produce reports on the effectiveness of enhanced 360° probe drilling to promote further adoption of these drill assemblies. With more machines utilizing 360° drills, the designs and operation can be further refined to impact boring operations less and reduce damaged or stuck machines. Robbins recommends that machines come equipped with 360° probe drilling with an array suited to the expected geology, and most importantly, that the operators schedule regular probe drilling and pre-grouting as needed into the boring operation to keep projects boring smoothly.

# Hard Rock TBMs—The Experience of Brenner Basis Tunnel, Construction Lot Mules 2-3, for the Evaluation of Risk Prediction Tools

**F. Amadini** ▪ Systra–Sws
**A. Flor** ▪ Systra–Sws
**D. Baliani** ▪ Webuild
**F. Cernera** ▪ Sapienza University of Rome
**M. La Morgia** ▪ Sapienza University of Rome
**A. Mei** ▪ Sapienza University of Rome
**F. Sassi** ▪ Sapienza University of Rome

## ABSTRACT

The risk of TBM jamming because of squeezing rock mass or excessive caving (instability of the cavity) conditions is always a focal point while boring long and deep tunnels with shielded TBMs. Mechanized excavation provides a large amount of real-time data, for each machine's parameters. Several analyses have been executed to assess the correlations between different sets of TBM parameters for the exploratory tunnel. Instead of analyzing the full dataset, a different approach is proposed, with data processed sequentially to simulate the real excavation situation. The probability distributions are recalculated together with their statistical parameters while advancing. This approach is extended also to the training of a Machine Learning model able to foresee the behavior at future rings based on past rings (using a Recurrent Neural Network). The predictions show a promising first step toward creating practical tools that can assist contractors and designers to correlate and predict TBM excavation data. These tools will assist to predict and quantify high risk situations that could arise during excavation.

## INTRODUCTION

### Project Description

Brenner Basis Tunnel (BBT) lot Mules 2-3 project is under construction by the Joint Venture led by WeBuild, Ghella with Pac and Cogeis. Project layout is widely described in literature (Citarei et al. 2020) and include 65 km of complex underground excavations. Focusing on mechanized excavation northbound, three double shielded TBM's were used, for a total excavation length of approximately 14 km each. The exploratory tunnel TBM (Ø 6.8 m) preceded the excavation of the mainline tunnels, performed by the other two largest TBM's (Ø 10.7 m), and it is the focus of the current study. The geomechanical context is particularly complex and heterogenous, characterized by the presence of fault zones, poor rockmass conditions, and squeezing rock masses.

While advancing with the exploratory tunnel, some exceptional shield jamming events occurred. Record of said events allows us to match them with TBM data. In December 2021, the Exploratory Tunnel TBM, named Serena, completed her race at the Italian/Austrian border.

#### *Geological Overview*

The Brenner Basis Tunnel crosses the central part of the Eastern Alps and encompasses the tectonic nappes involved in the collision between the European and the

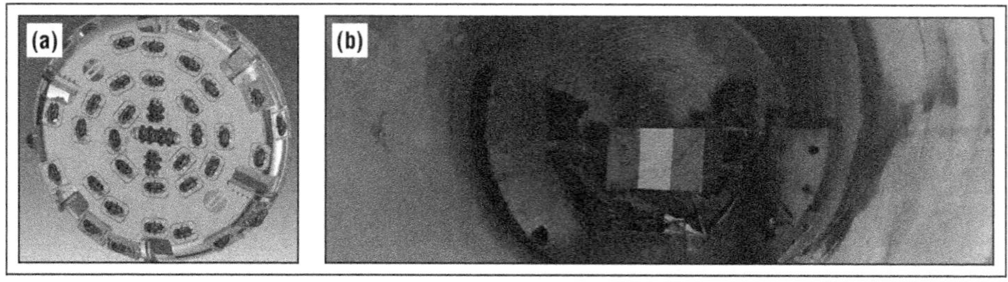


**Figure 1. (a) 3D view of the cutterhead by Herrenknecht (b) Final excavated tunnel face**

African plates, thanks to the presence of the huge antiform fold coinciding with the Tauern Window.

The Tauern Window is divided into two different nuclei, the Tux at the North and the Zillertal at the South and is constituted by a core of so-called Central Gneiss with its Mesozoic cover sediments.

Upward, the Central Gneiss and its cover are wrapped by the complex of the Penninic nappes, mainly calcschists and ophiolites. These are rocks of oceanic origin, over thrusted above the subpenninic nappes during the subduction phase. At both edges of the Tauern window, there are the Austroalpine nappes, once constituting the Adriatic (African) continental border (Figure 2).

According to the geological finding obtained from the excavation of the exploratory tunnel, is possible to summarize and discretize the encountered lithology as per the Table 1.

### *Geomechanical Data*

Geomechanical data are collected during geo-structural surveys carried out at the tunnel face from the openings in the cutter head. The collected rock mass parameters are mainly RMR, GSI, and UCS. Figure 3 and Figure 4 show the distribution of the RMR and UCS along the geological profile crossed by the Exploratory Tunnel (ET), the East Main Tunnel (EMT), and the West Main Tunnel (WMT) all driven northbound. The RMR for different tunnels at same chainage can be very different.

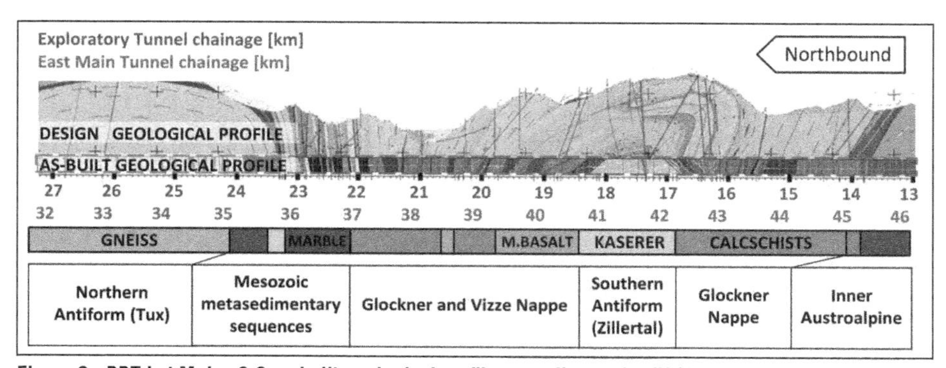


**Figure 2. BBT Lot Mules 2-3 as-built geological profile according to the JV (Pescara et al 2020)**

Table 1. Lithologies encountered along the alignment of the Exploratory Tunnel

| Start pk, m | End pk, m | Lithological simplified label | Lithological extended description |
|---|---|---|---|
| 13+085 | 13+788 | Paragneiss | Paragneiss |
| 13+788 | 13+928 | Amphibolite | Amphibolite |
| 13+928 | 16+807 | Calcschists | Calcschists, ophiolites, marbles |
| 16+807 | 18+200 | Kaserer | Kaserer Form. with phyllites, micaschists, quartzites, prasinites and chlorite-bearing schist |
| 18+200 | 19+690 | Metabasalt | Metabasalt with marbles, calcschists and serpentine schists interbedded |
| 19+690 | 20+421 | Calcschists | Calcschist, grey marbles and schists |
| 20+421 | 20+511 | Meta basalt | Metabasalt with marbles, calcschists and serpentine schists interbedded |
| 20+511 | 21+890 | Calcschists | Calcschist, grey marbles and schists |
| 21+890 | 23+013 | Quartzite/Marble | Carbonate quartzites and calcschists, marbles, prasinites, evaporites, metaconglomerates and meta-arkose |
| 23+013 | 23+310 | Kaserer | Kaserer Form. with phyllites, micaschists, quartzites, prasinites and chlorite-bearing schist |
| 23+310 | 23+925 | Quartzite/Marble | Quartzites and graphitic schist |
| 23+925 | 27+217 | Gneiss | Central gneiss and pre-granitic basement |

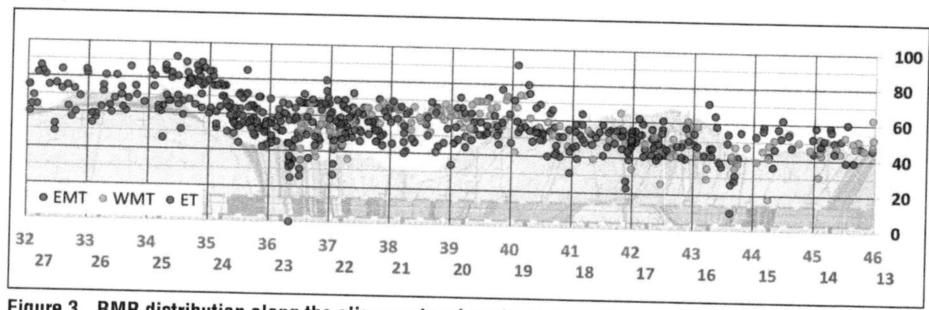


Figure 3. RMR distribution along the alignment and geological profile according to the JV

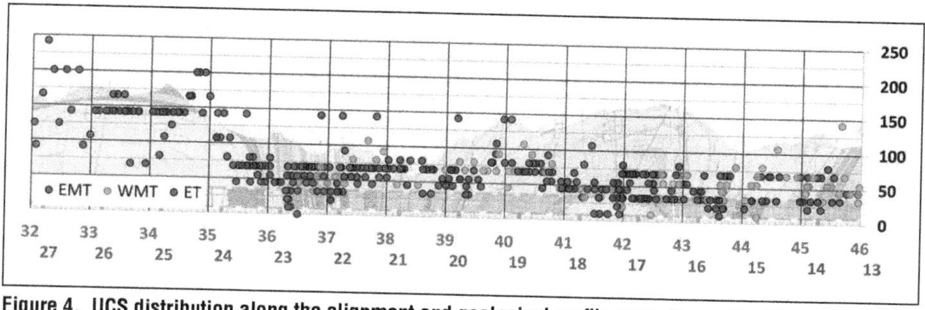


Figure 4. UCS distribution along the alignment and geological profile according to the JV

RMR distribution for main macro lithologies is given in Figure 5, where a simplified splitting of a more complex model is introduced for the purpose of this paper. The wide and highly variable distribution within the same lithology underlines the great heterogeneity of the rockmass.

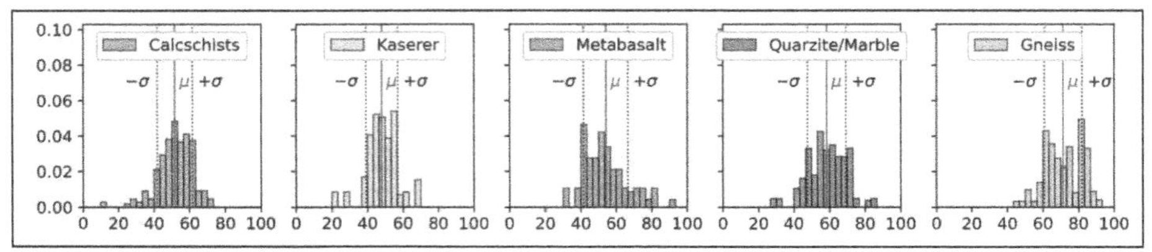


**Figure 5. Statistical RMR distribution split among macro lithologies (ET, WMT & EMT overall values)**

## ANALYSIS OF TBM KEY PARAMETERS

Key Performance Indicators (KPI) are recorded continuously during TBM excavation through an automatic acquisition system at intervals of 3 seconds. KPI trend is well explained in Maggio et al. (2022). In this paper, the analyzed parameters and the presented output are aggregated for a single excavated ring to manage the massive amount of data recorded from the machine.

Hereafter are listed some of the KPIs relevant for the understanding of the machine behavior:

1. Thrust Force (TF), kN
2. Torque (T), MNm
3. Rotational Speed (RPM), r.p.m.
4. Advance Rate (AR), mm/min
5. Penetration Rate (PR), mm/rev
6. Specific Energy (SE), MJ/m$^3$

Other relevant parameters are the ones obtained from the machine monitoring system such as the quantity of excavated muck material, shield pressure cells, and extensometers for shield-rock gap measurement. The main characteristics of TBM are listed in Pescara et al. (2020).

The statistical distribution shown in Figure 6 allows us to appreciate an empirical correlation between the RMR and the main KPI, mainly applicable to this project. Part of the geomechanical information is therefore intrinsically contained within the same machine parameters. Having completed the as-built geological profile for ET (Pescara et al 2020) all the criticalities encountered during excavation are available for post-processing and interpretation. Hereafter we focus on the possibility to detect signals of potential risk for TBM excavation and trying to quantify it a few meters ahead. Due to the high heterogeneity of the rockmass most of the risks raised suddenly.

### *"Moving Statistics" and Current Distribution*

For each KPI, we can populate the distribution sequentially, thus simulating a real excavation situation. If we consider the statistical parameters of the last 1,000 rings, we may be able to guess that an anomalous situation is arising by comparing the most recent values to the distribution sampled up to that point. The following statistical analysis is based on this concept.

We, therefore, define the *Current Distribution* for a KPI as the statistical distribution of its most recent measurements. In this case the last 1,000 measurements.

### *Evaluation of Standard Score on multiple TBM parameters*

Once we have collected the TBM key parameters and the Current Distribution for each of them, we can calculate the real-time deviation from the mean value. The greater the distance from the mean value, the higher the risk that TBM is crossing unexpected geostructural conditions.

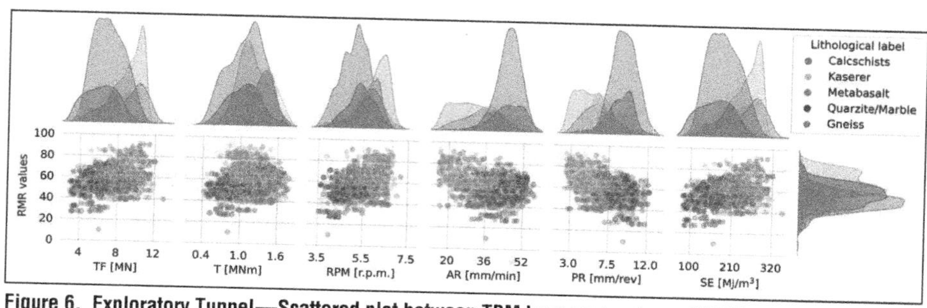


**Figure 6. Exploratory Tunnel—Scattered plot between TBM keys parameter and RMR values (<2σ cut-off)**

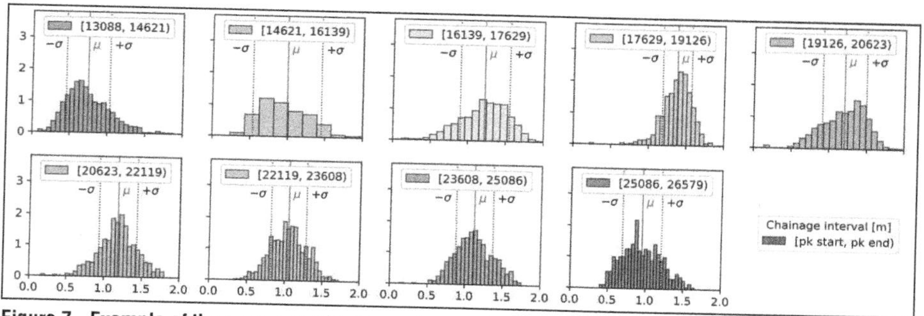


**Figure 7. Example of the same quantity (torque) current distribution, for different tunnel stretches**

For each KPI *x*, we define the Standard Score *z*, as in equation (1):

$$z = \frac{|x - \mu|}{\sigma} \quad \textbf{(EQ 1)}$$

with μ = mean; σ =standard deviation of the KPI Current Distribution.

It is important to stress that z is a function of the tunnel position (whether expressed in "rings" or in chainage meters). Since we are interested in anomalous behavior, we can disregard the standard score values lower than two (i.e., all TBM parameters measurement within 2 standard deviations from their *current* mean).

For each KPI we have then a Standard Score function of the chainage (or the ring):

$$z' = \begin{cases} \left|\frac{x - \mu}{\sigma}\right| & \text{if } |x - \mu| \geq 2\sigma \\ 0 & \text{otherwise} \end{cases} \quad \textbf{(EQ 2)}$$

The Sum of Standard Scores (SSS) can then be defined, by adding the standard score for each KPI. In Figure 7 we can see how the Sum of Standard Scores varies along the alignment. The background has color-coded accordingly to the level of caving recorded during excavation and judged by the JV geologist. Spikes of the SSS matches most of the zones of higher caving level (global instability) encountered during the excavation or emphasize the change of lithology.

This is the first candidate for a *monitoring quantity*, that can hint to future critical behavior. By looking ad Figure 8, it can be noticed that often there is a sort of *build-up* to the spike of our Score. This lead us to believe that some of the critical situations recognized by human intepretation can be noticed statistically with some degree of confidence with this back-analysis.

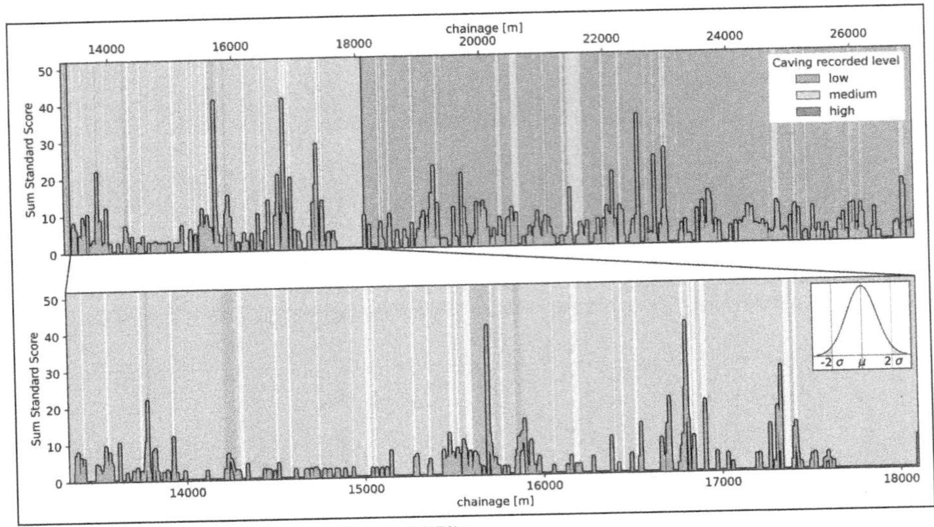


**Figure 8. Sum of Standard Scores (≥2) for all KPI's**

The goal is to predict each KPI trajectory in the next couple of rings and progressively see how past prediction diverge from real measurements, signalling possible problems. Next chapter will show the results with quantities prediction 4 rings before with Machine Learning application.

## MACHINE LEARNING METHODOLOGY

This work is based on following concept: Machine Learning (ML) techniques train their models on 'healthy' behavior to develop a model of 'normality' from which anomalous behavior may then be inferred (Sheil 2021).

Moreover, we deviate from the norm by dynamically generate a moving training set to better reflect the fact that the statistical distribution changes while advancing with the excavation. This is a consequence of the Current Distribution concept described in the previous paragraph.

### *State of Machine Learning Applied to TBM data*

With the increase of computational capabilities and thanks to the availability of enormous amounts of data, Artificial Intelligence methods, in particular Machine Learning (ML), have been proposed over the years to analyze TBM performances. For a recent review of the ML implementations related to the prediction of the penetration rate refer to Gao et al. (2021).

### *Architecture and Pre-Processing*

Our choice of architecture is dictated by the need for a multistep-ahead prediction. We opted for a type of Recurrent Neural Network (RNN): a Long Short Term Memory (LSTM) architecture, Hochreiter and Schmidhuber, (1997)

We want to predict the future value for each relevant KPI (see Chapter 2) at any given ring Figure 9(a). We then treat every quantity (TF,T,RPM,AR,PR & SE) as a feature to be predicted, and for each of them we construct a Neural Network (Figure 9(b)) that

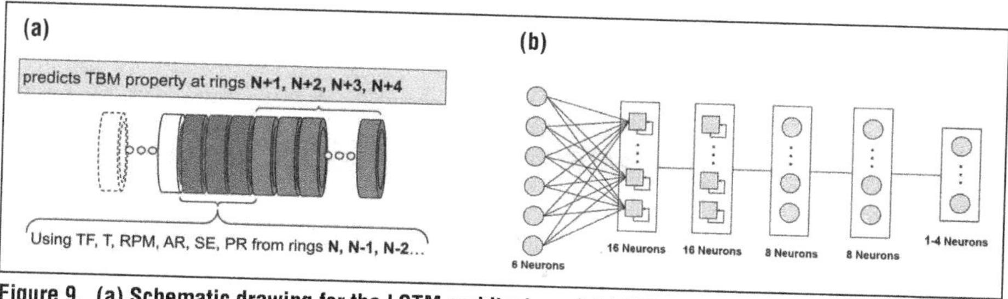


**Figure 9. (a) Schematic drawing for the LSTM architecture; (b) LSTM network architecture 4 neurons when predicting a sequence of values (in our case: the next 4 rings)**

uses the knowledge of the past 12 rings to predict the next 4 rings. Every ring we have the forecast next 4 values of the relevant quantities, i.e., 6 arrays of 4 values.

The pre-processing of the data has been done first by removing outliers (all values outside six standard deviations from the mean), by applying a Power Transform, in this case the Yeo-Johnson transform (Yeo, 2000) and then a zero-mean, unit-variance normalization.

### *An Alternative Approach: Progressive Retraining*

We believe that geological information is encoded in TBM readings, and we have shown how much the changing in geology alters the machine parameters. Contrary to the paradigm of training a prediction model on a large dataset before the application, we try a different route. While the TBM advances, we retrain the model incorporating all recent rings and, implicitly, the changes in geology.

We set aside the geological information only temporarily, aware of the fact that for an adequate understanding of the TBM behavior one cannot ignore the analysis of the rock itself.

We start with ring 200 with a small distribution, and then the weights of the LSTM neural network are retrained every 50 rings. The training set grows until it is composed by 1000 rings. From there on, at each retraining cycle, a past portion of the dataset is dismissed, so that only the last 1000 rings are considered, and the dataset undergoes the pre-processing procedure described in paragraph 0. There is no change to the hyperparameters at each retrain.

For each ring of simulated advancement, we apply the most recent version of the algorithm and predict the next 4 rings, storing the result.

## EXPERIMENT ON TBM PARAMETERS PREDICTION

The results reported in this section are mostly related to the prediction of Torque and, to a lesser extent, Penetration Rate. For these two KPI's, we managed to better tune the architecture and hyperparameters, hence obtaining the absolute best results.

### *Prediction*

The prediction versus real value of Torque and Penetration rate are respectively in Figure 9 and Figure 10. The overall trend is the same in both prediction and real value, while extreme values are, as expected, rarely reached by the prediction.

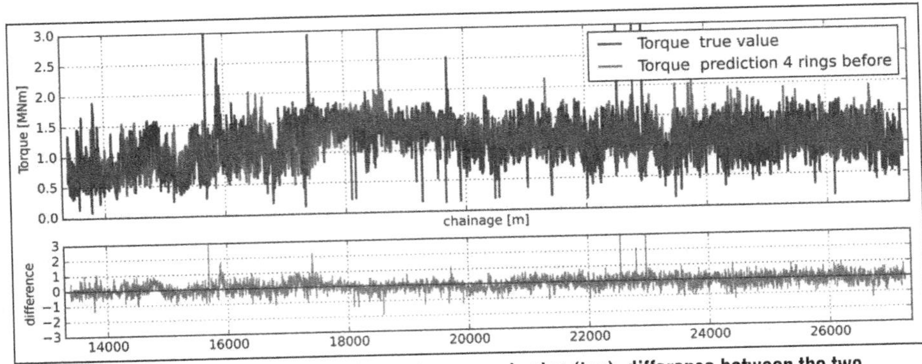


**Figure 10. Prediction Torque compared to real measured value (top), difference between the two (bottom)**

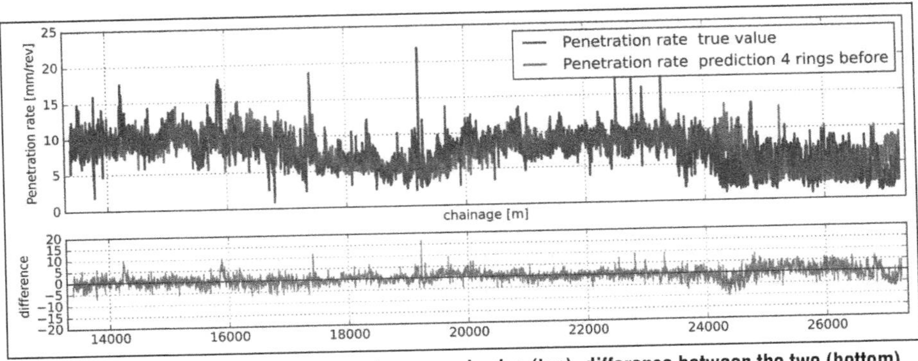


**Figure 11. Prediction PR compared to real measured value (top), difference between the two (bottom)**

Because of our hybrid train/test procedure, the usual metrics such as determination coefficient ($R^2$) or the Mean Absolute Error (MAE) are referred to smaller sets of values pairs. In Figure 12 the evaluation metrics, calculated on the Current Dataset before the new retraining cycle, even if calculated in the worst-case scenario, are still acceptable.

The commonly used metrics for the goodness of a predictive algorithm are summed quantities and provide a global and averaged interpretation. We are far more interested in the local reliability of the predictive algorithm, treating large discrepancies as positives if they happen to match a critical event.

In Figure 13 gives a closer look to Torque prediction in a smaller section of the tunnel. Similarly, to the figures in section 0, the background has been colored differently according to the risk measured (afterwards) in that point. The discrepancy between prediction and true value explodes in correspondence to the two elevated risk zones (red background). The other points show a reliable prediction, with some regions with very low error.

### *Machine Learning Discussion*

The choice to retrain the algorithm frequently, with an ever-changing training dataset that "follows" the advancement is a direct answer to the following consideration: the data that are continuous and objective are the Tunnel Boring Machine readings, while

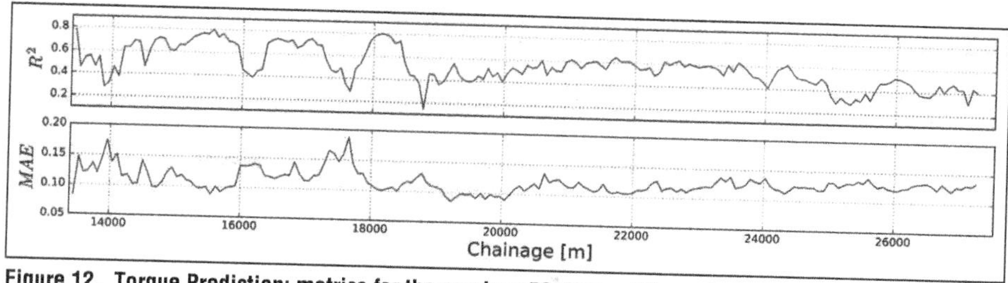


Figure 12. Torque Prediction: metrics for the previous 50 rings, at the end of each cycle (before retraining). Above the determination coefficient ($R^2$), below the Mean Absolute Error (MAE)

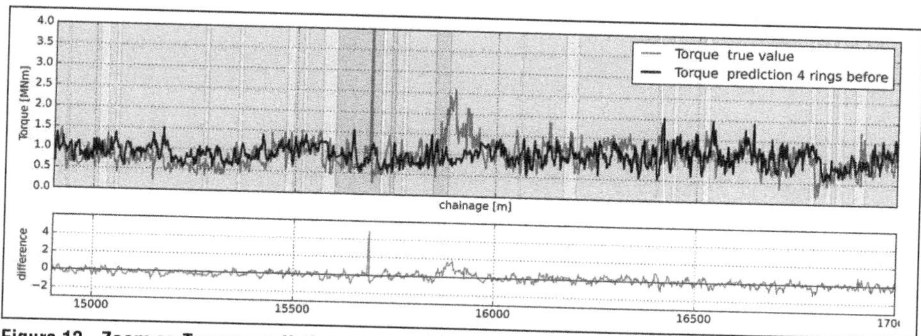


Figure 13. Zoom on Torque prediction on a smaller interval. The color-codes in the background indicates three different zones of caving risk recorded: with green no risk, yellow mild risk, and red high risk

the geological information is either discrete and subjective. It is nonetheless undeniable that the machine data are heavily affected and therefore indicators underneath geological variations. Hence the idea to disregard data that are "too far away" in the process of retraining since they are not representative of the TBM state.

By choosing a multistep-forward prediction (4 rings) we also avoided the "echoing effects," or naïve prediction mentioned in (Erharter 2021), since our algorithm does not predict only the next ring, but the next 4, making harder to the neural network to simply replicate the last result.

The results of these work are only preliminary and a lot of hyperparameters tuning is needed to obtain satisfactory results for all the KPI's and test on different datasets is certainly needed. But the overall paradigm of this self-contained progressive retraining seems a particularly good candidate for future applications in mechanized excavation.

## CONCLUSION

TBM excavation provides a significant amount of data to be analyzed. This wealth of information data is becoming increasingly object of studies by industry experts, who are discussing regarding the best computational approaches for the application of Machine Learning.

The analysis conducted on practical cases highlights how this anomalous behavior that precedes actual or potential blocking phenomena is sudden, causing the machine parameters to deviate significantly from regular behavior within few meters.

The combination of the Standard Score of the KPI in real-time allows to quickly visualize the superimposition of the effects that can lead to an actual or potential block.

From a practical point of view this visualization capability that combines Standard Score and an LSTM architecture-based prediction of the KPI deviation from "healthy behavior," approximately 4 rings in advance, could assist contractors and designers in the decision-making process by providing dynamic alert limits for TBM driving parameters.

Considering the fundamental importance of the time factor and the decisions that precede a potential block phenomenon in long and deep tunnels (WG no.17, 2017), exceeding a certain critical threshold could activate processes aimed at reducing the risk of blocking.

For the next generation of smart guided TBM, the following possible countermeasures could be introduced timely and in a semi-automatic way based on the proposed study:

- Dynamic variation of the cutterhead overcut (squeezing);
- Shield lubrification by bentonite injections for friction reduction (squeezing);
- Dynamic reduction of bucket openings in the cutterhead (caving);
- Cutterhead rotational speed reduction (caving);
- Stop timely TBM for ground improvement and treatment (face instability);
- Alert belt maintenance teams for the evacuation of excess excavated muck (caving);
- Supply timely TBM with heavily reinforced segmental lining (squeezing/caving);
- Shift modification to overcome the criticality encountered (squeezing/caving).

It remains indisputable that in a context of high complexity as mechanized excavation, human decisions cannot be replaced entirely (in the medium / long term) by Artificial Intelligence and Machine Learning. And yet, despite not having obtained exhaustive and definitive results, we believe this study represents a significative starting point toward the definition of practical predictive tools to aid human decisions. The research activity continues by the authors for future technical papers and practical applications on next TBM excavation projects to increase the reliability of these predictive tools.

## ACKNOWLEDGMENTS

The authors want to thank BBT, the owner company of the project, for the permission to publish the results of this study.

Special thanks go to WeBuild and Ghella for the possibility to collect and analyze all the TBM parameters recorded and the geological information and for the experience gained on this project with experienced team of professionals. The authors also thank Massimo Secondulfo for his contribution to this paper.

## REFERENCES

Citarei, S., Secondulfo, M., Buttafoco, D., Debenedetti, J. & Amadini, F. 2020. BBT, construction lot Mules 2-3. Production management and site logistics organization. *WTC 2019, May 3–9, 2019, Naples, Italy.*

Erharter, G.H. & Marcher,T. 2021, On the pointlessness of machine learning based time delayed prediction of TBM operational data, *Automation in Construction, Volume 121.*

Gao, B., Wang, R., Lin, C., Guo, X., Liu, B., Zhang, W., 2021. TBM penetration rate prediction is based on the long short-term memory neural network. *Underground Space 6, 718–731.*

Hochreiter, S., Schmidhuber, J., 1997. Long short-term memory. *Neural computation* 9, 1735–1780.

Maggio,G., Voza.A., & Egger.H. 2022. Key machine parameters for classification of rock mass in TBM excavation in BBT. *WTC2022, Copenhagen 2–8 September 2022.*

Pescara, M., Spanò, M & Della Valle, N. 2020. BBT, Lot Mules 2-3. Management of data gained by the pilot tunnel drive for the twin main tubes. *WTC 2019, May 3–9, 2019, Naples, Italy.*

Sheil, B. 2021. Discussion of “on the pointlessness of machine learning based time delayed prediction of TBM operational data” by G.H. Erharter and T. Marcher. *Automation in Construction Volume 124.*

Working Group no.17. 2017, ITA Report no.19—*TBM Excavation of Long and Deep Tunnels Under Difficult Rock Conditions*. Avignon: Longrine

Yeo I.K. and Johnson R.A., A new family of power transformations to improve normality or symmetry.” *Biometrika,* 87(4), pp. 954–959, (2000).

# New Diversion Hydropower Plant—Nedre Fiskumfoss

**Mads Aniksdal** ▪ Skanska Norge AS

## ABSTRACT

New Nedre Fiskumfoss is a construction project in North Trøndelag, Namsen River.

The purpose of the contract is construction of a new hydropower plant, that will phase out the existing power plant in the same waterfall.

The contract consists of a tunnel system with cross sections from 16 m$^2$ to 150 m$^2$, powerhouse and transformer in rock caverns, 6 shafts and major works with inlet and outlet below river level.

Technical challenges consist of demanding design on rock extraction, cofferdams, solid constructions with challenging design and demanding rebar work, as well as restrictions and unpredictable water flow in a national salmon river.

## PREFACE

This manuscript is written for the Rapid Excavation and Tunneling Conference and aims to provide insight into the development of a complex river power plant, which will phase out an existing hydro power plant that produces power as construction of the new is ongoing.

## INTRODUCTION

The purpose of this paper is to present New Nedre Fiskumfoss, a new diversion hydropower plant in Norway.

**Figure 1.**

New Nedre Fiskumfoss Kraftverk is located in the North of Trøndelag. The project site is bounded between the road E6 and the river Namsen, which present a minimal space for rigging and operations.

The superior goal is to phase out the existing power plant and increase the degree of utilization of the energy from the river. The existing power plant had the first unit in operation in 1946. Due to the age of the existing power plant there was a need for improvement, in the form of rehabilitation or a new plant. At the same time as the lifespan of the existing power plant is coming to an end, new technology makes better use of the same natural resource.

After a pre-project with interaction with the contractors in the project, the Client decided to build a new power plant next to the existing one. By using the same location for the outlet, new concession from The Norwegian Water Resources and Energy Directorate was not required.

Namsen is a national salmon watercourse. These watercourses are a part of a scheme to give a selection of the most important salmon stocks in Norway special protection and priority. In the national salmon waterways, new interventions and activities that could harm the wild salmon are basically not permitted.

Before the development of the first power plant, the salmon did not get past the waterfall. Now the power plant can be passed in salmon steps, that was put into trail operation in 1975. The salmon steps are still in operation and preserved during construction of the new power plant. As a supplement to the salmon steps, the project has built an arrangement for migration of smolt which runs in a separate tunnel from the intake and past the outlet construction. This is to provide safe passage for smolt to bypass the waterfall on its way back to the sea.

The project is considered important to ensure further utilization of already developed waterways, and to ensure security of power supply. The Client received in 2022 an award from the Norwegian association for rock blasting techniques on behalf Norwegian hydropower plants, with encourage to more hydropower development and not least the upgrading of the many existing hydropower plants in Norway.

Since 1971, it has been extracted a total of over 60 million cubic meters of rock to ensure Norway's clean energy.

1,739 hydropower plants stand for 90% of Norway's power generation and have a total capacity of 33,403 MW at the start of 2022. Norway is today one of Europe's leading producers of hydropower.

## Builder and the Main Contractors

The client for the project is NTE Energi AS, which has production facilities for hydropower in several municipalities in Trøndelag. The company is also a co-owner of hydropower plants in Nordland and in Sweden. NTE Energi currently produces renewable energy equivalent to the needs of 500,000 people. The company has several large energy projects, both under construction and planning. NTE's production of renewable energy was somewhat higher than normal production because of good inflows throughout the year 2021. Production ended at 4,123 GWh. Normal production was 3,894 GWh in 2020. The storage inventory at the end of 2020 was 1,726 GWh, compared to 1,375 GWh at the end of 2019.

The civil contractor for the project is Skanska Norge AS. Skanska Norway is one of the country's largest and leading contractors and project developers and has been present in the Norwegian market since 1906. In Skanska Norge, there are approximately 3,800 employees, and around 200 ongoing projects across the country at any given time. Based on the company's global environmental expertise, Skanska aims to be the first choice when it comes to green projects. In Norway, Skanska accounts for 36% of all BREEAM-certified buildings and has the country's first and only CEEQUAL-certified project on site. Skanska Norway has a long and proud tradition as a significant player in the development of hydropower plants in Norway and considers Nedre Fiskumfoss to be a demanding and complex project that highlights our position in the marked.

The contractor for electromechanical deliverance is Andritz Hydro in Norway. Overall, they are 160 employees. In 2006 Andritz Hydro AS became part of the Andritz Group. Andritz Hydro is one of the globally leading suppliers of electromechanical equipment and services for hydropower plants. With over 180 years of experience and an installed fleet of more than 471 GW output, the business area provides complete solutions for hydropower plants of all sizes as well as services for plant diagnosis, refurbishment, modernization and upgrade of existing hydropower assets. Pumps for irrigation, water supply and flood control as well as turbo generators are also part of this business area's portfolio.

Norconsult has assisted the client with the design of a new power station. Norconsult is one of the leading consulting engineering companies in power generation, energy supply and water resources planning with 80 years of experience in Norway and 50 years of international operations. Norconsult has worked on international projects in more than 150 countries over the past 50 years. They are both a sought-after partner for Norwegian companies abroad, as well as for local companies where we have a presence.

The cooperation between the parties in the project has been characterized by good constructive cooperation, common goals, and trust.

Figure 2 is a drone photo of the project site showing the river Namsen, the existing dam and power plant, the new intake and cofferdam.

**Table 1. Key Information**

| Project Name | Nye Nedre Fiskumfoss Kraftverk |
|---|---|
| Localization | Trøndelag, Norway |
| | |
| Client | Nord-Trøndelag Elektrisitetsverk (NTE) |
| Consulter | Norconsult |
| Contractor – Civil | Skanska Norge AS |
| Contractor – Electromechanical | Andritz Hydro AS |
| | |
| Maximum devouring ability | 300 $m^3/s$ |
| Installed power | 2*46MW |
| Turbines | 2*Kaplan |
| Annual production | 375 GWh |
| | |
| Construction time | 2019–2023 |

Figure 2.

Figure 3.

## The Project—New Nedre *Fiskumfoss* Hydro Power Plant

The project involves challenging and complex construction work, both in the open and underground, and fits well into Skanska's hydro power portfolio. In addition to the challenging design and construction, the surroundings of the power plant add increased complexity, being built between the river Namsen and the road E6, and the train line which passes in the immediate vicinity (Figure 3).

When the scope of work has been carried out, external factors have had a considerable impact.

The temperature on the project location has varied from below -30 degrees Celsius to over +30 degrees Celsius. It has at times been challenging to carry out work due to the low temperatures together with water from the river, especially at the bottom of the waterfall at low temperatures.

Namsen is a huge river with a large and varying water flow and melting ice, which can present challenges when carry out work in the river.

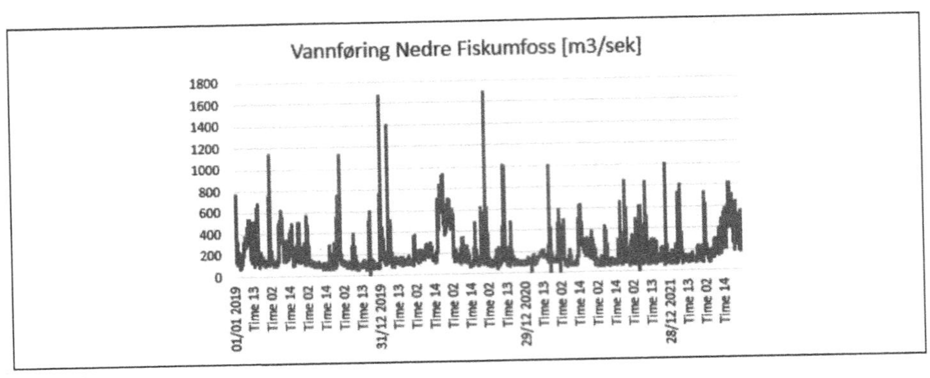


**Figure 4.**

The project has the following restrictions for dates when it is not allowed to work in the river:

- Smolt migration: 15 May–15 June
- Fishing period: 01 June–31 August
- Spawning time: 01 October–10 November

Progress for the project has been challenging and characterized by the complexities that the surrounding nature and climate have provided, unforeseen events, changes in the scope of work, restrictions, and many internal dependencies in the execution. With only one access road into the power plant logistics, transport plan and sequences have been very important to enable concurrency to minimize build time.

Skanska's scope of work:

- Preparatory work
  - New measuring house, Water, wastewater, drainage, electric power technical cables, access road to the outlet
- Extraction of rock
  - Tunnels, shafts, caverns, cut blasting and blasting of cofferdam with underlying rock
- Concrete work
  - Portal, Tunnel systems, Intake, Power Station and transformer hall, GIS-building, Outlet, Prefab
- Building complementary works
  - Steel work, painting, injection, doors, locks, fittings etc.
- Cofferdams
  - Project planning and establishment
  - Foundation Work

The design for the project has been communicated through a building information model from the client, which is processed and divided into "work packages" specially adapted to Skanska's production team. The information is then made available on selected places onsite, through BIM-stations. The project has been executed without physical blueprints.

### Organization

Through almost 4 years of production, the organization on the project has varied somewhat based on type of work and concurrences. At peak production Skanska's project crew consisted of 12 white-collar 80 blue-collar.

On power plant projects in Norway we use to have a rather slim organization of white-collars, which works well due to a flat structure and greater mandate with skilled blue-collar. Our perception is that this improves cooperation and utilization of knowledge on the project.

The white-collar mainly work onsite 4 days a week, except the foremen that follow the same rotation as the blue-collar. The rotation for blue-collar varies somewhat depending on field of expertise and will be presented later in the paper.

## Extraction of Rock

### Tunnels and Caverns

The tunnel system consists of several different tunnels and caverns on different levels. The tunnels have a cross-section varying from 16 $m^2$ to 150 $m^2$ and are excavated using the drill-and-blast method. In total the tunnels, shafts, and caverns sum up to 135,000 $m^3$ of hard rock to be excavated.

The blue-collar work 12/16-day rotation under the tunnel excavation, with 12 days at work and 16 days off-work, working night the first four days and rotating to daytime on the fifth day. Each shift consists of a drill-rig operator, a drill-rig serviceman, a machine operator, and an apprentice. The workers have high level of knowledge and experience, with a clear mandate. Due to the varying radius, inclination and cross-section is it a key factor that the workers are versatile and efficient, with a high initiative.

Table 2 shows the main machinery used for excavation of tunnels and penstocks:

4-axle trucks were used for transport of rock out of the tunnel system to a temporary storage, then to be transported out of the facility on semi-trailers. For the cavern and shafts from the transformer hall to the tailrace, tophammer drill rigs and excavators was used.

**Table 2. Main Machinery**

| **Drilling Rigs** | **Type of Equipment** |
|---|---|
| Epiroc Boomer XE3 | Tunneling face drilling rig |
| Epiroc Boomer E2 | Tunneling face drilling rig |
| Epiroc Boomer 282 | Tunneling face drilling rig |
| **Loaders** | |
| Cat 980M | Loader w side tipping bucket |
| Sandvik LH410 | Underground loader |
| **Excavators** | |
| Hitachi LC290 | Excavator w demolition hammer and tiltable cabin |
| **Service Machinery** | |
| Mercedes 1324 | Bulk emulsion carrier |
| Mercedes Axor 1824 | AMV working platform |
| Volvo L110 | Service loader |

Extraction of the tunnel system started in September 2019. The rock mass in the area for excavation consists of dioritic gneiss with a transition to quartz dioritic gneiss with elements of amphibolite layers. The rock mass integrity gives good conditions for excavation with moderate need for rock support. Scaling, both mechanical and manual, rock bolts (CT-bolt) and shotcrete was used continuous. Split tunnel face with reduced length was only used on the first blasts in the access tunnel. Remaining tunnel excavation was full-faced. The tailrace is 150 $m^2$ and was due to its height split in the horizontal direction. During the excavation it was no need for grouting. High pressure grouting was only carried out in the penstock and tailrace to avoid leakage into the powerhouse.

The excavation work has been on critical line of progress and gives several significant dependencies. To start the concrete work for the powerhouse and intake construction, all the excavation work except the last part of the tailrace tunnel had to be completed. Further, as the tailrace and outlet are well below the river level, the cofferdam downstream and the tailrace hatches had to be operative.

The first milestone was completing the heading in the powerhouse cavern and a temporary access tunnel through a lower level in the powerhouse cavern to the penstocks. This priority was made to establish the concrete girders for the overhead crane, and simultaneously excavate the penstocks. Due to the height in the powerhouse cavern, the first level was excavated without loading all the rock. The rock was compacted and leveled as a foundation for the concrete work, and fully loaded after completion of the concrete girders (Figure 5 and 6).

The remaining part of the powerhouse cavern was taken out successively. The cavern is 41,8 meters high, 57,3 meters long and have a width which measures 18,6 meters. On this diversion power plant it's no access through the penstock due to the design, which make the logistics a bit more challenging. Tophammer drill rigs and excavators was used, supplemented with tunnel rig for rock- and construction bolts drilling and a shotcrete robot from AMV. In this phase it's important to get everything done before the height difference is too big and ensure that everything is done correct. This of course also applies to rest of excavation.

Continuously during excavation, the profile has been scanned, and integrated in the BIM-model. The tunnel face is scanned using a profiler from Bever Control, which is

Status start of concrete work

Fully cast with the formwork on

Figure 5.

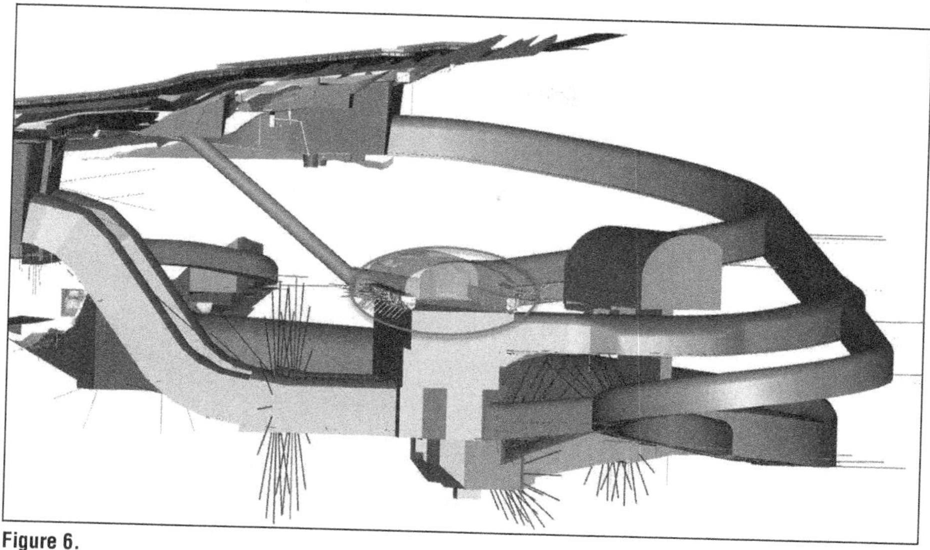

**Figure 6.**

mounted on the tunnel rig. This gives the operator immediate control over whether the profile is large enough. Then the survey engineer scans the profile, which is secured with shotcrete and rock bolts, using a MultiStation from Leica (MS50). The MultiStation is a combined scanner/total station, which also can be used to check the profile on local areas with a laser pointer. The point cloud from the scan can be imported into countless processing tools. We have used the Norwegian-developed Gemini, from the supplier Volue. Here the point cloud is processed and triangulated into a surface model which is directly implemented in the project's BIM-model.

**Figure 7.**

Below the scan which is implemented in the BIM-model is shown in Figure 8.

The last tunnel completed was the tailrace, due the mentioned dependencies. Figure 9 is a picture from the preparations for the outlet channel excavation.

### *Shafts*

The project has 6 shafts in total, excavated with three different methods. Raise borer, tophammer drill rigs and tunneling face drilling rig was used. Figure 10 shows all of the shafts, except the one from the transformer to the GIS-building (gas insulated switchgear).

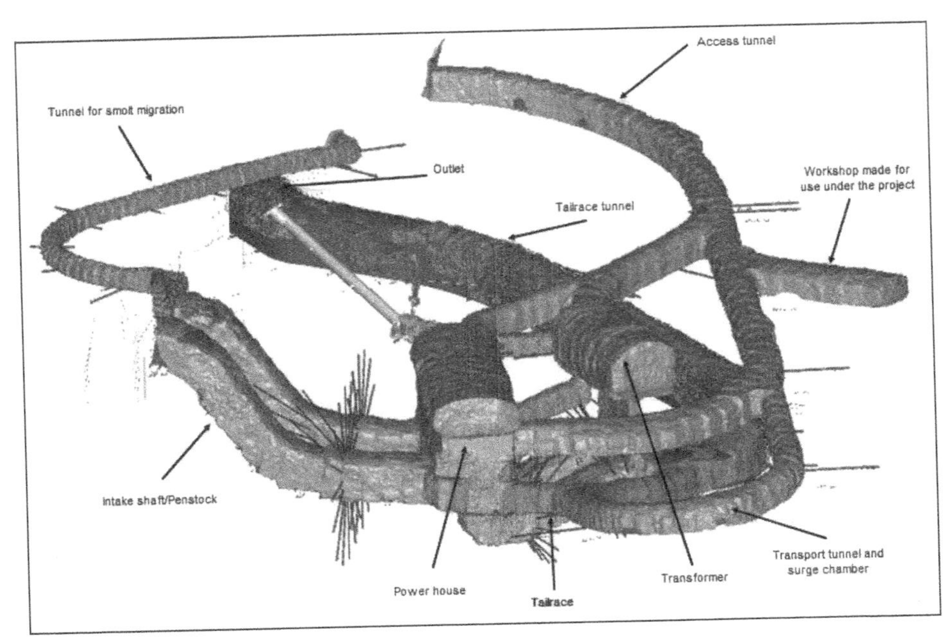


Figure 8.

Figure 9.

The shaft from the transformer to the GIS-building was excavated using a raise borer, with pilot hole and reamer head with diameter 2.8 meter. The shaft is approximately 40 meters.

Excavation of the 2 shafts from the transformer cavern to the tailrace was done using a tophammer drill rig. These shafts each have a length of 12 meters, width of 2 meters

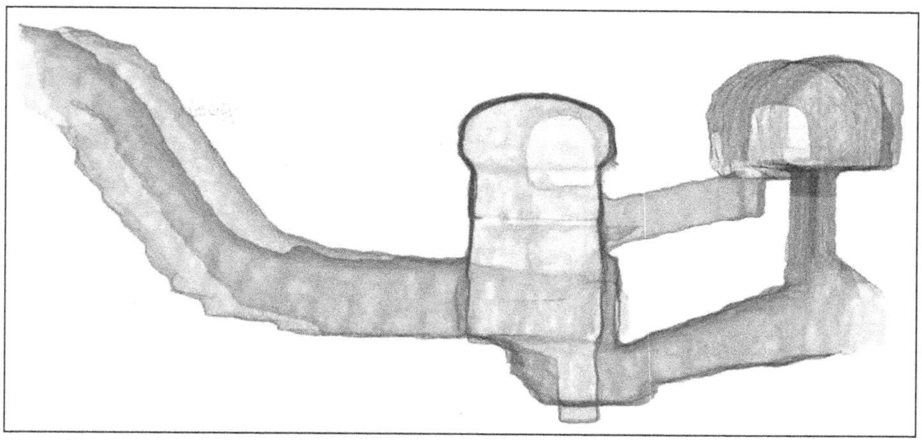

**Figure 10.**

**Figure 11.**

and a height of 12,5 meters (down to the tunnel heading below). For rock inspection it was used a truck with winch that lowered a work basket from the top. Rock bolts was drilled and installed form the same basket, using an Atlas Copco BBC 16W (pusher leg rock drills) for drilling (Figure 11).

One of the most interesting executions was excavation of the penstocks. The cross-section for these shafts is 75 $m^2$ each, with an inclination on 55 degrees. Due to the design, it was not enough space to use a tophammer drill rig. A Epiroc boomer XE3 tunneling face drilling rig was used for the extraction, done from below. This method ensured progress and was also beneficial for the surrounding work. The figures in Figure 12 show the shafts measures and a tunnel rig modeled in scale for illustrating the size and length of the shaft compared to the equipment.

To increased range, the road was raised 3,5 meters and the shaft was excavated full faced with length up to 7–8 meters each blast.

Figure 13 shows one of the penstocks, oriented towards the powerhouse cavern.

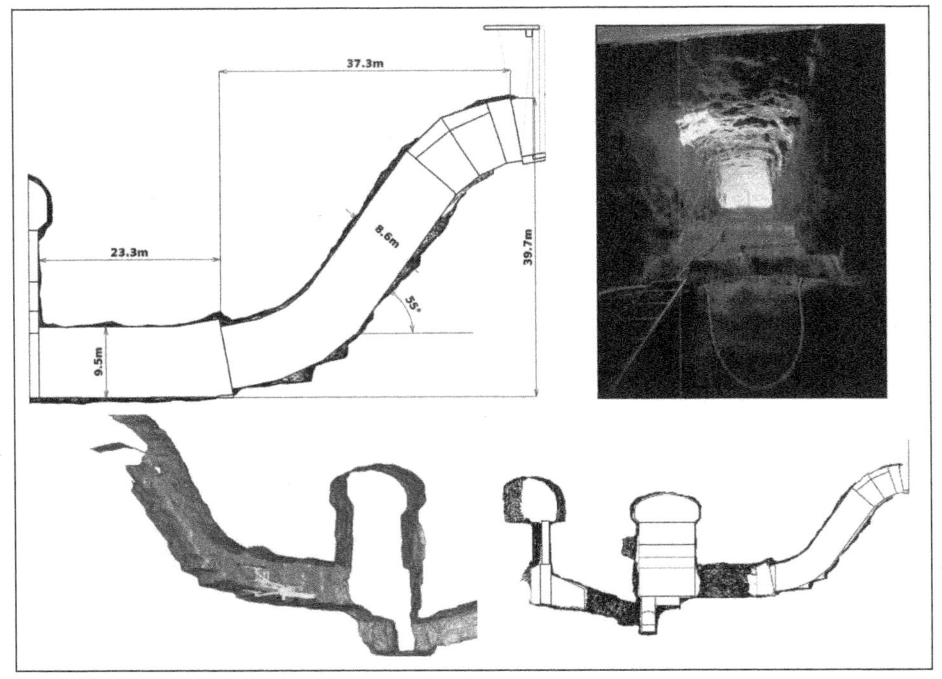


Figure 12.

Figure 13.

**Figure 14.**

## Concrete Work

The concrete work on the project is particularly challenging and consist of massive constructions with a challenging design and demanding rebar work. In total 21,500 $m^3$ concrete has been casted and 1,825 tons of rebar have been installed. Formwork on the project consisted of conventional formwork, systems from Doka and premade conventional formwork used in some places with demanding areas with chrome and radius. Due to the design of the tunnel complex, most of the concrete work on the project was accessed simultaneously giving a high peak of production (Figure 14).

This picture is taken from one of the penstocks, oriented towards the powerhouse cavern, showing the turbine construction.

The blue-collar that works with concrete have a 12/9 rotation, with 12 days at work and 9 days off-work. Each shift consists of a blue-collar supervisor with varying numbers of workers. Normally the concrete workers only work daytime, but at the peak of production they worked nights the first 4 days with rotation to daytime on the fifth day.

In Skanska Norway we have very experienced concrete workers that have worked with hydro power projects many projects in a row. Great communication and cooperation between white collar and blue collar characterize the day-to-day life on the project and is key to fully utilize the knowledge available on the project.

Construction of the powerhouse cavern was the main part of the concrete work and the most important towards completion of the project. Agile collaboration between the different contractors is key. Skanska and our main side contractor Andritz hydro have long history of good cooperation on different projects, it's all about the people.

In the cavern the turbine walls are the most impressive construction, with heavy rebar work and specially adapted formworks. The construction which are usually built with steel, are here made in concrete due to their size.

A criterion of success when building a powerhouse is carefully planned logistics, as everything in the cavern has to be lifted in and out by the overhead crane. Figure 15 shows three pictures from construction of the turbines.

The main concrete work in the cavern started in the fall of 2020 and was ready for the main electromechanical assembly in the spring of 2022. Figure 16 shows two pictures from December 2023 where the electromechanical assembly is well underway:

Both tailrace and penstock have concrete linings, with various issues in the execution. The tailrace tunnels had a demanding inclination with an increasing cross-section. Due to the statistics two levels of Doka system was needed, complemented with pre-made conventional formwork (Figure 17).

The shafts shown in the picture to the left, taken from the tailrace, leads to the transformer cavern. The transformer cavern, which is show on the right picture, includes constructions for transformers and arrangement for handling of the tailrace hatches.

For the outlet construction it was used climbing formwork. The construction was dependent on completion of the cofferdam downstream and excavation of the tailrace.

**Figure 15.**

**Figure 16.**

Drilling for the excavation of the outlet channel was done before the concrete work started. The temperature was at its worst -30 degrees, with wind from the open tunnel system. It was cold (Figure 18).

The intake construction was the largest surface construction. After completion of the cofferdam upstream and the excavation, both surface and shafts, was completed, the concrete work was started. The construction is mainly below the water level in Namsen and is where the water diversion of the new power plant starts. The construction is over 50 meters long and over 15 meters high. Over the intake its 4 small hatches that leads to the tunnel for smolt migration. A Liebherr 280EC-H 16 Litronic tower crane was used to serve the concrete worker (Figure 19).

**Figure 17.**

**Figure 18.**

**Figure 19.**

## Cofferdams

The upstream cofferdam was critical for security and to make the project possible. Skanska Norway designed and built the cofferdam, with a subcontractor to manage the pile work. Rock from the access tunnel was used to establish the area. The cofferdam is dimensioned for a Q500 flood and an ice load of 150 kN/m. The pile wall was made with PU32 (U-type) with bitumen in the locks. Steel pipes was mounted on the wall to make drilling for high pressure rock grouting possible. Between the pile wall and the bedrock chemical grouting was used. The cofferdam was functional for 2 years and was without leakage (Figure 20).

The intake channel was excavated without water pressure on the intake due to safety measures, meaning that the cofferdam was functional until the last blast. To achieve this the area under the pile wall had to be excavated in one blast. The construction was secured before the excavation work started with steel frames existing of HEB 300 beams, spacers and blasting mats. To reduce the size and impact from the last blast, excavation of rock inside the cofferdam was carefully done at first (Figure 21).

Around 60 blue-collar was involved in this work, that was ongoing 24/7 in teen weeks in the winter and spring of 2022. The machinery that was used was:

1. Excavators: Hitachi 690, CAT 352, Hitachi 670 LR, Volvo ECR 145
2. Dumpers: Volvo A45, Volvo A25
3. Surface drill rigs: Epiroc T35
4. Pipe drill rigs: Nemek 814, Nemek 407

The rock from the first excavation was used to establish the area outside the cofferdam, to access the entire profile of the channel. Casting pipes was then mounted

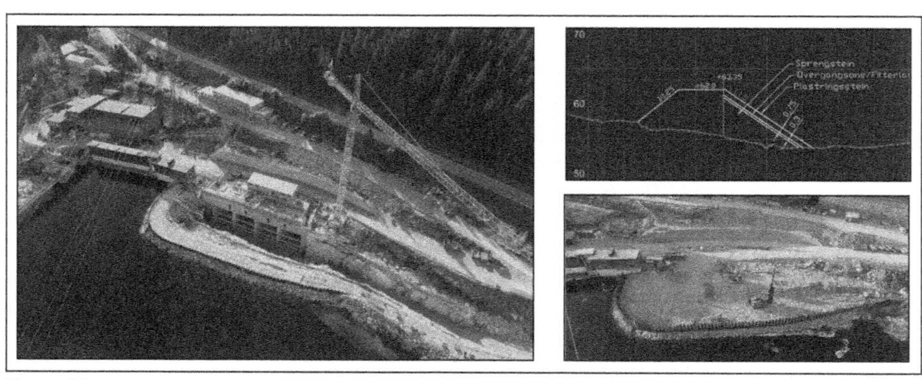


Figure 20.

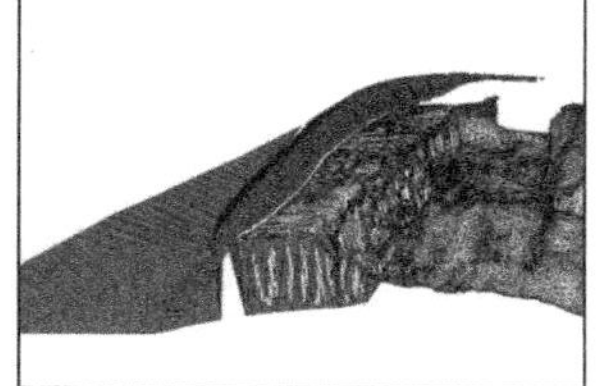

Figure 21.

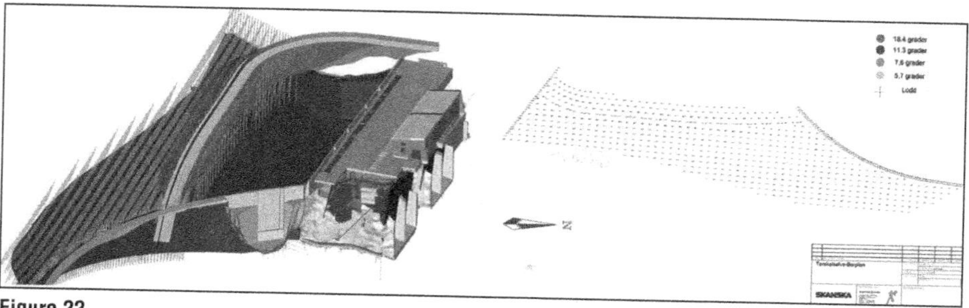

Figure 22.

Figure 23.

though these masses, to access the underlying rock. In total 824 holes was drilled, and 13,4 tons of explosives was used in the final blast. Such complex operation close to infrastructure require careful planning and continuous control. Figure 22 shows the drilling plan and the orientation of each hole.

After the drilling was finished and every hole was checked, the charging could start. Over 300 blasting mats were used in the operation. Everything went as planned. The first pictures in Figure 23 show drilling of casting pipes and the second shows the site during the blast. The two last pictures show loading of rock.

The downstream cofferdam was something special. Its placed right under the waterfall, with unlocated bedrock. Before we could start on the work a new access road from the main road to the outlet had to be made in steep terrain with need for stabilizing measures in the road cut. Establishment of the cofferdam started in the fall of 2020. During the work the existing power plant had to stop, due to the same point of outlet. This resulted in continues water over the dam and increased time pressure due to lost earnings. Stop time due to restrictions and the surrounding conditions was regular.

The area was at first filled with suitable masses to prepare for installation of the drilled RD pile wall (RD 406 × 10,0 mm) with rock and soil anchors in five levels. The deepest RD pile was over 25 meters long. Simultaneously as the pile wall was installed, surface rock excavation and establishing the tunnel for smolt was done. Figure 24 shows are some pictures of the processes.

A lot of resources were put into the grouting work to seal water leaks, with variable results. The cofferdam had to be approximately 16 meters deep. The further we excavated the masses, the higher the water pressure and leakage got. With large pump capacity and a concrete wall inside the cofferdam we manage to continue the loading inside and work with rock and soil anchors. For internal reinforcement HEB 1000 beams where used. The only access in and out of the cofferdam was eventually with a 220-ton mobile crane, that lifted all machines and equipment (Figure 25).

Eventually we manage to complete the cofferdam, with the right level inside. The tailrace hatches were closed, and the tunnel could be completed. After the tunnel was completed the outlet construction was established. Now it was time for excavation of the outlet channel and disassembly of the pail wall. As preparation the beams for the outlet construction was installed and covered with blasting mats. The pile wall corners of the east wall were blasted simultaneous as the rock excavation. Here there is only one attempt. Ice and cold made the charging problematic. Due to the large leakage the water pumps were the last installation that was removed. This was done to reduce risk of damage the structures and hatches. All the charging work had to be done form a basket (Figure 26).

After the blast it was used a great deal of resources for loading and clearing up the east pile wall. The south and the north wall were permanent (Figure 27).

Overview before start-up

Large water flow

Ongoing tunnel excavation and work on pile wall

Figure 24.

Figure 25.

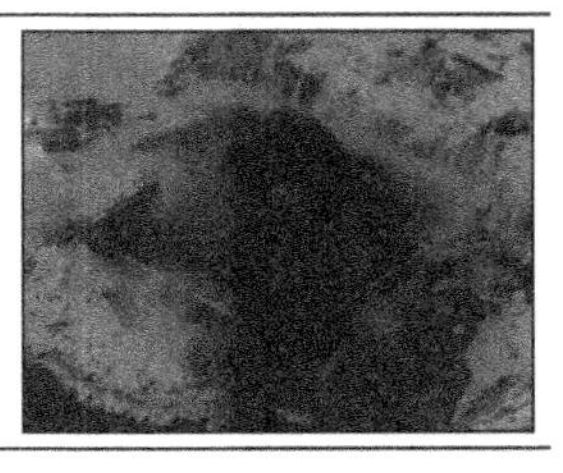

Figure 26.

Figure 27.

## SUMMARY

This has been a sustainable project with countless of challenges that have been resolved due to skilled professionals and good cooperation between the project participants.

## REFERENCES

Gullfeiselen for 2022 til norske vannkraftverk ved NTE • Byggeindustrien

Kraftproduksjon–Energifakta Norge

https://nte.no/om-nte/organisasjon-og-eiere/energi/

https://www.skanska.no/hvem-vi-er/skanska-i-norge/

https://www.andritz.com/hydro-en/about-andritz-hydro/locations/jevnaker-norway

https://www.norconsult.no/prosjekter/nye-nedre-fiskumfoss-kraftverk/

https://www.norconsult.com/about-us/about-norconsult/?targetId=cbody&method=insert#Global+presence

https://www.nve.no/om-nve/nves-utvalgte-kulturminner/kraftverk/nedre-fiskumfoss/

https://snl.no/nasjonale_laksevassdrag

# New Water Supply—Oslo

**Silje M. Tinderholt** ▪ Skanska Norge AS
**Ingunn Opland** ▪ Skanska Norge AS

## ABSTRACT

Drill and blast project in Norway's capital. The contract consists of 7 rock caverns (350–950 m$^2$), 9.2 km tunnels, and four shafts.

With 110 blue-collar and 31 white-collar, the project produces an average of 16–21,000 m$^3$ hard rock per week, with a total of 19 tunnel faces. Weekly, nine grouting rounds for water control, 800 m$^3$ shotcrete, and 1,200 bolts are required, and 60 trucks leave the site pr. day.

Challenges include blasting in densely populated areas, planning and coordination to maintain high production, obtaining a CEEQUAL certification of very good, and transporting rock on Oslo's main highway.

## INTRODUCTION

The capital of Norway, Oslo, is one of the fastest-growing cities in Europe. Today 90% of Oslo's inhabitants get their water from one primary source, Maridalsvannet. This makes Oslo's water supply vulnerable to climate change, pollution, and sabotage. The department of emergency response states that loss of water is not unlikely. Because of this vulnerability, the Norwegian Food Authority has demanded that Oslo County have a reliable second water supply by January 1, 2028 (Oslo Kommune, 2022).

The new water supply consists of a 19 km intake tunnel, from Holsfjorden to the water treatment plant at Huseby. From Huseby and Stubberud tunnels are built and connected into the existing distribution net, transporting water to the users (Oslo Kommune, 2022). Figure 1 shows the construction plan of the new water supply.

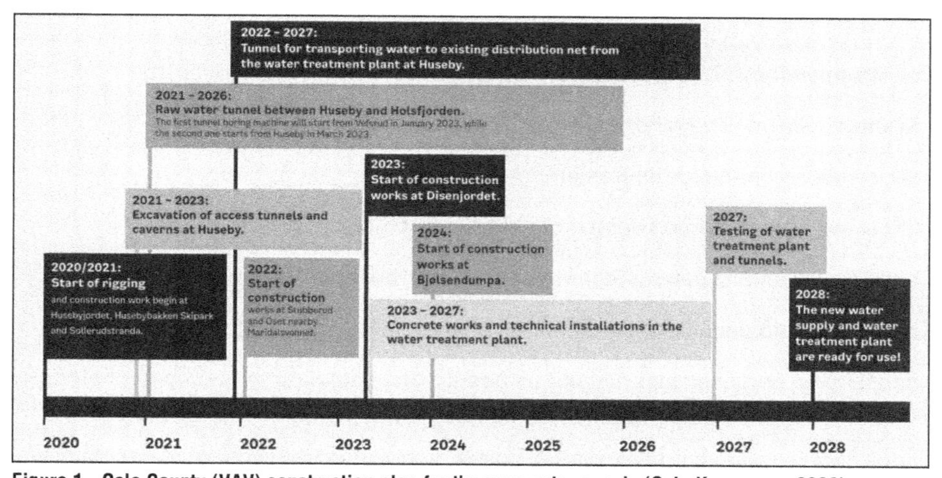


Figure 1. Oslo County (VAV) construction plan for the new water supply (Oslo Kommune, 2022)

Skanska Norge AS acquired the contract "E08 Preliminary works" excavating the new water treatment facility in the heart of Oslo. The contract consists of 8 rock caverns, 9.2 km tunnels, and 4 shafts. The caverns are positioned on three levels making this a complex tunnel system. At completion the project will have excavated and transported 1.05 million cubic meters of hard rock.

Oslo county aspires to reduce emissions from all construction sites. By 2030, their goal is to have reduce the county's total emissions by 95% (Oslo Kommune, 2022). Combined with Skanska's high climate and environmental ambition, the aim is to complete the project with a low climate footprint. One of the tools to reach this goal is using machines powered by electricity and biodiesel. The project will be certified through BREEAM Infrastructure Version 6 (BRE Group, 2022), aiming for a qualification of Very Good.

## KEY INFORMATION

- **Construction client:** Oslo County, the water and sewage department west (VAV)
- **Contract value:** 2.88 billion NOK/ 278 million USD
- **Construction start:** September 2020
- **Date of completion in contract:** April 2024
- **Estimated date of completion:** August 2023
- **Climate and environment:** BREEAM Infrastructure of Very Good. The aim is to complete the project using machines powered by electricity.

The contract sum of 2.88 billion NOK, plus options of 141 million NOK, was the largest construction contract ever signed in Skanska Norge at signing date. The estimated end contract sum 2.4 billion NOK.

## ORGANIZATIONS

The presentation of the project organization and machinery is based on the time of peak production, which lasted for approximately a year, from August 2021 to September 2022.

### Project Crew

To reach project goals, its essential to have the right people. Motivated and skilled workers are key. Skanska's project crew consisted of 110 blue-collar and 54 white-collar.

Figure 2 shows the project's organization chart. Green and orange boxes indicate the roles that make up the leader group. In this group, there is one leader per discipline, to make sure that all relevant information is accounted for in the decision-making process. The project is organized with six primary disciplines: HSE, outer environment, contract, survey, and production (tunneling, concrete, and groundworks).

The majority of the white-collar have a 4/3-rotation, working 12-hour days, Monday-Thursday. This rotation enables the white-collar to have a continuity and therefore better control over the activities in the project. Because of the complexity of the project, twelve foreman work in pairs. The foreman organizes the daily activity in the tunnel and have the same work rotation as the blue-collars. The line of communication between the white-collar and blue-collar is very open, making the information flow effective. This leads to an including work environment.

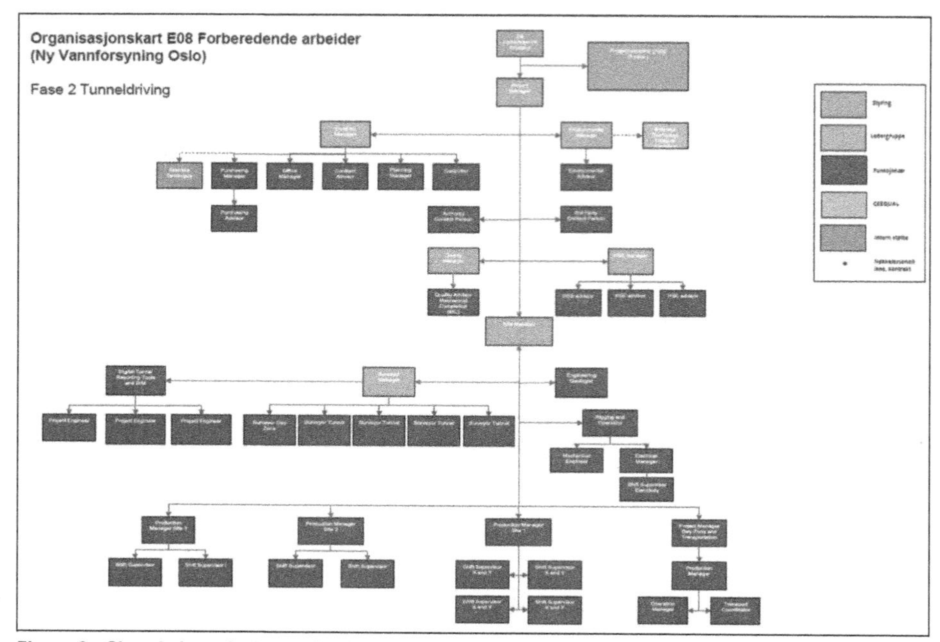


**Figure 2. Skanska's project organization chart (white-collar)**

The blue-collar work 12/16-day rotation, meaning 12 days at work and 16 days off-work, working night the first four days and rotating to daytime on the fifth day. Skanska Norge experience show that the 12/16 rotation increases production by approximately 25-30%. For the workers, it is the preferred rotation as they get more work hours when at work, and therefore more spare time at home. In total, one shift of 12 days gives 116.5 work hours and 10.5 hours paid lunch break. For each rotation, the worker uses 7.5 hours of their holiday, giving a total of 135.5 hours per rotation. In Norway every employee has the right to 5 weeks holiday, where 3 weeks must be uninterrupted.

Each shift consists of a drill-rig operator, a drill-rig serviceman, a machine operator/loader, and an apprentice. These four workers handle the entire excavation cycle. The low number of workers is possible due to their complementary and versatile skillsets. This is partly because of their education where the aim is to learn and understand all aspects of the excavation cycle, specializing to a specific role after the end of the apprenticeship. The emphasis on the entire excavation cycle allows the workers to assist each other with varied tasks, and step in to increase efficiency. Electricians and mechanics support the shifts. Four shifts (of 16 workers) go on rotation operating four drilling rigs. Because of the large amount of pre grouting, dedicated shifts (two-three blue collar) operate the grouting rig.

On an average week, the blue-collars work on 19 different tunnel faces, produce 16–21,000 cubic meters of hard rock, perform nine round of pre-grouting for water control, and install 1200 rock bolts ranging from four to six meters. Versatile workers give a streamlined production, Figure 3 shows the amount of blasted rock versus the number of work hours per week.

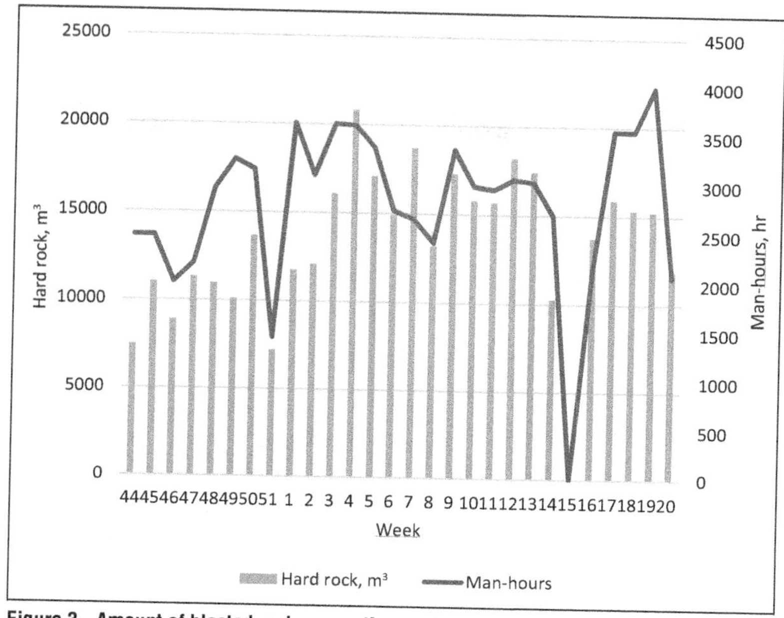


**Figure 3. Amount of blasted rock versus the number of work hours per week**

## Machinery

To achieve the goals with respect to environment and production, Skanska depend on suitable equipment. Five Epiroc Boomer XE3 rigs are in production, four in drilling, and one in maintenance rotation (Epiroc, 2022). The purpose of the drill-rig is to drill holes for blasting, grouting, and rock support, Figure 4. All the drill rigs are equipped with RHS (Rod Handling System). In essence, RHS is an automatic method of handling the drill rods during injection drilling (Epiroc, 2022).

One of the access tunnels has a smaller cross-section, not large enough for the EX3. Instead, a Boomer E2 drill-rig is used in this tunnel (Epiroc, 2022). This tunnel is three kilometers long with two access points, making it possible to excavate from both sides. Both the XE3 and E2 are fully electrical during drilling, creating favorable work conditions, and reducing greenhouse gas emissions.

Figure 5 shows a fully electric AMV FL70 front loader used in the caverns (AMV AS, 2022). The AMV loader is developed by Skanska and AMV. In the smaller tunnels an ITC SA 315 Superloader is used, this machine is powered on electricity when loading (ITC SA, 2022). The project has two AMV front loaders and two ITC Superloaders. In addition to the electrical loaders, the project use five wheel-loaders.

For grouting (water control), the project have two AMV Grouting Units which are container-based (AMV AS, 2022), and two grouting units which are skid based (AMV AS, 2022).

Scaling is paramount for safety; large tunnel faces demands both mechanical scaling and manual scaling (manpower) to prevent rock-fall for the newly blasted tunnel surface. Mechanical scaling uses a fully electric scaling machine, Cat 326. In total the project has three scaling machines.

Figure 4. Two EX3 and a L110 service loader bolting and drilling one the heading of B3

Figure 5. FL70 front loader

## EXCAVATIONS METHOD

The tunnel system is excavated using the drill-and-blast method (Berggren, Nermoen, Kveen, Jakobsen, & Arild, 2014). The caverns are located on three different levels, with access tunnels connecting them. Table 1 shows the dimensions of the caverns. A1 and A2 are located one the upper level, B1, B2, B3 and B4 in the middle and C1 (TBM assembly hall) and C2 on the lower level. In total 1.05 million cubic meters of rock mass is to be excavated.

Skanska proposed some alterations to the access tunnels as part of the negotiation process after the contract was signed. Figure 6 shows the tunnel system which was presented in the tender documents. Because the water treatment plant is seen as a vital infrastructure the overview of the tunnel system as its being built, is exempt from

**Table 1. Dimensions of the caverns in the tunnel system**

| Cavern | Average Area, m² | Largest Cross Section, m² | Width, m | Height, m | Length, m | Hard Rock to Be Excavated, m³ |
|---|---|---|---|---|---|---|
| A1 | 360 | 406.29 | 23.6 | 18.8 | 141.75 | 50,683 |
| A2 | 336 | 335.74 | 20.6 | 17.8 | 138.05 | 46,349 |
| B1 | 744 | 825.21 | 27.6 | 38.2 | 139 | 102,778 |
| B2/ B3 | 497 | 641.13 | 27.6 | 25.4 | 202.30 | 84,677 |
| B4 | 498 | 606 | 26.6 | 24.3 | 268 | 133,362 |
| C1 | 274 | 346.95 | 21.6 | 16.3 | 230.1 | 62,770 |
| C2 | 193 | 193 | 14.4 | 15.6 | 62 | 11,810 |
| Small access tunnel | 24.97 | 35.08 | 4.5 | 6 | 3,053 | 84,938 |

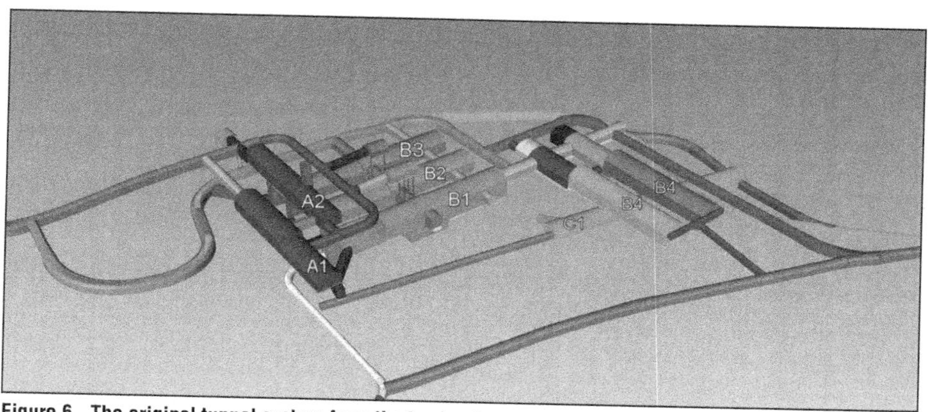


**Figure 6. The original tunnel system from the tender documents before changes were made**

the public. The first milestone of the project was completing the TBM (tunnel boring machine) launch and assembly tunnel C1. To reach this milestone 1,000 or 485 meters of access tunnel from the sites had to be excavated first. Because of neighbors' disputes, the nearest access point was delayed several months, and the changes made in the negotiation process were crucial to reach the milestone.

The client has made several incremental changes throughout the construction period to optimize the final design of the tunnel complex, and enhance the buildability for the contractor Skanska. These changes are possible because of the excellent relationship, trust, and constructive discussions between the client VAV and Skanska.

B1 is the largest cavern in the facility with a cross-section of 825 m². It is 139 meters long, 38 meters high, and has a width of 27 meters, Figure 7. This cavern is excavated in five sequential steps. The two lowest levels are excavated simultaneously as the heading from a lower access point. Then two rounds of benching connect the heading to the lower levels. After the heading is finished, a shaft is wire sawn and blasted from a higher-positioned tunnel into the cavern.

All blast rounds are full-faced, meaning that the tunnel faces are never split vertically. Figure 8 shows a drill-plan for a blast round in B1, it also includes the bolt drill plan. As a safety precaution the work height is restricted to 10 meters, meaning that tunnel faces are occasionally split in the horizontal direction, leaving a bench. The XE3 is also used when benching. Full-faced blasting is made possible using an electronic

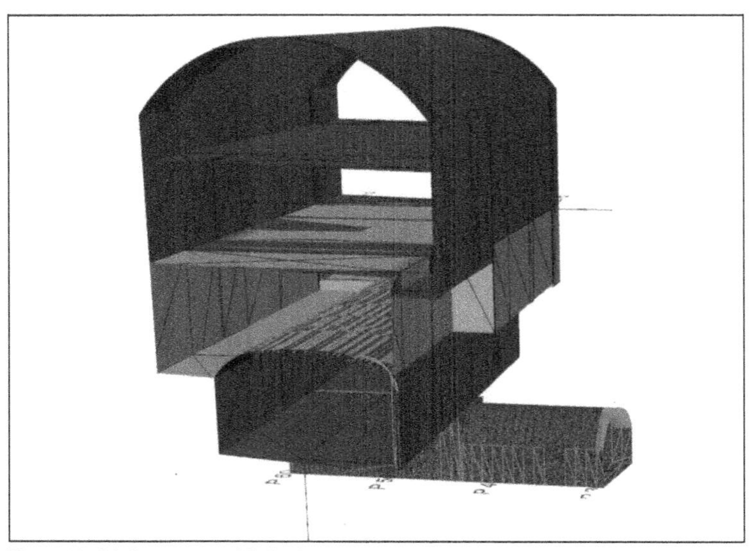

Figure 7. B1 the cavern with the largest cross-section, the horizontal planes separate the five different excavations steps

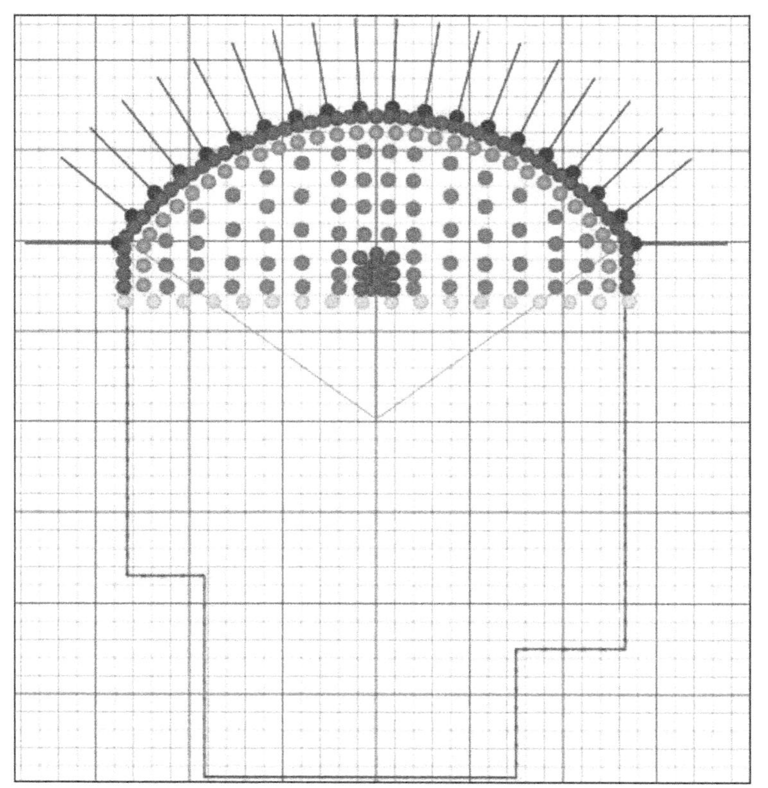

Figure 8. Drill-plan for blasting of heading in cavern B1 (including rock bolts drill pattern for permanent support), one square equal one meter

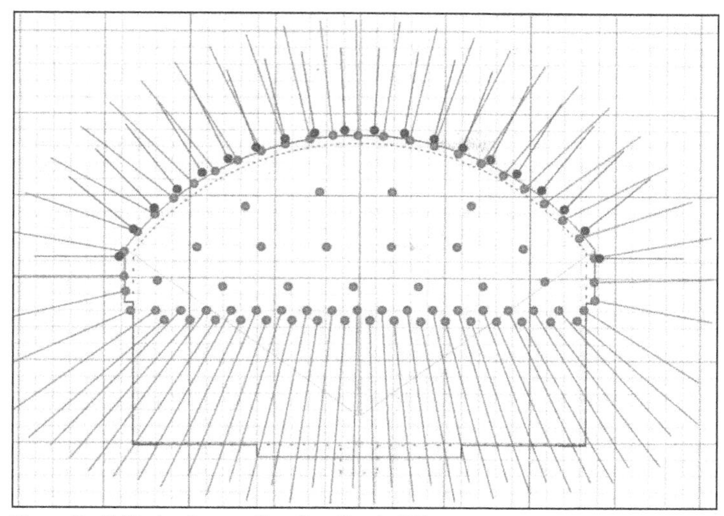

**Figure 9. Drill plan for grouting (heading in B3)**

blast system, which enables a hole-by-hole detonation with millisecond precision. In densely populated areas such as Oslo, electronic blast systems are preferred as it allows better control over blast-induced vibrations. All tunnel blasts use an ammonium nitrate emulsion bulk explosive.

The full-faced tunnel and cavern excavation are favorable because it gives an effective excavation cycle. This technique also minimizes the need to work under several temporary tunnel faces, because the rock support can be installed as a continuous arch.

The entire tunnel system is systematically pre-grouted. Cement (industrial or micro) is injected into the rock mass ahead of the tunnel face with a pressure of 30–60 bar. Injection is a measure for minimizing water leakage. The overlap between the injection rounds depend on local conditions, varying between 15–25 meters. The holes used for grouting are 29 meters long. Figure 9 shows a drill-plan for grouting in cavern B3. In the caverns, cement is injected into the bench level so that the lower level can be excavated without stopping because of grouting.

Four shafts are excavated between the three levels, dimensions shown in Table 2. The shafts are to be lined with concrete using one-sided forming. To minimize the use of concrete, VAV engaged the sub-contractors Kjell Foss and DW-teknikk through Skanska to excavate the shafts with a diamond-impregnated wire saw and then conventional drill-and-blast. A blasted surface is more uneven than a sawn surface, thereby the sawn surface reduces concrete usage. As a safety precaution, only three sides are sawn before the rock masse is belated. The fourth side is sawn after the shaft is excavated. As part of the drill pattern in the shaft, a contour is drilled towards the remaining intact wall to minimize the blast damage.

Ventilation of the tunnel system is split into two phases, before and after break-through between the two different construction sites. In the first phase a fan transports air from outside the tunnel via a duct. The fan delivers 120 $m^3$ air/sek, which at most was split between 15 tunnel faces. The size of the duct varies between 2,400 to 1,200 mm in diameter. In phase two, a stream of fresh air between the sites is created using

Table 2. Dimensions of shafts

| Name | Horizontal Cross Section, m² | Height, m | Volume, m³ |
|---|---|---|---|
| B-Z1 | 79 | 22 | 1953 |
| A-X1 | 108 | 26 | 2586 |
| A-X2 | 108 | 26 | 2790 |
| A-Y1 | 81 | 37 | 3167 |

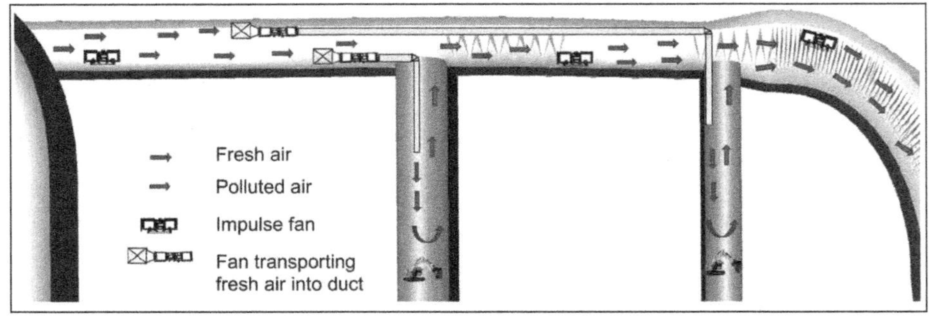


Figure 10. Ventilation of tunnel system Phase 2, fresh air is transport into the work area from the fresh air stream (blue arrows)

impulse fans mounted in the roof of the main tunnel. Air is then transported to the tunnel face using fans and ducts collecting air for the fresh-air stream, Figure 10.

## ROCK SUPPORT

The geological survey, conducted before the project started, indicated several weakness zones. Experience from projects close by in the Oslo geology confirms the conservative description of the rock mass integrity. Therefore, the project description included several mitigating solutions, such as lattice girders and split tunnel face to minimize blast impact on the rock stability.

The size of the tunnel system located in a limited area raised concerns regarding the local and global stability of the rock mass. Therefore, the client through Skanska engaged SINTEF (one of the leading industrial consultant and research institutions in Norway) to simulate the entire construction in a 3D model (SINTEF, 2022). The goal was to investigate how the excavation of rock mass would affect the stress system around the tunnel system and potential rock displacement.

Furthermore, SINTEF designed a surveillance strategy to monitor stress and displacement in the rock mass as the construction work progressed. The measurements are used to calibrate the model, incrementally refining the 3D model. The monitoring strategy consists of extensometers and 2D doorstoppers. Six extensometers are installed in total, one in the rock-mass above the largest caverns, one along an access tunnel between two caverns, and two in connection with shafts. The 2D doorstoppers are installed next to the caverns' extensometers, three total. Figure 11 shows the measuring equipment installed to monitor cavern B2, the same principle is used for B1 and B3. The measuring equipment is placed in concern areas with higher displacement and, or stress, identified through the 3D simulation.

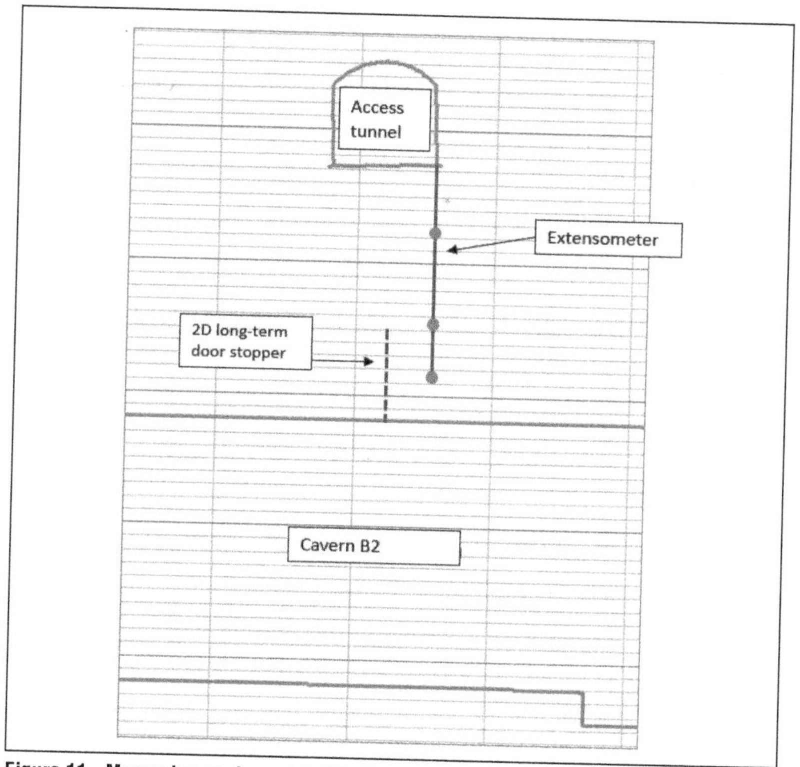


**Figure 11. Measuring equipment in B2, 2D long-term door stopper installed from the roof of the cavern measuring stress and extensometer installed form an access tunnel above the tunnel heading measuring displacement in three locations**

The excavation of the access tunnel showed that the rock mass was more competent than anticipated. As a result, rock bolts and sprayed concrete replace the cavern's heavy rock support, and the tunnel faces can be excavated as one. Both the in-situ measurements and the numerical model verified that local and global stability is ensured without the use of heavy rock support.

Compared to the contractual amount, the reduction in rock support is significant. Table 3 shows the estimated percentage of contractual amount installed on completion.

Less heavy rock support has a direct impact on the construction time and cost. As a result, the project will most likely be finished 8 months ahead of the contractual completion date.

**Table 3. Estimated percentage of contractual amount installed on completion**

| Rock Support Type | Contract Amount | % Off Contractual Amount On Completion |
|---|---|---|
| Shotcrete | 97,020 $m^3$ | 40 |
| Rock bolts | 101,410 | 60 |
| Grouting | 20,856,500 $m^3$ | 63 |
| Steel (heavy rock support) | 1,126,800 kg | 1 |

## ENVIRONMENT

Both the client and Skanska has high ambitions when it comes to climate and environment, and the aim is to complete the project with zero impact on the local and global environment.

Skanska offered BREEAM Infrastructure certification to the client as part of the negotiation process, aiming for a completion grade of Very Good. Skanska has completed several BREEAM Infrastructure evaluated projects throughout the last year, and experience shows that BREEAM Infrastructure gives a good foundation for reducing the environmental impact a construction site imposes.

Electrical-fueled machines are vital when it comes to minimizing $CO_2$ emissions. By substituting from a combustion powered loader to the AMW FL70, the project estimates a reduction in emissions of 77 tons carbon dioxide equivalents (kg $CO_2$-eq) per 100,000 $m^3$ rock. All people transport, both on the construction site and to and from sleeping accommodations, is performed with electrical vehicles. The remaining machinery that is not electrically powered, runs on HVO100, Hydrated vegetable oil, ensuring an 80–90% reduction in greenhouse gas emissions compared with standard fossil diesel (Circkle K, 2022).

Water is a limited resource in the Oslo region, and in the summer months of 2022, all inhabitants were asked to minimize their water consumption (Pettrém, 2022). The project treats and reuses all water used in the tunnel to avoid contributing to this water shortage. Water is transported out of the tunnel into a water treatment plant. In the water treatment plant, particles are sedimented, and the chemical character is regulated before it is sent back into the tunnel.

Fine rock particles from drilling and blasting result in large amounts of mud. In Norway, mud particles from tunneling are categorized as a waste product. Depositing the mud containing 80–90% water is a high project cost. To reduce the water content in the mud, Skanska built a filter press, Figure 12. The machine separates the fine particles and water, resulting in "mud" with a water content of only 20%, and the mud can then be deposited at a lower cost. The water is then treated in a water treatment plant and discharged into the sewer system. Skanska is simultaneously exploring the possibility of reusing the fine particles in the mud rather than handling it as waste. One option is to reuse the fin-milled particles in concrete.

A big part of the environmental focus is connected to the relationship between the project, and the local residents and community. Several mitigating measures have been implemented to reduce the noise and visual impact on neighbors:

- All machines must have white noise back-up alarms
- Changing windows on facades facing the construction site
- Noise-reducing screens
- Modern tunnel fans with low noise impact
- Facade cleaning

All wheels are washed before leaving the facility onto the public roads to reduce dust from the construction site. The vehicles drive over a cattle grid installed over a water-filled ditch, and along impressions in the asphalt filled with water. As an additional mitigating measure, a street sweeper cleans the roads on the construction site and the public roads in the nearby vicinity.

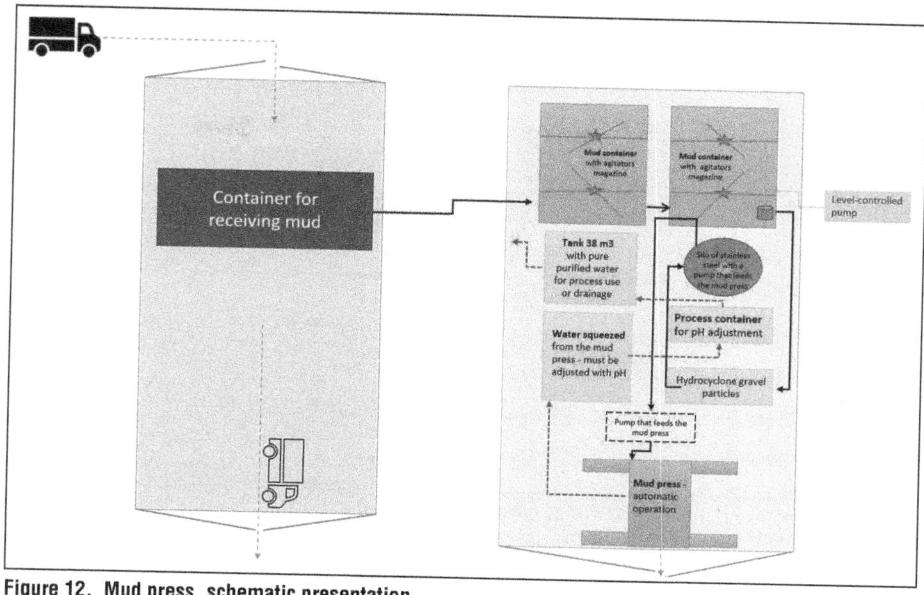


**Figure 12. Mud press, schematic presentation**

Informing and communicating with the local community about project progress, and how stages of the construction might impact them, is vital. The client has a good team of neighbor contacts that handle all incoming feedback; good and bad. They also arrange information meetings and distribute updates of the project status on social media.

## DIGITAL TOOLS

In recent years there has been a significant evolution in the construction industry with regards to digitalization.

Implementation of digital tools and workflows has led to increased production efficiency, time savings regarding documentation, and democratization of data, thus opening for data-driven decisions to further improve efficiency, HSE and cost reduction.

It is essential to have good reporting and documentation tools, especially when the production volumes are as substantial as they are for Skanska Norge. One key success factor for the project has been the utilization of Ditio and DRIV.

### DRIV

DRIV is a Skanska in-house developed software led by Skanska Digital in collaboration with Skanska's tunneling department. It is developed to give all relevant users a complete overview of the tunneling work tailor made to the user's roles and responsibilities. The solution is built on top of the Microsoft Azure framework (Microsoft, 2022).

The basis for the solution is to capture all data regarding tunnel production via APIs (Tyson, 2022) (Application Programming Interface) and manual input. All work processes are digitized, and both client and contractor can monitor the progress in real-time with updated construction plans and quantities produced. The solution consists of 4 main modules and 10+ microservices/app's, Figure 13 shows the dataflow in DRIV.

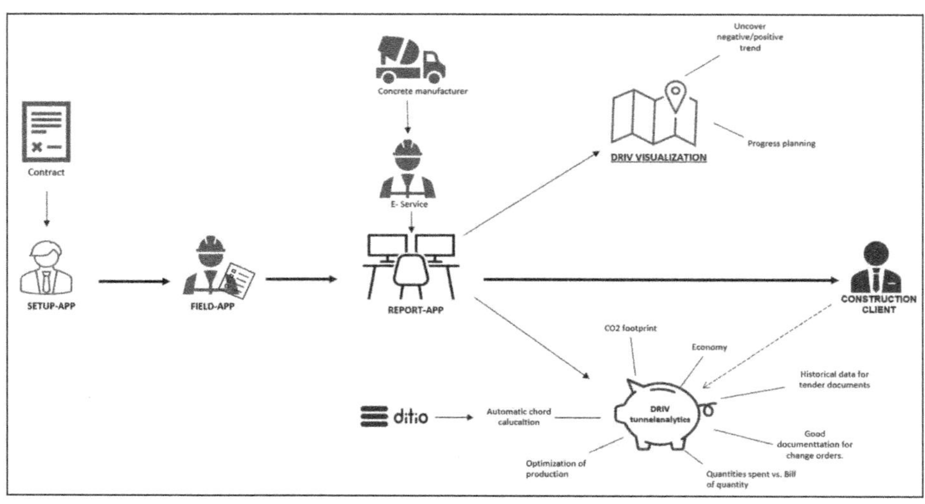


Figure 13. Dataflow in DRIV from SETUP to construction client

## The Modules

1. SETUP is the starting point for all projects. In this module, the projects are established, process codes from the contract are inputted and populated, tunnels, tunnel faces, and site areas. This module manages user rights, project-specific information, and connection to the shotcrete supplier.
2. FIELD is where large amounts of data are collected. The field module is the interface for blue-collar workers and has both offline and online support. This makes it easy to document the work while being in the tunnels with variable wifi coverage.
3. REPORT is the primary user interface for supervisors and production managers. All data registered in the FIELD module is transferred to the REPORT module. This is where the supervisors check and approve data. The production manager creates quantity reports in this module listing all produced quantities in a defined timeframe. These reports are then digitally transferred to the client.
4. INSIGHT consists of several different dashboards and apps, gathering for internal and external users allowing a complete overview of progress and activities. The solutions are also connected to execution plans and geolocated data (GIS data). A progress map is rendered based on the FIELD data merged with Skanska's GIS solution. This progress map is of great value to project management for task such as forward planning and monitoring negative and positive trends. With the help of this INSIGHT map, it is possible to see if the individual tunnel face or project is ahead or behind plan in terms of days.

A dashboard in real-time shared with the client shows produced quantities that have been verified and approved by all parts.

## DITIO

The software DITIO is used for man-hour registration and NMR (Near Misses report) reporting. All data is linked to the WBS structure, thus adding an extra dimension to the dataset.

Skanska Norge is the majority shareholder of Ditio with 40% stake and has been involved in the development and helped shape the solution. DITIO consists of a mobile application for field use and a desktop solution.

Each project worker, including subcontractors, checks in and out of the construction site daily via the field app on their mobiles. DITIO gives the project a complete overview of the location/subproject for each person on the project.

DITIO provides the project management with a real-time overview of the man-hours worked on the project. DITIO also handles NMR and checklist reporting. With the help of GPS coordinates, the HSE advisors can easily find their way back to the location where the incident occurred and provide rapid feedback to the person who notified them through the App. The solution complies with the current General Data Protection Regulation (GDPR).

The solution also consists of a News Feed and information center. This ensures fast and efficient distribution of project news and relevant safety information. The flow of communication is seamlessly in one place and immediately when users have downloaded the App and gained access.

All information from DITIO is available to other relevant applications through an API. NMR and mass transport data will flow into the inspection solution, GISIS, that Skanska uses to monitor production and achieve drawing-free production on civil projects.

## CHALLENGES

The scope of the New Water Supply project is large both in contract sum and workforce, with the added complexity of being situated in a densely populated area.

In a project of this scale, some challenges are to be expected. The main challenges the project has faced during the construction time is transportation, planning both for the day-to-day activity, but also to consider stakeholders, and blasting in densely populated areas.

To work efficiently and safely at the same time, excellent planning, tools, and logistics are needed. One of the main bottlenecks in the production line is identified as the transportation of the hard-rock. On one hand, it is important to maximize production for each day, on the other hand, it is necessary to have a predictable production such that subcontractors can arrange for the necessary resources to transport the muck/rock mass. On average, during the project, a truck leaves the site every 3 minutes and drives out on one of Oslo's main highways, driving 41.2 km to the muck deposit.

Many tunnel faces demand good organization, planning, and coordination to maintain high production. An essential part of success was choosing the right people for the job.

Such as a transport coordinator, an intermediary between subcontractors and Skanska, who ensured sufficient communication between the two parties and have been a key to success with alleviating the transport bottleneck. A resource that controls all the trucks, which tunnel face to drive to and when. The main goal of the transport coordinator was to get the rock out efficiently after the blasting. Because of the high level of temporarily employed truck drivers, a sole proprietorship and language barrier, this was a difficult task, but also one of the essential tasks on the project.

It was important in the planning and excavating phase to consider the surroundings and other stakeholders. To be able to succeed it was important that people showed understanding for the job and accept the temporary negative impact. To reduce the negative impact, several mitigating measures was implemented.

- To prepare the neighbors, a message was sent out half an hour before the blast went off.
- Time restrictions for blasting between 11 PM to 7 AM
- In certain areas, drilling and scaling were also restricted at night.
- Monitoring of blast-induced vibration on infrastructure nearby.
- Visual inspection of surface when grouting
- Traffic controller at pedestrian crossings
- Stop in transport between 07:30 AM to 08:30 AM in consideration to pedestrian going to school and work

In some places, the overburden was less than 4 meters, so it was essential to plan the blasts with respect to the induced vibrations. At one point, one of Oslo's main roads and railway lines was situated in the blasting area. The blasting had to occur at night when the fewest road users and trains passed the intersection between the blasting area and transportation network. The blasts were performed with reduced length and split tunnel faces. It took 9 weeks to pass this intersection of 102 meters.

One of the main access points is located directly underneath a house, Figure 14. Making the first 30 meters of the tunnel challenging both regarding the neighbors and the stability of the house. The neighbor was offered an alternative accommodation but refused to move. This resulted in a lot of coordination as the residents had to be evacuated for every blast. To monitor the building stability, 21 survey points were installed in front of the house and on foundation wall. They were monitored 24/7 by a multistation.

## SUMMARY

At 85% completion grade, it looks like the project will deliver eight months before the date of completion in contract, at a lower cost for the client, and with a BREEAM Infrastructure grade "Very Good," despite the challenges encountered during the

Figure 14. House located on the top off the access point, red open circles indicate the position of the prism and orange dots the position of the tilt sensors, the curved red line shows the tunnel portal

construction phase. This is made possible because of the skilled and versatile workforce working in the tunnel. At the same time, the day-to-day production in the tunnel was only possible with the coordinating and organizational efforts of white-collars and the constructive relationship with the client.

## REFERENCES

AMV AS. (2022, 11 18). *AMV FL70 Hybrid*. Retrieved from AMV AS Loading: https://www.amv-as.no/loading.

AMV AS. (2022, 11 18). *AMV Grouting Unit Conatiner Based*. Retrieved from AMV AS: https://www.amv-as.no/grouting-unit-container-based.

AMV AS. (2022, 11 18). *AMV Grouting Unit Skid Based*. Retrieved from AMV AS: https://www.amv-as.no/grouting-unit-skid-based.

Berggren, A.-L., Nermoen, B., Kveen, A., Jakobsen, P., & Arild, N. (2014, 05 *). Excavations and support methods. *Norwegian Tunneling Society*, pp. 35–43.

BRE Group. (2022, 11 18). *BREEAM Infrastructure Vesion 6*. Retrieved from BBRE Grop: https://bregroup.com/products/ceequal/the-ceequal-technical-manuals/ceequal-version-6/.

Circkle K. (2022, 12 13). *Circle K*. Retrieved from Circle K Milesbio HVO 100: https://www.circlek.no/bedrift/drivstoff/milesbio%C2%AE-hvo100.

Epiroc. (2022, 11 18). *Boomer E*. Retrieved from Epiroc: https://www.epiroc.com/sv-se/products/drill-rigs/face-drill-rigs/boomer-e.

Epiroc. (2022, 11 18). *Boomer XE/WE*. Retrieved from Epiroc: https://www.epiroc.com/sv-se/products/drill-rigs/face-drill-rigs/boomer-xe-we.

ITC SA. (2022, 11 18). *ITC 315 SL*. Retrieved from ITC SA: http://itcsa.com/index.php/en/machine-range/itc315-sl.

*Microsoft*. (2022, November 15). Retrieved from well-architected: https://www.microsoft.com/azure/partners/well-architected#well-architected-benefits.

Oslo Kommune. (2022, 11 18). *Slik bygger vi ny vannforsyning*. Retrieved from Oslo Kommune (Vann og avløp): https://www.oslo.kommune.no/vann-og-avlop/ny-vannforsyning-oslo/slik-bygger-vi-ny-vannforsyning/#gref.

Pettrém, M. T. (2022, Mai 06). Vannmangel i Oslo: Har Skrudd av fontener og toner ned vårvaksen. *Aftenporsten*. Retrieved from https://www.aftenposten.no/oslo/i/5GdWGW/vannmangel-i-oslo-har-skrudd-av-fontener-og-toner-ned-vaarvasken.

SINTEF. (2022, 11 18). *About SINTEF—Applied research, technology and innovation*. Retrieved from SINTEF: https://www.sintef.no/en/sintef-group/this-is-sintef/.

Tyson, M. (2022, November 15). *Infoworld*. Retrieved from what-is-an-api-application-programming-interfaces-explained: https://www.infoworld.com/article/3269878/what-is-an-api-application-programming-interfaces-explained.html.

# Record-Setting Tunnel Boring Below Lake Ontario at the Ashbridges Bay Outfall Tunnel

**Doug Harding** ▪ Robbins

## ABSTRACT

The 3.5 km long Ashbridges Bay Outfall in Toronto, Ontario, Canada was a challenging drive set below Lake Ontario. After a remote machine acceptance due to the global pandemic, an 8 m diameter Single Shield machine launched in March 2021 from an 85 m deep shaft and began its bore in shale with limestone, siltstone and sandstone. During excavation, the TBM and its crew bored a city-wide record of 30 rings in one day, or 47 m of advance. This paper will cover the unique project, from TBM acceptance through to launch, tunneling in difficult conditions, and completion in 2022.

## INTRODUCTION

The Ashbridges Bay Treatment Plant Outfall (ABTPO) project in Toronto, Ontario, Canada involves construction of a new 3.5 km long TBM-driven outfall in sedimentary rock. The completed outfall connects to fifty (50) in-lake risers to enable efficient dispersion of treated secondary effluent over a wide area of Lake Ontario, making this project the largest outfall in the country. The project for the City of Toronto improves the city's shoreline and Lake Ontario's water quality by replacing a 70-year-old existing outfall.

Outfall components (see overview map in Figure 1 and Figure 2) consist of an 85 m deep, 16 m diameter onshore shaft constructed adjacent to the shoreline; a 3,500 m

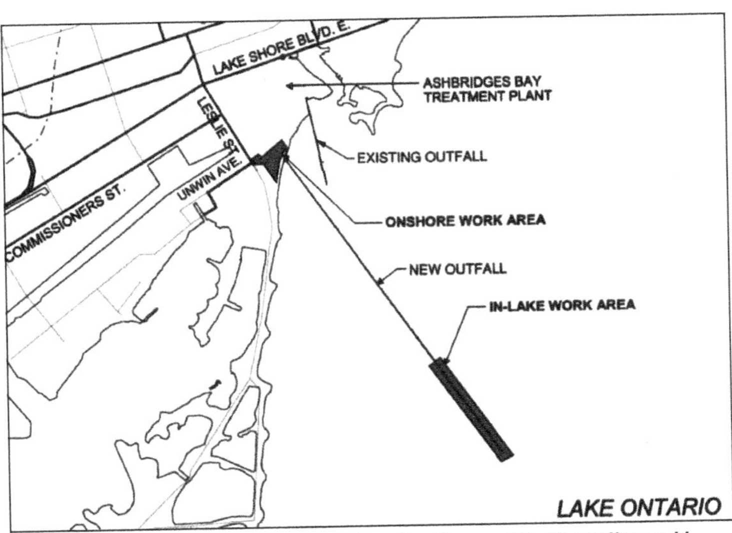


Figure 1. Ashbridges Bay Outfall map of works. (Image: http://tunnelingworld.com/project-show-case/ashbridges-bay-wastewater-outfall-tunnel/)

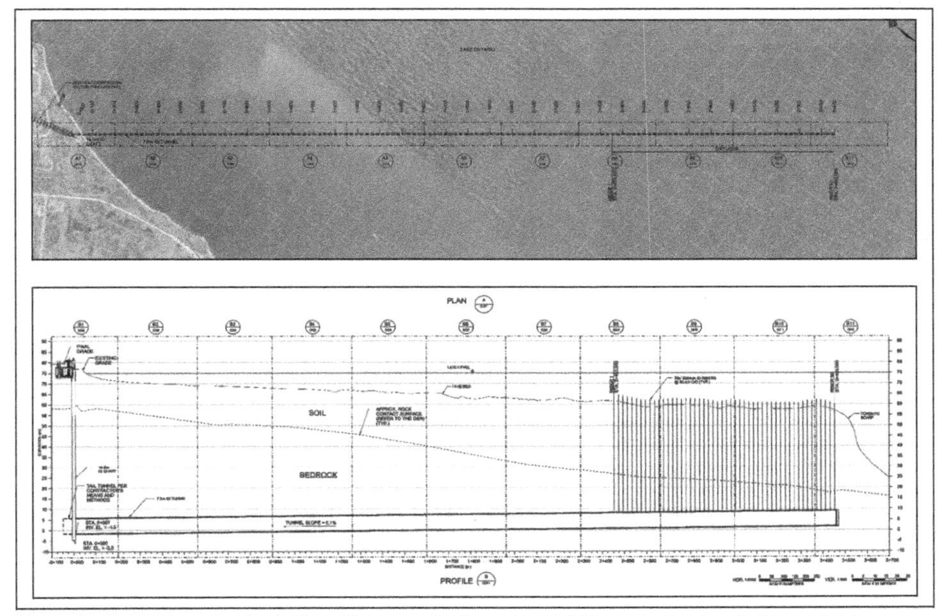


**Figure 2. Ashbridges Bay Outfall profile—running tunnel and diffuser shafts at end of tunnel drive**

long, 8 m diameter segmentally lined tunnel constructed through rock beneath the lakebed; and fifty (50) 1,000 mm diameter risers with 830 mm diameter ports installed in line with the tunnel and extending from the tunnel horizon to the lakebed at equal spacing along the diffuser (Solecki et al., 2022).

The project also includes the construction of a new effluent conduit that will convey treated and disinfected effluent from the ABTP to the new outfall for dispersion into Lake Ontario through the risers. Construction of the ABTPO project commenced in 2019 and is anticipated to be completed by the end of 2024. The main parties involved for the City of Toronto project include consultants Hatch/Jacobs/Baird and Contractor JV Southland/Astaldi. In 2022, the shaft and starter/tail tunnel were excavated, the tunnel was completed by TBM tunnelling, and all 50 risers were pre-installed from the lakebed. Work remaining on the project includes completing all tunnel riser connections, final cast-in-place lining for the shaft and starter tunnel, and conduit connections with the treatment plant.

## Geology

The tunnel alignment (see Figure 3) is situated entirely within the Georgian Bay Formation Shale (GBFS) which is described as a greenish to bluish grey non-calcareous shale. The shale is a fissile rock with widely spaced vertical or inclined jointing and closely spaced sub-horizontal bedding planes interbedded with limestone, siltstone and sandstone. As per the GBR, the average UCS values along the tunnel alignment baselined that the GBFS on average is "weak," according to the ISRM Classification System. In addition, specific groundwater baselines during TBM tunneling were stated in the GBR as follows: (1) 100 L/min transient inflow and (2) 15 L/min steady state conditions measured after 48 hours from initial excavation (Solecki et al., 2022).

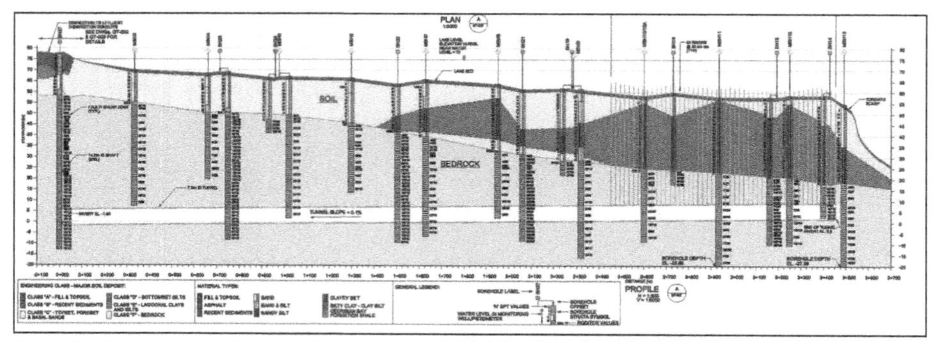


**Figure 3. Geology along the tunnel alignment—Ashbridges Bay Outfall profile**

## Project Sequence

The outfall construction sequencing involved, among other elements, sinking a shaft adjacent to the shoreline and then mining a tunnel through rock beneath the lakebed below Lake Ontario. The most important construction sequencing constraint required all risers to be installed, grouted, and tested for leaks before tunneling within 100 m of any riser (see Figure 4). Another constraint involved probing ahead of the TBM to assess the potential for water inflows. If flows were encountered during this probing, pre-excavation grouting would be performed to improve rock quality and limit water ingress into the tunnel. Following TBM mining, connections to the pre-installed risers would be made prior to outfall tunnel flooding.

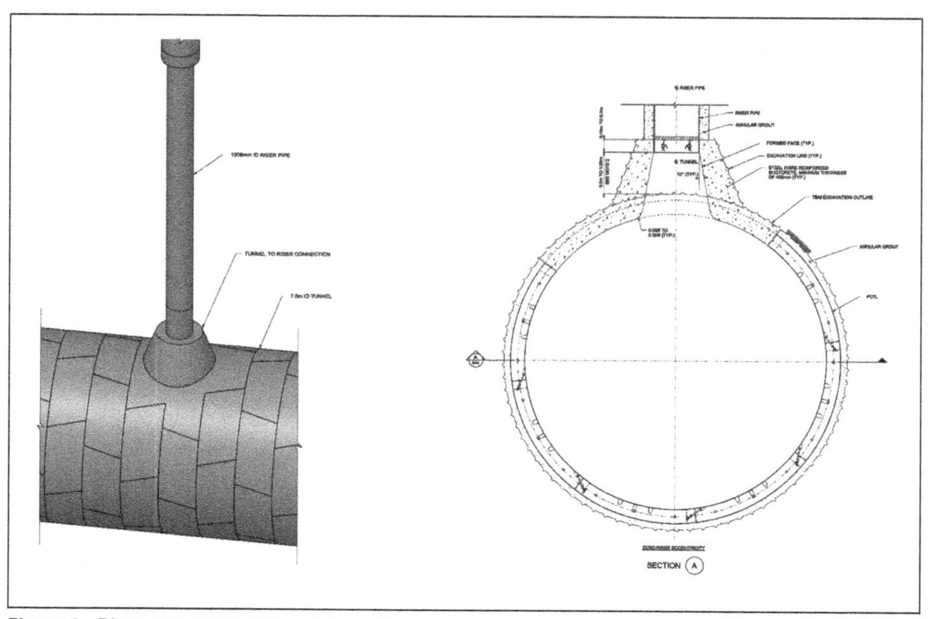


**Figure 4. Riser connections to main tunnel**

## TBM PROCUREMENT AND REBUILD

A Single Shield TBM type was decided on due to the requirement for a single-pass Precast Concrete Tunnel Lining. Other minimum TBM & tunnel requirements including TBM advance probing, two-component annular grouting injected from the TBM tail-shield, gas monitoring, refuge chamber and provisions for TBM pre-excavation grouting, Class 1, Division 2 compliance, CSA and Canadian Electrical and Safety compliance and certification were also specified to manage the anticipated conditions and potential project risks.

Per the contract design the tunnel would be a "dead end tunnel" and the TBM was to be abandoned after completion of the 3.5 km long tunnel drive. There were provisions in the TBM design to remove the usable components should there be time in the schedule to do so.

Upon contract award, the contractor procured a 7.95 m diameter Single Shield, High Performance (HP) Tunnel Boring Machine (TBM) manufactured by Robbins. The tender specifications allowed for use of a rebuilt TBM, and one was selected that was previously used at an 8.7 m excavated tunnel diameter in a Crossover XRE TBM configuration for both rock and soft ground excavation at the Túnel Emisor Poniente (TEP) II project located in Mexico City, Mexico (see original configuration in Figure 5) Design of the machine allowed both excavation in rock "open mode" and then converted to EPB "closed mode." The machine was converted to EPB mode for its final section of tunnel.

Robbins began remanufacturing and modifying the TBM for Ashbridges in the Robbins Mexico facility located just outside Mexico City. The Ashbridges machine incorporated new TBM shields and a 17-inch disc cutterhead as well as utilizing components from the previous project in Mexico. Robbins USA and Robbins Mexico worked together to perform the refurbishment work, which occurred directly before and during the height of the COVID-19 pandemic.

Tender specifications required that if a rebuilt machine was to be utilized, as a minimum the following major components had to be installed as "brand new": the main bearing, cutterhead electric drive motors, cutterhead gear drives and main bearing sealing system. In addition, the remanufactured machine had to comply with ITAtech Guidelines on Rebuilds of Machinery for Mechanized Tunnel Excavation—ITAtech Report No. 5, May 2015 and a new machine warranty had to be provided.

Additional machine features were added to comply with the TBM specification, and as the project was determined to be in "potentially gassy" conditions included the

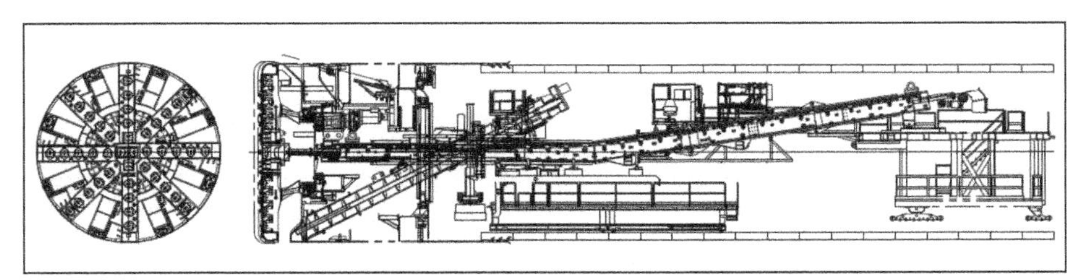

**Figure 5. The TBM in EPB Mode after completion of the Túnel Emisor Poniente (TEP) II project**

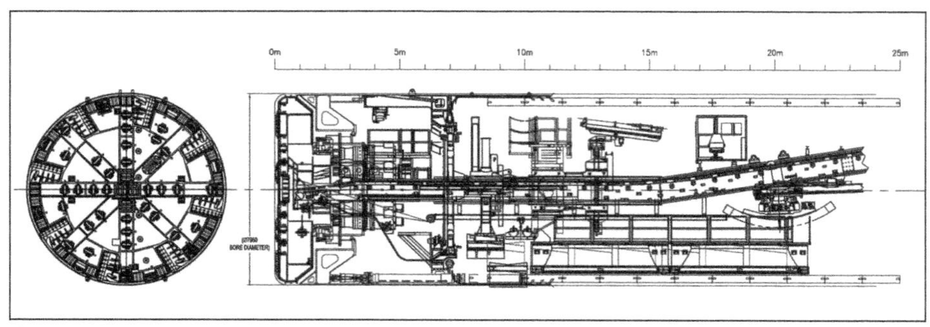


Figure 6. TBM configuration for the Ashbridges Tunnel project

| GENERAL SPECIFICATIONS | |
|---|---|
| TRC PROJECT NUMBER | 15889 |
| MACHINE NUMBER | SS234-280-4 |
| **SEGMENT LINING** | |
| DIAMETER (OUTER / INNER) | 7,660 mm / 7,000 mm |
| NOMINAL LENGTH / THICKNESS | 1,500 mm / 330 mm |
| NUMBER OF SEGMENTS IN RING | 6 |
| MAXIMUM SEGMENT WEIGHT | 5,400 kg |
| **CUTTERHEAD** | |
| BORE DIAMETER (NEW CUTTERS) | 7,950 mm |
| CUTTER TYPE | Ø17 in, BACKLOADING |
| CUTTER QUANTITY | 49 |
| BUCKET QUANTITY | 8 |
| WATER SPRAY NOZZLES | 6 |
| **MAIN BEARING** | |
| DIAMETER, TYPE | 3,910 mm, 3-AXIS |
| RATED L10 LIFE | +12,000 HOURS |

| MAIN THRUST | |
|---|---|
| MAXIMUM THRUST FORCE | 46,694 kN @ 345 bar |
| EXCEPTIONAL THRUST FORCE | 56,845 kN @ 420 bar |
| RECOMMENDED OPERATING THRUST | 6,464 kN |
| THRUST CYLINDER STROKE | 2,500 mm |
| STROKE SENSOR QUANTITY | 4 |
| **ARTICULATION** | |
| ARTICULATION CYLINDER FORCE | 23,683 kN @ 345 bar |
| ARTICULATION CYLINDER STROKE | 250 mm |
| STROKE SENSOR QUANTITY | 4 |
| **CUTTERHEAD DRIVES** | |
| TOTAL POWER | 7 x 330 kW = 2,310 kW |
| NOMINAL CUTTERHEAD SPEED | 0 - 7.1 RPM |
| MAXIMUM TORQUE | 3,410 kNm @ 0 - 6.5 RPM |
| EXCEPTIONAL (BREAKOUT) TORQUE | 5,116 kNm |

Figure 7. TBM

machine being fitted with gas monitors and supplied to Class 1, Division 2 specifications. In addition, due to the concern that water might leak into the tunnel from Lake Ontario, extra capacity for probing and grouting was built in around the machine shield periphery and through the face of the machine. Onboard drilling and grouting capabilities gave 360-degree ground detection capabilities as well as capacity to treat the ground if water was detected (see Figures 6 and 7).

In addition, remote data monitoring was enabled on the machine, for performance and drilling data, that was accessible from any location. Robbins monitored this data and worked with the contractor to troubleshoot any issues encountered during boring.

## Virtual Factory Acceptance Testing

By the time the assembly team reached the factory acceptance testing phase, much of the world was on Covid lockdown and travel was restricted between the U.S., Canada, and Mexico where the teams were based. Because of this a virtual TBM factory acceptance test was performed online for the first time ever, where personnel walked through the machine while it was in Mexico and showed each component performing the required testing (see Figure 8).

The virtual testing required coordination by all involved—consultant Hatch provided a virtual testing platform via Onsite Librestream software and Microsoft Teams, while Robbins and Southland/Astaldi secured fast broadband connections and set

Figure 8. The Ashbridges TBM during the virtual machine acceptance in Mexico

up multiple video feeds on the TBM for concurrent testing capabilities. The virtual testing also allowed more individuals including those from the project owner to take part in the test.

## TBM LAUNCH AND PERFORMANCE

Once the machine had been shipped to Canada from Mexico via ocean freight, it had to be hoisted and lowered in pieces down the 14 m diameter × 85 m deep shaft, which was located on the shores of Lake Ontario. TBM Assembly (see Figure 9 and 10) then commenced in the starter tunnel which was constructed to facilitate the TBM at shaft bottom.

Figure 9. Starter tunnel at shaft bottom

Robbins conveyor systems were supplied and included the 3.5 km long tunnel conveyor, vertical conveyor installed in the shaft, and a stacker conveyor on the surface for muck storage and haulage—the system would need to move over 174,000 cubic meters of material to build the tunnel. A large grout plant was also built at the surface to support the significant grouting operation behind the TBM tail shield while placing segments. Each ring required well over 5 cubic meters of grout material per ring (see Figure 11).

At the same time that the launch and initial tunneling were underway, crews were working to install the 50 in-lake risers from barges on the surface of Lake Ontario. Due to the weather conditions in the Northeast, there was a limited time to install the risers before winter and storm season set in.

Figure 10. Cutterhead being lowered down shaft

Figure 11. Project site at the surface with continuous conveyor configuration

## Encountering High Groundwater Inflows

The machine was officially launched in March 2021, and began boring in good ground conditions. The machine achieved a city-wide record of 30 rings in one day, or about 47 m of advance. The machine and crew surpassed a previous best day of 21 rings at a project with similar specifications (see Figure 12).

About 2.5 km into the tunnel, the machine hit its first section of significant groundwater. This was followed by another section about 3.2 km in. Both sections had high enough water inflows (>400 lpm) to significantly damage the pre-cast concrete lining. In one instance a sustained water inflow in excess of 100 lpm was measured per hole (see Figure 13). The holes were left to drain and there was no evidence of reduced flow over a period of time. It was therefore contemplated that water inflows could have a direct connection to the lake above the tunnel alignment. At these locations,

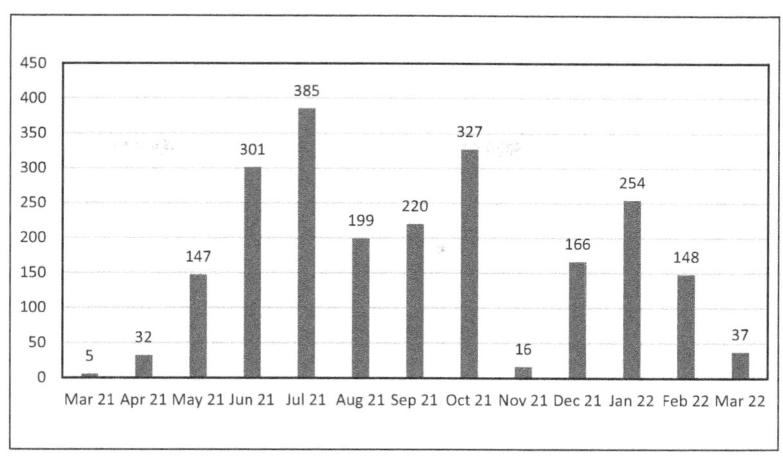


**Figure 12. Total rings erected during tunneling (each ring is 1.5 m long)**

**Figure 13. Significant, sustained water inflows from the rock formation above the tunnel crown (Solecki et al., 2022)**

the ground conditions consisted of approximately 15 m of GBSF rock cover above the tunnel and 35 m of overburden deposits. The distance from tunnel springline to lake level was approximately 70 m. In some holes, packers were installed to confirm groundwater pressures ranging between 1.7 to 3.1 Bar. When these conditions were first encountered, damage to the pre-cast concrete lining occurred in the crown of the tunnel consisting of structure cracks (>25 mm in thickness) and movement to the rings (steps & lips).

### *Solutions and Remediation Work*

Once the high-water inflows and initial pre-cast concrete lining damages were encountered, the contractor was directed to temporarily stop TBM mining to allow the designer to assess the conditions and integrity of the tunnel. It was determined that additional segmental lining repair was necessary consisting of crack repairs, channel installation within the crown, and rock bolt installations. The rationale for this required repair was to stabilize the rings to ensure axial load transfer was restored in the ring. Two channels per segment (C150×12.2), nominally spanning between the 10 o'clock to 2 o'clock position, were installed to temporarily stabilize the segments. Depending on the damaged segment and the orientation of the installed ring, up to 6 bolts per segment (1.8 m in length and 32 mm OD injectable anchors using TPH TD Solid Seal TX Resin) were installed.

Once the segments were repaired, the focus was then shifted to ensure that the annular gap between the lining and excavated rock was fully filled. Various check holes completed and numerous gaps (voids) in the shoulders and crown of the tunnel were discovered. The re-use of the two-component cement grout used during TBM mining was unlikely to be able to withstand the water inflows and pressure. Implementing an alternative PCTL annular grouting method was deemed necessary.

After contemplating several products that would meet the project requirements, a polyurethane chemical grout was selected as the preferred annular backfill grout. This chemical grout was a closed cell polyurethane foam used for filling voids, sealing larger volume leaks, and stabilizing the ground. Field trials were performed to select the necessary dosage of accelerant. The chemical grout was injected at closely spaced ports (⅜" diameter holes) around the periphery of the pre-cast concrete lining to completely fill the annular gap.
Ultimately, the following methods were successfully adopted to get through the sections of groundwater and to mitigate risk during the rest of tunneling:

1. Install a chemical grout collar (see Figure 14) around the first ring immediately outside the TBM shield. In total approximately 50–70 drill holes were installed around the complete ring, with approximately 2 to 4 liters of chemical grout injected per hole.
2. Mine and build the next ring while carefully examining movements and damages to nearby rings. During TBM mining, continue with injecting two-component annulus grout through the TBM shield while increasing the B-component accelerant.
3. Pre-install channel supports on the ring built within the TBM shield. It was determined that the pre-installation of channel supports prior to the ring leaving the TBM mitigates segment damage, and damage to lips and steps in the event that the lining moves further due to lack of annular confinement.

**Figure 14. Chemical grout collar radial injection ports to stabilize ring along with steel channel**

4. Monitor pre-cast concrete lining damage and have material and equipment available to repair segments as needed (crack repairs, rock bolts, additional channels, etc.).
5. Perform additional check holes and proof grouting using both the two-component grout and chemical grout. Ensure all segments are stable and the annular gap is full prior to mining the next ring set.
6. Repeat the pattern until the TBM successfully mines through the encountered ground feature.

### Breakthrough and Beyond

Upon breakthrough in March 2022, the machine shields and other components were left in place at the end of the tunnel, as there was no way retrieve them through the lined tunnel. Along with the good excavation rates in areas of good ground, cutter wear was low as the rock hardness was relatively weak; in fact only eight cutters were changed during the course of boring.

Work remained after the initial excavation: the contractor was required to excavate through the segmental lining and rock to connect the tunnel to the bottom of each of the 50 risers. Several mobile, working gantries are being used in this process with platforms to support specialized demolition equipment, as well as equipment to support the ground and perform final shotcrete lining in those 50 riser areas.

## CONCLUSIONS

Before and during tunneling, the Ashbridges Bay Outfall tunnel met with challenges—these challenges were met head-on by the team and innovative solutions developed. Whether a new way of virtually testing a TBM, or a solution for sections of water inflows, the project team quickly reacted to the situation by working closely together to develop a plan. The successful implementation of the plans demonstrates that challenging and adverse conditions can be safely and efficiently overcome by working closely as one diverse team.

## ACKNOWLEDGMENT

The author thanks Joe Savage for his contribution to this paper.

## REFERENCE

Solecki, A., Waher, K., and Kramer, G. 2022. TBM Tunnelling Challenges and Managing High Water Groundwater Inflows on the Ashridges Bay Treatment Plant Outfall Project. *Proceedings of the Tunnelling Association of Canada (TAC) 2022 Conference*.

PART 

# Health and Safety & Sustainability

*Chairs*

**Micheal Lang**
Frontier-Kemper Constructors Inc

**Kaveh Talebi**
Jay Dee Contractors Inc

# Carbon Footprint Reduction in Montreal Blue Line Extension

**P. Jarast** ▪ AECOM
**M. Bakhshi** ▪ AECOM
**V. Nasri** ▪ AECOM

## ABSTRACT

The Montreal metro Blue Line Extension toward the northeast consists of the construction of five new underground stations. The project includes 6 kilometers of the tunnel which will be constructed using a Tunnel Boring Machine (TBM). In line with the commitment to integrating sustainability best practices, Envision verification is pursued this project. The Envision reference framework was developed to cover all the sustainable development aspects of an infrastructure project and each phase of its life cycle (planning, design, construction, operations and maintenance, and end-of-life). As a part of this endeavor, low-carbon concrete and shotcrete mixtures are proposed for the linings of TBM and SEM tunnels as well as stations and auxiliary structures. The $CO_2$ emission of the low-carbon concrete and shotcrete mixtures were compared with their typical mixtures and the total $CO_2$ emissions reduction related to the tunnel lining is determined. The study is of strategic significance for achieving emission reduction in the tunnel industry.

## INTRODUCTION

Carbon footprint analysis is becoming more and more popular in every industry due to increasing concerns about global warming and greenhouse gas (GHG) emissions. The construction industry is a major producer of $CO_2$ emission. Huang et al (2018) reported that in 2009, the total $CO_2$ emission of the global construction sector (which includes building and infrastructures) was 5.7 billion tons, contributing 23% of the total $CO_2$ emissions produced by global economic activities. In the U.S., the construction processes generate the third highest greenhouse gas (GHG) emissions among industrial sectors. It is therefore important that this sector significantly reduces its consumption of energy and materials. One of the most relevant activities within this sector is tunneling construction. The significant impact of the construction industry largely arises from the embodied carbon in the primary construction materials—cement and steel and utilizing various types of high energy-consuming equipment. Research shows that more than 80% of the $CO_2$ emissions in the construction phase of a tunnel are attributable to the construction materials cement and steel. Thus, reducing the need for cement and steel can make a significant contribution to reducing $CO_2$ emissions. This comes along with a significant reduction in construction costs.

Several studies have focused on the calculation of emissions related to tunnel construction (Li et al. 2011, Miliutenko et al. 2012, Huang et al. 2013, Fremo 2015, Huang et al 2020, Lee et al 2016, Xu et al. 2019).

This paper estimates the portion of produced carbon emission of the project related to tunnel lining. Using a low-carbon concrete mixture instead of a typical one, the saving in $CO_2$ emission is evaluated. The comparison includes a baseline concrete mixture with ordinary Portland cement (OPC) and rebar reinforcement with a low-carbon concrete mixture with fiber reinforcement. The results of this study can be used as a reference for $CO_2$ emission calculation of tunnel projects. It also helps to have a better quantitative idea about the carbon emission of the tunnel liner.

## PROJECT INFORMATION

The PLB project consists of 5 stations and a TBM tunnel. The tunnel is approximately 4 km long. It is 15 m deep on average and will be excavated through rock.

In the baseline design a arc shape section with Cast in Place (CIP) concrete lining using Ordinary Portland Cement (OPC) concrete mixture was proposed. To reduce carbon emissions, an alternative case was suggested. In this case, while a similar section was assumed, CIP lining was replaced with low-carbon shotcrete lining.

TBM bored tunnel is another option instead of arc shape CIP section. For this case, an external diameter of 9.3 m with 7+1 precast segmental lining system and 1.8 m width was proposed. For this case, changing from an OPC concrete mixture into a low-carbon concrete mixture for prefabricated segments is suggested.

In the following detail of tunnel geometry and concrete mixtures for baseline and low-carbon cases are presented.

### Baseline Design

#### *Tunnel Geometry*

Figure 1 shows the tunnel geometry in the baseline design. It consists of three components. 75 mm temporary shotcrete, 40 cm CIP lining, and 32 cm thick invert.

#### *Concrete Mixture Designs*

The baseline concrete mixture consists of ordinary Portland cement (OPC) with no slag and silica fume. Table 1 presents the mixture design for temporary shotcrete, CIP arc, and invert.

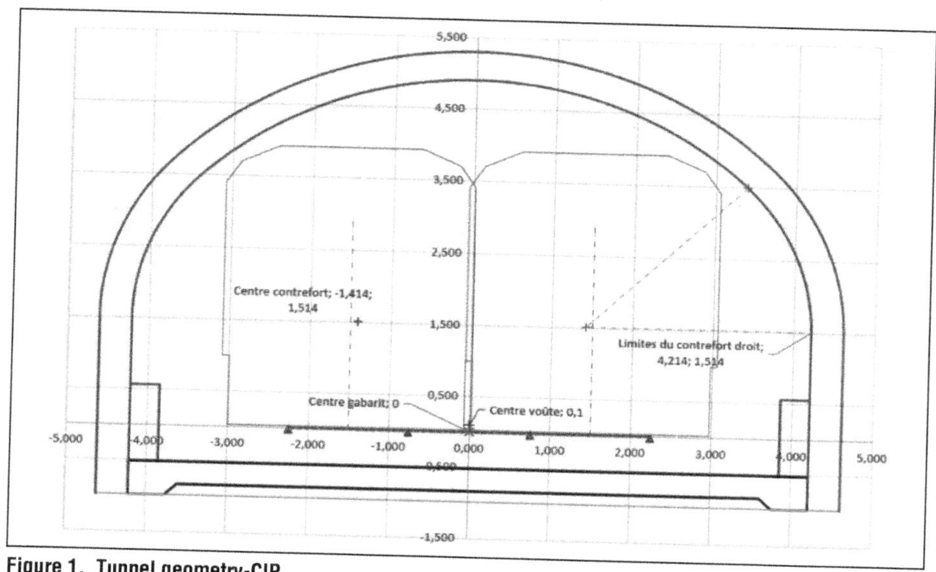


**Figure 1. Tunnel geometry-CIP**

### *$CO_2$ Emission Calculation*

The $CO_2$ emission of a material can be calculated using the following formula:

$$\text{Mass of } CO_{2eq} = CO_{2eq} \text{ Factor Material quantities} \quad \textbf{(EQ 1)}$$

$CO_2$eq Factor is the amount of equivalent $CO_2$ per kg of material from Environmental Product Declarations (EPDs). In this paper, EPA report to Congress EPA 530-R-08-007, Life 365 version 2.0, and Bath University's Inventory of Carbon & Energy (ICE) was used as the reference.

Table 2 summarizes the detail of $CO_2$ emission calculation for the unit volume of baseline concrete mixture designs. The quantities of materials are picked from Table 1.

Based on the geometry of the tunnel presented in Figure 1, the volume of shotcrete, CIP arc, and invert per 1 m of the tunnel can be calculated. Having these quantities together with the information presented in Table 2, the total $CO_2$ emission per 1 m length of the tunnel can be estimated. Table 3 summarizes this calculation.

In the calculation for the shotcrete volume, the thickness of the shotcrete is adjusted for over-excavation tolerance. Considering drill and blast as the excavation method, the over-excavation tolerance would be 0.5 m and so the average adjusted thickness can be calculated as:

$$t_{adjusted} = 75 + 500/2 = 325 \text{ mm} \quad \textbf{(EQ 2)}$$

**Table 1. Baseline concrete mixture design**

| | Cementitious Material, kg/m³ | Portland Cement, kg/m³ | Aggregate, kg/m³ | Water/ Cement | Rebar, kg/m³ | Fiber, kg/m³ | Admixtures, kg/m³ | Total, kg/m³ |
|---|---|---|---|---|---|---|---|---|
| Temporary shotcrete | 475 | 475 | 1,430 | 0.38 | 20 | — | 4.5 | 2,108.6 |
| CIP arc | 415 | 415 | 1,800 | 0.41 | — | 62 | 3.94 | 2,451.1 |
| Invert | 415 | 415 | 1,786 | 0.41 | — | 76 | 3.94 | 2,451.1 |

**Table 2. $CO_2$ emission of baseline concrete mixture designs**

| | | Baseline Temporary Shotcrete | | Baseline-CIP Arc Concrete | | Baseline-Invert Concrete | |
|---|---|---|---|---|---|---|---|
| **Components** | **$CO_2$ eq Factor*** | **Mass/m³, kg** | **$CO_2$ eqs/m³, kg** | **Mass/m³, kg** | **$CO_2$ eqs/m³, kg** | **Mass/m³, kg** | **$CO_2$ eqs/m³, kg** |
| Portland Cement | 0.92 | 475 | 437 | 415 | 382 | 415 | 368 |
| Admixtures | 1.67 | 4.5 | 7.5 | 3.94 | 6.6 | 3.94 | 6.3 |
| Aggregate | 0.006 | 1430 | 8.6 | 1,800 | — | 1,786 | — |
| Steel bar | 1.85 | — | — | 62 | 114.7 | 76 | 140.6 |
| Fiber | 0.86 | 20 | 17.2 | — | — | — | — |
| | | **Total** | 470.3 | | 513.9 | | 539.7 |

**Table 3. Total embodied carbon footprint per 1 m length of the tunnel-baseline design**

| | Thickness, mm | Volume, m³ /1 m tunnel | $CO_2$ eq/m³, kg | $CO_2$ eq /1 m length, ton |
|---|---|---|---|---|
| Temporary shotcrete | 325 | 5.9 | 470.3 | 2.8 |
| CIP arc concrete | 400 | 7.03 | 513.9 | 3.61 |
| Invert concrete | 320 | 2.854 | 539.7 | 1.54 |
| | | **Total** | | 7.9 |

## Low-Carbon Concrete Design

As an alternative solution, the temporary shotcrete and CIP arc were replaced with reinforced shotcrete. In the following, similar calculations presented for baseline design in section 2.1 is presented.

### *Tunnel Geometry*

Figure 2 shows the tunnel geometry for low-carbon. In this design, the CIP is replaced with reinforced shotcrete. The components consist of 50 mm temporary shotcrete, 100 mm permanent shotcrete lining, and 32 cm thick invert. The shotcrete was reinforced with 3D 65/35BG fiber and the invert has the same mixture design as reinforced shotcrete.

### *Concrete Mixture Designs*

To reduce the carbon emission while having the same strength properties, Portland cement was replaced with limestone Portland cement and 27% of the Cementous material was replaced with slag and silica fume (22% slag and 5% Silica fume which is considered moderate Supplementary Cementitious Materials (SCM)). This mixture was proposed for both shotcrete and invert. Table 4 presents this concrete mixture design.

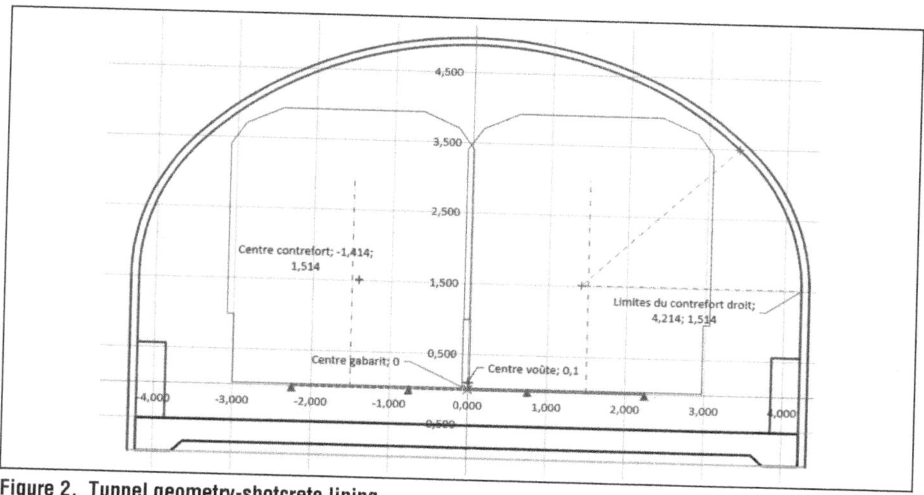


**Figure 2. Tunnel geometry-shotcrete lining**

**Table 4. Shotcrete mixture design-SCM**

| | Cementitious Material, kg/m$^3$ | Portland Cement, kg/m$^3$ | Slag, kg/m$^3$ | Silica Fume, kg/m$^3$ | Aggregate, kg/m$^3$ |
|---|---|---|---|---|---|
| Shotcrete-SCM | 475 | 346.75 | 104.5 | 23.75 | 1,430 |

| | Water/Cement | Rebar, kg/m$^3$ | Fiber- Fiber-3D 65/35BG, kg/m$^3$ | Admixtures, kg/m$^3$ | Total, kg/m$^3$ |
|---|---|---|---|---|---|
| Shotcrete-SCM | 0.38 | — | 40 | 4.51 | 2128.6 |

Table 5. $CO_2$ emission of SCM concrete mixture design

| Components | Shotcrete Mixture-SCM | | | |
|---|---|---|---|---|
| | $CO_2$ eq Factor | Mass/m³, kg | $CO_2$ eq/m³, kg | % Replacement by Mass |
| Portland Limestone Cement | 0.85 | 346.75 | 295 | |
| Slag | 0.1466 | 104.5 | 15 | 22% |
| Fly Ash | 0.093 | 0.0 | — | 0% |
| Silica Fume | 0.014 | 23.75 | 0.3 | 5% |
| Admixtures | 1.67 | 4.51 | 7.5 | 1% |
| Aggregate | 0.006 | 1430 | 8.6 | |
| Steel bar | 1.85 | — | — | |
| Fiber-3D 65/35BG | 0.86 | 40 | 34.4 | |
| | | Total | 361 | |

Table 6. Total embodied carbon footprint per 1 m length of the tunnel-Low carbon design

| | Thickness, mm | Volume, m³ /1 m tunnel | $CO_2$ eq/m³, kg | $CO_2$ eq/1 m Length, ton |
|---|---|---|---|---|
| Temporary shotcrete + Shotcrete lining | 160 | 1.8 | 361 | 0.65 |
| Invert concrete | 32 | 2.854 | 361 | 1.03 |
| | | | Total | 1.68 |

### *$CO_2$ Emission Calculation*

Using the same methodology as presented in *Baseline Design: $CO_2$ Emission Calculation*, the embodied carbon footprint for the unit volume of SCM concrete mixture is calculated. Table 5 summarizes the detail of this calculation. The quantities of materials are picked from Table 4.

Based on the geometry of the tunnel presented in Figure 2, the volume of shotcrete, CIP arc, and invert per 1 m of the tunnel can be calculated. Having these quantities together with the information presented in Table, the total $CO_2$ emission per 1 m length of the tunnel can be estimated. Table 6 summarizes this calculation.

In the calculation for the shotcrete volume, the thickness of the shotcrete is adjusted for over-excavation tolerance. Using the road header for excavation together with robotic shotcrete and real-time in-situ 3D laser scan, the over-excavation tolerance would be 0.02 m and so the average adjusted thickness can be calculated as:

$$t_{adjusted} = 150 + 20/2 = 160 \text{ mm} \quad \textbf{(EQ 3)}$$

Comparing the total $CO_2$ emission of this case with the baseline, the total saving is 79%.

## TBM Tunnel with Prefabricated Segmental Lining Design

### *Tunnel Geometry*

Figure 3 shows the tunnel geometry of the TBM segmental lining design. It consists of 7+1 segments 1.8 wide.

### *Concrete and Grout Mixture Designs*

**(a) Concrete design.** The baseline concrete mixture consists of ordinary Portland cement (OPC) with no slag and silica fume. To reduce the carbon emission while having the same strength properties, Portland cement was replaced with limestone Portland cement and 27% of the Cementous material was replaced with slag and silica fume (22% slag and 5% Silica fume). Table 7 presents these concrete mixture designs.

**(b) Backfill/Infill Concrete.** The baseline backfill/infill concrete consists of Ordinary Portland Cement (OPC) with no slag and silica fume. In the carbon-reduced case, 50% of the Portland cement was replaced with slag. Table 8 presents the backfill/infill concrete mixtures for these two cases.

**(c) Tail-Void Backfill Grout.** The tail-void backfill grout design is presented in Table 9. The same design is used both for the baseline case and the carbon-reduced case.

### *$CO_2$ Emission Calculation*

Using the same methodology as presented in *Baseline Design: $CO_2$ Emission Calculation*, the embodied carbon footprint for the unit volume of OPC and SCM concrete mixtures and grout mixture is calculated. Table 10 to Table 12 summarizes the detail of these calculations. The quantities of materials are picked from Table 7 to Table 9.

**Table 7. OPC and SCM concrete mixture designs**

| | Cementitious Material, kg/m³ | Portland Cement, kg/m³ | Portland Limestone Cement, kg/m³ | Slag, kg/m³ | Silica Fume, kg/m³ | Aggregate, kg/m³ |
|---|---|---|---|---|---|---|
| Segment-OPC | 475 | 475 | — | — | — | 1,430 |
| Segment-SCM | 475 | — | 346.75 | 104.5 | 23.75 | 1,430 |

| | Water/Cement | Rebar, kg/m³ | Fiber- 4D 80/60BGP, kg/m³ | Admixtures, kg/m³ | Total, kg/m³ | |
|---|---|---|---|---|---|---|
| Segment-OPC | 0.38 | 80 | — | 4.5125 | 2,168.59 | |
| Segment-SCM | 0.38 | — | 40 | 4.5125 | 2,128.59 | |

**Table 8. OPC and SCM backfill/infill concrete mixture designs**

| | Cementitious Material, kg/m³ | Portland Cement, kg/m³ | Slag, kg/m³ | Aggregate, kg/m³ | Water/ Cement | Water, kg/m³ | Admixtures, kg/m³ | Total, kg/m³ |
|---|---|---|---|---|---|---|---|---|
| Baseline concrete (OPC) | 220 | 220 | — | 1950 | 0.8 | 176 | 2.09 | 2,348.09 |
| Carbon reduced concrete | 220 | 110 | 110 | 1950 | 0.8 | 176 | 2.09 | 2,348.09 |

**Table 9. Tail-Void grout mixture design**

| | Cement + Bentonite+ Stabilizer, kg/m³ | Portland Cement, kg/m³ | Bentonite, kg/m³ | Stabilizer, kg/m³ | Water/ Cement | Rebar, kg/m³ | Admixtures, kg/m³ | Total, kg/m³ |
|---|---|---|---|---|---|---|---|---|
| Baseline-Grout | 325 | 325 | 42.5 | 5 | 2.55 | 0 | 3.0875 | 1205.59 |

Table 10. $CO_2$ emission of OPC and SCM concrete mixture designs

| | Baseline Concrete Mixture (OPC) | | | | Moderate SCM Concrete Mixture | | |
|---|---|---|---|---|---|---|---|
| Components | $CO_2$ eq Factor* | Mass/ $m^3$, kg | $CO_2$ eq / $m^3$, kg | % Replacement by Mass | Mass/ $m^3$, kg | $CO_2$ eq / $m^3$, kg | % Replacement by Mass |
| Portland cement | 0.92 | 475.0 | 437 | | — | — | |
| Portland limestone cement | 0.85 | — | — | | 346.8 | 295 | |
| Slag | 0.1466 | 0.0 | — | 0% | 104.5 | 15.3 | 22% |
| Fly ash | 0.093 | 0.0 | — | 0% | 0 | 0 | 0% |
| Silica fume | 0.014 | 0.0 | — | 0% | 23.8 | 0.3 | 5% |
| Admixtures | 1.67 | 4.5 | 7.5 | 1% | 4.5 | 7.5 | 1% |
| Aggregate | 0.06 | 1,430 | 8.6 | | 1430 | 8.6 | |
| Steel bar | 1.85 | 80.0 | 148.00 | | — | — | |
| 4D 80/60BGP | 0.88 | — | — | | 40 | 36.4 | |
| | | Total | 601 | | | 363 | |

Table 11. $CO_2$ emission of OPC and SCM backfill/infill concrete mixture designs

| | Backfill/infill (inside the tunnel) OPC concrete mixture-Baseline | | | Backfill/infill (inside the tunnel) OPC concrete mixture- 50% slag | |
|---|---|---|---|---|---|
| Binder | $CO_2$ eq factor* | Mass/$m^3$, kg | $CO_2$ eq/$m^3$, kg | Mass/$m^3$, kg | $CO_2$ eq/$m^3$, kg |
| Portland cement | 0.92 | 220.0 | 202 | 110.0 | 101.2 |
| Slag | 0.15 | 0.0 | 0 | 110.0 | 16.2 |
| Admixtures | 1.67 | 2.1 | 3.5 | 2.1 | 3.5 |
| Aggregate | 0.006 | 1950 | 11.7 | 1,950 | 11.7 |
| | | Total, kg/$m^3$ | 218 | | 133 |

Table 12. $CO_2$ emission of tail-void concrete mixture design

| | Tail-Void backfill OPC grout-Baseline | | | |
|---|---|---|---|---|
| Binder | $CO_2$eq factor* | Mass/$m^3$ (kg) | $CO_2$ eq/$m^3$ (kg) | % Replacement by Mass |
| Portland cement | 0.92 | 325 | 299 | |
| Bentonite | 0.093 | 42.5 | 4 | 9% |
| Stabilizer | 1.31 | 5 | 6.6 | 1% |
| Admixtures | 1.67 | 3.1 | 5 | 1% |
| | | Total, kg/$m^3$ | 315 | |

Table 13 and Table 14 summarize the detail of the total embodied carbon footprint of the tunnel based on the geometry of the tunnel and unit $CO_2$ emission presented in the previous tables for baseline mixture design and SCM mixture design.

Based on Table 14, the total carbon footprint of the two cases, baseline design, and carbon reduced design is calculated as below:

$$Total\ CO_2\ \text{eq/1 m tunnel}_{Baseline\ Design} = 6.7 + 2.2 + 1.4 = 10.3\ \text{ton} \quad \textbf{(EQ 4)}$$

$$Total\ CO_2\ \text{eq/1 m tunnel}_{Carbon\ Reduced\ Design} = 3.6 + 1.4 + 1.4 = 6.4\ \text{ton} \quad \textbf{(EQ 5)}$$

As is seen, by using the carbon-reduced concrete mixture for segments and backfill/infill concrete, the total carbon footprint per 1 m length of the tunnel is reduced by 38%.

Table 13. Total embodied carbon footprint of the tunnel- Segmental lining

| | Baseline Concrete Mixture (OPC) | | | | | | | | |
|---|---|---|---|---|---|---|---|---|---|
| | Ring Width, m | Tunnel Length, m | D-ex, m | D-in, m | Ring Volume, $m^3$ | Total Concrete Volume, $m^3$ | $CO_2$ eq/ $m^3$, kg | $CO_2$ eq/1 m Tunnel, ton | Total $CO_2$ eq, ton |
| 40 cm thickness Rings-OPC | 1.8 | 4086 | 9.3 | 8.5 | 20.1 | 45698.1 | 601 | 6.7 | 27,469.9 |
| 35 cm thickness Rings-SCM | | | | 8.6 | 17.7 | 40210.5 | 363 | 3.6 | 14,592.6 |

Table 14. Total embodied carbon footprint of the tunnel—Backfill/infill and tail void backfill grout

| | Volume, $m^3$/1 m Tunnel Length | Tunnel Length, m | Total Volume, $m^3$ | $CO_2$ eq/ $m^3$, kg | $CO_2$ eq/1 m Tunnel, ton | Total $CO_2$ eq, ton |
|---|---|---|---|---|---|---|
| Backfill/infill with OPC concrete mixture | 10.25 | 4086 | 41881.5 | 218 | 2.2 | 9,113 |
| Backfill/infill with OPC concrete mixture- 50% slag | 10.25 | 4086 | 41881.5 | 133 | 1.4 | 5,550 |
| Tail-Void backfill grout-Baseline | 4.45 | 4086 | 18195.8 | 315 | 1.4 | 5,725.5 |

## CONCLUSION

This paper presents the detail of embodied $CO_2$ emission calculation. As is shown, in CIP tunnel lining by improving the mix design (using limestone Portland cement and 27% supplementary cementitious material) together with using real-time in-situ 3D laser scan, $CO_2$ emission can be reduced by 80% (from 7.9 $CO_2$ kg per 1 m length of the tunnel to 1.7 $CO_2$ kg per 1 m length of the tunnel). For prefabricated segmental lining, by improving the mix design and using steel fibers instead of steel rebars, the $CO_2$ emission can be reduced by 38% (from 10.3 $CO_2$ kg per 1 m length of the tunnel to 6.4 $CO_2$ kg per 1 m length of the tunnel).

Considering the scale of tunnel projects, this reduction in $CO_2$ emission is huge and should not be ignored or underestimated. It should be noted that this emission estimation is a part of the total $CO_2$ emission during the construction phase and produced $CO_2$ by construction equipment and transportation should be added to it to have a complete estimate.

## REFERENCES

Fremo, O., 2015. Life Cycle Assessment of the Byasen tunnel in Trondheim, Norway. Master Thesis. Norwegian University of Science and Technology, Department of civil and transport engineering. Trodheim (Norway), 103pp.

Huang, L., Bohne, R., Bruland, A., Drevland, P., Salomonsen, A., 2013. Life Cycle Assessment of Norwegian standard road tunnel. In: *The 6th International Conference on Life Cycle Management in Gothenburg*.

Huang, L., Drevland, P., Bohne, R., Liu, Y., Bruland, A., Manquehual, C.J., 2020. The environmental impact of rock support for road tunnels: *The experience of Norway. Sci.* Total Environ. 712, 136421.

Huang, L., Krigsvoll, G., Johansen, F., Liu, Y., Zhang, X., 2018. Carbon emission of global construction sector. *Renew. Sustain. Energy Rev.* 81(Part 2) pp. 1906–1916.

Lee, J., Shim, J.A., Kim, K.J., 2016. Analysis of environmental load by work classification for NATM tunnels. *J. Korean Soc. Civ. Eng.* 36 (2), 307–315.

Li, X., Liu, J., Xu, H., Zhong, P., 2011. Calculation of endogenous carbon dioxide emission during highway tunnel construction: A case Study. *International Symposium on Water Resource and Environmental Protection, May 2011, Xian (China),* pp. 2260–2264.

Miliutenko, S., Akerman, J., Bjorklund, A., 2012. Energy use and greenhouse gas emissions during the Life Cycle stages of a road tunnel—*the Swedish case norra lanken. Eur. J. Trans. Infrastruct.* Res. 12 (1), 39–62.

Xu, J., Guo, C., Chen, X., Zhang, Z., Yang, L., Wang, M., Yang, K., 2019. Emission transition of greenhouse gases with the surrounding rock weakened—A case study of tunnel construction. *J. Cleaner Prod. 209*, 169–179.

# Steep TBM Challenge for the Limberg III Pump Storage Power Plant Project

**Karin Bäppler** ▪ Herrenknecht AG

## ABSTRACT

Hydropower produces almost two-thirds of the world's renewable electricity generation and delivers a major contribution on the ambition of the Paris Agreement and Sustainable Development Goals (SDG) with a range of benefits to society and environment. These include clean and flexible generation and storage in addition to reduced dependence on fossil fuels and avoidance of pollutants.

The Limberg III PSP is one of the European projects that addresses the implementation of the UN SDGs. The challenge of the project is the construction of a 770 m long pressure shaft to be excavated with a TBM at an inclination of 42°.

## INTRODUCTION

Austria has set the goal of covering one hundred percent of its energy needs from renewable sources by 2030. The Limberg III pumped storage power plant is intended to provide an important contribution to achieving this target. The project is located near the town of Zell am See (Kaprun municipality), southwest of the city of Salzburg in Austria. The pumped storage power plant expands the capacity of the existing Kaprun power plant group and will produce up to 480 megawatts of climate-neutral electricity by 2025. At the same time, the new pumped storage power plant is intended to ensure that energy can be efficiently temporarily stored and, when necessary, fed into the Austrian and Central European grid within a short time.

The Limberg III pumped-storage project is like Limberg II which began operation in 2011, designed completely underground between two existing storage reservoirs Mooserboden (final level 2,036 m) and Wasserfallboden (final level 1,672 m) and is intended to increase the overall output of the Glockner-Kaprun power station group.

Upon completion, Limberg III will be a power plant whose design is tailored very specifically to the future needs of the energy transition. Special machine sets with variable speed-controlled pump-turbines will be able to react extremely quickly and flexibly to the increasing need for balancing and control energy in the grid. With the increasing amount of energy generation from intermittent sources, such as wind or solar, pumped storage projects like Limberg III are essential to provide energy balancing to achieve grid stability for a safe and affordable electricity supply.

The focus in this publication is on the construction of the 770 m long pressure shaft by means of mechanized tunnelling technology. A 5.8 m-diameter Gripper TBM is used to excavate the pressure shaft with a steep uphill slope of 42° (90%) in calcareous slate.

## PROJECT OVERVIEW AND TUNNELLING REQUIREMENTS

The Limberg III project is a 480 MW pumped-storage power project (PSPP) located in Salzburg, Austria. It has been developed by Verbund Hydro Power (VHP), the biggest hydroelectricity provider in Austria.

Limberg III is a twin of the Limberg II power plant, which was built between 2007 and 2011. Like this, it is built between the high mountain reservoirs Mooserboden (2,036 m) and Wasserfallboden (1,672 m). The project includes the construction of a 3km long and 7.3 m-diameter headrace tunnel along with a 770 m-long and 5.8 m-diameter pressure shaft and a riser. It also involves the development of the Mooserboden inlet and outlet structures, inlet tunnel, valve chamber with access and drainage, pressure tunnel, surge tank, and distribution pipelines.

The pressure tunnel and most of the inlet tunnel of Limberg III run parallel to the headrace tunnel of Limberg II at a distance of about 40 meters. The lower and upper surge chambers of the two surge tanks and the pressure shafts are also parallel.

**Figure 1. Limberg III, project layout with 42° (90%) inclined pressure shaft [1]**

The paper focuses on the construction of the 770 m-long pressure shaft by means of a Gripper TBM. This is technically very challenging due to the steep gradient of 42° (90%) posing highest demands on the safety and performance of the TBM in use.

The 42° inclined pressure shaft has an inclined length of 584.62 m and the riser shaft to the surge tank Oberkammer is 193.38 m long. Both shafts are excavated with a Gripper TBM with an excavation diameter of 5.8 m. The overburden of the pressure shaft is similar to that of the headrace tunnel with a maximum overburden of 428 m and average overburden of about 400 m. The predominant rock type is calcareous slate with predicted rock strengths (UCS) in the range of 80 to 100 MPa. As part of the preliminary exploration for Limberg III, rock samples from the exploratory drilling were examined in the laboratory and CAI values up to 2.6 (very abrasive) and quartz contents of 20-30% were determined. The quartz content was taken into account in terms of occupational safety (suitable dust masks, etc.) and with regard to disc cutter wear.

A consortium of Porr Bau GmbH, G. Hinteregger & Söhne Baugesellschaft m.b.H (50%), Marti AG and Marti GmbH (50%) was awarded the contract for the construction of the headrace and access tunnels, passages, and shafts, along with the excavation of caverns for the Limberg III pumped storage power project in April 2021.

## TBM LAYOUT AND PROJECT-SPECIFIC DESIGN FEATURES

In close cooperation with Marti, Herrenknecht planned and developed the Gripper TBM that is in use for the construction of the 42° steep pressure shaft for Limberg III. The TBM is designed and tailored to this steep slope. A variety of interfaces had to be considered, both in the planning and design phase and also on the jobsite. Safety requirements have had the highest priority throughout.

**Figure 2. Limberg III gripper TBM Ø5.8 m for 42° (90%) inclined pressure shaft construction**

The TBM is designed with an excavation diameter of 5.8 m and a nominal torque of 1,755 kNm. The cutterhead is fitted with 17-inch backloading disc cutters, 25 single and 4 double discs. Wear protection is provided by grillbars, protection wedges and hardox plates on the face and in the gauge area. The cutterhead design meets the highest safety standards and with the backloading system the disc cutters can be replaced in the protected area inside the cutterhead so that workers do not have to work in front of the tunnel face where they could be exposed to the geology and possible rockbursts.

The steep tunnelling demands a reliable solution to prevent the Gripper TBM slipping back. Three clamping systems ensure that the machine is braced in two levels at all times and in all operating conditions. The gripper bracing directly behind the cutterhead is the first clamping level followed by a double anti-reverse lock located on the first back-up, with full backup redundancy of the available bracing levels for the 80 m long TBM having a total weight of 1,000 tons. While the first anti-reverse lock is rigidly connected to the back-up, the second moves along with the advancement in an automated, hydraulic manner so that the back-up moves consistently with the TBM. The operating cycle of the TBM includes advancement, standstill and re-gripping processes and in all these operating conditions at least one anti-reverse lock is always securely clamped against the rock. Always having at least two of three locking systems independently and thus absolutely safely braced against the rock significantly increases safety for personnel, machine and structure in all operating stages. Thus, any slipping back of the machine can reliably be prevented. The anti-reverse locks work mechanically on the principle of a self-locking toggle lever (automatic mechanical wedging) ensuring reliable bracing of the machine against the rock even in the event of a power failure or failure of hydraulic systems.

According to the expected rock behavior, the 770 m long tunnel section is divided into excavation classes in accordance with the Austrian Standard ÖNORM B 2203-2. The project tender documents specify excavation classes and the appropriate tunnel support measures based on the predicted rock types and its geomechanical behavior. If

Figure 3. Installation of rock support in the L1 area—two drill rigs for anchor drilling

the actual tunnel support measures need to deviate from the support standards then they are determined on site by mutual agreement between the client and the contractor.

According to the excavation classes, rock support measures need to be installed in the L1 area immediately behind the TBM and in the L2 back-up area or in the L3 tunnel working area only in the case of unexpected geological conditions. The time necessary for installing the tunnel support is dependent on the support classes. The length of working area L1 is defined as a 30 m distance from the tunnel face, working area L2 as a distance of 50 m behind the working area L1. The contractor is free to shorten the distance of tunnel support installation to the face for mechanical or construction reasons. About 85% of the pressure shaft required a rock support composed of the installation of anchors (3N° at 3 m ccs) in the L1 area and a 50 mm shotcrete support in the L2 area. Two drill rigs are provided on the TBM to install rock anchors in the L1 area with an anchor length of 3,000 mm (extendable to 4,000 mm) over the upper 180 degrees of the tunnel.

The cutterhead has an electrical main drive (Ø3,000 mm) and has an installed power of 1,750 kW (5 × 350 kW). The rock at the tunnel face is composed of calcareous slate and is cut by 17-inch disc cutters. The excavated rock chips are taken by the buckets and are further transported through an opening in the bottom of the cutterhead to a muck chute. From there the muck is further channelled to an intermediate storage in the starter cavern.

An overhead monorail is used to transport consumable supplies the TBM. On the TBM back-up there are locations for the consumables and intermediate storage locations for auxiliary and operating materials.

## ASSEMBLY AND LAUNCHING SITUATION

In September 2019 Marti AG awarded the contract for the design and manufacturing of the Gripper TBM and back-up system for the Limberg III project to Herrenknecht AG.

The factory acceptance took place in the TBMs manufacturer's headquarter in Germany at the beginning of February 2022 and was then directly transported to the jobsite in Kaprun, Austria by trucks.

Figure 4. Gripper TBM assembly in horizontal position short before 50 m vertical curve radius

The first parts of the Gripper TBM and back-up arrived on site at the end of February 2022. The TBM assembly started in a horizontal position in an assembly cavern of 45 m in length, 14 m in width and 11.5 m in height. It was completely assembled in the shaft base cavern in two months. During assembly, the machine was repeatedly moved forward through the starter cavern on steel beams by a pushing system. In the limited assembly area, the next back-up could then be assembled under the crane runway in the rear assembly cavern.

At the end of the horizontal section, a ramp with a vertical radius of 50 m and about 40 m long was built through which the TBM could be pushed into the profiled starting tunnel. The starting tunnel had a diameter of 5.83 m and a length of 22 m.

To launch the TBM the grippers were braced in the profiled starting tunnel with the anti-reverse locks still outside the starting tunnel. The gantries were pushed in the starting position using the cylinders of the back-up displacement system and the external power pack.

The TBM was launched on May 15, 2022 to excavate the 780 m long and 5.8 m diameter pressure shaft.

After breakthrough the TBM will be retracted through the constructed pressure shaft using strand jacks back to the assembly cavern.

## SITE EXPERIENCE TO DATE

Full TBM operation started in May 2022. The geological conditions were relatively good so far and only isolated rock falls needed to be secured with anchors and wire mesh support. Steel arches have not been required. After about 350 m of tunnelling, the geology turned to be slightly worse. Voids were also encountered with larger quantities of water inflow. The rock support in these sections was composed of anchors and wire mesh over the 120° in the crown area using also profiled foil to seal against the water inflow. Steel arches and shotcrete were not required.

## CONCLUSION

The construction of the 770 m-long steep inclined pressure shaft for Limberg III in Kaprun with a gradient of 90% complements Limberg 1 and Limberg 2 power plants that are already in operation with an additional power of 480 MW. The project then supports as an essential power bank for a safe, clean and affordable supply of electricity in Austria and will make an important contribution to avoiding power cuts. With

**Figure 5. Tunnel support using anchors, wire mesh and profiled foil**

the additional power, up to 100 new wind turbines or 100,000 home PV systems can be substituted or supported in seconds.

The project is another milestone in the construction of steep inclined shafts for hydropower projects using proven mechanized tunnelling technology. To date, the contractor Marti AG has successfully carried out steep inclined pressure shafts with tunnelling equipment out of Germany including the two pressure shafts for the Linthal 2015 project with inclined TBM drives of 85% and a 90% steep pressure shaft for the Ritom Hydropower project, both projects in Switzerland.

## REFERENCES

Marti Tunnel AG, Pumpspeicherwerk Limberg III, Kraftwerksgruppe Kaprun (AT), https://www.marti-tunnel.ch/de/Documents/PDF_Referenzen/Limberg%20III_d.pdf.

PART 

# International Projects

*Chairs*

**Boris Veleusic**
Michels

**Steve Kramer**
Cowi North America

# Bad Bergzabern Bypass Tunnel—NATM Tunneling Through Vineyards

**Richard Gradnik** ▪ BeMo Tunnelling GmbH
**Pafos Busch** ▪ BeMo Tunnelling GmbH
**Ralf Plinninger** ▪ Dr. Plinninger Geotechnik

## ABSTRACT

This paper presents the details of the Bad Bergzabern tunnel project. This tunnel is the main part of the B427 federal road bypass, intended to improve the regional traffic infrastructure and to reduce the traffic load in the historic city centre of the health resort Bad Bergzabern. Located in the south-western part of Germany, the project area is situated at the eastern margin of the so-called Palatinate Forest, a low mountain region of mainly triassic sediments with a complex and variable geological history.

The German federal government, represented by the office of mobility of Rhineland-Palatinate, awarded the contract to BeMo Tunnelling, a specialized Tunnelling company with a long history of innovation and success in NATM tunnelling. The contract includes the construction of two separate tubes with a length of approximately 1,44 km (0.9 mi)—the road (main) tunnel with a cross section of 101 $m^2$ (1,090 sq ft) and a parallel rescue tunnel with a cross section of ca. 23 $m^2$ (250 sq ft.) Both are connected by six cross passages. Tunnels and cross passages are excavated according to the principles of the New Austrian Tunnelling Method (NATM).

Due to BeMo Tunnelling's technical expertise and capabilities, the project team was able to engineer a speed-up concept that saves money and time for the construction team and client. This paper presents the innovative approach BeMo Tunnelling made to optimize the execution of the tunnelling works and the challenges the project team is facing with regard to a complicated geological situation, countermeasures against an increasing effect of climate change and volatile prices for construction materials.

## PROJECT BACKGROUND

The project area is located near the city of Bad Bergzabern in the state of Rhineland-Palatinate, Germany. Bad Bergzabern, a medium sized town with approximately 8,500 inhabitants, is situated about 5.5 kilometers (3.5 mi) northeast of the French border and about 40 kilometers (25 mi) northwest of the city of Karlsruhe (Figure 1). Morphologically, the project is located at the eastern margin of the so-called Palatinate Forest, at the transition to the plain of the Rhine Valley. The Palatinate Forest features a multi-national hilly region with mountain ranges reaching up to 670 meters (2,200 ft). Due to occurrence of thermal water, that is brought to the surface by wells from a depth of approximately 450 meters (1,500 ft), Bad Bergzabern enjoys a high reputation in health tourism and was officially rewarded the prestigious title of a "health resort." This, and the high regional popularity of the Palatinate Forest as a recreation area in south-western Germany, lead to a high traffic volume on the federal road B427, which interferes with the interests of residents and guests. Therefore, the decision of constructing a local bypass tunnel was made by the federal ministry of traffic and digital infrastructure and the project was included in the federal transport infrastructure plan to be under traffic by 2030 at the latest.

Figure 1. Location of Bad Bergzabern within Germany (left) and alignment of the Bergzabern tunnel (dotted red line) west of the city center (right)

In 2020 the tender procedure was started and the potential bidders were provided the necessary technical information. Two award criteria were defined, on the one hand dominantly the best price and on the other hand the optimization of the construction schedule. Thanks to careful processing and the implementation of innovative concepts, BeMo Tunnelling was able to convince in both categories and was awarded the construction contract of the Bad Bergzabern bypass in February 2021. The contract has a total value of 71 million euro (incl. VAT).

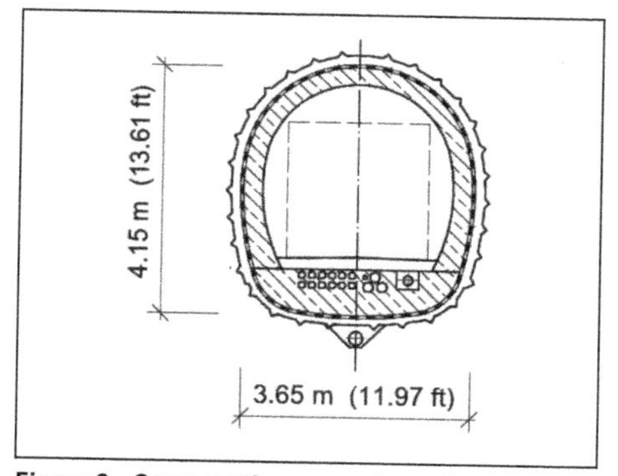


Figure 2. Cross section rescue tunnel

## THE BAD BERGZABERN TUNNEL

The tunneling part of the project consist of two tunnels, a 1.44-km-long (0.9 mi) main tunnel and a parallel rescue tunnel of smaller diameter. Both tubes are connected by six cross passages. The maximum overburden above the tunnel is around 120 m (400 ft). In addition, the contract includes the construction of two km (1.25 mi) of road, water tanks for fire defense and two electrical and mechanical equipment buildings (see Figure 1).

The main tunnel has a length of 1.44 km and is designed as a two-way traffic tunnel with a break out cross section of 101 m$^2$ (1,090 sq ft) (see Figure 3).

Breakdown bays are located 0.4 km (0.25 mi) in the tunnel from both portals. For these the tunnel section is widened up to 193 m$^2$ (2,080 sq ft). The rescue tunnel with an originally planned section of 15–30 m$^2$ (160–320 sq ft) is connected to the main tunnel by 6 cross passages with an excavated cross section of 13 m$^2$ (140 sq ft). The maximum distance between the main tunnel and the rescue tunnel is 12 meter (40 ft).

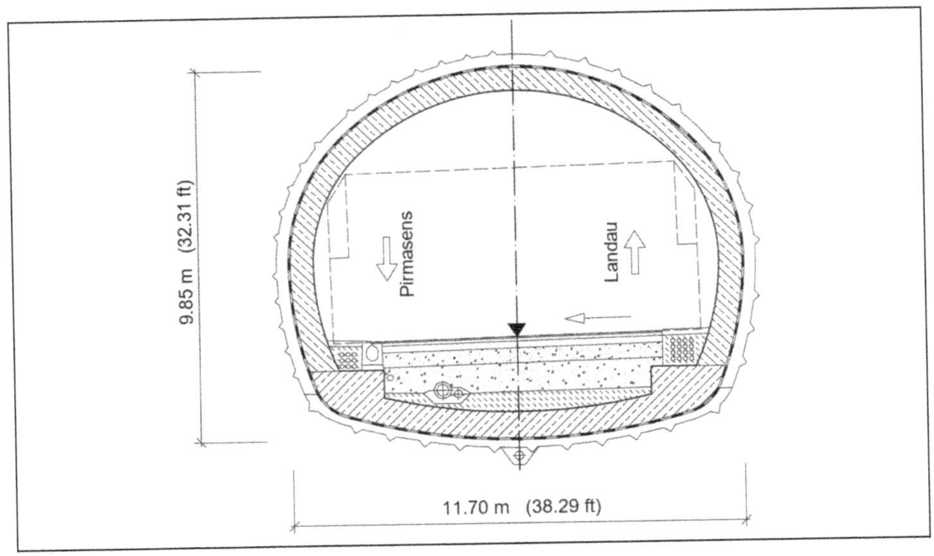


**Figure 3. Cross section main tunnel**

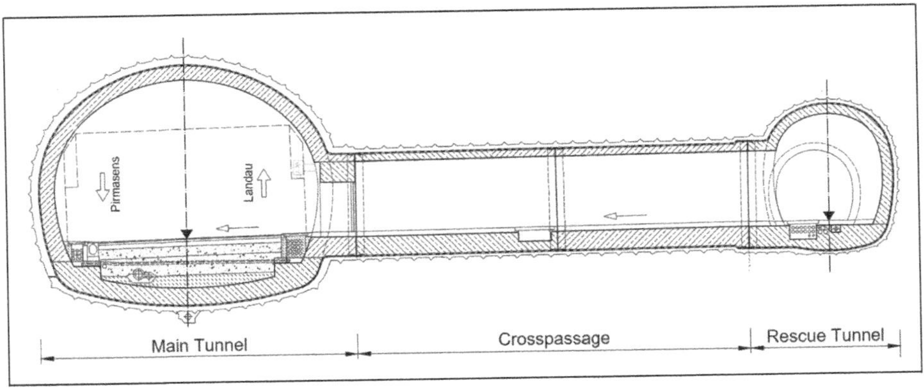


**Figure 4. Longitudinal section through the crosspassage**

## GEOLOGICAL OVERVIEW AND GEOTECHNICAL PROBLEMS

As a result of the complex regional geological history, the geological and geotechnical conditions in the Bad Bergzabern tunnel are highly heterogeneous and variable (Figure 5). In the portal areas, younger quaternary soil deposits (mainly silt, sand, gravel) are encountered, which feature isolated horizons of ground water.

After a short section of marl and limestones of the so-called "Muschelkalk" Formation, the two tubes encounter sandstones and siltstones of the so-called "Buntsandstein" Formation. These sediments have been deposited in a braided river system in the Triassic age, about 250 Mio. years ago. Later, during the opening of the nearby Upper Rhine Graben valley in the tertiary age, the formation was tilted, shifted by numerous faults and partly decomposed by migrating hot fluids. Recent weathering processes have further contributed to the alteration and deterioration of the rock mass.

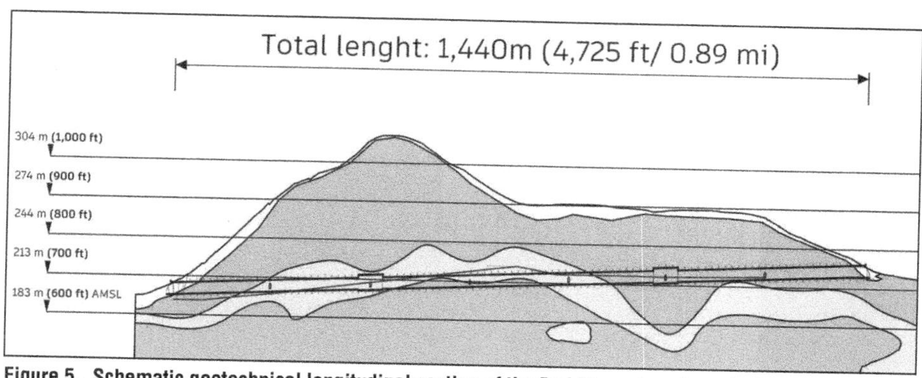


**Figure 5. Schematic geotechnical longitudinal section of the Bad Bergzabern tunnel, showing the different rock quality of decomposed sand (orange), intermediate sandstone and sand formations (yellow) and hard sandstone (green)**

The highly variable, small-scale changes in the degree of decomposition and strength represents one of the main challenges for tunnel excavation. When highly weathered or decomposed, the sandstone shows no or little strength and cohesion and provides only very limited unsupported standup-time. To make things worse, the key properties of the rock mass are structurally not at all linked to decisive elements, like bedding or discrete faults, so that the succession of softer and harder rock within the tunnel advance is more or less unpredictable. This requires a close observation and documentation of the rock mass conditions and frequent changes of excavation method and support classes.

Within the bedrock formation and partially above the gradient of the tunnel, a consistent ground water level exists, which is influenced by precipitation and climate and thus can rise and fall temporary within some feet in height. During excavation, a combination of decomposed sandstone with soil-like behavior and water inflows bears the potential risk of loose sand flowing into the opening. Regarding the long-term operation of the final tunnel, a full waterproofing of the tunnel is necessary as a result of the ground water level and the potential of the water to sinter the drainage.

In order to project-specifically classify different types of rock and rock mass, a system of homogeneous "Ground Units" (German term: "Homogenbereiche") has been established according to the actual German Standards for Subsurface Works. These Ground Units consider the specific properties of soil and rock and are intended to characterize larger volumes of rock with similar technical behavior, regarding for instance excavation, transport, and finally treatment and deposal. During the tunnelling works, the actual Ground Unit is jointly determined and documented by the Engineering Geologists employed on site on behalf of the client and contractor.

The following sections give a concise overview over the different Ground Units encountered at the Bad Bergzabern project.

## Ground Units B1 to B4—Gravel and Hillside Debris

In the portal areas quaternary soils such as sandy gravel or slope debris occur. These soil deposits vary in grain size from gravel to clay. The ground is unstable and needs additional support in order to remain stable. Without immediate support of the face with shotcrete, rock bolts, and grout injection, the face, and thereby the tunnel itself, would immediately collapse.

**Figure 6. Typical appearance of decomposed triassic sandstone/sand of Ground Unit X1. Left image: Sequential and partial mechanical excavation in the crown section of the main tunnel. Right image: Although originally gained in form of a hand specimen, the decomposed sandstone can easily be crushed by hand**

Under these conditions the face has to be stabilized with shotcrete and bolts to prevent a local failure and a pipe roof to prevent a global failure has to be installed. The pipe roof consists of 50 feet self-drilling pipes with a diameter of 140 mm (5.5 in). To create a supporting arch (canopy), the pipes have to be grouted with a cementitious grout mixture. In addition, the side and top have to be supported by systematically installed rock bolts. The excavation starts with the top heading, which is divided in two parts, followed by the bench and the invert A temporary invert of shotcrete has to be installed to ensure distribution of the vertical and horizontal forces by a completed ring. Due to these efforts, the heading in the unstable ground proceeds slowly with an advance length of only three feet per excavation round.

## Ground Unit X1—Loose Sandstone and Sand

Ground Unit X1 describes a type of completely decomposed sandstone / sand with soil-like behavior and non-existent Unconfined Compressive Strength. However, the quartz sand still has a relatively high density and is still highly abrasive. The behavior of the ground is similar to Ground Units B1 to B4. The face needs immediate support with shotcrete and rock bolts to avoid a collapse.

The face has to be stabilized with shotcrete and rock bolts in the face. Additionally, an umbrella of spiles has to be applied. The excavation then starts with the top, while the supportive wedge has to be spared, and continues afterwards with the bench and the invert. In the larger cross-section of the main tunnel, stability issues require excavation in smaller sections of only some square-feet and immediate sealing with shotcrete. As in Ground Units B1 to B4, after each stage of excavation a temporary shotcrete invert has to be installed to create a resistant support ring.

## Ground Unit X2—Weak and Loosened Sandstone

Ground Unit X2 describes weak sandstone series with an Unconfined Compressive Strength of below 15 MPa. The face usually consists of weaker and harder parts, representing a mixed face condition. This results in a more complicated excavation workflow, since some parts of the face are weak enough to be excavated by use of a ripper bucket, while other parts have to be excavated by use of a roadheader or jackhammer. However, the high quartz content of the sandstone implies the risk to generate harmful respirable dust which limits the use of roadheaders in these rock conditions.

Depending on the actual conditions on the face, the stability measures vary significantly. If the face shows a high amount of decomposed rock, the same procedure as in Ground Unit X1 is applied, which means rock bolts in the face, a spile umbrella and construction of a temporary invert. If the face shows more homogeneous rock conditions, less effort has to be put in stabilizing the face. The rock bolts then can be substituted by a supportive wedge of loose rock in the centre part of the face. Also, the length of advance naturally increases with the occurrence of more solid rock to around four feet .

### Ground Unit X3—Alternating horizons of Weak and Hard Sandstone

Ground Unit X3 consists of alternating horizons of weak and hard sandstone. The horizons are strongly bonded together and the rock is generally speaking of a better quality than in Ground Unit X2. As a result of mixed face conditions, both excavation by drill-and-blast and mechanical excavation is applied. As in Ground Unit X2, the stabilization measures are dependent on the actual composition of the face.

### Ground Unit X4—Hard Sandstone

This Ground Unit contains fresh, high strength sandstone, where the only suitable excavation method is drill and blast. Since this Ground Unit includes only rock of high strength and good rock mass quality, the stabilization measures can be divided in "must" and "can" measures. "Must" measures contain shotcrete covering of the walls and rock bolts in the side and top walls. Sealing the face with shotcrete and piling an umbrella of spiles are "can" measures that have to be coordinated with the engineering geologist and geotechnical engineer of the client.

## EXECUTION OF WORK

### Muck-Out Concept

#### *In the Bidding Process*

To place an economical offer, BeMo Tunnelling had to plan and simulate the different stages of the tunnelling works beforehand in order to identify critical work stages and to value risks. Especially the small section of the rescue tunnel included some challenges . There was no possibility to work safely with more than one big machine in the tight space. For the loose rock heading, BeMo decided to use an ITC 120, a combined tunnel heading and loading machine. The ITC machine is equipped with an excavation bucket and a conveyor belt for high-speed mucking. The excavated material is directly loaded to a dumper via the conveyor belt. For the rescue tunnels sections where drilling and blasting is performed, the mucking shall be done with a standard wheel-loader, due to restrictions with the bolder size processable with an ITC machine. The loader then transports the material via the next cross passage to the main tunnel and dumps it there for the final muck-out with larger machines. This concept proceeds the works applying a so called "Rottenbetrieb." That means a normal level of machine equipment with a higher than normal level of personal. The effect is a higher utilization of the machines and an increase in productivity in comparison with a standard level of personal and machinery. But in comparison to a work program with twice the machinery and twice the personal, the productivity of this concept is less than 100%.

#### *In Execution*

In the execution phase BeMo Tunnelling brought an acceleration concept to the attention of the client. This optimized concept included a slightly enlarged cross section of

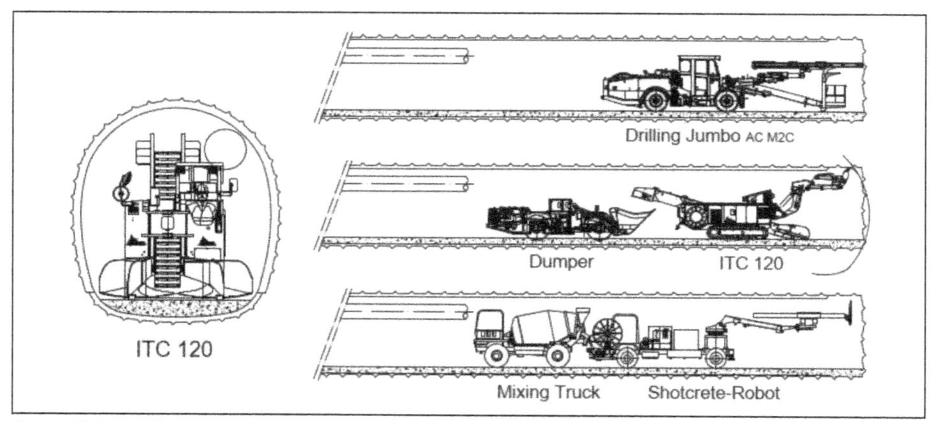


Figure 7. Tunnel works cross section for mucking and shotcrete logistics bidding process

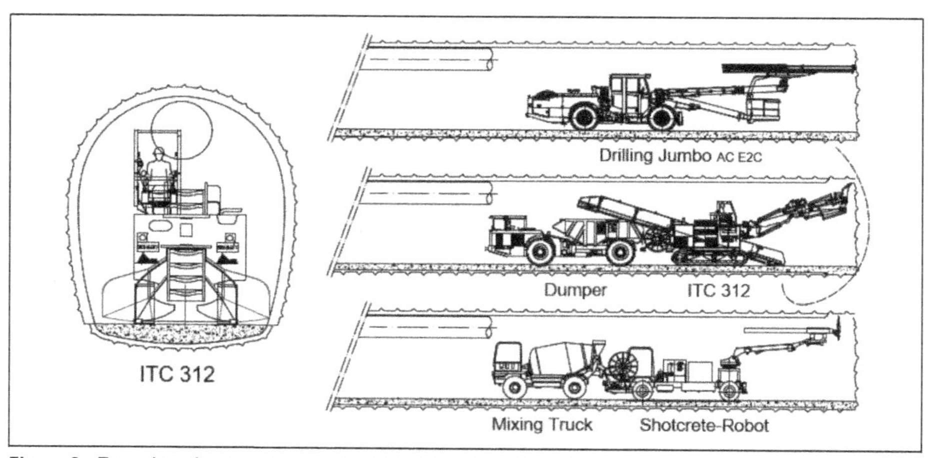


Figure 8. Tunnel works cross section for mucking and shotcrete logistics execution process

the rescue tunnel to make the use of larger machines technically possible. With the client's approval BeMo Tunnelling was now able to use a heavier and more powerful tunnel heading and loading machine, the ITC 312 tunnel excavator (Figure 9). Also, for the drill and blast sections an improvement of the performance given was possible.

With the improved concept dumping the material to the main tunnel by using a wheel-loader is no longer necessary, the material can directly be loaded on a dumper inside the rescue tunnel and transported to the surface without interfering with the operations in the main tunnel.

## Further Advantages of the Enlarged Rescue Tunnel

The enlarged rescue tunnel makes parallel works in the main tunnel possible. Usually, the top heading is finished before starting the excavation of the bench and invert, since the invert of the top heading gets used as a temporary road to transport concrete and rebar to the face—and excavated rock off the face. With the enlarged section of the rescue tunnel it`s possible to supply the top heading through the rescue tunnel and

**Figure 9. ITC 312 tunnel excavator in the Bad Bergzabern rescue tunnel**

the cross passages. This allows for the invert and bench heading to be started with the top heading still being active, since there is no interruption in the supply chain for the top heading. The acceleration in construction is determined to be more than six months, saving the client substantial time-related costs, and taking pressure off the project schedule.

## Sustainable Concrete Supply

The hydration of the cement content in the applied shotcrete is an exothermal reaction that sets free high amounts of thermal energy, what leads to an expansion of the hardening concrete. When the reaction slows down and the hardened concrete shrinks, cracks occur. Especially voluminous structural elements suffer from this effect, due to high temperature differences from the inside to the outside. The cracks can affect the functionality and stability of the applied concrete lining. To counteract this, the concrete gets cooled before installation ("Precooling"). Therefore, cooling down the components like water, cement, and aggregates can either be done individually before mixing or during mixing. Cooling down all components individually leads to high financial and logistical efforts; hence preference is given to cool down the mixture through the water itself and not all the components beforehand. A widespread technique is to add liquid nitrogen to the water. Liquid nitrogen is a by-product in the production of liquid air, that is widely spread used as a coolant across all industries. It has a temperature below –320 °F (77 K), is colorless and enables a low-threshold handling on site. Only an isolated tank is necessary for storage . The liquid nitrogen is then added to the uncooled water to cool it down to the needed, individually calculated temperature. This technique is wide-spread and there is a high level of application experience.

The increasing effects of climate change make the socio-ecologic responsibility of the construction industry obvious. BeMo Tunnelling has recognized this area of tension early and has been working towards a more sustainable approach on construction for a long time. To gain experience with eco-friendly and innovative technologies, the project team at the Bad Bergzabern tunnel project and BeMo's mechanical engineering department have chosen to put the current concrete cooling technique using

**Figure 10. Aerial view of the construction side with the city of Bad Bergzabern behind. The plain visible in the background is the Rhine Valley Graben, the Palatinate Forest rises at the left margin of the photo**

liquid nitrogen to the test. As a result, the decision was made to substitute the existing system by a new, fully electric, water-cooling system. For the new system, the periodic delivery of new liquid nitrogen by conventional diesel-powered trucks is made obsolete, thus benefiting the global climate and the local residents by reducing emissions. The experience and data so far look promising, although the final evaluation must be awaited to issue a recommendation for future projects.

## CHALLENGES FACED

### Dewatering

In February 2022 after approximately 0.42 km (0.26 mi) of excavation, the preceding rescue tunnel unexpectedly encountered decomposed sandstone with soil-like behavior in combination with significant water inflows and a ground water level above tunnel crown. In this situation, the potential risk of loose sand flowing into the tunnel had to be considered (Figure 11). Tunnelling works were immediately stopped in mutual consent of all project participants and additional measures were discussed and finally designed to be able to restart the excavation in a safe way.

With the support of the entire team at BeMo Tunneling, a suitable drainage system was quickly planned and implemented on site. Among others, the decisions included the specific experience, the company had gained during the construction of the 3.5 km-long Kramer tunnel (2.2 mi) near Garmisch-Partenkirchen, Germany, between 2019–2022.

The mechanical engineering department of BeMo Tunnelling provided the necessary machinery equipment on site in a timely manner, while the construction site team carried out immediate work preparation. The technical office, supported by the building maintenance department, was meanwhile able to elaborate a detailed design, which could be approved by the client within a few days. As a result tunnelling works were resumed after less than a week of downtime.

Figure 11. Small-scall flowing of non-cohesive sand (Ground Unit X1) in the face of the rescue tunnel in February 2022

Figure 12. Impression of underground dewatering measures in the Bergzabern tunnel: Left image: Drilling of AT76 drainage boreholes by use of tunnel boomer and rotary percussive drilling equipment in the rescue tunnel. Right image: Drilling of larger diameter vertical drainage well from the crown section of the main tunnel using a HUETTE drill with cased rotary drilling method

The adapted execution plan for the rescue tunnel included an enlargement of the cross-section of four rounds (≈4 m/13 ft) with at least eight drainage boreholes, each approximately 12 m (40 ft) long, using the DSI AT76 3" pipe system, which could be drilled and installed by use of the tunnel boomers tophammer equipment (Figure 12, left image). In order to enable the groundwater level to be lowered below the invert, boreholes were partially drilled downwards and then connected to a vacuum pump system.

When similar conditions were encountered in the much larger main tube, the drainage measures were adapted and effective dewatering was successfully achieved by use of a system of one or more sub horizontal (inclined 30°) 12" drainage pipes, 24 m (80 ft) long ahead of the face and vertical 12" wells with a spacing of approximately 30.5 m (100 ft) and a depth of 14 m (46 ft), drilled from the invert of the crown section (Figure 12, right image). For these longer and larger diameter drillholes a HUETTE special drill rig for cased rotary drilling was employed. Additionally, the DSI AT76 3" pipe system is applied also in the main tunnel, where necessary.

By use of these drainage systems, rescue tunnel and main tunnel could successfully and safely be driven under the given groundwater conditions.

## Material Prices and Availability

Since the beginning of the covid-crisis in 2020 prices for construction materials and for operating goods such as diesel fuel or electrical energy increased strongly. This leads to difficulties and uncertainty in the estimation and bidding process, as the development of prices in long terms can't be predicted. Being a contractor and having handed in fixed prices, makes taking the risk unforeseeable, drastic price developments after the submission of the bid a tough challenge. Public clients have recognized this problem area and implemented a compensation adjustment mechanism based on the development of prices published by the German Federal Office of Statistics.

For this project, the mechanism was also implemented, but is only limitedly effective, as it only covers the steel price and has some further disadvantageous restrictions regarding trigger limits or the difference between the theoretical official prices and the actual prices demanded by the suppliers. This, and the unforeseeable effects of the Russian attack on Ukraine in March of 2022 on the global markets, lead to a problematic situation.

Figure 13 shows the development of the mentioned official price index regarding diesel fuel and heating oil, reinforcement steel and electrical energy. Important milestones of the early project phase are marked on the timeline, as well as a general trend for the price development. In the binding offer BeMo Tunnelling made, prices of goods included in the calculation were as of September 2020. When the project was finally assigned in March 2021 and BeMo Tunnelling was able to close contracts with

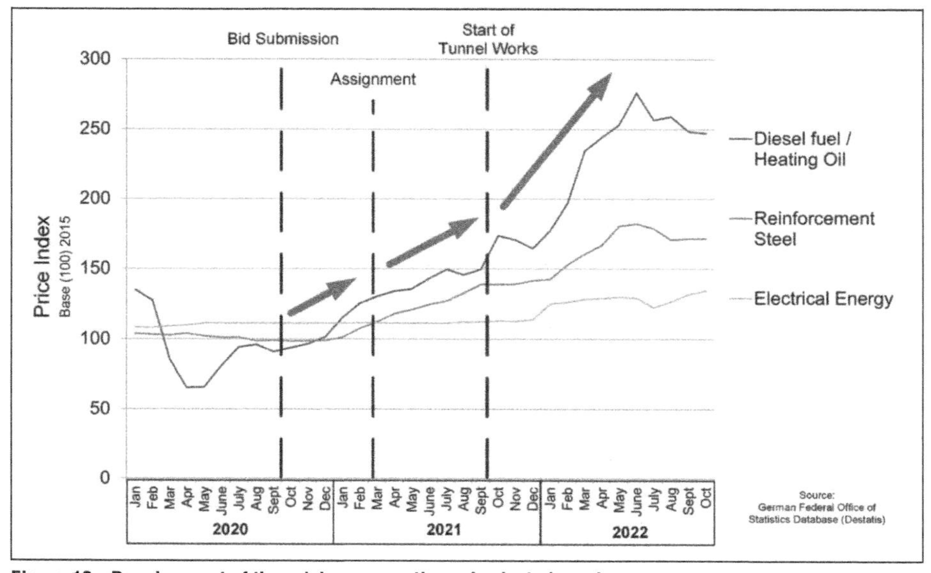


**Figure 13. Development of the pricing versus time of selected goods**

suppliers and subcontractors, prices for fuel, steel and other goods have increased drastically. Until the start of the tunnel works, and still ongoing today, Diesel and steel prices increased even more. In the European market, there are no fixed prices with fuel suppliers, prices for every filling are set in the moment of refuelling based on the market prices. This leaves the project team with the tough challenge of having uncertain, rising prices every day while still having mostly fixe*d billing prices to the client.

But not only the increasing prices are problematic, also the current availability of materials is concerning. As of June 2022, 45% of the German construction companies active in the field of infrastructure and civil engineering complain about shortages of critical and essential goods. This development began with the economic recovery phase after the COVID crisis in 2021, where international supply chains were still interrupted and production facilities weren't able to adjust their capacities to the again increasing demand. The situation was exacerbated by the impacts of the Russian attack on Ukraine, since a considerable share of raw materials used as construction materials such as steel or bitumen are imported from Ukraine or from, now sanctioned, Russia.

As a consequence, the project team takes action to absorb the effects of this development: supplier prices are renegotiated, the billing process and the stockkeeping are optimized. BeMo Tunnelling is confident, that handling the exceptional situation in an economical way is possible.

## CONCLUSION

The construction of the Bad Bergzabern tunnel provides several significant challenges. Difficult and highly variable geological and hydrogeological conditions required a variety of measures to ensure a safe and economical tunnel heading. While in hard and stable rock mass conditions drill and blast tunnelling is possible with only few stabilization measures, tunnelling in loose soil deposits and decomposed sandstone require pipe roofing, combined with a sequential and partial mechanical excavation with additional rock-bolts, intense shotcrete stabilization of the face and eventually additional drainage measures ahead.

The variable rock conditions require permanent adaption and frequent changes between drill and blast and mechanical excavation. Mixed-face conditions generate the need for a combined heading, where only parts of the face can be blasted, while other parts need to be excavated mechanically. Besides the geological challenges the engineers of BeMo Tunnelling were able to speed up the construction by turning in a sophisticated speed-up concept that includes largening the rescue tunnel to move logistical boundaries and forcing points. This results in significant savings in time and money for the client. Out of awareness for climate protection a new electrical concrete cooling system does now substitute a high number of periodical truck rides bringing in liquid nitrogen to the construction site for concrete cooling. The exact savings and the effectiveness of the system is to be reviewed in the future, but the intermediate status is positive. Apart from environmental concerns, the project team faces the impacts of global crises like the rising material and goods prices, as well as a material shortage, due to problematic global supply chains.

Nevertheless, the project team and the involved departments at BeMo Tunnelling meet the challenges and do their best to successfully drive the project forward. The project will get handed over to the client in June 2025.

# Design and Construction of REM Tunnels and Underground Structures

**Verya Nasri** ▪ AECOM
**Agustin Rey** ▪ SNC Lavalin

## ABSTRACT

Once completed, the Montreal Réseau Express Métropolitain (REM) will be the fourth largest automated transportation system in the world. The REM represents construction costs of approximately 7.0 billion Canadian dollars. The project consists of 67 km of twin tracks over four branches connected to downtown Montreal. The project includes 26 stations with 3 underground stations in downtown Montreal. One of the underground stations was built using the NATM method with thin permanent shotcrete initial and final liners separated by a sprayed on waterproofing membrane. The 2 other underground stations were built with the cut and cover approach including one using permanent secant pile wall with permanent walers and struts as the support of excavation and the final perimeter wall of the station. The project also includes the rehabilitation and enlargement of the Mont Royal Tunnel. This 100-year-old double track tunnel is about 5 km long. The REM also consists of 3.6 km new TBM tunnel connecting downtown to the Montreal International Airport through saturated soft ground and karstic rock. By the time of RETC, the construction of REM underground structures will be completed. This paper presents the design and construction aspects of the underground structures of this mega project.

## INTRODUCTION

The Réseau Express Métropolitain (REM) is an electric and fully automated, light-rail transit network designed to facilitate mobility across the Greater Montreal Region in Canada (Nasri, 2022). This new transit network will be linking downtown Montreal, South Shore, West Island, North Shore and the airport (Figure 1). The project consists of 67 km of twin tracks over four branches connected to downtown Montreal. The REM system will connect with existing bus networks, commuter trains and three lines of the Montréal metro (subway). Once completed, the REM will be one of the largest automated transportation systems in the world after Singapore, Dubai and Vancouver.

The project includes 26 stations with 3 underground stations in downtown Montreal. One of the underground stations will be built using the NATM method and the two others with the cut and cover approach. The project also includes the rehabilitation and enlargement of the Mont Royal Tunnel. This 100-year-old double track tunnel is about 5 km long. The REM also consists of 3.6 km new TBM tunnel connecting downtown to the Montreal International Airport through saturated soft ground and karstic rock.

The project is currently under construction by a joint venture of SNC Lavalin, AECON, Dragados, EBC, and Pomerleau and the final design was performed by a joint venture of SNC Lavalin and AECOM. Its construction will be completed at the end of 2024. To deliver this major project, several underground works are undertaken. This paper presents the major underground developments of the REM and the solutions used for the successful achievement of the underground construction objectives which include ensuring safety and stability of the opening during construction and for its full-service lifetime, minimizing impact and disturbance to the surrounding environment, meeting Owner's technical requirements, and minimizing cost, duration and risk of underground construction.

## GEOLOGICAL CONTEXT

Montreal geology consists of a variety of sedimentary horizons dating from the Precambrien, Cambrien and Ordovicien periods. The main associated lithologies are limestone and shale. Intrusive rocks dating from the Mesozoic/Cretaceous period are also encountered throughout Montreal, intersecting the sedimentary packages.

The strata are generally relatively sub-horizontal layers of sedimentary rock. However, events such as faulting, folding, glaciation and isostatic movement have shaped the strata differently in certain region of the island. Faults are generally considered inactive in the region.

Solutions that were put forth for the successful completion of the REM underground works were selected to best match project constraints and local ground conditions. Located in different parts of the city, intersecting different strata, a total of four different types of underground works are undertaken.

## DEEP UNDERGROUND STATION

To connect the deeply sitting REM track to an existing metro station located closer to the surface, an underground station accessible by an approximately 70 m deep vertical shaft is designed (Edouard Montpetit Station, EMP). Figure 2 shows the 3D model of the station with the main entrance shaft, the side platforms constructed by enlarging the existing Mont Royal Tunnel, concourse tunnel, ventilation tunnels and shafts and vertical circulation tunnels and shafts. Figure 3 shows the excavation of the main entrance shaft and tunnels. This station configuration was optimized in order to

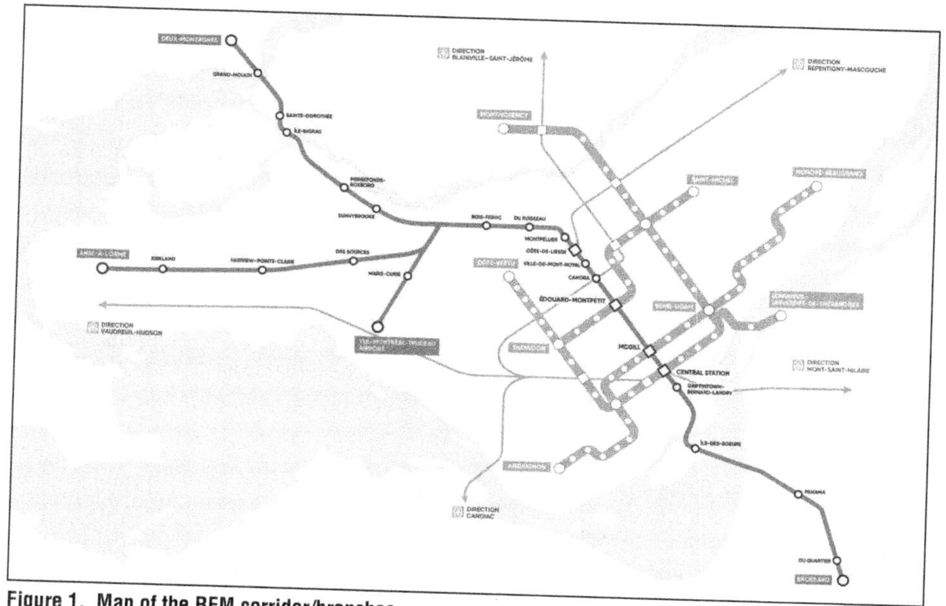


**Figure 1. Map of the REM corridor/branches**

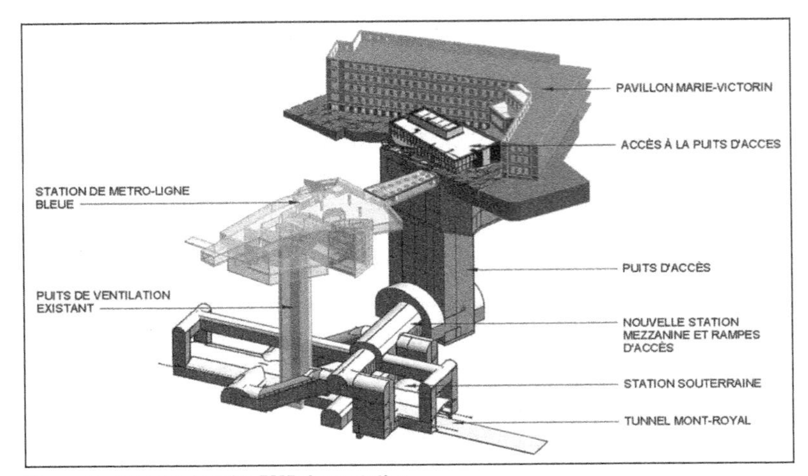


**Figure 2. 3D model of the EMP deep station**

**Figure 3. Excavation of EMP Station entrance shaft and tunnels**

minimize the rock excavation volume. With practically no overburden present in the area, the interchange station is almost entirely located within the Trenton formation which consists of interbedded limestone/shale packages and argillaceous limestone. The station is also located near the intrusive Mont Royal formation which consists of gabbro, monzonite and breccia, resulting in a significant number of hard dikes and sills in the area of interest.

In this particular area, probably resulting from the contact metamorphism due to the close proximity of the Mont Royal intrusive, the sedimentary rock package shows hard rock properties, with uniaxial compressive strength (UCS) varying between 125 and 180 MPa and Young's modulus varying between 75 and 88 GPa. Because of the nature of the work, the quality of the rock and the lower initial cost of the technique, controlled drill and blast method was selected to sink the deep vertical shaft and excavate the underground station.

The excavation took place in a densely populated area with major infrastructure in the near vicinity. Among those infrastructures is one of the main University of Montreal pavilions. This building houses classrooms and laboratories including a recently completed state of the art and highly sensitive acoustic laboratory located within only a few meters of the excavation. Hence, several engineering control mechanisms are put

Table 1. City of Montreal vibration limits in mm/s

| Building Type | Frequency (Hz) | | |
|---|---|---|---|
| | <10 Hz | 10–50 Hz | >50 Hz |
| 1-Commercial | 20 | 20–35 | 35 |
| 2-Residential | 5 | 5–15 | 15–20 |
| 3-Historical | 3 | 3–8 | 8–15 |

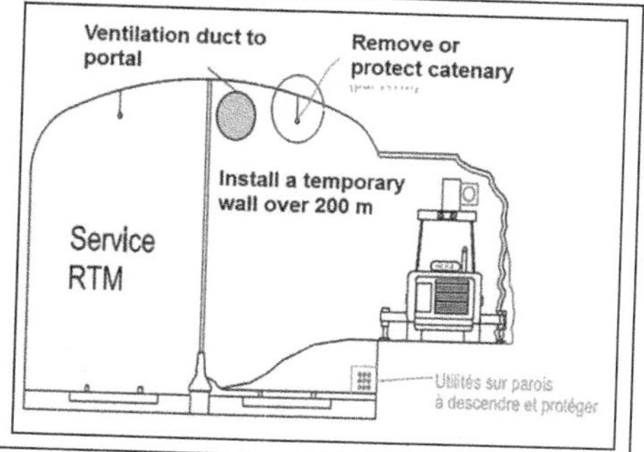


Figure 4. Enlargement of MRT to EMP Station side platforms

in place to minimize impact and disturbance including line drilling technique along the full entrance shaft excavation perimeter and for its entire depth, use of a maximum blast round length of 2.5 m, blasting sequences and patterns designed for low impact (specific blast hole, loading and delay patterns). Table 1 shows the City of Montreal allowable peak particle velocity for different types of building. Type 1, 2 and 3 are for commercial, residential and historical buildings respectively. The line drilling is performed using a DTH drill with the holes diameter of 140 mm placed at 250 mm center to center. The holes center is put at 200 mm from the excavation line to account for the vertical deviation of the drilling operation. To complement this effort, a comprehensive monitoring plan, counting over 150 instruments, was used.

Permanent CT bolts and shotcrete reinforced with steel fiber (Dramix 3D) were used as both initial and final liners for all shafts and tunnels. A layer of 5 cm of flashcrete is applied first for safety, the bolts are installed, spray on waterproofing membrane (BASF Masterseal 345) is added, and then another 5 cm of steel fiber reinforced shotcrete is applied. The liner is designed for a 125-year service life per the contract requirements. During development of the underground excavation, rock mapping is undertaken after each exposure of the final wall. Permanent rock bolting pattern is adjusted based on the ground condition and the need for additional rock bolting is assessed on site to ensure the overall stability of the excavation. In addition to durability, the shotcrete mix is designed specifically for the cold weather application in Montreal.

The EMP Station is built within the existing double track Mont Royal Tunnel (MRT). The side platforms are built by enlarging the existing tunnel. Figure 4 shows the side platform enlargement allowing one track to remain in operation.

## MONT ROYAL TUNNEL REHABILITATION

The Mont Royal Tunnel is a railway tunnel in operation since 1918, third longest in Canada, which connects the city's Central Station (Gare Centrale), located Downtown Montreal, with the north side of the Island of Montreal and Laval, passing through

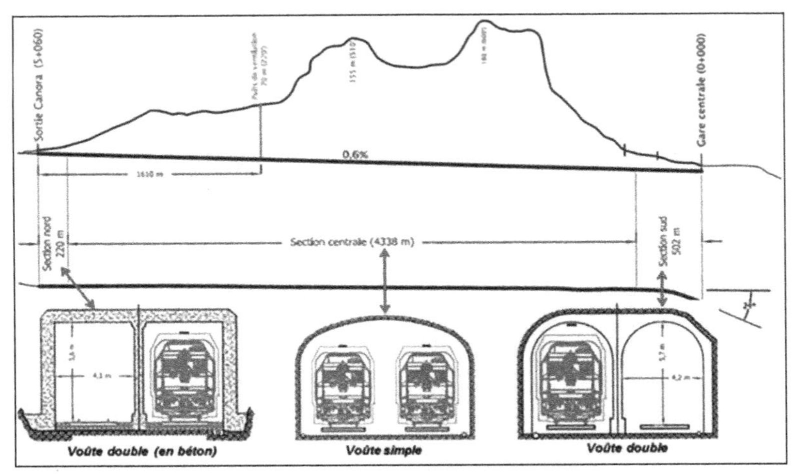


**Figure 5. Existing Mont Royal Tunnel**

Mount Royal (Figure 5). The REM project will use this existing double track horseshoe tunnel (5060 m long, 9.6 m wide and 4.4 m high, and a constant 0.6% grade) and two of the project stations will be built inside this tunnel by enlarging it from a double track tunnel to side platform station at the location of these stations. To accommodate the new track system and to ensure the tunnel is to current safety standards complying with NFPA 130 fire life safety requirements, the existing tunnel conditions were assessed. Various solutions including the installation of a center wall and boring a parallel egress tunnel on one side of the existing tunnel and connected to the existing tunnel through cross passages at regular spacing were evaluated. Adding a center wall to the existing double track tunnel was selected as the preferred solution.

To accurately evaluate the current conditions and define the tunnel enlargement needs based on the new train envelops, a high-resolution laser and optic scanning was performed by Dibit. Using this information, the current conditions of the existing tunnel and accurate clash analyses and interfaces requirements assessment was performed. Results from such analyses were used for the development of the optimal solution to minimize the volume of enlargement excavation.

## CUT AND COVER STATIONS

To connect the financial district of Montreal to the REM tracks located approximately 15 m below grade, a major station is planned on McGill Avenue (Figure 6). In this area, due to previous underground works for building the MRT and the Montreal Metro Green Line, the first 15 m of ground consists of backfill. The rock located below this layer of backfill was observed to belong to the Tétrauville formation, which consists of an interbedded limestone/shaley limestone.

Given the local stratigraphic column and low water table, the soldier piles and lagging wall solution is selected as the support of excavation for this station. Drilled soldier piles were socketed into the rock and steel fiber reinforced shotcrete were used for the lagging. The soldier piles and shotcrete lagging support of excavation walls were considered as the permanent station walls. Once the rock is reached, controlled drill and blast took place to cut the rock to the design level.

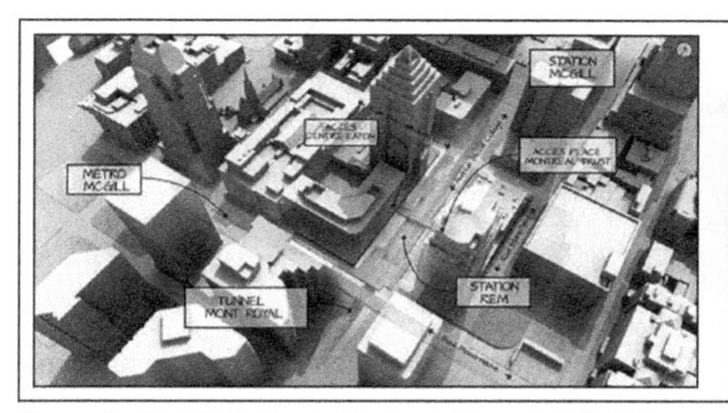

**Figure 6. McGill Station in Montreal Financial District**

The station is located between two major high-rise buildings and is connected to shopping centers inside these buildings. An existing underground commercial passageway connecting these two high-rise buildings is just above the station and were kept in place during the station construction. The station is at the intersection of McGill Avenue with two of the most important streets in Montreal and therefore the maintenance of traffic and utility relocation were among the main challenges of this station construction. In addition, the site limits of this station were at the edge of the adjacent buildings and the existing metro tunnel and therefore robust design of support of excavation in soil, controlled drill and blast in rock, and comprehensive instrumentation and monitoring program were required to prevent damage to the neighboring structures. The station new tracks were installed at the location of the existing Mont Royal Tunnel tracks and therefore the excavation had to be sequenced in a way to keep the existing tunnel liner in place as long as possible in order to minimize the existing tunnel closure.

The Technoparc Station, located at the northeast corner of Alfred-Nobel Boulevard and Albert-Einstein Street, is an underground station on the Airport branch. A method of construction using secant piles for the sidewalls was selected as the preferred structure for this station that serves as both temporary and permanent support of the excavation. Wide-flange permanent roof beams act as struts during excavation. This method is used for the approach ramp, the cut-and-cover tunnel structures and the Technoparc Station. The secant piles are drilled to rock and socketed into it. The excavation was drained during the construction and the final structures were tied down to the bedrock for buoyancy control during the permanent condition. Minimum 100 mm shotcrete liner is installed on the inside face of the secant pile wall. Sprayed on water proofing system is integrated in this liner which provides the water tightness required for the permanent condition.

The secant pile wall was designed to resist the lateral earth and hydrostatic pressure. To support lateral pressure, two levels of permanent struts were provided at station area. These struts and their walers were integrated as the main structural elements of the station carrying the lateral ground loads and the vertical loads of the station (Figure 7). 0.5% verticality was assumed for the construction tolerance of these secant piles. The wide flange section, the main pile reinforcement, installation tolerance was considered 25 mm horizontal and 1 degree rotation at top of the pile.

## AIRPORT TUNNEL

The REM connects downtown Montreal to the airport, requiring the development of an entirely new underground tunnel that runs below the international airport airstrips. The overall underground stretch consists of approximately 3.6 km. Along the tunnel alignment, bedrock elevation varies significantly, and a constant grade is observed at

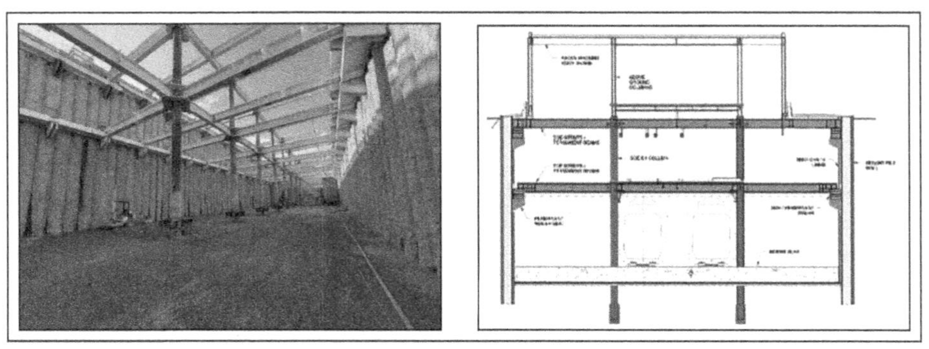

**Figure 7. Permanent secant pile at Technoparc Station on the airport branch**

surface, resulting in an overburden thickness varying between 12 and 20 m. The overburden consists of layers of backfill, granular material and glacial till, going from grade to bedrock. The bedrock consists of interbedded limestone/shaly limestone, belonging to two different formations: the Tétrauville formation and the Montreal formation.

Within the Tétrauville formation, two different members are expected to be intersected by the underground works. The upper horizon would consist of a good quality micritic shale with UCS values varying between 75 and 185 MPa, Young's modulus varying between 25 and 65 GPa and Cherchar abrasivity index varying between 0.8 and 1.8. The lower horizon would consist of softer shaly limestone with UCS values varying between 60 and 80 MPa, Young's modulus varying between 35 and 40 GPa and Cherchar abrasivity index varying between 0.3 and 0.6. Within the Montreal formation, only the Rosemont member is expected to be intersected. This member is expected to be of good quality with UCS values varying between 55 and 145 MPa, Young's modulus varying between 45 and 65 GPa and Cherchar abrasivity index varying between 0.8 and 1.8.

To align the REM system with the Airport station designed by others, the REM double-track descends from surface to approximately 40 m below grade. Given the significant constraints related to developing a tunnel underneath an international airstrip, the main portion of the underground works were performed using a hybrid tunnel boring machine (TBM). From the surface, the REM starts its descent and enters a cut and cover underground station (Technoparc Station). Heading out of the station, the cut and cover portion continues and widens over a 13 m length at about 125 m away from the station to serve as the TBM launch pit. At this point, the tunnel invert is at approximately 14 m below surface, in the overburden soil.

The hybrid TBM was launched in the overburden material and continued its descent and progress in this material for about 300 m. The ground was improved at the break-out over the first 10 m of the drive to allow watertightness and TBM control at the launch. Once the TBM reached the bedrock, it continued its descent over a course of about 300 m, down to 40 m below grade. From there, it progressed at this constant elevation, totalizing approximately 2.7 km of excavation within the rock. The hybrid TBM was able to progress within the overburden loose material in Earth Pressure Balance (EPB) mode and in open mode during its progression through competent rock.

As the TBM was advancing, precast segmental lining was installed, ensuring the stability of the opening and the safe development of the tunnel. Routine probing ahead of the face was also performed to assess ground mechanical and hydraulic conditions prior to advancement. Depending on ground conditions, pre-excavation grouting performed ahead of the face was required to improve mechanical and hydraulic properties of the ground before the TBM excavation.

## AIRPORT TUNNEL SEGMENTAL LINER DESIGN

Based on the geotechnical challenges and evaluation of cost estimate and risk mitigation, it was concluded that shielded TBM with one-pass lining system is the preferred alternative for the REM airport tunnel. Precast segmental lining was installed using an Erath-Pressure Balance (EPB) TBM as the initial and final lining for the main tunnel. With 6.478 m as the internal diameter of the tunnel, 300 mm was selected as the thickness of segments which is the common value used in practice for this size of tunnel. Lining thickness was verified during the design procedure. Optimized length of the ring (segment width) was selected as 1,700 mm. As shown in Figure 8, a 6+1 rhomboidal system assembled ring by ring was selected for REM airport tunnel lining. It consists of a trapezoidal reverse key segment slightly larger than other five full-size rhomboidal segments (3420 vs. 3210 mm, respectively along centerline) and one small trapezoidal key segment in a ring, slightly smaller than ⅓rd of ordinary segments (882 mm length along centerline). Ring segmentation into seven afore-mentioned segments results in segment slenderness/aspect ratio (segment curved length-to-thickness) of 10.7-11.4 which is near the maximum ratio used in the world for FRC segments. This results in a smaller number of segment and joints, stiffer segmental ring, reduced production cost as well as less hardware for segment connection, less gasket length and a smaller number of bolt pockets where leakage can occur. More importantly, the construction speed can increase significantly. In addition, advantages of using a rhomboidal system include staggered longitudinal joints, continuous ring building and compatibility with a dowel type connection in circumferential joints, which results in a faster ring assembly process comparing to rectangular systems.

Universal rings were selected for this project assembled from rings with circumferential joints inclined to the tunnel axis on both sides. One of the main advantages of this ring system over other systems (e.g., left/right rings) is using only one set of forms for segment production. Longitudinal and circumferential joints were designed as completely flat joints which are advantageous for load transfer between the segments and the rings compared to other types of joints. Also, flat joints have been proven to have a superior sealing performance. Bolt connection was designed for longitudinal joints and dowels were chosen for connecting rings in circumferential joints as they require less work for the construction of the segment form and less manpower in the tunnel as the insertion is automatically performed by the erector when the segment is positioned. For the first time in North America, a new dowel system, SOF-FIX ANIX 60 ASY, was used as the connection device in circumferential joints. This dowel is bolted by hand in the socket of the segment to be installed and then it is pushed by the erector in the socket of previously installed segment. This increases the precision of the dowel installation and reduces the ring offset.

The gasket type for sealing joints between segments was designed as fiber anchored gasket. Fiber-gasket system has been successfully implemented for the first time in the world by the Designer in South Hartford Tunnel, CT which is considered as latest innovation in EPDM gasket design (Bakhshi and Nasri, 2017). This new technology offers additional pull-out resistance comparing to conventional glued gasket system and has several advantages over anchored gaskets such as reduced risk of incorrect installation and reduced risk of air entrapment in the anchorage area. Gasket Profile is DATWYLER M 802 07 "type South Hartford" providing watertightness under the maximum expected groundwater pressure of 140 psi (9.6 bar). This gasket profile

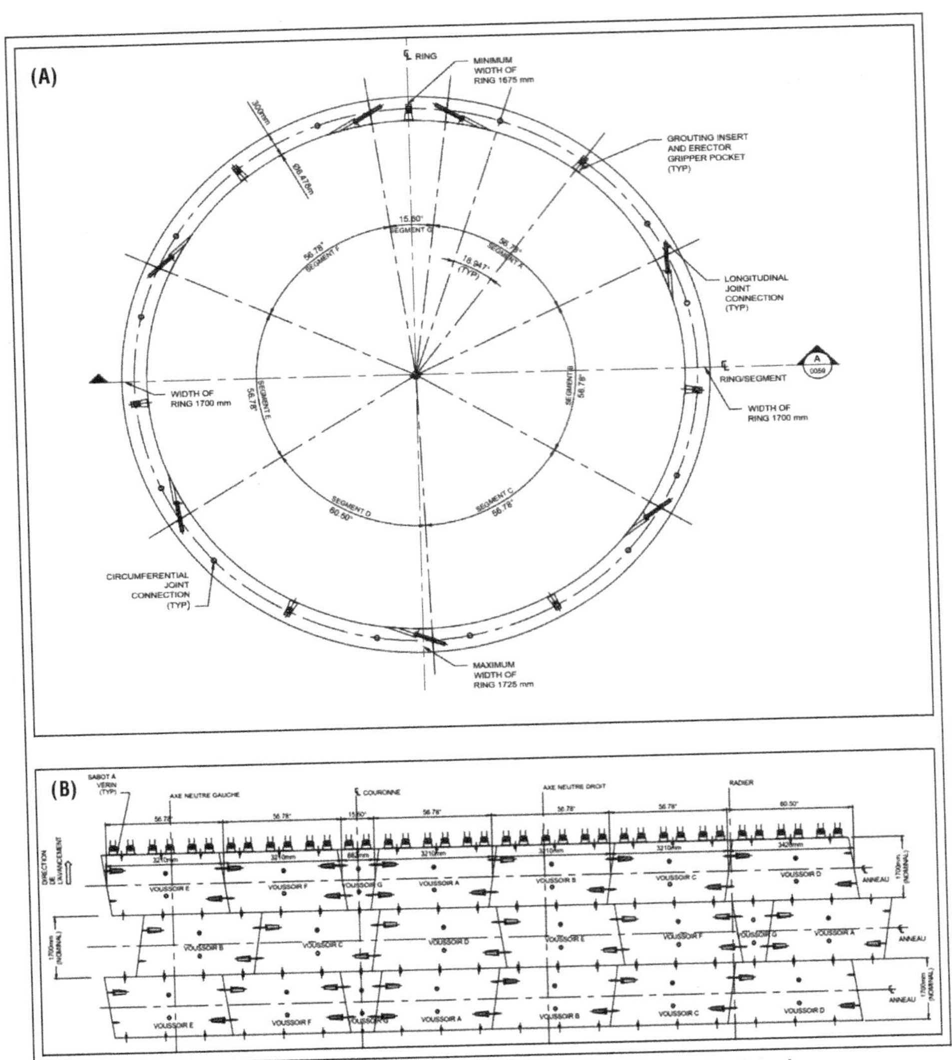


Figure 8. REM airport tunnel segmental ring: (A) section; (B) developed plan on intrados

guarantees watertightness for 1.5 times maximum working water pressure considering a combination of gasket differential gap of 0.2" (5 mm) and bearing surface offset of 0.4" (10 mm).

From the structural design point of view, the analysis and design of segments complies with the latest guideline (ACI 544.7R, 2016) in order to satisfy the intended objectives of the project during construction and service life of the tunnel (Bakhshi and Nasri, 2016). The design was carried out using the load and resistance factor design (LRFD) method. Accordingly, during segment production and construction, concrete segments should be able to withstand stripping (demolding), storage and handling loads, thrust force of jacks needed to drive the TBM forward during excavation, and contact/backfill grouting. After installation, segments should be able to resist loads imposed by the surrounding rocks and hydrostatic pressure. Segments were designed for final service stage and verified for production, transient and construction stages loadings.

Precast concrete tunnel segments for REM could be made of reinforced concrete (RC) or fiber-reinforced concrete (FRC). However, FRC segments were superior for crack control and were more cost effective, and therefore, were preferred solution to be used in this project. Minimum compressive strength of concrete at 28-day and at the time of stripping (demolding) was recommended to be 60, and 14 MPa, respectively. To achieve a high early-age strength segment and high performance and durable concrete lining, maximum water cement ratio was designed as 0.40 with the addition of silica fume in the level of 5 percent of cementitious materials. Minimum cement content was specified as 350 kg/m$^3$ and maximum aggregate size as 19 mm while chloride ion penetrability had to be less than 700 coulombs following ASTM C1202. Cold-drawn wire ASTM A820 type I steel fibers, with a minimum tensile strength of 1,800 MPa was specified for segment reinforcement. Other specified characteristics of fibers included double hooked ends, a minimum length of 60 mm, a maximum diameter of 0.75 mm and an aspect ratio (length/diameter) of 80. Minimum required fiber content for segments reinforced with steel fibers was recommended to be 42 kg/m$^3$ (71 lb/yd$^3$) in order to obtain a minimum first-peak flexural strength (LOP) and residual flexural strength ($f_{R,3}$) of 3 MPa, at the time of segment stripping (demolding) according to EN 14651. Specified FRC strength class at the time of stripping (demolding) was type "3d" according to fib Model Code 2010. At 28 days, minimum first-peak flexural strength (LOP) and specified residual flexural strength ($f_{R,3}$) at 28 days was considered 5 MPa, according to EN 14651. Specified FRC strength class at 28 days was, therefore, type "5d" according to fib Model Code 2010. FRC segments were designed following ACI 544.7R (2016) using these parameters and verified versus results of different performed analyses.

Note that the Designer, for the first time in the US, designed precast concrete tunnel segments with double hooked-end high-strength steel fiber (Dramix® 4D 80/60 BG). Same design was used for the first time in Canada for REM airport tunnel segmental lining. This new type of steel fiber satisfies the serviceability requirements by limiting time-dependent effects of creep on crack opening and more significantly guarantees ductility requirements with conventional fiber dosage rates by providing an ultimate bending moment higher than the cracking bending moment. This is especially important for the loads applied on segments during production, storage, handling and transportation when segment is subject to pure bending loads.

## BIM FOR REM TUNNELS AND UNDERGROUND STATIONS

Building Information Modeling (BIM) system was implemented to build a full virtual tunnels and underground stations complex that incorporates existing facilities and infrastructure, utilities, ground profile and geotechnical characteristics, proposed tunnels and underground stations architecture, structure, electrical, mechanical, ventilation and FLS components. BIM system was used to support Virtual Design Construction (VDC) process by visualizing completed tunnels and underground stations and their components and systems enabling the design team to resolve identified potential conflicts and issues prior to construction. In addition, BIM system was used to assist quantity take-off estimation by providing a realistic project cost estimate and minimizing human errors and inaccuracies during cost estimation procedure. The developed BIM will also provide significant assistance to the Owner with operation and maintenance during the service life of the REM tunnels and underground stations. Figure 9 demonstrates components of virtually built REM tunnels and underground stations. This process can be described in the following steps.

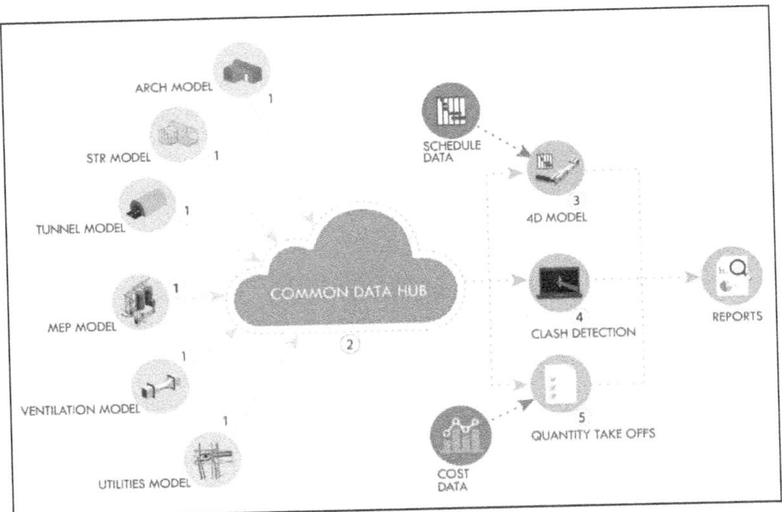


**Figure 9. BIM implementation process for REM tunnels and underground stations**

1. BIM Models: Building an intelligent 3D model-based process to more efficiently plan, design, construct, and maintain the REM infrastructure system.
2. Common Data Hub: Utilizing a common data hub to coordinate and communicate more effectively and resolve issues quickly with all involved disciplines such as architectural, structural, electrical, mechanical, etc… (Figure 8).
3. 4D Model: Adding time factor, construction schedule, to 3D model to visualize planned construction sequences.
4. Clash Detection: Detecting hard and soft clashes. Hard clash detection refers to finding the conflict between two objects occupying the same space. Soft clash detection refers to finding the conflict between objects that demand certain spatial and geometric tolerances or buffer zones with other objects within their space for access, insulation, maintenance, safety, etc...
5. Quantity Take-offs: Performing object based calculated quantity take-offs based on provided units.

Figure 10 shows the BIM model for Edouard Montpetit Station.

## TUNNEL VENTILATION DESIGN

The purpose of the Tunnel Ventilation System (TVS) in transit tunnels is to perform two primary functions. First function is to maintain environmental control during normal and congested operation within the tunnel such that the electromechanical equipment in the tunnel environment can function within their operational temperature and humidity limits. The second function is to provide tenable environment along the evacuation path during a fire emergency in the tunnel.

In the REM Project there are two tunnels, the existing Mont Royal Tunnel (MRT) in the Deux-Montagne branch and the new Airport Tunnel in the Airport branch. The Owner technical specification requires that a mechanical longitudinal ventilation system be provided in order to sweep the smoke and heat from a tunnel fire along the tunnel to one side of the fire and maintain tenable environment as defined by NFPA 130 on the other side of the fire, where the tunnel occupants will be evacuating the tunnel. The longitudinal ventilation system must provide sufficient air quantity to produce the

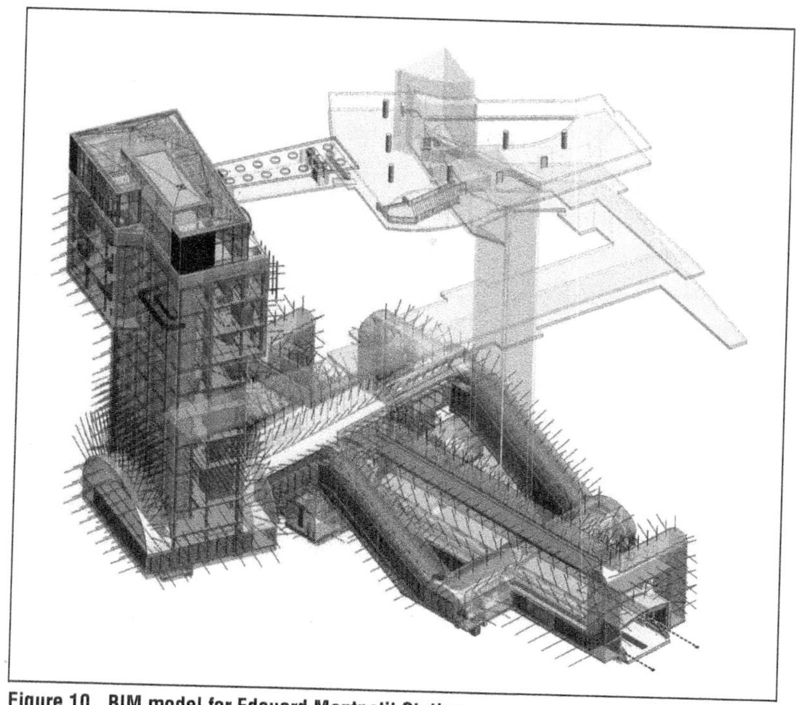

Figure 10. BIM model for Edouard Montpetit Station

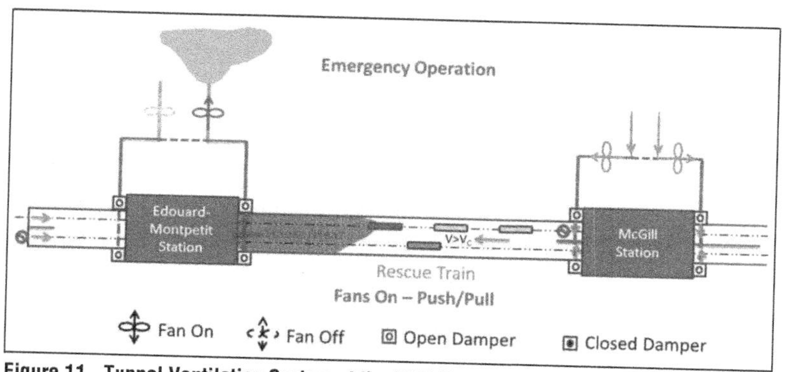


Figure 11. Tunnel Ventilation System at the REM Mont Royal Tunnel, configuration during fire emergency

Critical Velocity upstream of the fire to prevent back layering of smoke over the top of the evacuating tunnel occupants. Based on formulas provided by NFPA 130 the quantity of air to generate the Critical Velocity for the REM tunnels were calculated to be 130 $m^3/s$ for the MRT and 60 $m^3/s$ for the Airport Tunnel. This airflow is provided by fan plants in stations at the end of each tunnel, i.e., stations Edouard Mont-petit (EMP) and McGill within the MRT and Technoparc Station and the ventilation shaft at the southern end of the Airport Tunnel. This ventilation shaft is housed in the airport parking area. Figure 11 shows the schematic representation of the TVS configuration for MRT under fire emergency operation.

**Figure 12. BIM model for Edouard Montpetit Station**

The same TVS can be configured to provide ventilation during normal and congested operations. During normal operations the piston effect of the running trains is expected to create sufficient circulation within the tunnel to maintain the required temperature and humidity control. Therefore, fans can remain off and piston relief is achieved through bypass ducts.

During congested operations when trains are moving at reduced speed through each tunnel the heat rejection from the trains can increase the tunnel air temperature above the operating temperature of the electromechanical equipment operating within the tunnel environment such as the air conditioning condenser units on board the train. In this case the fans can operate in push-pull mode like the fire emergency configuration and cool the tunnel by passing fresh cool air over the trains and removing the heated air from the tunnel at the location of the fan plant operating in exhaust.

## ARCHITECTURAL DESIGN

The architectural signature of the stations is based on three main themes: movement, identity and transparency. A language expressing the kinetic movement of the train and the landscapes that follow one another is integrated through the various components of the network. The identity in the signature is constructed through a set of common elements and by considering the context introducing the notion of variability. Finally, transparency is implicitly reflected in the stations by design guidelines such as natural lighting, security and the relationship with the context (Figure 12).

## CONCLUSION

This paper presents the design and construction aspects of the underground elements of the REM mega project in Montreal. It discusses the details of one of the underground stations built using the NATM method and the two other stations constructed with the cut and cover approach. It also explains the rehabilitation and enlargement of the existing Mont Royal Tunnel. It describes as well the specifics of the airport TBM tunnel. Several innovative and cost-effective design and construction methods used for the REM project have been discussed in this paper.

## REFERENCES

Nasri, V., De Nettancourt, X., and Rey, A. (2022), Construction of REM Project in Montreal. World Tunnel Congress 2022, Copenhagen, Denmark, September 2–8, 2022.

Bakhshi, M., and Nasri, V. (2017), Design of Steel Fiber-Reinforced Concrete Segmental Lining for the South Hartford CSO Tunnel. Rapid Excavation Tunneling Conference 2017, San Diego, CA, June 4–7, 2017, pp. 706–717.

ACI 544.7R. 2016. Report on Design and Construction of Fiber-Reinforced Precast Concrete Tunnel Segments. American Concrete Institute (ACI).

Bakhshi, M., and Nasri, V. (2016), ACI Guideline on Design and Construction of Precast Concrete Tunnel Segmental Lining. World Tunnel Congress 2016, San Francisco, USA, April 22–28, 2016.

# Formwork Solutions for the Final Lining of the Kramertunnel, Germany

**Rainer Antretter** ▪ BeMo Tunnelling GmbH

## ABSTRACT

The city of Garmisch-Partenkirchen in southern Germany is located at a main transit road to Tyrol in Austria and suffers of a very heavy traffic load. Therefore, the Kramertunnel as a by-pass tunnel is built to release/mitigate the traffic through the city centre. The 3,600 m long single tube road tunnel is designed with a parallel but smaller emergency tunnel and with cross passages, fitted with single- and double-sided emergency bays, a vent shaft, and an intermediate ceiling to separate fresh air supply from the road traffic. The tunnel, concrete lined along its entire length provided the challenge to find smart formwork solutions for quick set-up, repositioning, setup and pouring. Three particularly specific sections are discussed in this paper.

## OVERVIEW OF TUNNEL WITH SPECIAL POURING SECTIONS

The Kramertunnel provides a variety of different designs of the main tunnel section. There are flat invert sections, deep and round shaped invert sections, and sections with and without intermediate ceiling. In addition, these different designs are executed with different types of waterproofing, from umbrella style waterproofing to a pressure-resistant membrane waterproofing to watertight concrete in combination with waterstops. One section with high water pressure even requires a combination of pressure-resistant membrane waterproofing with watertight concrete and waterstops.

The tunnel as shown in Figure 1 is of a single tube counter traffic running tunnel section in which the contractual requirement foresaw the inner lining to be poured by hit and miss mode. This meant that one vault block had to be poured and pouring of the following block had to be missed. The following block had to be poured again, and the next had to be missed again, so that a lining gap between two poured block was created. The skipped blocks could only be poured some days later by use of a second formwork, which followed the first and closed the gaps, after the concrete strength of the adjacent blocks has increased. The idea behind of such a pouring sequence is to prevent cracks in the young block during pouring caused by formwork deformations and transmission of mechanical pressure of the tail sheet of the formwork onto the fresh concrete of the next block. Cracks happen in case of daily pouring rhythm if the concrete of the previously poured block is only 24 hours old or even younger

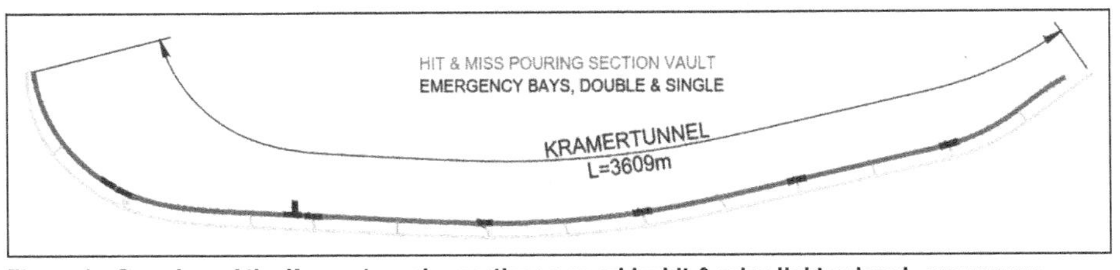


**Figure 1. Overview of the Kramertunnel—sections poured by hit & miss light colored, emergency bays dark**

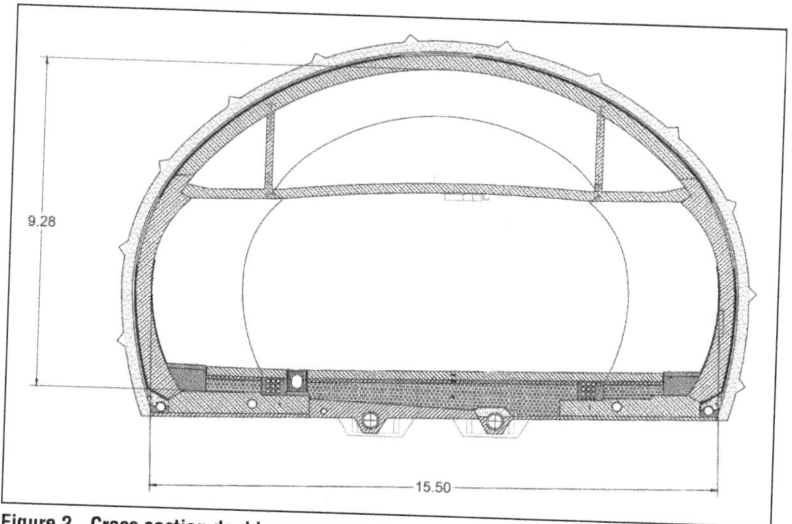


Figure 2. Cross section double emergency bay

(the stripping strength in Germany is specified to at least 2 N/mm$^2$, which is generally reached 24 hours after end of pouring). The risk of cracks after some days of concrete strengthen is significantly lower.

Due to the required pouring sequence with so called forerunner blocks and follower blocks in the vault, it was necessary to choose an invert formwork with special properties. The invert blocks could not be poured using the hit and miss system since an invert formwork able to pour two blocks per day would have been too expensive. Such an invert formwork would require a main beam length of 65 m with a costs factor of 2.5 to a normal invert formwork able to pour one block per day.

The wide emergency bays are designed double-sided which meant the cross section is 2.50 m wider either side of driving direction, thus, in total 15,50 m wide over a bay length of 55 m. Hand in hand with the enormous width for the bay cross section the tunnel height was also elevated in order to form an appropriate shape of the wide tunnel, shown in Figure 2.

Figure 3 shows a one-sided emergency bay which is designed with one breakdown lane in each driving direction, but these are offset in the longitudinal tunnel direction.

Up to the abutment for the intermediate ceiling, the tunnel geometry is the same for both types of emergency bays. This represented an important fact for selecting the formwork system.

## PARTICULAR FEATURES OF THE INVERT FORMWORK

Because the concrete invert also represents the driving surface for supplying all lining sites with concrete and reinforcement steel, it always had to be passable to the very front (invert formwork location). It was therefore necessary to pour the invert in block-by-block sequence before the vault formwork could follow moving on rails which are placed at the footings of the invert in order to be able to line the vault by the hit and miss sequence.

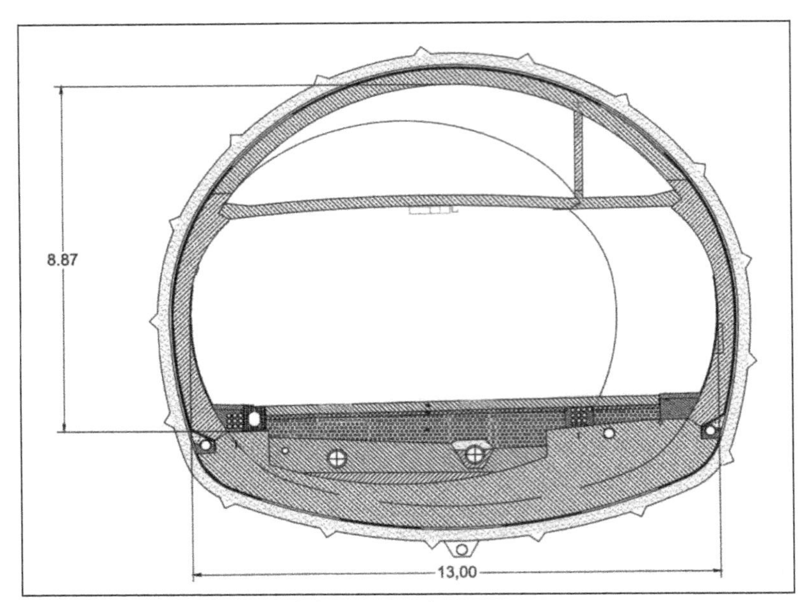


Figure 3. Cross section one-sided emergency bay

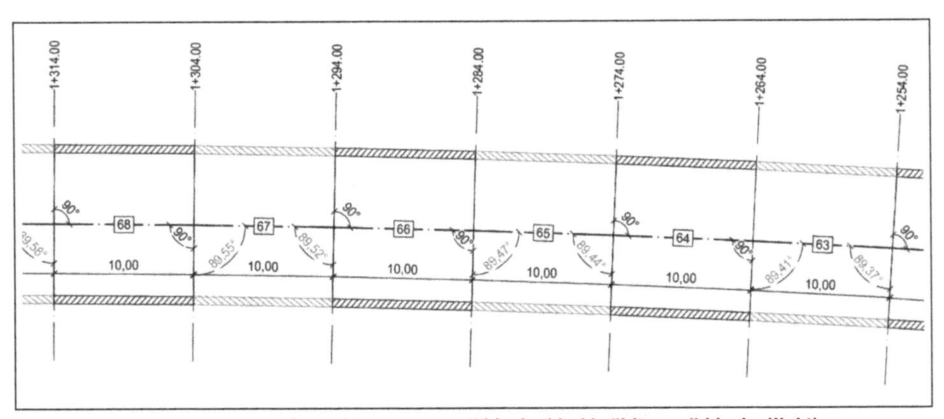


Figure 4. Formation of block joints with "miss pour" blocks (dark), "hit pour" blocks (light)

This meant that each forerunner invert block could be formed and poured with right-angled block joints which is common and quick. Every intermediate block, however, in which the vault must be poured using hit and miss sequence had to be executed in areas with a curved longitudinal tunnel axis with block joints that were not at right angles, but less than 90°, see Figure 4. The stop-end of the invert formwork had therefore to be an adjustable one in order to be able to form the skewed block joints, so that the water stop could be placed at the intended position and connected at the vault, (Figure 5). With the help of the surveyors on site, each block was formed with different block joints angles.

In order to achieve flexibility for the many different angles of the block joints, the solution was to design the stop-end with two hinges and an expansion joint for length

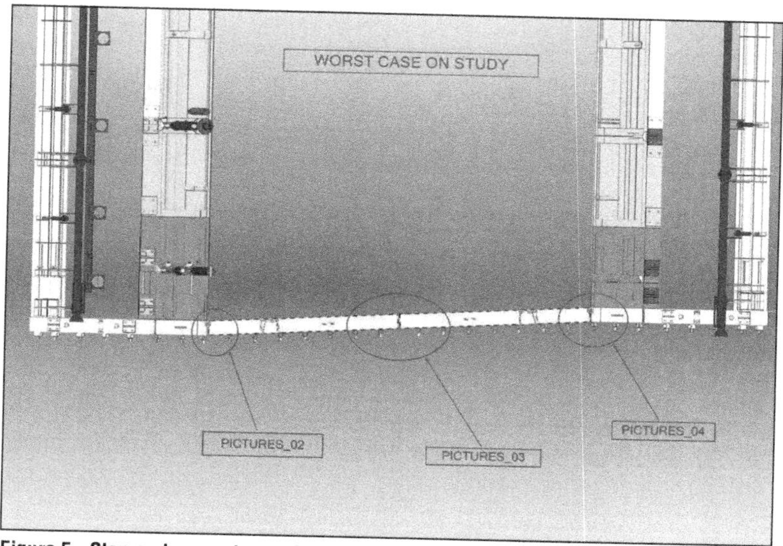


**Figure 5. Stop-end geometry on the intermediate block joint at worst case**

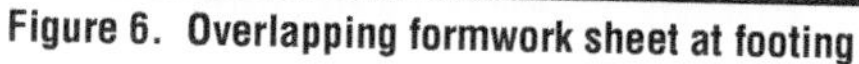

**Figure 6. Overlapping formwork sheet at footing**

**Figure 7. Formwork sheet inside attachment**

compensation at the tunnel axis. A certain part of the length of the tunnel was designed for a water pressure of 5 bar, which means that the precise installation of the water-stop was an important criterion for water pressure tightness. Therefore, the kinks in the stop-end were to be executed slightly rounded to achieve gradual transition of the direction within tolerable skew.

The invert footings and the steep inclined part of the round shaped invert were left right-angled to the tunnel axis (dark-colored formwork parts in Figure 5). These panels were fitted with formwork sheets that could be moved longitudinally to ensure length compensation for each block with different inner and outer chord length. In the same way the vertical panels for the footings were fitted, see Figure 6 and Figure 7.

## Practical Use of the Invert Formwork

During execution, the entire stop-end adjustment system proved reasonable and simple and quick to handle. Manual post processing of the water-stop in the kinked sections was necessary and executed right after finishing the pour when the concrete around the water-stop was still easy to remove around the hollow structure at the joint. The target of one pour per day could be achieved. Important was therefore, that stripping of the stop-end formwork parts took place already several hours after the pour to conduct post processing works like cleaning and coating of the joint, closing the reinforcement gap and preparation for advancement of the formwork.

Formwork advancement was done as the first step at the beginning of the next day before the consecutive pour.

Figure 8 shows the advancement of the invert formwork with the walking structure already extended over the block to be poured and the removed stop-end formwork parts of the footings on each side of the poured invert block.

Pouring was done by truck based concrete pump with a low unfolding tunnel mast, see Figure 9. Within the open part of the round shaped invert a concrete hose through the structure placed the concrete into the invert. In this section poker vibrators and a round shaped vibration bar was used, see Figure 10.

**Figure 8. Invert formwork during advancement into next block**

**Figure 9. Car concrete pump with mast**

**Figure 10. Invert pouring between footings**

## CHALLENGES AND SOLUTIONS FOR THE EMERGENCY BAYS

Within the Kramertunnel four double-sided and four single-sided emergency bays were to be built. Due to the time schedule, it was necessary to pour the emergency bays without interruption, so that the intermediate ceiling formworks could follow without interruption of the daily pouring rhythm. The intermediate ceiling formworks started later than emergency bay lining and the construction schedule envisaged that the six intermediate ceiling formworks—which were able to pour two blocks per day—should not hit the emergency bay formwork, which could only pour one block every two days. Therefore, a flexible solution for the bays was required, which should be able to move from one bay to the other quickly. At the same time, the emergency bay formwork should be able to be used for all cross-section types, both double-sided and single-sided. Furthermore, the one-sided emergency bay formwork should also be able to be used for the opposite lane.

After in-depth calculations of the costs and the estimated time required preference was given to a split formwork system. Instead of a single vault formwork for the huge double-sided cross section a single formwork for the sidewalls and a single formwork for the cap was chosen. The decisive factor was the time advantage determined by the flexibility of separating the side wall formwork from the vault formwork. One of the main reasons for the decision not to choose a huge single formwork for the entire cross section was that the time and effort required for dismantling and reassembling the heavy formwork elements of the side wall and the toe, as well as for reassembling these sections after relocation, was high and fraught with uncertainty because the site team was not familiar with the required type of work, especially the lifting techniques. Although more expensive in terms of formwork costs, it was decided to pour with a side wall formwork (shown in Figure 11) up to above the intermediate ceiling abutments and to cast the cap later using the existing arch formwork, which had to be adjusted for the enormous height of the double-sided bay. An advantage of splitting the side wall formwork and the cap formwork was the possibility to use it also for the single-sided bays since the tunnel geometry up to the was the same for both types of emergency bays.

**Figure 11. Side-wall formwork in double-sided cross section**

## DETAILS OF THE SIDE-WALL FORMWORK

However, the side-wall formwork also entailed the major expense that it could not be bolted into the rock at its upper end at a height of 6,25 m because the tunnel waterproofing membrane was not allowed to be penetrated to rule out later leaking damage. The structural requirements in this respect were underestimated in the cost calculation and turned out to be more expensive than expected. At sections with flat invert bolts had to be drilled and installed into the rock with a length of 4 m and in sections with flat invert and reinforced invert lining with a length of 3 m. Bolting was possible in the deep invert area since there had no waterproofing membrane to be installed and in sections with deep invert, the bolts only were installed in the reinforced concrete slab, not penetrating the membrane. The bolting procedure was extensive and expensive.

To transfer the forces from the concrete pressure and the torque into the invert, 15 Dywidag bolts with diameter 26 mm, shown in Figure 12, distributed over the formwork length of 12 m (not symmetrical) were required, with tensile forces of 250 kN within the formwork and 350 kN at the stop-end.

Despite the stable supporting structure and thus the weight of the existing carrier and side formwork of 82 tons, structurally determined deformations of the formwork at the intermediate ceiling support of 22 mm had to be accepted. The pouring speed was set to 1 m/hr and a concrete pressure of 45 kN/m$^2$ was assumed as a result of the deformation calculation is shown in Figure 13. From this knowledge, deformation during pouring was prevented by placing the formwork with additional deformation play to the general superelevation which was 25 mm.

Figure 14 shows bolting with Dywidag bolts and Figure 15 shows the joint with the cap geometry. A lot of positive experience was gained using this method during all pouring operations with regard to the actual geometry created, and no difficulties were encountered during the subsequent lining connection with the cap formwork, or during pouring of the intermediate ceiling sections.

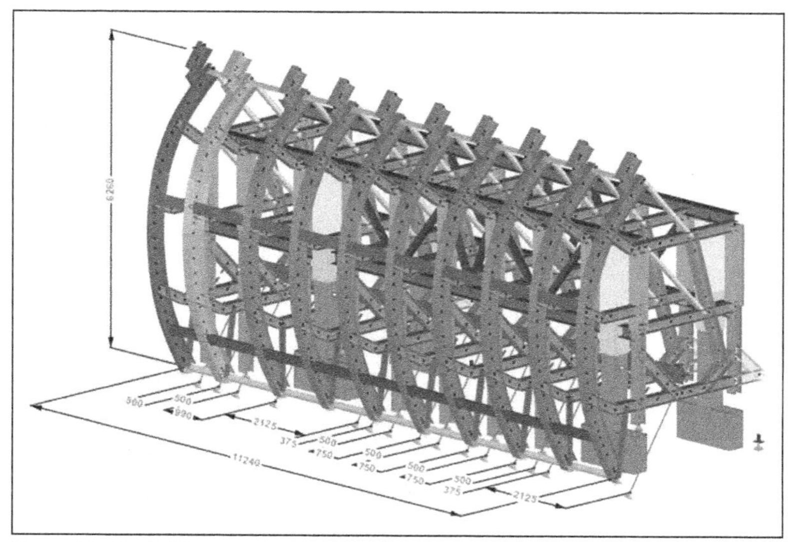


**Figure 12. Toe bolting spacing of side-wall formwork**

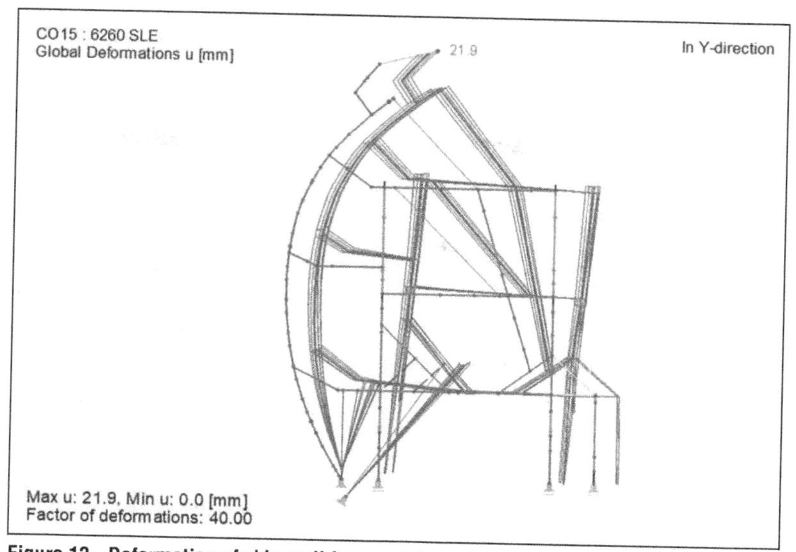


**Figure 13. Deformation of side-wall formwork from structural analysis**

**Figure 14. Toe bolting of side-wall formwork**

**Figure 15. Joint with cap ring and support**

Pouring was carried out applying a truck concrete pump with a tunnel boom, generally via hose connections on pumping ports, to eliminate the risk of failures (honey combings). For the approx. 75 $m^2$ large formwork with a length of 12 m was 48 air vibrators on the formwork structure were used.

After completion of all 4 blocks on one side of the double-sided emergency bay were completed, the formwork was turned by 180° using steerable driving units directly on the invert slab. The formwork with its max. length of 12 m had to be reversed nine times in order to perform such a rotation, see Figure 16. The stop-end working platforms and the tail sheet had to be dismantled for this purpose. The process worked very well and

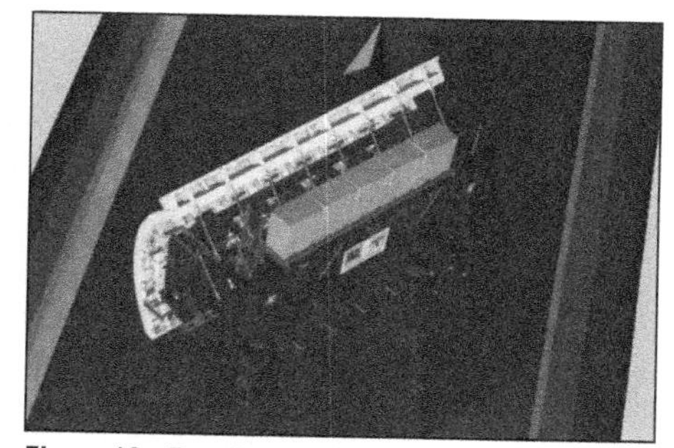

**Figure 16. Rotation of side-wall formwork**

Figure 17. Relocation through running tunnel

quickly from the start, so that only half a shift with the pouring crew was needed for the rotation.

### Turning and Relocation of Side-wall Formwork

For the process to move the formwork to the next emergency bay, it was necessary to attach an additional carrier structure with rail driving units. This was fitted in such a way that the opposite side of the existing rails with the track gauge of 7.50 m could be used, on which all the formwork equipment in the tunnel moved, shown in Figure 17. The relocation process, from one emergency bay to the next over a distance of 500 m, took a total of four days. Disassembly and reassembly of the platforms and the counterweights could be carried out relatively quick within one day, but assembly and disassembly of the bulky carriage structure, as well as the advance operation itself, took one and a half days each with double shifts.

## SPECIAL FEATURES OF THE CAP FORMWORK FOR EMERGENCY BAYS

The cap formwork for the big section of the double-sided emergency bays had to be operated at a height of 9.20 m above the invert slab with a span of 13 m. The formwork skin itself did not touch the invert to transfer loads into. It had to be folded so that it could pass through the poured profile of the running tunnel, and it also had to be usable for the single-sided emergency bays without major modifications. In addition, the formwork had to stay as close as possible to or below the budgeted costs, and it had to meet the low estimated costs for the relocation operations.

An existing vault formwork with a cross-section in the order of a highway tunnel from a different project was available at the plant yard of one of the JV partners. This formwork formed the basis for the considerations to convert it for use at great heights. Fortunately, formwork elements of the enlargement section of this completed tunnel were also available. This meant that the main components did not have to be manufactured from scratch but could be rebuilt. This resulted in significant savings and explains the peculiar appearance of the carrier of the formwork and the additional vertical supports which had to be installed.

## First Design Steps

Trying to use as many used parts as possible, a first project sketch was made and demonstrated both that vault of the cap can be formed and the folding of the formwork can be realized in such a way that the formwork fits through the running tunnel, even if tightly, see Figure 18. With a new formwork element on the left of the center line in Figure 18 (light colored) and the existing formwork structure adjusted to the desired geometry (cut at the right location of the rib, element parts slightly bent forming a "V-opening," the "V" filled with fitting sheets and welded together again) the main steps of adaptation should be done, was the idea.

The first structural calculations then showed that these changes were by no means all that was needed, but that a number of reinforcing measures were still required and locking of the double-cranked carrier legs became necessary.

The first static calculations showed that these changes were by no means everything that was needed A number of reinforcing measures were also required in addition, with the most expensive being a complete second auxiliary supporting structure made of standard formwork parts, and a locking system for the double-cranked carrier legs. Furthermore, due to the geometry change, adjustments had to be made to the stiffening structures of the formwork elements. In total, new and adapted parts of approx. 21 tons had be added to the formwork, shown in Figure 19.

Adaptations of the existing respectively new working platforms with an additional weight of five tons were necessary in addition so that in the end the entire cap formworks weighed about 180 metric tons.

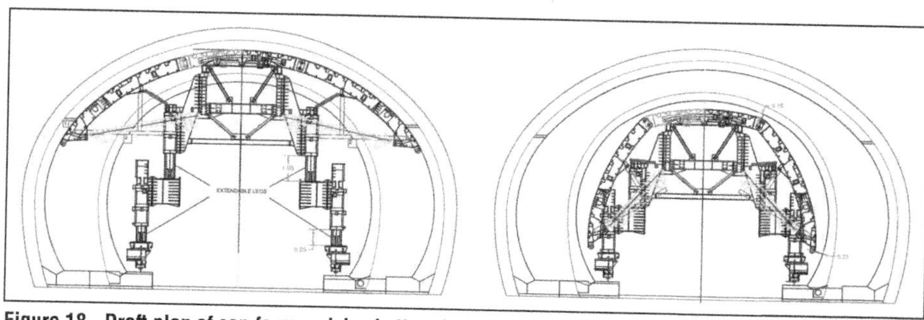

**Figure 18. Draft plan of cap formwork in shuttered and in relocation position**

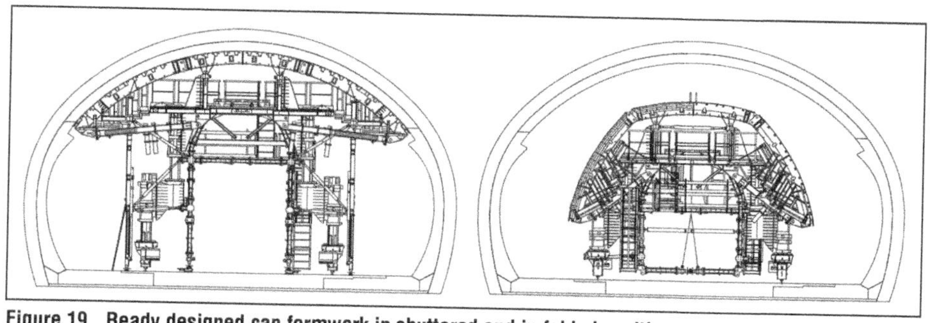

**Figure 19. Ready designed cap formwork in shuttered and in folded position**

Figure 20. Dywidag bolts and bracket

Figure 21. Counter plate in rear rib of formwork

## Final Design

After the formwork had been reinforced according to the structural calculations, a longitudinal anchorage had to be considered. The necessity arose from the concrete pressure at the stop-end since the formwork did not extend to the invert and the stop-end forces could not be transferred into the invert via friction. The stop-end started at the height of 6.25 m and ended at 9,75 m above the invert resulting in a pouring height of 3.50 m and inducing a longitudinal force of 220 kN. This force had to be dealt with at the top, since the vertical supports were not dimensioned for that. Hydrostatic concrete pressure was assumed.

The solution was found in Dywidag bolts which transferred the tensile force to anchor brackets which were bolted into the forerunner block, which is shown in Figure 20 and Figure 21. In total only two parallel anchor points were required.

## Connection To Side-Wall Structure

Since the connection at the side wall joint was just above the support of the intermediate ceiling, it was necessary to find a pressing system for the overlapping sheet that worked perfectly without the need for manual correction, because the possibility of reaching the joint by hand was almost non-existent. A press-on construction was considered consisting of a plate that ended freely at the bottom and could be pressed on at the bottom edge by means of disc springs with slight pretension. A strip of soft rubber was glued to the outside of the end sheet to ensure tightness, which was compressed and sealed perfectly. The max. deformation from the structural analysis showed only 9.5 mm at the overlapping area.

If the formwork deformed itself due to the concrete pressure, the slight pretension of the disc springs was sufficient to compensate for the deformation. The system worked splendidly, a drawing is shown in Figure 22, the already adjusted slight superelevation of the overlapping tail sheet after assembling shows Figure 23.

## Practical Experience with Cap Formwork

The experience with the cap formwork in the double-sided emergency bays corresponded to the ideas on the drawing board—the five blocks including the fitting block required for each bay could be completed in 8 to 10 days. The reinforcement had already been completed except for the gap closure required for the carpentry of the stop-end formwork, which is depicted in Figure 24, and so one block could be poured on average every second day.

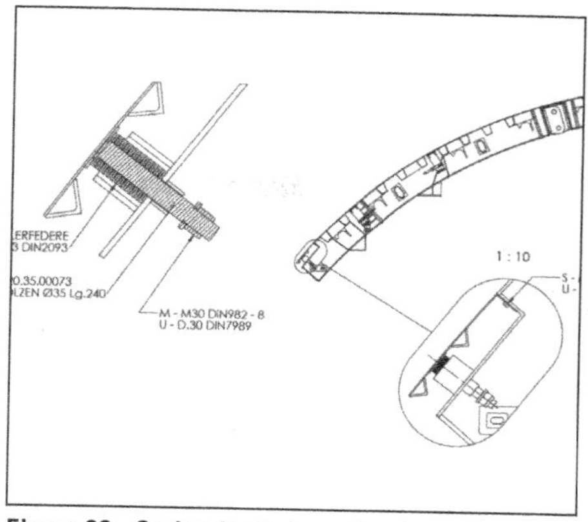


Figure 22. Spring-loaded overlapping tail sheet

Figure 23. Overlapping tail sheet superelevated

Figure 24. Stop-end on cap formwork

Figure 25. Pouring with the cap formwork

It must be mentioned that the emergency bays were constructed with a tunnel horizontal radius, just like the running tunnel. Here, another challenge lay in the execution of the pass block in the radius. The solution was a detailed and tricky block layout with the fitting block on the same direction of the axis as the previous poured block. This section—significantly longer than 12 m—had to be within the final lining tolerances. Figure 25 shows the pouring process.

Switching between the different formwork lengths could be done outside the critical pouring program whenever the formwork was relocated to the following bay. This process was subcontracted to the Formwork supplier and agreed in advance. Due to timely and careful work preparation, the conversion work on the formwork length could always be completed according to the program and it took an average of 3 days per action.

## Relocation of Cap Formwork

Not only was the tight fit of the cap formwork inside the running tunnel cross section a challenge, but also the passage in folded condition in curved sections of the tunnel. The clearance between the formwork and the final lined running tunnel in the straight section was just 200 mm on both sides. Therefore, each passage had to be checked on the drawing board with its radius and the formwork length. The consequence would have been the dismantling of formwork elements, which had to be avoided as far as possible due to the construction program and the costs. Figure 26 shows the cap formwork passing through a straight section of the lined running tunnel.

## Single-Sided Emergency Bays

At the time of writing of this article (Dec. 2022), pouring the side walls of the four single-sided emergency bays was still in progress but almost completed. The cap formwork for this cross section is significantly smaller and will be converted in a few weeks. The reinforced vault with side the walls partly poured is shown in Figure 27. The finished bulkhead wall of this cross section can be seen behind the truck concrete pump.

The conversion of the cap formwork from the double-sided to the single-sided cross section is about to start, so there is no experience available yet about the effort and use during pouring.

**Figure 26. Relocation of the cap formwork in the running tunnel cross section**

**Figure 27. Side wall pouring of single-sided emergency bay, bulkhead behind truck pump visible**

## CONCLUSION

The tight inner lining program at the Kramer Tunnel required special solutions in order to construct the special cross-sections quickly, so that the construction of the inner lining and intermediate ceiling of the running tunnel could be carried out uninterrupted with two vault formworks. Since the running tunnel had to be poured in hit and miss mode, a special solution for the invert formwork was required to respond to the angled block formation for the "miss" blocks. This was solved by means of a highly flexible stop-end formwork, which proved its worth in practical use at a reasonable extra effort of working hours.

The special cross-sections of the emergency bays also required a flexible solution, a formwork concept that could be quickly converted and relocated to avoid the risk of the intermediate ceiling formworks, running into the slower emergency bay formworks.

With the division of side walls and cap into individual formworks, the desired flexibility and speed of the pouring sequence could be achieved. Many different combinations of cross sections, concrete thicknesses, tunnel sealing systems, block lengths and curved tunnel sections had to be managed on top.

Since tunnelling work was also taking place during the lining operations, the solution to select steel formworks that could be transferred at any time and the division into side walls and cap proved to be the right ones.

# Nice—Tramway T2: A Success Story in Unusual Ground Conditions

**Guillaume Roux** ▪ Bessac, Saint-Jory
**Raoul Fernandez** ▪ CBNA/Bouygues
**Bernard Catalano** ▪ Bessac Canada/Bessac Inc

## ABSTRACT

The Thaumasia Joint-Venture, composed of Bouygues Travaux Publics, Soletanche-Bachy and Bessac, completed the construction of the underground section of line 2 of the Nice tramway, which comprised a 3.2 km twin track tunnel and four underground stations, in a particularly complex urban and geotechnical context.

The density of the surface structures, as well as the presence of many historic buildings on its route made the construction of the tunnel even more sensitive, with a settlement threshold of Half an Inch.

The joint venture faced a complex subsoil, varying from one extreme to another from highly plastic clay layers to fibrous peat and to cobbles with very high permeability, with artesian ground water, and with unforeseen anthropogenic obstacles in the tunnel section

Due to the diversity of the problems encountered and their variability over short distances, the construction of the underground section of the Nice tramway, in its very sensitive urban context, undeniably constituted an unprecedented challenge.

## PROJECT DESCRIPTION

### Award

Following an international prequalification and an open tender with four shortlisted JV teams, on November 7, 2013 the Metropole of Nice awarded to JV Thaumasia the construction of the underground section of the second tramway line from the Grosso Street open pit, to the Segurane Street open pit. This JV was composed of Bouygues Travaux Publics, Soletanche-Bachy and Bessac . The selection for this Design Bid Build contract was based on 50% price and 50% technical proposal. This contract amount was 284 M Euros (US$ 330 M) with an anticipated completion in June 2017.

Metropole of Nice Cote d'Azur selected Egis to be the construction manager. Other sections of the second tramway line (from Grosso to the airport, from Segurane to the Port) were awarded in different packages.

### Scope

The contract included:

One 3.2 km long tunnel drive in 27' ID bored by a pressurized face TBM . Thaumasia selected a slurry pressure balance TBM (see Figure 1).

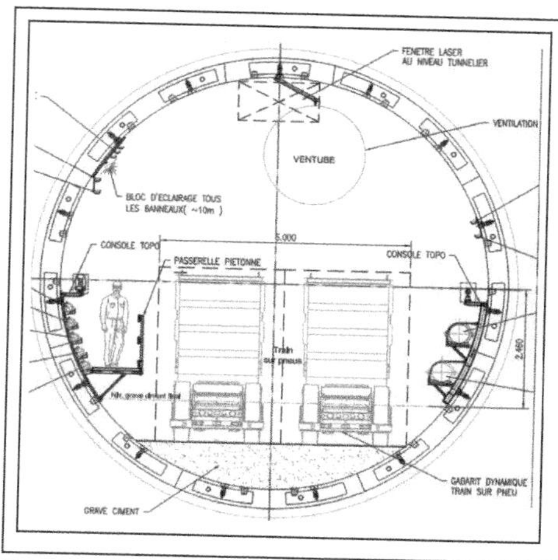

**Figure 1. Transversal section of the tunnel during construction**

- 2 open pits for the launch and retrieval of the TBM
- 4 underground stations, 55 m long average, 32 m deep constructed with diaphragm walls , built before, during and after TBM crossing: Allowing the concomitant construction of the station versus TBM tunneling was a key aspect of the project for its on-time delivery
- Internal civil structures
- Tunnel mechanical & electrical system.

## Geology

The first tramway line, built in 1991, was built at grade, without any underground sections. There are very few underground structures in Nice. The subsurface condition is highly convoluted mostly due to its geographical location: flanked between the Alps mountains and the Mediterranean Sea as a coastal city. The city's gravel beaches are indicative of the outwash geological context of the project.

Even at conceptual design, a number of geotechnical concerns were raised regarding the technical feasibility of underground alignment. But, as the first tunnel (at grade) had such a huge economic toll on the local businesses at ground level, a subsurface solution was imposed.

After extensive ground investigation involving 357 boreholes totalizing more than 50000 feet of drilling..., the extreme variability of the subsurface situation (see Figure 2) became clear: from highly plastic clay facies through fibrous peat to underground coarse gravelly channels that are cohesionless and with extreme permeability ($K > 10^{-3}$ m/s), changing from one extreme to another within a very short distance (i.e., a few meters of excavation).

Moreover, the hydrogeological analysis identified clear water with and without artesian conditions as well as brackish water, with both conditions affecting the confinement efficiency and the treatment process.

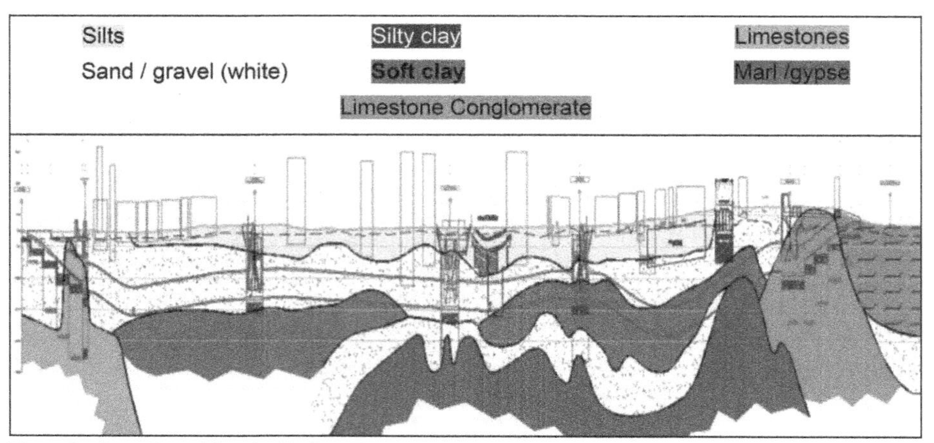


Figure 2. Complicated longitudinal profile

## Identified Challenges

As the project is located in an urban area, 75% of the tunnel alignment feature buildings located within the settlement through of the project. In addition, all those above surface structures are more than half a century old without any reliable as-built building drawings. Because of these two situational concerns, the owner has imposed a 10 mm (3⁄8") settlement threshold.

Combining everchanging ground conditions, an inability to improve ground from the surface, a sensitive surface living habitat, and a local community exacerbated by previous projects, not to mention a devastating terrorist attack, this project encountered multiple intense challenges.

# CHALLENGES DUE TO CONTEXT

## Restricted Site Staging in the Core of the Old City

At tender stage, the tunnel project was supposed to be excavated from Grosso pit , located in the middle of the city towards the city's port (see Figure 3). In order to limit nuisances to the residents due to hundreds of mucking trucks daily, the project team decided to reverse the tunneling direction to launch close to the port, so that the excavated muck could be transported by barges.

The drawback of this solution was launching from the very narrow Segurane Street , where the 75 feet deep, 145 feet long pit took the entire 46 feet wide street, which is constrained by the castle-steep hills and 6-story high old buildings, so that residents could see the bottom of the shaft from their balconies. Only a 4-feet wide sidewalk was maintained (See Figure 4).

This close proximity to inhabitants, when a project runs 24 hours a day imposed to implement the following drastic measures to maintain good relations with the stake holders

- The shaft was fully covered at night, even during tunneling. A motorized roof was opened in daytime to supply the tunnel and stockpile enough required material for the night operation (50 segments, 45 m$^3$ of grout, consumables…)

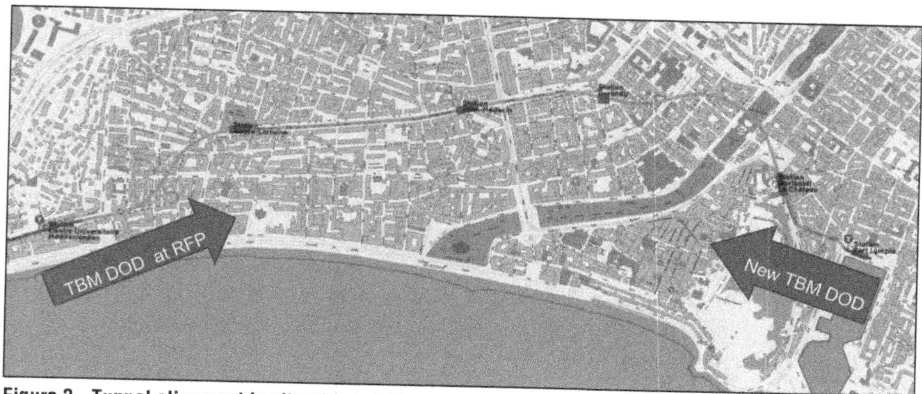


**Figure 3. Tunnel alignment in city with 4 stations**

**Figure 4. Segurane Street Pit: TBM launching and working shaft**

- The bottom of the shaft was equipped of a 40000 lb overhead crane hung under the concrete beams . This lifting device supported the TBM by loading the Multiple Service rubber tire vehicle
- All equipment on surface to be electrically supplied and fully enclosed (including elevators)
- Maintain employee sensibilizations to respect a low noise level policy
- All slurry pipelines, transporting the excavated muck were buried all the way to the 2,000 $m^3$/h separation plant (STP).

As Segurane Street was too narrow and too short to accommodate such a large STP, it was decided to stage it in the harbor itself (at the Quay Cassini). The purpose of this plant is to:

- Separate mechanically granular solids from the outlet slurry
- Extract dissolved solids from the recycled slurry
- Treat the discharged effluent
- Improve inlet slurry rheological properties by adding fresh mud and additives,
- Evacuate separate solids to the barging points.

Figure 5. Before and after STP installation in the middle of the Old Harbor

As this quay is an historic monument within a protected touristic district, the STP had to be acoustically enclosed (for the residents) in a 120-meter-long building validated by both a nationally accredited architect and the chamber of commerce and industry. As the two agencies can carry antagonistic goals, coordination was long and complex. One of the compromises was to drape the building wall façades with historical renderings in order to ease its visual integration in this protected landscaping (see Figure 5). The project team had to develop the layout of the plant without impairing: traffic and activity from : road, old harbor, cruise ship and pedestrian tourists.

The spoil was transferred in daytime only from the STP to a 2,500 tons moored barge in the port, thanks to fully enclosed 1,,000' long conveyor belt with a 800 tons/hour capacity. The 14" ID slurry pipe, linking the TBM site (located in Segurane Street) to the STP (located on the quay), was buried under the streets for aesthetic purposes and for noise mitigation.

All those measures, anticipated and taken at early stages of the project, allowed the project to operate 24 hours a day, 6 days a week, even during peak tourist seasons.

## Old Sensitive Buildings in a Historic Protected District

The slurry TBM was launched from Segurane Street, towards Garibaldi square, passing its first underground station and connecting to the first tramway (constructed at grade in the nineties). Then, the TBM passed under a 300-foot-wide cut & cover, constructed to channel a seasonal river, when the rainstorms hit the city and the surrounding mountains, in the direction of the second underground station under Durandy Square. Then the tunnel alignment follows Dubouchage Boulevard towards the third underground station at the corner of Jean Medecin Avenue From there, the tunnel remains under Victor Hugo Boulevard through the fourth underground station called Alsace Lorraine, until it reaches the Grosso open pit where the TBM was dismantled after a 10,500 foot journey in the heart of the old city.

The ground cover varied from 15 feet at Grosso pit to 100 feet under an underground parking lot.

The entire alignment is surrounded by old buildings up to 200 years old with very little as-built records. During the first section of the tunnel, up to Dubouchage Boulevard, 40% of the tunnel passed directly under buildings. As most of them are deemed vulnerable by their location, the following measures were taken:

- Intensive preconstruction survey
- Identification of the most sensitive habitations by reviewing any existing drawings (where available), and by visual inspection inside the underground caves
- Settlement analysis for the most critical buildings
- Auscultation and assessment on a case-by-case basis, the risks and the mitigation to put in place
- Several buildings have received consolidation measures (from foundation repair to external support such as micropiles)
- Permanent motorized auscultation thanks to Sixense system, connected in real time with the TBM cabin, with total stations, inclinometers, vibration sensors and automatic piezometers measurement.

## Logistical and Operational Obstacles

As this is an old city: streets are very narrow, which is neither practical for semi-trailers nor comfortable for site staging. The delivery and assembly of the 31.7' slurry TBM elements required intense coordination as the site could only store two elements at a time, so, each item had to be assembled in the pit thanks to an on-site 140-ton gantry crane, the day after delivery, at the risk of facing a traffic jam on site without any ability to clear it. The TBM delivery involved over a hundred oversized transports, arriving at night. In spite of these challenges—night delivery, noise restriction, lack of storage, synchronized operation—the machine was assembled in three months, which is in line with the usual duration in a normal setting for this size of TBM.

Despite being identified at the start of the project, the problem of site storage impacted the project throughout its entire duration. The jobsite, in Segurane Street, could accommodate only 12 rings on the surface and six rings in the shaft for the night shift. Therefore, segment delivery had to be adjusted every day, depending on the TBM's progress, to avoid keeping the semi-trailers on site. If the truck was parked on site, all other deliveries (fly ash, cement, rails, consumables, etc.) had to be suspended.

By selecting a slurry TBM, the excavated spoil could be evacuated by barge for disposal at sea, instead of 100 dump trucks daily in the city's tourist area. However, the logistics for the 2,500-ton barges was also complicated to manage as the barge had to observe restricted time in the harbor with loading patterns, had to be filled as quickly as possible (due to cost), and had to maintain a steady rotation with 2 barges. A total of 200 trips were required to transport the 450,000 tons of spoil. As the barges could be loaded only in daytime due to noise regulations, four spoil basins were constructed in one of the STP buildings so that the TBM could operate at night. The 2000 tons capacity spoil basins were emptied in daytime by an excavator reloading the conveyor belts.

As the site was delayed by several geological events (see next chapters), the project team decided to save on the project's critical path by building the tunnel invert during tunneling (see chapter 4.2). It created another logistical challenge as the site had to receive 250 $yd^3$ of gravel and sand and 60 $yd^3$ every day.

## GEOLOGICAL CHALLENGES

### A Highly Variable Geology

As mentioned above, Nice's underground conditions feature multiple surprises and challenges, and the project team discovered this the hard way: it was mandatory to always adapt the operating parameters to variable ground conditions.

Slurry confinement was preferred for the following reasons (in order of criticality):

- Capability to stage the storage of the muck away from the launch shaft and bring it closer to the barging point
- Environmentally more friendly due to reduced noise and fewer pollutants in the spoil
- Improved confinement quality under sensitive buildings
- Due to geological constraints, during more than 50% of the project, the TBM will face silty cohesionless sand, with some coarse gravely sections. The remaining horizons were composed of limestones, clay, and conglomerate. This variability is demonstrated by the amount of ultrafine diluted in the slurry (see Figure 6).

#### *First Concern*

From the beginning of the drive, the amount of highly plastic clay (PI 20–25) and its tendency to clog caused several delays as several hyperbaric interventions at 2.5 bars minimum were required to manually unclog the cutter head. In spite of 12.5 tons/hour effective outcome, the triple filter press (extracting the ultra fines from the slurry) at the STP became the limiting factor of the TBM's progress.

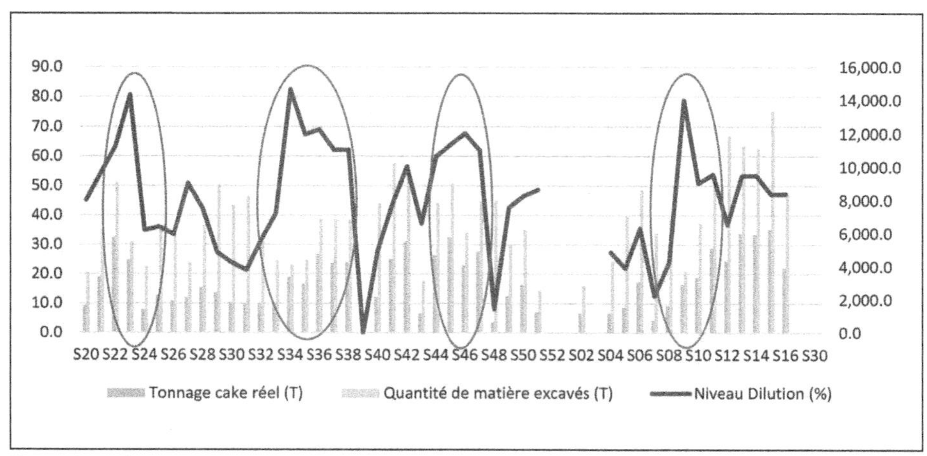


**Figure 6. % dilution per week, quantity of excavated soil per week**

### Second Concern

After a few thousand feet of excavation, between the first underground and the second underground station, the TBM encountered a large amount of fibrous peat. Beyond affecting the slurry properties, these organic fibers disrupted the efficiency of the hydro cyclones and the vibrating dewatering screens at the STP. Furthermore, downstream of the STP process, the conveyor belts faced multiple jamming events.

### Third Concern

Upon arrival at the second underground station, while on an uphill trajectory, very soft clay disrupted TBM guidance so badly that the TBM had to pass through the station diaphragm wall in a «crab» orientation, which completely blocked up the rear shield. The resulting high friction damaged the steel rear brushes to point that they had to be replaced on the spot in a risky location facing the risk of large water ingress. A second campaign of maintenance was then programmed in a dedicated improved ground plug inside the third underground station.

## Unusual Hydro-Geology

After 5,000 feet of excavation, 600 feet away from the third station under SOE construction, the cutter head was clogged up again by plastic clay, but this time, a higher amount of coarse sand and gravel was noticed at the outlet of the STP scalpers/trommel. Hyperbaric interventions were required to allow the TBM to resume its journey once again.

As the cutter head chamber was pressurized with compressed air (3.0 bars versus 2.7 bars of hydrostatic pressure) to allow «divers» to unblock the cutting tools, the compressed air traveled through the channel created by the coarse gravel until pressurizing and rupturing enclosed underground water pockets close to the slurry wall panels under construction. It led to the collapse of a freshly excavated panel during its concreting. As proof of the disruptive action of the compressed air, all piezometers within a 600-foot range started to blow water up to 6 feet in the air for the following 24 hours. The hyperbaric intervention had to be interrupted and the TBM had to resume its journey with a partially clean cutterhead until finding a less permeable location.

As the amount of plastic clay reduced after the third station, another large gravel underground channel was encountered. The TBM lost 35,000 $ft^3$ of slurry in this channel within few hours, while attempting to maintain the confinement pressure. When the TBM restarted, despite the information provided by a borehole a few feet away (which pointed out silty sand), an excessive amount of cobbles up to several inches in size was coming out of the STP scalpers. Maintaining face stability quickly became impossible with multiple blockages at the front chamber suction point and at the scalpers. Multiple attempts to boost the slurry properties to control the ground intake failed due to the large amount of ground water in the channel. The TBM, the slurry network and the STP suffered greatly due to these intense conditions. Although the confinement pressure never dropped below the hydrostatic pressure, the raveling excavated face led to ground losses, which were reported by compiling the various instrument readouts, and then several inches of surface settlement in the middle of Victor Hugo Boulevard was observed. Surface compassion grouting was later implemented to control the extent of the disruption.

## Anthropogenic Obstructions in the Last Tunneling Stretch

As if this was not enough, two more unexpected geological events kept the team on its toes.

At 500 feet before breakthrough, the TBM encountered an anthropogenic obstacle, most likely a poorly backfilled old water well. The TBM was progressing well in steady silty sand for weeks when the excavated face collapsed (due to sudden fluctuation of the compressed air bubble) during a slurry pipe extension. When the TBM attempted to restart, the TBM extracted a large number of gravels, bricks, steel elements, and construction debris. The cutter head was blocked. A collapse to the surface in the middle of the boulevard rapidly appeared.

In three weeks, the following measures were implemented to resume excavation:

- Non setting mortar injection by sonic drilling grout around the front shield and in front of the cutter head, acting as a curtain for the following stage
- Jet grouting injection above the front shield and the cutter head
- Jet grouting around the whole destabilized and decompressed area
- Over-Injection of bentonite slurry around the shield and in the excavation chamber
- Regular flushing and recycling of the excavation chamber to identify any cementitious pollution
- Installation of two additional cutter head motors to boost the installed unblocking power by 20%

The cutter head restarted after multiple cleanings and flushing without inducing any additional ground loss.

Then, 50 feet before breakthrough in Grosso open pit, another unidentified obstacle was encountered which resulted in another a surface collapse (with the site boundary) : the cover above the cutterhead was only 20 feet, when an old brick uncemented wall was touched by the TBM: another non-archived, unidentified obstruction buried under few feet of backfill. All the wall structural elements blocked the slurry suction point in the excavation chamber. Similar measures, as described above, were implemented in record time, thanks to the shallowness of the drive and the availability of the procedure and the equipment. After five days of work and hyperbaric intervention, the TBM could complete its journey (Figure 7).

**Figure 7. Cutter head lifting in Grosso pit after retrieval shaft being pumped out**

# TIMELINE CHALLENGES

## Crisis Control Mode

All stakeholders involved in the project knew the challenges ahead of this tunnel construction project. Everyone's concerns were justified as they went beyond the usual level of exposure in the tunneling industry.

Figure 8. MSV supplying TBM rolling on invert and invert ballast construction at the same time

The owner, the construction manager, the owner's technical advisor, and the EOR all managed to react promptly by being available to address the incidents when they occurred and by bringing in support to resolve it for the project's sake and for the neighbors. Pragmatic solutions were also considered before raising any contractual concerns, so that the project was not derailed.

As the problems multiplied along the path, the owner ran another extensive risk assessment analysis. It led to the implementation of preventive measures when there was any doubt: new boreholes followed, when required by some ground improvement operation (e.g., inert mortar injection in high concentration of gravels).

As the TBM was delayed by those unfortunate events, the owner understood that the project had to mobilize more equipment to remain on schedule: for instance, a fourth filter press was procured in order to cope with any excessive amount of clay.

### Tunnel Invert Construction During Tunneling

As part of the project's scope, the tunnel was to receive a 5-foot thick ballast below the final concrete slab. Thaumasia proposed the owner to initiate the construction of this section of the work during tunneling instead of doing it after TBM dismantling. This change of methodology saved two months on the project's critical path.

New equipment (trucks, tremie, silos), new personnel, and new materials formula were mobilized despite the congested site and complicated tunnelling work. (See Figure 8)

## CONCLUSION

Far beyond any common tunnelling projects, the construction of the single bore, twin tracks underground section of the Nice second tramway line pushed the limits of technical feasibility of transit projects in an urban environment due to its subsurface conditions and the above exceptional superstructures. This paper has shown that tunneling can be highly challenging, with its variety of challenging incidents, but solutions can be developed to meet and overcome them.

The success of the project relied first, on a highly professional and technical project team ready to cooperate and work together; second, on a contract that took into account the project challenges; and lastly, strong preparation at every stage of the project to anticipate risks.

Figure 9. Tunnel at hand over stage and tramway testing at Grosso pit

This project also demonstrates that large diameter tunnels (twin track, single bore) can be successfully completed in congested areas under with tight schedule, without annoying close neighborhood and impairing the social and economic life of the highly dense city

This section of the project was officially open to the public, 5.5 years after the award (See Figure 9).

# TBM Tunnel Offshore Connection

**G. Peach** ▪ Project Manager, Mott MacDonald (Doha, Qatar)
**H. Vigil Fernandez** ▪ TBM Manager, PORR (Doha, Qatar)

## ABSTRACT

The Musaimeer pump station and outfall project located in Doha Qatar, was designed to collect and manage ground and stormwater and then discharge the treated waters offshore. The project constructed a long outfall tunnel using an EPB TBM with segmental lining, each tunnel ring being 1.35 m in length with an internal diameter of 3.7 m.. The diffuser structure was a 6-arm structure, measuring 294 m by 40 m connected to a central manifold in turn connected to the TBM constructed tunnel via a riser shaft drilled from an off-shore barge. This technical paper will discuss the planning, methodology and construction practices employed to connect the TBM tunnel to the diffuser structure 10.2 km offshore.

## INTRODUCTION

The Musaimeer Pumping Station project is governed by the Qatar Construction Specifications (QCS, 2014). The Public Works Authority (PWA) have engaged in two contracts. Firstly Mott MacDonald Ltd., which is the Project Management Consultant (PMC) company, and HBK-PORR JV, which is the Design, Build, Operation and Maintain Contractor for the construction of the MPSO Project, which is the final receiver of the upstream drainage system and located at the end of the Abu Hamour Tunnel, immediately south of the Hamad International Airport (HIA). The outfall tunnel extends from the pump station, 10.2 km offshore discharging the flows into the Arabian Gulf. The discharging will be performed through a vertical riser shaft and a marine outfall diffuser field. Figure 1 shows the project location.

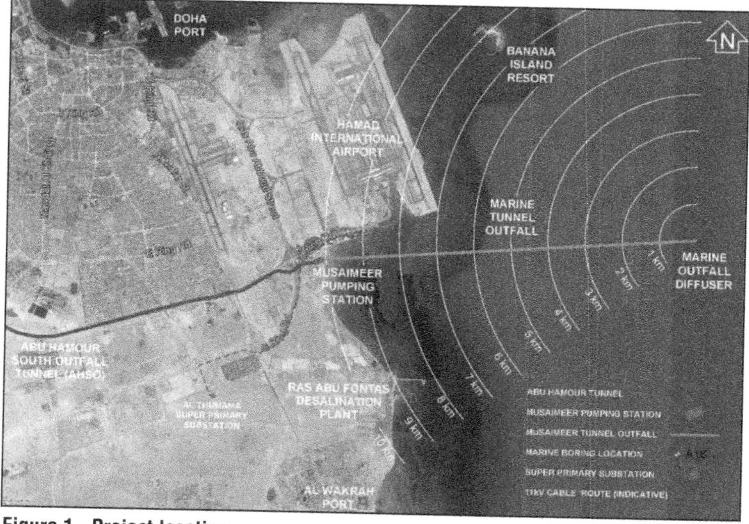


**Figure 1. Project location**

## DESIGN

The project three main components consist of a dump station, 10.2 km out fall tunnel and a diffuser bed structure attached to the end of the outfall tunnel. The system operates by the pumps lifting the water inflows up from the lower level –48 m to a ground level outfall chamber which has a top of structure level of +14 m. The inflows then pass through this structure and into a drop shaft which connects to the outfall tunnel, the connection of the drop shaft and outfall tunnel is the lowest point in the outfall tunnel. The outfall tunnel then rises at 0.005% along the alignment until it reached the riser shaft which connects to the diffuser bed system. Thus the system is a gravity operated system. Figure 2 illustrates the overall project hydraulics.

The marine diffuser field structure is shown in Figure 3. The central manifold is located directly on top of the riser shaft and is a concrete structure measuring 4 m × 4 m × 4 m and weighing 120 Tons. Connected the main manifold are 6 radial arm structures which either directly connect to the manifold or are connect by short secondary manifolds. Each arm is 147 m long and is separated from the adjacent arm by 20 m. This makes the entire structure outline 294 m × 40 m. The secondary manifolds and 6 arms are manufacture out of High-Density Polyethylene (HDPE) pipe and are of various diameters 2000,1400, 1200 and 710 mm to ensure an even flow throughout the various arms under varying flow rates from the pump station. The discharge from the structure to the sea is via 84 duck bill valves located equal spacing along the 6-arm structure.

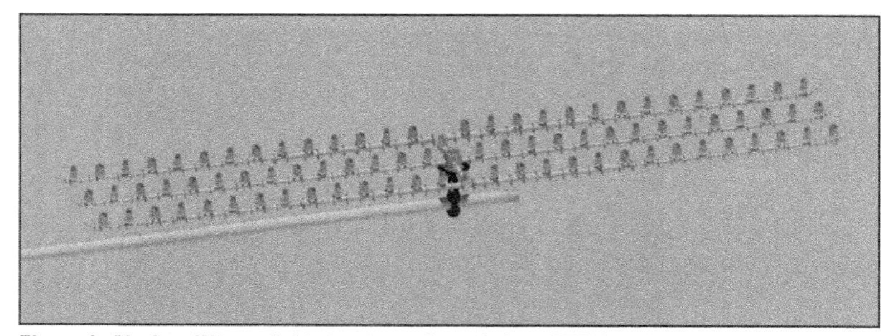

Figure 3. Marine diffuser field structure, riser shaft and TBM outfall tunnel

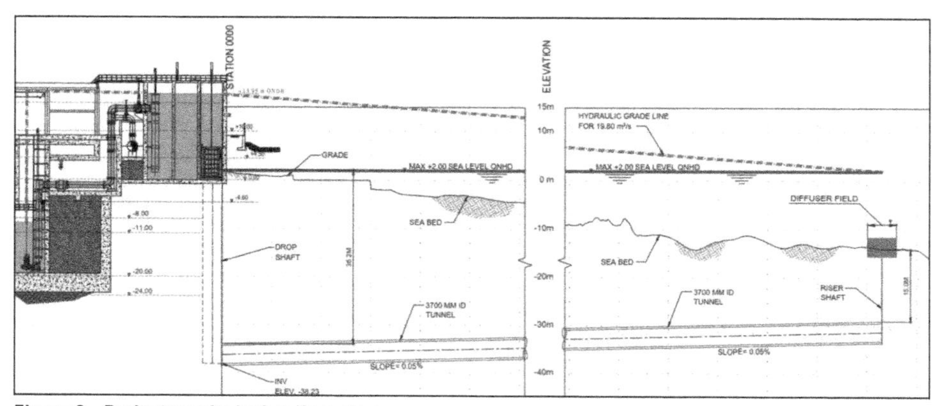


Figure 2. Project gravity hydraulic system

## DREDGING AND MATERIAL STOCKPILING

The dredging was carried out by a Cutter Suction Dredger (CDS), which consists of a cutter mechanism mounted upon a barge and is capable of being lowered and raised from the seabed allowing excavation of the diffuser field pocket, which was 300 m long, 60 m wide and 5 m deep. The total matter excavated was 65,000 $m^3$. The excavated material was placed in a stockpile located 50 m away and transportation of material was via a floating discharge pipe. All excavated material was later reused for back filling purposes, significantly reducing the environmental impact of the marine works.

## RISER SHAFT CONSTRUCTION

The riser shaft is the structure which connects the TBM outfall tunnel to the marine diffuser bed. The invert of the outfall tunnel is 14.5 m below the seabed and the depth of the sea at this location is 15.5 m. The specialist Contractor engaged for the section of the work by HBK PORR JV was MIC (W.L.L). The first stage in the construction is carried out by a dredger which on average removes 2.5 m of the loose sediments located on the seabed. Near the riser shaft a further 2 m is excavated to ensure competent strata is exposed which will improve the drilling operation required to construct the riser shaft.

With the dredging complete the next stage of the construction is carried out from a jack-up rig which is located direct adjacent to the centre line of the riser shaft. A cantilever frame is then installed to the side of the jack-up rig, this is the location the drilling equipment will be installed. The drilling rig or Bottom Hole Assembly (BHA) is shown in Figure 4. The BHA is lowered down to the correct location and commences to drill the rise shaft, shown in Figure 5 and 6. The BHA can dill various diameter holes from 4 to 5.3 m by use of extending hydraulically activated arms to suit the desired diameter of shaft. In this case the first section was drilled at 4 m diameter and then the arms are extended out, as shown in Figure 7 to the 5.3 m settings. The 5.3 m section is the location through which the TBM will excavate, giving a clearance of 500 mm each side of the TBM, which has an outer diameter of 4. 3 m.

Figure 4.

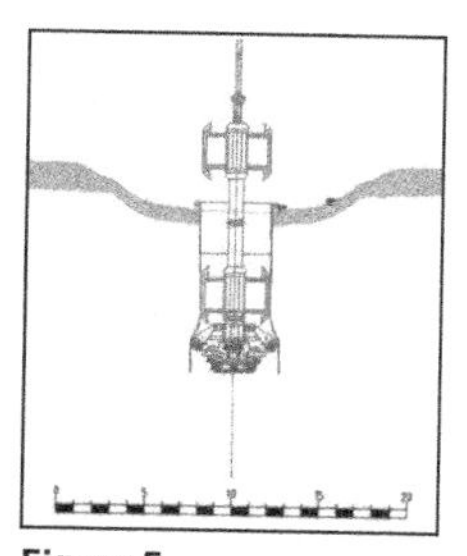

Figure 5.

On completion of the riser shaft drilling the BHA is then removed, and a three-dimensional survey carried out to determine the exact dimensions and position of the riser shaft, shown in Figure 8. The next stage shown in Figure 9 is to install a steel casing into the top 3 m of the riser shaft, to provide stability and prevent collapsing of the shaft inwards. Four Concrete manifold foundation blocks are

Figure 6.

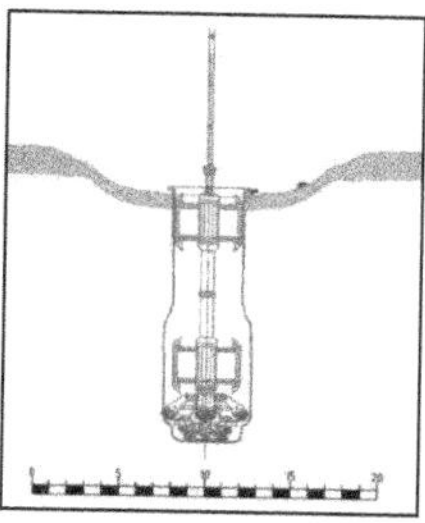

Figure 7.

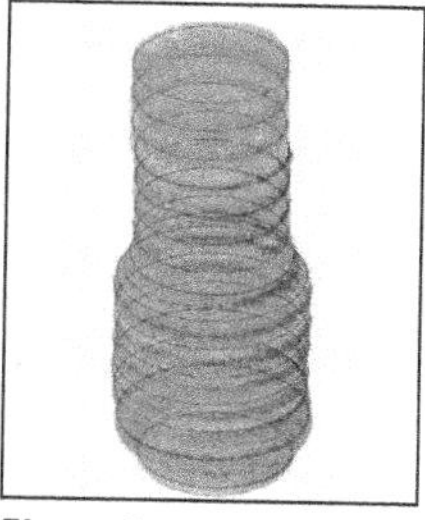

Figure 8.

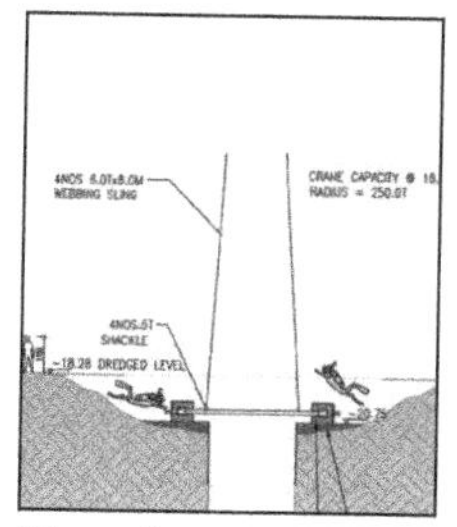


Figure 9.

then positioned and levelled into the correct position and connected to the steel casing, these activities are shown in Figure 9 and 10. A manifold concrete base slab (Figure 11) is then positioned on top manifold foundation blocks as shown in Figure 12.

Figure 10.

Figure 11.

The riser shaft internal lining protects downwards from the base of the concrete manifold to 500 mm above the TBM outfall tunnel and is made from Glass Reinforced Plastic (GRP). This is a cylindrical shape with a fixed end at the lowest end and a removable flange at the other end. A temporary steel support frame is installed within the GRP cylinder, refer to Figure 13. The GRP cylinder and steel frame are then installed into the riser shaft as shown in Figure 14 and positioned centrally within the riser haft and connected to the base slab. At this point the riser shaft is still full of sea water. Figure 15 shows the temporary kentledge which is then placed on top of the whole structure to prevent uplift during the grout placement.

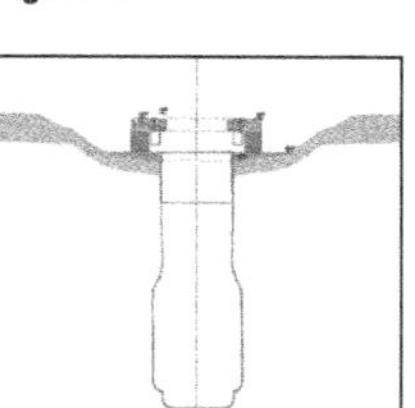

Figure 12.

Figure 13.

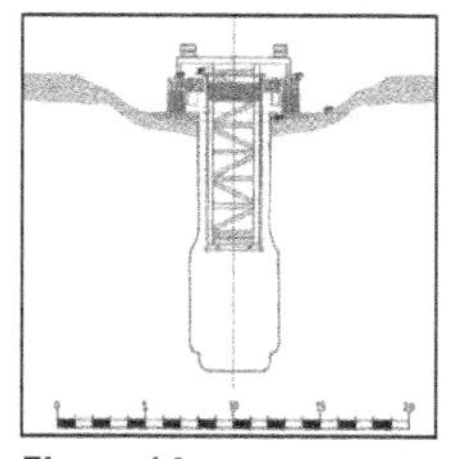

Figure 14.

Figure 15.

The void between the GRP cylinder and the whole of the lower portion is then installed with a grout mixture which fills the void in a controlled manner, slowly displacing the seawater. The grouting is shown by the blue color in Figure 16.

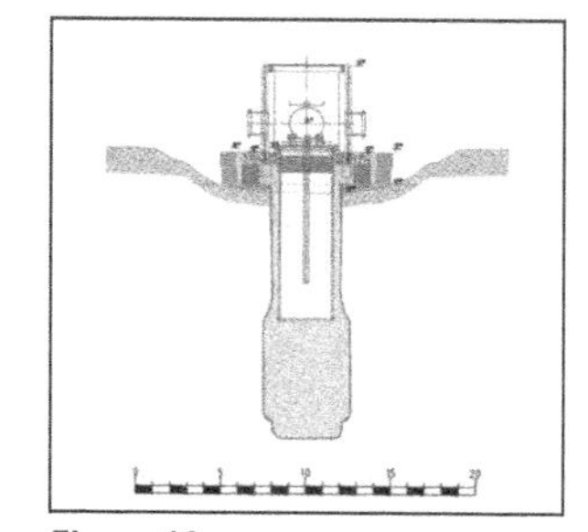

Figure 16.

When the grout has cured the kentledge is removed along with the internal steel frame and the upper flanged is secured and sealed to the top of the GRP cylinder. Both top and bottom ends of the GRP cylinder have a series of installed valves which allow venting of air or water. The upper valves are then used in connection with compressed air to remove all the seawater within the GRP cylinder and leave only air within the GRP cylinder.

The central manifold which collects the combined ground and storm water from the outfall tunnel and then distributed evenly to the diffuser structure, was constructed on shore, and weighed 120 tons The concrete manifold is shown in Figure 17 and is then lowered and positioned on top of the riser shaft and connected to the manifold base slab as shown in Figure 18.

Figure 17.

In due course all the diffuser field structure is connected piece by piece to the central manifold until the full diffuser bed structure is installed and covered with back fill material and then a final layer of scour protection installed and covering the entire diffuser bed structure, this shown in Figure 18. At this point the riser shaft structure is ready to receive the TBM constructing the outfall tunnel.

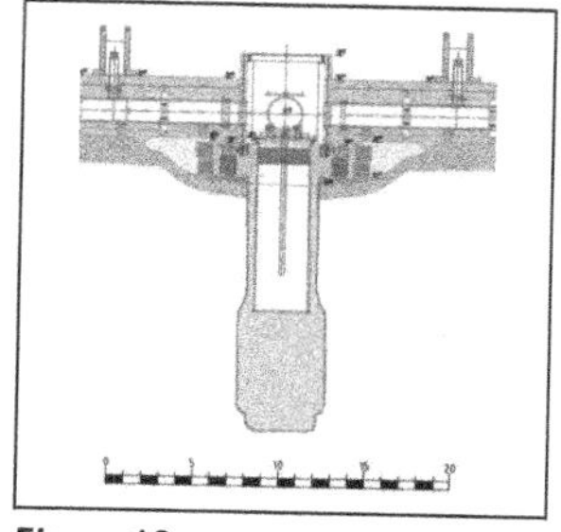

Figure 18.

## PREPARATION FOR OUTFALL TUNNEL CONNECTION

The programme for the overall project required the diffuser bed structure to be completed prior to the arrival of the TBM, this was achieved. The TBM excavated to within 5 m of the riser shaft shown in Figure 19 and depressurizes the cutterhead and monitors the situation for any water inflow over 8 hours. This was to test if the riser shaft was watertight to ensure the future connection works to be carried out in a safe manner. In this case there was no water inflow and thus the TBM excavates through the lower portion of the riser shaft and stops 25 m beyond the riser shaft as shown in Figure 20.

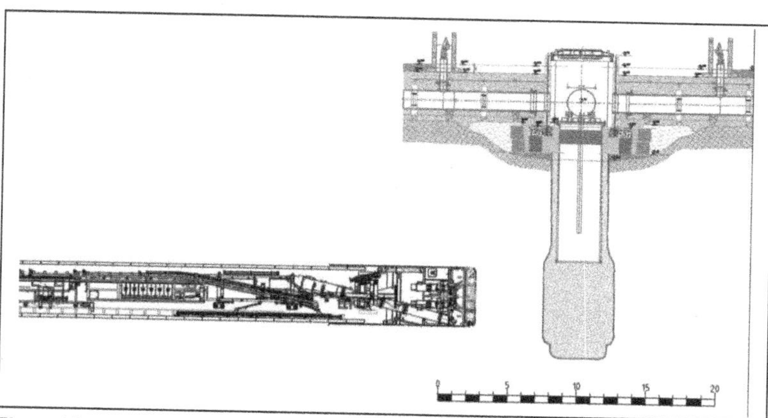


Figure 19.

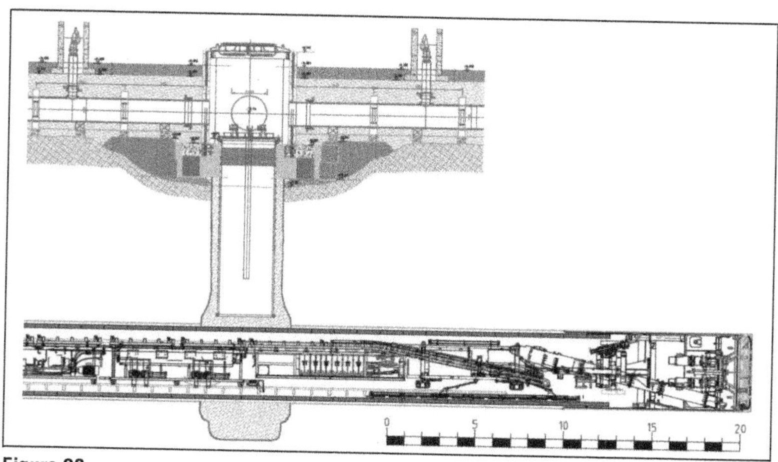


Figure 20.

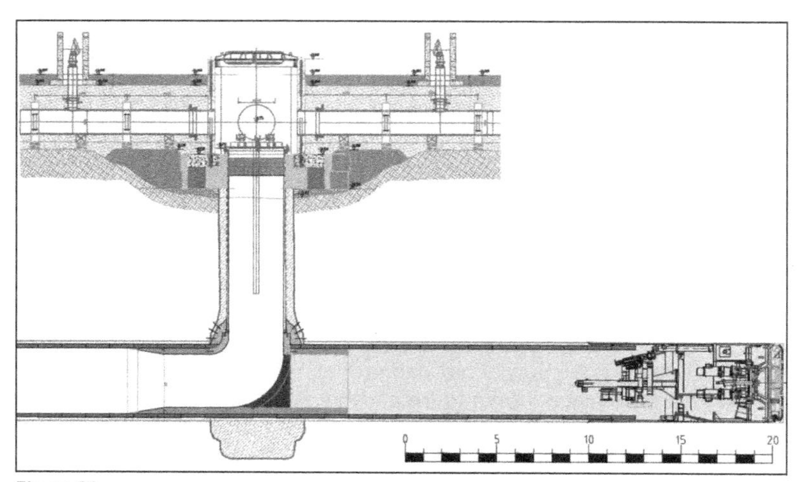


**Figure 21.**

The TBM and associated plant and equipment is then dismantled and removed back through the outfall tunnel and at the pump station location removed to the surface. The 25 m tunnel beyond the riser shaft is filled with concrete, refer to Figure 21, and then there follows very intricate and complex series of construction activities requiring precise logistical planning and execution to complete the connection of the outfall tunnel to the riser shaft and installation of the permanent works all of which is carried out 10.2 km offshore.

## TBM Approach to the Riser shaft

As the TBM approached the riser shaft a series inspections were planned checks with the aim to inspect the conditions of ground and water ingress conditions, the geological and geotechnical conditions of the surrounding rock mass, the alignment of the TBM against the riser shaft position and the conditions of the grout block. The checks involved excavating to a redetermined position and depressurizing the face for 8 hours and then measuring water inflow if any at this location. At chainage 10+146.5 m TBM will stop for the 1st inspection (See Figure 22).

Based on the monitoring assessment during the stoppages, it allowed the TBM crews to open the chamber in atmospheric conditions, the outcome of the inspections/interventions can be summarized as follow:

- Water inflow from the tail shield, around 0.6 l/sec.
- During the cutterhead and cutting tools inspection at the first stoppage, few tools were change, while the remaining stoppage no replacements were required.
- The face mapping carried out showed full Simsima Limestone with high RQD value, face was wet with not water inflows.
- The second stoppage confirm the interface riser shaft-outfall tunnel, being in this case the beginning of the grouting plug. The face mapping also confirms the good conditions of the bounding between grout and rock; while water inflow was stable inline with the amount of water evaluated during the first stoppage.
- During the third and last stoppage, the cutterhead was found surrounded by grout plug of the riser shaft, confirming the correct alignment of the riser shaft

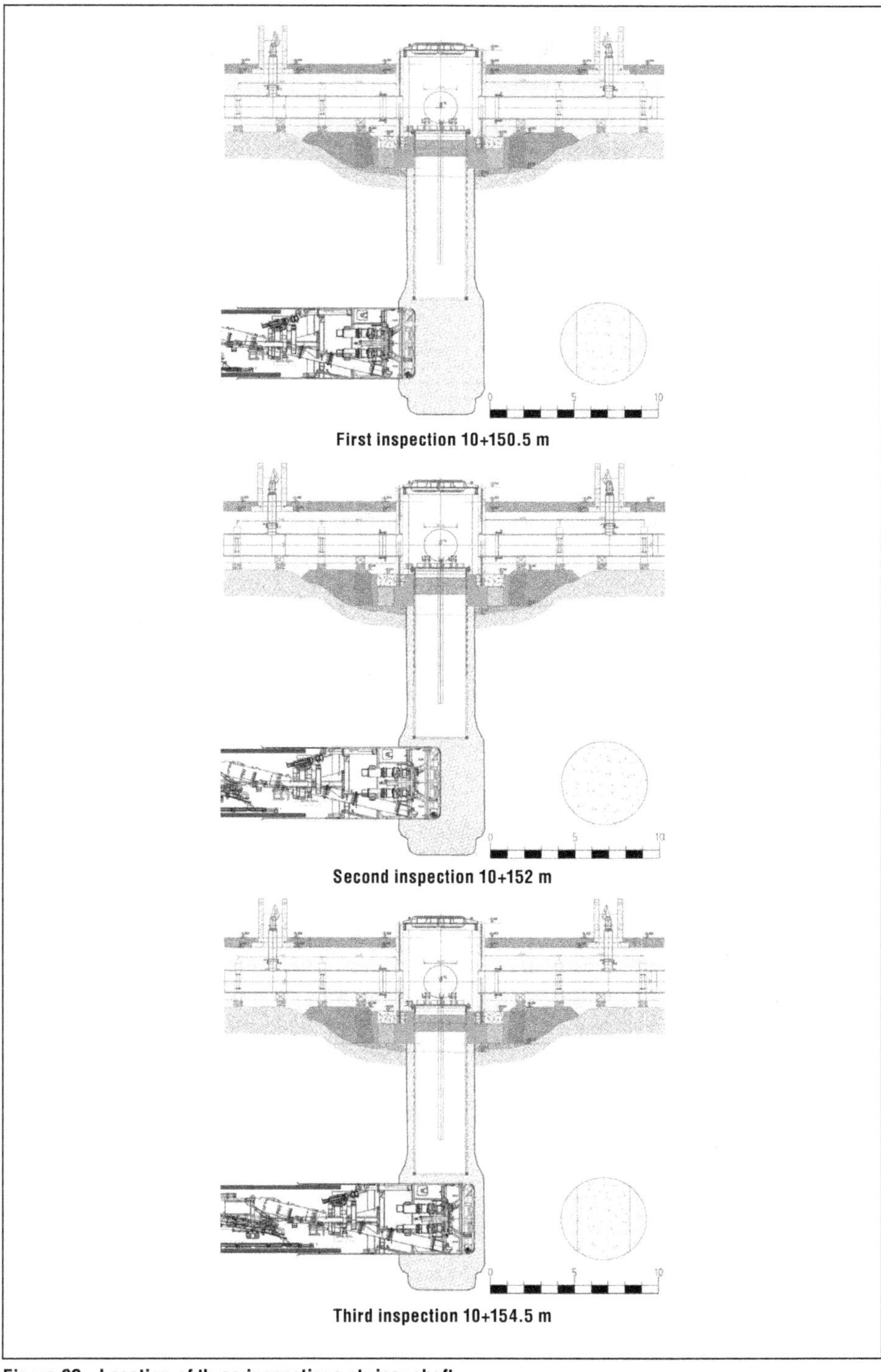


Figure 22. Location of three inspections at riser shaft

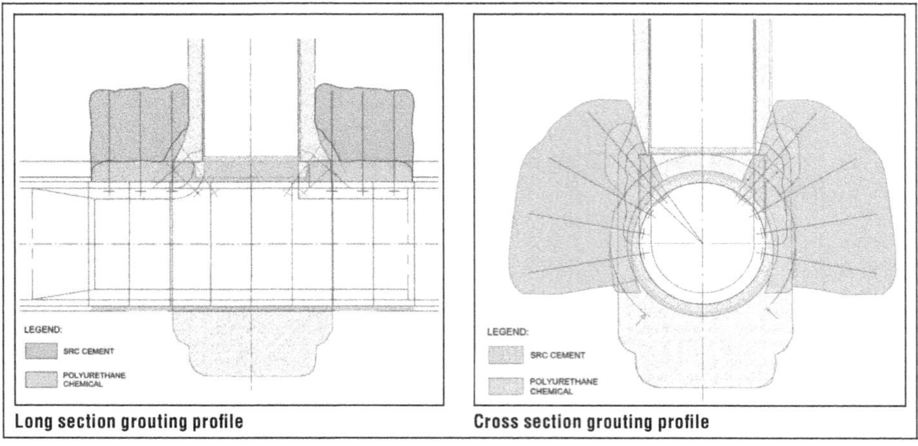


Long section grouting profile

Cross section grouting profile

**Figure 23. Grouting works around riser shaft**

with the outfall tunnel. The high strength value of the grout plug was also confirmed and the water inflow was minimal.

## TBM Sealing and Removal

Upon completion of the successful third inspection the TBM was then driven a further 33 m and stopped. The TBM cutter head and excavation chamber was then filled with grout and the front of the TBM sealed up. All the TBM gantries and associate equipment was then removed back through the 10.2 outfall tunnel.

## Grouting Works and Drainage for the Riser Shaft

In order reduce the possibilities to a minimum water inflow amount during the necessary excavation activities to connect the outfall tunnel with the riser shaft several activities related to grouting were carried out once the TBM was disassembly and the surveyor confirmed the location of the riser shaft. The two stage strategy followed was:

- Secondary grout using SRC cement.
- Chemical injection and rock improvement using polyurethane.

The secondary gout activities were applied to cover 10 rings length i.e., 2 rings before the beginning of the grout zone, 4 rings between beginning end of the grout zone, and 2 rings after the end of the grout zone, with this type of grouting the gaps between the rings and annular grout would filled and the possibility of water inflow minimized.

The chemical injection strategy followed was dividend in 4 different stages with different drill pattern, drill length and diameter, type of chemicals, pressures and stops criteria (See Figure 23).

The length of injection holes were considered variable from 0.4 m up to 3.5 m length around the outfall tunnel and riser shaft. The total quantities on material injected through 415 injection holes are summarized in 16,093 kg of OPC cement combined with 5,412 kg of 2 components of PU chemicals (See Figure 24).

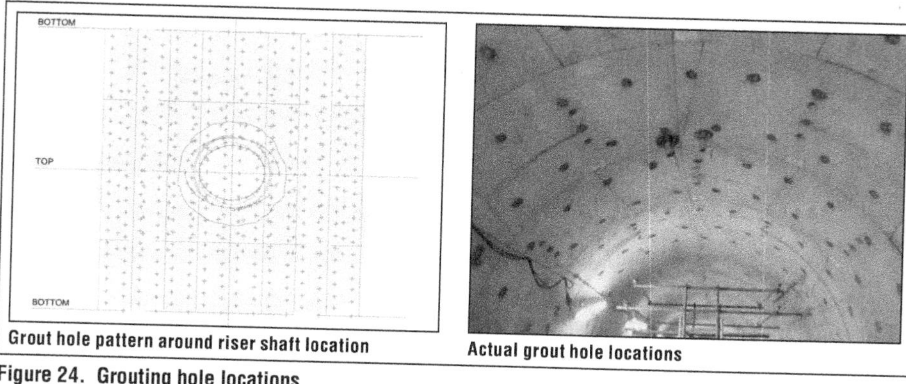

Grout hole pattern around riser shaft location
Actual grout hole locations

Figure 24. Grouting hole locations

## INNER LINING WORKS

The next stage of works to be carried out in the connection area was the installation of the permanent cast in situ internal lining. This permanent lining is constructed to create the necessary support to the segmental tunnel lining, during the excavation opening and connection construction activities to the invert of the riser shaft This part of permanent lining is extended either sides of riser shaft, reinforcing the tunnel and maintain a window on the riser shaft invert , to allow for the connection works to proceed. the stages of pouring concrete are shown in Figure 25.

Reinforcement used was Stainless Steel Grade 1.4436 or equivalent in line to Detail Design requirements and Project Durability Assessment Report (See Figure 26).

Single sided shutter was used for the casting of the inner lining. The concrete quality was Self-Compaction Concrete (SCC) C50/60 mix design was developed for the permanent structures considering the required time to reach the pouring destination, this period is expected to be 4 hours. The concrete was transferred from the mixer trucks to the bottom of logistics shaft via mobile pump directly into the concrete cigars by using the TBM rolling stock and finally pump the concrete into the shutter by static pump.

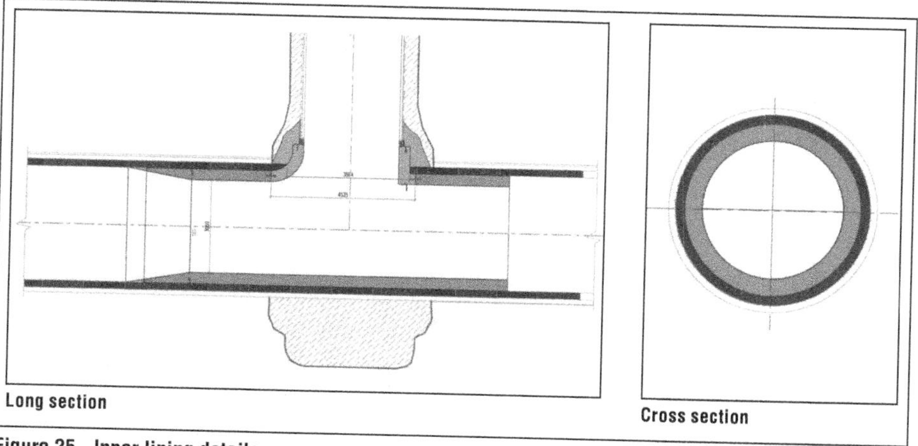
Long section
Cross section

Figure 25. Inner lining details

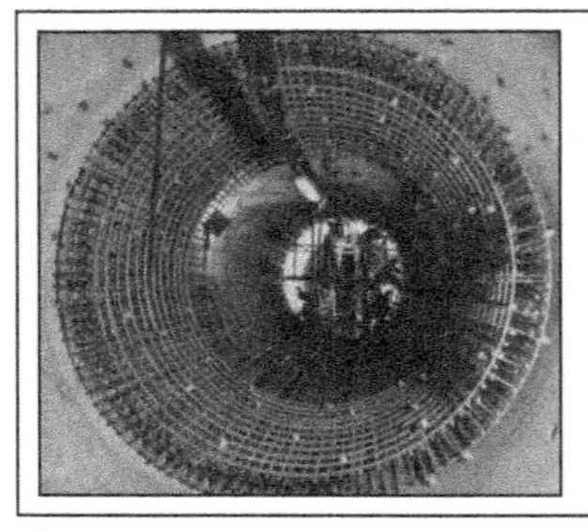  

**Figure 26. Inner lining reinforcement details**

## SEGMENT REMOVAL AND CONNECTION WORKS

A total of 11 segments were demolished and removed (some partially some as whole) at the location of the riser shaft inverts and outfall tunnel roof. All exposed segments were supported one to each other, on the same ring, with a steel structure consisting of horseshoe shape HEB120 and horizontal UPN120 beams.

The exposed roof segments were removed/demolished creating an opening from the centre of the riser shaft and then expanding the area towards the laterals. The arrangement for the removal/demolition is as follows:

1. Step 1, 2, 3, 4 will be done by coring drill with different diameters from 25 mm up to 200 mm; step 1 and 2 with the depth of concrete segment approximately while steps 3 and 4 will penetrate into the grout plug 100 to 200 mm. Coring bits diameters might experience small variation as per bits size availability. After every step, water inflow was observed and evaluated.
2. Steps 5, 6, 7 was the breaking of the segment and penetrate the grout plug towards the riser shaft.
3. Steps 8 and 9, excavation continue to locate the steel flange of the riser shaft invert.
4. Steps from 10 to 21 are the enlargement of the opening in stages (diameter and depth wise) as shown in the figure to stabilize the excavation following the philosophy of NATM (New Austrian Tunnelling Method).

Once the excavation was completed, the next step was to erect the scaffolding platform to cut the manhole flange provided for the purpose and the bottom of the riser shaft GRP pipe and then proceed with the cutting and lowering of the steel flange cup (7 ton), shown in the photography in Figure 27.

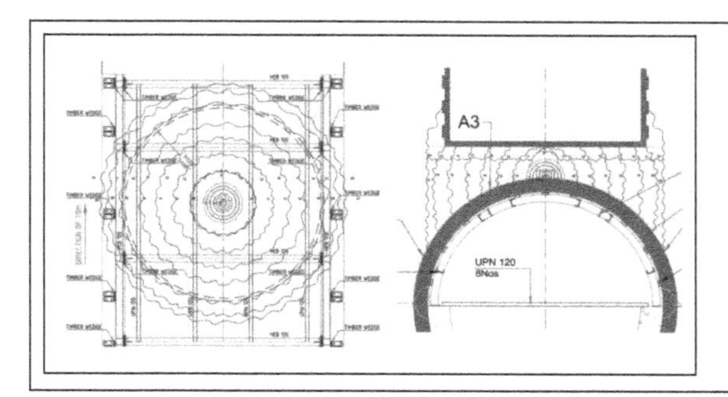

**Figure 27. Details of the excavation to connect the outfall tunnel and riser shaft**

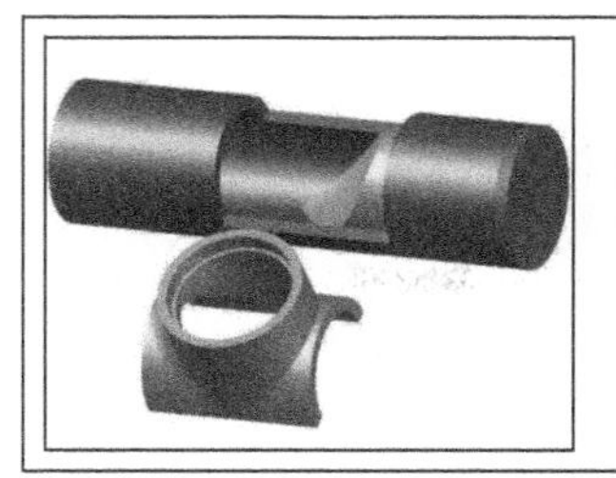

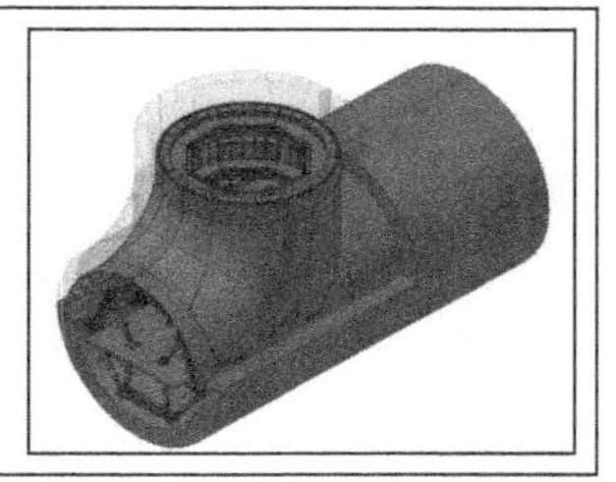

Figure 28. Details of the connection to the tunnel and riser shaft

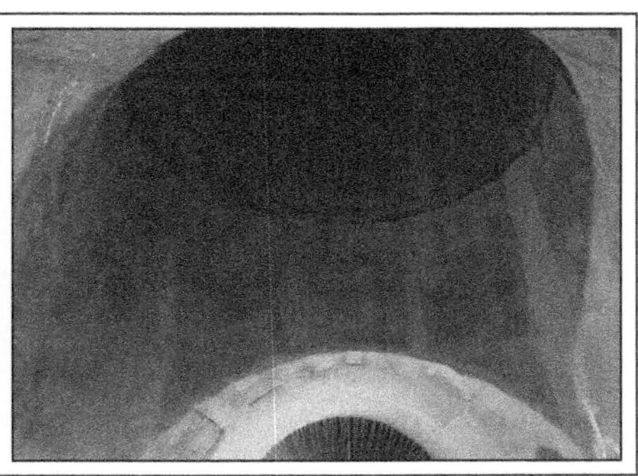

Figure 29. Concrete collar formwork and final finish

After the removal of the flanges, two out of the three main components of the project were physically connected. The activities continue with the installation of the water bars around the GRP pipe and reinforcement installation of the collar as shown in the Figure 28.

The concrete collar must be placed in one step to ensure the absence of construction and movement joints, a special shaped shutter was developed. For that a BIM model was generated not only facilitate the re-design of the formwork considering the design tolerances and deviation, the model was able to be fed directly with the data from the wriggle survey but also to simulate the transportation and assembly in a very confined space without lifting equipment, this saved time in mobilizing the formwork to the riser shaft location. Details of concrete connection collar are shown in Figure 29.

## HYDRAULIC BEND

On completion of the concrete collar the final step in completing the works was the backfilling of the 25 m TBM excavated beyond the riser shaft location. Figure 30 shows the details of the 25 m TBM tunnel being backfilled and the formwork for the hydraulic bend.

The back fill of the TBM was done using annulus grout since it is easy to transport, easy to pump and adjust the gel time to easy to allow flow into all areas of the TBM tunnel. The final stage of the connection works was the construction of the of the hydraulic bend/wall itself, for that again the BIM model was used to design an specific and unique shutter shape. In this case it is only mass concrete but stainless-steel dowels were required on the annulus to improve the shear effect of the water pressure

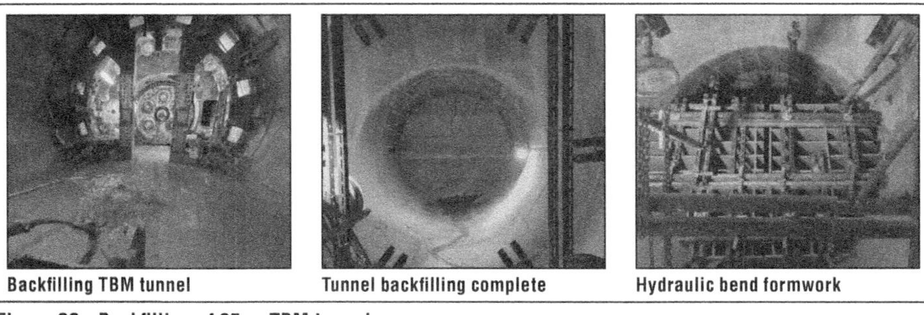

Backfilling TBM tunnel | Tunnel backfilling complete | Hydraulic bend formwork

Figure 30. Backfilling of 25 m TBM tunnel

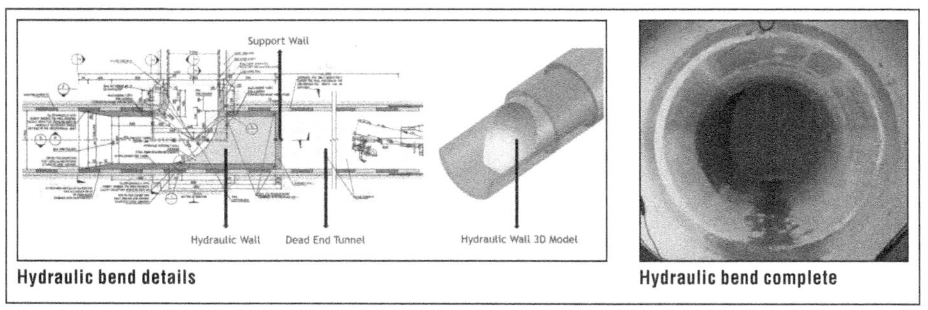


Hydraulic bend details | Hydraulic bend complete

Figure 31. Details of the excavation to connect the outfall tunnel and riser shaft

against the wall during the operation stages. The final details of the hydraulic bend and the actual completed structure are shown in Figure 31.

## CONCLUSIONS

This was the first of three possible outfall projects planned by Ashghal for the long-term management of rising ground water level and storm events, which will bring significant benefits to the population of Doha and surround areas.

There were many lessons learned from the project and these have all been categorized and been considered by Ashghal for future projects.

The project was a significant success and appreciation is extended to all parties involved in coming together and working to achieve the successful completion of this project.

## REFERENCES

Karagkounis, N., Sayers, K., Latapie, B., and Mulinti, S.R. Geology and geotechnical evaluation of Doha rock formations ICE Geotechnical Research UK, Volume 3, Issue 3. pp. 120–121.

Sadiq, A.M., Nasir, S.J., " Middle Pleistocene Karst Evolution in the state of Qatar, Arabian Gulf. Journal of cave and Karst studies, pp. 137–139.

Poulos, H.G., A review of geological and geotechnical features of some Middle Eastern countries—Innovative Infrastructure solutions 2018, Springer, pp. 6–8.

PART 

# Microtunneling and Trenchless Tunneling

*Chairs*

**Leah McGovern**
STV Inc

**Jay Sankar**
Amtrak

# Curved Microtunneling to Reduce Disruption in City Environment

**Daniel Cressman** ▪ Black and Veatch
**Allan Rocas** ▪ Black and Veatch
**Aswathy Sivaram** ▪ Black and Veatch
**Tatiana Chiesa** ▪ Metro Vancouver (formerly City of Toronto)

## ABSTRACT

The City of Toronto plans to construct a 1,350 mm consolidation sewer along the East Don Roadway (EDR). The project includes installation of approximately 260 meter of 1,350 mm diameter concrete pressure pipe through microtunneling methods. The sewer alignment crosses underneath a busy 5-lane arterial road and then follows the very narrow East Don Roadway road allowance. This paper discusses the significant challenges overcome to design the Don Roadway sewer using slurry microtunneling methods through mixed face conditions below the groundwater table in a congested road allowance. A curved microtunnel was ultimately required to avoid conflicts with existing utilities, condominium tiebacks and a bridge structure.

## BACKGROUND

To improve water quality in Lake Ontario and the Don River, the City of Toronto is embarking on an extensive wet weather flow (WWF) management project to capture, store and treat existing WWF outfalls. To achieve this objective a system of diversion structures, consolidation sewers and storage conveyance tunnels is being constructed. Once complete, the WWF project will provide approximately 600,000 cubic meters ($m^3$) of WWF storage in tunnels along the Don River and Lake Ontario shoreline. An existing WWF outfall at Queen Street has been identified as one of 11 outfalls along the Don River to be connected to the WWF storage tunnel, referred to as the Coxwell Bypass Tunnel (CBT). The East Don Roadway (EDR) consolidation sewer is being constructed to divert and collect flow from the existing Queen Street wet weather flow (WWF) outfall into the CBT.

The East Don Roadway (EDR) consolidation sewer picks up flow from the existing Queen Street Outfall, traverses south underneath the East Don Roadway crossing Eastern Avenue to the Vortex Drop Shaft which drops flow 50 meter vertically to the CBT. As detailed in Figure 1, the majority of this EDR consolidation sewer will be installed by means of microtunneling, and a section of the consolidation sewer will be installed with traditional open-cut methods. The microtunnel length has been broken up into two (2) sections; the first drive being installed between MH-04 and MH-03 to cross Eastern Avenue, and the second drive being installed between MH-02 and MH-01 to install the consolidation sewer beneath the EDR. The proposed alignment of the 1,350 mm EDR consolidation sewer is detailed in Figure 1.

The project area, depicted in Figure 1, is located in Toronto, Ontario, Canada, just east of the downtown City core. The project area is bound by Queen Street East to the north, the Don Valley Parkway on-ramp off of Eastern Avenue to the south, the multilane Don Valley Parkway highway to the west and new mix use commercial and residential condominium high rise development to the east.

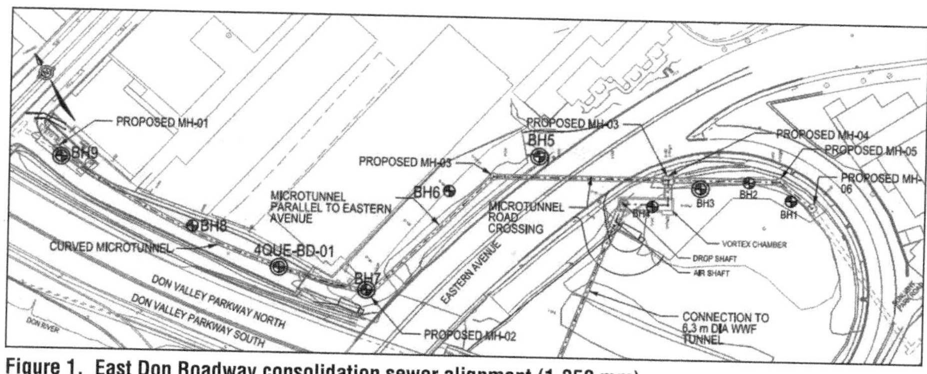


**Figure 1. East Don Roadway consolidation sewer alignment (1,350 mm)**

Eastern Avenue is classified as a major arterial road in the City of Toronto. It is a multilane roadway with three (3) eastbound lanes and two (2) westbound lanes. Eastern Avenue transitions to a bridge within the project area to cross over the Don Valley Parkway and Don River and facilitate traffic movement in and out of the Downtown Toronto core.

EDR is considered a local road however there is a fair amount of traffic as it is being utilized by the new adjacent commercial car dealerships as well as the new residential condominium developments. Within the right of way of EDR there is also a plethora of existing utilities including high pressure oil pipelines. Moreover, during construction of the new commercial and residential buildings off of EDR, tie-backs were left in place within East Don Roadway road allowance.

The congested nature of the project site and risks associated with construction of the EDR consolidation sewer in close proximity to the high-rise developments and a major commuter bridge were considered in detail in selection of the EDR consolidation sewer means of construction and detailed design. This process will be discussed in more detail herein.

## SUBSURFACE CONDITIONS

A geotechnical investigation was performed between April and May of 2020 by Wood Environment and Infrastructure Solutions (WOOD) to obtain information on the subsurface conditions along the 4SUN/4QUE Consolidation Sewer Alignment. The investigation consisted of advancing nine (9) boreholes (BH-1 to BH-9) and the installation of four (4) monitoring wells (BH-3, BH-5, BH-7, and BH-9). These boring locations are shown in Figure 1 and Figure 2, in plan and subsurface profile along the project alignment respectively.

The microtunnel alignments, between MH-02 and MH-01 and MH-03 and MH-04, are anticipated to be excavated solely in overburden soil baselined as Engineering Soil Class A Fill Material and Engineer Class D Clayey Soil.

The Class A Fill Material is anticipated to be encountered in <10% of the microtunnel alignment in a section of tunnel underneath the East Don Roadway from Station 0+080 to 0+120. The Fill Material has been placed by man-made processes with random and broad compositions which can include man made material such as debris; cobbles/boulders, concrete, pavements, glass, ashes, wood and or other organics.

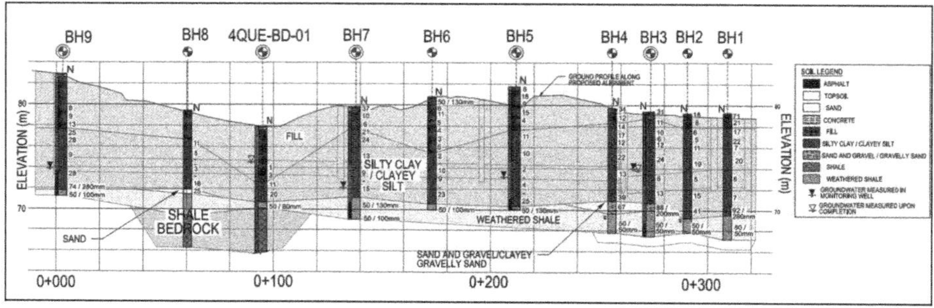


Figure 2. Anticipated subsurface conditions

The behavior of the fill materials is unpredictable and can vary from running to raveling to firm. Tunnelling through this zone with appropriate ground support will be critical as there is an existing 750 mm × 375 mm hydro duct bank directly above the tunnel alignment in this location.

The Class D Clayey Soil is anticipated to be encountered in >90% of the microtunnel alignment, all reaches of the tunnel other than the section of tunnel underneath the East Don Roadway from Station 0+080 to 0+120. The Clayey Soils are broadly graded, low to intermediate to high plasticity soils ranging from clayey silt to silty clay to clay with up to 50% coarse grain particle size. Cobbles and boulders are anticipated. This material is anticipated to be firm upon initial exposure, but will transition to slow-raveling over the course of a day on account of fissuring and the variability in these cohesive soils. Sand and gravel components of the material can be abrasive to cutting tools. Interbedded layers of high permeability sand and granular material are anticipated to be encountered.

## CRITERIA AND CONSTRAINTS

### Hydraulic Criteria

In the future, the EDR consolidation sewer is to be connected to divert flow from two outfalls to the WWF Tunnel: (1) the existing Queen Street Outfall sewer, just upstream of the microtunnel terminus shaft (MH-01) and (2) the existing Sunlight Park Road Outfall Sewer, south of the microtunnel alignment. The connection just upstream of the microtunnel alignment feeding into the future Queen Street Outfall sewer needs to be completed at the existing sewer invert elevation of 73.633 m to maintain adequate flow. A minimum gradient of 0.25% is to be maintained throughout the new EDR consolidation sewer in order to maintain adequate sewer velocities. An increased gradient or drop in the vertical alignment could be utilized after diverting flow from the existing Queen Street Sewer although this would require excavation of deeper shafts and a deeper vortex approach channel to the main drop shaft.

Flow from the 4SUN/4QUE sewers will collected into the vortex chamber, and then flow down a drop shaft into a deaeration tunnel, followed by a 13,200 mm diameter adit tunnel at a 0.2% slope, to connect into the CBT at an invert elevation of 28.04 m.

### Existing Constraints

As noted in the background, the East Don Roadway is congested with buried utilities. Some of the existing utilities within the roadway include Toronto Water infrastructure such as sewers and watermains. Additionally, there are several highly sensitive concrete structures such as concrete duct banks and concrete chambers, which are used to house electrical and communication infrastructure. A rare type of infrastructure which is located within East Don Roadway are three (3) parallel high pressure oil pipelines owned by Sun-Canadian, Imperial Oil and Trans-Northern respectively.

In order to mitigate conflicts with existing utilities a comprehensive subsurface utility engineering investigation (SUE) was carried out during detailed design to map out the location of the buried infrastructure. T2 Utility Engineers was retained by Black and Veatch to conduct the utility investigation which included a combination of both a desktop study in the review of existing record drawing and as well as an on-site field investigation. Through the field investigation, a total of twenty-nine (29) tests pits were carried out on EDR to expose buried infrastructure which could potentially conflict with the proposed microtunnel shaft and the microtunnel drive itself.

A new condo development was recently constructed on the east side of EDR, south of Queen Street East, and north of Eastern Avenue. This development is known as Riverside Square and was constructed by Streetcar Developers. During construction, steel tiebacks were placed in the secant pile wall around the perimeter of the excavation for temporary ground support. These tiebacks were left in place following completion of construction; they begin approximately 3.5 m below grade and were inserted into bedrock at a 45-degree angle.

The section of the EDR consolidation sewer underneath the East Don Roadway, MH-01 to MH-03, will be adjacent to the Riverside Square development. At its closest location, between MH-01 and MH02, the pipe is approximately 4 meters laterally from the edge of the building footprint, meaning interference with the tiebacks needed to be considered. The microtunnel boring machine used for construction of the sewer alignment would not be able to easily excavate through the steel tiebacks. Due to these factors the tiebacks needed to be removed from the ground or avoided completely. If feasible, the alternative to avoid the tiebacks was preferred to avoid disruptive and expensive work at surface. Considering the hydraulic criteria to tie in the existing Queen Street Outfall at an invert elevation of 76.63 m and maintain a gradient of 0.25% a shallow vertical alignment, a curved microtunnel was identified as a potential solution to excavate above the existing tiebacks with minimal clearance. Figure 3 and Figure 4 detail the tunnel alignment and the proximity of the tunnel to the existing tieback and utilities in the critical tunnel section along the East Don Roadway.

## DESIGN

### Alternatives Considered and Preferred Solution

#### *Shaft Location*

The positioning of shafts and the alignment of the EDR Sewer was driven by the numerous utility interferences around the microtunnel reception shaft, MH-01, and the potential conflicts, tiebacks, along the tunnel alignment.

The position of MH-01 was determined first, as it faced the most complex challenge in terms of fitting around utilities and the requirements of other stakeholders. The maintenance hole in this location was desired to meet the following objectives:

- Have minimal impact on all the utilities within East Don Roadway (power, water, gas, oil, sewage)
- Maintain 600 mm horizontal clearance from all Toronto Hydro infrastructure

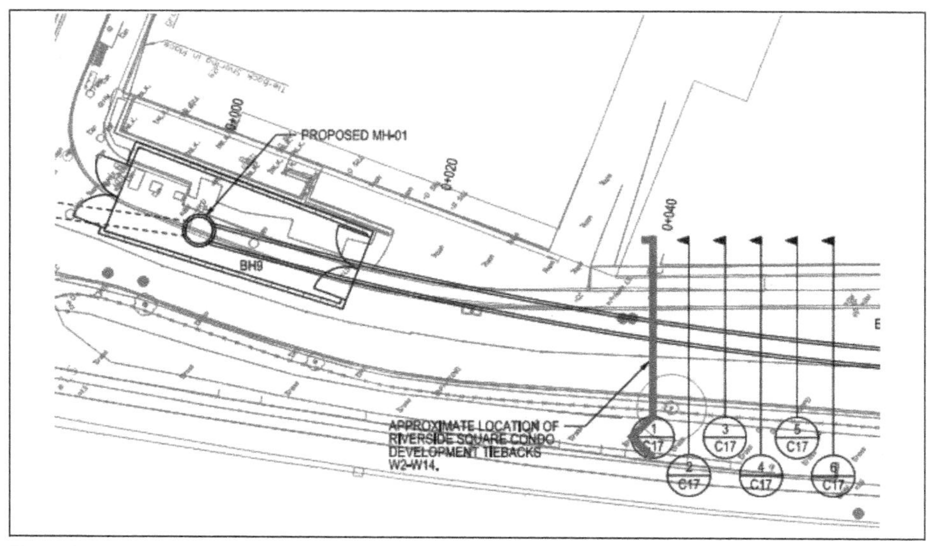


Figure 3. EDR consolidation sewer alignment under East Don Roadway

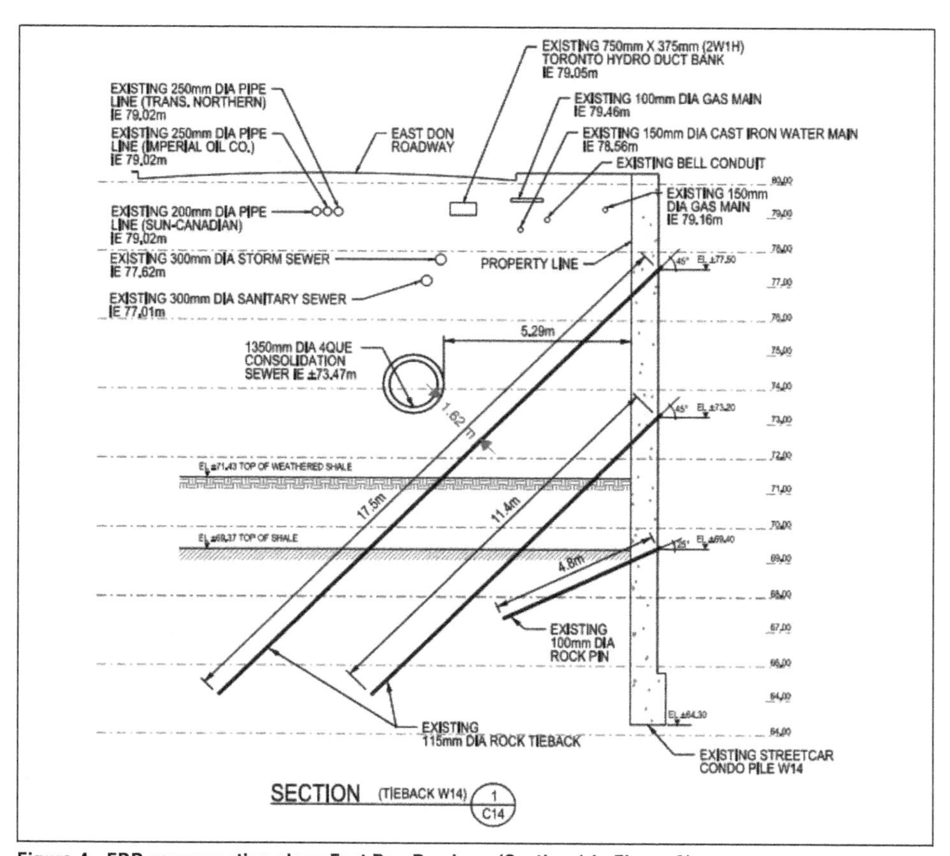


Figure 4. EDR sewer section along East Don Roadway (Section 1 in Figure 3)

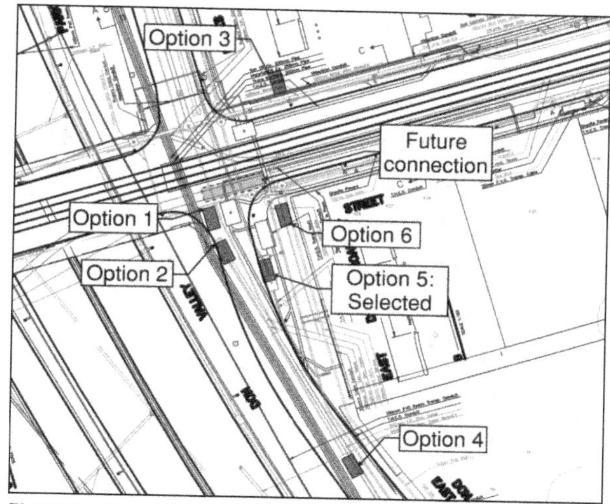


**Figure 5. Alternatives considered for MTBM termination points (MH-01 location)**

- Maintain 5 m horizontal clearance from all high-pressure oil pipelines
- Be positioned as close to Queen Street East as possible to make construction of the future connection to the 4QUE outfall easier
- Avoid impacts to the retaining wall separating East Don Roadway and the northbound DVP
- Maintain driving access to the alleyways and parking garages connected to the condos on the east side of the road

Numerous options were considered, as shown in Figure 5, with a MH-01 location slightly south of the Option 5 location in Figure 5 eventually being selected due to its ability to satisfy the required objectives. The selected location of MH-01 impacted a fibre optic, Bell Canada line, but it was found that other utility impacts were minimal. Avoiding all utility impacts was not feasible at the MH-01 location.

### *Tunnel Alignment*

Once the location of MH-01 was identified, several microtunnel alignments were considered to convey flow from MH-01 to MH-04 and the vortex drop structure. From a long list of alignments, four alignments were identified as feasible and needed for more detailed analysis of the associated cost and risk. The tunnel alignments considered to connect MH-01 to MH-04, are detailed in Figure 6 and summarized in the bullets below.

- Alternative #1 is shown above. The Alternative #1 alignment split the sewer alignment following EDR into three sections; (1) An approximately 140 meter microtunnel drive from MH-02 placed just north of Eastern Avenue to MH-01, (2) An approximately 60 meter section of cut and cover excavation on the North side of Eastern Avenue between MH-02 and MH-03 and (3) A second tunnel drive of approximately 80 meters from MH-03, underneath Eastern Avenue, to the MH-04 Vortex Drop Shaft connection location. The Alternative #1 alignment included the section of open cut on the North side of Eastern Avenue to decrease the length of tunnel and provide a more direct alignment underneath the busy 5 lane Eastern Avenue.

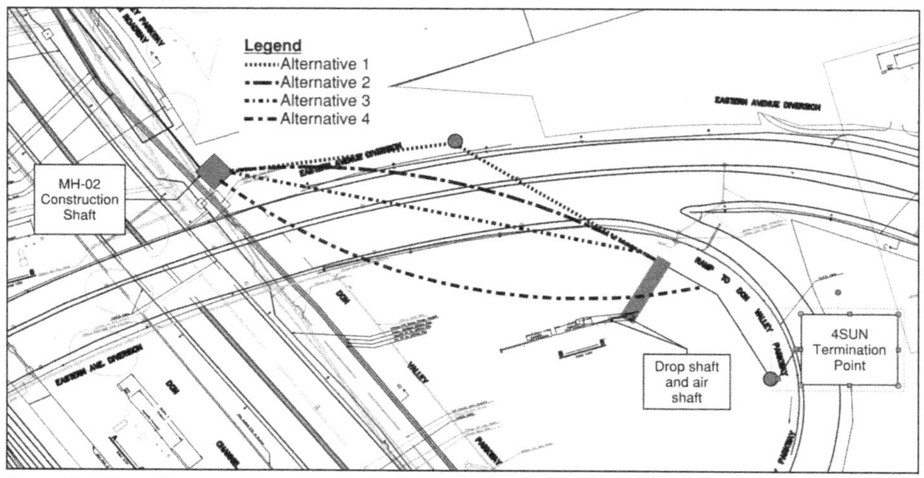


**Figure 6. Alternatives considered for the EDR tunnel alignment**

- Alternative #2 is shown above. The Alternative 2 alignment utilizes a 200 meter curved microtunnel from MH-02, underneath Eastern Avenue, to the MH-04 Vortex Drop Shaft connection location. The curve was considered to increase separation between the tunnel alignment and the Eastern Avenue Bridge Piers and remove the section of open cut excavation.
- Alternative #3 is shown above. The Alternative 3 alignment is very similar to the Alternative 2 alignment, however the 200 meter curve is omitted from the microtunnel alignment from MH-02, underneath Eastern Avenue, to the MH-04 Vortex Drop Shaft connection location and a straight pipe alignment is utilized.
- Alternative #4 is shown above. The Alternative 4 alignment would tunnel the EDR from the MH-04 Vortex Drop connection directly to MH-01 using a long curved microtunnel. This option provided most direct route to the terminus location and the added benefit of removing the requirement for both MH-02 and MH-03. However, this option had potential to conflict with the piles supporting the Eastern Avenue Bridge Pier.

In parallel to development of the alignments analysis was completed to determine the preferred method of excavation, open cut or trenchless, for each alternative. For this analysis, each section of sewer was examined independently.

- The length of sewer proposed between MH-01 and MH-02 runs under East Don Roadway. As noted previously, this road is extremely congested with existing utilities, which would require significant relocations and/or support of these existing utilities to allow for an open cut method. The use of a MBTM in this stretch presented significant advantages in the ease of constructability, thus it was selected.
- The length of sewer proposed between MH-02 to MH-03 has less significant utility interferences compared to its upstream counterpart. Either an open-cut or trenchless method could be used by the Contractor, as there is no significant advantage to either.
- The final section of sewer from MH-03 to MH-04 or underneath the Eastern Avenue in Alignment #2, Alignment #3 and Alignment #4 was required to be constructed using a microtunnel. Eastern Avenue is a 5 lane arterial road

providing access into Downtown Toronto and onto the Don Valley Parkway, an open cut method would result in a long closure of the road, which would not be acceptable to the City of Toronto.

The preferred tunnel alignment and means of construction was identified to mitigate the significant risk associated with performing construction underneath the East Don Roadway, in close proximity to the Eastern Avenue Bridge and across Eastern Avenue. While the estimated capital cost varied slightly across the alternatives it was not a driving factor in identifying the preferred alternative.

The Alternative 1 tunnel alignment was selected as provided the most effective risk mitigation measures:

- The section of the EDR from MH-01 to MH-02 would be constructed by microtunneling methods to avoid disruption to existing utilities, the traffic and the public with cut and cover excavation. The section of tunnel could be constructed to avoid existing tiebacks with minimal clearance, 0.5 m, and avoid the need to remove tiebacks in advance of the tunnel excavation (refer to Figure 3 and Figure 4). To achieve this clearance a curved microtunnel pipe with a radius of 3,000 m was utilized for the section of tunnel along EDR between MH-01 and MH-02. This curved profile was required in order maintain adequate clearances from existing utilities, while also maintaining adequate horizontal and vertical clearances from the Riverside Condo development tiebacks that occupy that length of the road. A straight microtunnel run would not have been feasible in this area due to these constraints. Figure 3 shows the curved pipe plan, with the tieback interferences outlined in red. By utilizing a curved pipe, the alignment was able to achieve vertical separation over the top of the tiebacks, so they did not need to be removed from the ground in advance of tunneling.
- The section of the EDR from MH-02 to MH-03 would be constructed by open cut methods. Adding in MH-02 and MH-03 prevents the potential to conflict with driven wood piles installed to support the Eastern Avenue Bridge pier. Considering the nature of constructing battered piles the feasibility of threading the tunnel between the piles was not desired and separation between the bridge pier and open cut alignment was preferred. Addition of MH-03 allowed the tunnel alignment across Eastern Avenue to not only be moved away from the Eastern Avenue Bridge pier but also shortened the length of the tunnel drive.
- The Section of the EDR from MH-03 to MH-04 at the Vortex Drop Structure would be constructed by microtunnel methods to avoid disruption to traffic on Eastern Avenue.

## Design

Following the selection of the Alternative 1 alignment the tunnel and pipe design was advanced through detailed design.

### *Tunnel*

The tunnel excavation method was specified to mitigate ground loss and surface settlement while excavating through challenging conditions which could present themselves as a risk to the project; running conditions in the Fill and lenses of sand and gravel material in the Clayey Soil. In consideration of the risk of encountering the subsurface conditions, described above, and the sensitive infrastructure present above the tunnel, the tunnel is designed to be constructed with a slurry microtunnel boring machine (MTB). A rigorous assessment of the anticipated volume loss performance of the MTBM, the potential for ground settlement and the extent of the zone of influence (ZOI) was completed to analyze the potential for tunnel induced impacts on utilities and structures. This analysis concluded that with a conservatively assumed volume loss of 1% there is minimal settlement predicted to occur over the centerline of the tunnel, with a significant decrease as the extent of the ZOI is approached. The extent of excavation induced settlements along the microtunnel alignment has been estimated and defined. Although the movement of utilities and structures within the project vicinity are expected to be minimal, a construction monitoring instrumentation program has been designed in order to verify design assumptions and mitigate the potential for ground movements to go undetected during construction (refer to Sivaram et al., 2021).

In addition to analyzing the anticipated volume loss and ground movement to achieve the Alternative 1 alignment the MTBM will need to achieve a 3,000 meter radius curve with tight alignment control. Black & Veatch experience in the completion of projects in the Toronto Area, specifically the YDSS Forecemain Twinning project in Newmarket, ON, informed the team that this radius would be achievable. However, the requirement for an experienced MTBM operator and the use of a precision guidance system, VMT or alternative on the MTBM was specified.

### *Pipe*

The pipe selected to construct the microtunnel drive is a concrete microtunnel jacking pipe. Specifically, it will be a C300 concrete pressure pipe in accordance with AWWA C300. C300 pipe was chosen as it can sustain any potential surcharge pressures in the overall system and the higher jacking forces required by tunnelling on a curve. Moreover, the joints in C300 help prevent any future ground water inflow and infiltration better than the standard reinforced concrete pipe. C300 concrete pressure pipe was chosen over C301 as it is more robust, and its life cycle cost is less expensive as C300 does not have prestressed wires. Damaged prestressed wires in C301 would ultimately compromise the pipe structure and would likely warrant a pipe replacement.

## CONCLUSION

The EDR sewer is to be constructed with a curved microtunnel alignment to avoid building tiebacks and utilities and to mitigate the project risk associated with utility conflicts and stakeholder impacts. This paper presents the steps taken to identify the preferred alignment through challenging conditions and risk mitigation measures implemented through the design of the project to mitigate potential impacts to utilities, structures and third-party stakeholders.

Currently the EDR microtunnel design is completed with bid award and construction anticipated to follow. Construction is anticipated to begin in the first quarter of 2022.

## REFERENCES

Sivaram, A., Cressman, D., Keller J., and Tatiana, C. (2021) Design of the 4SUN/4QUE Microtunnel Sewer in Toronto. North American Society of Trenchless Technology, Orlando, FL 2021.

# Northeast Boundary Tunnel Project: First Street Connector Tunnel and Mount Olivet Road Diversion Sewer Design and Construction

**Jeremiah M. Jezerski** ▪ Brierley Associates Corporation
**Basilio Giurgola** ▪ The Lane Construction Corporation
**Filippo Azzara** ▪ The Lane Construction Corporation
**Russell H. Lutch** ▪ Brierley Associates Corporation
**Federico Bonaiuti** ▪ The Lane Construction Corporation

## ABSTRACT

The Northeast Boundary Tunnel Project consists of a 23 ft diameter, 27,000 ft long, CSO tunnel, multiple underground connections, and significant surface works within an urban corridor. The NEBT alignment depths range from 60 to 140 ft and includes seven shafts ranging in depth from 77 to 155 ft with diameters varying from about 19.5 to 56 ft, some as in-line connections and others required adits for connection to the main tunnel. This paper focuses on the design and construction of two major underground connections which are the First Street Connector Tunnel and the Mount Olivet Road Diversion Sewer.

## INTRODUCTION

The Northeast Boundary Tunnel (NEBT) Project is a major component of the Consent Order driven DC Clean Rivers Program undertaken by DC Water. Upon project completion, it will relieve chronic flooding along Rhode Island Avenue NW, and near the intersection of Mount Olivet Road NE and West Virginia Avenue NE. The NEBT is 23 ft inside diameter by approximately 27,700 ft long and ranges in depth from about 60 to 140 ft. The project alignment passes beneath a dense urban corridor, rail infrastructure, and historical landmarks. Some of those are the RFK Stadium Complex, Langston Golf Course, National Arboretum, Mount Olivet Cemetery, New York Avenue, Amtrak Rail Yard, WMATA Red Line, CSX rail bridge, and beneath a significant length of Rhode Island Avenue through Northwest DC. The project includes nine major sites, eight of which include a shaft connection to the NEBT and seven shafts constructed as part of this contract. The shafts range from 76 to 154 ft in depth with diameters varying between 19.5 to 56 ft. Five of the shafts were connected via adits and two were inline.

The completion of the First Street Tunnel (FST) occurred in 2015, approximately three years prior to commencement of the NEBT project. Connection of the FST to the NEBT via the First Street Connector Tunnel (FSCT), was a critical component of the NEBT Project and was perceived to carry a significant portion of the project risk. Once FST tunneling was completed the TBM was abandoned along the FSCT alignment, requiring many TBM components to be removed underground through the NEBT after excavation of the FSCT. An important aspect to the work associated with the FST was that during the majority of the NEBT project duration, the FST was required to be in service and outages were only allowed for short durations to finalize the connection works.

The Mount Olivet Road Diversion Sewer (MOR-DSWR) connects two sites along Mount Olivet Road NE near the West Virginia Avenue NE intersection Due to the need to convey flow from and existing 7 ft diameter combined sewer to the NEBT and the distance from it, an approximately 730 ft long by 10 ft diameter diversion sewer was required to be constructed along Mount Olivet Road. To mitigate surface disturbance, the new diversion sewer was designed to be constructed using an EPB TBM.

DC Water elected to utilize Design Build project delivery for NEBT. A joint-venture comprising Salini Impregilo and Lane's S.A. Healy (SIH) is leading the Design/Build Team with Brierley Associates (Brierley) serving as the Prime Designer and Engineer-of-Record. Brierley was tasked with progressing the 30 percent design documents prepared during the bidding phase to final "Released for Construction." Brierley also manages the various design subconsultants including: traffic engineering/control, construction impact assessments, risk management, and concrete durability specialists. Design work began in September 2017 and construction will be completed in 2023, two years ahead of the Consent Decree schedule.

## FIRST STREET CONNECTOR TUNNEL

The FSCT is approximately 45 ft long and has a finished inside diameter of 18.5 ft. The tunnel begins at the terminus of the existing First Street Tunnel (FST) and ends at the intersection with the NEBT. The plan alignment is illustrated in Figure 1. As previously indicated the FST TBM was abandoned at the end of the FST Tunnel where connection to the NEBT was required to be made. The FST was stopped approximately 10 ft short of NEBT at its centerline to allow the NEBT TBM to pass and the connection to be made. As part of the FST TBM termination plan, the shield, cutterhead, main bearing, and the mid shield support frame were abandoned in place. Therefore, after excavating the FSCT up to the abandoned TBM from the NEBT, the TBM and its abandoned components were required to be disassembled and removed through the FSCT and NEBT.

The overburden depth is approximately 60 ft with groundwater head of about 40 ft above the crown. The FSCT lies completely within the Potomac KP (PTX) formation and primarily within the G3A and G4 soils. G3A soils are silty or clayey sands and G4

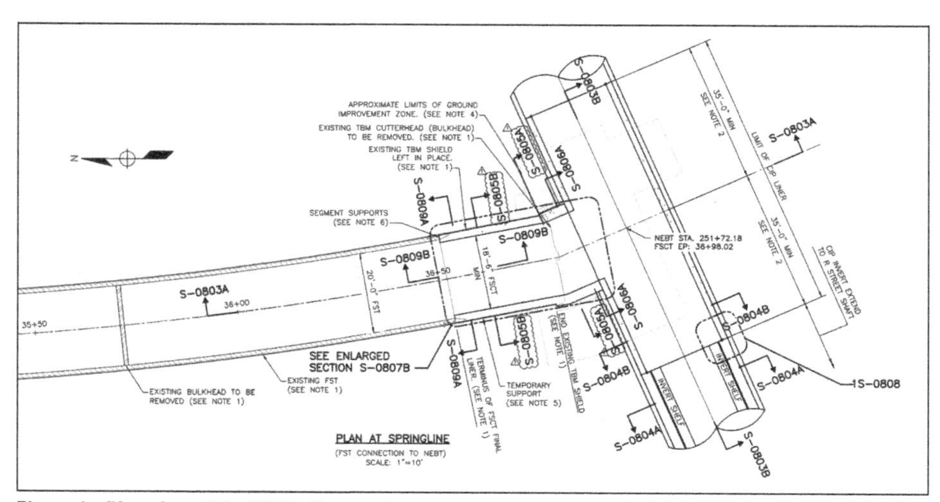


**Figure 1. Plan view of the FSCT alignment**

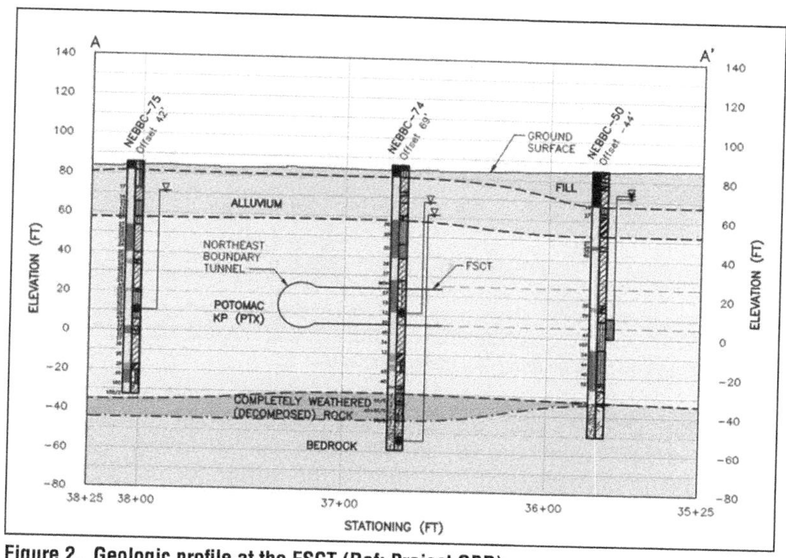


Figure 2. Geologic profile at the FSCT (Ref: Project GBR)

are sandy soils with trace of fines and gravels. The geologic profile is illustrated in Figure 2 from the Project Geotechnical Baseline Report (GBR).

## Design of the First Street Connector Tunnel

The FSCT is approximately 45 ft long from the terminus of the First Street Tunnel (FST) segmental liner to the intersection with the NEBT CIP liner. Approximately 30 ft of the FSCT was constructed inside the abandoned FST TBM. The remaining 15 ft was constructed through jet grout improved ground and utilized initial ground support consisting of shotcrete and lattice girders. The Design-Build Contract required that a 70 ft long, 18 inch thick, cast-in-place (CIP) liner be installed within the NEBT centered about the location of the FSCT connection to reinforce the precast segmental lining and mitigate potential for strength and serviceability issues over the 100-yr design life of the system.

Given that a CIP liner was required to be constructed inside the NEBT at the FSCT connection location the design philosophy adopted by the team was to utilize that liner to support the opening in the short-term, during construction, and the long-term, final condition. Evaluation of the CIP liner and the temporary support of excavation system was designed using the three-dimensional (3D) finite element analysis (FEA) software Midas GTS NX (MIDAS). MIDAS is a comprehensive finite element analysis software package that is equipped to handle the entire range of geotechnical applications including excavation of the tunnel system, seepage analysis, and consolidation analysis.

For the short-term condition, during construction of the FSCT, the model included the presence of a 44 ft by 44 ft jet grout plug that was required to be installed per the Contract requirements. The jet grout ground improvement was relied upon to improve the strength and deformation characteristics of the surrounding soil during the FSCT excavation and initial support installation sequence. Later, the jet grout plug dimensions were increased to 52 ft by 52 ft due to differing interpretations of the Contract requirements. However, the increased ground improvement zone was never relied upon in design as it had been previously demonstrated as being unnecessary. Overall,

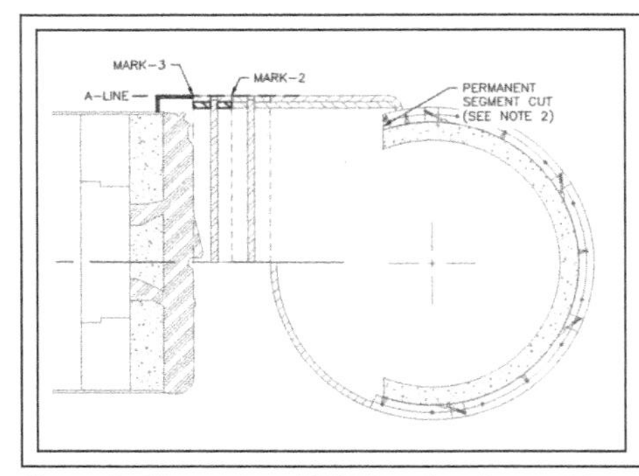


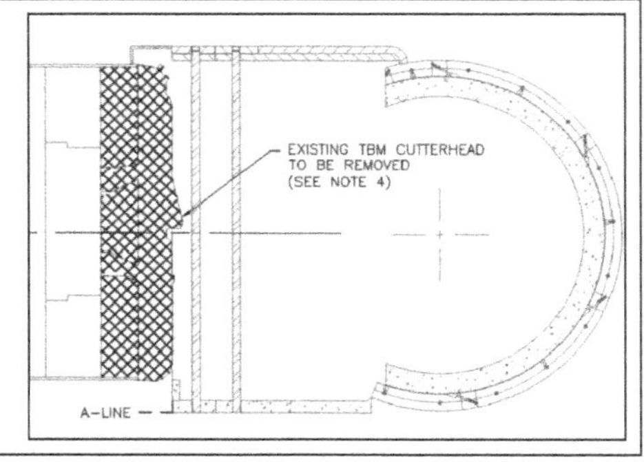


**Figure 3. Illustration of FSCT top heading and bench excavation**

the purpose of the 3D FEA was to evaluate the stress condition at the intersection of the NEBT and the FSCT within the 18 inch thick CIP liner within the NEBT, the stress state within the jet grout mass, and to determine the design forces and moments for the temporary shotcrete liner.

The soils were modeled using the Modified Mohr-Coulomb soil constitutive model, which defines the elastic region as nonlinear elastic. For this material model the power-law is used to obtain the elastic volumetric stress-strain relationship until the soil shear strength is exceeded, and then perfectly plastic deformation occurs at a constant shear stress for failed elements. While the FEM analyses are not intended to provide accurate determination of deformation, they are considered to provide reasonable approximation of deformation behavior that attempts to simulate steady-state conditions at each major construction stage. One important difference between standard Mohr-Coulomb model and the modified Mohr-Coulomb model is assigning an "unloading modulus." The unloading modulus is a critical feature considering the fact that the excavation of the FSCT involves unloading response.

The excavation was simulated with a top heading and bench as shown in Figure 3. The top heading was excavated in four stages, and the bench was excavated in three stages. The difference in stages was primarily due to supporting the crown at initial breakout. The top heading was excavated up to the abandoned FST TBM cutterhead prior to excavating the bench. The temporary support shotcrete liner was modeled as 12-in thick, two lattice girders were installed in the short distance between the NEBT and the FST TBM to maintain the tunnel profile and provide robustness in the case of inadequate ground improvement or unexpected ground behavior but were not considered in the analysis. To optimize construction the top heading excavation was designed to proceed through the first two excavation rounds with only 6-inch-thick shotcrete application. Then during excavation of the third round the final 6-inch application of shotcrete was placed as shown in Figure 4.

Figure 5 illustrates the model results for the temporary shotcrete liner and invert slab from the 3D FEA. An average load factor of 1.4 was applied to all results and the thrust moment interaction diagram was generated based on the design requirements presented in ACI-318 for plain concrete including a 0.6 strength reduction factor.

As part of this analysis, the resulting stresses in the 18-inch-thick CIP liner to be constructed within the NEBT were evaluated to verify that the liner would adequately support the break-out and subsequent excavation. The NEBT segments and the 18-inch-thick CIP liner were both modeled using shell elements. Figure 6 through

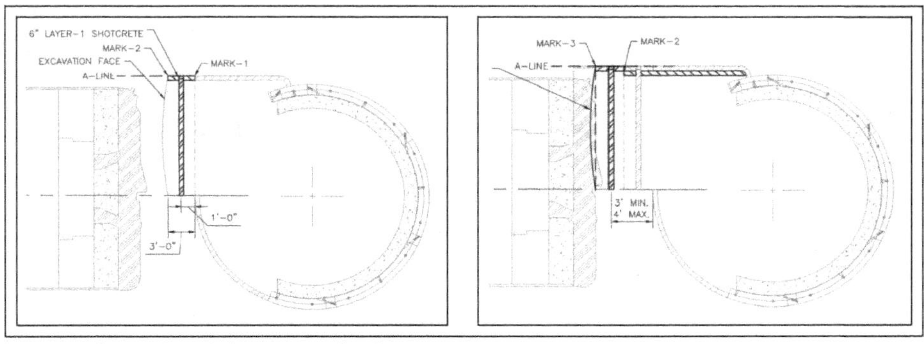


**Figure 4. Illustration of shotcrete application stages**

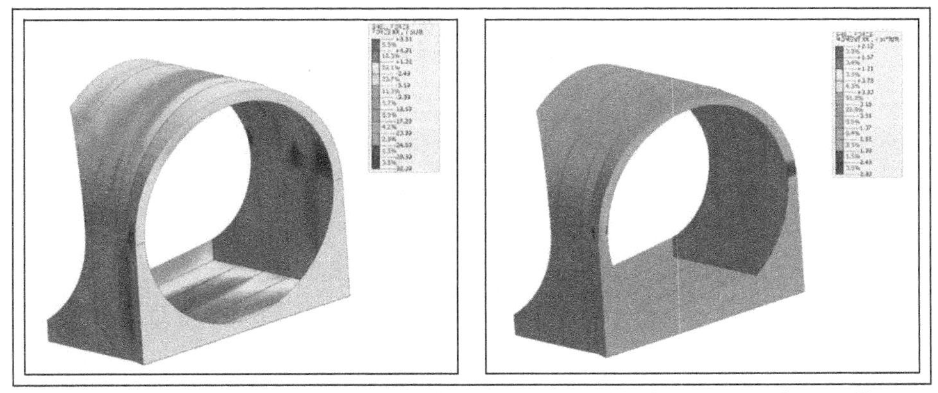

**Figure 5. Contour plots of hoop force (left); and hoop moment (right) in the temporary shotcrete liner**

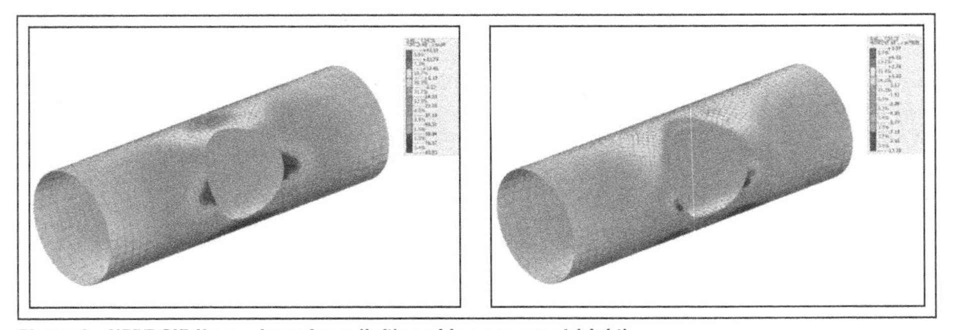

**Figure 6. NEBT CIP liner—hoop force (left); and hoop moment (right)**

Figure 8 show the contour plots of resulting forces and moments in the CIP liner. The resulting forces and moments were used to ensure that the requirements of ACI-350 were satisfied for the resulting temporary condition.

As part of the design, the ground improvement was checked for the ability to support the ground assuming installation of a homogeneous jet grout mass. The jet grout mass was modeled as 44 feet by 44 feet in section along the length of the excavation portion of the FSCT with 5 feet of overlap of the abandoned TBM cutterhead and front

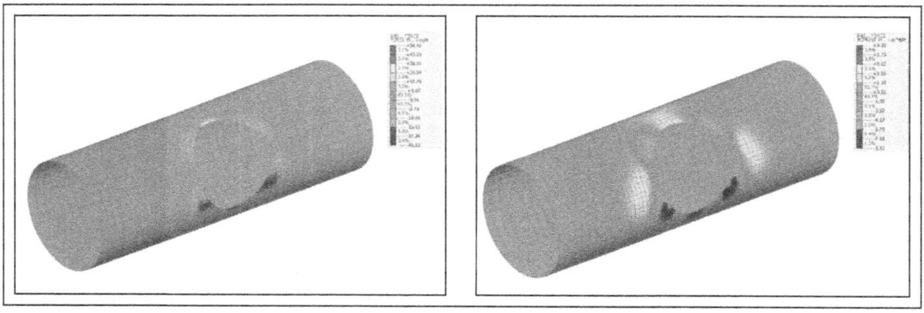

Figure 7. NEBT CIP liner—longitudinal force (left); and longitudinal moment (right)

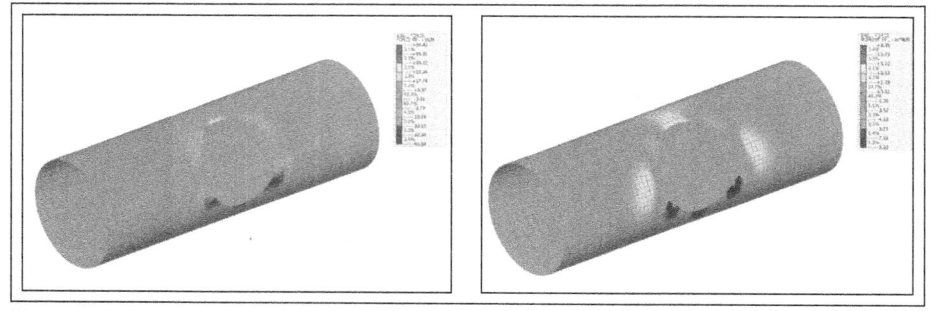

Figure 8. NEBT CIP liner—out-of-plane shear (x–z plane—left); and (y–z plane—right)

shield 5 feet. The jet grout was assumed to have a minimum unconfined compressive strength of 250 psi. Since it is known that the jet grout ground improvement zone will distribute the stresses around the excavation opening as an arch, hand calculations were performed prior to modeling to estimate thickness requirements and to act as a check the model results. Those calculations were performed assuming the jet grout would act as a thick-walled cylinder, which typically yield good results as compared to more sophisticated models. Those calculations indicate a maximum compressive stress in the jet grout of about 175 psi. However, this calculation assumes plane strain conditions, ignores the contribution of the temporary support, and also ignores three-dimensional effects from the presence of the abandoned FST TBM cutterhead and shield and the presence of the NEBT.

Figure 9 illustrates the maximum compressive stress in the jet grout plug for the design groundwater level case analyzed. The compressive stress is generally less than 175 psi for low ground water levels with the exception of stress concentrations at the inner 90 degree edges of the jet grout block, but decrease rapidly with distance from the corners. These concentrations should not be considered when evaluating the global stability of the ground improvement mass.

After completion of the excavation, the cutterhead, main bearing, main bearing frame and mid-shield support frame were to be removed. As part of the design, it was shown that the abandoned FST TBM shield had adequate capacity to support the surrounding ground and allow removal of the cutterhead and interior support frames. The shield varied in thickness between about 1.5 inches and 2-⅛ inches. Various buckling calculations were performed for the steel shield, such as Free Tube Buckling, Rotary Symmetric Buckling, and single lobe buckling for unstiffened pipe (Berti 1998). Those

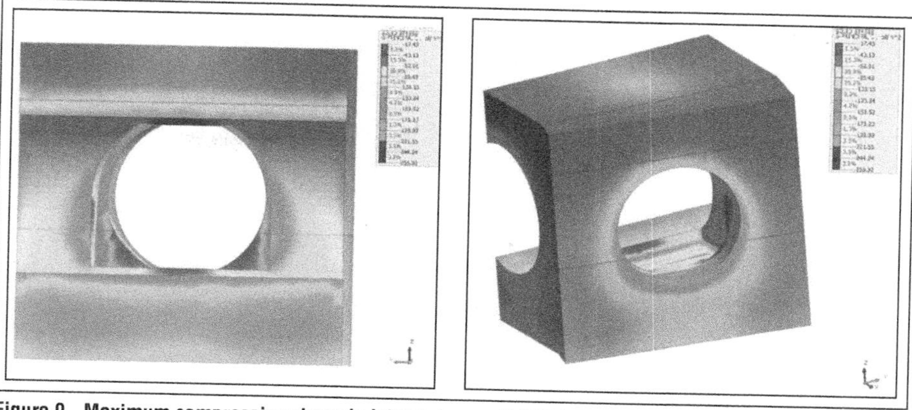

Figure 9. Maximum compressive stress in jet grout mass (LGWL)

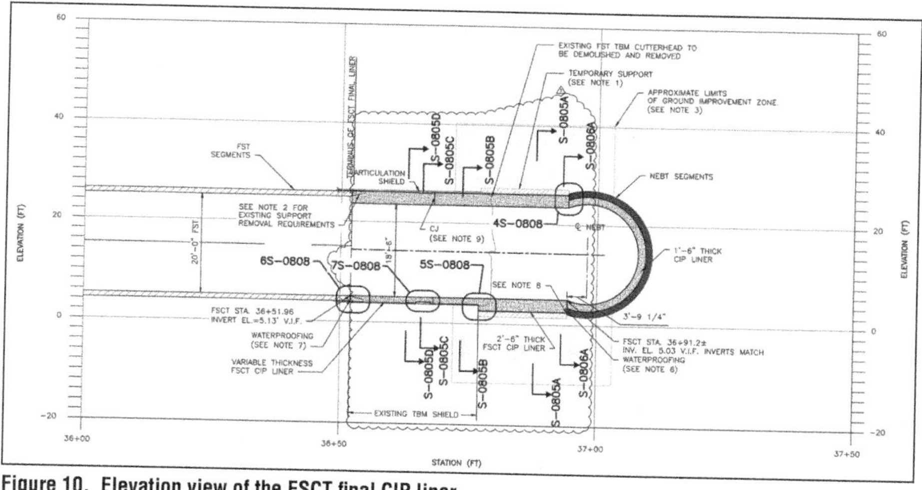


Figure 10. Elevation view of the FSCT final CIP liner

calculations demonstrated a safety factor of 2.3 conservatively considering that steering gap was not fully grouted. To mitigate risk associated with the work, a sequence was established to perform a sequential removal of the frames while constructing the final FSCT CIP liner. It was elected to remove the cutterhead and lower portion of the main bearing frame, leaving the crown section in for rigidity, and constructing the lower half of the liner between the NEBT and the mid-shield support frame, then remove the upper portion of the main bearing frame and complete that section of liner. Then the mid-shield support frame was removed, and the liner was completed up to the existing FST segments. The elevation view is shown in Figure 10.

## Construction of the First Street Connector Tunnel

To facilitate the construction of tunnel connection and mitigate the risk related to the underground excavation, ground improvement zones have been designed to resist the earth and hydrostatic loads during the Sequential Excavation Method (SEM), considering no dewatering was allowed. All the utilities crossing the jet grouting zone of influence were instrumented and monitored during the jet grouting operations. The

**Figure 11. FSCT excavation of top heading**

jet grouting layout had to be designed to avoid impacting utilities as well to reach the minimum ground improvement area.

The connection tunnel from the NEBT to the abandoned TBM towards the FST was constructed using the principles of the Sequential Excavation Method SEM. Tunneling of the First Street Connector Tunnel (FSCT) was performed from the NEBT through the jet grout block, due to access and construction restrictions at the FST. The excavation was performed with a remote controlled Brokk 400 excavator equipped with a demolition hammer and supported by a regular mini excavator used for mucking of the excavated material. The shotcrete setup used prefabricated dry-mix in bulk-bags and manual application by ACI certified nozzlemen.

After breaking through the NEBT segments partial excavation of the FSCT top heading was performed to assess the quality of the jet grouting, then top heading excavation was completed. A photograph of the breakout from the NEBT is shown in Figure 11. Given that the NEBT and FSCT were nearly the same size, the initial top heading excavation was primarily overhead. Some water ingress, about 8–10 gpm, developed at the interface between segmental grout and jet grout. Visual observation and frequent monitoring of the water inflow did not indicate any ground loss or increasing flow rate. The water was channeled into a hose which was grouted after completion of the shotcrete shell.

As the excavation progressed and approached the FST TBM, it was evident that the jet grouting quality and extents were sufficient to prevent any leakage that required significant remediation. Monitoring of convergencies in the NEBT and the freshly excavated FSCT was performed as an essential part of each SEM excavation. Convergence monitoring did not indicate any significant movement.

As the FST TBM cutterhead was exposed, during top heading and bench excavation, demolition and removal was performed. However, prior to cutting any portion, the bentonite-grout bulkhead located within the excavation chamber, placed during FST TBM abandonment, was removed. Chain falls and come-alongs were installed through the cutterhead openings to safely secure and lower each piece as it was cut and removed.

Following the cutterhead dismantling, the main bearing and its support frame was removed in accordance with the design sequence. Initially, the main bearing frame was cut to only 17 ft from the invert, so that the top remaining portion of the frame to

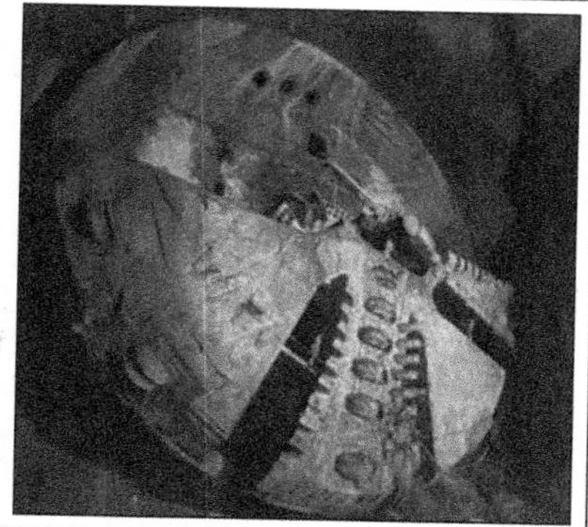

Figure 12. Cutterhead top portion demolition (left); and cutterhead bottom portion exposed (right)

provide support to the shield to mitigate convergence and potential for buckling.

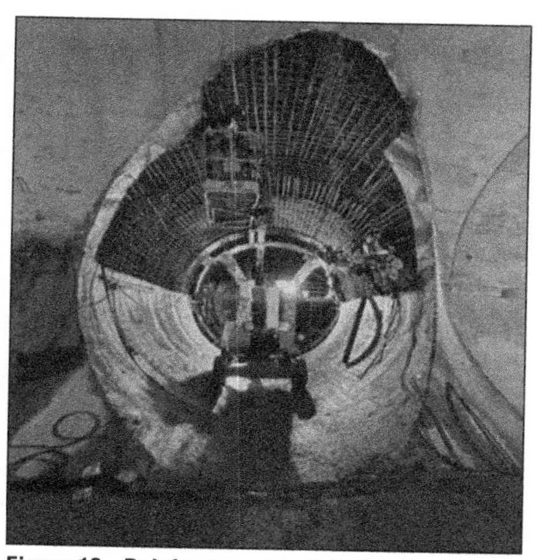

Figure 13. Reinforcement installation for FSCT pour #1B

Due to the geometric restrictions, a requirement to remove all TBM components within 1 inch from the shield was set to provide adequate space for the cast-in-place liner. Bentonite and grout ports on the TBM shield were supposed to have been grouted in place during TBM abandonment. During removal, it was discovered that they were not properly grouted and presented risk for water or material inflow. The bentonite and tail void grout ports were then sealed with fast setting polyurethane resin and cut flush with the internal shield surface prior to moving forward with CIP liner works. A different approach was taken for probe drill ports due to the size and shape of ports and proximity to the articulation shield. The approach was to locally adjust the concrete liner reinforcement to maintain the probe ports intact, as their complete removal would have constituted a significant risk to groundwater inflow and loss of ground due to their size.

The construction of the 18 inch thick reinforced CIP liner at FSCT was performed in four pours. The wooden formwork chosen was sectional and reused for both invert and crown pours. Given the geometry and relative size, the formwork had to be transported out of the tunnel to be flipped, brought back in, and then installed for the subsequent pour.

Pours #1A and #1B were completed first and extended from the CIP liner already poured into the NEBT to the FST TBM Shield articulation joint. Then, pours #2A and #2B extended from end of pour #1A and #1B to the FST segmental liner. Pours are illustrated in Figure 14. Due to the geometry of the pour and the reinforcement configuration, a self-consolidating concrete (SCC) was selected to be used. Concrete was placed trough ports and windows installed in the formwork to control pouring

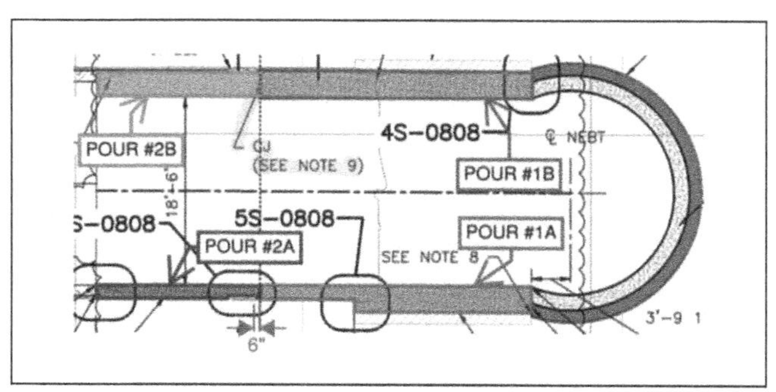


Figure 14. FSCT CIP Liner longitudinal section with pouring sequence

activities. Each pour was performed sequentially with the appropriate frame removal, as described in the design section.

## MOUNT OLIVET ROAD DIVERSION SEWER

The Mount Olivet Road Diversion Sewer (MOR-DSWR) connects two sites along Mount Olivet Road NE near the West Virginia Avenue NE intersection. The intersection of Mount Olivet Road and West Virginia Avenue represented both a local low spot for construction of inlets to reduce flooding potential as well the location of an existing large diameter sewer that would be connected to the NEBT for CSO diversion during storm events. Given that NEBT alignment was approximately 800 ft from the location of the Diversion Sewer, the Contract required that the connection be made via an Earth Pressure Balance (EPB) Tunnel Boring Machine (TBM) to mitigate disturbance along Mount Olivet Road. The total length of the excavated tunnel was just under 700 feet. The plan and profile are illustrated in Figure 15.

Cover along MOR-DSWR varies from 31 ft at the west end of the alignment and up to 46 ft at the east end of the alignment. The tunnel was excavated entirely through the Patapsco/Arundel Formation (P/A) of the Potomac Group Soils (KP). The tunnel invert was within 10 to 20 feet of the underlying Patuxent Formation (PTX) for approximately one half of the alignment. The western end is generally comprised of coarse grained soils under significant head. DC Water soil groups within the MOR-DSWR tunnel horizon were expected to vary between G1 and G2 soils, which are characterized as high and low plasticity cays and silts, respectively.

### Design of the Mount Olivet Road Diversion Sewer

The MOR-DSWR was designed to be a one pass system using 120 inch diameter HOBAS pipe. The EPB TBM selected would utilize pipe jacking methods requiring construction of a launching frame capable of jacking along the 730-foot alignment. The launching frame was required to react off the temporary support of excavation system for the adjacent drop shaft, which was constructed as an unreinforced slurry wall shaft. At the time of construction, the permanent shaft liner was complete, with the exception of the shaft lid. However, at the elevation of the MOR-DSWR a large block-out in that liner was present for the future connection of the approach channel within the launching excavation. Therefore, the launching frame was designed to span over the block out and transmit the thrust load to the slurry wall at the sides of the excavation and limit loading of the permanent structure at the opening. The plan and profile of the excavation and launching frame are shown in Figure 16 and Figure 17. During the

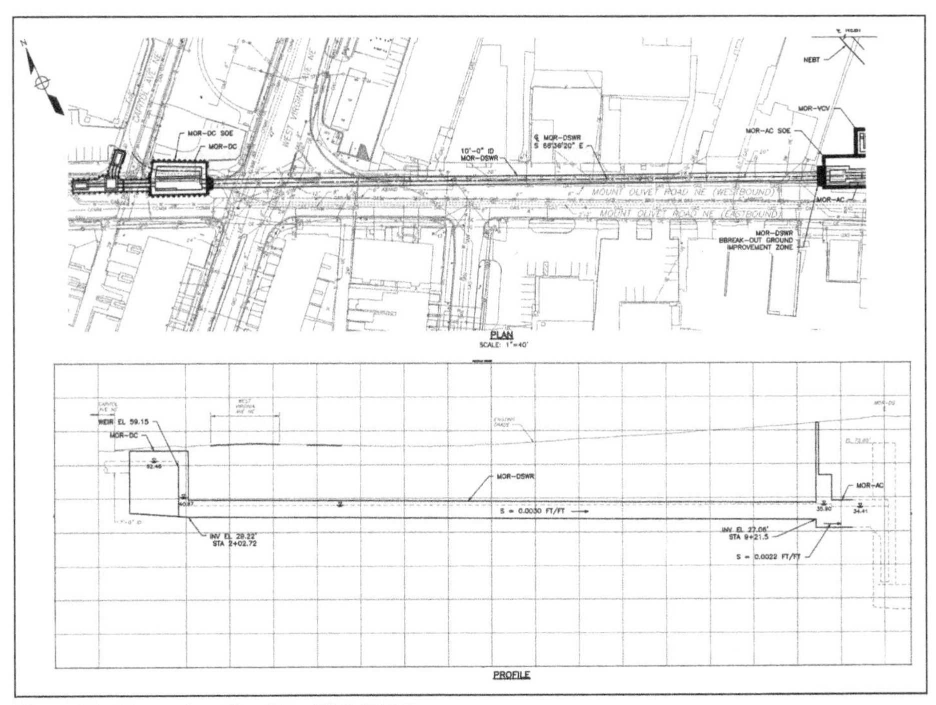


Figure 15. Plan and profile of the MOR-DSWR

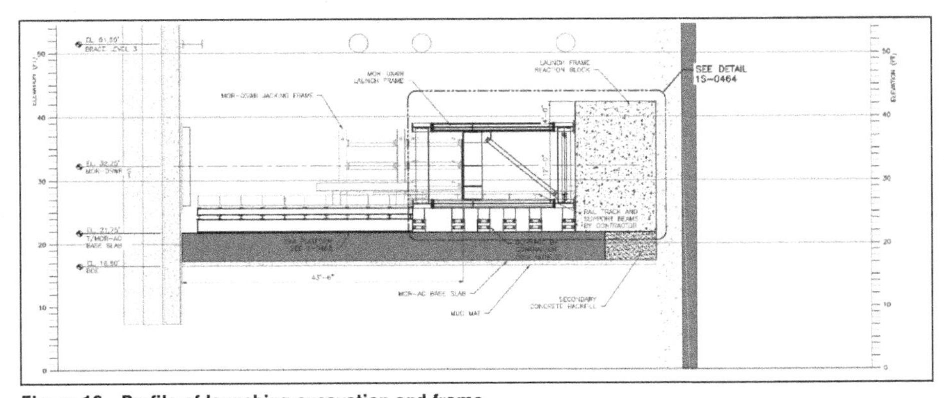


Figure 16. Profile of launching excavation and frame

design of the frame, it was decided to limit the frame to an allowable jacking capacity of 1,200 tons to keep frame member to reasonable sizes.

Calculations for suitability of the HOBAS pipe were performed and provided by the manufacturer's engineering team. Those calculations indicated a maximum allowable jacking load of 1,900 tons. Additionally, Brierley performed calculations for effects of eccentric loading in accordance with International Standards Organization (ISO) 25780 (2011). Those calculations demonstrated that as long as deflection change between pipe sections were maintained to satisfy a closed joint condition then the

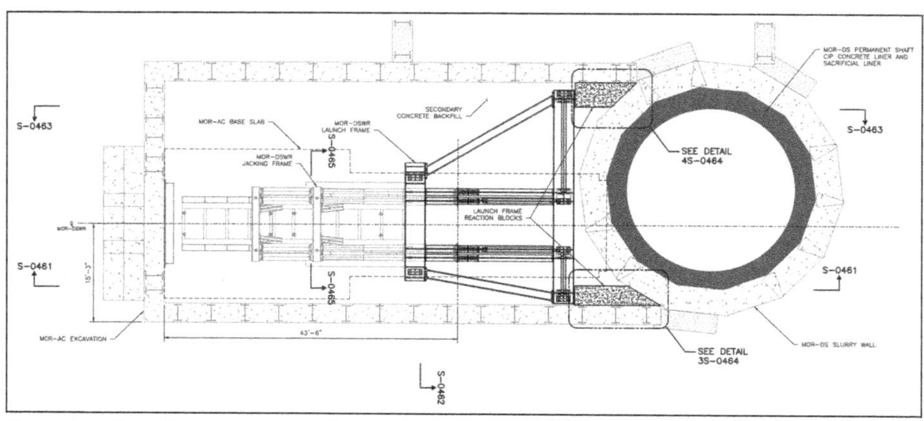


**Figure 17. Plan of launching excavation and frame**

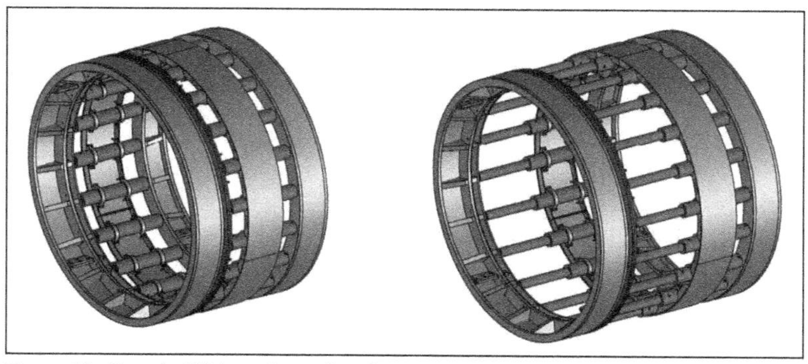

**Figure 18. Sample IJS from Herrenknecht**

reduced allowable jacking capacity was maintained above the 1,200 ton jacking frame design capacity.

To estimate the anticipated jacking loads and the potential requirement for an intermediate jacking station (IJS), an analysis for loading based upon the work of Bennett and Cording (Bennett 1999) was performed, as well as calculations based on empirical data. These calculations resulted in estimated jacking loads between 1,400 and 1,700 tons considering reasonably effective lubrication. With excellent lubrication techniques and materials, the jacking load could be as little as half of that estimate.

Given the estimated jacking loads, the limiting factor on the tunneling operation could be the allowable capacity of the jacking frame. Therefore, an IJS was procured prior to tunneling. The IJS was required to be installed within the center 240-ft of the alignment with the additional requirement that if prior to the installation of the IJS the jacking loads on the main jacking frame reach 70% of the allowable pipe jacking capacity, the IJS would have been immediately installed.

The IJS was provided by the TBM manufacturer, Herrenknecht (HK). A schematic of a similar IJS, as provided by HK, is shown in Figure 18. The outer casing was designed by Brierley. The casing was stainless-steel with an outside diameter of 126 inches, equal to the outside diameter of the HOBAS pipe, with a thickness of 0.993 inches.

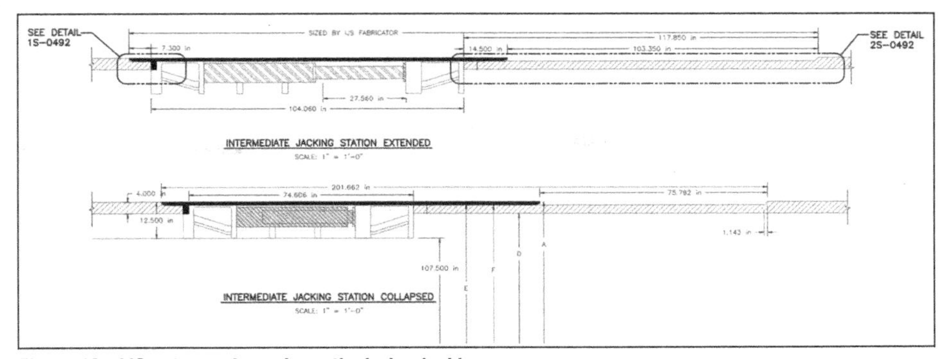


Figure 19. IJS outer casing schematic during jacking

Figure 19 illustrates the position of the IJS, HOBAS and outer casing in the extended and collapsed positions. Figure 20 and Figure 21 illustrate the leading and trailing edge of the casing. As shown the casing has a beveled edge such that when the casing is pushed over and the pipe is pushed into the casing, it does not damage or affect the setting of the gasket.

The EPB face pressure requirements were calculated using two different methods of analysis, which were based of the work of Anagnostou & Kovari (Anagnostou 1996) and Terzaghi (Terzaghi 1943). The Anagnostou & Kovari (A & K) method was performed twice, representing two different ground soil/water conditions in the event that the top of the closely underlying PTX soils was encountered. The analysis considered excavation in clayey soils indicated an optimal EPB pressure of about 1.1 bar and the case where the PTX would be encountered a EPB pressure of 1.5 bar would be required. The Terzarghi Silo analysis was performed for comparison and estimated a face pressure approximately 10% greater than the analysis that considered excavation within the PTX formation

Temporary construction loads were considered as it relates to the capacity of the HOBAS pipe, particularly TBM gantry loads and loading from the muck cars used for transporting material through the pipe. The gantry was installed within the sections of HOBAS pipe as advancement progressed. Those gantries were positioned on load distribution rails as shown

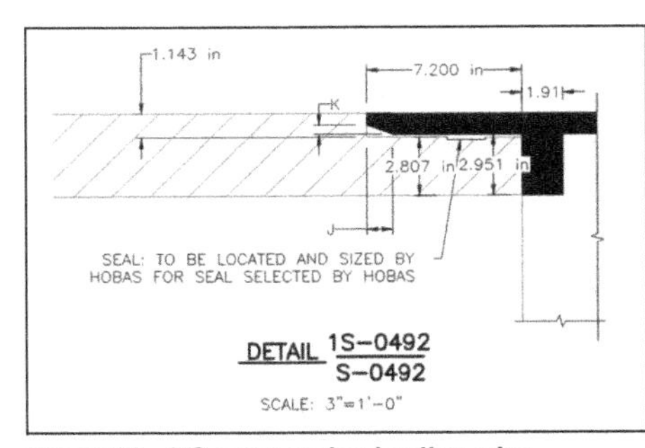


Figure 20. IJS outer casing leading edge

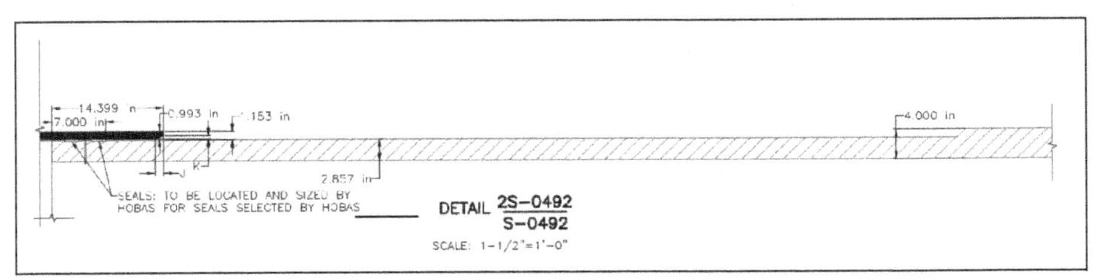


Figure 21. IJS outer casing trailing edge

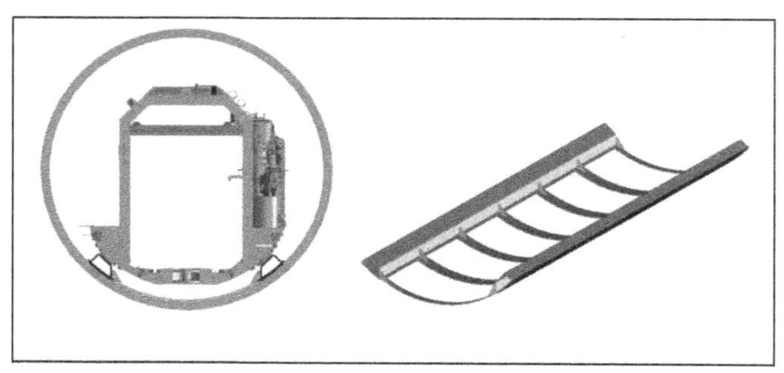

Figure 22. Gantry loading configuration and rails

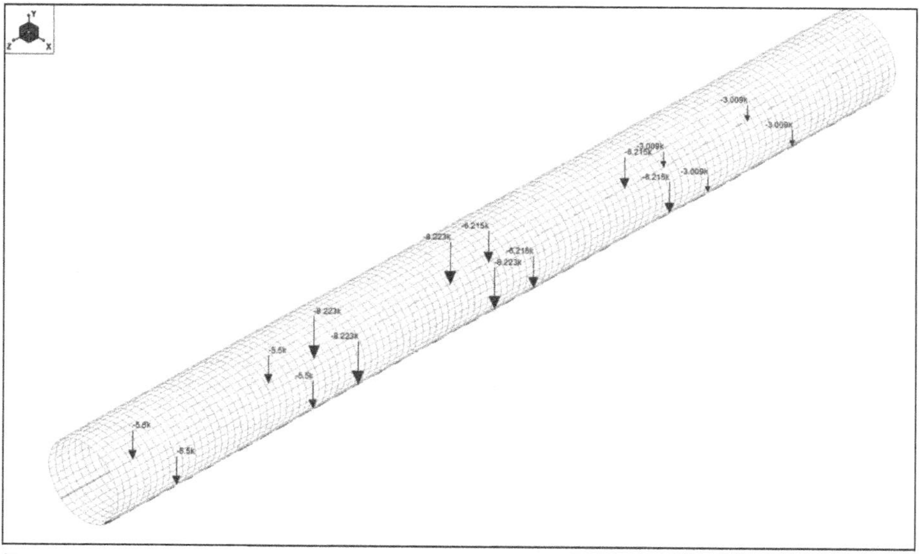


Figure 23. HOBAS Pipe RISA modeling

in Figure 22. These rails were HSS 10×6×⅜ connected with rolled plates to maintain position. To demonstrate acceptability of this loading condition, , a three-dimensional model was developed using the commercially available structural analysis software RIAS-3D. The model geometry is shown in Figure 23. The model indicated a maximum ovality of 0.4%, which is below the allowable limit of 1.5%.

Saddles were designed to spread the load across the invert of the HOBAS and maintain contact pressures less than 108 psi, which is the allowable contact pressure provided by HOBAS. Three alternate saddle configurations were provided, as shown in Figure 24, to allow selection of the saddle to be determined based on price and logistical type evaluations. As sections of HOBAS pipe are set in place the invert was lined with ¾" rubber horse stall mats to provide protection from damage due work activities and the saddles.

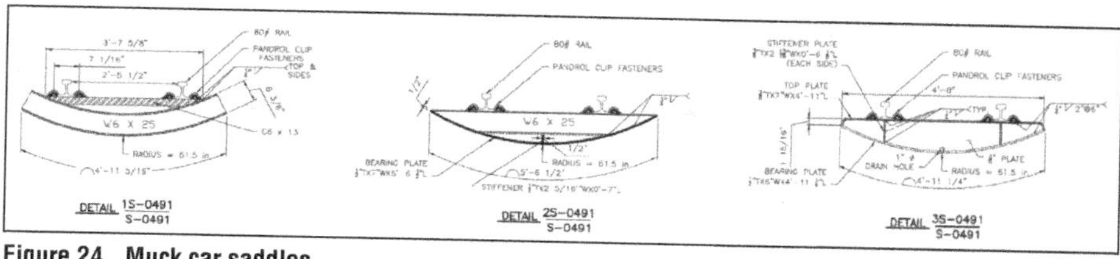


Figure 24. Muck car saddles

## Construction of the Mount Olivet Road Diversion Sewer

The 11.4 ft diameter EPB TBM, named "Tala" was used to construct the 720 ft long (straight) MOR-DSWR tunnel over a two-month period. The machine was a Herrenknecht EPB 2600 single shield TBM and was selected due to its ability to tunnel through unstable soils and below the water table. A rendering of the EPB TBM is shown in Figure 25.

The cutterhead was outfitted with 12-inch-diameter disk cutters, which excavated the foam conditioned ground in the excavation chamber under pressure. From there, the soil was directed through the screw conveyor and onto a belt conveyor that drops the soil into muck carts. The muck carts were rolling on steel muck cart rails on wooden and steel rail saddles adjusted to fit the curvature of the pipe segments.

As previously described, the launching frame, as shown in the rendering presented in Figure 22, was designed to spread the load away from the opening in the drop shaft behind it. The launching frame was configured to allow the muck carts to travel within its limits so that the carts could be lifted up through the top of the launching frame and through the support of excavation bracing system. Machine advancement utilized six (6) 100 ton jacking cylinders. Once the machine was advanced approximately 20 feet, HOBAS pipe was installed in 20 foot lengths.

To reduce friction around the pipes, bentonite injection stations were installed every 30 ft. Upon tunnel completion the BIS units were removed and to allow the grout placement within the annular space.

The position of the EPB TBM along with its tendency and direction was displayed on a monitor in real time using a laser system installed at the entrance of the tunnel. Tunnel alignment and excavation direction were determined through the VMT guidance system (U.N.S), which provided the machine operator detailed information on bearing, pitch, roll relative to the planned path. Tunneling commenced with an

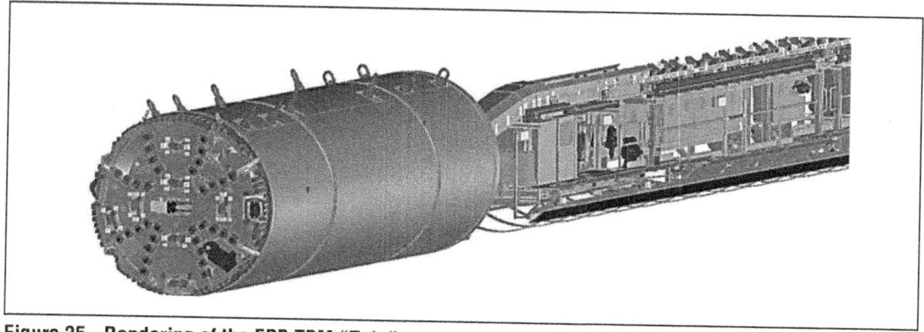

Figure 25. Rendering of the EPB TBM "Tala"

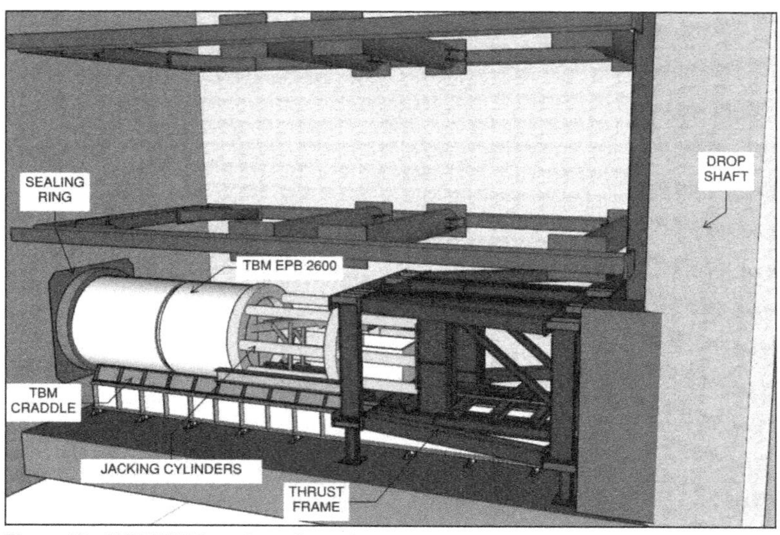


Figure 26. MOR TBM launch configuration

umbilical connection, which was disconnected after the installation of 4 pipes to install 4 TBM gantries to follow the TBM activity. The gantries were advancing on steel temporary gantry support rails, shown in Figure 27, which were installed on rubber mats against the pipes surface.

Figure 27. MOR TBM gantry support rails installation

For the TBM launch, the machine was set on a steel cradle precisely leveled and aligned with the tunnel orientation. A sealing ring, as shown in Figure 28, was installed at the interface with the Cement-Bentonite walls to maintain the TBM advancing pressure and prevent uncontrolled inflows of soil, bentonite, and groundwater during the breakout process due to the annular gap between the tunnel excavation and TBM shields.

The TBM started mining using a thrust force of 200 ton, which increased up to 383 ton at a jacking distance of about 220 ft. At this point, an intermediate jacking station, shown in Figure 29, was installed. Although the maximum thrust frame capacity was estimated to be 1,200 ton, the intermediate jacking station was installed to reduce the overall load on the main jacking frame and to provide a contingency in the case higher jacking loads than expected occurred during the TBM advancement. The intermediate jacking station allows the use of small hydraulic cylinders to push forward a smaller number of pipes and TBM, while reducing the number of pipes to be jacked by the main jacking station. During the advancement to the arrival at the MOR-DC site, the excavation was completed through mainly fat clay, which caused issues conditioning the soil and maintaining the belt conveyor with its optional operation speed. No considerable, ground movements, pipe gaps, or

Figure 28. MOR TBM and sealing ring

Figure 29. Intermediate Jacking Station (IJS)

pipe damage were recorded throughout the tunneling operation. The maximum jacking force reached was slightly greater than 600 tons as a result of heavy focus on the lubrication system and techniques.

## SUMMARY AND CONCLUSIONS

The interpersonal chemistry between the designer and contractor that developed during the tender phase of the NEBT evolved into a friendly, respectful, and collaborative relationship that was the key to the constructability of the design and the successful implementation of that design. The team took a collaborative approach to the development of key design criteria, limitations based on space and desired techniques such that the work could be performed with little to no changes as it progressed.

## REFERENCES

Anagnostou, G. and Kovari, K. 1996. Face Stability Conditions with Earth-Pressure-Balanced Shields. *Tunneling and Underground Space Technology.* 11(2): 165-173.

Bennett, R. D., and Cording, E. J. (1999). "Jacking loads associated with microtunneling." Proc. 3rd National GeoInstitute Conf. Geo-Engineering for Underground Facilities, G. Fernandez and R. Bauer, eds., Geotechnical Spec. Publ. No. 90, American Society of Civil Engineers, New York.

Berti, D.J., Stutzman, R., Lindquist, E.S., and Eshghipour, M., 1998. Buckling of Steel Tunnel Liner Under External pressure. *ASCE Journal of Energy Engineering*, (Dec), 55-89.

DC Clean Rivers Project Division J—Northeast Boundary Tunnel (NEBT) *Geotechnical Baseline Report.*

International Organization for Standardization. (2011). *Plastics piping systems for pressure and non-pressure water supply, irrigation, drainage or sewerage—Glass-reinforced thermosetting plastics (GRP) systems based on unsaturated polyester (UP) resin—Pipes with flexible joints intended to be installed using jacking techniques.* (ISO Standard No. 25780:2011). https://www.iso.org/standard/43221.html

Terzaghi, K. 1943. *Theoretical Soil Mechanics.* New York: Wiley.

Giurgola B, Bonaiuti F., Banov P, Fuegenschuh N, (2022). *Ground Improvement Works and Construction of Six SEM Adits as Part of the Northeast Boundary Tunnel Project in Washington, D.C. Proceedings RETC 2021,* pp. 763–772.

Giurgola B, Banov P, Fuegenschuh N (2022). *Ground freezing & Jet-grouting. A hybrid solution for Tunnelling in urban environments.* ITA-AITES World Tunnel Congress, WTC2022.

# Missouri River Intake Screen Structure and Tunnel: Overcoming Underground Challenges to Build Vital Infrastructure

**Ryan Ward** ▪ Michels Trenchless Inc.

## ABSTRACT

The Missouri River Intake Screen Structure and Tunnel (MRISST) is one phase of the Red River Valley Water Supply Project (RRVWSP), a vital piece of infrastructure in the state of North Dakota to transport water to drought-laden communities in the eastern half of the state. The scope of work consists of constructing a cofferdam in the Missouri River, mining a 1,600-foot by 74-inch tunnel from a secant pile shaft near the riverbank into a cofferdam, and erecting a 40-foot Y-shaped vertical pipe structure in the cofferdam to support the intake screens. Construction personnel worked in the river through the North Dakota winter to construct the cofferdam and contended with high hydrostatic pressures and raveling sands, gravels, cobbles and boulders while mining the tunnel. Diligent planning and meticulous execution were essential to overcoming the challenges encountered below the Missouri River.

## INTRODUCTION

The Red River Valley Water Supply Project (RRVWSP) is a nearly 150-mile-long water pipeline being built in North Dakota to convey water from the Missouri River to the Sheyenne River. The water will be used as a backup supply during times of drought for agricultural and industrial purposes. The project is entirely state-funded and will directly and indirectly reach approximately 50 percent of North Dakota residents. While conceptual planning has been ongoing for decades, permitting and design work began around 2017, and construction began at the end of 2020. The pipeline is slated to be complete by 2029 (GDCD, 2022).

The Missouri River Intake Screen Structure and Tunnel (MRISST) is one of the first phases of the broader pipeline project. The project scope consists of the in-river structure for the intake screens and more than a quarter mile of 74-inch pipeline installed by microtunneling to convey the river water from the intake screens to a pump station. The project used a design-bid-build contracting method.

The geology in the project area consists of glacial alluvium with sands, gravels, cobbles and boulders. These challenging subterranean conditions were a cause for concern from the planning phase on. Similar ground conditions had caused multiple failed microtunneling runs of the same diameter in the project vicinity.

## DESIGN/ENGINEERING

Construction of the complete intake system was broken into three contract packages. Contract 1 included the Missouri River Intake Pump Station (MRIPS) wet well, a 44-foot diameter by 63-foot-deep secant pile shaft that would serve as the microtunneling launch shaft and eventual wet well for the pumping station. Contract 2 included the buried foundation structure for the future intake screens, approximately 1,660 feet of 74-inch piping to convey water from the intake screens in the river to the wet well, and a final concrete liner of the wet well. Lastly, Contract 3 will include the final electrical and mechanical fit out of the MRIPS.

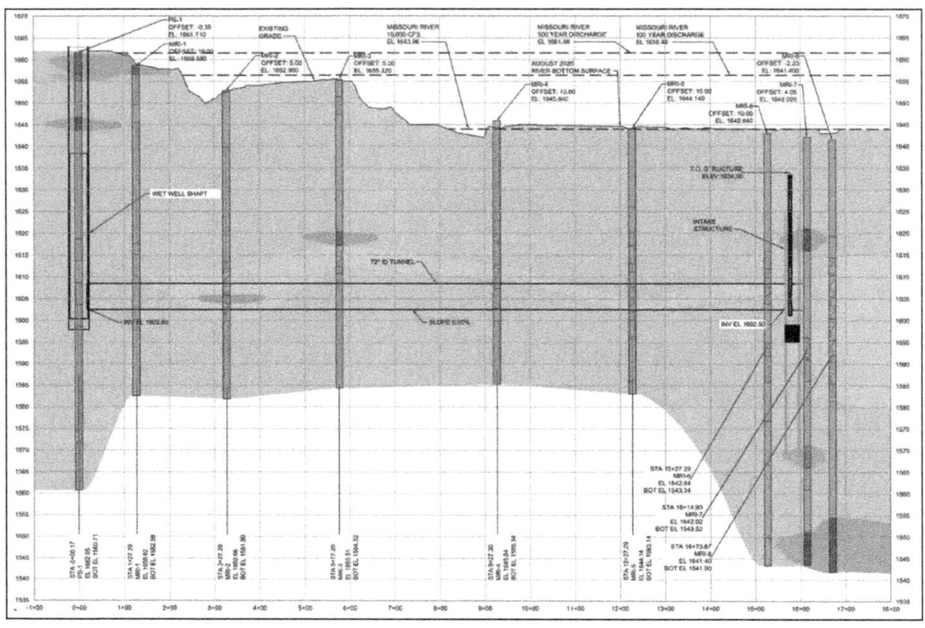


**Figure 1. Interpretive geotechnical profile (Black & Veatch, 2021)**

The MRISST contract bid package contained a Geotechnical Data Report (GDR) and Geotechnical Baseline Report (GBR). The reports were based off nine bore logs, five within the river and four on land. The GBR baselined 62 percent sands, 35 percent gravels, 3 percent clay. The GBR also baselined cobbles and boulders up to 24 inches without specifying a quantity. The sands and gravels were classified in accordance with the Tunnelman's Ground Classification as flowing and raveling. Lastly, the water table was baselined at 5 feet above river elevation, corresponding to approximately 48 feet of hydrostatic pressure at typical flow elevations (Black & Veatch, 2021).

The tunnel alignment connects state-owned property on the riverbank to the chosen location for the intake screens. The engineering team located the intake screens adjacent to a United States Army Corps of Engineers armored shoreline, which was constructed to avoid further erosion of the natural shoreline. The armored shoreline has caused a deep channel with high flow velocities to develop, leaving a favorable location of an environmentally resilient water intake (Encapera, 2022).

In response to the geological conditions of the chosen alignment, a fully shielded slurry microtunnel boring machine (MTBM) was specified for the trenchless installation. This equipment would allow the excavation of the tunnel to be completed remotely without personnel regularly working in the tunnel.

Rather than a reinforced concrete pipe product, which had proven detrimental in other nearby microtunneling projects, 1-inch-thick steel pipe was specified. Concrete pipe can be susceptible to axial point loading during jacking operations and normal point loading from cobbles and boulders rolling along the pipe. Steel pipe is more forgiving in this regard. Following the microtunnel boring machine, four 10-foot sections of pipe were required to improve steerability. The remainder of the tunnel alignment consisted of 20-foot pipe sections to increase production rates. Permalok's latest T-7 joint was specified in order to improve productivity rates while maintaining the ability to withstand the hydrostatic pressure and prevent infiltration of water and soil.

## CONSTRUCTION (AND CHALLENGES)

### Environmental Considerations

Pursuant to the project's United States Army Core of Engineers permit, mitigations were required to protect threatened and endangered species. The list of project-specific species included the piping plover, interior least tern, whopping crane, pallid sturgeon and northern long-eared bat. The mitigation measures included a prohibition on marine work in the river from April 15 through June 1, a prohibition on tree clearing for sitework outside of the months of January and February, and a prohibition on clearing and grubbing in June and July. Daily bird monitoring was required from April 1 through July 15, and September through November. The jobsite would be shut down if any of the protected birds were observed within the jobsite buffer during those monitoring periods. The project's schedule and work sequencing were developed around the associated monitoring and black-out periods. Specifically, around-the-clock tunneling operations were precluded during the bird-monitoring windows in case birds were observed in the project vicinity and necessitated a prolonged shutdown.

### River Work

While permanent access to the MRIPS site was constructed prior to the launch shaft excavation, Michels needed to pioneer an additional quarter mile of temporary gravel road to access the river. From the riverbank, a temporary bridge structure was erected to access the cofferdam and intake structure location. The temporary bridge was constructed by driving eight 42-inch pipe piles, two located on shore and six located within the river. Two pairs of 64-foot-long plate girders supporting timber mats yielded a 128-foot-long by 40-foot-wide work deck from which the cofferdam and intake structure were constructed. Bridge construction and piling was hampered by discarded armor stone from the original construction of the armored shoreline and required field modifications to its positioning. The temporary bridge structure was utilized in lieu of a barge or marine-based work surface to minimize costs. Additionally, the bridge structure enabled the construction of the cofferdam to continue safely through the winter months despite significant ice buildup in the river.

Following the completion of the bridge, sheet pile driving for the cofferdam began. Pile driving operations contended with wind speeds in excess of 50 miles per hour, windchill temperatures below negative 40 degrees Fahrenheit, and river ice 2 feet thick. Sheets were driven more than 65 feet below mudline elevation through sands, gravels, cobbles and boulders. After sheet piling, the cofferdam was excavated to support the retrieval of the MTBM and reinforced to withstand the hydrostatic loads while the excavation was dewatered. Due to the required depth of the cofferdam for retrieval of the microtunnel boring machine, the internal reinforcement bracing needed to be installed in phases as the cofferdam was excavated. The bracing of the completed cofferdam consisted of four levels of 12-inch ring beams. Approximately 1,200 cubic yards of soil was excavated.

Multiple issues were encountered during excavation of the cofferdam. First, boulders up to 2-feet in diameter were removed from the cofferdam. Second, during the installation of the lateral reinforcement beams below water, divers noticed that some of the sheet pile interlocks had separated during driving operations. Finally, after excavating down approximately 30 feet, a large volume of soil migrated into the cofferdam,

Figure 2. Cofferdam and temporary access bridge

leaving it only excavated approximately 15 feet below the original mudline. The project team's assessment of these three events and observations was that the boulders and cobbles present below ground caused the sheet piles to split while they were being driven, and that these gaps in the sheet piles allowed for a large migration of soil during excavation. The gaps between sheets were closed off by welding steel plate over them as the excavation continued. Despite the numerous challenges, excavation and reinforcement were completed, a 10-foot-thick plug was tremie poured, and the cofferdam was successfully dewatered, allowing for construction of the tunnel exit portal, retrieval of the microtunnel boring machine, and erection of the screen structure piping.

## Tunnel and Intake Piping

A quarter mile away at the MRIPS microtunnel launch site, microtunnel operations began with the setup of equipment and the connection of fluid, hydraulic, electrical and communication utilities. A skinned-up Herrenknecht AVN1500TB was utilized along with a Herrenknecht jacking frame rated for 1,100 tons of push force. Three intermediate jacking stations were custom built for the project with Permalok's T-7 snapping pipe and Herrenknecht's hydraulic rams. Two Herrenknecht lubrication injection pumps were plumbed on the surface to provide lubrication to the pipe string. The solids control equipment consisted of a separation plant with four Derrick shaker units and a big-bowl centrifuge. The tunnel was guided by VMT's hydrostatic water leveling and gyroscope system. Due to the likelihood of excessive moisture and humidity in the tunnel and the length of the tunnel drive, a laser guidance system was deemed insufficient for the tolerances required.

In the bottom of the MRIPS microtunnel launch shaft, a concrete thrust reaction block was poured behind the jack frame to distribute jacking loads and prevent any damage to the secant pile structure. Also in the launch shaft, the tunnel entry portal was constructed. The launch portal was a top hat system connected to the secant pile walls with rock bolts and cased in concrete. The portal window consisted of a pair of seals and adjustable fingers to prevent ground water and soil from infiltrating into the shaft. In the cofferdam, the exit portal was constructed by the cutting the steel sheet piles and pouring a low strength concrete block in the tunnel alignment. The exit portal window also contained a rubber seal and adjustable fingers to prevent the infiltration of water and soil into the cofferdam after completion of the tunnel.

Figure 3. MRIPS and microtunnel launch shaft work site

Twenty-four-seven mining operations began slowly with the MTBM advancing through nearly 10 feet of secant pile concrete. Once into virgin ground, however, the MTBM advanced quickly through the anticipated sands and gravels. The slurry muck circuit utilized a bentonite-based slurry to convey the excavated material to the separation plant on the surface. The gel strength and viscosity properties of the slurry provided sufficient face support in the running ground conditions throughout the tunnel drive. The only significant deviation in slurry fluid properties came in the days after mining through the concrete secant piles. The alkalinity of the concrete particles in the slurry caused the pH of the slurry to elevate to a point that the bentonite no longer reacted properly with the water. As a result, the slurry, despite a high unit weight and plenty of bentonite, did not have the high viscosity and gel strength properties desired. After introducing an environmentally friendly pH neutralizer, the slurry returned to its desired properties.

Also at the beginning of the tunneling drive, when jacking loads and friction on the pipe were minimal, the tunnel boring machine was exposed to substantial hydrostatic and lateral earth pressure loads pushing the MTBM back into the launch shaft. A hydraulic "pipe brake" was required to grip the pipe and hold it in place while the jack frame was retracted during pipe changes.

Mining operations continued through the running ground at instantaneous advance rates of nearly 200 millimeters per minute. Ground conditions were largely in line with that of the Geotechnical Baseline Report. Measured volumes of excavated soil indicated approximately 47 percent gravel compared to the 35 percent baselined. Additionally, small seams of coal were present in the tunnel alignment. While the coal had no adverse effects on the advancement of the MTBM, residue from the coal did slicken the separation screens and inhibit the efficient recycling of the slurry.

Near the midway point of the tunneling drive, the shaft crew noticed a small tear growing on the launch portal seal. As mining and pipe advancement continued through the seal, the tear grew to the point that mining operations were halted in order to prevent the total failure of the seal and flooding of the shaft and tunnel. An entirely new launch window with a rubber seal and adjustable fingers were rushed to the job site, fitted around the pipe, and secured to the existing portal, enabling mining operations to continue with a sealed shaft.

**Figure 4. Herrenknecht AVN1500TB**

Jacking loads proved not to be a significant issue during tunneling, largely due to a well lubricated pipe string and geology that reacted favorably to steering. Breakout forces only exceeded 600 tons at the very end of the tunnel drive, and average running jacking forces peaked around 400 tons at the end of the tunnel drive. As a result, the intermediate jacking stations were not utilized during regular mining operations. A minor but noticeable increase in jacking loads occurred following the downtime to replace the entry portal seal, but this increase regressed over the subsequent 100 meters.

Mine-through into the cofferdam occurred 24 days after the MTBM was launched. The pipe string was advanced to its final position, the cofferdam was dewatered, and the MTBM was removed from the shaft. Following removal of the MTBM from the cofferdam and utilities from the tunnel, the tunnel line was grouted with a cementitious grout. Additionally, the exit portal was grouted with a two-part chemical grout due to the volume of water infiltration. Once the tunnel line was grouted in place, and infiltration was plugged at both portals, construction of the riser pipe structure in the cofferdam took place. The riser pipe structure was made up of a 40-foot vertical "Y-shaped" pipe of 74-inch diameter that would eventually both support the intake screens and convey the river water to the tunnel.

## Concrete, Ancillaries and Final Completion

Following completion of the tunnel and intake piping, the cofferdam was partially filled with a controlled low strength material and finished with a reinforced concrete slab below the preexisting mudline elevation. The cofferdam sheet piles were cut above the finished concrete and removed from the river, allowing sedimentation to restore the riverbed to its preexisting condition. The temporary bridge structure and access road to the river were removed, and the site restored.

A cathodic protection system was included in order to protect the steel pipe structure over its design life. The system included 20 magnesium anodes and one zinc anode.

Figure 5. Intake structure riser pipe

At the time of writing this paper, a final concrete lining was partially poured in the MRIPS microtunnel launch shaft. The completed liner will consist of 2-foot-thick reinforced concrete walls around the perimeter of the cofferdam, as well as a "T-shaped" divider wall splitting the wet well into three bays. Hollow core concrete planks will be installed across the top of the wet well and divider walls to cover the structure until the final pumping fit out is complete.

## CONCLUSION

The Missouri River Intake Screen Structure and Tunnel was a technologically challenging piece of water supply infrastructure for the State of North Dakota, but it was needed to provide environmental resiliency as the state continues to grow. The project design implemented industry best practices and advanced the use of newer technologies. Challenges abounded during construction, from boulders and cobbles to high hydrostatic pressures and harsh North Dakota winter weather. Diligent planning and creative solutions yielded a successful project that will benefit the residents of North Dakota for decades to come.

## SOURCES

Black & Veatch, AE2S. 2021. *Geotechnical Baseline Report for Missouri River Intake Screen Structure and Tunnel, McLean County, North Dakota.*

Encapera, V., Sivaram, A., Kovar, K. 2022. *Design and Construction of the Missouri River Intake, Screen Structure and Tunnel Project*. North American Society for Trenchless Technology 2022 No-Dig Show TA-T3-04.

Garrison Diversion Conservancy District—Red River Valley Water Supply Project. 2022. "About the Project." www.rrvwsp.com.

# The Versatility of Tunnelling and Trenchless Methods for Sustainable Grid Construction

**P. Schmäh** ▪ Herrenknecht AG
**M. Peters** ▪ Herrenknecht AG

## ABSTRACT

The development of renewable energies and a sustainable power grid are the challenges for future energy supply. Underground cables will replace the vulnerable overhead lines, because of their safety benefits. Due to the public environmental awareness, smart tunnelling and trenchless solutions are required for inner-city and cross-country installations of underground cables, as well as for crossings and the landfall sections to connect offshore wind farms to the transmission grid. Methods from the tunnelling and pipeline industry provide a high flexibility in the planning of alignments, including versatile tunnel concepts and the installation of protective pipes with E-Power Pipe and Direct Pipe.

## INTRODUCTION

The field of power transmission contains several aspects. On the one hand, outdated networks need to be renewed and expanded. On the other hand, the implementation of renewable energies such as offshore wind power require powerful electricity grids in order to secure future energy supply. The fulfilment of the conditions for the energy transition is increasingly becoming the driving force for the industry. Existing networks have to be expanded, new routes have to be developed to transport electricity from offshore to onshore, over land and below obstacles. The trend is going towards underground network installations for environmental protection and public acceptance. Trenchless technologies answer this development and are getting more and more common in cable and pipeline installations. Furthermore, new trenchless construction methods and the further development of existing technologies now offer suppliers, grid operators and construction companies a wide range of methods for realizing projects. Over long and short distances, over land or when crossing obstacles, economic aspects, environmental protection and the concerns of residents and owners can be reconciled.

In future, the development of renewable energy sources such as offshore wind power will not only serve to supply electricity. In addition, solar and wind energy will be used to produce so-called green hydrogen. This can be used as a storage medium. It can be transported relatively easily, thus allowing the supply of industry as well as end consumers in all areas. The intelligent use of district heating also plays a key role. Smart district heating concepts are becoming increasingly interesting for businesses and municipalities, not only from an ecological but also from an economic point of view. The basis in each case is an appropriately designed grid and pipeline network that can keep pace with the requirements of the decades ahead.

## TRENCHLESS INSTALLATIONS OF UNDERGROUND CABLES

### Application Fields in Transmission Grid Construction

The landfalls of export cables coming from offshore wind farms are usually located in sensitive coastal areas where the operation of heavy construction equipment is not possible due to environmental protection requirements. When also dyke structures have to be crossed, this sets further increased demands on the safety aspects of respective trenchless installation methods. As the preservation of water quality has the highest priority, the use of drilling fluids must be strictly monitored. This mainly concerns the HDD method, which is not covered in detail in this paper. Even if HDD presents a common technology for the installation of steel casings for subsequent pull-in of the export cables, the alternatives with controlled microtunnelling systems such as Pipe Jacking or Direct Pipe need to be carefully investigated. In any case, the drilling works are carried out from onshore to offshore, whereas the protective pipe is inserted from the seaward side in a second step.

**Figure 1. Application range of trenchless methods from offshore to onshore and transmission grid construction**

Similar safety considerations also play a role where rivers have to be crossed along the route of the transmission grid. According to the width of the river and possible coverage, it is usually advised to design the cable crossing as a tunneled casing to provide a functional discharge of the cable heat emission by a tunnel ventilation system additionally installed in the large diameter tunnel. The heat generated by the cables and the magnetic field also play an important role in route planning over land. Depending on the temperature and the required limit values of the magnetic field at the surface, the depth and spacing of the cables (for separate installation e.g., with E-Power Pipe method) must be decided. Finally, these parameters have a deep impact on the required corridor width, planning of the alignment and the resulting approval process. Especially for crossing agricultural land, further aspects like route accessibility, the separation of soil layers, recultivation measures and restrictions to use bentonite further determine the possibilities for open-trench and the various trenchless methods.

In inner-city conditions, where more and more cables are placed underground to gain space on the surface and to protect the lines from external impacts, trenchless is mostly imperative. Due to restricted surface above ground or as cable tunnels often go deep and existing infrastructure can only be crossed under.

Figure 2. Grown power cable structures in inner-city areas

## Other Applications for Energy Tunnels

It is not only green electricity generation that makes a significant contribution to a successful energy transition, but also the intelligent use of industrial waste heat. Energy suppliers, municipalities and the industry are working on sustainable district heating concepts. These require a strong networking of all stakeholders and the construction of appropriate transmission and distribution networks. Tunnelled solutions are predominantly considered, when such district heating lines have to cross rivers or densely built-up industrial areas.

Coastal industrial facilities and power plants use seawater for cooling. New energy storage concepts with hydrogen require water for the electrolysis process: water is divided into hydrogen and oxygen with the help of electrical energy. The hydrogen can be stored as an energy source and converted back into electricity when needed. To ensure a safe installation and sufficient water supply volume, such water intakes are usually designed as tunnels to host the water lines.

## Advantages of Trenchless and Tunnelling Technologies

Reliable, quick, safe and environmentally friendly installation of onshore grids, realization of cable landfalls and crossings of natural and structural obstacles will play an important role for a high level of public acceptance and for the fastest possible approval procedures. On some sections, an open-cut method is the most efficient installation solution. With these, however, it is usually not possible to safely cross beneath existing obstacles above and below ground, such as traffic routes or bodies of water, without disrupting life above ground. Only trenchless methods can do this. They can also be used where the concerns of landowners and environmental protection must be especially taken into account.

Today, trenchless technologies cover the entire geological spectrum, with high flexibility in terms of diameters, drive lengths and depths of the network sections to be installed. Complementing proven methods from pipeline and tunnel construction, additional technologies have been developed that show their strength where the long-established methods reach their limits in terms of feasibility.

**Table 1. Overview of tunneling machine types according to diameter and installation length**

| | Tunnelling | Direct Pipe® | E-Power Pipe® | HDD |
|---|---|---|---|---|
| Installation of **cable / casing** | **Indirect** Cables in tunnel | **one-step** steel casing | **two-stage** HDPE single casings or bundle, steel casing | **multi-stage** Cable bundle or steel casing |
| Diameter | **> 10"** Ø tunnel (ID) | **24" – 60"** | **10" – 28"** < 36" with backreaming | **10" – 60"** |
| Max. installation length | **33,000 ft** | **6,500 ft** | **6,500 ft** | **16,400 ft** |

* The information in this table is intended as an initial guideline; the parameters may vary depending on the project

## Trenchless Technologies Overview

A remarkable number of trenchless installation methods has been developed over the past 50 years. Basically, these installation methods originate from the fields of pipeline installation and tunnel construction. Successive further development in terms of diameter range, geological suitability for use, installation accuracy and the variety of pipe materials to be laid have made these techniques increasingly economical and versatile. Additional process and technology developments (E-Power Pipe®) were aimed at new applications, such as the laying of underground cables, where completely new requirements were placed on the construction methods to be used from the subsequent operation of the pipelines.

Selection of the appropriate trenchless method depends largely on the specific project parameters. The pipeline can be installed directly, as it is the case with Horizontal Directional Drilling (HDD) or the Direct Pipe® method, or casing tunnels are created by means of pipe jacking or segment lining, into which the pipeline is inserted in a second step. Table 1 provides a simplified comparison of the common methods and some key parameters.

# SEGMENT LINING

Segmentally lined casing tunnels can host a large variety of utilities: from steel pipelines for oil and gas, for water and sewage, to electric cables and telecommunication lines. Accessibility of the tunnel during construction and operation offers a high degree of flexibility for inspection and maintenance to guarantee operational safety, reliability and durability of installed utilities.

## Safety Standards for Small Diameter Segment Lining

During the last decades, safety aspects in mechanized tunnelling gained more and more importance and safety standards are consequently improved: starting from basic topics such as Personal Protective Equipment (PPE) and space requirements, including refuge chambers and detailed rescue statements. From a European point of view, the latest tunnelling safety standards are summarized in DIN EN 16191:2014, which is currently being reviewed to even improve health and safety conditions by taking into consideration the lessons learnt. The latest DIN published in 2014 tightened regulations in regards of machine diameter and accessibility, also considering several other aspects as rescue systems and fire protection. This lead to confined space conditions in the machine, especially where segment lining logistics have to be considered. The upcoming reviewed version of DIN EN 16191:2014 will most probably define larger

**Table 2. Minimum access cross sections according to DIN EN 16191:2014**

| ID in mm | | 2600 – 3500 | 3500 – 6000 | ≥ 6000mm |
|---|---|---|---|---|
| Access height x width | ❶ | 1.0 x 0.45m | 1.4 x 0.45m | 1.9 x 0.45m |
| Exception for obstacles on less than 4m length | | 0.7 x 0.45m | 1.0 x 0.45m | |
| Min. access cross section | ❷ | 0.6m² | 0.8m² | 1.2m² |
| Min. walking width, with railing | ❸ | 0.3m clear | 0.3m clear | 0.3m clear |

minimum diameters as the required walkways for access and rescue purpose will consequently be further enlarged.

The logistic is the main key for a good overall performance on a segment lining TBM. A California crossing inside the tunnel is often needed for segment lining tunnels to allow trains to pass in order to reduce delays due to the travel times of the train. On the California crossing two trains can pass each other and an escape route along the California needs to be maintained at all times. The segment length is generally not less than 1,000 mm to keep the number of joints and segment seals to a minimum and to improve production. Furthermore, the smallest standard locomotive has a width of 1,000 mm. Considering a California crossing with two trains of 1,000 mm in width, sufficient escape route and enough space for a ventilation duct and the necessary tunnel lines, an inner diameter of 3000 mm becomes the preferable size for segment lining tunnels with a certain length.

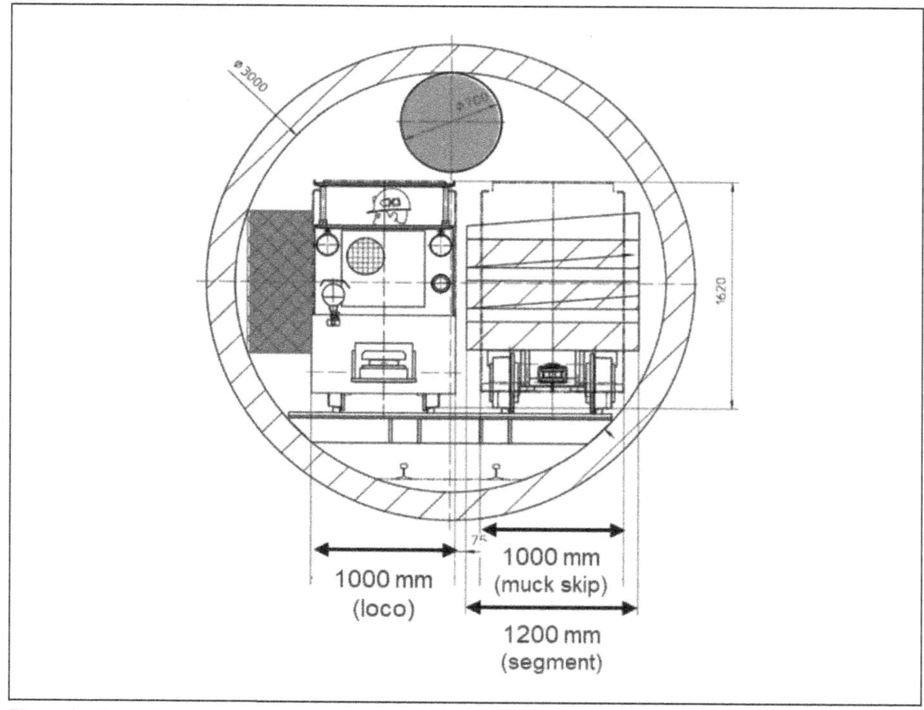


**Figure 3. Exemplary cross section and logistic concept of tunnel ID 3000 mm**

Figure 4. EPB 3000 for London Power Tunnel (LPT 2)

### Reference Projects

In grid construction segment lining casing tunnels are mainly considered for large tunnel structures in inner-city areas to host cables and other utility lines. In order to overcome either challenging soil or alignment conditions or to enable installations in great depths to avoid existing underground infrastructure, numerous accessible casing tunnels are currently in construction in European cities like Paris, London and Berlin. The preferred minimum diameter of these segmentally lined cable tunnels is ID 3,000 mm and larger. This minimum tunnel size is not only recommended for safety reasons as described above, but also a reasonable size to implement access ways and ventilation for discharge of the heat emissions in the tunnel once in operation.

An AVND 3000 machine is currently in operation in Berlin to excavate a 6.7 km long cable tunnel for grid operator 50 Hertz. Furthermore, a total of 3 EPB Shields are excavating 26.5 km of ID 3,000 mm tunnel in 4 drives for National Grid's LPT 2 project in London.

## PIPE JACKING

Pipe jacking is often a more economical alternative to segment lining, particularly in accessible diameters or when shorter casing lengths have to be realized. In addition, pipe jacking can also be used in the non-accessible diameter range. Thus, pipe jacking is possible in diameter ranges from 250 mm to approximately 4 m, whereby the maximum diameter is limited by the logistics of the pipes to the construction site. The jacking pipe materials should be pressure resistant, as due to the process the complete pipe string is moving when pushed from the starting point through the ground. The most commonly used pipe materials are concrete and vitrified clay

### Pipe Jacking versus Segment Lining

The specific project design and surrounding conditions such as diameter, tunnel length, alignment and ground conditions are crucial in selecting the most suitable lining method. For Microtunnelling, less personnel is required and due to remote-control from the surface the level of safety is considered as a major benefit, compared

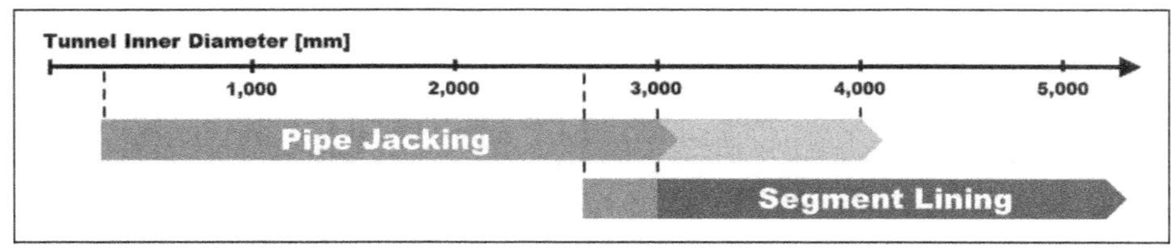


**Figure 5. Applicable diameter range of pipe jacking and segment lining**

to segment lining. Especially for large diameter and long-distance pipe jacking, the contractor´s experience in dealing with state-of-the art pipe jacking features such as volume-controlled bentonite lubrication, separation and navigation plays a key role for project success. In terms of tunnel diameter, the handling of the jacking pipes is considered as the most limiting factor of what is feasible in pipe jacking. Recent milestones set by contractors in all parts of the world demonstrate the current trend towards larger diameters and longer drives in pipe jacking.

In return, segment lining offers a high degree of flexibility concerning the planning of tunnel routes. Long drives and tight curved alignments are possible due to less friction compared to pipe jacking. Combined lining methods allow a high degree of flexibility whether to start tunnelling with pipe jacking in small launch shaft and switch to segment lining later on, or to use a push module to change from pipe jacking to segment lining in case of difficulties on long pipe jacking drives.

## Reference Project Cable Tunnel Swissgrid, Switzerland

In the canton of Wallis in Switzerland, a Herrenknecht AVND microtunnelling machine for pipe jacking is currently in operation to build a 1.2 km long tunnel for extra high-voltage underground cables (EHV line). The national transmission grid company Swissgrid will connect the Nant de Drance pumped storage power plant to the national electricity grid by this new section leading from the 20 m deep launch shaft in Le Verney to the target at La Bâtiaz substation. Along the route, the 3 m outer diameter (2.5 m inner diameter) tunnel will cross the Drance River, a motorway, SBB railway lines, a sewage canal of Emosson power plant and a gas pipeline.

**Figure 6. Contractor CSC Costruzioni, a subsidiary of Webuild, lowers the AVND 3000 into the launch shaft**

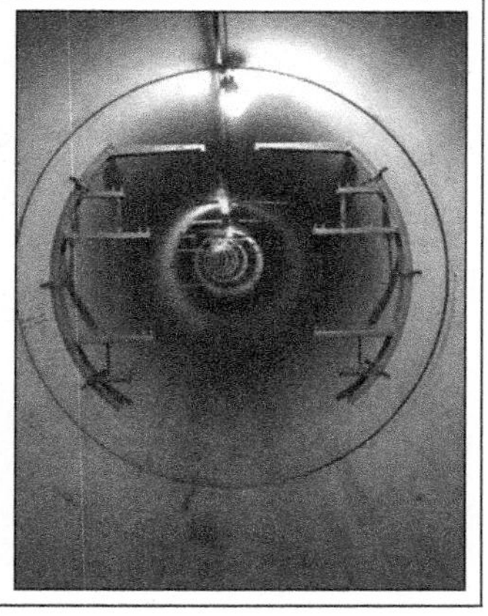

**Figure 7. AVND 2500 for pipe OD 3600 mm in launch shaft (left) and completed tunnel prepared for cable installation (right) - © Epping Rohrvortrieb GmbH + Co KG**

Initially planned as a small segment lining tunnel, the client could be convinced of the advantages of pipe jacking, carefully evaluated by a close cooperation of the consulting engineers, the experienced contractor and the tunnelling equipment supplier. The ground conditions in a formerly flooded area with gravel, boulders and sand and silt lenses with a groundwater pressure of up to 1.5 bar required a customized design of the cutting wheel and continuous face support by the Mixshield mode of the AVND machine. A complete set of tunnelling equipment, separation plant, interjacking stations and volume-controlled lubrication system with mixing unit has been delivered onsite. Due to the quite abrasive ground conditions, hyperbaric intervention for face access is needed for maintenance of the cutting wheel.

### Reference Project Cable Tunnel Legden, Germany

An even larger cable tunnel has been completed beginning of 2022 in Legden, Germany, where grid operator Amprion has built a 180 km long transmission line to bring wind power from the North Sea coast southwards. An AVND 2500 with extension kit for an outer diameter of 3,600 mm used concrete jacking pipes to execute two pipe jacking drives of 1,297 m and 813 m length. A total of 12 cables will be installed in the 2.1 km tunnel section, which will be accessible for maintenance when operating.

## DIRECT PIPE®

Over the past 15 years, the Direct Pipe® method developed by Herrenknecht for the trenchless installation of prefabricated steel pipelines has established itself worldwide. It combines the advantages of microtunnelling with the Pipe Thruster technology to enable trenchless installation of pipelines in difficult ground conditions while reducing the risks typically associated with HDD. Direct Pipe® allows excavation of the borehole and simultaneous trenchless installation of a prefabricated and tested pipeline in a single continuous step.

Typically, Direct Pipe® is used to safely cross rivers. Thanks to further technical development and growing popularity among clients and contractors, the range of applications for Direct Pipe® has steadily been expanded in recent years. Today, Direct Pipe® is also increasingly used for pipeline or cable landfalls. Here, the AVN tunnel boring machine can be decoupled from the pipeline at its target point in the seabed

**Figure 8. Direct Pipe for one-step pipeline installation**

and recovered. At the same time, the method has also evolved in terms of installation length and diameter. In New Zealand, in 2020 a world record was set with the installation of a 2,021 m long wastewater pipeline into the open sea. New developments in machine technology, for example the jet pump and the cutterhead design, now make it possible to use Direct Pipe® even with small diameters starting at 24".

## Reference Projects

The world premiere for Direct Pipe in 2007 was the installation of a 48" (1,219 mm) steel pipeline to host cables and a water pipeline in Worms, Germany. The Rhine River had to be safely crossed on 1,522 ft (464 m) with a minimum overburden of 10 ft (3 meters) only, through mixed ground of silt, sand and gravel. Since then, the Direct Pipe method has been used in more than 200 projects worldwide, in particular for crossing key sections along a pipeline route. 90 percent are below rivers and other bodies of water, but also under traffic routes occasionally.

### *Beatrice Offshore Windfarm Connection, UK*

In 2017 the Direct Pipe® method was first used for a shore approach near Portgordon, Scotland to connect the Beatrice Offshore Windfarm to National Grid´s substation on the mainland. With Direct Pipe®, even under the adverse weather conditions on the Scottish North Sea coast, two 48 inch cable conduits of 440 meter length each were safely installed beneath the coastline.

The main reason for choosing Direct Pipe® in this case was to reduce the impact on the flora and fauna of the environmentally protected coastal area. Not only the reliable avoidance of bentonite leakage favored Direct Pipe®, but also the wind farm operator's requirement to install the cables with the smallest possible overburden to ensure efficient cooling by the sea water. Since the seabed is covered with several meters of coarse gravel at this point, this would have meant a high frac-out risk for HDD, so that a successful installation would only have been possible at a greater depth. With Direct Pipe®, on the other hand, the desired shallow depth could be achieved without endangering the environment.

Figure 9. Recovery of the direct pipe MTBM from the seabed

After completion of the second drive, the 28 tons and 18 meter long MTBM was recovered from the seabed in one piece. With the help of a specially developed remote-controlled recovery and disconnect module it was separated from the pipeline – the MTBM was securely sealed against incoming water, while the casing pipe was flooded under controlled conditions in advance. The machine was then loaded onto a jack-up rig with salvage crane and hauled away by boat.

## E-POWER PIPE®

The expansion of the power grid in Germany, over long distances as underground cables, set new demands on trenchless methods. The near-surface, precise individual installation of the non-pressure-resistant HDPE cable protection pipes was not possible with existing methods. With E-Power Pipe, Herrenknecht has developed a new two-stage process that enables precise, safe and reliable installation of cables even at shallow depths. As a result, small-diameter pressure-resistant and non-pressure-resistant product pipes can be quickly and safely installed underground close to the surface over long distances of more than one kilometer. In the small-diameter range, it is now possible to realize drives up to 10 times longer than before.

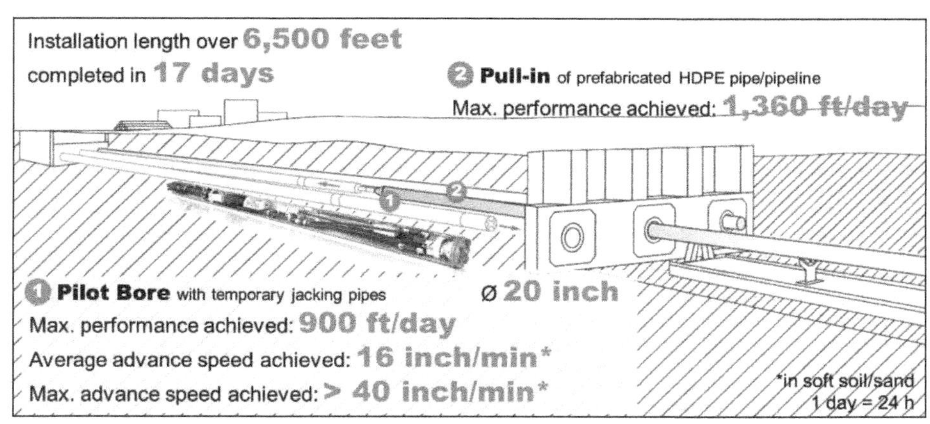


Figure 10. The E-Power Pipe technology and milestones achieved in underground cable installation

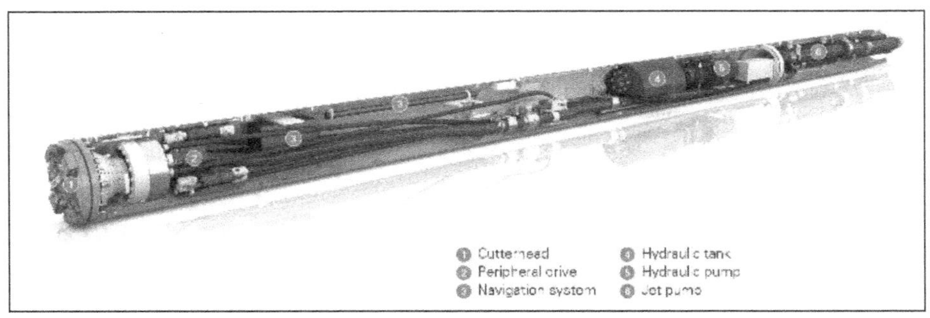


**Figure 11. The jet pump integrated in AVNS350XB Slurry microtunnelling system**

A jacking frame installed in a launch pit pushes the tunnel boring machine through the ground along the specified alignment. Specially designed reusable steel jacking pipes house all the supply lines for the TBM and the material discharge. After breakthrough at the target point, the TBM is separated from the steel jacking pipes and a pull head is installed instead. The prefabricated product pipe is connected via the pull head to the steel jacking pipe still in the borehole and pulled in by retracting the steel jacking pipes with the jacking frame. The borehole remains mechanically supported the whole time. During insertion, if necessary, the product pipe is mechanically and thermally connected to the ground with the addition of backfill material.

## The Role of the Jet Pump Innovation

The AVNS350XB tunnel boring machine with an excavation diameter of 20" (505 mm) is the key component of the E-Power Pipe method. With this slurry-supported micro-tunnelling machine, drives are possible even under ground water level and in changing soil conditions from silty soft soils to medium hard rock 22.000 psi (150 MPa). The fully remote-controlled tunnel boring machine is equipped with a jet pump as the slurry pump as well as an integrated hydraulic power pack so drives of more than 3,000 ft (1,000 m) in length can be realized.

With the integration of the jet pump into the well-known technologies of Slurry Pipe Jacking (AVNS) and HDD, the range of these trenchless technologies has been expanded again by going longer with smaller diameter pipes and pipelines and to lower the frac-out risk in HDD significantly. The new AVNS slurry machine concept featuring the jet pump presents a milestone in slurry microtunneling pushing the boundaries towards the feasibility of longer drives in small diameters. Thus, several applications and technologies benefit from the development of one single tool. E-Power Pipe® with AVNS and the jet pump as a core component presents a new trenchless alternative for the installation of HDPE pipes, e.g., for cable casings in grid expansion. As a further development, the jet pump system for Direct Pipe® extends the application range in smaller diameters down to 24", to install steel pipes in one step safely and economically.

## Reference Projects

The efficiency of the E-Power Pipe method has already been demonstrated in practical use in several projects. Following the successful pilot project in 2017, the E-Power Pipe method was used in further construction projects for the expansion of the grid in Germany. With its remarkable advance rates, high planning reliability and environmentally friendly construction method, E-Power Pipe scores well with grid operators,

**Figure 12. Breakthrough of AVNS600 machine after pilot drill and pullback of protection pipe bundle (6,000 ft length)**

**Figure 13. Breakthrough of the AVNS350XB after pilot drill with temporary steel jacking pipes**

contractors and landowners. The current record installation length of 6,000 ft has been achieved beginning of 2022 in the Netherlands, where a HDPE pipe bundle for underground cables has been pulled in a 20" borehole.

### *SuedLink Cable Connection to Substation, Grossgartach, Germany*

The SuedLink project is one of the major HVDC cable links in the transmission grid as a connection between the offshore wind farms in the North and Baltic Sea and the consumer centers in Southern Germany. The project comprises two DC links with a total transmission capacity of 2 gigawatts and a total length of 702 km and 558 km respectively. In order to connect the Grossgartach substation to the transmission grid, the E-Power Pipe method was used to install cable protective pipes underground. Between the Brunsbüttel substation near Hamburg in the north and Grossgartach near Stuttgart in the south, this represents the first completely prepared section to host the SuedLink cables lateron.

In the period from January to March 2021, three cable protection pipes of 455 m length each were installed by Implenia using the E-Power Pipe method on behalf of the transmission grid operator TransnetBW. The three boreholes to pull-in the cable protection pipes were drilled through clayey soil in only 2.5 m to 5 m depth, in a 3D curve alignment with radii of 500 m vertically and 788 m horizontally.

## CONCLUSION

This paper has given an overview of the different technical alternatives for trenchless respectively tunnelling solutions for the trenchless installation underground cables. For the successful implementation of the energy transition, as a consequence to the Paris climate agreement of 2015 to which 195 countries and institutions have committed themselves by ratification, environmentally-friendly mechanized solutions for the construction of required underground infrastructure are indispensable. In the coming decades, one focus of trenchless construction will be the expansion of transmission grids in order to connect households and industry to alternative energy sources throughout the country. In this context, not only the expansion of renewable energies but also the laying of underground cables in the most environmentally-friendly way possible plays a decisive role. Small-diameter tunnelling solutions and installation methods from the pipeline industry make an important contribution to this and will get a greater share in future. In order to provide a feasible solution to any kind of project design, e.g., for sensitive river crossings in transmission grid construction and challenging offshore—onshore connections in coastal areas, existing technologies are being further developed and improved. Lessons learned in small diameter tunnelling and the expertise in trenchless pipeline installations lead to the development of new trenchless methods, such as the E-Power Pipe® technology, which opens up new application fields and sets new standards in the trenchless industry.

## REFERENCES

DIN EN 16191:2014, (2014) Tunnelling machinery—Safety requirements; English version EN 16191:2014, DIN Deutsches Institut für Normung e.V., Berlin, Germany.

PART 

# New and Innovative Technologies

*Chairs*

**Kevin Smyth**
Frontier-Kemper Constructors Inc

**Bade Sozer**
McMillen Jacobs Associates

**Glenn Larose**
Jacobs

**Sergio Moya**
Frontier-Kemper Constructors Inc

# Applying Automation and Machine Learning for Tunnel Inspections

**LiLing Chen** ▪ Arup
**Yung Loo** ▪ Arup
**Fabio Panella** ▪ Arup
**Tristan Joubert** ▪ Arup
**Michael Devriendt** ▪ Arup
**Nasir Qureshi** ▪ Metrolinx
**Ahmad Ali** ▪ Metrolinx

## ABSTRACT

Tunnel inspections have traditionally been carried out manually by inspectors writing up observations and taking photos of defects. The results from the inspection and the defects observed are dependent upon the rigor of the inspectors and may be subject to repeatability and consistency issues and is often time consuming with elevated health and safety risks. This paper will discuss work that is being carried out in developing and implementing an innovative hardware and software solution integrating machine learning to automate the process of capturing objective tunnel condition information, offering cost and programme savings as well as health and safety improvements

## INTRODUCTION

The Eglinton Crosstown West Extension (ECWE) is a proposed rapid transit line in the Greater Toronto Area that extends the current Eglinton Crosstown Line (ECLRT) further west from Mount Dennis Station to Renforth-Eglinton Station. The extension will stretch approximately 9.2 kilometers from Mount Dennis to Renforth Drive, incorporating seven new stations, and interchanges with the Mississauga Transitway at Renforth Gate. The project is split into 4 different contracts with the Advance Tunnel Contract 1 (ATC1) currently under construction. The ATC1 will consist of an approximately 6.5-kilometer long twin bored tunnel of 5.75 m internal diameter with 9 cross passages running from Renforth to Scarlett area, as shown in Figure 1.

Arup has been appointed as the Technical Adviser to Metrolinx and Ontario Infrastructure and Lands Corporation, collectively, the Contracting Authority (CA) and is providing design compliance review and site construction oversight services to the CA. With support from the CA on implementation of digital initiates for the project, Arup has proposed, amongst other digital works developed for the project, to develop and implement an automated tunnel inspection system for the ATC1, which is currently in the first phase of the development for a pilot run. The current phase of work is to develop the tool to undertake photogrammetric survey of the tunnel and the development of a virtual model of the tunnel to allow users to view, review and compare defects.

This paper will provide an overview of the survey methodology and benefits it will provide in different stages of the project, as well as the development of the inspection tool and potential future expansion of this to bring additional values to the asset owner over the long term asset life.

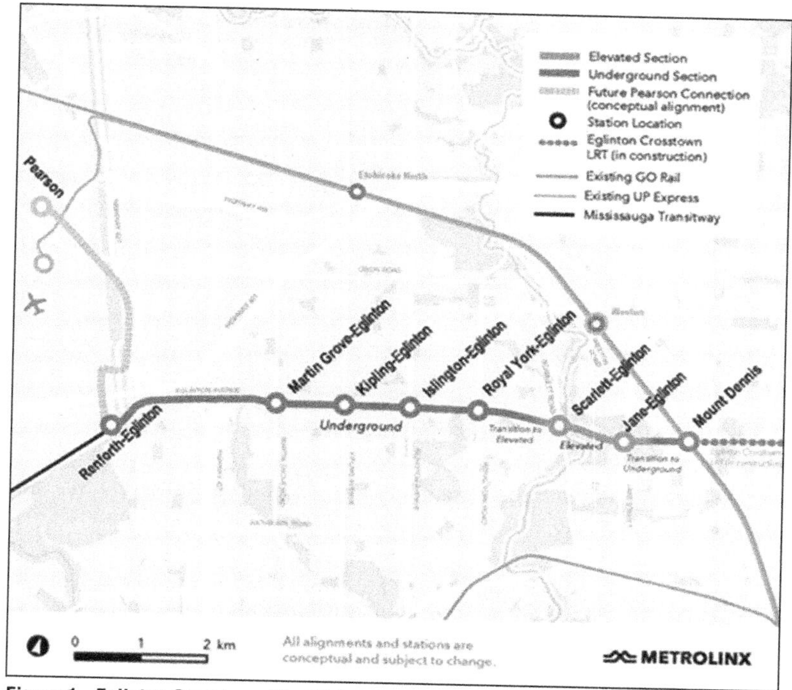


Figure 1. Eglinton Crosstown West Extension Project

## PROJECT BACKGROUND

The tunnel to be constructed under ECWE ATC1 contract consist of twin bored tunnel of approximately 6.5km long running between Renforth Gateway and Scarlett Road. The tunnels are 5.75 m internal diameter with 275 mm thick precast concrete segments. The ATC1 contract was awarded to Westend Connectors (WEC) and design and construction for the ATC1 contract started in May 2021. Both TBM tunnels are progressing and are 39.8% and 28.6% complete as of end of November 2022. The tunnels are expected to breakthrough in early 2024 and be handed back to the CA in 2025. Typical tunnel cross section and tunnelling progress are shown in Figure 2.

## AUTOMATED TUNNEL INSPECTION SYSTEM

At different stages of the project, the automated tunnel inspection is intended to provide information to serve different purposes and bring benefits to the users and asset owners. The automated tunnel inspection system that will be used for the survey comprises of a camera array system for automated capture of hi-resolution imagery at speed, and a web-based platform for visualization and interrogation of the inspections.

### Construction Stage

As part of the Technical Advisory services to the CA, Arup is providing construction oversight to the on-going construction works for the ATC1 contract. Site inspectors carry out daily tunnel walks and undertake visual site inspections. Site observation notes on defects and photos are taken by the site inspectors and are then manually logged in a deficiency log afterward for tracking and site record. Locations and defect descriptions largely depend on inspector's notes and further review the photos

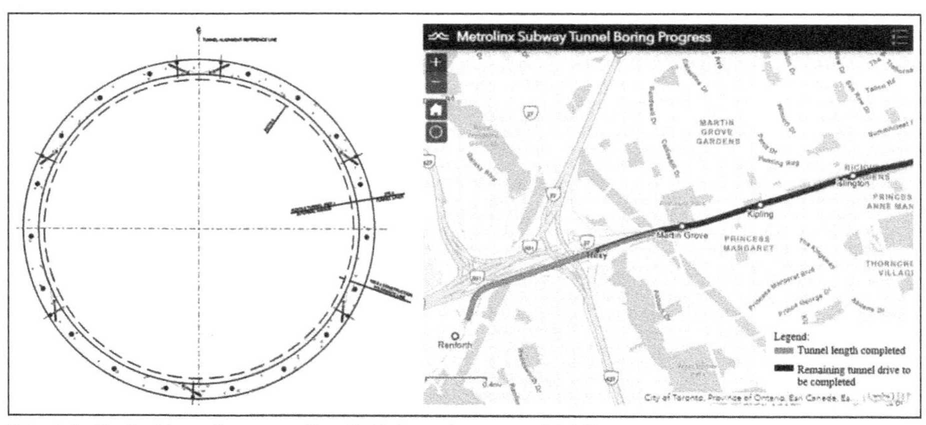


**Figure 2. Typical tunnel cross section (left); tunnel progress (right)**

afterwards when logging into the deficiency log. The survey taken by the automated tunnel inspection tool would supplement the current manual tunnel inspection process to allow:

- Virtual tunnel environment for review of defects observed visually on site.
- Dashboard platform providing a centralized location for summary of defects and user-friendly features for defect reviews.
- Identification of additional defects missed by the manual visual inspection.
- Supplement the visual inspection to provide an overall tunnel conditions baseline at tunnel breakthrough.

## Project Handover

Prior to handover of the tunnel asset to the CA, Project Co is to ensure all defects are repaired and accepted by the CA. The tunnel inspection tool can be utilized at this stage to:

- Provide the common digital platform for relevant parties and stakeholders to review inspection data and make decisions on repair and acceptance.
- Validate the condition of the as-built tunnel against the design intent and required specifications.
- Provide a baseline of condition in the tunnel for future record, audit and comparison.

## Operational and Maintenance

Following tunnel commissioning and fit-out, the historic collected data from construction and hand-over will provide the baseline for comparisons over time against upcoming general and targeted inspections. This will provide rich visual data across the lifetime of the asset's operation and maintenance. The tunnel inspection tool can be utilized at this stage to:

- Provide a digital repository of visual data of tunnel condition over time.
- Enable the asset manager and tunnel inspector to effectively identify and record issues in the tunnel.
- Enable quick and repeatable data capture of the tunnel to be collected consistently over time.

- Enable the asset manager to be proactive rather than reactive in identifying potential emerging issues based on historic objective and consistent survey data collected. It should be noted that the civil asset condition often impacts on other tunnel systems, for example groundwater ingress/leakage affecting track and M&E systems.

## METHODOLOGY

With the support from the CA on extending digital initiatives, ECWE ATC1 will be one of the first tunnel projects in Toronto to envision implementing automated tunnel inspections. The development for ECWE utilizes Arup's *Loupe 360* tunnel inspection service and digital platform engine (https://www.arup.com/services/tools/loupe-360). This builds on technology and knowledge developed with public and private sector asset owners predominantly in the UK, where the technological, commercial, health and safety benefits of automated tunnel inspections is being demonstrated.

The development of the automated tunnel inspection system is carried out in several phases to meet project specific requirements. It consists of an initial stage on proof of concept for the pilot run on tunnel inspection survey; further development with the CA on additional project specific requirements including incorporation of machine learning for construction and handover phase for tunnel baselining; and the final phase explores on program-wide asset management requirements implementation for tunnel maintenance and management.

The current phase of the work is at the initial proof of concept stage and the automated tunnel inspections system comprises a physical hardware system and a web-based digital platform to view and interrogate the captured data to meet project needs.

### An Industry Shift Change

The combined hardware and software setup provides a step change approach to how tunnel inspection is conventionally undertaken and automated. It moves forward all aspects of the approach:

1. A change from manual inspection to automated inspection. This speeds up the data capture process, removes personnel from the tunnel creating improvements in health, safety and welfare, enabling engineers to focus on less manual intensive and prioritized tasks based on the visual outputs.
2. Use of data rich computer vision data and analytics to support decision making. 360 imagery provides a fully comprehensive view of the tunnel, enabling objective analysis and baselining over time for comparison. This removes the reliance on inspector specific subjectivity and descriptive differences, which can always be compared back to the underpinning data.
3. Digitalization of assets, documentation and analysis. Representation of the asset and the inspection records within the digital platform enables easy and user-friendly access to a wide user base (on a secure access controlled basis), and enables reporting and review of the tunnel environment to be a lot more accessible than within the conventional approach which relies on traditional documentation and reporting. As more surveys, layers and datasets become available, these can be iteratively built on within the platform and deeper analytic insights can be found through overlapping data analysis.
4. Machine learning. Using machine learning to automatically detect defects and inventorying of tunnel furniture enables a dataset of tunnel health metrics to be quickly developed across the tunnel, especially over long kilometer extents of repeat tunnel structures. The automatic insights to flag up and

identify issues enables a first pass assessment of the distribution and relative severity of deficiencies. The machine learning approach enables a move from preventative to predictive maintenance regimes which ultimately leads to better informed understanding of the tunnel condition and the ability to intervene, plan and monitor the ongoing health of the asset.

## Hardware Design

The physical hardware system comprises a camera array system and a frame support onto the tunnel multi-purpose service vehicle (MSV). The camera array system comprises multiple GoPro cameras that are clustered together on a 3D printed mount, specifically designed to ensure 360-degree field of view coverage of the tunnel, and benefits from utilizing low-cost commercially accessible and upgradable hardware. The cameras and mount are affixed onto the MSV by attachment onto an aluminum frame designed modular system, to ensure stability and centering within the tunnel. The setup is also powered by an on-board battery and independent memory storage. The build is modular in design to allow future-proofing for future technology sensor development and additions, including: live geospatial location tracking, thermal vision and sound recordings, aiding the inspection process. Should the in-tunnel communications network allow, live transmission of visuals and data storage in the cloud is possible in the future (Figure 3).

The frame system is designed to be compatible with the equipment available on site considering various site constraints. The frame needed to meet the requirements of the inspection, adhere to best practice design standards and ensure compatibility and improvement to the site survey teams workload and conditions. The development of the frame system was coordinated among different parties, including the CA, Project Co, Arup's UK team, project site team, and the project tunnelling team in the North Americas. The tunnel is currently under excavation using the TBM, some of the main site constraints that are taken into consideration in development of the frame include:

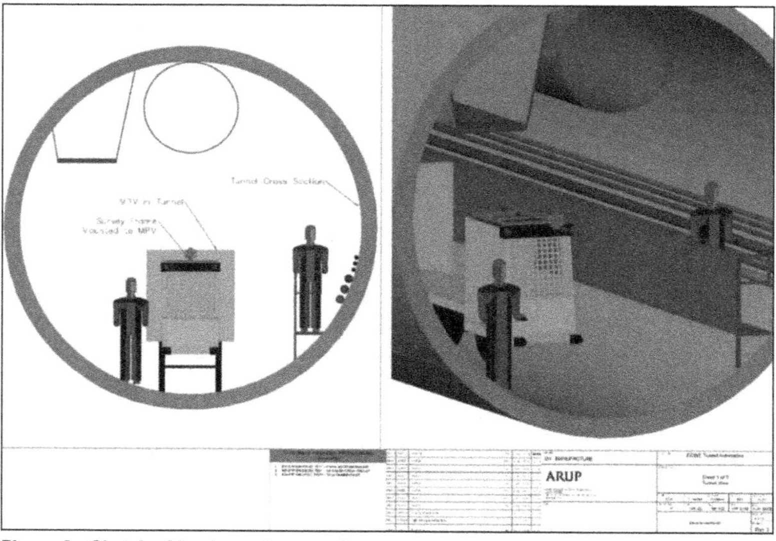


**Figure 3. Sketch of hardware frame setup**

- Utilization of existing equipment used by Project Co for frame attachments.
- Space limitation imposed by temporary systems within the tunnel, including ventilation duct, conveyor belt system, temporary sidewalk for cameras array mounting.
- Lightweight for ease of lifting and handling during setup.
- Limitation on type of power supply for the inspection system within tunnels.

One of main considerations in development of the frame is to not have any, or, minimize any disruptions to the tunnelling works and make use of the existing equipment current in service in the tunneling activities on site. The teams collaborated on the project specific frame design development, types of equipment and vehicles currently in use for the tunnel construction and concerns on utilizing the different equipment for attachment were reviewed with Project Co's inputs. The MSVs currently in use on transportation of the tunnel segments or workers within the tunnel were identified as the most suitable method for the frame attachment and carrying out the survey within the tunnel. A 3D virtual model of the MSV was developed based on the specification provided by Project Co and on-site imagery captures as well as on-site measurement verifications. Overall size of the frame was then designed to fit.

The hardware frame is designed in such a manner that adjustments can be made to ensure the equipment is always in the optimal position, i.e., tunnel center point and cantilevered over rails with sufficient clearance from temporary systems within the tunnel. Positioning the frame and subsequent equipment above the MSV drivers cabin is important for several factors: improved image quality, reduced impact to site survey team working conditions and avoiding existing equipment within the tunnel.

The use of aluminum material minimizes the total weight of the frame, thus allowing the frame to be easily mounted and detached from the MSV for site surveys and equipment maintenance. A combination of adjustable feet, an anti-split rubber mat and ratchet straps improve the overall friction between the frame and MSV; ensuring the frame remains in position during normal survey conditions and no permanent modification is required to the MSV.

As the existing MSV's power supply was intended for its own operation powering, external power for the tunnel inspection tool is required for the camera array system. With coordination with Project Co, an on-board battery attached to the MSV will be used.

## Software Platform

To visualize the captured imagery and to enable the tunnel engineer to analyze and interrogate the data in an effective manner, a user-friendly web-platform has been developed to serve as a virtual survey environment. This enables the user to review the full data outside of the tunnel environment and inspect the relevant parts of the tunnel in more detail as required. The platform is built within Arup's *Loupe 360* tunnel inspection platform, with expanded functionality to take into account specific needs and requirements for ECWE. The platform comprises development and integration of various tools and features:

- Overview management dashboard. This provides an overview of tunnel information, 360 visuals summaries, tunnel health metrics and mapping layers.
- 360 viewer: High resolution imagery viewer with manipulation toolsets including side-by-side comparison mode, geospatial localization tab, asset information tabs.

- Mapping pane: 2D map including tunnel alignment layers, and map styles e.g., streetmaps, satellite imagery.
- Deficiencies log: Upload functionality for integration of daily inspection deficiencies logs and point-and-shoot images database. Integration of inspections database with 360 viewer and geospatial querying.
- Machine learning: Automatic detection of selected defects and tunnel features

Specific nuances of these toolsets have been developed through user interviews and workshops with the ECWE teams to ensure functional usability and prioritization of the feature development pipeline (Figures 4 and 5).

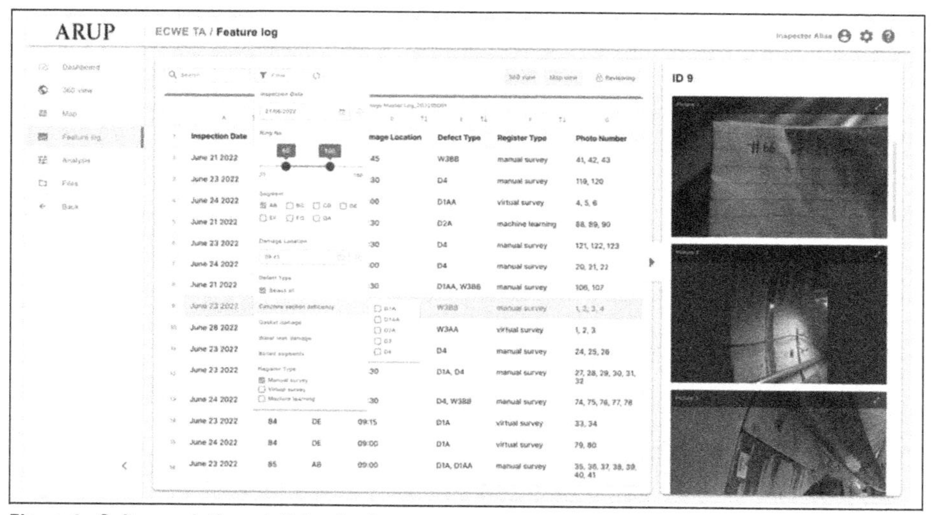


**Figure 4. Software platform deficiencies log**

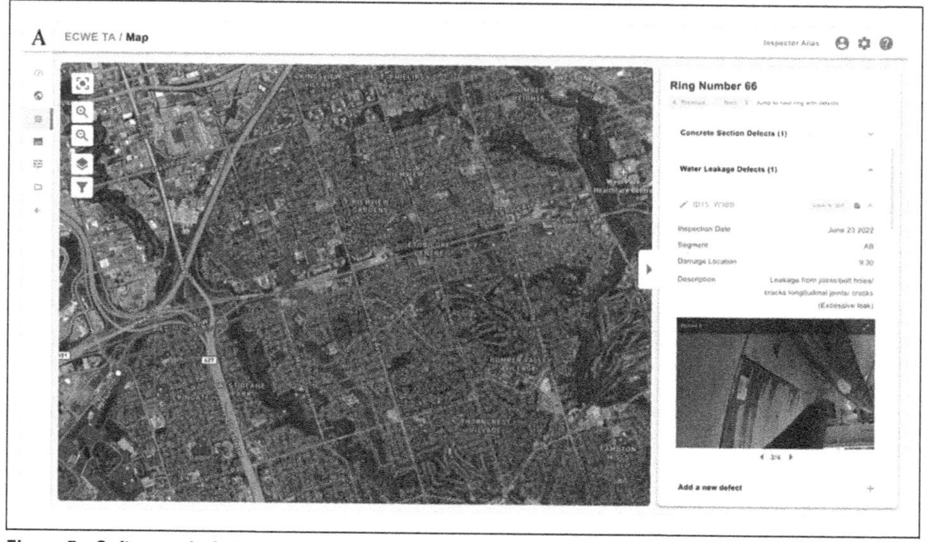


**Figure 5. Software platform mapping pane**

## Machine Learning (ML)

The application of ML for transport civil engineering asset management purposes is seeing increasing uptake and interest, especially in light of the productivity and quality benefits demonstrated by early adopters and the success that ML application to other industries and sectors has shown. The development of ML within the automated tunnel inspection tool is focused around automatic defect detection and change detection. This will ultimately aid the engineer in identifying issues in the tunnel, understanding spatial and temporal changes in the tunnel, and making better informed decisions on clear visual data.

Since early 2010s, with the publication of ImageNet (Russakovsky, et al., 2015) and thanks to increasing computational power, Deep Learning (DL) has attracted more research as well as industry applications. Both ML and DL are algorithms that give the computers the possibility to learn directly from data, without being explicitly programmed. However, DL is a subset of ML. The core of DL algorithms is the convolution operation (DL algorithms are also called Convolutional Neural Networks—CNNs). The main three properties of a CNN are parameter sharing, sparse interaction and equivariant representation (Goodfellow, Bengio, & Courville, 2016). Thanks to these property CNNs, differently from ML algorithms, can learn from unstructured data such as documents, images, and text. DL is being applied to the image defect detection problem on this project. This section describes the automation of visual inspection of tunnels with DL-based image understanding tasks

By applying DL to ECWE's asset inspection, this will aid in demonstrating that accurate defect and inventory detection in tunnel environments can be effectively automated with DL approaches and enables ECWE to develop predictive analytic capabilities in their inspection regimes that can aid proactive inspection and maintenance plans. We highlight the importance of the data collection design phase to maximize the detection output. Finally, we describe our approach to map images and defects in the tunnel's local coordinate system.

The process to deliver a ML based application for ECWE, from ideation to deployment is described in the next section.

### *Problem Statement*

The ATC1 construction works requires the tunnels to be progressively inspected and surveyed to ensure that construction progress matches the design intent. Traditionally, the record of deficiencies and tunnel condition are manually noted, recorded and compared. With kilometers of repeat tunnel structures being inspected and visual changes being observed on a daily basis, automation and use of computer vision technology to aid these traditional processes is becoming more possible and accessible with the development in new ML based technology computer vision stacks. Identification and understanding of the specific asset owner's needs and cost-benefits needs to be accounted for too. The primary objective of the ML aspects of the automated tunnel inspection tool is image understanding of the tunnel environments with the aim of detecting structural defects like cracks and water ingress. Among the multiple ML approaches to perform this task (including image classification, object detection and image segmentation) object detection is the approach of choice for this application as sufficient geometric information can be identified (bounding box class, location, and dimensions—see Figure 6) and it is inexpensive in terms of data labelling and required computational power.

**Figure 6. Example of DL defect detection on equirectangular image**

### *Data Collection*

To be sure that the ML algorithm performs as expected it is important to train the ML model on images representative of the dataset used in production. Well validated datasets and pre-trained algorithms that achieve high quality results do not widely exist for concrete tunnel linings and lowlight tunnel environments. Arup has built up a dataset of labelled data in these tunnel environments and builds off this to develop and train models applicable for the ECWE environment. As the base imagery collected is panoramic imagery, much of the training dataset includes panoramic images too, to match the format of the data collected. As more data from the ECWE in-tunnel surveys are collected over time, this will serve to increase the pool of data for the ML training to be developed across to improve the accuracy and precision of predictions.

The training is supported by a process of manual labelling for object detection purposes which consists in defining the coordinate of the bounding box containing the feature of interest and the respective label. Although this is a highly repetitive and time-consuming process, labelling accuracy is important especially on new datasets, and it is important that a significant portion is labelled and reviewed by engineers familiar with identifying and recognizing these defects.

### *Model Training and Evaluation*

To be sure that the ML algorithm performs as expected it is important to train the ML model on images representative of the dataset used in production. Well validated datasets and pre-trained algorithms that achieve high quality results do not widely exist for concrete tunnel linings and lowlight tunnel environments. Arup has built up a dataset of labelled data in these tunnel environments and builds off this to develop and train models applicable for the ECWE environment. As the base imagery collected is panoramic imagery, much of the training dataset includes panoramic images too, to match the format of the data collected. As more data from the ECWE in-tunnel surveys are collected over time, this will serve to increase the pool of data for the ML training to be developed across to improve the accuracy and precision of predictions.

The training is supported by a process of manual labelling for object detection purposes which consists in defining the coordinate of the bounding box containing the feature of interest and the respective label. Although this is a highly repetitive and time-consuming process, labelling accuracy is important especially on new datasets, and it is important that a significant portion is labelled and reviewed by engineers familiar with identifying and recognizing these defects.

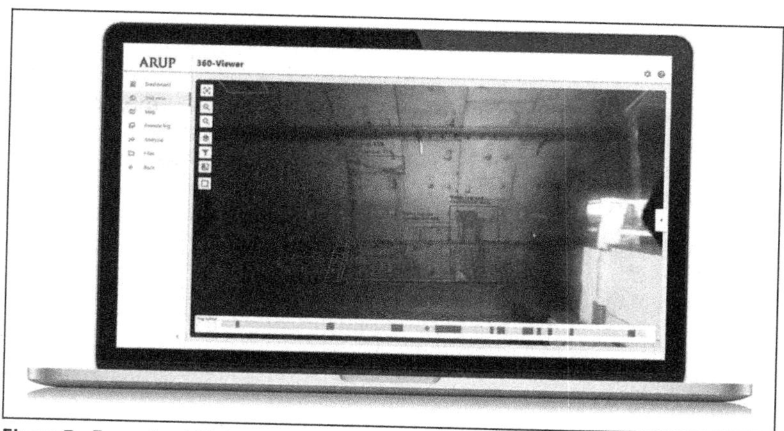


Figure 7. Example of DL-based condition monitoring visualization in Loupe 360

### *Defects Tracking for Condition Monitoring*

For the next step, the panoramic video, collected with the hardware, will then feed to the object detection algorithm for defect and fire equipment detection and counting. The algorithm will analyze each frame of the video as an independent image. It means that there is no correlation between two detections of the same object when it appears in two consecutive frames. This could negatively affect the defect/inventory counting from overlapping frames. To overcome this limitation, combination of YOLOv5 algorithm with the DeepSort algorithm will be considered. DeepSort (Wwojke, Bew ulus, 2017) combines the Kalman filter (estimating the velocity and motion) and appearance descriptors per each bounding box to track uniquely identified objects.

Finally, for defect and inventory mapping it is essential to identify the camera position in the tunnel coordinate system. Considering the linear nature of tunnels, thus the impossibility to have any loop closure, a bespoke pipeline will be considered to be implemented for in tunnel mapping. It will be based on the detection of distinctive natural features of the tunnel lining, for example segment joints, and tracking their position to the centre of the equirectangular image of the tunnel. This way, the number of rings inspected can be counted. This would be the same approach used by engineers to locate themselves when performing a traditional visual inspection.

Figure 7 is a screenshot of an example of final deep learning-based condition monitoring. Detected objects are mapped in the tunnel's local coordinate system giving the possibility to identify, with an easy to understand red-amber-green visualization, the areas that require more attention.

## FUTURE DEVELOPMENT AND CONCLUSION

The strength of the system grows with collection of repeat data over time. DL algorithms for automated defects and inventory detection represent a fast and repeatable approach to automate visual inspection of large infrastructure. As additional surveys are undertaken, additional training will be able to be integrated to improve even more the detection performance. Future works will implement ML for the tunnel inspection system and automate data labelling from the deficiencies log that is provided at the handover of the new infrastructure to the asset owner, and future asset inspection logs and reports generation.

In the progression from coarse to fine inference, image semantic segmentation represents the last evolution of DL approaches, providing a per pixel classification. This approach would benefit especially the detection of cracks giving the possibility to define length, width, and orientation of the detected cracks. Even though extensive research is available on the topic (Panella, Lipani, & Bohm, 2022), extensive training datasets for tunnel lining inspections are still missing. Future effort can be made to create an engineer-validated dataset for semantic segmentation of cracks, and other defects at a per pixel classification level, and attuned to specific deficiency categories and monitoring and prediction of its change over time.

The development of the automated tunnel inspection system for ECWE is at an early stage on ATC1. Following the pilot trials in early 2023, collection of the on-site data, and integration into the software platform, it is anticipated that further iterative development will enable the system to be embedded into business-as-usual processes for inception during the TBM drive progression, through handover and onto the operational period. Adoption of new technology and systems requires changes in human processes and operations, so embedding technology adoption change needs to be accompanied with processes which enable this to happen. As other drives and Metrolinx assets will operate similar systems, scaling the innovation to other assets would benefit consistency of approach across assets. As the projects progress, integration with ECWE's future operational and maintenance asset management systems, sensors and processes may also be possible, which can help build up a richer picture of the tunnel's health, building up different data sets across the structure that will be useful for operators, maintainers and cross-discipline teams.

## REFERENCES

Boehm, J., Panella, F., & Melatti, V. (2019). *FireNet*. Retrieved from http://www.firenet.xyz/

Everingham, M., Van Gool,, L., Williams, C., Winn, J., & Zisserman, A. (2012). Visual Object Classes Challenge 2012 Dataset (VOC2012). *Academic Torrents*.

Goodfellow, I., Bengio, Y., & Courville, A. (2016). *Deep Learning*. MIT Press.

Panella, F., Lipani, A., & Boehm, J. (2022). Semantic segmentation of cracks: data challenges and architecture. *Automation in Construction*.

Russakovsky, O., Deng, J., Su, H., Krause, J., Satheesh, S., Ma, S., ... Fei-Fei, L. (2015). ImageNet Large Scale Visual Recognition Challenge. *International Journal of Computer Vision (IJCV)*, 211–252. doi:10.1007/s11263-015-0816-y.

Wojke, N., Bewley, A., & Paulus, D. (2017). Simple Online and Real-time Tracking With a Deep Association Metric. *International Conference on Image Processing (ICIP)* (pp. 3,645–3,649). IEEE. doi:10.1109/ICIP.2017.8296962.

Zou, Q., Zhang, Z., Li, Q., Qi, X., Wang, Q., & Song, W. (2018). DeepCrack: Learning Hierarchical Convolutional Features for Crack. *IEEE transactions on image processing*.

# A Case Study from India on TBM Driving Under Low Overburden

**Debasis Barman** ▪ MICE

## ABSTRACT

TBM (TBM) mining in urban areas particularly under low overburden is a big challenge to design and excavate. During excavation of the Lucknow Metro project, in Uttar Pradesh, India the metro authorities encountered a design stretch where the TBM had to cross a sewer canal crossing of 15 m length under overburden of less than 1 m. Not only the stability of TBM due to uplift was of concern but also the stability of both banks. Most structures on the canal banks were non-engineered structures whose stability due to vibration of TBM was also of concern. Due to these risks, a method was proposed that met stability requirements for both ground and superstructures. The design was practical and cost effective to construct, and a collaborative exercise between the client, contractor and designer that took approximately 18 months to implement.

## INTRODUCTION

The Lucknow Metro Rail project phase 1A (North–South corridor) comprises a total 22.8 km of metro network connecting 22 stations from CCS Airport to Munshipulia. A total 3.67 km is an underground section (package LKCC-06) starting near Charbagh Metro station and ending near K. D. Singh Babu Stadium Metro station.

The underground package consists of two ramps, three underground stations, twin bored tunnel of internal diameter 5.8 m and five cross passages along the alignment. The respective stretch belongs between Hussainganj station & south ramp TBM drive.

### Description of the Area

Haider Canal is in the Hussainganj area of Lucknow, Uttar Pradesh, India. The area is a very congested, and densely populated. There is limited access to any investigation and construction sites. All buildings along the banks of canal are mostly brickworks with no proper foundation for even two storied buildings. Most buildings constructed were lacking of proper engineering design & execution technology. Materials are often of inappropriate quality. Both banks of the canal were developed by dumping locally available fill material and daily used garbage's had been dumped from time to time. The initial site assessment predicted the following risks during execution

- Mostly due to dumped fill on banks of canal there were stability of building's lying on banks due to TBM vibrations driving below buildings.
- Ground cover available from the tunnel crown was estimated to be less than 1 m. This presented a high risk of flotation along with sinking of the TBM at the canal bed due to soft ground conditions.

## REVIEW OF EXISTING INFORMATION

### Ground Characterization

During the tender stage two boreholes had been drilled in the Haider canal banks, of 25 m depth each, as shown in Figure 3. Additional new boreholes, of 15 m depth, were recommended to be drilled to better define the Haider canal bed ground conditions, as shown in Figure 3.

The boreholes on the Haider canal bed were of major considerations in the present studies as they replicated real ground conditions. The results of lab tests and bore hole log charts indicated that the strata at the site was dominantly cohesive soil structure. The cohesive type soil comprised of either silty clay soil, of low to medium plasticity and compressibility, or clayey silt soil, of low plasticity and compressibility belonging to CL, CI, ML group (62 to 99% material finer than 75 microns). The only non-cohesive type soil is found to comprise of sandy silt, ML type, (having 67% fines). Standard Penetration blows as recorded in the filled-up soil zone presented up to 3.00 m depth below ground level. In both boreholes SPT values ranged from 6 to 7 till 3 m depth. Beyond 3 m depth, SPT values increased from 8 to 31. No water table was encountered in any of the boreholes during investigation. Table 1 depicted the soil strength parameters as obtained from lab tests on samples obtained from boreholes conducted in Haider canal bed,

Figure 1. Haider Canal (Depicting poor ground conditions on banks of canal along with structures lying on bank)

## Existing Building Condition Assessment

Building condition survey were conducted on the buildings lying on tunnel influence zone and further building damage assessment analysis were done on the same. Building damage assessment were done as per Burland et al. (1977).

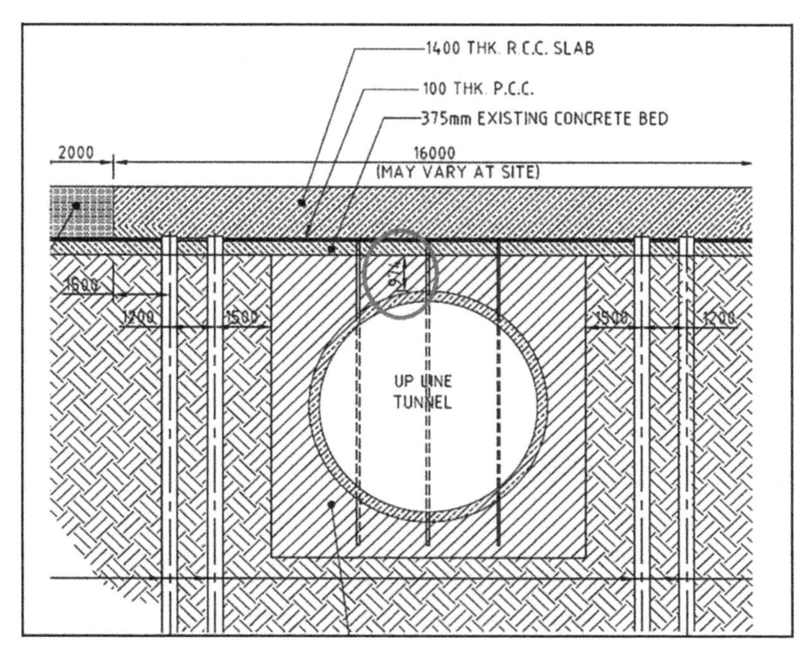


Figure 2. Section view showing existing ground cover above tunnel

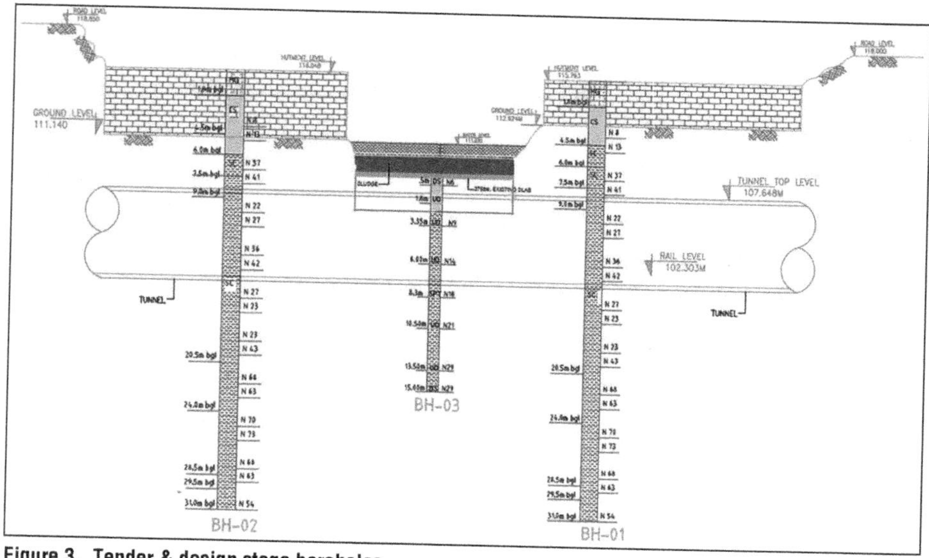


Figure 3. Tender & design stage boreholes

Table 1. Ground characterization and soil parameters

| Thickness of Soil Layer, m | Description of Ground | SPT/Blows | Soil Stiffness, MPa | Cohesion, kPa | Friction Angle, degree | Unit Weight, kN/m$^3$ |
|---|---|---|---|---|---|---|
| 3 | Fill | - | 3 | 0 | 26 | 17 |
| 4 | CL (Low compressible clay) | 10 | 9 | 30 | 10 | 19 |
| 5 | CI (Intermediate compressible clay) | 24 | 22 | 100 | 13 | 20 |
| 3 | CL (Low compressible clay) | 28 | 25 | 140 | 13 | 21 |

Boscardin and Cording (1989) theory. Analysis results predicted that most of the structures lied in *severe,* the most critical category of building damage assessment and were required to be evacuated prior to site activities.

During site visit it was found that certain buildings were tilting towards the Haider canal and cavities were located on the floor of the structure. Cracks had developed on building walls in due course of time and were anticipated to widen during TBM driving. One of the structures had a column projecting towards the canal bank which had the chance of getting dislodged due to vibration of construction equipment's and machine movement.

## Identification of Possible Problems

As per site visits and review of existing information, the following risks were identified for TBM excavation:

- Damage, or partial collapse, to buildings lying on both banks during TBM mining;

**Figure 4. Existing building conditions along canal banks (Representing dilapidated conditions of buildings lying on canal bank prior to machine drive)**

- Significant movement of ground lying on banks and possible failure of the bank slopes as the ground is mostly dumped fill;
- Drop of face pressure, or loss of ground at face, during mining due to weak ground conditions; and
- Ingress of water from cutter chamber as mining progressed under the canal, leading to possible flooding inside the TBM and rings built

## TBM SAFE MINING OPTIONS

### Preliminary Option

During tender design studies, it was interpreted from ground investigation report that soil below the canal bed level of thickness 2 m–2.5 m would be loose due to erosion of water, so the zone might need to be improved through which TBM would pass. Also, flotation of tunnels would occur due to low ground cover above tunnels. Looking at the scenario, it was proposed that TBM driving zone to be confined within a solid concrete mass block so that weight of the block resists the upward thrust due to flotation. Both the TBM were proposed to be confined within unreinforced concrete blocks of dimension 12 m × 12 m × 12 m. The concrete block was proposed to be prepared by interlocking bored piles of 1.2 m diameter. Piles were proposed to be bored in such a way by leaving an overlap of 100 mm between two piles. Each pile was proposed to be terminated at least 3 m below tunnel invert level. All piles were proposed to be of cast in-situ concrete grade M10 (concrete cube strength of 10 MPa) so that TBM can cut without difficulties.

During design studies of the solution the following risks were identified by me which might be critical during real execution,

Maintain stability of the concrete block as a whole during TBM cutting the block was one of major concern, as a chance the whole block might move ahead.

Number of total piles to be bored were 300 no's and being of big diameter i.e., 1.2 m, it was difficult to execute as heavy capacity pile driving rig was difficult to be placed on such soft ground

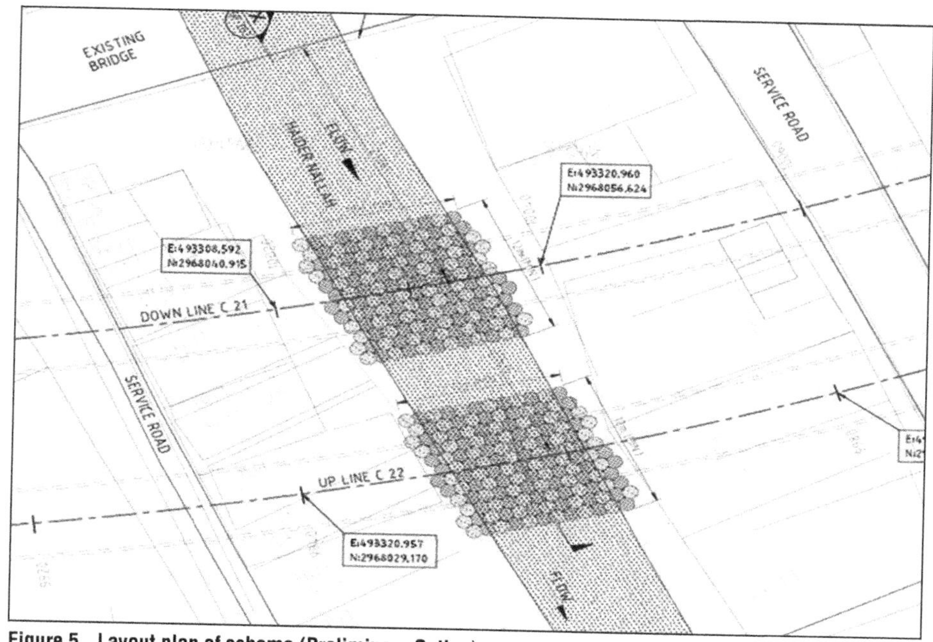


Figure 5. Layout plan of scheme (Preliminary Option)

Efficiency of the whole scheme would depend on the tolerances of each pile bored and such efficient agency was not available locally in the area.

Though the solution was acceptable in design checks but above possibilities during execution and installation of 300 bored piles would take a significant amount of time & costs in terms of initial preparations. Being from designer side, I proposed client to explore alternative options in terms of cost feasibility and ease of execution.

## Design Option

To counteract flotation issues, another option of dead weight increase on the ground surface above TBM was proposed by me. As per my initial studies I found that a 375 mm thick brick slab was lying over canal bed. As the cushion between brick slab and TBM was less than 1 m, an additional thick concrete slab of 1,400 mm was proposed to be constructed over existing slab as additional dead weight to counteract the flotation during tunnel mining process. Prior to construction of the slab, a number of 400 mm diameter cast in-situ concrete piles were bored for a depth of 20 m at a spacing of 3D (D = pile diameter) leaving a 2.5 m clear space from TBM both sides. The piles were supporting the thick slab and transferring load to deep soil stratum. I interpreted that dead load of slab over the ground would try to compress the ground above tunnel crown to some extent which would release the pore pressure in clays to some extent. This would reduce the chances of consolidation settlement of ground to some extent prior to the tunnel mining process. The ground beneath the canal bed is very soft cohesive soil and additional load like slab weight would impose increase in pore water pressure which were required to be drained prior TBM driving.

## DESK STUDIES

### Design Studies

The bored tunnel I checked for flotation possibilities considering the dead weight of slab would counteract the uplift forces. Both FOS in temporary & permanent conditions were respectively less than 1.1 & 1.5.

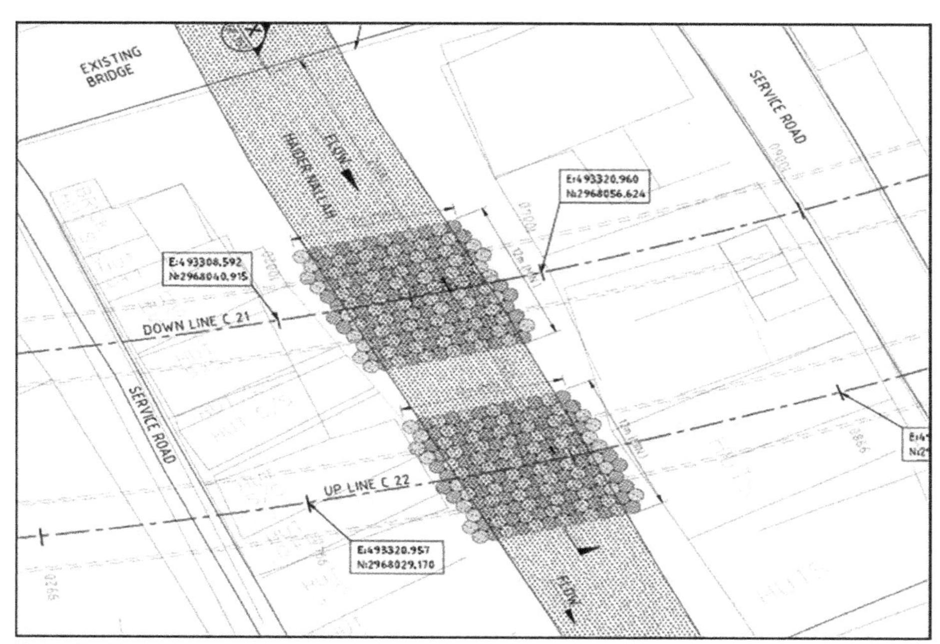


Figure 6. Plan of scheme (final option)

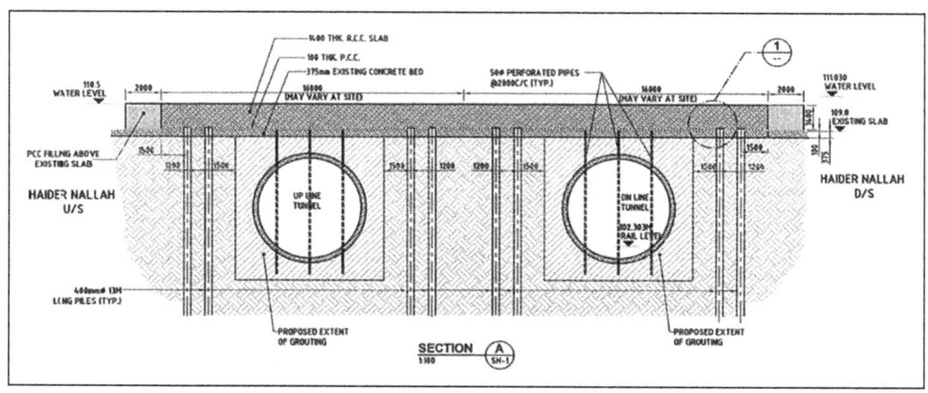


Figure 7. Elevation of scheme (final option)

## Bored Pile Capacity Determination

The bored pile of 400 mm diameter with depth 13 m, I checked for vertical capacity as per available geotechnical information's. The design capacity was 400 kN for a single pile considering FOS of 2.5 as per Indian standards.

## FEM Analysis to Check Tunnel Ring Movements and Associated Structures

Numerical analysis was performed using Rocscience, Inc.'s RS2 software to check the stability of the composite system behavior during execution.

The soil layers were modelled as per thickness encountered during investigation. Tunnel liners & Piles were modelled as plate elements with respective properties provided. The existing slab & thick slab were taken in considerations too. Soil zone around the tunnel between outer boundary piles were modelled with increasing stiffness taking in considerations of grouting. Since in RS2, TBM volume loss (%) can't be directly provided, equivalent relaxation (%) were simulated. A volume loss of 0.2% were adopted taking in considerations of the low cover above tunnel & dilapidated conditions of the building lying on tunnel alignment.

The following steps were modelled in RS2 to simulate the whole process of construction starting from slab casting to TBM mining:

- Setting up original conditions prior to construction with all boundary conditions
- Casting of bored piles and recording associated movements
- Casting of 1,400 mm thick slab in two stages and recording associated movements
- Resetting the all above displacements to zero
- Mining of 1st TBM volume loss (%) simulated
- Cast-in situ lining simulation of 1st tunnel completed
- Mining of 2nd TBM volume loss (%) simulated
- Cast-in situ lining simulation of 2nd tunnel completed
- Recording forces in all structural elements

FEM analysis predicted a maximum of ~60 mm movement due to slab dead weight imposed on slab and a minimum 5–6 months period to dissipate 75%–80% of ground pore water pressure. Accordingly, the TBM duo driving were planned. During TBM driving 5–6 m movement were predicted at slab bottom which were very negligible considering slab weight. Forces imposed on concrete slab & pile were within allowable limits due to TBM movement.

## Checking of Superstructures Movement and Building Damage Assessment

In order to carry out building impact assessment, buildings adjacent to the proposed alignment were classified in terms of potential risk of damage based on categories proposed by Burlandet et al. (1977) and Boscardin and Cording (1989) which showed

**Table 2. Ground support parameters**

| Type of Material | Stiffness, GPa | Cohesion, MPa | Friction Angle, ° | Unit Weight, kN/m$^3$ |
|---|---|---|---|---|
| Grout | 15 | 0.5 | 40 | 20 (assuming as lean concrete/backfill material) |
| Cast In-situ segments | 35 | — | — | 25 |
| Concrete Pile | 30 | — | — | 20 |

**Numerical Analysis Predictions**

| Stage of Execution | Maximum Settlement of Concrete Slab, mm | Pile Toe Movement, mm | Remarks |
|---|---|---|---|
| 900 mm slab Casting (1st stage) | 47 | | Consolidation settlement of slab in stages |
| Pile casting | 52 | 5 | |
| Full Slab 1.4 m casting (2nd stage) | 61 | 20 | |
| Additional settlement due to tunneling | 66 | 22 | |

that these categories of damage are related to the magnitude of the maximum tensile strain induced on the structure. The damage classification table is presented in Table 3.

A staged approach was adopted to assess the degree of damage to the buildings due to ground movements from the station excavation.

Stage 1 included the settlement analysis and preparation of contours along the alignment. Empirical approach was used to carry out the analysis. Any structure where the predicted settlement was less than 10 mm and the predicted ground slope was less than 1/500 were not subject to further assessment. This stage was primarily used to carry out an analysis to predict settlement, determining the influence zone and identify the critical buildings in influence zone. There were 27 buildings around Haider canal

**Table 3. Building Damage Classification (after Burland et al., 1977 and Boscardin and Cording, 1989)**

| Risk Category | Description of Degree of Damage | Description of Typical Damage and Likely Form of Repair for Typical Masonry Buildings | Approximate Crack Width, mm | Maximum Tensile Strain, % |
|---|---|---|---|---|
| B-0 | Negligible | Hairline Cracks. | Less than 0.1 mm | Less than 0.05 |
| B-1 | Very Slight | Fine cracks easily treated during normal redecoration. Damage generally restricted to internal wall finishes. Perhaps isolated slight fracture in buildings. Cracks in exterior brickwork visible upon close inspection. | 0.1 to 1 mm | 0.05 to 0.075 |
| B-2 | Slight | Cracks easily filled. Redecoration probably required. Recurrent cracks can be masked by suitable linings. Exterior cracks visible: some repainting may be required for weather tightness. Doors and windows may stick slightly | 1 to 5 mm | 0.075 to 0.15 |
| B-3 | Moderate | Cracks may require cutting and patching. Tuck pointing and possibly replacement of a small amount of exterior brickwork may be required. Doors and windows sticking. Services may be interrupted. Weather tightness often impaired. | 5 to 15 mm or a number of cracks greater than 3 mm | 0.15 to 0.3 |
| B-4 | Severe | Extensive repair involving removal and replacement of sections of walls, especially over doors and windows required. Windows and door frames distorted. Floor slopes noticeably. Walls lean or bulge noticeably. Some loss of bearing in beams. Services disrupted. | 15 to 25 but also depends on number of cracks | Greater than 0.3 |
| B-5 | Very Severe | Major repairs required involving partial or complete reconstruction. Beams lose bearing, walls lean badly and require shoring. Windows broken by distortion. Danger of instability. | Usually greater than 25 mm but depends on number of cracks | Greater than 0.3 |

**Table 4. Damage Category of buildings combined for construction impact and building conditions vulnerability**

| Section | Number of Buildings | Damage Category | | | | |
|---|---|---|---|---|---|---|
| | | B-0 Negligible | B-1 Very Slight | B-2 Slight | B-3 Moderate | B-4 Severe |
| Haider canal area | 27 | 4 | 3 | 15 | 3 | 2 |

area. Almost all of them were small hutments and in poor condition. Hence, all buildings were taken for further assessment keeping in mind their critical nature.

In stage 2 analysis all buildings as identified within influence zone from stage 1 analysis were considered for assessment. The assessment was based on the limiting tensile strain approach by Mair et al (1996), wherein damage risk would then be carried out with reference to the classification of Burland et al. (1977) and Boscardin and Cording (1989), as presented in Table 3. Buildings with Degree of Damage classified as 'Slight' or better are considered acceptable and further structural assessment would not be required. In the Stage 2 assessment, the Greenfield surface ground movement estimated above was used to compute the damage risk to the building due the proposed construction.

In stage 3 analysis, buildings lying in the category of moderate & above had been identified and stage 3 analysis were done based on settlement derived from numerical analysis. Numerical analyses using soil-structure interaction was performed. Vertical & horizontal displacement as derived from the FEM analysis were superimposed with respect to building position and limiting tensile strain were evaluated for each building and risk categorization were done with the improved results.

There were 4 buildings identified near Haider canal which were falling in damage category B-3 moderate or above. These buildings were already in critical category as per building condition survey data hence proper monitoring arrangements were made prior to excavation,

## CONSTRUCTION SCENARIO

### Retrofitting of All Superstructures Within Influence Zone (Prior to TBM Drive)

To evaluate the impact on all superstructures lying within influence zone of both TBM, building impact assessment was carried out and classified in terms of potential risk of damage based on categories proposed by Burlandet et al. (1977) and Boscardin and Cording (1989) criteria. A total of 27 buildings around Haider canal area were identified for analysis and almost all of them are small hutments in poor condition. Based on damage category buildings were classified for minor repair/major repair/partial demolition along with continuous monitoring by instruments. The following recommendations, were implemented in site on buildings depending on its existing health,

- Major part of the buildings was temporarily supported using steel props. All the cavities as found in building floors were filled with cement slurry immediately. The cavities had been originated due to lack of proper foundation beneath the buildings as well due to soft subsoil and not been compacted prior to construction. In due course most of the buildings bottom slab got cracked and created voids. All voids were filled with cement grout in pressure range of 2–3 bar so the voids get filled and create a compacted surface for all buildings.

**Figure 8. Building damage assessment results and predictions**

- All backfilling works conducted by cement slurry were checked against water tightness and extent of fill along the voids in the soil mass. This were done by injecting water at high pressure through existing holes and possibilities of water leakage on the bank side or in other portions of buildings were explored.
- All ground lying on foundation of all buildings were properly regarded and compacted. Voids in the ground were filled with lean concrete.
- External edges of all buildings projected on the bank side were supported by propping as per availability of space.
- Total 1,462 bags (73.1 ton) of cement were used to fill the cavity for both side of the houses along the canal
- Very critical category buildings as identified by building damage assessment discussed previously were temporarily evacuated till TBM mining & ring building were not completed.
- Monitoring activities were continued as per design proposal and recordings interpreted for any type of movement in regular intervals.

## Regrading of Slopes on Both Banks of Canal (Prior to TBM Drive)

Soil mass on both banks were regarded to stable slope and supported with shotcrete & wire mesh. A retaining wall were constructed along the banks till influence zone of TBM on both sides and backfill with lean concrete completed behind wall.

## Casting of Slab and Installation of Bored Cast-In Situ Piles

The following steps were proposed to ensure that slab-pile monolithic system was built in canal bed without harassing the canal flows

After all building structures were supported both externally and internally and ground of the both banks were regarded, bored piling operations were initiated.

To prepare a dry surface for working, canal was diverted on one side using temporary bunds.

Holes were punctured slightly higher than pile diameter and bored piles constructed using percussion drilling technique. Reinforcement from piles were projected to be connected with slab in future.

All piling operations was completed, reinforcement placing & casting of the half slab commenced.

The total thickness of 1,400 mm for slab were executed in two stages of 500+900 mm.

Completion of half slab, water was diverted on the slab and the other half was made dry to repeat the above steps.

Finally, the whole canal flows were restored to its original condition.

## TBM Mining Parameters Fixation

The following recommendations as per discussion with client & contractor, were proposed to be implemented during TBM drive

Figure 9. Retrofitting of buildings

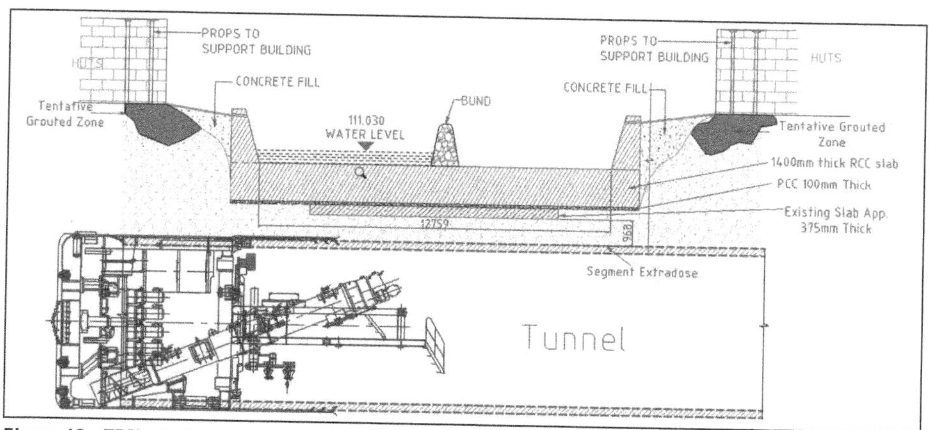


Figure 10. TBM mining arrangement

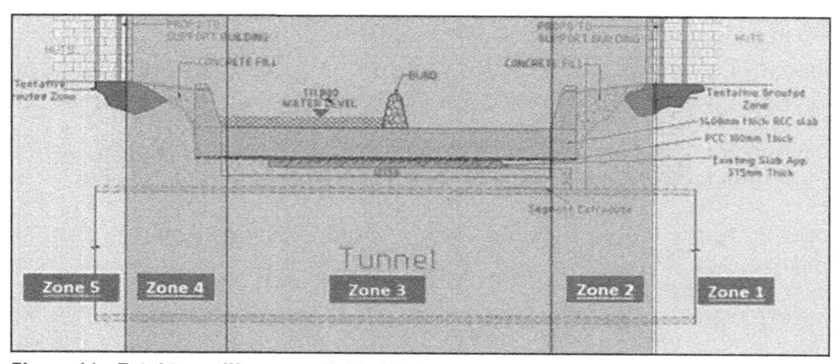


**Figure 11. Total tunnelling zone divided in 5 parts**

While ring building wherever face pressure dropped down, bentonite slurry was proposed to be injected to maintain the face pressure.

In case of water ingress from the screw conveyor from cutter chamber, the bulkhead gate was proposed to be closed to minimise the tunnel flooding from the canal water.

Heavy duty dewatering pumps to be mobilized to prevent flooding inside tunnel

Enzan system as in-built in TBM to be followed to monitor excavation volume and keeping volume loss (%) at minimum.

Following TBM Face Pressure were recommended for all five zones depending on position of TBM

1. Zone 1: 0.7 to 0.8 bar
2. Zone 2: 0.7 to 0.8 bar
3. **Zone 3: 0.5 to 0.6 bar (canal crossing)**
4. Zone 4: 0.7 to 0.8 bar
5. Zone 5: 0.7 to 0.8 bar

## Monitoring of Ground and Superstructures

For building structures monitoring the following instruments were recommended as per results of building damage assessment,

1. Building settlement marker
2. Tilt plate
3. Crack meter
4. Bi Reflex target

For Ground Monitoring of bank slopes along tunnel alignment the following instruments were proposed,

1. Ground settlement marker
2. Multiple point borehole extensometer
3. Piezometer

Table 6 shows monitoring frequencies along with proposed instruments both for ground as well as for buildings were proposed and implemented in site during execution

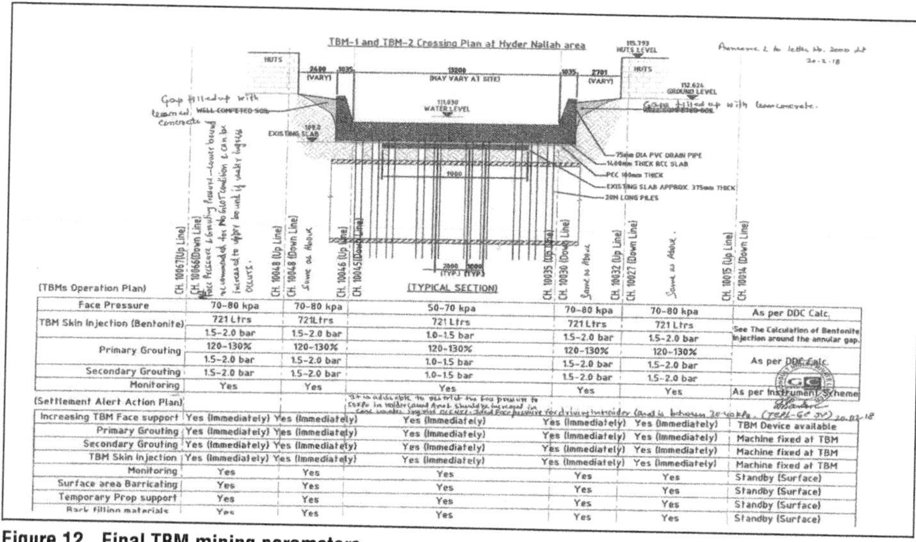


(TBMs Operation Plan)

| | | | | | | |
|---|---|---|---|---|---|---|
| Face Pressure | 70-80 kpa | 70-80 kpa | 50-70 kpa | 70-80 kpa | 70-80 kpa | As per DDC Calc. |
| TBM Skin Injection (Bentonite) | 721 Ltrs | 721Ltrs | 721 Ltrs | 721 Ltrs | 721 Ltrs | See The Calculation of Bentonite injection around the annular gap. |
| | 1.5-2.0 bar | 1.5-2.0 bar | 1.0-1.5 bar | 1.5-2.0 bar | 1.5-2.0 bar | |
| Primary Grouting | 120-130% | 120-130% | 120-130% | 120-130% | 120-130% | As per DDC Calc. |
| | 1.5-2.0 bar | 1.5-2.0 bar | 1.0-1.5 bar | 1.5-2.0 bar | 1.5-2.0 bar | |
| Secondary Grouting | 1.5-2.0 bar | 1.5-2.0 bar | 1.0-1.5 bar | 1.5-2.0 bar | 1.5-2.0 bar | |
| Monitoring | Yes | Yes | Yes | Yes | Yes | As per Instrument Scheme |

(Settlement Alert Action Plan)

| | | | | | | |
|---|---|---|---|---|---|---|
| Increasing TBM Face support | Yes (Immediately) | Yes (Immediately) | Yes (Immediately) | Yes (Immediately) | Yes (Immediately) | TBM Device available |
| Primary Grouting | Yes (Immediately) | Yes (Immediately) | Yes (Immediately) | Yes (Immediately) | Yes (Immediately) | Machine fixed at TBM |
| Secondary Grouting | Yes (Immediately) | Yes (Immediately) | Yes (Immediately) | Yes (Immediately) | Yes (Immediately) | Machine fixed at TBM |
| TBM Skin Injection | Yes (Immediately) | Yes (Immediately) | Yes (Immediately) | Yes (Immediately) | Yes (Immediately) | Machine fixed at TBM |
| Monitoring | Yes | Yes | Yes | Yes | Yes | Standby (Surface) |
| Surface area Barricating | Yes | Yes | Yes | Yes | Yes | Standby (Surface) |
| Temporary Prop support | Yes | Yes | Yes | Yes | Yes | Standby (Surface) |
| Back filling materials | Yes | Yes | Yes | Yes | Yes | Standby (Surface) |

Figure 12. Final TBM mining parameters

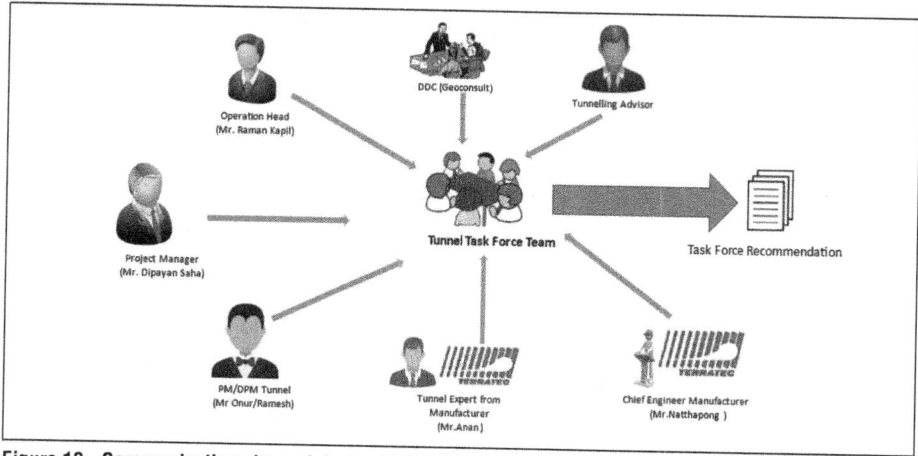


Figure 13. Communication channel during TBM mining

**Table 5. Summary of instruments**

| Sr. no. | Instrument Type | Quantity (Nos.) |
|---|---|---|
| 1 | Building settlement Marker (BSM) | 66 |
| 2 | Ground Settlement Marker (GSM) | 22 |
| 3 | Pavement settlement marker(PSM) | 7 |
| 4 | MPBX | 2 |
| 5 | Piezometer | 1 |
| 6 | Tilt Plate | 7 |
| 7 | Crack Meter | 2 |
| 8 | Bi-Reflex Target | 10 |

Table 6. Frequency of Monitoring of instruments for tunnelling works

| Type of Instruments | One Month Prior to Tunnel Construction | During Excavation | | After Completion of Tunnel |
|---|---|---|---|---|
| | | Position of TBM | Monitoring Frequency | |
| Ground settlement monitoring point/ vibrating wire piezometer/ Inclinometer/ extensometer | Weekly | 50 m behind the machine and 25 m in front of cutter head | Daily | Monthly |
| | | Between 50 m to 100 m behind the machine | Twice a week | |
| | | 100 m to 300 m behind the machine | Weekly | |
| | | 300 m behind the machine | Nil | |

Table 7. Frequency of Monitoring of instruments for existing buildings

| S. No | Types of Instruments | One Month Prior to Excavation | Risk Category B1 to B2 | Risk Category B3 to b4 | After Completion of Tunnel |
|---|---|---|---|---|---|
| | | | During Excavation | | |
| 1 | Crack Meters | Weekly | 24 Hrs. Interval | 12 Hrs. Interval | Monthly |
| 2 | Building Settlement Markers | Weekly | 24 Hrs. Interval | 12 Hrs. Interval | Monthly |
| 3 | Tilt meters | Weekly | 24 Hrs. Interval | 12 Hrs. Interval | Monthly |
| 4 | BI Reflex Targets | Weekly | 24 Hrs. Interval | 12 Hrs. Interval | Monthly |

## Team Communication Process

A dedicated tunnelling task force team was formed between the TBM crew and respective senior personnel from contractor including consultant representative. The function of the team was to make a coordinated approach to the problem with inputs from different levels and to generate a common understanding. The team used to monitor all activities related to tunnel boring machine movement on shift basis and provided their inputs based on TBM parameters recorded along with ground & structures monitoring records. Overall, the inputs of the team were very useful and valuable on successful completion of the operations. During real execution, designer representative was mobilized in site during both TBM drive to provide his inputs based on ground monitoring & TBM record obtained. As per designer observations, machine speed, face pressure & grout pressure were modified to avoid blowing of ground & minimize TBM tail end volume loss.

## Safety Preparedness During TBM Mining

- While ring building if face pressure drops down, bentonite slurry was injected to maintain the face pressure.
- Mining parameters was followed strictly and jointly with designer representative while tunnelling operation under Haider Nallah, to avoid any surface disturbances.
- In case of water ingress inside TBM, the bulkhead gate was closed to minimise the tunnel flooding from the Nallah Water.
- Tunnel team kept one more backup plan to dry out the Nallah if required. Total 9 dewatering pumps of capacity from 10 HP to 35 HP were mobilized in site.
- Excavation volume was monitored by ENZAN system of the TBM at TBM operator Cabin and at the Surface at survey Cabin

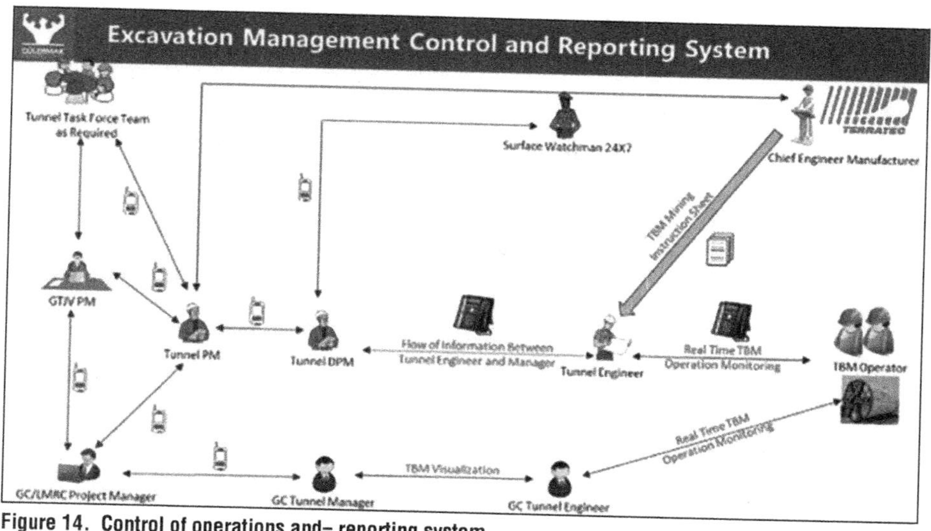


**Figure 14. Control of operations and= reporting system**

- Apart from monitoring the ENZAN system the excavation volume was measured by the muck weight by using muck bucket and gantry crane, and ring wise measurement was done
- While tunnelling under the nallah, watchman was kept round the clock around the influence zone of TBM
- Two numbers of additional 10 HP pumps wills be placed in the tail skin area of TBM to pump out the water from the tunnel.

## HSE Measures During TBM Mining

- 24 Hrs. Ambulance at TBM site.
- Rescue basket kept ready at site with slings and master ring attached.
- Two full time first aider will be deputed at TBM site during Nala Crossing.
- Continuous Gas Monitoring and record keeping.
- Emergency torch light availability inside TBM 06 Nos.
- Stretchers and first aid boxes will be kept ready inside TBM.
- Refresher training to all workers "How to use Emergency Escape Breathing Device."
- Evacuation Mock Drill before Nallah Approaching.
- Ensuring fire curtain sprinklers are in working condition.
- Sufficient Nos of fire extinguishers inside TBM.
- Ensuring public addressing system in working condition and connected with Shaft, Surface and First Aider.
- Deputation of Sr. Safety Manager in each shift during Nala Crossing.

## TBM Mining Parameters (Prior to Mining)

All building settlement points recorded movement from 2 mm to 4 mm.

The following machine records were obtained which were quite close to recommended:

**Table 8. TBM parameter record from Enzan during mining (down line) under Haider Nallah and adjacent building**

| | | | | | | | | Date | 23-02-2018 | |
|---|---|---|---|---|---|---|---|---|---|---|
| | Net Stroke | Thrust Force | Cutter Torque | Earth Pressure, kPa | | | | Grout Volume, m³ | | Grout Pressure |
| Ring No. | mm | kN | kN–M | Right UP | Right Mid | Left Up | Left Mid | Liquid A | Liquid B | Bar |
| 109 | 1,351 | 14,617 | 911 | 53 | 74 | 61 | 69 | 3.623 | 0.241 | 2.5 |
| 110 | 1,393 | 17,076 | 892 | 94 | 134 | 99 | 96 | 3.332 | 0.219 | 2.5 |
| 111 | 1,443 | 18,172 | 793 | 107 | 112 | 106 | 114 | 3.349 | 0.223 | 2.5 |
| 112 | 1,459 | 17,879 | 783 | 104 | 120 | 107 | 101 | 3.349 | 0.212 | 2.5 |
| 113 | 1,442 | 17,409 | 896 | 106 | 125 | 115 | 102 | 3.294 | 0.210 | 2.5 |
| 114 | 1,444 | 16,214 | 1,136 | 78 | 78 | 88 | 90 | 3.100 | 0.227 | 2.5 |
| 115 | 1,419 | 15,551 | 1,029 | 88 | 123 | 85 | 67 | 3.287 | 0.226 | 2.5 |
| 116 | 1,428 | 13,763 | 1,155 | 85 | 102 | 88 | 86 | 3.303 | 0.218 | 2.5 |
| 117 | 1,407 | 13,606 | 1,215 | 78 | 106 | 89 | 90 | 3.348 | 0.225 | 2.5 |

**Table 9. Down line tunnel drive between South cut & cover and Hussainganj**

| Activity | Dur. | Planned Start | Planned Finish |
|---|---|---|---|
| **Main Drive First Stage (rings No. 1 through No. 50)** | **10** | **03-02-2018** | **12-02-2018** |
| Install Working platform in Shaft area and Y Switch | 2 | 13-02-2018 | 14-02-2018 |
| Connect all service lines (water, air and grouting) and electrical cables & re-check all systems | 1 | 15-02-2018 | 15-02-2018 |
| **Main Drive (rings No. 51 through No. 318)** | **33** | **16-02-2018** | **20-03-2018** |
| Reaching Haider Nalla (From R51 to R124) | 9 | 16-02-2018 | 24-02-2018 |
| Crossing Haider Nalla (From R125 to R144) | 3 | 25-02-2018 | 27-02-2018 |
| Completion of Main Drive (From R145 to R318) | 22 | 28-02-2018 | 20-03-2018 |
| Removal of Tunnel rail, walkway and water pipe | 5 | 21-03-2018 | 25-03-2018 |

1. Face Pressure—0.3 bar (min) to 1.0 bar (maximum)
2. Thrust Force—18,000 kN (maximum)
3. Cutter Torque—1,500 kN-m (maximum)
4. Grout Volume—130% of excavation volume
5. Grout Pressure—2.0 to 3.0 bar

To minimize the surface settlement above the shield , bentonite slurry was pumped in the gaps between the excavated soil and the shield

**Note:** Mixing Ratio (Water 1000:Bentonite 50 kg)
Used Bentonite = 36 kg/Ring

## Ground and Superstructures Settlement During TBM Mining

Building settlement point showed a maximum movement of 4 mm which is less than predicted value of 7.5 mm. Also ground settlement markers showed a maximum value of 5 mm which was within tolerable range as per design.

## CONCLUSIONS

The case study is a classic example of TBM driving under low overburden in urban areas. Lessons learnt after completion were selection of appropriate methodology along with their feasibility in terms of execution, coordination between all parties involved, early contractor mobilization. Detailed risk analysis and proposal of mitigations provided the opportunity to take care of contingency measures during execution. TBM mining parameters monitoring along with monitoring of movement for surface and superstructures lying on tunnel alignment and interpretation of the records provided confidence to both designers as well to execution team.

## ACKNOWLEDGMENTS

I convey my regards to Mr. Dipayan Saha, PM (Afcons) for keeping trust on me since conceptual design till execution completion. Also, cooperations from Mr. Ashish Dwivedi (Gulemark engg.), Mr. Sukanta Mondal (Afcons infrastructure pvt ltd) were quite appreciated. Finally, I thank all my ex-employer Geoconsult India colleagues for their co-operations to make project successful.

## REFERENCES

1. Loganathan, Nagen. 2011, An Innovative Method for Assessing Tunnelling-Induced Risks to Adjacent Structures: PB 2009 William Barclay Parsons Fellowship Monograph 25.

2. Marco D. Boscardin, Edward J. Cording, Building response to excavation-induced settlement, Journal of Geotechnical Engineering, Vol. 115, No. 1, January, 1989.

3. Jardine, F.M. 2003, Response of buildings to excavation induced ground movements, Proceedings of the international conference held at Imperial College, London, UK, on 17–18 July 2001, CIRIA.

4. Poulos, H.G., Pile settlement zones above and around tunnelling operations, Australian Geomechanics Vol 41 No 1 March 2006, 81–90.

5. Selemetas, D, Standing, J.R., Mair, R.J., The response of full-scale piles to tunnelling, 2006 Taylor & Francis Group plc, London, UK.

# The Woodsmith Project—Construction of the Access Shaft at Lockwood Beck Using Innovative Blind Boring Technique

**Andrew Raine** ▪ ARUP
**Peter Stakne** ▪ STRABAG UK
**Callum Fryer** ▪ STRABAG UK
**Craig Sewell** ▪ STRABAG UK
**Carmen Hu** ▪ ARUP
**Cameron Dunn** ▪ ARUP

## ABSTRACT

The Woodsmith Project is located in north-eastern England within the North York Moors National Park. The strata contains the world's largest known high-grade Polyhalite resource of around 2.66 billion tonnes. The Project was initiated by York Potash Limited (YPL), a subsidiary of Sirius Minerals plc., but in March 2020, a take-over was completed by Anglo American, one of the largest mining companies in the world.

Polyhalite is an evaporite mineral and contains essential nutrients required for plant growth. The Polyhalite will be extracted via two deep mine shafts and transported underground from the mine site to a Materials Handling Facility on a Mineral Transport System (MTS), which comprises a high-capacity conveyor belt system located within a 4.9 m ID tunnel.

In 2018, STRABAG were awarded with the design and construction of the MTS tunnel and associated infrastructure, including the high-capacity conveyor system. The scope of work consists of 300 m of open cut, 100 m SCL tunnel, 36.5 km TBM drive (originally anticipated as three separate TBM drives) along with a network of underground caverns at the Woodsmith Mine. The construction of two 380 m deep intermediate access shafts was later awarded as Amendments to the Contract.

The tunnel is constructed using a Herrenknecht Single Shield TBM. As a result of how the various elements of the project have developed over time, the first TBM drive has been extended to ~29 km making it the worlds longest single heading TBM Tunnel when complete. The option of completing the entire tunnel length with a single heading TBM drive is currently being assessed by Anglo American and STRABAG.

This unprecedented tunnel length requires two intermediate shafts to provide essential access and services, the first intermediate shaft is at Lockwood beck (LWB) and the second is at Ladycross (LDX). The LWB shaft has been constructed utilising the innovative Blind Boring technique never used before in the UK. This paper will explain the methodology and technical risk management associated with this technique.

The authors believe that the collaborative approach between all contract partners is and continues to be the important contributor for the success of this project.

## INTRODUCTION

The Woodsmith Project is located in north-eastern England within the North York Moors National Park. The strata contains the world's largest high-grade Polyhalite resource of 2.66 billion tonnes (JORC), around seven percent of it within the project's area of interest (Figure 1). The rapidly increasing population of the world requires more

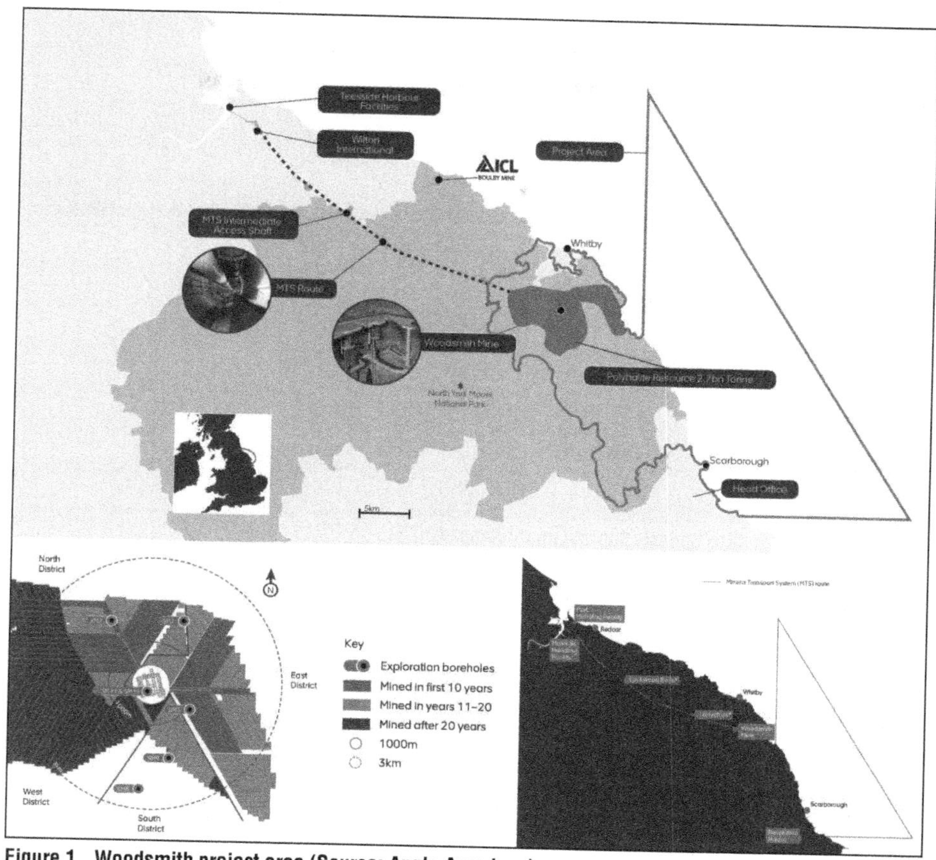


**Figure 1. Woodsmith project area (Source: Anglo American)**

food to be grown on less land available for agriculture. POLY4 is Anglo Americans flagship multi-nutrient fertiliser product. Made from Polyhalite, it contains four of the six macro-nutrients and many micronutrients that are essential to plant growth. It allows farmers to maximise their crop yield, increase quality and improve soil structure with one simple product.

From the Woodsmith Mine, the mineral will be extracted via mine shafts 1,550 m below ground and transported via the MTS to the Materials Handling Facility in Teesside where it will be processed. Due to the location within the National Park, the project is designed to minimise the environmental impact and carbon footprint. The 37 km underground MTS as shown in Figure 2, is an integral part of this concept and was a mandatory requirement for obtaining the Mining Licence and Planning approvals from the Country Parks Authority by the Client. The original concept had been a transportation pipeline which was abandoned due to the negative environmental impacts, the innovative new project construction plan is fundamentally more environmentally friendly.

In 2018, YPL awarded STRABAG with the contract for the design and construction of the MTS tunnel, associated infrastructure and fit out works. The scope of work comprises 300 m of open cut, 100 m SCL tunnel, three TBM drives totalling 36.5 km

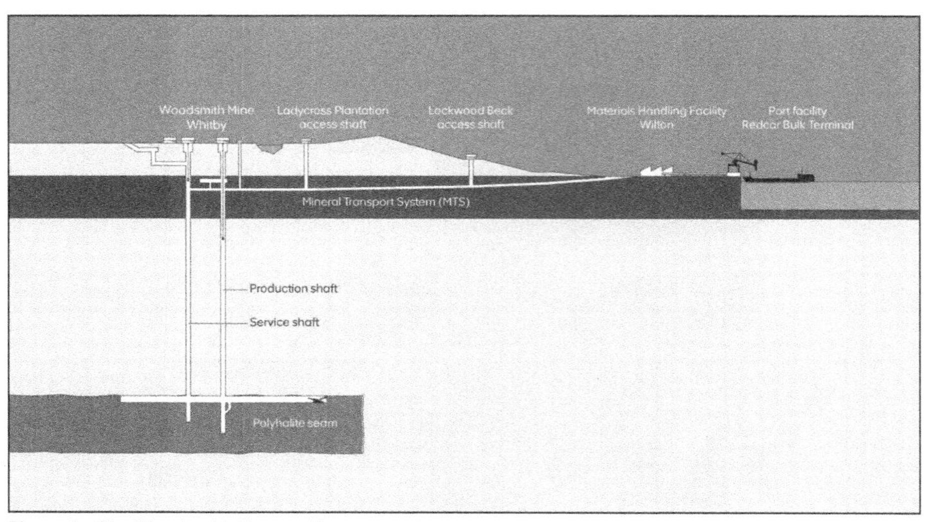


**Figure 2. The Woodsmith Project (Source: Anglo American)**

of segmentally lined tunnel of 6 m bored diameter (4.9 m ID), underground caverns, a permanent conveyor belt system, tunnel utilities and maintenance railway system.

In July 2020, Anglo American awarded the construction of the 380 m deep intermediate access shaft at LWB to STRABAG, who subsequently awarded the shaft design works to ARUP. In parallel STRABAG conducted a detailed constructability review, considering the history of shaft sinking works associated with the local ironstone mines and all available Geotechnical Investigation data collected in the area.

Various methods of shaft construction were investigated including ground freezing, conventional shaft sinking and Blind Boring method. At the culmination of this study, the Blind Bore method was selected as the preferred method due its overall lower risk profile, efficiency, safety, sustainability, and schedule benefits.

In September 2020, after a worldwide competitive procurement phase, STRABAG engaged Shaft Drillers UK (SDUK), a subsidiary of Shaft Drillers International (SDI) from the USA as the specialist Contractor to deliver this highly complex and innovative scope of work.

It is the first time such a shaft sinking method has been delivered within the UK.

## BACKGROUND AND GENERAL DESCRIPTION OF GEOTECHNICAL SITE SETTING

### Lockwood Beck Intermediate Shaft, Adit and Future Cavern

In September 2019, YPL announced that the required funding for the next phases of development was in doubt. At this time, the ongoing construction of a 9 m diameter shaft at LWB (being delivered by others) to support TBM #2 was placed on hold to limit cost expenditure. The existing 9 m diameter shaft was approximately 60 mbgl when abandoned.

With the TBM progressing five months ahead of schedule and the LWB Shaft works on hold, Anglo American and STRABAG agreed that TBM #2 would be omitted and TBM

#1 would continue to ~29 km. As part of this strategic change to tunnelling methodology, the shaft at LWB was no longer required for TBM mobilisation and the diameter could be significantly reduced.

The intermediate shaft is located along the route of the MTS tunnel at LWB, 12.8 km from the Wilton Site with a 21 m long adit connection between the shaft and the tunnel. This shaft will be used for services including ventilation, power, grout supply and personnel access during the construction phase and will be retained in the permanent case for maintenance access to the tunnel.

During the construction of the adit, the TBM is parked at TM 18.6 km. TBM cutter head repairs and critical maintenance are carried out in parallel to the adit constriction works.

The connection of the LWB shaft to the MTS tunnel will provide a second means of tunnel access and egress, allowing the TBM to advance safely to the next intermediate shaft location at LDX. It is considered that the construction of the LDX Shaft will mirror that of LWB.

Following completion of the TBM drive, a cavern will be constructed at LWB 365 m below ground which will house the MTS conveyor booster drives and other critical infrastructure.

## Logistical Challenges

As the original plan considered three individual TBM drives of ~12.5 km, the extension to a 29 km (and potentially 36.5 km) TBM drive requires significant redesign of tunnel logistics and service provisions. All bulk materials such as tunnel segments continue to be supplied by locomotives from Wilton, however upgrades of essential services such as ventilation, power, material conveyor and grout systems are implemented to extend the single drive capability to approximately 21 km.

Beyond 21 km, ventilation, grout and power will be supplied from the intermediate shaft at LWB and subsequently LDX, and additional booster stations are installed at pre-determined locations to increase the capacity of the temporary conveyor system.

As logistics are based on a single rail, fixed crossings are required at intervals of approximately 6 km with an additional California crossing located behind the TBM. TBM crews will also access the tunnel at LWB, and later from LDX, to reduce journey time to and from the TBM. Due to the long distances and respective travelling time, special man-riders to improve comfort and efficiency have been designed and procured.

## Geotechnical Setting at Lockwood Beck

The predominant geology within the tunnel alignment is relatively uniform, consisting of Glacial Till at the start and the low permeability Redcar Mudstone with Ironstone layers of up to 200 MPa UCS. The geotechnical conditions along the route have been evaluated using a combination of existing geological data, geophysical surveys, surface mapping and 12 boreholes along the tunnel alignment (Figure 3). Several fault zones have been identified along the 36.5 km long tunnel drive. Besides the faults, the main geotechnical risks include the possible interference with old mine works, water inflows, gas and mixed face conditions in the presence of ironstone.

Specifically, at LWB there are four main fault intersections, highly flowing underground water courses and changeable and challenging geological and geotechnical conditions.

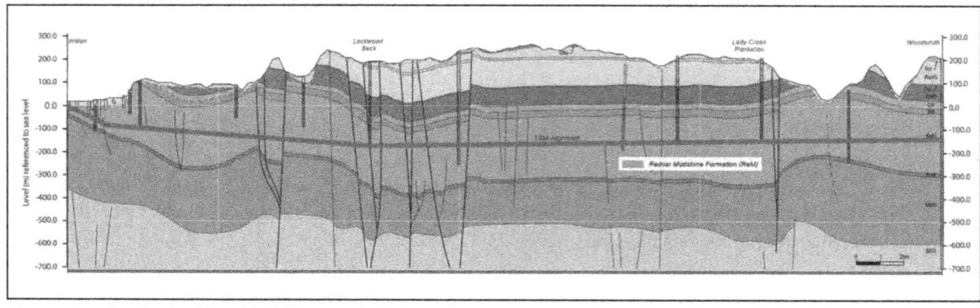


**Figure 3. Geological profile (Source: STRABAG)**

## TECHNICAL SOLUTION TO ACHIEVE AN OPTIMIZED RISK MANAGEMENT APPROACH

### Shaft Requirements

A feasibility investigation was carried out by STRABAG to determine whether the existing shaft would be suitable to continue excavation down to MTS level (365 m deep) to provide ventilation supply to the TBM. Several key aspects of construction and design were reviewed, including the construction programme and the geotechnical risks associated with the shaft excavation. 320 m of excavation was still required to reach shaft bottom, 107 m of this would be through the Ravenscar group, which was expected to be poor quality rock, containing a number of water bearing features. Control of inflows from these water bearing features for conventional shaft sinking was a key concern.

The review concluded that extension of the existing shaft using a traditional shaft sinking method would introduce significant delay to the TBM operations and additional risk, as such it was not a feasible option.

STRABAG then investigated an alternative shaft excavation method, namely 'the Blind Boring method', in order to meet the challenging construction programme whilst consideration given to the difficult ground conditions expected during the shaft excavation.

To determine the feasibility of the Blind Boring method, a space proofing review was one of the critical steps to be taken. The minimum requirements for the shaft were established as hoist access for ~18 persons, tunnel ventilation, grout and mechanical service line supply and power supply. The space proofing exercise determined the minimum internal diameter allowable for the shaft, ~3.2 m. The Blind Bore method soon became the most viable and preferred option.

The methodology concept and construction detailing were provided by SDUK, the design for the shaft sinking works was carried out by ARUP on behalf of STRABAG, with ILF as the independent CATIII checker. Under the UK Construction Design Management (CDM) regulations STRABAG takes the role of both Principal Contractor and Principal Designer.

### New Shaft Location

Several attempts had already been made at ground treatment at the original shaft location, which were unsuccessful in stemming water inflow from the upper part of the shaft coming from the Ravenscar group. At the time of award to STRABAG, the original shaft at 60 m deep was flooded with recorded inflows up to 400 m$^3$ per day.

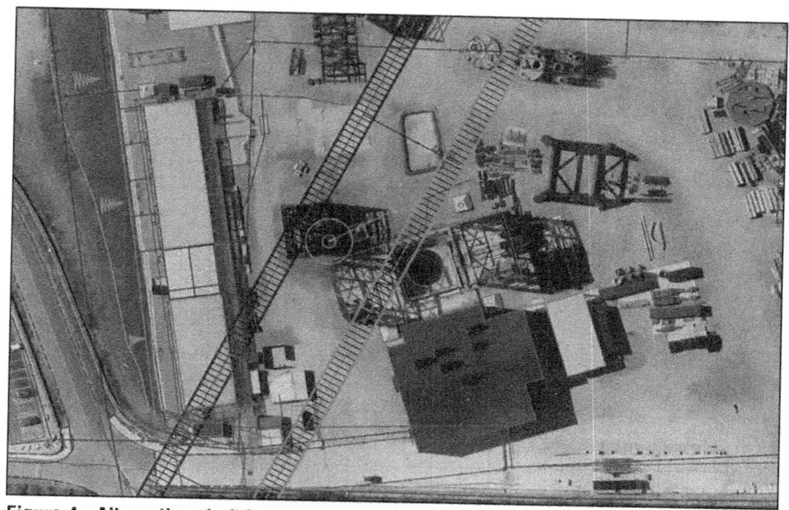

**Figure 4. Alternative shaft location at Lockwood Beck with shaft to tunnel offset shown (green)**

The client had expended significant resources at the beginning of the Project on an extensive suite of downhole geophysical logs including shallow and deep resistivity, natural gamma and acoustic BHTV. These were very useful and led the team to re-evaluate the shaft position, after a series of test bores were sunk and a full desk study conducted on the local geology and seismic traverse information, the shaft position was relocated.

The change in shaft location also meant that to optimise the connection of the shaft to the tunnel, and to maintain the appropriate safety clearance distances, the tunnel alignment was shifted with a 7,500 m curve radius maintained.

## ENGINEERING AND DESIGN

ARUP evaluated the rock block size to be encountered along the LWB shaft applying a discrete fracture network (DFN) modelling approach using the FracMan software, to avoid potential risks associated with blockage of the spoil pipe. Based on the ground investigation data provided, the generated DFN models yielded results demonstrating that the shallower formations are more likely to produce smaller blocks given the bedding plane and subvertical joint fracture intensities, while the deeper formations are more likely to produce larger, more massive rock blocks. Rock block volumes could potentially vary significantly at depth producing blocks with volumes between 0.04 $m^3$ to 6 $m^3$. For the deeper formations this range was estimated between 10 $m^3$ to 14 $m^3$. The results of the DFN modelling provided an informative understanding of the potential rock block size based on persistence, fracture intensity and joint orientation, and allowed for planning mitigation measures of potential blockages during excavation. STRABAG concluded that this alternative method will not only allow for the ventilation connection to be constructed in time and the associated geotechnical risks are considered to be manageable.

Preliminary designs were provided by SDUK and detailed Design of the shaft lining was conducted by ARUP., The design had to consider the complexity of the excavation and casing installation as well as the high water pressure of up to 36 bar in the operational case. A detailed structural assessment was performed to capture the global behaviur of the steel casing using Oasys GSA and the local effects at the ring stiffener connecting to the casing using LS-DYNA. For structures like LWB shaft where a thin steel lining is the main supporting element subject to high external pressure, the effect of geometric imperfections was one of the critical considerations in design. In addition to the permanent state structural capacity design, ARUP completed extensive studies to ensure sufficient capacity of the lining for the various temporary phases. This included evaluation of the liner structure during installation with an assessment of hydrostatic pressure from external water and grout and internal water used to overcome buoyancy. Internal fill levels could then be monitored on site to ensure both the liner and the installation hoist would not be overloaded.

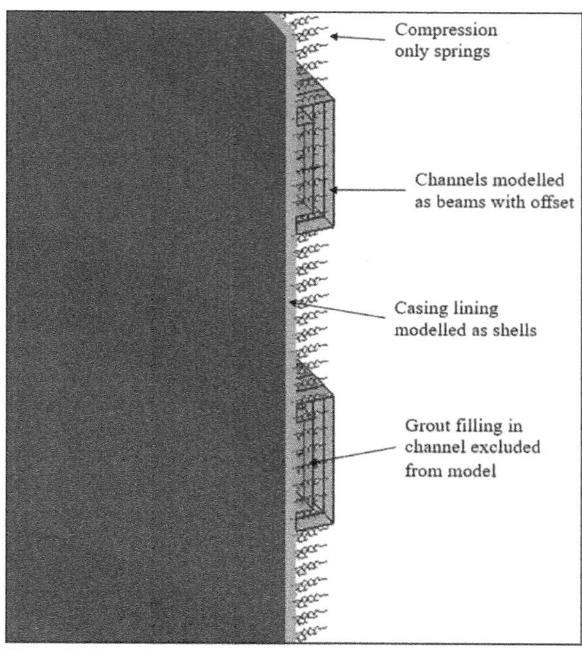


**Figure 5. Overview of the GSA model (Source: ARUP)**

Due to the construction of adit, particularly the installation of the opening of the shaft lining, ARUP also completed an impact assessment. This included design of reinforcement to support the opening, considering the differential pressures of groundwater between the lining and the adit.

STRABAG engaged ILF Consulting Engineers to perform the independent CATIII design check.

Other essential design items include the 21 m long connection adit and shaft head house which was designed by STRABAG and their in-house designer Zentral Technik. The head house and surrounding infrastructure includes the landing level of the double deck Alimak, access lobby and an emergency rescue winch. Surface loads, such as those from the head house facility and rigs, is taken by the secant pile foreshaft and applies no loads onto the steel casing and are therefore not included in the loading conditions for the casing design. The adit is designed to house the tunnel ventilation fan, fire doors as well as electrical and mechanical services.

## SITE ESTABLISHMENT AND PREPARATION

The infrastructure established at LWB, which was designed and installed to support traditional shaft sinking was more than sufficient to support the new Blind Bore scope, with ample mains water, power, and compressed air supply available. However, some

Figure 6. Drone image of the shaft setup including 10,000 m$^3$ lagoon and cuttings pit

additional works were still required to prepare the site for Blind Boring. The specific site establishment requirements were defined by SDUK and delivered by STRABAG.

Additional boreholes were drilled into the superficial materials within the proximity of the shaft to identify the depth of suitable bedrock in which the surface equipment would be founded. Once confirmed, an end bearing secant piled wall was designed and constructed to allow safe excavation of the foreshaft to –20 m. 27 interlocking concrete piles working in hoop stress toed into bedrock, providing a watertight, structurally sound retaining structure to enable excavation by BROKK and mucking skip.

A 10,000 m$^3$ lagoon and wedge pit was designed and constructed adjacent to the proposed shaft to provide a sufficient basin for successful water recirculation. The lagoon was excavated and lined with a geomembrane. A concrete structure was constructed with legato blocks to encourage circulation of the water within the lagoon, providing adequate suspension time for solids to fall from the drill water. A cuttings pit provides a confined cell to capture cuttings whilst allowing the water to flow into the lagoon. The cuttings were constantly removed by an attendant excavator to maintain capacity throughout the drilling operation.

## PRE-EXCAVATION GROUTING CAMPAIGN

The pre-grouting phase of Blind Boring is integral to the methodology and is key to mitigating risk. At LWB this was made increasingly important due to geological challenges including known faulting and high groundwater flows in water bearing strata within the upper 180 m. For the Blind Boring methodology to be a success, the reamed 3.65 m diameter hole, 390 m deep, must be able to maintain a body of water and must not collapse when supported by drilling water only.

The principal objectives of pre-grouting are therefore twofold. Firstly, to improve stability and minimise risk of sidewall collapse achieved by consolidating faults/shear zones, joint/foliation planes and low strength rock. Secondly, to create a curtain of low permeability strata around the shaft perimeter so the shaft can maintain a body of water, essential for reverse circulation and control of buoyant loads lifted by the drilling rig.

The pre-grouting regime was self-performed by SDUK using a RD20 drill rig. A series of vertical holes were drilled in a radial pattern around the outside of the shaft. Working in a top-down approach to overcome groundwater conditions and stabilisation challenges, cementitious grout was pumped through tremie pipes until yield pressure in that zone was achieved. Feedback from the drilling operation and results from permeability testing are examined by vastly experienced members of the SDUK team and are ultimately used to determine success. Pre-grouting must remain an iterative, results based, dynamic approach without cost or programme pressure. Due to the known geological issues, the forecast programme and grout consumption more than doubled, with 34 holes drilled and almost 900T of cement pumped—a significant quantity in a small area indicating the geotechnical and hydrogeological challenges. Results from water loss tests showed significant reduction in permeability and ability to hold water.

## PILOT HOLE AND MICON SYSTEM

The directional control of the main auger relies upon a stinger protruding from the hemispherical reaming head. The stinger follows a pilot hole which is predrilled to full shaft depth. SDUK drill the pilot hole using the same RD20 drill rig used for the pre-grout regime. The verticality of the shaft is critical to maintaining minimum offsets from the tunnel and to allow installation of an Alimak man riding lift. Directional drilling techniques were therefore implemented, and in this case the system specified was Micon's Rotary Vertical Drilling System (RVDS).

A common problem of the pilot hole is the precision. Common reasons for deviations are steeply dipping stratification with differing rock hardness. On the Uttendorf II power station project (in Austria 1982), a directional drilling tool was used for the first time, positioned immediately behind the pilot bit. The use of such a drilling tool enabled drilling with a precision of less than 0.5% of the boring length.

With the technological developments of the drilling tool (RVDS) a precision of about 0.1 m over 500 m (0.002%) can be achieved. This self-steering tool delivers, in addition to the technical functional data for the tool, the current location (inclination and direction) of the drill to the control panel of the boring machine. A turbine with coupled generator and hydraulic pump, driven by the flow of flushing water in the drill-string, serves energy and operates the tool. The inner pulser sub is the module of the RVDS carrying the components which ensure the data transmission to the surface, store the hydraulic oil for the hydraulic pads / ribs and generates the electric and hydraulic power. The measured and processed data are transmitted from the RVDS-Tool to the surface, to the control panel of the boring rig by positive mud pulse technology. A

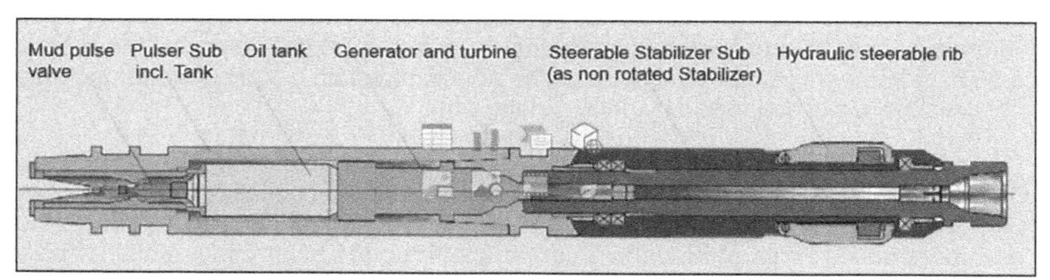


**Figure 7. RVDS tooling (Source: Micon)**

conical piston reduces the flow through the RVDS and thereby creates and in-crease of the backpressure over the RVDS. The change in the pressure is measured on the surface outside the bore hole. A defined sequence of changes in the pressure creates a signal which is measured and de-coded outside the bore hole and displayed on a computer at surface.

## REAMING, LINER INSTALLATION AND GROUTING

Following completion of a thorough pre-grouting campaign, install of pilot hole, construction, and excavation of foreshaft and RC drill pad, shaft reaming can commence.

The bottom hole drilling assembly, a 3.65 m diameter hemispherical dome dressed with disc cutters, is suspended from the drilling rig which has been designed and manufactured by SDUK, having been enhanced over their 40 years of operation. The rotary table provides the torque required to turn the hemispherical bit, and kentledge loaded on the cutter head provides the downward pressure necessary to engage the disc cutters into the development face to facilitate the cutting action. Following setup of equipment, all operations are controlled by a small specialist drilling team on surface.

Reverse circulation is used to remove the cuttings generated at the cutter head. Compressed air is injected down through the drill string to create a less dense fluid and thus pressure differential between the water column in the shaft (heavy) and the water in the drill string (light). This creates the lift necessary to bring cuttings to surface, which are then discharged into the cutting lagoon. The 10,000 m$^3$ lagoon provides sufficient time for suspended solids to fall from the water, reducing water density before it is recirculated back to the shaft.

Additional drill strings are added from surface and the process is repeated until target drill depth is achieved. Top up water is added form surface to account for minor losses into the strata.

**Figure 8. Surface setup of drilling rig**

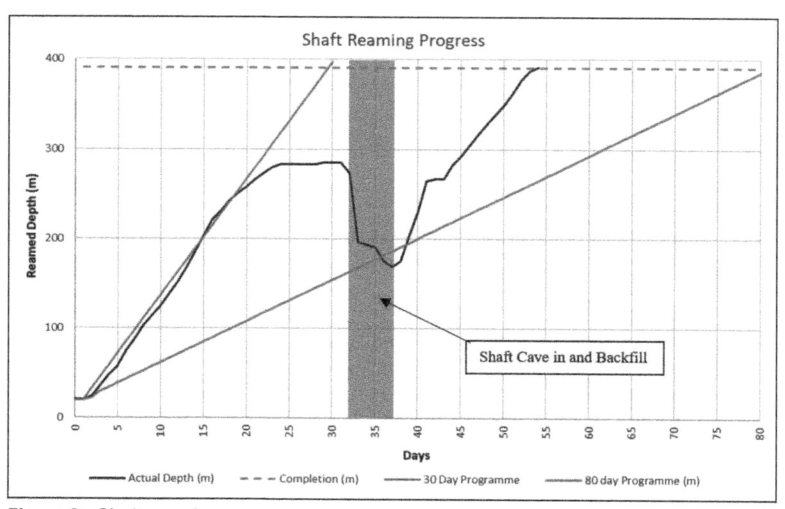


**Figure 9. Shaft reaming progress**

The shaft was reamed in 53 days, a hard average of 7.4 m/day, 27 days ahead of the anticipated programme duration of 80 days. This included overcoming a construction challenge of side wall caving within the shaft. During pre-grout works, a potential fault was identified, intersecting the shaft at 180 mbgl. The zone of strata had comparably high grout take and slaking was evident within down hole calliper surveys. During shaft reaming, it was noted that ground began to cave within this zone. As per mitigation plans and a collaborative approach between all parties on site, the shaft was swiftly backfilled, a grout infill poured and reaming recommenced, consequently achieving full depth excavation without any further issue.

Following reaming completion, the drill tooling is removed, and the shaft is subsequently lined. Another significant advantage of Blind Boring is that the permanent, watertight, shaft lining can be installed immediately following shaft excavation in a single phase, without requiring access within the shaft.

A 1,000T steel liner was installed in 43 No., 9 m long, 3.2 m diameter sections. SDUK oversee this process with the joint welding being performed by a local fabrication company, Industro Solutions. The steel sections, also fabricated locally in Teesside by Francis Brown Ltd, range in thicknesses of 15 mm to 30 mm and are built with up to 10No external stiffener rings depending on their depth and subsequent ground pressure. The intrados of the shaft lining remains smooth.

The bottom of the initial casing section is equipped with an engineered and fabricated sealed bulkhead. The bulkhead allows the completed section of casing to become buoyant in the flooded shaft. Subsequent lining sections are lifted and welded with a full penetration butt weld at ground level. The welds are Ultrasonically tested before being lowered into the ground. A detailed buoyancy programme, structurally assessed by ARUP, is manged by SDUK, filling the intrados of casings to provide ballast and maintain the weight on the drilling rig. The proficiency of the grout curtain is essential, any significant water losses have the potential to result in the entire operation failing catastrophically.

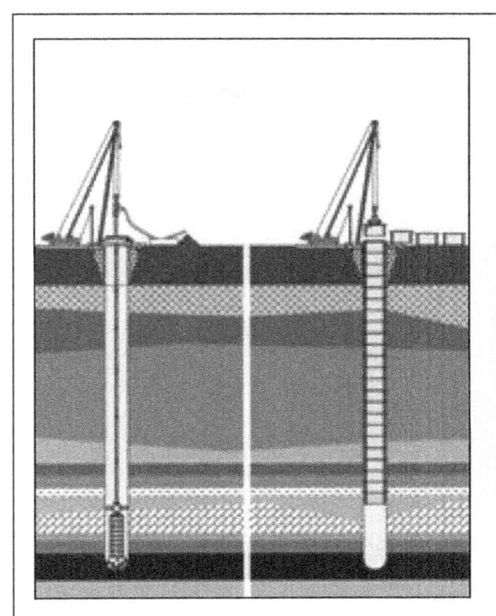

**Figure 10. Schematic of casings installed top down and picture during install**

To stabilise the liner within the hole, the annular space between reamed shaft and extrados of the steel liner is grouted. A neat OPC and water mix is tremied in stages from bottom up by SDUK. Once grout strength is achieved, a submersible pump is lowered from surface and the water within the shaft removed and the shaft handed over for fitout. Surplus water from the anulus, shaft intrados and drilling fluid was treated for suspended solids and pH and then batch released into the local watercourse under permit. Cuttings were refined and removed from site and the lagoon backfilled. Construction of the shaft was completed to programme, in 287 days.

## SHAFT FITOUT

Upon completion of the shaft, underground access was granted to personnel. Shaft fitout installations included an 18-person double car ALIMAK lift, mechanical services for air, grout and water, 11 kV and 3.3 kV power supplies and tunnel ventilation. Installations were prefabricated off site and lowered into the shaft using a combination of winches and crawler cranes. Prior to the ALIMAK lift car installation, all access was gained from a working basket, lifted by a primary and secondary (rescue) crawler crane. The Liebherr cranes were fitted with state-of-the-art Programmable Logic Controller (PLC) functionality meaning they could perform like shaft winders and adhere to strict legislation including Mines Regulations 2014.

Following the hierarchy of control for hot works, the requirement to weld services within the shaft was minimized, with alternative engineering solutions identified. For the majority of fixings, Hilti X-BT threaded fasteners were used to provide a mechanically fused connection to the steel liner. Fixtures were lowered and fixed to the liner.

The 21 m long adit (shaft to tunnel centre) connecting the shaft to the tunnel was constructed using conventional excavation, rock bolting and fibre reinforced sprayed concrete lining (SCL). At the connections to the shaft and the tunnel, the structures were

Figure 11. Up shaft view of shaft, electrical cable, ALIMAK mast, ventilation, and mechanical pipework

Figure 12. (left) Opening within shaft lining, picture taken from within adit, (right) Alimak landing levels within headhouse prior to cladding

supported using steel jamb frames, welded to the shaft lining. The adit separates the tunnel from the shaft and acts as a fire compartment. Surface ventilation downcasts to create a fresh air base within the shaft/adit whilst the main tunnel fan, positioned in the adit, draws air down the 3.2 m diameter shaft and pumps it to the face of the TBM.

At surface level, the shaft top is equipped with a head house with four floors, standing approx. 14 m high. Access to the ALIMAK is available on two levels and the third and fourth provide access to emergency winch equipment, used for ALIMAK rescue. Strict controls are implemented at surface to eliminate dropped object risks and communications are now interlinked to the tunnel.

There was no requirement for working underground until the shaft construction was complete and fit-out commenced. Throughout all construction phases, works were planned and executed safely and efficiently, with no accidents or incidents being recorded.

## CONCLUSION

The success of the LWB shaft construction is based on the collaborative and innovative approach between all contract partners and this culture continues to be an important contributor for the success of this world class, highly complex project.

Geotechnically, the LWB shaft is located at four main fault intersections, highly flowing underground water courses and challenging geological and geotechnical conditions was encountered and expected during the construction. As explained in this paper, the Bling Boring Method is implemented by keeping the entire shaft filled with water while installing the steel casing, to provide support for the casing by 'floating' it during installation. The amount of water added to the shaft was adjusted proportionally as new casing segments were added. The careful planning and consideration in design of the sequence of works was vital for the successful construction of the LWB Shaft.

Additionally, another major challenge for this very long MTS drive are the safety provisions, ventilation requirement, tight construction programme and the evacuation strategy. The construction of LWB Shaft and its adit connection to the MTS tunnel means it provides these essential requirements in time to support the continued operation of the TBM towards unprecedented distances.

The Blind Boring shaft install was completed without safety incident, a record achieved with great teamwork between STRABAG, SDUK, ARUP and Anglo American as well as several local supply chain partners. By using a small, highly skilled workforce with no requirement for underground access during construction many of the traditional risks were eliminated at source. Minimal noise and visual impact as well as ~50% less raw material used when comparing to traditional shaft sinking provided an environmentally positive solution; and a reduced programme, workforce and limited equipment resulted in a highly cost-effective sinking method.

Overall, the design and construction (including pre-excavation grouting and foreshaft) of the LWB shaft has been completed within 2.5 months and 11 months, respectively. In comparison with the traditional shaft sinking programme, this alternative shaft sinking method allowed STRABAG to improve upon the overall construction programme by at least 18 months.

# Risk-Based Design Study and Innovative Ventilation Strategy for a High-Gradient Short Tunnel

**Juan Pablo Muñoz** ▪ ILF Consulting Engineers Austria GmbH
**Maximilian Weithaler** ▪ ILF Consulting Engineers Austria GmbH
**Juan-Carlos Rueda** ▪ ILF Consulting Engineers Austria GmbH
**Harald Kammerer** ▪ ILF Consulting Engineers Austria GmbH
**Reinhard Gertl** ▪ ILF Consulting Engineers Austria GmbH

## ABSTRACT

The Guillermo Gaviria Echeverri tunnel and its access roads project in Antioquia, Colombia is a mega infrastructure project that includes a road network with the longest tunnel in Latin America and several short tunnels, the latter being the focus of this study. The design of the project required careful consideration of several special factors, such as the tunnels' steep slopes of up to 5%, high fire load of 100 MW, bidirectional operation, and short lengths that make it difficult to place ventilation equipment. The combination of all these parameters cause a challenging design problem for the ventilation systems and smoke control of the short tunnels.

To address this challenge, a risk assessment study was conducted to support the design of a new and innovative ventilation strategy that deviates from relevant guidelines and standards, like the Austrian RVS [1][5][6] or the German EABT [8] in order to provide a customized solution for this particular project. This strategy resulted a cost-effective solution, that allowed a reduction in the number of required jet fans and at the same time reaches or reduces the risk level compared to traditional ventilation strategies.

## INTRODUCTION

The Guillermo Gaviria Echeverri tunnel, it's access roads and connecting short tunnels are part of the mega project "Prosperity Roads." The project includes 18 tunnels, 30 bridges and several kilometers of new roads. This paper is based on Tunnel 14, one of the four short tunnels with similar characteristics for which the ventilation system was designed. Tunnel 14 is located at an average altitude of 1368 m above sea level, with a total length of 919 m and a continuous longitudinal gradient of 5%. The tunnel includes one rescue tube that connects the traffic tube directly to the exterior, and similar to the other tunnels in the project, will be operated under bidirectional traffic.

An initial and preliminary ventilation design specified 24 jet fans along the tunnel to reach the air velocities required for the ventilation according to the regulations. However, a detailed analysis of this initial design concluded that the ventilation configuration, which follows the guidelines for this tunnel with its special characteristics, was both overdesigned and inefficient, and it may lead to unnecessary investment, operation and maintenance costs. It was necessary to review the ventilation configuration and adapt it to the specific tunnel characteristics, including the high longitudinal gradient, the short tunnel length and the bidirectional traffic operation. However, to implement that review it was necessary to deviate from the guideline's requirements. Many tunnel safety guidelines and regulations define only a general framework for tunnel design approach, often not taking specific characteristics of an individual tunnel into account. This is especially noted for tunnel ventilation control which are only evaluated with respect to the traffic operation type (unidirectional or bidirectional traffic). However, ILF experience has shown that for some tunnels with specific characteristics it can be reasonable to deviate from these general regulations to improve the ventilation system and implement different guidelines like the RVS 09.03.11 [6]. Further, the recommendations of PIARC allow a tunnel design to deviate from the guidelines when a quantitative and comparative risk analysis proves that the tunnel design solution, that deviates from the guidelines, shows a risk profile equal or lower than the tunnel designed according to the guidelines.

## INITIAL SITUATION

Tunnel 14 is a short tunnel in the road network on the project "Prosperity Roads" in Colombia. It is a 919 m long tunnel with a constant slope of 5% and with a mean elevation of 1,367.95 m above sea level. Furthermore, the cross section of the traffic tube is a horse shoe profile with 60.7 m$^2$ and a perimeter of 30.1 m.

The initial studies and design for the ventilation of Tunnel 14 were completed in April 2017. The objective of this initial design was to analyse the tunnel, its characteristics and to propose a ventilation system design according to the main design guidelines such as the PIARC [2], EC-Directive 2004/54/EC [4] and the NFPA 502 [7], as well as to local guidelines such as the "Colombian Handbook for design, construction, operation and maintenance for Road Tunnels," among others. Since this initial design phase until now the geometry of the tunnel has remained the same. The ventilation configuration of the initial design specified 12 locations of jet fans, with 2 fans in each location, and a power of 41 kW per fan (see Figure 1—initial design).

After the re-analysis of the initial design and considering the particular characteristics of the tunnel and its boundary conditions, a preliminary conclusion was that the number of jet fans seemed unnecessarily high, and given the short length of the tunnel, this would force the spacing of fan sets to be reduced. This arrangement largely reduces the efficiency of each jet fan, as in order to reach its optimum performance they should be installed with a minimum of 100 m from each other, and in this tunnel the jet fans were evenly distributed and separated only by 60 m as an average. The loss of efficiency would lead to a higher number of jet fans installed and a larger power supply system, which will result in additional CAPEX and OPEX costs. A more thorough re-analysis was conducted to re-evaluate the number of fans required. It was concluded that the initial design was the result of the application of the guidelines to the ventilation design, and although it was correct from the perceptive perspective, it was an inefficient ventilation system.

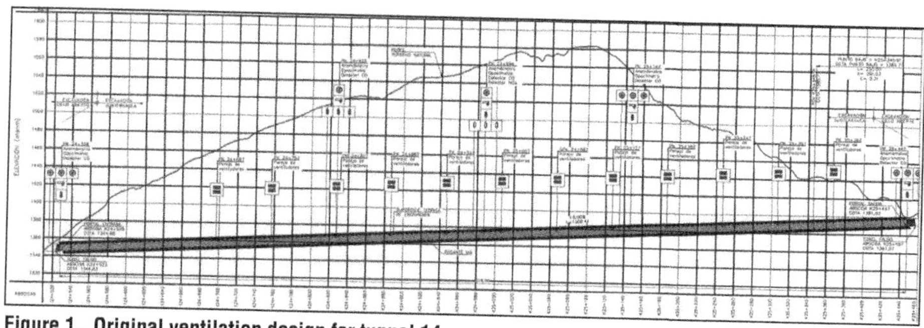

**Figure 1. Original ventilation design for tunnel 14**

During an incident, both the RVS 09.02.31 [7] and the EABT [9] suggest a ventilation strategy to maintain the air flow towards the same direction it was at the time of the incident, making it easier in regular tunnels to ventilate the smoke in the direction of one of the portals. However, it is not the best approach to take for a tunnel with such a high gradient as Tunnel 14. In case of higher temperatures, the buoyancy of the fire is a key factor to take into consideration and that changes the most critical scenario from accelerating the smoke to decelerating it depending on the location of the fire. When following the ventilation strategy proposed by the guidelines, the amount of jet fans needed for changing the direction induced by the buoyancy of the fire (i.e., in order to keep the original ventilation direction) increases, especially as the fire location gets further into the tunnel. At more remote locations from portals, the thrust required to overcome the fire bouyance and change the direction of the induced air flow to follow the original air direction grows considerably.

After analyzing the behavior of the air flow in the tunnel during an incident, and determining that the triggering factor for the large number of jet fans was the buoyancy of the fire causing the traditional ventilation to be inefficient for this particular scenario, it was decided perform a quantitative risk assessment study in order to determine a suitable alternative ventilation strategy. The goal of adapting the ventilation strategy was to overcome the complications generated by a large buoyancy due to high heat release rates and a steep gradient.

In general terms, guidelines offer a general approach to the design challenges, making it always a good practice to review each tunnel individually, especially if its geometry diverges considerably from the average tunnels considered in developing the guidelines.. The decision of deviating from the guidelines required a totally new optimized ventilation design and strategy. This was accomplished via the analysis of different scenarios to determine the best control strategy, and subsequently a detailed quantitative risk assessment to be used as an decision-making tool. This permitted a comparison between the perceptive ventilation configuration and the optimized ventilation design and strategy, which deviates from the guidelines, to evaluate if the improved ventilation design meets the risk level of the perceptive design.

## GUIDELINES

The guidelines deemed more appropriate for the design of this tunnel were the EABT-80/100 R2 [8] as a main guide, next to the RVS 09.02.31 [6], RVS 09.02.32 [1], RVS 03.01.11 [5], which are mentioned in the "Colombian Handbook for design, construction, operation and maintenance for Road Tunnels" [9], and the EC-Directive 2004/54/EC. As a more broad and general guideline the PIARC [2] will also be considered for the fresh air demand, emission data and traffic calculations.

The EABT-80/100 R2 [9] was taken as a main guideline instead of the NFPA 502 [8], since the Columbian Handbook uses the same criteria as the EABT in term of tunnel ventilation. Furthermore, a comparison to the RVS guidelines was performed, since these guidelines are mentioned as "equivalent" for several scenarios in the Colombian Handbook.

When considering this difference in guideline philosophy, it is clear that a risk analysis is needed if an innovative ventilation strategy is to be considered. The PIARC [4] establishes a control strategy for longitudinal ventilation which maintains the main ventilation direction previous to the incident and states that any deviation from this strategy must be only implemented after a risk analysis is completed. The EC-Directive 2004/54/EC [5] also states that a risk analysis is needed to determine if the right amount of safety measures are being taken for tunnels with gradients higher than 3%, or bidirectional tunnels.

## RELEVANCE OF TUNNEL SLOPE WITH RESPECT TO DECISION ON VENTILATION SYSTEM

### EC-Directive

In the European Directive 2004/54/EC [5] the guideline specifies some restriction for longitudinal ventilation. Clause 2.2.3 in the Ref 5, states "*In tunnels with gradients higher than 3%, additional and/or reinforced measures shall be taken to enhance safety on the basis of a risk analysis.*" Reference 5 does not restrict the implementation of a longitudinal ventilation system for tunnels with greater slopes, but does require the implementation of a risk assessment during the design

### National Guidelines

In many European countries, there are national tunnel design guidelines in addition to the EC-Directive. The RABT or EABT-80/100 [9] is a set of German national technical guidelines relevant for planning of tunnel structures and equipment (including ventilation systems).

These national guidelines are continuously updated, implementing new findings and developments. In the 2019 version, Reference 9 states that the application of longitudinal ventilation systems for tunnels longer than 400 m, with gradients higher than 3%, must come with additional and/or reinforced measures in order to improve the safety. For this, a safety assessment (risk analysis) must be done.

## TECHNICAL ASPECTS AND VENTILATION CALCULATION

In addition to regulatory aspects, some technical specifications have to be considered during the design of the ventilation system for bidirectional road tunnels.

### Fresh Air Demand

The fresh air demand depends on the emissions of the vehicles inside of the tunnel. This depends on the hourly traffic and the length of the tunnel. The amount of fresh air itself is no longer a restriction for any ventilation system, especially for shorter tunnels such as Tunnel 14. However, guidelines like the PIARC, NFPA, EABT and RVS set thresholds that have to be met and considered during the design.

### Pressure Loss

During both normal operation and incidents, there are several factors that affect the air velocity in the tunnel, and those factors may impact the ventilation systems in positive or negative ways depending on the scenario. In case of Tunnel 14, the higher pressure losses for the ventilation system to overcome are the buoyancy of the fire, the meteorological pressure difference and the losses by fire. Most significantly, the higher pressure due to the high slope of the tunnel.

### Calculation of Ventilation System

To determine the appropriate number and specifications of jet fans, and to develop an alternative ventilation strategy, it is necessary to identify the critical fire location within the tunnel. This can be achieved by running multiple fire scenarios and analyzing the results to observe the location where the system will have a higher requirement of thrust and air velocity.

During the simulations, it is clear that the buoyancy of the fire caused by the high heat release rate and steep slope increases the air flow velocity as the fire site gets further from the lower portal into the tunnel. This is the reason why the previous design, which followed traditional ventilation strategy, calculates such a large number of jet fans for the tunnel. That is, given the case that the ventilation direction before the incident is towards the lower portal, the buoyancy will induce a large air flow towards the upper portal, making the ventilation according to the guidelines extremely challenging and increasing the amount of jet fans required.

The key to making a more efficient design is locating an inflection point, which is a location where the system can change the direction of the smoke to both portals and also slow down the air flow if necessary. In the event of an incident, this inflection point will be the location where the ventilation system requires the most thrust to safely ventilate the tunnel. For Tunnel 14, the inflection point location is at 300 m from the lower portal (Portal Entrada), which is between the first quarter to the first third of the tunnel.

Due to the way that the buoyancy affects the air flow in different fire scenarios, it was determined that the most effective way to optimize the ventilation system is to always ventilate towards the lower portal if the fire starts within the first 300 m of the tunnel, from the lower portal. This is because the buoyancy in this section does not accelerate the air to a velocity that cannot then be slowed down or have it's direction changed. For the remaining portion of the tunnel, the ventilation system should always be directed towards the upper portal, using the jet fans thrust to either slow down the effect of the buoyancy or turn the flow around with the assistance of it.

As a result, the tunnel divided into two different sections and the ventilation strategy then is set as follows, depending in the location of the fire:

- Lower Section: If the fire starts in any place within the first 300 m from the lower portal, the ventilation system will either accelerate, decelerate or change direction of the air flow in order to direct it towards the lower portal.
- Higher Section: If the fire starts in any other place out of the Lower Section zone, the ventilation system will either accelerate, decelerate or change direction of the air flow in order to direct it towards the higher portal.

After the initial determination of the most critical scenario, the ventilation strategy and a rough calculation of the number of jet fans needed for the critical scenario, a more detailed design was done to determine the specific number of fans and location needed.

The detailed calculations show the need of 8 jet fan locations including the systems requirements and the redundancy considerations. Giving a total of 16 unidirectional jet fans installed with the main blowing side towards the lower portal.

The comparison between the original design and the review with the new and innovative ventilation strategy shows a reduction of 33% in the amount of jet fans, and between 13% and 27% reduction in the installed power needed.

## RESULTS OF RISK ASSESSMENT STUDY

Taking in consideration the high slope, the bi-directional traffic operation and the modification of the ventilation strategy from the guidelines, a quantitative risk assessment study was performed for Tunnel 14. For that, the Austrian Tunnel Risk Model TuRisMo (defined by the RVS 09.03.11 [6]) was applied, following the principles defined by PIARC [4] for risk assessment studies.

The risk assessment study was performed for the proposed alternative as well as the ventilation strategy according to guidelines. The risk evaluation was carried out using a relative approach. The risk value of Tunnel 14 with the proposed ventilation layout and equipment was compared to the risk value of the "reference risk profile," defined by the initial design following the guidelines regarding the ventilation strategy as dictated by the PIARC.

The fire consequence analysis considered a large number of simulation scenarios. Transient 1D/3D smoke propagation simulations are identified as most demanding in terms of computational resources, therefore the number of simulations was limited to 324 case combinations between traffic volume and symmetry, fire size, location and delay time; giving a total of 40,000 simulations.

The model differentiated between three types of fire events for risk analysis:

- Primary: During normal traffic flow, a vehicle has an accident or a break-down and catches fire.
- Secondary: An accident or a break-down causes a traffic jam. A vehicle hits the rear end of the vehicle queue and catches fire.
- Tertiary: Traffic overload causes a traffic jam. A vehicle located in the traffic jam has a technical defect and catches fire—not relevant in this study, since the average daily traffic is very low, traffic jam is not expected

Figure 2 illustrate the results of the quantitative risk analysis for Tunnel 14, showing the expected risk values for the fire risk of the two investigated ventilation controls.

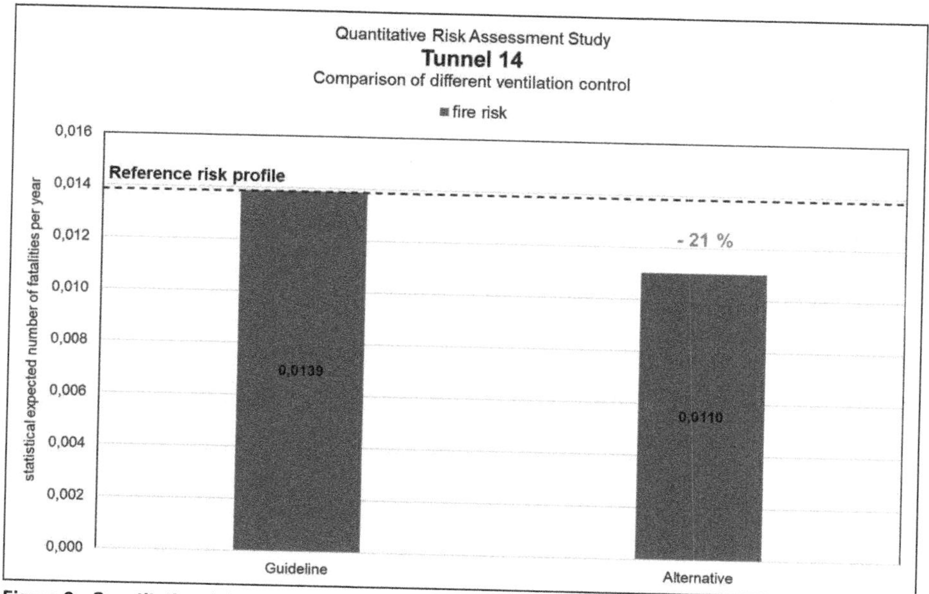


Figure 2. Quantitative risk assessment study—final results

The results can be interpreted as follows: The previous design following the guidelines defines the reference risk profile. In order to be able to implement the proposed ventilation strategy, the risk of the alternative should be the same or lower than the reference. As can be seen in the Figure above, the alternative not only can achieve the same risk level but even lowers it by 21%, which means that it can be implemented without any additional risk mitigation measures. In addition, the alternative ventilation control mode is able to compensate the increase in risk due to the high tunnel gradient and can be seen as a risk-mitigation measure for this specific characteristic.

## SUMMARY AND CONCLUSIONS

A risk assessment study was performed in order to investigate and quantify how the variation in the ventilation strategy affects the system in comparison with the initial design. The study compared the risk of a system design to comply with the classic longitudinal ventilation control as stated by the guidelines, with a system designed with an innovative ventilation control strategy that deviates from the guidelines, and use the results as a decision-making tool for the ventilation design.

The results of both the ventilation design and the risk assessment show that the innovative ventilation strategy is not only more efficient from a technical point-of-view, by lowering the power consumption in at least 13% and the costs related to the acquisition, installation and operation in around 33%, but is also a better option from a safety perspective, lowering the risk on up to a 20% from the reference profile.

The results obtained from the quantitative risk assessment study as well as the ventilation study shown that the proposed alternative ventilation system with the strategy deviation from the guidelines improves significantly the initial ventilation configuration in Tunnel 14. Therefore, that is the system that will finally be implemented and operated.

Following the successful results of this design approach on the first tunnel, the same studies were applied to the remaining three tunnels, showing similar conclusions with the selected ventilation strategy, meaning this innovative strategy both reduced the amount of jet fans required and maintained or even increased the safety in comparison with the reference risk profile.

## REFERENCES

Österreichische Forschungsgesellschaft Straße, Schiene, Verkehr, Arbeitsgruppe „Tunnelbau“, Arbeitsausschuss „Betriebs- und Sicherheitseinrichtungen“, Tunnel—Tunnel Equipment—Ventilation Systems—Fresh Air Demand, RVS 09.02.32, Issue 06/01/2010.

World Road Association (PIARC), Technical Committee C4 Road Tunnel Operation, “Road tunnels: Vehicle emissions and air demand for ventilation,” 2012.

World Road Association (PIARC), Technical Committee 3.3 Road Tunnel Operations, “Risk analysis for road tunnels,” 2008.

World Road Association (PIARC), Technical Committee 3.3 Road Tunnel Operations, “Current practice for risk evaluation for road tunnels,” 2013.

Directive 2004/54 / EC of the European Parliament and of the Council „on minimum safety requirements for tunnels in the trans-European road network“; Brussels, 29.04.2004.

Österreichische Forschungsgesellschaft Straße-Schiene-Verkehr, "RVS 09.03.11 Tunnel Risk Analysis Model," 2015.

Österreichische Forschungsgesellschaft Straße-Schiene-Verkehr, "RVS 09.02.31 Tunnel Ventilation," 01.06.2014.

National Fire Protection Association NFPA 502, Standards for Road Tunnels, Bridges, and Other Limited Access Highways.

Forschungsgesellschaft für Straßen- und Verkehrswesen, Arbeitsgruppe Verkehrsmanagement, Empfehlungen für die Ausstattung und den Betrieb von Straßentunneln mit einer Planungsgeschwindigkeit von 80 km/h oder 100 km/h, 2019, EABT 80/100.

Manual para el diseño, construcción, operación y mantenimiento de túneles de carretera para Colombia, 2021 Edition, 12.02.2021.

# Lessons Learned: Implementing BIM for the Chiltern Tunnels for High Speed 2 in the UK

**Kurt Zeidler** ▪ Gall Zeidler Consultants
**Yuan Le** ▪ Gall Zeidler Consultants
**Tomasz Kecerski** ▪ Gall Zeidler Consultants
**Vojtech Ernst Gall** ▪ Gall Zeidler Consultants
**Dominic Reda** ▪ Gall Zeidler Consultants

## ABSTRACT

High Speed 2 (HS2) mandated that Building Information Modelling (BIM) processes be implemented during design and construction. Gall Zeidler Consultants (GZ) was responsible for developing the BIM Models for the portals, tunnels, and cross passages of the of the 16 km-long Chiltern twin-bore Tunnel. In addition, GZ developed the GBR for the Chiltern Tunnels and incorporated the relevant information into the Project's BIM framework. This paper presents the implemented BIM framework, describes the modelling approach, and discusses the advantages and challenges of implementing BIM, particularly how BIM improved interface coordination and how BIM was used in feasibility assessments of different construction options.

## INTRODUCTION

HS2 is a new high-speed rail line that will be the backbone of the UK's rail network. Phase One of this project will connect London to Birmingham between 2029–2033. In July 2017 the Align joint venture, consisting of Bouygues Travaux Publics, Sir Robert McAlpine, and Volker Fitzpatrick, was awarded the contract to complete the design and begin construction of the C1 package. The C1 package consists of 21.6 km (13.5 mi) of high-speed rail infrastructure that includes a 3.5 km (2.2 mi) viaduct, the 16 km (10 mi) twin-bored Chiltern tunnel, and five vent shafts accommodating both intervention and tunnel ventilation facilities.

The Chiltern Tunnel is the longest tunnel on the HS2 route between London and Crewe and will carry passengers under the Chiltern Hills. The Chiltern Tunnel consists of a twin TBM tunnel, 5 intermediate shafts, 40 Cross Passages (CPs) and 2 portals, one each at the South and North ends of the tunnels. The two Tunnel Boring Machines (TBMs), Florence and Cecilia, that are being used to create the twin bore Chiltern Tunnel were launched in the summer of 2021. As of the writing of the paper, tunneling under the Chiltern Hills is currently ongoing.

Gall Zeidler Consultants (GZ) is working on tunnel design and serving as a geotechnical specialist supporting Align's design partner Align D, consisting of Ingerop and Jacobs. GZ's main scope of works for the tunnels includes the design of 38 out of the 40 Cross Passages and the adits associated with each shaft.

Following the UK Government's Building Information Modelling (BIM) Mandate, HS2 is being delivered using BIM. BIM, as defined by the BS EN ISO 19650-1:2018 is the:

> *"use of a shared digital representation of a built asset to facilitate design, construction and operation processes to form a reliable basis for decisions."*

In other words, BIM is a process that involves the generation and management of project and asset information using digital representations of physical and functional characteristics of structures and facilities over their entire life cycle. At the time of the commencement of the HS2 project, the UK BIM mandate required that all projects be implemented to a "BIM Level 2," meaning (Innovative UK, 2017):

> *"Projects will use intelligent, data-rich objects in a managed 3D BIM environment. All parties working on a project are able to combine their BIM and design data to collaborate and share information through the use of a common data environment (CDE). The CDE enables users to carry out checks against data validation strategies to make sure they are on target."*

The definitions above are, however, necessarily vague, as they need to broadly describe BIM as a concept. They do not provide the average engineer, project manager, or CAD technician with a concrete understanding of how BIM is implemented in a tunneling project. In current practical international usage, BIM is often used as an umbrella term to describe the use of any number of digital tools, such as, but not limited to, 3D modelling, computational design, visualization, clash detection, 4D/5D modelling and information management used to improve design, project delivery, asset management, and collaboration (ITA WG 22, 2022). This contribution is therefore intended to provide a concrete example of BIM implementation for a tunneling project by giving an overview of how BIM was, and is being, implemented for HS2's Chiltern Tunnel.

It should finally be noted that the implementation of BIM in tunneling differs somewhat internationally. Some countries have concrete standards regarding BIM while others do not. BIM in the UK is largely regulated by the BS EN ISO 19650 series as well as the PAS 1192 series of documents, and this contribution should be viewed with this frame of reference in mind.

## CONTRACTUAL FRAMEWORK

### Employer's BIM Requirements

The Employer's BIM Requirements form the basis of BIM implementation for the HS2 Project. The BIM requirements differentiate between a Project Information Model (PIM) and an Asset information Model (AIM). The PIM refers to the BIM implementation during the design and construction phases of the project and the AIM refers to the BIM framework in place after handover and during asset management. The BIM Requirements also state the goals of the PIM and AIM. These goals are for BIM to:

- Provide a consistent approach to the ways in which data requirements are defined and how data is procured,
- Provide a consistent data-driven approach to the production, management, and delivery of the Project Information Model (PIM) that is structured such that it can be efficiently shared and reused,
- Provide assurances and validation of the quality, integrity and completeness of the design and construction through the utilization of an up to date and validated PIM,
- Provide an improved assessment of health and safety and risk impacts through the utilization of an up to date and validated PIM,
- Provide greater value for cost planning, estimating and carbon management through the utilization within an up to date and validated PIM,

- Provide a validated and verified Asset Information Model (AIM) for operational use,
- Achieve target capital delivery during design and construction by eliminating waste,
- Enable information governance and information security, and
- Deliver best value through innovation.

The Employer's BIM requirements further constrains the JV to internal HS2 standards (CAD, Health & Safety, Sustainability, etc.).

## BIM Execution Plan

While the Employer's BIM requirements define which BIM goals are to be met by the project, they do not define how these goals are to be met. The BIM implementation, therefore, falls under the purview of the JV, provided that HS2 standards are followed. The Employer's BIM requirements dictate that the actual tools and methods used are to be set forth in a BIM Execution Plan (BEP). While the BEP is seen as a "living document," and may be modified due to changes in the Employer's BIM requirements, the BEP forms the contractual framework that describes the BIM implementation process for the PIM as well as AIM.

Among other items, the BEP describes the information delivery plan, the management of the modeling process the Common Data Environment (CDE) used to store and transfer the project/asset information, and a description of the BIM deliverables.

## Digital Delivery of the Design Information

An integral part of BIM is the data management process. Specifically, to ensure conformance to ISO 19650, all data should be stored, exchanged, and managed within a Common Data Environment (CDE). A CDE is, per the ISO 19650 Series an:

> *agreed source of information for any given project or asset, for collecting, managing and disseminating each information container through a managed process.*

With the additional following note being provided for clarification:

> *A CDE workflow describes the processes to be used and a CDE solution can provide the technology to support those processes.*

At its core, a CDE is often a cloud-based file storage system, and many software programs used to develop a CDE allowing an owner to define workflows which regulate when information is uploaded, edited, checked, or approved (ITA WG 22, 2022). For the HS2 C1 project, ProjectWise was used as CDE to store and manage all project information, including all models, drawings, design reports, specifications, and geotechnical information. It was also used as a tool for configuration management and deliverable submission.

To ensure a coordinated workflow for the delivery of all the BIM models, each model and drawing uploaded to the CDE were assigned to a unique reference number and added to the Task Information Delivery Plan (TIDP) and Master Information Delivery Plan (MIDP) by the Project Manager. Essential attribute information such as Seed file, Primary Asset and Sub Asset ID were stored/assigned to ensure uniqueness. A unique script was developed to import the information from TIDP and MIDP and generate the placeholders on CDE. This ensured the proposed deliverables created on

ProjectWise being consistent with the logs on TIDP and MIDP. It also eliminated the risk of creating redundancy and losing control of the CDE.

The workflow (shown in Figure 1) for this project was based on requirements presented in BS 1192:2007+A2:2016. Establishing this concrete workflow streamlined document/model/submittal management and distribution and ensured a minimum quality of design information. Every placeholder has an embedded property called "state" and can be approved by person with appropriate authority to move it to next state. The first state was always "Work in Progress" and it was used to produce the design work. After the work was done the placeholder was approved by the originator which triggered an automatic background process called CAD-QA to ensure the placeholder was compliant with project standard. When successful, the placeholder was moved to state 'Checking'. At this state the next person in the design process, a discipline engineer, was responsible to check the design. Once the design was successfully checked and approved, the placeholder then moved to the state 'Package Shared', where it became visible to other disciplines in the design. This ensured that only checked designs were shared with others. This checking loop was performed on a weekly basis. The next State for the placeholder was "Reviewed." Only the project manager or other individuals with "Reviewer" authority were able to approve the "Package Shared" placeholder. The last state of the process was either Team Shared or "Client Shared." "Team Shared" was used for those submissions which were only submitted to the Contractor, i.e., the Align JV, while Client Shared included the submission to HS2.

The workflow process included built in automatic procedures. In the "Package Shared" state incorporated an automatic model creation of 3D models, and in the "Approved for Team Shared/Client Shared" state PDF renditions of the drawings were automatically created.

## CREATION OF 3D MODELS

A 3D model was created for each structure of the Chiltern Tunnel system. In contrast to simple 3D models, these BIM objects included metadata or information beyond the simple 3D geometry of the object. The metadata includes information such as asset IDs, materials, design life, etc, that allow the user/Asset Manager to query the BIM model much more effectively. An example of a BIM model showing the 3D model including the metadata stored on the object is provided in Figure 2.

Before creating the 3D models, 2D cross sections were developed to ensure that all spaceproofing requirements were met. The 2D Models were also used as a basis for all structural calculations to ensure that the base structural concept was sound before beginning modeling. Finally, the 2D sections were additionally used to verify the 3D models as part of the QA process by comparing cut sections in the 3D model with the original 2D concepts.

To effectively and efficiently create the BIM models for the Chiltern tunnels and define HS2 metadata for each element, several techniques were used. The following sections provide a description of three of the primary techniques to do so: DataGroup Catalog Profiles, Feature Modelling, and Generative Components.

### DataGroup Catalog Profiles

To model all the components more easily along the Chiltern Tunnel alignment, a Catalog of elements was created for each space proofed 2D cross section. This allowed implementation of HS2 Metadata before creating models, as much of this data

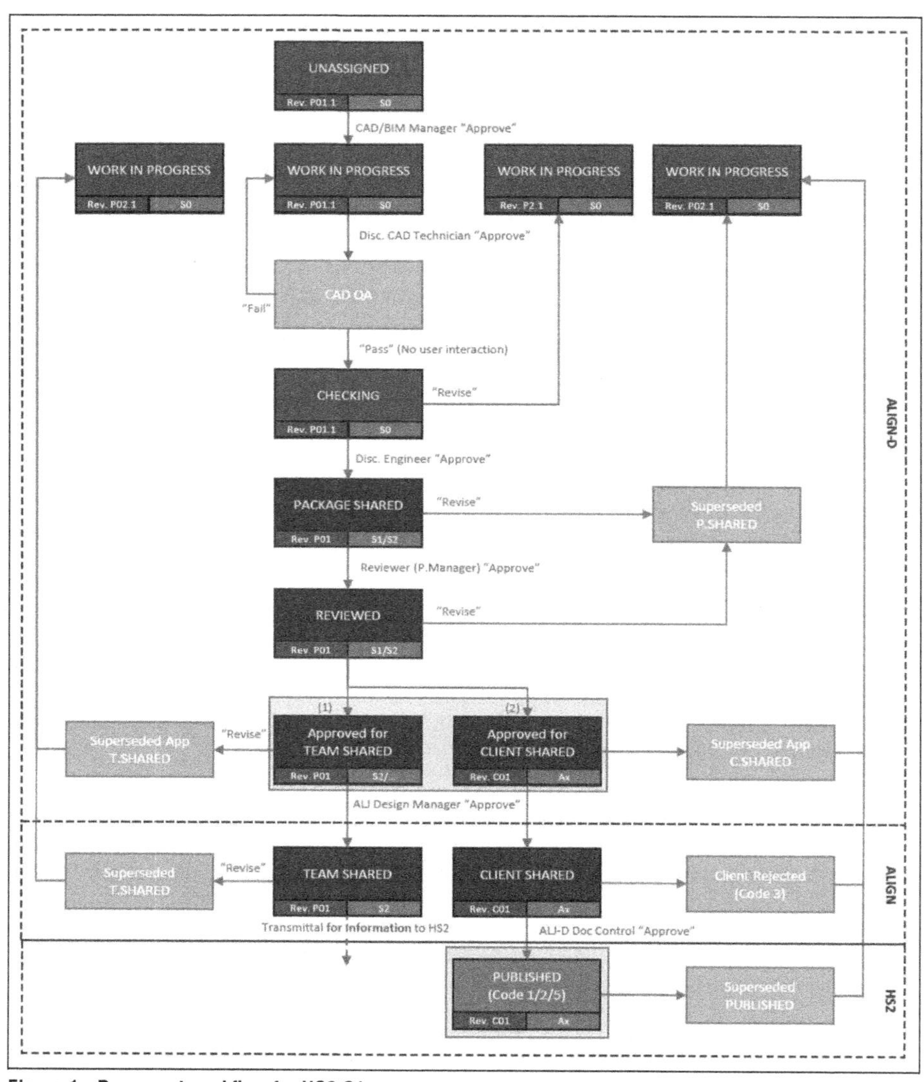


**Figure 1. Document workflow for HS2 C1**

was standardized. This allowed a modeler to simply pick up a required shape from the Place Profile library and extrude it along either the alignment or any predefined setting out line. Not only does this technique simplify the modelling process, but it minimizes mistakes made from manual creation/insertion objects along 3D setting out lines that may follow complex 3D paths. Even if the BIM object required further modification of its Metadata, any Metadata could be easily edited from inside the model after the model was created.

## Feature Modelling

Feature modelling is an AECOsim modelling technique that is used to create complicated 3D objects while maintaining a detailed history of the changes made to the model. This technique was used for the Cross Passage collar transition and TBM tunnel walkway transition at door frame locations (Figure 3). The disadvantage of feature modeling is, however, that no properties from the DataGroup Catalog as discussed above are implicitly embedded into the objects. Hence HS2 metadata must be added to objects in a different way. To simplify and expedite this process, HS2 metadata were added to feature elements by Boolean operation. Boolean operations allow the user to combine objects, create differences between two objects, or create an intersection between two objects that can inherit the properties of both objects. By intersecting a feature modelling object with a DataGroup Catalog object, such a wall or slab, which both contain metadata, the feature object was able to inherit the required properties.

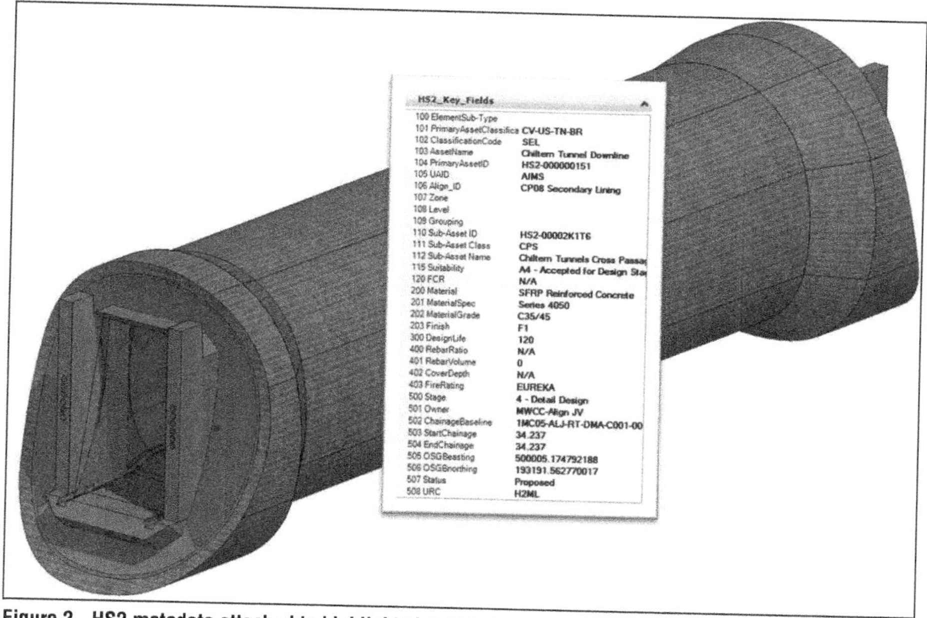


**Figure 2. HS2 metadata attached to highlighted element**

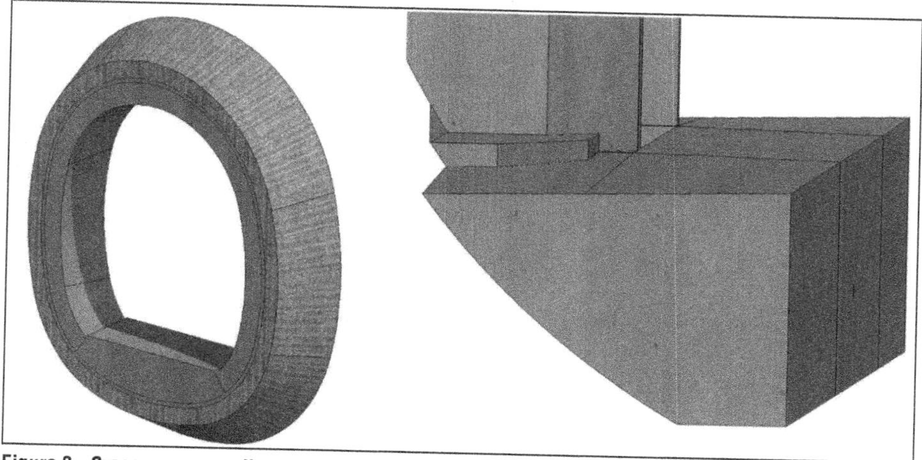

**Figure 3. Cross passage collar transition and TBM walkway transition at door frame**

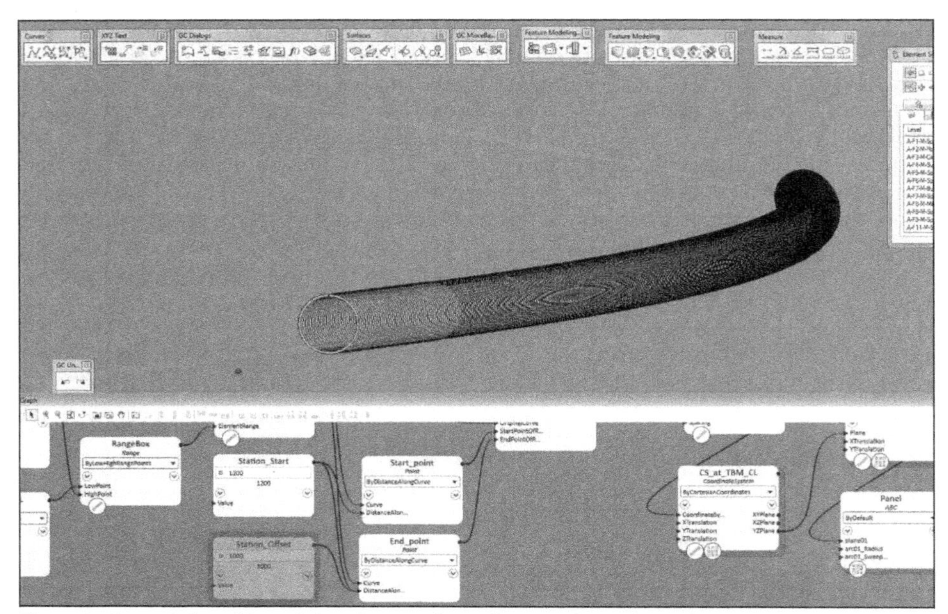


**Figure 4. Computation of tunnel ring element along alignment for station 1200 to 2200**

## Generative Components for TBM Ring Model

Modelling a 16 km long twin-tube TBM tunnel provided a set of unique modeling challenges. The DataGroup technique only proved useful in creating an extruded element to a maximum length of 200 m, such as for the CPs. After 200 m, the extrusion function no longer functioned properly. As such, to model the main TBM tunnel with the DataGroup technique, the tunnel alignment would have had to be divided into a series of 200 m long sections with individual setting out lines to extrude longitudinal elements like walkways, slabs, or the tunnel lining.

The TBM tunnel lining is/will be comprised of 2 m wide pre-cast concrete segmental rings. To create more than 8,000 segmental rings over the entire length of tunnels more efficiently, the Generative Components (GC) software was used to build the TBM model. This technique allowed for the creation of a set of rules, which govern the automatic model creation. In doing so, the full 16 km length of alignment could be used as setting out line, and a pre-defined set of rules was used to replicate a single ring object to create the 8,000+ rings with precision and control. This process is shown in Figure 4.

Another benefit of GC was the ability to automatically input data into each element. It should be noted, however, that this software feature did not perform as anticipated, as data cannot be replicated into reporting schedules. Nevertheless, feedback was provided to the Software developer who is now working on resolving this issue.

## BIM USE CASES

The following sections describe some of the BIM use cases that were implemented in the HS2 C1 project.

### Design Interface Coordination

One of the biggest advantages of using 3D models is to coordinate the design between several disciplines. For example, the Chiltern Tunnel structural design has interfaces with shafts, portals, utilities, and mechanical and electrical design packages. As per the BIM requirements, each discipline developed their own 3D models for the design. The design coordination models could then be easily prepared by bringing different 3D models from different disciplines into a "federated model." A federated model is described in the ISO I9650 Series as a model that "...comprises information models from different lead appointed parties, delivery teams or task teams." By sharing these federated models across all packages on ProjectWise, engineers were able to study the coordination models and identify interface issues. With the aid of the federated models, the interface issues could be visualized and discussed in the regular interface meetings as mandated in the BEP.

One very powerful example of interface coordination is the utilities passing through the cross passages. The design of the Cross Passages adopted innovative design approach of using Reinforced Concrete (RC) door frames inside the TBM tunnels to provide support to the opening on the TBM segmental linings. With the door frames structurally connected to the TBM linings at each CP location, the route of the services and utilities running along the TBM tunnels needed to by-pass the door frames or feed into the CPs per another method.

After the 3D models of the TBM and CPs had been created, they were shared with the tunnel M&E team and drainage team for design coordination. The other designers could use them to create their own 3D models for their design elements such as cable ducts, fire mains, drainage channels etc. All 3D models were then shared and referenced in a federated design coordination model (Figure 5). Thus, designers could always see the latest design developments from other design disciplines and conduct necessary design coordination. The responsibility was also clear as each discipline was responsible to keep its own model correct and updated. When design coordination required recesses and openings to be modelled on the tunnel structures, solid blocks were modelled by the proposing party to represent cut-out in the concrete. The tunnel designer would then use the solid blocks to cut their structure models

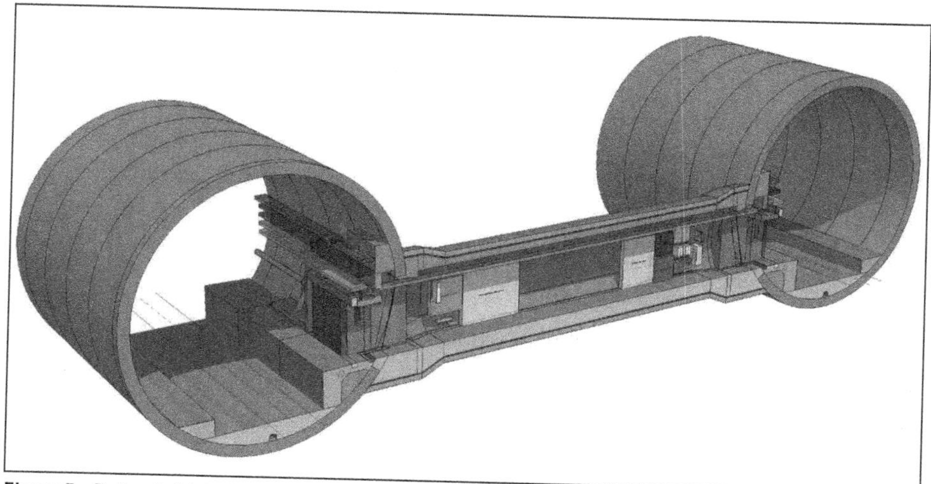

**Figure 5. Federated interface model for CP8**

upon the agreement between the designers. To ensure the coordination was carried out smoothly, it was important that a RACI (Responsible, Accountable, Consulted, Informed) chart was agreed upon between different parties on their modelling scopes.

## Use of 3D Model for Construction Method Assessment

The designers used BIM to help the contractor select suitable construction methods. Benefitting from the ECI (Early Contractor Involvement) approach, the contractor had been involved in the design development from an early stage. As an integrated design team, 3D design models were available to the contractors to help them understand the design approach and assess the feasibility of different construction methods.

In the Scheme design stage, the proposed pre-support design for CP excavation was to install pipe arch canopy tubes (as shown in Figure 6) from the TBM tunnel and provide crown support throughout the excavation of the CP. After adopting the door frame support approach, a 3D federated model of the TBM tunnel, CP and door frame was jointly reviewed by the designer and the contractor. It was identified that the pipe arch canopy tubes approach was no longer desirable due to the heavily reinforced RC frame around the opening. An alternative design of grouted pre-support spiles was therefor considered. Although 3D modelling of temporary works was not required by the Client, a full 3D model of the pre-support spiles system was conducted for space proofing and feasibility study purpose (Figure 7). The proposal was reviewed and subsequently accepted by the contractor, benefiting from 3D model.

## Use of 3D Model to Produce Design Drawings

3D models were used to prepare the design drawings. Sections and views on the drawings were generated by creating dynamic views (D-views) at the specific locations

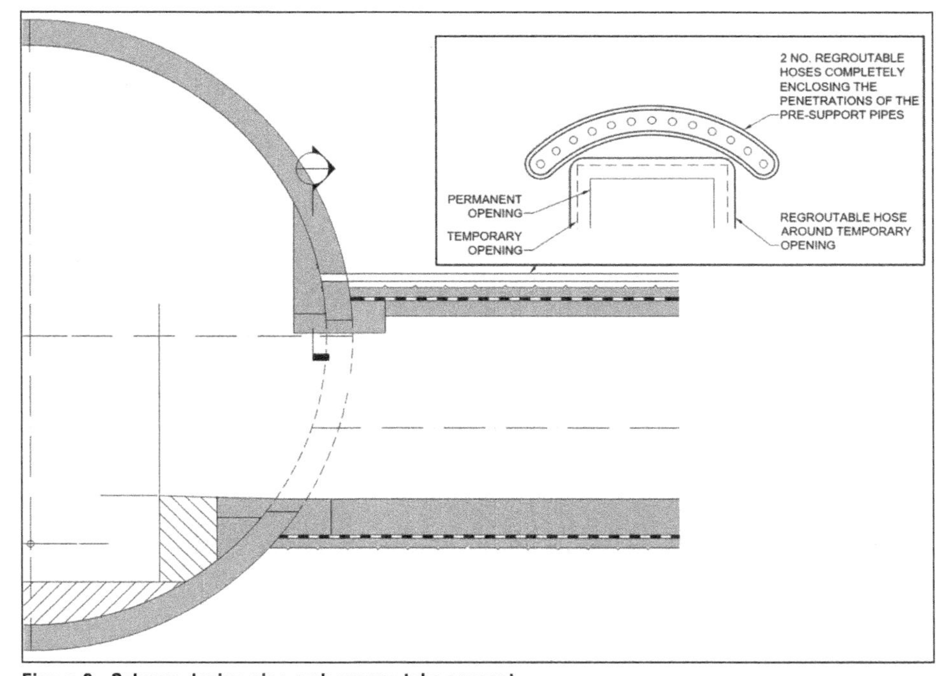


**Figure 6. Scheme design pipe arch canopy tube concept**

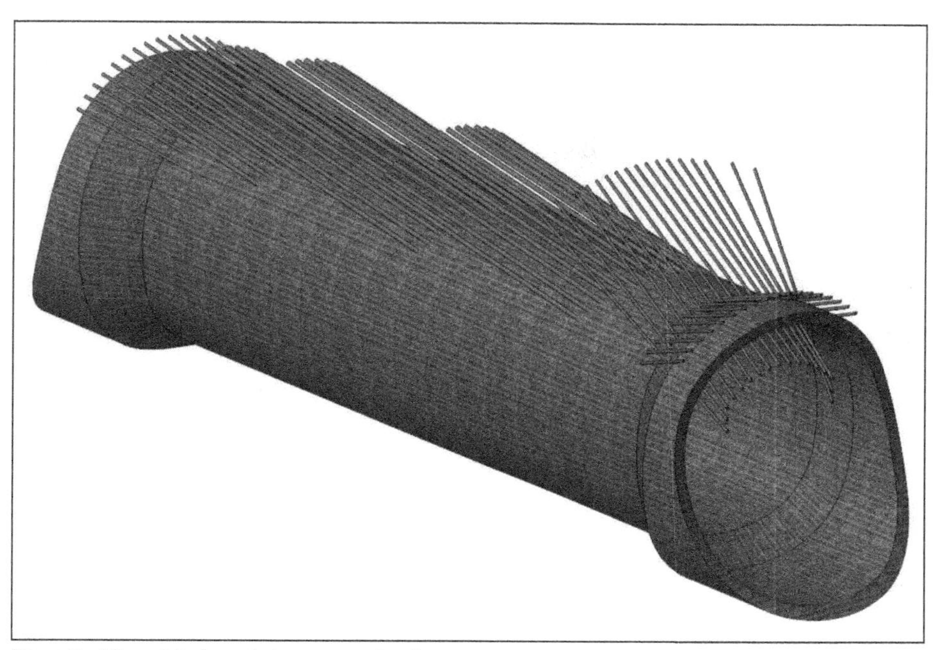

**Figure 7. 3D model of grouted pre-support spiles**

on the 3D models. These D-views were saved in a separate D-view model and referenced to the drawing sheet with annotations and notes required to complete the drawings. Figure 8 shows the D-views set up in a CP 3D model with the highlighted D-view displayed in the small window.

Sometimes line types and hatchings needed to be manually adjusted at the D-views to improve the clarity of the drawings. An element from D-views could be selectively shown to fit the drawing purpose. For example, only the primary lining was shown in the excavation sequence drawings while the secondary lining was turned off from the D-view as it was not relevant for an excavation sequence drawing.

Once the drawings were set up with the correct D-views, it was much more efficient to update the drawings when the 3D model was updated as the design developed. CAD technicians only needed to refresh the D-view model to pick up D-views from the most up to date 3D model, and the sections and views in the drawing were updated accordingly. While this was very helpful in producing design drawings, the number of D-views increased as more drawings were produced. This led to issues with locating the correct D-views. Draftsmen spent too much time to find the correct D-view for a drawing which subsequently affected the efficiency of drawing production. Had a coherent naming system for the D-views been used from beginning, this issue could have been avoided. Superseded or unused D-views should have been deleted or renamed accordingly to avoid the development of too many D-views. A good D-view name should contain detailed location information and which drawing it was used to help the draftsmen select the correct the D-view during drawing production.

Although 3D models should have enough details so that the correct details were reflected on the 2D drawings, it was important that a balance between insufficient details and too many details was maintained during the production of 3D models.

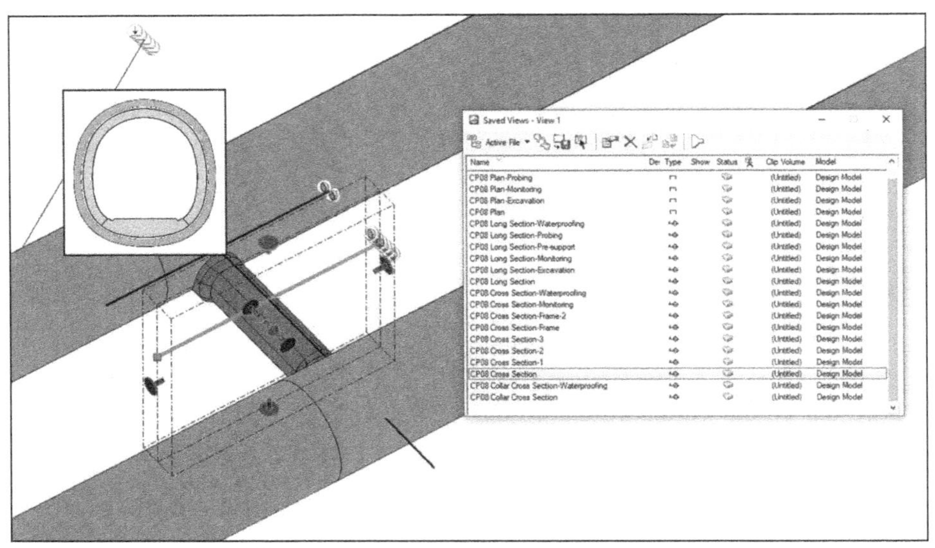


**Figure 8. Dynamic view of CP cross section**

Models with insufficient details would not provide the contractor enough information to understand the designer's intent. Models with too many details were slow to model and could be confusing and distracting from the focus subject

One of the design requirements from the drainage team was to provide 1:100 transverse slopes across the invert slab of shaft adits to drain any surface water into the side channels. Including this feature in the model would not add any additional value to the model, instead it would increase the complexity of the 3D model. the end it was decided that invert slab was modelled as flat, not including the slope feature. A note was added to the cross section drawing to highlight the 1:100 fall design requirement rather than include this in the 3D Model.

### Use of 3D model to Store As-Built Information

As discussed in the introduction, the Chiltern tunnel is currently in construction. To this end, BIM implementation has progressed from the BIM setup through modelling, and the existing models are currently being updated with as-built information. The current stage of the project requires the CAD/BIM team to support the construction team. The Site survey team is providing precise coordinates of CP positions, and CPs are remodeled using the new coordinates. Laser scanning and cloud modeling is being used to assess the accuracy of constructed structures, as well as forming the basis for updating the as-built models. This contribution will provide a description and lessons learned of the initial stages of the BIM implementation through to the current as-built model updating phase.

## CONCLUSION AND LESSONS LEARNED

The use of full-scale BIM for a large scale and complex multidisciplinary tunneling project has been pioneered on the HS2 Chiltern Tunnels Project in the UK. While there were many challenges throughout the different design stages which required collaboration between the developer and the designer, many lessons were learned, and the BIM system evolved and improved as the project progressed. Nevertheless, the BIM implementation for HS2 has proven that a mature BIM system can increase design production efficiency, reduce construction risks and project cost, and add value to a large tunneling project.

In addition to the topics discussed in this contribution above in detail, the following are a few additional key lessons learned during the BIM implementation in the HS2 C1 project:

- Contractor involvement at early stage is important to allow design to be suited to used construction method and to avoid costly or unnecessary changes on later part of the project.
- The established workflow needs to be taken seriously and maintained during the project.
- The CDE needs to be used by all involved parties (e.g., design files should not be allowed to be sent via email)
- MIDP and TIDP are the only source of truth and should be used to create placeholders on CDE.
- A neutral platform/file format (e.g., IFC) should be used as an output to allow all involved on the project to have access to design regardless of software used.
- 3D Modeling needs to be overviewed by design engineer to make sure the design is represented correctly. This requires Engineering staff to learn new review tools.
- Metadata input into models needs to be correctly understood and input with overview by design engineer.
- When a project consist large number of 3D models, which are used also for drawing production, it is important to setup section cut systems from the 3D model in a way, it allows the efficient introduction of changes.
- All 3D models need to be developed to the same coordinate system to allow for smooth coordination.

## REFERENCES

Innovative UK. 2017. Creating a Digital Built Britain: what you need to know. www.gov.uk/guidance/creating-a-digital-built-britain-what-you-need-to-know.

International Tunneling Association Working Group 22: Information Modelling in Tunneling (ITA WG 22). 2022. ITA Report No. 27: BIM in Tunnelling—Guideline for Bored Tunnels—vol 1. The International Tunnelling and Underground Space Association/Association Internationale des Tunnels et de l'Espace Souterrain (ITA/AITES). https://about.ita-aites.org/.

# Mechanized Tunneling—New Trends in TBM Development

**Werner Burger** ▪ Herrenknecht AG
**Marcus Lübbers** ▪ Herrenknecht AG

## ABSTRACT

TBM development in recent decades has been dominated by increased face pressures, larger diameters and multi-mode functionality paving the way to realize projects deemed impossible before. In addition, the trend towards digitalization does not stop at the tunnel portal. A shortage of skilled personnel and the ever-present wish to increase workplace safety, quality and system performance automation, operator-assist or even autonomous systems have become key development targets. The paper will present the current state of the art in light of these special conditions.

## DRIVING FACTORS FOR TBM DEVELOPMENT

For decades the development targets in mechanized tunneling circled around more difficult ground conditions, higher face pressures, larger diameters, greater tunnel length and finally higher production. Much has been achieved in this respect that changed the world of mechanized tunneling making tunnel projects feasible that were deemed impossible before. This developmental trends will continue, however, in a rapidly changing world the TBM industry cannot ignore other general developmental trends. Whilst increasing workplace safety is and has been a permanent target, sustainability and shortage of personnel shifted into the focus of TBM development. Intelligent data management, artificial intelligence and automation entered the underground world resulting in a new approach to industrialized mechanized underground excavation, enabling further steps towards increased safety, quality and economy.

Different to many other industries that can provide large R&D budgets and build numerous different prototypes for extensive real size testing and improvement programs, in the TBM world the options for real size testing are extremely limited. There is normally no such opportunity like a test tunnel. New developments need to be applied on a real project under real performance and schedule requirements. Such conditions have been present for the classical TBM development in the past as well and could be successfully tackled by a "controlled risk" approach and close cooperation between owner, contractor and TBM supplier.

## AUTOMATION

As in other industries, the tendency for automation has also reached the field of mechanized tunneling. To a large extent the TBM operation can be looked at as an industrial process. Of course, the ground is an important factor in this process, as it is for example on open gripper TBMs where the rock support method has to and can be adapted to the prevailing ground conditions. On shield machines with segmental lining ground support installation however, is largely independent of the existing ground and face conditions.

For the operations within the TBM and gantry area, two tendencies are present:

- Fully automated subsystems
- Operator-assist systems

Besides the ground support installation, the supply of ground support materials and consumables, the discharge of excavated muck and the maintenance of the equipment are frequent ongoing activities.

Applying tunnel conveyors for dry mucking systems or slurry circuits for closed mucking systems provides continuous muck discharge and the required manual operations are limited to belt structure or pipe extension. Similar frequent services extension work is required for all other supplies from the portal like power, water, compressed air and ventilation that are essential for the TBM operation.

## Segment Handling

The supply and installation of segment rings is "the" frequent operation in mechanized tunneling using precast segmental lining. Therefor the approach towards automation of individual steps along the entire supply and installation process has moved into the focus for economical as well as for quality and safety reasons. Especially for long tunnels where thousands of rings need to be installed an investment in fully automated or operator assist systems can become beneficial.

Whereas automation in segment production (precast plant) and storage yards has been introduced for some time, segment handling operations within the TBM moved into the focus in recent years.

Fully automated subsystems can include all operations for segment handling starting with unloading from the transport vehicle up to the point where the erector picks up the segment and even the ring placement itself.

However, when introducing automated handling systems the segments themselves as well as the loading configuration or segment stack on the supply vehicle have to be suitable for automation. Whereas the segments itself are high precision precast elements that may already incorporate means like bar codes or similar to be identified automatically increased requirements on geometrical accuracy of the segment stacks including the dunnage placed between segments and the stack position on the supply vehicle become necessary.

Since no human operator is involved any more in segment handling within the gantry area the previous possibility for a final visual check of the individual segments before being supplied to the erector is no longer possible. The visual check and subsequent correction measures typically includes the detection of damages caused during transport in the tunnel, correct position of the segment seals and other installation or contamination of the segment surface by dirt or snow / ice that could cause malfunction of the vacuum pick up.

In order to avoid disruption of the ring erection by detecting such issues at a late stage by the erector operator during the ring erection automated handling systems should include at some stage of the handling process also automated inspection systems that could be realized by "intelligent" cameras. Also, the integration of an automated cleaning device for the segment intrados is advisable to eliminate the risk of vacuum plate malfunction due to segment surface contamination (Figure 1).

**Figure 1. Automated segment transfer sequence in gantry**

Even a fully automated system still needs to provide the option for safe manual operation including the possibility to reverse the transport direction and bring segments back out from their installation position in the tailskin to the transport vehicle and finally out of the tunnel.

Several 10 m TBMs incorporating automated segment handling systems have been supplied to jobsites and started operation in 2022 as a first application.

## Ringbuild Assistance

For the ring erection process sensor technologies provide the possibility to display additional information on the erector operators control panel for the ring erection itself. Such information can include ring build clearance or the relative position of the currently to be placed segment to it neighbor segments.

Measuring and providing such geometrical information to the erector operator electronically can in consequence eliminate the need of an assistant ring builder climbing up and down the erector platforms and performing manual measurements. Thus this can eliminate safety risks and reduce the crew size of the ring build team (Figure 2).

## Automatic Ring Erection

The next step beyond providing such information to the erector operator is the fully automated ring erection aiming towards a configuration where no personnel need to be present in the ringbuild area.

Ring erection is by nature an operation including the handling of heavy precast concrete elements in an area that can, especially in smaller diameter TBMs, be considered as a confined workspace. Difficult access and limited visibility are typical challenges to be overcome in this area, therefore safety aspects are the major driving factors for automated ring erection besides quality and economy.

The main component of the ring build system to be automated is the erector, a specialized manipulator for handling the heavy concrete segments.

Like for all automation processes, it is essential for the system to know the exact position of all components involved. Therefore, all moving parts of the erector, the thrust

Figure 2. Erector suction plate equipped with sensor devices

cylinders and the segment feeder need to be equipped with position measuring systems. The recorded sensor data can be also available for internal documentation and "Building Information Modelling" systems as required.

For an automated ring build process, not only the exact position of all erector functions and other moving parts is required. It is also crucial to detect the exact segment position and orientation on the segment feeder for picking it up. Where optical sensors are used, unrestricted view of the optical sensors is essential. For this reason, for example, attention must be paid to unnecessary exposure or pollution of the sensors during cleaning and maintenance work.

First automatic ring erection systems for a 3 m and a 8 m TBM have been developed and tested with real size test arrangements at the Herrenknecht facility in Schwanau. A first field prototype is now installed on a 3 m EPB TBM being operated under actual site conditions (Figure 3).

Whenever automated systems are installed physical barriers with interlocked access doors need to be installed preventing personnel from entering the area where automated handling operations are performed. As consequence such locations within the gantry or ring erection area are only man accessible during manual operation mode or for maintenance. The currently ongoing revision of the European safety standard for tunnel boring machines EN16191 will address specific requirements for automated systems.

It is also obvious that automated systems require a larger amount of sensor technology and more computerized control systems. Especially the harsh environment in a tunnel heading needs to be addressed for component selection and design as well as an appropriate qualification of the on-site maintenance personnel.

## DATA MANAGEMENT

Many processes take place simultaneously during the operation of a tunnel boring machine. Each individual process is monitored by numerous sensors and controlled

**Figure 3. Vacuum suction plate of a field prototype for automatic ring erection on a 3 m EPB**

by various actuators. All process variables and control actions together with parameters and fault messages are recorded in a data management system over the entire duration of the project. With an average of 2000 sensors recorded every second and a project duration of 2 years, significantly more than 100 billion data points must be recorded and saved. If several or more complex machines are used and with longer project times, this number can quickly double. This huge amount of data from the machine processes alone is enriched by data from the periphery, geotechnical engineering and other construction-related processes.

Furthermore, data management must be able to cope with both structured data and unstructured data. Structured data comprise time series, metadata for sensors or aggregated values for individual processes, whereas images, reports or geotechnical data are described as unstructured data. A suitable data management system must be able to process and save the different data types and formats. In addition, the data needs to be interconnected and relationships and references established.

Data recording systems of lower performance have been around in the industry for almost two decades but only in recent years intelligent real time data processing and availability on-site and off-site have become available instruments for operator-assist or semi- and fully automated systems and processes.

In summary such processed data present a huge value related to operator-assist systems and performance optimization for the ongoing operation itself, for remote control and supervision as well as for documentation and creating additional knowledge for future operations and developments.

A second important aspect of operational data processing is related to TBM maintenance. A maintenance software can monitor the data and check the processes on the TBM. Possible malfunctions are recognized at an early stage and the necessary steps can be initiated before an unplanned event occurs. Thanks to automated evaluations of, for example, the operating hours of the components, necessary maintenance is recognized at an early stage and reported to the user. This creates planning reliability for the user, as the necessary components or personnel can be ordered or planned in time and the entire maintenance history of the equipment can be documented automatically.

A specific use case for a predictive service is, for example, the condition monitoring and cutting tool change forecast for a cutterhead. This is about the condition monitoring of the cutting wheel or the cutting tools in connection with the real existing ground conditions. The aim is to provide an early prognosis for necessary tool changes based on the current and past tunnelling conditions and to adapt these dynamically with the mined distance.

TBM rebuilds or remanufacturing of TBM subassemblies and components can have a major effect on reducing the carbon footprint of the equipment. The availability of the full operational data and maintenance history of an entire TBM or its subassemblies is key information to evaluate the remaining in-service life and / or the required steps of remanufacturing to establish a high quality and reliable product for the next project.

## CONCLUSION

Automation and intelligent data management have entered mechanized tunneling and will further change the underground work environment and support the tendency towards more sustainable solutions. First automated systems are out in the field. There will be lessons learned resulting in further improvements and developments towards a safe and more efficient operation and maybe a more appealing place to work.

## REFERENCES

T. Weiser, J. Tröndle, 2022. Digital construction—data management in mechanized tunneling, World Tunnel Congress 2022.

# Mobile Solid-State Lidar for Construction Quality Assurance

**Steve Miller** ▪ Schnabel Engineering, LLC
**Travis A. Shoemaker** ▪ Schnabel Engineering, LLC
**Adam Saylor** ▪ Schnabel Engineering, LLC

## ABSTRACT

Lidar scanning has become an increasingly popular tool for documenting underground excavations; however, high costs of traditional mechanical electrical lidar hardware, software, and training have limited the frequency at which lidar scans can be performed. Additionally, traditional scanning can be time consuming because multiple scan positions are necessary. With Apple's incorporation of low-cost mobile solid-state lidar on select smartphones and tablets, lidar is becoming more tractable for everyday construction documentation similar to construction photo collection with smartphones. Here, we describe benefits of mobile solid-state lidar and describe several practical applications for use in underground construction, such as documenting the working face, over-blast, and geologic features of tunnels.

## INTRODUCTION

Field exploration, construction of tunneling projects, and tunnel inspections require engineers and geologists to collect data representing the site and the completed work. The purpose of this data is to inform the design of underground structures or ensure that the constructed work meets the design intent or contractual requirements. Although this purpose has not changed throughout the history of tunneling, the tools to collect the data has. Over the last several decades, the engineer's toolbox has shifted from primarily analog or manually operated data collection tools, such as theodolites for measuring distances and analog cameras for collecting photos, to digital measurement devices. Lidar, or light detection and ranging, has been an increasingly popular digital measurement tool because lidar can collect accurate and detailed 3D models in less time than traditional surveying techniques. Lidar is also particularly attractive for tunneling applications because it does not require consistent or strong lighting of the scanning area. To date, one of the largest drawbacks of lidar has been the high cost of hardware, software, and training; however, new technologies are breaking down this barrier. Apple begun incorporating a solid-state lidar sensor on select consumer-grade devices starting with the release of the iPhone 12 Pro and the iPad Pro (2020). Although these sensors have significant limitations, the introduction of this technology impacts the tunneling industry in two ways: (1) Mobile solid-state lidar allows for immediate 3D model creation by any tunneling engineer at low costs without special training. (2) This technological advance follows a trend of additional digital sensors becoming available on all-in-one devices, such as smartphones and tablets. This paper will describe the mobile solid-state lidar available on current Apple devices, summarize the currently available literature on its accuracy, precision, and resolution, and provide several examples of its application to tunneling.

## MOTIVATION

Over the last couple of decades, lidar has become nearly common place on construction sites. Despite these applications and growing popularity, cost, and access to lidar remain relatively large barriers to everyday use of lidar in tunneling applications. The costs of lidar have historically been very high in comparison to traditional surveying and measurement equipment. For example, surveying total stations typically cost on the order of a few thousand dollars, whereas tripod-mounted survey grade lidar typically costs on the order of tens or hundreds of thousands of dollars. Because these lidar units are expensive in comparison to tape measures, levels, and total stations, lidar scanning has traditionally been a specialty service. Therefore, lidar scanning on construction sites is typically limited to only most critical data collection. For example, lidar can be used to measure rock tunnel excavations immediately before and after shotcrete applications to calculate shotcrete thicknesses and volumes. These operations require careful coordination to ensure that lidar scanning crews are available at the correct time so that construction operations aren't delayed. Because access to traditional survey-grade lidar is limited and the costs are high, there is a motivation to consider alternative lidar tools. Solid-state mobile lidar has become a relevant topic to engineering applications because of its ease of use, portability, and low cost.

The first lidar scanner available on consumer-grade devices came to market with the introduction of the iPad Pro and iPhone 12 Pro in fall of 2020. These devices begin at a cost of approximately $800 USD. Prior to this, entry level commercial handheld lidar started at around $50,000 USD. In addition to over a 700% price entry level lidar cost reduction, mobile device lidar becomes even more practical when it is no longer a single purpose specialty device but is embedded in a multi-purpose device that many engineers carry with them regardless of whether they intend to perform lidar scans. As an example, the authors identified that approximately 20% of Schnabel Engineering employees as of fall of 2022 use an iPhone that is already equipped with the lidar sensor. This number is expected to grow as newer devices are manufactured and become more mainstream. For these users, there is essentially no additional hardware cost to perform mobile lidar scans.

This evolution is not dissimilar from the transition from film, to digital, and then mobile devices for photography. Using film photography, an engineer in the field had to be very deliberate about when they would take photos and how many they would take. Digital photography removed the development process and consumable film, but also required a single purpose device that was typically not network connected and still required the user to physically return to an office with the photos. Smartphones revolutionized this process. By default, engineers in the field started carrying a network connected mobile device, allowing them to take a virtually unlimited number of photos, and instantly transmit them anywhere allowing for real-time feedback. Solid state lidar embedded in mobile devices allows for the same ease of use and practicality. Without specialty equipment or advance planning, an engineer can take a 3D scan and transmit the data back to the office in just a few minutes.

## LITERATURE REVIEW

As a general class of sensors, lidar encompasses several methods to measure the distances between a sensor and a feature by aiming or directing emitted light pulses at the object and measuring the time required for the light to return to the sensor. This technique is often referred to as Time-of-Flight (ToF) and the measured time can be converted to a distance using the speed of light. Because the specific information about the Apple's lidar sensor is proprietary, information from the manufacturer is limited. Additionally, Apple's marketing materials identify the primary intents of the sensor as improving photo focus and augmented reality (AR) capabilities, and the marketing materials do not provide measures of accuracy. Several researchers claim that Apple's lidar sensor is likely solid-state (Murtiyoso et al. 2021, Spreafico et al. 2021).

Solid-state lidar can be used in harsh environments because it does not have moving mechanical parts (Raj et al., 2020). Additionally, solid-state lidar can be produced on a single chip, and therefore, it can be produced more cost effectively at scale.

To the authors' knowledge, there have not been any papers published describing the application of mobile device equipped solid-state lidar in tunneling applications. Therefore, this section reviews literature relating to the device's accuracy, precision, and data collection speeds as employed on applications different than tunneling.

Vogt et al. (2021) studied the ability of the lidar to scan small components and found that lidar could not reasonably scan a Lego brick with a volume of approximately 6 $cm^3$. Shoemaker et al. (2022) performed a proof-of-concept study using the lidar as an alternative method for performing soil density testing. The researchers excavated holes in compacted file, weighed the excavated soil, scanned the hole with mobile lidar, and calculated the hole's volume and the soil's density. This density was compared to nuclear density gauge testing. The researchers concluded that the lidar method compared well against nuclear gauge testing when the excavated hole size was greater than approximately 0.15-$ft^3$ (4,200 $cm^3$), but the lidar accuracy was inadequate for smaller samples.

Luetzenburg et al. (2021) evaluated the lidar at a larger scale by scanning a coastal cliff and beach with approximate length, width, and heigh of 130 m, 15 m, and 10 m, respectively. The scans reportedly took approximately 15 minutes to collect per scan. The Apple lidar has a range of approximately 5 m, so the researchers used a selfie stick to elevate the device to obtain coverage of the entire area. The researchers compared scans collected with 3D Scanner App, which uses the iPhone's lidar sensor, and EveryPoint, which uses the iPhone's lidar and camera photos, to Structure-from-Motion Multi-View Stereo (SfM MVS) reference models. The multi-scale model-to-model cloud comparison (M3C2) between the lidar point clouds and the SfM MVS reference were performed using the Cloud Compare software. The average difference between these two datasets was approximately 0.1 m with a standard deviation and RMS of 0.68 m and 0.69 m, respectively. Luetzenburg et al. repeated scans of 10-m long by 15-m wide by 10-m tall on a portion of a coastal cliff and beach six times using the 3D scanner app. The average M3C2 between one reference scan and the other five scans was 0.02 m. The researchers also compared smaller objects within the scans and found that the percent precision increased with increasing size of the object. In particular, objects smaller than about 10 cm had precisions ranging from 40% to 100%, whereas objects larger than 10 cm had precision of at least 85%.

Tavani et al. (2022) created 14 independent models of a folding rule with the ruler extended between 20 cm and 150 cm using the 3D Scanner App with an iPad 12 Pro. In each model, the model length of ruler was compared to the actual ruler length. All measurements had an error of 1 cm or less, except one measurement with an error of 3 cm. These researchers also evaluated the amount of scanning drift by scanning a 30-m wide vertical wall. The scan drifted perpendicular to the wall from the end of the wall to the middle of the wall by approximately 30 cm, and the wall scan was curved in plan view. Tavani et al. 2022 also tested Pix4DCatch, which is a structure from motion (SfM) photogrammetry application and cloud processing platform that utilizes the iPad lidar during data collection. The researchers compared the Pix4DCatch model of an outcrop against six GPS survey points and found that the photogrammetry model was miscalled by about 5%. This result brings in to question the value of performing SfM photogrammetry when a lidar sensor is available.

Labedz et al. (2022) used reference metal spheres located approximately 270 mm apart to measure the accuracy of the lidar scanner. The center-to-center distance between the spheres were calculated for 18 successful scans and the average error and standard deviation were 1.41 mm and 4.06 mm. These researchers also evaluated the accuracy of the lidar device for four applications: (1) a 99.2 mm by 40.8 mm by 90 mm plastic figurine, (2) a bronze statute 678 mm by 458 mm by 754 mm, (3) a room of dimensions 690 cm by 525 cm by 275 cm in an academic building, and (4) a portion of a building facade with dimensions 1,740 cm by 123 cm by 1203 cm. Scans with the iPad Pro collected with the Scaniverse App was compared to photogrammetry models created using a DSLM camera and Agisoft Metashape software. Both data techniques were verified using scans by a Konica-Minolta scanner, Creaform Academia 3D structural slight scanner, or a laser scan. CloudCompare was used to calculate the RMS distance between the photogrammetry models and the iPad lidar scans and the reference models. The two methods produced similar accuracies at room and building façade scales (experiments 3 and 4), but the iPad produced significantly less accurate results for the figurine and statue (experiments 1 and 2) at smaller scales. Another key outcome presented by Labedz et al. 2022 is that iPad lidar scans were repeated several times because portions of the scanned objects were often missed in the scanning operations. Lastly, Lapedz et al. (2022) pointed out that the iPad scans had resolutions several order of magnitudes coarser than the photogrammetry models.

Diaz-Vilarino et al. (2022) compared iPad lidar scans of rooms to a BIM model created using data from a Faro Focus x330 laser scanner. The researchers found that the scans compared reasonably well for one room but were significantly different for the second room of two-room scans. For example, the iPad measured the interior wall thickness between two rooms as about 40 cm, when the actual wall thickness is about 10 cm. This highlights the significant scanning drift associated with moving the scanner from one room to the next. Diaz-Vilarino et al. (2022) also calculated the local precision by comparing the RMSE of point distances to fitted planes on planar surfaces. For indoors scans, the RMSE on 6 planar walls was 0.53 cm for the iPad and 0.28 cm for the Faro Focus. For outdoor scans, the RMSE on 6 planar segments was 0.56 cm for the iPad and 0.17 cm for the Faro Focus.

Katrinia et al. (2022) performed iPad lidar scans of caves for archaeologic documentation. The researchers compared three scanning applications: 3D Scanner App, EveryPoint, and SiteScape. The researchers recommended the SiteScape application because it produced scan density nearly ten times that of the other two apps and had less systematic error. The researchers concluded that the iPad lidar is suitable for documenting caves for archaeology purposes.

In summary, the literature shows that the accuracy and precision of the mobile solid-state lidar device depends on the particular application. For small models, the lidar has an accuracy on the order of 1 cm, which is also the reported accuracy of the 3D Scanner application. However, at larger scales, the accuracy depends on the scanning route. When scanning is performed between different corridors, such as moving between rooms, significant drift in the scanning data is observed. The precision of the lidar scanning for small scans on indoor and outdoor walls was found to be approximately 0.5 cm, which will not limit the scanners use for most tunneling applications. The resolution of the mobile solid-state lidar scans is much lower than traditional mechanical-electrical lidar scans and SfM photogrammetry models. However, the mobile lidar scans have been shown to have adequate resolution for many applications, such as archeological documentation of cave walls and architectural documentation of building facades. The mobile solid-state lidar scanning is rapid in comparison to traditional lidar scanning because the mobile lidar scans continuously and is not mounted at several tripod locations. The mobile lidar system also processes in a matter of minutes on several applications, such as 3D Scanner and Polycam, whereas SfM photogrammetry applications, such as Pix4D Catch, require photos to be uploaded for cloud computing, proceeded on the cloud for several tens of minutes, and then downloaded back to the mobile device or laptop. Lastly, none of the iOS lidar scanning applications identified in the literature automatically register the scans to regional or global coordinate systems. Geolocation of the mobile lidar scans can be performed using reorientation and translation in secondary post-processing software using surveyed tie points visible in the scans.

## FIELD APPLICATIONS FOR QUALATY ASSURANCE

Construction documentation and tunnel inspections were considered in the present study as potential tunneling related mobile solid-state use cases A short description of both use case is described and a summary of lessons learned from all use cases is provided in the next section. A lidar-equipped iPad Pro was used for both applications with the Polycam scanning application.

### Construction Documentation

The mobile solid-state lidar was tested during the construction of a 300-ft long, 10-ft wide by 10-ft tall horse-shoe shaped drill and blast tunnel. The construction documentation process followed typical construction observation practices but was also supplemented with the lidar scans. The following aspects of the construction process were included in the field testing of the mobile lidar:

- Rock bolt measurement
- Tunnel overbreak scan
- Working face documentation
- Tunnel clearance dimensions

#### *Rock Bolt Measurement*

The mobile solid-state lidar allowed the on-site tunnel inspection personnel to make scans quickly and measure distances between rock bolts to ensure compliance with the initial support in the tunnels. Figure 1 shows the scan of the tunnel and rock bolts and the distance measured with the app. The tunnel inspector was able to quickly use the scan to measure the distance between multiple rows of rock bolts for compliance with the design, which included a 4-ft bolt spacing. The scan was taken from the ground level and required less than a minute to scan, with little to no interruption to the construction of the tunnel. The scan can then be fully evaluated off site to ensure the bolt spacings are compliant with the design requirements.

#### *Tunnel Overbreak Scan*

Scanning was also used to aid in the calculation of the fill required between steel sets and the tunnel wall in an area that experienced overbreak. The scan and measurements were performed in a few minutes (Figure 2). This calculation provided an estimated volume required for backfill. This measurement is important for estimating the required amount of material and documenting quantities for contractual payment. This type of measurement can be used for any type of excavation measurement such as test pits, trenches, and voids beneath slabs, and can quickly capture the data for use by both the contractor and the engineer.

Figure 1. Rock bolt scan

### *Tunnel Working Face Documentation*

Another use case was for construction documentation was mapping of the working tunnel face. A scan was performed on the tunnel face after the tunnel was supported and the face has been cleaned off (Figure 3).

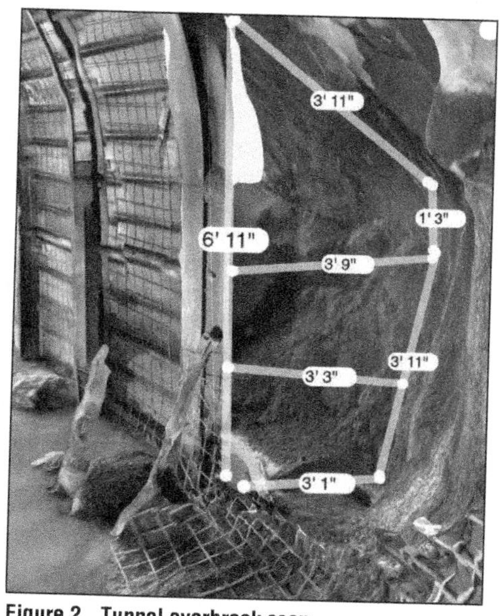


Figure 2. Tunnel overbreak scan

This lidar tunnel scan does not replace traditional methods of face mapping in projects, but it does provide supplemental information to recently exposed rock in tunnels. In the scan, measurements can be taken to show that the tunnel was excavated to the required 11-feet width and height in this location. The recently exposed area can also be mapped in 3D from the safety of where tunnel support has been installed. The results can be exported to a 3D PDF model that can be shared easily with stakeholders to allow them to observe information unclear in traditional photos. This information can be used to supplement traditional face mapping, which is performed by hand with visual observation. Because mobile lidar scanning is rapid, it can be done by any engineering staff with a mobile device with lidar and the Polycam app.

### *Tunnel Clearance Determination*

The project used in testing had three vertical 20-inch shafts that were drilled down from the surface to intersect the tunnel. The shafts were designed for vertical turbine pumps with specific clearance distances from the tunnel walls that were required for the pump operation. Using the mobile lidar, the position of each of the shafts was able

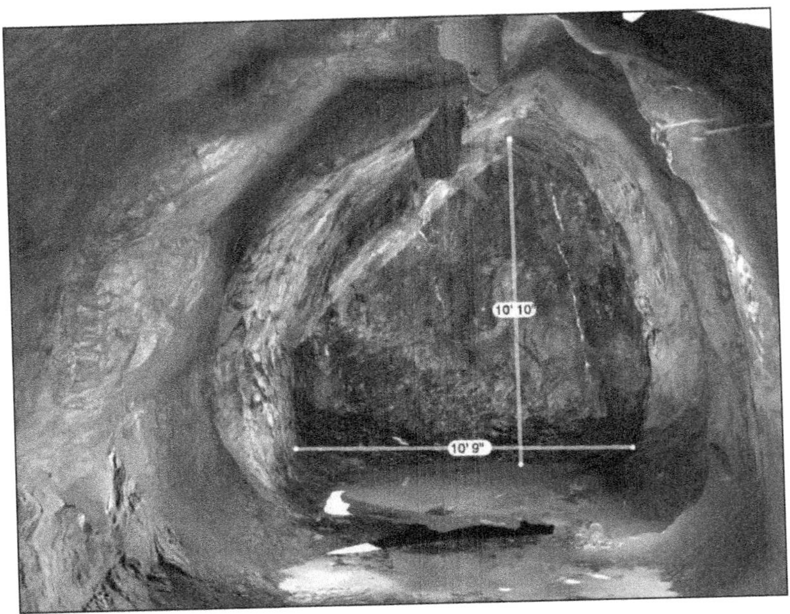


Figure 3. Tunnel working face scan

to be viewed within 5 minutes to determine if there were any potential issues. The scan was viewed in a Polycam's "blueprint" mode, which showed the tunnel locations in plan view (Figure 4). The scan was able to show the locations of the three shafts in proximity to the tunnel walls. The scan helped the engineering staff confirm that clearance issues were unlikely, and the tunnel contractor continued with the tunnel liner installation. The clearances were also confirmed using a traditional survey of the tunnel.

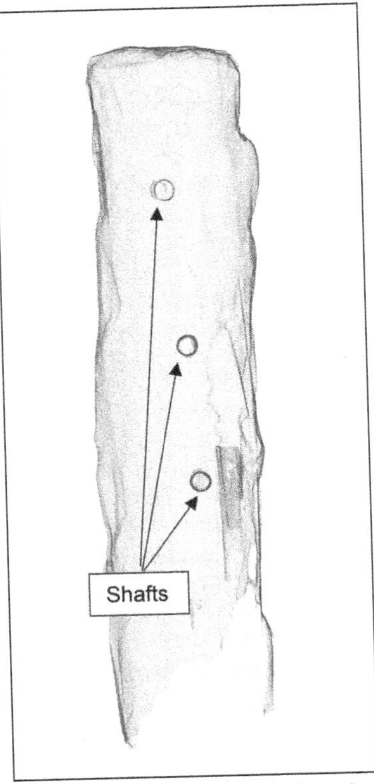


Figure 4. Blueprint scan of the tunnel

## Tunnel Inspections

Mobile solid-state lidar scanning was used during several abandoned mine tunnel inspections. The underground mines were undergoing remediation and were inspected and mapped as part of the remediation process. Mine maps are typically available to the inspectors. However, mine maps can have inconsistencies, and they may not have the latest geometric information. During the inspection process of one mine, several areas of the mine were scanned. Scans of the entryway and an ore chute are presented in Figure 5 and Figure 6, respectively. for the scan provided dimensions of the openings as well as layouts of the tunnels, which were compared to the mine drawings without the need to perform a complete

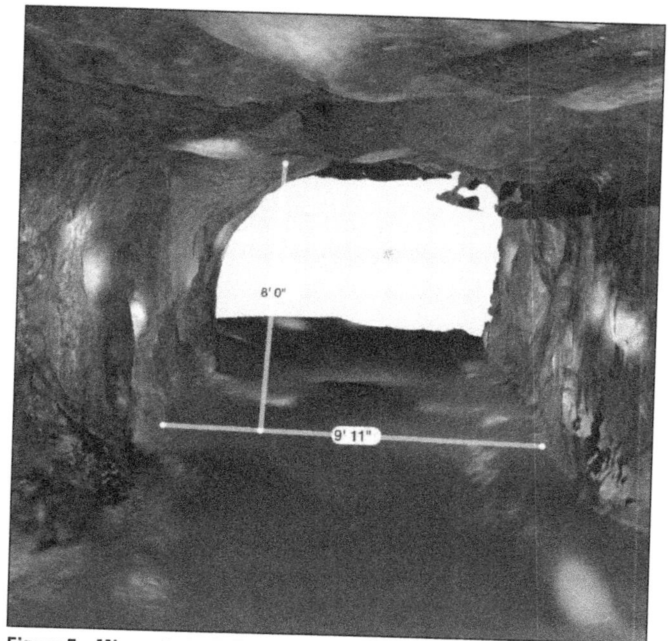


Figure 5. Mine entryway scan

Figure 6. Mine ore chute scan

traditional survey. These scans were collected quickly, and the data was analyzed at a later time, which reduced the time required to be in the mine. The scans, when combined with photos, notes, and sketches, help to provide information that can be used to validate existing data on the tunnels, which is essential to the remediation process.

## FIELD APPLICATION LESSONS LEARNED

The field applications informed the authors of several lessons for performing mobile solid-state lidar scans in tunnels. These lessons about the limitations of the mobile lidar are divided into the following categories: scanning range, scan sizes, and scan level of detail.

### Scanning Range

The mobile lidar sensor has a limited range of approximately 5 meters or 16.4 feet. This distance is substantially lower than traditional lidar scanners and limits the size of structures that can be scanned. However, the range of the lidar sensor does not limit longer scans, which can be collect by moving the scanner. Nonetheless, additional time is necessary to move the scanner to capture all aspects of the scanned features. One key limitation is that tunnel roofs in tunnels taller than 20 to 22-ft cannot be scanned by a person holding an iPad in their hands. It may be feasible to use a selfie stick or other extension in a tunnel to scan the roof of taller tunnels; however, this was not tested in the present study.

### Scan Sizes

Scans of large structures are feasible, but the scans can suffer from drift associated with the collection. Drift in lidar collection is common with traditional methods, although newer tradition lidar has ways to correct the drift both in the field or during post processing of the data. The mobile lidar scanning software available to the authors does not currently correct for the drift in scans, and the drift is often more pronounced with a mobile lidar scanner than traditional lidar scanning methods. During the field testing, a 300-ft long straight tunnel was scanned with mobile lidar. The collected lidar data incorrectly showed a curve in the tunnel alignment (Figure 7). There is a horizontal lateral drift of 15 feet along the alignment of the tunnel scan, which represents approximately a 1-foot drift for every 20-ft scan. Drift was not observed when taking longer or larger scans during all of the field testing. The total drift that is associated with mobile lidar scanning has not been fully evaluated as part of this testing and should be considered as part of future evaluations. While the overall tunnel scan may be unusable for global mapping of the tunnel alignment, the scans can be

Figure 7. Tunnel scan with drift

Figure 8. Steel liner tunnel scan

useful for developing localized analysis and documentation. For example, cross sections still represent an approximation of the tunnel outline despite the drift. Additionally, the scan can be used for confirming the number of rock bolts installed in the tunnel.

### Scanning Details

The mobile solid-state lidar typically did not capture high detail of items that are less than approximately 4 inches. This is consistent with information found in the literature review on limitations associated with small object sizes. The color and texture of small items was captured as part of the scanning process, and the texture can be seen in textured meshed of the lidar scans. The mobile lidar scanning tends to generalize these features when models are produced, because the details cannot be captured. A steel liner plate train tunnel was scanned (Figure 8) and the details of the liner places were not able to be captured in the scan. The flanges and ribs of the plates extend approximately 4 inches from the outside plate. The result is that the overall shape of the tunnel is captured, but the details of the liner plate flanges and ribs are generally not captured in high enough detail to provide accurate measurements. These scans can, however, be used for basic observations and limited analysis.

## SUMMARY AND CONCLUSIONS

Field testing showed that there are currently practical applications for use of mobile solid-state lidar technology in heavy civil and underground construction. It is important to understand that this technology currently cannot replace traditional practices in underground construction or non-mobile lidar scanning. However mobile lidar can supplement field work by efficiently providing high-level scans of a project site. Mobile lidar should not be used for precise measurements but is very useful when rapid approximations are necessary. Although the lidar accuracy is on the order of inches for small structures with flat and/or linear features, the lidar struggles to accurate record intricate structures and scans taken over a large area. Scans over large areas can suffer from drift, which current mobile software cannot correct.

If a picture says 1,000 words, a lidar scan can tell the story of 1,000 pictures. These mobile lidar scans are quick and cheap for the information they provide. As mobile solid-state lidar technology continues to improve and is added to additional mobile devices, it will become easier to make use of this technology for heavy civil and underground projects. As cameras on phones advanced, their adoption as the go to digital solution for photos became the standard. Scanning with mobile devices will follow a similar trend and will start to become mainstream as the applications for scans become more apparent.

## RECOMMENDATIONS FOR FUTURE WORK

The currently available mobile solid-state lidar hardware has limited accuracy and range compared to traditional lidar scanners, and the current mobile device software is limited in overall functionality. As hardware and software improve, additional field uses will be tractable. Software can be designed to automatically calculate voids, identify and geotag items for mapping, such as rock bolts or other key project features, and possibly calculate material evaluations such as the Rock Quality Designation (RQD) of rock cores. One immediate recommendation is to evaluate the mobile lidar's ability to capture the details of rock and concrete cores, and Figure 9 provides a scan of an example concrete core. As phone processing power continues to increase, mobile lidar scanners will incorporate more functionality that is typically available in traditional lidar scanners and software, making mobile lidar a more powerful tool for construction use. Items that are currently being explored are the use of geotags that can be used to register mobile scans together that are taken at different time intervals. This can be used to stitch together multiple scans quickly and efficiently into one scan. Geotags should also be considered as potential tie points to tie mobile lidar scans into project or global coordinate systems. Access to geolocated models opens the doors to multi-temporal scan analysis, including change detection and deformation monitoring. Eventually, it may become practical to overlay prior construction scans overtop of reality to compare present and past conditions using augmented or mixed reality.

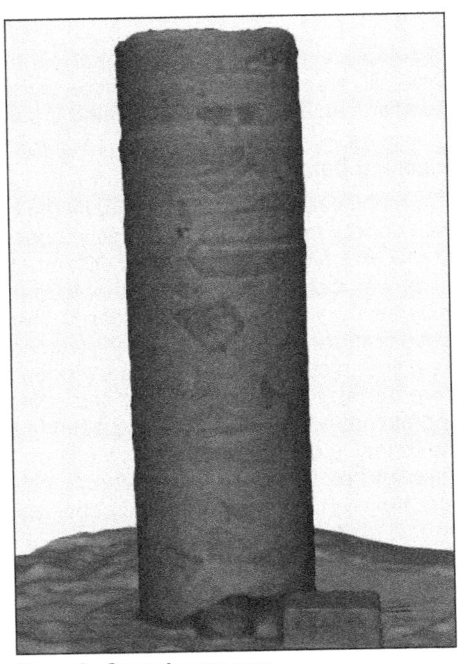

**Figure 9. Concrete core scan**

## REFERENCES

Díaz-Vilariño, L., Tran, H., Frías, E., Balado, J., and Khoshelham, K. 2022. 3D mapping of indoor and outdoor environments using Apple smart devices. *Int. Arch. Photogramm. Remote Sens. Spatial Inf. Sci.* XLIII-B4-2022, XXIV. ISPRS

Congress (2022 edition), 6–11 June 2022, Nice, France. https://doi.org/10.5194/isprs-archives-XLIII-B4-2022-303-2022.

Kartini, G. A. J., Gumilar, I., Abidin, H. Z., and Yondri, L. 2021. The comparison of different lidar acquisition software on iPad Pro M1 2021, *Int. Arch. Photogramm. Remote Sens. Spatial Inf. Sci.*, XLVIII-2/W1-2022, 117–120, https://doi.org/10.5194/isprs-archives-XLVIII-2-W1-2022-117-2022, 2022.

Łabędź, P., Skabek, K., Ozimek, P., Rola, D., Ozimek, A., Ostrowska, K. 2022. Accuracy verification of surface models of architectural objects from the iPad lidar in the context of photogrammetry methods. *Sensors*. 22(21):8504. https://doi.org/10.3390/s22218504.

Luetzenburg, G., Kroon, A., and Bjørk, A. A. 2021. Evaluation of the Apple iPhone 12 Pro LiDAR for an Application in Geosciences. *Sci Rep* 11, 2222. https://doi.org/10.1038/s41598-021-01763-9.

Murtiyoso, A., Grussenmeyer, P., Landes, T., and Macher, H. 2021. First assessments into the use of commercial-grade solid state lidar for low cost heritage documentation. *Int. Arch. Photogramm. Remote Sens. Spatial Inf. Sci.* XLIII-B2-2021. XXIV ISPRS Congress (2021 edition). https://doi.org/10.5194/isprs-archives-XLIII-B2-2021-599-2021.

Raj, T., Hashim, F., Huddin, A., Ibrahim, M., and Hussain, A. 2020. A survey on lidar scanning mechanisms. In *Electronics*, 9(741), doi:10.3390/electronics9050741.

Tavani, S., Billi, A., Corradetti, A., Mercuri, M., Bosman, A., Cuffaro, M., Seers, T., and Carminati, E. 2022. Smartphone assisted fieldwork: Towards the digital transition of geoscience fieldwork using LiDAR-equipped iPhones, *Earth-Science Reviews*, 227(103969). https://doi.org/10.1016/j.earscirev.2022.103969.

Shoemaker, T. A., McGuire M. P., and Penzone S. 2022. Soil density evaluation using solid-state lidar. Geo-Congress 2022 : Advances in Monitoring and Sensing; Embankments, Slopes, and Dams; Pavements; and Geo-Education. https://doi.org/10.1061/9780784484067.009.

Spreafico, A., Chiabrando., F., Teppati Losè, L., and Giulio Tonolo, F. 2021. The iPad pro built-in lidar sensor: 3D rapid mapping tests and quality assessment. *Int. Arch. Photogramm. Remote Sens. Spatial Inf. Sci.* XLIII-B1-2021. XXIV ISPRS Congress (2021 edition). https://doi.org/10.5194/isprs-archives-XLIII-B1-2021-63-2021.

Vogt, M., Rips, A., and Emmelmann, C. 2021. Comparison of iPad Pro®'s lidar and TrueDepth Capabilities. *Technologies* 9(25), https://doi.org/10.3390/technologies9020025.

# Remote De-Tensioning of Tieback Anchors After Structural Completion

**Sean Peterfreund** ▪ Delve Underground
**Grant Finn** ▪ Delve Underground
**Ty Jahn** ▪ Condon-Johnson & Associates

## ABSTRACT

For urban underground development involving tieback anchors for excavation support, authorities having jurisdiction typically require tiebacks to be de-tensioned before project completion. This paper updates "A Proactive Approach to Tieback Anchor De-tensioning" (by Peterfreund and Finn, published in the 2017 RETC Proceedings), which discussed a new method employed in a cut-and-cover transit station project. Since publication of that paper, 237 tiebacks were successfully and rapidly de-tensioned from using a small-diameter drill to cut strands from the surface. To the authors' knowledge, this was the first intentional large-scale use of a drilling method for this purpose.

## INTRODUCTION AND BACKGROUND

The use of tieback anchors for temporary excavation support is common practice in soft ground applications across the world. They are flexible and economical, and easily procured. However, once the anchor is no longer required, authorities having jurisdiction typically require tiebacks to be decommissioned (de-tensioned) before the project is completed. One reason for this is a perceived risk of sudden or catastrophic failure if future adjacent activities unexpectedly encounter the post-tensioned strands. The 2017 paper, "A Proactive Approach to Tieback Anchor De-tensioning (Peterfreund et al.), discusses these risks in more detail, while presenting an alternative double soldier pile arrangement. That design allowed access to the tieback strands from the surface after completion of the permanent structure. This paper focuses on the planning and execution of the drilling program to cut those tiebacks. The work was conducted in 2019 over a total of three weeks, shortly before substantial completion of the project.

### Status Quo for De-tensioning Tiebacks

There are generally two options for de-tensioning tiebacks where a permanent waterproof wall is required. The first requires staggered sequencing of the waterproofing membrane, slabs, and walls. After a portion of the waterproofing membrane is in place, a slab is cast, acting to prop the ground loads permanently. Then one or more rows of tiebacks are de-tensioned, the wall membrane installed, permanent perimeter wall constructed, and the subsequent floor slab placed, before repeating the sequence. The second method (illustrated in Figure 1) is to build the entire permanent structure at once, incorporating temporary windows, or block-outs, in the perimeter walls and membrane to allow access to the tieback heads for de-tensioning at a later date. After de-tensioning from the head, this method requires patching the waterproofing, reinforcing the window, and constructing formwork before pouring back the concrete to fill the window.

Source: Delve Underground

**Figure 1. Typical basement wall construction with tieback windows**

### Recap of Innovations at U District Station Design as Discussed in Previous Paper

The design of the soldier piles at Sound Transit's U District Station (part of the Northgate Link light rail extension) in Seattle, Washington, provided a third option for the tieback de-tensioning dilemma. These piles were double-wide flanges welded together by face and backing plates at discrete sections along the length of the piles. The stiffener plates at the tieback head locations were only located on the outsides of the webs, resulting in a completely clear conduit up the center of the entire pile height. Once placed in the ground, the conduit was backfilled with controlled density fill (CDF) along with the rest of the bored hole (Figure 2). This allowed for the opportunity to insert a drill into the top of the pile and cut through each of the tieback anchors. Each pile had up to six rows of tieback anchors, and each tieback consisted of up to 12 strands. Rather than sequence the slabs and walls or construct windows in the waterproofing and conduit, the general contractor built the permanent structure with minimal consideration given to the tiebacks.

## PROOF-OF-CONCEPT FOR REMOTE TIEBACK DE-TENSIONING

### Perceived Risks and Safety Considerations

In a traditional sense, tieback ground anchors for basement wall construction are typically de-tensioned at the head from inside the basement. Methods typically employ a tradesperson with an oxy-acetylene torch focused on the wedge anchors for each strand (Figure 3). Heating the strands allows them to slip through the wedge anchors, releasing the tension. This traditional method requires the tradesperson to be present at the tieback anchor head and is therefore at risk from flying debris or potential erratic response of the strands during the de-tensioning process.

From an owner's perspective, the energy released during tieback de-tensioning can cause concerns for impacts to a completed basement wall and the integrity of the

Source: Delve Underground

**Figure 2. Double-wide flange soldier pile arrangement with CDF**

Source: Delve Underground

**Figure 3. Typical de-tensioning process for tiebacks (note protection measures for waterproofing)**

waterproofing system. In their 2010 paper, "Investigating the Risks Associated with Allowing Temporary Tiebacks to Remain Stressed," Smith et al. performed a demonstration project in which one of the stated aims was to observe the response of a concrete basement wall to instantaneous de-tensioning of a tieback. Their tieback de-tensioning method centered around exposing the unbonded length and using either an excavator or hydraulic shears to cut the tieback strands. Smith et al. (2010) found that the risk of damage to both the basement wall and the waterproofing system was low. They noted that the risk of water ingress associated with de-tensioning a tieback after basement wall construction was complete was much lower than the risk associated with patching waterproofing at conventional tieback de-tensioning block-outs. This suggests that where de-tensioning is required, post-basement construction tieback de-tensioning is preferred.

The method presented in this paper and discussed above aims to de-tension tiebacks with less of an impact to the surrounding area compared to excavating a trench, as this is not always possible, especially when multiple rows of tiebacks are installed.

## Preproduction Testing

There were several unknowns prior to starting the production work on de-tensioning of the tiebacks, including:

- Which down-the-hole drilling technique would be most efficient
- Whether drill tooling would get stuck down the hole after the tiebacks were de-tensioned
- How to verify that all the tieback strands were cut
- Whether the cut tieback strands would prevent drill spoils from coming up the drill hole

The space between the two W21x111 welded-together soldier piles was 11.6 inches, with an outside diameter of 2 inches for the tieback strand bundle. Drill holes 8- and 9⅝-inches in diameters were evaluated based on the diameter required to ensure all of the tieback strands were severed. A 9⅝-inch-diameter drill hole was selected because it would increase the drill bit to strand overlap to a minimum of 2.8 inches versus 1.2 inches using an 8-inch-diameter drill hole. This overlap mitigated the risk of the drill hole drifting completely off center to one side of the soldier pile web (Figure 4).

Multiple down-the-hole drilling methods were considered prior to selecting two drilling techniques to conduct a test program in Condon-Johnson & Associates' yards prior to mobilizing to the jobsite. This was done to determine if it was feasible to drill through tieback strands and which method would be the most efficient. The two methods used were (1) open hole with carbide button bit attached to a down-the-hole hammer and (2) a cased duplex drill hole with a junk mill casing shoe. The test program consisted of

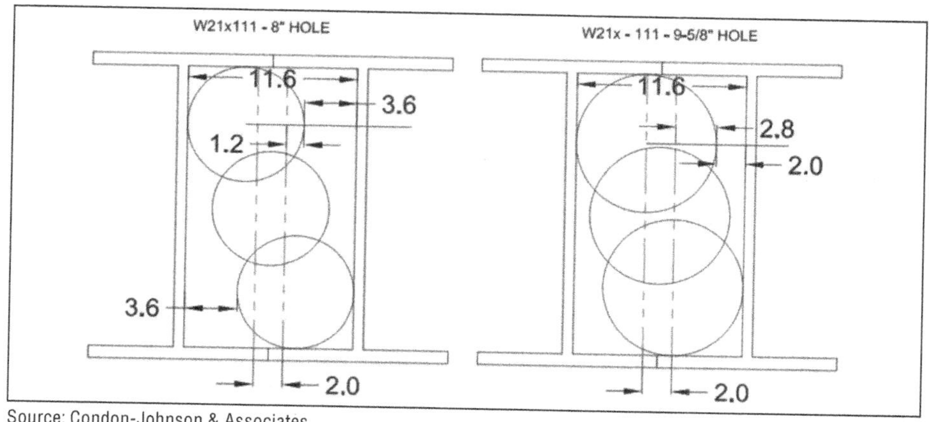


Source: Condon-Johnson & Associates

**Figure 4. Tieback de-tensioning design for space inside of soldier piles, with circles representing potential drill bit locations**

Source: Condon-Johnson & Associates

**Figure 5. Down-pile view from testing (1 of 2) view of CDF, cut PVC sheath, and cut tieback strands**

double W21x111 beams welded side by side to replicate the soldier piles with a nine-strand tieback locked off on both sides of the beam.

The open hole drilling method proved to be successful in drilling through all of the strands and de-tensioning the tieback. Breaking the strands required little time and effort as they were locked-off, and because the 270 ksi steel is relatively brittle. Drilling through the tieback strands resulted in a "rat's nest" of wires that was a major concern for the bit getting stuck down the hole when trying to extract the drill string back out of the drilled hole. The drill bit appeared to wallow out an area around the tieback strands because of the low strength of the CDF and the drill bit bounding around down the hole at the tieback location (Figures 5 and 6).

The cased hole drilling method also proved to be successful in drilling through the tieback strands to de-tension the tieback anchor. The majority of tieback strands were observed to be cleanly cut by the casing cutting shoe but discharging the cut strands fully out of the casing seemed to be an issue. Ultimately, open hole drilling using a down-the-hole hammer was selected as the drilling method because of its increased efficiency resulting from not having to handle casing.

## Production Work

A Klemm 806 drill rig was used for all production tieback de-tensioning work on site, with a reduced crew size of three personnel because casing was not required during drilling. One mini excavator and an air compressor were also used. The area around the top of each soldier pile needed to have clear access for the drill rig to be able to drill vertically down each soldier pile, along with an area for the chuck tenders to load the drill string. On-site constraints limited the drill rig size and will be a limiting factor in the future on other projects.

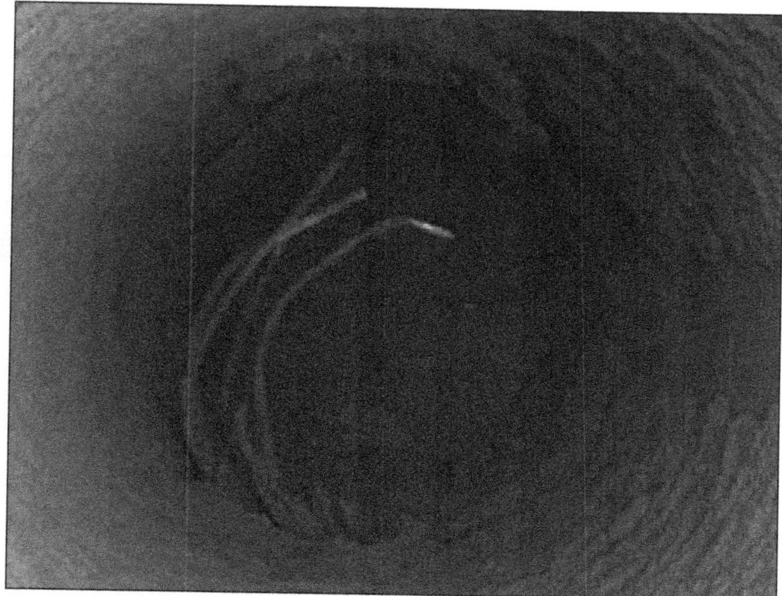

Source: Condon-Johnson & Associates

**Figure 6. Down-pile view from testing (2 of 2) of CDF and cut tieback strands, noting inside of steel conduit at bottom**

During the production work, there was no visible impact from inside the station that drilling through tiebacks to de-tension them had an impact on the structure's permanent concrete walls. Additional cracking or water intrusion was not observed during or after tieback de-tensioning.

The production work was completed in two separate phases because the de-tensioning work had to be timed and coordinated with on-site work. This did not impact overall production work on site, except for the additional time to remobilize to the jobsite. In total, approximately three weeks of active drilling was required.

## Results and Production Rate

Production rate on the drilling program to de-tension tiebacks was faster than expected. Once a tieback was encountered, it took approximately 3 to 4 minutes to cut through the plastic casing and strands. The typical de-tensioning of a soldier pile required 60-feet of drilling to fully sever all the tiebacks in that pile. The 237 tiebacks were de-tensioned over the three-week timeframe.

There was no noticeable difference in drill behavior for the top row of tiebacks (which included a long de-bonded zone) versus any other row. The drill operator felt that the work did not present any unusual challenges and that the CDF was critical to the design for keeping the bit from wandering. Bit wear was normal for CDF, and a total of two bits were used.

## Verification and Backfilling

Initial discussions included whether to lower a camera down each drilled hole to verify that all the tieback strands were de-tensioned. Although potentially feasible, this

Source: Condon & Johnson Associates

**Figure 7. QL-8 9⅝-inch hammer drill bit wear near end of drilling program**

verification method was not selected because the camera could get caught on the tieback strands, even if the cased hole drilling method was used. It was agreed that having a minimum 2-inch overlap on each side of the outside of the polyvinyl chloride (PVC) sleeve was sufficient to verify that all the strands were de-stressed. This was simply based on the fact that a locked-off strand could not deform enough to be pushed to the side while staying stressed, and that it would then do the same when the drill bit was removed from the drill hole. If resistance was observed during extraction at a tieback location, the drill string would be moved up and down twice at the tieback location to verify that the strand was, in fact, not still locked off. It should be noted that this verification method was not required during production work because all tieback strands were cut and de-tensioned during the initial drilling of the hole.

To ensure each row of tiebacks was de-tensioned, the drilled hole was advanced a minimum of 3 feet below the tieback row elevation. The drill depth could not be verified with a drop-tape but was instead measured by counting the number of drill rods added to the drill string. This proved to be a reliable and accurate method for measuring the drilled hole length.

It was very noticeable when a tieback was encountered, as the drill mast would bounce around because the drill bit had to break through the strands and the drilling penetration rate would significantly decrease. Verification that a tieback was encountered was also confirmed as pieces of the PVC no-load zone would be ejected to the surface in the drill cuttings. (If a greased and sheathed strand were used, then typically PVC no-load zones are not incorporated, so this may not be a reliable way to verify a tieback was encountered.) Drill rig action and penetration rate were the most reliable methods to verify a tieback was encountered.

The drill holes were initially planned to be backfilled with grout, but concerns were raised that the grout could travel out the head of the anchor into the waterproofing system. This was a concern because of the fluid nature of grout coupled with the head pressure exerted on the grout, with up to six rows of tiebacks being backfilled at once. Pea gravel was proposed and selected as the preferred backfill material because of its self-compacting nature and cost-effectiveness. There were no issues with the pea gravel acting as a water conduit, as there was waterproofing installed on the station walls. Furthermore, the bays between the soldier piles were lagged with timber boards, which would also allow water to freely travel from the subsurface materials to the waterproofing membrane. Thus, a pea gravel-filled hole would not introduce a new water path that was not already envisioned and catered for in the design of the waterproofing system.

## OPPORTUNITIES FOR ENHANCEMENT

### Technical Enhancements

Remote de-tensioning on this project was only made possible through a concentric tieback connection and the use of side-by-side wide flange sections, which formed a vertical conduit. Double-wide flange vertical elements were selected in this particular case to provide sufficient wall strength and stiffness while keeping the pile drilled hole diameter as small as reasonably practical to avoid existing adjacent infrastructure. However, double-wide flanges represent a premium over conventional support of excavation walls, which typically comprise either reinforcement cage or single-wide flange vertical elements within a drilled hole.

For regions where reinforced concrete drilled shafts are the primary vertical elements in a support of excavation earth retention system, introducing a thin-walled hollow structural section (HSS) within the cage could facilitate remote de-tensioning. The vertical conduit would have hollow structural section stand-off attachments at each tieback location, which would be recessed behind the vertical reinforcement to allow access for tieback installation at predetermined depths. These enhancements are the subject of Delve Underground patents granted in the United States (US Patent No. 10240315) and pending in Australia (AU Patent Application No. 2021221381).

For conventional single-wide flange vertical elements with eccentric tieback pockets, the thin-walled HSS vertical conduit may be welded behind either the front flange or back flange and fabricated as part of the pile assembly. When detailing the connections at the tieback pockets, care must be taken to not interrupt the vertical conduit with web stiffener plates.

These modest adjustments to conventional support of excavation systems offer the advantage of remote de-tensioning.

Specifications for support of excavation systems with remote de-tensioning should require a mock-up or test program before fabricating production elements so that the drilling contractor can select the correct tooling, prove out the size of the vertical conduit, and confirm quality control measure for verifying tieback de-tensioning.

### Contractual Environment

Traditional design-bid-build contracts for permanent underground facilities do not typically include the design of temporary support-of-excavation systems. General contractors would normally subcontract the design and installation of such systems to specialty firms, and those firms tend to win the work based on lowest price. Reconfiguring a pile to allow for remote de-tensioning requires additional material and labor, which would adversely affect the cost and installation schedule of a support-of-excavation system. The benefits of remote de-tensioning are only realized during permanent structure build-out (schedule) and long-term performance of the facility (no wall block-outs/waterproofing patches). The use of a remote de-tensioning system is therefore particularly attractive where a single entity is responsible for at least design and construction of a permanent underground structure, including the temporary support-of-excavation, where the capital cost of a temporary earth retention system is not considered in isolation. Projects procured using public-private partnerships (P3) contracts therefore offer a viable opportunity to adopt temporary support of excavation systems configured for remote de-tensioning.

## CONCLUSION

The project update presented herein demonstrates that remote de-tensioning of multi-strand ground anchor tiebacks is viable using open hole down-the-hole hammer drilling with a carbide button bit through a vertical steel conduit. Correctly coordinating the size of the vertical conduit with the diameter of the drill bit, together with monitoring of drill advance rates, provides a reliable measure of tieback de-tensioning verification.

The post-drilled conduit may be backfilled with pea gravel, which is a readily available, low-cost drainage material that is simple to handle and place.

Modest adjustment to the configuration of traditional temporary support-of-excavation systems can facilitate tieback de-tensioning from the ground surface and the construction of permanent underground works without regard to the location of tieback pockets or construction sequence of the permanent structure. Remote de-tensioning also eliminates the need for waterproofing patches and block-outs within basement walls.

The system presented in this paper offers benefits in terms of overall schedule reduction, site safety, and enhancements to permanent structure durability compared to conventional tieback de-tensioning methods. These benefits may be best realized in P3 contracting environments, where a single entity is responsible for design and construction of the underground facility.

## REFERENCES

Peterfreund, S., and Finn, G. 2017. A proactive approach to tieback anchor de-tensioning. In *Proceedings of the 2017 Rapid Excavation and Tunneling Conference*. San Diego, CA: Society of Mining Engineers, Inc. 852–857.

Smith, M., Flangas, L., and Ciani, D. 2010. Case history: Investigating the risks associated with allowing temporary tiebacks to remain stressed. In *Proceedings of the 2010 Earth Retention Conference*. Bellevue, WA: American Society of Civil Engineers. 188–195.

# Structural Underpinning an Airport Terminal to Mitigate Tunneling Risk—Atlanta Plane Train West Extension Project

**Thomas Hennings** ▪ Delve Underground
**Daniel Ebin** ▪ Delve Underground
**John Murray** ▪ Delve Underground
**Robert Gould** ▪ Mott MacDonald (formerly Guy F. Atkinson Construction)
**Ryan Smith** ▪ Keller North America

## ABSTRACT

The Atlanta Plane Train Tunnel West Extension was excavated at Hartsfield-Jackson Atlanta International Airport using Sequential Excavation Methods (SEM). At its shallowest point, tunnel cover to bottom of Domestic Terminal footings was limited to 6 feet in soft ground. To mitigate risk of excessive building movement due to tunneling-related activities, the columns near the alignment were directly underpinned. Underpinning was accomplished using a combination of grouted mini-piles, installed under low headroom conditions, that supported steel framing spanning over the tunnel excavations. This paper focuses on design and construction of the underpinning system used to mitigate risk from tunneling.

## PROJECT INTRODUCTION

The Atlanta Plane Train Tunnel West Extension Project at the Hartsfield-Jackson Atlanta International Airport includes an extension of the tunnels to the west to create a tail track that will allow a quicker turn-around of trains. This will result in an increase in capacity of the existing Automated People Mover (APM) system (DOA, 2017; Aldea Services, 2018).

The project is being delivered using the progressive design-build method. Clark-Atkinson-Technique, Joint Venture (CAT JV) is the design-builder for the project. Delve Underground (Delve) is the lead designer and the tunnel and shaft designer for the project. The shaft and tunnel works, including final lining, are being carried out by Guy F. Atkinson Construction (Atkinson), a joint venture partner and a subcontractor to CAT JV.

The tunnel was to be excavated using Sequential Excavation Methods (SEM). Prior to tunneling beneath the existing Domestic Terminal, the terminal building columns, supported on shallow footings required some form of temporary alternative support or ground improvement to mitigate the risk of tunneling-induced movement. CAT JV elected to develop an alternative scheme to support the terminal columns using an underpinning framing system supported on minipiles. These piles supported jacked steel frames that were structurally connected to the existing steel columns. Underpinning was selected as an alternative to a ground improvement concept that was presented in the bridging documents and is further discussed below. A reference design for the underpinning works was developed by Delve, which took the design to a 35% level to prove-out the underpinning approach and develop performance requirements for the final design. CAT-JV then appointed Keller North America (KNA) to complete the design and execute construction under a Delegated Design agreement—the

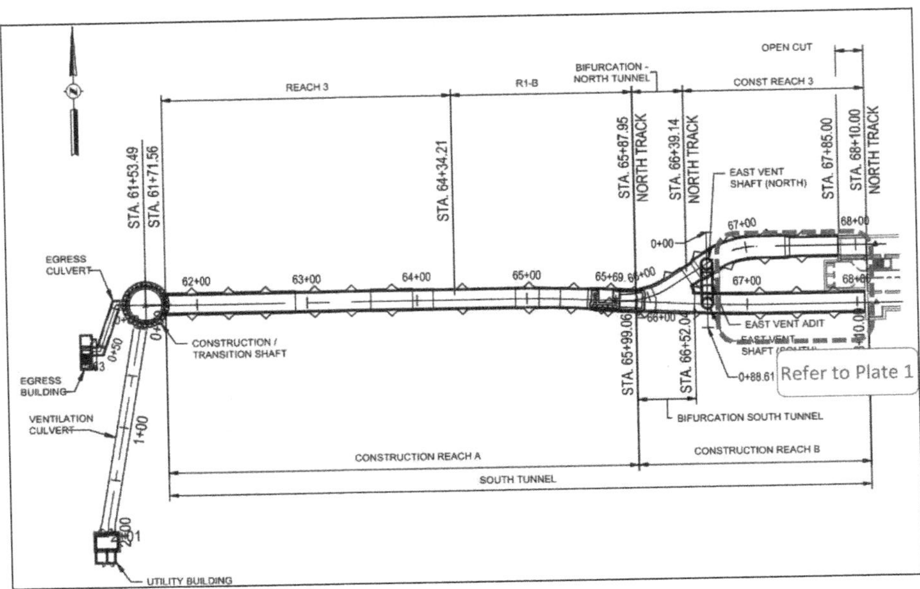


**Figure 1. General layout of the APM tunnel west extension**

**Figure 2. Existing Domestic Terminal building and apron area above the new west extension tunnels**

transfer of design responsibility to a specialty contractor for the final design and construction of typically more complex or unique works.

Figure 1 presents the tunneling and other subsurface works for the APM west extension. Figure 2 presents the domestic terminal building and apron area above the new west extension tunnels.

## REASON FOR UNDERPINNING

The project bridging documents included a ground improvement program relying on permeation grouting to improve the strength of the soil overlying the rock in the area underneath the Terminal foundation footings. Specifically, a layer of saprolite ranging from 0 feet to approximately 20 feet thick created a mixed face zone below 19 Terminal footings. The tunnel in this area has only 6 to 8 feet of soil cover between the tunnel crown and the Terminal footings. These footings immediately overlie the north and south ribs of both the North and South Tunnels that are being constructed as part of the project. Although the project bridging documents incorporated permeation grouting for supporting the ground above the tunnels, the Basis of Design Report provided with the bridging documents included data indicating that 70 percent of the soils in the mixed face area would be ungroutable due to high fines content. The design-build team was concerned with the feasibility of tunneling through the saprolite relying solely on permeation grouting, which would have limited effectiveness.

As a result of these concerns, the team agreed that an alternatives evaluation was warranted. During the preconstruction phase, potential mitigation approaches for the mixed face tunnel zone were evaluated. Three potential approaches were identified including: underpinning of the terminal foundations, jet grouting, and compensation grouting. Based on geotechnical and structural evaluations of all options, as well as logistical considerations for maintaining airport operations, underpinning of the terminal foundations was determined to be the most reliable method for mitigating the risk that the terminal structure would not settle during tunnel excavation. This approach was not as cost effective as compensation grouting, which could be staged but only executed if a trigger settlement was reached. However, the risk of relying on a reactive approach with some uncertainties with respect to effectiveness was not considered acceptable to the design-build team or owner. The preliminary design for the underpinning, including design criteria, was provided to the underpinning design-build contractor to complete the final design and to install the underpinning system. Figure 3 presents a longitudinal section through the south tunnel showing the Domestic Terminal above and the proximity of the terminal spread footings to the crown of the new tunnel. The geologic profile is superimposed onto the tunnel profile showing the mixed faced tunneling conditions.

## UNDERPINNING DESIGN

### Reference Design

Development of the original concept considered physical constraints including existing apron level space arrangements, support of excavation at the west elevator lobby (WEL) and TCE/PDS Rooms, and alignment and spatial extent of both the north and south SEM mined tunnels.

Consideration was made for permanently underpinning the terminal columns but was rejected because it would have required significant excavation disturbing sensitive high secure areas that required continuous reliable operation and could not be relocated.

Re-use of the existing terminal foundation system was desirable because ground settlement due to tunneling was expected to be less than 0.75-inches. Design assumed undisturbed subbase conditions throughout construction. Undisturbed subbase conditions were actively maintained by requiring duplex method of pile drilling. The footings were pre-cored prior to pile installation. Cores were located far enough from the footing center such that moment demand was less than capacity considering loss of reinforcement. The piles were jacketed with a bond breaker through the footing core holes to ensure the footings would move freely during ground settlement. Depending on the extent of ground settlement and condition of the existing anchor bolts, design included a method of retrofitting the existing base plate system if necessary.

Four existing terminal columns are supported-on pier/footing combinations. These include columns at K11, L11, L12 and M12. This complicated things since these columns were technically not acting independently. Due to the proximity of L12 to the

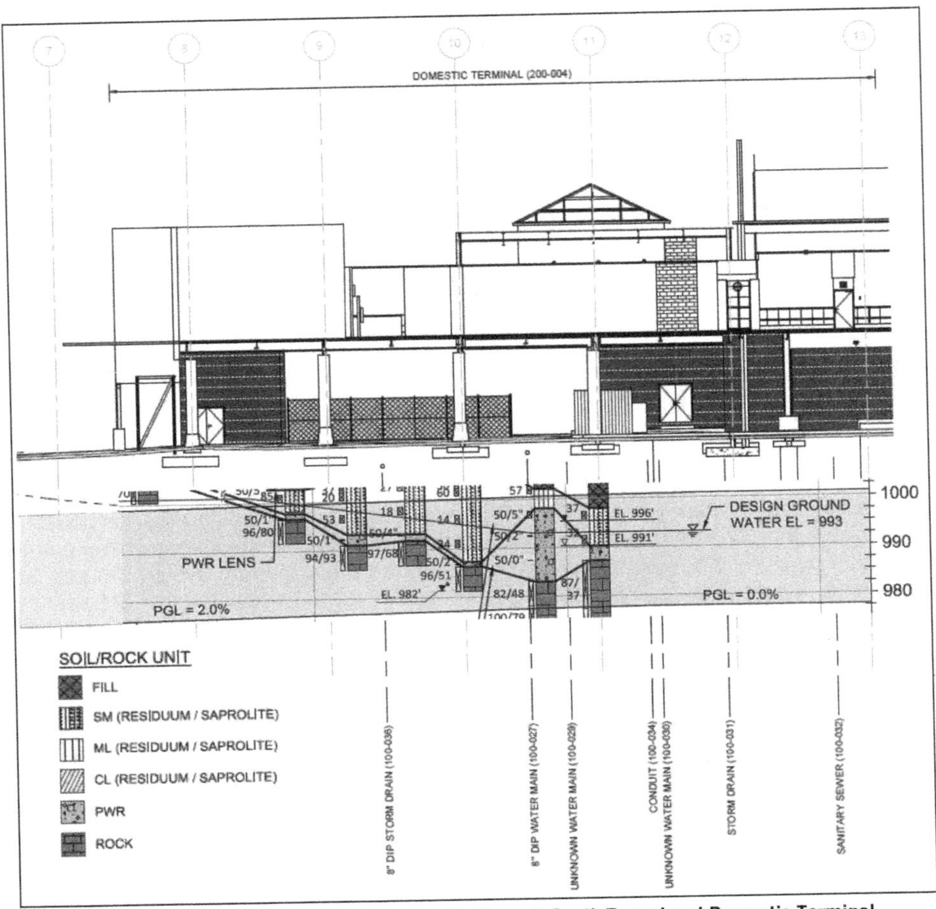


**Figure 3. Longitudinal section with geologic profile through South Tunnel and Domestic Terminal**

new WEL structure, the existing footing would need to be replaced by a new narrower footing 3'-6" wide by 18'-0" long.

Delve and Thornton Tomasetti (TT) developed building loads at column bases using record information made available from past construction. Delve developed the reference design, with TT support, of underpinning system based on constraints, anticipated loads, and serviceability needs to a 35% level. Keller North America (KNA) would take on the role of EOR for the underpinning design and advance the reference design to IFC.

The reference design underpinning arrangement included seven column pairs linked by main beams (four piles total for each pair of columns), three columns (L8, L12 and M12) supported by eccentric framing (four piles each column), two columns (M8 and J12) supported by concentric framing (four piles at M8 and eight piles at J12). L12 was to be supported on four piles, all extending into the tunnel profile. Intent was to trim piles at tunnel crown after initial lining completely installed and shotcrete reaching full compression strength. K12 was not underpinned because it had been previously been adjusted to rest on a foundation wall constructed during Phase 1 of the Terminal

Redevelopment Program .work performed in the 1990s. Several piles were* dual purpose, performing as underpinning piles as well as SOE piles (two of the four J12 and L12 piles adjacent to the WEL and TCE/PDS Room excavations).

## Delegated Design

Under a delagted design agreement, KNA developed the design to 60%, 100%, and Issued-for-Construction (IFC) levels between May 2020 and October 2020. The underpinning design consisted of two distinct elements: the minipiles and the underpinning steel frames. The minipiles consisted of 9-⅝-inch outer casing with a 7-inch inner casing. Minipile were grouted with 5,000 PSI grout. For the tension piles (tie-down anchors), 1-¼-inch threaded bar was used in lieu of the 7-inch casing and a tie down assembly installed above the transfer beam.

Figure 4 presents the plan of the underpinning configuration developed by KNA during the delgated design phase. The underpinning frame design developed three basic types and included frames for column pairs as well as some individual columns. KNA finalized the underpinning frame design into three different types, summarized in Table 1.

During the delegated design phase, KNA and CAT-JV re-configured some underpinning arrangements including the following:

1. J12: The permanent piles were maintained, but the subsequent reloading stage that involved the transfer to a permanent flat jack that was locked off and encased in concrete was eliminated. This meant there was only one transfer to the underpinning frame and the frame was incorporated into the permanent structure of the back of house PDS Room. Therefore, only four piles were required.

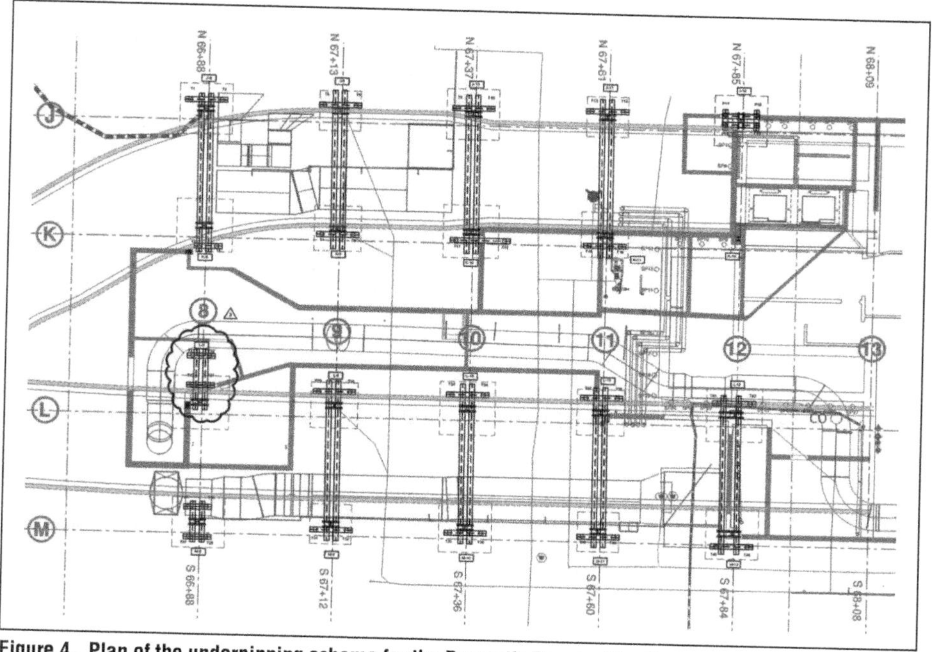


**Figure 4. Plan of the underpinning scheme for the Domestic Terminal footings above the tunnels**

Table 1. Summary of underpinning frame types

| Type | Description | Column (Pairs) | Notes |
|---|---|---|---|
| Type 1 (Unbraced) | Bridge beams connecting two micropiles at a column location with transfer beams spanning between the bridge beams of two columns | J8-K8<br>J9-K9<br>J10-K10<br>J11-K11<br>L9-M9<br>L10-M10<br>L11-M11 | |
| Type 1 (Braced)<br>(with lateral bracing) | As above, but with lateral bracing. | L12-M12 | L12 will be load transferred to new Apron level framing using W30 beams in a cantilevered configuration. This is because the existing footing has been demolished for the new West Elevator Lobby. |
| Type 2 | Eccentrically loaded transfer beams using two compression piles and two tension ('tie down anchors') | L8 | This configuration was required because it was not possible to utilize transfer beams between L8 and M8 due to critical airport MEP infrastructure that could not be relocated. |
| Type 3<br>(all with lateral bracing) | Concentric framing around one column | M8<br>J12 | J12 installed with permanent mining piles with underpinning frame incorporated into the permanent structure |

2. L12 and M12: L12 was originally designed as a concentric frame around the column and M12 was designed as an eccentric frame with two piles and two tie down anchors. This was converted to a column pair arrangement to eliminate the L12 piles within the tunnel profile and the eccentric framing at M12. During the construction phase, to accommodate Contractor concerns related to removal of the existing footing at L12 and replacing it with a new narrower footing discussed previously, CAT-JV opted to utilize the new framing design at Apron level and design and install a beam crossing through the WEL space to permanently support the column, eliminating the need for local SOE support and significant demolition in a tight space.

### Design Considerations for the Load Transfer Back to Existing Footings

For load transfer from column footings to underpinning frames, Delve provided KNA with not-to-exceed lift and load limits. The goal was to lift columns off footings to ensure load had been fully transferred but a load limit was required so as not to overstress the existing columns. Load limits were conservatively kept at or below 75% of the column buckling capacity assuming full height unbraced length from top-of-footing to Boarding Level. It became apparent that the load to lift ratio did not appear to act linearly. Following load transfers, it was difficult to access the area and assess whether full lifts of columns were achieved. Knowing the amount of load that was transferred into the underpinning system, it was clear the majority was transferred. After evaluation, it was determined that the majority of load from each footing was transferred to the underpinning system and mining of the north and south tunnels could commence. In most cases, load limits were reached prior to visual confirmation that the column base was lifted off the footing.

## UNDERPINNING CONSTRUCTION

The underpinning elements consisted of minipiles that supported the underpinning frames. The underpinning frames consisted of structural steel beams and other steel elements (including lateral bracing members for some underpinning frames), hydraulic jacks and spray applied fire protection. Additionally, the underpinning system also comprised extensive real time monitoring which included in-ground vibrating wire strain gauges installed in some of the piles, optical survey targets on the existing columns that were monitored automatically using automated motorized total stations (AMTS) and survey marks made onto the columns for manual survey verification as required, thus providing a level of redundancy in settlement data should the automated monitoring system go offline during critical load transfer activities.

Construction of the underpinning system involved six main stages:

**Stage 1:** The concrete protection (column jackets) for the existing columns was removed to a height of 8 feet above the existing grade. This enabled the welded connections for the connector channels which would house the hydraulic jacks.

**Stage 2:** Minipiles were installed, either two or four per column depending on the type of underpinning frame. Minipiles (and tie-down anchors for L8) were socketed 8 feet into competent rock below tunnel invert elevation.

**Stage 3:** Excavation to the top of the footing was carried out to expose the column baseplate connection and the nuts were loosened.

**Stage 4:** The underpinning frame was installed, including the temporary hydraulic jacks

**Stage 5:** Nuts were backed-off and, with the support of the real time instrumentation, the columns were lifted and the load transferred to the underpinning frame and minipiles. When the appropriate load was applied or the maximum lift had been achieved, the frames were locked off by tightening nuts on the threaded rods and the hydraulic jacks were removed and used for the next column.

**Stage 6:** The underpinning frame and all exposed steelwork was coated with spray applied fire proofing.

The following Figure 5 and Figure 6 show a Type 1 (braced and Type 2 underpinning frame.

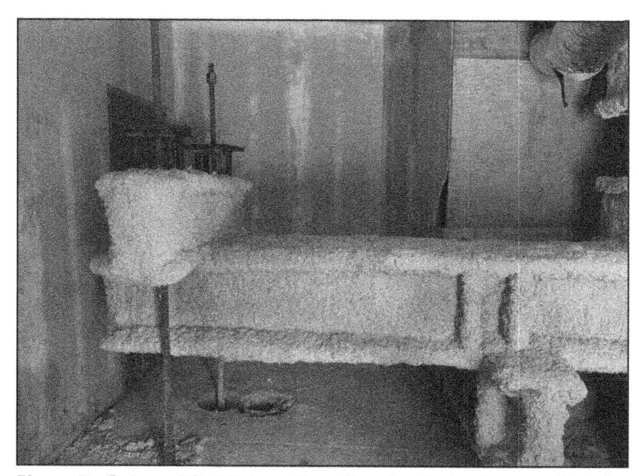

**Figure 5. Column L8 utilizing a Type 2 underpinning frame with tension piles (tie-down anchors)**

Many of the minipiles were installed within existing equipment rooms at Apron level. This necessitated the use of small drilling rigs with umbilical power packs that could fit through double doors and could work in low headroom areas less than 10-feet high.

Figure 6. Column M12 paired with L12, utilizing a Type 1 underpinning frame with lateral bracing

## CONSTRUCTION MONITORING AND MINIPILE PERFORMANCE

During minipile drilling, the geotechnical instrumentation that was already installed was monitored to determine any movements to the terminal structure. Monitoring generally comprised of the automated reading of Structural Monitoring Points (SMP) that were optical prisms installed on the columns and read by an AMTS unit. Data was uploaded to an instrumentation data management system (IDMS) with user access via a web portal. Overall, the effects of the drilling could be observed in the monitoring data, but the data remained below the response criteria indicating there was no ground loss below the footings that could cause settlement from the minipile drilling. Duplex drilling was mandated, which ensured the minipile casing was advanced as the hole was advance, minimizing the loss of material from the outside of the casing.

For the load transfers, it was necessary to ensure first t*hat all surround instrumentation to a column, or column pair, being transferred was operational and stable data had been obtained for 14 days prior to the load transfer and data was available for 48 hours prior to the load transfer. During the load transfer, the AMTS was switched to manual control and the SMP points monitored at 15-minute increments during the transfer. The load and stroke distance of the hydraulic jacks were also measured. Criteria were provided by Delve for both the maximum lift or the maximum load and the jacks were locked off for whatever criteria was reached first. When either of the criteria were reached, the jacks were locked off and the existing column welded to the connector channels. Hydraulic jacks were then removed.

Due to the advancing excavation east of the bifurcation, CAT-JV was able to work with Delve so that the tunnel could be advance to just west of the Terminal as the load transfer work was progressing, so long as the load transfers has already taken place at least two gridlines ahead of the advance tunnel face.

During the tunnel excavation, some minipiles that were immediately adjacent to the tunnel were exposed during the tunnel excavation. These were incorporated into the shotcrete initial lining but did not pose a structural concern to the underpinning of the terminal structure and this was considered in the design by assuming an unbraced length equivalent to the height of the tunnel.

One major interface with the underpinning and tunnel excavation was that drill and blast excavation of the rock in the tunnel was in very close proximity to the piles and monitoring of the column movements was closely observed. Blast events could be seen in the data due to the structure response to blasting, but movements remained within the response criteria. Mechanical excavation was required to remove the rock within 5 feet around a minipile in the tunnel horizon.

## RELOADING

Prior to footing re-load, baseline conditions are to be re-established by re-leveling underpinned columns targeting angular distortions at all locations to be at or below 1/10000. Prior to the re-leveling activity, SMP data in the vertical (Z) direction will be averaged over a two week period and used to determine adjustments required to meet the target angular distortion limit. The target of 1/10000 or less is selected as a goal with the understanding that it may not be achieved at some or even all locations but would improve the odds of meeting the 1/1000 limit at the conclusion of the work. It also became apparent during the original transfer of load from footings to underpinning system that the relationship between the lift and load was not linear and not easily predicted. It was decided that reinforcing certain columns depending on the magnitude of the re-leveling lift would be prudent to increase their capacity. The high locked-in column load will be of most concern when the columns are resting back onto their footings and the full unbraced length was again realized. Chapter 6 of AISC Design Guide 15 was followed to perform the design. Two details were developed depending on column orientation relative to the underpinning arrangement. Eight columns were identified as needing a capacity increase. Three columns along 8-Line were reinforced with the Type 2 detail, and the remaining with Type 1. The re-leveling activity and footing re-load is scheduled for January 2023. Planned column re-leveling and footing re-load sequence is as follows:

1. Prior to footing reload, re-level underpinned columns with a goal of achieving angular distortions equal to or less than 1/10000
2. Restore grout pad below column base plate
3. Re-load footing
4. Monitor vertical movement for one week
5. If vertical movement is within acceptable limits and is stable, remove underpinning framing
6. If vertical movement is unacceptable or has not stabilized, repeat Steps 1 thru 4

## REFERENCES

Aldea Services. 2018. *Basis of Design Report (BODR), Plane Train Tunnel West Extension Phase 1, Hartsfield-Jackson Atlanta International Airport*. Prepared for Materials Managers & Engineers, Inc. (2MNEXT).

Brockenbrough, R.L and Schuster, J. 2018. *Design Guide 15, Rehabilitation and Retrofit, Second Edition*. Chicago, IL: American Institute of Steel Construction.

City of Atlanta, Department of Aviation (DOA). 2018. *Domestic Terminal, PlaneTrain Tunnel West, Extension Phase 1. Basis of Design Report Volume 1,* Post RFP Revision #2, March 23, 2018.

# TBM Trailing Sandwich Belt High Angle Conveyor

**Joseph A. Dos Santos** ▪ Dos Santos International LLC

## ABSTRACT

The largest TBM (Tunnel Boring Machine) tunnels are typically along modest slopes, but not always. Occasionally high slope angles are required beyond the capabilities of conventional open trough trailing conveyors. A 2019 project required a large TBM to ascend to an incline angle of 25°. A Sandwich Belt high angle conveyor could easily solve the high angle conveying problem but such a system must include all of the features of the conventional trailing conveyor, particularly the ability to extend with the TBM advance. The writer responded to the challenge and developed a TBM trailing Sandwich Belt high angle conveyor with all of the needed features. The invention (U.S.A. patents awarded October, 2021 and November, 2022) however goes beyond the 2019 project requirements and extends to any high angle incline.

This writing first presents the broad Sandwich Belt high angle conveyor development, then its use in various construction and tunneling projects prior to this TBM Sandwich Belt trailing conveyor invention.

Finally we present the new TBM Trailing Sandwich Belt high angle conveyor with all of its features and implications.

## INTRODUCTION

### What Are Dos Santos Sandwich Belt High Angle Conveyors?

This article deals predominantly with the Dos Santos Sandwich Belt High Angle Conveyors, a technology developed by this writer more than 40 years ago. The first commercial installation went into operation in Wyoming, USA in 1984. This is to clarify that when referring to Dos Santos Sandwich Belt high angle conveyor systems in this writing I am referring to the work of J. A. Dos Santos since 1979 while at the various companies of employment:

- Dravo Corporation, Pittsburgh, PA USA (1975–1982): During the period 1979–1981, under a US Bureau of Mines study; the writer developed the sandwich belt high angle conveyor technology, rationalized in the conventional conveyor technology. This also produced the landmark publication "Evolution of Sandwich Belt High Angle Conveyors", a writing that is complete in defining the theory and design rules and in the conceptualization of the designs that went on to commercialization.
- Continental Conveyor and Equipment Company, Winfield, AL USA: The HAC Systems from 1982–1997.
- Since the founding of Dos Santos International: The DSI Snake Sandwich and GPS (Gently Pressed Sandwich) High Angle Conveyors, as well as the invention of the Adder Snake, and now the TBM Trailing Sandwich Belt High Angle Conveyor, from 1997 until the present.

## Sandwich Belt High Angle Conveyors

Development of the Sandwich Belt high angle conveyor concept has come a long way since its first introduction in the early 1950s. Over the approximate thirty year period until 1979, significant advances were few and only came in spurts. Such advances did not build on previous developments. Rather, they were independent developments which soon reached their technical limitations.

The latest significant development of this technology, beginning in 1979, is the first to take a broad view of the industries to benefit from high angle conveying and of all previous developments. As a result these latest developments know few technical limitations, address a broad range of applications and offer a forum for continued logical development or evolution.

## Sandwich Belt Principle

Dos Santos Sandwich Belt high angle conveyors represent logical evolution and optimization of the sandwich belt concept. The sandwich belt approach employs two ordinary rubber belts which sandwich the conveyed material. Additional distributed force on the belt sandwich provides hugging pressure to the conveyed material in order to develop sufficient friction at the material-to-belt and material-to-material interface to prevent sliding back at the design conveying angle.

Figure 1 is a realistic model of the belt sandwich. An ample belt edge distance assures a sealed material package during operation even if belt misalignment occurs. This model also illustrates the interaction of forces within the sandwich. The applied or induced hugging load is distributed across and along the carrying belt sandwich. Of that hugging pressure, only the middle pressure hugs the material load while the outer pressure merely bears against the material free edges of the belt. Both belt surfaces apply their frictional traction on the material. From this model one can calculate the required material hugging pressure that will ensure the material does not slide back due to the tangential gravity forces. This is expressed by Equation 1:

$$Nm \geq \frac{Wm}{2}\left(\frac{\sin\alpha}{\mu} - \cos\alpha\right) \qquad \textbf{(EQ 1)}$$

where: $\mu = \mu_m$ or $\mu = \mu_b$, whichever is smaller

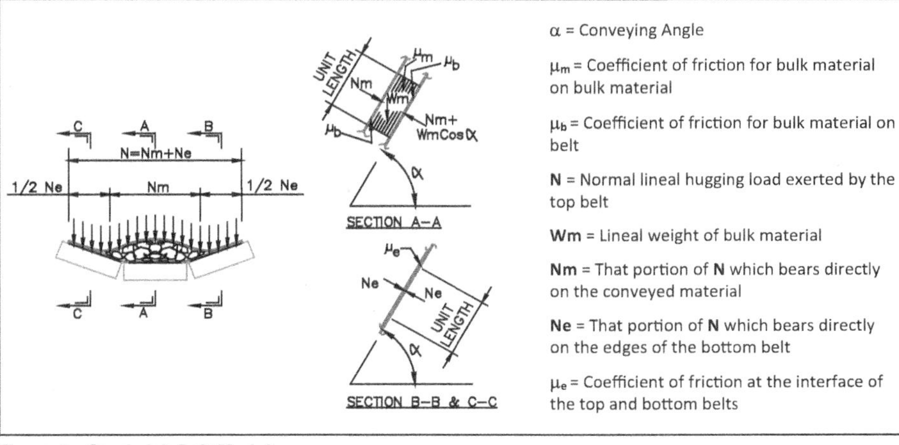


**Figure 1. Sandwich Belt Model**

## Dos Santos Sandwich Belt High Angle Conveyors

When investigated anew in the late 1970s, it was clear that the sandwich belt concept offered the greatest potential for a cost effective, operationally appropriate high angle conveying system to address the broad needs of the mining and bulk materials handling industries.

Following the extensive study of past Sandwich Belt conveyors, the governing theory and constraints, and development of the governing design criteria, a broader scope effort was undertaken in 1982 to develop the first Sandwich Belt high angle conveyor to meet these needs. The resulting Dos Santos Sandwich Belt high angle conveyors are truly evolutionary in judiciously selecting and advancing desirable features while avoiding the pitfalls of the past. They conform entirely to the governing theory, to the constraint equations and to the development criteria.

These Sandwich Belts fulfill all established operational requirements. The profiles can conform to a wide variety of applications.

### *Advantages of Sandwich Belts*

Dos Santos Sandwich Belt high angle conveyors offer many advantages over other systems including:

- *Simplicity of Approach:* The use of all conventional conveyor hardware, for very high availability and low maintenance costs
- *Virtually Unlimited in Capacity:* The use of all conventional conveyor components permits high conveying speeds. Available belts and hardware up to 3000 mm wide make possible capacities greater than 10,000 t/h.
- *High Lifts and High Conveying Angles:* High lifts to 300 m are possible with standard fabric reinforced belts. Much higher single run lifts are possible with steel cord or aramid fiber belts. High angles up to 90° are possible.
- *Flexibility in Planning and in Operation:* Dos Santos Sandwich Belts lend themselves to multi-flight conveying systems with self-contained units or to single run systems using externally anchored high angle conveyors. A system may be shortened or lengthened or the angle may be altered for the requirements of a new location.
- *Belts are Easily Cleaned and Quickly Repaired:* Smooth surfaced rubber belts allow continuous cleaning by belt scrapers or plows. Smooth surfaced belts present no obstruction to quick repair by hot or cold vulcanizing.

# BACKGROUND

A brief background of the Sandwich Belt high angle conveyor development is presented here. For a deeper, more detailed understanding of the technology the reader is referred to the 1982 landmark writing "Evolution of Sandwich Belt High Angle Conveyors"[1] a comprehensive presentation on the development.

## Sandwich Belts of the 1950s

The concept of elevating bulk materials at high angles using the sandwich belt concept was first introduced in 1951. That introduction did not produce any lasting success. It did produce a simple mathematical model that facilitated calculation of the hugging pressure needed to develop the material's internal friction so that material slide back did not occur when operating at the design steep incline. The sandwich model of Figure 1 was inspired by the original model but it is a more realistic representation of a Sandwich Belt conveyor.

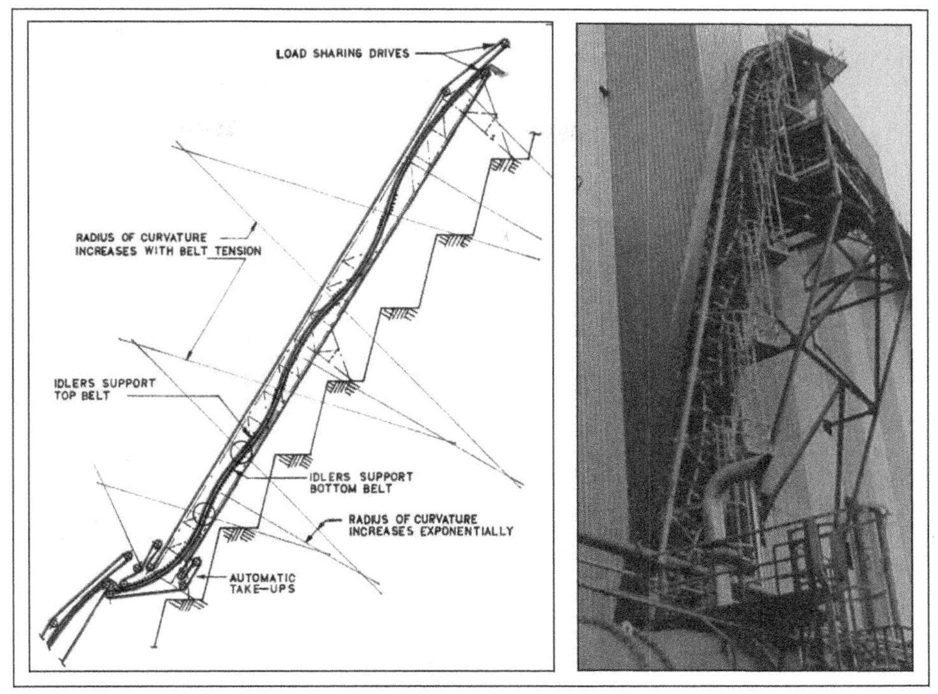


**Figure 2. Snake Sandwich, from concept (left) to reality (right)**

## Loop Belts of the 1970s

Commercial success in sandwich belt high angle conveying was actually first achieved in the 1970s with the Loop Belts, sandwich belt elevators of C-shaped profile developed by Stephens Adamson of Canada. The Loop Belt was and is the vital element of any conveyor based self-unloading ship system. Such self-unloading ships have had great success achieving unloading rates above 10,000 t/h with wide Loop Belts to 3,048 mm (120"). Being strictly of C-profile, Loop Belts could not be adapted to the general high angle conveying path which is predominantly along a straight incline.

## Dos Santos Sandwich Belts of the 1980s

It was the success of Loop Belts that inspired the Dos Santos Sandwich Belts of the 1980s. These expanded the Loop Belt capabilities by producing endless elevating profiles that could take the most direct and/or conforming path between the loading and discharge points. The Dos Santos development was broad, rationalizing the sandwich belt high angle conveyor technology in the conventional conveyor technology. Two Dos Santos inventions came out of this development, the Snake sandwich belt and the mechanically pressed sandwich belt. Whereas the former is a technological extension of the Loop Belts of the 1970s, the latter could be considered additionally the technological extension of the Cover Belts of the 1950s. Figure 2 shows the Snake sandwich high angle conveyor concept and reality in the DSI Snake. Figure 3 shows the mechanically pressed sandwich high angle conveyor concept and reality in the DSI GPS (Gently Pressed Sandwich). Both systems utilize the basics of the Snake sandwich. An additional variation of the mechanically pressed sandwich allows for the use of steel cord belts in a very high single flight system. Such a system utilizes

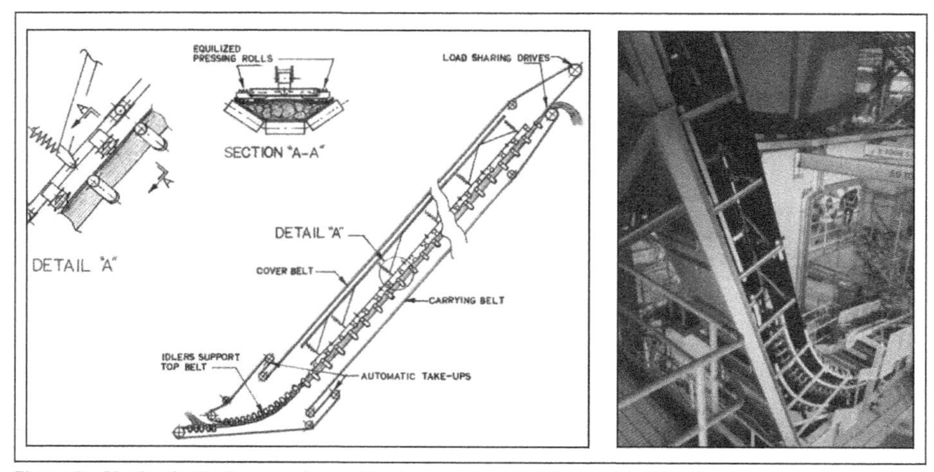


**Figure 3. Mechanically Pressed Sandwich, from concept (left) to reality (right)**

a large uplift type transition curve that intrudes into the open pit. Very high single lifts exceeding 300 meters are possible.

Both Snakes and GPSs typically utilize the same transition features in increasing the conveying angle from the low angle at the loading point to the ultimate high angle. A transition curve is also typical from the high incline angle to the discharge. The former utilizes two convex curves after the top belt joins the bottom belt forming the sandwich. The first convex curve is carried along closely spaced inverted troughing idlers. The second reverses the curvature and is carried along closely spaced upright idlers to the straight high angle incline. An inflection zone between the two curves transitions the sandwich smoothly from the inverted troughing to the upright troughing configuration. The Snake differs from the GPS only along the straight incline. The former continues utilizing alternating convex curves with inflection zones between the curves. The latter utilizes a straight high incline with fully equalized pressing rolls applying a gentle hugging pressure onto the top belt which in turn hugs the bulk material onto the lower troughed belt.

## Select Sandwich-Belt Installations in Construction and Tunneling

Dos Santos Sandwich Belt high angle conveyors are well established in the various industries with the first commercial unit beginning operation in 1984. Since then more than 100 units have gone into operation throughout the world. Installations have been in many diverse industries. Here we present only installations in construction and tunneling. Table 1 presents these units each with a brief technical summary.

DS 005 (shown in Figure 4.) was the first commercial Dos Santos Sandwich Belt high angle conveyor for construction and it was the first vertical installation. The system was part of the Los Angeles metro expansion of the late 1980s. It was located downtown at the corner of 7th and Flower Streets. The excavation in this area was an open cut with timber lagging on steel beams to support the busy city street above while excavation proceeded below. Material movement was by LHD (load-haul-dump) with FELs (front end loaders) to a grizzly covered hopper. The hopper loaded the tail of the high angle conveyor through a vibratory feeder. The high angle conveyor elevated the excavated earth continuously from under the street to a truck loading bin above. Though it was

**Table 1. Select sandwich-belt high angle conveyor installations**

| DS # | Application/ Location | Material/ Rate, t/h | Ang, ° | Elev, m | Lgth, m | Width, mm | Speed, m/s | Top/Bott, kW | Year |
|---|---|---|---|---|---|---|---|---|---|
| 005 | Construction/ L.A., CA, USA | Excav. Earth/ 272 | 90 | 32.3 | 42.5 | 914 | 1.6 | 22.4/22.4 | 1988 |
| 037 | Tunneling/ Illinois, USA | TBM Muck/ 1266 | 90 | 70 | 83.8 | 1372 | 3.6 | 186/186 | 1993 |
| 108 | Tunneling/ Paris, France | TBM Muck/ 800 | 90 | 24.6 | 33.5 | 1400 | 3.0 | 75/75 | 2018 |
| 109 | Tunneling/ Paris, France | TBM Muck/ 800 | 90 | 26.3 | 35.2 | 1400 | 3.0 | 75/75 | 2018 |
| 116 | Tunneling/ Singapore | TBM Muck/ 800 | 45 | 32.8 | 90.1 | 1200 | 2.0 | 75/75 | 2022 |

designed to load directly into the bin, for reasons of truck access and traffic flow, a connecting conveyor was added in order to locate the bin further away from the intersection of the two busy streets. The surge capacities of the hopper below and the bin above allowed independent discontinuous excavation and truck loading without interrupting the continuous elevating of the high angle conveyor. The system was designed to begin operation during the early excavation, requiring only 25 meters of lift. Then it was extended down as the depth increased, in 1219 mm increments until reaching the design maximum depth and the corresponding design lift of 32.3 meters.

**Figure 4. DS 005 represents two firsts in one Sandwich Conveyor; first vertical, first in a construction project**

Valuable lessons learned during this early project included:

1. Though the project was of short duration the duty was harsh
2. Oversized material consisting of large rocks and large clumps easily passed through the hopper's grizzly which consisted of parallel 51 mm wide bars spaced at 203 mm
3. Such material was too large for the 914 mm belt width and corresponding equipment
4. Though not detected by the layman, continuity of hugging lapse along the vertically straight elevating section prompted a review and revision of the continuity of hugging criteria

Though the system completed its task successfully, we decided that future Sandwich Belt high angle conveyors for such projects would use wider belts (not less than 1,200 mm belt width), thicker damage resistant wear covers and rubber disc center rolls at the idlers along the transition curves in order to soften the indents of the large lumps as they are radially pulled into the troughed transition curves.

DS 037 was the first Sandwich Belt installation to elevate tunnel muck from a TBM (tunnel boring machine). This was part of the Chicago TARP (Tunnel and Reservoir Plan) project. The entire excavating and muck haulage system consisted of the TBM, a trailing conveyor below ground, the vertical Sandwich Belt conveyor system to lift the material to the surface and finally a transfer and stacking system consisting of a grasshopper conveyor and radial stacker. As with subsequent systems the major equipment (drives and take-up systems) was located on the surface, at the head end so it could be easily accessed and serviced. Only the intermediate structure and tail pulleys were located underground. Additionally the intermediate structure served as the support and guidance for the trailing conveyor's return belt strands that were elevated to its belt storage unit located on the surface, 70 meters above.

From the first Sandwich Belt application at a TBM (DS 037) in 1993, many years passed before the next such application. In 2005 three (3) vertical Sandwich Belt high angle conveyors were used to elevate TBM tunnel muck at the extension of London Heathrow airport. These were HAC's by Continental Conveyor, USA.

From the time period 1993 until 2018 (25 years) there were notable successes with Sandwich Belt high angle conveyors elevating tunnel muck from TBMs but their use did not escalate to dominance in this industry. Pocket belt systems continued to dominate this duty despite their known difficulties in completely discharging very sticky materials. Engineers of the Paris Metro expansion project recognized the distinct Sandwich Belt advantage that the smooth surfaced rubber belts could discharge the muck completely and could be scraped clean. This knowledge led them to specify the exclusive use of Sandwich Belt high angle conveyors for the high angle elevating duties of the Paris Metro Expansion project.

DS 108 and 109 were both part of the Paris Metro expansion project. These units and the project were presented in detail at the Rapid Excavation and Tunneling Conference in 2019 and published in the proceedings. Here we describe these units only briefly.

Both units were built according to the Standard Muck Lifter-1 design which is illustrated in Figure 5 with technical features summarized in the adjacent table. In the Standard Muck Lifter design we recognize the typically short life of any tunnel construction project. Accordingly, it is designed to be versatile and adaptable to a range of tunnel projects. This Standard Muck Lifter-1 has a nominal design rate of 800 t/h. It can be installed in a C-profile or in an S-profile. It has a muck lift range from 17.07 meters to 40.54 meters. The height range is achieved by increasing or decreasing the straight vertical length. It is worth noting that all drive and take-up equipment is located at the head end terminal for easy access and servicing at the surface. Thus the vertical and loading sections are kept simple for easy handling in the shaft.

The two SML-1 units of the Paris metro project are shown in Figure 6. DS 108-unit-P1 (at left) has a lift of 24647 mm while DS 109-unit -P2 (at right) has a lift of 26323 mm. The difference is one GPS (Gently Pressed Sandwich) module with its net length of 1676 mm.

DS 116 is shown in Figure 7. It is a tailored design for the specific requirements of the Changi Airport project in Singapore. It features 1,220 mm wide belts running at 2.0 m/s for an elevating rate of 800 t/h. It began operation at the beginning of 2022.

## PATENTED TBM TRAILING SANDWICH BELT HIGH ANGLE CONVEYOR

After the many installations that received the tunnel muck from conventional TBM trailing conveyors, beginning in 2019 Dos Santos developed the TBM trailing Sandwich Belt high angle trailing conveyor. Two USA patents were awarded to Dos Santos International. The first patent was awarded on October 19, 2021. A second patent

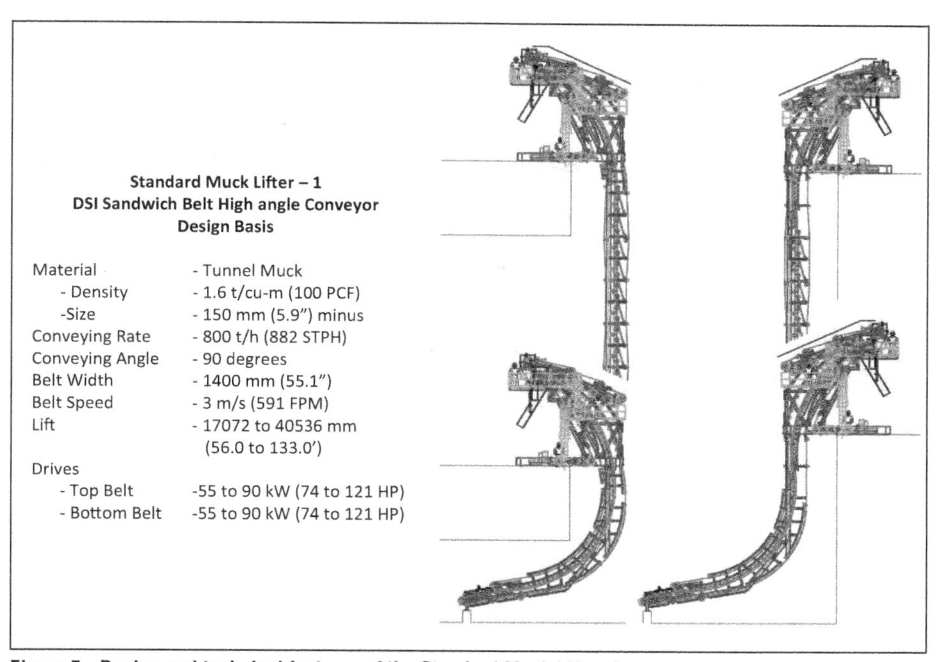


**Figure 5. Design and technical features of the Standard Muck Lifter-1**

**Figure 6. Paris Muck Lifter-1 units; (left) P1 –24.647 m lift, (right) P2 –26.323 m lift**

that broadened the invention and expanded the claims was awarded on November 1, 2022. Joseph A. Dos Santos is the inventor. The invention is to haul bulk material away from continuous mining machines and, more particularly, from a TBM (tunnel boring machine) along any high angle incline.

In a typical mining operation such as tunneling, it is well known to use a continuous mining machine, such as a TBM, to advance the mine face continuously without interruption. The rock material or muck must be hauled away continuously from the face. The conventional, open troughed extendable trailing conveyor (see Figure 8) has traditionally fulfilled this function as typical tunnel excavations are not along steep inclines.

Occasionally a tunnel must be excavated along a steep incline exceeding 15 to 18°. A TBM is capable of such steep excavation but the traditional open trough conveyor

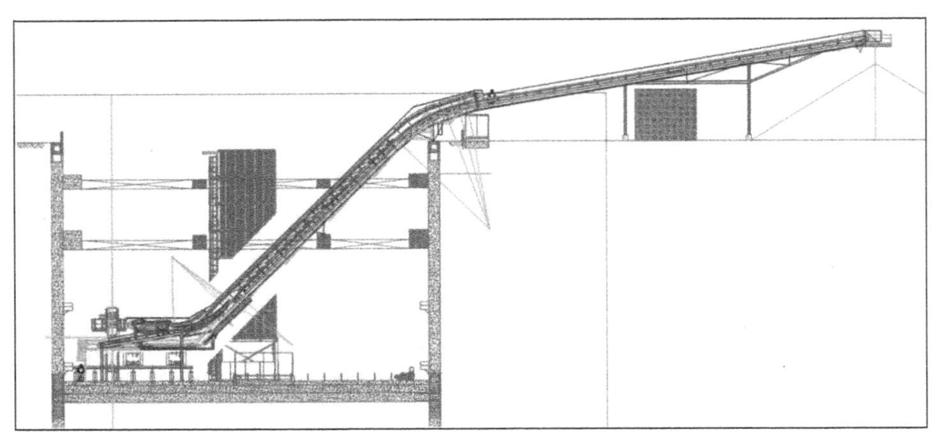

Figure 7. DS 116 is elevating and stacking TBM tunnel muck at the Changi Airport project in Singapore

cannot serve the muck clearance function as the muck would slide down on the belt at such angles. Dos Santos Sandwich Belt high angle conveyors are well developed and well known to have the operating characteristics of conventional open trough conveyors but they also have the capability of conveying the bulk at any high angle. This is done by sandwiching the bulk between two belts, hugging the material continuously (without lapse) in order to develop its internal friction which resists any tendency to slide down. The DSI invention extends the high angle capability to the TBM trailing conveyor.

Figure 8. Conventional TBM trailing conveyor at the Paris Metro project

In 2019, the challenge was presented to Dos Santos International. A major tunneling project in Australia required the excavation to begin along a horizontal path then to increase in slope to 25°. The 25° slope continued 1163 meters then again reduced to a lower, near horizontal angle along the remainder of the tunnel. Necessity being the mother of invention, DSI embraced the challenge. Some key elements of the invention were determined in the requirements of this project but not all. The requirements are summarized:

1. The TBM trailing conveyor system must begin at the launch of the TBM and must continue to haul the muck to the discharge point that is established at the launch.
2. The loading point must be on the TBM trailing deck.
3. The system must extend in length while in operation as the TBM advances the tunnel excavation.
4. This requires a belt storage unit at the launch for both the top belt and for the bottom belt of the sandwich conveyor.

5. This also requires that conveyor structure is added safely at the tunnel, from the TBM trailing deck as the Trailing Sandwich conveyor extends with advance of the excavation.

Whereas the use of the belt storage units (one at each belt as shown in Figure 10) is relatively straightforward, as is the loading at the TBM trailing deck (see Figure 11), the transition from the low loading angle (onto the bottom belt) to the sandwich, then to the high angle is not. Indeed it is unprecedented. Traditionally the transition from the sandwich entrance to the high angle has been gradual along a troughed convex transition curve. Such a curve is not possible within the space of the straight tunnel.

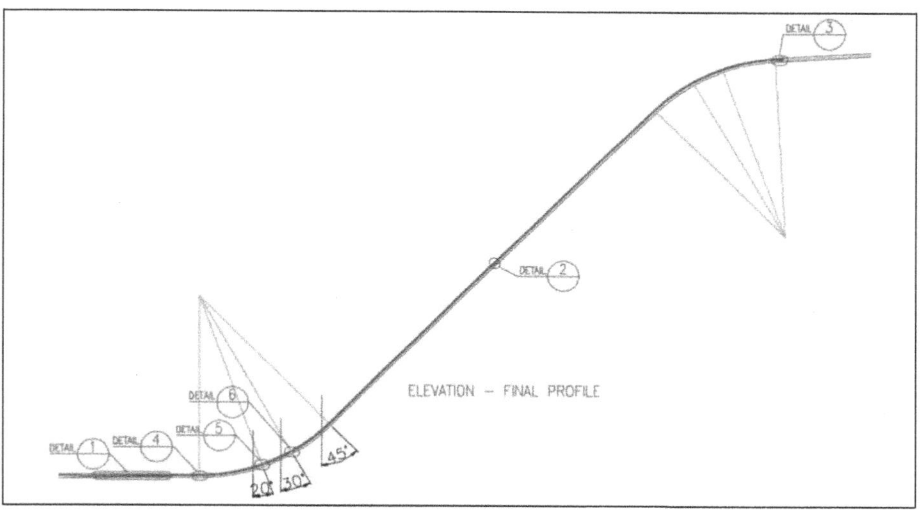


**Figure 9. Basis tunnel profile. Incline angle is illustrated at 45° but it can be any high angle to 90°**

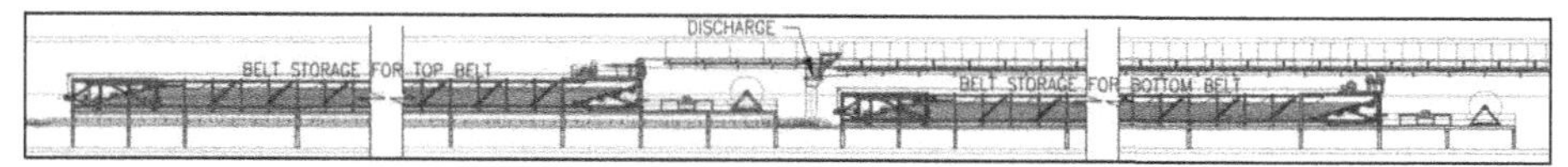


**Figure 10. Belt storage units, one at each belt, pay out belting as the TBM advances and the Sandwich Belt Trailing Conveyor lengthens**

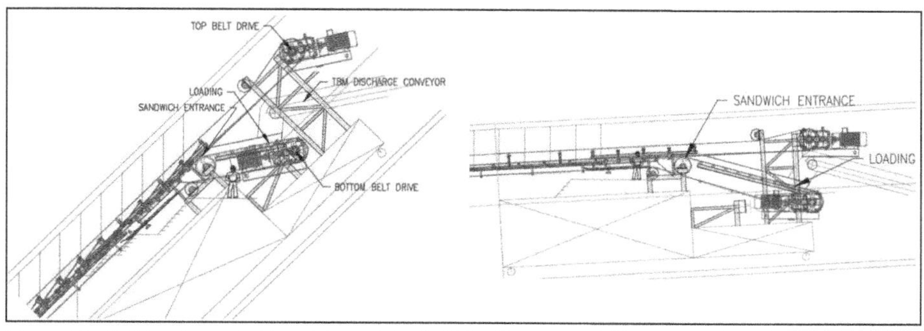


**Figure 11. Tail loading and drive station at the TBM trailing deck. TBM is excavating along the high incline at left and along a low angle at right**

The transition is thus abrupt. The skirted, flat bottom belt with the bulk load travels into the sandwich over a large diameter bend pulley that immediately deflects the belt line to the high angle. The top belt joins the bottom belt at the high incline covering the material stream immediately, imposing only a gentle hugging pressure onto the bulk material.

The challenge of the 2019 Australian project inquiry prompted the invention, but vision of the broader requirements, including much higher tunnel angles, formed the basis for the broader invention.

Most challenging of the broader higher angle requirements is the need for an interim hugging pressure as the permanent high angle structure is added at the tunnel from the TBM trailing deck. This is because the conveyor is running without lapse so the material hugging must also be without lapse. This challenge does not occur at conveying angles up to about 30° as the contact and mere weight of the top belt stabilizes the bulk load and develops its internal friction. This also means that at high angles to approximately 30°, GPS (gently pressed sandwich) type pressing rolls are not required and the mere top belt suffices.

The interim hugging is accomplished with a weighted blanket over the top belt for conveying angles of approximately 30 to 55° and torsionally sprung air plenums for incline angles above 55°. In either case, the interim hugging system is secured to the high angle structure of the TBM trailing deck and creates a smooth upper surface that the permanent GPS pressing rolls can roll over and onto the top belt as they assume the permanent hugging function. Figure 12 presents a series of six slides that depict adding the permanent conveyor structure as the TBM advances, leaving the tunnel anchored hanging structure behind.

The invention promises to revolutionize haulage from high angle tunnel excavations affording all of the benefits of TBM trailing conveyors that were not previously available at high incline angles.

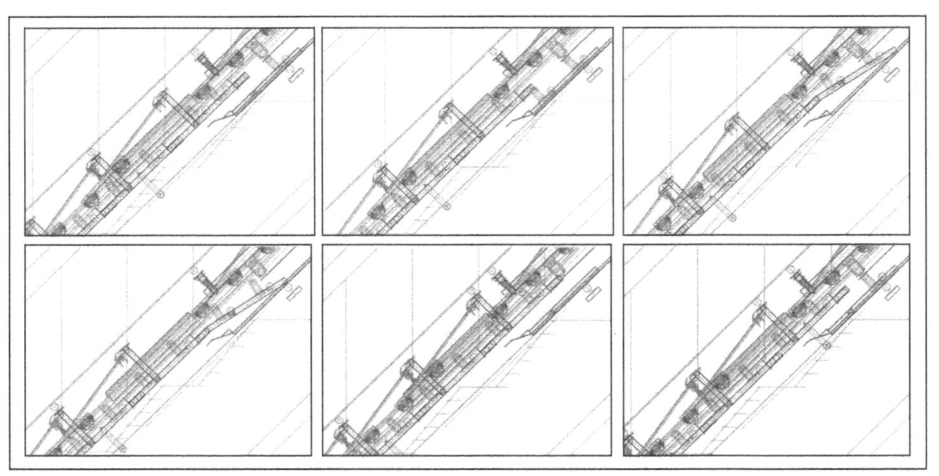

**Figure 12. Sequence, adding chain hung Sandwich conveyor structure at the tunnel from the TBM trailing deck**

## CLOSING

The modern Sandwich Belt high angle conveyor technology was developed by the writer from 1979 to 1981, then commercialized beginning in 1982. The record is one of success. The first application in a construction project came early in the commercialization, in 1988, only the fourth commercial installation. The first use elevating TBM muck came in 1993 at a significant tunneling project in Chicago, USA. From that point, over the years there were other significant units elevating TBM muck. A significant breakthrough happened in 2017 when the engineers of the Paris metro expansion project specified the exclusive use of Sandwich Belt high angle conveyors at the shaft, for elevating the TBM muck to the surface. The significance of the specification is that they knew that the muck would be sticky and they recognized the distinct Sandwich Belt advantage. The use of all conventional conveyor equipment including smooth surfaced rubber belts ensured that the material would discharge completely and the belts could be scraped clean with standard belt scrapers.

The latest development, starting in 2019, the patented TBM Sandwich Belt high angle trailing conveyor takes the muck clearance function to the trailing deck of the TBM when large tunnels must be excavated at any high angle. With this expanded capability the future looks bright for the continued success of Sandwich Belt high angle conveyors in construction and tunneling projects.

## REFERENCES

Dos Santos, J.A. and Frizzell, E.M., Evolution of Sandwich Belt High-Angle Conveyors. CIM Bulletin. Vol.576, Issue 855, July 1983, pp. 51–66.

Dos Santos, J. A., Sandwich Belt High Angle Conveyors Exclusively at Paris Metro Expansion-2019. Presented at Rapid Excavation and Tunneling Conference-2019, published in Proceedings, 10 pages, 17,18 June 2019, Chicago, IL, USA

dos Santos, M.J., Adder Snake: Low-Angle to High-Angle with no Transfers, IMHC, Presented at Beltcon 20, 2019, Johannesburg, South Africa, published in the conference proceedings.

Conveyor Equipment Manufacturers Association (CEMA). Belt Conveyors for Bulk Materials. Conveyor Equipment Manufacturers Association, Naples, FL, USA, 7th ed., 2013.

Dos Santos, J.A., (1998). The Cost/Value of High Angle Conveying. Bulk Solids Handling, 18, No. 2. 253–260.

Dos Santos, J.A. (2000). Theory and Design of Sandwich Belt High Angle Conveyors According to The Expanded Conveyor Technology. Society for Mining, Metallurgy & Exploration. Salt Lake City, February 28–March 1, 2000, Salt Lake City.

# Using the Digital Geologist to Count Cobbles at the RSC7 Tunnel

**Shane Yanagisawa** ▪ Osprey Engineers LLC
**Dan Preston** ▪ ClaroVia Holdings LLC
**Don Deere** ▪ Deere & Ault Consulting Engineers
**Robert Marshall** ▪ Frontier-Kemper Constructors Inc.

## ABSTRACT

Geotechnical Baseline Reports often estimate the number of cobbles that will be encountered in terms of volume or weight. These estimates are practically impossible to verify in the field. ClaroVia Holdings in conjunction with Frontier Kemper Constructors Inc. developed Digital Geologist 2.0 to classify and count cobbles discharged from an EPB TBM at the RSC7 project in Burbank, CA. The system worked by using two stereoscopic laser sensors with a global shutter to scan and count cobbles at 12 frames per second. Results of the effort are revealed along with lessons learned and paths for improvement.

## INTRODUCTION

Cobbles are an important consideration when reviewing any Geotechnical Baseline Report for a soft ground EPB TBM tunnel. How many times have you read a GBR that says, "For Reach *Y*, Cobbles are anticipated to be *X*% of the total excavated volume of material"? Engineers use this language to better define the ground and set a baseline. Contractors review this language closely when they are bidding on the work or when they are wondering whether the ground presents a differing site condition.

Defining the cobble quantity is one thing, counting cobble is another. A field determination of the cobble content can be done by weighing the muck cars and then separating out the cobbles using a bar screen. The cobbles are then weighed and the cobble percentage by weight of muck is calculated. Conveyor belt scales are not accurate enough when one is trying to measure cobbles to an accuracy of 1%. If the baseline given is by volume, then the cobbles are weighed, and the volume calculated based on a sample cobble specific gravity determined laboratory test. Sorting out and weighing cobbles on a continuous basis is not practical or economical.

Face mapping by a Registered Engineering Geologist can be done, but not continuously and the interpretation and mapping can be subject to dispute. The geologist must be ready to go in the cutterhead when the TBM stops for cutterhead inspection. Sometimes the exposed face is unstable and the TBM must advance ahead a few more rings in hopes of better ground. If the entry is being done under compressed air, the geologist must go in and out through a compressed air lock with the rest of the crew. Clearly, a better way is required to quantify cobbles.

## THE PROJECT AND THE PROBLEM

The Los Angeles Department of Water and Power (LADWP) contracted with Frontier-Kemper Constructors, Inc. (FKCI) to complete the River Supply Conduit Improvement Upper Reach–Unit 7 (RSC7) Project. Tunnel Segment 10 of the project was excavated by a 4.12 m (13.51 ft) diameter Earth Pressure Balance Tunnel Boring Machine (EPB TBM) advancing 3,584 m (11,754 ft) or 2,602 tunnel liner

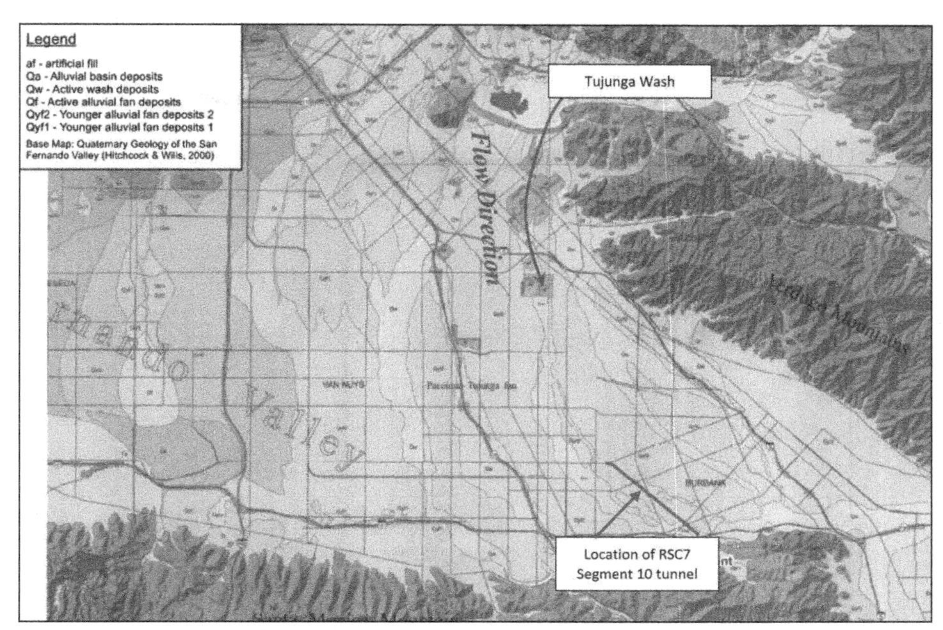


**Figure 1. Location of the RSC7 tunnel in the San Fernando Valley**

rings 1.37 m (4.5 ft) wide. The TBM holed through on March 15, 2022, 597 calendar days after launching on July 26, 2020, and 236 days behind the baseline schedule completion date.

The tunnel traverses the Tujunga Alluvial Fan (Figure 1) and encounters a combination of Wash Deposits and Alluvial Fan Deposits. These deposits are labeled in the contract documents as Young Alluvium. The Young Alluvium is described as sand with relatively large percentages of gravel with some silt. Within the fan deposits are storm channels or channel lag deposits that are known to contain coarse gravel, cobbles and boulders. These channels, as is the main channel of Tujunga Wash, are oriented generally parallel to the tunnel drive.

During the tunnel drive, FKCI thought that the tunnel alignment could be following an old storm channel of the Tujunga Wash infilled with cobbles and large gravel. As a result of these geologic conditions, FKCI believed that the 76 to 305 mm (3 to 12 inch) cobbles encountered were substantially more than the 7% by weight of excavated materials baseline given in the Geotechnical Baseline Report (GBR) and constituted a Differing Site Condition. FKCI began documenting the number of cobbles by having a Certified Engineering Geologist map the face and assess the percentages of cobbles. Mappings were done at 10 locations starting at Station 89+87 and ending at Station 49+85, a span of 4,002 ft. These mappings yielded an average cobble content by weight of 22%. A claim was submitted to the Owner alleging a Differing Site Condition (DSC) based on the number of cobbles mapped but the claim was denied.

## DEVELOPING A TECHNICAL SOLUTION TO COUNT COBBLES

FKCI engaged CVH to adapt and install a system that could classify and count coarse gravels and cobbles by direct three-dimensional measurement as they moved up the conveyor belt immediately after discharge from the TBM screw auger.

**Figure 2. The guts of the Digital Geologist 2.0**

The goals of the Digital Geologist were to:

1. Capture 100% of the mined materials with a direct measurement system and a High Definition (HD) image
2. Using the direct measurement data collected, generate a 3-dimensional light coded depth map using state of the art technology developed by Intel
3. Provide a reliable method to count cobbles and large gravel
4. Convert the cobble and gravel volumes to weights
5. Calculate % by weight of cobbles as a % of weight of the mined materials by ring advance

Digital Geologist 2.0 was assembled from commercial off-the-shelf technologies to be a data generation and data logging system configured to operate as a real-time "digital sieve." The underlying technology was based on hardware and software developed by Intel and sold as Realsense D435i. DG 2.0 was based on a Class 1 laser projector and 2 global shutter infrared imagers in a stereo solution offering high quality millimeter accurate depth measurements for a variety of applications. The DG 2.0 also had a single high-definition imager for validation. These images could be used by personnel to validate output data produced by the Realsense imagers.

The Realsense D435i imagers scan an object and transform it into voxels. Voxel is analgous to a "pixel". While a pixel represents a picture element, a voxel represents a volume element, a value on a grid in three-dimensional space.

The height of each voxel against the baseline depth or height map yields a column 1 mm by 1 mm. When multiplied by height it equals the volume of the column in cubic millimeters (Figure 3). When filtered by rules and integrated across the image a 3-dimensional polygon(s) emerge with volume in cubic mm. In the images below (Figure 4) is an example taken by the DG 2.0 at RSC7. On the left is a light encoded 3-dimensional rasterized image of the high definition (HD) image on the right. The left image is a visual representation of a 3-D point cloud

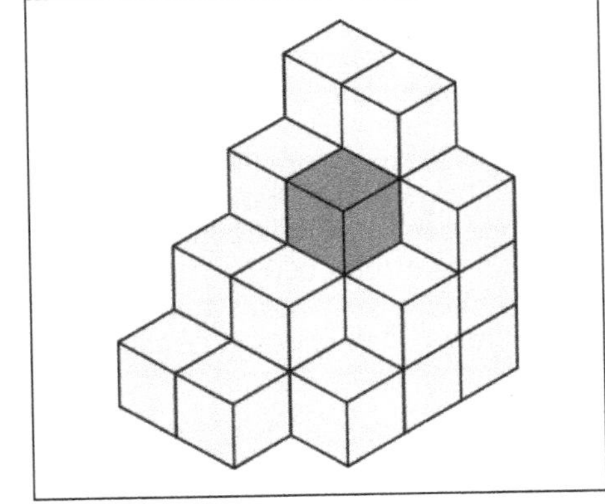

**Figure 3. Voxel object**

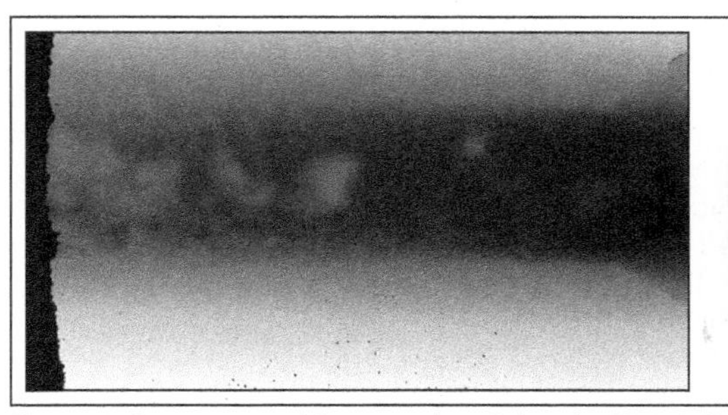
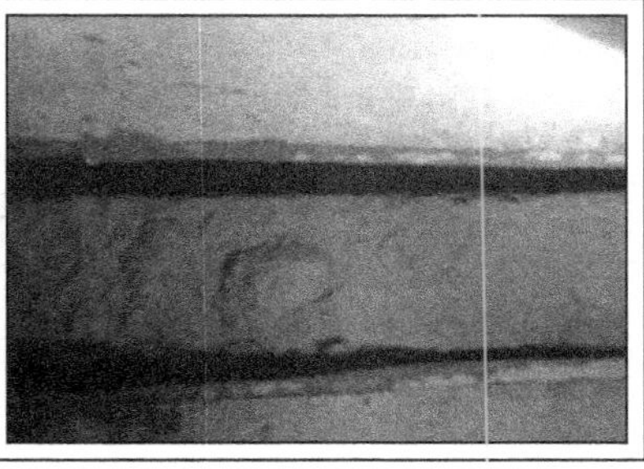

**Figure 4. Image created from voxels at left and a high-definition pixel image at right**

in a 2-D image. The voxels are 1 mm square boxes color coded to a shade of gray that represents elevation; or simply stated, it is a 3-D image presented in a 2-D coded format. The blur in the image is a function of resolution of the voxels 1 mm square.

The Realsense 435i generally operates as a direct measurement system where the geometry of parallax in the stereoscopic imagers generate a distance forming a heightmap as represented by a point cloud in the form of a light coded rasterized image. The right image in Figure 4 is the visual image typical in conditioned soil/muck.

The imagers were set to scan and capture 100% of the materials discharged from the TBM screw auger onto the conveyor belt. Frame rates were set high to ensure that subsequent analysis would be accurate. The images had to be stitched together so that no material was missed or counted multiple times. The images were broken into "slices" that represented a portion of the material as the belt moved across the imager's field of vision. The counting algorithm flagged material that had been measured in previous frames. Rather than erroneously adding the value to the volume total, the overlapping portion of these slices were averaged to limit the effect of any noise in the bulk volume data.

There were three embedded controllers located in the DG 2.0 housing configured for the data collection and logging needs, external communications, web interface and real-time analysis and reporting. There was also an interface to the TBM PLC for signaling from the PLC. An analog 4–20 ma signal transmitted belt speed information and a fiber connection for data collection, movement, and storage.

## MEASURING A COBBLE

The hardware used to assemble the DG 2.0 can be bought by anyone. The key technologies developed by ClaroVia are the patented algorithms used to reliably sift through the massive amounts of data generated by the stereoscopic sensors. The sensors produced numerical data which was input into a huge matrix and manipulated by algorithms after the data was uploaded from the jobsite to CVH's offices. The algorithms accounted for noise in the data, scanned the belt, and located and classified the cobbles by size. During post-processing, the DG 2.0 was able to take raw imaging data and rerun the data with different filters to classify the rock in different ways. The operation of the DG 2.0 is described below in non-technical terms.

One of the challenges was the presence of conditioned ground (Figure 5) defined as ground to which a soil conditioner has been added during the TBM excavation process to produce a more plastic mass of material. The ground conditioner was made of water, foaming agent, and air. The conditioned ground was opaque, and the instruments selected were not configured to image below the surface. Any cobble or large gravel masked by the conditioner was not counted.

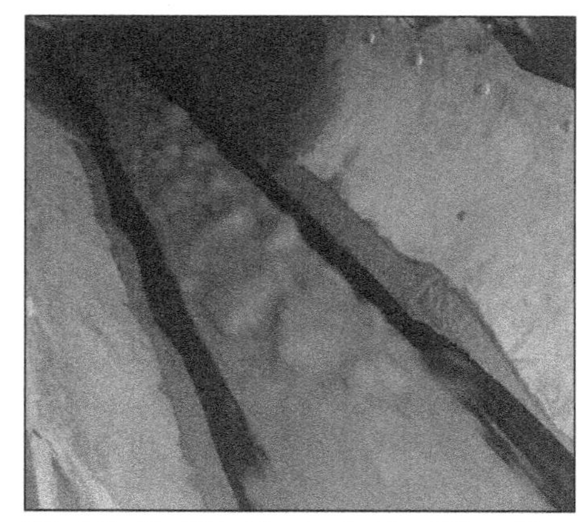

Figure 5. Conditioned ground with cobbles whizzing by on the TBM conveyor belt

Prior to starting a TBM advance cycle, the empty running belt was scanned to determine belt position relative to the stereo scanner resulting in a baseline measurement of ±1 mm. The conveyor surface was assumed in later calculations to be 10 mm above the measured surface to account for any bounce in the running belt. The bounce was determined to be nearly zero, but the 10 mm cut-off was kept as an added measure of conservatism and became the baseline in the calculations.

When a cobble passed under the DG 2.0, the profile of the muck was determined by direct measurement and differenced from the baseline taken at the start of the advance. The minimum threshold for a 3-inch cobble was set at 78 mm (3.07 in). The DG 2.0 can be configured to make volumetric determinations based on one of three configurations:

- Cylindrical
- Spherical
- Reflection

DG 2.0 was configured to implement the most conservative configuration which is the reflection method shown in Figure 6. Measurements made above the muck are reflected and inverted. When the baseline is subtracted, any material projecting below the baseline is eliminated. Any part of the cobble below the conditioned muck outside the reflected volume did not contribute to the volume determination.

If there was no muck masking the cobble, the reflection line was made at the last part of the cobble visible to the scanner. Again, the cobble 3D image was inverted about

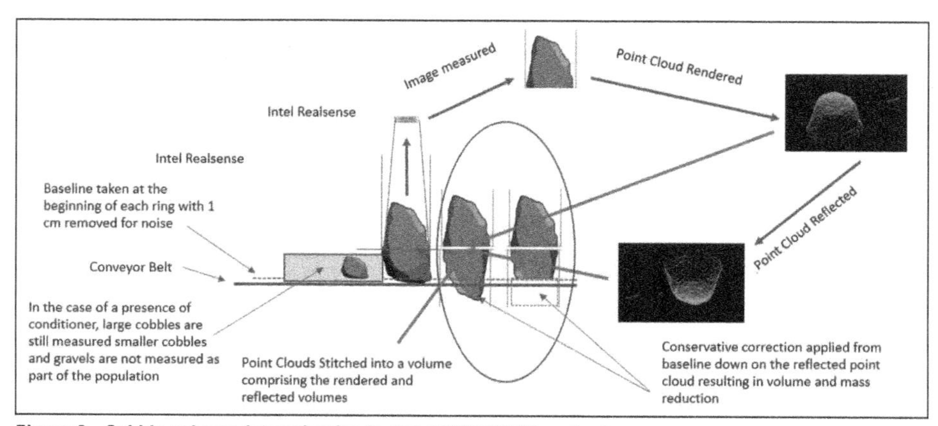


Figure 6. Cobble volume determination by the REFLECTION method

the reflection line and the part projecting below the baseline was cut off. Cobbles tended to lay with the widest, heaviest, part of the cobble on the conveyor belt surface and the reflection method would not count this part of the cobble if it was below the muck line.

## COMMISSIONING THE DIGITAL GEOLOGIST FROM RINGS 1450 TO 1573

The DG 2.0 was mobilized, calibrated, and tested from Rings 1450 to 1573. During the commissioning period between Rings 1450 and 1573, an extensive validation exercise was done using known object dimensions on the belt.

The DG 2.0 was mounted directly over the conveyor belt about 60 cm (2 ft) behind the TBM screw auger discharge and used multiple 3-dimensional direct measurement systems for volumetric determination (Figure 7). The direct measurements were taken at 12 frames per second to classify cobbles over 76 mm (3 in) and large gravel over 38 mm (1 ½ in). The scanned cobble volume was calculated. A density derived from lab data of cobble samples was then applied to the cobbles to compute a weight.

Since the cobbles and boulders were baselined in the GBR as the percent weight of excavated material, the total weight of muck removed from the tunnel also had to be known. The contract specifications required that FKCI keep and maintain a Mass Balance Sheet that tracked the weight of material excavated for each ring. FKCI tracked the weight of all the tunnel muck discharged from the beginning of tunneling activities. The total weight from each 4.5 ft advance of the TBM was measured by a remote reading below-the-hook load cell on the shaft crane that measured the weight of every muck car. Adjustments were made for the amounts of water, bentonite, and soil conditioner added to arrive at weight for each tunnel liner ring advance.

A human technician using images from the high definition camera spot checked the objects analyzed by the DG 2.0. After reducing and analyzing the first 50 ring excavation cycles, we identified challenges that impaired some of the data. The challenges identified and resolved included:

### Belt Speed

The reported belt speed from the TBM was not its actual speed. The reported speed was actually the belt set by the TBM PLC and sent to the conveyor belt drive. As the belt loaded and moved materials, the belt speed shifted up to 0.5 m/sec which resulted in either an over sampling or under sampling. This issue was solved by analyzing the

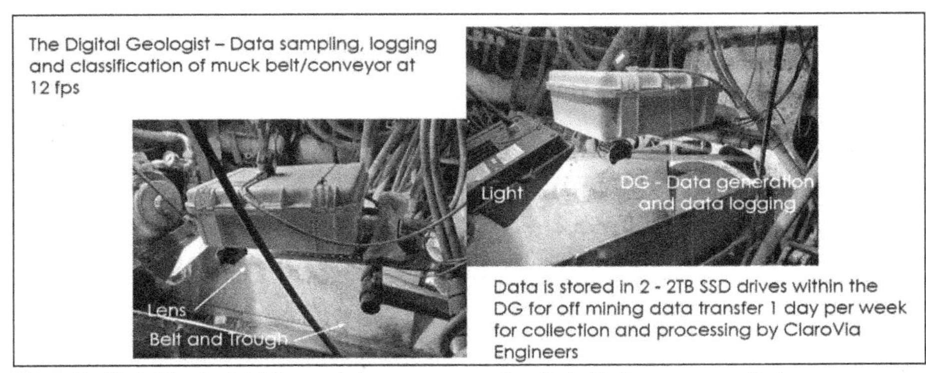


Figure 7. Digital Geologist 2.0 mounted over conveyor directly behind screw auger discharge

data to determine belt speed to ensure a sliced and stitched belt images represent a 100% sampling without over or under sampling.

### Image Data Corruption

Relying on miners to clean the imager lenses between rings was only a partial solution. The DG 2.0 was located close to the screw discharge and mud would often splash on the lenses during TBM advance. Specks of mud on the imager lenses would result in false readings of depth on some voxels gathered by the 3-D imagers. Post processing could average the adjoined voxels to detect any unexpected jumps in volume for adjoining voxels, if these were detected, the voxel was flagged, and these anomalies were eliminated and corrected. Sometimes the imager lenses were covered by too much mud to render a usable image. Although enough data was collected to meet statistical requirements, the mud problem persisted to the end of the project

### Low Voltage Supply

Blocks of data missing, or incomplete data were correlated to error logs generated by the image processor. These errors were identified as "under volt and over temperature" shutdowns and each event required 15 min auto restart. Some low voltage events were correlated to welding repairs done on the TBM during a maintenance period. An uninterrupted power supply was installed ahead of the DG 2.0 to eliminate damage from low voltage events.

### Raw Data

The data collected and transferred was a constant and time-consuming element of the post processing. Two Network Attached Servers were installed; one located onsite, and one at CVH. Ultimately, the stored raw data was in the range of 20 terabytes of data for 1028 rings or 19 gigabytes of data per ring. This data was transferred to the CVH lab for analysis. The data included image data, raw depth images, baseline belt data, differenced images from the baseline, HD images and metadata.

### Constrained Processor Resources

Up to 36 hours of processing on a desktop computer were required to analyze all the stereoscopic images for a single TBM liner ring advance of 1.4 m (4.5 ft) due to the immensity and complexity of the image data. Even though the results of the analyses were good, running multiple analyses to ensure repeatability took another 1 to 2 weeks per ring cycle per run. A dedicated data processing machine was built that analyzed and reduced the data processing to 3–4 hours for each TBM ring advance

## DIGITAL GEOLOGIST IN FULL OPERATION

The Digital Geologist 2.0 was in full operation for the last 4,633 ft. of the tunnel from Stations 58+04 to 11+71 (Ring 1574 to 2602). Station 11+71 and Ring 2602 are at the terminal end of the tunnel. Following the data collection phase, post processing analysis was conducted according to ANSI/ASQ methods Z1.4 and Z1.9. Please note that in this discussion the words "strata" and "stratum" are used in a statistical sense, not a geological one. Initially, a random sampling methodology was planned. The muck from all the 1,028 TBM ring advances would be scanned, and a random sampling would be used for analysis. Statistical theory says that with a population of 1028 rings, 379 samples were required to have a confidence interval of ±4%. Computer image processing was a slow and expensive process so not all viable data was processed.

However, the data collection was impacted by several events that rendered a portion of the data unusable. Because the DG 2.0 was close to the discharge of the screw auger, mud would occasionally splash on the digital imager lenses that were not immediately cleaned. There was a low voltage event that caused the DG 2.0 to shut down. Sometimes the number of images taken was insufficient.

CVH was able to objectively shift analytical methods from random sampling across the population to a stratified random sampling. This was done by applying objective filtering to form the data strata. The strata were collectively and mutually exclusive in that every element in the population was assigned to one and only one strata. The data from these 1,028 rings was divided into 5 strata:

Strata 1—Unusable Data 311 rings or 30.25%
  a. 269 ring data with obstructed measurements (e.g., mud on lens)
  b. 42 ring data with low voltage—over temperature shutdowns
Strata 2—Insufficient Data—138 rings or 13.4% with less than 10,000 images per ring available to process
Strata 3, 4, 5—Viable Data—579 rings or 56.3% random samples including
Strata 3 Viable Data with <7% high density materials by weight
Strata 4 Viable Data with >7% high density materials by weight
Strata 5 Viable Data with >>7% high density materials by weight

The data processed for 416 rings (out of 579 viable rings) represented a sample size of 40.3%. Data processing was stopped after 416 rings as the criterium for a 4% confidence factor was met. The remaining 165 rings could still be analyzed if necessary. The 416 ring data sets were selected at random and no cherry picking of data was done.

## DATA ANALYSIS RESULTS

Upon completion of the analysis, the results determined that the population of data yielded an arithmetic mean of cobbles greater than 78 mm (3.07 in) by weight of the excavated materials calculated to be 11%, with a confidence of 95%, and a confidence interval of 3.71%. What this means is that any ring not analyzed within the total population of 1028 rings will have a 95% confidence that cobble volumes will be at the arithmetic mean of 11% with a confidence interval of 3.71% around the arithmetic mean (or ±1.86% margin of error (which is ½ the interval). For example, if a typical ring had a dry soil weight of 32,658 kg (72,000 lbs), we would have a 95% confidence the weight of cobbles in excess of 78 mm (3.07 in) would weigh 32,658 kg × 0.11 = 3,592 kg ±1.86% (or ±67 kg). This yields an uncertainty of 133 kg represented as a range of 3,525 to 3,630 kg or 10.8% to 11.2%. The populations represented by these numbers are in the Strata of 3, 4 and 5. Strata 5 included the sample population of interventions and cleanouts but also some very heavy rings during normal TBM advance.

In addition to cobble volumes greater than 76 mm (3 in), the DG 2.0 was able to quantify large gravel visible above the conditioner, or in the absence of conditioner. The mean value of the large gravel 38 to 76 mm (1½ to 3 in) by weight of the excavated materials was 2% of the excavated material detected at the same 95% confidence and confidence interval of 3.71% around the mean. Only large gravel visible on the conveyor belt was counted. The majority of gravels in this size range were probably not visible and not counted as a result of being buried in the muck.

Figure 9 shows the calculated percentages of cobble for individual ring cycles. Only the 416 valid data points are shown. The data is shown with the TBM drive from right to left. The heavy dark line is the 20-ring moving average. The horizontal line represents

Figure 8. Layer of cobbles and coarse gravel in the tunnel face

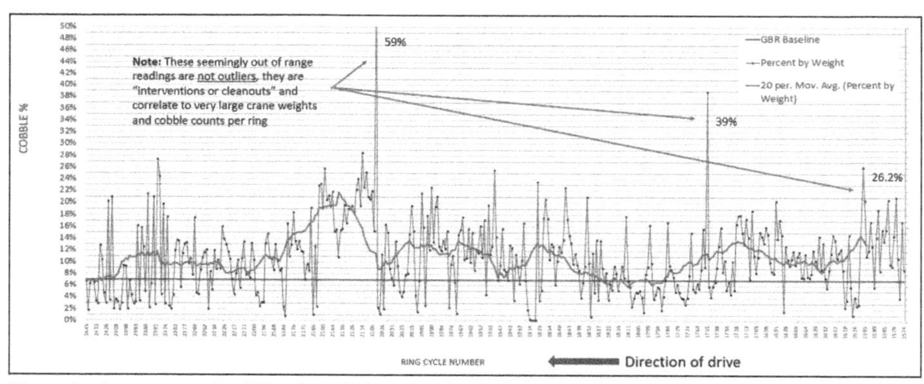


Figure 9. Percentage of cobbles by weight sorted by liner ring number

the cobbles and boulders 7% by weight guideline from the GBR. Within the analyzed data sets of the 1028 ring cycles, there was observed a low of 0.2% and a high of 52% weight of cobble to weight of the mined materials within the ring cycle. Outliers and all suspect values were evaluated. The 3 large peaks identified above are real and coincide with interventions and they also coincide with an extreme muck car ring weights. The graph clearly shows the high variability of the cobble content.

Figure 10 pie chart shows that 69% of the rings sampled exceeded the 7% GBR guideline.

The arithmetic mean of 11% does not adequately describe the ground. As was shown in CVH's report, there was a wide range of cobble percentages in the data with a maximum of 52% and a minimum of 0.2%. This information would not have been available but for the Digital Geologist 2.0.

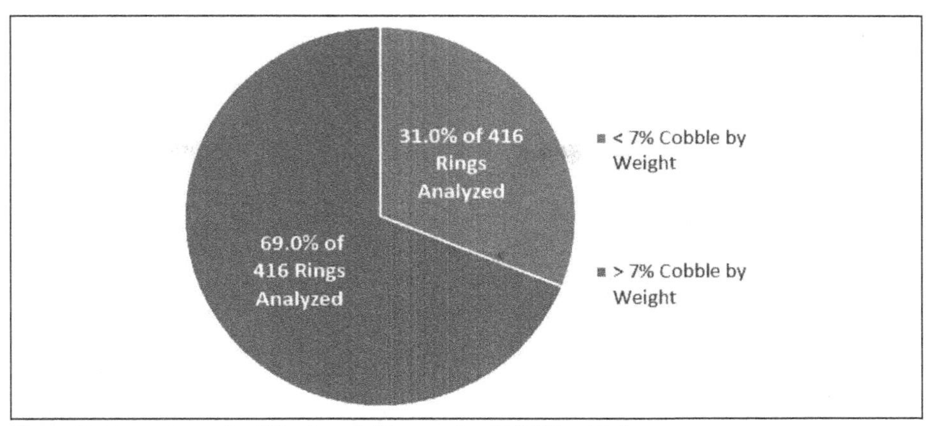


**Figure 10. Percentage of cobbles by weight exceeding GBR 7% guideline**

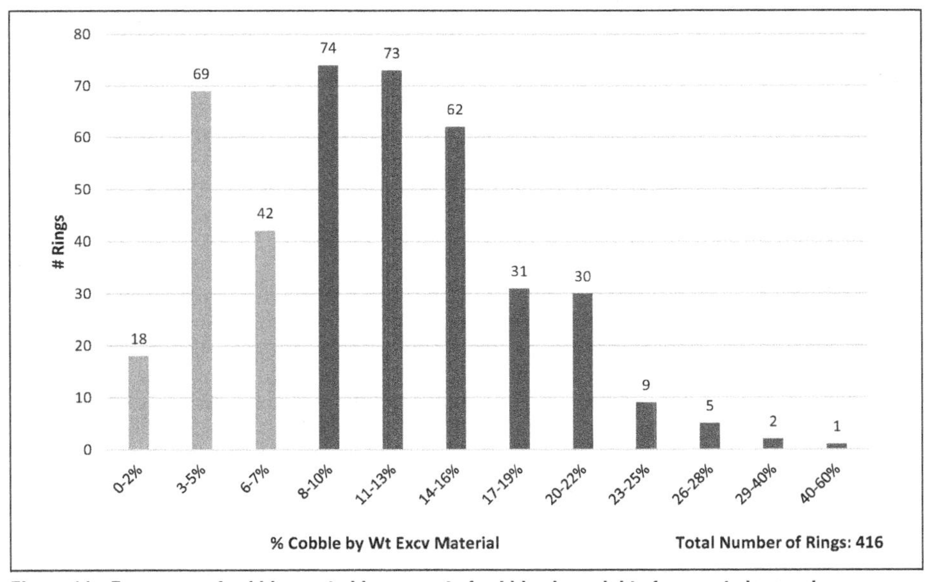


**Figure 11. Frequency of cobbles sorted by percent of cobbles by weight of excavated ground**

Because the DG 2.0 measured every visible cobble in the 416 ring dataset, we can also classify the cobbles by size. Figure 11 shows the percent cobbles by excavated weight sorted into categories based on the frequency of detection by ring. For example, the green bar that the far left says that there were 6 sample rings where the cobble weight was between 1 and 3% of the excavated weight.

The green bars on left show there were 129 sample rings where the cobbles were 7% or less of the excavated materials by weight. The red bars show 287 Rings with cobble weight greater than 7%. The largest occurrences shown were for 209 rings where the cobbles were between 8 and 16% by weight. A significant number, 70 rings or 17% of the total were in the 17 to 25% range.

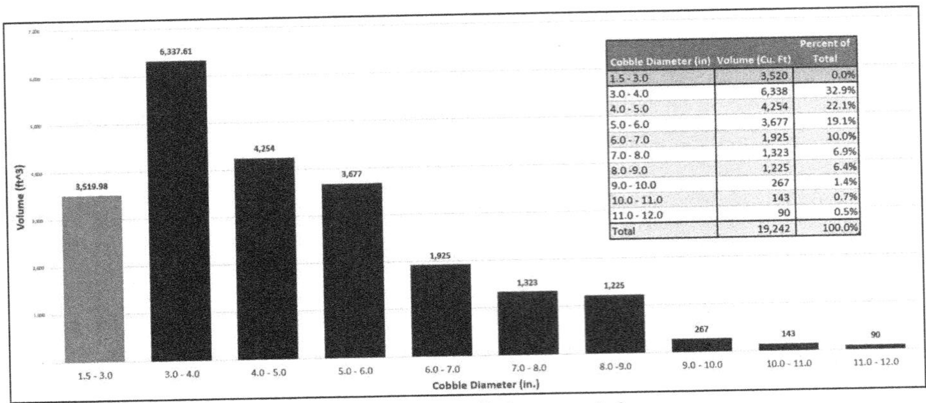


| Cobble Diameter (in) | Volume (Cu. Ft) | Percent of Total |
|---|---|---|
| 1.5 - 3.0 | 3,520 | 0.0% |
| 3.0 - 4.0 | 6,338 | 32.9% |
| 4.0 - 5.0 | 4,254 | 22.1% |
| 5.0 - 6.0 | 3,677 | 19.1% |
| 6.0 - 7.0 | 1,925 | 10.0% |
| 7.0 - 8.0 | 1,323 | 6.9% |
| 8.0 -9.0 | 1,225 | 6.4% |
| 9.0 - 10.0 | 267 | 1.4% |
| 10.0 - 11.0 | 143 | 0.7% |
| 11.0 - 12.0 | 90 | 0.5% |
| Total | 19,242 | 100.0% |

**Figure 12. Large gravel and cobbles by volume for 416 rings sampled**

Figure 12 shows the volumes of the large gravel and cobbles sorted by size counted during the excavation of 416 rings. The large gravel is not included in the 557 m$^3$ (19,242 cf) total shown in the table. The size distribution of the cobbles is the normal distribution one would expect to find in an active stream bed. The quantity of large gravel is low because the DG 2.0 can only count the visible large gravel. Most of the gravel is hidden under the conditioned muck. The amount of gravel exceeds what was indicated by the contract geotechnical documents. If we make a projection for the large gravel that follows the charted size distribution for the cobbles, the amount of large gravel would be about 2.5 times what is shown or about 5% by weight of excavated materials.

The dimensional accuracy of the measurements was ±1 mm and the reflection method of volume assessment was conservative. Any part of a cobble from 0 to 10 mm above the conveyor belt was not included in the volume calculation. Larger dimensions of the cobbles hidden by the muck were not counted. This may explain why the DG 2.0 averaged 11% cobbles by weight of excavated material while the geological mapping averaged 22% over the span of 10 face mappings. Another explanation is that there were more cobbles in the area assessed by face mapping.

The DG 2.0 also counted large amounts of gravel in the 38 to 76 mm (1½ to 3 in) range although only the visible gravel was counted. The first 10 mm above the belt was excluded, therefore the counts for gravel are very conservative.

## NO RELIABLE CORRELATION BETWEEN COBBLE COUNT AND MUCK CAR WEIGHT

Since we weighed all the muck cars with good accuracy, we wondered if there was a correlation between muck car weight and cobble count. We did not find a reliable correlation. Variations in the surrounding soils matrix density overshadowed the cobble density. Actual soil densities at the TBM reception shaft varied from 1,842 to 2,210 kg/m$^3$ (115 to 138 pcf) with an overall density of 1,986 kg/m$^3$ (124 pcf). Consider the chart in Figure 13 that shows theoretical bulk densities for the excavated soil as the percentage of cobbles by weight is varied. For the calculations, an initial density of 2,210, 1,986, and 1,602 kg/m$^3$ (138, 124, and 100 pcf) were used. For reference, compact dense clean sand has a density of about 1,602 kg/m$^3$ (100 pcf) and wet concrete typically has a density of 2,322 kg/m$^3$ (145 pcf).

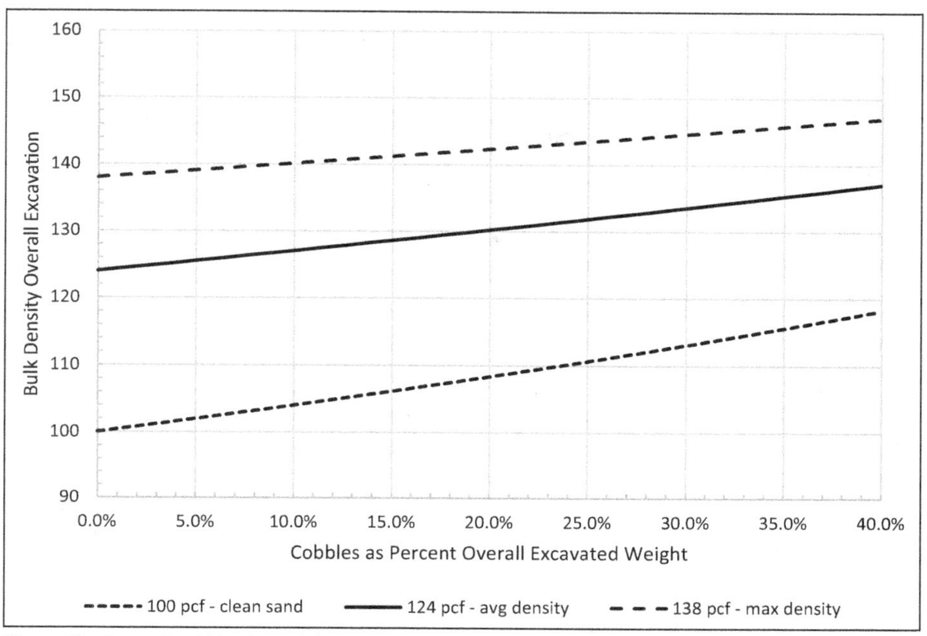


**Figure 13. Percent cobbles by weight versus bulk density of excavated ground**

The graph shows that bulk soil densities did not vary much with cobbles variations that would clearly constitute as DSC if they occurred. Using the average density of 1,986 kg/m$^3$ (124 pcf), a 10% variation in cobble count by weight only results in a 48 kg/m$^3$ (3 pcf) change in overall bulk density. Even a lighter soil such as clean sand shows only a 64 kg/m$^3$ (4 pcf) change with a 10% change in cobble count.

Face mappings showed that soil matrix was always in different layers, never a completely uniform face. The face often showed sand layers interlaced with extremely dense layers of cobbles, large gravel, and sand. The point is that variability in geology resulted in swings in bulk density that were not necessarily related to cobble density. The cobbles must be tracked by screening, face mapping, or electronic means such as the Digital Geologist.

## CONCLUSION

The findings of the Digital Geologist were used to prepare a second claim presented to the LADWP. This claim was also rejected and is now going through the dispute resolution process, so we are not able to comment on every aspect of DG 2.0's performance. Looking ahead to the future FKCI and CVH intend to use this technology on future EPB projects with improved technology and gain industry acceptance for its use.

The RSC7 tunnel was an ideal candidate for counting cobbles using the DG 2.0 optical technology. The ground was largely absent of plastic fines that would produce a tunnel muck that could blanket the cobbles and make them uncountable to optical technology. Clay soil was nearly non-existent, so the DG 2.0 didn't have to worry about distinguishing cobbles from clay balls. For the most part, the cobbles stood out clearly on the conveyor belt. For the next project CVH is looking into the use of millimeter wavelength radar to count cobbles from above and under the conveyor belt.

On the next project, we intend to have a more rigorous program to verify the percentages of cobbles by comparing them to a field sieve analysis of the tunnel muck. This program will also establish how conservative the Reflection Method used to establish volumes of cobbles is compared to the actual cobbles encountered.

Future projects will have all data analysis and reduction done in real-time onboard the Digital Geologist eliminating the need for post-processing and providing more immediate feedback to site management. The Communications Management Unit will present the data volumes via a local web interface. Reports will be generated and electronically transferred as configured.

Geotechnical Baseline Reports would do better to specify the cobbles in terms of volume rather than as a percentage of the excavated weight. This would be a more direct measurement that would eliminate the step of calculating the weight of all the material excavated from the material. In ground known to have cobbles and boulders, Hunt (2017) recommends the use of Boulder Volume Ratio methods which screens larger samples than possible with conventional borings to determine cobble and boulder volumes. As stated by Hunt (2004), a focused, boulder-sensitive subsurface investigation and proper baselining are essential information to communicate risks to the contractor and allow better tunnel risk mitigation

## REFERENCES

Hitchcock, C.S., and Wills, C.J., 2000, Quaternary Geology of the San Fernando Valley, Los Angeles County, California, Map Sheet MS-050, California Division of Mines and Geology.

Hunt, S.W., Tunneling in Cobbles and Boulders, 10th Annual Breakthroughs in Tunneling Short Course 2017, Chicago, IL.

Hunt S.W., 2004. Risk Management for Microtunneled Sewers, Proceedings of Collection Systems 2004: Innovative Approaches to Collection Systems Management, Water and Environment Federation, Alexandria, VA, Paper 9D, F2004.6, 15p

# Using GIS Application for Inlet/Outlet Tunnel Geologic Mapping for Ground Support (CH Reservoir Case Study)

**Nadav Bar-Yaakov** ▪ Stantec
**Gregory Raines** ▪ Stantec
**Eric Zimmerman** ▪ Stantec
**Cory Bolen** ▪ Stantec

## ABSTRACT

The Chimney Hollow Reservoir Project broke ground in August 2021, constructing the tallest Asphalt Core Dam in North America, intending to improve reliability of drinking water supply to Northern Colorado communities. Aiming to expedite and streamline the geological mapping process in the tunnel, Stantec's teams have developed a mobile and web-based GIS application for digital mapping input and rock mass evaluation, allowing real time decision making for assigning tunnel support class and resulting in a faster more efficient process. The following article presents and demonstrates the GIS mobile and web application-based process, the mapping results and review additional uses of the database created.

## THE CHIMNEY HOLLOW PROJECT

As Northern Colorado's growth continues, the region will need collaborative projects that not only support meeting its future water demands, but also protect the environment and wildlife, help maintain local food production, and preserve the region's quality of life. The Chimney Hollow Reservoir Project, which is the main component of the overall Windy Gap Firming Project, is an example of such a project. This effort is a collaboration between nine municipalities, two water districts and a power provider, working together to build Chimney Hollow Reservoir. The new reservoir will be located just west of Carter Lake in southern Larimer County and will store 90,000 acre-feet of water when built—slightly smaller than neighboring Carter Lake (112,230 acre-feet). The project will provide a firm yield of 30,000 acre-feet annually, which will meet a portion of the future demands of the participants. The reservoir connects to existing infrastructure by storing water diverted into Windy Gap's system, which pumps water into Lake Granby and uses Colorado Big Thompson (C-BT) Project facilities for delivery to participants. This project will not take away water from agriculture or other users but would utilize the water rights associated with the existing Windy Gap Project, which has been delivering water since 1985. Once built, Chimney Hollow Reservoir will provide the additional storage needed to make the Windy Gap Project a more reliable delivery system for the residents who currently depend on it, as well as their children and grandchildren.

After receiving final approval from the federal government in 2017, preconstruction work began in December 2019 and included non-site activities. Barnard Construction, the construction firm hired for the project has started the onsite construction in August 2021 and is expected to last four years. The construction cost is being contracted at $485.4 million.

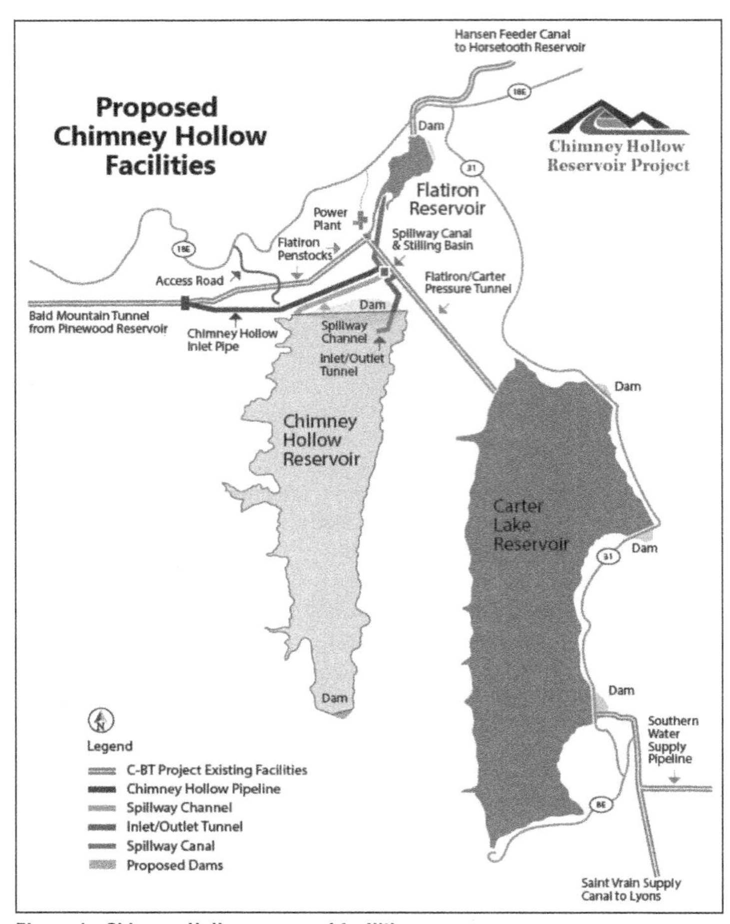


**Figure 1. Chimney Hollow proposed facilities**

## The Inlet/Outlet Tunnel

The Chimney Hollow Reservoir Inlet/Outlet (I/O) works are designed to be used for first filling of the reservoir, and for the day-to-day reservoir filling/water distribution during the reservoir operation period. The I/O tunnel is designed to house a 72" steel conveyance pipe and is comprised of the following elements: Down- Stream Tunnel, Valve Chamber, Grouting Chamber and Upstream tunnel. General information about the different tunnel elements is presented in the Table 1.

## The Geological/Geotechnical Setting

The Chimney Hollow Valley strikes roughly north-south and exhibits characteristic Front Range hogback topography with a relatively gentle western slope and a steep eastern slope. The tunnel is located in the Fountain Formation (locally around 1,150 ft thick) on the eastern side of the valley, which typically consists of interbedded sandstone, claystone, and shale sedimentary rocks at the site, dipping shallowly (typically from 10° to 35°) to the east. Along the steep escarpment on the eastern side of the valley, the sedimentary Fountain Formation is overlain by the Ingleside Formation (hard sandstone) at the top of the hogback.

**Table 1. General information- Chimney Hollow I/O tunnel**

| Tunnel Element | Total Length, ft | Section type | Excavated Diameter, ft | Inner lining |
|---|---|---|---|---|
| Downstream tunnel | 670 | Horseshoe/circular | 25/27 | 24"/18" concrete lining |
| Valve Chamber | 50 | Circular | 30 | 24" concrete lining |
| Grouting camber | 30 | Horseshoe | 11 | Concrete backfill |
| Upstream Tunnel | 1,200 | Horseshoe | 11 | Concrete backfill |

**Table 2. Discontinuities information- Chimney Hollow**

| ID | Discontinuity Type | Dip/Dip Direction, ° | Discontinuity Spacing |
|---|---|---|---|
| B-1c | Bedding Primary | 15/091 | Moderately close to wide |
| J-1c | Joint Secondary | 71/273 | Moderately wide to wide |
| J-2c | Joint Tertiary | 79/322 | Wide to Very wide |
| J-3c | Joint Secondary | 85/213 | Wide to Very wide |
| J-4c | Joint Secondary | 22/268 | Very wide |

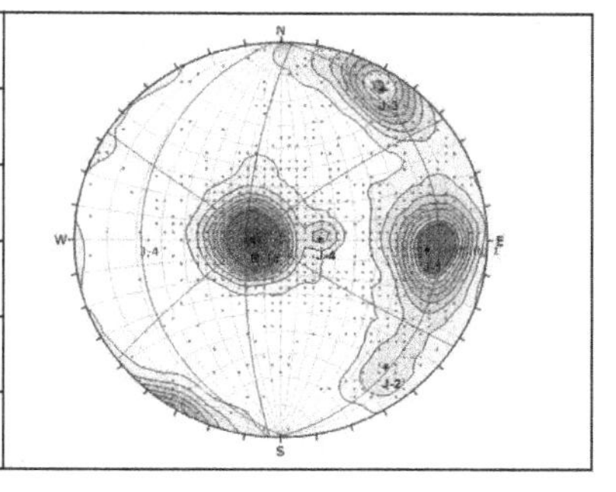


During the design phase it was envisioned that Carter Lake will likely have a major impact on the groundwater levels along the tunnel, and that groundwater inflow during construction has been almost exclusively through discontinuities in the rock mass.

Studies done during design stages suggested that the probability of seismic shaking at the site during the construction period is estimated to be extremely low.

The Discontinuities sets mapped during the site investigation are presented in the Table 2.

### Initial Support Design

Excavation of the tunnel was done using the drill and blast method, where the initial support design, included 4 support types as a function of the rock mass quality, as presented in Table 3.

## DEVELOPMENT OF THE GEOLOGICAL MAPPING APPLICATION

### Requirements for Geological Mapping in Chimney Hollow

The geological mapping in dam projects is essential for validating design assumption regarding foundation conditions, fill production, dam performance and identifying adverse conditions that may affect the stability/safety of the dam during the lifetime of the project. Identifying such conditions may results in additional works done to eliminate/reduce the risk involved (e.g., Stitch Grouting of discontinuities). Additionally, geological mapping of conventional tunneling operations is inherent to the tunneling cycle and has to support real time decision making for initial support implementation.

During the preparation period of Stantec's resident engineer team, it was clear that geological mapping of the entire dam foot print (1,035,208 Sq Y), the quarry operation (12 million CY of fill) and 2,000 ft of tunnel , will create a significant work load on the team, and may create friction points with the contractors team. As a result, Stantec's team was looking for ways to streamline the geological mapping process.

Table 3. Initial Support types- Chimney hollow I/O tunnel

| Ground Support Type | Support Components | Q Value* | Rock Mass Condition† |
|---|---|---|---|
| 1 | Rock bolts and welded wire fabric in crown | Greater than 2.0 | Massive, Moderately Jointed |
| 2 | Rock bolts, welded wire fabric and shotcrete | 0.2–2.0 | Moderately Blocky and Seamy |
| 3 | Rock bolts, welded wire fabric and shotcrete | Lower than 0.2 | Very Blocky and Seamy |
| 4‡ | Steel ribs with lagging of shotcrete or steel | — | — |

* NGI Q system hand book , (2015).
† Terzaghi (1946).
‡ Used at the portals and as a contingency.

## The Geological Mapping Process

Generally, the geological mapping is being done on a clean section of dam foundation or tunnel face, and serves as a hold point for construction works, either rock fill placement or initial support installation. The main objectives of the on-site geological mapping are:

- Geo-locate the mapping section (tunnel alignment, foundation coordinates, etc.)
- Identify rock types.
- Identify discontinuities, shear zones, faults and their attributes (fill, alterations, etc.)
- Identify ground water conditions
- Crate a local geological map
- Assess rock mass quality and initiate the associated construction action (e.g., grouting for foundation mapping, initial support type for tunneling).

Additional back-office actions are associated with geological mapping as specified below:

- QA/QC inspection of the mapping results (sign off)
- Filing the mapping results, creating a geological data base
- Producing geological maps and profiles as part of project deliverable
- Producing statistical analysis of the mapping results (e.g., rock mass distributions for baseline confirmation).

Traditionally, the on-site mapping is done on paper, and later is filed as is or reformatted and then filed as a hard copy. In later years, the field copies would be reformatted to an electronic file (either scanned or as a designated electronic form).

## Characterization of the Application

Preliminary meetings between the resident engineer team members, and Stantec's GIS team members has led to the following characterization of the tunneling on-site application objectives:

- Reduce on site mapping time
- Standardization of geological description
- Produce a geological map and georeferenced it
- Calculate Q Barton values for rock mass quality assessment

- Auto calculation of numerous values to and formulas to minimize time and calculation errors and the need for additional devices such as calculators, etc.
- Collect pictures/files for later days reference
- Allow QA/QC process to mapping results
- Reduce post mapping activities (re-formatting, filing etc.) to the minimum
- Creating a Database of the mapping results for project hand-over
- Data security
- Report producing capability
- Data analysis capability
- Paperless activity

## The Tunnel Geological Mapping Application

The abovementioned characteristics has led the team to develop a mapping application using ESRI Survey123 GIS environment. The system is divided to two main components, the field mapping application (Data acquisition), installed on a handheld device (ipad mini 6th generation) and the online dashboard (Data base access) (Figure 2).

### *The Field Application*

The field application is divided to 11 subsections, as specify below:

1. *Survey information*—This section used for general data input: location/ georeferenced information, geologist, time and date and general comments
2. *Field map*—Providing tunnel section template (according to survey station) for a free hand sketch using Apple pencil (Figure 3).
3. *Lithology*—used for rock data input using set of scroll-down lists for description standardization and assessing RQD for Q Barton calculation
4. *Discontinuities*—used to input full description of discontinuities geometry, roughness and alternation, and the associated Q- Barton coefficients (Jr, Ja, Jn).
5. *Groundwater and seepage*—describing the tunnel in- flow and assigning value to Jw coefficient.
6. *NGI-Q Barton calculation*- used to assign SRF coefficient ad calculating the Q-Barton value, and assessing the rock mass quality.
7. *Tunnel support*—used to assign and document the initial support type used for the mapped section.
8. *Over break and fall outs*—allow free text description of the overbreak and observed fall outs.
9. *Reference Library*—used to present all relevant design material/ reference material (e.g., NGI Q system hand book) to allow access on site.
10. *Photo and video*—this section allow for Photographs taking, and annotate them for documentation of the excavated face, additionally, an option for video file attachment, for documentation of any dynamic phenomena (water inflow, spalling etc.). is available.
11. *Review and sign off*—this section allow for QA/QC sign off, mainly used in the back office.

During mapping, the ability to copy previous mapped section will carry over all the information except of the data from sections 1, 2 and 10, while all other sections can be edited. By doing so, mapping time reduced dramatically.

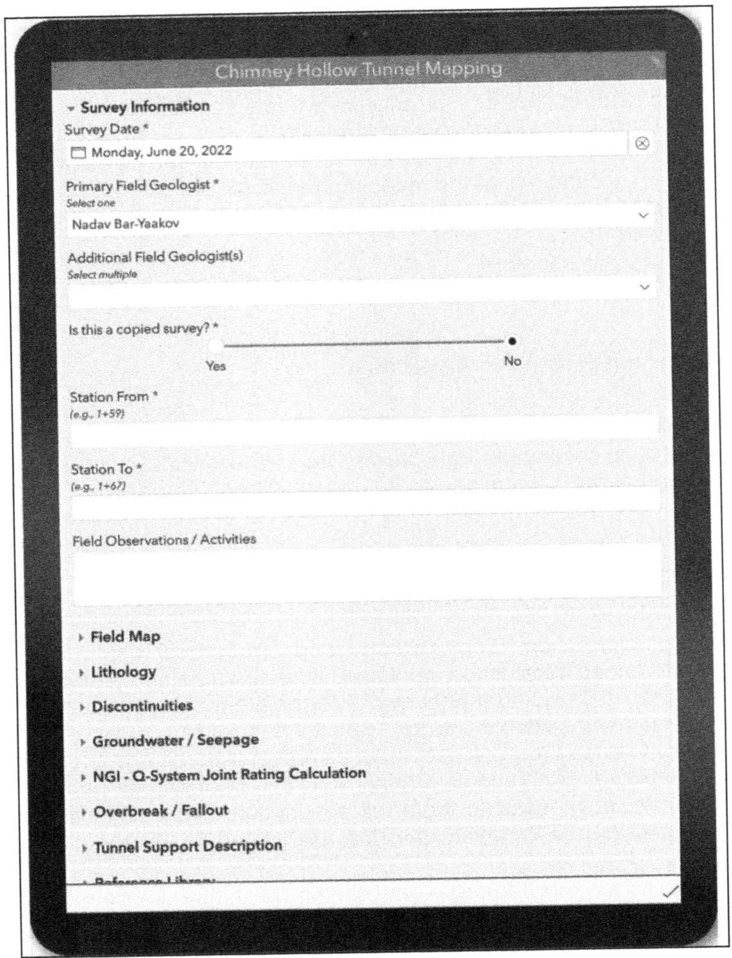


**Figure 2. View of the mapping application GUI**

### *Field Mapping Using the Application*

During excavation of the Chimney Hollow I/O tunnel, geological mapping of the face was integrated into the tunneling construction cycle just after mucking of the excavated materials. The mapping was done by the resident engineer team and was supported by the tunneling contractor's team. The geological mapping was done parallel to the contractor's survey (alignment and rock face scan), which allowed for surveying of the geological contacts, and accurately represent them in the geological face map. After mapping, calculating Q-Barton values and evaluation of other relating factors, initial support type was assigned by the resident engineer according to the design assumptions. At that point, a short discussion regarding the results and support type was held between the engineer and the contractor's team resulted in an agreed support type to be installed, and a field form sign-off for documentation. Using the application resulted in net mapping time of less than 10 minutes, minimizing the interference to tunneling cycle (Figure 4).

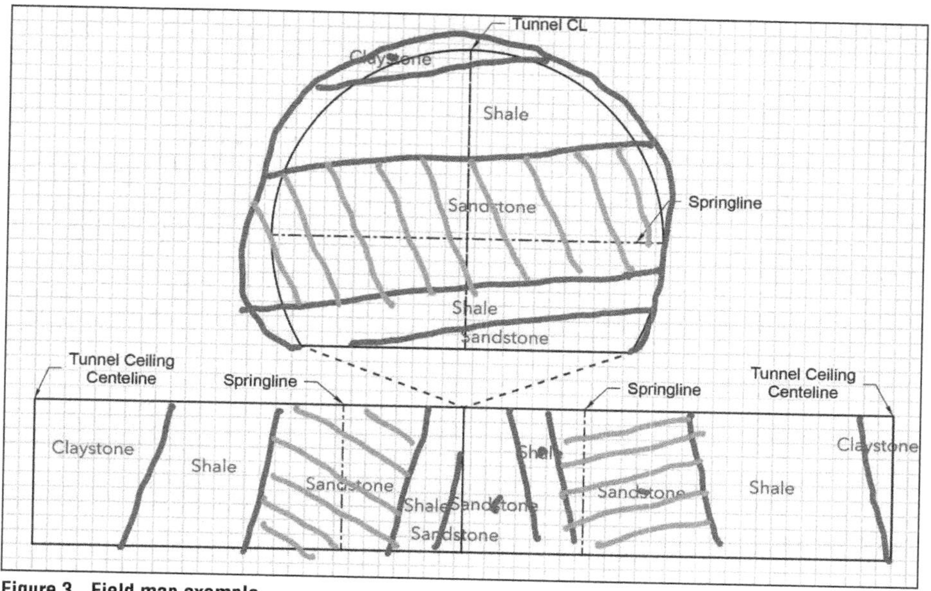


**Figure 3. Field map example**

**Figure 4. Field mapping in progress**

### *The Online Dashboard*

The online dashboard provide access to all data that has been gathered using the mapping application, and to additional information gathered in the project (Drone images and survey information).

The dashboard allow for review and sign-off the mapping forms, as well as editing them, present the overall progress of the tunnel, and allow filtering to find different forms. Since the on-line application is accessible to any authorized user, the data flow from the project site to any other involved parties is seamless.

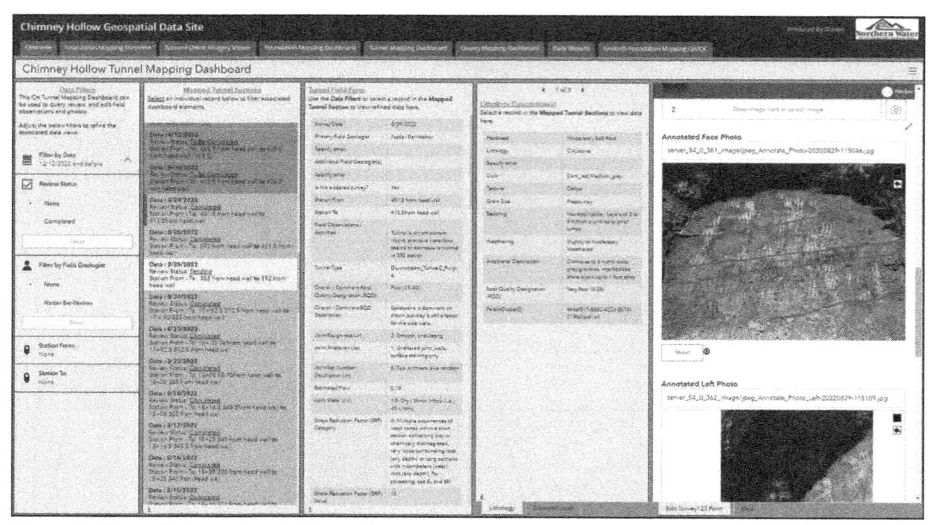


**Figure 5. View of the on-line dashboard**

Additional features available are the mapping sign-off color coding and the report generator. On later stages, the database and the dashboard interface will be transferred to the client as part of the project's documentation deliverables (Figure 5).

## SUMMARY

The implementation of the tunnel mapping application as presented above resulted in the following benefits:

- Reducing mapping time and construction cycle interference
- Increase data availability
- Reduce post mapping activities and human errors during those activities
- Allow real time QA/QC process for off -site personal
- Paperless activity

Total time saving related to the mapping activities comparing to previous projects was assessed to be 75%.

The GIS platform used allows the implementation of additional features that will increase productivity and documentation quality, and may be developed in future projects:

- Incorporating scan and surveying results to create tunnel digital twin
- Introduce augmented reality features for off-site staff

Generally, the utilizing of the GIS platform in the Chimney Hollow project is considered to be very successful and has resulted in similar implementations in other projects.

## REFERENCES

NGI 2015., Using the Q-system, Hand book, Oslo, Norway

Terzaghi, K., 1946. Rock Defects and Loads on Tunnel Supports. Harvard Univ., Graduate School of Engineering

# Virtual Master Rings—Replacing a Tradition

**Dieter Loh** ▪ VMT GmbH
**Florian Werres** ▪ VMT GmbH
**Mathias Knoll** ▪ VMT GmbH

## ABSTRACT

When a new tunnelling project starts, it has been a vital tradition for decades to cast and erect master rings horizontally on a flat surface to demonstrate the compliance with dimensional tolerances. Some tenders ask for the erection of two or even three rings upon each other or repeated master ring erections at intervals throughout the project. Especially for large diameters, this is a challenging and risky task for the workforce in the segment factory. A truck-mounted crane is often necessary, and in some cases it can take up to a week to complete a master ring.

The concept of virtual rings has the potential to replace this tradition by assembling master rings digitally instead of physically. This is based on a best-fit approach, considering the 3D coordinates from laser tracker measurements of segments or segment moulds.

Building a virtual ring, a sub-millimeter accuracy can be achieved, the alignment of neighboring circumferential and ring joints observed, and the fit of bolt holes checked. The traditional method, on the other hand, is discussed controversially despite the significant amount of work and risks—particularly as the horizontal position of the ring and absence of gaskets do not simulate the reality in the excavated tunnel.

This paper will highlight the technical aspects of virtual rings, as well as their pros and cons, illustrated by a case study from London Thames Tideway where a total of 30 physical master rings could be replaced by virtual rings.

## PERFORMANCE SPECIFICATION

The reasoning for specifying very tight geometrical tolerances in standard specifications issued by associations like the International Tunnelling Association (ITA) or the British Tunnelling Society (BTS) are:

- Geometric sensitivity to inaccuracies and distortions of individual segments
- High load effects from earth, water, and grouting pressure on the tunnel lining
- Jacking forces during mining
- The load transfer takes place only in limited areas (partial surface load)
- Damage cannot always be detected (for example, on the outer side of the ring)
- Repairing of damaged segments are costly and time-consuming

Tapered rings enable the tunnel lining to accommodate alignments that contain curves or to compensate for misalignment of the tunnel due to site-specific problems.

The use of reinforced segments necessitates precise segments, and great care in their installation is needed.

## MASTER AND TRIAL RINGS

### Traditional Erection of Master/Trial Rings

Quite frequently, the erection of master / trial rings is a basic requirement, and the geometrical tolerances are defined in the project's specification. This procedure, including measurements, is complex and time consuming.

It takes anywhere from several days up to a full week to build a master ring and one or two additional trial rings, requiring huge man-power and additional logistics.

This does not represent the physical ring build situation in the TBM due to a different orientation, the absence of installed gaskets, no pressure from jacking forces etc.

A certain number of master and trial rings must often be built according to the requirements at the beginning and throughout the segment production, which multiples the efforts and costs for the contractor. Moreover, this risky task does not always comply with health and safety standards in place for the onsite personnel.

### Measurement of Master/Trial Rings

Metrology companies perform mould and segment 3-D measurements as well as master and trial ring 3-D measurements using a laser tracker system. Portable laser trackers are used to measure large parts, jigs, moulds, assemblies, and machines on-site quickly and easily. They measure 3-D coordinates by tracking a reflector that the user moves across the object being measured. These measurements can be compared against nominal CAD data to provide information to the onsite personnel, allowing the team to move forward with confidence or to make the appropriate adjustments if required.

These measurements are performed using a premier portable metrology software for large-scale applications, which has been adapted for this specific task. Specified measurement profiles are used for collecting the data and specified measurement plans (programmed scripts) are used for an immediate and fast semi-automated calculation and reporting procedure.

The results show an unrealistic deviation range from –9.4 mm to +7.8 mm. The ring is strongly deformed ellipsoidal, and the ring build workers confirmed significant physical difficulties to erect this ring using cranes, chain blocks and belts.

Furthermore, only the master ring's intrados and extrados cylinders could be measured. The adjacent and hidden joints cannot be controlled.

Looking at these unsatisfying measurement results, a Virtual Ring Build was introduced as an alternative concept to a physical erection.

### Alternative Concept of a Virtual Ring Build

Instead of building a physical ring, the alternative concept proposes to measure individual segments or moulds followed by an assembly of these results in the virtual world.

The advantages of a Virtual Ring Build include:

- Quality control for a complete ring
- Avoiding the master / trial ring construction and its inaccuracies due to physical limitations

- Realizing and eliminating possible physical ring build problems caused by geometrical effects prior to the segment transportation into the TBM
- Prove that segments from different mould sets can be combined within one ring, if this is not allowed in the project's specification
- Prove that segments exceeding the minor sub-millimeter tolerance can be combined within one ring and still meet the ring build tolerances; ultimately decreasing the production's reject rate
- Saving time and costs
- Avoiding accidents of onsite personnel

The individual segment measurements from the trial ring in Figure 1 were used for a Virtual Ring Build and show more accurate results compared to the physical trial ring. The deviations now range from –1.9 mm to +1.5 mm, approximately five times better than in the real world.

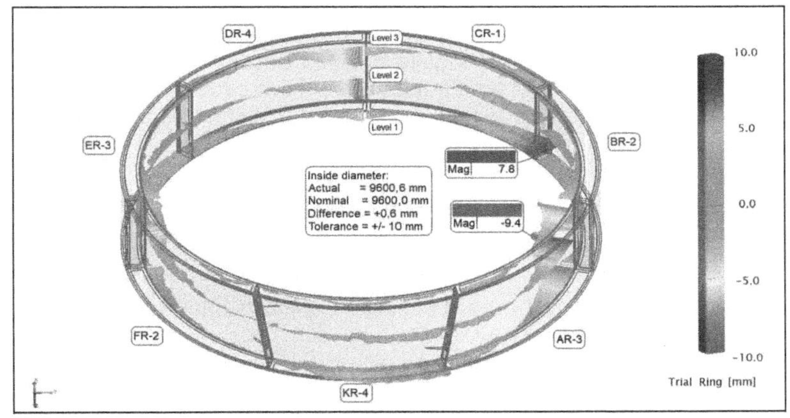


**Figure 1. Trial Ring 3-D measurement with large deviations on the intrados caused by physical ring build difficulties**

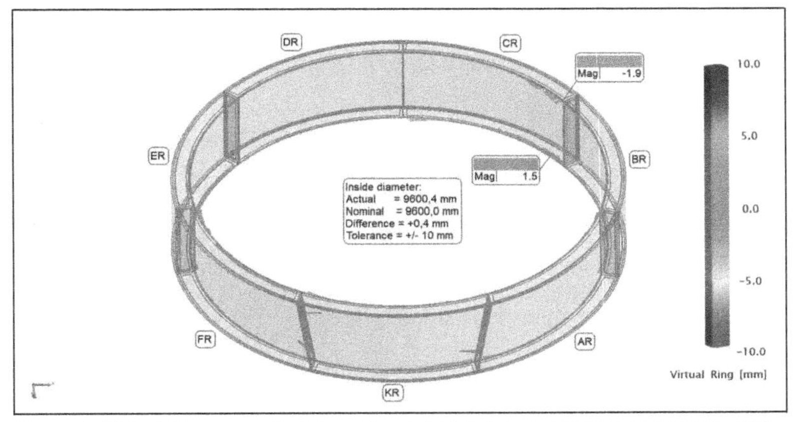


**Figure 2. Virtual Ring assembled from Figure 1 trial ring segment measurements with small deviations on the intrados, eliminating the physical limitations**

### Conclusion

The Virtual Ring Build successfully assembles individual segment measurement results mathematically and is unaffected by atypical physical difficulties and limitations which do not reflect the real situation in mechanized tunnelling. Thus, it is a practical solution to prove the ring build quality.

## TECHNICAL PROCEDURE

The moulds or segments 3-D measurement results are combined mathematically for adjacent moulds or segments to present the assembly of the ring ("Virtual Ring Build"). This is realized through so-called mathematical relationships, which is a powerful method with a number of uses: dynamic inspection, alignment, virtual fit-up, real-time assembly and optimization.

Relationships are not physical objects by themselves but a defined link between objects. In the simplest case, a relationship calculates the distance between two related entities and re-calculates this distance whenever either of the related items are moved. This linking process is quite efficient and can be expanded to include many points and many objects simultaneously—in this case, many point groups from the individual measurements and many objects from all the segments nominal CAD.

### Mathematical Calculations

The following calculations are performed for a virtual ring:

- Relationship fitting of the measured intrados points to the nominal intrados cylinder
- Relationship fitting of the measured front and back points to the nominal circumferential joint
- Relationship fitting of the measured left and right points to the corresponding adjacent actual longitudinal joint
- Using constraints of only positive residuals in the adjacent longitudinal joints, as neighboring concrete joints physically cannot overlap

### Results and Reporting

The resulting 3-D deviations to the nominal ring as well as the resulting gaps and laps at the adjacent joints are presented as a graphical output, showing variations from theoretical measurement.

The following example shows the effect of overlapping on neighboring longitudinal concrete joints before (Figure 4) and after the Virtual Ring Build (Figure 5).

### Consideration in Tunnelling Tenders

To date, 3D mould and segment measurements are a common best practice, but the possibility of building virtual rings is not yet published by the industry nor specified in a remarkable portion of segmental lining tenders.

The format and contents of a Virtual Ring Build report are not defined and standardised. However, the graphical and numerical layout described in *Results and Reporting* was accepted by a dozen clients across the globe.

To allow contractors for virtual ring building, the industry needs to reflect this possibility in the tenders by changing the corresponding chapter as part of the segmental lining specification.

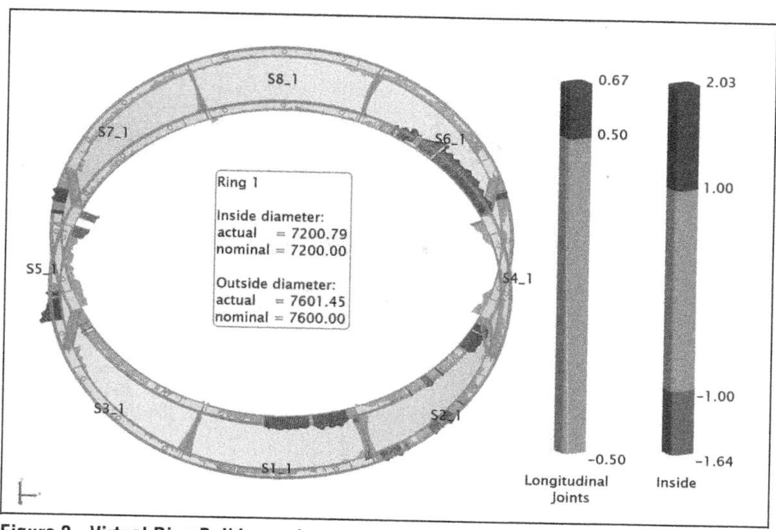


**Figure 3. Virtual Ring Build sample report**

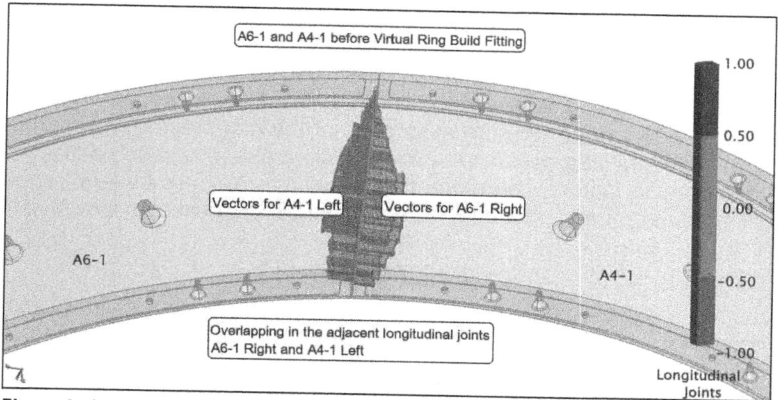


**Figure 4. Impossible overlapping on neighboring concrete joints before the Virtual Ring Build**

## CASE STUDY THAMES TIDEWAY TUNNEL

London relies on a 150-year-old sewer system built for a population less than half its current size. As a result, millions of tonnes of raw sewage spills, untreated, into the River Thames each year. Tideway is upgrading London's sewer system to cope with its growing population. The 25 km tunnel will intercept, store, and ultimately transfer sewage waste away from the River Thames.

Starting in Acton, west London, the sewer tunnel (the Thames Tideway Tunnel), will travel through the heart of London at depths of between 30 and 60 meters, using gravity to transfer waste eastwards. The works are divided into three sections, each of them being built by a different contractor.

One chapter specified in the precast segment requirements of this large infrastructure project contained a mandatory clause to erect physical trial rings at intervals not

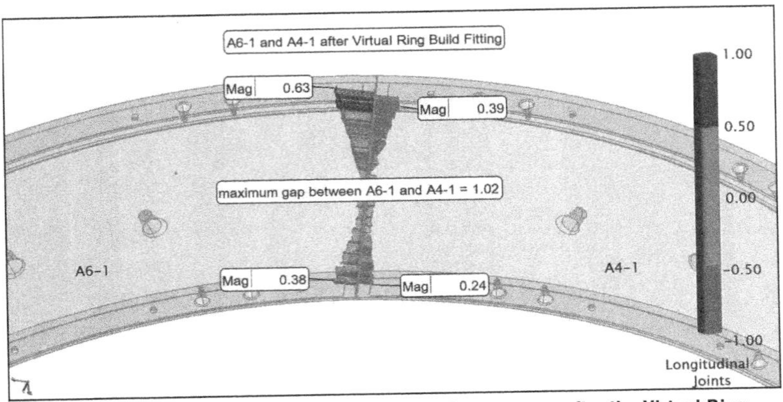


**Figure 5. The resulting 3-D deviations do not overlap anymore after the Virtual Ring Build; the remaining gaps are of interest**

exceeding 0.5% of the segment production and compare these against the master ring. Applying this clause on the C415 East Section main works with a total production of around 39.000 individual elements, the contractor was asked to build and erect 30 trial rings during the precast production.

This procedure had a large negative impact on the contractor's project budget and limited benefit for the client, which created the willingness to investigate alternative solutions. After a technical seminar, both parties explored the possibility of replacing all physical by virtual rings with the exemption of the first master ring. Furthermore, results from dozens of segment measurements were already available due to regular quality checks in the factory. Therefore, the decision to submit a change request was taken by the contractor and unanimously approved by a project manager's instruction.

## REFERENCES

Fast, accurate measurements for large-scale applications. Accessed 26 May 2021 (2021), https://www.faro.com/en/Products/Hardware/Vantage-Laser-Trackers.

Bala Muralikrishnan, Steve Phillips, Daniel Sawyer (2015) "Laser trackers for large-scale dimensional metrology: A review." Precision Engineering Volume 44, April 2016, Pages 13–28.

Nod Clarke-Hackston, Manfred Messing, Dieter Loh, Rainer Lott (2006) "Modern high precision high speed measurement of segments and moulds." Tunnelling and Underground Space Technology 21(3):258–258.

Bazalgette Tunnel Limited (2021). London's super sewer. Accessed 14 May 2021, https://www.tideway.london/.

# Wireless Optical Displacement Sensor for Convergence and Divergence Monitoring

**Raphael Victor** ▪ Senceive Corp, USA

## ABSTRACT

The construction and modification of tunnels brings a requirement for precise measurement of convergence and divergence to safeguard the structure, maintain safety and provide the assurance needed to maintain efficient progress.

Established methods include manual measurement using tape extensometers and automated methods using photogrammetry and automated total stations. Drawbacks include the need for frequent access, power supply, cabling, and cost.

The paper describes how the development of reliable optical displacement sensors (ODS) has changed the landscape. ODS sensors are connected to a long-range wireless mesh communications network via a solar-powered gateway outside the tunnel, with data instantly transmitted to the internet. A sensor and its reflector target can be installed in 20 minutes and is maintenance free for a decade. Tunnel movement is measured to sub-millimeter precision with repeatability of ±0.15 mm. Integration of a triaxial tilt sensor allows slope distance to be converted to horizontal and vertical changes and allows rotational movement to be determined.

Applications include new-build and long-term structural health monitoring. Case studies will be given, including the refurbishment of rail tunnels in Spain where the ODS measured movement during track lowering in a situation where no other automated system was considered viable.

## INTRODUCTION

Automation of monitoring in general appears undeniably tempting exactly until that moment when prices are discussed. That's when suddenly manual surveys seems not so outdated anymore.

Obviously this perspective is rather short sighted if automation is applied where it can exploit its advantages, i.e.,

- In areas where access is limited, difficult or dangerous
- In projects where high recording rates (e.g., shorter than one day) are required
- Where long term observations (e.g., longer than one year) are desired or
- Where repetitive procedures can efficiently free human resources for decision requiring tasks
- Any combination of the above

The nature of tunnelling more or less represents all of the above. Monitoring works are not only repetitive but due to the limited space this environment provides could also hamper advance for tunnels under construction or general operation for existing tunnels with significant impact on schedule and cost.

Common parameter in tunnel monitoring is convergence as observation of changes of the tunnel profile perpendicular to the alignment. This can be conducted as life cycle monitoring of aging masonry tunnels as well as temporary observation of deformation induced by adjacent construction measures or even advance and construction of the tunnel itself.

In all of the above cases wireless condition monitoring (WCM) seems to be an intelligent solution with regards to costs, ease of installation and efficiency. Optical displacement sensors (ODS) are a typical component of a modern WCM-system representing unrivalled simplicity in convergence monitoring.

## CONVERGENCE

### Tunnel Deformation

Excavation of a tunnel or close to an existing tunnel cause volume loss and thus a disturbance of the surrounding medium. In order to re-establish the initially lost equilibrium, depending on the material, the ambient environment starts to rearrange with the tendency to fill the void represented by the tunnel or cavity.

Tunnel profile design and lining, respectively support are designed to deflect the outside loads in order to maintain stability and establish a new final equilibrium with the surrounding ground. Only if this precondition is reached operation of the tunnel is safe and support of the overburden, for instance bearing urban infrastructure, is verified.

Therefore it is essential to observe tunnel deformation as component of a project's holistic monitoring concept.

Depending on the tunnel profile, surrounding regime, overburden, etc. the tunnel can experience various deformation (see Figure 1). As convergence we consider the displacement of a point towards or way from the interior, i.e., the centre point of the profile area. For a circular profile this would be the perpendicular direction to the outside tangent.

For efficient convergence observation it is crucial to estimate the pattern of deformation in advance to distribute sensors appropriately and yield representative results. Segmental lined tunnels will deform in the joints as the segments itself can be considered while masonry lined tunnels might deform non homogeneously and require more thorough consideration of placement of the sensors.

### Common Observation Techniques

The strategy of continuous observation depends on a variety of criteria, not the least those already mentioned in the introduction above. Generally speaking monitoring is the observation of change and thus the comparison of a in-situ current state with an

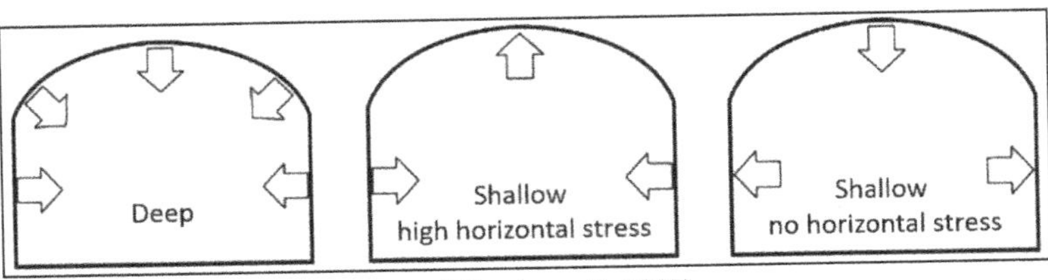


**Figure 1. Scenarios of convergence (Erlandsson,O. 2020)**

initial state some time in the past, represented by a zero- or baseline measurement. In addition the parameters observed need to be identical and comparable in both epochs which might require some information about the environmental conditions. In case of geometric observations like convergence, the location of the observed spot must be identical our at least be restored with sufficient precision. The latter becomes increasingly significant if the tunnel undergoes renovation or the initial survey is carried out when only preliminary lining is installed.

First and foremost question would be the observation frequency. If annual observation repetition is required and access is not an obstacle automatization is hardly economical. Manual observations can be conducted using Geodetic Total Stations, tape extensometers (see Figure 2) or in modern days even terrestrial laser scanning.

For total station observations the tunnel profile is equipped with reflectors, ideally prisms, that are observed from the total station's location by horizontal and vertical angle and slope distance (Figure 3) which are converted into local coordinates at mm accuracy level. The linear distance between these coordinates represents the observation parameter. Depending on the spacing between the profiles and the visibility along the line of sight, a number of profiles can be observed from one standpoint before moving onwards.

In order to observe convergence by tape extensometers eye bolts will be installed at the opposing ends of the intended chord and the extensometer temporarily hooked into them. The reading can then be taken from the micrometer. Tape extensometers are easy to use and provide high accuracy up to ±0.2 mm. The measurement will require some effort and support such as cherry pickers etc. In any case the measurement tape or wire has to be removed after measurement as it would impede tunnel works or hamper operation.

Laser Scanning covers the tunnel wall with an enormous density and can provide valuable information particularly in case of inhomogeneous deformation or irregular surfaces. However, identification of discrete points might be difficult and the extraordinary

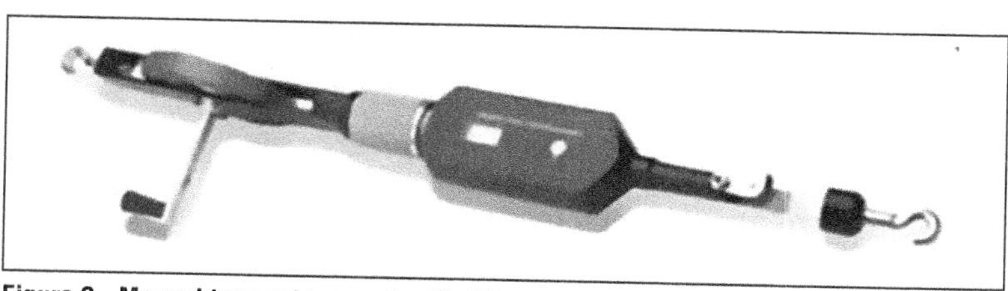

**Figure 2. Manual tape extensometer (Soil Instruments)**

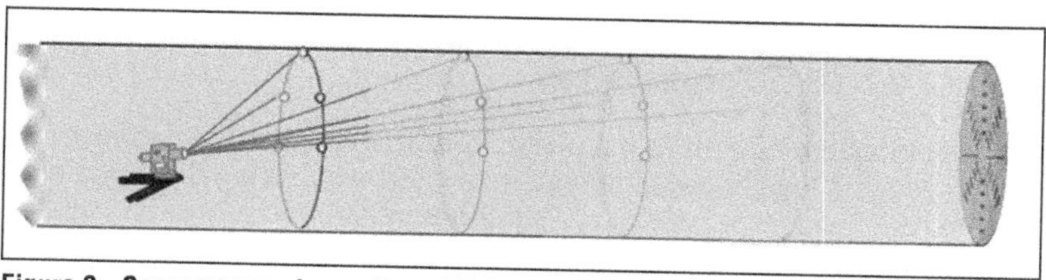

**Figure 3. Convergence observation utilizing total stations**

volume of data requires some effort in data processing and most probably some manual intervention.

All the above methods require qualified personnel operating the survey equipment, working themselves "through the tunnel" by observing section after section. This sequential strategy limits the observation frequency significantly and makes short observation intervals <1 day unrealistic or at least highly inefficient.

## Automation of Observation

Automation is a valid strategy to overcome the observation obstacles encountered by manual surveys and initially mentioned in the introduction. In order to realize automation the survey equipment needs to be permanently installed. Hence, tape extensometers are already disqualified as they would interfere with tunnel and construction operations. Theoretically laser scanners could be installed but the hardware cost simply prohibit the permanent installation of multiple instruments. Thus, Robotic Total Stations (RTS) remain as conventional means of deformation observation. Here by far the biggest share of hardware investment lies in the totals station itself. In order to minimize these costs multiple convergence sections would be observed from one location.

As efficient as this appears with increasing distance dust, fuel emissions and other sources of haze blur the visual line and therefore decrease the quality of observation (Figure 3). Furthermore, besides physical obstruction by traffic, machinery etc., the further away the section the closer aligned reflectors appear in the RTS's field of view, which makes it harder for the automatic target recognition to distinguish them. To make things worse, the RTS itself will most likely by located in the unfavorable position on a bracket mounted to one side of the tunnel. Hence, alternatives might be required.

In recent years miniaturization led to the development of devices, so called *Nodes*, that combine sensor, data transmission and power supply in one very compact housing. The keyword here is *Wireless Condition Monitoring* (WCM). The simplicity of those sensors' handling promote them for complicated environments.

## Wireless Observation Constellation

Wireless Condition Monitoring systems essentially consist of three components: (i) the sensor-node, (ii) the gateway, and (iii) the data management platform.

Compact design and low power consumption of the node allow easy installation of the field equipment even by none-experts without maintenance required for up to 10–15 yrs depending on the sensor. The nodes establish communication between an implemented sensor, like a MEMS tilt sensor and/or a laser based optical distance sensor ODS (Figure 4), or a connected external sensors (potentiometric crack meters, VibratingWire Strain gauges, PT100 temperature sensors etc.). The data is submitted either via 868/915 MHz LoRaWAN based signal or Mesh 2.4 GHz Wifi-frequency to a gateway. There it is intermediately stored and submitted either via cellular network or ethernet landline to a remote central server.

Wireless communication minimizes installation effort and reduces the risk of damage and vandalism causing loss of data that cabled systems occasionally experience. The long battery life reduces the necessity of maintenance almost to zero and allows redeployment on multiple projects as well as long term life cycle monitoring.

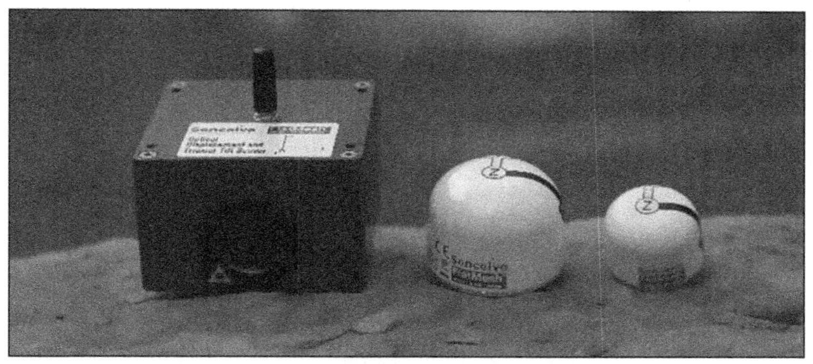

Figure 4. Wireless 3-axes tilt nodes, left with integrated optical distance sensor (ODS)

Intelligent systems allow remote system modification according to the stage of works, related expectations of the development of changes or requirements of contingency measures etc.

Last but not least WCM allows easy extension of the system by gradually adding more and even different sensors, communicating within the same platform. No extra cables have to be pulled. Hence, even relocation according the construction stage or replacement is simple.

Eventually the data management platform provides access for multiple users with adequately defined user roles. The data is stored and further processed if required. Integrated alarming and messaging prevent that breach of predefined threshold values or system malfunction goes unnoticed.

## PROJECT EXAMPLES

### Costa Blanca—Martorell Tunnels—Spain

As part of the ambitious Mediterranean Rail Corridor the Spanish Rail Infrastructure Operator ADIF decided to renovate three disused several hundred meters long tunnels between the Catalan Martorell and Castelbisbal to suffice modern requirements. Main task of modernization was to strengthen the lining and lower the track in order to allow electrification.

By lowering the invert the overall tunnel geometry was altered and it had to be closely observed whether the redistribution of loads would jeopardize structural integrity. Continuous monitoring was instructed in order ensure a safe work environment and provide in situ information for rapid structural response to potential deformation.

The obligation to carry out permanent observations made automated systems mandatory. However, the expected ambient conditions, specifically dust did not allow for long distance optical observations and therefore excluded total station observations ex ante.

The contractor Dragados and its assigned monitoring consultant INSTOP opted for a Senceive™ WCM system utilizing laser distance sensors (ODS) in a triangular arrangement, i.e., two on the spring line and one in the crown (Figure 5). Although also relying on optical observations the distances observed in this configuration are much shorter than in any applicable total station concept.

Specific diligence was spend on the Martorell tunnel as the city of Martorell with some high rises is located directly above. Installed in 25 m intervals Costa Balance tunnel featured 30 ODS profiles, Martorell tunnel 27 and Castelbisbal tunnel 35.

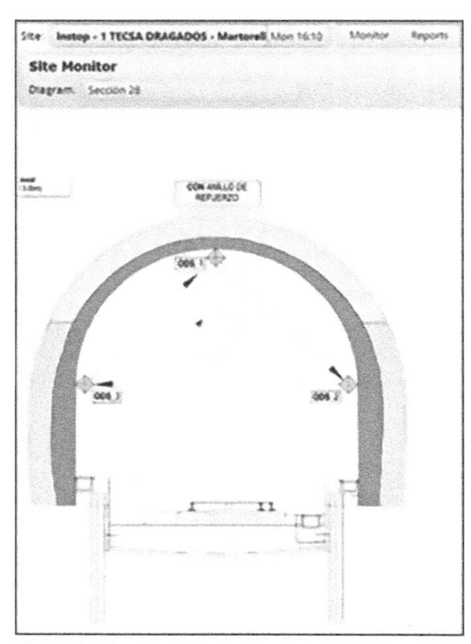

Figure 5. Section diagram screenshot from the data management platform

Achieving the nominal submillimeter accuracy not only depends on the environmental conditions, i.e., the medium the laser beam has to travel through, but also the quality and orientation of the target. Aiming at rough and/or oblique surfaces decreases the accuracy as the incidence point might shift inadequately. Therefore specific targets were installed to provide a reflective, smooth surface at 90° incidence angle (Figure 6).

The ODS nodes themselves are equipped with 3-axes MEMS tilt sensors. Thus rotational movements at the node itself are observed as well without the node being restricted to horizontal orientated installation (Figure 7). Besides the challenging environment one more reason for WCM was the tight schedule which only left a short window for installation. Wireless systems are easy to install, even by none specialists. At four sections per hour an entire tunnel installation was completed within one working day.

Figure 6. ODS-target; in the background ODS equipped with retro target for redundant total station observation

Pre-configuration and wireless character allow for effortless extension of the system, even to observe other parameters. Costa Blanca and Martorell tunnel were additionally equipped automated longitudinal settlement observation. The system consists of fixed length beams daisy-chained with cardanic wall mounts (Figure 7) with wireless tilt nodes fixed to the beams. Via trigonometric functions any longitudinal tilt can be converted into vertical displacement by the known length of the beam. The accumulation of the incremental vertical displacements along the chain yields longitudinal settlement or upheaval. Furthermore above Martorell tunnel a number of rod extensometers were installed to observe vertical ground movement each at three levels (3 m/6 m/9 m) communicating its readings to the solar powered gateway via multi-channel vibrating wire nodes (Figure 8).

**Figure 7. ODS pointing towards a target located at the tunnel crown; 3-axes tilt sensors allow installation in arbitrary orientation; simultaneous operation of tilt beams for observation of longitudinal settlement/upheaval**

As the components of the system were located relatively far apart from each other the long range GeoWAN™ system was chosen as communication platform. Each tunnel was equipped with its own gateway serving a system with the furthest node 1.2 km away. Eventually gateways also received signals from the nodes of the further away other tunnels up to 8 km away. However, to maintain redundancy the multi gateway constellation remained.

**Figure 8. 3× rod extensometer connected to a LoRaWAN 4-channel vibrating wire node**

The system was delivered preconfigured. Therefore immediately after installation the proficiency of the system could be tested online without further need to configure the hardware.

The renovation of the track including the three tunnels intermediate galleries etc. was completed after 3 months (see Figure 10) with a monitoring system of more than 400 nodes in place for about 4 months. After demobilization the components will be redeployed on one or more subsequent INSTOP projects.

**Figure 9. Martorell Tunnel before (left) and after (right) renovation**

## Chipping Sodbury Tunnel—UK

A similar situation as in the previous case was encountered in the Chipping Sodbury Tunnel near Bristol/UK. The 1902 completed and thus more than 100 years old structure is 4,000 m long and features a brick-lined arch-profile of 8.4 × 6.4 m (w × h). In order to comply with modern requirements clearance had to be increased by lowering the track 150 mm for a new alignment design including electrification.

As well a tight schedule forced for automated monitoring to maintain continuous operation without interference of survey works. The monitoring consultant AECOM already gained experience in similar renovation projects and could therefore utilize the before successfully proven technology of WCM. Again tunnel geometry, obstructions by traffic and construction as well as harsh atmospheric conditions generated by dust and humidity prohibited total station observations.

Unlike in the above Spanish tunnels Chipping Sodbury Tunnel provided cellular signal reception throughout the entire length as initial checks confirmed. Therefore Senceive's communication platform Flatmesh™ could be employed. Due to the larger band width of the 2.4 GHz signal the system provides optimized synchronization and extremely short message pulses for high observation frequencies. Due to meshed and thus redundant signal path structure high reliability is achieved.

Hence, establishing communication was rather straight forward by simply installing gateways in regular intervals to gather the data of adjacent nodes.

As stated before, 3-axes tilt sensors can be installed in any orientation without the necessity of prior levelling. Therefore the initial approach was to install daisy chained tilt-beams within the tunnel section representing deformation arrays (Figure 10).

AECOM preferred miniaturized Senceive Nano+ MEMS tilt nodes (Figure 4, middle) for compactness and their gained trust in the IP68 performance under the expected

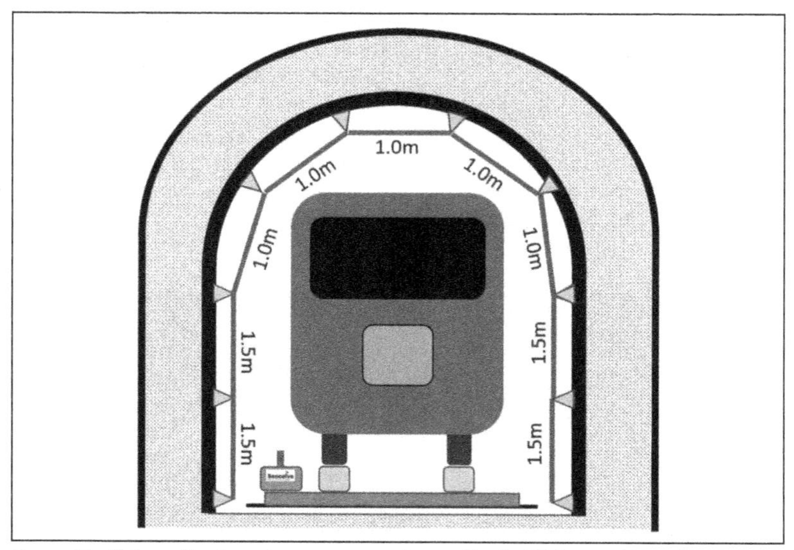


**Figure 10. Schematic view of convergence observation by tilt beam chains; lower left: track monitoring by high-G tilt node mounted to sleeper**

wet and dusty ambient conditions. In addition shape arrays require one gateway per chain while the wireless tilt nodes' signal of several vicinal profiles can be gathered by one gateway which reduces hardware requirements considerably.

Eventually however, it was not possible to install beam chains throughout the entire profile as the beams, cutting curvature like chords, provided considerable obstruction for operations in localized "pinch points." Hence, individual tilt beams were installed only at crucial locations (Figure 11), still employing more than 200 nodes throughout the tunnel. Where convergence observation was required mere tilt nodes were substituted by ODS, pointing to the opposite tunnel side respectively (Figure 12).

Permanent online access to the data management platform allowed remote temporary deactivation of the inherent alarming during blockage of the line of sight by construction works in order to avoid false alarming.

As track lowering was conducted one after the other the remaining track was observed by rail specific vibration resistant tilt nodes (Figure 10). 843 high-G tilt nodes were installed, spaced 3 m within the immediate area of impact and further apart outside.

**Figure 11. Vertical tilt-beams supplementing horizontal beams chains, Nano+ nodes provide compact design**

**Figure 12. Convergence observation by ODS mounted to tilt beam**

In addition it was crucial to ensure proper operation of the tunnel drainage, realized by a 6ft culvert underneath. As it was expected that significant impact on this structure would result in detectable deflection of the invert and thus the track, culvert monitoring was also conducted by tilt-node-based trackbed observations.

## CONCLUSION

Both presented projects did not encounter significant deformation or even structural hazards. However, works were conducted under the umbrella of risk mitigation by meticulous observation of structural changes and deformation and thus confidence for utmost progress of works.

Automated systems proved capable of encompassing the entire workflow from data acquisition via transfer and processing towards multi-user data visualization and thus, provision of the observed parameters to all involved disciplines without delay.

The obtained information can be fed into the "observational method," i.e., the feedback loop suggested by Peck (Peck R. 1969), confirming or refuting the design assumption in a "learn-as-you-go" procedure. This allows timely modification of processes in order to optimize construction progress and reduce the hazard risks.

One additional advantage of the presented system solutions is that the initial installation ("ab initio" system) can be expanded to multi-sensor, multi-purpose systems by successively adding sensors and thus take into account changing conditions and requirements ("best-way-out" method).

The systems' simplicity promotes the common and efficient utilization of automated methods in physical and commercial risk mitigation, even at early stages by making "deformation monitoring an integral part of decision making in the design-construction-supervision-maintenance system" (Kavvadas M.J. 2005).

## REFERENCES

British Tunnelling Society (2011), "Monitoring Underground Construction—A best practice guide," British Tunnelling Society and Institution of Civil Engineering, Reprint 2014), ISBN 9780727741189.

Erlendsson, Olof (2020) "Comparison of Tunnel Convergence Measurement Methods"; Degree Project in the Built Environment, KTH Royal Institute of Technology, School of Architecture and the Built Environment, Stockholm, Sweden.

INSTOP (2021) https://www.youtube.com/watch?v=hSOjfC8eGho.

Kavvadas, M.J. (2005) "Monitoring Ground Deformation in Tunneling: Current practice in Transportation Tunnels." Engineering Geology Volume 79, Issues 1–2, 2005, pp. 93–113, ISSN 0013-7952.

Peck, R.B. (1969) "Advantages and Limitations of the Observational Method in Applied Soil Mechanics" 1969, Géotechnique, pp. 171–187, V 19, N 2, 10.1680/geot.1969.19.2.171.

Soil Instruments (2021) https://www.soilinstruments.com/products/extensometers/digital-tape-extensometer/.

PART 

# Pressurized Face Tunneling

***Chairs***

**Veronica Monaco**
Jacobs

**Dan Dreyfus**
McMillen Jacobs Associates

# Design and Construction Considerations for the Pawtucket CSO Tunnel

**Kathryn Kelly** ▪ Narragansett Bay Commission
**Victor Despointes** ▪ CBNA
**Stephane Polycarpe** ▪ CBNA
**Vojtech Ernst Gall** ▪ Gall Zeidler Consultants
**Chris Feeney** ▪ Stantec
**Brian Hann** ▪ Barletta Engineering

## ABSTRACT

The Pawtucket Tunnel is a 2.2-mi, 30.2-ft diameter CSO storage tunnel in Rhode Island. The primary tunnel is built using a dual-mode open-EPB TBM capable of sealing the face within 120 seconds to manage poor ground and water ingress which is launched from within a SEM starter tunnel. The main tunnel connects to four adits along its alignment, three of which are SEM tunnels and one being a MTBM tunnel. This contribution describes the selection process, potential risks, and requirements for the TBM, as well as the design of the main tunnel, the adits, and the adit-bored tunnel connections.

## INTRODUCTION

The Narragansett Bay Commission (NBC) owns and operates Rhode Island's two largest wastewater treatment facilities, the Field's Point Wastewater Treatment Facility (FPWWTF) and the Bucklin Point Wastewater Treatment Facility (BPWWTF). NBC also operates 112 miles of interceptors, nine outlying pump stations, 26 tide-gates, 65 combined sewer overflows (CSOs), and a septage receiving station. In addition, NBC owns and operates a 62-million-gallon-capacity CSO storage tunnel, a tunnel pump station and numerous CSO diversion facilities.

NBC embarked on a three-phase CSO control program in 1998, aimed at lowering annual CSO volumes and reducing annual shellfish bed closures in accordance with a 1992 Consent Agreement (CA) with the Rhode Island Department of Environmental Management (RIDEM). Phases I and II of the Program, which focused on the Field Point Service Area in Providence, were completed in 2008 and 2015, respectively. The program succeeded in lowering annual CSO volumes and reducing annual shellfish bed closures to levels that are in keeping with a 1992 CA between NBC and RIDEM.

The third and final phase (i.e., Phase III) is focused primarily within the communities of Pawtucket and Central Falls. The implementation strategy prioritizes water quality benefits, while also limiting the financial impact on rate payers. The primary elements include the Pawtucket Tunnel and ancillary underground components including a launching shaft, receiving shaft, drop/vent shafts, adit tunnels, and tunnel pump station. The tunnel system provides volume to store all contributing overflows during a storm event up to the three-month storm for subsequent pump out and treatment. The required minimum storage volume to achieve the defined hydraulic criteria (i.e., no overflows for the three-month design storm, and no more than four overflows per year per outfall for typical year storms) is 58.5 million gallons (MG).

The Pawtucket Tunnel is a rock tunnel, 115-ft to 155-ft below the ground surface, located adjacent to the Seekonk and Blackstone Rivers in Pawtucket, RI. The tunnel is approximately 11,600-ft in length with a 30-ft inside diameter (see Figure 1). The tunnel is mined with a Herrenknecht dual mode (open/EPB) TBM. The machine is fitted with a hard rock cutterhead and a conveyor to extract muck from the face when it is operated in open mode or a screw conveyor when operated in earth pressure balance (EPB) mode. The lining is made of universal double taper rings of seven fiber reinforced segments, thickness 14-inches and 6.6-ft in length. Each segment joint both radial and circumferential is fully gasketed to achieve the water tightness criteria.

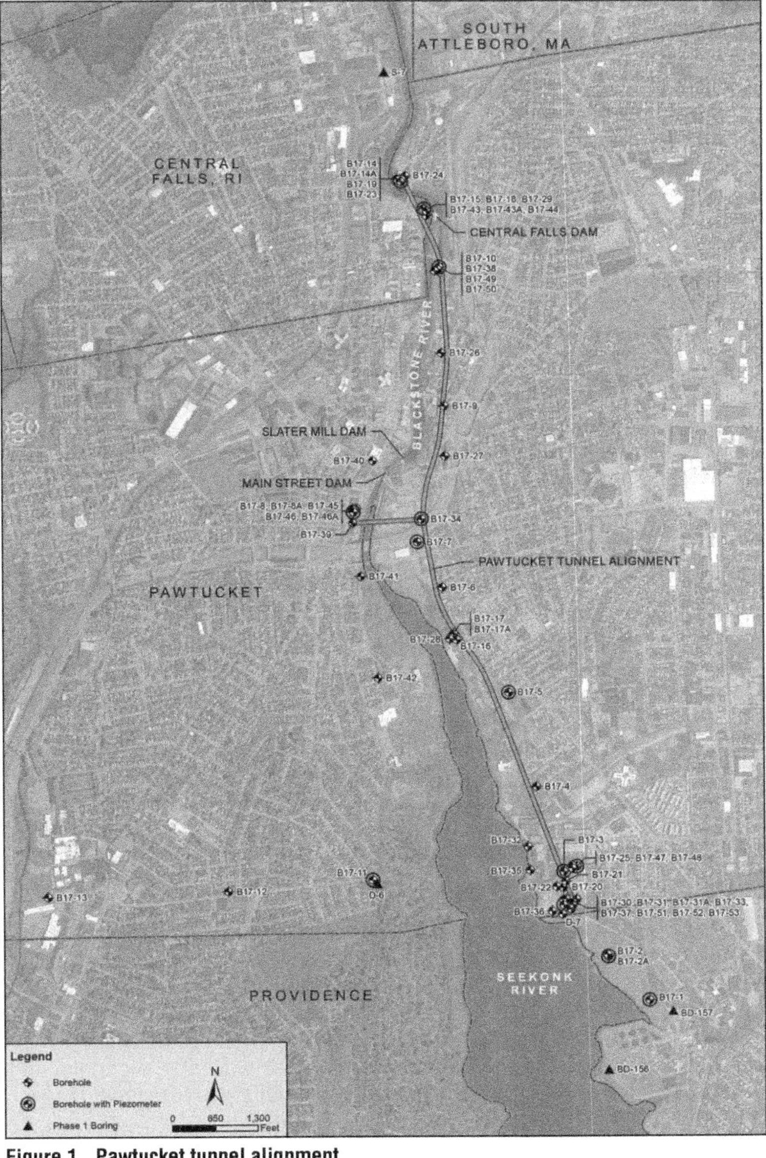


Figure 1. Pawtucket tunnel alignment

NBC is under a CA with RIDEM, which defines the implementation schedule for the Phase III CSO Program. The final construction completion date is for a fully operational tunnel system. The tunnel, tunnel pump station, and near surface projects must be completed and operational prior to the dates specified in the CA. Near surface structures planned at drop shaft sites and the mechanical fit-out of the tunnel pump station will be constructed by others under separate contracts.

NBC selected a design build delivery to meet the regulatory schedule and manage risk. The objectives included the following:

- Establish a collaborative relationship between NBC, Program Manager, and Design-Builder to deliver a quality Project on time and within the defined Project budget.
- Maintain a safe, injury-free work site(s).
- Minimize impacts to the community through close coordination with NBC and community stakeholders.
- Control costs to reduce economic burden to ratepayers.
- Manage risks to successfully deliver the Project.
- Minimize disruption to customers and NBC operations.
- No impacts to environmental water quality and/or BPWTF performance during construction.
- Maximize participation of Rhode Island based engineering firms, contractors, subcontractors, suppliers, and construction trade labor.
- Protect and enhance water quality in Narragansett Bay and its tributaries by providing safe and reliable wastewater collection and treatment services to customers at a reasonable cost.

The project team for the Pawtucket Tunnel Design Build include Stantec/Pare, Program/Construction Managers; CBNA/Barletta JV, Design Builder; and AECOM in association with Gall Zeidler Consultants, Engineer of Record.

## GEOLOGY

The geologic history of the bedrock is complex. The rock originates from fluvial deposits, which characteristically are not laterally continuous. Rock types range from conglomerate to coal with sandstone and siltstone as the predominant rock types. Sequential episodes of tectonic deformation have superimposed structural features including folds, faults, and joints.

The Pawtucket Tunnel is constructed mainly in the Rhode Island Formation, a Carboniferous-age sedimentary rock comprised of sandstone and siltstone, with lesser amounts of conglomerate, shale, and coaly deposits. Bedrock is moderately folded and faulted. Bedrock is overlain by a thin layer of glacial till and thick layers of glaciofluvial deposits. Glacial deposits are comprised mainly of sand, gravel, and silt, and are overlain by granular fill.

The project area is within the Narragansett Basin, which is approximately 55 miles long, 15 to 25 miles wide, and is made up of several thousand feet of non-marine sedimentary rocks that have been folded, faulted, and slightly to moderately metamorphosed. Sedimentary rocks of the Narragansett Basin have been divided into five formations, of which the Wamsutta Formation and the Rhode Island Formation occur within the tunnel horizon. Bedrock is overlain by glacial deposits consisting of till and glaciofluvial deposits as well as man-made fill deposits near surface.

The Rhode Island Formation is widespread and consists mainly of gray sandstone and siltstone, with lesser amounts of gray to black shale, conglomerate, and coaly rock. These sediments were deposited in medial to distal alluvial fan environments, and are comprised of meandering stream deposits, bank and flood plain deposits, and swamp deposits. Sediments of the Rhode Island Formation generally are finer-grained than those of the Wamsutta Formation. Rock types are laterally and vertically discontinuous, a characteristic of the environment in which they formed.

The Wamsutta Formation occurs in the northern part of the Narragansett Basin, where it underlies and partially interfingers with the Rhode Island Formation (Skehan, Rast, and Mosher 1986). The Wamsutta Formation consists of a sequence of red conglomerates, sandstones, and shales up to 3,000 feet thick. Volcanic fragments are common. Conglomerate in the Pawtucket quadrangle contains boulders as large as 4 feet in diameter (Quinn 1971). These sediments were deposited in an alluvial fan environment, and are composed of braided stream deposits, with related crevasse splay deposits and flood plain deposits. Bedrock of the Wamsutta Formation is present along the northern 200-ft to 400-ft of the tunnel.

## PROJECT CRITERIA/RISK MANAGEMENT

### Project Criteria

Stantec developed Project Criteria, including Base Technical Concept (BTC) Design, to support the procurement. The BTC design included preliminary design drawings, Geological Baseline Report (GBR), Geotechnical Data Report (GDR), and Environmental Data Report (EDR). The Design Builder (DB) was responsible to developing final design documents for project elements to comply with project criteria for each of the project elements. The project elements included bored tunnel, large diameter shafts, suction header, drill and blast adit tunnel, and drop/vent shafts. The design was completed in January 2022 to comply with dates in the CA.

The primary functionality of the tunnel is to provide storage of combined sewage during wet weather events to control CSOs. The project criteria require the tunnel provide a storage capacity of 58.5 MG to store CSO generated during the 3-month design flow. The hydraulic criteria are prescriptive specifying vortex style drop shafts with hydraulic capacity to convey peak flows from the 2-year storm.

Durability criteria requires precast concrete segmental lining, internal components, permanent bolts, and associated inserts to be designed for a minimum of 100-year service life. Service life for reinforced concrete is defined as total duration to initiation of steel corrosion plus five years of propagation duration of carbon steel. Design and durability analysis considers pH level, chloride content, sulfates and other contaminants in soil, rock, ground water, and future wastewater. Analysis and testing confirmed exposure conditions for service life modeling.

Key project criteria for the Pawtucket Tunnel included the following:

- Diameter 30-ft ID
- Slope 0.001 ft/ft
- Minimum rock cover of two tunnel diameters. *Note: CBNA/Barletta raised tunnel elevation by 25-ft, during the Alternative Technical Concept process.*
- Maximum allowable groundwater infiltration 1 gpm per 1000 ft of tunnel and 0.1 gpm at any one segment

## Risk Management

NBC identified the following top five (5) risks: worker and public safety; TBM mechanical failure; damage to third party building, structures, or utilities due to ground movement during excavation; delay in production or delivery impacting project schedule and milestone completion, and differing site conditions. The procurement commenced in December 2019 prior to risks/issues associated with COVID, supply chain delays, and macroeconomic inflationary pressures. The risks associated with bored tunnel focused on ground conditions, which were baselined in the GBR for Bidding (or GBR-B): methane, water inflow, low strength rock, and fault at northern alignment.

The GBR-B identified construction considerations attributed to how predominant types of ground are expected to behave in response to construction. During preliminary design, the vertical profile provided two diameters of sound rock cover over the crown in an area of low rock near tunnel station 146+00. The horizontal alignment and shaft locations accommodated property acquisition constraints and operational preferences. As previously noted, the JV submitted an alternative technical concept (ATC) to raise the tunnel elevation, thereby reducing the amount of rock cover.

Pawtucket area coal has the capacity to retain high amounts of methane, but geologic history combined with mining and drilling experience indicates that the actual methane content of coal is low. Methane can be adsorbed on organic surfaces of coal and is released by a reduction in hydrostatic pressure. The potential exists for discharge of methane from carbonaceous rock, so excavations in rock were classified as "potentially gassy."

RMR values along the tunnel range from about 30 (poor) to 70 (good). Rock mass quality generally improves from south to north. For an excavated diameter of 33.5-ft, immediate collapse in the crown is indicated for most of the alignment. To mitigate the risk of roof fallout, the project criteria require tunnel excavation with a shielded TBM and concurrent installation of a permanent lining of gasketed precast concrete segments. The single-pass lining approach minimizes groundwater discharge volumes. This approach mitigates problems associated with localized deposits of weak mudstones and coal, such as loss of bearing on gripper pads, and a soft invert.

A potential fault crosses the tunnel alignment under the river at the northern end of the alignment (see Figure 2). The GBR-B baselined the fault to be vertical, composed of a major slip zone 3-ft wide, surrounded on each side by a damage zone comprising a network of low-throw faults trending subparallel to the major fault. Each damage zone is estimated to be 60-ft in width, for a total width of the fault zone of 123-ft. To mitigate the risk of excessive inflows while tunneling under the river, it is expected that probing and grouting ahead would be needed. The baseline conditions were conservative based on the inability to locate the contact between the two formations during the preliminary geotechnical investigation.

The JV successfully located the contact, which indicated a contact zone (i.e., no fault) and no fractured water bearing features. The JV conducted additional investigations from the surface to revise the baseline within the reach in the GBR-C. In addition, a probe at the base of the receiving shaft was drilled in December 2022 with two horizontal probes confirming the contact between the Rhode Island and Wamsutta formations. At the time of this article, one of the two probes was completed, confirming the following: no apparent faulting along the contact between the two formations, two formations appear to be interfingered, and no significant groundwater inflows. The JV

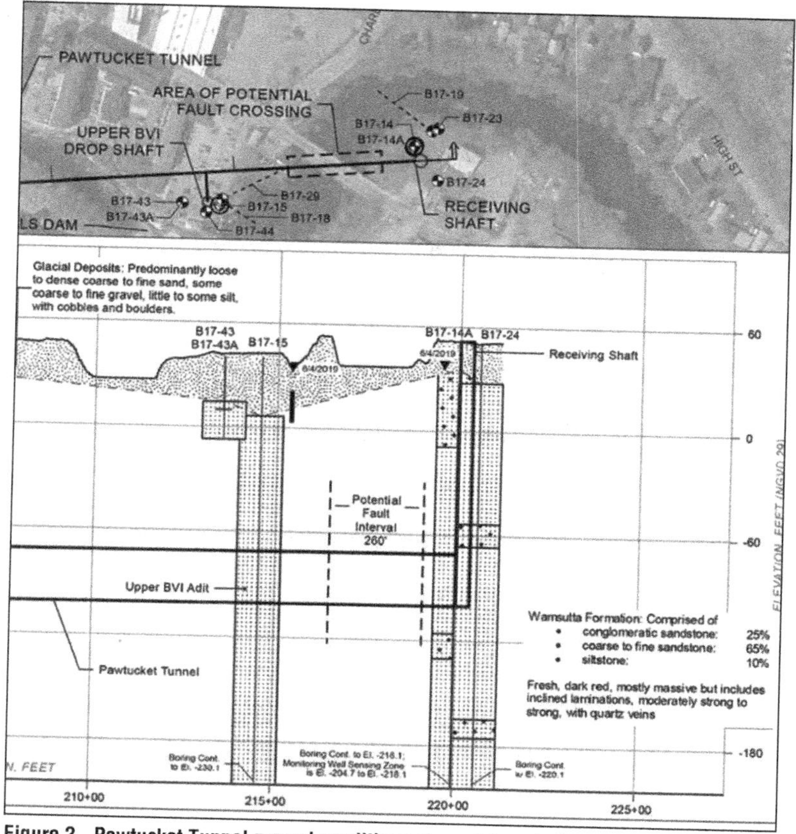


**Figure 2. Pawtucket Tunnel ground conditions at receiving shaft**

updated the GBR-C accordingly, identifying TBM operation through this zone to be in open mode.

Groundwater inflow was an identified risk due to flow rates encountered during construction of the Providence tunnel. A secondary risk was impact to NBC wastewater treatment facility from elevated loadings of total suspended solids, pH, and impact to sludge production. Efforts to reduce tunnel dewatering flows became important for both points.

Estimated flush flows from the heading of Providence Tunnel were typically 100 gpm with brief intervals ranging up to 200 gpm (Kaplin, Peterson, and Albert 2009). These heading inflows ranged from about ½ to 1 times the steady state inflow.

The GBR-B baselined the main tunnel drive to cross one or more shear zones. Such locations of high fracture density would produce high inflow rates over a short distance of up to 100-ft. The baseline for localized heavy inflows is set at a sustained inflow rate of up to 130 gpm per 100-ft of tunnel, to occur at one interval up to 100-ft long, at an unidentified location south of station 208+00. It also baselined the tunnel drive to cross a fault zone under the river. As a baseline, sustained inflows of up to 400 gpm per 100-ft of tunnel may occur within the fault zone. The fracture network related to the fault zone is expected to be directly connected to the river.

## BORED TUNNEL CONSTRUCTION AND TBM DESIGN

The JV selected Herrenknecht from Germany as the TBM supplier. Through negotiation and risk allocation, the Project Team agreed on the use of a TBM with open mode/EPB capabilities which presents the advantage of controlling muck management, minimized maintenance and optimized production time. The TBM contract was secured during the tender time and triggered when the Notice to Proceed was provided on December 18th, 2020. The TBM was manufactured in the Herrencknecht factory in Germany and despite the disturbance due to the COVID pandemic, the Factory Acceptance Test was completed ahead of schedule in January 2022. The TBM delivery time frame went from end of April to end of June 2022.

The 33.8-ft diameter TBM has a 46-ft long shield and a 300-ft long trailing gear, comprised of four gantries to supply power to the shield, and connect the utilities and logistics. The shield is composed of three main sections with one active articulation. This enables operators to achieve the minimum required curvature of 1,000-ft. The structure and seals are designed for 72 psi of pressure.

The 33.8-ft diameter cutter head is designed for hard rock and protected against wear. The cutterhead has a total of sixty four 19-inch cutter discs for high efficiency and reliability in the anticipated conditions. It can be configured for either EPB mode or open mode. Transition between the two modes is approximately a week.

The machine is fitted with a conveyor belt in open mode and a screw conveyor in EPB to expel muck from the cutting chamber. The TBM can be isolated in less than ten minutes to control flow at the heading for elevated inflows and/or face instability.

Mining has commenced in open mode and will be the primary mode for the Pawtucket tunnel. EPB mode will be used when confinement pressure is required to ensure the stability of the ground and/or to control water ingress. The conformation of the contact between the two rock formations at the northern river crossing gave credence to the selection of mining in open mode. The JV confirmed the contact did not have any fractured zones with water bearing features. Mining in open mode increases production rate and eliminates the need for soil conditioners.

The mode would transition to EPB due face instability and/or elevated groundwater inflow. The change of mode is anticipated to be executed in a week of work mainly in the cutting chamber to re-configurate the cutter head. In open mode, the muck is collected and lifted to a conveyor fitted in the center of the shield; these buckets would be removed to allow the extension of the screw conveyor to collect the muck in the bottom of the excavation chamber. The bulkhead will then be closed by retracting the conveyor belt and the machine will be able to operate in the confinement mode.

A dual mode TBM is advantageous to the project balancing production and risk management, During favorable ground conditions, JV will mine in open mode achiever higher production. In poor ground conditions, mining in EPB reduces the need for probe drilling and pre-grouting, which would be required in a single open mode TBM (Figure 3).

The TBM began mining in November 2022 from the launch shaft on the main site at the Southern end of the project area (Figure 4). The launch shaft is 62-ft diameter and 155-ft deep. Soil excavation is within a cofferdam made of secant piles and in the rock using drill and blast techniques. At the base of the shaft excavation (120-ft), a starter tunnel of 230-ft and a tail tunnel of 60-ft were excavated in two stages: a top heading and bench. The remainder of the shaft was then completed after the starter tunnel and

Figure 3. TBM in Herrenknecht factory in January 2022

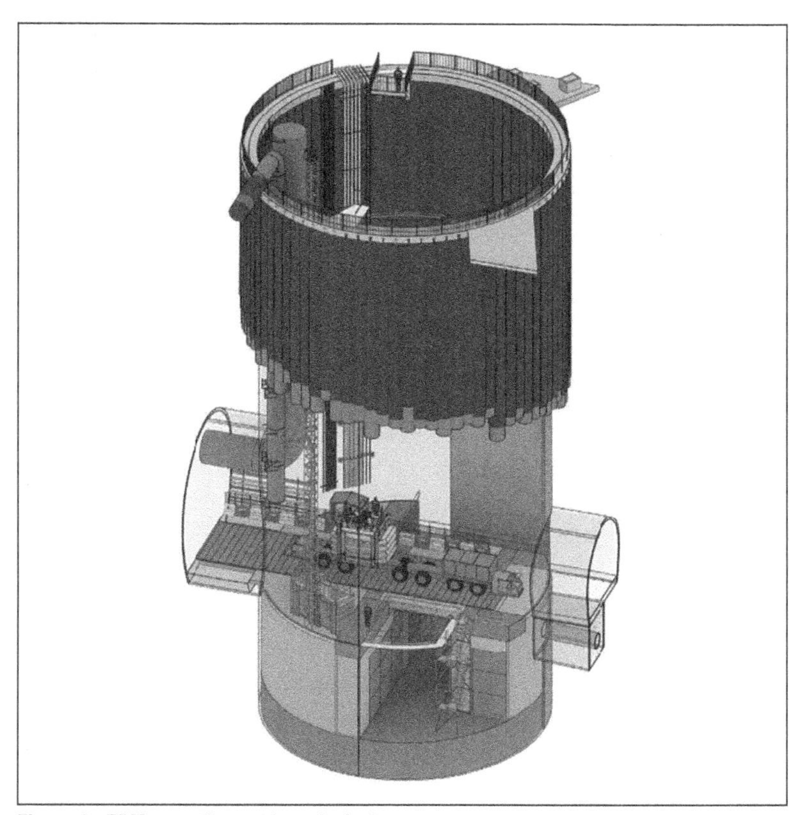

Figure 4. TBM operations at launch shaft

tail tunnel were excavated. The bottom of the shaft was backfilled, and a trench was kept below the tunnel invert to install the bottom part of the vertical conveyor. This sequence of works has avoided remobilizing the drill-and-blast team for deepening the shaft to its final level and brings benefit to the general schedule of works.

To leave sufficient time to complete the preparation works in the shaft and starter tunnel to receive the TBM, it was decided to pre-assemble the TBM in surface. The 1,665-ton machine was delivered in 82 packages of about 20 tons and for the five heaviest ones just under 100-tons. The pre-assembly on the surface made possible to lower parts into the shaft up to 375-tons. The mobilization of a 600-ton crawler crane for six weeks was necessary to lower the three elements of the shield and the cutter head. The back-up gantries are pre-assembled in half section not limited by the weight but by the size to fit inside the 62-ft diameter launch shaft (Figure 5).

**Figure 5. 375-ton TBM front shield lowered with the 600-ton crawler crane in the launch shaft**

Surface logistics at the main site provides the necessary support to the TBM operation. A crawler crane of 350-ton capacity handles the segments and the services inside the shaft. It is assisted by a loader to unload the segments from the delivery trucks. There is sufficient space on top of the shaft to store up to sixty rings, more than a week of TBM production. The 350-ton crawler crane is also used for the TBM assembly. It is used primarily for the pre assembly of the gantries (Figure 6 and 7).

Tunnel muck is transported from the TBM by a chain of conveyors at a rate of 1,250 tons per hour. The muck is transported out of the shaft by using a vertical bucket conveyor. At the surface, the muck is dropped by a radial stacker in a stockpile accumulating 15,000 cubic yards (the average weekly quantity excavated by the TBM).

The rest of the surface installation comprises of the following:

- Mortar and Cooling Plant
- Electrical Substation
- Water Treatment Plant
- Main Offices
- Mechanical and Electrical Workshop

## SEM TUNNEL CONSTRUCTION

The Pawtucket project contains several mined tunnels (i.e., main drive, adits). In a similar fashion to the contract form and TBM selection, the mined tunnels were designed with the goal of minimizing risk. However, in contrast to TBM tunneling, mined tunneling is performed under open face conditions without the aid of a protective shield. As

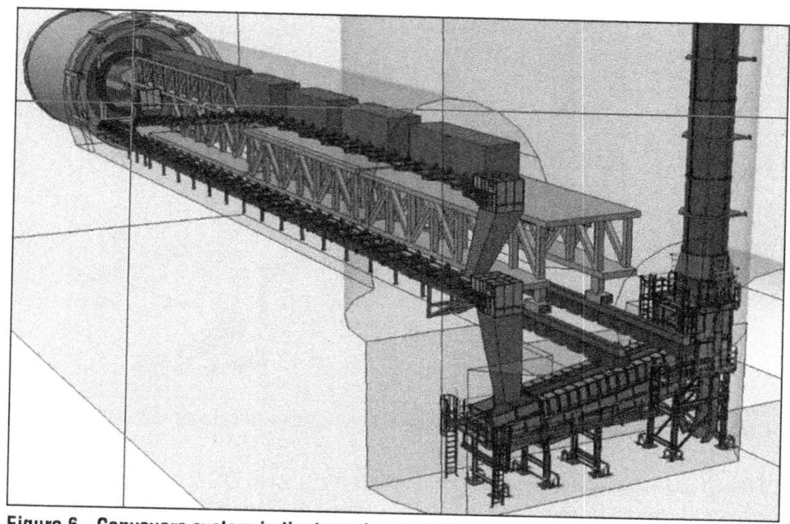

Figure 6. Conveyors system in the tunnel and shaft

Figure 7. Main site aerial picture during TBM assembly

such, to minimize risk, the Sequential Excavation Method (SEM) is employed using separate support classes depending on the expected rock behavior. The specifications also call for rock excavation and support meetings (RES meetings), in which representatives from the Engineer of Record (EOR), NBC, and JV decide the appropriate support measures based on the existing rock conditions. The design build delivery allows for a close interaction between the engineer and contractor during tunneling. This close interaction allows for efficient adjustment of rock support to ensure that what is installed is both safe and economical.

The largest mined tunnel along the alignment is the starter tunnel. The starter tunnel extends northwards from the launch shaft in the direction of the main tunnel and houses the TBM during launch. To further extend the launching area for the TBM, a shorter tail tunnel has been constructed between the launch shaft and the future pump shaft. In addition to these larger mined tunnels, The Pawtucket tunnel will be connected to four drop shafts by tunnel adits along its alignment. The adits, listed from

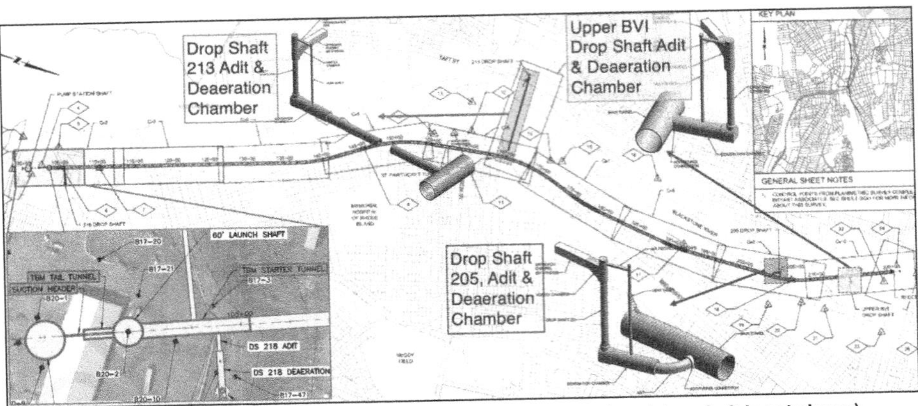


Figure 8. Layout of the Pawtucket Tunnel and connecting structures (receiving shaft is not shown)

south to north, are referred to as the Drop Shaft (DS) 218, DS 213, DS 205, and Upper Blackstone Valley Interceptor (UBVI) adits. The DS 218 adit connects to the SEM starter tunnel portion of the alignment and the DS 213, DS 205 & UBVI adits connect to the segmentally lined portion of the main tunnel alignment. The location of the adits, the tail tunnel, and the starter tunnel are shown schematically in Figure 8. At the time of writing this contribution, the starter tunnel and tail tunnel have been fully excavated.

The JV has proposed to excavate the OF-213 adit from the access shaft to the main spine tunnel by micro tunneling to mitigate the geological risks inherent to drill and blast mining initially anticipated in the BTC. The adit is over 1,000-ft in length and crosses under the Seekonk River. Additional ground investigations from the surface were limited by accesses not granted into private properties. The choice of a confined microtunnel TBM prevent the use of pre-excavation grouting and secure the schedule against geological conditions.
The JV has decided to utilize Hobas fiber reinforced plastic (FRP) jacking pipe for the 1,000-ft adit. FRP pipe was selected to attain the 100-year durability requirement for the project as well as achieving pressures needed on the gaskets. The fiberglass pipe is 96-inch in nominal diameter with a pipe wall thickness of 3-inch. The pipe joint is a flush bell and spigot joint which has an allowable safe jacking load of about 1,150-ton and a maximum allowable hydrostatic pressure up to roughly 70 psi.

## Starter and Tail Tunnel

The starter tunnel is ca. 220-ft long with a design excavation width at the springline of 37'-2" and a height of 36'-11 ¾", making it the largest horizontal excavation by area along the tunnel alignment. Based on the geotechnical borings, and as discussed in the previous section on geology, the starter tunnel was expected to pass through primarily siltstone and sandstone with the potential for shale seams and coaly deposits. The rock mass was expected to have a median RMR of around 40. In addition, the contract requirements include the stipulation for a "contingency" support class be developed, in case worse-than-expected conditions should occur.

To provide greater flexibility in this relatively short tunnel, the design includes two support classes plus the contingency support class. The general support class (SCI) is intended for a rock face with a mapped RMR of 40+, a heavier support class, SCII, is intended for a rock mass with an RMR of 30–40, and the contingency support class, which is also referred to adverse ground support, is for ground/rock with an RMR of below 30, in which soil-like behavior may be encountered. Both SCI and SCII feature rock dowels with fiber reinforced shotcrete, and the contingency support class also features a closed invert. Excavation was carried out by drill and blast and the excavation sequence was subdivided into a top-heading and two benches, to match excavation levels in the connecting shafts and tail tunnel. As mentioned above, the stand-up times predicted in the GBR for the large excavations were not high. As such, all support classes stipulate that a 2-inch minimum flashcrete layer is to be applied immediately after excavation and rockbolting to ensure stability.

Rock conditions at the shaft-tunnel interface were as expected, with the greatest tunneling risk being a graphitic shale seam located at the excavation crown. The rock quality did, however, significantly improve along the length of the tunnel, with RMR values at the face consistently exceeding 50 deeper in the rock. This represented, contrary to expectations in the GBR, stable rockmasses with sufficient stand-up times. As such, the excavation procedure was modified. A third support class, SCI-A was developed with a single top heading/bench, the longest round length, and the lightest support. Figure 9 describes the support classes, their intended application range, and depicts the lightest (SCI-A) and heaviest Support Class (Contingency Support).

The tail tunnel is 65' long with a design cross section of 24'-4" width × 22'-2" height. Like the starter tunnel, three ground support classes were developed, GSI (RMR>40), GSII (30<RMR<40), and a contingency ground support class (RMR<30) for adverse ground conditions. During excavation, the rockmass was found to have higher RMR values than expected, partly because the rock mass was dipping away from the excavation. As such, it was decided to excavate the tail tunnel full-face to improve logistics. This decision was supported with experience gained from mining the starter tunnel. No significant instabilities were observed during construction of the starter tunnel top heading. Because the starter tunnel top heading has an area similar to that of the tail tunnel, it was assumed that no major instabilities would be observed in the tail tunnel even with full-face excavations.

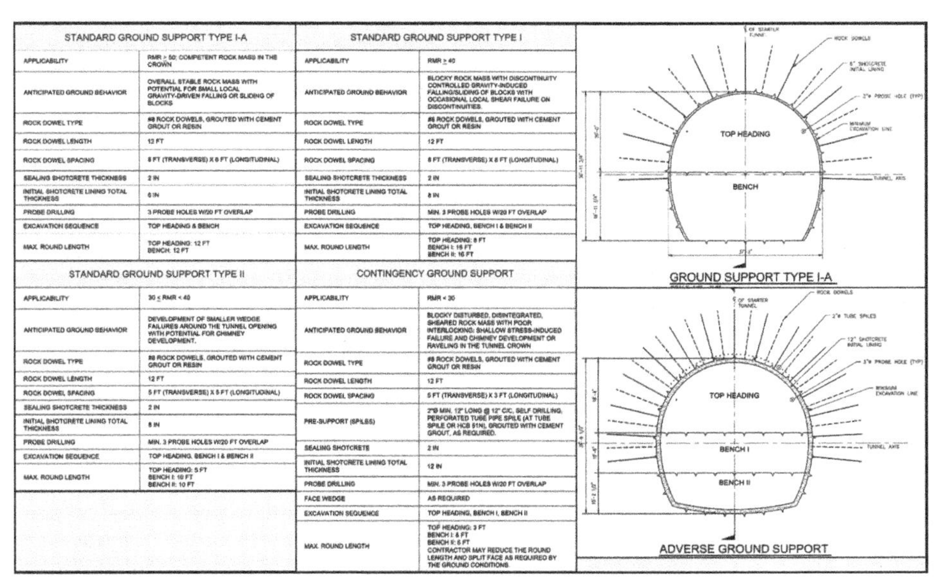

| STANDARD GROUND SUPPORT TYPE I-A | |
|---|---|
| APPLICABILITY | RMR > 50; COMPETENT ROCK MASS IN THE CROWN |
| ANTICIPATED GROUND BEHAVIOR | OVERALL STABLE ROCK MASS WITH POTENTIAL FOR SMALL LOCAL GRAVITY-DRIVEN FALLING OR SLIDING OF BLOCKS |
| ROCK DOWEL TYPE | #8 ROCK DOWELS, GROUTED WITH CEMENT GROUT OR RESIN |
| ROCK DOWEL LENGTH | 12 FT |
| ROCK DOWEL SPACING | 8 FT (TRANSVERSE) X 8 FT (LONGITUDINAL) |
| SEALING SHOTCRETE THICKNESS | 2 IN |
| INITIAL SHOTCRETE LINING TOTAL THICKNESS | 6 IN |
| PROBE DRILLING | 3 PROBE HOLES W/20 FT OVERLAP |
| EXCAVATION SEQUENCE | TOP HEADING & BENCH |
| MAX. ROUND LENGTH | TOP HEADING: 12 FT<br>BENCH: 12 FT |

| STANDARD GROUND SUPPORT TYPE I | |
|---|---|
| APPLICABILITY | RMR ≥ 40 |
| ANTICIPATED GROUND BEHAVIOR | BLOCKY ROCK MASS WITH DISCONTINUITY CONTROLLED GRAVITY-INDUCED FALLING/SLIDING OF BLOCKS WITH OCCASIONAL LOCAL SHEAR FAILURE ON DISCONTINUITIES. |
| ROCK DOWEL TYPE | #8 ROCK DOWELS, GROUTED WITH CEMENT GROUT OR RESIN |
| ROCK DOWEL LENGTH | 12 FT |
| ROCK DOWEL SPACING | 8 FT (TRANSVERSE) X 8 FT (LONGITUDINAL) |
| SEALING SHOTCRETE THICKNESS | 2 IN |
| INITIAL SHOTCRETE LINING TOTAL THICKNESS | 8 IN |
| PROBE DRILLING | MIN. 3 PROBE HOLES W/20 FT OVERLAP |
| EXCAVATION SEQUENCE | TOP HEADING, BENCH I & BENCH II |
| MAX. ROUND LENGTH | TOP HEADING: 8 FT<br>BENCH I: 16 FT<br>BENCH II: 16 FT |

| STANDARD GROUND SUPPORT TYPE II | |
|---|---|
| APPLICABILITY | 30 ≤ RMR < 40 |
| ANTICIPATED GROUND BEHAVIOR | DEVELOPMENT OF SMALLER WEDGE FAILURES AROUND THE TUNNEL OPENING WITH POTENTIAL FOR CHIMNEY DEVELOPMENT. |
| ROCK DOWEL TYPE | #8 ROCK DOWELS, GROUTED WITH CEMENT GROUT OR RESIN |
| ROCK DOWEL LENGTH | 12 FT |
| ROCK DOWEL SPACING | 5 FT (TRANSVERSE) X 5 FT (LONGITUDINAL) |
| SEALING SHOTCRETE THICKNESS | 2 IN |
| INITIAL SHOTCRETE LINING TOTAL THICKNESS | 8 IN |
| PROBE DRILLING | MIN. 3 PROBE HOLES W/20 FT OVERLAP |
| EXCAVATION SEQUENCE | TOP HEADING, BENCH I & BENCH II |
| MAX. ROUND LENGTH | TOP HEADING: 5 FT<br>BENCH I: 10 FT<br>BENCH II: 10 FT |

| CONTINGENCY GROUND SUPPORT | |
|---|---|
| APPLICABILITY | RMR < 30 |
| ANTICIPATED GROUND BEHAVIOR | BLOCKY DISTURBED, DISINTEGRATED, SHEARED ROCK MASS WITH POOR INTERLOCKING: SHALLOW STRESS-INDUCED FAILURE AND CHIMNEY DEVELOPMENT OR RAVELING IN THE TUNNEL CROWN |
| ROCK DOWEL TYPE | #8 ROCK DOWELS, GROUTED WITH CEMENT GROUT OR RESIN |
| ROCK DOWEL LENGTH | 12 FT |
| ROCK DOWEL SPACING | 5 FT (TRANSVERSE) X 3 FT (LONGITUDINAL) |
| PRE-SUPPORT (SPILES) | 2"Ø MIN. 12' LONG @ 12" C/C, SELF DRILLING, PERFORATED TUBE PIPE SPILE (AT TUBE SPILE OR HCB 51N), GROUTED WITH CEMENT GROUT, AS REQUIRED. |
| SEALING SHOTCRETE | 2 IN |
| INITIAL SHOTCRETE LINING TOTAL THICKNESS | 12 IN |
| PROBE DRILLING | MIN. 3 PROBE HOLES W/20 FT OVERLAP |
| FACE WEDGE | AS REQUIRED |
| EXCAVATION SEQUENCE | TOP HEADING, BENCH I, BENCH II |
| MAX. ROUND LENGTH | TOP HEADING: 3 FT<br>BENCH I: 6 FT<br>BENCH II: 6 FT<br>CONTRACTOR MAY REDUCE THE ROUND LENGTH AND SPLIT FACE AS REQUIRED BY THE GROUND CONDITIONS |



**Figure 9. Starter tunnel support class types I-A through to contingency support; The heaviest (contingency/adverse support) and lightest Ground Support type I-A are also schematically depicted**

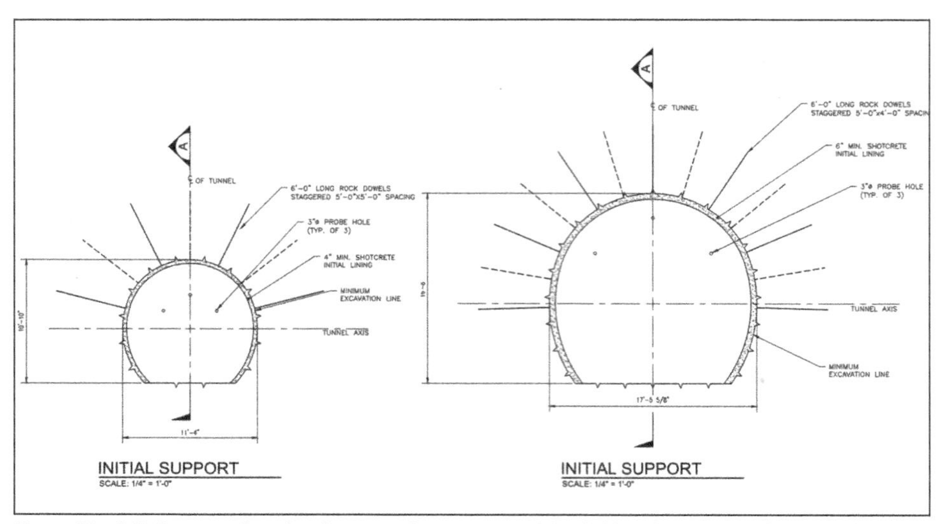


**Figure 10. Adit Cross section showing ground support type 1 (on left) and Deaeration Chamber Cross section showing ground support type I (on right)**

The final lining of the starter tunnel will be cast-in-place fiber reinforced concrete cast flush with the inner diameter of the TBM tunnel. Fiber reinforced concrete was selected not only out of economy, but to increase robustness of the lining against corrosion, further minimizing risk that significant damage would occur to the as-built-structure over time. The tail tunnel will be back filled.

## SEM Adits

In contrast to the tail and starter tunnel, the SEM adits are much smaller in cross section. The SEM adits excavations are 11'-4" in width by 10'-10" in height and are of varying length. The adits are "preceded" by a slightly larger deaeration chamber. The deaeration chambers are immediately below the drop shafts. When water is collected in the drop shafts, it flows first through the deaeration chamber and then into the adit. The deaeration chambers all have an excavation cross section of 17'-5 ⅝" width by 16'-8" height and are all ca. 90-ft long.

Because of the smaller size of the deaeration chambers and the adits, a single support class, in addition to the contractually mandated contingency support class, was deemed sufficient. In addition, the rock quality along the tunnel alignment generally improves as the alignment stretches north, and it is therefore assumed that the adits and deaeration chambers will be built in a more stable rock mass to that of the starter & tail tunnels. The cross section of this support class, SC1, is shown in Figure 3 for both the adit and deaeration chamber.

## CONCLUSION

NBC selected design build delivery for the Pawtucket Tunnel to manage risks and align design responsibilities with the DB team. The primary risk inherent to underground construction was understanding the geologic properties along the alignment and its influence on construction. The GBR is an important tool in risk management and risk allocation between the owner and the JV. At time of tender, Stantec baselined the ground conditions in the GBR-B defining properties of soil and rock along the alignment utilizing accepted statistical analysis of the data. The GBR-B baseline comprised the geological data, the interpretation of existing geotechnical conditions, and the expected ground's response to construction.

The design build delivery system offered an opportunity for the NBC and JV to jointly engage in the process of risk management while final design was on-going. The JV had an opportunity to engage their design team by providing interpretations of GBR-B in developing their approach to design and construction. The JV identified gaps in data and performed supplemental investigations. NBC and the JV then negotiated the terms of the baseline to formulate the GBR-C. The GBR-C reflected a joint understanding of the interpreted baseline conditions that was consistent with the design, the planned equipment, and the construction means and methods.

The supplemental investigation of the contact between the Rhode Island and Wamsutta formations and the absence of a fractured water bearing feature is an example of the two step GBR process having a net benefit to the project. The tunneling approach was optimized based on the results of this investigation. As such, the JV elected to use a dual mode TBM with the primary operating mode being the open or unpressurized mode. The JV engaged their design team to refine the interpretation based on the supplemental data, which culminated in a revised GBR-C. Stantec and NBC actively participated by assessing the appropriateness of the update and preventing a re-allocation of risk through the process.

The close coordination with JV and engineer team on ground conditions during SEM afforded efficient decision making in appropriate rock support. The underlying objective is for construction to proceed in a safe, efficient manner to maintain schedule. Analogous to the TBM selection process, the investigations and interpretations described in the GBR-C were used to develop an efficient SEM excavation scheme, which has proved successful.

At the time of this article, the TBM mining has commenced in November 2022. The 11,600-ft, 30-ft diameter tunnel is anticipated to be fully mined in January 2024. Final lining and site restoration is scheduled for December 2024 to allow the follow-on pump station fit-out team to install the mechanical, structural, and electrical systems for the tunnel pump station to become fully operational by December 2026.

The purpose of the paper was to highlight a few examples of the evolving design process inherent to design build delivery. The evolution leads to innovation through refinement linking selected equipment, material, and construction methods to design.

Open and effective communication have been key to successful project delivery, initiated by NBC leadership. The project has endured numerous challenges and risks from outside forces (i.e., COVID, supply chain delays, inflation) and typical project specific challenges. The JV and design team are constantly refining design and approach to manage the challenges. NBC, JV, Program Manager, and design team each understand the role in maintaining the success of the project.

## ACKNOWLEDGMENTS

The authors would like to thank the leadership and support of the Narragansett Bay Commission, Stantec, CB3A, and GZ Consultants for permission to publish this paper. Special appreciation is due to Laurie Horridge, Jim McCaughey, and Rich Bernier of NBC; Raoul Fernandes and Pete Williamson of CBNA/Barletta: Mellissa Carter of Stantec: and Thomas Klingenberg and Dominic Reda from GZ Consultants.

## REFERENCES

Kaplin, J. L., J. P. Peterson, and P. H. Albert. 2009. Underground construction for a combined sewer overflow system in Providence, Rhode Island. In *Proceedings, 2009 Rapid Excavation and Tunneling Conference.*

Quinn, A. W. 1971. *Bedrock geology of Rhode Island.* Geological Survey Bulletin 1295. Washington, D.C.: U.S. Govt. Printing Office.

Skehan, J. W., N. Rast, and S. Mosher. 1986. Paleoenvironmental and tectonic controls of sedimentation in coal-forming basins of southeastern New England. In *Paleoenvironmental and tectonic controls in coal-forming basins of the United States.*

# Cohesive Soil Conditioning Practice for Earth Pressure Balance Tunneling

**Mike Mooney** ▪ Colorado School of Mines
**Rakshith Shetty** ▪ Colorado School of Mines
**Diana Diaz** ▪ Colorado School of Mines
**Vitaly Proschenko** ▪ Colorado School of Mines

## ABSTRACT

Cohesive clay soil poses a significant risk to earth pressure balance tunnel boring machines (EPBM) due to unstable face pressure, clogging potential, high cutterhead torque, poor muck flow, low advance rate, etc. This paper details the performance measures desired when EPBM tunneling in clay soil, and the scientifically-supported conditioning methods to transform in-situ clay to a muck with such characteristics, giving particular attention to ground conditions. The paper then examines TBM data from multiple clay soil projects in the context of these performance measures to demonstrate the influence of cohesive soil conditioning on EPBM performance.

## INTRODUCTION

Effective earth pressure balance tunnel boring machine (EPBM) operation involves using the excavated soil as a support medium to counterbalance in-situ pore water pressure and lateral effective stress at the face in order to prevent tunneling-induced deformation and damage to overlying infrastructure. Effective EPBM operation also requires efficient processing of the excavated material through the cutterhead openings into the excavation chamber, into the mouth of the screw conveyor, up the screw conveyor and along the belt conveyor to a temporary muck pile, whereafter, it is transported (typically by truck) for disposal. Conditioning fluids such as foam, water, bentonite slurry, polymer, etc. are introduced in up to three distinct locations—face, chamber and screw conveyor—to facilitate smooth and consistent face pressure support and efficient muck processing.

EPBMs have long been well-suited for tunneling through fine-grained cohesive soils. In soft to medium soft marine clay environments, EPBMs progress very efficiently without the use of any conditioning. In most areas in North America, however, fine grained cohesive clays and silts, as well as claystones and siltstones that behave as a fine grained cohesive soil when excavated, require conditioning to decrease shear strength, improve workability, and in many cases to reduce stickiness and clogging potential.

This paper reviews the established fundamentals cohesive soil conditioning per the literature and examines the influence of cohesive soil conditioning on EPBM performance on multiple recent projects.

## DESIRED COHESIVE MUCK PROPERTIES

Early studies on EPBM tunneling linked desirable cohesive soil properties to consistency and undrained shear strength as defined in the soil mechanics literature. Based on field experience, Maidl (1995) suggested a desired range for the cohesive soil's

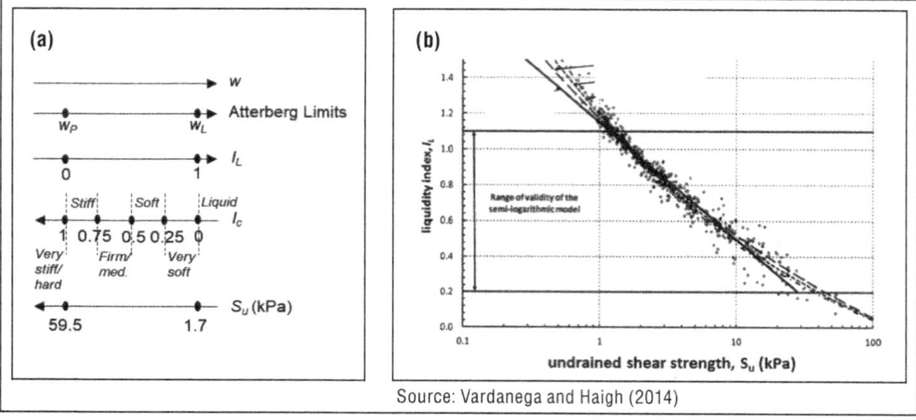


Source: Vardanega and Haigh (2014)

**Figure 1. (a) Schematic representation of the inter-relationship between Atterberg limits, liquidity and consistency indices, and undrained shear strength, (b) Relationship between liquidity index and undrained shear strength**

consistency index $I_C$ (Equation 1) between 0.40–0.75 to provide both face pressure support and workability through the EPBM. $I_C$ provides a normalized metric that is defined by a cohesive soil's native (in-situ, before excavation) gravimetric moisture content $w$ in relation to its liquid limit $w_L$ and plastic limit $w_p$. The corresponding $I_C$ values are also widely used to characterize the cohesive soil's consistency from very stiff/hard to viscous liquid for defining the general behavior (Figure 1a).

$$I_c = (w_L - w)/I_P \quad \text{where} \quad I_P = w_L - w_P \qquad \text{(EQ 1)}$$

In a state of the art review on soil conditioning, Milligan (2000) indicated, without reference or data, that the undrained shear strength $S_u$ of a conditioned cohesive soil should be in the range of 10–25 kPa. The recommended $I_C$ range in Maidl (1995) can be related to the recommended $S_u$ range in Milligan (2000) using published results from a wide range of cohesive soils in the soil mechanics literature. There is a relatively unique relationship between $I_C$ and $S_u$ that holds across cohesive soils. Vardanega and Haigh (2014) examined the relationship between liquidity index ($I_L$ = 1 – $I_C$) and $S_u$ from more than 600 fall cone tests using a database of 101 soil samples across 12 countries. The relationship used to estimate $S_u$ is shown in Figure 1b where $S_u$ = 1.7 kPa at $w_L$. The relationship can be rewritten in terms of $I_C$ per Equation 2 and is valid over the $I_C$ range –0.1–0.8.

$$S_u = 1.7 * 35^{I_c} \qquad \text{(EQ 2)}$$

Per Equation 2, $S_u$ varies from 7.1–24.5 kPa over the $I_C$ range of 0.4–0.75, thus generally equating the combined Maidl (1995) and Milligan (2000) recommendations.

There are deviations in the literature regarding desirable cohesive soil properties for EPBM tunneling. A case study paper commenting on the St. Clair project indicated that $S_u$ = 80–100 kPa and soft to stiff consistency was ideal for EPBM tunneling (Jancszek 1997). Thewes and Budach (2010) suggest an $I_C$ range of 0.6–0.7 to provide sufficient pressure gradient in the screw conveyor. Conversely, Merritt (2004) reports very good screw conveyor behavior with $S_u$ as low as 4 kPa. This would equate to $I_C$ near zero.

Regarding clogging potential in EPBMs caused by cohesive soils, a common reference is Hollman and Thewes (2013), where they state strong clogging potential occurs when the plastic index $I_P > 0.2$ and when $I_c$ is in the range 0.5–0.75, whether native or conditioned. Medium clogging potential occurs in the cohesive $I_c$ range of 0.75–1.0, and little clogging potential occurs for $I_c < 0.5$, according to Hollman and Thewes (2013). Laboratory testing by Zumsteg et al. (2013) showed the highest stickiness was observed in the $I_c$ range 0.4–0.75, corroborating the clogging potential suggestion of Hollmann and Thewes (2013), but also showing that the greatest clogging risk occurs in the desired $I_c$ and $S_u$ range recommended by Maidl (1995) and Milligan (2000).

The change in $I_c$ of the muck (Equation 3) can be determined from the native dry density $\rho_d$, water density $\rho_w$, and the liquid injection ratio $LIR$, defined as the volumetric ratio of liquid injected to excavated ground (Equation 4).

$$I_c = -\left(\frac{\rho_w}{\rho_d}\right)\left(\frac{LIR}{I_P}\right) \qquad \textbf{(EQ 3)}$$

$$\text{Conditioned muck } I_c = \text{Native } I_c + I_c \qquad \textbf{(EQ 3a)}$$

$$LIR = V_{Liquid}/V_{Exc} \qquad \textbf{(EQ 4)}$$

An important point to consider is that the $\Delta I_c$ and the final or conditioned $I_c$ (Equation 3a) is calculated and represented as a homogeneous average throughout all the excavated muck. In reality, the conditioned material exists at the interfaces between slices and blocks of the cohesive soil scraped and cut by the cutterhead scrapers and knife bits. The dimensions of the slices and chunks depend on the penetration rate $PR$ (mm/rev) and the soil's consistency. The minimum dimension (the slice thickness) will be determined by dividing the $PR$ by number of scrapers per orbit, and typically ranges from 5–20 mm on the projects considered here. There is insufficient time and pressure for liquid to penetrate into the largely unaltered slices and chunks. To this end, the interface $I_c$ is what governs the muck behavior, and this final interface $I_c$ will be considerably lower than what is reflected by the aggregated conditioned $I_c$ using Equation 3. Estimating the interface $I_c$ requires knowledge of the slice and chunk size distribution, which is unknown.

## USE OF FOAM AND POLYMER IN COHESIVE SOIL

Foam was primarily introduced in the 1980s for EPBM tunneling in cohesionless soils as a way to both, reduce hydraulic conductivity and increase void ratio, and thereby decrease stiffness and strength, and to increase compressibility for smooth face pressure (Mooney et al. 2017). Foam has been recommended for clay soils (EFNARC 2005, Milligan 2000) but the rationale is far less documented. The hydraulic conductivity of clay does not need modification like in cohesionless soil. The strength and workability are altered by decreasing $I_c$, which is generally achieved by adding liquid. Foam can increase the compressibility of clay muck for more smooth face pressure distribution. EFNARC (2005) suggests a foam injection ratio $FIR$ of 30–80% for clayey soils and recommends foam expansion ratios $FER$ in the range of 5–30, though for all soils and not just clay. $FIR$ and $FER$ are defined in Equations (5) and (6), and illustrated in Figure 2. The Shield Tunnel Association of Japan (2020) suggests that if the purpose of soil conditioning is only to prevent adhesion the standard $FIR$ should be 30% and recommends $FER$ of 8 for cohesive soil. These ranges and recommendations are offered without basis.

$$FIR = V_{Foam}/V_{Exc} \qquad \textbf{(EQ 5)}$$

$$FER = V_{Foam}/V_{Foam\ Liquid} \qquad \textbf{(EQ 6)}$$

In addition to providing compressibility, foam is used as an economically and mechanically more efficient way to distribute liquid (to decrease $I_c$) and/or added polymer to affect a change in clay surface properties.

The primary mixing environment on EPBMs is the cutterhead tool gap, wherein a finite number of nozzles (typically 1 nozzle per m of cutterhead diameter) inject foam or liquid. The collection of scraping tools provides an inefficient mixing environment for cohesive soil given (a) that mixing is not their primary purpose, (b) the low permeability of cohesive soil and (c) spatially infrequent nozzles. The injection of larger amounts of fluid in foam form compared to just a smaller volume of liquid gives a better chance of fluid distribution.

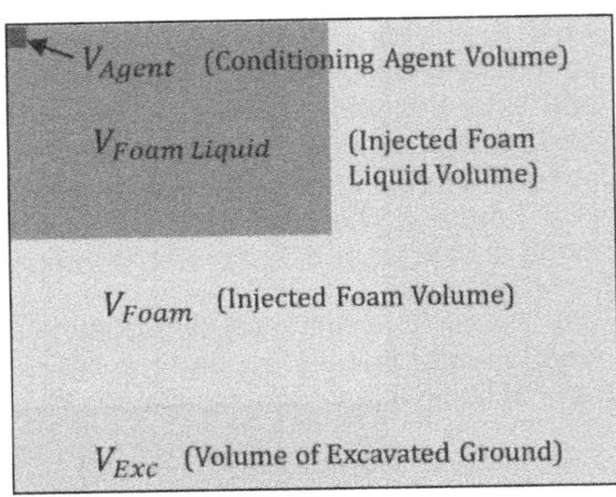


**Figure 2. Schematic showing the volumetric proportions excavated ground, foam, foam liquid mixed with compressed air to create foam, and the conditioning agent(s) that are diluted in water to create foam liquid**

Polymer is sometimes added directly (diluted with water) or in the foam liquid solution, primarily to target stickiness and adhesion of clay onto metal surfaces that causes clogging. Clay particles carry surface charge, typically anionic, that leads to adsorption of free water, including the water in foam bubble walls. These so-called anti-clay polymers, typically anionic, arrest the clay's adsorption of free liquid. The anti-clay polymer provides stability to foam bubble walls, preventing wall thinning and foam bubble destruction that releases air in the excavation chamber. The anionic polymer, whether in foam or liquid form, further serves to increase the electrostatic repulsive force between clay particles, therefore reducing shear strength and improving workability for face pressure control and efficient excavation.

Despite the stated foam stabilization capability of anti-clay polymers, many EPBM projects have reported significant foam destruction and air void accumulation in the top of the excavation chamber (e.g., Borgui 2006, Borgui & Mair 2008, Langmaack 2002, Langmaack & Feng 2004, Alavi et al. 2013, Mori et al. 2017). This has driven some recent projects to use foam with very low air content, for example, *FER* = 1.5 in Seattle clay (McLane 2014), *FER* = 1.5 in London clay (Borgui and Mair 2008), average *FER* = 3 in Toronto hard clay (Boone et al. 2005), and *FER*≈ 1 in very stiff with medium-high plasticity clay of Lambeth group (Borgui 2006).

## ANALYSIS OF PROJECT DATA

### Projects Overview

EPBM data from four US projects is included in the analysis. Project A, B, and D are west coast projects while project C is an east coast project. Only reaches where the EPBM mined through fully water-saturated cohesive soil are included in the study. Liquid and plastic limits, in-situ moisture content and native $I_c$ are summarized in Table 1. In all cases, the native cohesive soils are firm to stiff and require conditioning to reduce $I_c$ for efficient EPBM tunneling. The soil conditioning quantities are described by the *LIR*, *FIR*, *FER* and foam liquid injection ratio *FLIR*. Anti-clay polymer was used

**Table 1. Summary of EPBM projects analyzed**

| Project | A | B | C | D |
|---|---|---|---|---|
| EPBM D (mm) | 6440 | 6440 | 8009 | 6556 |
| Soil description (USCS) | Lean and high plastic clays (CL and CH). | Lean and high plastic clay (CL, CH); pockets lean and high plastic silt (ML and MH) | High plastic clay (CH); pockets of lean clay (CL) and clayey sand (SC) | Low and high plasticity silt and clay (ML/MH and CL/CH) |
| $w_L$ | 31–96 | 40–56 | 50–93 | 31–76 |
| $w_P$ | 21–47 | 22–31 | 19.5–26 | 23–40 |
| $w$ | 24.9–43.2 | 21.6–37.1 | 19.4–26.4 | 22.9–62.9 |
| In-situ (native) $I_c$ | 0.6–0.8 | 0.8–1.0* | 0.9–1.1 | Ave = 0.95 |
| Conditioned muck $I_c$ | 0.2–0.7 | 0.4–0.8 | 0.8–0.9 | 0.2–0.6 |
| *LIR* (%) | 3–30 (Ave=11)† | 2–20 (Ave=7) | 6–26 (Avg=10) | 6–25 (Avg=19) |
| *FIR* (%) | 10–79 (Ave=39) | 6–52 (Ave=23) | 32–102 (Ave=49) | 20–46 (Ave=34) |
| *FER* | 2–19 (Ave=9) | 1–36 (Ave=4) | 2–73 (Ave=5) | 1–13 (Ave=2) |

* Two outliers, one with very high $I_c$ value of 2.7 classified as OH soil, and another one with low $I_c$ of 0.55 have been ignored in from a total of 15 borehole data.

† *LIR* of Project A is the sum of foam liquid injection ratio and other liquids (water/bentonite slurry) added.

in the foam of each project. The resulting conditioned $I_c$ is also shown. Recall that interface final $I_c$ will be lower than the reported conditioned $I_c$.

## RESULTS

Cohesive soil conditioning effort is shown in the following figures together with EPBM key performance indicators (KPIs) including penetration rate *PR*, advance rate *AR*, cutterhead torque $T_{chd}$, cutterhead thrust force $F_{chd}$ or net thrust force $F_{net}$ as a proxy, excavation chamber pressures, top and bottom chamber pressure gradients, screw conveyor gate opening, and screw conveyor torque normalized by rotation speed $T_{sc}/N_{sc}$.

Figure 3 shows twin EPBM tunnel data from Project A through water–saturated clay. Both tunnels were excavated with the same EPBM. The native $I_c$ ranged from 0.8 to 0.6 (ring 450 to ring 650). The conditioned muck $I_c$ resulting from direct and foam–based liquid injection ranged from 0.7 to 0.2, with noted differences in conditioned $I_c$ as well as *FIR* and *LIR* between the parallel tunnels. *PR* reached 50 mm/rev during both tunnel drives. More consistent *PR* = 50 mm/rev was achieved with greater *LIR* from tunnel 1, ring 600 onward and throughout tunnel 2, in both cases with conditioned $I_c$ in the range 0.2–0.4.

The data from the first and second halves of tunnel drive 1 suggest an important relationship between conditioning and KPIs. Considerably higher *FIR* and *LIR* was used during the second half of tunnel drive 1. $F_{net}$ was decreased considerably with higher conditioning, leading to a significant $T_{chd}$ reduction. The $F_{net}$ decrease caused a reduction in *PR* from 50 to 25 mm/rev. An increase in *FIR* from 20% to 50% did not lead to any improvement in *PR*. Only after liquid–only *LIR* was introduced around ring 590 did the *PR* increase back to 50 mm/rev while maintaining the lower $F_{net}$. The achievement of considerably lower $F_{net}$ and $T_{chd}$ was realized by increasing the *LIR* and lowering the conditioned $I_c$ to 0.2–0.4. The increase in *FIR* from 20% to 50% was insufficient by itself to improve the *PR*. The reduction in $F_{net}$ and $T_{chd}$ is valuable because it reduces tool wear.

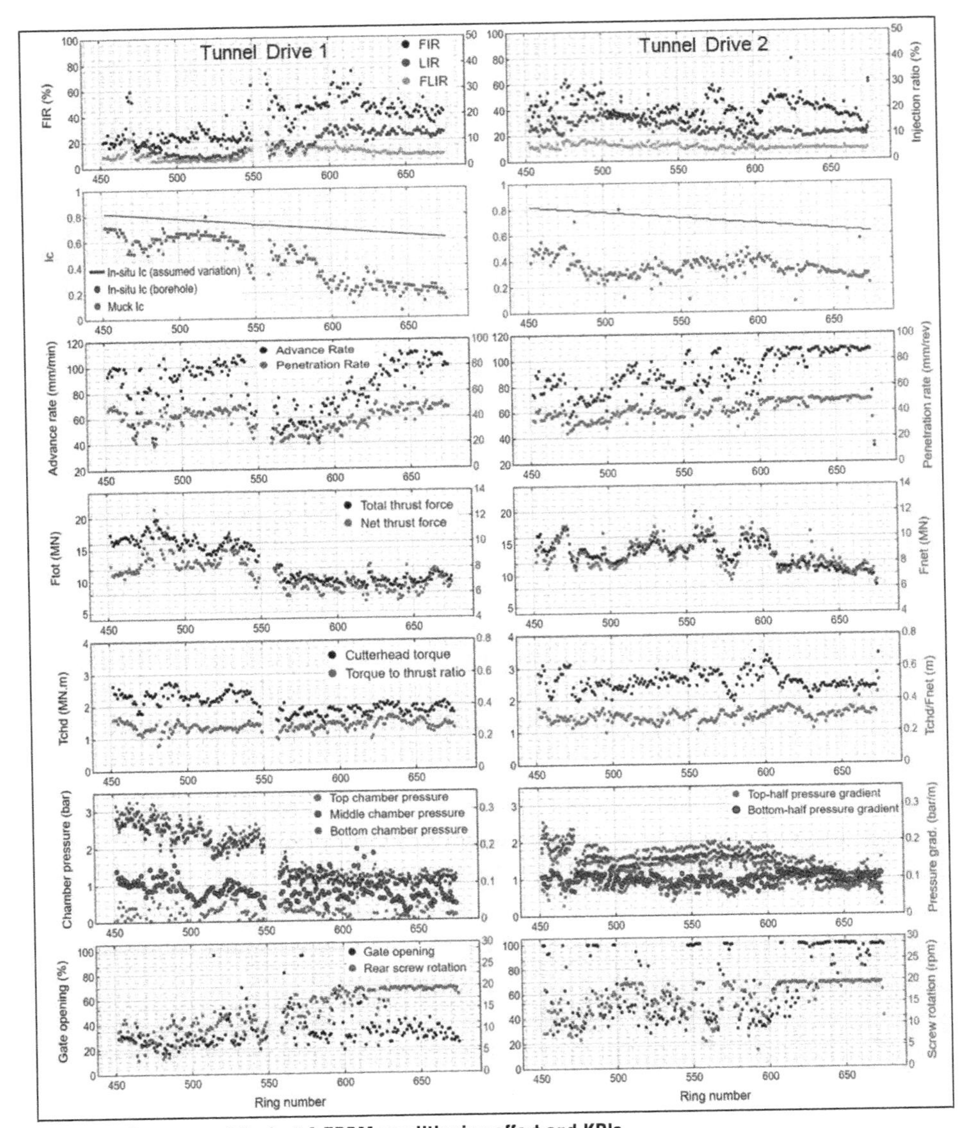


**Figure 3. Summary of Project A EPBM conditioning effort and KPIs**

Project B data (Figure 4) is shown over three 700–800 ft-long stretches through water–saturated clay. The native $I_c$ averaged 0.9 (stiff consistency). *FIR* and *LIR* varied considerably over these three stretches bringing the conditioned $I_c$ to different levels as shown in Figure 4. Stretch 2 experienced much greater *PR* (65 mm/rev) and *AR* (120 mm/min) than stretches 1 and 3, and arguably as efficiently as an EPBM can perform. In this middle stretch, the chamber pressure response is very smooth with little fluctuation. The chamber pressure gradient or apparent density (not shown) was less than 0.15 bar/m and is considered ideal. *FIR* during this stretch was 20–30% and *LIR* was 3–5%. The conditioned $I_c$ remained in the 0.7–0.8 range. The screw

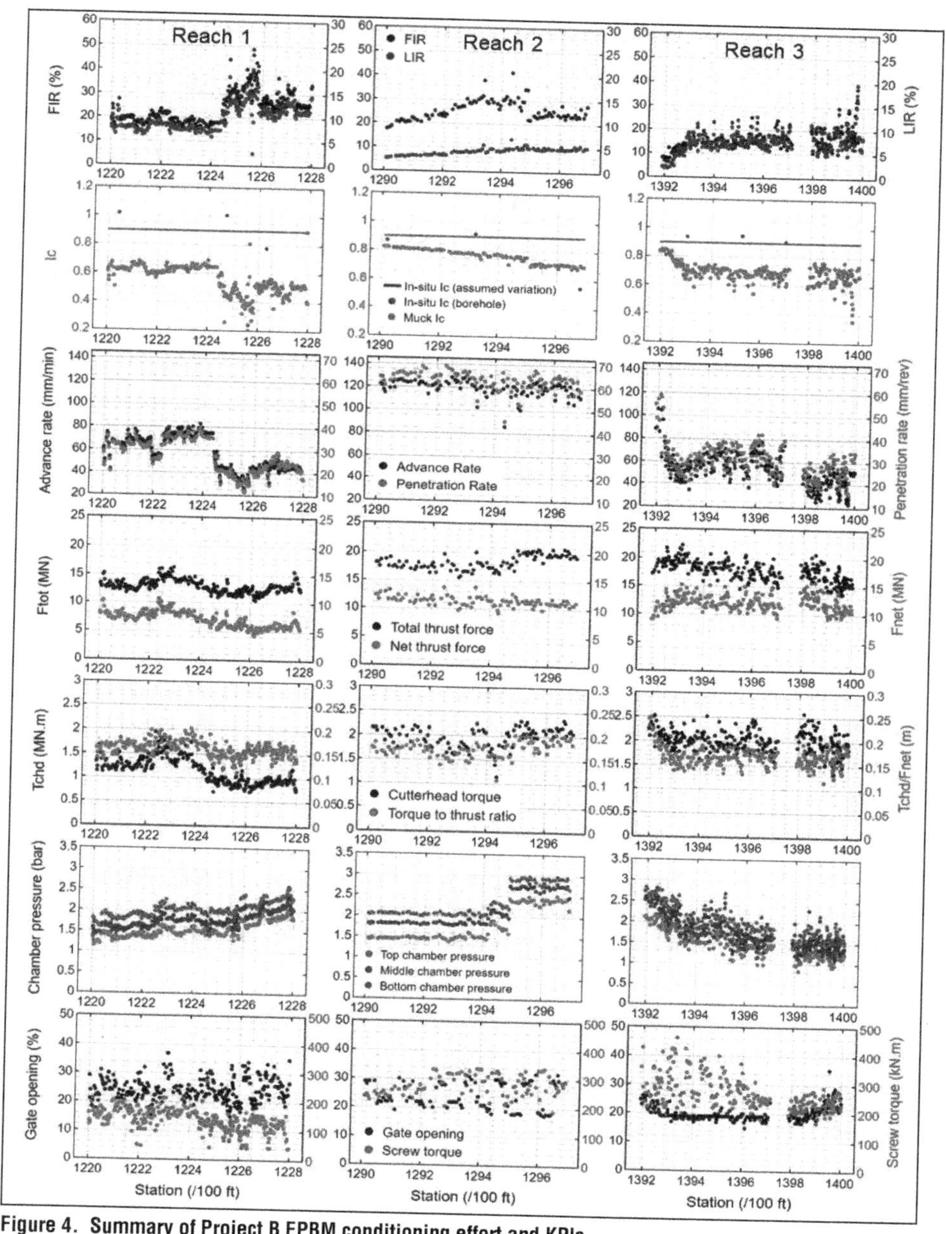


**Figure 4. Summary of Project B EPBM conditioning effort and KPIs**

conveyor gate was operated only 20–30% opened and the low normalized screw conveyor torque conveys the low undrained shear strength.

The EPBM exhibited smooth chamber pressure during stretch 1 and more variable erratic chamber pressure during stretch 3. The higher *FIR* (20–40%) during stretch 1, compared to *FIR* = 10–20% during stretch 3, is the likely reason for this. Further, however, the upper chamber pressure gradient was often below 0.1 bar/m indicating the presence of an air void at the top of the chamber. This air void often forms when EPBM tunneling through clay and is typically managed by purging the chamber (Alavi et al. 2013).

There is a notable increase in *FIR* and *LIR* midway through stretch 1 that decreases the conditioned $I_c$ from 0.6 to 0.4 on average. The *AR* and *PR* decrease almost by a factor of two as a result and $T_{chd}$ decreases accordingly. One might be led to conclude this increase in foam and liquid is the reason for the reduction in *PR* and $T_{chd}$. However, both chamber pressure and the cutterhead thrust (net thrust force) decreased as well, and this has a significant influence on *PR* and $T_{chd}$.

Data from 750 rings (4,500 ft) of Project C EPBM tunneling is presented in Figure 5. LIR of 6–12% and FIR of 35–60% was used along this stretch. As a result, the native $I_c$ of approximately 1.0 was reduced to mostly 0.8–0.90 with a 50 ring stretch at the end with conditioned $I_c$ of 0.7–0.8. The EPBM advanced at 30–60 mm/min with *PR* = 15–35 mm/rev. While there are stretches where higher *PR* coincide with lower $I_c$, there are also instances of the opposite. Linear regression of *PR* vs. conditioning parameters *FIR*, *LIR*, and $I_c$ produced no statistical correlation. Therefore, there seems to be no influence of conditioning within the ranges used.

$T_{chd}$ varied from 4–6.5 MN-m throughout; however, local fluctuations occur much more frequently than changes in conditioning and therefore cannot be attributed to conditioning. Chamber pressure response illustrates that the top and springline pressures are similar. The top gradient tends towards zero suggesting the persistent presence of an air bubble in the top of the chamber. This presents a disadvantage of foam conditioning in clay soil described in Section 3. Lower chamber gradient varies from 0.1 to 0.2 bar/m. The normalized screw conveyor torque to rotation speed $T_{sc}/N_{sc}$ does decrease with decreasing $I_c$ as would be expected with $S_u$ decreasing.

Project D data from 700 rings of excavation through clay soil is presented in Figure 6. The native $I_c$ of nearly 1.0 was reduced to the range of 0.3–0.6 with *LIR* = 15–20% and *FIR* = 30–40%. Very low air foam was used, i.e., *FER* = 1.5–2. *LIR* and *FIR* values are very constant along these 700 rings; a gradual increase in *LIR* from ring 450 to 700 creates a gradual reduction in $I_c$ from 0.6 to 0.3. To this end, changes in EPBM KPIs are not due directly to changes in conditioning.

*PR* varies from 15–40 mm/rev with the highest *PR*s occurring when $I_c$ is 0.45–0.50 and lower *PR*s occurring when $I_c$ is around 0.3. This observation does not suggest causation and there are many parameters that vary and may be the primary reason. The chamber pressures exhibit greater spread (and gradients that reflect this) from ring 550 onward when conditioning was the highest. Much of the $T_{chd}$ increase from ring 500 onward can be attributed to the increased $F_{chd}$ that itself could be reducing the *PR*. The normalized $T_{chd}$ /$F_{chd}$ shows a decrease over the same ring 450–700 range where $I_c$ is decreasing.

## SUMMARY

Most cohesive soil formations in North American EPBM tunneling require some degree of conditioning. The native $I_c$ of the projects evaluated here ranged from 0.9–1.1, indicating stiff consistency around the plastic limit. The main objectives behind conditioning are to enable consistent and smooth face pressure control and to improve material processing through the cutterhead openings, chamber, and screw conveyor.

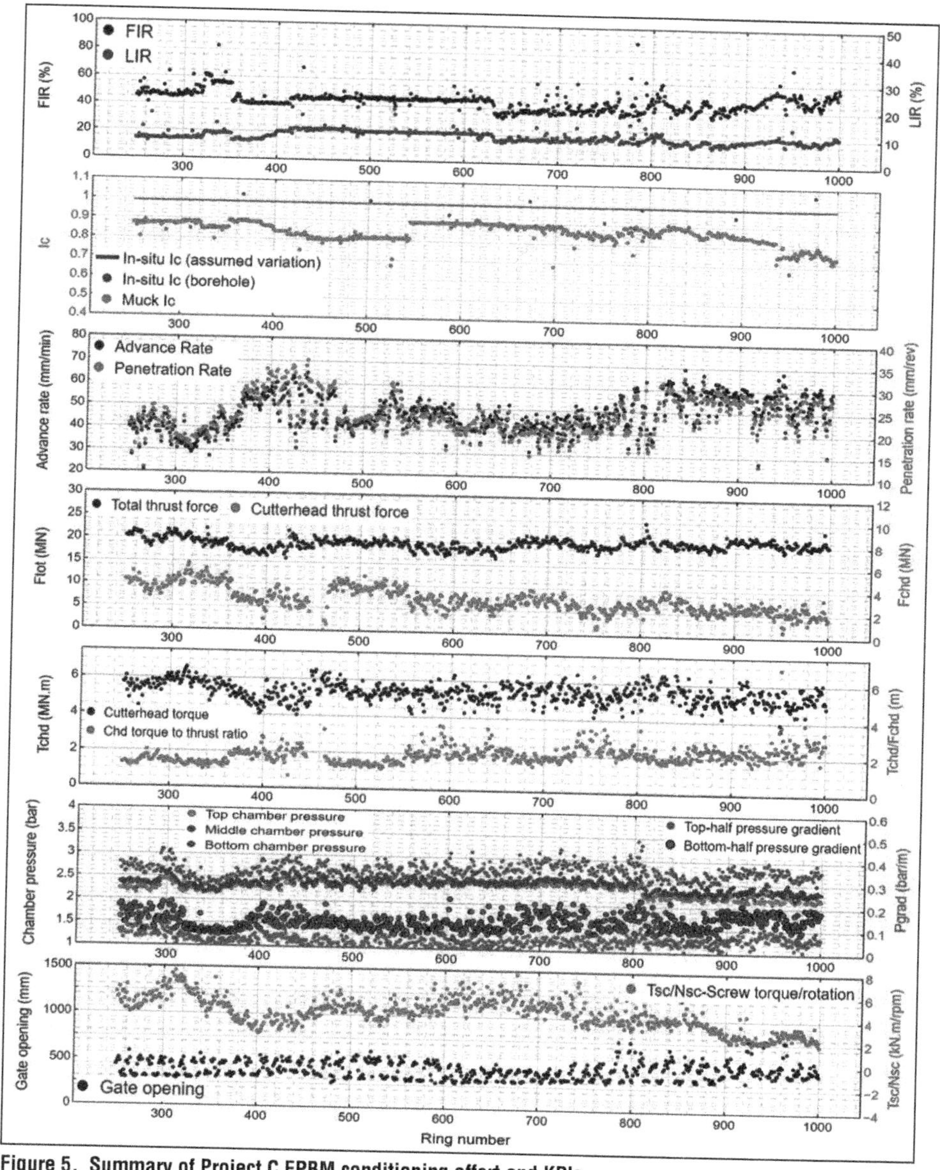


**Figure 5. Summary of Project C EPBM conditioning effort and KPIs**

Resulting KPIs include lower cutterhead torque, increased penetration rate, and smooth chamber pressure.

Both liquid (water + chemicals) and foam (liquid + air) are used to condition cohesive soils, and were used in the four projects examined. *LIRs* and *FIRs* varied considerably across and within the projects, as did the resulting conditioned $I_c$ values—ranging from 0.2–0.9. Good EPBM performance was observed across this $I_c$ = 0.2–0.9 range. There was no clear decisive $I_c$ range within which KPIs were clearly better across all

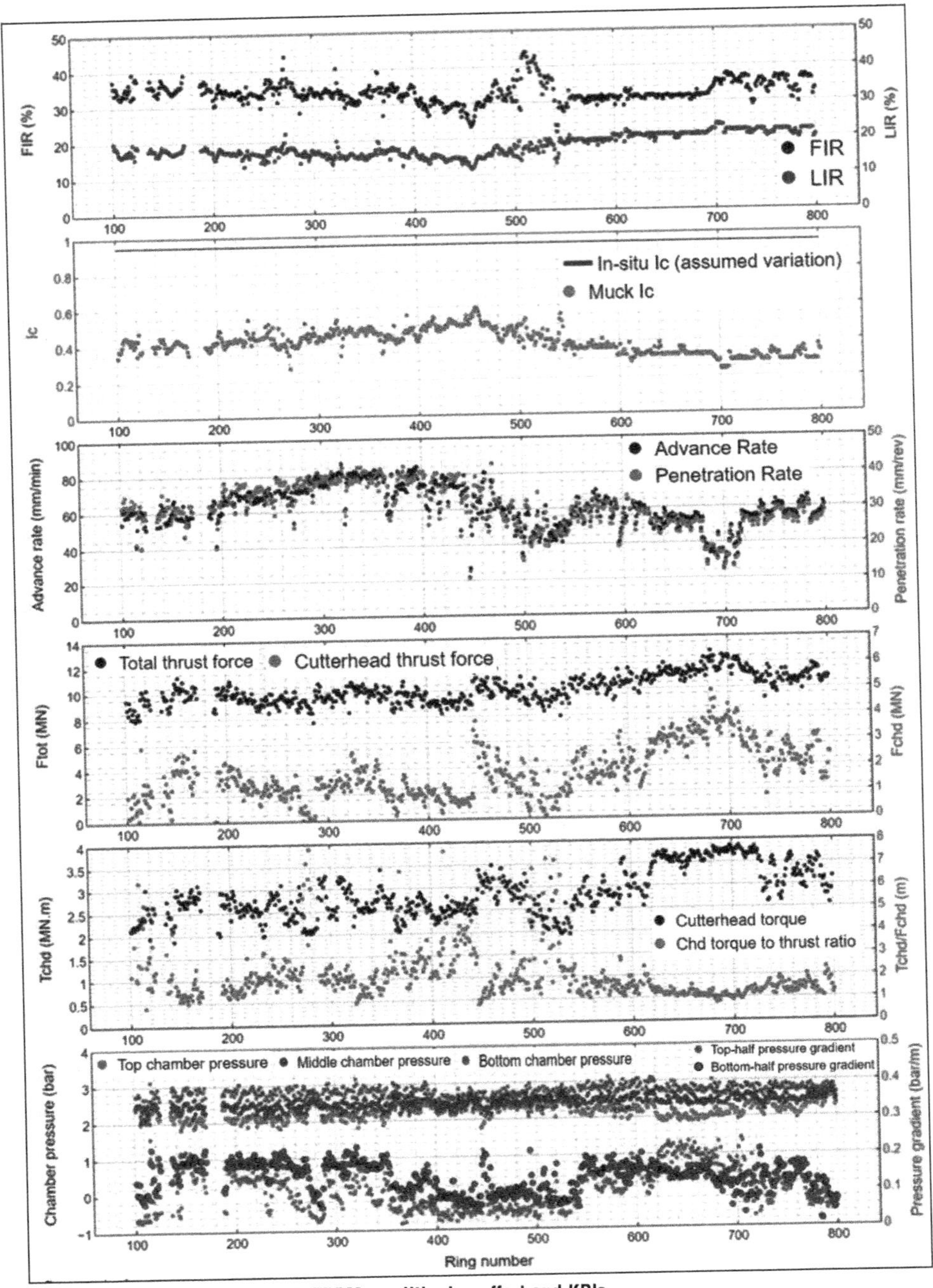


Figure 6. Summary of Project D EPBM conditioning effort and KPIs

projects. It is worth noting that it is difficult to isolate the influence of conditioning on KPIs because there are so many interrelations between the KPIs.

There were numerous cases where *LIR* was found to influence KPIs. Higher *FIR* (above 35–40%) did not translate into improved KPIs, at least in the data examined. In fact, many cases of upper chamber pressure gradient revealing an air bubble were evident, and therefore, the foam is causing a problem. Further, most projects operated with the screw conveyor gate only partially open, conveying the challenge that exists to dissipate pressure within the conditioned and homogeneous muck. Finally, it is difficult to compare across projects because EPBM configurations (e.g., design opening ratio, cutting tools, location and quantity of foam injection nozzles, etc.) and diameters are different as is the nature of the clay soils.

## REFERENCES

Alavi, E., Diponio, M.A., Raleigh, P. & Hagan, B. 2013. Keeping the Chamber Full: Managing the Air Bubble in EPB Tunneling. *Proc. Rapid Excavation and Tunneling Conference*. pp. 1,065–1,073.

Boone, S., Artigiani, E., Shirlaw, Nick, Ginanneschi, R., Leinala, T. & Kochmanova, N. (2005) Use of ground conditioning agents for Earth Pressure Balance machine tunnelling. *Proc. AFTES International Congress, Chambery;* 313–319.

Borgui, Xavier 2006. *Soil Conditioning for Pipe-Jacking and Tunnelling.* Doctoral dissertation, University of Cambridge, UK.

Borgui, F.X., & Mair, R.J. 2008. The effects of soil conditioning on the operation of earth pressure balance machines. *World Tunnel Congress, 2008—Underground Facilities for Better Environment and Safey, India.*

EFNARC. 2005. Specification and Guidelines for the use of specialist products for Mechanised Tunnelling (TBM) in Soft Ground and Hard Rock. *European Federation for Specialist Construction Chemicals and Concrete Systems, Surrey, UK, Part 2: EPBM, Foam.* Available at: http://www.efnarc.org/publications.html

Hollmann, F.S., and M. Thewes. 2013. Assessment method for clay clogging and disintegration of fines in mechanised tunnelling. *Tunnelling and Underground Space Technology* 37:96–106. doi: 10.1016/j.tust.2013.03.010.

Janscez, S. 1997. Modern shield tunnelling in the view of geotechnical engineering: A reappraisal of experiences. *Proc. 14*th *ICSMFE, Hamburg:* 1,415–1,420.

Langmaack L. 2002. Soil conditioning for TBM chances & limits. *AFTES Underground works: living structures.*

Langmaack L. & Feng Q. 2005. Soil conditioning for EPB Machines: Balanced of Functional and Ecological Properties. *Proc. Of the Int. World Tunnel Congress and the 31*st *ITA General Assembly, Istanbul, Turkey,* 7–12 May 2005.

McLane, R. 2014. Automatic Soil Conditioning through clay. *Proc. North American Tunneling,* 127, 195–200.

Maidl, U. 1995. *Erweiterung der Einsatzbereiche der Erddruckschilde durch Bodenkonditionierung mit Schaum* (in German). PhD Thesis, Department of Civil Engineering, Ruhr University.

Merritt A.S. 2004. *Soil conditioning for earth pressure balance machines*. PhD Dissertation, University of Cambridge, UK.

Milligan, George. 2000. Lubrication and soil conditioning in tunnelling, pipe jacking and microtunnelling: A state-of-the-art review. British Geotechnical Association Meeting, Institution of Civil Engineers, London, 8 November 2000.

Mooney, M.A., Wu, Y., Parikh, D. and Mori, L. 2017. EPB Granular Soil Conditioning under Pressure. Keynote Paper. *Proc.* 9th *Intl. Symp. Geotechnical Aspects of Underground Construction in Soft Ground*, IS–Sao Paulo, April 4–5, 2017.

Mori, L., Alavi, E. and Mooney, M.A. Apparent Density Evaluation Methods to Assess the Effectiveness of Soil Conditioning, *Tunnelling & Underground Space Technology*, 2017, 67, 175–186.

Shield Tunnel Association of Japan 2020. Rheological Foam Shield Tunneling Method-Technical Materials (in Japanese), August 2020. Available at: https://shield-method.gr.jp/wp/wp-content/uploads/doc_tec_rf.pdf.

Thewes, M., and Budach C. 2010. Soil conditioning with foam during EPB tunnelling. Geomechanics and Tunnelling 3 (3):256–267. https://doi.org/10.1002/geot.201000013.

Vardanega, P. J., and S. K. Haigh. 2014. The undrained strength—liquidity index relationship. *Canadian Geotechnical Journal* 51:1,073–1,086. doi: 10.1139/cgj-2013-0169.

Zumsteg, R., M. Plotze, and A. Puzrin. 2013. Reduction of clogging potential of clays: a new chemical applications and novel quantification approaches. *Geotechnique* 63 (4):276–286.

# Innovative TBM Launching in Urban Areas

**Fabrizio Fara** ▪ Lane Construction Corporation

## ABSTRACT

The Ship Canal Water Quality Project (SCWQP) includes a 2.7-mile long, 18'-10" internal diameter tunnel which is located in West Seattle. Tunneling in urban areas and soft ground conditions (Till and Till-like soils in Seattle) is associated with the risk of settlements, especially in presence of groundwater above the alignment. The Designer foresaw the use of soil improvement (jet grouting) outside of the launch shaft to mitigate the risk of flowing ground and to provide a redundant sealing. The Contractor proposed and ultimately successfully executed a different approach by using a "Steel Bell System" or Pressurized Can as described in this paper.

## INTRODUCTION

The Seattle Ship Canal Water Quality Project is located in north Seattle and spans the neighborhoods of Ballard, Fremont, and Wallingford. The Project, executed in September 2019, is part of the Ship Canal Water Quality Program from Seattle Public Utilities (SPU), designed to control and treat the combined sewer overflow to the Ship Canal. The Tunnel will diverge all the water to a new pump station, where it will be treated prior to being discharged into the Ship Canal.

The Project consists of a 21 ft, 14,000 ft tunnel, five drop shaft structures, as well various diversion structures, ventilation facilities and an electrical building. The main Drop Shaft is called West Shaft (WSS) and is a 86' internal diameter secant pile structure from where the TBM was assembled and launched.

To mitigate adverse potential effects at the TBM launching shaft, such as soil losses and settlements, an improved soil block outside the shaft has been proposed by the designer. The improved ground was designed to provide redundancy in case the sealing ring typically installed at the tunnel launch portal would fail while the TBM is being launched. At the TBM launching typically there are considerable risks associated with this phase. In fact, at the WSS Shaft the geology near by the shaft includes presence of Till-like material which consist of gravel, sand and clay soil, as well cohesionless Sand and Gravel, CSG material. These materials would exhibit "flowing behavior" while they are under a groundwater head as showed in the Geotechnical Baseline report (Figure 1).

## TBM LAUNCHING METHODS

Several options for the TBM launching or breaking in methods are available from the literature and have been successfully implemented in past different worldwide projects. However, numerous factors related to the specific site location at the launching area are to be considered. The main goals for a TBM launching are the following:

- Tunnel soil face stability is ensured throughout the TBM launching process.
- No groundwater flow takes place toward the inside of the shaft. A groundwater flow inside the launching shaft might be insidious if it brings material along with.

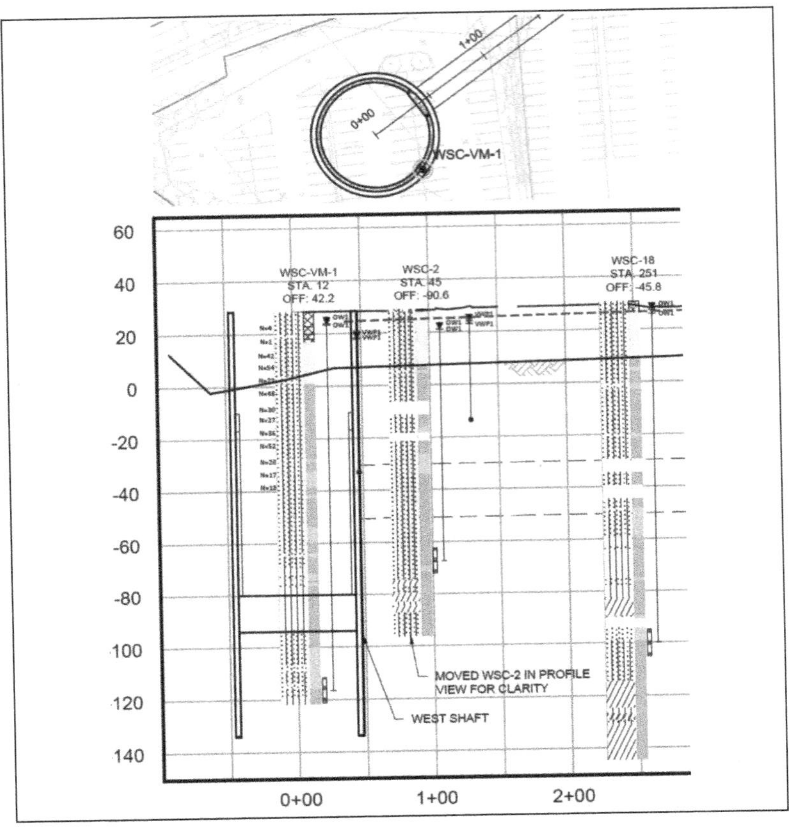


Figure 1. Geotechnical baseline report

If these two goals described above are not met, settlements and or volume losses may represent a safety risk other than causing significant damages especially if utilities or buildings are around the underground facility.

To facilitate the TBM Launching, the methods can be implemented are generally divided into two main categories: (1) Ground improvement near the shaft. (2) Mechanical barriers to limit potential soil losses. A general description of the various methods is provided on the following chapters.

## Ground Improvements

### *Jet Grouting Block Near the Shaft*

An improved ground block by the launching shaft would mitigate a risk of waterflow and settlements. Jet grouting uses a high-pressure 'jet' of either grout, water, air or a combination to erode soil whilst simultaneously injecting grout into the soil through a "jet monitor" (Figure 1). The specially designed drill stem and monitor are raised and rotated at the same time to combine the grout with the original soil to form "soilcrete." The final product is cemented round columns and is basically effective in any soil but not necessarily efficient in every soil (Figure 3). The Jet grouting activity usually requires the area above the block available for an extensive amount of time. In addition,

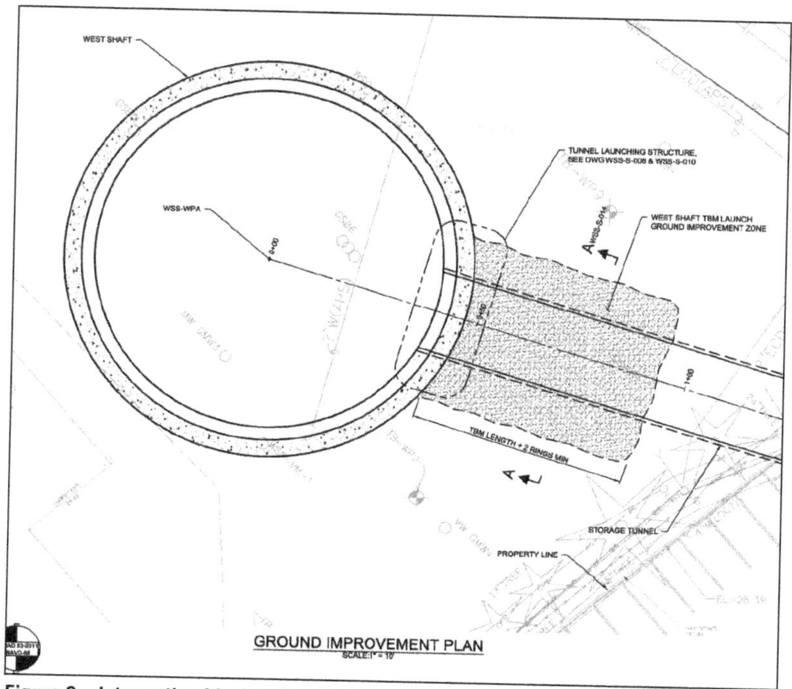


Figure 2. Jet grouting block by the shaft

a dedicated area is need for the spoil treatment and material hauling. In urban environments, available working areas could represent a limitation. Furthermore, a Jet grouting block requires an extensive quality control to ensure the improvement zone is entirely treated and do not present zone of untreated material that could compromise the TBM launching.

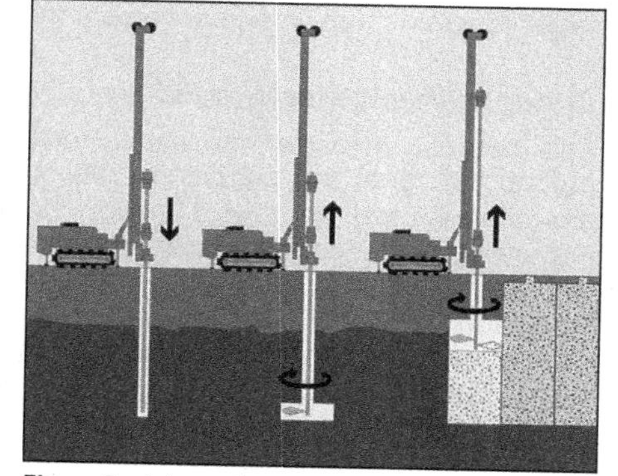

Figure 3. Jet grouting

### *Permeation Grouting*

Permeation grouting, also known as penetration grouting, is a common and conventional grouting method. It involves filling any cracks, joints or void in rock, concrete, soil and other porous materials. The idea is to fill a void without displacing the formation or creating any change in volume or configuration in the medium. This is typically done to strengthen the existing formation, creating an impermeable water barrier or both. Although this technique do not generate spoil material like the jet grouting, it has a limitation on its applicability. Typically, fine grain materials are not easily injectable and it requires an extensive quality control as well.

### *Artificial Ground Freezing (AGF)*

The ground treatment is carried out by a series of freezing plants and freezing pipes installed interconnected into the block object of the treatment through which a coolant

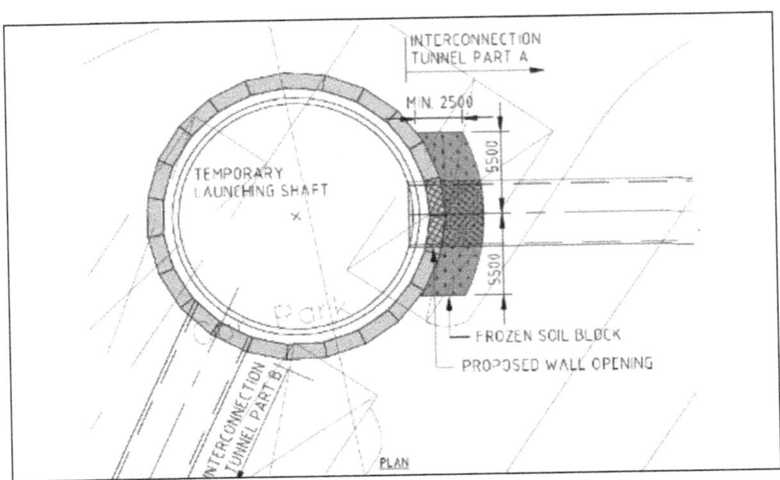


**Figure 4. Artificial ground freezing for TBM breakout**

fluid at very low temperature is driven. By thermal exchange, columns of frozen soil begin to form around pipes as the fluid flows through the network. Two are the main fluids used nowadays for freezing: brine (high-concentration salt solutions) and liquefied nitrogen (LIN). Circulating temperatures are –25 °C for brines and –196 °C for LIN. During the last two decades soil freezing with LIN has developed from an application uncertainty into a more standard and successful procedure for treating unstable soil and leakages. Although ground freezing is one of the most reliable techniques for ground improvements, it doesn't generate spoil and it easy to verify. However, it requires a minimum time to develop a frozen soil curtain (Figure 4).

### *Canopy Drilling/Pipe Umbrella Tubes*

This process involves horizontally drilling at the excavation face, generally into ground with poor soil conditions, whilst simultaneously installing steel tubes which are located above the tunnel crown. The tubes are injected with the goal of reinforcing the soil creating a retaining element to provide support and stability, prior to launching a TBM. The finished canopy (or umbrella) of tubes produces a zone of reinforced ground.

**Figure 5. Canopy drilling/pipe umbrella tubes**

This method usually is associated with other ground improvement technique such soil nailing or horizontal injection at the TBM face. For very loose soils this type of ground improvement may still carry risks since it doesn't stop a potential "soil flowing" behavior where there is a groundwater head pressure (Figure 5).

## Mechanical Barriers

### *Sealing Ring*

Figure 6. Sealing ring (entrance seal)

Also known as an Entrance Seal, this is a large steel ring placed against the entry wall of the tunnel, which contains a circular rubber seal that is held tight against the TBM shield to seal the annular gap created between the wall and shield (and also between the wall and dummy ring at a later stage). This ensures that there is no (or very limited) water seepage into the launching area. For the break-in it is a system that, at the same time, can stabilize the ground using the TBM itself, manage the ground water by closing the gap, minimize this gap by allowing the TBM to bore through the retaining wall and seal the gap by immediately filling of the annular void against the retaining structure. Usually used with other techniques (jet grouting or steel bell). As mentioned in the introduction, a sealing ring system itself doesn't provide a certain redundancy to minimize risks associated with the TBM breakout (Figure 6).

### *Bathtub Structure by Cement-Bentonite Walls*

Cement-bentonite walls consist of improved-ground panels which are installed in sequence and interlock together to create a watertight "bathtub" in front of the shaft. The soil inside the bathtub is then dewatered to ensure no soil migration take place while the TBM breakout from the shaft. The panel—bathtub structure represents a reliable solution to limit the extent of potential soil affected from the TBM launching. However the area above the tunnel entrance still need to be available for the entire duration of the bathtub works other than being free from utilities (Figure 7).

### *Steel Bell (False Tunnel)*

The steel bell also known as false tunnel consist of a circular steel structure installed inside the launching shaft. The final goal of the steel bell is to create a watertight temporary tunnel around the TBM, such the TBM launching can be performed using

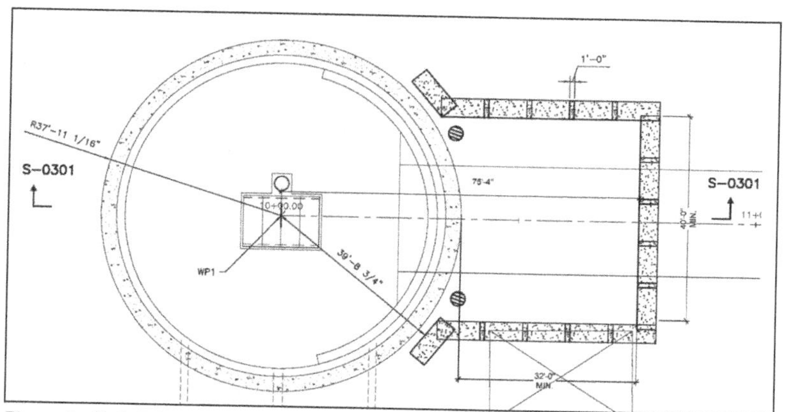


Figure 7. Bath-tub structure

Figure 8. Steel bell

the whole EPB pressures at the face. Once the steel bell is installed, the TBM is slid into the bell and to complete the setup, the first (dummy) ring behind the tunnel boring machine is erected. While the TBM advances the first 5 ft, reaching the retaining wall of the shaft, one or two additional dummy rings are installed and grouted by tail void grouting against the steel bell structure. Alternatively, the steel bell can be installed together with the sealing ring described above, to reduce the number of dummy concrete rings solely used for the TBM launching (Figure 8).

### *Use of the Steel Bell for the Ship Canal Water Quality Project TBM launch*

Lane Construction, a branch company of Webuild, successfully implemented in 2019 a steel bell structure as launching method for the Earth Pressure Balance TBM Mud Honey, avoiding the construction of ground improvement near the busy shaft. This system was integrated with pushing frame and sealing ring system and provided a means to create a regular EPB TBM tunneling conditions at the face and around the entire tunneling envelope even before breaking out of slurry walls and into the glacial deposited ground.

The Steel Bell consisted of one Front Element (left), four Typical Elements (middle) and a Connecting Ring (right), Figure 9. All Elements are supported underneath by integrated central supports as well as outer leg supports which are attached using bolted connection. The Connecting Ring provides the interface between the Steel Bell and other launching structures.

The Steel Bell is approximately 11 m (36 ft) long and 3.335 m (10.94 ft) ID radius and it is made by 2.2 m long sections. Each section consists of 6 steel curved pieces

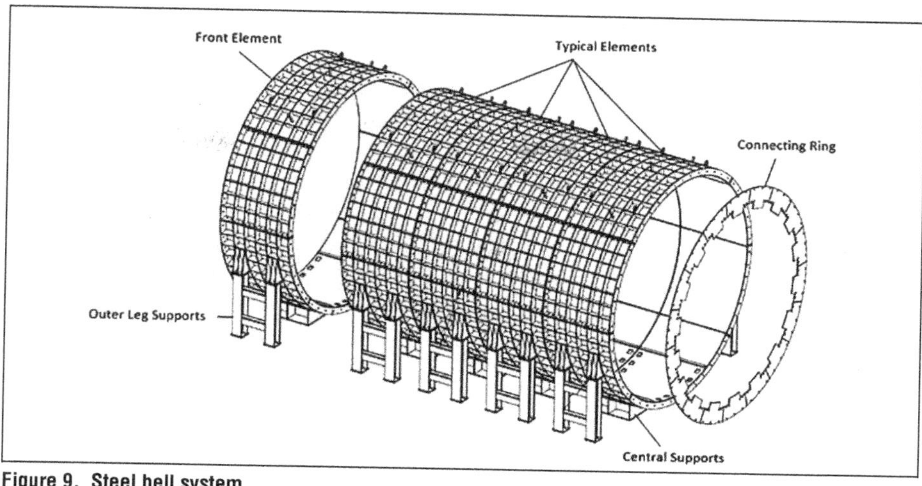


Figure 9. Steel bell system

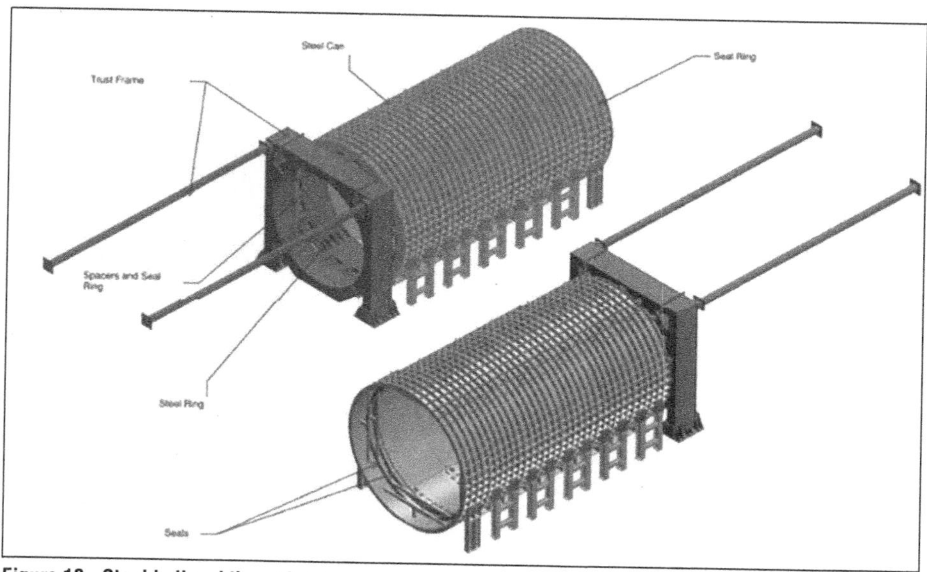


Figure 10. Steel bell and thrust frame

mechanically connected with bolts. The bottom portion of the bell is anchored to the shaft bottom slab and acts as a regular cradle (Figure 10).

The TBM is fully assembled behind the steel bell and then pushed into this latter by means of a hydraulic translation system. Then the steel bell is closed and sealed, ready to be pressurized (Figure 11).

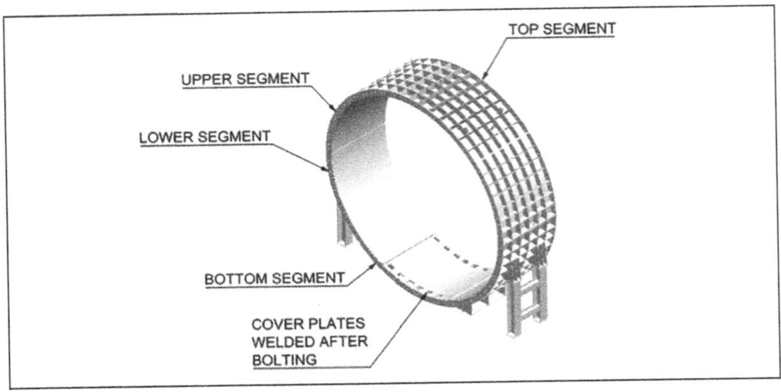


Figure 11. Typical section of the steel bell

## INSTALLATION SEQUENCE

The seal ring is the first part to be installed. The Ring is preassembled on surface and installed into a recess created in the final lining. Anchoring bolts are used to position the ring into the final lining and ensure proper connection to the structure.

Two lip seals are installed in the sealing ring along with an injection pipe to pump a sealing agent between the steel lips (Figure 12). The goal of the sealing agent is to ensure the pressure is maintained within the lips Datwyler P 90-326 was selected and used as a seal profile ensuring a working pressure up to 5 bar. The filler profile is pressed in simultaneously, locking the seal profile into position. When all but the last 75 cm of the starting seal has been installed, the remainder is cut to an extra length (+5 cm) creates pressure on the profile butt joint, providing and adequate seal. For added water tightness, adhesive is applied to the butting surfaces.

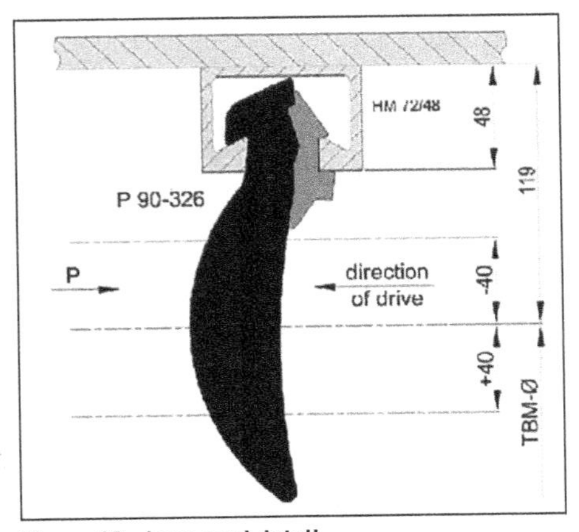


Figure 12. Leap seal detail

The 2 ft gap between seal ring and slurry wall shaft was closed applying shotcrete, except at the bottom portion which was grouted by gravity.

The 5 ft metal ring was erected behind the tail shield and provided the reaction needed against which a regular precast tunnel liner ring was erected for the first advance. The 5 ft steel ring ensured a better load distribution on the thrust frame with a plane and flat surface. The metal ring was installed using the TBM erector and the different sections were bolted together, additional plates were welded between steel sections as redundant safety measure. The connection ring in order to closes the gap between the metal ring and the steel bell was welded to provide efficient sealing (Figure 13).

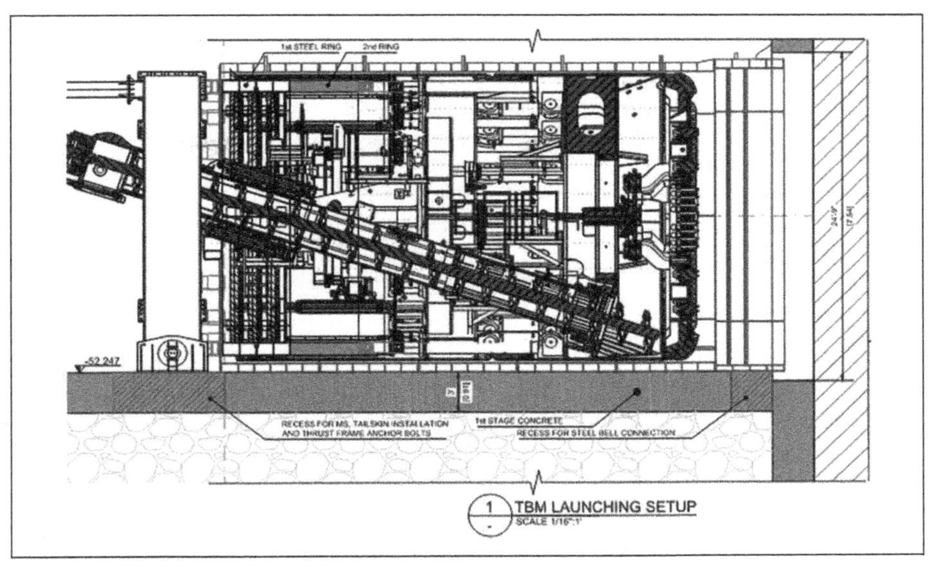


**Figure 13. TBM launching set up using the steel bell**

## DESIGN

The Steel Bell design was carried out starting from the following inputs:

- Diameter of TBM Shield
- Length of TBM Shield
- EPB and Grouting pressure required to advance the TBM.

The steel bell design was developed using a 3D FEM analysis. The inner diameter of the Steel Bell is equal to the cutting diameter of the TBM plus a pre-defined tolerance to prevent the TBM from making contact with the Steel Bell. The length of the Steel Bell was defined by the length of the TBM shield in addition to other launching structures and requirements for the number of dummy concrete rings to be installed prior to the tunnel face.

Due to the need to transport the steel bell to the launch site and subsequently move it to other launch sites or other projects, the Steel Bell elements are split into a number of segments as shown in the chapter above. The element sizes were dictated by transportation regulations and therefore tend to be around 2,200 mm in length (maximum). The steel bell segments consisted of an inner steel plate which had multiple stiffeners attached running both longitudinally (#4, #5, #6) and circumferentially (#2). The segment edge plates were used to attach each section to its adjacent segments/elements. (#1, #3, #7). Lifting hooks (#8) were provided for handling each segment for transportation and assembly/disassembly (Figure 14).

The number of stiffeners and their size depends on a few factors, principally the weight of the TBM and the grouting pressure required (which is associated with the hydrostatic pressure of the ground at the launch site). The design followed an iterative process, whereby the number of stiffeners and their size are initially approximated based on experience and preliminary calculations.

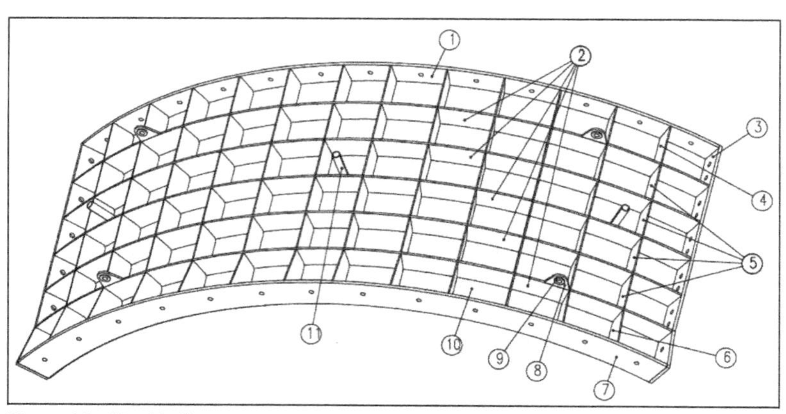


Figure 14. Steel bell segment

Once the dimensions were defined from the FEM model, then Shop drawing were developed and sent for manufacturing.

## DESIGN CALCULATION

The design calculations were based on the Eurocode which is based on Limit State design and defines two types of Limit States such as:

- Serviceability Limit States concern (SLS): the functioning of the structure or structural members under normal use; the comfort of people; the appearance and durability of the construction works.
- Ultimate Limit States concern (ULS): the safety of people, and/or the safety of the structure.

Boundary conditions were defined based on the TBM geometry and constraints. For example, where the Steel Bell was in contact with the ground, these contact points were fixed points. The loads were considered from the TBM weight and grouting pressure plus a safety factors. Once the geometry along with the boundary conditions was defined, then a 3D-Mesh was generated. As shown in Figure 15. Small red arrows represent grouting pressure applied all around the internal Steel Bell surface while pink arrows represent the TBM contact points and force applied from the weight of the TBM.

Finally, the red arrow in the center pointing in the downwards direction represents gravity and therefore the weight of the structure itself.

### Outputs

The main outputs consisted of stress distribution, strain and deformation around the steel bell. The software used to model the steel bell produced a visual representation of these outputs with a color scale, as shown in Figure 16, where was possible to identify the most stressed areas.

The main outputs were the maximum stress and maximum deflection in the structure under the maximum load conditions and were finally compared to the allowable factored required values confirming they met the requirements. The geometry was slightly adjusted to reduce stress concentrations avoiding a change on the geometry

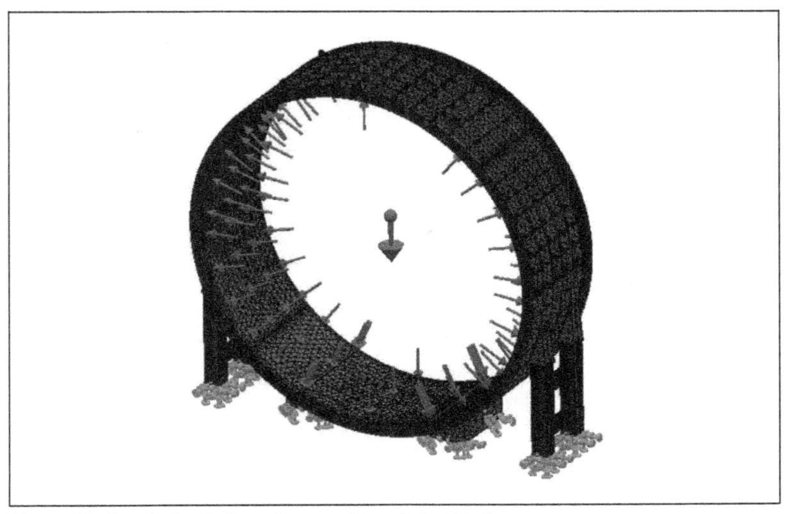

Figure 15. Inputs for design of one section of steel bell

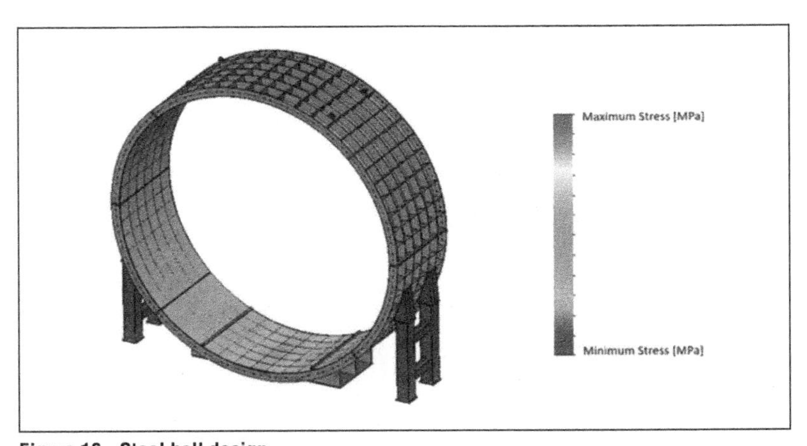


Figure 16. Steel bell design

which would have caused a migration of grout and therefore a possible depressurization of the steel can.

## STEEL BELL—CHALLENGES

While the use of the bell has allowed a significant save in terms of costs and time, the team at the Ship Canal Project has faced some challenges during the assembly of the steel can and during the launching of the TBM. Below are listed some of the issues main issues:

- Installation of the seal ring and lip seal: The team has planned to use the steel bell bottom as a cradle for the final assembly of the TBM, in this optic and to optimize the schedule it was considered to install the seal ring and the leap seal with the TBM positioned close to the tunnel access. This has resulted in a series of logistic issues, installing the ring has taken much longer than foreseen, particularly the positioning of the same and the later installation of the leap seal with the ring already in place. Accessing the seal ring area with a manlift, installing the leap seal itself (following the manufacturer guidelines) took a lot of effort and extra time (Figure 17).

Figure 17. Seal ring installation

- For the next application it is strongly suggested to assemble seal ring and lip seal on surface and preferably with the ring in horizontal position then only after having completed the installation the rong can be lowered into final position. It is suggested to leave enough space in front of the launching wall so to allow an easy and precise installation of the seal ring.
- Once the steel bell was pressurized a multiple number of leaks have been noted at different locations. The leaks were coming from the joints between the bell sections and from the bolts used to connect the same. The main reasons for this have been found in two main factors:
  1. Due to schedule constraints the FAT was executed at the factory only for the single sections of the bell, an FAT for the full system (including the seal ring and the closing ring) was never done (Figure 18).
  2. The compressible tape used between the bell sections was not strong enough to keep the pressure of the bentonite, Lane has opened few sections and add an additional layer without having any success. The team had to spend significant time welding all the seams of the steel bell, including all the bolts and nuts where the bentonite under pressure was finding a path to leak (Figure 19).
  3. A better sealing system has to be found to guarantee full water tightness of all the bolts, rubber washers to be considered.

Figure 18. Small misalignment between bell sections

Figure 19. Welding around bell joints and bolts

- The spacers used to connect the closing ring with the thrust frame were not adjustable, it has been necessary to do some packing at time which took time and lot of efforts. Use of adjustable spacers is suggested for the future uses (simple screw jacks will be of great help).

## GENERAL NOTES FROM THE LAUNCHING

Despite the loss of time due to the above-mentioned issues the system worked very well, the TBM head was filled with pe-gravel and bentonite, the steel pressurized at the design pressure of 2.3 bar, the excavation has started smoothly and without any significant issue. The leaks have been monitored closely, before to break through the slurry wall a final inspection of all the connections and bolts has been done and only at that time the decision to start mining into the ground has been made. There were no losses of pressure recorded during the launching, the presence of two injection valves on the steel bell has allowed a full control of the pressure inside the steel bell. The launching was successful, no settlements have been recorded in the area surrounding the shaft. When the TBM body and the first gantry used for the launch with the umbilical fully inside the ground and after having 20 nos. rings installed and grouted, the pressure on the steel bell has been released and the initial launching face considered completed. The steel bell structure has remained in place till the completion of the TBM initial drive (1,000 ft), this decision has been made mostly for schedule reasons, the team has planned to dismantle the steel bell just prior to start the final assembling so not to disrupt the production during the initial drive.

## STEEL BELL PROS

- The Pressurized steel bell system allows the TBM to counterbalance the ground pressure and ensure water tightness from the very initial phase of the launching without required construction of false tunnels, neither the execution of ground improvements
- Schedule is certainly one of the biggest advantages of this Method. Steel Bell/Steel Rings/Starting Seal can be manufactured off-site simultaneously as other work is being done to prepare for TBM launching. These steel structures can be ready to assemble at an exact date with few risks to timescales compared to ground improvement work. Installation of the bell required not more then two weeks at the Ship Canal project, excluding the extra time required for the leak's rectification. Ground improvements requires normally longest duration for their application and in many instances (like at The Ship Canal Project) due to the nature of the ground, will not guarantee a 100% of reliability.
- Logistic is always a big challenge particularly when the site areas are tight and activities can be performed only on a linear sequence. The original CPM schedule at the Ship Canal Project foreseen to have the JG works performed in parallel with the shaft excavation, this would have been extremely challenging considering the huge amount of space required by both those activities. The team has started looking at possible alternatives when realized that the logistic for the JG operations (including spoil management, area to be treated, equipment) was prohibitive while the surrounding are of the shaft was busy with the shaft excavation. The selection of the steel bell method has resulted in a good logistic optimization.
- Steel bell can be dismantled and reused for breakthrough purpose or as well future uses in different jobs

- The steel bell can be pressurized before starts mining the ground, this enables fine tuning and checking/inspections to take place before the TBM starts cutting the ground
- Concrete block and other ground reinforcements usually need to be demolished to allow platform infrastructure to be built. Steel bell will be removed, leaving space for platform construction.
- All the above advantages have a quite significant impact in terms of cost savings. Adding to the time saved for the ground improvement activities the bell avoids costs like pouring and dismantling (disposal) of eventual launching/receiving blocks, avoids all the costs related to the JG spoils disposal, replace the use of expensive ground improvement methodologies like the ground freezing. The Pressurized steel bell method is a costs advantageous methodology.

## CONCLUSIONS

At the Ship Canal Water Quality Project the TBM was completed in October 2021. The Steel Bell used to complete the TBM launching provided reliable results in terms of Safety, Quality and Schedule. An extensive area where the Jet grouting was planned to be installed was left available for other activities. The use of the steel bell was demonstrated to represent a valid alternative to the more traditional TBM launching methods. However, the TBM steel bell assembly was time consuming and can be definitely improved for next applications. The sealing between steel panels requires special attention since a grout migration through the joints can compromise the TBM breakout operations. Finally, an extensive jet grouting block was never installed and conclusively consisted in a time and economical saving without compromising safety and Quality of the TBM launch, providing a significant saving in the work schedule of the project.

## REFERENCES

https://www.bmcmicrofine.com.au/the-different-grout-injection-methods/.

TBM Shield Machines Break In And Break Out, International Tunnelling Symposium in Turkey: Challenges of Tunnelling (Tunnel Turkey, Istanbul) Istanbul, 02–03 December 2017, Nicolla Della Valle, Patricio García de Haro, Ignacio Sáenz de Santa María Gatón.

https://www.dsitunneling.com/fileadmin/downloads/dsi-underground-canada/dsi-underground-at-pipe-umbrella-en-ca.pdf.

MTR Shatin to Central Link—(Contract 1128), Temporary Works Excellence Award 2017 MTRC Contract 1128 SOV TO ADMIRALTY TUNNELS.

# Launch of an Earth Pressure Balance Tunnel Boring Machine in Short-Mode on a Congested Site from Narrow Deep Shafts for The Alexrenew Project in Alexandria, Virginia: A Case Study

**Luis Fernandez-Deza** ▪ Traylor Bros., Inc.
**James Hawn** ▪ Traylor Bros., Inc.
**Dustin Mount** ▪ Traylor Bros., Inc.
**Jean-Marc Wehrli** ▪ Traylor Bros., Inc.

## ABSTRACT

The RiverRenew Tunnel System project in Alexandria, VA forms part of a CSO scheme aimed at reducing overflows of combined stormwater and sewage into the Potomac River during storm events. A major piece of the project is a 12-foot internal diameter tunnel of approximately 12,000 feet length lined with a single-pass precast concrete segmental liner and excavated by an Earth-Pressure Balanced Shield TBM. The tunnel lies between 120 to 140 feet below grade in the clays of the Upper Potomac Formation with occasional dips of more granular, saturated soils of the Terrace Formation and the Alluvium. The TBM is launched from a figure-8 configuration Pumping/Screening Shaft located within AlexRenew's operational Water Resource Recovery Facility. This paper discusses the challenges encountered during the launch of the TBM in short mode within a highly congested site and from narrow deep shafts.

## INTRODUCTION

Traylor-Shea Joint Venture (TSJV) was awarded the RiverRenew Tunnel System Project in November of 2020. An overview map of the Project can be seen in Figure 1. The Project includes a total of four shafts and the 12-foot internal diameter Waterfront Tunnel to address combined sewer overflows at Alexandria's outfalls into the Potomac River and tributary waterways. TSJV utilized one Earth Pressure Balance (EPB) Tunnel Boring Machine (TBM) from Herrenknecht on the Project. The 4.472 meter (14 ft-8 in) cut diameter TBM is designed to a maximum operating pressure of 4.5 bar (65.3 psi) and an advance rate of 100 millimeters per minute. The TBM was launched from the Pumping and Screening Shafts (see green dot in Figure 1), mined through an intermediate "Outfall 002" drop shaft, and will be recovered from the "Outfall 001" drop shaft at the completion of mining. The tunnel is lined with a single pass fully gasketed and bolted precast concrete segmental liner.

Primary design responsibilities and design services during construction were led by Jacobs Engineering.

## GEOLOGY

The City of Alexandria is a historic neighborhood, which has been built up over centuries to its current elevation. The major geologic strata of the area are Fill, Marsh, Alluvium, Terrace, Upper Potomac formation, and the Lower Potomac formation.

The Tunnel alignment on the Project is primarily within the Upper Potomac soils when underneath existing land, which includes approximately the first 6,600 feet of tunnel alignment and last 1,800 feet of tunnel alignment. Underneath the Potomac River,

Figure 1. Project overview

Alluvium and Terrace layers will be encountered for approximately 3,000 feet. The tunnel alignment has been divided into six distinct reaches based on the dominant geology and intermediate shaft locations.

## Reach 1 and 2—Upper Potomac Formation

Reach 1 and 2 (approx. 6,600 feet) consist of very stiff to hard, fine-grained soils containing fat clays, lean clays, silty sand, and clayey sand. During TBM launch the Upper Potomac formation was found to be very firm. It has allowed for free-air interventions and has allowed for the final lining to be constructed before the ground starts to move.

## Reach 3, 4, and 5—Alluvium and Terrace

Reach 3 and 4 (approx. 3,000 feet) are underneath the Potomac River and will encounter the Alluvium and Terrace strata. The Alluvium strata consists of interbedded sedimentary deposits of clay, silt, and sand with varying amounts of gravel and organic material. The Terrace strata is generally underneath the Alluvium strata and consists of interbedded sand and gravel with varying amounts of fines. The Terrace is expected to contain cobbles and boulders, particularly at the bottom of the strata at the interface with the Upper Potomac strata. Mixed soil conditions will be encountered throughout Reach 3, 4, and 5, particularly at the interfaces between the Potomac formation, Terrace, and Alluvium strata. An excerpt from the geotechnical plan and profile of Reach 3 and 4 can be seen in Figure 3, which shows the Tunnel alignment at multiple layer interfaces.

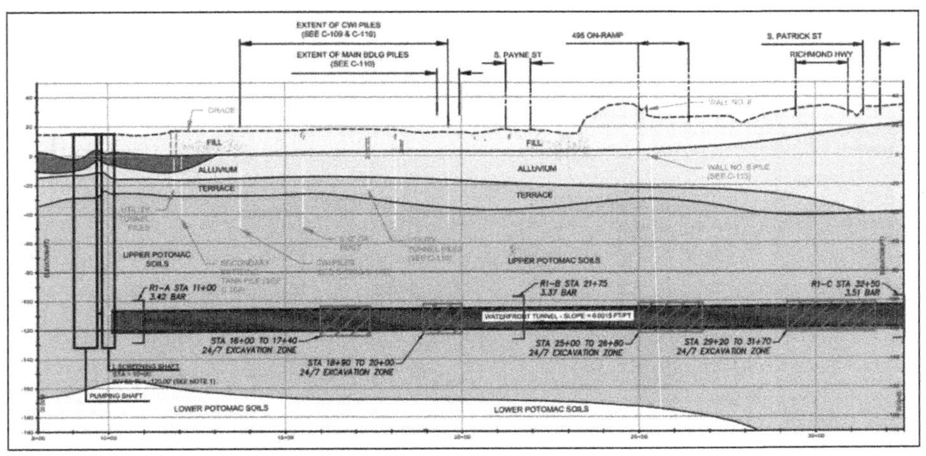


Figure 2. Reach 1 tunnel profile and geology

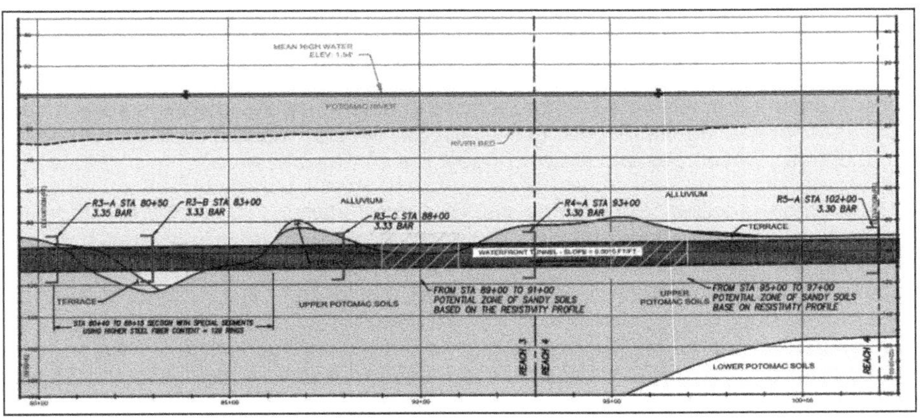


Figure 3. Reach 3 and 4 tunnel profile and geology

### Reach 6—Upper Potomac Formation

In Reach 6 (approx. 1,800 feet), the tunnel alignment returns to the Upper Potomac formation in similar soils as Reach 1 and 2.

## CONCURRENT ACTIVITIES ON SITE

### Near Surface Structure Construction

During the early stages of tunneling, there were multiple simultaneous construction activities being performed on the surface.

Keller was selected to install the support-of-excavation (SOE) for the near-surface structures (NSS) that connected to the Pumping Shaft. The slurry panel installation required more ancillary equipment, tanks, and more laydown area than originally anticipated. The entire NSS SOE operation took up roughly half of the available lay down area at the launch site. This work required TSJV to effectively plan to keep the tunneling progressing as much as possible. Meetings between TSJV and Keller took

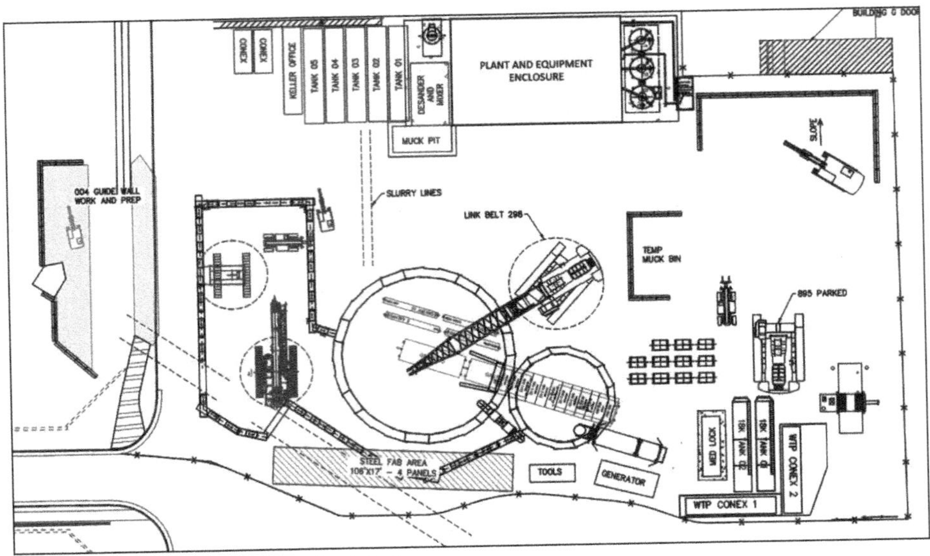


**Figure 4. Site layout during TBM short-startup**

place to ensure that the work was sequenced in a way that would always keep a haul route open for hauling out tunnel muck. Deliveries were scheduled in an efficient manner to optimize resources and to not congest the Plant access roads.

To begin more ancillary work during the TBM short startup, TSJV collaborated with the Owner to permit a long-term closure of one of the water treatment plant's access roads. This would re-route plant traffic through a small private road and to the plant's South gate. The Owner agreed, and this proved to help the operations, especially during concrete pour days. This road closure also required TSJV to provide permanent maintenance of traffic (MOT) personnel to facilitate construction traffic, water treatment plant traffic, and the occasional rogue civilian traffic. Additionally, the Owner requested supplement security measures, such as a security booth and more security cameras. In the spirit of partnering and to save schedule, TSJV complied with all requests.

Once the road closure was in place and the NSS SOE was complete, the roadway was utilized for construction of a 6-foot I.D. microtunnel boring machine (MTBM) approximately 30 feet below grade. The launch shaft featured sheet piles and the reception pit featured auger-cast tangent piles as vertical elements of SOE. The roadway was also used for compaction grouting for MTBM deep foundations, and the MTBM plant and equipment. A site layout during TBM mining, NSS SOE construction, and start of Auger-cast work can be seen in Figure 4.

## TBM SUPPORT SYSTEMS

### Backfill Grout

As previously mentioned, this tunnel project utilized an EPB TBM to excavate the tunnel and is lined with a precast segmental liner. The overcut, or annulus, of the TBM is backfilled using a two-component grout. To batch this grout, TSJV purchased a Sealcrete bi-component grout batch plant by Simem. The grout batch plant (GBP) was staged on the surface adjacent to the grout constituent storage tanks and housed within the plant and equipment (PE) enclosure. Due to space limitations, the GBP and storage tanks were tightly located, leaving just enough space to safely walk between components. The decision to enclose the GBP stemmed from the temperature sensitivity of the designed grout mix, with the local environment ranging from below freezing during winter months to over 100°F (38°C) during summer months.

A series of grout tests were conducted by TSJV, and the approved grout mix ratios were incorporated into the GBP controls to batch the correct ratios every time. Additionally, The GBP was wired into and controlled by the TBM programmable logic controller (PLC) to efficiently provide grout when needed. Grout constituent silos were outfitted with ultrasonic sensors that would alert the batch plant operator of low levels to restock constituents.

During the early stages of tunneling, the main site was projected to host several subcontractors simultaneously. The subcontracted work ranged from support of excavation install, micro-tunneling work, to compaction grouting for deep foundations. All these activities had the potential need for high water demand. In anticipation of this and the fact that the site only had one fire hydrant to supply water, TSJV made the decision to purchase a water buffer tank to keep a safe amount of water readily available for the GBP. This buffer tank would provide enough water to batch grout for five additional advances in addition to the 3,600-liter grout storage tank on the TBM.

## Compressed Air

The compressed air (CA) plant was also tightly staged adjacent to the grout batch plant and inside the enclosure. The compressed air plant supplies compressed air to the TBM & tunnel, and to other areas of the launch site from a dedicated auxiliary supply line. The CA plant was composed of Kaishan compressors, Mikropor dryers, air receiver tanks by Samuel Pressure Vessel Group, and a series of Mikropor air filters. Stemming from the CA plant, TSJV dedicated two 6-inch air supply lines to feed the TBM and tunnel. A third line was installed to provide air to the rest of the launch site. The receiver tanks were outfitted with two pressure relief valves each in case of a pressure build-up in the system. Also incorporated was a bypass pipe which fed directly from the compressors to the TBM supply lines in case work needed to be performed on any of the components within the system. Throughout the CA plant, several moisture discharge fittings were installed to expel moisture build-up in the piping. Prior to the supply lines, which raced to the tunnel, a bank of in-line filters were installed. This was done to ensure high quality breathing air in anticipation of hyperbaric interventions. In addition to the filter bank in the CA plant, TSJV also constructed a mobile filter bank engineered to be secured down to the multi-service vehicle decks, in case anything was to happen to the topside filters during a cutter head intervention.

**Figure 5. Plant and equipment enclosure**

## Cooling Water

The cooling water tower and water tank were located topside adjacent to the PE enclosure. Due to the corrosive nature of cooling water vapor, TSJV made the decision to enclose this system separate from the GBP and the compressed air plant. The cooling water system was composed of a hot/cold water tank outfitted with a level sensor coupled to an actuated valve that would open a water supply line if the water level reached too low. This water was conveyed by pumps through the cooling tower, supplied by Delta Cooling Towers, which in turn discharged the cooled water to the cold side of the storage tank. The cold water would then be gravity fed to two pumps in parallel, powered by Baldor motors, which would send the water through the tunnel to the TBM heat exchangers. The cooling water piping was also outfitted with Endress Hauser thermophant sensors to ensure adequate water temperature monitoring was achieved.

### TBM Utility Trench

Due to space constraints on site, TBM utilities were installed in a trench underneath the future muck hauling route. The utilities were routed around the location of the tunnel muck bin, which was installed during short startup. They were buried with sufficient depth to avoid damage from surface traffic and to stay below the frost line. TSJV elected to bury power conduit in a separate utility trench. This trench also had to cross under the future haul route. Due to the anticipated traffic flow and the fact the TBM 35-kV power line was contained in this trench, TSJV elected to encase this trench in concrete.

### Hyperbaric Operations

The maximum earth pressure expected for this tunnel drive was 3.9 bar and multiple cutter head interventions were planned throughout the tunnel. To support potential hyperbaric operations, the TBM was outfitted with a two-stage man-lock, a material lock, and a top side medical air lock. The medical air lock purchased was a 7806D by Reimers Systems and had a capacity to treat a crew of six up to 5.1 bar. The medical air lock was tightly staged within the swing radius of the tunnel service crane, adjacent to the water treatment tanks. To operate the hyperbaric equipment, TSJV partnered with Ballard Marine Construction. Ballard was contracted to provide training on hyperbaric operations and contracted to perform hyperbaric interventions if one were to be needed above 3.5 bar.

## LAUNCH TECHNIQUE SELECTION

During the bidding process, it was evident that there would be multiple construction activities in progress during the launch of the TBM. Added to the minimal space available at the launch site, TSJV elected to pursue a multistage short-mode launch. TSJV elected to design most of the trailing gear of the TBM to be capable of partial disassembly. The gantries were designed to have all, if not most of their respective essential equipment on one dedicated side. The other side of the gantries would be dedicated to being a walkway for personnel. This design concept allowed TSJV to optimize the number of half-gantries to be staged at the bottom of the launch shaft and to minimize the footprint needed on the top side of the site. With the exception of the service crane and a few precast segment stacks, the TBM launch was fully located at shaft bottom, allowing site laydown for partnering subcontractors to progress their scope of work. Although the half-gantry concept minimized the footprint of the TBM for launch, it would also generate a need for lengthy umbilical hoses. After much analysis, TSJV decided on the rebuild locations and procured the required lengths of umbilical hoses and cables to reach those end points.

## SHORT-START UP CHALLENGES

### Shaft Geometry

The TBM launch site includes two shafts in a figure-8 configuration with an interconnect approximately 137 feet below grade. The Pumping Shaft (PS) is the bigger of the two, with an inside diameter of 72 feet. The Screening Shaft (SS) has an inside diameter of 40 feet. The support of excavation method elected by TSJV was diaphragm walls. The PS was enclosed with 4-foot thick slurry panels, while the SS was supported with 3-foot thick slurry panels. The top of slab of the PS reached a depth of approximately 137 feet while the SS was excavated further down to install the permanent base slab, for a top of slab depth of 150 feet. Additionally, a safe-haven or "bathtub" was constructed at the tunnel break-out. This bathtub was enclosed by slurry panels filled with flow fill. Due to the small shaft diameter, the TBM assembly and launch was done in multiple stages.

### Work Platform

TSJV elected to construct the SS permanent base slab prior to launching the TBM. This required an elevated steel platform to be engineered and installed to make the launching elevation the same between both shafts. The SS steel work platform included two main configurations. The first configuration would be tailored to accommodate the launch of the TBM. This required very precise installation to get the shields in the correct location for the start of the tunnel. The second configuration would be tailored to facilitate full production mining. The work platform was composed of three subassemblies: The runway (middle) section, left wing, and the right wing. The middle section was suited to have the TBM shield cradles bolt on to the runway lateral beams for a fast and efficient installation. The work platform included bolt hole patterns that would allow for the subsequent installation of starter-ring supports. The work platform also contained a designated location to allow for the installation of a rack and pinion construction hoist, allowing personnel to travel up and down from the surface, as needed. Staged underneath the working platform are the base structures for the proposed vertical conveyor and belt systems.

**Figure 6. SS work platform**

When the TBM advances far enough to allow for full TBM installation, plus 150 feet for the construction of the permanent conveyor belt transition, the work platform will be transitioned to the second configuration to facilitate full production mining. This change will encompass the removal of the false rings, the TBM cradles, the tunnel eye-seal, and the installation of the cleated belt for the vertical conveyor system. When this work is complete, custom-made crane mats will be flown down and installed in the runway portion of the work deck, allowing surface space in the SS shaft to service the tunnel, and releasing the PS shaft for future permanent works.

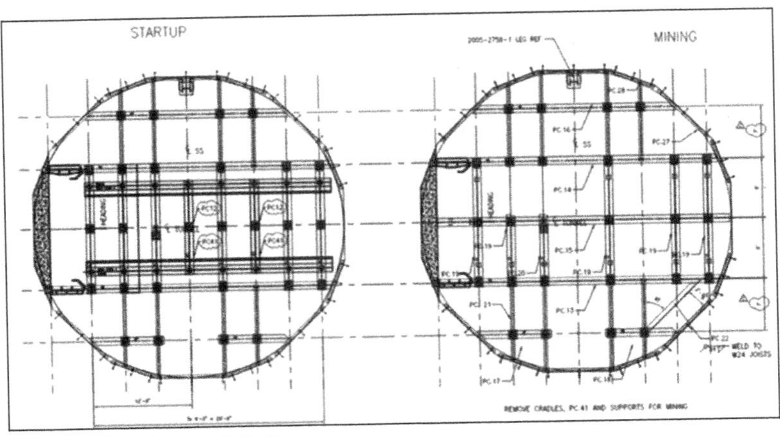


Figure 7. Two stage work deck

## Shield Assembly

The EPB TBM's front, mid, and tail shields arrived on site in August 2022 overlanded by a Goldhofer transporter in a 6-deck-6 configuration, provided by W.O. Grubb. The Goldhofer modules were utilized to successfully navigate through the tight residential streets of old-town Alexandria and to satisfy axle load requirements of the Virginia Department of Transportation (VDOT). The decision was made to transport the TBM shields during the middle of the night, to minimize impact to the public. Once the shields arrived at the launch site, inside of AlexRenew's water treatment facility, the shields were unloaded using a Grove GMK7550 all-terrain crane. To minimize required work on the TBM shields once on site, TSJV chose to have the front shield transported in one piece with the cutter head installed and the complete mid-shield installed with

Figure 8. TBM shields on thrust slab Stage 1

the main beam. Immediately afterward, TSJV began preparing the shields to be lowered down the launch (PS) shaft.

Once the bottom of the pumping shaft and screening shaft were reached and ready to receive the TBM shields, TSJV mobilized a Manitowoc MLC-300 crawler crane with variable positioning counterweights (VPC). The TBM shields were lowered as such: The front shield with the cutterhead installed, and the mid-shield with the erector main beam installed. The mating of the shield articulation cylinders occurred at shaft bottom with the aid of a separate hydraulic power pack used to manually extend the cylinders. Due to the size constraint of the launch shaft, the two front TBM shields needed to be inched forward enough into the screening shaft (SS) to facilitate the installation of the screw and the tail shield. TSJV accomplished this by engineering a bolt-on push beam that would fasten to the top of the steel cradles. A hydraulic power pack was used to manually extend two bottom thrust jacks on the TBM and advance the machine forward until the next location of bolt holes were reached. Once the screw conveyor was installed and secured, the tail shield was lowered and mated in a similar fashion.

Prior to lowering and mating the first gantry, the TBM shields needed to be pushed half-way into the screening shaft to allow the installation of the trailing gear cradles. To achieve this assembly sequence, TSJV engineered a 2-stage thrust slab. The first stage would include the embedment of components of the thrust frame, and the necessary rebar and concrete to resist and transfer the thrust loads anticipated during launch. In addition, a rectangular block out was constructed in the first phase to accommodate setting the TBM shields at the necessary elevation. The second phase of the thrust slab would include filling in the rectangular block out and the installation of dowels to tie this secondary slab to the first. This allowed the installation of the trailing gear cradles at the proper elevation and allowed the construction of the complete thrust frame once the tail shield had passed it. Once this step was complete, the first gantry was lowered and mated to the main beam of the TBM.

**Figure 9. Trailing gear installation on thrust slab Stage 2**

## Staged Launch

The planned launch sequence required the trailing gear decks be separated, and the essential half-gantries staged down in the PS. The remaining structure of the trailing gear was stored at a neighboring site until needed. The staged launch consists of six primary phases to achieve full production mining:

- Phase 1
  - Install TBM shields (front, mid, and tail) and screw conveyor
  - Push TBM shields into the SS shaft
  - Mate Gantry 1 & umbilical lines
  - Install temporary belt & drive

  - Test & Commission
  - Advance machine approximately 50 feet
- Phase 2
  - Mate Gantry 2 & umbilical lines
  - Relocate temporary belt & drive
  - Test & Commission
  - Excavate & tunnel approximately 60 feet
- Phase 3
  - Mate Gantries 3 & 4. Extend umbilical lines
  - Relocate temporary belt & drive
  - Test & Commission
  - Excavate & tunnel approximately 114 feet
- Phase 4
  - Mate Gantries 5, 6, 7, and 8. Extend umbilical lines
  - Test & Commission
  - Excavate & tunnel approximately 112 feet
- Phase 5
  - Mate Gantries 9, 10, 11, and 12. Extend umbilical lines
  - Test & Commission
  - Excavate & tunnel approximately 56 feet
- Phase 6
  - Mate Gantries 13, and 14
  - Excavate & tunnel approximately 236 feet
  - Remove & reinstall all tunnel utility lines
  - Reconfigure SS work deck
  - Install permanent belt structure & vertical conveyor
  - Test & Commission
  - Begin full production tunneling

## Umbilical Management

To facilitate the multistep launch, umbilical hoses and cables were required to service the TBM and its essential equipment. The distances the TBM needed to advance between rebuild stops ranged from 50 to 240 feet. This meant that the required lengths for each phase needed to be in the shaft and readily available for tunnel advancement. TSJV had to manage thousands of feet of cables and hoses for launch and to do this, TSJV engineered a festoon system. This festoon system included steel cable reels hoisted by beam trolley chain falls. These beam trolleys were mounted on W-beams which were mounted on the slurry panels approximately 20 feet above the shaft bottom. This festoon system allowed for less amount of cable and hoses on the floor and the lifting and lowering capabilities of the cradles allowed for unreeling of the cables as the TBM advanced.

**Figure 10. Umbilical management system**

## Segment Delivery

By the time the TBM shields were installed, and the first gantry mated, there was limited space at the back of the PS shaft to lower and deliver the precast tunnel lining segments to the heading. For this stage of the launch, TSJV utilized a custom-made segment cart that was designed to receive and transport tunnel segments to the TBM feeder. During this phase, the TBM segment quick unloaders were located on Gantry 2, and there was not enough space in the shaft to utilize this machinery. Instead, TSJV used a beam trolley hoist to lift the segments from the cart and transport them onto the segment feeder, one segment at a time. The segment cart was designed to run on a modular bolt-on rail system, that was synchronized with the TBM advances. As the TBM advanced, a new section of rail would be installed, allowing the segment cart to keep up with the machine. This method was used until Phase 2 of the launch, when there was enough room to mate Gantry 2, and utilize the multi-service vehicles (MSVs) to receive and transport the tunnel segments.

## Temporary Belt Extension

Another critical aspect of this launch was the installation and jumping of the temporary belt head drive. TSJV engineered and utilized a temporary belt head drive to power and facilitate mucking operations. The head drive was envisioned to be re-installed throughout the launch, so a bolt-on frame was designed to facilitate easy assembly and disassembly. This head drive was operated by a hydraulic valve stack and the TBM PLC and received its fluid power from the hydraulic pump of the TBM. For the first three

**Figure 11. Temporary belt drive and segment cart**

phases, the head drive assembly would need to be relocated to the back of the last trailing gear of that respective phase, for example, Gantry 4 during Phase 3. With these jumps, it was also necessary to install additional hydraulic hoses to power the valve stack and install belt roller structures on the top of the gantries to support the additional belt that would be installed. For receiving the muck at the belt discharge area, a muck box with modified feet was engineered to securely sit on the MSV deck. The MSV operator drives the box up to the discharge area to fill the box, then backs the MSV up, to a location in the PS shaft where the service crane can hoist and dump the box. This method was utilized until the complete TBM was buried and the permanent belt conveyor system was installed in the SS shaft.

## CONCLUSION

With adequate planning, coordination, and efficient use of site staging area, TSJV was able to successfully launch an EPB TBM from a congested site in multiple short-mode configurations. The pre-design of the trailing gear to be capable of partial disassembly proved to provide the minimal footprint needed to launch the machine. In addition, the procurement of umbilical hoses and cables allowed the progression of the TBM to advance to key stages along the tunnel to minimize downtime due to building trailing gear. With the growing number of cities requiring tunnels, and a growing population, the lessons learned from this launch will be invaluable to future launches as lay-down areas continue to shrink.

# Unique Umbilical Launch of a Slurry TBM in LA

**Nick Karlin** ▪ Dragados USA

## ABSTRACT

Dragados USA is constructing the LA Effluent Outfall Tunnel, a 7-mile, 18 ft internal diameter effluent discharge tunnel, utilizing a 21.6 ft (6.585 m) Herrenknecht Mixshield TBM. This paper highlights the means and methods used for TBM launch from a 55 ft diameter shaft excavated in wet with diaphragm wall support of excavation. This included SEM excavation of jet grout blocks for starter and tail tunnels, temporary concrete structures and steel installations, TBM sealing system, umbilical management for TBM hydraulic, electric and bentonite circuits as well as complex segment delivery logistics. The many launching challenges were overcome thanks to the innovative approaches from the project team.

## LAUNCHING SHAFT LAYOUT AND EARLY WORK

The contractual design of the starting shaft for the tunnel is a 55 foot internal diameter diagraph wall shaft. Additionally, the contract stipulated a ground treatment program to extend from the launching shaft in direction of the tunnel drive of sufficient length to provide encapsulation of the TBM plus three segmental rings in the ground prior to cutterhead exiting the treated area.

After review of several possible arrangements for TBM assembly and launching, it was decided to use this treated ground zone to construct a starter tunnel to assemble the TBM into prior to launching. This enabled Gantry 1 to be placed in the shaft for the start of tunneling which includes both P2.1, the first slurry pump which draws from the excavation chamber, as well as the operators cabin. Due to the fact that the contract already required ground treatment for TBM length plus 3 rings there was not a significant increase to the quantity of required jet grout for this approach.

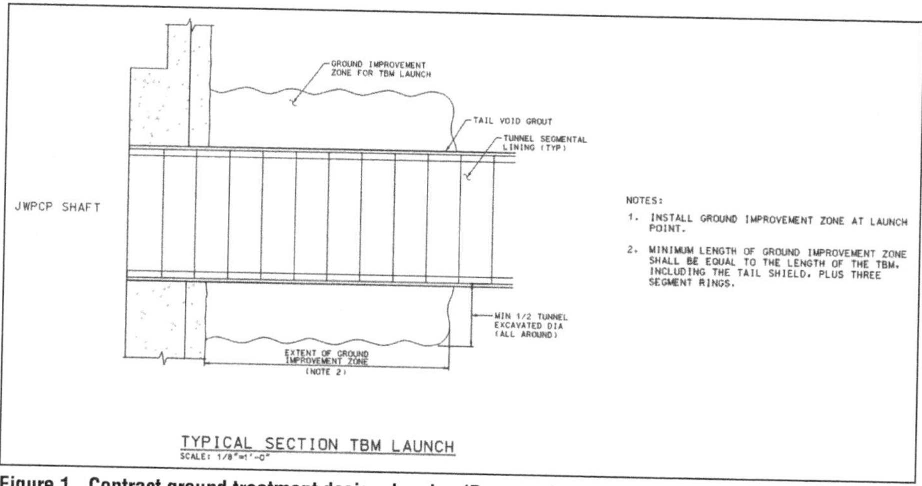


Figure 1. Contract ground treatment design drawing (Parsons & McMillan Jacobs, 2018)

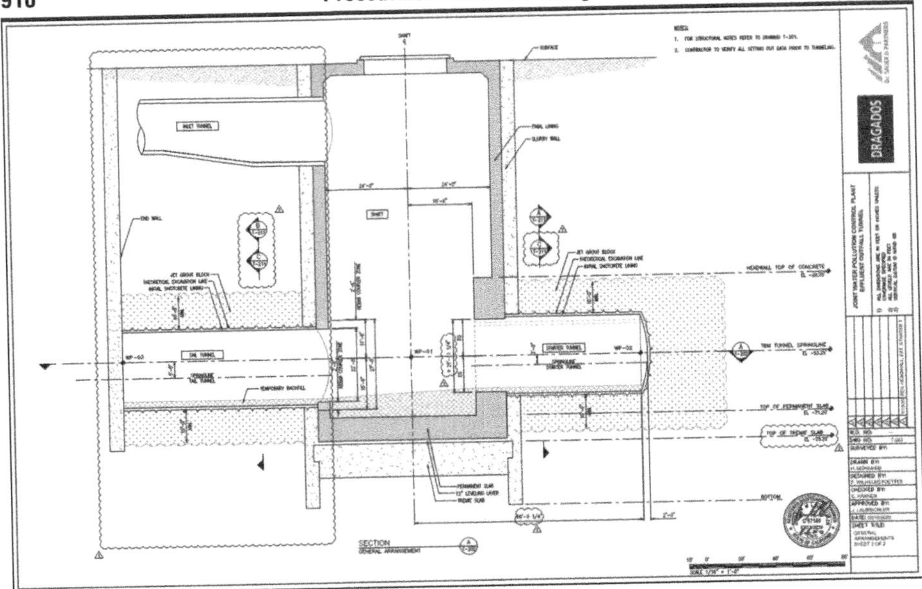


**Figure 2. Dragados shaft configuration and ground treatment**

In addition to the starter tunnel, Dragados also elected to perform jet grouting for the construction of an SEM tail tunnel. The tail tunnel was partially necessary to enable the launching of the TBM with Gantry 1, however its primary purpose was for the logistics of the overall tunnel drive. Given the tunnels 7 mile length double ring MSVs were purchased, and in order to efficiently load a vehicle of this length construction of a tail tunnel was required. After developing the full scope of the early ground treatment and tunneling, Dragados engaged the services of Dr. Sauer & Partners for design of the SEM starter and tail tunnels and to serve as the EOR during construction.

After reviewing options for several arrangements of the tail tunnel it was realized that there would be issues using only jet grout ground treatment to construct the required tail tunnel length. The jet grout required a 10' unexcavated plug at the end of the tunnel to resist the ground and hydrostatic loads. This 10' plug could not be installed due to existing overhead 34.5 KV transmission lines on the project site. As such, the 10' end plug was optimized down to just 40" by installing a diaphragm panel end wall. As diaphragm panels were already the support method for the shaft, all the equipment was onsite to perform this scope without additional mobilizations or delays.

The final configuration for the launching shaft included starter and tail tunnels excavated to lengths of 39.4' and 48.25', respectively, as measured from inside face of shaft wall.

## CONCRETE WORK & STEEL STRUCTURES

The overall TBM launching concept was to construct a launching barrel to seal the TBM within to contain the slurry and maintain the face pressure for launch. This concept is commonly seen implemented with both EPB and Slurry TBM by utilization of a steel launching barrel, however given the job site conditions the launching barrel would be constructed with a combination of shotcrete and cast in place concrete, both within the SEM starter tunnel and within the shaft. Minor steel elements to seal the cast in place concrete to the launching frame would also be required to ensure complete seal of the TBM for proper ground support.

Figure 3. Completed TBM launching barrel

After careful review of construction tolerances, as well as TBM shield diameters the concrete launching barrel diameter was set at 6.685 m, as compared to a cutting diameter of 6.585 m this left an annular space around the cutter head of 50 mm or approximately 2."

With construction diameters set, formwork was procured. The starter tunnel would be finished out with a cast in place invert and smooth shotcrete crown for 30', whereas the 6' closest to the shaft would be cast in place full 360 degrees to ensure complete sealing of joints between starter tunnel and diaphragm walls. This full 360 degree concrete placement was carried out into the shaft by an additional 6' to fully encapsulate the TBM. This additional concrete placed out into the shaft was finished as a flat face head wall with a height of 28'-9" and a width of 35'. The weight of this concrete was considered in the thrust frame design to act as ballast helping to resist the uplift reaction of the vertical member of the classic A-Frame style thrust frame. The thrust frame was secured into a temporary structural concrete shaft invert slab. This slab had a thickness of 4'-6" with a double mat of #10 rebar at 1' centers, as well as hair pins in critical locations such as the raker corbels and thrust frame column anchorages.

To further expedite TBM assembly and launching, two separate steel support structures in addition to the thrust frame were developed and procured by Dragados. The first was a fairly standard set of cradles for assembly of TBM shields. The height of these assembly cradles was designed to match the elevation of the temporary shaft concrete as well as the starter tunnel invert so that no vertical translation of the TBM would be required during the assembly process. In addition to the shield assembly cradles, Dragados procured a segmental false steel invert to cross the shaft and extend into the tail tunnel. The fabrication of the false invert was again coordinated with the shaft and tail tunnel concrete elevations to expedite field installation. The

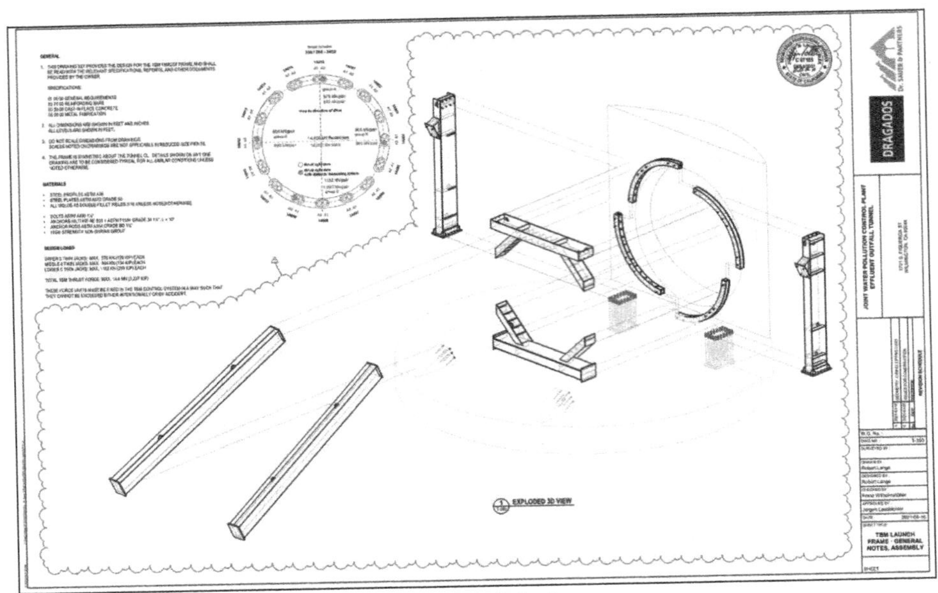


**Figure 4. Thrust frame general arrangement (exploded view)**

false invert allowed for easy assembly and rolling of Gantry 1 during the initial tunneling, as the standard installed bogey wheels could be used, and they would transition smoothly into the segmentally lined tunnel. The false invert also facilitated later access of people and equipment into the tunnel for the startup tunnel drive, as it could accommodate the TBM jump rail for later gantry install, MSV traffic, etc.

## TBM SEALING AND UMBILICAL MANAGEMENT

As with the initial launch of any pressurized TBM, control of face pressure in the initial launching configuration is of critical importance as any failures can lead to face instability and intrusion of ground or water into either the TBM or the shaft excavation area. For this reason, Dragados developed and installed a 3 level sealing system to contain the 1.8 bar initial launching pressure. The first seal was a conventional Bullflex bag, installed within the cast-in-place starter tunnel. The Bullflex bag was inflated with grout to close the gap between the cast in place launching barrel and the tail shield. The next layer of sealing consisted of a pair of rolled steel plates, one affixed to the concrete headwall and one sandwiched between the first ring and the thrust frame, these plates were then connected by an Omega Seal to allow for minor movement or distortions while maintaining a designed seal pressure of 4bar. The final component of the sealing system was use of the installed TBM bi-component grouting system to pre-fill the first 18" of PCTL and omega seal area prior to pressurization of the slurry circuit. To aid in this pre-advance grouting a double row of temporary exclusions plates were welded to the lower 90 degrees of the tail shield so that the entire circumference of the tail shield would have exclusion plates installed. In addition to aiding the pre-pressurization grouting, these plates were intended to help maintain grout/bentonite separation while the TBM shield passed through the starter tunnel where it was elevated on 2"×2" slide rail. Once the TBM entered native ground and began to drag on the shield extrados the expectation was that these plates would wear/break off with time after having served their intended purpose.

**Figure 5. TBM shields on assembly cradles, jacking into position within starter tunnel**

The initial launch phase of the TBM required the shield to advance with just Gantry 1 for a distance of 350 tunnel feet. This would allow enough length in the tunnel to install all critical hydraulic and electrical gantries underground. To facilitate this 350' advance and connect the TBM and G1 to the remaining six critical gantries on surface would require a nominal umbilical length of 150–180 m depending on connection location to surface gantry and termination location underground. Several different systems and solutions were required to manage the variety of TBM utilities.

The primary umbilical of the TBM, which consisted off all electrical, control, and hydraulic systems were housed on a custom designed derrick system. This derrick system consisted of two fixed boom, hydraulic winch driven derrick cranes with each crane holding a 5' diameter drum with powered hydraulic rotation and normally closed brake. Powered rotation and braking were elected to be installed by the contractor for safety as the cables/hoses would be suspended in the shaft for several months, with varying tension loads as the TBM progressed. The ability to maintain even loading on the two drums and prevent unintended sliding or rolling of the drums was seen as critical to the operation for both safety and logistics.

For the TBM slurry circuit a separate umbilical system was designed by Dragados. This consisted of support brackets and a continuous trolley beam which was secured to the PCTL using the erector. 14" diameter × 20' long bentonite feed and slurry return

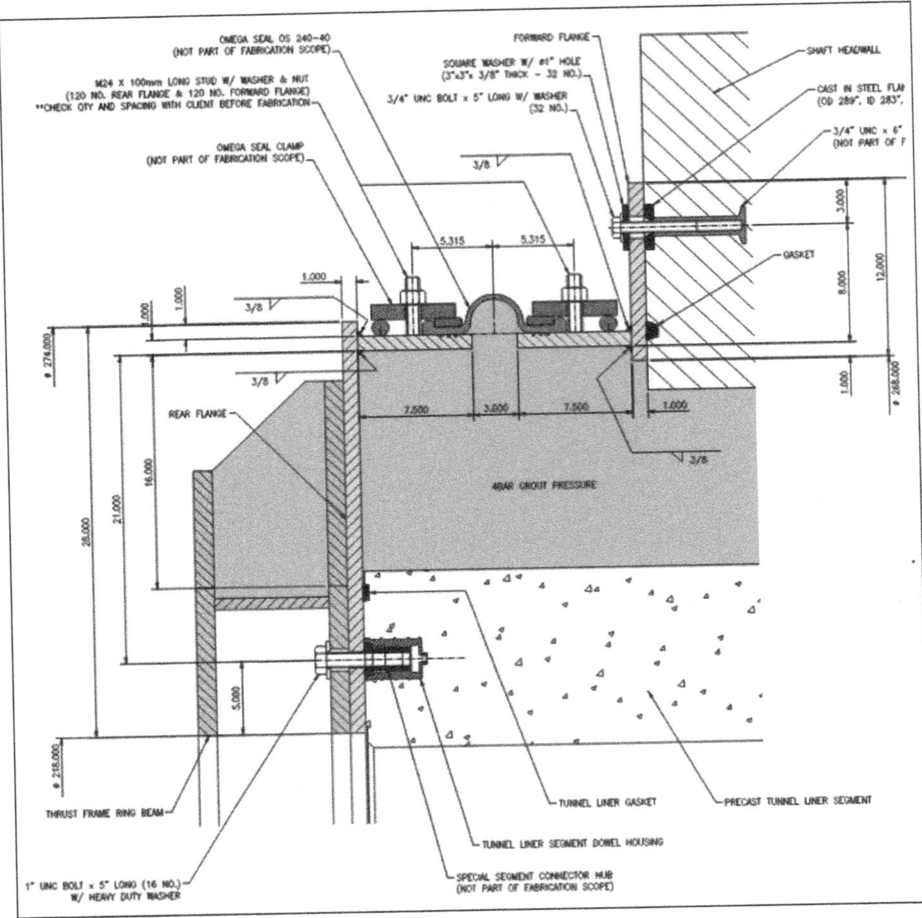


**Figure 6. Omega seal components and initial grouting zone**

lines were then paired up with riser clamps and affixed to beam trolleys to roll along the continuous trolley beam in the tunnel. A 50' × 10" diameter hose was maintained on each side of the slurry circuit in the shaft to provide the flexibility required for pipe extensions between the rolling tunnel pipe string and the fixed pipe mounted to the shaft wall.

For the TBM cooling water, grout, and compressed air conventional drag hoses were used behind the TBM with pipe installed in the lower quarter arch of the tunnel.

## SEGMENT DELIVERY AND GANTRY MODIFICATIONS

The shaft diameter and initial launching arrangement of the TBM and gantry one required three phases of segment delivery logistics to continually feed the heading. Phase 1 delivery was utilized to construct the first 9 rings of tunnel. For this phase a special top section of the rear half of Gantry 1 was fabricated with a large opening to allow segments to be directly lowered to the feeder. For this initial phase the segment hoist was not installed in the gantry, as it would have obstructed the clear window to the feeder.

Figure 7. Umbilical systems in tunnel during initial TBM launching

Figure 8. Segment delivery for ring one

Phase 2 segment delivery occurred once the clear window for direct loading of the segment feeder became obstructed by the thrust frame. For Phase 2 there was not yet enough space in the shaft to lower an MSV, so an additional intermediate solution was required. A light C-Channel track was fabricated with conventional shaped, but small gauge rail ties and heavy duty carts outfitted with segment dunnage to transport half stacks of segments. At this stage the open top rear section of gantry one was removed, and the standard top was installed, complete with segment hoist to load the segment feeder from the rail carts. Additionally, an air tugger was installed on the

Figure 9. View from tail tunnel towards heading

hindside of the segment feeder pushing snout to facilitate in the pulling and towing of segment carts. This second phase was used for an additional 10 rings, which was far enough for the entirety of gantry 1 to clear the shaft and allow for lowering of an MSV.

With the lowering of the MSV, the third and final stage of modified segment delivery for the initial launch sequence began. This phase would be used for the remaining 260' of tunnel which had to be driven with only G1 on the full umbilical set up to enable the space underground for installing G2-9 in the tunnel. This last configuration consisted of a modified MSV in which the chassis had be reconfigured to be as short as possible and the heading end cab was removed. These changes were necessary so that the MSV could deliver segments directly to the segment hoist at the back of G1, as the designed gantries with quick unloaders were not installed for the initial launching.

## CONCLUSIONS AND COLLABORATORS

The launch of a pressurized TBM in shallow urban settings always requires attention to detail and project specific considerations to ensure success. At the LA Outfall Dragados utilized several underground construction techniques along with both traditional and novel approaches to civil and mechanical structures required for a successfully TBM launch. This would not have been possible without good cooperation within the project team and the assistance of our key collaborators below.

Dr. Sauer & Partners—Design of Starter & Tail Tunnels, Thrust Frame, and Structural Concrete
Big River Engineering—Omega Seal Steel Ring Assembly & Arrangement Drawings
Kelley Engineered Equipment—Design & Supply of Umbilical Derricks
M&L Engineering—Supply of TBM Cradles & False Invert
Pini Group—Geotechnical Review & Support
Herrenknecht—TBM Supplier

PART 

# Project Planning

*Chairs*

**Colin Sessions**
Jacobs

**Tony Cicinelli**
Kiewit

# Developing Project Cost Estimates in Volatile Markets

**Connor Langford** ▪ Mott MacDonald
**Murray Gant** ▪ Metro Vancouver
**Ken Massé** ▪ Metro Vancouver

## ABSTRACT

For public organizations, preparing accurate project cost estimates is a critical component in capital project planning. Project delivery methods provide a structured approach to moving projects through the design and construction lifecycle, and typically use cost estimates as key inputs at each stage to determine if the project will advance. As design stages can occur years or even decades before construction contracts are awarded, estimates need to make reasonable assumptions about how the work may be executed (means and methods) as well as potential future market trends so that appropriate values for material pricing, labour costs, escalation, and owner's costs can be included.

This process is challenging enough during stable markets, as historic trends can often be misleading and require a suitable understanding of economic factors to use appropriately. When you mix in the market volatility that has been observed since 2020, developing accurate project cost estimates becomes even more challenging. In cases such as this, having a structured approach to cost estimating that includes an adequate risk-based reserves cannot be overstated as it provides the basis for ongoing project decisions.

This paper explores some of the historic and more recent market trends that have impacted key aspects of project cost estimates for capital project planning. It also provides recommendations on how to develop cost estimates for infrastructure projects, with examples from Metro Vancouver, which plans for and delivers regional-scale services to 2.6 million residents. A brief commentary on the use of escalation clauses in contracts is also presented.

## INTRODUCTION

With infrastructure projects becoming bigger, more complex, and more costly, owners are facing increased scrutiny and pressure to deliver their projects within budget. To do so, owners need reliable project cost estimates early in the project lifecycle to start seeking funding from both internal and external sources. As the project advances, these cost estimates will also be the bottom line against which stakeholder and public trust will be measured.

Developing project cost estimates is easier said that done, as several risk factors complicate the process (Figure 1). While most professionals focus on project execution risks, such as technical challenges that occur during design and construction, market factors can be one of the most difficult to quantify and can have one of the largest impacts on the project's bottom line. This was true during stable markets, however when you mix in the market volatility that has been observed since 2020, the uncertainty around labour and material prices as well as escalation costs has skyrocketed.

**Figure 1. Typical risk factors associated with an underground construction project**

To address this, project teams need to work collaboratively to develop a structured process for estimating project costs. This includes understanding how uncertainty in cost estimates change as the project progresses towards construction, as well as developing a set of reasonable assumptions to base the cost estimate on. Developing adequate risk-based reserves is also a critical component as this improves the predictability of project outcomes. This provides decision makers and project team members with a more accurate assessment of the ultimate project cost and a realistic confidence level in achieving that target.

This paper explores these concepts and provides insights from one of Canada's largest regional service provider, Metro Vancouver.

## OVERVIEW: METRO VANCOUVER

Metro Vancouver is a partnership of 21 municipalities, one Electoral Area and one Treaty First Nation that collaboratively plans for and delivers regional-scale services to the 2.6 million residents currently living in the region. Its core services are drinking water, wastewater treatment and solid waste management. Metro Vancouver also regulates air quality, plans for urban growth, manages a regional parks system and provides affordable housing.

Metro Vancouver's funding and financing objective is to minimize the annual household impact on regional ratepayers and promote fairness amongst the users paying for the facilities that Metro Vancouver builds. Accurate project cost estimating is a key input to this objective. To determine the household impact, the schedule for each project is used to forecast the annual cashflow for all capital expenditures across the corporation. The resultant household impact is then calculated for upcoming budget years, and if the impact is above the target, more scrutiny is applied to both the scope and schedule of each project. Decisions are made to either reduce scope, stretch out the schedule, defer the project (or portions of it) or seek other smoothing mechanisms to avoid rate shock with respect to the impact on households and ratepayers.

## PROJECT COST ESTIMATE COMPONENTS

Several elements are included in a typical project cost estimate. These are summarized here, including challenges that are identified and need to be considered during development.

### Construction Cost Estimate

The construction cost estimate is prepared by the engineer and is intended to provide an accurate estimate of the contractor's bid price. At the outset of a project when information is limited, the estimate is often prepared using unit rates and using similar projects as baselines. As the project develops into the later stages of design, a bottom-up estimate is prepared that incorporates key information from the various contract documents, such as the drawings, technical specifications, measurement and payment items, and documents such as the Geotechnical Baseline Report (GBR).

The estimate itself consists of a few different components that can be arranged in several different ways, depending on the preferences of the cost estimating team:

- **Direct Costs**: costs associated with constructing the specific project elements called out in the contract documents. This includes labour, material costs, major equipment purchases (such as a tunnel boring machine), subcontractor costs, power and fuel etc.
- **Indirect Costs**: costs borne by the contractor which are necessary for the execution of the project, but which are not 'chargeable' to the other major cost elements. This can include items such as general mobilization and demobilization, temporary site services, supervision, equipment maintenance, servicing, weather protection and site security.
- **Contractor Markup**: to reflect the contractor's profit margins. This varies depending on the type of work being done (i.e., tunnel excavation, chamber construction, pipe installation etc.) and how busy the market is at the time the procurement documents are issued.
- **Contractor Risk**: a contractor contingency amount to cover their productivity and construction risks as well as missed items. Similar to the markup, this amount will vary based on the type of work.
- **Insurance & Bonding**: to cover the insurance and bonding requirements laid out in the contract documents.
- **Construction Escalation**: an estimate of the escalation in both direct and indirect costs over the life of the project. This plays an important role in tunnelling projects, as they often run for several years with major purchases occurring throughout.

The following subsections identify some of the recent challenges being experienced in properly estimating these numbers, in particular due to market factors over the past two years.

#### *Labour*

Figure 2 shows a historical comparison between hourly wages in the United States (U.S.) economy and the 'core' Consumer Price Index (CPI) since 2007 (note that 'Core' CPI-U for all urban consumers omits distorting price movements occurring among food and energy items). As can be seen, prior to 2020, these two items have been closely linked. In 2020, hourly wages increased significantly relative to the core inflation rate. This was primarily due to significant layoffs associated with lower-paid workers at the outset of the global COVID19 pandemic, with many essential workers being kept on and receiving additional incentive and danger pay.

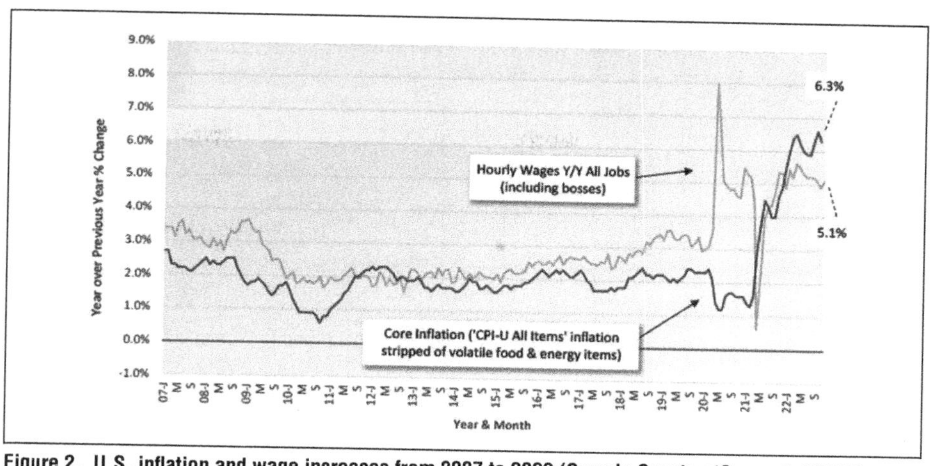


Figure 2. U.S. inflation and wage increases from 2007 to 2022 (Canada ConstructConnect, 2022b)

Since 2020, these two items have moved back into close alignment, however they are significantly higher than what they have been historically. In November 2022, a core inflation rate of +6.3% stood beside an 'all jobs' wage hike of +5.1%. This increase over historic trends is likely due to the critical construction labour shortage in North America right now, resulting in wages being driven higher. How these trends will continue to evolve over time has a significant impact on the construction costs in both Canada and the U.S. as skilled U.S. workers are often used on both sides of the border.

### *Fuel*

Fuel prices have been changing drastically over the last few years when compared to historic trends (Figure 3). After the collapse of oil prices during the global COVID19 pandemic in 2020, the future of the oil industry was unclear, leading to a start in fuel prices rising. After the Russian invasion of Ukraine, international trade sanctions were levied against the Russian oil industry, which is one of the top producers of oil and natural gas for the global market. This subsequently led to a drop in supply and a further increase in prices (Forth & Fromberger, 2022), leading to significant year-over year energy-related costs for diesel (+63.8% in March 2022) and gasoline (+61.5% in March 2022).

As the underground infrastructure industry still relies heavily on diesel and gasoline for site power and backup power, these price increases have had a significant impact on construction costs. Regardless of the size of the project, contractors are considering how to deal with this increased financial risk when bidding on contracts. One outcome is including higher fuel and energy costs to manage this uncertainty, therefore increasing the expected construction costs.

### *Materials*

Increases in U.S. the cost of construction materials started in earnest for many key items in 2020 at the start of the global COVID19 pandemic (Figure 4). This was driven by uncertainty in the market as well as considerable issues with material sourcing and transportation. These increases have not only impacted final project materials, such as steel pipe and concrete, but also temporary materials such as lumber used for concrete forming.

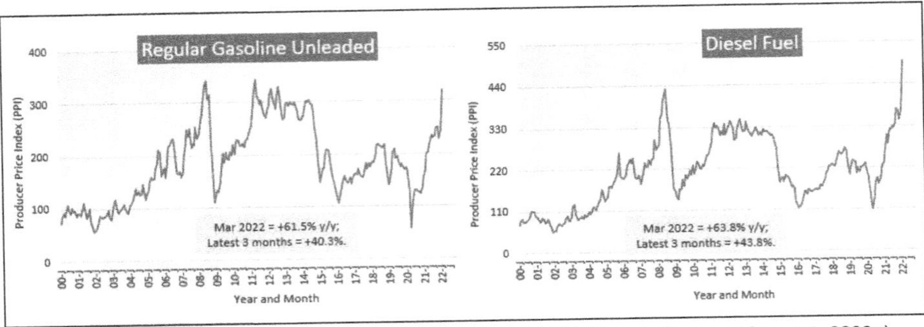


**Figure 3. U.S. gasoline and diesel costs between 2000 and 2020 (Canada ConstructConnect, 2022a)**

As material costs make up a considerable portion of construction costs, these trends are having a significant impact on project budgets. This is shown by the increase in the 'final demand construction' index from the Producer Price Index (PPI) data, otherwise known as the 'bid price' index, since the start of 2021 (Figure 5). This increase is closing the gap with material cost increases, indicating that bid margins by contractors are starting to stabilize (despite the "spikey" material cost changes) signalling the emergence of a somewhat healthier construction marketplace.

### *Escalation*

In addition to the challenges identified in the previous sections in selecting an appropriate basis for labour, fuel and material costs for the base project cost estimate, perhaps a greater issue is determining how these rates will be changing over time to accurately estimate escalation. As project cost estimates are often needed years or even decades before construction contracts are awarded, estimators are forced to interpret historic trends and utilize guidance from economists to forecast these costs well into the future. While this process has always had an element of uncertainty in it, the "crystal ball" often used by economists and estimators has gotten even more foggy since the global COVID19 pandemic.

Looking back at Figures 2–4, it's clear that in addition to the year over year increases resulting in higher construction costs, many show a "spikey" up and down behavior (i.e., labor, steel and lumber products), which creates a great deal of uncertainty when estimating project budgets. Attempting to project these trends into the future while the market is still correcting itself is fraught with issues and leads to several different interpretations of project costs. Will the trend continue? Will it correct? Is another global market correction pending? The answer to these questions can have a considerable impact on construction cost estimates and it seems every economist has a different answer.

As such, the most useful method to account for escalation is likely a combination of projecting recent trends and carrying suitable risk-based project reserves—this allows for a confidence-based assessment of the project cost estimate. This is discussed further in the *Risk Reserves* section.

### *Market Competition*

The timing of project procurement relative to how busy the rest of the industry is can have a significant impact on the construction cost and therefore is of considerable importance to owners. Putting a request for proposals (RFP) or tender package out

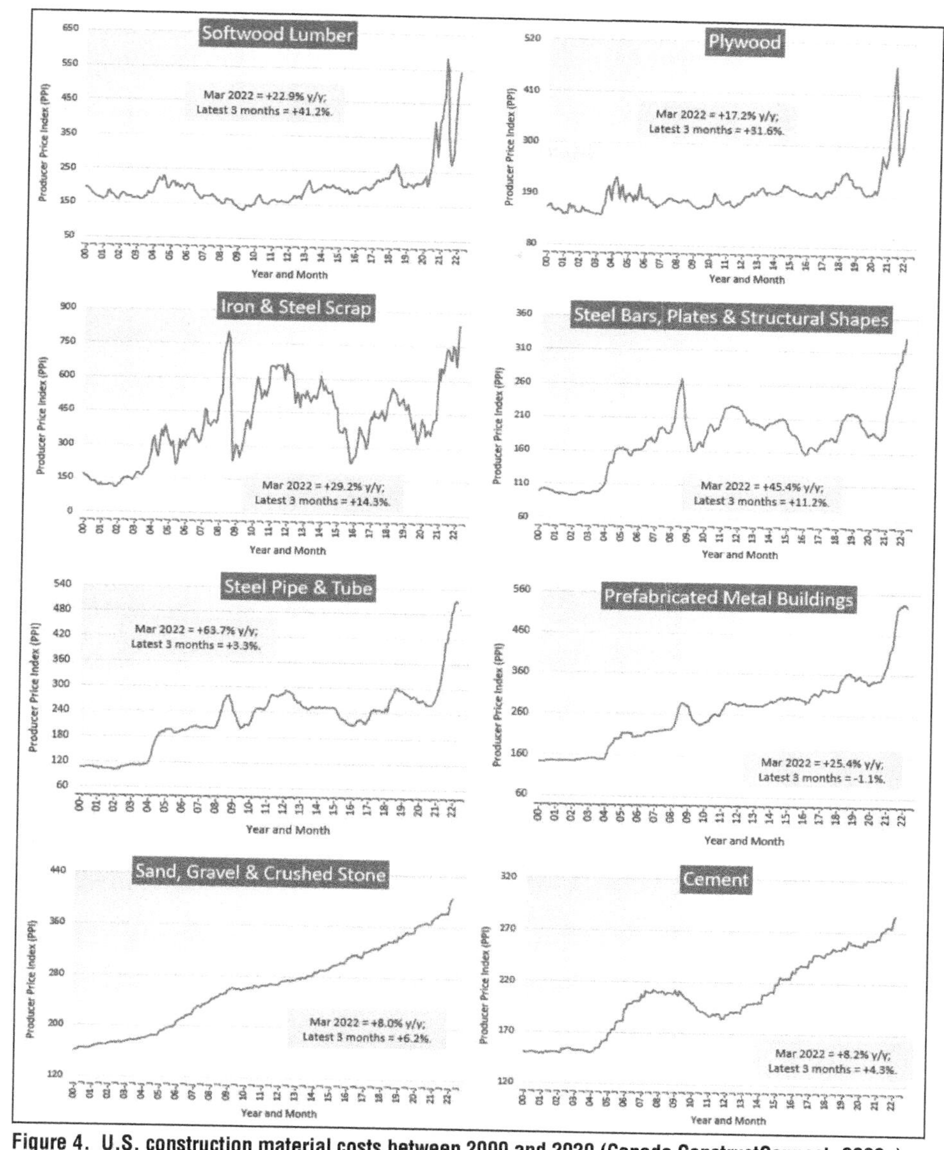


**Figure 4. U.S. construction material costs between 2000 and 2020 (Canada ConstructConnect, 2022a)**

on the street at a time when the market is quiet, typically results in market competition—multiple bids are received, and contractors are motivated to put together a competitive price to win the work. This is the ideal place to be in as an owner as it provides confidence to the funding agencies and boards that the preferred contractor truly stood out relative to the other submissions. Conversely, when the market is "hot" and saturated with other infrastructure projects, interest to take on more work may be low, resulting in few bids and potentially higher prices. A smaller number of bids also draws more intense scrutiny from governing/approving bodies, which could result in schedule delays if a rebid is required.

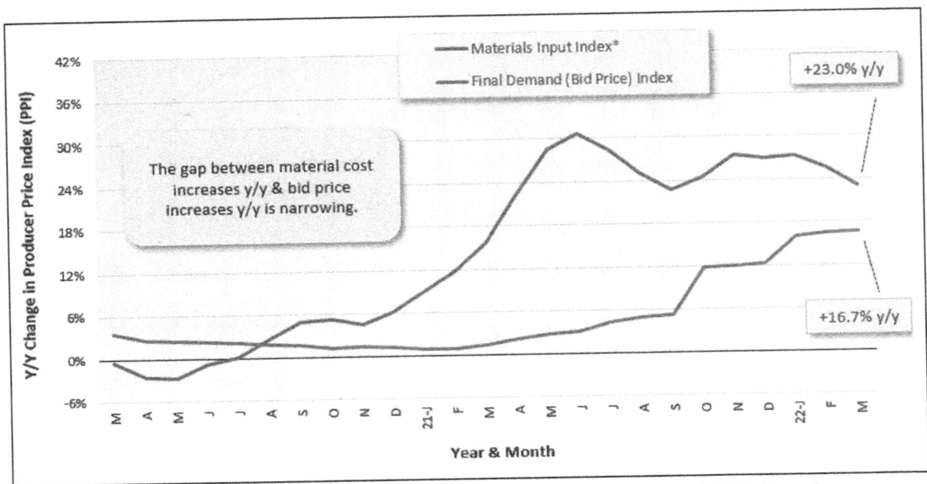


**Figure 5. U.S. construction bid prices versus material costs between 2000 and 2020 (Canada ConstructConnect, 2022a)**

As the current market conditions in the underground construction industry are quite "hot," owners are forced to compete with other projects for the best contractors. This is often done by breaking projects into several smaller elements to reduce contractor risk, providing more beneficial contract terms, and/or accepting higher construction costs. Whatever approach is selected, it will need to be reflected in the construction cost estimate so the confidence in this value can be assessed appropriately.

## Owner's Costs

Following the development of the construction cost estimate, the owner's costs need to be considered. Owner's costs include costs associated with:

- Land-related costs such as site acquisition, rights-of-way, and associated legal fees;
- Efforts required to acquire permits and associated permit fees;
- Internal staff costs, including owner oversight and overhead costs for the project team and any legal staff;
- Owner-procured services such as community engagement, First Nations consultation, archaeologists, and construction management;
- Any internal construction costs that are not covered by the contractor (such as disinfection and tie-ins on water mains etc.); and
- Owner consultants not related to technical design.

For smaller projects, these costs are typically well understood by the owner and therefore there is limited uncertainty associated with them. For larger and more complex projects, however, owners often do not have sufficient resources in house. As such, they need to pull from a resource constrained market, often at salaries significantly higher than their own historic target rates, thereby increasing staff costs on a project and putting even further pressure on consultants who they rely upon to help deliver their projects.

### Risk Reserves

Following development of the construction cost estimate and schedule, a reserve amount is developed. There are typically two "buckets" of reserves that often have different names, depending on the region of the project, the owner, and the funding sources:

- **Contingency reserves**: this amount is meant to cover project risks that can be identified and quantified as part of the risk management process (also referred to as "known unknowns").
- **Management reserves**: this is typically an additional amount to account for impacts associated with unidentified risks (also referred to as "unknown unknowns"). The amount is often a function of the owner's risk tolerance and therefore varies from project to project.

Some owners will only focus on one of these—as an example Metro Vancouver only considers contingency reserves for projects. Regardless, developing a total reserve amount that meets the owner's needs requires a good understanding of risk tolerance and the desired confidence level in that number. This is not a trivial exercise and involves all members of the project team to work together to confirm the desired outcome.

## CONSIDERATIONS FOR COST ESTIMATES

Several tools are available to manage the challenges with developing a project cost estimate. These are described in the following subsections.

### Estimate Accuracy

As part of the cost estimate development, the accuracy range of the estimate needs to be considered. This range provides an indication of the degree to which the final cost for a given project could vary from the estimated cost, due to factors such as the level of project definition, the purpose of the estimate (e.g., end use), the level of effort (e.g., number of hours) for the estimate, and the estimating methodology used. Accuracy is presented as a plus/minus range around the mean estimate applied after the application of the mean contingency value. According to the Association for the Advancement of Cost Engineering (AACE), this accuracy range provides a 90% confidence level that the actual cost outcome would fall within this range (e.g., P90 level, or 9 times out of 10 the Project cost will fall within this range). As the degree of project definition increases, the expected accuracy of the estimate typically improves, resulting in a narrower range of values. This is based on the premise that as more design and project details are available, project estimates will become more accurate. Similarly for the construction schedule, it is important to use expected productivities in the base schedule, as opposed to slower productivities that include potential risk impacts. Variability around these rates can be noted (i.e., range of possible advance rates), however they should be used as inputs to the risk assessment.

To standardize the estimate uncertainty, Metro Vancouver has developed a Project Cost Estimating Framework that prescribes what accuracy levels are to be quoted for each design stage (Figure 6). Metro Vancouver also includes a segregated line item for estimating uncertainty in their owner's costs, which allows for estimating assumptions as the site and oversight model becomes further defined throughout the lifecycle. This ensures that an uncertainty amount is carried in the cost estimate (rather than

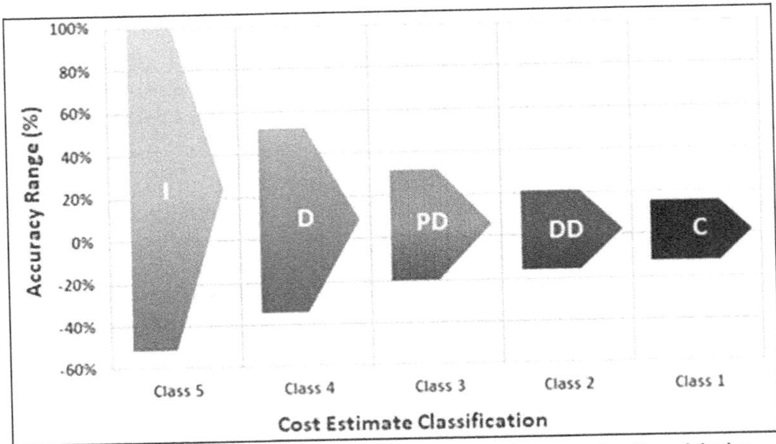


**Figure 6. Metro Vancouver cost estimate accuracy guidelines as a function of design level**

just quotes) and allows for consideration to be given to the level of scope definition, complexity of the project, and novelty of the nature of work or delivery model. By including this number, it provides added confidence in the overall project cost estimate at each stage.

## Stage Gate Process

One of the key issues with developing cost estimates is determining when they are to be prepared, what each estimate is used for, and who will review and provide comments. One of the methods for accomplishing this is through a structured stage gate process. Stage Gating is a complete, structured and transparent process for review and oversight of project key deliverables at each stage of the project lifecycle. It includes key milestones at which certain project-specific requirements need to be met before receiving approval to progress to the next stage. A Gate Review is where decision makers (i.e., the Gate Panel) review progress and assess readiness for the project to go forward. If successful, the Gate Review provides the project team with authorization to move the next stage.

For example, once the project definition stage is complete, approval is sought before moving into the subsequent stage of preliminary design. Likewise, approvals are sought before proceeding with detailed design and before proceeding with construction. Cost estimates are a key component of each Gate Review, and the estimates become progressively more accurate at each gate as the project gains certainty in terms of scope, schedule, and risk.

Metro Vancouver has been developing such a process since 2020 when their Board expressed an interest in undertaking a review of Metro Vancouver's capital project delivery practices to provide value for residents. Metro Vancouver's stage gate framework comprises 5 gates for major projects (see Figure 7). At each gate review, the deliverables on which the review is based includes a cost estimate for the project in its entirety as well as a cost estimate for the subsequent stage.

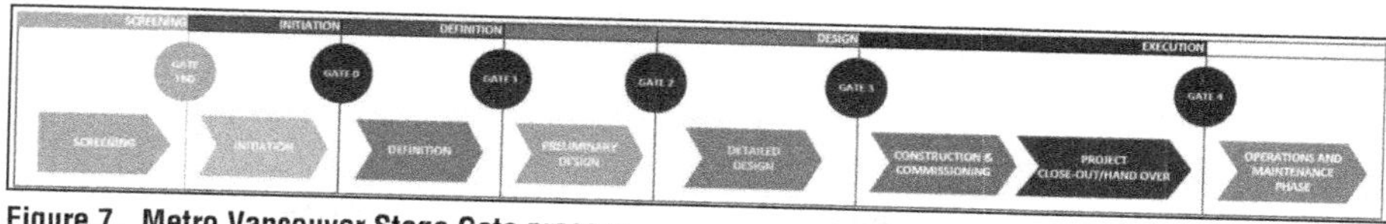


**Figure 7. Metro Vancouver Stage Gate process**

A summary of the Stage Gate process is provided below:

- At **Gate 0**, after a project is initiated, approval is sought to begin project definition. This gate review requires an AACE Class 5 Cost Estimate for the whole project and an AACE Class 3 cost estimate for the Project Definition Phase.
- At the conclusion of the Project Definition stage, the project is at **Gate 1** where approval is sought to proceed to the Preliminary Design stage. An AACE Class 4 Project Cost Estimate and an AACE Class 3 Preliminary Design Cost Estimate is required at this gate.
- **Gate 2** is where the project seeks approval to enter the Detailed Design Stage, and it requires an AACE Class 3 Project Cost Estimate as well as an AACE Class 3 cost estimate for Detailed Design stage.
- **Gate 3** is the final investment decision as this is when the panel decides whether to approve the expenditure of funds for construction, which is where the bulk of the project budget lies. At this gate review an AACE Class 2 project cost estimate is required.
- **Gate 4** is approval to handover the project to Operations and Maintenance staff and approval of project financial close-out. Cost estimates related to the project are not required for this stage. Although, throughout construction (and all prior stages), annual cash flow projections are required for the purpose of calculating the corporation's annual budget and the resulting household impact.

Utilizing such a process adds structure to the cost estimating framework and ensures that a financial basis is provided to advance a capital project to the next stage.

## Risk Reserves

Another key item to consider when developing a project cost estimate is the development of risk-based reserves. While the construction cost estimate provides the "most likely" cost estimate, assessing an appropriate reserve amount discretely considers all risks being tracked by the project team. This point is critical, as carrying allowances for risks within the cost estimate and schedule could lead to double counting of risk when the reserve amount is calculated.

The reserves can either be a function of the construction cost estimate (i.e., a percentage of the construction costs) or determined through a more rigorous risk process involving the analysis and assessment of project risks via a probabilistic quantitative risk assessment (QRA). While a percentage may be useful at the outset of the project when the scope is being confirmed and risks are not as well understood, often a probabilistic QRA is needed at the later design stages, particularly for complex projects.

The outcome of a probabilistic QRA is a distribution of risk loaded costs/schedule completion dates that are associated with different confidence levels (e.g., P50, P75, etc.), as shown in Figure 8. This information can then be used to determine an appropriate amount for reserves based on a confidence level. This desired confidence level is often a function of various factors:

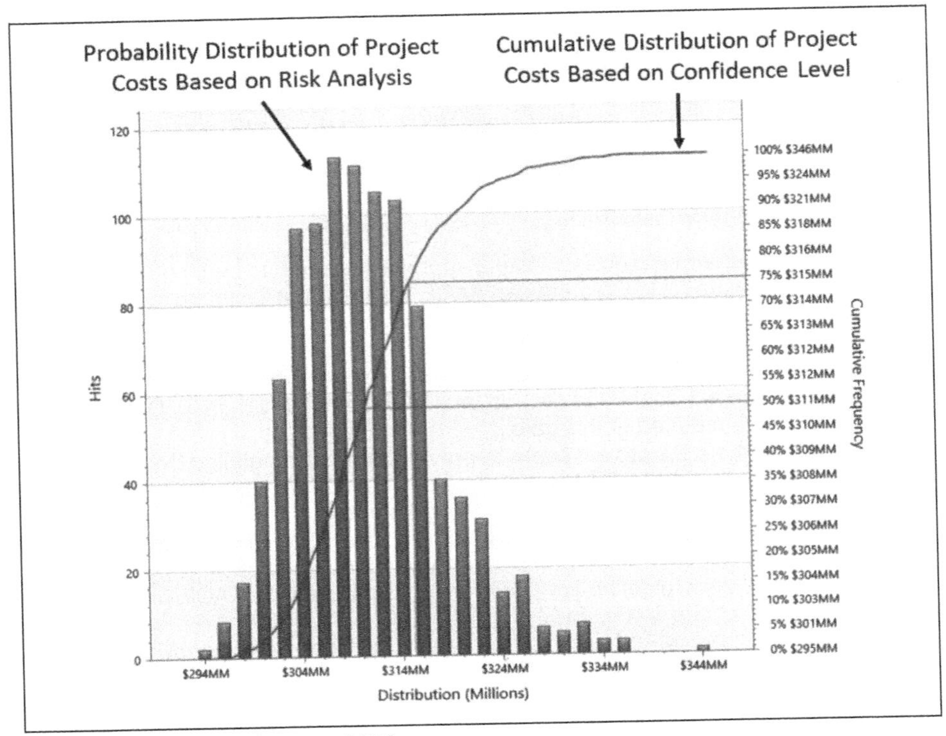


**Figure 8. Example output from a cost QRA**

- **Risk tolerance of the owner:** higher confidence levels are typically desired by public clients and governments as cost certainty is often a key project objective.
- **Stage of the project:** a higher degree of confidence is typically required as the project gets closer to the procurement stage as project budgets need to be more firmly defined at this stage.

Metro Vancouver has been a strong proponent of using QRAs to determine contingency reserves for their major projects. Engineers will work with Metro Vancouver staff to identify and capture risks for the full project lifecycle—from design through procurement, construction, and operations—which are then inputted into a project risk register. With respect to market risks—Metro Vancouver has taken a hands-on approach to reviewing the estimate base case and looking at risk scenarios to address the risk of increased escalation and market factors that could result in an increase in the contractor's bid price. The risk register is then used to prepare a cost and schedule QRA to better understand the range of possible project costs (including construction costs and owner's costs). The selected value is based on a confidence level selected by Metro Vancouver that considers the overall project complexity. This process provides a more defendable project cost estimate when requesting funds from the Board and at each Gate Review stage.

## PROCUREMENT CONSIDERATION—THE ESCALATION/DE-ESCALATION CLAUSE

In addition to the cost estimating approach detailed in the previous section, another tool that has been gaining exposure over the last two years to address rapidly changing market conditions is the "Escalation/De-Escalation Clause." This is a provision in a contract that calls for adjustments in fees, wages, or other payments to account for fluctuations in the costs of raw materials or labor. The escalation clause shifts the burdens of increasing material and labor costs from the contractor to the owner, while a de-escalation clause does the opposite if costs come down. The clause can be triggered in several different ways, ranging from a simple approach (defining that an adjustment will be made if an increase in the cost of materials exceeds a certain percentage threshold) to a more complicated set of formulae based on different indicators such as material pricing or an identified cost index in a specific geographic area.

While escalation clauses can be helpful for owners because contractors will be more comfortable submitting lower bids, meaning more contractors will bid on jobs, they can also present challenges because the prices that are being submitted aren't being guaranteed. As Metro Vancouver is focused on providing cost certainty at the outset of a project, an escalation clause has not been considered at this time, however, this may be re-assessed in the future.

## CONCLUSIONS

Developing an accurate project cost estimate is critical for infrastructure projects, however recent market factors have made this more challenging. Several tools have been identified in this paper to improve the effectiveness of developing a project cost estimate, with specific examples provided from Metro Vancouver, which provides regional-scale services to the 2.6 million residents in the area. These include:

- A structured approach to defining estimate accuracy and including it in the overall project cost estimate.
- Developing a Stage Gate process that allows for a structured review and approval process at various stages over the project lifecycle.
- Utilizing an effective risk management program to develop appropriate contingency reserves that discretely consider risks and uncertainties, thus providing a confidence-based approach to budgeting. This process includes quantifying market risks based on a review of historic trends.

## REFERENCES

Canada ConstructConnect. 2022a. "Construction Material Costs—Spikes Everywhere," dated May 3, 2022. Available from: https://canada.constructconnect.com/canadata/forecaster/economic/2022/05/construction-material-costs-spikes-everywhere [last accessed 13 December 2022].

Canada ConstructConnect. 2022b. "Among Sectors, U.S. Construction Workers Rank 4th for Hourly Wage Hikes," dated December 7, 2022. Available from: https://canada.constructconnect.com/canadata/forecaster/economic/2022/12/among-sectors-u-s-construction-workers-rank-4th-for-hourly-wage-hikes [last accessed 13 December 2022].

Forth, C. & Fromberger, T. 2022. "The Cost of Urbanization—The Effects of Global-Scale Events on the Canadian Tunnelling Market" In: Tunnelling Association of Canada (TAC) 2022 Conference, Vancouver BC, November 2–4, 2022.

# Digital Work Preparation Tool for Underground Construction Tasks—Planning and Optimizing Projects

**Julia Herhold** ▪ Hochschule RheinMain

## ABSTRACT

The demand for sustainable and profitable/efficient projects within the tunneling and other underground engineering sectors grows rapidly. Construction tasks play a key role within the project conduction, whether as primary (tunneling), secondary (e.g., mine securing) or tertiary processes (e.g., workshops). The construction tasks of primary processes create the main structures, such as tunnels, which resemble the main project goal. For the secondary and tertiary processes, the purpose of the task shifts. The main goal is to provide structures, such as the securing of the cavity and workshops, that enable the project and enhance the efficiency of the primary conduction underground.

To plan and optimize the conduction of construction tasks underground, whether they are identified as primary, secondary or tertiary tasks, the work preparation can be used. This method, known from the above ground construction, can be adapted and used underground. The approach of a digital work preparation tool combines, connects and adapts the main tasks "choice of procedure," "logistics" and "site planning." In addition, the digital tool is connected with digital models (construction, site, surroundings). With the adaption and forward projection of the work preparation, underground constructions (e.g., tunneling) can provide a more sustainable and efficient outcome.

## CURRENT SITUATION

Companies in the underground construction, mining and raw material extraction industries are increasingly under the pressure to perform economically to sustain their position on the global market. Profitability, efficiency and productivity are the basic conditions to be an active participant on the global market as well as a valid business partner and contractor for underground/tunneling construction projects or mining operations.

The completion of large-scale as well as common construction projects above ground has been under the pressure to minimize costs and time whilst maximizing the potential outcome, in the meanings of quality and economic gain, for more than a few decades now. Whereas the construction industry has to fight the competition on the market with efficient project outcomes, the tunneling as well as the mining industry not only have to meet challenges of the market, but face the set influences of the underground surroundings to efficiently conduct construction tasks underground. The tunneling industry is under high pressure considering time, money and quality, while the projects are mainly conducted under political influences as well as the public eye. Therefore, the primary tasks, the tunnel constructions processes, need to be efficient and optimized.

On the side of industrial processes underground, mining and raw material extraction operations can be named. The ongoing exploitation of existing raw material deposits

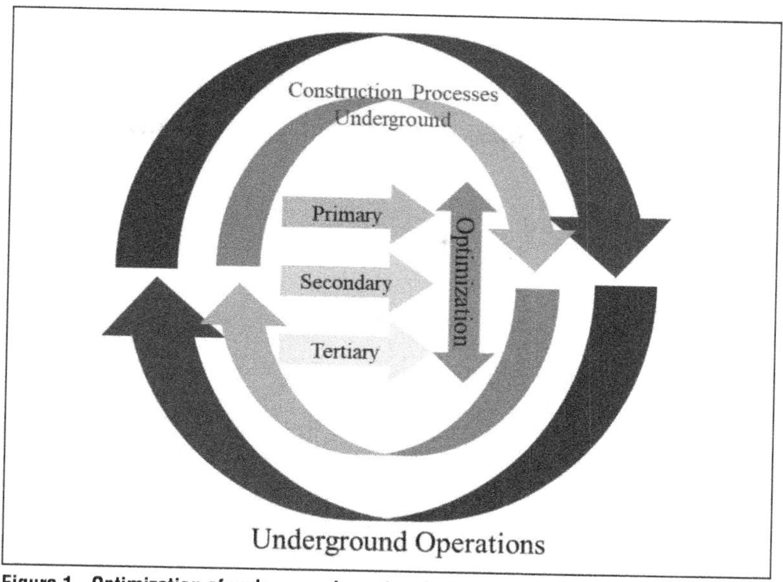


Figure 1. Optimization of underground construction processes

to satisfy the constantly increasing raw material demands world-wide, led to the predicament that economically mineable deposits have decreased dramatically. Even tough mineral deposits do still exist on a global level, near-surface deposits decrease rapidly. Therefore, the extraction of raw materials within mining operations will be relocated further away from the surface in greater depths.

With the growing distance from the surface to the mineral deposit, the extraction costs will increase many times over. With that in mind, the extraction costs will increase whilst the economic efficiency will decrease. The extraction of certain deposits will no longer be profitable with the common approach. To further extract raw materials efficiently even in greater depth, companies have to change their practice in different ways. That leads to the fact, if the deposits are in the depths, where the extraction processes will not be economically, other (support) processes have to be in order to guaranty an economical corporate process itself.

Whether or not the project on hand identifies the ongoing construction processes as primary (e.g., tunneling), secondary (e.g., mine securing and development) or tertiary processes (e.g., workshops in mines) the optimization of these processes is crucial for the efficiency, productivity and profitability of ongoing underground processes as shown in Figure 1.

## APPROACH

The fundamental approach to optimizing underground construction processes lies in the restructuring, digitization and adaptation of the work preparation, which is already familiar from surface construction. Before restructuring and digitization can take place, the interdependencies of the elements must be recognized. By taking a closer look at the individual elements, the dependencies and sequence of planning can be recognized. Construction method selection as well as construction logistics planning represent primary processes that are significantly influenced by project and construction conditions as well as the environment or construction site. The construction site setup

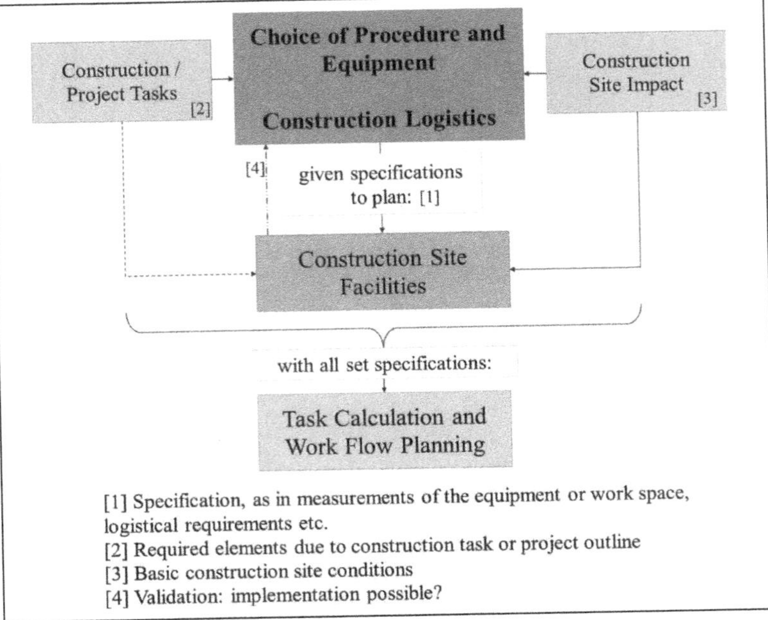


**Figure 2. Interacting elements of the work preparation**

in turn builds on the specifications from the process selection as well as the logistics planning. This is an iterative process. Costing and scheduling are based on the specifications made in the process selection, logistics planning and construction site setup. The dependencies are shown in Figure 2.

With this principle, a concept for a digital tool can be developed that enhances work preparation and focuses on connections and interdependencies. In the following, this concept as well as its elements will be explained with regard to underground construction work.

## UNDERGROUND CONSTRUCTION PROCESSES

### Construction Trades Underground

The conduction of construction work underground takes place in various trades with different construction tasks. The various trades, for which construction work can be conducted underground, are shown in Figure 3.

The underground construction trades include a variety of construction tasks and projects. Civil engineering as well as special civil engineering are the counterpart to construction engineering (above ground). Construction tasks within the trades of civil engineering or special civil engineering are usually conducted underground or subsurface. Within the shaft engineering, construction tasks usually result in underground cavities, that are elongated as well as vertical or angled. They are not only used to overcome the difference in altitude, but also as connecting ways for example. In some cases, they are only built as support constructions during the construction process of the main construction (e.g., tunnel or pump storage plant).

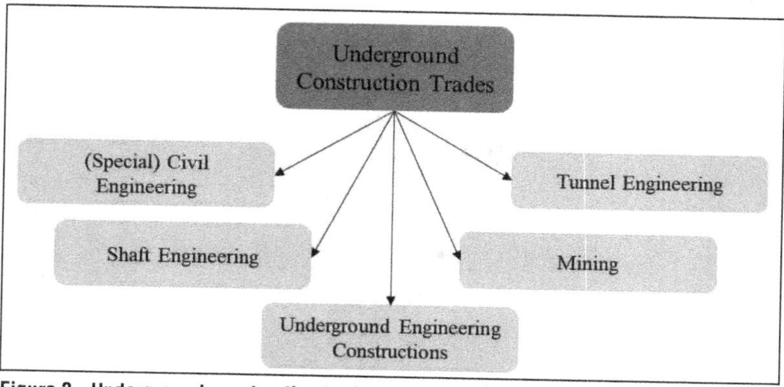


**Figure 3. Underground construction trades**

Whereas tunnels, in comparison to shafts, are elongated, leveled or slightly sloped underground cavities, with two openings (to surface). Tunnels can be excavated with different techniques. It is common and necessary to secure the excavated cavities with support constructions, such as shotcrete. The tunnel constructions can fulfill diverse tasks, such as intake of traffic routes, smaller accessible profiles for water supply and waste water management, access and supply constructions for underground facilities or within mining operations.

The mining industry excavates raw material deposits in the areas of surface mining as well as underground mines. The global demand for raw material shall be satisfied with the extraction of raw materials and minerals of deposits all over the world. For the extraction itself different mining techniques and technology can be used. The conduction of construction work during the mining process can be seen as support or secure processes to ensure the outcome or provide security. [Buja, 2001; Maidl 1994; Weyer 2015]

## Construction Tasks Underground

Various types of construction tasks are required to build a structure. This applies to the execution of construction tasks above ground as well as underground. Regardless of the subdivision into primary, secondary or tertiary processes, there is no differentiation of construction tasks. Identical construction tasks are found in all three processes. Furthermore, it can be stated that these construction tasks are similar to those above ground. It can therefore be stated that underground construction work does not differ from that above ground, but that the environmental or surrounding conditions create the differentiation and challenge. Common construction tasks include reinforced concrete, steel and shotcrete constructions as well as brick work and in rare cases even wood work. Figure 4 shows a selection of constructions and structures in which the construction tasks listed above were applied.

Figure 4 also shows, that the construction tasks can be anything from main processes to supply processes. For example, the reinforce concrete work for the pump storage plant was a primary process. The main project goal was to build the pump storage plant, therefore reinforced concrete tasks needed to be conducted. On the other hand, the concrete work was used as a tertiary process to build a workshop underground. This workshop was not essential for the mining process, but it made the ongoing processes more efficient by providing repair services underground. The steel work shown in Figure 4 can be seen as a typical secondary process. It provides a structure, which

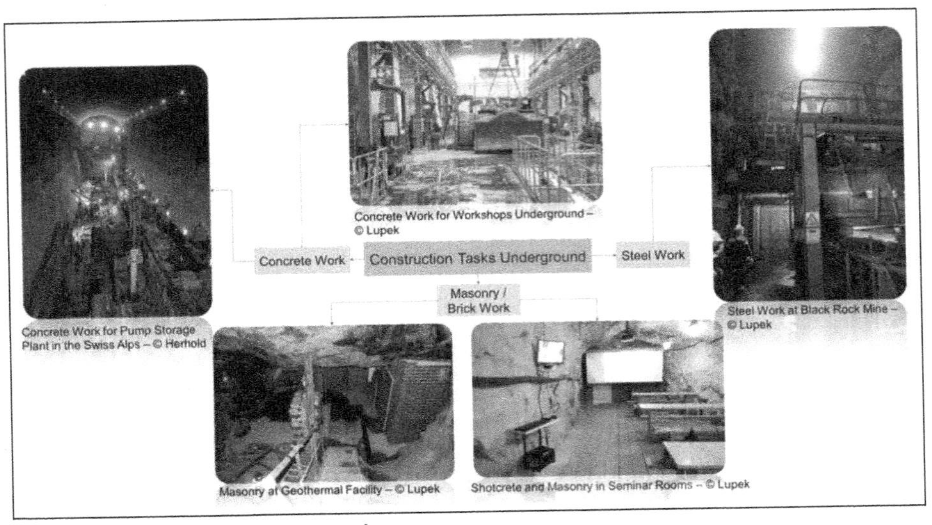


**Figure 4. Construction task underground**

is needed to transport raw material and waste rock above ground, but does not count as a mining process.

## Fundamental Conditions Underground

When construction tasks are conducted in another surrounding than above ground, such as in tunneling or underground mining operations, the conditions of the current environment change significantly. The given influences of the sub-surface areas have an important impact on the actual conduction of the construction work underground. Every action, which has to be taken underground, has to consider the existing influences.

The conditions above ground under which the conduction of civil engineering construction takes place, are mostly well known and considered within the given data. This changes when we shift to the underground areas, where the influences depend highly on the different operating areas and the tasks, which therefore have to be completed. It can be said, that the planning and preparation of construction tasks or their conduction can only be done efficiently, when all influences and conditions of the underground area are considered in the used data as well as the work preparation tool.

Common influences of the underground work space are known to be:

- Restricted space conditions
- Elongated ways or large distances (it is possible, that the distance from shaft to the underground construction site grows during the process of mining or other processes)
- Mediocre visibility due to the lack of natural sunlight (and in some cases dust)
- Mediocre visibility due to narrow and bendy ways
- Boundaries to technical opportunities
- Negative impact on work conditions, leads to decreasing performances
- Critical point within the infrastructure is the transfer from above ground to underground

It has to be considered, that the conduction of construction work underground and the calculation of the construction tasks stand under different influences than above ground. Therefore, the data and the work preparation technique need to be adjusted to the present conditions and environment. To achieve a close to reality calculation and opportunity to control the processes. This paper is planning to find an approach to adjust the recent work planning skills to the conditions underground, so it can be used for the construction tasks conduction in sub-surface area. The focus of this paper is drawn towards the construction logistics, the choice of equipment and procedure as well as the task calculation under adjusted data, including the underground influences on performance and effort. [Buja, 2001; Maidl 1994; Weyer 2015; Girmscheid 2013]

## ELEMENTS OF THE WORK PREPARATION

### Introduction

The planning and preparation of construction tasks is a common tool used for decades in the area of civil and building engineering above ground. With the given data (literature or experience), the future construction occurrences can be predicted closely to reality and then be coordinated during the process to maintain the status-quo. With the clear structures of the processes, the actual conduction as well as the construction process will be efficient.

The work preparation is usually conducted in 5 different planning elements (shown in Figure 5), which are linked to one another. Therefore, every assumption that will be made in one element of the work preparation, will influence other elements. The consequence is that the work preparation can be seen as an interpolating process throughout the project preparation and the construction itself. [Krause 2011; Girmscheid 2015]

The individual elements are explained in more detail below. The focus will be on the choice of construction method and construction logistics.

### Choice of Equipment and Procedure

The choice of the applied and used technology/procedure and equipment can be made in different ways. On one hand the choice can be made from a financial point of view and on the other hand by including additionally further criteria. It is common that for every construction task, different procedures and technology can be used. Therefore,

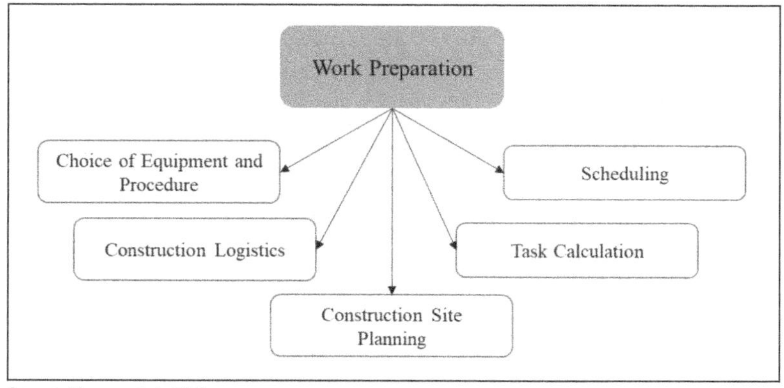


Source: Krause et al. 2011

**Figure 5. Element of the work preparation**

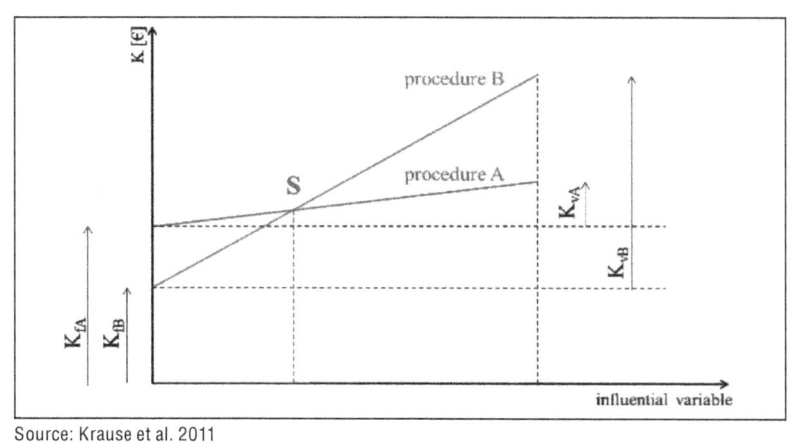


Source: Krause et al. 2011

**Figure 6. Procedure costs**

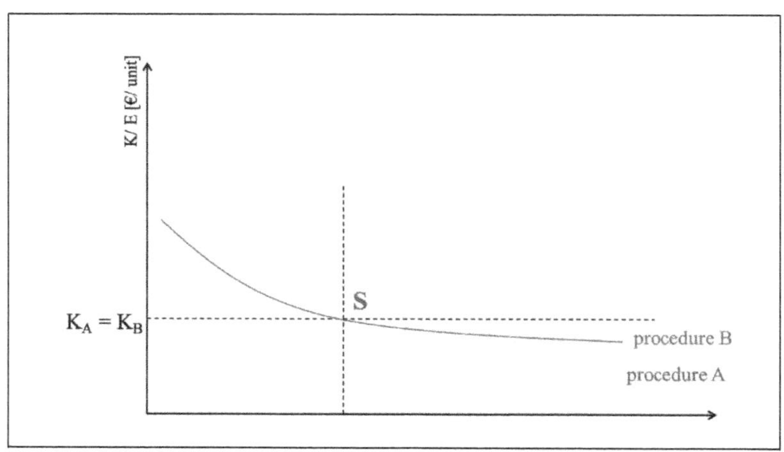


Source: Krause et al. 2011

**Figure 7. Procedure costs per unit**

the most fitted procedure with the equivalent equipment needs to be chosen under the consideration of costs and other criteria (technical, geometrical, naturally, lawfully).

If the selection of the procedure and equipment occurs under financial considerations, the costs of the procedures can be compared as shown in Figures 6 and 7. [Krause 2011]

Furthermore, the costs of each procedure can be connected to units (Figure 7) which have to be conducted, e.g., costs per $m^3$ of concrete.

## Construction Logistics

The construction logistics are divided in several logistics that occur on the field or throughout the construction process. The concatenation of the logistic elements, enable the construction process to function. Every element has its own tasks, which are necessary to operate the construction site. The logistic can be divided in: [Koch 2011]

- Supply logistics → ensure that all commodities are transported and provided to the construction site. The responsibility of the supply logistics ends with the limit of the construction site
- Construction site logistics → means all logistical tasks to move goods and elements on the construction site from one point to another. It includes distribution, provision and storage
- Disposal logistics (in the understanding of return or retrace) → some of the goods, materials, equipment as well as waste have to be disposed or returned away from the construction site. This type of logistics includes all logistical streams away from the construction site. It can be seen as the counterpart to the supply logistics
- Information logistics → is the logistical part, which provides information for every single logistical task or stream; the information logistic can be seen as an informative part of logistical chain

The logistical processes have a clear goal to provide the right goods, at the right time, in the right amount and quality, at the right place to the optimal costs. This can be seen as the logistics policy. Furthermore, a newer system has evolved: Just in time (JIT). The goal of just in time is to provide the goods only when needed and with that, decrease the need for stockings. [Krause et al. 2011; Ruhl, Binder, Motzko 2015; Schach, Schubert 2009; Schach, Schubert 2010]

## Construction Site Facilities

Construction site facilities are the fundamental elements of every construction site. Without these the conduction of the construction tasks would not be possible. The facilities itself and their configuration on site vary depending on each construction site and its circumstances. The construction site facilities can be summarized in elements, shown in Figure 8. [Krause 2011; Schach, Otto 2011]

## Calculation and Work Flow Planning

The task calculation is the result of performance and effort values (under the specific construction site impacts) and the amount of construction tasks that need to be done. The performance and effort values are highly dependable to the influences and circumstance of each construction site and construction task.

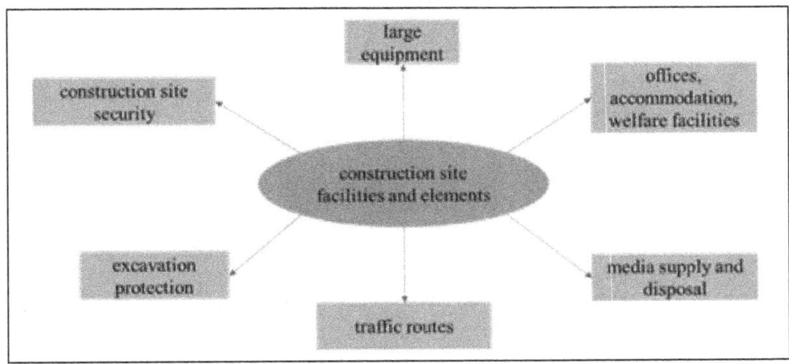


Source: Schach, Otto 2011

**Figure 8. Superordinate elements of construction site facilities**

The work flow planning is the result of the concatenation of the different tasks under the consideration of their duration.

In addition to the work flow planning the cost calculation is another task of the work preparation. It includes a continuous cost planning by creating and updating a working calculation by not only looking backwards on the executed construction processes, but also looking forward and giving a cost prognosis on ongoing and future construction processes. [Krause 2011]

## DIGITAL WORK PREPARATION TOOL

### Connected Elements of the Work Preparation Tool

In the course of updating and digitizing the work preparation, six elements are defined in the concept. These are based on the classic elements of work preparation and are shown in Figure 9.

The project manual contains basic information about the project, such as contractors, contractual regulations, project content, and other specifications and information. The two models are to depict the structure as well as the construction site/environment, picking up necessary project information, displaying it and matching it with basic processes. Within the framework of the database, necessary data are to be collected, processed and made available for the current project as well as future projects. This data includes effort and performance values as well as technical characteristics of equipment and machinery. With the help of the matrix for process and equipment selection, possible processes for the execution of the construction work are to be examined, taking into account all conditions and influences, and the best possible procedure within the set construction work, construction site and present influences can be determined. The logistics network plan shall depict all logistical flows on the construction site as well as to and from it.

The elements are closely interconnected; a constant flow of information between them is necessary for planning and conducting the project. In the following the elements

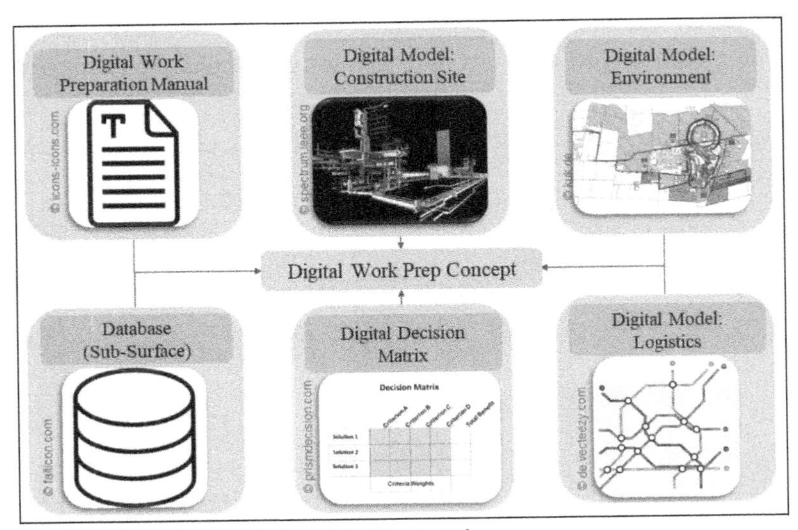


Figure 9. Elements of the digital work prep concept

"digital decision matrix," "logistical net plan" and "construction site facilities" will be explained in more detail.

## Digital Decision Matrix: Choice of Procedure and Technology

The goal of this matrix is to find the most fitted equipment and procedure for the construction task(s) under the given influences of the construction site. The matrix uses the essentials of the usual methods of selection, but focusses especially on the terms and criteria, which have to be fulfilled. Other than the common comparison, this approach grades every single criterion concerning its importance in the first step. In the second step every technique will be judged on how well it fulfills the set criterion.

This allows the approach to narrow down possible techniques and equipment in a first wave with so called exclusion criteria as shown in Figure 10.

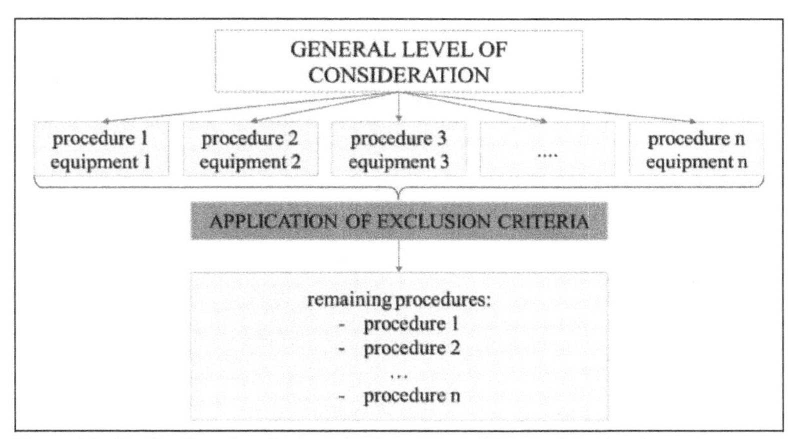


**Figure 10. Application of exclusion criteria on potential procedures/equipment**

Furthermore, in the next level the remaining techniques and equipment can be judged on how well they fit the criteria. In the end of the evaluation process, the best fitted techniques and equipment will be ascertained.

This approach offers the possibility to include every single criterion in the selection process to narrow down the best fitted choice. The criteria will not only be considered, but also rated with the value of its importance. This enables the user to highlight certain criteria as import or less important, and the selection process to bring out the optimal solution.

The approach is shown in Figure 11.

## Logistic Net Plan

To find the perfect logistical concept for the project, the circumstances of the construction site, the surroundings as well as the construction tasks and the equipment need to be considered. In the first step, the site and its surroundings need to be mapped in a scheme. This can be used to draw in the necessary logistical processes, which are necessary to guaranty the construction site operations.

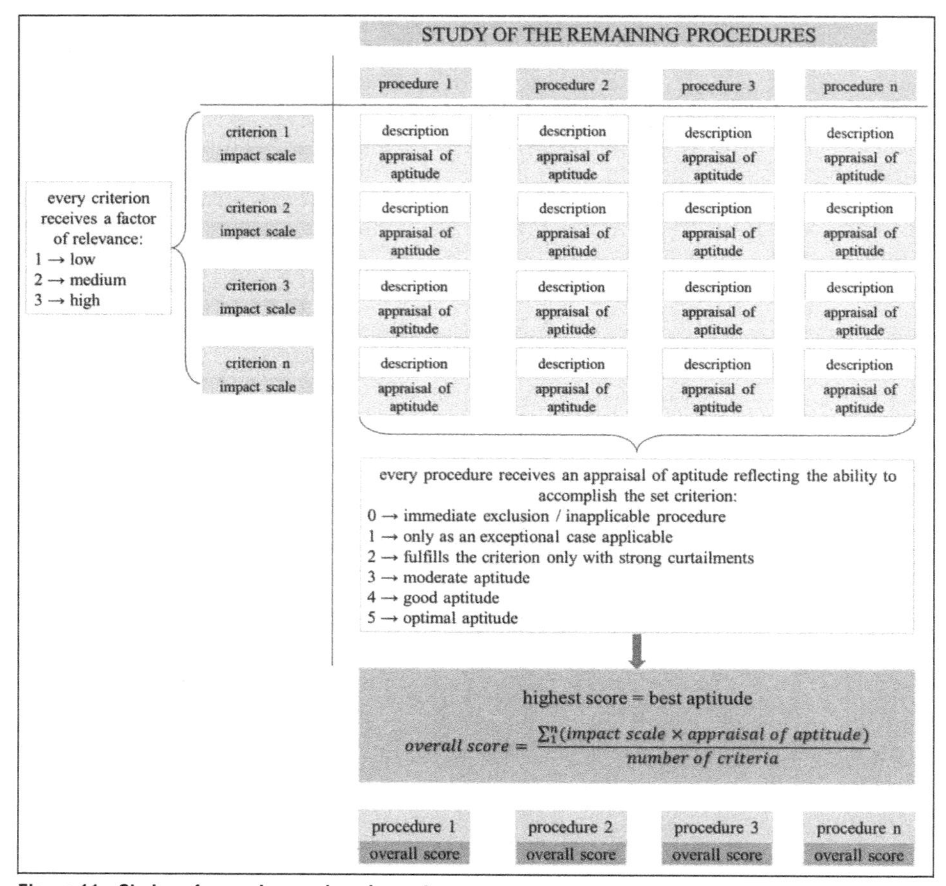


**Figure 11. Choice of procedure and equipment**

The following step includes the dimension of the construction site and its surroundings. It is crucial for the logistical planning to know the distances between the different points of the construction site as well as the dimension of an access tunnel or shaft for example.

Especially the construction and dimension of the connection above ground to underground is crucial for the logistical concept of the site. Every single item has to pass the connecting way; it is the most frequented point of the site.

In the last step of the logistics planning the scheme will be provided by information concerning the properties of the construction site infrastructure and ways. This is necessary for example to decide whether the necessary equipment can be provided underground or not. Or in another case, if the chosen equipment can be supplied sufficiently under the given circumstances of the construction site. Therefore, the information concerning the diameter of the access tunnel, which is also used to secure the supply of the construction site, will be provided with information about his diameter. It then will be cross checked with the necessary dimensions for the equivalent supply or the dimensions of certain large equipment.

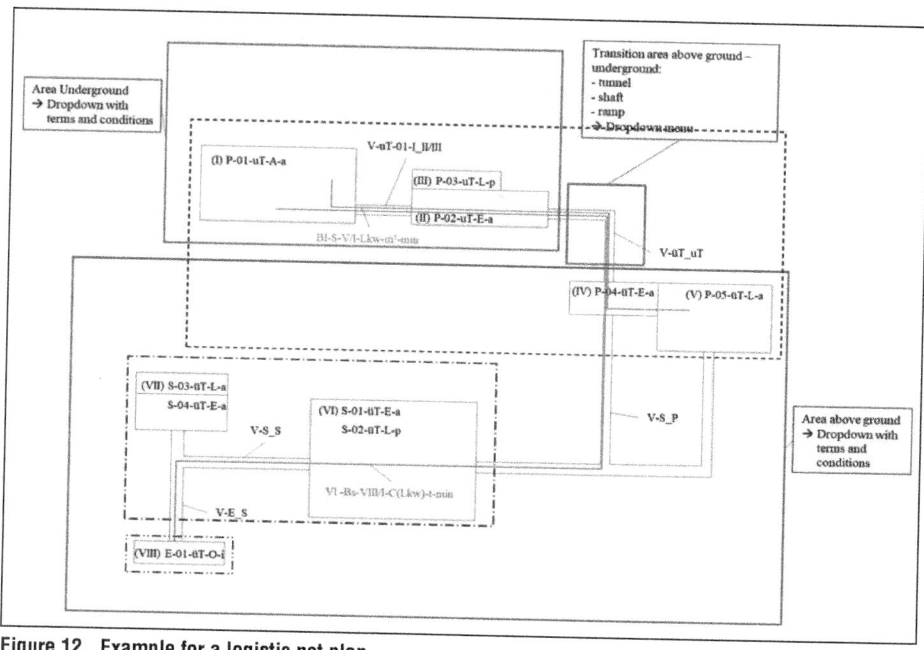


**Figure 12. Example for a logistic net plan**

The logistics can be seen as the instrument, which can ensure the sufficient supply of the construction site and guarantee efficient processes and utilization of the (transportation) equipment.

An example for a logistic net plan is shown in Figure 12.

## Construction Site Facilities

The research on underground construction sites has shown that the most common problem is the limited amount of space. Therefore, the setup of the construction site can have a crucial impact on the productivity. Under the consideration of the choice of equipment and technology as well as the logistical options under the given constraints of the construction site, the segmentation of the construction site can be done.

This paper follows the approach that most frequently needed and highly time-critical elements shall be installed as close to the work spot as possible. Installations, which are important for the work process, but are specious, need to be installed on large areas, closely available to the site. Other facilities, such as offices, accommodations or canteens can be installed with a distance to the construction. These facilities do not have an impact on the actual construction process, the only impact results from the time loss within the shift. The greater the distance between accommodations/ offices and the actual work places, the lower the actual work time on site (shift time – travel time).

The arrangement of the construction site facilities is directly connected to the choice of equipment and technology as well as the logistics. The optimization of the construction and corporate processes is therefore only possible, if every element is considered.

## CONCLUDING REMARKS

### Summary

The work preparation and task calculation are common planning tools used for the building engineering above ground. With the different elements it is possible to plan the task close to reality and organize the construction site under the consideration of the construction site circumstances.

This approach shall be transferred to the application on construction processes in an underground environment. As it has been made clear, the original concept cannot be simply assigned to the sub-surface area. The elements, especially the data as well as the ground concepts for choice of technique and the logistics need to be adjusted. The paper focusses mostly on the logistics and the choice of equipment and procedure. The adaptation of the work preparation tool to an underground area will provide opportunities and challenges. But with the implementation of the work preparation the enhancement of efficiency in corporate processes is highly likely

### Future Research

Until the work preparation tool can be applied to its full extend, several tasks need to be completed. One of them is the revision of the existing data under the consideration of changed circumstances. The sub-surfaces conditions and their impact have to be valued and transferred to the existing data.

Furthermore, the connections between the elements of the work preparation need to be researched. The connections and dependencies of the elements have a significant impact on the planning and possible enhancement of efficiency.

Eventually, the opportunity to transfer the work preparation tool, especially for the underground operations, into a digital system, is given. Building information modeling (BIM) could be one of the solutions. Yet again, the system has to be adapted to the underground circumstances. Another problem might be the digital image of the underground facilities.

## REFERENCES

Buja, H.-O. 2001. Handbuch des Spezialtiefbaus. Geräte und Verfahren. Werner, Düsseldorf.

Girmscheid, G. 2013. Bauprozesse und Bauverfahren des Tunnelbaus. Ernst, Berlin.

Girmscheid, G. 2015. Angebots- und Ausführungsmanagement-prozessorientiert. Erfolgsorientierte Unternehmensführung. Springer Vieweg, Berlin [u.a.].

Hiester, M., Pries, M., and Hanau, A., Osebold, R. 2017. Lean versus BIM. Interaktion von Lean Construction und Building Information Modeling. BauPortal, 01/2017, 66–70.

Koch, S. 2012. Logistik. Eine Einführung in Ökonomie und Nachhaltigkeit. Springer Vieweg, Berlin.

Krause, T., Martin, J., Kuhlmann, W., Pick, J., Olk, U., Schlösser, K. H., Streit, W., and Winkler, N. 2011. Zahlentafeln für den Baubetrieb. Praxis. Vieweg + Teubner, Wiesbaden.

Maidl, B. 1994. Konstruktion und Verfahren. Handbuch des Tunnel- und Stollenbaus/ von Bernhard Maidl; 1. Glückauf, Essen.

Ruhl, F., Binder, Florian, and Motzko, C. 2015. Baulogistik in Planung, Ausschreibung und Bauausführung. BauPortal, 2, 12–16.

Schach, R. and Otto, J. 2011. Baustelleneinrichtung. Grundlagen–Planung–Praxishinweise–Vorschriften und Regeln. Praxis. Vieweg + Teubner, Wiesbaden.

Schach, R. and Schubert, N. 2009. Logistik im Bauwesen. Wissenschaftliche Zeitschrift der Technischen Universität Dresden 58, 1-2, 59–63.

Schach, R. and Schubert, N. 2010. Baulogistik als Wettbewerbsfaktor. Die Effizienz auf Baustellen erhöhen. GS1 network, 3, 7–13.

Weyer, Jürgen. 2015. Einführung in den Bergbau für Nebenhörer. Einführung. Vorlesungsskript, TU Bergakademie Freiberg.

# Flood Resilience for San Francisco

**Eva K. Fernandez** ▪ Delve Underground
**Renée L. Fippin** ▪ Delve Underground
**Paul Y. Louie** ▪ San Francisco Public Utilities Commission
**Derek S. Adams** ▪ San Francisco Public Utilities Commission

## ABSTRACT

The Folsom Area Stormwater Improvement Project is part of SFPUC's flood resilience efforts under the Sewer System Improvement Program. This 4,000-foot-long, 12-foot-inside-diameter tunnel will help reduce the risk of flooding in San Francisco's low-lying Inner Mission neighborhood. The project is in a high seismic hazard zone and extends through a complex geologic corridor that includes full-face soft, compressible Bay Mud, Colma Sands, and full-face rock with several stretches of mixed-face conditions. Additional challenges include removal of 99 piles beneath an existing box sewer, tunneling under commuter rail tracks, and an underground tie-in. This paper examines these challenges and how they are being addressed.

## BACKGROUND

The Inner Mission neighborhood from 18th to 10th Streets is a low-lying area whose combined sewers drain a densely developed area of over 4,000 acres. The drainage area generally extends from near Cesar Chavez Street to the south, the edge of Golden Gate Park to the west, and to Pacific Heights at its northern edge, as shown in Figure 1. The highest reaches of the drainage area, just northwest of Twin Peaks, reach some of the highest elevations in San Francisco. The elevation drops approximately 900 feet before flattening out in the vicinity of 17th and Folsom Streets.

Flooding is not a new issue to the Inner Mission neighborhood. The area coincides with what had until the mid-1800s been Mission Creek, a navigable waterway surrounded by marshland. As the City grew, this naturally low-lying area was filled in and developed. Mission Creek is now covered by development and has been incorporated into the sewer system.

Local flooding was first documented in 1899 (Grunsky, et al.). Old photographs believed to be taken around 1890 to 1910 show flooding after the marsh area was built out with roads and structures in the neighborhood that is now the Inner Mission, shown in Figure 2.

In the last decade, multiple storms have caused flooding to properties. The extent of the damage depends on storm magnitude, but the properties experiencing the flooding are often the same because of local topographic low points.

The Alameda Street Wet Weather Conveyance Tunnel (AWWCT) is one portion of the San Francisco Public Utilities Commission's (SFPUC) mitigation plan to reduce the risk of this flooding.

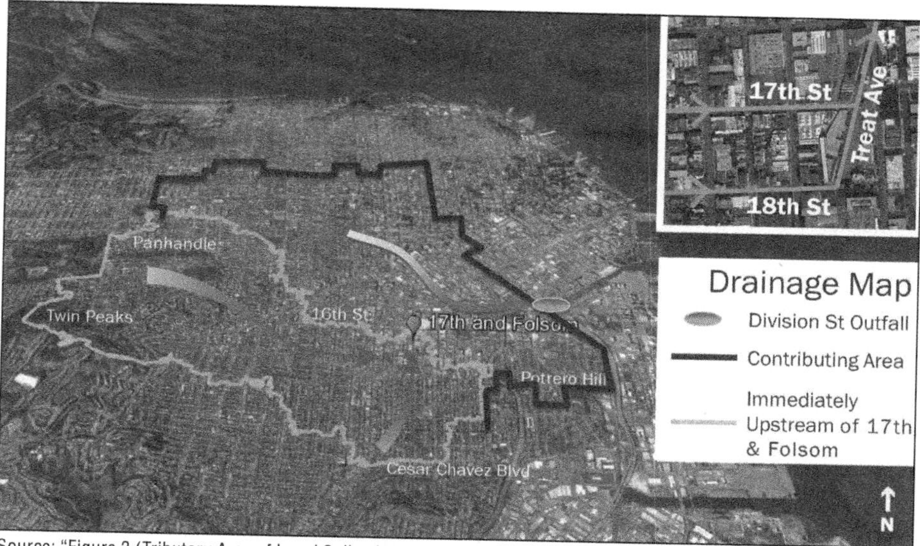


Source: "Figure 3 (Tributary Area of Local Collection System)," in Folsom Area Stormwater Improvement Project, Needs Assessment and Alternatives Analysis Report, 2017

**Figure 1. Project drainage area**

Source: SF Public Library Archives (from Grunsky, et al. 1899).

**Figure 2. Archival photographs (1890–1910) depicting historical flooding in vicinity**

## INTRODUCTION

The AWWCT is an approximately 4,000-foot-long, 12-foot-wide ID conveyance tunnel that will accept and convey stormwater to existing downstream wet weather infrastructure with the purpose of improving reliability of the stormwater system in the flood prone areas described. It is designed for a 100-year service life. See Figure 3 for project location.

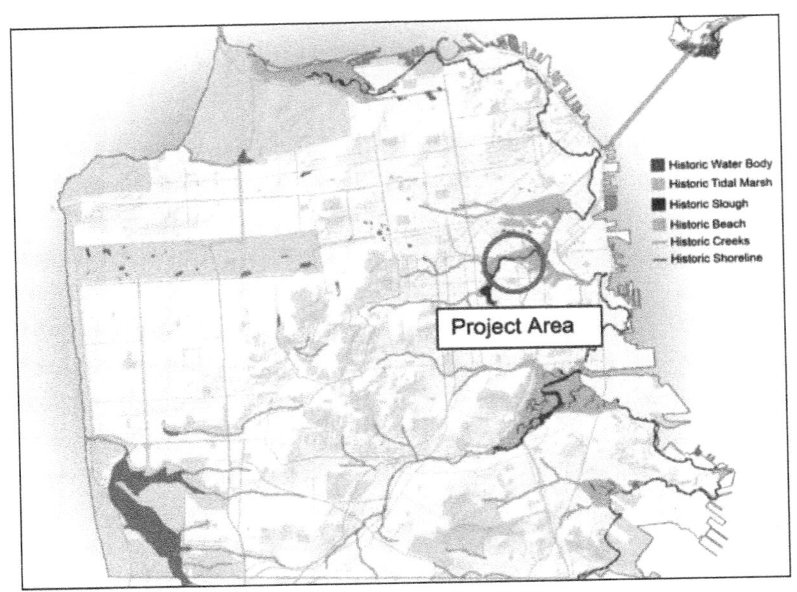


**Figure 3. Project location**

Lower-lying areas in the vicinity of 17th and Folsom Streets experience up to several feet of flooding during certain rain events. The new tunnel and associated infrastructure will manage stormwater to meet the San Francisco Wastewater Enterprise's level of service within the project area. This level of service is to manage flows from a 3-hour storm delivering 1.3 inches of rain.

The major design components of the AWWCT project include:

- Approximately 4,000 feet of tunnel
- Three shafts: Florida Street Shaft, Caltrans Launch Shaft, and De Haro Rotation Shaft
- Underground tie-in to the existing Channel Consolidated Transport/Storage (CCTS) Box
- Connection to a new upstream sewer box with passive flow control in the Florida Street receiving shaft
- Retrofit of four bents of US Highway 101 (US 101) viaducts to remove piles within the tunnel path
- Rebuild of the existing four cell Division Street Sewer top-down and removal of piles
- Portions of segmental lining fully surrounded by soft marine clay

## GEOLOGY

The AWWCT is located in the Coastal Ranges geomorphic province. This province consists of northwest-trending mountain ranges and valleys that are subparallel to the San Andreas Fault (CGS 2002). Near San Francisco, the northern and southern ranges are separated by an incised valley that is present-day San Francisco Bay. Rocks of the Franciscan Complex make up the bedrock and have eroded to form a series of hills and valleys that dominate the topography of San Francisco.

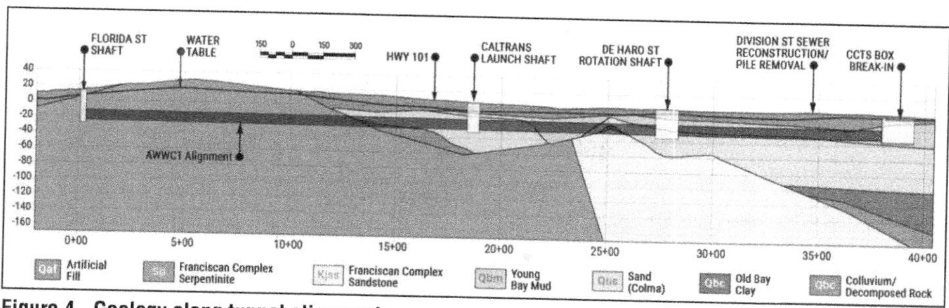


**Figure 4. Geology along tunnel alignment**

The Franciscan Complex includes a variety of sedimentary, igneous, and metamorphic rocks. The tunnel alignment is within the Hunters Point shear zone. The mélange within the shear zone can be subdivided into an upper serpentine matrix mélange and lower sandstone-shale matrix mélange.

## Site Investigations

Site investigations included advancing 13 borings and 6 cone penetration tests (CPTs) along the AWWCT alignment to depths of up to 115 feet. Additional field testing was also completed, which consisted of shear vane, slug testing, and P-S logging. Historical geotechnical investigations adjacent to the alignment as well as data from similar projects were also reviewed and used for characterization purposes as appropriate.

As is typical in San Francisco, the Franciscan Complex undulates across the project alignment with approximately 1,300 feet being entirely in rock and approximately 2,700 feet in sediments, with occasional protrusions of Franciscan reaching the alignment. See Figure 4 for the geology along the project alignment.

Subsurface materials consist of relatively young (Quaternary) alluvial, colluvial, and marine estuarine deposits and bedrock of the Franciscan Complex. Bedrock at tunnel depth consists of moderately strong, hardened blocks or nodules of serpentine within a weaker, sheared serpentine matrix. Blocks of sandstone and shale are also present.

The quaternary sediments consist of:

- Artificial Fill comprising a variety of soil types and debris (e.g., brick, wood, concrete)
- Young Bay Mud comprising a soft, compressible, plastic marine clay
- Colma Formation comprising medium dense to dense coarse-grained soil deposits within localized interbeds of stiff clay and sandy clay
- Old Bay Clay comprising stiff marine clay

Groundwater elevation varies across the alignment and roughly follows the profile of the ground surface. The elevation ranges from El. 0 (Old City Datum) in soft ground to El. 10 to 20 feet in bedrock perched within fractures. Groundwater at the shaft in Florida Street is expected to be 8 to 17 feet below ground surface (El. 0) and El. –5 at the CCTS tie-in. Groundwater is subject to tidal fluctuations because of the proximity of the waterfront.

## Excavation Methods

The selected excavation method for the AWWCT is via a pressurized face tunnel boring machine. The method was selected because of the tunnel length, ground conditions,

Table 1. Geologic units and typical properties

| Geologic Units | Properties | | | |
|---|---|---|---|---|
| | Saturated Unit Weight, pcf | Drained Friction Angle, degrees | Cohesion, psf | UCS Hard Blocks, psi |
| Artificial Fill | 120 | 30 | 0 | — |
| Franciscan Complex—Serpentinite/Sandstone | 145 | 31 | 550 | 4,500 |
| Young Bay Mud | 100 | 28 | 0 | — |
| Colma Sand | 130 | 40 | 0 | — |
| Old Bay Clay | 110 | 30 | 0 | — |
| Colluvium—Decomposed Rock | 115 | 30 | 0 | — |

* UCS = Unconfined Compressive Strength.

ground cover, and environmental considerations, as well as the documented successful use of TBMs in the widely variable San Francisco geology. The inside diameter of the tunnel is 12 feet minimum with a lining thickness of approximately 10 inches.

While the most likely type of TBM will be an earth pressure balance machine given the observed ground conditions, it is likely that the project will allow for contractor selection of exact means. Microtunneling was considered; however, the required internal diameter of the tunnel is a minimum. This diameter is on the upper end of common microtunneling technologies, and the length and size of pipe jacking do not have precedence.

## PROJECT CONSIDERATIONS

### Launch Location and TBM Drives

The most conventional and convenient tunnel drive, especially for a 4,000-foot-long tunnel, would be from a launch shaft to a reception shaft. However, in a dense urban environment like San Francisco, property must be used that is available from amenable landowners and with a reasonable temporary easement agreement. For the AWWCT project, this translates to launching the TBM from a central location, which is between two elevated highways: the northbound and southbound Interstate 80 (I 80) freeway. This would create two tunnel drives. At this time, SFPUC is working with the California Department of Transportation (Caltrans) to define the analysis of the launch shaft and tunnel with I-80 bents and piles. The area surrounding the launch shaft contains parking lots, some of which include highway bents to work around. The total launch shaft staging area is estimated to be just over 2 acres.

The first TBM drive will head westward just over 1,800 feet straight up Alameda Street to Florida Street, where it will be received into a shaft that serves as a connection point for upstream sewer works designed as a separate project. This will be the first drive because the TBM can be retrieved and recovered, whereas complete reception and recovery are not possible for the eastward drive. See Figure 5 for an illustration of the TBM drives.

The westward drive will start in soft soil, Colma Sand, but be primarily within the serpentinite of the Franciscan Complex. The primary obstacle with the westward drive is the Caltrans foundation piles described below.

The eastward drive is about 1,900 feet long heading down Alameda Street and onto Berry Street. The eastward drive is primarily within soft soils—Colma Sand and then

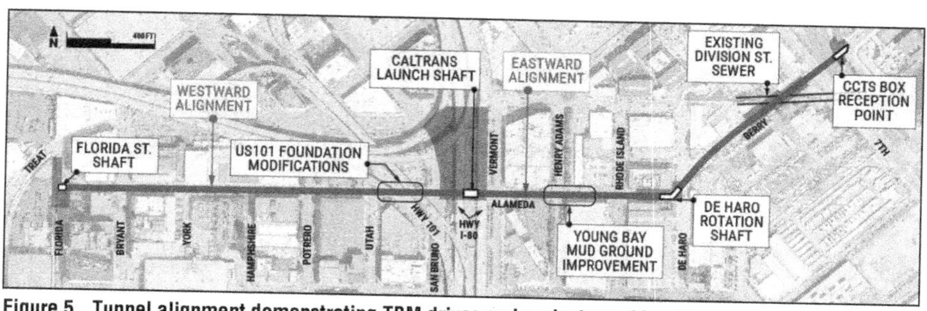


Figure 5. Tunnel alignment demonstrating TBM drives and project considerations

into Young Bay Mud. The primary obstacles with the eastward drive are Young Bay Mud below the tunnel invert, a tight turn onto Berry Street, a 40-foot-wide box sewer on piles, a three-track commuter rail crossing, and an underground reception.

## US 101 Foundation Modifications

The AWWCT alignment passes beneath four bents of the Central Freeway Viaduct, which connects US 101 and I-80. The Central Freeway Viaduct curves over the alignment on elevated overpasses supported on pile-founded bents. In the 1990s, the piled foundations were modified as part of a seismic retrofit program throughout the State of California following the 1989 Loma Prieta earthquake. This retrofit placed several piles within the Alameda Street boundaries. Of these, four piles, one at each of the four bents, conflict with the tunnel alignment.

Upon discovery of the conflict, SFPUC worked with Caltrans to determine the best path forward. This included several alternative reconfigurations of the alignment between the piles including splitting of flows. This proved difficult since the San Francisco collection system operates by gravity and the elevated viaducts have numerous bents crossing the alignment at an angle. In addition, a new alignment along an alternate street was considered but opened the project to additional unknowns. With Caltrans' concurrence, it was decided to keep the AWWCT alignment as originally proposed and to perform a highway foundation retrofit to remove the conflicting piles and replace them with new piles compensating for any capacity loss.

The retrofit calls for the removal of four piles and the addition of 10 new 24-inch-diameter cast-in-drilled-hole piles socketed 5 feet into rock with thickened pile caps. Construction would be staged to ensure stability of the highways and will require low overhead work to accomplish both pile installation and pile removal.

At this time, the foundation retrofit will be part of the overall AWWCT contract. Figure 6 shows the reconfiguration of the foundation under US 101.

## Young Bay Mud Below Tunnel Invert

Between Vermont Street and Rhode Island Street, the TBM will pass through a 220-foot-long deep valley of Young Bay Mud. Young Bay Mud is a soft highly compressible marine clay and silt that was deposited in the San Francisco Bay by rising sea levels through the Holocene epoch. It includes silty clay, lean clay, sandy lean clay, elastic silt, silt, and clayey sand. Sampling during field logging of the adjacent borehole indicates an SPT "N" value of less than one and weight of hammer. Test results indicate an undrained shear strength of approximately 500 psf, a liquid limit ranging between 120 and 33, and a plasticity index between 73 and 3.

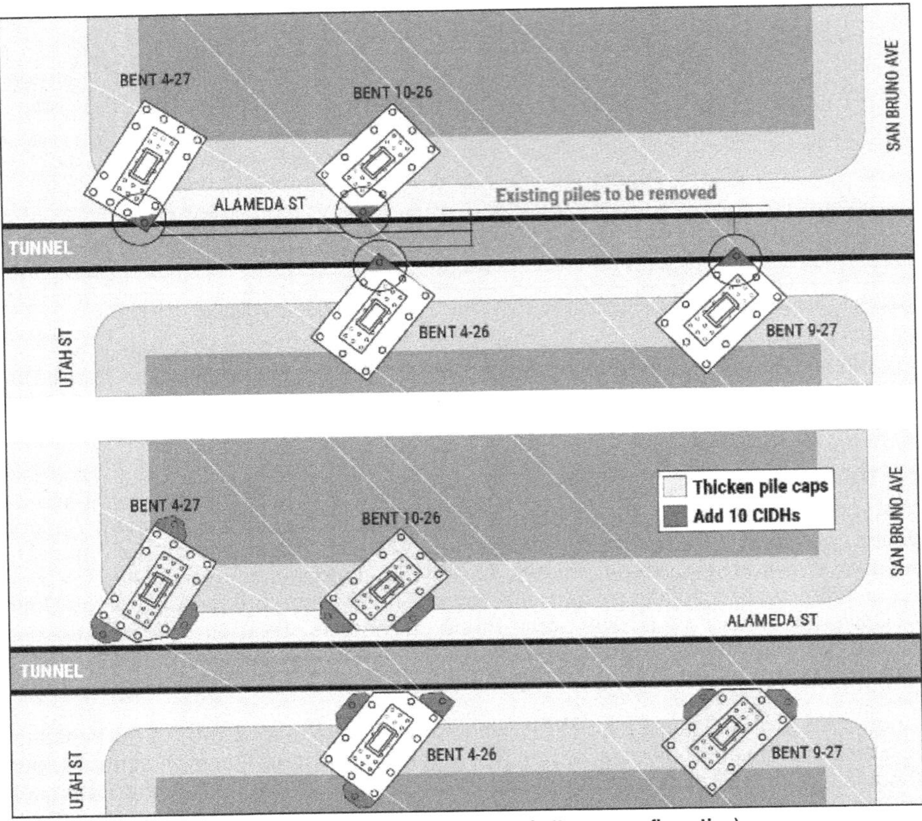


Figure 6. US 101 foundation reconfiguration (top: Existing; bottom: reconfiguration)

Of concern is the presence of the Young Bay Mud below the tunnel invert. To prevent the TBM from losing vertical alignment (e.g., diving downward by unacceptable amounts) in the weak Young Bay Mud as well as long-term settlement of the sewer, ground improvement will be incorporated along this stretch. The most appropriate ground improvement method will be left to the contractor to select; however, jet grouting of Young Bay Mud has been successfully performed for the Sunnydale Auxiliary Sewer project in 2011 and for Islais Creek Contract E in 1999. Both projects required tunneling through YBM and strengthening of the strata.

A separate 250-foot stretch of thin, 1- to 2-foot-thick Young Bay Mud is projected below the invert where the TBM will pass below the three-track commuter rail and into the final CCTS reception box. Treatment from the ground surface is viewed as not feasible because of the railroad service requirements. For this area, a three-dimensional finite difference model was developed in FLAC3D (Itasca 2019) to assess the potential for settlement. The model indicated a settlement less than 0.2 inch, which is viewed as tolerable.

## TBM Rotation

Where the tunnel alignment reaches the eastern end of Alameda Street at De Haro Street, the alignment continues on a diagonal eastward up Berry Street to its final terminus. In San Francisco, the cost of a subsurface easement and the time it takes to negotiate such an easement drive much of the decision to stay within the City right-of-way in almost all situations. At this corner, staying within the right-of-way is not possible without a 300-foot radius turn. Given the limited experience with a small radius in the United States, instead it was elected to incorporate an intermediate shaft at this intersection to allow for rotation of the TBM and relaunch along Berry Street. This option allows for a more reasonable curve for the TBM to navigate. This intermediate shaft will be decked the majority of the time and only serve for rotation purposes. Spoils handling and segment transportation will continue to be managed from the Caltrans Launch Shaft.

## Division Street Box Sewer

The four-compartment Division Street box sewer is an approximately 2,700-foot-long major component of the SFPW wastewater infrastructure, delivering a peak 21 million gallons per day (MGD) of dry weather flows and 1,560 MGD of wet weather flows. This four-cell box sewer crosses above the AWWCT alignment along Berry Street. Each of the four cells is approximately 10.0 feet wide by 11.5 feet tall and interconnected, sharing a side wall with the adjacent cell. Three original cells were constructed in 1909 as a single concrete structure, and a fourth cell was added in 1968. The original three cells are supported on wooden piles, and the newest cell is supported on 10.75-inch steel pipe piles filled with concrete. Pile spacing as indicated on historical design drawings is between 3 and 6 feet where the Division Street sewer crosses the alignment in Berry Street.

The tunnel alignment crosses the Division Street box sewer at an approximately 45-degree angle, making the crossing approximately 80 feet long (see Figure 7). At this location, the bottom of the concrete boxes is about 5 feet above the crown of the sewer and over 100 piles interfere with the tunnel alignment. The soils at this location are Fill and Young Bay Mud. Several concepts to address the pile/tunnel conflict were considered. These included underpinning of the existing box and hand-mining below

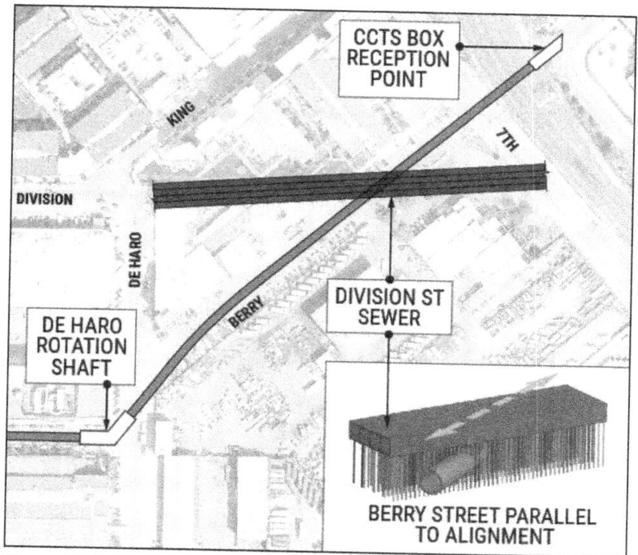


Figure 7. Division Street sewer crossing

the box to remove piles. This method was complicated by the fact that the pile rows are diagonal to the alignment. The spacing of piles diagonal to the alignment is 3 feet and the rows are offset, making it difficult to maintain support of the structure during pile removal

Ultimately, a top-down reconstruction of the box has been selected with a new foundation system being reestablished outside of the tunnel limits. The top-down reconstruction will need to take place during the dry weather season (April to October), when flows are manageable by bypass pumping.

## Segmental Lining Within Young Bay Mud

The AWWCT is the first segmentally lined tunnel known to be constructed completely within You1ng Bay Mud (i.e., surrounding all sides). Approximately 300 feet of the tunnel will be in these geotechnical conditions with another 700-foot stretch having some Colma Formation in the lower 1 to 2 feet. Site response analyses were performed using DEEPSOIL Version 7.0.30 (Hashash et al. 2020). The site response analyses for these critical locations demonstrated shear strains within the soft marine clay to be around 1%, which is significantly higher than typical segmentally lined tunnels. These site response analyses were used as inputs to the soil structure interaction (SSI) and racking analyses for the segmental lining design.

As a pseudo-static analysis, all soil units and the tunnel are simulated as elastic materials based on the linear elastic constitutive model in FLAC3D (Itasca 2019). Strain compatible modulus is estimated for each soil layer based on corresponding free field shear strain provided through the site response analyses, and the soil properties were based on project-specific data.

Within the model, the segmental lining is represented by elastic structural liner elements with an interface between liner elements and the surrounding soil. The length of each segmental lining ring is assumed to be 4 feet in longitudinal direction. Circumferential joints between two adjacent segmental lining rings are simulated using compression only link elements to prevent penetration of segments into each other. To account for the effect of radial or longitudinal joints on the stiffness of segments, the equivalent stiffness has been determined based on the method proposed by Muir Wood (1975).

The model configuration at the location of maximum shear strain of 1.0% is shown in Figure 8. The FLAC thrusts and moments induced in the segmental lining were used in the structural design of the segmental lining, resulting in a 10-inch, 7,000 psi reinforced concrete section, and the evaluation confirmed the joints do not open during shaking.

Soil structure interaction modeling was also used to evaluate the impact of differential settlement in the sands below the tunnel, transitions from soil to rock, and at the tunnel's connection to existing and future structures confirming detailing of the flexible connections.

## CCTS Break-in

The AWWCT alignment terminates at an existing belowground structure—the Channel Consolidated Transport/Storage (CCTS) Box. This terminus occurs within the City right-of-way, which is overlapped by a Caltrain franchise agreement area containing four commuter rail tracks. The CCTS Box has an existing chamber that was constructed for the purpose of accommodating future connections to the box. There is a nearly full height wall separating the chamber from the active wastewater system.

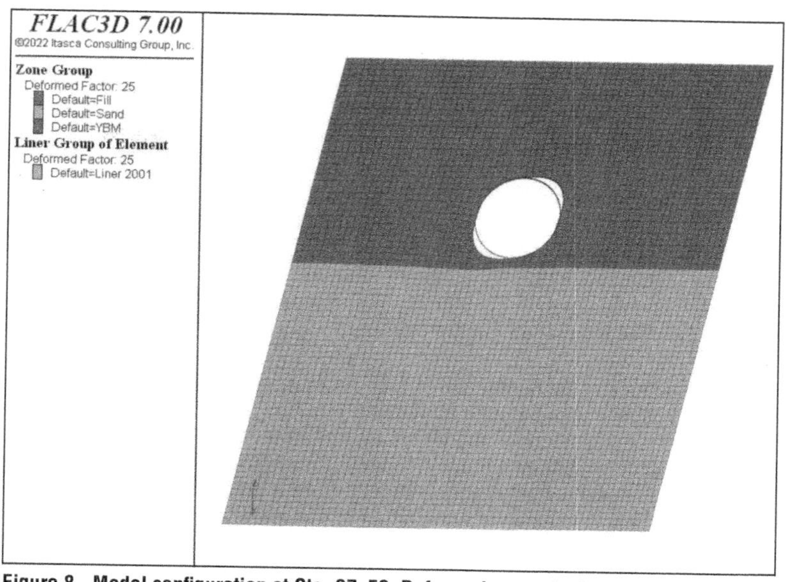


**Figure 8. Model configuration at Sta. 37+50: Deformed geometry is exaggerated (mag factor = 25)**

The chamber is only a few feet wider than the TBM but long enough to allow for complete entry of the machine. The chamber is currently strutted crosswise at an elevation interfering with TBM reception.

To allow for TBM retrieval, the chamber will be backfilled, struts removed, and configured for TBM reception. At this time, it is assumed that the TBM shield will be left in place and the cutterhead retrieved from a small opening to be created in the box roof. The contractor will be allowed to propose an alternate to the retrieval scheme within the confines of the contract documents and permitting.

## SUMMARY

The 17th and Folsom area is historically prone to flooding. Several hydrologic and hydraulic factors contribute to flood challenges in the area:

- **Local topography:** 17th and Folsom and vicinity form a naturally low-lying area.
- **Rainfall/runoff from upstream areas of the drainage basin:** The large drainage basin combined with the steep terrain can lead to significant flow in the collection system in a short period (within 15 minutes) during large storms, including some short storms with high rainfall intensity.
- **Conveyance capacity:** System flows surpass the combined carrying capacity of the underground sewer and overland streets.
- **Land settlement:** Because the area is built on a historical creek and landfill, settlement and subsidence of land in this area has potentially resulted in lower property elevations.

A key component in providing flood resiliency for this impacted neighborhood is the 4,000-foot-long Alameda Street Wet Weather Conveyance Tunnel. While the tunnel length is short, the project faces a number of complexities within this short stretch. This includes over 100 piles within its path supporting critical existing infrastructure,

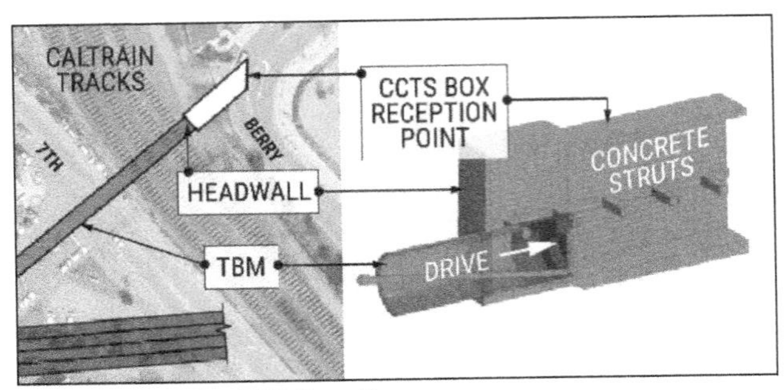


**Figure 9. CCTS Box TBM Reception**

rock and soil ground conditions, soft clay tunneling, a rail track crossing, and an underground reception. The removal of piles—for US 101 and for the Division Street Sewer—will require significant infrastructure improvements prior to the launch of tunneling activities. Soft clay tunneling will require ground modification in some stretches. The underground reception will require thoughtful reconfiguration of the reception box to ensure stability as well as to avoid impact to the adjacent commuter rail tracks.

## ACKNOWLEDGMENTS

The authors would like to acknowledge our joint venture partner Stantec Consulting, Inc., which performed the field investigation and advanced hydraulics and is leading the SSI and foundation modification work for the Caltrans area. We also would like to acknowledge San Francisco Public Works, which is designing the new conveyance infrastructure that will tie into the new AWWCT and lead the City-wide hydraulics.

## REFERENCES

California Geological Survey (CGS). 2002. California Geomorphic Provinces, Note 36.

Grunsky, C.E., Manson, Marsden, and Tilton, C.S. 1899. *Report Upon a System of Sewerage for the City and County of San Francisco*. San Francisco, CA: Britton & Rey.

Hashash, Y.M.A., Musgrove, M.I., Harmon, J.A., Ilhan, O., Xing, G., Numanoglu, O., Groholski, D.R., Phillips, C.A., and Park, D. 2020. *DEEPSOIL 7, User Manual*. Urbana, IL: University of Illinois at Urbana-Champaign.

Itasca Consulting Group. 2019. Fast Lagrangian Analysis of Continua in 3 Dimensions (FLAC3D), Version 7.00. Minneapolis, MN: Itasca Consulting Group.

Muir Wood, A.M. 1975. The circular tunnel in elastic ground. *Geotechnique* 25(1):115–127.

# Planning for Sound Transit's Proposed LINK Light Rail Expansion in Seattle Washington

**Matthew Preedy** ▪ Sound Transit
**Dirk Bakker** ▪ Sound Transit
**Anthony Pooley** ▪ Sound Transit
**Mike Wongkaew** ▪ HNTB
**Raghu Bhargava** ▪ HNTB

## ABSTRACT

Sound Transit's ST3 program will add 62 miles of light rail to the Puget Sound area, including connections to the neighborhoods of West Seattle and Ballard. These connections include considerations for both elevated and tunnel alternatives in West Seattle and Ballard, as well as a new tunnel and six underground stations in the densely developed downtown Seattle. This paper describes the conditions and constraints along the corridor, and the latest development in the ongoing process to plan the underground guideways and stations.

## INTRODUCTION

Sound Transit's West Seattle and Ballard Link Extensions (WSBLE) Project is part of the 62-mile ST3 regional transit system expansion approved by the voters in November 2016. As shown in Figure 1, the WSBLE project consists of two separate Link light rail extensions: one from West Seattle to south of downtown (SODO) and the other from SODO, through the downtown core, to Ballard.

The West Seattle Link Extension will add 3.8 miles of light rail service from SODO extending southwest to West Seattle's Alaska Junction neighborhood with four new stations. The West Seattle Link Extension will eventually connect to the Everett Link Extension.

The Ballard Link Extension will add 7.1 miles of light rail service from SODO extending north and west through downtown Seattle to Ballard. The Ballard Link Extension will include nine stations and a new downtown Seattle light rail tunnel from SODO to South Lake Union and Seattle Center/Uptown. The Ballard Link Extension will eventually connect to the Tacoma Dome Link Extension.

Sound Transit's project development process includes three phases prior to the final design. These phases ensure readiness for engineering and compliance with the requirements of Federal Transit Administration (FTA), National Environmental Protection Act (NEPA) and State of Washington Environmental Protection Act (SEPA):

- Phase 1 (2017 through early 2019) included alternatives development, environmental scoping, and identification of preferred alternatives and other alternatives for detailed study.
- Phase 2 (mid 2019 through early 2023) includes the preparation of the conceptual design of the alternatives and a Draft Environmental Impact Statement (DEIS) for public comment. Upon reviewing the comments from the public, business and community organizations, tribal organizations,

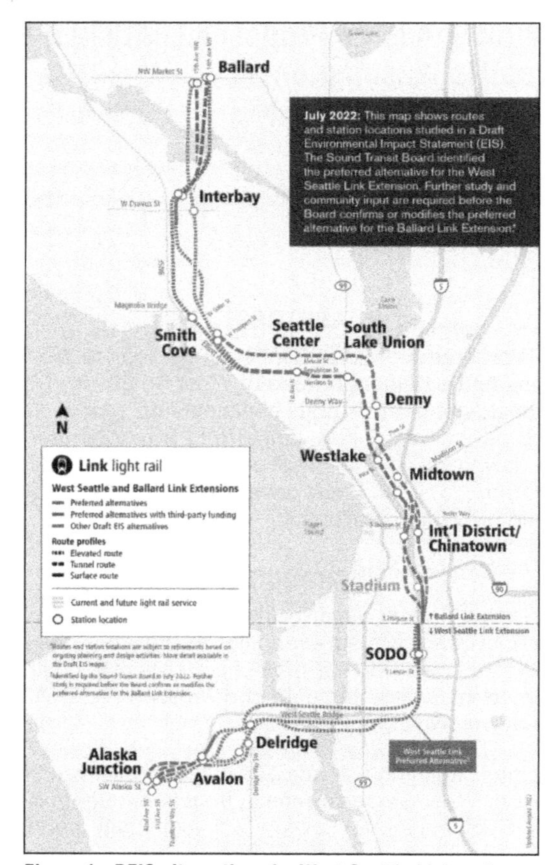


**Figure 1. DEIS alternatives for West Seattle Link and Ballard Link Extensions (Sound Transit, 2022)**

and agency partners, the Sound Transit Board will confirm or modify the preferred alternative.

- Phase 3 (starting in early 2023) will include the preparation of the Final Environmental Impact Statement (FEIS) and preliminary engineering (PE) of the preferred alternative.

*Preedy* et al. (2021) summarized the work performed in Phase 1 and the conceptual engineering work performed in Phase 2 of the underground section in downtown Seattle. This paper summarizes the subsequent project development through December 2022 and expands the discussion to include the underground alternatives in West Seattle and Ballard.

## RECENT PROJECT DEVELOPMENTS

Conceptual engineering of the alternatives was completed in early 2021. Subsequently, comparative cost estimates and evaluation of the potential impacts and benefits of the alternatives on the natural and built environment were prepared to inform the evaluation of the alternatives. Targeted refinements of the conceptual design were also made in 2021 to explore opportunities to reduce risks, impacts and right of way acquisition.

Sound Transit and the Federal Transit Administration published the DEIS for the West Seattle and Ballard Link Extensions for public comment in January 2022. Sound Transit received nearly 5,200 individual communications, each containing one or more comments, during the 90-day public comment period. After reviewing the DEIS and the comments received, the Sound Transit Board identified the preferred alternative for the West Seattle Link Extension in July 2022. The Board requested further studies and community engagement in ten areas from Chinatown-International District through Downtown, Interbay and Ballard before confirming or modifying the preferred alternative for the Ballard Link Extension. Studies are to be finalized in the first quarter of 2023.

The preferred alternative for the West Seattle Link Extension includes a cut-and-cover Alaska Junction Station, a retained-cut Avalon Station, and 0.35-miles of twin guideway tunnels between the two below grade stations. North of West Seattle, the alignment is mostly elevated in Delridge and Duwamish segments and at grade in SODO segment where the new line will connect with the existing Central Link. Further description of the underground work in West Seattle is provided in the next section.

The Sound Transit Board has established an overall project budget of $14.1 billion (in 2022$) and the delivery timelines for the two extensions as shown in Figure 2. With the preferred alternative identified for the West Seattle Link Extension, and the identification of a preferred alternative for the Ballard Link Extension anticipated by the end of Q1 2023, the preparation of the Final Environmental Impact Statement and preliminary engineering will begin in 2023 with the final design expected to be completed in 2026. Construction for the West Seattle Extension is expected to begin in 2026 and be completed in 2032.

As part of project development, Sound Transit has advanced the linear schedule and the preliminary contract packaging plan for the project in the third quarter of 2022, to reflect identification of the preferred alternative for West Seattle Link Extension.

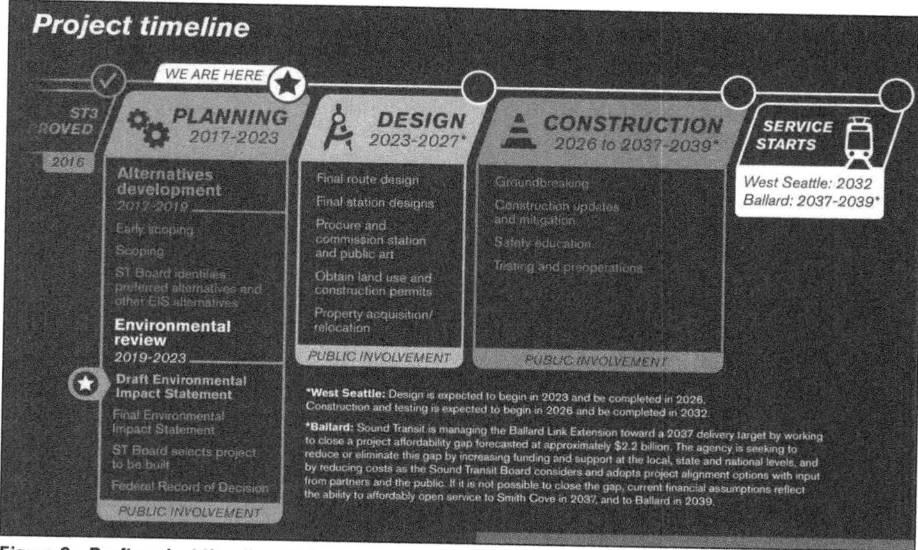


**Figure 2. Draft project timeline (July 2022)**

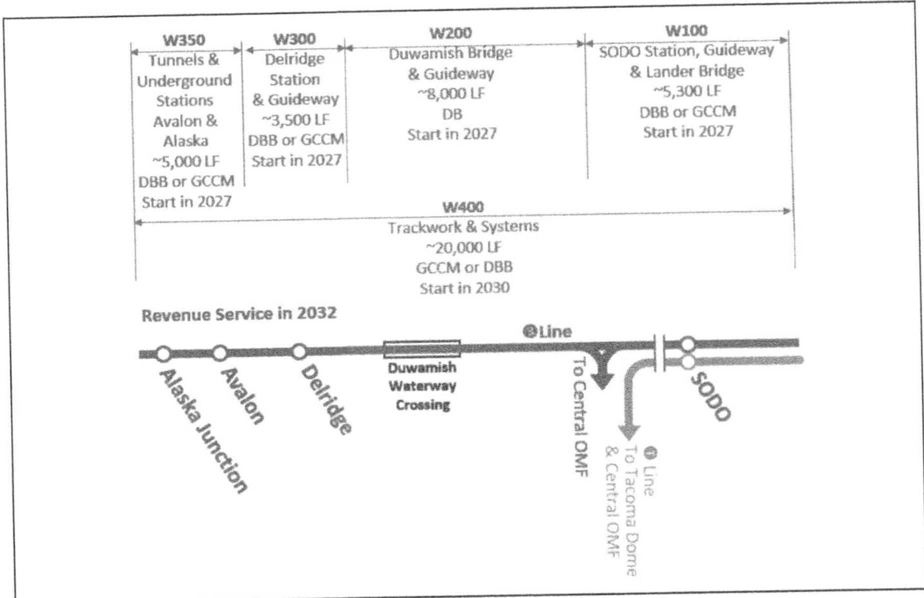


Figure 3. Preliminary contract packaging breakdown for West Seattle Link Extension (July 2022)

An extensive industry outreach program was initiated, to obtain feedback on the preliminary procurement and contract packaging approach for construction and professional services. Project representatives attended several conferences and spoke at Sound Transit's annual Contracting Expo event. Two WSBLE "Open Houses" were held, aimed at construction contractors and professional service providers, to provide more detail on the project and the preliminary procurement and packaging plan. As a follow-up to the Open Houses an extensive series of one-to-one meetings were held with a wide range of interested consultants and contractors to obtain more specific feedback on the preliminary packaging of the work and the proposed delivery methods.

Figure 3 shows the preliminary contract packaging approach for the West Seattle Link Extension, which remains under review and subject to further adjustments. In the figure, the abbreviations DBB, DB, and GCCM refer to design-bid-build, design-build, and general contractor-construction manager delivery methods, respectively.

## TUNNEL SECTIONS IN THE PROJECT

The underground sections for both extensions of the WSBLE project are based on center platform stations and twin guideway tunnels between the stations. Figure 4 shows the general arrangement cross sections of the guideway tunnels. It is anticipated that the twin tunnels will be constructed using tunnel boring machines (TBM) although the sequential excavation method (SEM) is considered for the short tunnel in West Seattle. Cut-and-cover, SEM and retained-cut methods have been considered for station construction. The selected construction method and configuration for each station depend on the site-specific conditions, such as the relationship between the station platform and entrance(s), the platform depth, surface and subsurface obstructions, and the ground condition.

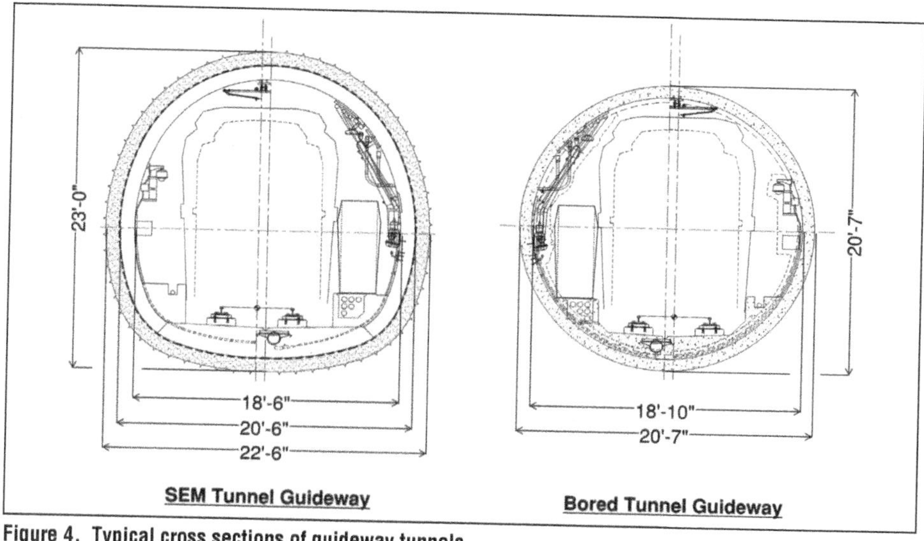


Figure 4. Typical cross sections of guideway tunnels

## West Seattle Tunnel Configuration

The preferred alignment for the West Seattle Link Extension was identified in July 2022 by the Sound Transit Board. As illustrated with a pink line in Figure 1, the alignment begins at South Holgate Street in the SODO area and ends south of Alaska Junction at 41st Avenue Southwest and Southwest Hudson Street intersection. At South Holgate Street, the alignment is mostly at grade or in shallow retained cuts for about 900 feet. It then transitions to the elevated (bridge) section for about 1.5 miles, running nearly parallel to and south of the existing West Seattle Bridge, and continues elevated on long-span bridge structures overcrossing the east and west Duwamish Waterways before entering the Delridge neighborhood in West Seattle. After the elevated station at Delridge, the guideway alignment transitions to a short section of at-grade guideway then into a retained cut structure at Avalon Station. At this station, lids will be constructed at 35th Avenue Southwest and Fauntleroy Way Southwest to maintain the surface traffic over the retained cut. The alignment continues underground with twin tunnels and terminates at the Alaska Junction station, which will be cut-and-cover straddling the intersection of 41st Avenue Southwest and Southwest Alaska Street. Figure 5 provides the plan and profile views of the underground section of the West Seattle Link Extension.

The length of the guideway in twin tunnels between the two stations is close to 1,800 feet. Although the tunnels can be constructed by either the sequential excavation (SEM) or a tunnel boring machine (TBM), the shorter tunnel length lends itself to the SEM method with less onerous equipment and logistic needs. The main portal and construction staging area for the tunnels will be located adjacent to the Avalon Station. A site of approximately 3.25 acres is identified as part of the DEIS footprint to support the construction of tunnels, station and retained cuts.

For the emergency egress from the twin tunnels, the Sound Transit and National Fire Protection Association (NFPA) standard 130 require either cross passages at a maximum spacing of 800 feet along the tunnels or a fully enclosed, fire-rated point of safety

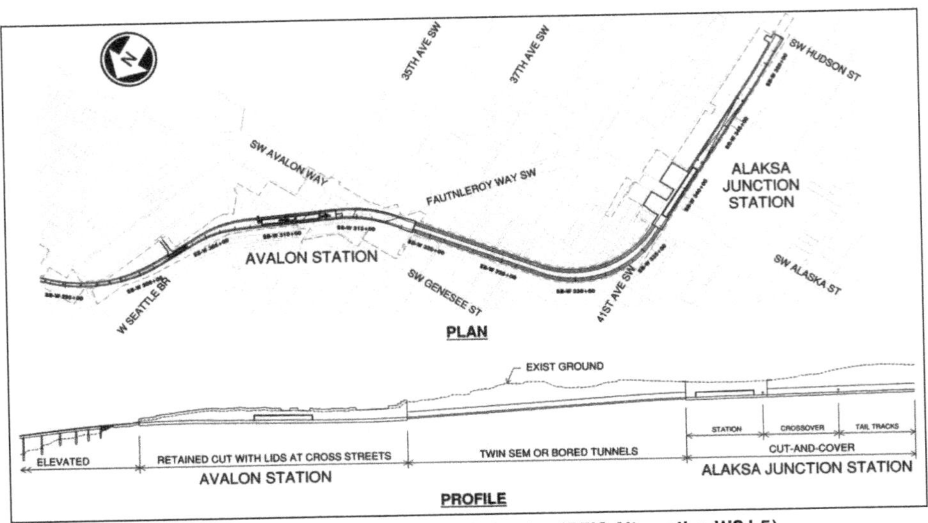


**Figure 5. Underground section of West Seattle Link Extension (DEIS Alternative WSJ-5)**

within 2,500 feet of walking distance from any point in the tunnel. Either of the two means of egress can be adopted for the West Seattle tunnels.

## West Seattle Geotechnical Conditions

Geotechnical investigation for the tunnel and station structures will be performed as a part of Phase 3 GI program discussed on page 7. Based on limited number of borings done for previous projects in the vicinity of the alignment, the surficial geology in West Seattle between Avalon and Alaska Junction typically comprises of strata of Vashon recessional outwash (Qvro, ESU2-RGD) and Vashon recessional lacustrine sediments (Qvrl, ESU 3-RCS) and occasional areas of fill and modified ground (Hf, ESU1-ENF). These soils were deposited as the glacial ice retreated. The ESU2-RGD consists of clean silty sand, gravelly sand, and sandy gravel with cobbles and boulders. Its consistency ranges from loose to very dense. The ESU3-RCS consists of fine sand, silt, and clay and its consistency varies from dense to very dense and soft to hard. The Fill (ESU1-ENF) consists of various materials but through this segment it generally consists mainly of dense or stiff soils if engineered but very loose to dense or very soft to stiff if un-engineered.

The subsurface geology consists of Vashon Till, Pre-Vashon lacustrine deposits and Pre-Vashon glaciomarine drift. Vashon Till (Qvt, ESU4-TD) consists of gravelly, silty sand and silty, gravelly sand with cobbles and boulders. It is very dense in consistency. Pre-Vashon lacustrine deposits (Qpnl, ESU7-CCS) are fine-grained lake deposits consisting of fine sandy-silt, silty fine-sand and clayey silt with scattered to abundant fine organics. It is dense to very dense or very stiff to hard in consistency. Pre-Vashon glaciomarine drift (Qpgm, ESU8-TLD) till-like deposit consists of a variable mixture of clay, silt, sand, and gravel with cobbles and boulders and locally contains scattered shells. It is very dense or hard in consistency.

The thickness of the ESU3-RCS (recessional lacustrine) and ESU2-RGD (outwash sediments) that overlie the glacially overridden Pre-Vashon sediments range from approximately 30 feet in the vicinity of the West Seattle Junction Station to approximately 4 feet in the vicinity of the Avalon Station, based on borings done for previous projects in the vicinity of the alignment (Shannon and Wilson, 2004; Altinay and Associates, 1926). The Pre-Vashon ESU7-CCS and ESU8-TLD underlie the project alignment at the total depth of all the borings advanced within this segment.

Groundwater levels could range from 40 to 80 feet depth in the few borings where groundwater depths were measured (Shannon and Wilson, 2004). Shallow, discontinuous perched groundwater may be encountered within the surficial granular soils near their contact with the underlying lower permeability Vashon Recessional Lacustrine deposits, Vashon Till, and Pre-Vashon sediments.

There are no identified liquefaction hazards along the underground section in West Seattle (City of Seattle Potential Liquefaction Area Map, 2019). There are no slopes greater than 40 percent adjacent to the underground alignment in West Seattle and no identified existing or potential landslide hazards in the vicinity of the tunnel section of the alignment.

## Downtown Tunnel Configuration

The proposed twin tunnel alignment will consist of approximately 3.3 miles of guideways with approximately 21-foot-7-inch outside diameter transit tunnels. The preferred alignment (DT-1 alternative), as identified in the DEIS but subject to further studies as directed by Sound Transit Board in July 2022, extends from the south portal located adjacent to 6th Avenue South between South Massachusetts Street and South Edgar Martinez Drive, to the north portal near the intersection of 3rd Avenue West and West Republican Street. The twin tunnels will pass through six stations, including the Chinatown-International District Station (connecting to the existing International District Station), Midtown Station, Westlake Station (connecting to the existing Westlake Station), Denny Station, South Lake Union Station, and Seattle Center Station.

Before confirming or modifying the preferred alignment for the Downtown Tunnel, further studies as directed by the Sound Transit Board are being performed to maximize community benefits and passenger experience while minimizing impacts and costs. These are described below and indicated geographically in Figure 6:

- Explore alternative station locations, station configurations and alignment in Chinatown-International District
- Explore opportunities to reduce station depths and improve passenger access and experience
- Explore hybrid alignment connecting Westlake/5th Avenue station (DT-1 alternative) to Denny/Terry station (DT-2 alternative) to South Lake Union/Harrison station (DT-1 alternative)
- Explore opportunities to provide station access from north and south sides of Denny Street to the Denny/Terry station (DT-2 alternative)
- Explore hybrid alignments connecting South Lake Union/Harrison station (DT-1 alternative) to Seattle Center/Mercer station (DT-2 alternative) to either DT-1 or DT-2 north portal location
- Explore westerly shifting of Seattle Center Republican station (DT-1 alternative)
- Explore shifting the north portal of DT-2 alternative south to Mercer Place

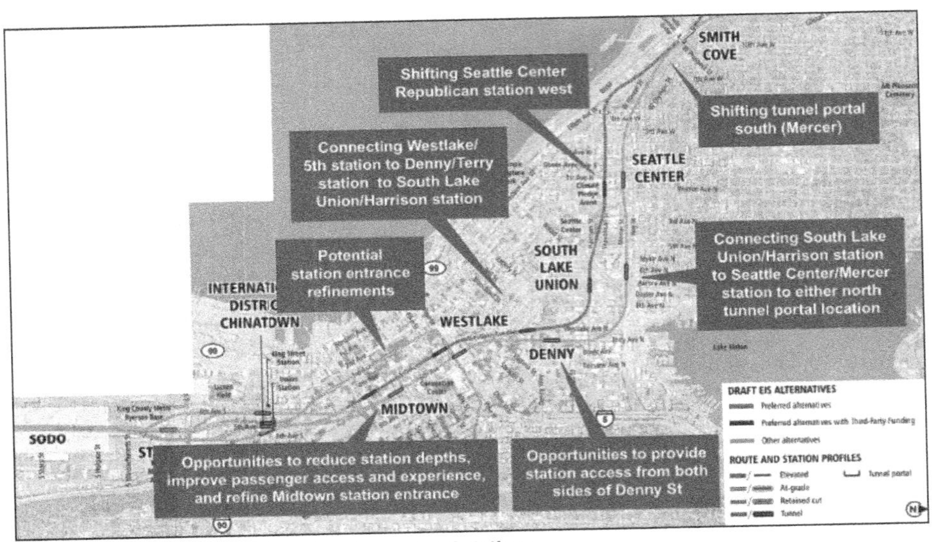


Figure 6. Further studies of downtown tunnels and stations

## Downtown Tunnel Site Constraints

The profile depth and horizontal alignment of the downtown tunnels are constrained by topography, existing surface and subsurface obstructions, right of way width constriction, and other site-specific conditions. The existing topography and subsurface obstructions constrain the profile depth of the downtown tunnels and the corresponding platform depth of the proposed stations. Starting at the southern end in the Chinatown-International District area, the existing grade elevation is low. The native ground was at elevation +20 feet NAVD88. This was subsequently filled to elevation +47 feet NAVD88 in the last century as the city expanded through reclamation of the low-lying land. The ground surface rises steeply to the Midtown area (elevation +180 feet NAVD88) in less than 3,000 feet of distance. This combined with the need to locate tunnels in soils with competent strength and sufficient ground cover and the upper desirable limit of 5% sustained gradient of track profile resulted in relatively deep tunnels and stations. Left-in-place tiebacks of the support of excavation used for constructing deep multi-level basements of several high-rise buildings (notably, the Columbia Tower) are major obstructions to construction of tunnels using the tunnel boring machine (TBM). This further restricts the ability to raise the tunnel profile in the Midtown. To the north of Midtown, the existing Westlake Station is oriented east-west across the project's north-south route requiring the new tunnels to go under the existing station with sufficient clearances under its drilled shaft foundations. Between Westlake and South Lake Union, the major constraints are deeper than usual, including large gravity sewers, fiber optic cables, complex network of numerous utilities, and the SR-99 tunnel portals. Also, several high-rise buildings with deep basements and left-in place tiebacks have been constructed in this area in the last 10 to 15 years.

The relatively narrow public right-of-way (ROW) width and the presence of existing buildings with deep basement adjacent to the ROWs constrain the horizontal alignment of the downtown tunnels and the location and configuration of the stations. The ROW width of many streets in downtown Seattle is 66 feet. Between Chinatown-International District and Midtown, the presence of battered pile foundations of the existing Sound Transit Chinatown-International District Station and buildings of the Union Station, the pile foundation of the retaining wall of the 5th Avenue South and the enlarged abutment of the Yesler Bridge further constrict the available width for construction of major underground elements of the Project. North of Westlake, the ROW width remains narrow (66 feet). The presence of residential and commercial buildings with basements, garages and business operations at the street level further constrain the use of more typical construction techniques and require innovative approach to construct underground stations and ground improvement.

In addition to the physical constraints, the tunnel profile depth and horizontal alignment as well as the potential construction methods for the stations are further constrained by the presence of buildings in the South Lake Union and Seattle Center areas with receptors that are sensitive to noise and vibrations induced during construction and subsequent train operations. The sensitive receptors include medical and research laboratories, recording studios, and performing art and cultural venues.

## Downtown Tunnel Geotechnical Conditions

South of South Main Street, through the Chinatown-International District and the South Portal at the 6th Avenue South and Massachusetts Street, the existing ground surface is relatively flat with native ground elevation varying between 18 to 22 feet above mean sea level. Several major streets are constructed on retained fill or structures supported on piles. The tunnels, stations and associated underground connections will be constructed mainly below (except at some locations partly in) the recently deposited post-glacial Holocene loose to medium dense sands, and soft to firm clays and silts (normally consolidated soils). These soils are underlain by glacially overridden and over-consolidated Quaternary dense to very dense sands, glacial, very stiff to hard silts and clays, and glacial Till and Till-like soils.

The groundwater conditions are anticipated to consist of a shallow, unconfined aquifer 3 to 10 feet below native ground surface formed within the granular normally consolidated soils over less permeable clays and silts. Seasonal and tidal variations could result in groundwater levels close to the native ground surface as well as in the retained fill material above it.

Where underground structures are within or in close proximity to the recently deposited soils and below the groundwater, ground treatment will be required to mitigate liquefaction and potential settlement impact to the nearby buildings and WSDOT I-90 bridge structures. Ground treatment will also increase the strength of the ground for safe and stable construction, especially for critical tunnel activities like break-in and break-out of the TBM from the station box.

North of South Main Street, the existing ground surface varies greatly with the highest elevation at Midtown, gradually descending through Westlake and becoming gentler and flatter in South Lake Union, rising again through the Seattle Center, and then descending to the north end close to Elliott Avenue at the western end of Republican Street. In this reach, the two miles of tunnels, five stations, and associated underground connections will be constructed in glacially overridden and over-consolidated Quaternary dense to very dense sands, glacial, very stiff to hard silts and clays, and glacial Till and Till-like soils. The glacial and interglacial soil units are typically of limited lateral extent and grade laterally into, or are inter-layered with, or may contain blocks of material from other stratigraphic units. In many instances, these materials are mixed and interbedded sheared and intact soils with contacts that are not easily defined. This complexity in subsurface geological and groundwater conditions presents difficult tunneling conditions, with specific challenges for the sequentially excavated mined caverns, connection adits and cross passages between bored tunnels.

## Interbay and Ballard Tunnel and Underground Stations

In the DEIS, the Interbay-Ballard segment included alignment alternatives that use tunnel as well as high level fixed and medium height movable span bridge types to cross Salmon Bay. The preferred alternative studied in the DEIS, with Ballard station located at 14th Avenue and Market Street, did not show a substantial cost difference between a tunnel and an approximately 150-foot high, fixed span bridge crossing over the Salmon Bay east of the existing 15th Avenue Bridge.

As part of the Further Studies directed by the Board, which are geographically depicted in Figure 7, the North Interbay station location as well as the preferred location of the Ballard station at either 14th or 15th Avenue are being studied for adjustments in location and configuration refinements. This includes a shift in the preferred DEIS alternative tunnel portal location in North Interbay and the adjustment of the alignment from the North Interbay Portal to either the 14th or 15th Avenue station locations in Ballard. The Further Studies do not include re-assessment of the Bridge crossings. It is anticipated that the Salmon Bay crossing will be a tunnel.

The length of Interbay-Ballard section guideway in twin tunnels is approximately 1-mile, of which about 1,700 feet is below Salmon Bay. The 1,100-foot-long section in the middle of bay has greater depth below the bay's mudline with potentially higher groundwater head on tunnels. Taking into consideration the length of tunnels, presence of softer soils at and adjacent to the south and the north banks of the bay and groundwater ingress and sediments under the high-water level of the bay, the tunnels will be constructed utilizing a closed face and fully pressurized tunnel boring machine (TBM). The main portal and construction staging area for the tunnels will be located at the North Interbay portal, where an area of 3.5 acres is planned/proposed to establish the site and staging to support construction of the tunnel and retained cut station and guideway.

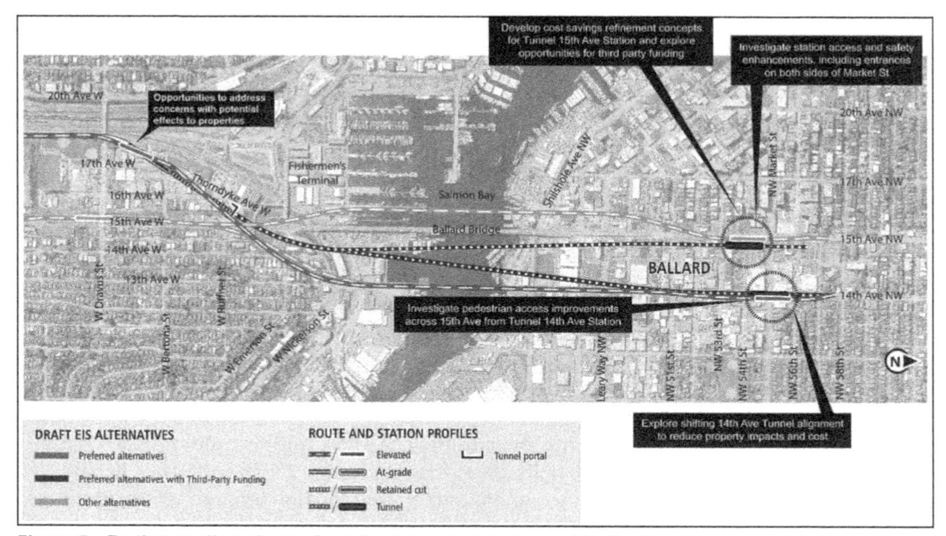


Figure 7. Further studies of tunnels and stations in Interbay and Ballard

## CURRENT PROJECT ACTIVITIES AND NEXT STEPS

A three-phase geotechnical exploration (GE) program was established in 2018 to gather geotechnical data appropriate to the three levels of project development process. The Phase 1 GE consisted of 9 borings and tests; these in conjunction with existing borings along the Project route provided information for technical concepts and feasibility studies. The Phase 2 GE consisted of 46 borings and testing to support conceptual engineering and the DEIS phase of the Project. The testing comprised in-hole seismic logging, installation of vibrating wire piezometers and pumping tests. A suite of laboratory tests to measure engineering, strength properties, including abrasivity and chemical analyses, were carried out utilizing soil samples from depths relevant to the Project's structures. The two pumping tests with a suite of six borings each were undertaken at the Midtown and the Westlake stations to understand hydrogeology of the two sites, analyze and assess potential connectivity of the deep confined aquifer with the perched groundwater in the sand layers at or above the depth of the proposed SEM station caverns and cut-and-cover entrance shafts, occurrence of artesian groundwater heads, and to estimate rate of pumping and associated groundwater drawdown should dewatering measures be considered to support the excavation of the mined caverns proposed for the Midtown and Westlake Stations.

The Phase 3 GE program for the entire Project's route (West Seattle to Ballard though the Downtown Seattle) will consists of about 200 to 250 borings at an average interval of 300 feet. The borings will be more closely spaced at station locations, cross passages, in areas of variable and challenging ground conditions and where presence of existing structures sensitive to construction exists which may require ground treatment, underpinning or structural protection designs. The Phase 3 GE also includes borings and field logging to identify and delineate subsurface obstructions mentioned in the historical reports and records as well as additional pumping tests to further understand hydrogeology. The main objective of the three-phase GE program is to incrementally and ultimately provide sufficient number of borings and testing data to complete the Project's final environmental impact statement, the preliminary engineering, and support the preparation of the construction procurement documents.

### Next Steps

In addition to the Preliminary Engineering of the Tunnel segments, Sound Transit (ST) will continue to update the linear project schedule based on constructability assessments, risk assessments, value engineering considerations, real property acquisitions and discussions with third parties as it relates to utilities, maintenance of traffic, dewatering and other constraints that affect tunnel design and construction. ST will also further define the scope and limits of Early Works Contracts such as major utility relocations needed prior to excavation and tunneling. Following industry outreach activities ST will re-assess its Preliminary Construction Packaging and Procurement as a basis for Program Management for the Final Design, Construction and Start-up Phases of the project It is anticipated that a phased Program Management approach will be in place by the end of 2023.

## REFERENCES

City of Seattle, 2019, Maps of Landslide Prone Areas.

King County, 2019, Liquefaction Susceptibility Maps.

Preedy, M., Bakker, D., Pooley, A., Wongkaew, M. and Bhargava, R., 2021, Methodologies for Sound Transit's Proposed New Tunnels for LINK Light Rail

Expansion, *Proceedings of the 2021 Rapid Excavation and Tunneling Conference*, Las Vegas.

Shannon and Wilson, 2004, Report Addendum No. 095-1, Geotechnical Data Report (GDR), Seattle Monorail Project (SMP), Seattle, Washington.

Sound Transit, 2022, West Seattle and Ballard Link Extensions Draft Environmental Impact Statement.

# Planning for State-of-Good Repair, Hazard Mitigation, and Resiliency for San Diego County Water Authority's 2nd Aqueduct Water Supply Infrastructure

**Mahmood Khwaja** ▪ CDM Smith
**William Brick** ▪ CDM Smith
**Paul Taurasi** ▪ CDM Smith
**Anjuli Corcovelos** ▪ San Diego County Water Authority, San Diego, CA

## ABSTRACT

California's San Diego County Water Authority provides a safe and reliable water supply to more than 3.3 million residents through twenty-four member agencies. The 2nd Aqueduct pipelines, a critical infrastructure consisting of three large-diameter pipelines, are susceptible to natural threats including streambed erosion, channel migration, and seismic hazards within the 3,840-foot-long Moosa Canyon segment of the alignment. A state-of-good repair, hazard mitigation, and resiliency project is progressing to ensure uninterrupted water supply. For each pipeline, a segment will be replaced by tunneling. A high-level project development approach is presented, focusing on alternative selection, environmental and geotechnical considerations, and operations and maintenance constraints.

## PROJECT BACKGROUND

The 2nd Aqueduct is part of the San Diego County Water Authority's (Water Authority) aqueduct system that delivers water to twenty-four member agencies in San Diego County. The 2nd Aqueduct pipelines consist of Pipeline No. 3 (a 72-inch steel welded pipe), Pipeline No. 4 (a 90-inch prestressed concrete cylinder pipe [PCCP]), and Pipeline No. 5 (a 96-inch PCCP) that pass through Moosa Canyon (Figure 1). The three pipelines are aligned within an approximately 100- to 200-foot-wide permanent right-of-way (ROW) that extends from southern Riverside County in the north and to Lower Otay Reservoir in the south. All the pipelines are buried, and substantial portions in this area are constructed of PCCP.

Over the last several years, the pipelines have begun exhibiting vulnerabilities including steel pipe pitting and bend yielding in some steel segments, as well as circumferential cracking and joint separation in the concrete pipes at the bends. In addition, over the decades since the pipelines were installed, high intensity storms and increased urbanization of watershed have caused erosion of the soil cover and a scouring of the streambed to expose sections of the pipelines.

## SITE AND PROJECT DESCRIPTION

The project site is within Moosa Canyon, a southeast to northwest trending canyon that contains Moosa Creek. Three subsurface pipelines presently extend through the canyon in a roughly north to south orientation. Pipeline 3 is a steel pipe that is 72 inches in diameter, Pipeline 4 is a PCCP that is 90 inches in diameter, and Pipeline 5 is a PCCP that is 96 inches in diameter. The ground surface elevation in Moosa Canyon is approximately 270 feet above mean sea level (MSL). Steeply ascending hillsides are

Figure 1. Moosa Canyon study area

Figure 2. Canyon bottom view

Figure 3. Canyon slope view

located to the north and south of Moosa Canyon with elevations of approximately 650 and 600 feet MSL, respectively. See Figure 2 and Figure 3 for canyon views.

## CURRENT VULNERABILITY

A condition assessment was conducted in 2019 to determine the probable cause of damage and distress. The soil erosion and scouring has led to an overstressing of the pipelines where the surrounding soil is not fully supporting the pipelines longitudinally with friction, particularly at the joints with high angle bends. The pipe overstressing is further exacerbated when the local alluvial soil is saturated and displaced. In the event of an earthquake, the alluvial soil has the potential to liquify, which would result in the structural support of the pipeline being compromised for long lengths, greatly increasing the risk of significant damage or pipe rupture. To strengthen the pipelines, several layers of carbon fiber were installed in Pipelines Nos. 4 and 5. However, this provides only a temporary solution. Another temporary mitigation includes creek stabilization measures to reduce the impacts from scour.

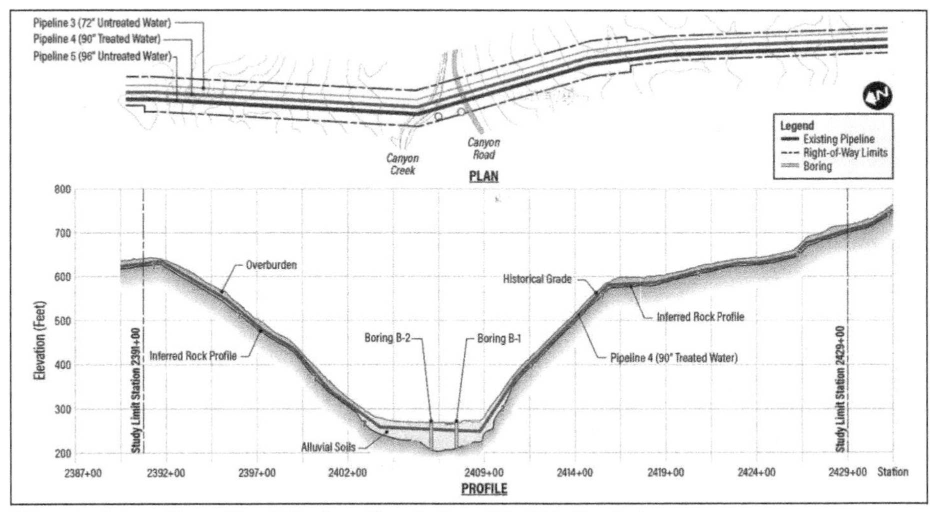


**Figure 4. Schematic profile of Pipelines Nos. 3, 4, and 5 within the study area**

The 2nd Aqueduct pipelines are critical to regional water supply. A failure of any one of the three pipelines would be a major disruption to the Water Authority's 24 member agencies. Previous and current studies have shown that the pipelines are highly vulnerable due to continued scouring of the streambed crossing, the liquefaction and lateral spread due to seismic events, and instability of the Moosa Canyon slopes. A schematic view of the profile is shown in Figure 4.

To assist with developing a plan to address current vulnerabilities it was necessary to understand the potential natural hazards and the subsurface conditions. A high-level geotechnical investigation and evaluation was conducted to support the coarse and fine screenings of alternatives with the understanding that a project focused geotechnical investigation would be developed when the project moves into the next phase involving final design. Elements of the geotechnical investigation and evaluation for the project are briefly described.

## GEOTECHNICAL INVESTIGATION

### Field and Subsurface Exploration

To support the conceptual engineering effort and screening process, subsurface exploration consisting of drilling, logging, and sampling of two small-diameter borings (B-1 and B-2) was executed. The purpose of the borings was to evaluate subsurface conditions and to collect soil samples for geotechnical laboratory testing.

The borings were drilled to depths up to 75 feet, with the approximate locations of the exploratory borings shown on Figure 4.

### Laboratory Testing

Geotechnical laboratory testing of representative soil samples included the performance of tests to evaluate in-situ moisture content and dry density, gradation, Atterberg limits, consolidation, shear strength, and soil corrosivity.

### Geophysical Survey

Seven seismic refraction survey lines, each up to 345 feet in length, were used to evaluate the approximate depth to bedrock and the general rippability along the pipeline alignment. The seismic line profiles generally indicated surficial soils, sill soils, colluvium, and alluvium underlain by granitic rock. In general, the seismic refraction surveys performed on the slopes of the canyon indicate that the moderate ripping conditions with some difficult ripping to be expected. The seismic refraction survey performed in the bottom of the canyon indicated moderate ripping conditions in the upper 20 to 30 feet; difficult ripping should be expected below 30 below ground surface (bgs).

## GEOLOGY

### Regional Geology

The project site is situated in the western portion of the Peninsular Ranges Geomorphic Province. This geomorphic province encompasses an area that extends approximately 900 miles from the Transverse Ranges and the Los Angeles Basin south to the southern tip of Baja California (Norris and Webb, 1990; Harden, 2004). The province varies in width from approximately 30 to 100 miles and generally consists of rugged mountains underlain by Jurassic metavolcanic and metasedimentary rocks, and Cretaceous igneous rocks of the southern California batholith. The portion of the province in western San Diego County that includes the project site generally consists of uplifted and dissected coastal plain underlain by Upper Cretaceous-, Tertiary-, and Quaternary-age sedimentary rocks.

The Peninsular Ranges Province is traversed by a group of sub-parallel faults and fault zones trending roughly northwest. Several of these faults are active: Elsinore, San Jacinto and San Andreas faults located northeast of the project site and the Rose Canyon, Coronado Bank, San Diego Trough, and San Clemente faults located west of the project site (Figure 5). Major tectonic activity associated with these and other faults within this regional tectonic framework consists primarily of right-lateral, strike-slip movement. The Temecula segment of the Elsinore Fault Zone, the nearest active fault system, has been mapped approximately 10.3 miles east of the project site.

### Site Geology

Geologic units encountered included fill, alluvium, and granitic rock and are consistent with the geologic units previously mapped by Kennedy and Tan (2007). Additional units, such as residual soils and colluvium may be presented, but were not encountered during the geotechnical investigation. The current geotechnical investigation program supported previous findings from the boring data at the site (SCST, Inc. 2016 and San Diego County Water Authority, 1957). The regional geologic map of the site is shown on Figure 6. A geologic cross section is included as Figure 7, which includes a preliminary interpretation of the subsurface units along the alignment based on the project specific subsurface investigation conducted.

#### *Fill*

Fill was encountered in boring B-1, consisting of silt and gravelly sand.

#### *Alluvium*

Alluvium was encountered below the fill strata in B-1 and at ground surface in B-2 for majority of the boring depth. As encountered, the alluvium generally consisted of

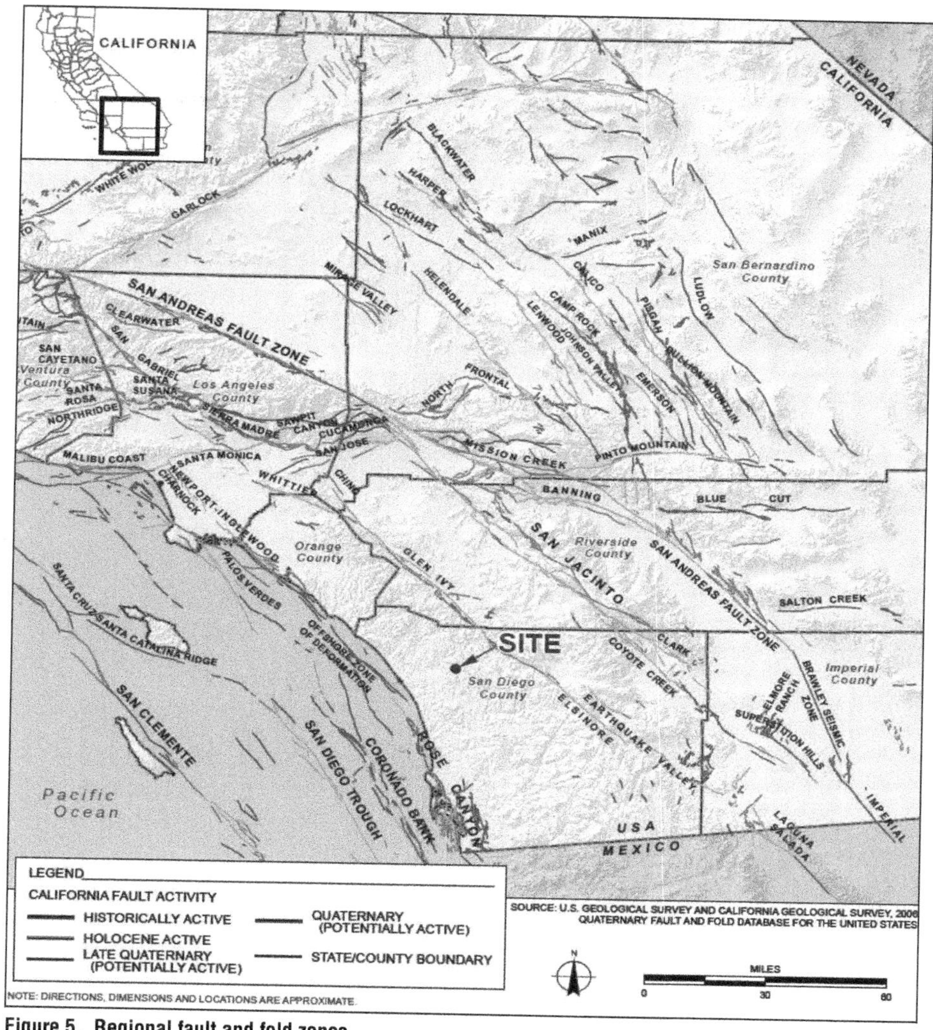


**Figure 5. Regional fault and fold zones**

sands and silty sands, with evidence of sandy clay within the alluvium in B-1. Scattered amounts of gravel were observed within the alluvium. Although not encountered in our borings, cobbles and boulders may be encountered in the alluvium.

### *Granitic Rock*

Granitic Rock was encountered in both borings below the alluvium. As encountered, the granitic rock generally consisted of yellow, gray, and black, wet, moderately weathered, coarse-grained granitic rock. Granitic rock outcrops were observed along the ascending hillsides to the north and south of Moosa Canyon.

### *Groundwater*

Groundwater was encountered in both borings and is estimated to be at relatively shallow depths and consistent with observation well readings being monitored by

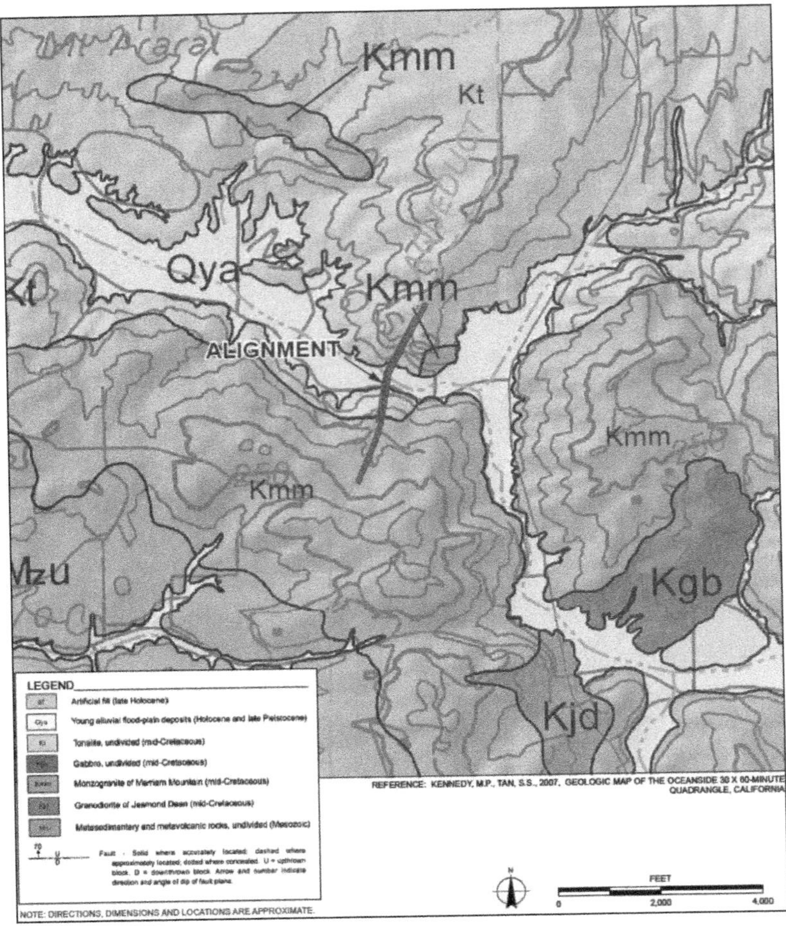


**Figure 6. Project site geologic map**

Valley Center Municipal Water District. These observation wells are in proximity of the project location.

## Faulting and Seismicity

The numerous faults in southern California include active, potentially active, and inactive faults. As defined by the California Geological Survey, active faults are faults that have ruptured within Holocene time or within approximately the last 11,000 years. Potentially active faults are those that show evidence of movement during Quaternary time (approximately the last 1.6 million years), but for which evidence of Holocene movement has not been established. Inactive faults have not ruptured in the last approximately 1.6 million years. The approximate locations of major active and potentially active faults in the vicinity of the site and their geographic relationship to the site are shown in Figure 5.

The site is a seismically active area, as is the majority of southern California, and the potential for strong ground motion is considered significant during the design life of the proposed project infrastructure. However, no known faults are mapped underlying

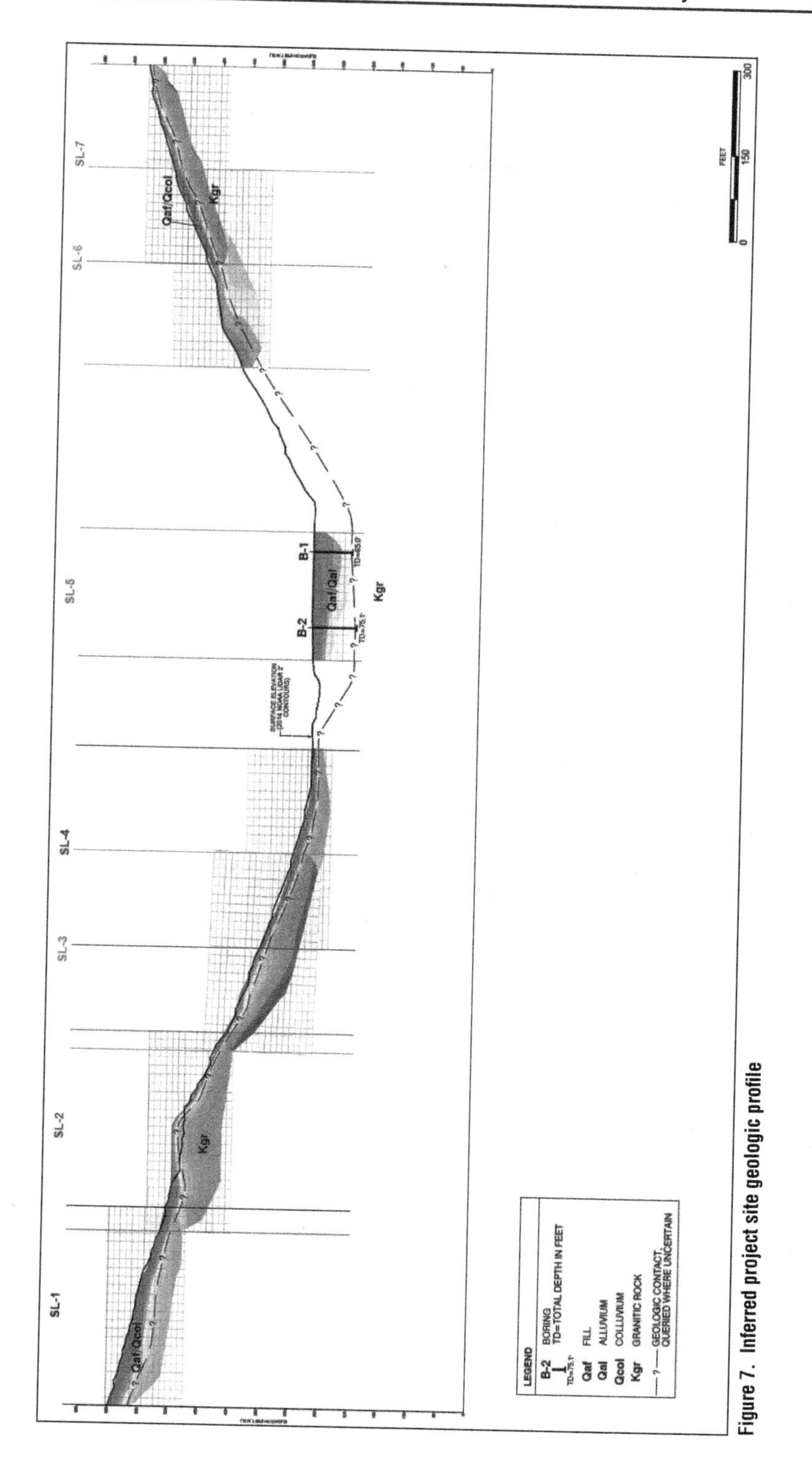


Figure 7. Inferred project site geologic profile

**Table 1. Principal active fault**

| Fault | Distance to Site | Maximum Moment Magnitude |
|---|---|---|
| Elsinore (Julian Segment) | 10.3 | 7.4 |
| Elsinore (Temecula Segment) | 10.3 | 7.1 |
| Newport-Inglewood (Offshore) | 17.2 | 7 |
| Rose Canyon | 17.6 | 6.9 |
| Elsinore (Glen Ivy Segment) | 25.6 | 6.9 |
| Coronado Bank | 33.5 | 7.4 |
| Earthquake Valley | 34 | 6.8 |
| San Jacinto (Anza Segment) | 34.7 | 7.3 |
| San Jacinto (San Jacinto Valley Segment) | 36.8 | 7 |
| San Joaquin Hills | 37.2 | 7.1 |
| San Jacinto (Coyote Creek Segment) | 38.5 | 7 |
| San Jacinto (Clark Segment) | 40.6 | 7.1 |
| Palos Verdes | 43.6 | 7.3 |
| Elsinore (Whittier Segment) | 45.9 | 7 |
| South San Andreas (South San Bernardino) | 51.6 | 7 |
| San Jacinto (San Bernardino Valley) | 51.8 | 7.1 |
| Elsinore (Coyote Mountain Segment) | 52 | 6.9 |
| San Jacinto (Borrego Mountain) | 55.7 | 6.8 |
| Pinto Mountain | 60 | 7.3 |
| South San Andreas (North San Bernardino) | 60.9 | 6.9 |

the project site. The site is not within a State of California Earthquake Fault Zone (EFZ) (formerly known as an Alquist-Priolo Special Studies Zone). The nearest known active fault is the Julian segment of the Elsinore Fault Zone, located approximately 10.3 miles east of the site.

Table 1 lists selected principal known active faults that may affect the site.

Based on this information, during design development, the project will likely use seismic parameters associated with the closest known active fault, the Julian segment of the Elsinore Fault Zone

### *Ground Motion*

The project area has a high potential for experiencing strong ground motion considering the proximity of the site to active faults capable of producing a maximum moment magnitude of 6.0 or more. The 2019 California Building Code (CBC) specifies that the Risk-Targeted, Maximum Considered Earthquake (MCER) ground motion response accelerations be used to evaluate seismic loads for design of buildings and other structures. The MCER ground motion response accelerations are based on the spectral response accelerations for 5 percent damping in the direction of maximum horizontal response and incorporate a target risk for structural collapse equivalent to 1 percent in 50 years with deterministic limits for near-source effects. The horizontal peak ground acceleration that corresponds to the MCER for the project area was calculated as 0.48g using the 2021 Structural Engineers Association of California (SEAOC)/Office of Statewide Health Planning and Development (OSHPD) seismic design tool.

The 2019 CBC specifies that the potential for liquefaction and soil strength loss be evaluated, where applicable, for the mapped Maximum Considered Earthquake Geometric Mean (MCEG) peak ground acceleration (PGAM) with adjustment for site class effects in accordance with the American Society of Civil Engineers (ASCE) 7-16 Standard. The MCEG PGA is based on the geometric mean PGA with a 2 percent probability of exceedance in 50 years. The PGAM was calculated as 0.52g using the 2021 SEAOC/OSHPD seismic design tool (web-based).

### *Ground Rupture*

At present, there is no evidence of active faults that cross the project vicinity. Therefore, the potential for ground rupture due to faulting at the site is considered low. However, lurching or cracking of the ground surface because of nearby seismic events cannot be ruled out.

### *Liquefaction*

Liquefaction is the phenomenon in which loosely deposited granular soils with silt and clay contents of less than approximately 35 percent and non-plastic silts located below the water table undergo rapid loss of shear strength when subjected to strong earthquake-induced ground shaking. Factors known to influence liquefaction potential include composition and thickness of soil layers, grain size, relative density, groundwater level, degree of saturation, and both intensity and duration of ground shaking. According to the County of San Diego Seismic Department of Planning and Land Use (2008), the project site is in an area that is prone to liquefaction particularly due to the alluvial soils. Accordingly, the liquefaction potential of the subsurface soils was evaluated using subsurface data from borings B-1 and B-2.

For purposes of our initial liquefaction evaluation, the historic high groundwater table was at a depth of 15 feet. The liquefaction analysis indicated that portions of the alluvium encountered below the assumed groundwater level to depths on the order of 60 feet are generally susceptible to liquefaction during the design seismic event.

### *Seismic Settlement*

The proposed improvements may be subject to several hazards, including liquefaction-induced settlement. To estimate the amount of post-earthquake settlement, the method proposed by Tokimatsu and Seed (1987) was used in which the seismically induced cyclic stress ratios and corrected N-values are related to the volumetric strain of the soil. The amount of soil settlement during a strong seismic event depends on the thickness of the liquefiable layers and the density and/or consistency of the soils.

Based on an assumed peak ground acceleration of 0.52g and an earthquake magnitude of 7.7, a post-earthquake total settlement of up to approximately 15 inches is estimated for the site, depending on the groundwater levels.

### *Lateral Spread*

Lateral spreading of ground surface during an earthquake usually takes place along weak shear zones that have formed within a liquefiable soil layer. Lateral spread has generally been observed to take place in the direction of a free-face (i.e., retaining wall, slope, channel), but has also been observed to a lesser extent on ground surfaces with very gentle slopes. Due to the site being situated adjacent to Moosa Creek, the site's susceptibility to lateral spreading was evaluated. Preliminary results indicate that the site could experience lateral deflections on the order of 14 feet in response to a design seismic event, equal to an earthquake magnitude of 7.7. The thickness of the alluvium is expected to decrease toward the margins of the canyon. The amount of lateral spread can be expected to decrease as the thickness of liquefiable soils decreases toward the margins of the canyon. In addition, analysis indicates that the amount of lateral displacement will decrease with distance away from the channel free faces adjacent to Moosa Creek.

### *Landsliding and Slope Stability*

Review of referenced geologic maps, literature and topographic maps, and subsurface exploration, landslides or indications of deep-seated landsliding were not noted underlying the project site. Preliminary deep-seated stability of the slopes that descend from the east and west property lines was performed on geologic cross sections. The ultimate shear strength parameters used in our stability calculations under static conditions for the different material types are presented in Table 2.

**Table 2. Strength parameters used in static slope stability analysis**

| Earth Material | Unit Weight pcf | Static Shear Strength Parameters | |
|---|---|---|---|
| | | Cohesion psf | Friction Angle, degrees |
| Alluvium (Liquified) | 120 | 300 | 0 |
| Residual Soil | 120 | 75 | 30 |
| Weathered Granitic Rock | 135 | 600 | 36 |

Slope stability evaluations under both static and seismic conditions assumed worst-case conditions with respect to groundwater. A factor of safety of at least 1.5 under static conditions is generally considered adequate as per the guidelines of California Geological Survey Special Publication 117A (CGS, 2008a) and accepted engineering practices. Preliminary results on the slope stability analyses (static conditions) yielded factors of safety greater than 1.5. The analyses of slope stability, when subjected to seismic loading, was also conducted. The strength parameters, for different soil strata, used under seismic conditions are presented in Table 3.

Preliminary results for slope stability with seismic conditions yielded pseudo-static factors of safety of greater than 1.0 for all scenarios, which is considered adequate for seismic conditions.

**Table 3. Strength parameters used in seismic slope stability analysis**

| Earth Material | Unit Weight, pcf | Seismic Shear Strength Parameters | |
|---|---|---|---|
| | | Cohesion, psf | Friction Angle, degrees |
| Alluvium (Liquefied) | 120 | 300 | 0 |
| Residual Soil | 120 | 75 | 30 |
| Weathered Granitic Rock | 135 | 600 | 36 |

## ALTERNATIVE SCREENING PROCESS

An alternatives screening evaluation process was conducted by the project team and the Water Authority to determine the best long-term solution to mitigate these vulnerabilities over a planning horizon of 75 years or greater. The evaluation began with a coarse screening approach that distilled several preliminary repair and replacement alternatives down to three preferred alternatives that then advanced to the fine

**Table 4. Coarse screening rehabilitation/replacement alternatives**

| Alternative No. | Alternative Component | | | | | | | | | | | |
|---|---|---|---|---|---|---|---|---|---|---|---|---|
| | Canyon Slopes | | | | Canyon Bottom | | | | | | | |
| | Spot Repair Pipeline No. 3 and Reline Pipelines No. 4 and 5 with Structural Steel Liner | Reconstruct Pipelines No. 3, 4, and 5 | Stabilize Pipelines | Rock Tunnel (One to Three Bores) | Spot Repair Pipeline No. 3 and Reline Existing Pipelines No. 4 and 5 with Structural Steel Liner | Reconstruct Pipelines No. 3, 4, and 5 Using Open Cut Excavation | Reconstruct Pipelines No 3, 4, and 5 on Utility Pipe Bridge with Deep Foundation | Reroute Pipelines No. 3, 4, and 5 Outside of Existing ROW (New Alignment) | Rock Tunnel (One to Three Bores) | Soft Ground Tunnel (One to Three Bores) | Deep Foundations to Support Pipelines in Place | Improved Streambed Stabilization Techniques |
| 1 | • | | • | | • | | | | | | • | • |
| 2 | • | | • | | | • | | | | | • | |
| 3 | • | | • | | | | • | | | | | |
| 4 | | • | • | | | • | | • | | | | |
| 5 | | • | | • | | • | | | • | | | |
| 6 | • | | • | | | • | | | | • | | |
| 7 | | • | | • | | • | | | | • | | |

screening stage. The coarse screen alternatives included four options for the slopes, and seven options the Moosa Canyon bottom. A total of seven alternatives were developed for evaluation based on the four and seven options for the slopes and the canyon bottom, respectively. These options are noted in a matrix form in Table 4.

The seven alternatives were evaluated as part of the coarse screening process, shown schematically in Figure 8.

The three highest ranking options were progressed to fine screening, these include:

### *Alternative No. 1—Reline Existing Pipelines in Canyon Bottom*

This alternative includes performing internal or external spot repairs to the steel Pipeline No. 3 in selected locations along the canyon bottom to address the areas of advanced degradation. This alternative also includes relining the PCCP Pipelines Nos. 4 and 5 with steel cylinders to provide a robust system capable of withstanding significant ground movement. This alternative also includes improved streambed stabilization, deep foundations to the bedrock to support the pipelines, and anchors at the horizontal bends to mitigate seismic and scour risks.

### *Alternative No. 2—Replace Using Open Cut Construction in Canyon Bottom*

Within the canyon bottom, the pipelines would be fully replaced with new piping using open-cut construction designed to mitigate seismic and liquefaction risks. The new pipelines would be either thicker-walled steel piping or specialized seismic-resilient piping, depending on future analyses. The new pipelines would be constructed using deep open-cut excavation at elevations below the 500-year scour depth of Moosa Creek but remain within the seismic liquefaction zone. The piping at the bottom of the canyon would likely be constructed adjacent to the existing pipe alignments while the existing water supply operations are maintained. The new piping would be supported on deep foundations or have special seismic steel pipes (SSSPs) designed for seismic resiliency. This alternative also includes improved streambed stabilization and anchors at the horizontal bends.

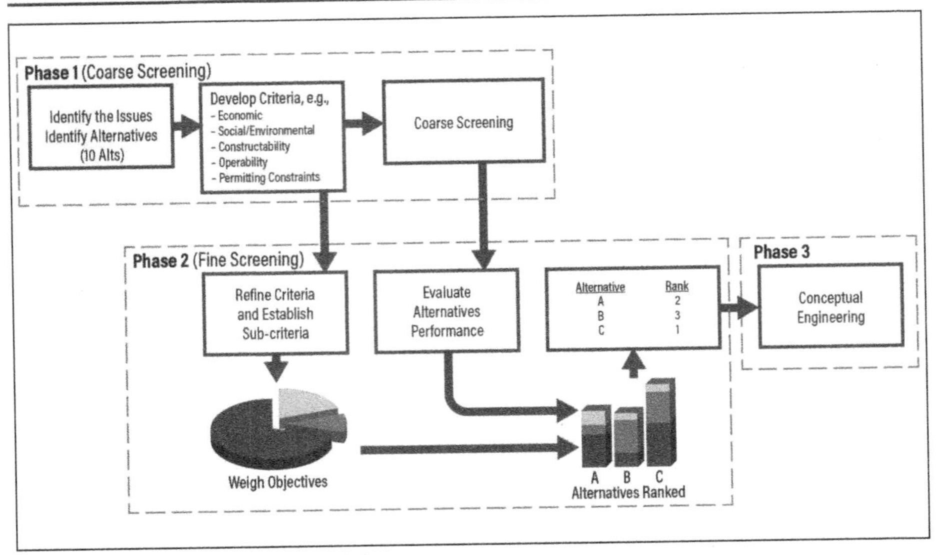


**Figure 8. Schematic representation of the screen processes**

### *Alternative No. 3—Replace with Tunnels in Canyon Bottom*

Within the canyon bottom, this alternative includes excavating, on each side of the canyon bottom, shafts to below the 500-year scour depth and constructing a tunnel between the shafts, using either conventional or mechanized tunneling methods. The depth of the tunnel would be just above the bedrock depth to mitigate liquefaction issues. Each pipeline would be evaluated for seismic and other slope stability concerns on the canyon slopes. Pipe stabilization measures would be used on the existing pipelines if necessary. Groundwater control measures for shaft and tunnel construction would be needed. For the shafts, lowering the groundwater table, or using an impermeable excavation support system with a grout plug at the bottom of the shaft, may be viable. The use of microtunneling would help mitigate construction risk associated with groundwater control and stabilize the weak ground below groundwater table. In this alternative, construction would remain outside of the creek channel. Therefore, no streambed stabilization or restoration would be needed, and the existing armoring would remain as-is with continued monitoring in the future.

## FINE SCREENING ACTIVITIES

The three alternatives were further analyzed and ranked using the following criteria:

### Reliability/Resiliency

Comparatively, Alternatives Nos. 2 and 3 were evaluated to be more resilient than Alternative No. 1 with respect to seismic risks, stream erosion concerns, and unbalanced thrusts forces within the canyon bottom. Further refinement of the evaluation ranked Alternative No. 3 as the most preferred with respect to resiliency and reliability.

### Project Costs

All alternatives require capital and life-cycle investments. A Class 4 cost estimate, according to the AACE International cost classifications, was developed for the three fine-screened alternatives. In all three alternatives, the option for the slope was the same; the cost comparison was essentially between the options for the canyon bottom. Based on the preliminary cost estimate, Alternative No. 3 had the lowest construction cost, as well as the lowest life-cycle cost.

### Operations and Maintenance

Since Alternative No. 1 was a rehabilitation option in place within the threat zones to mitigate the risk of impacts due to scouring of Moosa Creek bottom, Alternative No. 1 would require periodic, and potentially extensive, monitoring and repairs of the improved stream stabilization system. This would increase the overall maintenance and operational costs, without improving the reliability of the pipelines. Alternatives Nos. 2 and 3 reconstruct the pipelines within the canyon bottom which provides an opportunity to fully redesign the pipelines through the canyon bottom to optimize operational needs.

The increased depth of the pipelines associated with Alternatives Nos. 2 and 3 add to the challenges when performing draining, inspection, and/or repairs, as compared to Alternative No. 1.

### Social, Environmental, and Permitting Impacts

All alternatives require that some form of armored channel protection remain following completion of all construction. However, only Alternative No. 3 allows the existing channel armoring to remain as-is through and after construction. The environmental impacts, mitigations, and the permitting requirements for implementation are anticipated to be the least for Alternative No. 3.

Alternative No. 2 is expected to potentially have the greatest environmental impacts due to construction of shoring systems, open cut excavations, and pile supported anchor blocks, dewatering and treatment efforts, and additional stream stabilization enhancements. Alternative No. 2 requires complete removal of the existing armoring and three full trenches; it also involves the most dewatering. The level of future additional geotechnical investigations and engineering analysis is anticipated to be greater for Alternatives Nos. 2 and 3.

### Constructability

Alternative No. 1 offers the least constructability concerns but would require close coordination for extended water supply shutdowns for rehabilitation of the existing pipelines. Alternatives Nos. 2 and 3 would be constructed adjacent to the existing pipelines and require temporary system shutdowns to make the connections—potentially a shorter duration impact. Alternatives Nos. 2 and 3 are anticipated to have increased construction risk due to nature of the construction methods and will require adequate ground investigation to help mitigate and/or manage the risks.

The objective of this study was to identify a preferred alternative that best fulfills the Water Authority's goals for the project objectives described above. The weighting criteria is given in Table 5.

## EVALUATION OUTCOME

Notable outcomes of the fine screening evaluation are as follows:

Table 5. Summary of evaluation criteria

| Evaluation Criteria | | |
|---|---|---|
| No. | Description | Weighting |
| 1 | Reliability/Resiliency | 34% |
| 2 | Project Costs | 13% |
| 3 | Operations and Maintenance | 20% |
| 4 | Social/Environmental/Permitting Impacts | 20% |
| 5 | Constructability | 13% |
| | Total | 100% |

## Reliability/Resiliency

Alternatives Nos. 2 and 3 were felt to be more resilient to mitigating the seismic, stream erosion, unbalanced thrusts, and alluvial soils within the canyon bottom than Alternative No. 1, with Alternative No. 3 felt to be the best of the three.

Under the scenarios evaluated and assumptions utilized, the northern and southern slopes of Moosa Canyon were found to be just stable (pseudo-static factor of safety greater or equal to 1.0 is stable). The need for slope stabilization will be evaluated further in the future, and the slope stabilization techniques that were evaluated may be eliminated or modified as additional data becomes available.

## Project Costs

The Initial Construction Cost between the three alternatives ranged within a narrow band with a less than 6 percent difference between the least and the most expensive alternative.

Alternative No. 3 also had the lowest total life-cycle cost.

## Operations & Maintenance

Because the pipelines will remain within the potential scour depth of Moosa Creek, it was felt Alternative No. 1 would potentially require more extensive monitoring and repairs of the stream stabilization system, which would increase overall maintenance and potentially reduce the reliability of the pipelines.

Because Alternatives Nos. 2 and 3 allow for the pipelines to be rebuilt within the canyon bottom, these alternatives allow for the option of maintaining or increasing pipe size and reducing headloss through this reach. This allows for slightly better operations than Alternative No. 1, which would reduce the diameter of Pipelines Nos. 4 and 5 in the canyon bottom.

The increased depth of the pipelines associated with Alternatives Nos. 2 and 3 pose increased challenges when performing draining, inspection, and/or repairs, as compared to Alternative No. 1.

## Social/Environmental/Permitting Impacts

All alternatives require that some form of armored channel protection remain following completion of all construction. However, only Alternative No. 3 allows the existing channel armoring to remain as-is through and after construction.

The environmental impacts, mitigations, and the permitting requirements for implementation are anticipated to be the least for Alternative No. 3.

Alternative No. 2 is expected to potentially have the greatest environmental impacts due to construction of shoring systems, open cut excavations, and pile supported anchor blocks, dewatering and treatment efforts, and additional stream stabilization enhancements. Alternative No. 2 requires complete removal of the existing armoring and three full trenches; it also involves the most dewatering.

The level of future additional geotechnical investigations and engineering analysis is anticipated to be greater for Alternatives Nos. 2 and 3.

### Constructability

Alternative No. 1 does not require that new connecting sections of pipe be installed to join the new and existing pipelines, as is required for both Alternatives Nos. 2 and 3.

Alternative No. 2 is anticipated to require the most dewatering of the three alternatives.

Both Alternatives Nos. 2 and 3 are anticipated to have increased construction risk, as compared to Alternative No. 1, due to the potential for geotechnical conditions that would not be known until construction.

Alternative No. 3 will require that the vertical shaft and 90-degree bends associated with transitioning between the existing shallow pipelines and new deep tunnels be anticipated and designed correctly.

All three alternatives are anticipated to require approximately eight years for design, bidding, and construction. Is it anticipated that taking a pipeline out of service and/or bypassing can be scheduled to occur during low water-demand months (November—April) and for a manageable period. The results of the evaluation and weighted criteria scoring activity are presented in Table 6. The alternative with the highest score is Alternative No. 3—Replace with Tunnels in Canyon Bottom. This alternative is currently the Recommended Alternative to further evaluate in the subsequent Conceptual Design Phase of the project.

**Table 6. Ranking of alternatives**

| Alternatives | Ranking | Total Score* |
|---|---|---|
| Alternative No. 3 Replace with Tunnels in Canyon Bottom | 1 | 3.847 |
| Alternative No. 1 Reline Existing Pipelines in Canyon Bottom | 2 | 3.641 |
| Alternative No. 2 Replace Using Open Cut Construction in Canyon Bottom | 3 | 3.336 |

* (5) Fulfills the Objective with Added Value; (4) Fulfills the Objective; (3) Fulfills the Objective with Limited Challenge(s); (2) Fulfills the Objective with Challenge(s); (1) May Not Fulfill the Objective

## NEXT STEPS

All three alternatives were anticipated to require approximately eight years for project execution from planning through construction and commissioning. The alternative with the highest score was *Alternative No. 3—Replace with Tunnels in Canyon Bottom*. This alternative was selected by the Water Authority to move forward to the Design Phase of the project. The primary points that had the greatest influence in the selection of Alternative No. 3 as the preferred alternative included the following:

- The added value of risk mitigation associated with seismic hazards, stream bed erosion, and unbalanced thrusts forces within the canyon bottom by relocation outside the threat zones.
- Lowest construction and life-cycle cost.

- The option of maintaining or increasing pipe size and reducing headloss across the canyon bottom.
- Eliminates the need to work within or restore the creek channel and eliminates the need to replace or upgrade the existing channel armoring.
- Eliminates construction activities within Moosa Creek.

Overall, the study demonstrated that replacing the pipelines in the canyon bottom with tunnels will offer the Water Authority a resilient infrastructure repair solution. In addition, it will mitigate natural hazards for decades and provide reliability for future generations at the lowest total life-cycle cost.

## ACKNOWLEDGMENT

The authors would like to acknowledge the San Diego County Water Authority for the opportunity to present this paper for the planning phase of the project.

# Planning the Black Creek Tunnel Project

**Jeff F. Wallace** ▪ Black & Veatch
**Daniel G. Cressman** ▪ Black & Veatch
**Cary Hirner** ▪ Black & Veatch
**Malcolm Sheehan** ▪ City of Toronto
**Tony Cicchetti** ▪ City of Toronto
**Prapan Dave** ▪ City of Toronto

## ABSTRACT

The City of Toronto is currently undertaking the design of the Black Creek Sanitary Trunk Sewer Relief Project. The Project involves the design and construction of approximately 20 kilometers of tunnel through shale rock of the Georgian Bay Formation and soft ground soil consisting of glacial till, glaciolacustrine and glaciofluvial sand, silt and clay deposits. The 20 kilometer alignment has been split into four sections of tunnel (1) the 12.9 km, 3.0 meter diameter Keele Trunk Relief Sewer, (2) the 2.9 km, 1.5 meter diameter Keele Trunk Relief Sewer Microtunnel, (3) the 3.9 km, 1.8 meter diameter Jane Relief Sewer, and (4) the 0.5 km,1.5 meter diameter Combined Sewer Overflow Storage Tunnel. Additionally, one 40,000 $m^3$ wet weather flow (WWF) storage tank and numerous WWF connection shafts are to be constructed and connected to the Keele Trunk Relief Sewer to pick up WWF from existing outfalls currently discharging to Black Creek. The anticipated cost of the project is approximately $600 million with construction of the Project's Phase One Works, the Keele Trunk Relief Sewer and Jane Relief Sewer, scheduled to commence in 2025.

This paper provides an overview of the project in terms of the anticipated geotechnical conditions, the scheme developed for design and construction of the tunnel system and the associated schedule and procurement strategy. The consideration for planning of the tunnel system, the risk-based analysis and decision-making process are presented and form the basis for selection of a precast tunnel lining support method in the rock and soft ground.

## INTRODUCTION AND PROJECT BACKGROUND

The design of the Black Creek Sanitary Trunk Sewer (BCSTS) Relief Project is currently underway by the City of Toronto (City). This Project has been initiated to upgrade sanitary servicing in the Black Creek sewershed and reduce the frequency and severity of combined sewer overflow (CSO) events during wet weather flows (WWF) to improve the water quality in Black Creek and the Humber River. The Project includes four major components: the Keele Trunk Relief Sewer, the Keele Trunk Relief Sewer Microtunnel, the Jane Relief Sewer, and the Combined Sewer Overflow Storage System.

The existing BCSTS, as shown in Figure 1, was built in the 1960s and is approximately 15 kilometers (km) in length. The BCSTS services a sanitary drainage area of approximately 5,500 hectares (ha), with 20% of the drainage area having local combined sewers. In 2016, the BCSTS serviced a population of 351,000 with two-thirds of this population being residential (City of Toronto, 2016).

During wet weather events, such as heavy rainstorms, flow in the existing BCSTS exceeds the sewer's design capacity causing two different issues:

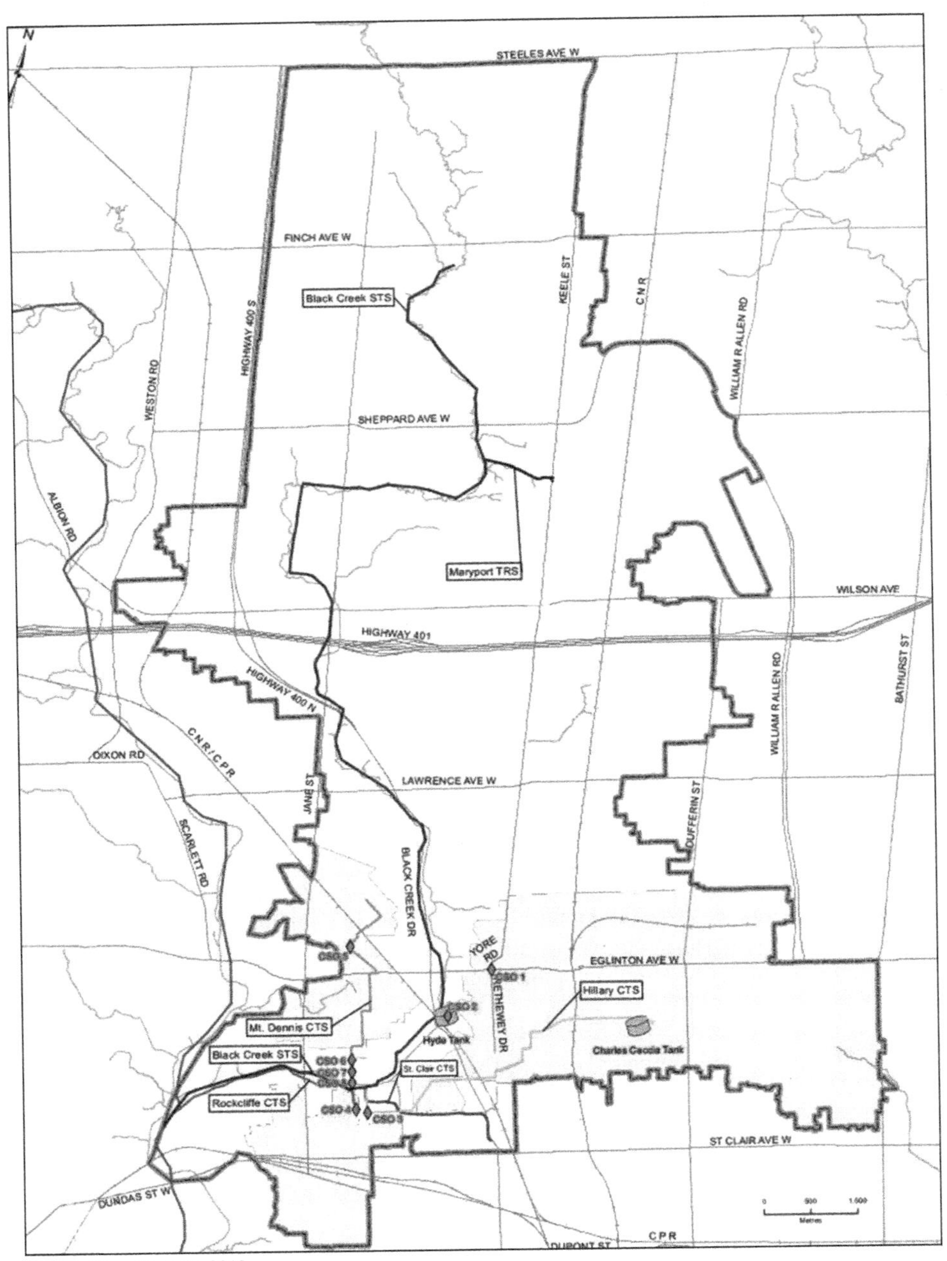


Source: City of Toronto 2016

**Figure 1. Map of the existing BCSTS and associated sanitary drainage area**

1. Flow is known to backup into the local sewer system and cause basement flooding
2. CSOs occur into the Black Creek watercourse at existing Combined Trunk Sewer (CTS) outfalls

A Municipal Class Environmental Assessment (EA) Study was initiated by the City in November 2015 to determine an EA solution to address the two issues identified above by providing sufficient sanitary servicing capacity to meet future populations needs and reduce CSOs to Black Creek. The EA solution served as the starting point for the Preliminary and Detailed Design of the Project which is currently underway.

## PROJECT COMPONENTS

The BCSTS Relief Project consists of four major components:

1. The Keele Trunk Relief Sewer (KTRS)
2. The Keele Trunk Relief Sewer Microtunnel (KTRS Microtunnel)
3. The Jane Relief Sewer (JRS)
4. The Combined Sewer Overflow Storage System (CSO Storage System)

These components are shown in Figure 2 as they currently stand at the point of writing this paper. As the design of the Project is currently underway, changes to the shown layout may occur prior to construction.

### Keele Trunk Relief Sewer

The KTRS has an inside diameter of 3000 mm and is approximately 12.9 km long. It collects flows directly from the existing Maryport Trunk Relief Sewer (TRS) and the existing BCSTS. However, it also will collect flows from the three other components of this Project (the KTRS Microtunnel, JRS, and CSO Storage System) and the proposed Downsview Lands Redevelopment project. Thus, this segment of the sewer has the largest diameter of the Project. The KTRS ultimately carries the flows to the Humber Sanitary Trunk Sewer (HSTS) through a gated connection that will limit flows into the HSTS based on the capacity available at the downstream Humber Treatment Plant (HTP). The alignment of the KTRS as shown in Figure 2 was heavily influenced by the locations of the required hydraulic connections and the highly congested nature of the Project area both above and below ground. The KTRS will be installed by a Tunnel Boring Machine (TBM) with a precast concrete tunnel lining. There will be nine construction shafts along the KTRS which will all ultimately function as maintenance access shafts or connection shafts.

### Keele Trunk Relief Sewer Microtunnel

The KTRS Microtunnel has an inside diameter of 1500 mm and is approximately 2.9 km long. It collects flow at a single location from the existing BCSTS. These flows are then carried and dropped into the KTRS. The alignment of the KTRS as shown in Figure 2 was selected to provide the required hydraulic connection, avoid conflict with the existing Toronto York, Spadina Subway and mitigate impacts to third party stakeholders. The KTRS will be installed by a Microtunnel Boring Machine (MTBM) with direct jacked reinforced concrete pipe along with a very short section of open cut at the most upstream end. There will be four construction shafts along the KTRS Microtunnel (plus one shared with KTRS) which will all ultimately function as maintenance access shafts or connection shafts.

## Jane Relief Sewer

The JRS has an inside diameter of 1800 mm and is approximately 3.9 km long. It collects flows at two locations from the existing BCSTS. These flows are then carried to the KTRS. The alignment of the KTRS as shown in Figure 2 was heavily influenced by the locations of the required hydraulic connection and the highly congested nature of the Project area both above and below ground. The KTRS will be installed by a Microtunnel Boring Machine (MTBM) with direct jacked reinforced concrete pipe. There will be five construction shafts along the KTRS Microtunnel (plus one

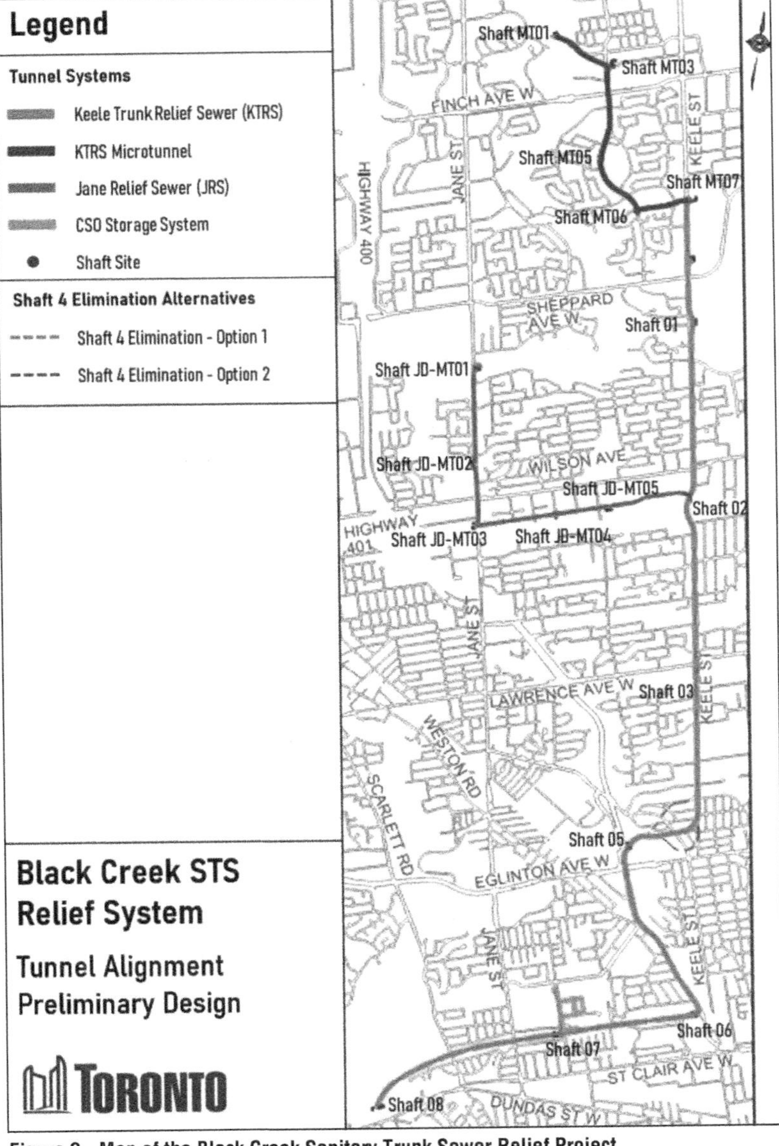


Figure 2. Map of the Black Creek Sanitary Trunk Sewer Relief Project

shared with KTRS) which will all ultimately function as maintenance access shafts or connection shafts.

### Combined Sewer Overflow Storage System

The CSO Storage System consists of two components: A 1,500 mm diameter and approximately 0.5 km long Storage Tunnel and a 39,000 $m^3$ CSO Storage Tank along with its connecting pipe. The Storage Tunnel collects flows from three existing CSO overflows that currently release excess flows into the Black Creek watercourse. The CSO Storage Tank collects flows from two existing CSOs, two existing CTSs and the existing Black Creek STS at varying locations along its connecting pipe. As shown in Figure 2, the CSO Storage Tank is connected to the Storage Tunnel by a single connecting pipe that also facilitates the connections to the existing sewer systems. The Storage Tunnel then transmits all flows from the CSO Storage System to the KTRS. The Storage Tunnel will be installed by a Microtunnel Boring Machine (MTBM) with direct jacked reinforced concrete pipe. The CSO Storage Tank and its connecting pipe will be installed by open-cut. There will be two construction shafts along the Storage Tunnel (plus one shared with KTRS) which will all ultimately function as maintenance access shafts or connection shafts.

## ANTICIPATED GEOTECHNICAL CONDITIONS

The Project area is located in the West Lowland of the Great Lakes (St. Lawrence lowlands) which are affected by the Pleistocene glaciation and covered by surficial deposits. The overburden varies from approximately 10 to 60 meters thick. North of Eglinton Avenue (as identified in Figure 2), the overburden mainly consists of Late Wisconsinan young tills. These tills are mainly clayey silt, silty clay, clay, sandy silt, and silty sand. South of Eglinton Avenue, the overburden mainly consists of modern lake deposits, glacial lake deposits, or older lakes deeper-water deposits. These river and lake deposits consist mainly of gravel, sands, silty sand, silt, and clay (Ministry of Natural Resources n.d.).

The bedrock in the project area consists of Upper Ordovician shale, limestone, dolostone, and siltstone of the Georgian Bay Formation at varying depths below ground surface. Buried valleys and steeply dipping bedrock surface are present along the alignment in the vicinity of Black Creek. The shale of the Georgian Bay Formation is typically low to moderate strength and exhibits swelling behavior. The limestone, dolostone, and siltstones are typically more competent in terms of strength, harness, and toughness. They also typically exhibit lower swelling potential compared to the shale.

To date, data from 33 boreholes is available along the KTRS and KTRS Microtunnel alignments. A comprehensive geotechnical investigation is underway for the Project to provide the subsurface information that will be used during preliminary and detailed design of the Project. This geotechnical investigation comprises of an additional 216 borings along the four Project alignments to be drilled over the course of two phases. The phase 1 investigation is anticipated to occur between May 2023 and October 2023 while the phase 2 investigation is anticipated to occur between November 2023 to April 2024. Based on the boreholes available to date, the following ground conditions are anticipated for each of the sewer alignments:

1. KTRS: hard rock consisting of shale, limestone, dolostone, and siltstone for the northern half of the alignment; soft ground consisting of silty clay, silt, silty sand, sand, gravel, clayey silt till, and silty clay till with potential for some mixed face conditions along the southern half of the alignment

2. KTRS Microtunnel: soft ground consisting of silty clay, clayey silt, silt, sandy silt, silty sand, and sand
3. JRS: soft ground consisting of silty clay, clayey silt, silt, sandy silt, silty sand, and sand for the upstream sections transitioning into hard rock consisting of shale, limestone, dolostone, and siltstone as it approaches KTRS
4. CSO Storage System: soft ground consisting of silty clay, silty sand, and silty clay till

## TUNNEL LINING SELECTION

As highlighted in the previous section, the ground conditions across the Project range from soft ground to hard rock. The varying ground conditions combined with the hydraulic requirements of the project do not allow for the use of single tunneling method and thus tunnel lining across all four Project components in a manner that would be cost effective.

In order to minimize costs and to increase clearance with the various below ground utilities already present in the Project area, a one pass microtunnel lining was preferred along the KTRS Microtunnel, JRS, and CSO Storage Tunnel. The easements and infrastructure along these three alignment also required horizontal curves along all drives. These curves and the nature of the flows to be carried drove the decision for direct jacked reinforced concrete pipe to be selected.

Unlike the microtunnels, the KTRS tunnel presents a relatively unique situation where the alignment is split roughly in half between soft ground and hard rock. Initially, in the EA study, two different TBMs along with two different pipe sizes were proposed to be mobilized to mine the soft ground and hard rock sections of the KTRS respectively. However, stakeholder coordination occurring during preliminary design added significant additional flows into the upstream end of the KTRS and resulted in the need to extend the KTRS upstream further than it was proposed during the EA. This provided an avenue for cost savings by utilizing a consistent lining approach for the entire KTRS sewer, a precast concrete tunnel lining with an inside diameter of 3,000 mm was selected for both sections of tunnel. The consistent approach to lining the tunnel has also allowed a reduction in the number of shaft sites and permitted the deletion of an especially challenging shaft in the immediate vicinity of a school.

## PROCUREMENT STRATEGY

As described above, the BCSTS Relief Project is comprised of four relatively large components. These four components are currently anticipated to be split across four main contracts with one early works contract as described in the following sections. The anticipated timing of the contracts are also described below.

### Contract E-1

Contract E-1 will encompass the setup of the TBM mining site for KTRS including site grading, utility relocation, environmental protection, and site grading in order to provide schedule acceleration for the KTRS Tunnel Contract and allow for increased involvement by local Contractors. Tunnel projects recently and currently in construction in the City of Toronto have been experiencing 1–2 year waits for utility relocations. Therefore, easement/property acquisitions and utility relocations at other shaft locations will also be advanced under this early works contract as dictated by a critical path analysis of the construction schedule.

## Contract 1

Contract 1 encompasses the KTRS from Shaft 8 to Shaft MT07. This contract encompasses the entire length of the 3,000 mm ID precast concrete tunnel lining to be installed by TBM and will also consist of eight (8) construction shafts (Shafts 01, 02, 03, 05, 06, 07, 08, MT07), six (6) maintenance hole structures (structures for Shafts 01, 03, 05, 06, 07, 08), and five (5) diversion structure/connections (DS-2, hand-mine to Maryport TRS, HSTS, two future redevelopment connections). Real time flow monitoring upgrades at existing BCSTS/HSTS connection (upstream) shall be installed, and real time remote monitoring at the connection of HSTS and KTRS, with manual override controls for emergency or contingency situations, shall also be installed under this contract. This will ensure that flows from the entire Project can be controlled based on flow capacities for both the HSTS and the Humber WWTP.

## Contract 2

Contract 2 encompasses the JRS from Shaft 2 to Shaft JD-MT01 installing 1800 mm ID RCP with a one-pass MTBM. This Contract will also consist of five (5) construction shafts (JD-MT01, JD-MT02, JD-MT03, JD-MT04, JD-MT05), six maintenance hole structures (structures for all JD-MT shafts as well as Shaft 02), and three (3) diversion/connections (DS-3, DS-4, Shaft 02).

## Contract 3

Contract 3 encompasses the CSO Storage Tank and an upsize of an existing combined sewer on Keele Street. In addition to the 39,000 $m^3$ CSO Storage tank, the contract scope will also include a diversion structure to divert flow from CSO 4 and also a maintenance and drop structure to divert flows from CSO 3, with overflows via existing CSO 3 if the tank is full. It will also have high water level diversion to reduce surcharge in Hillary CTS, Mt. Dennis CTS, Saint Clair CTS, and Rockcliffe CTS. After the CSO storage facilities are complete the existing combined trunk sewer flow control regulators at connections of Hillary CTS, St. Clair CTS, and Mt. Dennis CTS to BCSTS will be modified. The existing combined sewer on Keele St. from the flow control structure at Eglinton Ave W. to Juliet Crescent will be upsized to 1200 mm ID pipe over approximately 592 m to increase capacity, manage combined sewer overflows, and achieve one CSO per year or less.

## Contract 4

Contract 4 encompasses the KTRS Microtunnel from Shaft MT07 to Shaft MT01 and the CSO Storage Tunnel since both sewers are the same inside diameter and to be installed by MTBM. This contract will also include the following along the KTRS Microtunnel: four (4) construction shafts (Shafts MT01, MT03, MT05, MT06), five (5) maintenance hole structures (structures for Shafts MT01, MT03, MT05, MT06, MT07), and two (2) diversion structure/connections (DS-1, to KTRS at Shaft MT07). The CSO Storage Tunnel components of the Contract will also include two (2) diversion structures connected to the tunnel at Shaft ST01.

## Contract Phasing and Schedule

The above five contracts are currently anticipated to be grouped into two phases currently named the Phase One Works and Phase Two Works. While the preliminary design of both phases are concurrent, the detailed design and thus construction of the two phases will occur sequentially. The Phase One Works will encompass the KTRS and JRS (i.e., Contract E-1, Contract 1, and Contract 2). Contract E-1 is anticipated to begin in Q1/Q2 of 2024, Contract 1 in Q1 of 2025, and Contract 2 in Q1 of 2026. The schedule of Contract 3 and 4 will be further defined during detailed design of the Phase Two Works but is currently anticipated to begin in 2029/2030.

## SUMMARY

In summary, the Black Creek Sanitary Trunk Sewer Relief Project consists of four major infrastructure components, the KTRS, the KTRS Microtunnel, the JRS, and the CSO Storage System, that will reduce basement flooding issues in the Black Creek sanitary drainage area, reduce the number of CSOs that currently occur in the Black Creek watercourse, and provide increased sewer capacity for future population growth. These four major components include four tunnels: the 12.9 km, 3.0 meter diameter Keele Trunk Relief Sewer in rock and soil, 2) the 2.9 km, 1.5 meter diameter Keele Trunk Relief Sewer Microtunnel in soil, 3) the 3.9 km, 1.8 meter diameter Jane Relief Sewer in rock and soil, and 4) the 0.5 km, 1.5 meter diameter Combined Sewer Overflow Storage Tunnel in soil. These tunnels are currently anticipated to begin construction across multiple contracts with start dates ranging from 2024 to 2030 as described in detail above. Additional details of the contracts and details of the infrastructure will be made available once detailed design is finalized.

## REFERENCES

City of Toronto. (2016, April 28). Black Creek Sanitary Drainage Area Servicing Improvements Municipal Class Environmental Assessment Study Public Consultation Drop-In Event #1. Toronto, Ontario, Canada: City of Toronto.

Ministry of Natural Resources (not dated). Quaternary Geology, Toronto and Surrounding Area, Southern Ontario. Ontario Geological Survey Preliminary map P.2204. Ministry of Natural Resources.

PART 

# Risk Management

*Chairs*

**Joe Rigney**
Parsons

**Mo Magheri**
Kiewit

# 30 Years of Advances in Risk Management for Underground Projects

**John Reilly** ▪ John Reilly International
**Philip Sander** ▪ University of the Bundeswehr Munich
**Kevin Lundberg** ▪ RiskConsult GmbH
**Daniel Weinberger** ▪ RiskConsult GmbH

## ABSTRACT

Since the early days of risk management, developments in technology and guidelines (ITA, ITIG, OGG, UCA) have advanced risk management and probable cost estimating, which are now widely recognized. WSDOT's CEVP risk-based cost validation process is now established after 20 years. This paper traces these developments, summarizes advanced techniques, and presents examples of US and international projects with results. A particular case of using advanced risk evaluation to allow enhanced-capability TBMs to be considered in a competitive bid process is presented. Further advances, including the use of digital twin risk-based representation of complex projects, are described.

## INTRODUCTION

*"A plan, whatever it may be, must be made for the bad ground, it must be calculated to meet all exigencies, all disasters and to overcome them..."*
*—Marc Brunel, after the water break-ins of 1826–1828 at the Thames River Tunnel*

*We advocate for such plans—risk identification and risk response—to be made before construction starts!*

## UNDERSTANDING RISK—A SHORT HISTORY

As an introduction to recent US risk-management conferences, an overview "Update, A History of Risk Management," (Reilly 2018) was developed including content derived from Bernstein (1996). The overview included the following:

*"Documentation related to risk dates back to 1754 BC. The Code of Hammurabi was a well-preserved Babylonian code in ancient Mesopotamia, a collection of 282 rules and established standards for commercial interactions, which set fines and punishments with sections detailing events that could occur in construction and penalties for mismanagement of outcomes. Some historians believe the concept of risk arose through gaming from early times since people in ancient civilizations played games with dice and bones—games that evolved into chess and checkers over 2000 years ago. Subsequently, writings by Dante, Galileo, and many others indicate historical evidence that gaming gave rise to probability theory. In the 1600s the mathematicians Pascal and Fermat wrote to each other about games of chance—a correspondence that is believed to have led to modern probability theory."*

The relatively recent realization that project risks could be postulated, identified, quantified, and better managed is implicit in a series of publications (Reason 1990, 1997; Chapman 2003; Cooper 2005; Talib 2007; Vose 2009) and developments over the last 30 years, some of which are reported in this paper.

## RISK RELATED TO MEGAPROJECT COST ESTIMATING AND MANAGEMENT

In the mid-to late 1990s the International Tunnelling Association (ITA) and the American Underground Construction Association (AUA, now UCA) were concerned about poor megaproject cost and delivery performance. Reports showed cost overruns in the order of +50% to +200% of budget in a significant number of cases. A 2001 conference in Seattle sponsored by AUA presented cases of cost overruns up to +260% over budget (Stark 2001). Subsequently, a study of the out-turn costs of 258 infrastructure projects over 70 years confirmed the above findings and asserted that the problem of grossly inaccurate cost forecasts was chronic and had been so for 70 years (Flyvbjerg 2002). It noted that a key factor appeared to be "strategic misrepresentation."

One of the authors (Reilly), as President of AUA, had raised these issues with international and North American owners, managers, engineers, and contractors in order to identify key processes necessary for improved cost performance and management to budget. This work included a survey of international projects followed by presentations of the findings and recommendations in Boston, China, and Italy (Reilly 2000a, 2000b, 2000c). In these discussions, unanticipated risk events and their impacts on cost and schedule were acknowledged as key factors requiring further study and process implementation.

About that time, the Washington State Department of Transportation (WSDOT) was about to deliver a significant number of complex megaprojects. The Secretary, Douglas MacDonald, was concerned about the cost problem and requested evaluation of a better process to address cost issues and to better estimate the probable cost of complex transportation projects. This resulted in the development and implementation of the risk-based Cost Estimate Validation Process (CEVP) (Reilly et al. 2002, 2004). The CEVP process, now in service for 20 years (Reilly 2022), requires independent validation of base costs and the characterization of potential risks, which are then combined probabilistically with the validated base costs to produce a range of probable cost and schedule. Budgets are set, generally at a 60–80% probable cost level, depending on the owner's risk tolerance.

This led to formal procedures for the development of risk management plans, risk registers, risk requirements in construction specifications and the implementation of risk management related to the owner's risk management responsibilities, building on international guidelines and practices (WSDOT 2006, 2018).

## EARLY TIMELINE–RISK MANAGEMENT AND RELATED COST CONCERNS

Examples include:

- 1990s: Project risk workshops (e.g., LA Metro, 1995; Sir Adam Beck Tunnel, Niagara, 1997).
- 1993–99: Discussions regarding risk—AUA (now UCA), ITA, BTS and international conferences (ITA 1992). UK, Australia and New Zealand address risk as part of Alliancing with owner-engineer-contractor pain-gain provisions (e.g., Wandoo oil platform, Western Australia; Sydney Northside Tunnel Project [Henderson 1999]).
- 2001: AUA Conference on megaproject cost concerns, Seattle (Stark 2001).

- 2002: WSDOT develops the CEVP risk-based cost estimate validation process—leads to explicit risk definition and adoption of risk management as a required Agency practice (WSDOT 2006, 2018)
- 2003: WSDOT briefs FTA & FHWA on risk-based cost estimating. FTA and FHWA evaluate risk-based processes and adopt for specific projects (procedures and timelines vary).
- 2002–Present: Risk management processes gradually recognized and implemented by states and agencies—understanding grows as applications become routine and more professionals are engaged, conference topics routinely include risk. Integrated cost-risk-schedule models developed. Agencies develop risk management procedures and risk-based cost estimating. Codes of practice and industry guidelines for risk management are developed and used—ITA (1992, 2004), ITIG (2006, 2012), UCA Better Practices (Reilly 2008a), UCA Goodfellow and O'Carroll (2015), and ÖGG (2016).
- 2021: FHWA/NHI eBook course in probabilistic risk-based cost estimating.

## RISK MODELLING TECHNIQUES FOR COMPLEX TUNNELLING PROJECTS

Risk assessment methods and techniques have been developing steadily since they gained acceptance as a scientific field in the 1970s and 1980s (Aven 2016). Initially, and often still today, risk assessments were done with qualitative methods, where risks were estimated with descriptive words in terms of their likelihood and impact (e.g., low, medium, and high). Numerical methods, on the other hand, quantify the likelihood and impact in order to calculate the risk deterministically or probabilistically. Whereas the deterministic methods merely multiply single-point estimates for probability and impact, probabilistic methods allow for the inclusion of uncertainty by using distribution functions (usually three-point estimates such as the beta-PERT or triangular distribution) for certain parameters and running a high number of simulations to produce a range of possible outcomes. Probabilistic analysis can be used either to depict actual uncertainty and variation or to counter a lack of knowledge (Aven 2016).

Although the outdated idea of calculating risk as the deterministic product of its likelihood and impact is not entirely a thing of the past, a multitude of organizations have adopted the usage of probabilistic methods for risk analysis as they better reflect the inherent uncertainty of cost and risk estimates with ranges. But are these advancements sufficient in complex environments such as tunnelling projects, where inadequate modelling of cost and risk can result in estimate discrepancies that run into millions of dollars?

In very early project phases, simple probabilistic approaches usually suffice in representing even the largest infrastructure projects. However, as a project progresses, more sophisticated methods are needed to accurately convert the available information into reliable estimates for project cost and schedule. In recent years, more advanced probabilistic modelling techniques (e.g., event tree analysis (ETA), fault tree analysis (FTA), correlations and dependencies, etc.) have been developed and existing techniques have been refined in order to meet these requirements.

## EXAMPLES–TWO IMPORTANT US PROJECTS THAT USED RISK ANALYSIS

### Alaskan Way Viaduct Replacement Tunnel Project, Seattle

Famous for being the largest Tunnel Boring Machine (TBM) at the time (2013) at 57.5 ft/17.5 m in diameter, the machine stopped mining approximately 1,000 ft into the drive. Stoppage and damage to the machine required recovery of the TBM via a deep access shaft to remove and replace the cutting head, main bearing and seals-resulting in significant delay and added costs (Alaskan Way TBM 2013)

When the decision was made to use a single large-diameter TBM, a quick workshop was held to estimate the probable cost of the new project using risk-based CEVP principles. An 80% probability level of $2.2 billion was recommended, however the State Legislature approved $1.9 billion (approximately the 60% probability level). The owner's risk register was extensive and included two risks related to machine breakdown issues with a combined impact of approximately $125–180 million (most likely) and $150–220 (high range) but with probabilities of less than 5%. These risks did eventuate, which affected the agency and contractor, who brought litigation to recover costs. The court, however, ruled that these costs were not recoverable from the owner under the design-build contract provisions.

The final total program cost to the owner was a little over the authorized budget. The 60% CEVP was $3.1 billion and the actual cost was $3.2 billion (rounded numbers). The lesson here is that major risk events are foreseeable and can be quantified in cost and probability; however, risks that are of great consequence but low probability are a continuing concern (Talib 2007).

### SR520 Lake Washington Floating Bridge, Seattle

The longest floating bridge in the world, this structure links the east and west shores of Lake Washington, which is too deep for a conventional bridge or a tunnel. Seventy-seven prestressed concrete pontoons were linked together to form a 7,700 ft long floating structure.

Numerous CEVP evaluations were made of several options and the budget for the final preferred alternative was set according to the 60% probability level. During the bidding and initial construction of the east side approach, extremely competitive bids were received and the Legislature, as a result (and against recommendations related to potential risk), removed approximately $300 million from the budget. Subsequently, problems with pontoon cracking required expensive remediation, and funding had to be added back into the budget. The actual cost of the project closely met the 60% CEVP number of $3.1 billion for the lake crossing and the east side approach.

## CASE STUDIES—EUROPEAN

An advanced risk analysis process, the digital project risk twin process (Sander, Eßig, and Reilly 2021), explained in detail in the following sections, has been used successfully on the following large underground projects:

- Construction of the new subway line U5 in Hamburg (Hamburger Hochbahn AG), a 30 km-long subway tunnel estimated to transport 270,000 passengers daily that connects the northeast and northwest of the city to the city center and main train station.
- Construction of the Brenner Base Tunnel (BBT SE), a tunnel connecting Austria and Italy with a 64 km-long high-speed railway tunnel.
- Construction of the Koralm Base Tunnel (Austrian Federal Railways), a 33 km-long high-speed railway tunnel connecting the cities Graz and Klagenfurt.
- Intersection of the U2 and U5 metro lines in Vienna (Wiener Linien AG), including 10 new stations with over 6 km of new tracks.
- Extension of the U6 subway line in Munich (U6-PMG), a 1 km extension of the U6 connecting Planegg to the Munich metro system.

Experience in these projects has also shown that earlier inclusion of the risk management process allows the owner to better communicate with both the public and other official stakeholders. Most frequently these stakeholders are the regional and national governments and government agencies that are often responsible for providing the funds and budgets for large infrastructure projects. Especially in the early planning stages of a project, making use of risk management can help develop more realistic cost estimates by including the additional costs from risks and price escalation that the more common approaches do not always fully consider. Additionally, project success can be significantly improved by targeting countermeasures to risks that most frequently lead to cost spikes and large delays. This is best accomplished by including all of the internal project stakeholders (agencies, owner, contractor, and designer) in the risk management process.

## TBM ENHANCED CAPABILITY USING RISK EVALUATION

The particular case of using advanced risk evaluation to allow enhanced-capability TBMs to be considered in a competitive bid process was the subject of an article in TunnelTalk (Reilly 2021). This arose from the following question: Who should specify the TBM—the agency, engineer, contractor, or TBM manufacturer? Feedback to an earlier TunnelTalk article in 2020 strongly suggested that the contractor, if under a design-build procurement, should select the TBM. Concerns were also noted that some engineers may not be sufficiently experienced to specify the TBM. The text of the article included the following:

> *"In comments received, one TBM manufacturer reported that commercial terms and cost limitations can often preclude the supply of a machine with the tunnelling capabilities and risk mitigation measures that they believe are necessary to reduce risk and result in a successful drive with good cost and schedule performance.*
>
> *In this regard, it is recognized that the cost of the TBM, or TBMs, is about 5% of the overall project cost, but that it is a key determinant of project outcome. It is therefore foolish to constrain the cost of the TBM and thereby reduce or exclude prudent capability that may be required by the ground conditions. To avoid this dilemma, a sufficiently detailed risk analysis of the machine and the drive or drives should be undertaken to cover machine design, expected performance, issues that may be encountered, and anticipated ground conditions. Such a risk analysis should be made in order to inform a TBM design and price that balances machine capabilities to risk exposure. Armed with the analysis, risk mitigation measures should be adopted that produce benefit for the owner and the contractor.*
>
> *To allow this contractually, and maintain cost and value considerations, it may be that the contract price could be increased above a defined base by the estimated value of the risk mitigation measures. For example, the TBM procurement tender could include a competitive base price for a TBM that meets the owner's prescriptive and performance specifications, with additional monies then authorized to fund additional TBM capabilities related to the estimated value of risk reduction measures, if implemented. Under this concept, while overall initial cost to the owner might increase, overall value to the owner would also increase (and offset initial costs).*
>
> *In a contribution, it was reported that an owner and its contractor could be prepared to pay 10–15% more in capital purchase for a certain TBM if it*

*believes that a greater than 15–20% improvement of tunnelling production rates will be gained. This would require early-stage collaboration between the owner and the contractor and a contract that allows, and makes provision for, such early-stage collaboration.*

*Clearly, and generally speaking, any new product or innovation offered by a TBM manufacturer may only be recognized by the contractor as a means of improving the TBM manufacturer's chances of being selected by the contractor and/or the contractor's chances of being awarded the project and may not be linked to overall project gains.*

*A noted concern is that strong competition among TBM manufacturers may lead to a market where pricing dumping is a disruption strategy adopted by some TBM manufacturers to gain access to a new market.*

*There was feedback from those invited to comment that more problems may come from poor TBM operation rather than a TBM that is not suited for the project, in other words, not fit for purpose. This was challenged by other respondents who pointed to cases where there were problems with both the TBM and its operation. In this regard, one comment referenced a 2005 recommendation of the British Tunnelling Society that 'the correct choice of machine, operated without the correct management and operating control, is as bad as choosing the wrong type of machine for the project."*

## DIGITAL PROJECT RISK TWIN

Major projects involve a high investment level, a high degree of uncertainty, and long durations. In order to cope with these factors, it is advantageous to use a digital project risk twin (PRT) within which uncertainties and dependencies between cost and schedule can be transparently displayed and analyzed. The results of the analysis are then summarized in comprehensible dashboards.

These methods provide valuable information on uncertainties compared to deterministic methods and enable the validation of a project's budget. These advanced modelling techniques are put into practice with a digital project risk twin. The following sections describe PRT basics and structure.

A digital twin (DT) is a representation of a tangible or intangible object or process from the real world in the digital world. DT has become a commonly recognized term in large infrastructure projects as technological advances in modelling and simulation make it applicable. Figure 1 shows the elements of a DT for cost and schedule in infrastructure projects with the corresponding steps: the input/system integration of cost, schedule, risk, and budget data are gathered at the beginning (Figure 1[1]). In the PRT (Figure 1[2]), the data is linked by software. Simulation is used to perform an integrated cost and schedule risk analysis including uncertainties. The analysis results are processed and displayed in the form of individually tailored dashboards (Figure 1[3]).

The PRT is made up of several components. Figure 2 shows chronologically from top to bottom the process steps to create such a model, starting with basic cost and basic schedule planning, through risk analysis and risk mitigation, to the selection of an approach for escalation (Sander et al. 2022).

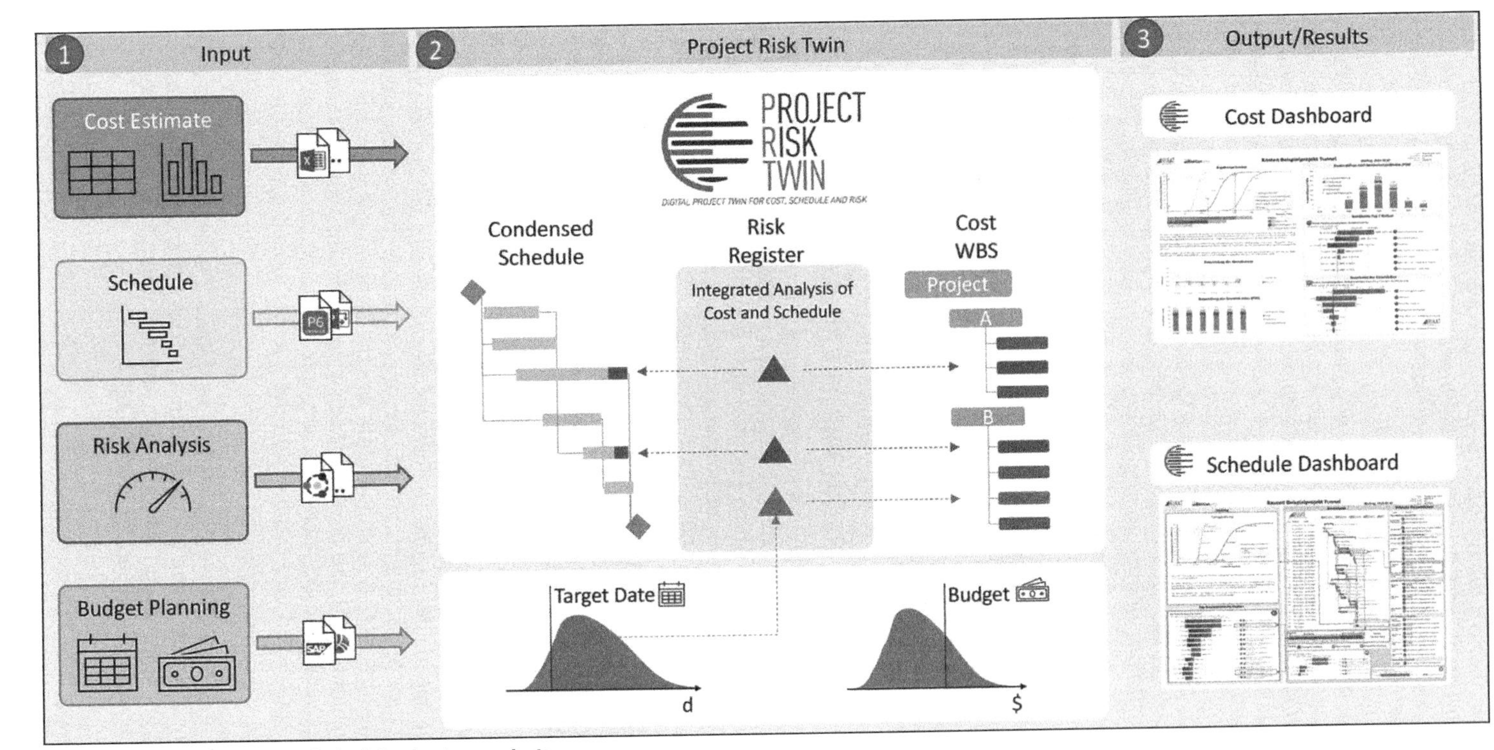


Figure 1. Digital project risk twin for infrastructure projects

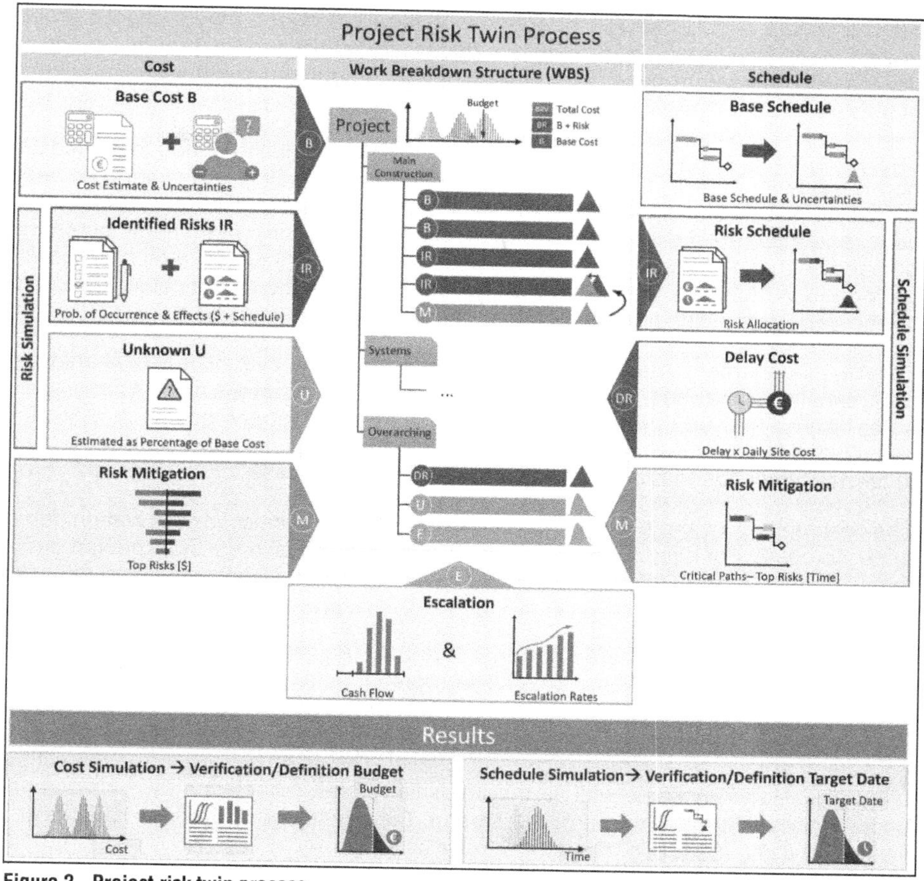


**Figure 2. Project risk twin process**

In the lower part of Figure 2, the results for cost and schedule are shown. Furthermore, a distinction is made between the sub-process for cost in the left part of the graphic and the sub-process for schedule in the right part of the graphic. In the middle, the work breakdown structure (WBS) is shown, which structures the recorded cost items hierarchically so that they can be analyzed individually.

At the beginning of the process, the base cost is determined, uncertainties for quantities and prices are evaluated, and a base schedule is created. It is recommended to have these base results validated by external experts. This is followed by the risk analysis—specific risk scenarios are identified, described, and quantified in terms of their probability of occurrence and impact on cost and schedule.

In order to take uncertainties into account, the evaluation can use three-point estimates (best, expected, and worst case). Both positive effects (opportunities) and negative effects (threats) are considered from the perspective of the two contracting parties: client and contractor. Risks that could delay construction completion will in most cases also trigger additional time-related cost. To calculate the cost of delay, the potential delay due to risk impact is linked with time-related costs taken from the base cost estimate.

All information is obtained in moderated workshops attended by the client's project team. Conditioning of the participants is necessary before the workshops. Important goals are to raise awareness regarding risk management and to reduce several forms of bias (such as optimistic bias).

In the second step of the risk analysis, unknown risk, consisting of non-identified and non-identifiable risks, is considered. Structured questionnaires are used to evaluate the project based on numerous factors such as maturity, complexity, geological conditions, and the market situation. The result is a project-specific percentage surcharge on base cost, which also takes into account the quality of the individual risk analysis performed. A higher degree of complexity—particularly unfavorable geological conditions—results in a higher surcharge to be applied.

For risk management, the impact of individual risks on the overall project is analyzed (e.g., sensitivity analyses, what-ifs, critical paths, etc.). Appropriate risk mitigation measures can be derived from this to mitigate high-impact cost and schedule risks. The risk mitigation measures can be quantified in advance by simulations and evaluated in terms of their cost-effectiveness. This provides a solid base for decision-making.

The cash outflow over the entire project duration is used to calculate escalation (future price increase). For long-running projects, high inflation cost must be expected due to the compound interest effect, which must be considered for both base cost and risks.

After the final simulation of the entire model, the results are transparently summarized by means of dashboards (Figures 3 and 4). Figure 3 shows the implementation for a tunnel construction project. The representation shown here uses the Risk Administration and Analysis Tool (RIAAT) software. In the left part of the picture, the WBS is shown with the cost components. To the right, the probabilistic results of all cost items are shown for the selected WBS level. The next window shows the schedule with assigned risks as well as the prediction of the completion date considering uncertainties. On the far right of the screen, the sensitivity of the individual risks is visualized using a tornado diagram (Sander 2022).

Figure 4 shows the results of the cost analysis of a tunnel based on a fictitious example. Here, for example, the predicted total cost (base cost, risk cost, and escalation), cash outflow according to cost components, and impact ranges of the individual risks and their sensitivity can be read off. As already described in part one, the S-curves (cf. Figure 4. top left) can be used to evaluate the robustness of the budget (Sander, Becker, Lammers, et al. 2021).

## SUMMARY—RISK-BASED COST ESTIMATING BENEFITS

From surveys and feedback after risk-based workshops we have identified the following:

- Many benefits were noted by participants—including improved project understanding and awareness of potential risk-based costs and better risk management practices.
- Earlier understanding of the project, better alignment to key goals and objectives.
- Ability to set budgets related to an agency's risk tolerance, considering the probability of success.
- Process is transparent to stakeholders—stakeholder communication is improved.
- Range of probable cost better aligns with stakeholder expectations.

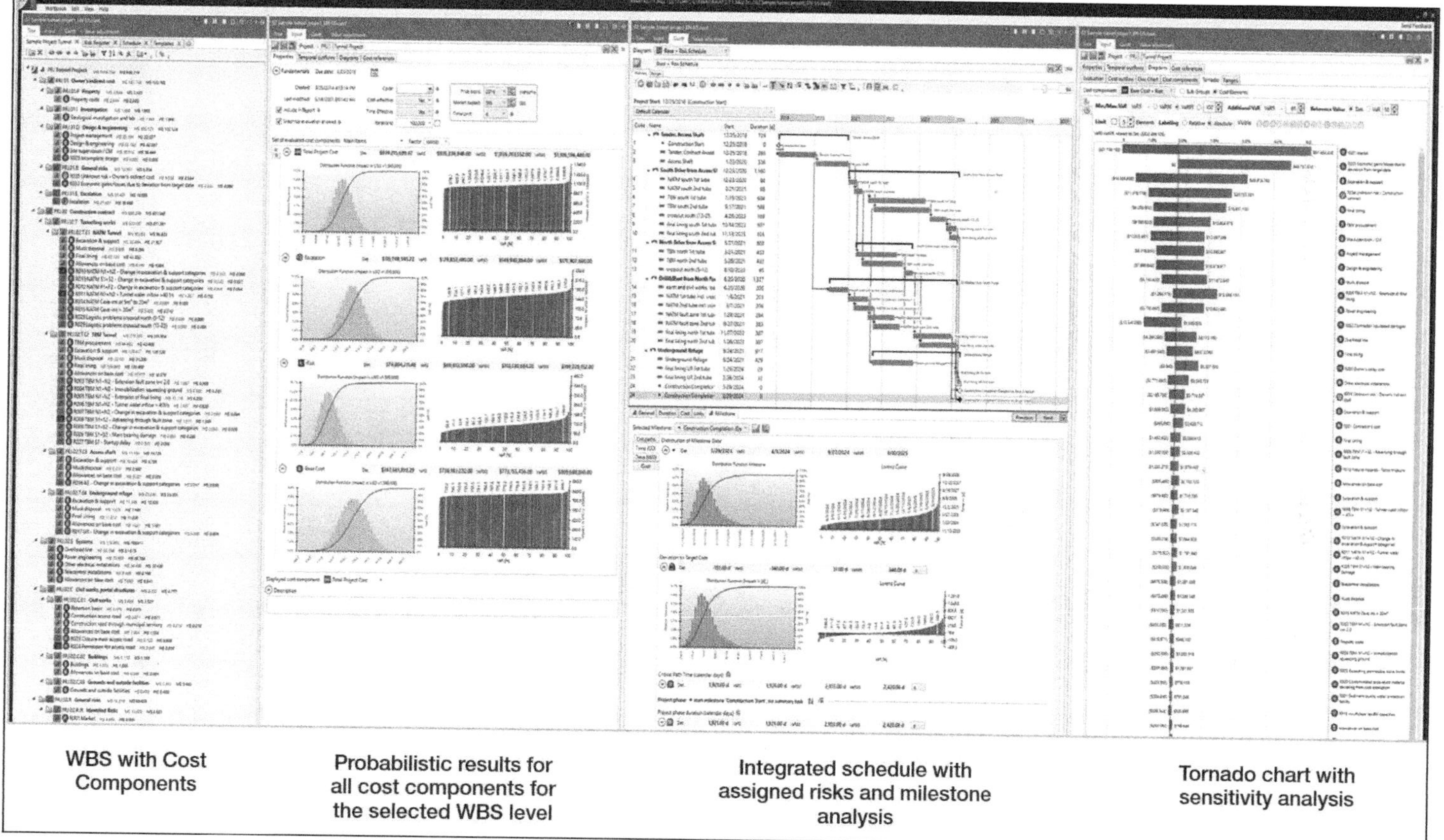


Figure 3. Sample main interface RIAAT software

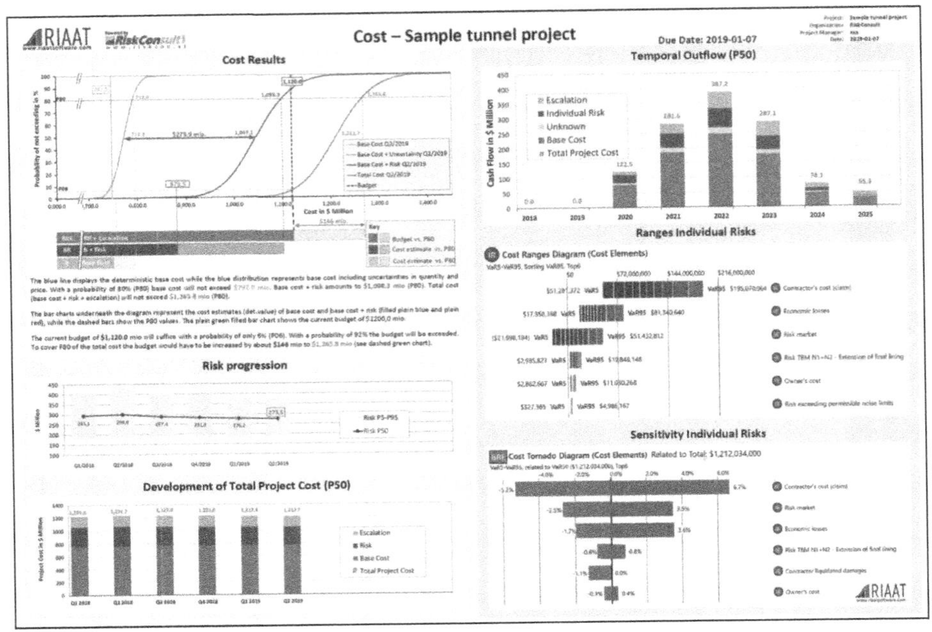


Figure 4. Dashboard for cost of an example tunnel

- Early CEVP-validated estimates counter an "optimism bias"—allows more realistic budgeting and scope management.

Examples of out-turn performance using risk-based cost estimates and processes:

- WSDOT Alaska Way—CEVP 60% $3.1 billion, actual $3.2 billion (rounded numbers)
- WSDOT SR520—CEVP 60% $3.1 billion for Lake crossing and east side equaled the actual cost
- WSDOT/ODOT evaluated outcomes for 28 smaller highway projects comparing projects with a cost-risk assessment (mini CEVP) vs. no cost-risk assessment. Change order costs were reduced from +10.3% to +3.9% and schedule changes were reduced from +18.2% to +8.2%

## REFERENCES

Aven, T. 2016. Risk assessment and risk management: Review of recent advances on their foundation. European Journal of Operational Research August.

AlaskanWayTBM2013(https://en.wikipedia.org/wiki/Bertha_(tunnel_boring_machine).

Bernstein, P.L. 1996. Against the Gods: The Remarkable Story of Risk. John Wiley & Sons.

Chapman, S., and Ward, S. 2003. Project Risk Management. John Wiley & Sons.

Cooper, D., Grey, S., Raymond, G., and Walker, P. 2005. Project Risk Management Guidelines—Managing Risk in Large Projects and Complex Procurements. John Wiley & Sons.

Flyvbjerg, B., Holm, M.S., and Buhl, S. 2002. Underestimating Costs in Public Works Projects: Error or Lie? Journal of the American Planning Association 68(3): 279–296.

Goodfellow, R., and O'Carroll, J. 2015. Guidelines for Improved Risk Management Practice on Tunnel and Underground Construction Projects in the United States of America. UCA.

Henderson, A. 1999. Northside Storage Project. Proc 10th Australian Tunnelling Conference, Melbourne. p. 57.

ICE. 1985. Risk and its Management in Construction Projects. Proceedings of the Institution of Civil Engineers Part I 78(3): 449–521 and discussion 80(3): 757–764.

ITA. 1992. Recommendations on Contractual Sharing of Risks, 2nd ed. TUST Vol. 7(1): 5–7.

ITA. 2004. Guidelines for Tunnelling Risk Management. International Tunnelling Association, Working Group 2, Eskensen, S. TUST 19(3): 217–237.

ITIG. 2006, 2012. Code of Practice for Risk Management of Tunnelling Works. January 2006 and May 2012.

ÖGG—Österreichische Gesellschaft für Geomechanik. 2016. ÖGG Guideline for the Cost Determination for Transportation Infrastructure Projects.

Reason, J. 1990. Human Error. Cambridge University Press.

Reason, J. 1997. Managing the Risks of Organizational Accidents. Ashgate Publications.

Reilly, J.J. 2000a. Tying it together—Policy, management, contracting, and risk, an overview. Proceedings of the North American Tunnelling Conference June 2000, Boston. pp. 159–168.

Reilly, J.J. 2000b. The management process for complex underground and tunnelling projects. Journal of Tunnelling and Underground Space Technology 15(1): 31–44.

Reilly, J.J. 2000c. Management of complex underground projects. Keynote Address, 5th Beijing Tunnelling Conference, November 9. Tunnels & Tunnelling International, Chinese Edition. pp. 3–9.

Reilly, J.J., McBride, M., and Dye, D. 2002. Guideline Procedure, Cost Estimate Validation Process (CEVP). Washington State Department of Transportation, January.

Reilly, J.J., McBride, M., Sangrey, D., MacDonald, D., and Brown, J. 2004. The development of CEVP®—WSDOT's cost-risk estimating process. Proceedings of the Boston Society of Civil Engineers, Fall/Winter.

Reilly, J.J. 2008a. Risk management. In Recommended Contract Practices for Underground Construction. Denver, CO: Society for Mining, Metallurgy, and Exploration, Inc.

Reilly, J.J. 2018. Update: A history of risk management. Presentation at the Risk Management Conference, Los Angeles, CA. November.

Reilly, J.J. 2021. TBM procurement, risk and technology advancement, part 2. TunnelTalk. April.

Reilly, J.J. 2022. CEVP—How it started, how it's going, recognizing founders. Virtual presentation, Seattle, WA. June.

Sander, P., Eßig, M., and Reilly J. 2021. RISK 3689: Integrated cost schedule risk analysis: Application of project risk twin process for major infrastructure projects using RIAAT (digital twin). AACE International.

Sander, P., Becker, C., Lammers, M., Upfoff, K., Brodehl, R., and Van Droogenbroeck, A. 2021. Risk management in major tunnelling projects—Part 2: Digital project risk twin. Application for the Construction of U5 East, Hamburg.

Sander, P., Spiegl, M., Reilly, J., and Burns, T. 2022. Digital project twin for quantitative cost, risk, and schedule assessment of capital projects. Australian Journal of Multi-Disciplinary Engineering.

Stark, C., 2001. Chart of 21 projects, cost and cost overruns compared to budget. Presented at the US Underground Construction Association Conference, March 19.

Talib, N.N. 2007. The Black Swan: The Impact of the Highly Improbable. Random House.

WSDOT 2006 Initial Alaskan Way Risk Management Plan.

WSDOT 2018 WSDOT Risk Management—current document see: https://wsdot.wa.gov/publications/fulltext/CEVP/ProjectRiskManagementGuide.pdf.

Vose, D. 2009. Risk Analysis: A Quantitative Guide, 3rd ed. Wiley & Sons.

# Ashbridges Bay Treatment Plant Outfall TBM Risk Mitigation—Prescriptiveness and Verification of TBM Fabrication

**Dan Ifrim** ▪ Hatch
**Andre Solecki** ▪ Hatch

## ABSTRACT

Tunneling could be a major challenge to tunneling contractors as tunneling operations can be seriously affected by the Tunnel Boring Machine (TBM) performance. Specifying the appropriate TBM methodology, selection of systems and features, is critical for managing tunnel risks and contribute to successful tunneling project. The paper analyzes tunneling records from the recently completed City of Toronto's Ashbridges Bay Treatment Outfall tunnel project and underlines the righteous selection of the TBM type and associated TBM systems in achieving good advance rates during tunnelling and managing project risks. The paper intends to draw attention of a few aspects of TBM technical specification importance and highlight the lessons learned from this process.

## INTRODUCTION

The Ashbridges Bay Treatment Plant Outfall (ABTPO) project by the City of Toronto involves construction of a new tunnelled outfall to convey treated secondary effluent (water) from the ABTP into Lake Ontario. The ABTP is one of Canada's largest and oldest wastewater treatment plants. This new outfall will allow cessation of operations of the existing outfall which is reaching the end of its service life and has limited hydraulic capacity. With only 1 m of driving head available at peak design flow under high lake water conditions, the new ABTPO has been designed to minimize headloss and allow the outfall to operate by gravity with a design capacity of 3,923 Million Liters per Day (MLD).

Outfall components (as illustrated in Figure 1) consist of an 85 m deep, 14 m internal diameter onshore shaft constructed adjacent to the shoreline; a 3,500 m long,

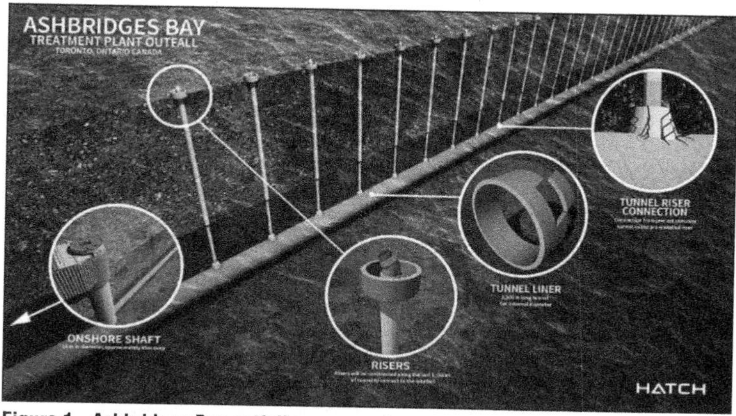


Figure 1. Ashbridges Bay outfall components

7 m internal diameter tunnel constructed through rock beneath the lakebed; and, fifty 1,000 mm diameter risers with 830 mm diameter ports installed in line with the tunnel extending from the tunnel horizon to the lakebed at equal spacing along the diffuser. The project also includes the construction of a new effluent conduit that will convey treated effluent from the ABTP to the new outfall for dispersion into Lake Ontario through the risers.

Construction of the ABTPO project commenced in 2019 and is anticipated to be completed by the end of 2024. The delivery of the construction contract followed a traditional design-bid-build procurement. Main team members on the project include:

- Owner: City of Toronto
- Consultant: Hatch with Jacobs/Baird
- Contractor: Southland Astaldi Joint Venture

## TUNNEL WORKS

The tunnel works started in the summer of 2020 and completed in the spring of 2022. The prequalified Contractor, a joint venture of Southland and Astaldi was the low bidder and awarded the project.

The tunnel was mined using a Shield Rock Tunnel Boring Machine (TBM). The tunnel specification allowed for a new or remanufactured TBM. The Contractor selected Robbins as supplier of a remanufactured TBM. The TBM launched from an onshore shaft of 16 m in excavated diameter, and adjacent fore tunnel of approximately 100 m long to facilitate assembly of the full TBM before launching.

## THE TBM

Upon contract award, the contractor procured a re-manufactured 7.95 m bore diameter single-shield high performance hard rock single shield Tunnel Boring Machine (TBM) from The Robbins Company (Robbins). The original TBM was an 8.7 m excavated tunnel diameter cross-over TBM utilized on the Túnel Emisor Poniente (TEP) II project located in Mexico City, Mexico. Refer to Figure 2 for a general arrangement of the TBM on utilized on the TEP II project.

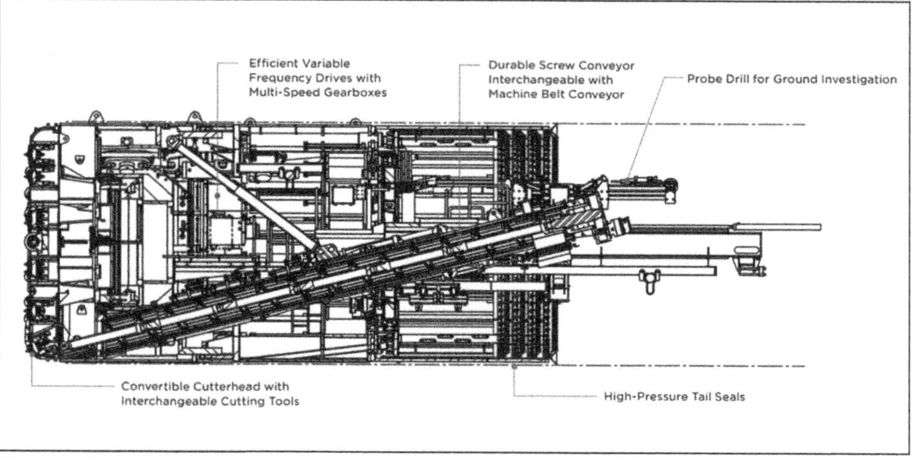


**Figure 2. Emisor Poniente crossover TBM general arrangement**

Figure 3. Remanufacturing site in Mexico City

## REMANUFACTURING PLAN

The existing TBM required extensive remanufacturing to fit ABTP Tunnel design specification requirements. These requirements include new Cutterhead and Shields, new Main Drive (Bearing, Seals, Motors and Gearboxes) a remanufacturing plan to reuse existing components such as Gantry, Belt Conveyors, Propulsion and Articulation Jacks, VFDs, Erector, Refuge Chamber, etc.

The plan included reconfiguring of the original multimode also known Crossover XRE (EPB Open Mode) TBM utilized on the Emisor Poniente project, into a Rock Shield TBM (operating only in open mode).

The ABTPO contract required that TBM re-manufacturing followed ITAtech Guidelines on Rebuilds of Machinery for Mechanized Tunnel Excavation (ITAtech, 2019). The remanufacturing plan was prepared by Robbins/SAJV and submitted to the Construction Management team for review and consideration. The plan was based on Robbins Quality Plan and included quality verification requirement throughout the manufacturing process, starting with the raw material verification, Finite Element Analysis, components re-manufacturing process, new components submittals and verification for compliance with the TBM technical specification requirements.

The remanufacturing location, as elected by Robbins, was in Mexico City and was located in the same construction site on the TEP II project were the TBM was retrieved from the project. Re-manufacturing on site saved the project time and cost by allowing the TBM to be remanufactured and shipped directly to the ABTPO project from TEP II project location. The main re-manufacturing site was complemented by various small shops that were performing part of the remanufacturing including the hydraulic cylinders plant.

The locations were not the typical American or Canadian shops and did not inspire a lot of thrust with respect to a successful remanufacturing process. The Contractor and Robbins efforts, along with the local Mexican shops proved to be up to the task and deliver a quality re-manufacturing process.

## TBM REMANUFACURING PROCESS

The remanufactured TBM was specified by the TBM technical specification "Work included in this section includes the procurement, manufacturing, testing, assembly, installation, operation, maintenance, abandonment of a new or fully re-manufactured/refurbished, shielded, rock Tunnel Boring Machine (TBM) capable of performing as specified in the Contract Documents."

The remanufacturing plan as initially drafted by Robbins was discussed with the project team and altered to mitigate risk and ensure early compliance for all TBM systems and plant.

The plan was affected by Covid-19 pandemic and some components were delayed or refurbished on the remanufacturing site in Mexico City.

## Cutterhead

Having in consideration the ABTP ground conditions, a new Rock Cutterhead design was produced and manufactured for the project.

## Main Drive

As per the TBM technical specification a new Main Drive including new Bearing, Gearboxes and Motors.

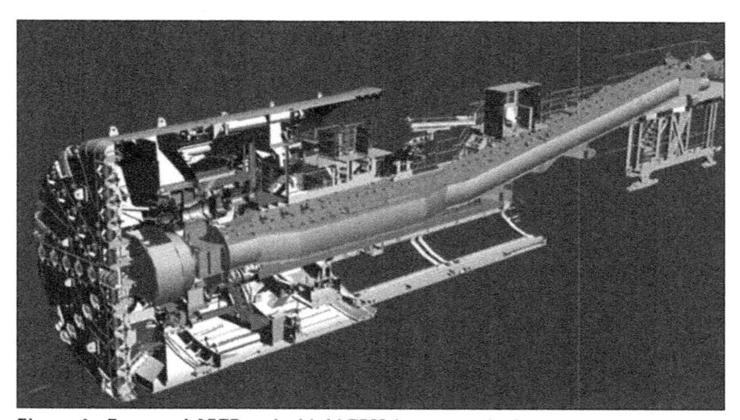

Figure 4. Proposed ABTP rock shield TBM (cross section)

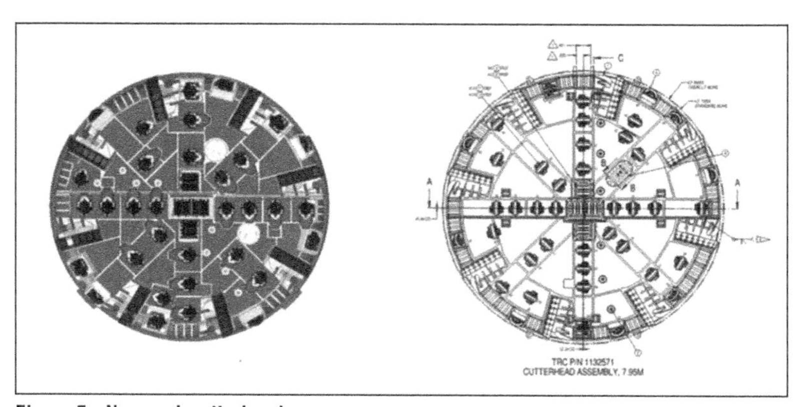


Figure 5. New rock cutterhead

The Contractor and Robbins submitted a detailed remanufacturing procedure identifying step by step operations and remanufacturing procedure with identification of the new components and provenience.

The main drive components were purchased from Italy (seals and Gearboxes) utilized a new main drive bearing manufactured by Liebherr in United States new and remanufactured Elin motors and a new ring gear manufactured in Turkey.

## Shields

Due to smaller tunnel diameter on the ABTPO Project (8 m OD) compared to the original TBM utilized on the TEP II project (8.7 m OD), new shields were required. The shields were designed by Robbins and validated through a Finite Element Analysis.

The manufacturing was realized in China and due to Covid-19 the CM team could not do any verification and rather rely on the Hatch China review of the manufacturing process.

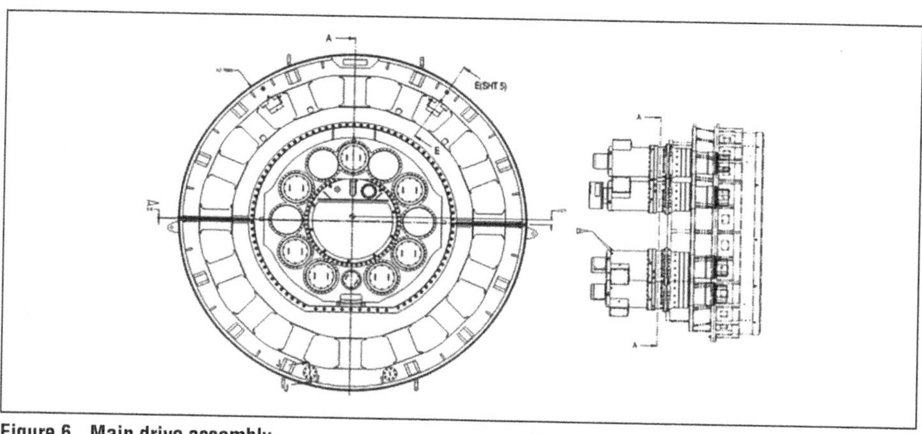

**Figure 6. Main drive assembly**

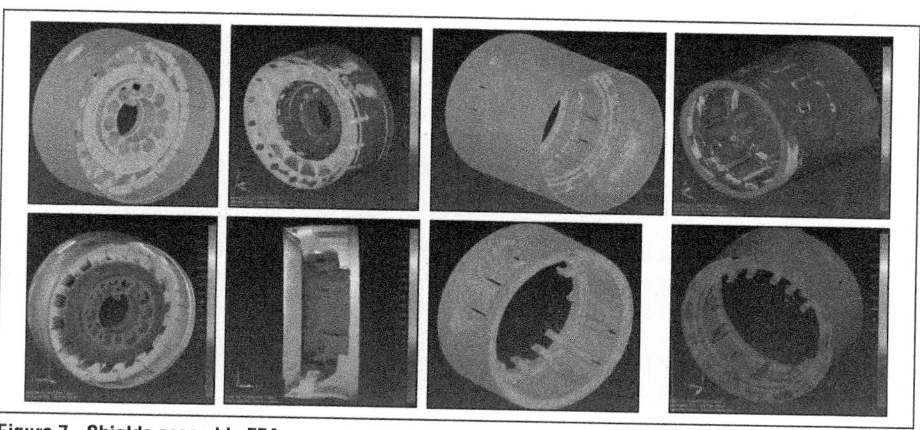

**Figure 7. Shields assembly FEA**

## Electrical System

The Substation transformers were tested per applicable standards (such as transformer turns ratio (TTR), insulation resistance, winding resistance, power factor, excitation, and over potential—Hi Pot) prior to cleaning.

The internal inspection is conducted to provide an assessment of the condition of bushing, gaskets, windings, and connections verification for resistance.

The transformers were tested after remanufacturing including insulation resistance, polarization, dielectric absorption, winding resistance, power factor insulation, excitation, over potential, etc.

All Motor Control Center (MCC) cabinets were remanufactured, fitted with new components MCC's and VFDs. All connections were tagged for ease of connecting on site to the appropriate motor's connections. Remanufacturing was carried by a local Mexican electric firm.

## Class 1 Division 2

The TBM required to be Class 1 Div.2 compliant due to tunnel gassy conditions. All electrical enclosures were redesigned and remanufactured to meet that criterion.

## Hydraulic System

The hydraulic system was remanufactured entirely (cylinders, pumps and motors) in Mexico City.

All components were initially tested and then disassembled to individual parts, verified, replacing worn seals and packers with new parts. Also, all hoses, connectors and manifolds were replaced with new.

## Segmental Lining

The tunnel as designed by Hatch, consisted of 7-piece ring (6+1). During the submittals period the Contractor requested a change in design (from 6+1 to 6-piece

**Figure 8. Electrical remanufacturing**

Figure 9. Remanufacturing of propulsion jacks

trapezoidal) to accommodate the configuration of the existing TBM propulsion system. The proposed configuration was adopted, and a design verification was performed by Hatch. The change was processed as a value engineering proposal with the benefits of reducing re-manufacturing cycle as well as reducing the ring erection time during tunnelling.

## Segment Erector

A ring segment erector with vacuum pickup system utilized on the Emisor Poniente project was adapted to ABTP project with small modifications such as the addition of spacers behind the rollers around the ring to address the larger diameter and modification of the vacuum pad for matching the ABTP segmental lining design.

## Grout Injection System

The grout injection system (a two-component type) was adopted from the Emisor Poniente project and was thoroughly re-manufactured and fitted with new components from the manufacturer (Verder). One mention, the remanufacturing took place in Mexico City and carried by Robbins skilled technicians with some support from the manufacturer. The system included two- Grout tanks and 5 sets of Component A and B peristaltic pumps.

## Probe Drill

The probe drill was tested and concluded that all parameters were in like-new condition and did not require major re-manufacturing, however the mounting was modified to match the new Shield. Throughout the tunnelling process the unit was used extensively and successfully.

## Emergency Refuge Chamber

A new refuge chamber was manufactured and supplied in conformance with the technical specifications and Class 1 Div.2 requirements.

## TBM Conveyor

The tunnel conveyor remanufacturing consisted of adapting the existing conveyor bridge to the new TBM, manufacturing of a side transfer conveyor and replacing all rollers and belts.

Figure 10. Two-component grout system with peristaltic pumps

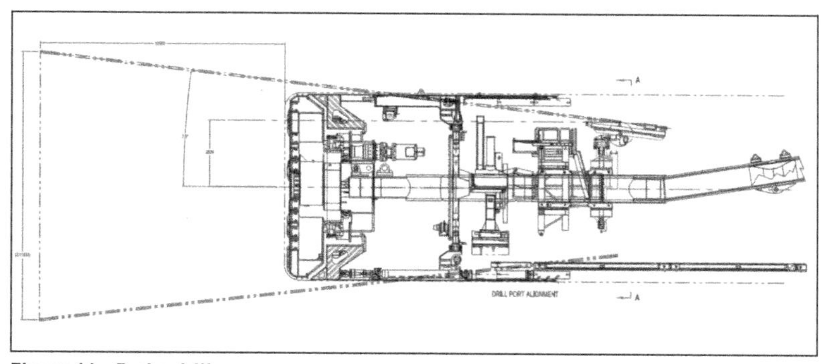

Figure 11. Probe drill

## TBM Transportation

The TBM was transported by sea from the port of Vera Cruz in Mexico to the port of Hamilton in Canada.

# TBM VERIFICATION AND ACCEPTANCE

## Verification Criteria

The TBM verification was somewhat strict and prescriptive. The prescriptiveness came from the challenges and specifics of the project along with requirements to prevent avoidable delays meet schedule.

Among the requirements, the followings are deemed important to share:

- The TBM main components shall be all new and of a recognized industry standard "best"
- The proposed TBM shall carry a full warranty by the manufacturer (original manufacturer-for rebuilt TBMs) for the duration of the project
- During the TBM fabrication, the TBM manufacturer shall collect and make available the followings:

Figure 12. Refuge chamber

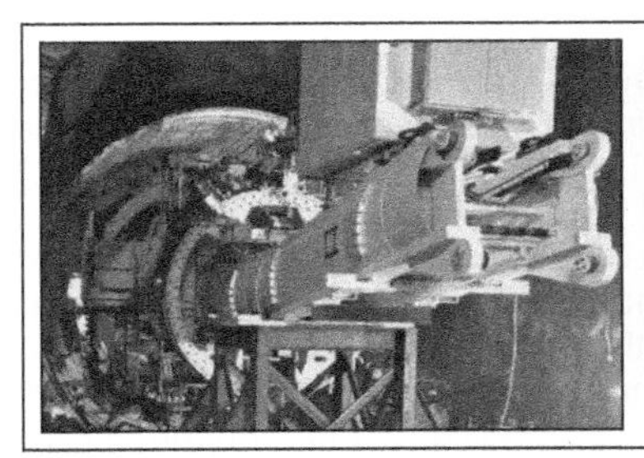

Figure 13. Images of remanufactured TBM conveyor

Figure 14. TBM unloading at Hamilton port

  - All materials Certificates of Origin and testing certificates issued by an ISO 17025:2005 certified body.
  - All fabrication geometrical tolerances and verifications.
  - All purchased components Certificates of Origin, data sheets and certifications.
  - All calibration certificates.
  - All internal assembling and testing verifications.
- TBM Test and Verification (FAT)
  - The TBM manufacturer must assemble and provide a demonstration test conducted at the factory in the presence of the Consultant before shipping to the Site.
  - The Contractor with TBM manufacturer assistance shall provide an additional demonstration test conducted onsite before launching the TBM for the tunnel drive.
  - The TBM manufacturer shall provide Finite Element Analysis for all structural components, shields and main drive assembly with simulated tunnelling loads.
  - A minimum of two test PCTL rings shall be produced and shipped to the TBM manufacturer for testing of the ring erector prior to the TBM acceptance test.
- TBM Manufacturer support:
  - A technical representative from the manufacturer: knowledgeable in the assembly, operation, maintenance and repair of TBM, to assist the TBM assembly onsite and be on site throughout the first 500 m of tunnel drive.
  - The manufacturer technical representative will require training all TBM operators during the first 500 m of tunnelling. This requirement does not reduce the qualification criteria for TBM operators specified elsewhere.
  - TBM manufacturer's representative proposed for site assistance with TBM assembly, launching, commissioning and tunnelling support shall be a senior technician, employee of the manufacturer, with a minimum of ten years' experience.

## Risk Mitigation

Prevention of avoidable delays and mitigating risk of TBM getting stuck under Ontario Lake were the drivers for mitigating risk.

Knowing that an eventual main drive failure is associated with months, maybe even over 1 year of delays and the impracticability of accessing the main drive bearing for replacement (constructing a cavern lateral to the tunnel to access the front part of the main drive) and cost of dropping a cofferdam in the lake made the quality of the main bearing paramount. Selection of the main drive bearing and proven record of supplying quality bearings to the industry provided the risk mitigation and comfort for the owner.

The new ring gear and pinions quality was also very important knowing that ring gear replacement would be as hard as replacing the main drive bearing. The ring gear was manufactured in Turkey while the pinions were manufactured in the US by verified suppliers, previously approved as preferred supplier by Robbins Quality Management System.

The CM team was able to verify compliance of the TBM for most of these items less some of the equipment and materials Certificates of Origin and testing certificates.

**Figure 15. Gear cutting and heat treatment**

Manufacturing verifications were carried away in Mexico City and in China to mitigate risk of components breakdown during tunnelling while ensure quality of the systems provided.

Just before the TBM was all assembled on site in Mexico City the Covid-19 travel and material movement delays and restrictions had an impact on the final assembly readiness for FAT. To mitigate any delays, the team quickly switched gears to complete the first-ever virtual TBM Factory Acceptance Test. Leveraging Hatch's experience with non-destructive testing and virtual inspections in some of the most remote locations, Hatch implemented a virtual platform using Onsight Libre stream software along with Microsoft Teams to provide the resources and technologies needed to pull off the test successfully.

They were however items that could not be fully tested or verified because of the site limitations in terms of power and services. All these items testing was finalized in Toronto at the Site Assembly Testing (SAT) and the TBM was accepted by the City of Toronto.

## TBM PERFORMANCE

Following TBM assembly and on-site testing, initial mining commenced in late March 2021 but was briefly halted to allow the initial PCTL installed within the starter tunnel to be fully grouted. Mining recommenced on April 19, 2021, and continued until completion of the total 3,353 m tunnel drive on March 2, 2022. See Figure 16 which shows the cumulative tunnel progress over time as well as the weekly totals. The total average advance rate (including all downtime) was approximately 14.9 m per working day. During TBM tunnelling, the working day consisted of two working shifts (12 hours each) per day, 5 days per week (2 × 12 hr shifts, 5 days per week).

Major downtime events during TBM tunnelling are demarked on Figure 16 with symbols A through F. The following summarizes these major event points:

A. TBM mining learning curve and planned stopped for grouting the initial PCTL rings installed within the starter tunnel.
B. Repair to TBM segment feeder cylinder, oil contamination purge, and installation of the tunnel conveyor booster station. Total downtime: 9 working days.

C. Methane gas infiltration standby and pre-excavation grouting due to observed groundwater inflows during advance TBM probing. Total downtime: 3 days.
D. Groundwater inflows and PCTL damages (Groundwater Inflow Zone 1). Total days impacted: 25 working days.
E. Conveyor drive motor replacement and year-end holiday site shut down.
F. Groundwater inflows and PCTL damages (Groundwater Inflow Zone 2). Total days impacted: 15 working days.

Events D & F were attributed to encountering significant water inflow which halted TBM mining and damaged the PCTL. The encountered conditions were attributed to differing site condition and mining was successfully completed by developing a re-mining plan for this special section. The details regarding these conditions, impacts to tunnelling and the re-mining plan are described in greater detail in another paper (Solecki et al., 2022). When excluding these two specific events (which together contribute up to 40 working days), the average tunnel advance rate for the entire tunnel drive was 17.7 m per working day. Figure 16 also demonstrates that the rolling weekly advance rate was drastically impacted with encountering the high groundwater inflows during tunnelling.

The maximum weekly TBM advance rate was achieved the week of October 15, 2021 with a total excavated distance of 167 m (111 PCTL Rings) which translates to a sustained daily rate of 33.4 m per working day. The project also recorded several days of exceptional daily productivity. The best day was recorded as 46.4 m (31 PCTL Rings) which translates to an average PCTL ring installed and excavated every 46 minutes. This productivity rate is understood to be a record setting advance rate for this type of TBM, size and ground conditions. Overall, there were 78 working days when the daily productivity rate exceeded 20 m/day, which equates to approximately 40% of the entire mining duration excluding the two major water inflow downtime events (see D & F above).

The impressive productivity rates are further reinforced that the contract required that advance pre-excavation probing be complete at all times ahead of the TBM. This

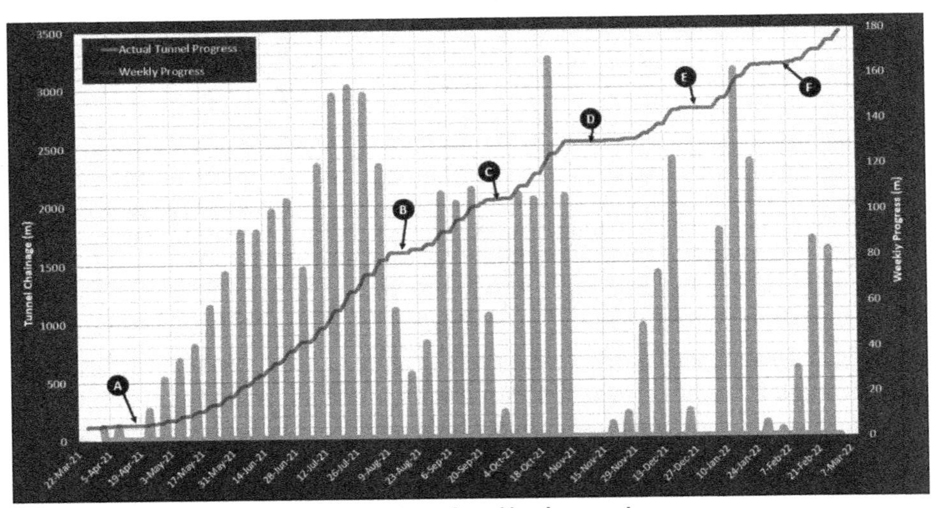


**Figure 16. TBM tunnelling overall productivity and weekly advance rates**

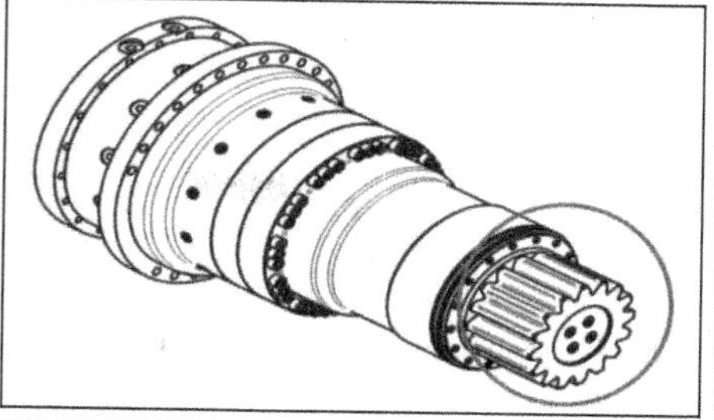

Figure 17. Damaged main drive pinion

essential meant that every 25 to 30 m, mining stopped to complete this probing which took several hours per day on average to complete.

## LESSONS LEARNED

What would we do different if we have the chance to go through the process again. First, we would provide more clarity on the technical specifications regarding the re-manufacturing process, and about certain items for which we faced time constrained decisions. These decisions however were proven to be right, and the specific deviations accepted did not affect the TBM productivity or generated any breakdowns or repairs that could not be solved by the Robbins technical team.

We would inquire for more detailed quality control on items that you typically take for granted such as manufacturing gearbox pinions at a factory in the United States. We have had the unpleasant surprise to witness a severe pinion breakdown at the first turn of the cutterhead on site. It appeared that the raw material (steel grade) and heat treatment were not verified during manufacturing at Robbins pinion supplier. After this failure the pinion cavity was cleaned and flushed with oil and a complete set of new pinions were manufactured and installed. This is a true lesson learned not to take anything for granted.

Like any tunnel in challenging ground conditions, we went through breakdowns, ground squeezing, ground water inflow, etc.; however, there was no problem or issue too big to be tackled and solved by Robbins field technicians.

One special mention for Robbins Field Assembly Manager, Alfredo Garrido that was proven to be on top of all matters, hydraulic, electrical, structural or final assembly and testing. Outstanding all together.

## REFERENCES

ITAtech Report. March 2019. Guidelines on Rebuilds of Machinery for Mechanized Tunnel Excavation.

Solecki A., Waher K., and Kramer, G.J.E., 2022. TBM Tunnelling Challenges and Managing High Groundwater Inflows on the Ashbridges Bay Treatment Plant Outfall Project, TAC 2022 Conference, Vancouver, Canada.

# Fire Damage Assessment of Reinforced Concrete Tunnel Linings

**Nan Hua** ▪ Mott Macdonald
**Anthony Tessari** ▪ University at Buffalo
**Negar Elhami Khorasani** ▪ University at Buffalo

## ABSTRACT

Fire hazards can cause severe and irrecoverable damage to reinforced concrete (RC) tunnel linings, threatening the serviceability and resilience of transportation networks. This paper first examines the existing methodologies that are used to classify damage to tunnel linings from fire. These approaches are then supplemented with the results of advanced modeling and inputs from industry experts to propose new fire damage/repair classifications with the associated thresholds. The results of the presented research can be used to guide fire risk assessment as well as the performance-based design of tunnel fire protection.

## INTRODUCTION

Fires and high temperatures can generate severe demands on a tunnel structure. The rapid rise of gas temperature in excess of 1,000°C inside a confined tunnel space and long fire durations due to limited fire-fighting access may lead to severe damages to the reinforced concrete (RC) lining. This damage can include cracking, heat-induced spalling, material deterioration, leading to reduced structural capacity. The repair work and long-term service interruption caused by tunnel fires can result in significant economic losses. However, there is no well-established guideline for post-fire damage assessment and repair procedures for tunnels.

According to a report published by SCOR Global P&C (Schütz, 2014), 177 tunnel fire events were recorded in 29 countries from 1966 to 2014. Numerous causes were reported, such as those resulting from electrical fire, car crash, bomb attack, derailment, collision, arson, etc. Among the 177 events, 28 were categorized as major events, resulting in more than 700 deaths, 1,000 injuries, loss of more than 500 vehicles, and financial loss of at least €1 billion (1.17 billion US dollars). Well-known examples include the 2008 Channel Tunnel fire, which involved several heavy goods vehicles on carrier wagons and damaged 650 meters of the tunnel lining with a cost of €250–286 million to resume service (Lu, 2015; Schütz, 2014). While a fire event would not necessarily cause collapse of a tunnel structure, the potential threat to the serviceability of the tunnel itself and the resilience of the overarching transportation system emphasize the need to establish a standardized and effective post-fire damage assessment method to guide repairs and restore functionality.

The primary goal of post-fire damage assessment is to identify proper repair methods to restore the structure's functionality (if demolition is not an option, as for most fire damaged tunnel linings). An efficient repair strategy requires quantitative estimations of the extent and degree of damage, such as the spalled/deteriorated depth of affected sections, the maximum temperature reached in fire-affected areas, and the strength loss of structural materials. By properly mapping the documented damage levels to defined categories, with guided repair methods corresponding to each damage category, engineers can effectively evaluate the repair work and predict the downtime after a fire event.

Aside from post-fire assessment, a damage classification system would also benefit performance-based design of tunnels for fire. With numerical modeling tools, tunnel fire protection systems can be optimized based on the performance requirements such as the acceptable downtime duration given a fire demand. Improving the structural response for probable disruptive events is key to enhance resilience.

This paper proposes a fire damage assessment framework for RC tunnel linings, which integrates advanced modelling with traditional inspection methods. The framework quantifies fire damage to RC tunnel linings in terms of surface discoloration, crack width, concrete spalling, sectional temperatures, strength loss of materials, and residual displacement. A damage classification system is discussed based on a collection of international guidelines and feedback from industry experts to facilitate a systematic determination of the necessary damage metrics and repair strategies. A discussion on incorporating the proposed framework into risk analysis and performance-based design is also included in this paper.

## FIRE DAMAGE ASSESSMENT FRAMEWORK

Figure 1 summarizes the proposed damage assessment framework for RC tunnel linings exposed to fire. It is suggested that practitioners utilize advanced modelling in three steps and combine the results with traditional post-fire inspection procedures to evaluate safety and obtain a full understanding of fire damage to a tunnel lining. The overall damage conditions obtained from the two paths in Figure 1 can be incorporated into a damage classification system to help determine the level of required repair work. The following paragraphs will introduce each step followed by a discussion of fire damage classification for RC tunnel linings.

### Traditional Post-Fire Damage Inspections

Although no detailed damage assessment framework is available for tunnel structures, there are some general guidelines for fire damage assessment and repair of concrete structures (Concrete Society, 2008; Federal Highway Administration, 2015; Fire Safety Committee of the Concrete and Masonry Industry, 1994; International Federation for Structural Concrete (*fib*), 2008). Generally, these documents recommend similar post-fire inspection methods, which can be summarized into three levels depending on the extent of damage.

#### *Visual Inspection (and Hammer Sounding)*

Visual inspection provides a recording of features such as concrete color, concrete spalling, cracking, surface crazing, distortion, deflections, and collapse. A small hammer may be used to conduct a tapping survey on concrete surfaces to detect concrete damage. According to the UK concrete society (Concrete Society, 2008), a "ring" from the hammer tap indicates sound concrete, while a "dull thud" indicates weak concrete.

The surface appearance of concrete provides engineers the first indication of which elements require cosmetic repairs, and which, if any, will need further assessment. The color of concrete can change due to high temperature, which is apparent upon visual inspection. In many cases concrete shows pink/red discoloration above 300°C (572°F) due to oxidation of ferric salts in aggregates, also indicating a strength loss of the material.

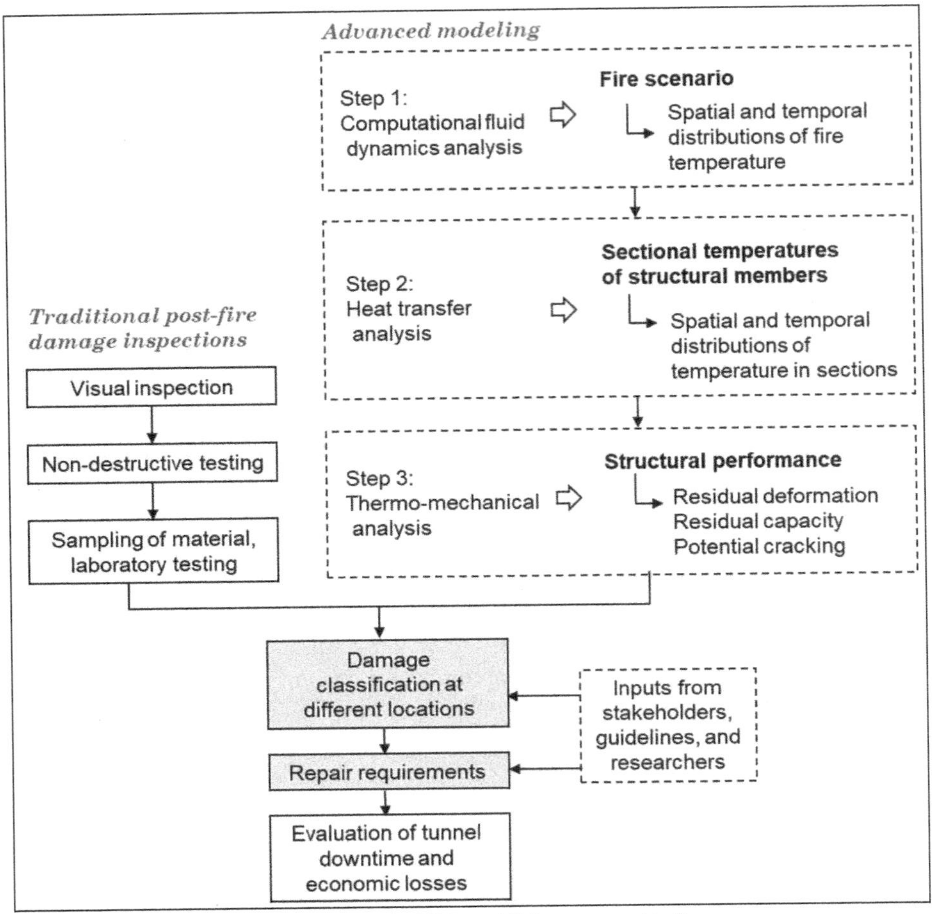


**Figure 1. Damage assessment framework for RC tunnel linings exposed to fire**

Heat-induced spalling is a major and common issue for fire performance of concrete structures, and is defined as the violent ejection of pieces from the surface of concrete elements when exposed to rising temperatures (Bisby et al., 2014). Spalling conditions range from localized popping-off of small and thin chips to dramatically breaking away of large pieces of concrete and exposing reinforcement. Spalling also enables the fire frontier to propagate deeper into the concrete and thus increases the deterioration depth. The visual inspection after the fire should include an assessment of spalling, documented as spalling depth across the affected area, as well as the total weight/volume of spalled concrete.

### *Non-Destructive Testing (NDT)*

Although visual inspections serve as the most direct assessment, numerous NDT techniques can be used to provide a more reliable assessment. Due to general availability and the ease of application, the most commonly used NDT methods for fire damage assessment are the rebound (Schmidt hammer) test and the ultrasonic pulse velocity (UPV) test (Albrektsson et al., 2011; Concrete Society, 2008; International Federation for Structural Concrete (fib), 2008). These methods can provide an estimation of concrete strength loss in situ but are largely influenced by other factors such as surface smoothness, size, and shape of the test specimens, and therefore their results are not always consistent.

NDT techniques for fire damage assessment are becoming more common as a result of technology developments across multiple disciplines. For example, digital image processing can be used to map cracks on a concrete surface after cooling (Felicetti, 2013), machine learning algorithms and image texture analyses can be adapted to automatically detect concrete spalling (Hoang et al., 2019), and short pulse radar method using electromagnetic waves can be used instead of the UPV method (Joakim Albrektsson et al., 2011).

### *Sampling of Material and Subsequent Laboratory Testing*

Samples of damaged material, together with undamaged references, may be removed for laboratory investigation. Concrete samples are typically obtained by drilling cores or by careful extraction of lump samples (Ingham, 2009). Concrete core samples can be taken for petrographic examination and compression tests. The petrographic analysis is mainly concerned with damage and property change of concrete at high temperatures, i.e., depth of microcracking, paste alteration (color change and strength softening), and carbonation.

## Advanced Modeling and Analysis

As an alternative or in addition to traditional inspections, predictive fire engineering tools, such as empirical equations or computer modelling, can be used to assess the fire severity in the structure (Ingham, 2009). Given the complexities involved in completing a thermal-mechanical analysis and the extensive factors influencing the structural behavior under fire, a three-step methodology utilizing advanced modeling (Figure 1) is proposed to assess the material damage and residual structural conditions for tunnel linings after fire. As an additional advantage, the mentioned numerical analysis methods are able to incorporate comprehensive parametric studies in a more economic way compared to traditional lab testing, and therefore enable damage-based performance prediction and optimized design for tunnel linings in fire risk.

### *Computational Fluid Dynamics (CFD) Analysis*

The first step in the methodology is to determine the fire scenario. A fire scenario describes the evolution of gas temperature over time during a fire event. A tunnel fire scenario is typically characterized by a fast heating rate and a high peak temperature. Due to the special geometry of the tunnel space, gas temperature varies significantly along the tunnel length and across its cross-sections during a fire event. Understanding the temporal as well as spatial distributions of temperature inside a tunnel is important for fire damage assessment. Modelling fire scenarios in tunnels requires an understanding of a number of factors including the tunnel geometry, fuel size and type, ventilation condition, fire spread pattern, the existence and activation of a deluge system, etc. CFD modelling tools, such as FDS and Ansys Fluent, can capture complex 3D fire scenarios (Ministry of Transportation and Public Works of the Netherlands, 1999). After a site survey to evaluate the cause of fire, location of fire, amount and type of fuel load, and the natural or active ventilation conditions during the fire event, CFD modelling can be used to replicate the fire scenario and estimate the fire severity, such as the maximum gas temperatures at different locations within the tunnel space and over time.

### *Heat Transfer Analysis*

Once a credible demand fire scenario is determined, temperatures within the lining sections can be obtained by completing a heat-transfer analysis. The analysis can be conducted using finite element software such as ABAQUS, ANSYS, and SAFIR (Franssen and Gernay, 2017), the results of which can be conveniently imported into the mechanical analysis in the next step. Given that most modern tunnel linings are made of thin shells, one-dimensional heat-transfer analysis across the thickness of the lining would satisfy the need. The greatest challenge in this step is to model concrete spalling and incorporate the process into a heat transfer analysis. Following an actual event, engineers/inspectors can measure the spalled depth of concrete and use a spalling-incorporated heat-transfer analysis (Hua et al., 2021) to obtain realistic sectional temperatures. The results can be used to characterize the residual material strength across the lining depth.

### *Sequentially Coupled Thermo-Mechanical Analysis*

Sequentially coupled thermal and structural analysis, performed with finite-element software packages such as Abaqus and SAFIR, can help determine the residual deformation and capacity of a damaged structure. This step takes time-dependent sectional temperatures of the structural elements as inputs and involves complexities such as large thermal gradients across sections, distinct material properties, residual strain and deformations, etc. Residual capacity is an important indicator for the serviceability condition, durability, and resilience of a structure. Determination of residual capacity after fire requires capturing the cooling behavior of the structure, which brings more complexity into the modeling effort (Kodur and Agrawal, 2016). Another challenge is to model the soil-structure interaction under elevated temperatures, for which supporting test data and relevant modeling verifications are very limited. Examples of successful modeling approach can be found in Hua et al. (2022). Given the involved complexities, this step can be applied to moderate/severe damage classifications only (as introduced in the next section) based on the results of damage assessment in the first two steps.

## Damage Classification and Repair Requirement

Two major standard documents, from the UK concrete society and the *fib* (Concrete Society, 2008; International Federation for Structural Concrete (*fib*), 2008), provide fire damage classification for building structures, where the one by the UK concrete society (Concrete Society, 2008) includes a categorized repair requirement for the damage classifications. Tables 1, 2, and 3 provide a summary of the available guided damage classifications and repair requirements.

The above two documents provide valuable fire damage indicators and grading criteria in the absence of fire damage guidelines in the US. However, not all details can be applied to tunnel structures. Also, using concrete color as an indicator is a fairly crude assessment, especially when the lining surface is partially or fully covered by soot. Most of the listed criteria are qualitative, and the evaluation would largely rely on the inspector/engineer's experience and judgement.

In Table 3, the repair requirement after fire varies from superficial repair to demolition. Likewise, these requirements are primarily defined for building elements, and may not be directly applicable to tunnel structures, especially the last class requiring demolition, as tunnels typically maintain stability after major fire events and would not reach a state that necessitates demolition.

**Table 1. Fire damage classifications of concrete structures according to International Federation for Structural Concrete (*fib*) (2008)**

| Class | Characterization | Description |
|---|---|---|
| 1 | Cosmetic damage, surface | Characterized by soot deposits and discoloration. In most cases soot and color can be washed off. Uneven distribution of soot deposits may occur. Permanent discoloration of high-quality surfaces may cause their replacement. Odors are included in the class (they can hardly be removed, but chemicals are available for their elimination). |
| 2 | Technical damage, surface | Characterized by damage on surface treatments and coatings. Limited extent of concrete spalling or corrosion of unprotected metals. Painted surfaces can be repaired. Plastic-coated surfaces need replacement or protection. Minor damages due to spalling may be left in place or may be replastered. |
| 3 | Structural damage, surface | Characterized by some concrete cracking and spalling, some deformation of metal surfaces or moderate corrosion. This type of damage includes also class 2 damages, and can be repaired in similar ways. |
| 4 | Structural damage, cross-section | Characterized by major concrete cracking and spalling in the web of I-beams and deformed flanges. |
| 5 | Structural damage to members and components | Characterized by severe damages to structural members and components, with local failures in the materials and large deformations. Concrete constructions are characterized by extensive spalling, exposed reinforcement and damaged compression zones. Mechanical decay in materials may occur as a consequence of the fire. Class 5 damages usually will cause the dismissal of the structure. |

**Table 2. Fire damage classifications according to the Concrete Society (Concrete Society, 2008)**

| Damage Class | Color | Crazing | Spalling | Exposure and Condition of Main Reinforcement | Cracks |
|---|---|---|---|---|---|
| 1 | Normal | Slight | Minor | None exposed | None |
| 2 | Pink/red | Moderate | Localized to patches | Up to 10% exposed, all adhering | None |
| 3 | Pink/red | Extensive | Considerable | Up to 20% exposed, generally adhering | Small |
| 4 | Whitish grey | Surface lost | Almost all surface spalled | Over 20% exposed, largely separated from concrete | Severe |

**Table 3. Repair classification adopted from the Concrete Society (Concrete Society, 2008)**

| Damage Class | Repair Classification | Repair Requirements |
|---|---|---|
| 1 | Superficial | Superficial repair of slight damage |
| 2 | General repair | Non-structural or minor structural repair restoring cover to reinforcement where this has been partly lost |
| 3 | Principal repair | Strengthening repair in accordance with the load-carrying requirement of the member. Concrete and reinforcement strength may be significantly reduced requiring check by design procedure. |
| 4 | Major repair | Major strengthening repair with original concrete and reinforcement written down to zero strength, or demolition and recasting |

In fire-exposed RC lining sections, distinct temperature gradients could develop due to the low thermal conductivity of concrete. As a result, a portion of the lining close to the exposed surface may experience high temperatures and severe damage, whereas the deeper parts of the lining may be unaffected, even after a major fire. Therefore, determining the depth of influence, i.e., the depth that heat travels within the lining thickness, is crucial for the damage assessment of tunnel linings. Several standards recommend 300°C as the threshold for discoloration and a reduction in the residual strength of concrete (Concrete Society, 2008; Federal Highway Administration, 2015;

Fire Safety Committee of the Concrete and Masonry Industry, 1994; International Federation for Structural Concrete (*fib*), 2008). Eurocode3 (European Committee, 2005) specifies that steel starts to lose yield strength upon reaching 400°C. According to a collection of test data compiled by Deshpande et al. (Deshpande et al., 2020), the bond strength between rebar and concrete decreases by up to 20% after being heated to 300°C.

Structural-level damage assessment of a tunnel lining shall focus on its post-fire serviceability and capacity (Kodur and Agrawal, 2016). The residual crown displacement reflects the residual strain and locked-in stresses within structural elements. For buildings, ACI 318 (ACI Committee 318, 2019) specifies the maximum permissible deflection of floor slabs as $l/360$, where $l$ is the span of the floor. However, deformation of tunnel linings is constrained by the surrounding soil/rock, and therefore the amount is largely controlled by the type of geomaterial. According to a study on the influence of geologic profiles on fire behavior of tunnels (Hua et al., 2022), the residual crown displacement of bored tunnel sections (with an internal diameter of 7.5 m) in soft ground ranges from 23 mm to 60 mm after 72 hours of exposure to an extreme standard fire, while the residual crown displacement of tunnel sections in rock is about 1 to 3 mm. In another study (Hua et al., 2021), the residual crown displacement of a cut-and-cover tunnel section (with an internal diameter of 8.2 m) was calculated as 7 mm after exposure to a passenger train fire.

This work proposes quantitative damage thresholds to assist with damage diagnosis, ensure safety, and save on repair costs through a proper assessment of structural conditions. Table 4 summarizes the damage classification thresholds associated with sectional temperatures and residual crown displacement, which can be documented using laser scanning technology or calculated using advanced modelling. The proposed damage classification table incorporates inputs from experts from the tunnel engineering industry. Note that some of the proposed thresholds, such as the residual crown displacements in Table 4 are based on tunnels with internal diameters ranging from 7–9 m in soft ground, and require further research and refinement. Tunnel segments diagnosed as Class D damage, where there are severe section losses and residual deformations, would require a complete structural analysis (Step 3 in the advanced modelling procedure) to evaluate the residual structural capacity and locked-in stresses within the sections.

The corresponding repair requirements for the defined damage classes are listed in Table 5. To implement the proposed framework, the affected tunnel lining sections shall be divided into a series of segments with a diagnosed damage class for each

**Table 4. Proposed fire damage classification as related to the results of advanced modelling and onsite measurements**

| Damage Class | Depth of Concrete Reaching Temperatures > 300°C | Temperature of Reinforcement | Crown Residual Displacement |
|---|---|---|---|
| A | $d$ < half cover depth | $T$ < 100°C | Negligible |
| B | half cover depth ≤ $d$ < cover depth | 100°C ≤ T < 300°C | $\Delta$ < 6 mm or $\Delta$ < ¼" |
| C | cover depth ≤ d < half section depth | 300°C ≤ $T$ < 600°C or partially exposed | 6 mm ≤ $\Delta$ < 13 mm or ¼" ≤ ½" |
| D | $d$ ≤ half section depth | $T$ ≤ 600°C or significantly exposed | $\Delta$ ≤ 13 mm or $\Delta$ ≤ ½" |

Table 5. Proposed repair requirements for each damage classification

| Damage Class | Repair Requirements |
|---|---|
| A | Superficial repair, such as surface cleaning and crack sealing |
| B | Concrete restoration |
| C | Concrete restoration and rebar replacement |
| D | Advanced thermal-structural analysis to determine residual strains, locked-in stresses, and residual capacity of the lining to determine if re-design is needed |

segment. The repair cost can then be estimated for each segment as well as the entire affected lining.

## DAMAGE-BASED RISK ANALYSIS AND OPTIMIZED DESIGN

The risk analysis and performance-based design (PBD) require a better understating of influencing factors on the structural response and considerations of uncertainties, which also allows for more innovative and cost-effective designs. The concept of PBD was first developed in earthquake engineering and has been adopted by other disciplines, such as wind engineering. The application of a performance-based fire design (PBFD) methodology, for which the benefits have been demonstrated for building applications (Khorasani et al., 2019), can be extended to tunnels to enable safe and economic solutions. The proposed damage assessment framework in this paper paves the way for PBFD of tunnel linings by incorporating the required repair work and downtime after a possible event when evaluating different solutions and optimizing the cost. The following paragraphs provide insights on key influencing parameters and uncertainties involved in the applications of advanced modeling to evaluate fire damage.

### Fire Scenarios

Understanding the spatial and temporal distribution of temperature inside a tunnel is important for fire damage assessment. The most common approach in engineering practice today is the application of standard design curves and deterministic procedures. However, defining a tunnel fire scenario involves a high level of uncertainty and requires inputs from a number of factors such as the tunnel geometry, variety of car materials, ventilation conditions, etc. Based on published test data, Hua et al. (2020) studied the uncertainties in the ventilation velocity, amount of fuel in terms of heat release rate (HRR), tunnel slope, ignition point, and ignition criteria for fire spread in a passenger train fire event. The research randomly sampled the said parameters and investigated 540 fire scenarios. The established fire temperature demands are shown in Figure 2a. According to the result, while high intensity fires (maximum HRR>40 MW) have a mean peak temperature of about 1,000°C, that of low intensity fires (maximum HRR≤40 MW) is less than 300°C.

### Concrete Spalling

Heat-induced concrete spalling has always been a challenge when predicting the fire behavior of structures because of its large uncertainty. Although thermo-hydro-mechanical theories for heat-induced spalling exist, there is no widely acknowledged numerical prediction or modelling method for this phenomenon. Hua et al. (2021) collected over 100 test data, proposed a simplified spalling model, and implemented the model in the heat transfer analysis to obtain a more realistic concrete damage. Figure 2b shows the predicted concrete damage distribution considering a mean

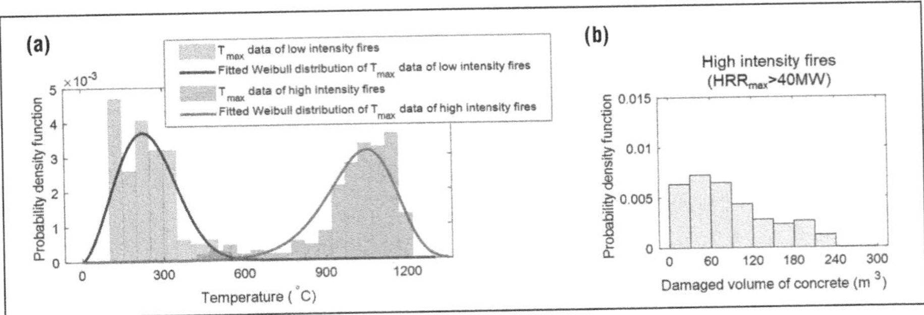


**Figure 2. Uncertainties in (a) fire scenarios and (b) concrete damage**

value of possible spalling rate. The calculated volume of damage was based on concrete depth reaching the 300°C-threshold as discussed in previous sections.

## Temperature Dependent Material Properties

Both thermal and mechanical properties of structural materials show large uncertainties under elevated temperature based on the observations from extensive published experimental data. Jovanovic et al. (2020) developed probabilistic models for thermal conductivity and specific heat of concrete. Qureshi et al. (2020) established probabilistic models for the temperature-dependent strength of concrete and steel. Shahraki et al. (2022) proposed a probabilistic model for the residual compressive strength of concrete to capture the uncertainties in material strength as a function of the maximum temperature history, after reviewing and analyzing 1,240 experimental datapoints.

The proposed advanced modelling framework for fire damage assessment, along with the above research efforts on quantifying the uncertainties, can be integrated within probabilistic risk-assessment methods to guide the design of fire protection for RC tunnel linings.

# CONCLUSIONS

This paper proposed a framework for fire damage assessment of reinforced concrete tunnel linings. The framework relies on both observation and measurements as well as numerical modelling of the thermal and structural response. Damage classes and the corresponding repair methods were indicated based on the depth of concrete exceeding 300°C, the maximum rebar temperature, and residual displacement. The involved uncertainties in the proposed advanced modeling framework were discussed at the end of the paper. This framework can be integrated with risk-assessment methods to optimize the fire design of tunnels with associated active and/or passive fire protection.

# ACKNOWLEDGMENTS

This work was supported by the CAIT Region 2 UTC Consortium, and the Institute of Bridge Engineering (IBE) at the University at Buffalo. Any opinions, findings, and conclusions or recommendations expressed in this material are those of the authors and do not necessarily reflect the views of the CAIT Region 2 UTC Consortium. The authors sincerely appreciate the valuable feedback received via questionnaire from our private industry and government colleagues. Their time, opinions, and comments are greatly appreciated.

## REFERENCES

ACI Committee 318, 2019. Building code requirements for structural concrete (ACI 318-19). American Concrete Institute, Farmington Hills, MI, USA.

Albrektsson, J., Flansbjer, M., Lindqvist, J.E., Jansson, R., 2011. Assessment of concrete structures after fire. SP Technical Research Institute of Sweden.

Bisby, L., Mostafaei, H., Pimienta, P., 2014. White paper on fire resistance of concrete structures. US Department of Commerce, National Institute of Standards and Technology.

Concrete Society, 2008. Assessment, design and repair of fire-damaged concrete structures, Camberley, UK.

Deshpande, A.A., Kumar, D., Ranade, R., 2020. Temperature effects on the bond behavior between deformed steel reinforcing bars and hybrid fiber-reinforced strain-hardening cementitious composite. Construction and Building Materials 233, 117337.

European Committee, 2005. Eurocode 3: Design of steel structures—Part 1–2: General rules—Structural fire design (EN 1993-1-2). European Committee, Brussels, Belgium.

Federal Highway Administration, 2015. Tunnel Operations, Maintenance, Inspection, and Evaluation (TOMIE) Manual, Washington DC, USA.

Felicetti, R., 2013. Assessment methods of fire damages in concrete tunnel linings. Fire technology 49, 509–529.

Fire Safety Committee of the Concrete and Masonry Industry, 1994. Assessing the condition and repair alternatives of fire-exposed concrete and masonry members, Skokie, Illinois.

Franssen, J.-M., Gernay, T., 2017. Modeling structures in fire with SAFIR®: theoretical background and capabilities. Journal of Structural Fire Engineering 8, 300–323.

Hoang, N.-D., Nguyen, Q.-L., Tran, X.-L., 2019. Automatic detection of concrete spalling using piecewise linear stochastic gradient descent logistic regression and image texture analysis. Complexity 2019, 5910625.

Hua, N., Tessari, A., Elhami-Khorasani, N., 2020. Quantifying uncertainties in the temperature-time evolution of railway tunnel fires. Fire Technology, 361–392.

Hua, N., Tessari, A., Elhami Khorasani, N., 2021. Characterizing damage to a concrete liner during a tunnel fire. Tunnelling and Underground Space Technology 109, 103761.

Hua, N., Tessari, A., Elhami Khorasani, N., 2022. The effect of geologic conditions on the fire behavior of tunnels considering soil-structure interaction. Tunnelling and Underground Space Technology 122, 104380.

Ingham, J., 2009. Forensic engineering of fire-damaged structures. Proceedings of the Institution of Civil Engineers—Civil Engineering, 12–17.

International Federation for Structural Concrete (*fib*), 2008. Fire design of concrete structures—structural behavior and assessment, Lausanne, Switzerland.

Joakim Albrektsson, Mathias Flansbjer, Lindqvist, J.E., Jansson, R., 2011. Assessment of concrete structures after fire. SP Technical Research Institute of Sweden.

Jovanovic, B., Elhami Khorasani, N., Thienpont, T., Chaudhary, R.K., Van Coile, R., 2020. Probabilistic models for thermal properties of concrete. Proceedings of 11th International Conference on Structures in Fire, Queensland, Australia, Nov. 30–Dec. 2.

Khorasani, N.E., Gernay, T., Fang, C., 2019. Parametric Study for Performance-Based Fire Design of US Prototype Composite Floor Systems. Journal of Structural Engineering 145, 04019030.

Kodur, V.K.R., Agrawal, A., 2016. An approach for evaluating residual capacity of reinforced concrete beams exposed to fire. Engineering Structures 110, 293–306.

Lu, F., 2015. On the prediction of concrete spalling under fire, Dept. of Civil, Environmental and Geomatic Engineering. ETH-Zürich, Zurich, Switzerland.

Ministry of Transportation and Public Works of the Netherlands, 1999. Evaluation of Memorial Tunnel CFD simulations Ministry of Transportation and Public Works of the Netherlands, Netherlands.

Qureshi, R., Ni, S., Khorasani, N.E., Coile, R.V., Hopkin, D., Gernay, T., 2020. Probabilistic Models for Temperature-Dependent Strength of Steel and Concrete. 146, 04020102.

Schütz, D., 2014. Fire protection in tunnels: Focus on road & train tunnels. Tech. Newsl. SCOR Glob. P&C.

Shahraki, M., Hua, N., Elhami-Khorasani, N., Tessari, A., Garlock, M., 2022. Residual compressive strength of concrete after exposure to high temperature: A review and probabilistic models. Fire Safety Journal, in press, 103698.

# Overexcavation Risk Management During Pressurized Face Tunneling in the Pacific Northwest

**Ulf Georg Gwildis** ▪ CDM Smith

## ABSTRACT

Overexcavation and ground loss events during pressurized-face TBM advance in soft ground can result in unplanned deformations above the tunnel and damage to existing infrastructure. Successfully managing the risk of this occurring is crucial for any project in an urban area to proceed. Common risk management approaches include a combination of establishing suitable ranges of TBM operational parameters and using geotechnical instrumentation for deformation monitoring. Another risk management tool is material flow reconciliation, which varies depending on the specified mining method and the equipment choices by the contractor. This paper compares recent and ongoing projects in the Pacific Northwest regarding overexcavation risk management.

## GROUND LOSS ESTIMATE FOR PRESSURIZED FACE TUNNELING

Design of tunnels within urban areas needs to limit surface deformations and minimize the impact on existing infrastructure to avoid damage, service disruption, and litigation. Where this cannot be achieved to the extent necessary by the tunneling approach alone, pre-tunneling ground improvement measures, foundation improvements, or replacement of existing structures may be required. To minimize deformation impact during the tunneling process, construction contracts of large soft-ground tunnel projects in urban areas typically specify pressurized-face mechanized tunneling, either with an earth-pressure balanced TBM (EPBM), a slurry-pressure balanced TBM (STBM), a hybrid of both, or by leaving the selection of the TBM type to the contractor. Estimating tunneling-induced ground deformation during the design by using either analytical methods or numerical modeling requires the assumption that ground loss can be limited to a certain percentage of the theoretical excavation volume. For most of the recent and current TBM projects a value of 0.5% or less is assumed. The actual ground loss that occurs during tunneling depends on various factors such as the face support pressure applied during TBM advance, with the exact target value falling under the contractor's means and methods. This target value needs to be within the range of the minimum pressure required for face stability and the maximum allowable pressure to avoid heave or a blow-out event, which is a risk factor especially in shallow overburden scenarios. The target pressure ideally is optimized for the project-specific subsurface conditions to allow effective and efficient mining advance without spending an excessive amount of time on cutterhead maintenance, which can be time consuming in abrasive ground, especially under high hydrostatic head condition.

For verifying that the design phase assumption of ground loss percentage is not exceeded and to manage the risk of overexcavation, continuous monitoring of the tunneling operation is required as the TBM advances. Modern TBMs record over 200 operational parameters in intervals specified or agreed-upon by the project parties, typically less than 30 seconds. Process control software allows easy visualization and sharing of the data. Other data sources also provide valuable information if they are integrated with the TBM operational data via synchronized time stamps. Geotechnical instrumentation is commonly installed at critical locations along the TBM drive.

However, to what extent the geotechnical instrumentation data applies beyond the specific instrument locations remains an open question, especially in highly variable subsurface conditions such as those provided by the glacial geology of the Pacific Northwest. Further, the delay of measuring the deformation effects of the tunnel excavation process can be significant, depending on the thickness and geologic makeup of the tunnel overburden. This limits the usefulness of this data source as the basis for decisions to apply mitigating measures and modify TBM operations. Other than geotechnical instrumentation, monitoring of the material flow of the excavated spoils provides data that can be tabulated on a per-ring basis. However, the data quality of this source depends on the specifics of the measuring equipment and the collected data require some interpretation. These seem to be some of the reasons why in the past this data source has not always been considered with the priority it deserves.

## EVOLUTION OF TBM GROUND LOSS MONITORING IN THE PACIFIC NORTHWEST

Table 1 lists selected tunneling projects whose experiences illustrate the evolution of overexcavation risk management during TBM advance in the glacial and alluvial deposits of the Pacific Northwest.

### Beacon Hill Transit Tunnel

The Beacon Hill Tunnel project included an underground station and twin light rail tunnels in glacial and lacustrine, glacially overconsolidated deposits. 18 months after completion of the drives by an EPBM the discovery of a sinkhole above the tunnel triggered a multi-phase exploration of ground loss effects and subsequent grouting program for remediation of the underground voids and loosened soil volumes that had been mapped (Figure 1).

Although the TBM had been equipped with two conveyor belt scales for muck weight measurements, the collected data was not relied on during construction. Post-construction review derived a ground loss volume of 2,520 m$^3$ (3,296 cy), which correlated within 2% of the injected grout volumes plus delineated disturbed soil

**Table 1. Selected TBM projects**

| | Project | Face Conditions | TBM Type* | # of Drives | Length, m | Diameter, m | Completion |
|---|---|---|---|---|---|---|---|
| 1 | Beacon Hill Transit Tunnel (Seattle) | Glacial Deposits | EPBM | 2 | 2*1,310 | 6.5 | 2008 |
| 2 | Brightwater Conveyance System East Contract (Seattle) | Glacial Deposits/ locally Alluvium | EPBM | 1 | 4,231 | 5.9 | 2008 |
| 3 | Brightwater Conveyance System Central Contract (Seattle) | Glacial Deposits/ locally Alluvium | STBM | 2 | 6,651 | 5.4 | 2011 |
| 4 | Brightwater Conveyance System West Contract (Seattle) | Glacial Deposits | EPBM | 1 | 6,424 | 4.7 | 2010 |
| 5 | Brightwater Conveyance System BT3 Completion Contract (Seattle) | Glacial Deposits | EPBM | 1 | 3,018 | 4.9 | 2011 |
| 6 | Alaskan Way Viaduct Replacement Tunnel (Seattle) | Glacial Deposits | EPBM | 1 | 2,825 | 17.5 | 2017 |
| 7 | Annacis Island WWTP† New Outfall System (Vancouver, BC) | Alluvium | STBM | 2 | 780 | 5.0 | On-going |

* EPBM = Earth Pressure Balanced TBM, STBM = Slurry Pressure Balanced TBM
† WWTP = Wastewater Treatment Plant

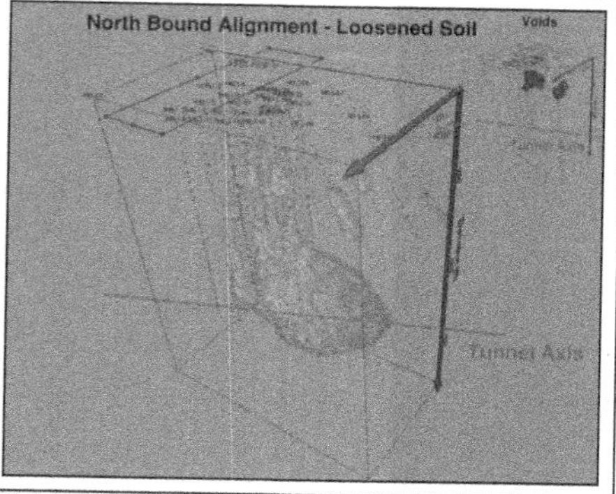


**Figure 1. Surface feature (left) and presentation of exploration results of areas of loosened soil and voids (right) above the Beacon Hill twin-bore tunnel**

volumes (Robinson et al. 2012). This demonstrated the value of muck weight measurements for monitoring of ground loss during TBM advance.

## Brightwater Conveyance System

The Brightwater Conveyance System included a 20 km long, deep lying tunnel alignment to gravitationally convey effluent from a new regional wastewater treatment plant serving the Seattle metropolitan area to a marine outfall into Puget Sound. The tunnel alignment was oriented in east-west direction perpendicular to the Pleistocene age glacial advances and retreats. This meant that its five TBM drives through a sequence of glacial and non-glacial, glacially overconsolidated deposits and locally through the alluvial filling of a valley carved deep into these deposits would encounter highly variable face conditions. These highly variable face conditions required a geotechnical baseline approach that provided a baseline of face conditions as percentage of the tunnel drive lengths without location specificity other than at the locations of the exploratory borings. This approach in turn made the tracking of the face conditions encountered during the TBM advances necessary, which for the pressurized-face drives required to derive the results from various data sources such as EPBM spoils or materials separated from the STBM return slurry as well as the TBM operational parameters. In addition, weight and volume estimates were used for checking trends of overexcavation. In the case of the STBM drives of the central contract, this monitoring was based on slurry density and flow measurements. Figure 2 shows a sinkhole and the related data analysis identifying insufficient operational control as cause. At the time of the sinkhole occurrence the TBM had been advanced through overconsolidated, non-glacial deposits, which were identified by face condition tracking as belonging to the geotechnically defined tunnel soil group of predominantly sandy soils. The nearest exploratory boring indicated the overburden to consist also of mostly granular soils with few fine-grained interlayers. This scenario offered an explanation to the fact that the sinkhole opened within a few hours after the recorded occurrence of overexcavation at a depth of 50 m (150 ft).

## Alaskan Way Viaduct Replacement Tunnel

The Alaskan Way Viaduct Replacement Tunnel with an excavation diameter of 17.5 m (57.5 ft) to house a double-deck highway was excavated by an EPBM mostly in glacially overconsolidated glacial and non-glacial deposits, which due to the general north-south direction of the alignment showed less variability compared to the Brightwater

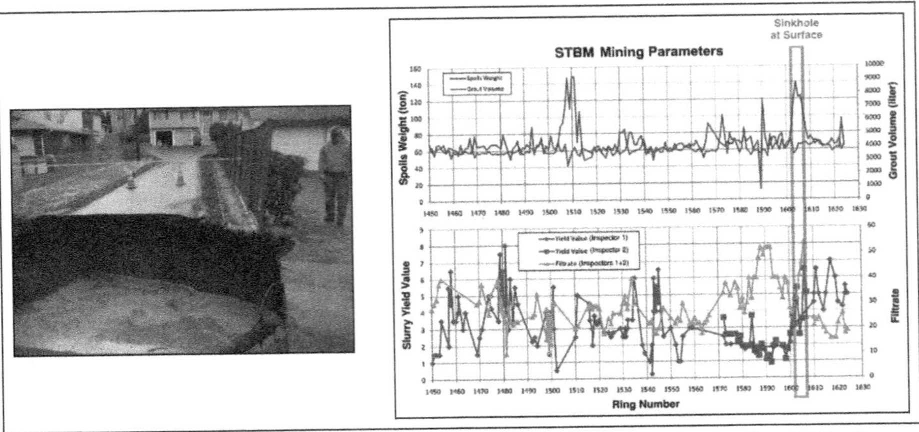


**Figure 2. Sinkhole (left) and presentation of TBM data evaluation results at that location (right) above the BT3 drive of the Brightwater Conveyance System tunnels**

tunnels. Undercrossing of the Seattle downtown area meant focus on ensuring that tunneling-induced deformations did not exceed the limits per design to minimize the risk of damage and disruption of existing infrastructure. An extensive instrumentation program to measure groundwater hydrostatic head, subsurface deformation around and above the TBM, and settlement of structures at the surface acquired data that was provided to the project parties including third-party infrastructure owners via a web-based geographic information system (GIS) software (Geoscope). Based on extensometer deformation measurements 1.5 m (5 ft) above the tunnel crown the ground loss was back calculated as ≤0.11% by volume (Cording et al. 2017). Such back calculation results are limited to the instrument locations, which for this project had a relatively tight average spacing of about 15 m (50 ft). This spacing still exceeded the 2-m (6.5-ft) length of an individual excavation step (liner ring) by almost an order of magnitude.

Given the high risk level when considering the large excavation volume and the project location, TBM operational parameters were also shared with all project parties to allow independent tracking of TBM performance and reconciliation of TBM spoils weight measured by belt scales with the theoretical weight under consideration of the theoretical excavation volume, face conditions per GBR profile, densities of the soil materials constituting those face conditions, and materials added during the tunneling process. Those materials included soil conditioners and bentonite, the latter injected into the annular space to maintain the pressure envelope around the TBM and into the excavation chamber to maintain face pressure during stoppages. The material flow reconciliation process also considered the tail void grout volumes and the volumes of secondary contact grouting through pre-installed ports in the liner rings. This process was continuously applied for trend analysis.

## Annacis Island Wastewater Treatment Plant New Outfall System

The Annacis Island WWTP New Outfall System includes two STBM drives through the alluvial deposits of the Fraser River delta at a hydrostatic head of about 3.4 bar at the tunnel invert. The river sands consist of poorly graded sands with fines content up to about 20% and include zones of higher fines content and with interbeds and lenses of fine-grained soils up to several meters in thickness. The saturated river sands generally behave as flowing ground under unsupported condition, resulting in the risk of significant ground loss in case of leakages and requiring focus on procedures managing this risk, especially during TBM launch and reception (Gwildis et al. 2022). To protect existing plant infrastructure and large warehouses of the industrial park on Annacis Island, material flow reconciliation on a per-ring basis was specified in addition to deformation monitoring. The specific material flow reconciliation procedure set up for this project was also used to identify and document changes in tunnel face conditions along the drives.

## MATERIAL FLOW RECONCILIATION APPROACHES

Material flow reconciliation compares the weight of the materials removed during a TBM advance increment ('push') with the theoretical excavation volume and weight of the in-situ soil under consideration of materials added in the process. The reconciliation approach selected for a specific project depends on the type of TBM system used, which in turn determines the equipment options for measurement of the material flow from which to derive weight and volume of the excavated materials (tunneling spoils). The following methods have been used at the selected projects for data collection and reconciliation calculations:

- **Method A:** EPBM systems can use belt scales to measure the spoils weight on the conveyor belt immediately behind the TBM's screw conveyor (Figure 3).
- **Method B1:** STBM systems can use flow meters and density sensors to measure slurry flow rate and slurry density in both the feed line to the tunnel face and the return line to the slurry treatment plant (STP) (Figure 4).
- **Method B2:** STBM systems can use belt scales at the STP to measure the weight of output from the desander units and centrifuges or filter press units as well as density measurements at slurry tanks (Figure 5).

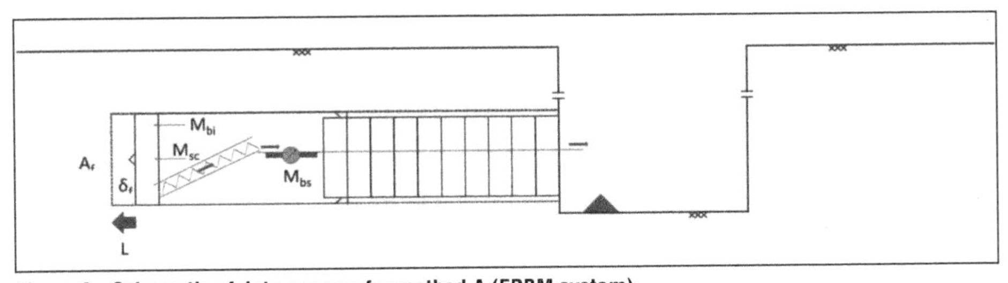


**Figure 3. Schematic of data sources for method A (EPBM system)**

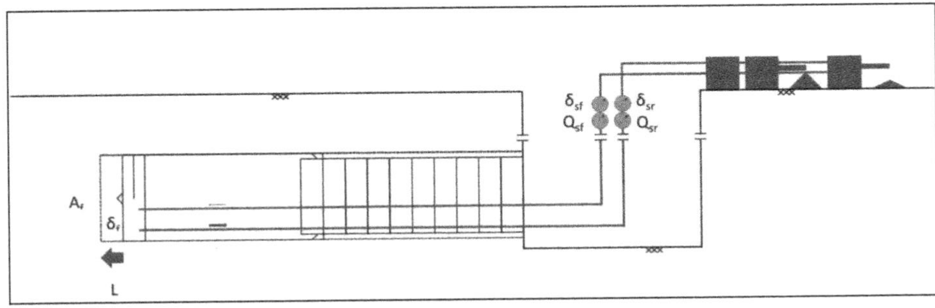


**Figure 4. Schematic of data sources for method B1 (STBM system)**

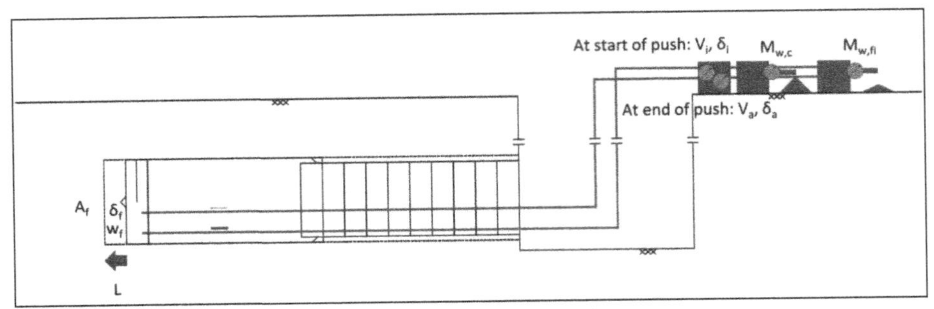


**Figure 5. Schematic of data sources for method B2 (STBM system)**

**Method A** for EPBM systems generates the reconciliation value (RV) by comparing the weight of the tunnel spoils as measured by the belt scales ($M_{bs}$) with the theoretical weight of the excavated geometry (tunnel face area $A_f$ times advance length L times average in-situ density $\delta_f$) and the weight of any materials added during the push such as soil conditioners $M_{sc}$ and bentonite injected into the excavation chamber or the annular space around the shield $M_{bi}$. This approach is summarized by Equation 1.

$$tRV = \frac{M_{bs} * 100}{(A_f * L * \delta_f) + M_{sc} + M_{bi}}, \%$$ **(EQ 1)**

The accuracy of the RV depends on various factors including the tolerances of the belt scale system and the quality of the estimate of the in-situ density of the tunnel face materials, which is derived from geotechnical exploration data and ideally checked by tracking the tunnel face conditions and testing of spoils samples. How the TBM is operated is another influencing factor. An example of operational decisions that impact the RV is modification of the earth pressure distribution in the excavation chamber by purging air volumes originating from soil conditioning. Another example is the practice of compacting the excavation chamber filling prior to a weekend stop by halting the conveyor screw extrusion before the end of a push while continuing with the TBM advance.

**Method B1** for STBM systems uses flow meters and density meters at the vertical slurry pipeline sections at the launch shaft. This method compares for each push the flow rate ($Q_{sf}$) and specific weight ($\gamma_{sf}$) of the bentonite slurry in the feed line to the tunnel face with the flow rate ($Q_{sr}$) and specific weight ($\gamma_{sr}$) of the material-loaded slurry in the return line from the tunnel face to the slurry treatment plant to determine the dry weight of the excavated material ($M_d$) (Bochon et al. 1999). This approach is summarized with Equation 2 (specific weight of water, $\gamma_w$):

$$M_d = \sum_{t=0}^{n} \left(\frac{\gamma_{sr} - \gamma_w}{1 - \frac{\gamma_w}{\gamma_{sr}}} * Q_{sr}\right) - \sum_{t=0}^{n} \left(\frac{\gamma_{sf} - \gamma_w}{1 - \frac{\gamma_w}{\gamma_{sf}}} * Q_{sf}\right), \text{kN}$$ **(EQ 2)**

The $M_d$ value is then used as input value to conduct reconciliation using the general approach summarized by Equation 1. The accuracy of the RV depends on the tolerances of the measuring equipment, the quality of the estimate of face condition and average in-situ density, and specifics of the operation of STBM and STP. Fully emptying the excavation chamber from the excavated material in the return slurry before the start of the next push is a prerequisite for obtaining reconciliation values of reasonable accuracy.

**Method B2** for STBM systems determines the dry weight of excavated materials at the STP, separately for coarse (granular) components ($M_{d,c}$) and for fines ($M_{d,fi}$). $M_{d,c}$ is determined by weighing the output of the desander units ($M_{w,c}$) and using the ancillary parameter $w_w$ (ratio of weight of water to the weight of wet solids in percent) as per Equation 3. $M_{d,fi}$ is determined via slurry tank level sensor and density meter by comparing slurry volumes and densities before ($V_i$, $\delta_i$) and after a push ($V_a$, $\delta_a$) as per Equation 4 (amount of water from the excavated ground added to the slurry as estimated from the moisture content of the in-situ face material, $M_{w,insitu}$).

$$M_{d,c} = M_{w,c} * (1 - w_w),\ t \qquad \text{(EQ 3)}$$

$$M_{d,fi} = \delta_a * V_a - \delta_i * V_i + M_{w,c} * w_w - M_{w,insitu},\ t \qquad \text{(EQ 4)}$$

If fines content or presence of fine-grained soils in the face are significant, output from filter press (or other separating units) can be measured by weight and a second reconciliation approach may be used to determine $M_{d,fi}$ values for comparison. This may be considered as a control procedure at longer intervals (several days, weekly). Materials such as lime that are added to facilitate the separation process by filter press need to be subtracted during the calculation. As for method B1, the combined $M_d$ value is then used to conduct reconciliation using the general approach summarized by Equation 1. The accuracy is determined by measuring equipment tolerances and the input values derived from geotechnical exploration data. It is influenced by events related to the STBM and STP operations such as clogging of pipelines, slurry transfers between tanks during a push, etc. Such events need to be recorded and taken into account. This reconciliation procedure does not consider water ingress or slurry losses.

The procedures as described do not take into account changes to the reconciliation value at the time the liner ring is built at the location of the preceding excavation step. Tail void grouting and secondary grouting for filling the annular space around the liner ring extrados can either be integrated into the reconciliation table or evaluated separately. Specific project conditions including tunneling behavior of the ground may determine preferences.

## RECONCILIATION DATA USE

Material flow reconciliation values summarize for each mining advance step the interaction of TBM operation and ground conditions encountered. The reconciliation value is a function of many variables. For understanding the accuracy of a specific value, detailed recording and reporting of the mining process is required. Calibration of the measurement equipment used for determining the spoils weight allows estimating measuring tolerances. Automated recording of the TBM operational parameters and detailed reporting of non-automated process steps as well as unplanned events provide the basis for better understanding differences between subsequent pushes in seemingly uniform ground conditions. However, the accuracy of each reconciliation value is also limited by the assumptions that need to be made regarding composition of the tunnel face and its average characteristics in terms of in-situ density and in-situ moisture content. These values must be estimated using the geotechnical exploration data per Geotechnical Data Report or baseline values per Geotechnical Baseline Report.

During pressurized-face tunneling the direct observation of face conditions encountered is not possible. However, in cases where location-specific prediction of tunnel face conditions along the alignment is not feasible and a non-location-specific baseline approach of baselining face conditions as a percentage of the overall drive length has been adopted, face condition tracking is required. This can be achieved by regular sampling, geotechnical laboratory index testing, and geotechnical and geologic classification of the spoils combined with the use of other data sources such as the TBM operational parameters (Gwildis et al., 2009). Where such a tracking effort is not specified, regular spoils samples may still be taken for other reasons, for example for testing to optimize soil conditioning during EPBM operations, in which case sample material would be available for checking assumptions on face conditions.

Figure 6 illustrates an example of material flow reconciliation for a tunnel reach in variable face conditions with the RV plotted over the alignment length by ring number. The left graph was generated using the average in-situ density for the materials in the tunnel face per exploration phase data, while the right graph shows reconciliation values for the same reach with adjustments of this geotechnical input parameter to reflect the data collected during the construction phase.

As is evident from this example, reconciliation graphs serve primarily for trend analysis. In soil types with some standup time under unsupported condition such as glacially overconsolidated till or lacustrine deposits, a ring location that seems to indicate overexcavation (positive RV value) may be followed by a ring that seems to indicate underexcavation (negative RV value). Plotting curves with moving averages over a small number of rings can help to obtain a clearer picture.

Use of the reconciliation graphs allows to evaluate the need for modifications of the mining process and allows to compensate for potential overexcavation by tail void grouting and secondary grouting in soil conditions with some standup time. Further, the reconciliation graphs can serve as a data record for assessing post-tunneling ground surface deformations as well as damages to existing infrastructure or claims thereof.

Under certain conditions, reconciliation graphs can also be used as a tracking tool for face conditions and to either verify the geotechnical baseline or identify a differing site condition (DSC). Figure 6 already provided some indication of the potential for this use of the data. If all other data sources indicate that changes of RV trends are not caused by changes in TBM operation and related factors, then these changes may reflect variation of face conditions.

Figure 7 provides the example of a STBM drive in alluvial sands with interbeds and lenses of fine-grained soils. Use of reconciliation method B2 allowed determining the mass dry weight of the spoils separated at the STP separately for the coarse granular materials weighted at the desander units and the fine-grained materials. The latter were estimated from slurry tank volumes and densities on a per-ring basis and also determined based on filter press output on a weekly basis. Differences in derived fines content provide the range for DSC evaluation.

## CONCLUSION

Mechanized tunneling in an urban environment requires to estimate tunneling-induced deformations during the design phase and to manage the risk of overexcavation and exceeding the estimated deformations during the construction phase. Risk management tools available to the owner/engineer include the ability to specify the requirements for the mechanized tunneling process regarding equipment, operation, data collection and reporting procedures. TBM operational parameters document the excavation process while measurement data from geotechnical instrumentation and survey points installed along the tunnel drive document the effects on the surroundings of the tunnel, depending on the subsoil conditions with or without much

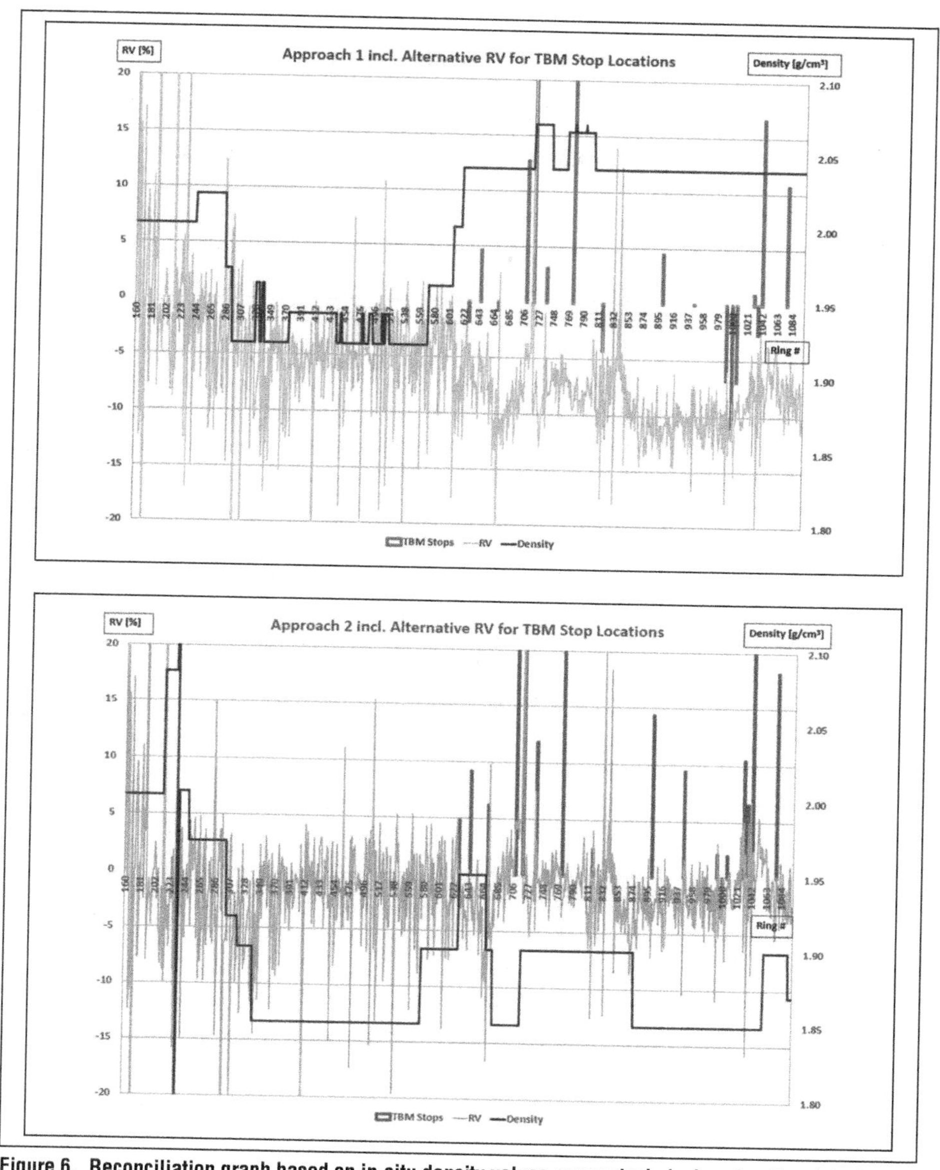


**Figure 6. Reconciliation graph based on in-situ density values per geotechnical exploration data (left) and with adjustment of in-situ density values to reflect changes in face conditions encountered**

delay. Pressurized-face tunneling does not allow direct observations of the geotechnical conditions at the tunnel face; however, data collected from the material flow of spoils during TBM advance can be used to better understand the face conditions and machine-ground interaction.

It is this author's conclusion that despite the inherent uncertainties of the data derived from the material flow of spoils, TBM material flow reconciliation has evolved as an essential risk management tool for mechanized tunneling projects in urban areas

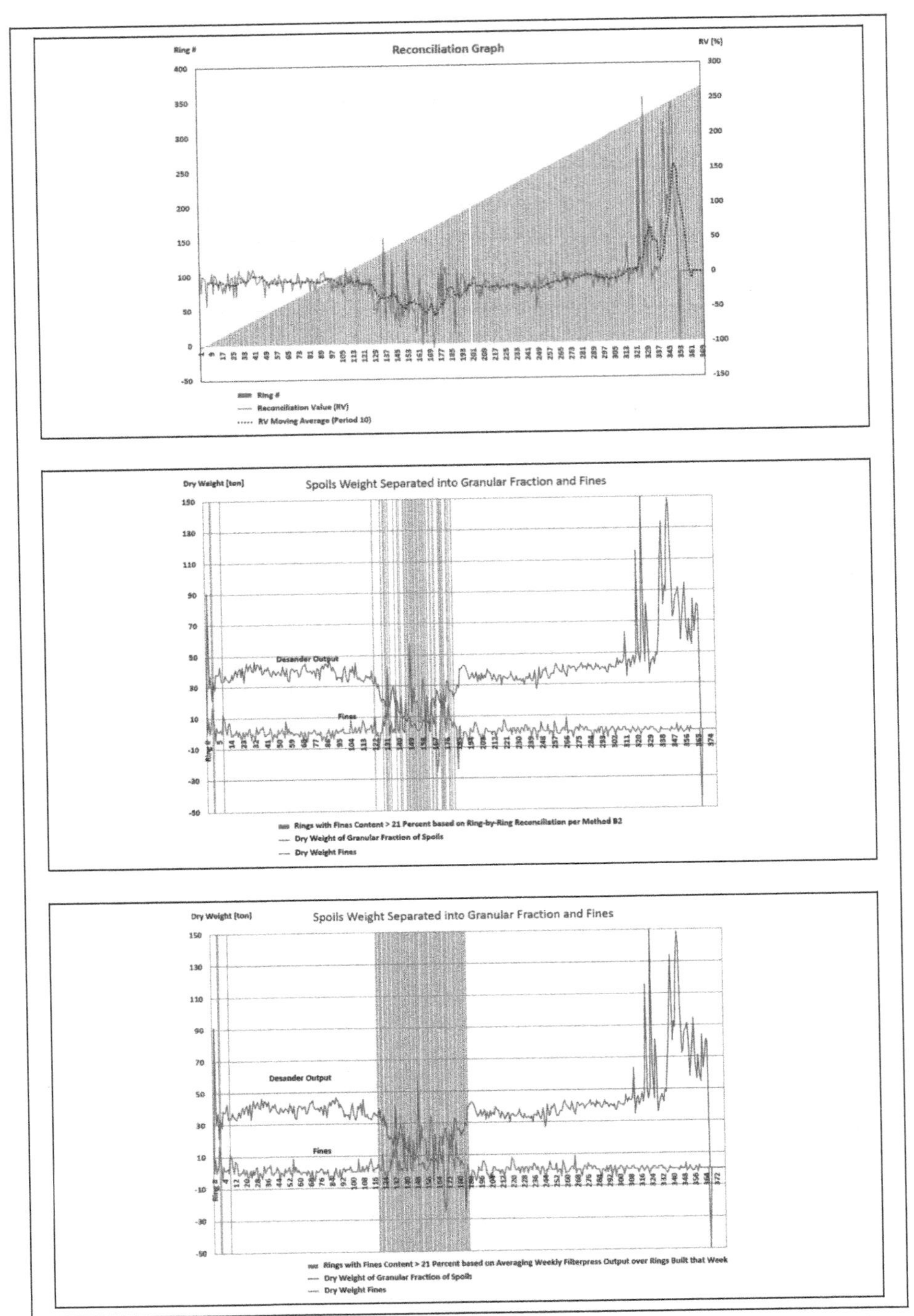


**Figure 7. Reconciliation graph (top) divided into desander output and fines with rings exceeding a fines content of 21% overlain as grey columns based on slurry volume and density changes (bottom left) and based on averaged weekly filter press output (bottom right)**

**Table 2. Use of Material Flow Reconciliation Data**

| Application | Method A<br>EPBM Belt Scales | Method B1<br>STBM Slurry Flow and Density Meters | Method B2<br>STP Discharge Measurements |
|---|---|---|---|
| Use for Detection of Overexcavation Trends | + | + | + |
| Use for Face Condition Tracking | –* | –* | + |

Legend: + applicable, – not applicable
* Additional tracking effort based on spoils sampling is required

that is indispensable for tracking, assessing, and pro-actively responding to the risk of overexcavation and resulting risk of surface impact and damage to existing infrastructure. Based on the project experiences from various TBM drives in the Pacific Northwest with its variable glacial geology and more uniform overlying Alluvium, the following general conclusions can be drawn:

- Material flow reconciliation provides a tool for tracking the excavation process on a per-ring basis to identify overexcavation trends for timely implementation of countermeasures and to provide a documentation that may be used for evaluating post-construction impact damages or claims.
- Setting up the data collection process for material flow reconciliation needs to consider the type of TBM system. For EPBMs the data collection focuses on conveyor belt scales. For STBMs two general approaches are available. Depending on specific project conditions, the degree to which equipment requirements and reconciliation procedures are specified varies. Some redundancy of measurement equipment or, in the case of STBM projects, the option of using two independent reconciliation methods may be considered for specific risk scenarios.
- Material flow reconciliation requires some understanding of the face conditions encountered during each push, either based on the geotechnical exploration data alone or supplemented by face condition tracking efforts that include taking spoils samples and using them for material characterization.
- In the case of reconciliation method B2 for STBM projects, where various types of measurement equipment and procedures are used to determine dry weight of granular materials and of fines separately, the material flow reconciliation data may also be used for the purpose of face condition tracking (Table 2).

## REFERENCES

Bochon, A., Rescamps, Y., Chantron, L. 1999. La détection des anomalies d'excavation au tunnelier a pression de boue: Mèthode mise au point sur le chantier EOLE. In *AFTES 1999.*

Cording, E., Nakagawa, J., McCain, J., Stirbys, A., Sowers, D., Vazquez, J., Painter, C. 2017. Managing Ground Control with Earth Pressure Balance Tunneling on the Alaskan Way Viaduct Replacement Project. In *Proceedings of the 2017 Rapid Excavation and Tunneling Conference.* 82–98. Englewood, CO: Society of Mining, Metallurgy & Exploration.

Gwildis, U., Maday, L., Newby, J. 2009. Actual vs. Baseline Tracking during TBM Tunneling in Highly Variable Geology. In *Proceedings of the 2009 North American*

*Tunneling Conference*. 250–262. Englewood, CO: Society of Mining, Metallurgy & Exploration.

Gwildis, U., Newby, J., Kennedy, E., Sidhu, A., Roux, G. 2022. Mechanized Tunneling in Flowable River Sands—The TBM Drives of the Annacis Island WWTP New Outfall System. In *Proceedings of the 2022 North American Tunneling Conference*. 687–695. Englewood, CO: Society of Mining, Metallurgy & Exploration.

Robinson, R., Sage, R., Clark, R., Cording, E., Raleigh, P., Wiggins, C. 2012. Conveyor belt weigh scale measurements, face pressures, and related ground losses in EPBM tunneling. In *Proceedings of the 2012 North American Tunneling Conference*. 65–72. Englewood, CO: Society of Mining, Metallurgy & Exploration.

# Risk Baseline Report: An Innovative Risk Management Approach for a Complex Underground Project

**K. Bhattarai** ▪ HNTB Corporation
**David J. Hatem** ▪ Donovan Hatem LLP

## ABSTRACT

The increasing popularity of alternative methods in delivery of underground infrastructure projects has given rise to innovative and collaborative contracting strategies to allocate and manage the unique risks associated with the work.

This paper presents approaches and recent examples of how risk has been dealt with using various alternative delivery methods, including Progressive Design Build (PDB) and Construction Manager/General Contractor (CM/GC). Based on the experience gained from these examples, recommendations for an innovative and collaborative risk baselining approach using contingency sharing are presented that should advance the overall goals of equitable risk sharing and compensation.

## INTRODUCTION

In recent years, there has been increasing interest by public agencies to evaluate and implement alternative delivery methods in underground infrastructure projects.

Several tunneling projects using alternative delivery methods have successfully been completed or are under active delivery process. Some of these projects, alongside their corresponding delivery methods, are:

- Westside CSO Tunnel Project in Portland, Oregon—Cost Reimbursable and Fixed-Fee Contract (modified CMGC)
- Alaskan Way Viaduct Tunnel in Seattle, Washington—Standard Design Build
- Anacostia River CSO Tunnel in Washington DC—Standard Design Build
- Lake Mead Intake 3 Project, Nevada—Standard Design Build
- Northeast Boundary Tunnel Project, Washington DC—Standard Design Build
- Los Angeles Metro Purple Line Section 3 Project—Standard Design Build
- SCVTA's BART Silicon Valley Extension Phase II San Jose, California—Progressive Deign Build

The evolution of alternative delivery methods has also led to continued development and adaptation of new ways of allocating, managing, and achieving equitable share/transfer of risks between Owners and Contractors, such as risk registers and the Geotechnical Baseline Report (GBR).

The GBR and the Differing Site Condition (DSC) clause have been a part of the construction contract documents for many years to improve risk allocation and facilitate the resolution of claims and disputes. The GBR has been instrumental so far in allocating geotechnical risks. The GBR, however, is not intended as a contractual tool to identify, allocate and manage *non-geotechnical* project risks.

The risk register is increasingly being adopted as an approach to identify, allocate, mitigate and manage risks. However, it has so far been adopted as a non-contractual tool in the risk management process. A risk register is generally developed during the design phase and updated as the project progresses. The risk is allocated, controlled, mitigated, and managed for each of the risk items and generally transferred to a party that is best qualified to manage that risk. A qualitative or quantitative analysis (QRA) or both are performed to assess and quantify, mainly, the impact of a risk in terms of cost and schedule.

## Compliance with Federal Requirements

Federal agencies have defined requirements for government agencies seeking project funding for their projects. The following documents provide guidance for risk management in Federally funded projects.

- Risk Analysis Methodologies and Procedures—Federal Transit Administration (FTA) (2004)
- Project Management Oversight Procedure 40 Risk Assessment and Mitigation Review—Federal Transit Administration (2022)

FTA requires fund seeking agencies to undergo rigorous risk management processes to obtain funding for Entry into Engineering and a subsequent project development phase. Systematic risk workshops are performed and the documents are provided to the FTA for review and approval.

## Typical Risks in Tunneling and Identification of Unique Project Risks

Some typical tunneling risks are:

- Temporary support failure in deep excavations and tunnels
- Settlement, deformation or collapse of buildings close to tunnel and excavations
- Ground different from what is anticipated leading to differing site conditions claims
- Inadequate ground investigations leading to inadequate structural design
- Choice of a TBM with inadequate features, not compatible to the ground encountered during excavation or alignment, leading to soil failures or sinkholes
- Bearing failure leading to a TBM being frozen in place at the heading
- A TBM with inadequate features, not compatible with a sharp alignment curve, leading to an out-of-tolerance liner ring building. Pitfalls—water leakage, inadequate liner structural capacity, deviation from the alignment.
- Uncontrolled groundwater drawing leading to settlement of properties close to the excavation—a source of protest, delays, and lawsuits
- Fractured/weathered rock and loose soil with a shallow cover

## Risk Impacts

Exposure to risks impacts both the cost and schedule of a project. During the risk management process in the project execution phase, both cost and schedule risks are generally quantified using both qualitative and quantitative methods.

One common practice is to estimate risk scores—likelihood (probability) times severity (impact)—before and after mitigation in the construction submittals. The scores are classified using three color codes—red (high), yellow (medium) and green (low). The scores would show how risks in terms of cost and schedule are reduced or mitigated by applying certain innovative design and construction techniques. A mitigated risk will show a low to very low risk score. During a risk workshop, a risk register is updated to review the status of the existing risk items and document the newly identified risks, corresponding risk scores, ownership, transfer, and mitigation of risks.

Risk scores for the schedule impacts are derived from five probability and five severity classes. The probability classes range from remote (low probability) to expected (high probability) while the severity classes range from a schedule delay from a week (negligible) to delays of more than 6 months (catastrophic). Similarly, the severity classes for cost usually range from cost impacts of less than $200k (negligible) to impacts greater than $20 million (catastrophic). The magnitude of severity, however, varies from a project to a project based upon how it is defined to suit the complexity and size of the project.

## RISK BASELINE REPORT—DESIGN AND CONSTRUCTION

Risk Baseline Report for Design and Construction (RBR-D&C), an innovative contractual tool, has been suggested to baseline and manage risks during design and construction (Bhattarai 2018).

It is recommended that the RBR-D&C be prepared using a three-step process. First, the Owner prepares Risk Baseline Report (Owner) during the preliminary/project design development phase and includes it as a part of an RFP contract requirements. Second, the Contractor (or Design-Builder) prepares Risk Baseline Report (Procurement) based on the RFP requirements. Third, a Risk Baseline Report for Design and Construction is collaboratively developed by and negotiated as a contract document between the Owner and the Contractor (or Design-Builder) (Bhattarai 2018).

The proposed RBR-D&C will include a Project Risk Register (PRR) and a risk management plan. The PPR will also include any other risks that the Owner identifies. The RBR-D&C will state and baseline the risks, which will form a contractual basis for payment, in case of exceedance of risk trigger levels.

The RBR-D&C will clearly state how each risk in the design and construction has been allocated and the ownership will be specified. Furthermore, it will also state how each risk is monitored and trigger levels for each risk will be defined.

The Contractor (or Design-Builder) and the Owner will jointly administer the RBR-D&C and the risk management process during the life of the contract. Risks may be revised, added, or eliminated as work progresses, as well as probabilities, severity, impacts, and mitigation plans.

The RBR-D&C will establish and clarify a contractual understanding of the risk (risk baseline) and how the risks will be allocated in the contract. The baseline will be defined in terms of risk type, ownership, risk scores, life of the risk items, and the milestone of their completion and contingencies associated with them. The payment will be based on the Incentivized Contingency Allocation Method.

Both the GBR and RBR-D&C should be contract documents and be legally and technically consistent. Moreover, the contract documents (inclusive of the RBR—D&C) should provide the basis to allocate and manage risks and resolve differing site conditions claims.

## PROCUREMENT METHODS AND RISK MANAGEMENT PROCESS

Increasingly, Owners are exploring and selecting alternative delivery methods that maximize collaboration in design development between them and the Contractors or the Design-Build Team. Progressive Design Build (PDB) and Construction Manager/General Contractor (CM/GC) are two such methods.

### Progressive Design Build (PDB)

In the PDB delivery method, the Owner, early in the design development phase, selects a Design-Builder using a qualification-based assessment who collaborates with and assists the Owner in developing an overall project design and construction process, project schedule, and cost. The Design-Builder will also support the Owner in developing a GBR, a project risk register and an overall risk management approach. A recent example of a tunnel project that is being procured using this procurement method is "SCVTA's BART Silicon Valley Phase II Extension Project."

In this project, the Owner VTA (Valley Transit Authority) prepared its own preliminary risk register prior to the RFP solicitation and required the bidders to state their approach in risk management, risk sharing, and allocations methods; their approach in using the risk register, maintain and manage the risk register, and control the cost impacts; the top five design risks and methods to mitigate these risks (VTA 2022).

The Joint Venture (JV) of Kiewitt, Shea and Traylor (KST) was selected in June 2022 by the VTA as the PDB Contractor. The PMT (the project management team, a JV of HNTB and WSP) is the Owner's risk management consultant. The risk management team is led by a dedicated Program Risk Manager who conducts the risk workshops and monitor project risks. The project is currently at Stage 1, the programming and design stage of the PDB contract. Stage 2 of the contract includes design completion, construction, start-up, and commissioning stages. During the Stage 1 level, the VTA is working with the PDB contractor to monitor contract risks plans to utilize quantitative analysis of identified risks to estimate project contingencies. A risk register could be incorporated into the Stage 2 contract documents.

### Construction Manager/General Contractor (CM/GC)

In the CM/GC method, the contract allows a Contractor to bring its expertise in construction and risk management early in the design process. A CM/GC Contractor works in synergy with the Owner, Engineer, and stakeholders in developing a risk management process that includes a project risk register and mitigation methods for those risks. A relatively new method of delivery in transportation tunnels, this approach is currently being evaluated for Austin Cap Metro Blue Line and Orange Line Subway projects in the City of Austin, Texas.

PDB and CM/GC, in different approaches, both represent efforts to increase collaboration of the Owner's and DB Team (in PDB) and the CM/GC (in CM/GC) in the design development process. These approaches are especially beneficial on major subsurface projects in providing the primary project participants with a mutual and transparent foundation in which to collaboratively assess technical challenges, pricing considerations, and risk identification, allocation, and contingency and mitigation strategies.

## COLLABORATION IN DESIGN DEVELOPMENT PROCESS: AN ALTERNATIVE TO CONVENTIONAL DESIGN-BUILD

There are multi-dimensional concerns for Contractors and Consulting Engineers, presented by certain prevailing procurement and contractual practices in conventional Design-Build (DB). At root, these concerns principally derive from mandates that a fixed price be contractually committed prior to sufficient clarity and comprehension of the reasonable and realistic expectations as to what is required of the DB team in the final design and construction approaches. Those concerns are exacerbated by the aggressive and imbalanced risk allocation obligations often assigned to the Design-Builder (Hatem 2022a; 2022b; 2022c).

The cumulative effect of these concerning practices and dynamics often produces both serious financial losses for Design-Builders and substantial professional liability "cost overrun" claims asserted by the latter against their Consulting Engineers, as well as the negative reputational impacts to those firms participating in DB public infrastructure projects (PIPs).

### Project Cost and Risk Realities

At root, the principal concerns with conventional DB approaches on PIPs primarily arise out of unrealistic expectations of project participants during the proposal phase as to the actual and inherent project cost ("project cost") and risks required to be reasonably assessed in the pricing, planning and execution of the design and construction of a project that meets the Owner's ultimate requirements. Simply put, the realistic project cost is frequently not captured in the fixed-price award. This consequence is significantly detrimental due to the inability of the majority of Design-Builder proposers to adequately define, capture and reasonably predict during procurement all of the relevant design and construction considerations, costs, and risks inherent and necessary to assess and price in order to achieve the Owner's ultimate requirements. On megaprojects, the risks of unrealistic project cost and overly optimistic risk assessments are elevated (Hatem and Corkum 2010, Hatem and Gary 2020).

The overarching question is when can sufficient understanding of project-specific design and construction approaches reasonably and realistically be known in a manner to adequately and realistically inform commitments as to contractual pricing and risk allocation terms.

On most complex DB subsurface and other infrastructure projects (and especially megaprojects), it is neither realistic, reasonable, nor fair to expect that such an understanding can or should be known or knowable at the time of DB contract execution. A recent study by Travelers Insurance Company provides compelling data to support and validate that observation (Travelers Study 2021).

The acute problems associated with procurement and contractual practices in conventional DB PIPs that (a) require a fixed price at the time of initial DB contract award and (b) mandate imbalanced risk allocation terms need to be corrected and a more sensible and collaborative path forward developed. In general, the solution should allow for deferral of contractual commitments as to the final price and risk allocation terms until the Design-Builder has had a reasonable opportunity to understand, assess, price and plan for the required project-specific design and construction approaches, and for the site, subsurface, and other relevant conditions and constraints (physical and political) in which those approaches will be executed (Hatem and Gary 2020; Hatem 2020, 2022a, 2022b; Stephenson 2022).

## PDB and CM/GC

PDB is a significant step in the right direction to correct some of these root causes and resultant problems in conventional DB PIPs by providing for early Design-Build Team involvement in the design development process that allows for meaningful and significant opportunities to achieve risk allocation balance (Gransberg and Molenaar 2019). PDB generally involves a process in which contractual commitments as to fixed cost and risk allocation terms are deferred by the Owner and Design-Builder until at least approximately 60 percent of design development has been achieved. Meaningful involvement, interaction and collaboration among the Design-Build Team, and the Owner in PDB should serve to improve their mutual understandings and transparencies of risk perceptions, and positively influence pricing and contingency realism and balanced risk allocation (Hatem 2022a, 2022b, 2022c). It is generally recognized that the advantages of PDB particularly on subsurface infrastructure projects, include the ability of the Owner team and Design-Builder team to be better informed and aligned as to both perceptions and realities of critical risk variables and contingencies—such as those involving evaluation of subsurface conditions and assessments as to final design approach and construction methodologies—prior to reaching contractual commitments on price and risk allocation terms (Hatem 2022a, 2022b).

There are other approaches to defer final price and risk allocation commitments in DB until the Design-Builder has had adequate time to evaluate relevant project factors and conditions. The Virginia DOT "scope validation" approach relating to the pricing and risk for subsurface conditions work, is noteworthy in this regard. Under that approach, the Design-Builder has a period of time following a limited notice to proceed within which to validate its pricing and risk assessments as to subsurface conditions prior to making final contractual commitments.

Both PDB and CM/GC are intended to address problems associated with premature fixed prices and imbalanced risk allocation in conventional DB (Richards 2021, Forsey 2021). Like PDB, CM/GC allows for the opportunity for meaningful Owner and CM/GC collaboration and shared input on all issues relevant to project-specific risks identified in a Risk Register, including final design and constructability issues. In CM/GC the Owner is typically responsible for the adequacy and suitability of the final design and the CM/GC is typically responsible for the means/methods.

The RBR-D&C approach is consonant with the PDB and CM/GC approaches and potentially an extremely beneficial tool in their implementation and collaboration in the design development and risk allocation processes.

The RBR-D&C approach can and should be applied—albeit adaptively, and, sensitive to the differing roles, risks and responsibilities of the project participants—in PDB and CM/GC. The RBR-D&C should certainly be classified and defined as a Contract Document, and the contents of the RBR-G&C should be conscientiously written in a manner that is consistent with other portions of the Contract Documents, including the GBR.

## Benefits of Design Collaboration on Subsurface Projects

Major subsurface work involves several critical interactions, interdependencies, and dynamics (IIDs) among ground conditions, final design approaches and construction means/methods (Hatem 2018; Hatem 2022b). Rarely does any particular portion of the Contract Documents address risk factors and considerations for *all* of those IIDs in a universal, integrated and consistent manner. The RBR-D&C provides an opportunity to achieve that objective and to contractually document the results of a collaborative effort of the Owner and other project participants thereby enhancing effective and transparent risk allocation. The IID factors reinforce that effective risk allocation on major subsurface projects depends upon integrated and consistent (contractual) understandings as to:

- Ground conditions as documented by data and their evaluations specific to relevant design and construction methodologies
- Permanent works design
- Construction means/methods

Especially in PDB and CM/GC, the ability of project participants—in a collaborative mode—to define risks relating to (a), (b) and (c) in an integrated manner is significantly enhanced; as well as their ability to document their mutual understandings contractually and collaboratively as to the allocation of those risks. The RBR-D&C provides an excellent mechanism to accomplish those objectives in an integrated, consistent, and consolidated manner. The key (and concern) is to draft the RBR-D&C in a manner that is conscientiously consistent with other portions of the Contract Documents.

The collaboration in the design and construction processes in PDB and CM/GC should promote and facilitate utilization of the observational method (Powderham and O'Brien 2021).

At 60+% level of design development on a major subsurface project—i.e., the (minimal) point at which the Contractor (in CM/GC) or the Design-Builder (in PDB) is typically expected to contractually commit to a fixed price and risk allocation terms—the following has transpired:

- The subsurface investigation and data evaluation is complete or substantially complete
- Deficient subsurface data is available to adequately inform final design development
- The final design is substantially complete
- There has been a reasonable opportunity to address and mitigate in final design and construction means/methods issues that have been identified in a Risk Register
- There is a sufficient basis and understanding to reliably inform the selection and design of construction means/methods, and equipment
- The proposer has a realistic and reliable basis upon which to plan and price (with appropriate contingencies) the permanent works and construction means/methods
- There exists an adequate, reasonably informed and realistic basis to negotiate and contract on relevant risk allocation terms
- An adequate contractual basis exists to facilitate resolution of subsequent differing site condition dispute

An RBR-D&C can—in conjunction with other provisions of the Contract Documents—provide an integrated and consolidated basis to collaboratively capture and contractually document relevant risk factors and considerations as to all of the preceding points.

Viewed in this context, the virtue and value of the RBR-D&C approach—especially in PDB and CM/GC—are the ability to effectively promote fairness, balance and transparency in pricing bases and risk allocation on major subsurface projects.

## INCENTIVIZED CONTINGENCY ALLOCATION METHOD

Some Owners, such as the Washington Department of Transportation (WSDOT), have added contract incentive clauses in contracts to allow for additional payments to Contractors to promote better contract delivery performance (WSDOT 2010).

In this incentivized contingency allocation contractual arrangement, the Owner's Risk Manager estimates the contingencies for the risks, generally by using QRA analysis. The Owner also establishes the contractual criteria for how this contingency is allocated and distributed.

To facilitate this contractual process, the risk contingency is presented and agreed between the Owner and the Contractor as a separate line item in the Bill of Quantities. The contingency fund is controlled by the Owner and defined as an allowance in the Contract. The Contract will allow the contingency reserve fund to pay for mitigating the risks that appear during construction.

The release of the contingencies, upon completion of the risk item, are generally done in the following ways (Bhattarai 2018).

- Divide contingencies for each major risk item in an equal proportion, if no contingency is utilized at the completion
- Distribute the remaining (residual) contingency fund if utilized during the construction under other agreed proportion

In this arrangement, the Contractor also signs an agreement with their designer/subcontractor to distribute the contingency for a risk item associated with the relevant design/construction element. This fosters partnership and creates a win-win situation at all levels of the construction organization.

The Owner's Risk Manager and the Contractor's Risk Facilitator jointly track the contingencies—for both the cost and schedule—regularly and discuss them in the project risk workshops and monitor their completion and mitigation status via the Project Risk Register (PRR) over the defined life of the risks.

A similar approach was successfully used in WSDOT SR 99 Bored Tunnel Alternative Design-Build Project. This contract included three main clauses:

- Shared Contingency Allowance
- Deformation Mitigation and Repair Fund
- Completion Incentive

As a part of this contract provision, WSDOT established shared contingency allowance fund. The contract allowed the D-B Contractor, per the "Deformation Mitigation and Repair Fund" clause, to receive 75 percent of any amount remaining in the Shared Contingency Allowance following Physical Completion of the Work and payments for change orders made to the Contractor. The contract also allowed WSDOT to retain the remaining 25 percent of the fund. This payment will be added to the total contractual payment by a change order and shall be paid at the same time as the Final Payment.

The completion Incentive clause compensated the Contractor for achieving substantial completion prior to the contractual Substantial Completion Deadline with an incentive of $100,000 per day, up to the maximum of $25 million (WSDOT 2010).

There are some situations where an incentivized contingency approach can result in the timely and cost-effective resolution of unanticipated field conditions. By establishing well-defined work categories, such as unanticipated subsurface conditions, unknown utilities and unexpected contaminated soil, and other unanticipated risks, the Contractor and Owner can establish agreed-upon contingency funds to be used if and when these conditions are encountered. In addition, to incentivize the Contactor to perform the required work in a timely and efficient manner, the balance of each contingency fund at the completion of the project is apportioned between the Contractor and Owner based upon an agreed upon split (e.g., Owner 60-percent and Contractor 40-percent) (Bhattarai 2018).

## INSURANCE AND SURETY—THE ELEPHANT IN THE ROOM

The insurance and surety industry forms an ineliminable partner in the underground construction industry as they insure against various inherent project risks. Well defined risks, quantification of risks, equitable risk sharing mechanism in the contracts, and an effective risk management process will provide clarity and basis for the insurers to objectively rate the project and insure various risk elements to various project parties, as applicable. Although it is not customary, it is important that the insurance industry is brought onboard during the project development, procurement, and construction phases. With their expertise, they greatly contribute to the risk management process during the development stage and the final contract negotiations.

## COORDINATED INSURANCE PROGRAM (CIP)

CIP Insurance Program (or wrap-up insurance program) insures the project parties for general liability, worker's compensation and employer's liability, excess liability, builder's risk, and in some cases professional liability and environmental liability. A well-defined and effectively executed RBR-D&C would bring clarity in risk identification, mitigation methods and equitable payment mechanism. This clarity would provide a strong basis for the underwriters to insure the projects, increasing potential for lowering insurance premiums.

## PROFESSIONAL LIABILITY

On major subsurface projects, project-specific professional liability (PSPL) insurance is essential to provide adequate risk transfer and protection for all project participants (Hatem 2010, Hatem and Gary 2020). Fixed price and imbalanced risk allocation procurement and contractual practices in conventional DB have elevated professional liability exposure and adverse claims experience resulting in a concerning reduction in the availability and capacity of PSPL insurance limits in the insurance markets (Hatem 2022c). PDB and CM/GC should improve the PSPL availability and capacity issues by addressing problems associated with the need to establish fixed price commitment prematurely relative to design development by deferring such commitments until significant design development has been achieved and providing a more balanced and informed context for contractual risk allocation decisions (Hatem 2022c). On major subsurface projects, the Joint Code of Practice for Tunneling Projects in the UK ("Joint Code"), jointly developed by the British Tunneling Society and the Association of British Insurers aims to improve both contractual risk allocation and insurance underwriting practices on those projects (Hatem 2020a, Hatem 2010). The Joint Code approach should have positive application in improving, adapting, and aligning those practices in the collaborative contracting approaches of PDB and CM/GC. More specifically,

Owners, Contractors and Consulting Engineers may utilize the Joint Code approach as guidance to align and collaborate in developing mutually acceptable processes for PDB and CM/CG implementation in a manner that comports with professional liability insurance underwriting requirements and expectations (Hatem 2022a, Hatem 2022b, Hatem 2022c).

## SURETY/BONDING/BUILDER'S RISK

The Owner's preference of adopting a standard D-B contract in procurement without proper equitable risk sharing protocols has contributed to diminishing returns to the Design Builders as well as increased risk exposure to the Surety/Insurance Industry leading to increased reluctancy in insuring projects by underwriting communities. This results in higher premiums for some critical projects. The adoption of RBR-D&C will increase transparency in risk exposure as well as equitable risk sharing among the project parties, thus enhancing potential for better project insuring and bonding.

## CONCLUSION AND RECOMMENDATIONS

Over the last decade alternative delivery methods have become more frequent choices of project delivery, especially for large projects. The increasing popularity of the PDB and CM/GC delivery methods comes with a growing interest by project owners to seek ways to minimize design and construction risks and deliver the project on cost and schedule.

Lessons learned from the execution of successful large infrastructure projects, especially conventional DB projects, show that the success depends upon understanding of risks from the project inception and formulating balanced risk allocation and effective risk management strategies through more collaborative design development and risk allocation processes.

The RBR-D&C provides a formal contractual risk allocation, contingency sharing, and effective risk management process for CM/GC and PDB project delivery methods. Thus, it is recommended as an effective risk allocation and management tool for projects procured under CMGC and PDB delivery methods as well as for other delivery methods.

For a successful and effective implementation of RBR-D&C as a contract document it is important that the report and process follows formal guidelines. The suggested check lists and a flow chart are provided in Figure 1.

### Suggested Guidelines (Check List) to Prepare a RBR—D&C

#### *Introduction*

- Project name
- Project organization, owner/owner organization, design team, contractor (PDB/CMGC contractor)
- Purpose, scope, and organization of report
- Description of past project experience in risk management, as applicable

#### *Project Description*

- Project location, type, and purpose
- Summary of key project features

- References to contract documents—general and supplementary conditions, contract specifications, contract drawings, GBR, GDR
- Risk Baseline Report—Design and Construction defined as a contract document
- RBR and GBR, references and domain of applicability
- Hierarchy of contract documents with reference to RBR—D&C
- Bill (Schedule) of Quantities
- Owner's risk management structure
- Contractor's risk management structure

### *Risk Management Plan and Process*

- Risk management plan/scope
- Contractor's risk management task force organization
- Risk management workshops—frequencies and risk register updates
- Risk partnering plan and process—owner and contractor
- Risk identification phase and process
- Risk evaluation phase and process
- Risk control phase and process
- Risk monitoring and reporting
- Risk action logs and reports (incl. software adopted for the purpose)
- Risk mitigation/reduction methods and process—design, construction, testing and commissioning, QA/QC process

### *Project Risk Register*

- Major risk items
- Risk characterization and breakdown structure
- Probability of occurrence and severity—cost, schedule, quality, health and safety, sustainability, as appropriate.
- Risk ownership and mitigation
- Quantitative/qualitative risk assessment—pre- and post mitigation
- Risk contingencies
- Risk trigger levels

### *Risk Contingencies, Bill of Quantities and Sharing*

- Quantification, scope, and agreement
- Reference to other relevant contract documents and applicable clauses of the contract

### *Risk Baselining and Payment Provisions*

- Methodology
- Register of baselines
- Payment provisions

### *Insurance and Surety*

- Scope, program, coverages, and limitations
- Reference to other relevant contract documents

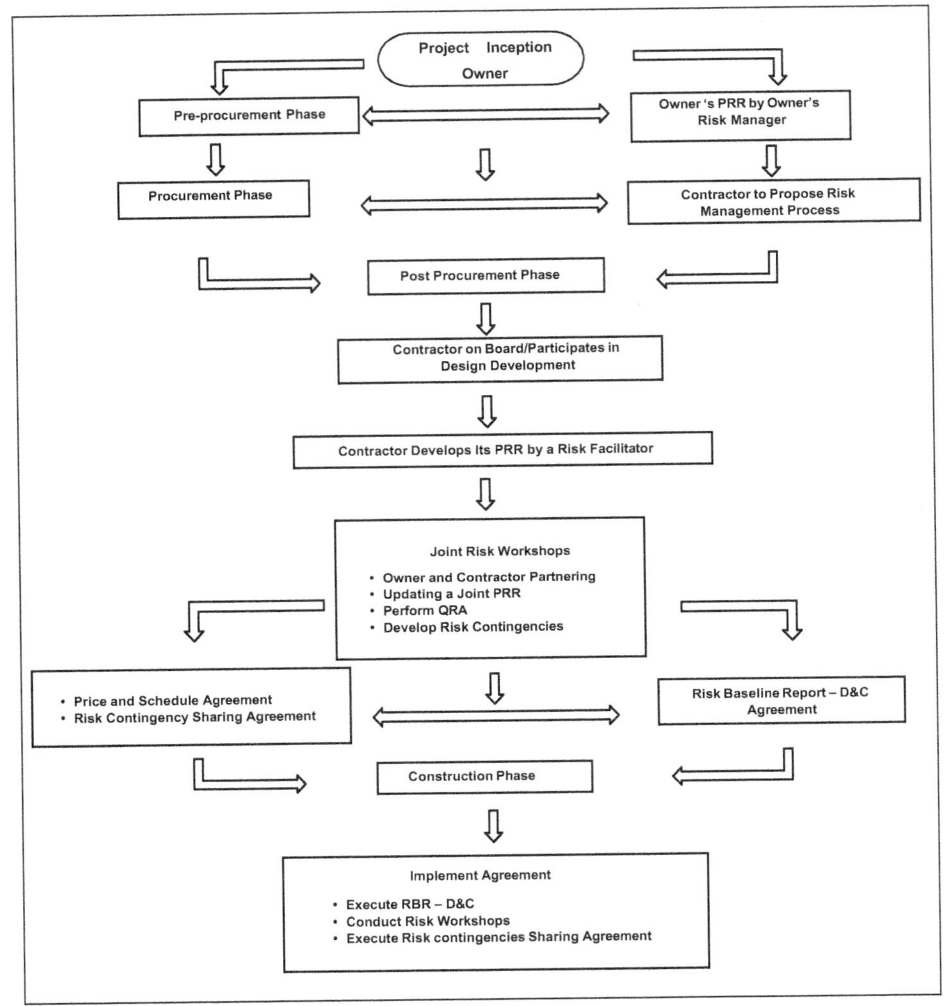


**Figure 1. Suggested risk management flow chart**

## REFERENCES

Bhattarai, K. 2018. Design, Construction, and Risk Management Strategies for Shallow Tunnels in Urban Settings. *Proceedings of North American Tunneling Conference*. pp. 687–687. Washington DC, USA.

District of Columbia Water and Sewer Authority. 2013. Requests for Proposals—Anacostia River Tunnel Project.

District of Columbia Water and Sewer Authority. 2016. Requests for Proposals—Northeast Boundary Tunnel Project.

Feroz, M., Moonin, E., and Grayson, J. 2010. Lake Mead Intake No. 3, Las Vegas, NV: A Transparent Risk Management Approach Adopted by the Owner and the Design-Build Contractor and Accepted by the Insurer. *Proceedings of North American Tunneling Conference*. pp. 559–65. Portland, Oregon.

Forsey J., Weatherall, M., Kehoe, J. 2021. Perfect Procurement. 9th *Int'l Society of Construction Law Conference*.

Gransberg, D., and Molenaar, K. 2019. Critical Comparison of Progressive Design-Build and Construction Manager/General Contractor Project Delivery Methods. Trans. Res. Rec.

Gribbon, P., Irwin, G., Colzani, G., et al. 2003. Portland Oregon's Alternative Contract Approach to Tackle a Complex Underground Project. *Proceedings of North American Tunneling Conference*. pp. 68–76. New Orleans, Louisiana.

Hatem, D.J., and Corkum, D, eds. 2010. Megaprojects: Challenges and Recommended Practices. *American Counsel of Engineering Companies*. Chapter 18, pp. 508–38 and 599. Washington.

Hatem, D.J. 2018. Subsurface Conditions and Design Adequacy Risk Allocation in Design Build: Dynamics, Interactions and Interdependencies. *Tunnel Business Magazine* (October 2018).

Hatem, D.J., and Gary, P. 2020a. Public-Private Partnerships and Design-Build: Opportunities and Risks for Consulting Engineers, Ch. 12, 12.5 12.3.2, 12.5, 12.6.2. *American Council of Engineering Companies*. 3rd ed. Washington.

Hatem, D.J. 2020b. Improving Risk Allocation on Design-Build Subsurface Projects. *Tunnel Business Magazine*.

Hatem, D. J. 2022a. Recalibrating and Improving Design-Build on Public Infrastructure Projects. *American Bar Association*.

Hatem, D.J. 2022b. Design-Build: Recalibrating Procurement and Contractual Approaches. *George A. Fox Conference*. New York.

Hatem, D. J. 2022c. Project-Specific Professional Liability Insurance on Design-Build and Public-Private Partnership Projects in North America: A Path Forward. *Donovan Hatem LLP*. Boston, MA.

International Tunneling Insurance Group. 2012. A Code of Practice for Risk Management of Tunnel Works. 2nd ed. A 3rd ed. is in progress.

Los Angeles County Metropolitan Transportation Authority. 2017. Request for Proposal—Westside Purple Line Extension Project Section 3.

O'Carroll, J., and Goodfellow, B. 2015. Guidelines for Improved Risk Management (GIRM) on Tunnel and Underground Construction Projects in the United States of America. Underground Construction Association of SME.

Powderham, A., and O'Brien, A. 2021. The Observational Method in Civil Engineering. Ch. 14, pp. 325–331 CRC Press 2021.

Richards, T., Bolland H., and Bradstreet, B. 2021. Buildability Risk Allocation and Mitigation. 9th *Int'l Society of Construction Law Conference.*

Southern Nevada Water Authority (SNWA). 2007. Requests for Proposals: Lake Mead Intake Number 3—Shafts and Tunnel.

Stephenson, A., and Suhadolnik, N. 2022. Getting Risk Right: Risk Allocation for Ground Conditions in Major Subsurface Projects. *39 The International construction Law Review.*

Travelers. 2021: Travelers Infrastructure Study A 17-Year Deep Dive into Heavy Civil Projects in North America.

https://www.constructionbusinessowner.com/resources/infrastructure-study-finds-inefficiencies-opportunities-design-build

Valley Transportation Authority (VTA). 2021. Request for Proposals to Provide Progressive design Build Services for Contract Package No. 2 Tunnel/Trackwork.

Washington Department of Transportation (WDOT). 2010. SR 99 Bored Tunnel Alternative Design-Build Project Confirmed Document.

Note: Additional references regarding implementation of PDB and CM/CG on subsurface projects may be accessed from https://donovanhatem.com/wp-content/uploads/2022/12/DH-Footnotes.pdf.

# Risk Mitigation of Natural Gas in Louisville MSD's Deep Bedrock, Ohio River Tunnel

**R. M. True** ▪ Black & Veatch
**Todd Tharpe** ▪ Black & Veatch
**Jonathan Steflik** ▪ Black & Veatch
**Jacob L. Mathis** ▪ Louisville and Jefferson County Metropolitan Sewer District (Louisville MSD)

## ABSTRACT

In June 2022, Louisville MSD completed a 6.5 km (4-mile), 6.1 m (20-foot) finished-diameter CSO conveyance and storage tunnel system located along the Ohio River in downtown Louisville, Kentucky. During construction, the Ohio River Tunnel alignment was expanded 2.4 km (1.5 miles) into areas where natural gas was detected. Without delaying construction, the project team developed and implemented a natural gas risk mitigation strategy to safely excavate the tunnel via TBM and install concrete lining. The risk mitigation strategy included the installation of gas-vent borings, increased tunnel ventilation, retrofitting tunneling equipment for "gassy" operations, and modifying the project specific HASP.

## LOUISVILLE MSD'S OHIO RIVER TUNNEL

### Project Description

As part of a Federal Consent Decree to address unauthorized discharges from combined sewer overflow (CSO) systems, Louisville MSD retained Black & Veatch to design and manage the construction of the Ohio River Tunnel (ORT), also known as the Waterway Protection Tunnel. The original 4.0-km (2.5-mile), 6.1-m (20-ft) finished diameter ORT and associated pump station, drop shaft infrastructure and interceptors provided an innovative solution to replace three large CSO storage basins in the Louisville area. Notice to Proceed for the ORT project was issued by MSD on November 8, 2017.

#### *Ohio River Tunnel Extension and Natural Gas Encounter*

In April 2018, due to favorable bid prices received for the ORT project, MSD requested Black & Veatch perform a subsurface investigation to evaluate the feasibility of extending the ORT approximately 2.4-km (1.5-miles) upstream from the ORT terminus to a separate CSO basin under construction. To verify subsurface soil and rock conditions for the proposed ORT Extension alignment, Black & Veatch performed a geotechnical subsurface investigation that included drilling several deep bedrock borings within the tunnel zone of the proposed extended alignment and at the relocated Retrieval Shaft site. Unlike the subsurface investigation performed for the original ORT design, Black & Veatch encountered natural gas during drilling operations. Following completion of the ORT Extension subsurface investigation MSD, Black & Veatch, and the ORT contractor, S-T JV (a joint venture between J.F. Shea and Traylor Brothers), worked together to finalize the design and safely complete the project.

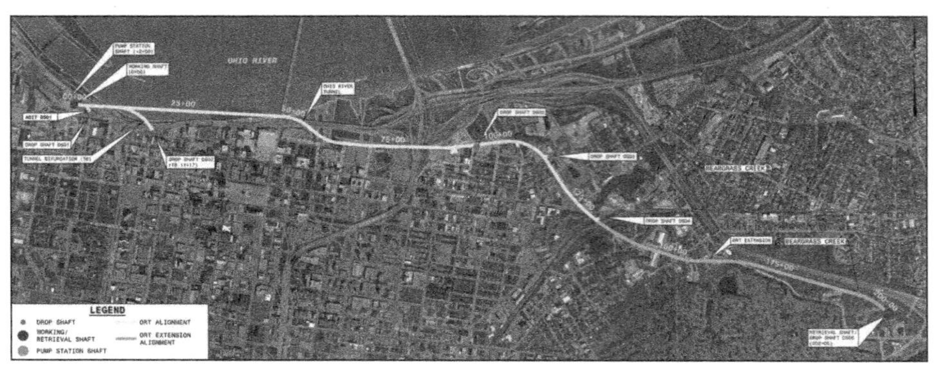


**Figure 1. ORT and ORT Extension alignments**

The extended ORT project included the following major components:

- An additional 2,147-m (7,045-ft) of tunnel, totaling 6,158-m (20,205-ft) of 6.1-m (20-ft) finished diameter tunnel, constructed approximately 55 to 67 meters (180 to 220 feet) below ground surface (bgs), providing a total of approximately 208,200-cubic meters ($m^3$) (55-million gallons (MG)) of CSO conveyance and storage capacity.
- Conversion of original Retrieval Shaft/baffle type drop shaft to vortex type drop shaft with adit connection to tunnel.
- Relocated Retrieval Shaft/drop shaft structure within terminated CSO basin site, thus eliminating a stand-alone pump station and associated infrastructure.
- A total of six vortex flow drop shafts with deaeration chambers and associated adits and tunnel connecting infrastructure.
- A Deep Dewatering Pump Station that conveys captured CSOs to MSD's collection system for treatment at the Morris Forman Water Quality Treatment Center (WQTC).

The original ORT alignment and ORT Extension alignment are shown on Figure 1. Despite the presence of natural gas along the ORT Extension alignment, construction of MSD's deep bedrock ORT was completed in June 2022 with no gas-related injuries to personnel and no significant delays to the project schedule. This paper documents the natural gas encounter during the ORT Extension subsurface investigation, the implementation of a modified risk mitigation strategy and its effectiveness specific to the handling of natural gas during project construction.

## RISK MITIGATION OF NATURAL GAS

Natural gas was encountered in April 2018 during the ORT Extension subsurface investigation. Black & Veatch re-evaluated and modified the existing project natural gas risk mitigation strategy to safely and efficiently complete construction. Some of the risk mitigation strategies are "proactive" avoidance strategies, while others are "reactive" mitigative strategies. Several preventative strategies were developed specifically for the exploration and handling of natural gas during tunnel and shaft construction.

### Hazard Identification and Exploration Through Subsurface Investigations

Prior to the development and re-evaluation of the project risk mitigation strategy, potential construction and safety hazards like natural gas were first identified and explored through geotechnical subsurface investigations. The following sections cover ORT project geology and contextualize the project team's mitigative approach to natural gas.

### *Project Geology, Subsurface Investigations*

The structural geology of the ORT project area has been influenced by tectonic activity associated with the Cincinnati Arch and the Illinois Basin, which resulted in local and regional faulting, regional downwarp, and deformation of the bedrock units. Bedrock units encountered during the ORT and ORT Extension subsurface investigations were Ordovician, Silurian, and Devonian in age and were explored by deep bedrock borehole drilling, water pressure testing, laboratory analyses of collected geological samples, and borehole geophysical surveys. Vertical and inclined borings were advanced through overburden soils using 108-mm (4.25-inch) ID steel casing and seated into top of rock. Nearly all vertical borings were drilled and sampled to top of rock (TOR). Soil was not sampled in inclined borings. Bedrock was cored to the full depth of the vertical and inclined borings with HQ tooling. Following coring operations, bedrock portions of select borings were water pressure tested using pneumatic packer assemblies with downhole pressure transducers to determine hydraulic conductivity of the respective rock units. Water pressure testing provided insight to potential groundwater infiltration into the ORT during construction which can also lend insight into gas infiltration into the tunnel bore.

### *Original ORT Subsurface Investigation*

For the original ORT design, Black & Veatch drilled 18 deep bedrock borings (~35–78-m (115–255-ft) deep) to explore geologic risks and characterize the bedrock along the project alignment. Of the bedrock units of Devonian age, the following lithological formations were encountered: New Albany Shale, Sellersburg Limestone, and Jeffersonville Limestone; of Silurian age: Louisville Limestone, Waldron Shale, Laurel Dolomite, Osgood Formation, and the Brassfield Formation; as well as the Drakes Formation of Ordovician age.

Natural gas was not encountered in any of the deep bedrock borings completed along the original ORT alignment. The subsurface geologic profile showing the original ORT alignment and deep bedrock borings is presented on Figure 2.

### *ORT Extension Subsurface Investigation*

To characterize the subsurface of the proposed ORT Extension alignment, an initial six deep bedrock borings, B-GE-1R through B-GE-6, were drilled from April to June 2018. Of the six borings, natural gas was encountered in four, B-GE-3 through B-GE-6. For these four borings, it was necessary to temporarily halt drilling operations due to natural gas encounters. These borings were vented to atmosphere until drilling operations could safely resume. However, only borings B-GE-3 and B-GE-4 were completed. Borings B-GE-5 and B-GE-6 were not drilled to completion due to continued venting of natural gas, and thus drilling efforts were discontinued.

Following the encounter of natural gas in the four borings, a subsequent subsurface investigation included the drilling of two additional deep bedrock borings, B-GE-7 and B-GE-8. Boring B-GE-7 was drilled within the footprint of the relocated Retrieval Shaft and encountered natural gas continuously at low pressures throughout the drilling process until completion.

Due to regional uplift associated with the Cincinnati Arch, the deep bedrock borings along the ORT Extension did not encounter the New Albany Shale, and Sellersburg

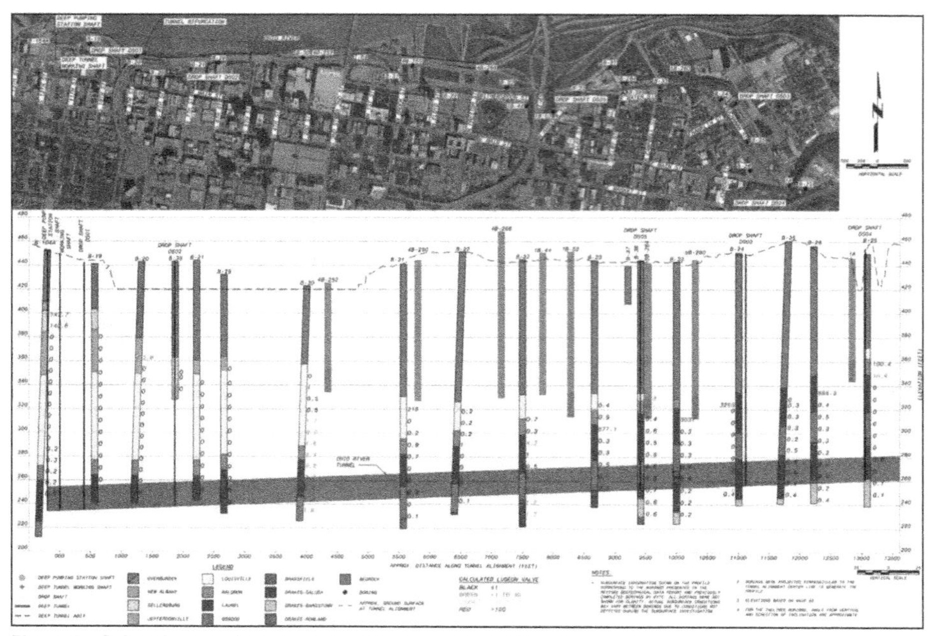

**Figure 2. Subsurface geologic tunnel profile for original ORT alignment**

and Jeffersonville Limestone formations of the Devonian age. However, the deep bedrock borings along the ORT Extension alignment did encounter the Bull Fork Formation and Grant Lake Limestone formations of the Upper Ordovician System, which were not encountered in the original ORT borings. The subsurface geologic profile of the ORT Extension portion of the deep bedrock tunnel and deep bedrock borings is presented on Figure 3.

Geophysical logging was performed to gain insight into the structural geology along the portion of the ORT Extension alignment where natural gas was encountered. Boring B-GE-3 was geophysically logged, as were B-GE-5 and B-GE-6 once gas venting subsided after approximately five weeks. Downhole tools utilized to log the borings included an acoustic televiewer, optical televiewer, single point resistance, spontaneous potential, three arm caliper, normal resistivity, fluid conductivity, and natural gamma. The acoustic televiewer tool utilizes an internal fluxgate magnetometer, which allows for the orientation (referred to as strike and dip) of downhole fractures to be determined. During the drilling of boring B-GE-5, natural gas was encountered between 38 and 39 meters (125 and 128 feet) bgs. The results of the televiewer logging indicated that the gas producing fracture, noted at that same depth, was dipping at an angle of 42 degrees from vertical, with a dip direction of 227.5 degrees. Additionally, once the collected geophysical data was corrected to an elevation datum, it became apparent that an inactive bedrock fault was situated between borings B-GE-3 and B-GE-5, indicating the possibility that the fault plane provided a conduit for the migration of natural gas from a deeper source.

## Implementation of Gas Venting Mitigation Strategy

Following the natural gas encounter and geophysical logging in the ORT Extension borings, the project team developed a gas mitigation plan to eliminate or dissipate

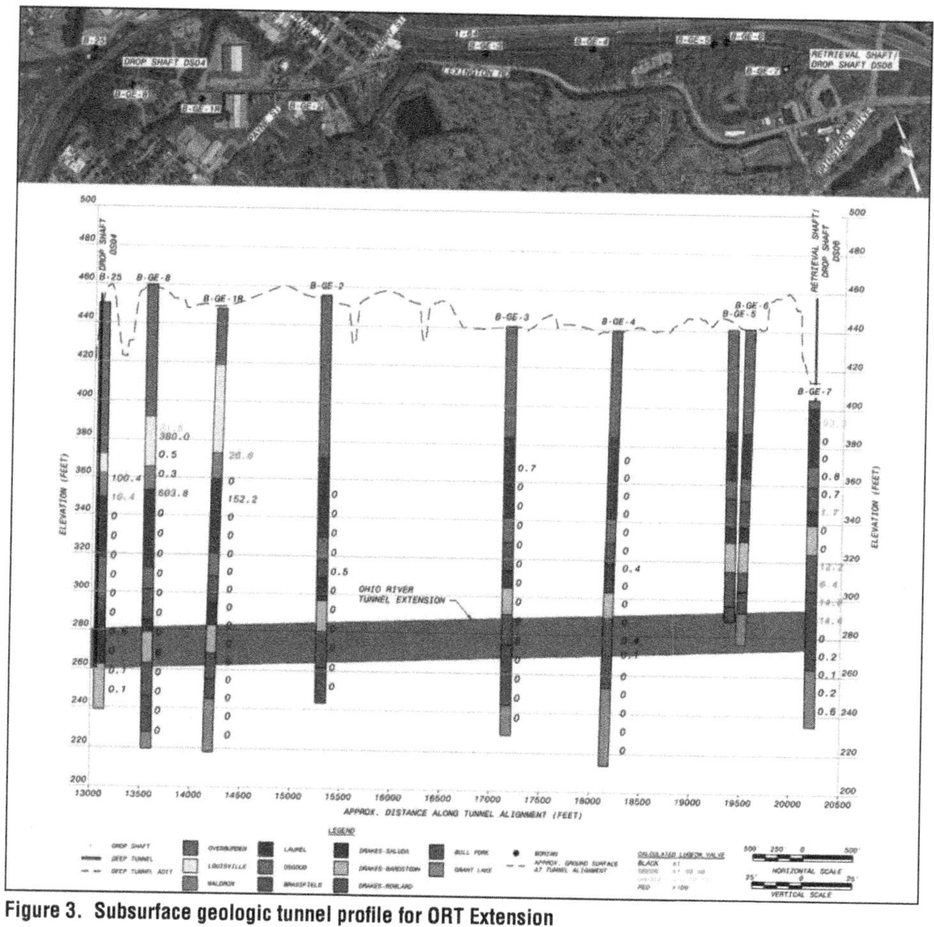

Figure 3. Subsurface geologic tunnel profile for ORT Extension

gas prior to TBM arrival. The plan included the drilling of 21 gas vent borings (GVBs), GVB-1 through GVB-21, that were completed along the ORT Extension alignment. The GVBs were completed with packer assemblies, gas vent piping, and an isolation valve at ground surface. Following geophysical borehole logging, gassy borings B-GE-3 through B-GE-6 were also converted to GVBs, totaling 25 gas venting media (GVM). The gas venting strategy was implemented following approval from the Kentucky Division of Oil and Natural Gas. A subsurface profile of the 25 GVM and ORT Extension portion of the deep bedrock tunnel is presented in Figure 4.

### *Characterization of Natural Gas*

Although, the primary purpose of the GVM was to facilitate the release of pressurized natural gas trapped within the underlying bedrock, the installation of isolation valves at land surface allowed for continued monitoring and sampling of gas. To evaluate the nature and extent of natural gas trapped within the bedrock, samples of in-situ gas were collected from select GVM and were sent to Isotech Laboratories, Inc. for analysis.

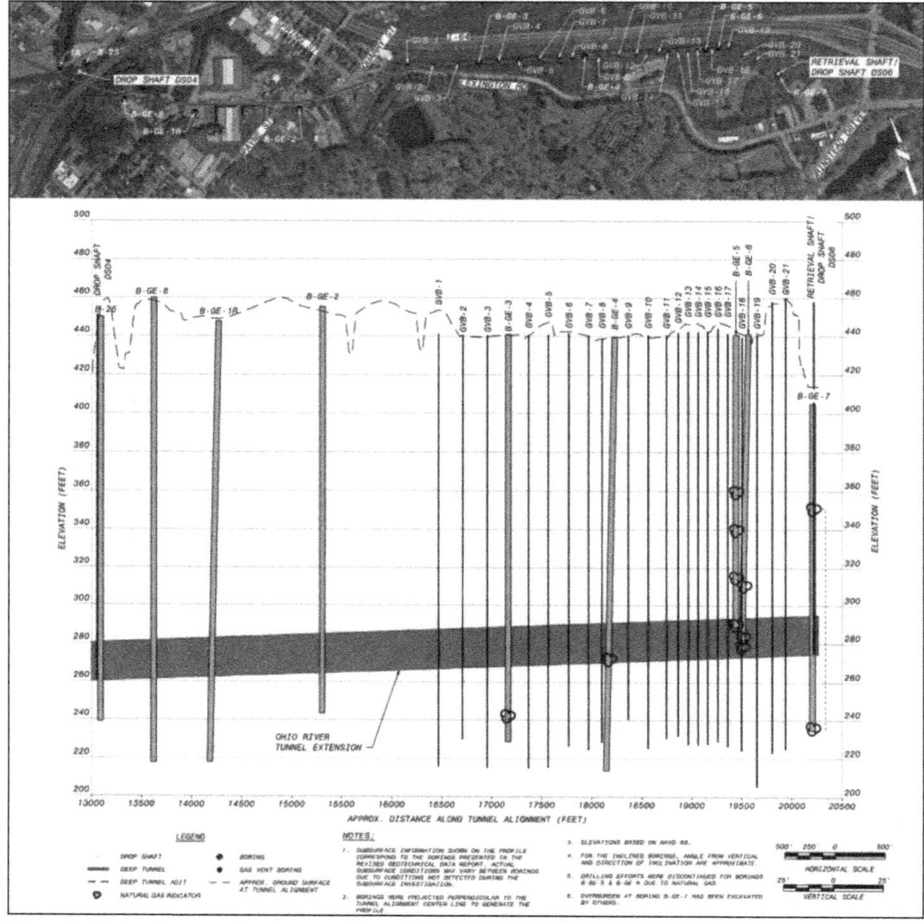


**Figure 4. Subsurface profile of gas venting media along the ORT Extension**

To determine the source of the gas (whether biogenic or thermogenic, and whether it could be associated with nearby gas storage facilities), the molecular and isotopic composition was analyzed. Through stable carbon and hydrogen isotope analyses of methane, in addition to the stable carbon of ethane and propane, gas chromatography analyses reflected a native thermogenic gas originating from a relatively immature source rock in the local area, most likely the New Albany Shale Formation. The analysis also determined that the ORT Extension gas samples were consistent with published results of gas samples from the New Albany Shale Formation in the eastern part of the Illinois Basin geologic feature.

The results of the gas sampling analyses indicated that, due to the proximity of the ORT project to the New Albany Shale Formation (which was encountered in only one deep boring at the westernmost portion of the tunnel alignment), gas was likely originating from this rock unit and traveling along deep bedrock bedding plane fractures, faults, and or fault splays to bedrock strata near the ORT Extension alignment, with some of the ORT Extension deep borings and GVBs contacting these subsurface features.

### *Abandonment of GVM*

The 25 GVM installed between August and October 2018 were actively venting gas during Working Shaft, Pump Station Shaft, and Starter/Tail tunnel construction prior to TBM assembly. Therefore, the GVM were open to atmosphere for approximately 20 to 22 months prior to abandonment in June 2020 prior to TBM arrival.

The Contract Documents originally required the contractor to abandon all GVM approximately 457-m (1,500-feet) prior to TBM arrival; however, this requirement was revised to 122-m (400-feet) to maximize gas venting. Because the GVM were open through most of the underlying rock column and likely intersected water bearing features, all GVM were required to be abandoned, rather than only GVM that directly intersected the tunnel bore. All GVM were abandoned to avoid groundwater and gas infiltration. The GVM were abandoned by pressure grouting the gas vent riser piping and then removing the surface piping and valve components.

## Additional Risk Mitigative Strategies

In addition to the installation of GVM, Black & Veatch re-evaluated and modified several risk mitigation strategies implemented for the original ORT design to account for the ORT Extension portion of the project.

### *Tunnel Classification, OSHA CFR 29 1926.800*

Many of the components of the risk mitigation strategy utilized for the ORT project stemmed from classifying the tunnel as "potentially gassy." Occupational Safety and Health Administration (OSHA) guidelines dictate whether a subsurface work area, such as a tunnel excavation, is to be classified as "potentially gassy" or "gassy." Per OSHA 29 CFR 1926.800, potentially gassy operations occur when the following conditions exist:

- "Air monitoring shows 10 percent or more of the LEL for methane or other flammable gases measured at 12 inches (304.8 mm) ±0.25 inch from the roof, face, floor or walls in any underground work area for more than a 24-hour period: OR" ((h)(1)(i))
- "The history of the geographical area or geological formation indicates that 10 percent or more of the LEL for methane or other flammable gases is likely to be encountered in such underground operations" ((h)(1)(ii))

The latter conditional requirement established by OSHA determined the original ORT project to be classified as "potentially gassy." In another deep bedrock tunnel constructed in the Louisville area, the Louisville Water Company's (LWC) Riverbank Filtration (RBF) Phase II Tunnel, natural gas accumulated (>10% LEL) during construction downtime when the tunnel ventilation system was inactive. Knowing about this regional natural gas encounter, and despite not encountering natural gas during the original ORT subsurface investigation, the project team elected to conservatively classify the ORT as "potentially gassy" which would influence risk mitigation strategies for safer tunnel construction.

After encountering natural gas during the ORT Extension subsurface investigation questions arose regarding whether the ORT should be conditionally re-classified as "gassy." Per OSHA CFR 29 1926.800(h)(2), reclassification was unnecessary as the ORT did not satisfy the three listed requirements: (i) exceedance of 10% LEL for three consecutive days; (ii) ignition of in-situ gases; or (iii) connection to underground areas that are actively classified as "gassy."

### *Tunneling Air Ventilation Requirements*

Because the original ORT was classified as 'potentially gassy', specific requirements pertaining to tunnel and shaft ventilation were incorporated into the Contract Documents per OSHA CFR 29 1926.800. At minimum, the original ORT ventilation system was required to be in full compliance to all design and operational ventilation requirements as stipulated by OSHA CFR 29 1926.800(k). In addition to OSHA requirements, the contractor was required to supply a ventilation system that could achieve at minimum 3.33x the OSHA required airflow velocity of 9.1 meters per minute (mpm) (30 feet per minute (fpm)) of 30.5-mpm (100 fpm) throughout the tunnel cross section in the event gas was encountered. Further, Black & Veatch required the tunnel ventilation system to be maintained and in operation continuously (24 hours per day, seven days per week). This requirement was included to prevent a 'system start up' encounter with natural gas like those referenced from the LWC RBF Phase II Tunnel.

For the ORT, the contractor used two Aerovent DDPRV 54R6 fans in parallel to supply airflow to underground working areas. Table 1 documents the ventilation system schedule for the original ORT, ORT Extension, and for concrete lining operations.

The ventilation system used for the construction of each phase of the ORT project exceeded OSHA CFR 29 1926.800(k) requirements, as well as project contractual requirements.

### *Tunneling Equipment and Safety and Health Plan Modification*

Risk mitigative strategies were added to the Contract Documents regarding the contractor's tunneling equipment and safety and health plan.

A cursory list of tunneling equipment fitted for "gassy" conditions as required by OSHA CFR 29 1926.800, as well as the Contract Documents is included below:

- Automatic shut-off capabilities for drilling/TBM equipment;
- Continuous combustible gas analyzers (CGAs) capable of sounding alarms and activating warning lights;
- Explosimeters capable of visual and audible alarms at various locations, specifically the TBM heading.

**Table 1. ORT ventilation system schedule**

| | ORT Construction Phase | | |
|---|---|---|---|
| **Ventilation System** | **Original ORT** | **+ ORT Extension** | **Post- Hole-Through/ Concrete Lining** |
| Tunnel length, m (ft) | 4,386 (14,061) | 6,433 (21,106) | 6,433 (21,106) |
| Design Input Flow, $m^3/s$ (cfm) | 28.3 (60,000) | 35.4 (75,000) | 29.3 (62,000) |
| Air Velocity, mpm (fpm) | 52.4 (172) | 65.5 (215) | 60.0 (197) |
| Percentage Greater than OSHA CFR 29 1926.800(k) minimum (9.1 mpm | 30 fpm) | 576% | 720% | 659% |
| Percentage Greater than Contract Required Minimum (30.5 mpm | 100 fpm) | 172% | 215% | 197% |

Likewise, the contractor's safety and health plan was updated to address natural gas encounters and evacuation procedures. A cursory list of components of the contractor's safety and health plan developed for the "potentially gassy" ORT is included below:

- Conditional evacuation and reentry procedures, including: the use of Brookville locomotives, designated trained personnel for reentry, and in-situ gas level verification protocol.
- Conditional ventilation system procedures per natural gas levels and location of encounter (i.e., heading, behind TBM).
- Specific Personal Protective Equipment (PPE) for safe handling and or avoidance of natural gas hazards (i.e., self-rescuers, handheld gas monitors headlamps and flashlights)

Even though natural gas was not encountered during the original ORT subsurface investigation, contract documents required that the contractor "*shall* anticipate that combustible and toxic gases will be encountered during tunneling operations and construction of the shafts" (Contract Specifications).

These requirements were considered risk mitigation strategies in that safety concerns, loss of schedule, and the costly retrofitting of tunneling equipment to accommodate "gassy" conditions would be avoided in the event of a natural gas encounter during construction. With these requirements already in the Contract Documents, no modifications were necessary when the contract was revised to include the ORT Extension.

### *Modifications to Project Schedule*

The original ORT contract included 100 hours in the project schedule for downtime due to natural gas encountered during construction. Because none of the deep bedrock borings drilled under the original ORT subsurface investigation encountered natural gas, the inclusion of downtime hours for gas mitigation and handling was a conservative approach that still met MSD's Consent Decree schedule requirement. After encountering natural gas during the ORT Extension subsurface investigation, an additional 120 hours of downtime were conditionally added to the project schedule, thus totaling 220 hours for the full length of the tunnel.

During ORT construction, natural gas was encountered three times; once during blasting efforts for the relocated Retrieval Shaft and twice during TBM excavation within the ORT Extension. The contractor did not claim use of downtime hours due to natural gas for the first two instances but did claim use of 86.25 downtime hours of the 220 total for the third, resulting in no loss of schedule due to encountering natural gas during construction.

### *Continued Sampling and Monitoring of Gas Venting Media*

During project construction, the contractor was permitted to perform additional monitoring and sampling of the GVM. The contractor followed the sampling procedure for in-situ gas sampling performed by Black & Veatch and collected gas samples for similar analyses performed by Isotech Laboratories, Inc. to compare composition of the gas over time. Testing results were similar to analyses performed 12 months earlier.

The contractor attempted to further characterize the nature of the natural gas by completing time-volume discharge testing on the GVM. Surface valves were closed for set times allowing gas to accumulate in the boring. Once pressures stabilized, the isolation valves were opened, and pressures and flow rates were recorded while discharging gas to 49-to-114-liter (13-to-30-gallon) bags (affixed to the gas vent piping). Results from the April 2020 testing, when compared to the November 2019 testing, reflected that shut-in pressures and venting rates had decreased in certain GVM, while other GVM that were previously dormant began to show small pressure increases. The contractor's specialty subconsultant suggested that these results indicated gas "accumulations in the vicinity of the vent wells are recharged by a gaseous phase slowly migrating up from deeper strata by intermittent buoyant flow through the fracture network of relatively low permeability." Consequently, no accurate prediction of the maximum gaseous phase discharge rate into the tunnel could be made due to the complexity of the geologic conditions.

Consideration was also given to solution phase gas infiltration. By projecting a potential gas concentration of 0.25 ml/ml of water at a groundwater inflow rate of 7,571 liters per minute (LPM) (2,000 gpm), which was the baseline for ORT Extension maximum steady-state flow, the estimated maximum gas inflow rate would have been 0.032 $m^3/s$ (66.8 cfm); an inflow rate that would have been sufficiently diluted with the existing ventilation system to well below the 10% LEL for methane.

### *Tunnel Boring Machine (TBM) Probing and Pre-Excavation Grouting*

The project team required TBM probing as another risk mitigative strategy. The primary concern influencing the use of TBM probing was potential groundwater infiltration, as projected through the water pressure testing of deep bedrock borings drilled during subsurface investigations conducted along the tunnel alignment. However, TBM probing was also an effective means of exploring unexcavated ground directly ahead of the TBM for natural gas. Fortunately, actual groundwater infiltration was significantly lower than projected inflow rates (approximately 10,978 LPM (2,900 gpm) and approximately 7,571 LPM (2,000 gpm) steady-state inflow for original ORT and ORT Extension, respectively). Due to lower groundwater infiltration, probing requirements were relaxed in select sections of the alignment; however, all parties agreed to a rigorous probing quality work plan that specifically addressed potential natural gas encounters. The quality work plan specifically addressed procedures to mitigate natural gas infiltration into the open tunnel bore. For example, to minimize groundwater and gas infiltration into the tunnel, an ultra-fine grout mix was utilized to grout gas and water-bearing features should they be encountered.

## NATURAL GAS ENCOUNTERS DURING CONSTRUCTION

Natural gas was encountered three times during construction of the ORT. The first encounter occurred during drilling and blasting efforts for the Retrieval Shaft on June 4, 2019. While removing blasting mats from the shaft following blasting operations, gas monitors located within the center of the shaft registered approximately 3 PPM $H_2S$. This natural gas encounter occurred within the same footprint where natural gas was encountered during the continuous drilling of deep bedrock boring B-GE-7 approximately nine months earlier.

The latter two natural gas encounters occurred within the ORT Extension portion of the excavated tunnel. On April 28, 2020, during TBM maintenance efforts, personnel discovered a methane ($CH_4$) reading of 26% LEL from bubbling process water in the rail pit floor while using a handheld gas monitor. Concurrently, the TBM rail pit atmospheric monitor registered a reading of 0.6 PPP H2S. While the encounter did not represent an immediate safety risk due to sufficient airflow ventilation, it did alert all parties to the presence of natural gas and the potential for encountering gas during the remaining alignment to be bored.

On August 26, 2020, approximately 183-m (600-ft) before TBM hole-through, the contractor encountered methane gas emanating from a probe hole at an excess of 60% LEL, which prompted evacuation of the tunnel per the approved safety and health plan. Upon reentry, trained personnel continuously monitored gas levels in all areas of the TBM, including: the tunnel face, within the cutterhead, and the interior and exterior of the two probe holes. On Friday August 28, 2020, two days after the initial gas encounter, it was decided to allow the ORT to ventilate over the weekend, with air monitoring verification on the succeeding Monday dayshift. The following Monday, August 31, 2020, after crews verified safe working conditions per the safety and health plan, TBM excavation resumed.

The natural gas encounter directly downstream of the ORT terminus was directly within the vicinity of the two deep bedrock borings B-GE-5 and B-GE-6 that were discontinued at tunnel zone due to natural gas venting nearly 26 months prior. It is believed that a geologic bedrock fault is located within the vicinity of this area that vents natural gas from underlying bedrock strata.

## LESSONS LEARNED

Important learning opportunities were developed during the construction of this deep bedrock tunnel with geologically complex ground and numerous construction risks such as construction through "potentially gassy" ground.

### Thorough Geotechnical Subsurface Investigations

As exampled by the ORT project, when investigating the geologic subsurface within projected tunnel alignments, the following considerations can be useful in the exploration and characterization of natural gas:

- Extensive review of geologic literature of the project area specific to the presence and characterization of natural gas.
- Geophysical borehole logging of select deep bedrock borings
- The implementation of a gas vent boring (GVB) or GVM for monitoring the project subsurface prior to and during construction. The GVB(s) can be fitted with surface valves to allow for groundwater and gas monitoring (both qualitative and quantitative).

### Additional Exploratory Borings and GVM

As the TBM progressed toward the ORT Extension portion of the tunnel alignment, the project team discussed the installation of additional GVM. Given the characterization of the subsurface within the project area, the following considerations influenced the decision to not install additional GVM:

- Geologic complexity of the subsurface relative to horizontal fracturing and the presence of vertical fractures and or a steeply dipping fault that would be costly and difficult to locate.
- Confidence in other risk mitigative infrastructure such as the ventilation system, safety and health plan, upgraded tunneling equipment, and revised probing quality work plan.
- Timing; relative to TBM positioning and progression rate, GVM may only be in place for a short time with variable venting rates, which would reduce their effectiveness.

## CONCLUSION

The risk mitigation strategy adopted by the project team for work in "potentially gassy" conditions enabled safe and efficient construction of the ORT project with minimal downtime due to natural gas. The risk mitigation strategy also allowed workers to construct the ORT with no gas-related injuries. The implementation of GVM along tunnel alignments prior to excavation can be useful in characterizing and dissipating natural gas. More effective gas dissipation may be expected for less geologically complex project areas (i.e., rock units with less vertical faulting, horizontal jointing, and fault splays). Gas venting media were used as a primary tool to mitigate safety and construction risks influenced by natural gas during construction of the ORT. Additional risk mitigation may be achieved through implementation of indirect strategies such as:

- Appropriately classifying tunnels for hazardous working conditions;
- Modifying project schedule;
- Upgrading tunneling equipment;
- Implementing an appropriately robust safety and health plan;
- Installing additional gas wells for monitoring and characterization of the subsurface
- A multifaceted risk mitigation approach can benefit all stakeholders and help ensure a safe, timely, economically viable and otherwise successful project.

The discovery of natural gas during the construction of the ORT Extension challenged the project team to adapt the project execution and implement risk mitigation measures, ensuring the safety of personnel and timely completion of this important infrastructure project. Because the project team was able to successfully mitigate risks during construction, the ORT met MSD's Consent Decree with EPA, provided surplus storage at a lower cost than the originally planned storage basins, while minimizing disruption to the community. At the surface sites, at both ends of the ORT project, MSD has partnered with Louisville Metro government and the Waterfront Development Corporation to create green spaces and extend public park areas, creating a community asset both above and 60-m (200-ft) below ground.

## REFERENCES

Black & Veatch. June 2019. Louisville and Jefferson County Metropolitan Sewer District.

Ohio River Tunnel—Tunnel and Shafts Package—Revised Geotechnical Data Report.

Black & Veatch. June 2019. Louisville and Jefferson County Metropolitan Sewer District.

Ohio River Tunnel—Tunnel and Shafts Package—Revised Technical Specifications.

Occupational Safety and Health Administration (OSHA). 2019. Underground Construction. CFR 29 1926.800.

PART 

# SEM Applications and Projects

*Chairs*

**Zeph Varley**
WSP USA

**Lisa Smiley**
Jay Dee Obayashi JV

# Construction of the Bypass Tunnel for the Upper Llagas Creek Flood Protection Project

**Clayton Williams** ▪ Mott MacDonald
**Dale M. Hata** ▪ Drill Tech Drilling & Shoring
**Glenn M. Boyce** ▪ McMillen Jacobs Associates

## ABSTRACT

The Upper Llagas Creek Flood Protection Project in Northern California encompasses portions of the communities of Morgan Hill and Gilroy, and unincorporated areas of Santa Clara County. The project provides for flood protection improvements along Llagas Creek and two of its tributaries, West Little Llagas Creek and East Little Llagas Creek. Llagas Creek is one of the tributaries of the Pajaro River and drains a watershed of approximately 104 square miles. The river system has a history of flooding the watershed communities, and the project's goal is to provide 100-year flood protection by creating a bypass infrastructure, including channels, cut-and-cover culverts, and a tunnel.

The project has been divided into several phases. This paper describes Phase 2A, which consists of approximately 2,150 linear feet of lined horseshoe-shaped bypass tunnel measuring 14 feet wide by 12 feet high. The bypass tunnel was sequentially excavated using a combination of roadheader and controlled blasting through colluvium and weathered to fresh greenstone. The tunnel was mined near residences, requiring strict noise and vibration controls.

Phase 2A also includes the construction of twin bypass concrete culverts, portal shoring, and dewatering systems. It features the use of a Design Review Board and partnering for project delivery. This case study discusses the bypass tunnel design, progress of tunnel construction, plans to complete the work, construction risks, and the software and procedural tools being used to manage the project, including ProjectWise construction management software.

## INTRODUCTION

### Project Team

The project team consists of the client, Valley Water, the City of Morgan Hill, McMillen Jacobs Associates as the tunnel designer, Mott MacDonald as the construction manager responsible for contract administration and quality assurance oversight, Flatiron West Inc. as the general contractor, and Drill Tech Drilling & Shoring as the tunneling subcontractor.

### Project Background

The first recorded instance of flooding in the area dates to 1937. Federal funding for flood control was initially approved by Congress in 1954, but issues with funding and property acquisition caused delays until recent years. Valley Water and Natural Resources Conservation Service (formerly known as the Soil Conservation Service) began construction of the Llagas Creek Project in the 1980s. Approximately 60% of the project had been completed by 1994. In 1999, due to lack of funding, the project was transferred to the US Army Corps of Engineers to secure state and federal funds needed for the completion of construction.

In 2009, many businesses and residences in downtown Morgan Hill were flooded under approximately one foot of water. In 2012, Santa Clara County residents voted to enact the 15-year Safe, Clean Water and Natural Flood Protection Program, where funding was identified for the Llagas Creek Project. Phase 2 of the Upper Llagas Creek Flood Protection Project will divert high flow waters into the bypass tunnel from West Little Llagas Creek as it approaches residential and commercial zones. High flows are diverted into twin bypass concrete culverts, which transition into the tunnel and back again into twin bypass concrete culverts until the high flow's confluence with existing Llagas Creek downstream.

## Project Overview

The construction contract was awarded to Flatiron West Inc. on May 12, 2021, and the Notice to Proceed was issued for a construction start of June 21, 2021. The major works of the construction contract are summarized in Table 1.

One primary features of the Upper Llagas Creek Flood Protection Project is the lined horseshoe-shaped bypass tunnel, which is being excavated using a combination of controlled blasting and roadheader methods. The tunnel extends from Hale Avenue to West Dunne Avenue, as shown in Figure 1.

The tunnel excavation dimensions are 15'-10" wide by 14'-5" high, with the final internal dimensions measuring 14'-0" wide by 12'-0" high. The cover above the tunnel ranges between 13 and 142 feet, with the smallest amount of cover occurring at approximately the midpoint of the alignment. As of December 15, 2022, the full length of the original tunnel (2,045 linear feet) has been excavated. A contractor-proposed change that increased the length of tunnel by approximately 125 feet, discussed in Section 3.5, is nearing completion as well.

**Table 1. Major works for the upper Llagas Creek flood protection project**

| Scope of work | Descriptions |
|---|---|
| Box culverts | Cast-in-place twin box culverts connect to West Little Llagas Creek.<br>Gas, electric, sanitary sewer, and water lines are rerouted or protected in place to prevent conflicts with construction.<br>Two access structures on either side of Nob Hill allow ingress and egress to and from the box culverts.<br>Transition structures connect the box culverts to the tunnel.<br>Excavation is dewatered as needed.<br>Excavation material is managed. |
| Hale Avenue Extension | New gas, electric conduits, sanitary sewer, and water lines are trenched and installed along the future Hale Avenue Extension.<br>Roads are constructed and graded. |
| Tunnel | Temporary shoring is installed at the launch portal with soldier piles and internal bracing.<br>Portal development includes clearing and grubbing, material disposal, protecting utilities in place, construction of a sound wall around the portal area, and installation of canopy tubes over the first 60 feet of the tunnel.<br>Tunnel is excavated with roadheader and drill-and-blast methods.<br>Initial ground support uses W6x20 steel sets and unreinforced shotcrete in weathered greenstone, and rock bolts and shotcrete in fresh greenstone.<br>Final lining is shotcrete with synthetic fiber reinforcement and a 12-inch-thick concrete invert slab with a single layer of welded wire fabric. |

**Figure 1. Tunnel alignment by reach**

## DESIGN AND CONSTRUCTION

### Tunnel Design

During the preliminary design phase, several different alternatives were evaluated, including the use of trenchless methods like microtunneling through Nob Hill. The issue with microtunneling was that much of the drive would be in hard rock and a single jacked pipe would not be large enough for the projected flow capacity needed to prevent flooding of the downtown area.

McMillen Jacobs Associates was brought onto the design team to design a larger tunnel. In evaluating the ground conditions and tunnel length, a conventionally driven tunnel with a classic horseshoe shape could be built at the invert elevations established by the cut-and-cover box culverts. As shown in Figure 2a, the tunnel has a finished width of 14 feet and a height of 12 feet. A tunnel of this size would accommodate the use of a roadheader to excavate in weathered ground conditions and to allow the use of controlled blasting in harder, stronger ground conditions. These finished dimensions provided enough capacity to accommodate the design stormwater flows of the project. Ideally the tunnel design would have two diameters of ground cover over the tunnel, which would provide ground arching and a stable tunnel.

The upstream end of the box culvert extended to the intersection of Warren and Hale Avenues, which was the start of a residential neighborhood. The downstream end of the tunnel was at West Dunne Avenue, where Nob Hill transitioned into a flat area. The original alignment would have connected these two locations with a straight tunnel under Nob Hill. The problem was that portions of Nob Hill was subdivided and owned as private parcels. The homeowners were not interesting in providing easements under their properties. The design team was forced to place the tunnel under the City's street rights-of-way.

The final tunnel alignment, shown in Figure 1, had a portal just short of the intersection of Warren and Hale Avenues. The tunnel continued down Hale Avenue to Nob Hill

Terrance, made a left turn under Nob Hill Terrace to Del Monte Avenue, then turned right on Del Monte and continued under Nob Hill, daylighting at West Dunne Avenue and into the box culvert.

One issue that developed with this alignment was the lack of ground cover over the tunnel within reaches along Hale Avenue (defined as Reach 1), at the corner of Nob Hill Terrance and Del Monte Avenue (defined as Reach 3), and at the downstream end of the tunnel (defined as Reach 5). The ground cover was a little as 13 feet in some cross sections. The ground cover of 13 feet provided only one tunnel diameter of cover.

Adding to the tunnel cross section under the residential streets were various utilities including sewers, storm drains, water mains, and gas mains. These utilities were installed by trenching, which took the available ground cover and subdivided the area into discontinuous zones with course bedding materials and trench backfill. The concern was that the trenches would hinder ground arching and tunnel stability.

A geotechnical exploration program was completed along the tunnel alignment. Borings found colluvium as the overburden material transitioning into extremely weathered greenstone. The weathering of the greenstone improved with depth. For the deeper tunnel sections (Reaches 2 and 4), the ground consisted of weathered and slightly weathered greenstone and the stability of the tunnel was not in question. For the shallower tunnel sections (Reaches 1, 3, and 5), the tunnel invert was in the weathered greenstone, but the crown was colluvium and extremely weathered rock.

To help enhance tunnel stability, the design team elected to include provisions in the contract documents for formational grouting from the surface. The plans called for the drilling of three grout sleeves per cross section in a staggered pattern (see Figure 2b). The center sleeve would lie along the tunnel centerline and outer sleeve would be five feet on either side. The plan was to solidify any colluvium into a rock-like mass and provide in-situ stability from the surface before the tunnel was excavated.

In addition to the surface grouting program, the contract documents included a geotechnical instrumentation program to monitor for surface and utility movement along the street rights-of-ways during active tunneling.

## Portal Area

The launch portal (North Portal) is located on a vacant residential lot at the intersection of Warren Avenue and Hale Avenue. The portal's invert elevation is roughly 27 feet below the road elevation. McMillen Jacobs Associates developed the portal shoring concept with a 30-foot-wide access ramp down to the portal along the bypass alignment, as shown in Figure 3. Once construction of the tunnel is completed, the ramp will be removed to make way for the cast-in-place box culvert that abuts the tunnel.

Drill Tech designed and constructed the shoring at the launch portal. The shoring system used soldier beams with up to two levels of internal bracing for support. The portal shoring will be installed in two phases, with the first phase installed to support the access ramp down to the portal for the duration of tunnel construction. The second phase will be installed after the tunnel is complete to support the ramp excavation required to make way for the box culvert. Phase 1 of portal shoring was completed in January 2022, and Phase 2 is planned for early 2023.

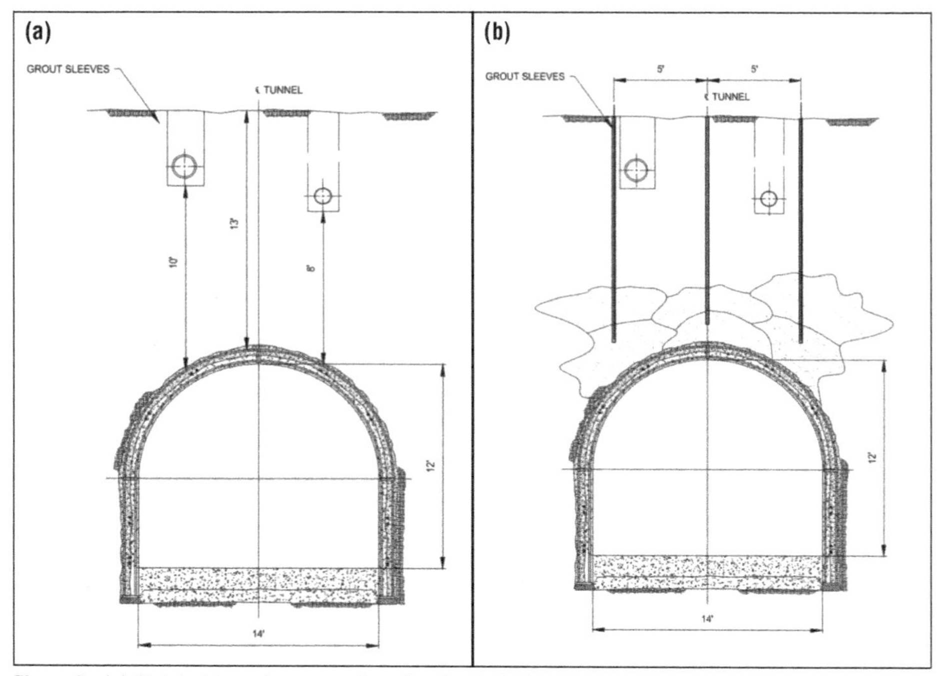


Figure 2. (a) Finished tunnel cross section showing typical low ground cover; and (b) grout injected from the surface to improve the ground in the tunnel crown

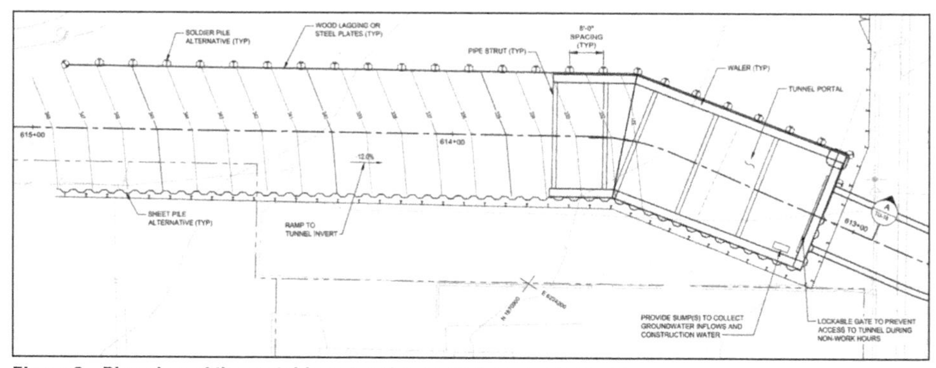


Figure 3. Plan view of the portal layout and ramp to the tunnel elevation

## Tunnel Construction

The tunnel was advanced with the Sequential Excavation Method (SEM), using mechanical excavation equipment and controlled blasting. Most of the tunnel excavation on the project has used an Alpine Miner 75 (AM-75) roadheader, except for 230 linear feet where rock mass strength warranted the use of controlled blasting. The AM-75 is powered by electric over hydraulics, capable of grinding (cutting) through soil and extremely weak to strong rock.

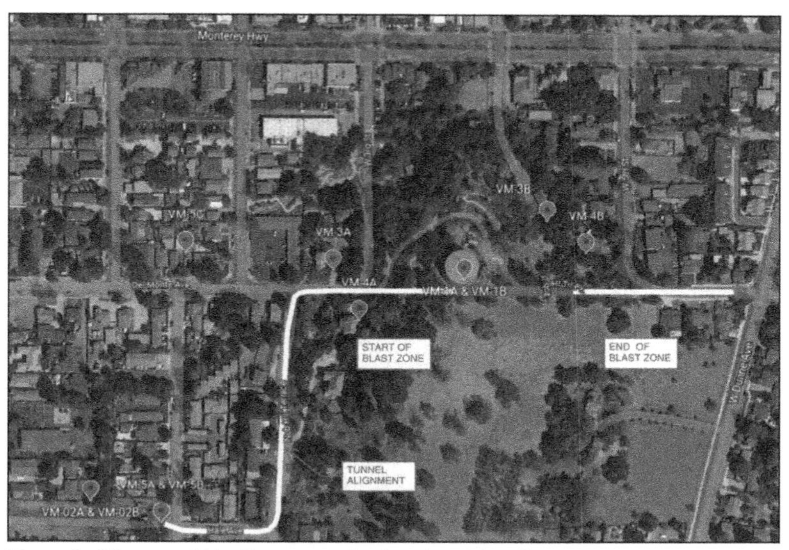


Figure 4. Site map of blasting zone and seismograph locations

For this tunnel size, the roadheader is the preferred excavation method in stable ground, as it is proven to be more productive and economical than blasting. For comparison, the AM-75 advances 20 to 30 feet per shift in stable ground, whereas blasting production is limited to a single round length per shift, in this case 8 to 10 feet. The cutting capacity for the AM-75 was established in fresh rock with an approximate unconfined compressive strength of 6,125 psi and rock mass rating (RMR) range of 39.5 to 61.5. At that point, roadheader pick consumption increased to more than 1 pick per cubic yard mined, and production dipped below 8 feet per shift.

Blasting was used when the rock mass strength exceeded the cutting capacity of the AM-75. Factors that influenced the switch to blasting included low production with the AM-75, equipment wear, pick consumption rate, and risk of damage to adjacent improvements. With the tunnel blocks away from downtown Morgan Hill, numerous residential and commercial structures are within a 1,000-foot radius of the tunnel. To protect these structures from blasting, the project required peak particle velocities measured at the nearest structures to be no more than 0.5 in/sec for 2.5 to 10 Hz, 0.05 × frequency in/sec for 11 to 40 Hz, and 2 in/sec for greater than 40 Hz.

Up to five seismographs were used to monitor blasting ground vibrations at a water tower and at structures closest to the tunnel heading. The seismographs were self-contained units able to upload data to a cloud server within minutes of blasting. Figure 4 shows the limits of blasting and seismograph locations denoted as "VM-xx."

Drill Tech utilized controlled blasting methods to regulate charges and control peak particle velocities at neighboring structures. Blasting utilized Dyno Nobel's Powermite 1-½"x16 emulsion cartridges with 200/5400 NONEL EZ Drifters, and both 17 ms and 42 ms NONEL EZTLs. The charge ranged from 6.5 to 22.7 lbs/delay. Powder factor ranged from 4.5 to 6. With this approach, blasting was successfully completed while maintaining ground vibrations below the peak-particle-velocity limit by a factor of 2. Figure 5 shows the plotted distance versus peak particle velocity data for all seismographs.

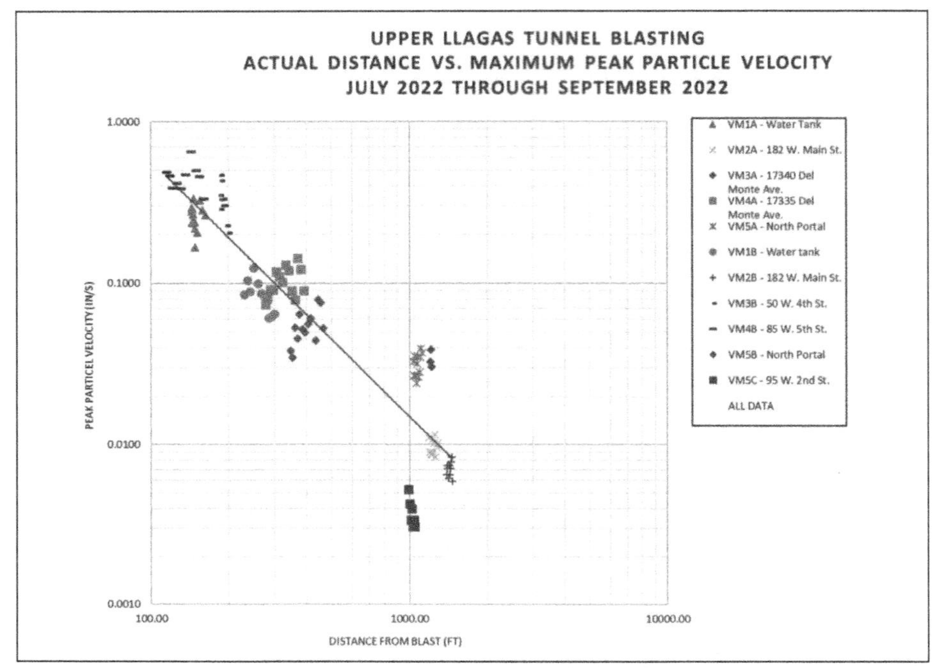


**Figure 5. Plot of distance versus maximum peak particle velocity**

### *Settlement Monitoring Program*

Segments of the alignment where the tunnel runs below active roadways with low cover are closely monitored by Municon West Coast for surface and utility settlement, in accordance with the project's tunnel geotechnical instrumentation and monitoring requirements. Surface settlement was monitored by Automated Motorized Total Stations, which routinely shoot arrays of prisms along the roadway and sidewalks.

Utilities were monitored for settlement using YieldPoint Digital Utility Monitoring Points, each equipped with Ackcio BEAM-AN dataloggers placed in a traffic-rated enclosure. This setup allows the system to collect readings multiple times a day, without the need for manual data collection. Data collected from the Automated Motorized Total Stations and Utility Monitoring Points is continuously uploaded to Municon's web portal where it is available to the project team for review. Deflection thresholds for surface and utility monitoring are shown in Table 2.

### *Control of Water*

A dewatering pump was progressed with the tunnel heading as it was excavated, pumping to a sump pump set in a concrete-lined area. This portal sump sends all water through two 10,000-gallon settlement tanks and treatment system with an automated data logging system. One aspect of the treatment system is to control pH. The maximum expected discharge for the winter months is expected to be 200 gallons per minute. Actual discharge from the tunnel during excavation was on the order of 5 to 10 gallons per minute.

Stream conditions and discharging conditions (whether the water reaches the creek, as opposed to filtering solely through the ground) dictate the water quality objectives

**Table 2. Copy of the maximum allowable movement and corrective action trigger levels from the project's technical specifications for the different types of instrumentation**

| Type of Geotechnical Instrumentation (movement direction) | Action Trigger Level | Maximum Allowable Movement |
|---|---|---|
| Settlement Monitoring Point (vertical) | ½ inch | 1 inch |
| Survey Reference Point (vertical) | ½ inch | 1 inch |
| Survey Reference Point (horizontal) | ¼ inch | ½ inch |
| Convergence Point (total distance) | 0.3% | 0.4% |
| Multiple Point Borehole Extensometer (vertical) | ½ inch | 1 inch |
| Inclinometer with Settlement Casing (horizontal and vertical) | ½ inch | 1 inch |
| Utility Monitoring Point (vertical) | ¼ inch | ½ inch |

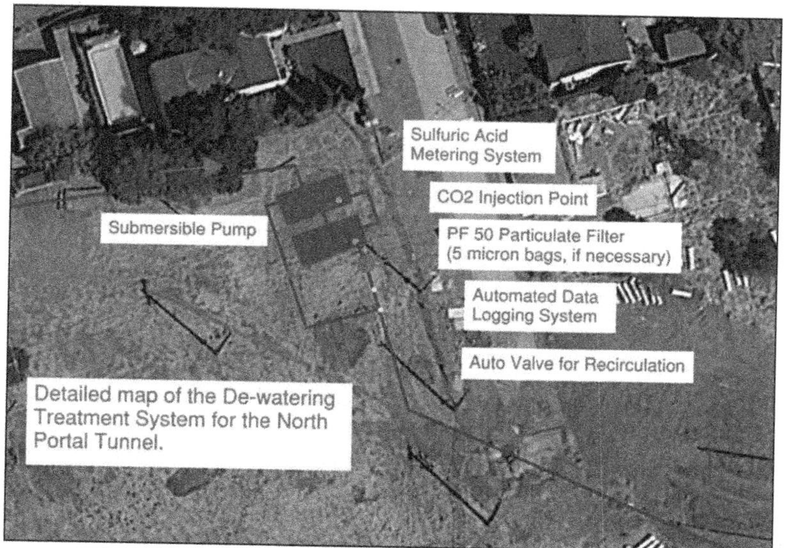


**Figure 6. Detail of dewatering filtration system (contractor submittal)**

for the tunnel discharge water. An electric auto valve monitors the pH of the water so only water with a pH between 6.5 to 8.3 pH is discharged. The capacity of the settlement tanks is kept below 60% during regular operations to allow for the additional capacity requirements during recirculation periods (see Figure 6). The discharge point is placed on grasslands, approximately 850 feet away and 26 feet in elevation above West Little Llagas Creek. If the discharged water flows to the creek, the discharged water cannot cause changes in turbidity more than 20% of the creek's upstream turbidity.

## VALUE ENGINEERING PROPOSALS AND CHANGES

### Surface Grouting

The contract documents require ground improvement by grouting above the tunnel crown. This would be carried out over Reaches 1, 3, and 5 from the surface for a total length of 535 feet. The purpose of this requirement would be to mitigate the risk of high settlement to existing utilities and enhance the development of ground arching above the tunnel crown in reaches with low ground cover. However, provisions in the contract documents allow delayed or cancelled implementation of the ground improvement based on the observed ground settlements from geotechnical instrumentation and monitoring during construction.

During the portal development, the ground conditions and behaviors were observed to be better than anticipated. Drill Tech proposed the use of spiles and canopy tubes as pre-support in lieu of ground improvement for tunneling in Reaches 1, 3, and 5. Elimination of the ground improvement operation would also limit the potential impact to the adjacent community and business.

With Drill Tech constructing the tunnel, having submitted a plan on how it would initially support the ground, the project owner and McMillen Jacobs Associates were open to eliminating the grouting from the surface and incorporating the use of canopy tubes and spiling from within the tunnel if necessary. This value engineered change was accepted.

Drill Tech proceeded with installation of canopy tubes using the Terraroc Symmetrix drilling system with 7.5-inch O.D. by 60-foot-long tubes. As the tunnel was advanced beyond the canopy tubes, ground conditions were consistently found to be more competent than anticipated and did not require spiling. Tunneling through these critical low-cover areas was successfully completed with maximum recorded utility and surface settlements of 0.13 and 0.26 inches, respectively. This value engineered change also mitigated a significant impact to the public by eliminating repaving in the public right-of-way.

## Initial and Final Support

Initial support types were designed in reference to various ground classes. Three ground classes were defined based on the available data from the project-specific geotechnical investigations to cover the anticipated range of ground conditions along the tunnel alignment. These three ground classes are summarized as follows:

- Ground Class I represents predominantly slightly weathered to fresh greenstone, moderately to highly fractured with intensely fractured local zones, rock quality designation (RQD) of 40% and higher with unconfined compressive strength of intact rock from 1,000 psi to 27,000 psi.
- Ground Class II represents highly to completely weathered greenstone with possible remnants of Class I boulder size clasts, intensely fractured, very weak to extremely weak, RQD of 60% and lower.
- Ground Class III represents overburden soil consisting of colluvium.

Three initial support types are defined in the contract drawings corresponding to the three ground classes defined above. These three support types are judged to be able to fully address the anticipated variations of ground conditions for maintaining tunnel stability and controlling ground movements during tunneling. Details of the three initial support types are as follows:

- Support Type I consists of rock dowels and fiber-reinforced shotcrete lining with variable thickness based on encountered ground conditions. Alternatively, steel sets spaced at five feet on centers with shotcrete or timber lagging as required may be installed in lieu of rock dowels and shotcrete lining.
- Support Type II consists of bar spiles pre-support as required, steel sets spaced at four feet on centers with shotcrete or timber lagging.

- Support Type III consists of bar or channel spiles pre-support, steel sets spaced at two to four feet on centers with shotcrete lagging to face of steel sets and curved invert struts.

The initial support elements for each of the support types as described above are the minimum requirements. The initial ground support was the design responsibility of the contractor. Drill Tech engineers estimated the vertical and horizontal earth pressures from the information provided in the Geotechnical Design Report and Geotechnical Baseline Report, then utilized the methods from Proctor and White's *Rock Tunneling with Steel Supports* to design the steel sets for Ground Class II and III ground conditions.

Steel sets were W6×20, spaced four to five feet on center depending on the ground conditions. Initial ground support design for Ground Class I ground utilized the methods from Evert Hoek's *Practical Rock Engineering*. Systematic bolting and unreinforced shotcrete were utilized where the roadheader was used. 7' long rock bolts and shotcrete with 4"×4" W2.1 Welded wire fabric were utilized in the drill and blast harder rock areas.

The final shotcrete liner for the tunnel required a total shotcrete thickness of 11 inches (initial plus final), and a minimum of 3 inches of reinforced final shotcrete. The reinforcement for the final liner was required to be either 4"×4" × D4.0 × D4.0 welded wire fabric, or steel fibers conforming to ASTM A820 Type I or II, with length ranging from ¾" to 1-½", and minimum tensile strength of 95 ksi. Shotcrete with steel fiber reinforcement was required to attain a minimum 28-day flexural toughness of 240 ft-lb, resulting in 1.5 inches of deflection per ASTM C1550, or a minimum residual strength of 400 psi at a maximum deflection of L/150 per ASTM C1609.

Drill Tech proposed the use of BarChip BC-54 synthetic macro fiber as an alternative material to steel fiber and welded wire fabric in the final lining shotcrete. Synthetic fiber is the preferred reinforcement method due to its numerous advantages, including worker safety, longer service life considering corrosivity, superior workability, integration with waterproofing membranes, and low carbon footprint from manufacturing. The product substitution was accepted by McMillen Jacobs Associates and will be incorporated into the upcoming final liner shotcrete work. Shotcrete will be dosed with 9 pounds per cubic yard of synthetic fiber, which has been demonstrated to meet the requisite 28-day flexural toughness through pre-construction testing.

## Tunnel Extension

The design of the South Portal presented multiple accessibility constraints. Constructing the box culverts would require temporary shoring down the centerline of a narrow residential road. The road has multiple overhead lines including high-voltage power lines and residential power and communications drops that inhibited crane access needed for shoring installation. Furthermore, the box culvert needed to be constructed in multiple phases to maintain access to driveways along the road per the project requirements. To resolve these constraints, the project team elected to eliminate the cut-and-cover work along the road by extending the tunnel 125 feet. This proposal presented its own challenges, as the tunnel would be advanced in shallow cover ranging from 6.5 to 11 feet in mostly Ground Class III material.

McMillen Jacobs Associates, as the engineer of record for the tunnel, evaluated the feasibility of this proposal. The evaluation focused on the initial support measures and final lining requirements for this extension section. The results of evaluation concluded that the proposal would be acceptable with the following provisions:

- Canopy tubes as pre-support should be installed from both ends of the extension. The length of canopy tubes should be long enough to cover the entire length of this extension as possible. Recognizing that Drill Tech's equipment would only allow the canopy tubes to be installed for a maximum length of 50 feet, spiles would be used in lieu of canopy tubes for the remaining section.
- The initial tunnel support consists of W8 × 31 steel sets with 8-inch-thick shotcrete lagging or lining with 6-inch-thick concrete invert. A layer of welded wire fabric is also required in the arch and sidewalls.
- To monitor potential damages to existing utilities, surface settlement monitoring points were installed along the tunnel centerline on Del Monte Avenue, on maximum 10-foot spacing. These are checked twice a day during construction of the tunnel extension to monitor ground movement.
- The final lining for the tunnel extension section consists of 14-inch-thick shotcrete (including the 8 inches installed as part of the initial support and W8 × 31 steel sets) and 24-inch-thick final invert (including 6 inches installed as part of the initial support). The final lining is reinforced with #6 bars spaced at 6 inches on centers transversely and longitudinally. The same size and spacing of rebars are required in the final concrete invert on both extrados and intrados.

A challenge posed by the tunnel extension was the installation of canopy tubes from inside the tunnel. Canopy tubes at this location were crucial to provide pre-excavation support for the tunnel and the sanitary sewer manhole directly above the crown.

To create access for drilling, Drill Tech proposed a modified tunnel section with a flatter crown, and a step down at the starting point of the canopy tubes, as shown in Figure 7. The step down had the added benefit of providing an additional two feet of cover in the tunnel extension. Canopy tubes were drilled from both ends of the extension with a custom fabricated drill (shown in Figure 8) mounted to an arch frame sized to match the layout of the canopy tubes. The arch frame setup acts as a drilling template, providing highly accurate placement and alignment of drilled holes. Canopy tubes utilized the Terraroc Symmetrix system and were successfully advanced to a depth of 50 feet with less than four inches of deviation in either direction at the tip.

## CONSTRUCTION MANAGEMENT TOOLS

### Construction Management Software

Valley Water required ProjectWise construction management software to be used for the contract, contract documents, correspondence, inspector daily reports, contract and change management documentation, submittals, RFIs, quality control documentation, and all other construction contract documentation. The project benefited from using the software in several ways:

- All parties (Valley Water, City, designer, subcontractor) and their subs can be given appropriate access and permission to the system.
- Web-based interface means that the project files can be accessed anywhere with an Internet connection.
- Double handling of documentation is reduced. Instead of having to submit hard copies or email in documents to one person, the contractor can submit directly to the system.

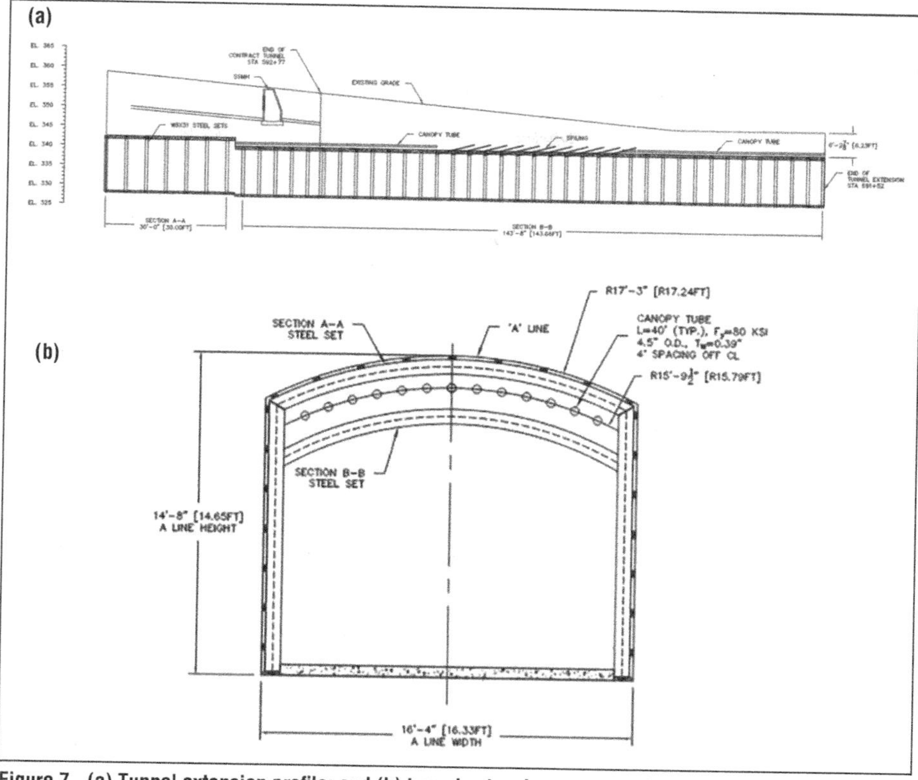


Figure 7. (a) Tunnel extension profile; and (b) tunnel extension cross-section at step down

Figure 8. Canopy tube drill at South Portal

### Partnering

Construction partnering is a proactive approach to risk management and mitigation, problem identification and solution, effective communication, and overall stakeholder collaboration. Partnering provides a framework for smoother problem-solving and communication among project team members across all parties.

The partnering consultant is responsible for providing a facilitator for the partnering meetings. Before any partnering meetings, the facilitator will reach out to Valley Water, the City, the construction manager, and the subcontractor to get input on any issues, actual or potential, that they would like brought up during upcoming meetings.

### Dispute Review Board

A Dispute Review Board (DRB) was required by the contract. A three-person board was jointly selected after Valley Water and the subcontractor provided a list of candidates. DRB members were selected based on their experience on other DRBs and their familiarity with the construction of projects with similar elements to this one. The DRB will be used to engage in dispute avoidance activities, such as fostering and conducting proactive discussion on project issues. Members of the DRB attend regular DRB meetings, perform site visits, and provide advisory opinions. The intention of the DRB is to attempt to resolve issues that may present themselves on the project that cannot be addressed at lower levels of the dispute resolution ladder established during partnering, to avoid escalating processes via the legal process.

## CONCLUSION

The completion of the tunnel for the Upper Llagas Creek Flood Protection Project will be a crucial milestone for the project's overall success. Tunnel excavation and initial support was the riskiest component of the project, given the tunnel's proximity to roads, structures, and utilities. The project team collectively adapted to varying site geology and constraints, and successfully implemented solutions to drive the successes of the project.

Major achievements included the elimination of surface grouting, completion of controlled blasting without ground vibration exceedances, and the tunnel extension. Various tools were used for construction management and allowed for a streamlined approach for discussing and mitigating these changes.

## ACKNOWLEDGMENT

The authors of this paper would like to acknowledge Xavier Irias and Glenn Hermanson of Woodard and Curran as the civil design consultants; Brett Mainer and Brian Harris of Drill Tech Drilling & Shoring; Yiming Sun of McMillen Jacobs Associates; Scott Hensler, Scott Ball, and Joseph Ang of Mott MacDonald; the Valley Water's project managers, project engineers, and water resources specialists; and the City of Morgan Hill's engineers for their assistance with the construction of this project and with the writing of this paper.

## DISCLAIMER

Valley Water and the City of Morgan Hill were not authors of this paper and assume no responsibility or liability for any possible errors or omissions in the content of this paper.

## REFERENCES

Geo-Logic Associates dba Pacific Geotechnical Engineering (GLA). 2019 Geotechnical Investigation, Nob Hill Tunnel, Upper Llagas Creek Flood Protection Project, Santa Clara County, California dated February 22, 2019.

McMillen Jacobs Associates. 2020 Upper Llagas Creek Flood Protection Project—Phase 2, Nob Hill Tunnel, Santa Clara County, California, Geotechnical Baseline Report, dated November 2020.

Renewed_SCW_Program_Report—Nov2020.pdf (valleywater.org): Safe, Clean Water and Natural Flood Protection Community Preferred Program Report.

Safe Clean Water 15-year Program.pdf (valleywater.org) | Safe, Clean Water and Natural Flood Protection General Information.

Upper_Llagas_Map_KC (valleywater.org) | Map of Reaches.

2012 E6: Upper Llagas Creek Flood Protection | Santa Clara Valley Water | Upper Llagas Creek Flood Protection General Information.

# Design and Planning of New Passageway Tunnel for Circulation Improvements at Grand Central—42nd Street Station, New York City

**Dominic Reda** ▪ Gall Zeidler Consultants, LLC
**Alfredo Valdivia** ▪ Gall Zeidler Consultants, LLC
**Vojtech Gall** ▪ Gall Zeidler Consultants, LLC

## ABSTRACT

The design and planning for the construction of a new passageway to improve access and circulation to better accommodate anticipated increased passenger flow within Grand Central—42nd Street Station. The passageway includes the break-in of an over 100-year-old station cavern with an innovative design implementing a protective structure above the existing track to allow for safe train operations and passenger circulation during construction. Construction of the new passageway follows the Sequential Excavation Method. This paper will present the innovative solutions that went into the preliminary design and planning considering construction access, site logistics, risk management, structural retrofitting, instrumentation and monitoring, and impact assessments.

## INTRODUCTION

To relieve congestion and overcapacity issues between the Lexington Avenue Line and the Flushing Line at Grand Central—42nd Street Station, which are expected to worsen with increased ridership into and out of the station, Metropolitan Transportation Authority Construction and Development (MTA C&D) is planning to implement several projects to improve access and passenger circulation for the interchange, including widening of existing stairways from the northbound 4/5/6 Line and the construction of additional stairs and passageway to the Flushing Line. This paper focuses on planning of the new passageway connecting the existing passageway with the Flushing Line at Grand Central—42nd Street Station.

Grand Central Terminal serves as a transportation hub for New York City and connects the Metro-North Railroad and New York City Transit (NYCT) systems. NYCT's Flushing Line at Grand Central Terminal is a highly used, multi-level underground structure mainly located underneath 42nd Street and thus, presented many challenges for constructing a new tunnel with minimal disturbance to station operation.

The selected procurement method for the Circulation Improvements at Grand Central—42nd Street Station is Design-Build (DB) with the selection process following a Request for Qualifications (RFQ) and Request for Proposals (RFP). Planning and preparation of the procurement documents included Parsons (Prime Engineer), Gall Zeidler Consultants (Sub-Consultant for Tunnel Engineer and Station Connection), Sowinski Sullivan (Sub-Consultant for Architecture), Mueser Rutledge Consulting Engineers (Sub-Consultant for Geotechnical Services), and NAIK Group (Program Manager).

The technical and contractual requirements for the project were included in the RFP documents. The RFP documents comprised of: Invitation to Bidders (Volume 1),

Design-Build Agreement & General Provisions (Volume 2), Division 1 General Requirements (Volume 3), Project Requirements & Design Criteria (Volume 4) and Reference Documents which included Reference Design Drawings, As-Built Drawings, Standard Drawings, Utility Drawings and Surveys, Environmental Drawings, Reports & Surveys, Geotechnical Data Report and Geotechnical Baseline Report (Volumes 5–10).

The RFP documents require the Design-Builder to perform the final design of the underground structures including temporary and permanent works in accordance with the Project Requirements and Design Criteria (PRDC).

A project specific PRDC was developed which included mandatory requirements related to: scope of work, codes & standards, loads & load combinations, design life, materials, durability, minimum spaceproofing & train clearances, mined tunnel excavation, initial rock support, shotcrete, tunnels & underground structures, watertightness & waterproofing, subsurface investigation, ground characterization & geotechnical parameters, instrumentation & monitoring, protection of existing structures, support excavation, pre-construction and post-construction condition surveys, noise & vibration and settlement monitoring.

## HISTORICAL BACKGROUND

The planning and construction of Steinway Tunnels under the East River—connecting Queens, New York with Manhattan, New York—dates back to the late 1800's. The construction of the tunnels was delayed and temporarily ceased due to difficult tunneling conditions, funding issues, and in some instances, tragic accidents. As a result, it took many years to complete the Steinway Tunnels. The tunnels under the East River were complete by mid-1907, followed by the completion of an approximately 150-ft long Station Cavern (see Figure 1 and Figure 2).

For roughly the initial 15 years, the Flushing Line at Grand Central—42nd Street Station operated with a loop at the current west end of the station (see Figure 3). Shortly after opening the Flushing Line to Manhattan the platform was extended approximately 450-ft to meet the ridership demand. This upgrade to the station included the construction of a mid-platform staircase. This station upgrade took place between circa 1914 and 1916.

Figure 1. Construction of Grand Central—42nd Street Station (courtesy of NYCT)

Grand Central—42nd Street Station operated with this configuration until circa 1927 when the Flushing Line was extended to Times Square Station. The station undertook another upgrade in circa 1951, again to meet increased demand with the construction of a second mid-platform staircase. The next upgrade to the station did not occur until circa 1995 with the construction of an elevator near the west end of the platform.

Figure 2. Completion of Grand Central—42nd Street Station prior to opening (courtesy of NYCT)

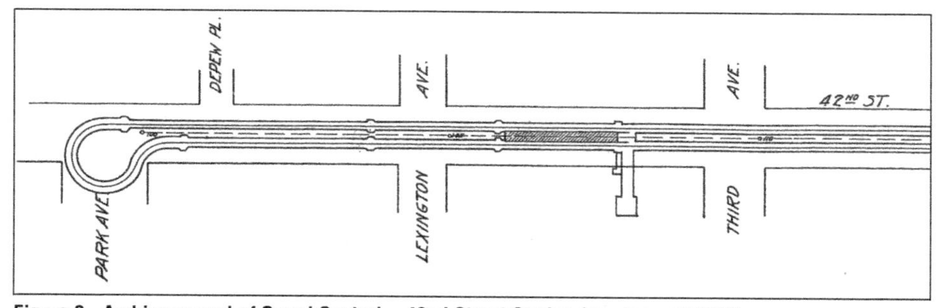


Figure 3. Archive record of Grand Central—42nd Street Station (courtesy of NYCT)

Detailed planning for the next major upgrade at Grand Central—42nd Street Station for circulation improvements began in early 2019. To enable construction of the new passageway tunnel, a vertical shaft has been included to provide construction access (see Figure 4).

## GROUND CONDITIONS

The regional geology (Bedrock) of the project location consists of interlayered schist, gneiss, granofels, and amphibolite of the Hartland Formation. Younger intrusive granitic pegmatite dikes, less than a meter in thickness, and sills, more than a meter in thickness, have been documented within the Hartland Formation. Soils atop bedrock consist of glacial till, and compact granular fill. The stratigraphy at the project location is shown in Figure 5.

A small ground investigation program, consisting of two (2) boreholes and a surface geophysical survey of the project site, was conducted for development of the local subsurface conditions at the project location. The primary aim of the investigation was determination of the bedrock topography and for the development of geotechnical

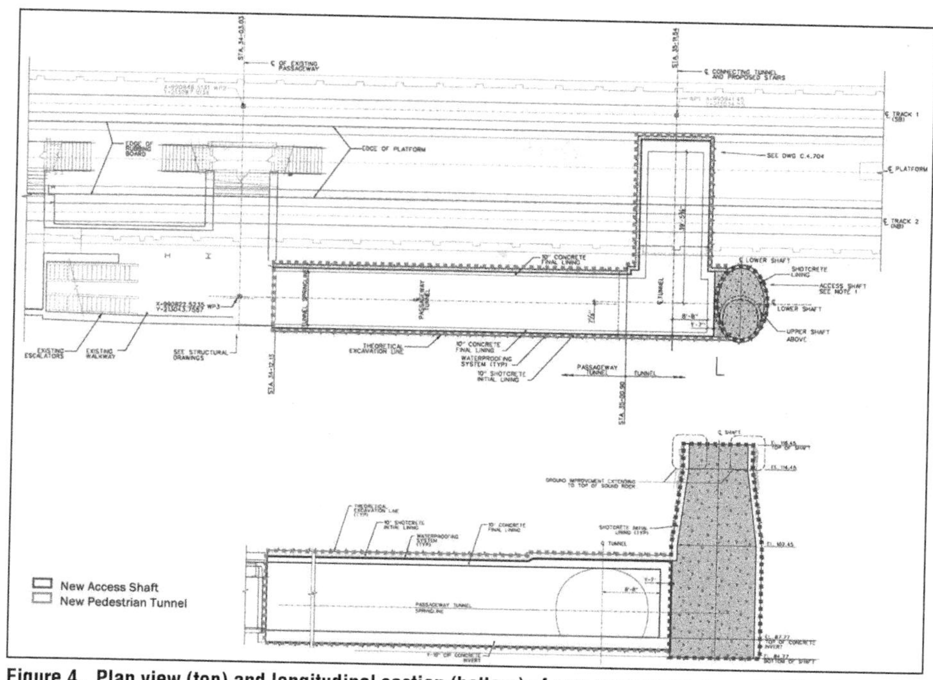


**Figure 4. Plan view (top) and longitudinal section (bottom) of new passageway tunnel at Grand Central—42nd Street Station**

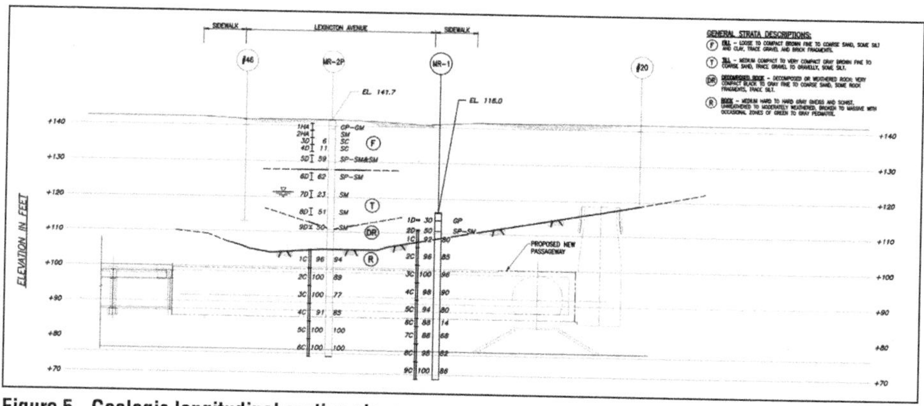


**Figure 5. Geologic longitudinal section along new passageway tunnel**

parameters for preliminary design. Groundwater monitoring established the groundwater elevation at +120.9 (approximately 11-ft above tunnel).

Bedrock foliation is well developed, generally dipping between 20° and 60° toward NW-WSW. A conjugate joint set is also present, with a strike similar to that of the foliation and dips between 25° and 66° ESE-SSE. Ungrouped localized joints are also present within the project area. Joint surface conditions ranged from rough or irregular to smooth and planar. Iron staining was typical on the joint surfaces with some altered surfaces, consisting of silty coatings or mineral coatings, also observed.

The bedrock topography was a critical element for consideration of excavation and support sequencing for construction of the new passageway. A combination of project/historic boreholes, geophysical surveys, and published geological mapping suggests a local linear depression of the bedrock surface, which results in apparently shallow cover (< 5 ft) along part of the passageway's alignment parallel to East 42nd Street.

## PROJECT CHALLENGES

Construction of a new passageway from within an active station presents a number of challenges.

The main project challenges considered during the preliminary design and planning of the project included: excavation and rock removal during construction, maintaining train operations and limiting track outages during construction, pedestrian access/egress during normal operations and emergencies, coordination of work with on-going and planned adjacent contracts, site access for personnel & materials, limited available space for construction laydown/staging areas, protection of existing structures in particular during tunneling and the breakthrough of the station cavern, unforeseen conditions (ground conditions, existing structures, systems and utilities).

Considering the surrounding infrastructure of the new passageway and the importance to limit impacts from noise and vibrations, alternative means-and-methods were required as opposed to more traditional drill-and-blast method for hard rock tunneling as was done during the initial construction of the station. Examples of alternative rock excavation methods included the use of rock splitters and hydraulic breakers.

Initially, project plans did not envision the need for a construction access shaft. Early plans for construction without a shaft were complicated due to the limited space at the existing passageway and work restrictions (construction laydown, access path, operation, work windows, track outages, etc.). The construction access shaft was introduced as an option as the project was developed. This not only helped construction logistics, but it also helped limit the number of track outages for nights and weekends which was a goal from early in project planning.

While it was agreed to further develop the shaft further, the project still needed to evaluate where the shaft could be located and how this would be accessed from the street surface. A trade-off study was performed to select an optimum location for the construction access shaft. Figure 6 shows the two preferred locations. The main difference between the two concepts is the location of the shaft; either before or after the tunnel turn towards the Flushing Line station cavern.

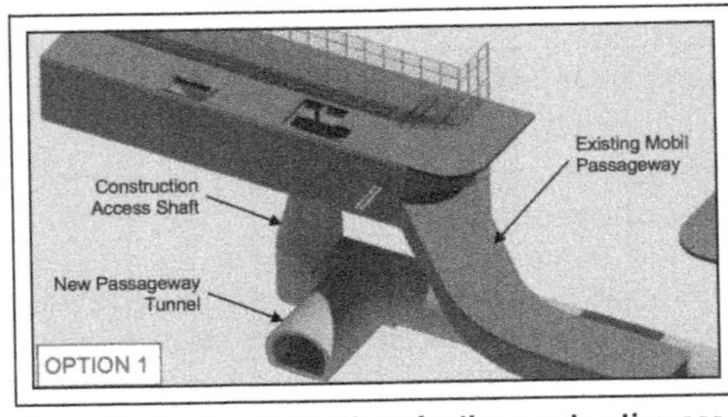


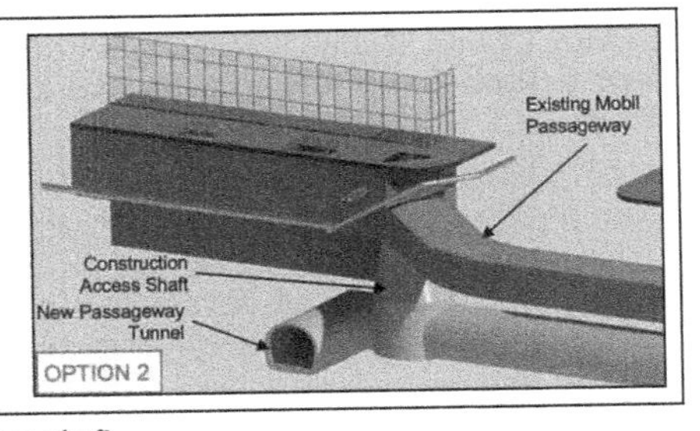


**Figure 6. Preferred locations for the construction access shaft**

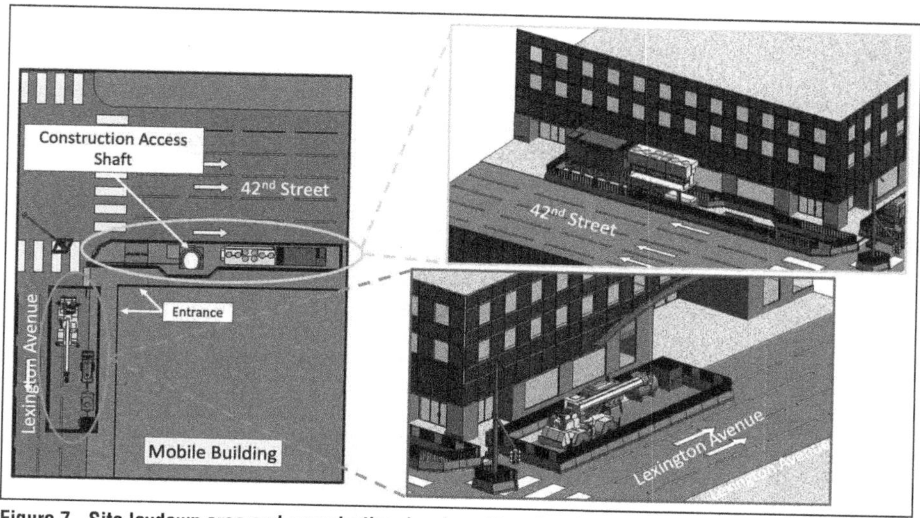


**Figure 7. Site laydown area and organization developed for planning purposes**

Each shaft location had its benefits, challenges, and risks. While Option 1 provided the largest access shaft diameter and most traditional shaft-to-tunnel interface, it required the most demolition for internal elements and impacted MTA's back of house. Alternatively, Option 2 had the flexibility to be constructed from within the existing Mobil Passageway or the from street surface, however, it required structural modification of the headwall at the interface with the Mobil Vault and Mobil Passageway Tunnel. Since there was not a clear benefit for one over the other, the preferred solution for the procurement was Option 1, however, the DB Team was given the flexibility to adjust the position of the construction access shaft.

As the project progressed with the shaft location based on Option 1, further planning at the street surface commenced. The shaft shown is located along 42nd Street outside the Mobil Building, near the southeast corner of 42nd Street and Lexington Avenue. To limit disruption to traffic at a busy intersection, a compact construction laydown area was considered split along 42nd Street and Lexington Avenue (see Figure 7). This configuration allows the site to maintain operation of both pedestrian and vehicle traffic during daytime hours along the sidewalk and street, respectively. During overnight construction hours, the two individual sites would be connected and expanded one additional lane along 42nd Street to allow for deliveries and removal of excavated rock. This concept was provided as part of the procurement documents to be further developed and finalized by the DB Team.

## DESIGN AND CONSTRUCTION CONSIDERATIONS

### Design Considerations

The new passageway extension consists of a construction access shaft, passageway tunnel and the connection tunnel to the existing station cavern. Tunnel design and construction for the passageway follows the Sequential Excavation Method (SEM) approach with installation of initial support following every excavation round. Strengthening of the existing station cavern arch around the new opening is enabled by installation of permanent supports using steel columns jacked against the arch before any tunnelling above the station cavern arch and any cutting/demolition of the arch commences. Temporary protective measures are required at the opening location to ensure safe operation of the station during construction.

To include flexibility in the DB procurement, the DB team had the ability to adjust the final position of the connection to suit their means-and-methods as well as optimize the project cost and schedule while considering associated risks for the ultimate selection. An isometric view of the new passageway tunnel is shown in Figure 8 with the maximum allowed length of the new passageway. The shortest possible location on the connection would coincide with the lowest rock cover above the tunnel.

Figure 9 shows the existing conditions at each end of the new passageway tunnel. The design for the new passageway has to consider, among others, train clearance to temporary and permanent support, installation of the column foundation, structural reframing at the existing passageway, relocation of suspended services and potential

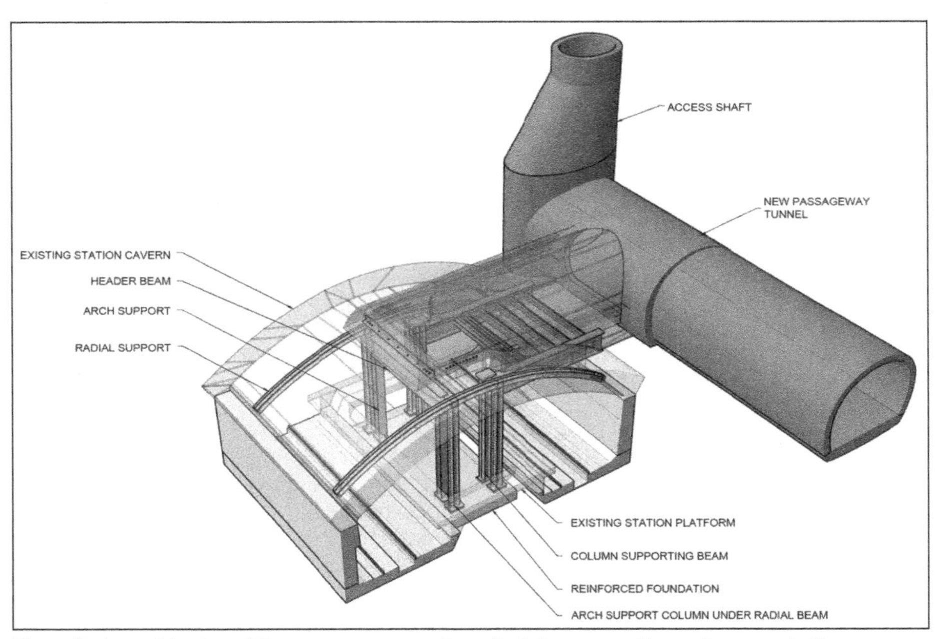


**Figure 8. Isometric view of the new passageway tunnel, station connection and access shaft**

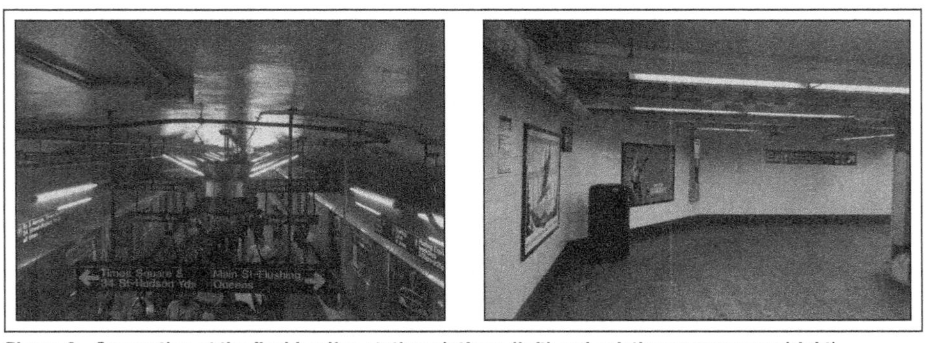

**Figure 9. Connection at the flushing line station platform (left) and existing passageway (right)**

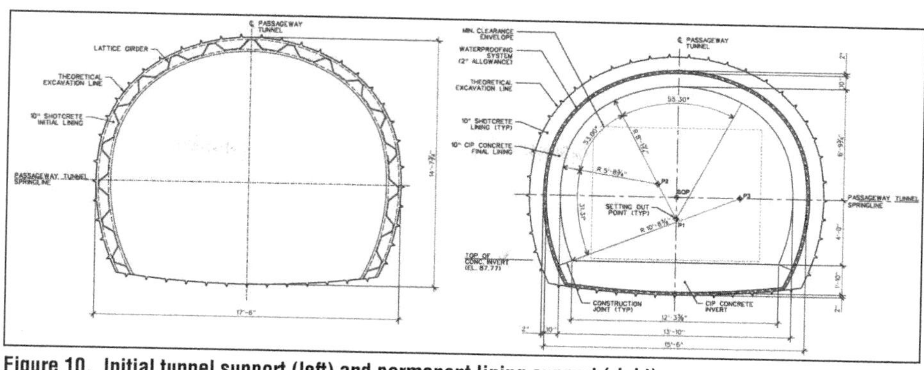


**Figure 10. Initial tunnel support (left) and permanent lining support (right)**

under platform services, rerouting of passengers, maintain staircase and escalator usage, and the platform boarding area.

Preliminary excavation and support sequencing for the tunnel envisioned a maximum full face 4-ft excavation round, followed by the installation of 10 inches of fiber-reinforced shotcrete and lattice girders (see Figure 10) with an allowance to adjust to rock bolts and reinforced shotcrete when sufficient rock cover is present to develop a reinforced rock arch. Meanwhile, at the low rock cover area, contingency measures include spiling ahead of the advance and pre-excavation grouting.

With the permanent opening frame installed and excavation completed, the break-in to the station cavern can proceed. Upon completion of the break-in, a waterproofing system is applied followed by the installation of a reinforced concrete final lining. The waterproofing system will comprise a geotextile and PVC membrane and will be terminated at the existing station cavern.

The preliminary design considered a comprehensive surface & subsurface instrumentation and monitoring program that included in-tunnel, geotechnical, and structure monitoring within the tunneling zone of influence. The project requires monitoring data to be evaluated on a continuous basis to facilitate adjustments of tunneling works, as needed, to minimize construction impact and enable execution of the works in a risk-managed approach. Considering the potential influence of the break-in on the Flushing Line Station Cavern, a real-time monitoring system is planned through the installation of continuous shape arrays. This monitoring approach was selected over a more traditional prisms target with automated total stations due to limited site distances and to minimize the requirement for maintenance of the monitoring system over the duration of the project. For instance, accessing the prisms for realignment or cleaning would require track outages, disrupting station operation.

To ensure the protection of existing structures and utilities, the Design-Builder is required to evaluate and prepare a damage assessment report documenting the impact of the work on existing overlying and adjacent structures and utilities. Pre-construction and post-construction condition surveys of foundations, interior and exterior of buildings, properties, railroad, utilities, appurtenances, and structures are also required within the tunnel zone of influence.

During the preliminary design and planning of the project, it was determined that the Design-Builder shall identify and develop mitigation measures for potential risks associated with the execution of the tunneling works. Per the project requirements, the Design-Builder shall consider the following:

a. Unforeseen ground conditions and low rock cover along the passageway and connection to the existing station cavern.
b. Under platform services and surface mounted not previously identify that require relocation and/or modifications.
c. Conflicts or additional work with above ground utilities and services that will need to be relocated if an access shaft is utilized by the Design-Builder.
d. Unable to perform additional supplemental geotechnical investigation due to access and/or schedule constraints of the project.
e. Impacts to the existing Station Cavern and Mobil Passageway due to tunnel excavation and structural modifications including service interruptions.
f. Delays to access potentially impacted properties within the ZOI to perform pre-construction condition survey and install instrumentation as part of the I&M program.
g. The need for additional staging and laydown areas.
h. Delays due to unable to efficiently remove excavated/muck material.
i. Delays due to unavailable/cancelled track outages and station closures.
j. Delays due to work by Adjacent Contractors.

Considering the project's location and scope of work, during the preliminary design and planning of the project it was established that the Design-Builder shall not begin mined tunnel excavation including demolition of existing structure until the following conditions (work restrictions) have been met:

a. Required submittals have been submitted and approved.
b. Pre-construction survey of existing structures, facilities, utilities and railroad infrastructure has been completed by Design-Builder and pre-construction documents have been provided.
c. Installation of geotechnical instrumentation for required monitoring program (subsurface, surface and building) has been completed and instrument baselines have been established.
d. Ground improvement, where required, to facilitate mined excavation has been completed and efficacy verified.
e. Approved temporary bracing (propping) to support existing tunnel has been installed where required to facilitate mined excavation. Unless otherwise noted, keep temporary bracing in place until final structural concrete has reached minimum specified design strength.
f. Prior to tunnelling above the existing station cavern and making any openings in existing cavern arch, the permanent structural steel system shall be installed to resist and transfer all vertical and lateral loads around the openings.
g. All issues related to health and safety have been met and submittals have been made in accordance with OSHA requirements, and other applicable codes and regulations of Federal, State of New York, and local agencies having jurisdiction.

The project requirements mandate that the Design-Builder prepares and submits for approval the following work plans associated with the tunneling work: Sequential

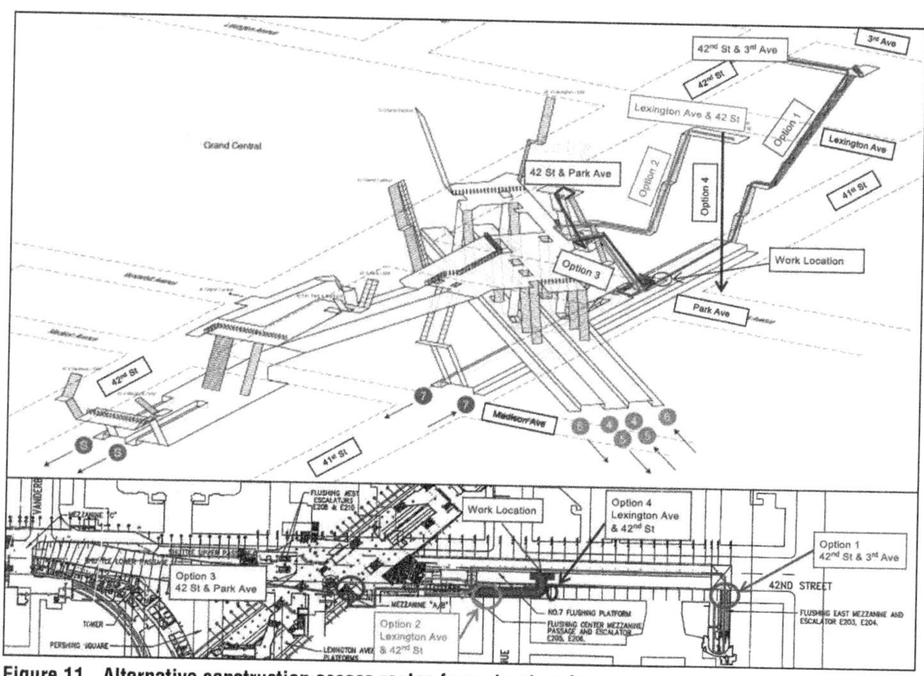


Figure 11. Alternative construction access routes from street surface

Excavation Method (SEM) work plan, Pre-Excavation Grouting work plan and Instrumentation & Monitoring work plan.

## Construction Considerations

While the construction access shaft was the ultimate decision for project planning, there were three (3) other alternatives without a construction access shaft reviewed as part of project planning. Each of these options provided access from different entrances to the Flushing Line at Grand Central—42nd Street Station (see Figure 11). The entrance to Option 1 at 42nd Street and 3rd Avenue connects to the east end of the Flushing Line platform. Meanwhile, Option 2 and Option 3 connected from different street access locations to the existing passageway via Escalator 205/206. Of the three (3) options, Option 3 proved to provide the most direct access for construction and logistics. However, each of these options were ultimately eliminated due to heavy disruption to station operation as they required significant laydown space and several nightly track outages.

Once the project ruled out these three (3) options, the shaft option began to be investigated and a choice had to be made to have the shaft extend to the street surface or to limit the shaft to the Mobil Passageway level (see Figure 12). As shown, there is limited working height within the Mobile Passageway for construction. For this location, it was anticipated that structural reframing was required to create additional working height by demolishing the base slab and potentially deepening of the original excavation profile. Even with this enabling works, all lifting of materials, muck and equipment would be limited to a small capacity gantry system. Ultimately, the decision was made to create a shaft from street surface outside of the Mobil Passageway.

Figure 12. Shaft access from street (left) versus Mobil Passageway (right)

## CONCLUSION

Since the original construction of the Grand Central—42nd Street Station, improved access and passenger movement necessitated the upgrading of the station. A number of transformations have occurred since the early 1900's with the next planned upgrade commencing in early 2023. The major difference between previous upgrades and this current plan is the limitation of construction access and the demand to maintain station operation throughout construction. As a result, preliminary studies were undertaken to find an optimized solution to carry out the construction of a new passageway tunnel. While a number of iterations for preliminary design and planning were required to have alignment from all parties involved, a clear priority to minimize disruption to station operation early on in the project helped guide the refinement of each alternative and ultimate selection of the construction access shaft. The introduction of a construction access shaft to project planning increased the magnitude and scope of the project, however, it simplified construction and logistics planning. The shaft allows for direct access for, among other, material deliveries, construction services, personnel, and muck removal.

The project challenges presented for the Circulation Improvements at Grand Central—42nd Street Station are common to station upgrades elsewhere in urban settings. This paper presented a unique solution for the construction of a new passageway tunnel and the connection tunnel to an existing station cavern with minimal disruption to the operation of the station.

## REFERENCES

Mueser Rutledge Consulting Engineers. (2021). Access Improvements at Grand Central Station GCT Flushing Line, New York, NY—Geotechnical Baseline Report & Geotechnical Data Report; Metropolitan Transportation Authority Contract CM-1550 Package 3, Parsons Transportation Group, June 12, 2021 including revised Drawings B-1 and GS-1 issued August 10, 2022.

IRT Flushing Line. 2012. https://www.nycsubway.org.

The Steinway Tunnels (1960). 2012. https://www.nycsubway.org.

# Excavation and Support of Cross Passages on Westside I

**Jeff Brandt** ▪ Traylor Bros., Inc.
**Kerwin Hirro** ▪ Traylor Bros., Inc.
**Peter Dietrich** ▪ BeMo Tunneling GmbH
**Norbert Fuegenschuh** ▪ BeMo Tunneling GmbH
**Mike Cole** ▪ Stantec
**Joseph DeMello** ▪ LA Metro

## ABSTRACT

This paper presents a successful case history of the ground improvement efforts and excavation of twenty-three cross passages between twin bore tunnels driven by EPB tunneling machines in Los Angeles, California on Section 1 of the Westside Subway Extension. Unique and complicated ground conditions local to the tunnel alignment include San Pedro sands, siltstone/claystone Fernando Formation, and zones of "asphalt impacted soils." To provide water cut-off and soil stabilization, jet grouting was performed prior to TBM excavation at the location of many of the cross passages. However, upon investigating the cross passages prior to excavation, the subsurface conditions required additional measures and SEM tools to ensure successful completion of the project.

## INTRODUCTION

The Los Angeles County Metropolitan Transportation Authority's (Metro) Purple Line Extension will add nearly nine miles of underground heavy rail to the Los Angeles Metro Rail system. The project is broken up into three different sections and upon their completion will connect the existing Wilshire/Western Station in Koreatown to the VA hospital in Santa Monica and include the addition of seven more subway stations. The goal is for all three sections to be fully operational prior to the 2028 Los Angeles Olympics.

LA Metro's Section 1 of the Purple Line Extension is a $1.6 billion, design-build contract in partnership with a joint venture of Skanska-Traylor-Shea (STS). The Section 1 segment ties into the existing Wilshire/Western station, providing connection to the existing D-Line and the LA Metro Rail System. Section 1 extends the system West 3.9 miles under the Wilshire Boulevard Corridor, passing directly adjacent to the renowned La Brea Tar Pits. The project includes two, twenty-two foot diameter tunnels and the construction of three underground subway stations. Along the tunnel alignment, smaller, sixteen foot diameter cross passage tunnels were included to provide emergency egress and facilitate tunnel support system function.

## GEOLOGY

Section 1 of the Purple Line Extension was divided into four separate tunneling drives with various soil conditions encountered throughout. The twin TBMs launched heading East from the Wilshire/La Brea Station and tunneled 9,570 feet toward the existing Western/Wilshire Subway Station, terminating at a retrieval shaft built for the reception and disassembly of the TBMs. After completion of Reach 1, the TBMs were disassembled, transported back to the Wilshire/La Brea Station, and re-assembled for launch to the West. Reaches 2, 3, and 4 connect La Brea station to Fairfax and La

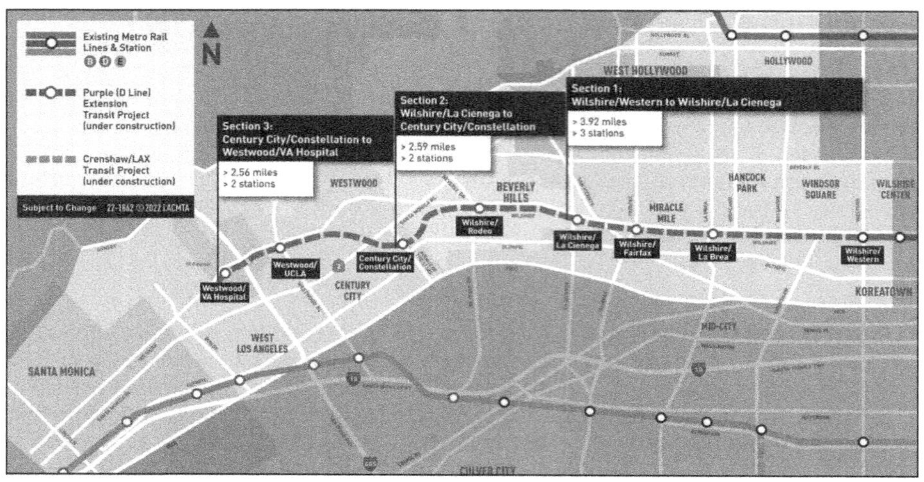


**Figure 1. LA Metro Purple (D-Line) Extension Transit Project (Metro.net)**

Cienega Stations before terminating the tunnels in Reach 4 approximately 550 feet West of the La Cienega Station. The tunnel profile depth of cover ranged between sixty to one hundred feet from the surface level and was situated within the San Pedro and Fernando Formations, with some sparse zones through the Lakewood Formation.

The project's Geotechnical Data Report (GDR) defined the San Pedro Formation as containing both fine-grained dense sand and silty sands with medium to course-grained sands, as well as the inclusion of some stiff silts. The San Pedro Formation also contained cobbles and gravelly sand layers, as well as shell fragments (Parsons Brinkerhoff, 2014). Core sample recoveries along the alignment were typically incomplete, indicative of low cohesion. Along with a high permeability, fast raveling to running conditions were to be expected. The Fernando Formation typically underlies the San Pedro Formation and contains stiff to hard siltstone and claystone, yielding Unconfined Compressive Strength (UCS) testing results between 25 to 135 psi and noted stand-up time tests beyond several hours. The Lakewood Formation would be sparsely encountered, characterized by poorly graded sands, silts, clays, and some clayey sand layers. Groundwater measurements for the project indicated that all the tunnel and cross passage profiles would be below the water table and to expect both perched and semi-perched groundwater conditions, as well as the presence of locally confined aquifers (Parsons Brinkerhoff, 2014).

Common to the Southern California Geological profile, the GDR was clear in anticipating potentially active and inactive faults in the vicinity of the alignment, posing potential difficulties with both hydrocarbons and gassy conditions. The Los Angeles Basin is home to many oil fields, and the Section 1 tunnel alignment passes through both the Salt Lake Oil Field and the South Salt Lake Oil Field according to the California Division of Oil, Gas, and Geothermal Resources (CalGEM, 2020) records.

The hydrocarbons, and other notable petroleum byproducts such as methane or hydrogen sulfide gases, were expected to migrate through any available pore space and along any geological fractures, particularly near the La Brea Tar Pits. In contrast to their gassy state, the hydrocarbons are also characterized along the alignment as zones of asphaltic and tar-impacted sand lenses or discrete pools and seeps of a viscous tar and oil (Parsons, 2019). The large void spaces in the coarse-grained

San Pedro Formation led to zones of "asphalt-impacted soils" as well as the zones between Fernando and San Pedro Formation where seeps could most readily travel. The project was initially classified by CalOSHA as Gassy due to the expected presence of methane, hydrogen sulfides, and Volatile Organic Compounds (VOCs) (California Code of Regulations, 2022). As the TBM tunneling encountered levels of gas beyond that in the initial GBR profile, the project was re-classified along Reach 2 and Reach 3 as Extremely Hazardous. These challenging geological conditions were all paramount in considering the approach for both the design and the execution of cross passage construction.

## CROSS PASSAGE DESIGN

As the Section 1 Design-Builder, STS was partnered with Parsons to complete the design and construction of the cross passages. The tunnel alignment of approximately 3.4 miles requires cross passages spaced at intervals between 750 and 800 feet, resulting in a total of twenty-three cross passages, designated as CPB-4 through CPB 26. The cross passages were designed with a typical interior dimension to accommodate the project defined systems layout and code compliance, ultimately requiring a minimum interior width of 10'-6" by 9'-6" high. The twin tunnels are typically separated by a pillar of material approximately 17' wide and thus resulted in the average interior space of the cross passages ranging from 13'-6" to 16'-1" long (Parsons, 2016).

The cross passages were initially designated by four different ground classifications which would then require different programs for both excavation and initial support lining of the cross passage, however as the ground treatment program was further developed, a fifth ground condition classification was developed. According to the Cross Passage Design Report, the ground classifications for the cross passage profile are as follows:

Ground Class 1 – Profile is within the coarse-grained San Pedro Formation
Ground Class 2 – Profile is at the intersection of both the coarse and fine-grained San Pedro Formation
Ground Class 3 – Profile is in a mixed face of both San Pedro and Fernando Formations
Ground Class 4 – Profile is entirely within the Fernando Formation
Ground Class 5 – Profile is within Tar-Impacted San Pedro Formation

All locations within the San Pedro formations were designed with the anticipation of localized depressurization and ground treatment would be required to limit ground water seepage to less than 50 gallons per minute. Jet grouting was assumed as the model for limiting the seepage and providing soil stability in all Ground Classes 1, 2, and 3. Initially, the cross passages within the tar-impacted sands, Cross Passage (CPB) 17, CPB 21, CPB 22, and CPB 23, were thought to be subject to ground improvement via jet grout, but the tar-impacted soils were not able to be washed thoroughly for treatment and the addition of Ground Classification 5 was necessary. The design for Ground Class 4 assumed that ground improvement was not required, only local relief wells may be necessary for sufficient gas and water depressurization (Parsons, 2016). Considerations were made at the design phase to move the cross passages into more favorable ground conditions, including even lowering the tunnel alignment in its entirety, to better locate cross passages within the preferred Ground Class 4 designation. Ultimately, Reach 1 included CPB 4 through CPB 15, Reach 2 CPB 16 through CPB21, Reach 3 CPB 22 through CPB 25, and Reach 4 CPB 26.

The twenty-three cross passages were excavated by the Sequential Excavation Method (SEM) after the completion of both TBM drives in any given Reach. The SEM method requires in-depth practical experience, extensive geotechnical understanding, and engineering, as well as skilled execution. To ensure successful excavation of the challenging ground conditions of the Section 1 project, STS employed the assistance of BeMo Tunneling GmbH (BeMo) at the early design phase as well as throughout the entire excavation operation, employing their services as the lead SEM Engineer for the project. Utilizing BeMo's unique and exceptional experience in the industry, the excavation of all twenty-three cross passages was planned to intricate detail including excavation sequencing, ground support measures, instrumentation and monitoring concepts, basic toolbox items, as well as contingent and emergency response procedures.

Prior to any excavation, pre-excavation probing was required to verify the ground conditions and determine the ground water conditions at the location, ensuring that the design parameters for the location were accurate. Any changes in condition were recorded in the probing log and shared with the Engineer of Record for review. With all information gathered and ground conditions confirmed, prior to any excavation a general Required Excavation and Support Sheet (RESS) was produced by the SEM Engineer and reviewed with the Tunnel Manager, the Engineer of Record, the Surveyor, and the Owner's Representative Tunnel Manager to agree on the process of excavation and ground support. The general RESS was documented and used as a standard for a Daily RESS sheet to be reviewed and signed off by all parties to document the previous excavation advance's ground conditions, support systems installed, and any changes that may be required that the SEM Engineer could require. With this system, the SEM process allows for broad adaptation of different "toolbox" measures to ensure that the excavation is completed safely.

Typical to many SEM designs, the cross passages include an initial supportive liner, a HDPE water and gas tight membrane with redundant grouting capabilities, and a final cast-in place concrete liner. The prescribed initial support for the cross passages on Section 1 was designed based on the system carrying a minimum soil load equivalent to that exerted by a height of soil above the opening or twice the diameter of the opening, whichever is smaller, and hydrostatic pressures (Parsons, 2016). Based on this analysis and the individual cross passage ground classification designation, a design was produced for the initial lining support utilizing a steel fiber-reinforced shotcrete layer either eight or ten inches thick with the inclusion of steel lattice girders sized and spaced as necessary.

## GROUND IMPROVEMENT FROM THE SURFACE

Cross passages with a designation of Ground Class 1, 2, 3, and 5 required some form of ground improvement due to the presence of soft ground or mixed-face conditions under full hydrostatic head (Parsons, 2016). Ground treatment options considered, but were not limited to, soil-cement mixing, chemical injection, jet grouting, ground freezing, and localized depressurization via deep wells. The tunnel alignment runs under Wilshire Boulevard, one of the most trafficked non-interstate roads in Los Angeles, and the long-term street closures required for depressurization or deep well operations proved to be extremely difficult to obtain, particularly in union with the Project's other traffic closures. A ground improvement program that could operate independent of the tunnel boring machines (TBM) schedule and operation was preferred. Further, as cross passage excavations could be performed as quickly as a matter of weeks or last a duration of months, a "one-and-done" treatment program that could be deployed, installed, and demobilized was required, eliminating the consideration of ground freezing and depressurization at many locations.

Through analysis of ground conditions, the amount and pressure of ground water present, restrictions on schedule, and applicability of installation from the surface compared to treatment from within the tunnel, and not without cost consideration, jet grouting was chosen by STS as the most preferred ground treatment solution. The San Pedro Formation was readily erodible by the jet grout method. Additionally, the jet grouting method allowed the installation of the jet grout block treatment ahead of the schedule of the advancing TBM tunnels, completely removing the schedule constraint of ground treatment from the cross passage excavation schedule once the tunnels were built.

A specialized jet grouting Subcontractor was selected to install the jet grout blocks along the alignment. The Subcontractor designed a jet grout program with 10-foot columns at 8 foot spacings on center, providing complete coverage of the entire cross passage excavation profile. The intent was to create a jet grout block with minimum dimensions of 40 feet wide and centered, at a minimum, along the centerline of the twin running tunnels for complete coverage. The jet grout blocks were designed to extend 10 feet above the crown of the cross passage and approximately 8 feet below the invert of the cross passage profile. The jet grout block design parameters for this project required a minimum compressive strength of 300 psi by U.C.S. testing methods (Parsons, 2016).

The Subcontractor performed a series of test blocks on the project prior to beginning their jet grout program with results that yielded acceptable results to the Engineer of Record (Parsons) in the Ground Classification Type 1 through 3. The testing in San Pedro Formation impacted with tar (Ground Classification 5), however, was not favorable. It must be noted that the jet grouting method was shown to be possible in the tar-impacted sands, but the cost and effort to make the block perform as specified was beyond the limits available to STS. Ultimately, thirteen cross passage locations were chosen to have the jet grout treatment. The remaining four locations in Ground Class 5 would require additional ground improvement efforts to ensure safe excavation.

## CROSS PASSAGE EXCAVATION

Prior to cross passage excavation, probe drilling was completed after completion of TBM mining to verify the ground conditions at each cross passage location. Shut-off valves were mounted to the intrados of the precast concrete tunnel liner and drilled through, retrieving the soil cuttings. The approximate flowrate was measured, and relevant photo and video documentation was taken before closing the valve to document pressure. Typically, probe holes were located at the center of the cross passage profile, as well as two locations to each side near the crown. However, as many of the probing results did not yield enough data to fully consider the condition of the cross passage ground condition, particularly in the jet grouted zones, additional probe drilling was performed with up to nine probes from each tunnel being performed to adequately map the ground. At many locations, the probing results indicated to STS that the jet grouting campaign did not improve the ground conditions to levels that would provide a safe excavation without further ground improvement.

Figure 2. Flowing sands encountered during probe drilling

Figure 3. Typical STS cross passage excavation setup

STS chose to begin cross passage excavation at locations within the Ground Classification 4 to properly train the excavation crews and thoroughly test the equipment and systems required for working in a gassy/extra-hazardous environment. After installation of the necessary propping support cages to allow removal of sections of the precast tunnel liner, excavation began with the top heading. The cross passage design drawings provided by the Engineer of Record (Parsons) and the SEM Engineer's RESS typically provided for a top heading excavation followed by the bench excavation. In the Fernando Formation and properly treated ground, the entire excavation and installation of the initial liner could be completed in roughly three weeks, working continuously for two twelve-hour shifts, four days per week. However, as soon became the challenge of the project, the remaining nineteen cross passages required much more effort and engineering to safely complete.

## Cross Passage Excavation in Incomplete Jet Grout Blocks

After successfully completing several of the Type 4 cross passages, CPB 14 was next but had to be skipped due to concerning probe drilling results. CPB 14 was a cross passage treated with jet grout, however the probing results yielded flowing, untreated sands and running water beyond 20 gallons per minute. While the Subcontractor responsible for the jet grout program and the Engineer of Record (Parsons) reviewed CPB 14 data and developed a remediation program, STS continued the cross passage excavation at CPB 15. Immediately upon removing the pre-cast tunnel liner, the Lakewood formation was encountered. Fine sand pockets intermingled with jet grout columns occupied the crown, with stand-up times in the range of less than one hour. However, when applying shotcrete to these sand pockets, the sand would begin to flow and the face would lose stability. An application of a low air-pressure, eggshell thin layer of shotcrete was required to provide initial stabilization of the ground before adding a layer of wire mesh anchored into the nearby jet grout to provide adhesion for the following layer of shotcrete. Unfortunately, as the excavation proceeded below spring line, the competency of the jet grout columns began to fail, and the columns did not overlap to provide improved ground conditions and effective water cut-of.

The lower half, or bench, of the excavation of CPB 15 yielded flows of sands and water with much higher pressure and volume. Pockets and veins of flowing sands with

**Figure 4. Sands Intermixed with Jet Grout at CPB 15**

**Figure 5. Sheet piles installed to secure running sands at CPB 15**

flow rates beyond 200 gallons per minute were encountered where the jet grout was missing. The largest inflows came from under the running TBM tunnels, boiling into the excavation from below where the flow could not be stopped without damaging the tunnel liner. Small-section sheet piles had to be driven between the cross passage excavation and the TBM tunnel liner utilizing the hammer of the mini excavator. The ground loss and intake of water during this procedure was extremely concerning to the project management team. Once the excavation was secured with a sealing layer of shotcrete, compensation grouting was performed and further attempts at stabilizing the ground began by cementitious and chemical injection.

After failed attempts of grouting with both ultra and micro-fine cementitious grout, a permeation chemical grouting campaign was developed with the jet grout Subcontractor and their chemical injection consultant. A two-part acrylate resin was injected into the cohesionless soil, creating a gelatinous mass that could stem or reduce water

Figure 6. Flowing silty sands at CPB 15 bench excavation

flows. As the pressure required to permeate the silts and sands was increased, the layers of shotcrete would rupture, and the entire cycle would need to be repeated until enough of a grout curtain was created before continuing. To prevent the shotcrete from rupturing, the invert was braced with breasting boards and shoring prior to applying additional layers of shotcrete. Drainage pipes were inserted to relieve water pressure away from injection zones but caused issues as the chemical could readily find this path of least resistance and would not penetrate the ground. Three inch diameter well points were strategically installed to use negative pressure to draw the chemical grout into desired areas of the excavation.

After grouting, additional verification probe holes were drilled before continuing excavation. If required, grouting campaigns would continue, often for several shifts before acceptable results were achieved. After grouting, excavation would have to continue using hand tools as the vibrations from larger equipment would fragment the grout curtain and the work would have to be repeated. Along with the permeation grouting, the RESS was updated to account for shorter round lengths, smaller pockets, inclusion of drainage mats, forepoling ahead of the excavation, additional probing, and ultimately, more layers of shotcrete.

In reflection, the experience gained at CPB 15 provided valuable insight for the future RESS planning, including permeation grouting programs, as well as reinforcing the importance of contingency and toolbox items being immediately accessible during excavation. CPB 15 was jet grouted and no flowing materials were encountered during probing. However, the bench excavation of this cross passage alone took nearly two months to complete. STS made the decision to re-probe all the previous cross passages and all future cross passages a minimum of nine times to gain further information before opening the excavation. Jet grouted locations were no longer considered to be safe enough for excavation without additional grouting efforts and the jet grout could not be relied on to have fully overlapping columns.

## Cross Passage Permeation Grouting Prior to Break-Out and Excavation

With the experience and lessons learned from CPB 15, a permeation grouting program from within the tunnels was deployed by the jet grouting Subcontractor at all jet grouted locations that did not meet their Specified Performance Criteria. Ultimately any probing that indicated non-overlapping columns of jet grout and produced flowing sands or water beyond four gallons per minute required some additional grouting effort. The Subcontractor developed a grouting program that consisted of a chemical grout curtain that was then reinforced with steel spiling, installed from both running TBM Tunnels to fully encapsulate the cross passage profile. Within the cross passage profile, a three foot grid in the interior was drilled and grouted to stabilize soils and cutoff any paths of flowing water.

The new ground improvement operation from within the tunnel proved to be extremely challenging for STS' schedule and was a main reason why this method was not considered initially when ground improvement options were considered. During the drilling and grouting operation, the equipment blocked the tunnel and strategic planning was required to ensure all other daily activities could still take place. Additionally, to supply the permeation grouting operation while supporting base-scope work led to large schedule issues and major constraints. Thousands of gallons of grout were injected into each of the cross passage locations requiring treatment and the campaigns could take nearly eight weeks to complete.

After both jet grouting and the additional permeation grouting was completed, STS could not consider a cross passage safe for rapid excavation. In locations where no flowing material was encountered in probe drilling, additional steps were taken in case problematic areas were missed during the investigation. Smaller pocket sizes were usually the first revision to the RESS. Instead of the two pockets seen in the Fernando, four or more pockets was typical. Two pockets would be excavated above the wedge, and a pocket on either side of the wedge for each leg of the lattice girder would complete the initial round. The two pockets above the wedge could be reduced in half, allowing the top sections to be handled and adequately supported first before moving on to middle and lower levels. These extra steps made the excavation take much longer than previously planned but allowed for safe and consistent progress.

## Cross Passage Excavation in Tar-Impacted Soils

In contrast to the permeable sands and flowing water of the cross passages in Reach 1, the introduction of tar-impacted soils in both Reach 2 and Reach 3 added another difficult challenge to overcome. The tar created cohesive sands that would slowly ravel but would quickly succumb to air-slack and transform into fast raveling, eventually collapsing in a similar manner as a pile of hot-mixed asphalt, even when simply being disturbed by the movement of the mini-excavator. The tar also limited permeability of the sands that all cement and chemical grout treatments proved ineffective at strengthening or binding the soils. Efforts were made to both cool and freeze the ground with some moderate success, but the application to the entire excavation profile was not feasible with the level of effort required. The tar-impacted soils had the additional quality of restricting the penetration of the shotcrete into the substrate, causing layers of shotcrete to remain susceptible to delamination, collapsing off the crown, face, and walls before a round of excavation could be completed.

The SEM Engineers modified the RESS to provide additional lateral and overhead support to the ground prior to excavation and forepoling prior to advance. A roof canopy of R51 hollow-steel bars and a second, offset row of R32 bars at one-foot centers were installed above the excavation profile from both tunnels and grouted with micro-fine cement. A curtain of bars was also installed along each side of the cross passage profile walls. There was concern that because the tar prevented permeation of the grout to form a full canopy, the bar would not hold back the running sands. This concern proved accurate; however, the spiling bars not only created a shorter arch span for the soils and increased stand-up time, but they offered additional anchoring connections for shoring and shielding plates during excavation which proved invaluable in preventing ground collapse.

A further addition to the RESS toolbox was to use face dowels through the face of excavation. Hollow fiberglass IBO rods were drilled from one running tunnel through

Figure 7. Ground support spiling curtain at a tar-impacted cross passage

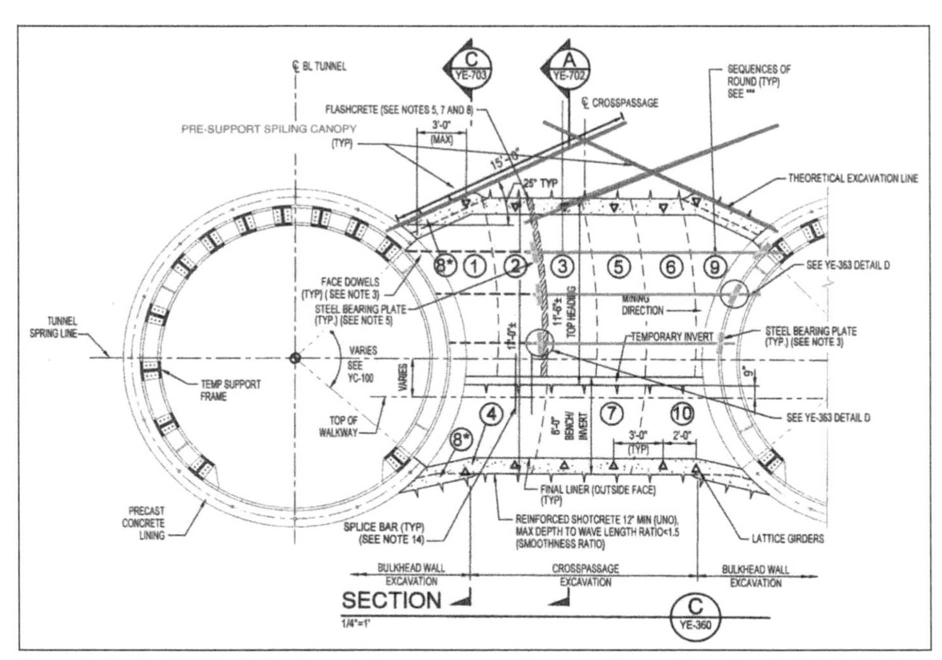


Figure 8. Typical RESS of a tar-impacted cross passage. Canopy spiles and face dowels highlighted for clarity (Parsons)

Figure 9. PB 17 small pocket excavation and face dowels with channel straps

Figure 10. Pre-advance spiling installation at CPB 17

the cross passage and into the opposite tunnel on three foot centers. The rods were anchored with steel plates and nuts in the tunnel opposite of advancing side, while steel channel, plates, and nuts were used to provide compression on the excavation side. As the excavation advanced in small pockets, the steel channels were removed, the excavation would advance, and then the channels were re-installed and tightened, compressing the face. The rods could then be cut back to the face of excavation to provide room for the next pocket excavation.

The constant movement and squeezing nature of the tar-impacted soil caused the shotcrete to crack and rupture, requiring additional bracing to the girders and shotcrete prior to advance. The extra weight of the top heading shotcrete and girders also caused the pore space within the tar-impacted soil below the excavation to collapse and the entire cross passages would sink. To anchor the cross passage in position, lateral spiling and shotcrete "elephant feet" were installed outside the limits of the excavation.

RESS revisions eventually included shortening the lattice girder spacings from three feet to eighteen inches and modifying the heading from top and bench headings into four headings: top, top-middle, middle-bench, and bench. Smaller hand excavated pockets, eighteen inches by eighteen inches, were required to eliminate the vibrations caused by the heavy mini excavator. Pockets could not be opened adjacent to one another or the vibrations from impacting the previously excavated round would damage the shotcrete and material would ravel through while the shotcrete cured. This methodical alternating of small excavation pockets, although time-consuming, eventually allowed the excavation of each face to proceed at a predictable, safe pace. A successful two-foot round, including excavation, top-heading girder installation, spile installation, and full shotcrete application would be completed in three shifts. For comparison, the entire top heading of a cross passage in Fernando soils could be completed in less than eight shifts, but a similar sized excavation in the tar-impacted sands required more than twenty shifts.

## CONCLUSION

Section 1 of the Westside Subway Extension had several challenging ground conditions to overcome to ensure a successful project. While preparing for flowing sands, gas, and tar-impacted soils posed a significant challenge, the utilization of the Sequential Excavation Method for the cross passage excavation allowed the tunneling crews to quickly adapt and respond to unknown and ever-changing scenarios. The installation of a ground improvement system could not be guaranteed in such challenging conditions and every possible solution to potential problems had to be considered and made available to safely complete the project. While pre-excavation planning and ground improvement methods were utilized to limit risk, it was seen on this project that additional tactics were required once the ground was exposed.

## REFERENCES

CalGEM. (2020). *California Geologic Energy Management Division (CalGEM)*. Retrieved from California Geologic Energy Management Division (CalGEM). Well Finder. Online geographic information system. https://maps.conservation.ca.gov/doggr/wellfinder/#/.

California Code of Regulations. (2022, September 23). 8 CA ADC T. 8, D. 1, Ch. 4, Subch. 20 Tunnel Safety Orders. *Division of Industrial Safety, Subchapter 20 Tunnel Safety Orders*. California, United States: Barclays California Code of Regulations.

Parsons. (2016, June). Cross Passage Design Report. *LACMTA DU5 Tunnel Cross Passages AFC Submittal*. Pasadena, CA, USA: Parsons.

Parsons. (2016, 02 29). Jet Grouting Ground Improvement Design Calculations. *Westside Subway Extensions Project, Extension 1 Design Calculation Report*. Pasadena, CA, USA: Parsons.

Parsons. (2019, 03 14). Tar-Impacted Soil Cross Passage Analysis—Tool Box Items. *AFC Submittal—Supplemental Information*. Pasadena, CA, USA: Parsons.

Parsons Brinkerhoff. (2014, November 3rd). Geotechnical Baseline Report. *LACMTA C1045 Westside Subway Extension Project, Section 1*. Los Angeles, CA, USA: Parsens Brinkerhoff.

*Purple (D-Line) Extension Transit Project*. (2022, 12). Retrieved from Metro.net: https://www.metro.net/projects/westside/.

# Kramer Tunnel—Construction of the Ventilation Shaft Project in Germany's Southern Alps

**Lukas Walder** ▪ BeMo Tunnelling GmbH
**Richard Gradnik** ▪ BeMo Tunnelling GmbH
**Roland Arnold** ▪ BeMo Tunnelling GmbH
**Raphael Zuber** ▪ Staatliches Bauamt Weilheim

## ABSTRACT

This paper presents the construction of the ventilation shaft for the single-tube, double-lane Kramer tunnel. The project is located in the Federal State of Bavaria, Germany. This road tunnel, which has been under construction since December 2019, is the key component of the western bypass of the town of Garmisch-Partenkirchen. Once completed, it will relieve the town's central district from enormous volumes of traffic.

The contractor is the joint venture named ARGE Kramertunnel consisting of BeMo Tunnelling and Subterra. The whole project will be completed and handed over to the project owner, Staatliches Bauamt Weilheim, in 2023. The opening of the entire bypass section including the tunnel to traffic is planned for 2024. The tunnel, with a total length of 3,609 m (2.24 miles) and an estimated cost of EUR 178 million incl. VAT (US-Dollar 201 million), is designed as a single-tube, double-lane structure with one traffic lane for each direction of travel. The safety issue that exists for single-tube tunnels is solved by a parallel escape tunnel passable for vehicles. For tunnel ventilation, an intermediate shaft has been designed with a depth of approximately 90 m (295 feet) and an inner diameter of 4.5 m (14.8 feet). All tunnels and ducts are excavated according to the principles of the New Austrian Tunnelling Method (NATM).

The geological and hydrogeological conditions, the change in the profile (from round to square geometry), and also the shaft being located at a high elevation presented a demanding technical and logistical challenge. The site arrangement with the pit bank was located in a cut, the slopes of which had to be stabilized with shotcrete and soil nails. This excavation was carried out from the fall of 2020 to the beginning of 2021 because construction operations had to be completed outside the nesting period of protected birds.

## PROJECT BACKGROUND

A variety of different tunnel projects lie within the Loisachtal (Loisach river valley) due to the two nearby federal highways B2 and B23 meeting there. These include the Farchant and Oberau tunnels, which are already open for traffic, the Kramer tunnel, which is still being constructed, and the Wank and Auberg tunnels.

The federal motorway A95 (Munich–Garmisch-Partenkirchen) connects the cities of Munich and Eschenlohe and was built between 1965 and 1982. To the south of Eschenlohe, the B2 serves as a continuation of the A95, heading south through Oberau and Farchant. At the intersection of the B2 and B23 lies the town of Garmisch-Partenkirchen, Germany's largest winter sports resort, which simultaneously hosts the main holiday region at the foot of the Wetterstein Mountains and the most vital traffic connection of the area.

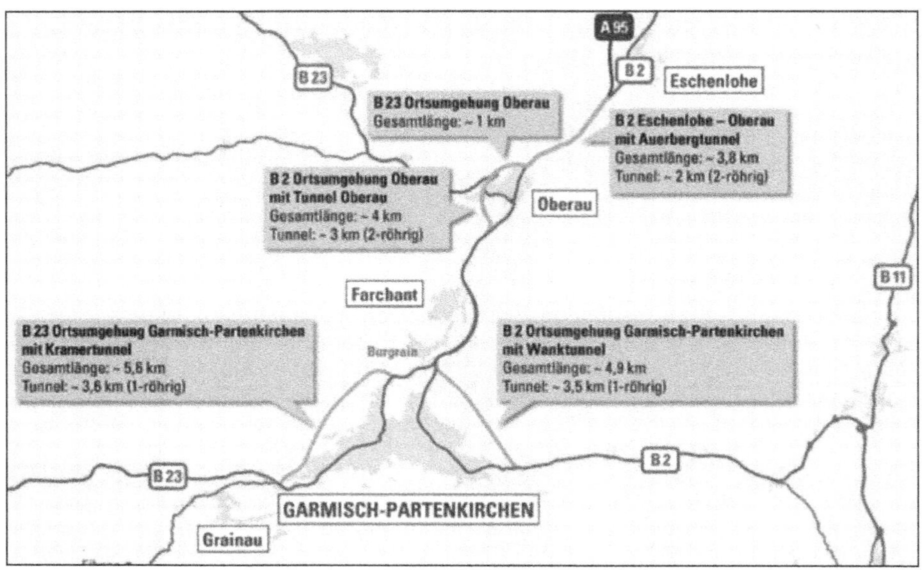


**Figure 1. Loisachtal tunnel chain[2]**

The southbound route continues along the B2 and leads through Mittenwald to Innsbruck, Austria and Italy, and is mainly used by southward holiday commuters. From Garmisch-Partenkirchen, the B23 leads west to western Austria and Switzerland, passing through Griesen. Because of the high proportion of leisure-related commuters and transportation, traffic figures for both highways increase significantly around holidays and towards weekends.

The relocation of the B23 through the Kramer tunnel and the relocation of the B2 through the Wank tunnel is expected to make the traffic stream much more efficient and significantly reduce the impact of traffic within Garmisch-Partenkirchen.

## KRAMER TUNNEL

The western bypass of Garmisch-Partenkirchen, which pierces the Kramerspitz (Kramer mountain) and gives the Kramer tunnel its name, begins south of the Farchant-Burgrain bypass. The route for the project begins in front of the planned northern tunnel portal along the route of the existing B23 highway. From the southern portal, the route continues to the southwest and reconnects to the B23 highway with two bridges over the Loisach river. The project has a total length of 5.56 km (3.5 miles).

## TUNNEL DESIGN AND CROSS-SECTION OF THE SHAFT

The driving tube of the Kramer tunnel is designed as a single-tube, two-way traffic tunnel with a total length of 3,609 m (2.24 miles). Breakdown bays are arranged at intervals of up to 600 m (2,000 feet). For this purpose, the clearance profile for the tunnel tube is widened on one or both sides by 2.5 m (8.2 feet). Cross passages are routed from the tunnel to the accompanying rescue tunnel at regular intervals. The single tube driving tunnel has been designed as a standard cross section according to RABT as RQ 11 t. This results in a total width of paved road of 7.5 m (25 feet).

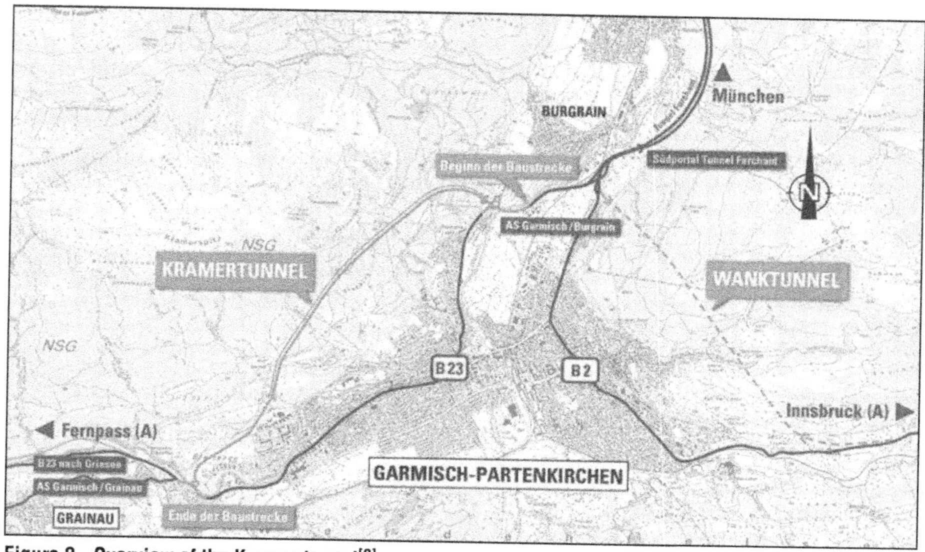


Figure 2. Overview of the Kramer tunnel[2]

In the Kramer tunnel portal areas, ventilation will be provided by axial fans, while closer to the middle, exhaust emissions will be discharged into the ventilation duct formed by the intermediate deck and, subsequently, routed into the above-mentioned cavern and ventilation shaft. The cross-section through the ventilation shaft for the initial (upper-most) 76.4 m (250 feet) is circular with a diameter of 5.50 m (18 feet) in the primary lining. At a depth of 76.4 m (250 feet), the circular profile transitions to a rectangular profile with dimensions of 5.50×5.40 m (18×17.7 feet). The transition area is 4.0 m-long (13 feet) and the rectangular profile itself continues for an additional 3 m (10 feet) down to the depth of 83.40 m (274 feet), where the excavation terminates at the level of the top heading of the future ventilation chamber (see Figure 3).

In the final state, next to the vent stack is located a service building with a direct access to the ventilation shaft. In this building is also placed the cage for inspection (see Figure 4).

## GEOLOGICAL CONDITIONS

The ventilation shaft is located in the northern tunnel section, approximately at km 1 + 615 (1 mile) and has a depth of approximately 90 m (295 feet). The subsurface conditions encountered in the course of the excavation can be summarized in three zones:

1. *Debris flow sediments:* Debris flow sediments can be found under a thin layer of holocene soil down to a depth of around 30 m (98 feet). These sediments consist of gray-brown, silty to very silty, sandy, slightly stony gravel with predominantly angular carbonate components. Occasionally blocks are embedded, which can reach a size of up to approximately 1 m$^3$. Thin, fine-grained layers (silt, sandy) and isolated soil horizons also appear within this strata.
2. *Alternating series of debris flow sediments and moraine deposits:* Between approximately 30–50 m (98–164 feet) depth, a "mixed face" sequence of debris flow sediments and moraine deposits is encountered. Within this section of the shaft, the above described debris flow sediments are interbedded

with glacial moraine deposits. The moraine deposits consist mainly of clays/silts with partly sandy or gravelly layers and also partly with stones.

3. *Triassic limestones and marlstones of the so-called Kössen Formation* are encountered from around 50 m (98 feet) to the final depth of the shaft. These marine sediments were deposited about 205 million years ago in the shallow shelf of the so-called Tethys ocean basin. They consist of predominantly thin to medium bedded, gray to dark gray limestones and gray- to black-colored

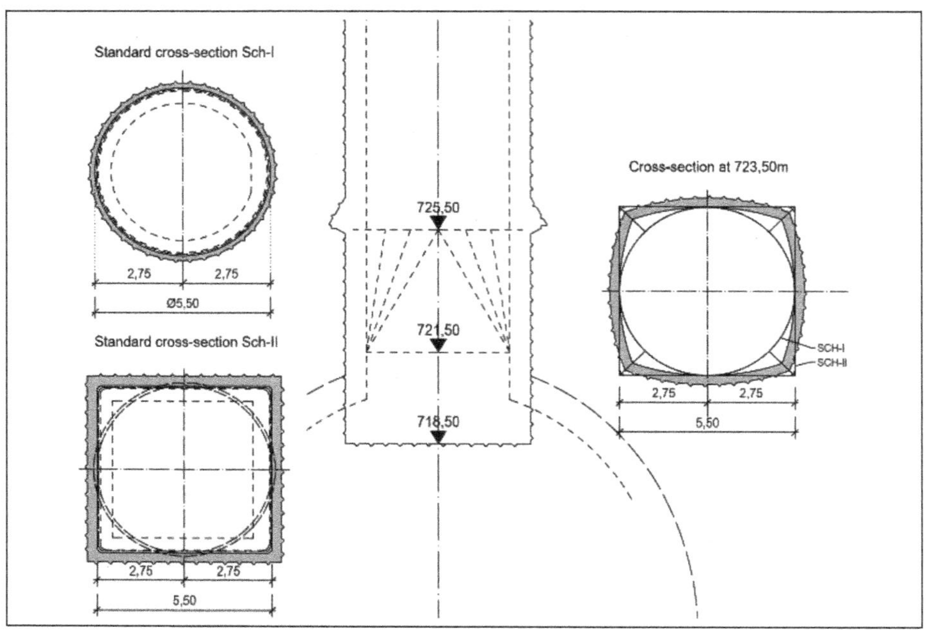


**Figure 3. Geometric changes in cross section of the shaft[3]**

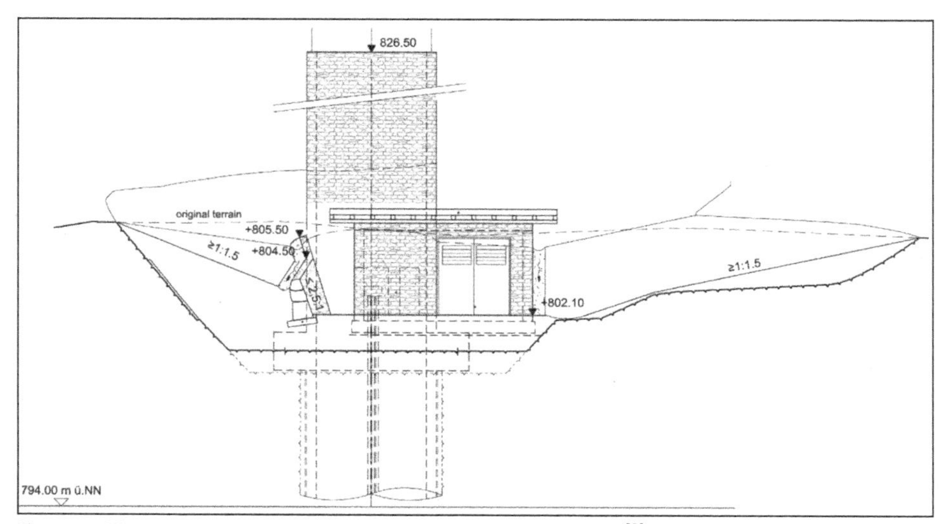


**Figure 4. View of the vent stack with service building in its final state[3]**

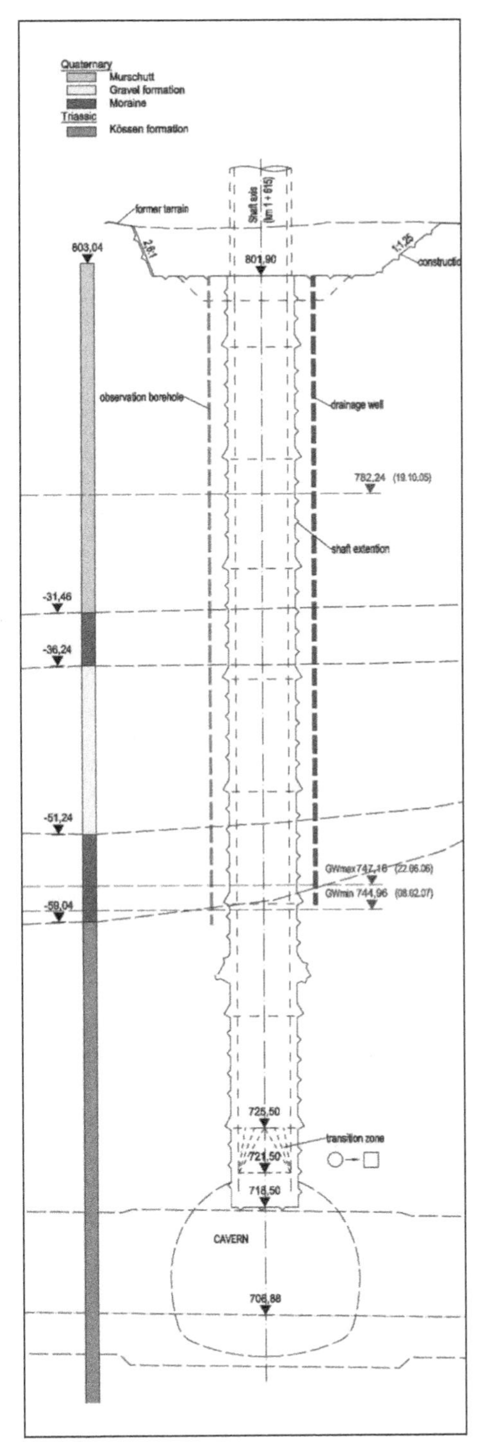


Figure 5. Shaft longitudinal section[3]

marlstones with moderately high to high compressive strength. The proportion of limestone and marlstone varies, so that the sequence is partly dominated by limestone and partly dominated by marlstone. The sequence of layers is folded internally in some areas and locally, slates are interbedded. The layer of slate show non-durable behavior and tend to soften and decay when exposed to water or the atmosphere.

The groundwater level is just above the level of the limestones and marlstones of the Kössen formation, at around 743–750 m (2,438–2,460 feet) above sea level.

## PREPARATORY WORKS BEFORE STARTING SHAFT EXCAVATION

### Access Road

The first phase of construction consisted of establishing the access road and to establish the site in the location of the future shaft. The roughly 600 meter-long (1,968 feet) access road branched off from a public road (Pflegersee Straße), which itself posed a challenge in terms of the logistics of muck and transport of construction materials. The approximately 750 m-long (2,460 feet) section of Pflegersee Straße before the populated area of the town is sloped at an average longitudinal gradient of about 10 percent (Figure 6).

### Construction Pit

The working area of the cut for the construction site arrangement itself has plan view dimensions of about 35×25 m (115×82 feet). It had to provide space for all material storage, space for intermediate stockpile of muck and for the water management system with a neutralization station. To achieve a relatively flat work surface, it was necessary to carry out extensive excavation in the northwestern area of the site arrangement. The adjacent 2.60:1 gradient slope was stabilized at 1.5 m (5 feet) height stages with 4.0 m-long (13 feet) soil

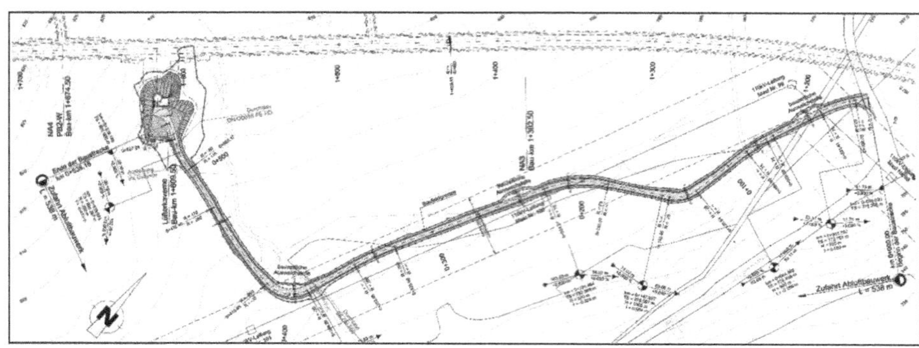

Figure 6. Map of the access road[5]

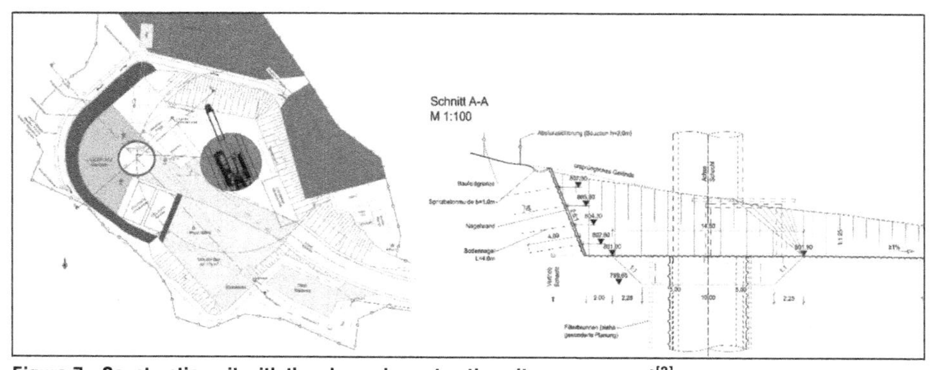


Figure 7. Construction pit with the planned construction site arrangement[3]

nails complemented by one layer of KARI welded mesh and a 15 cm-thick (6 inches) layer of shotcrete. The height of the wall stabilized with soil nails amounted to 7.0 m (23 feet). In the locations where it was not necessary to overcome great differences in the heights, the excavation slopes were stabilized at 1:1.25 gradient (Figure 7).

## Pre-Dewatering

The groundwater level in the solid rock fluctuates between 743–750 m (2,43–2,460 feet) above sea level. While drilling the borehole for the groundwater measuring point GAP 06/05 in the year 2005, another groundwater level at 782.24 m (2,566 feet) above sea level in the debris flow sediments was detected. Therefore, for pre-dewatering the layer of the debris flow sediments and the moraine, three drainage wells with a total delivery rate of 45 l/s (12 gallons/s) were considered in the tender design (Figure 8).

The three dewatering boreholes B1–B3 were designed to pass the entire layer of the debris flow sediments until they encountered the native rock of the Kössen formation at a distance of at least 4.5 m (15 feet) from the centre of the shaft at a regular radial spacing of 120°. A fourth borehole P1 was designed as a supplement to be used for check monitoring of the water table and to assess the success of its lowering.

The drilling works for the three dewatering boreholes started at 801.90 m (2,631 feet) above sea level. The initial borehole roughly 30 m (98 feet) in the debris flow sediments resulted in the grab drilling process with a diameter of 622 mm (24.5 inches).

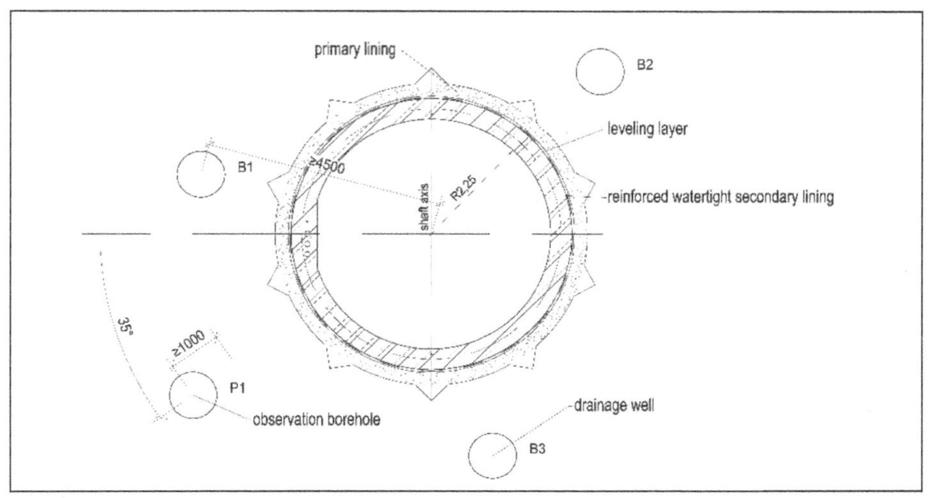


**Figure 8. Shaft cross-section with the locations of drainage and checking boreholes**

Then the process was changed to the rotary drilling method with a diameter of 500 mm (19.7 inches) from approximately 30 m to 55 m (98 to 180 feet) depth. The boreholes were cased from 13 m (43 feet) to approximately 50 m (164 feet) with DN 300 (11.8 inches) filter tubes packed with filtration gravel (Figure 9).

# SYSTEM BEHAVIOR (TUNNELLING CLASS)

## General

The system behavior for the shaft excavation and selection of the presupport was defined in the Geotechnical Report.

It is expected that the required support consists of a competent layer of reinforced shotcrete. Shaft abutments are located every 10 m (33 feet). As presupport in the debris flow sediments, it is necessary to use self-drilling grouted spiles.

## Tender Design—Execution Design

All the above-mentioned mandatory means of support have led to the tender design. Three excavation support classes were designed for the shaft excavation. The cross-sectional area (see Figure 10) is equal to about 27 $m^2$ (291 square feet).

Class SCH 6 with an excavation length of 1 m (3.3 feet) is used for the zone with debris flow sediments. The proposed primary lining is 200 mm-thick (8 inches) and reinforced with one layer of KARI Q188 welded mesh and a lattice frame. In addition, radial anchoring with four pieces of 2.5 m-long (8.2 feet) IBO R32-210 rock bolts is used for this class. As presupport in this class, self-drilling grouted spiles are designed for the stabilization of the excavation before application of shotcrete. In this class mechanical excavation is provided.

Class SCH 4 with a maximum excavation length of 1.5 m (4.9 feet) is used as a transition class for a more compact environment. A 150 mm-thick (6 inches) lining reinforced only with KARI Q188 welded mesh is used for this class. As radial anchoring

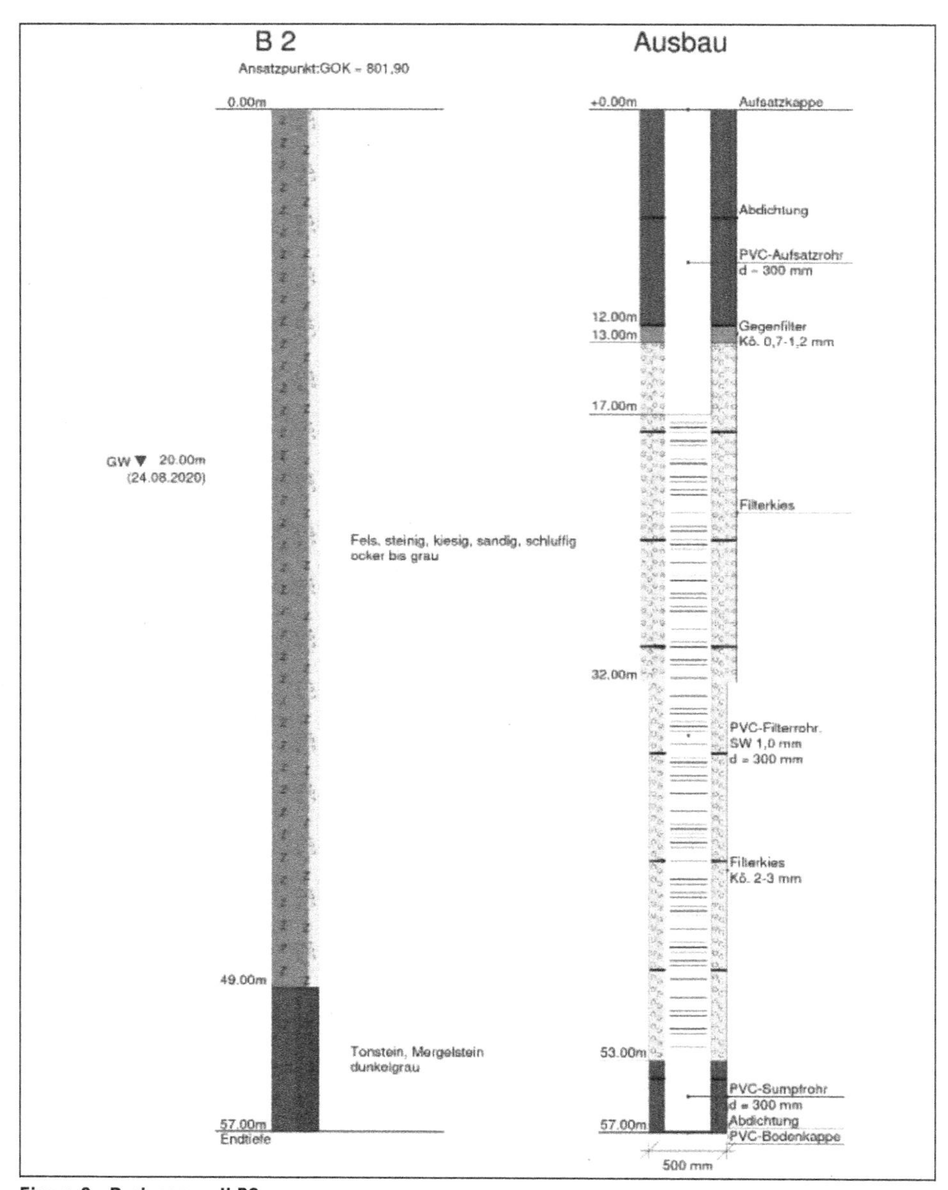


**Figure 9. Drainage well B2**

in this class, four SN rock bolts at 2 m-long (6.6 feet) are used. In this class combined rock breaking techniques (mechanical/blasting) are provided.

Class SCH 3 with a maximum excavation length of 2.2 m (7.2 feet) is designed as the optimal class for the best quality rock environment. The shotcrete lining is 100 m-thick (4 inches) and is also reinforced with KARI welded mesh Q188. Four pieces of 2 m-long

(6.6 feet) hydraulically expanded rock bolts are used for radial anchoring in this class. Rock breaking in this class is done by blasting.

During the execution of the design phases, the tender design was placed under review and prepared for execution. The assumptions of the tender design were confirmed and were included in the execution design (see Figure 11 and Figure 12).

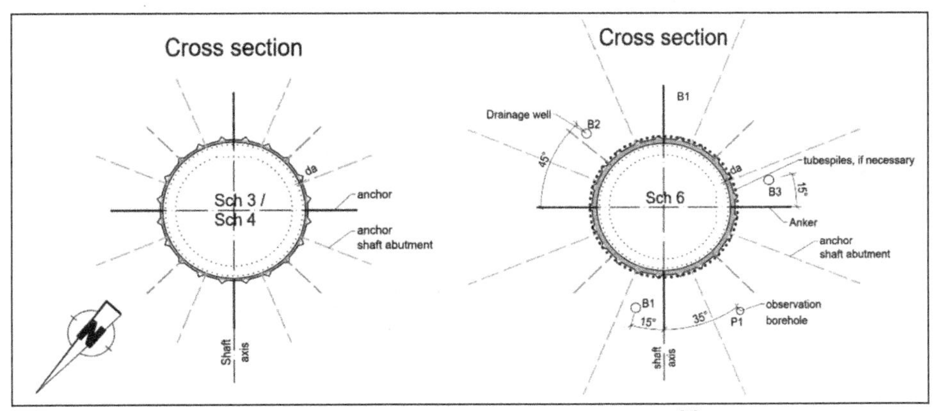


**Figure 10. Cross sections of excavation support classes SCH 3/SCH 4/SCH 6[3]**

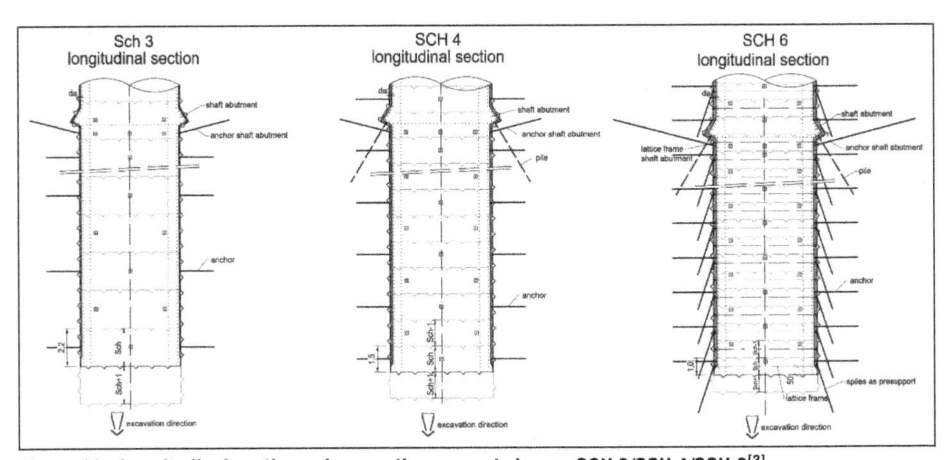


**Figure 11. Longitudinal sections of excavation support classes SCH 3/SCH 4/SCH 6[3]**

| Excavation support class | | | | Sch 3 | | | Sch 4 | | | Sch 6 | |
|---|---|---|---|---|---|---|---|---|---|---|---|
| excavation | | procedure | blasting | | | blasting/mechanically | | | mechanically | | |
| | | advance length | | L | 2,2 m | | L | 1,5 m | | L | 1,0 m |
| | | tolerance | | td+ü | 3 cm | | td+ü | 5 cm | ∅ | td+ü | 5 cm |
| | | | | ta | 10 cm | | ta | 10 cm | | ta | 10 cm |
| support | reveal | advance support | | - | - | | - | - | tubespiles 51mm, injected a≤0,25m; e=2,0m; if necessaey | l | 3,0 m |
| | | lining | | da | 10 cm | | da | 15 cm | | da | 20 cm |
| | | reinforcement | mountain side | Q | 188A | mountain side | Q | 188A | mountain side | Q | 188A |
| | | | air side | - | - | air side | - | - | air side | - | - |
| | | anchor | RR-Anker, 100 kN | l | 2,0 m | SN-anchor, 100 kN | l | 2,0 m | SB-anchor, 100 kN | l | 2,5 m |
| | | lattice frame | | - | - | | - | - | Typ S | e | 1,0 m |

**Figure 12. Excavation support classes for the shaft**

# EXPERIENCE GAINED FROM EXECUTION

## General

Because of the nesting season for local protected birds, the works on the ventilation shaft are restricted to the time from the beginning of July to the end of January. The work was started in the summer of 2020 by building the construction site arrangement in a hard-to-reach location in the middle of the forests stretching on the Kramer massif at an elevation of 800 m (2,625 feet) above sea level. The shaft excavation itself started in fall of 2020 and was successfully completed with the end of January 2021. The preparatory works before starting the shaft excavation were described above.

## Equipment for Excavation

In the preliminary phase, different methods were investigated for material transport in the shaft and lifting of muck. The first version investigated was according to TAS[4] (see Figure 13). This version contains a head tower with a cover plate for the shaft and two winches. One for the transport of persons and the other for loads.

Further, the local mining authority was consulted, who pointed out that for this kind of shaft with a depth of only 80 m (262 feet), the requirements of TAS do not apply. This opened up other possibilities. For the execution phase, in addition to the abovementioned system according to TAS, the following three alternatives were investigated:

- Tower crane
- Gantry crane
- Telescopic crawler crane.

The problem with the tower crane variant was the hook clearance. To swing over the trees at this place, a hook clearance of 40 m (131 feet) is necessary. Combined with the shaft depth of 80 m (262 feet), the required rope length is 120 m (394 feet). In windy conditions at a rope length of up to 120 m (394 feet) the use of this crane version is too difficult.

For the gantry crane variant, the space available was too small. Also, it is very dangerous when people are close to the rails of the crane.

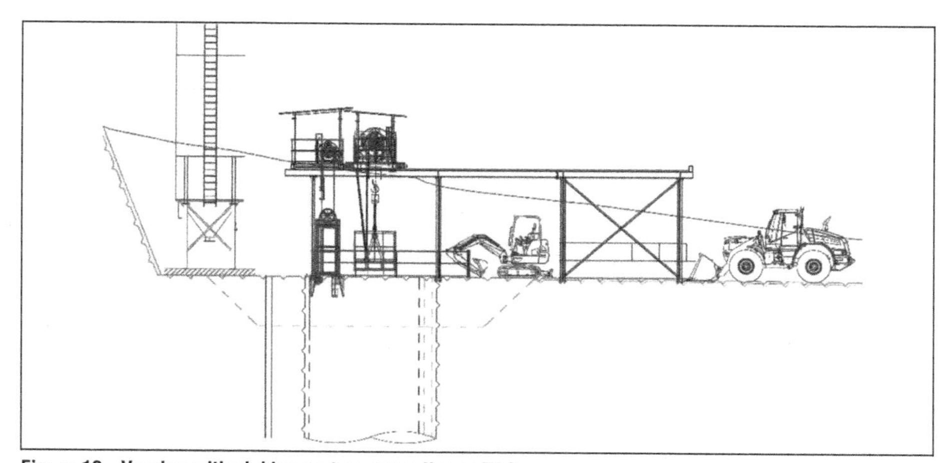

Figure 13. Version with sinking system according to TAS

Figure 14. A view of the site arrangement and the pit bank with the Sennebogen 6133e

Figure 15. Protective shield/roof in the shaft

The Sennebogen 6133e tracked telescopic crane (lifting capacity of 130 t), which is equipped with two lifting devices and two independent drives (electrical and diesel), became the reason this version with sinking system according to TAS could be abandoned. In the end the telescopic crawler crane was chosen because it was the more economic variant, and the time of delivery was convenient.

As mentioned above the transportation in the shaft and lifting of muck was carried out using a Sennebogen 6133e tracked telescopic crane. Transport of persons was ensured by a hanging cage and muck was transported by a suspended hoisting skip with a volume about 0.5 $m^3$ (17.6 cubic feet). For mechanical breaking and drilling, the excavator Takeuchi TB230 was used. A blowing ventilation system was installed.

For protection of the persons working at the shaft bottom, a protective shield (roof) was used to protect against falling objects during transportation of muck or material (see Figure 15).

## Start-Up Process of Excavation

The shaft head was constructed as a reinforced concrete structure (see Figure 16), which set the geometry of the work and provided, among other things, fall protection. The shotcrete primary lining directly linked to this structure with a rebend connection (see Figure 17). Geometric surveying of the excavation round was conducted by means of the Leica/Navigator system provided by the company of Leica/Amberg. Four lasers were installed at the pit bank for checking purposes and to continuously direct the excavation.

Figure 16. Start-up process

## Presupport with Steel Sheet Piles

Due to the experience of the drilling works for the three dewatering boreholes and the work near the surface, it was decided together with the client to execute the first meters of presupport with sheet piles. The provided solution of presupport with spiles was ultimately abandoned. Instead sheet piling using UNION steel piles was executed before each excavation round of the class SCH 6. In Figure 18 the process of sheet pile driving is shown, and in Figure 19 the sheet piles before and after excavation are shown.

Pumps in the three dewatering boreholes adjacent to the shaft were mostly sufficient for pumping ground water. During the shaft excavation, water was pumped from the shaft itself only occasionally.

## Shotcrete for Primary Lining

C25/30 J3 curve shotcrete was used for the primary lining. Damp concrete was mixed in a mobile batch plant established at the northern portal of the Kramer tunnel. It was subsequently transported to the pit bank and was deposited in a tipping storage box. From this storage, the mixture was distributed to ALIVA 262 spraying machine (see

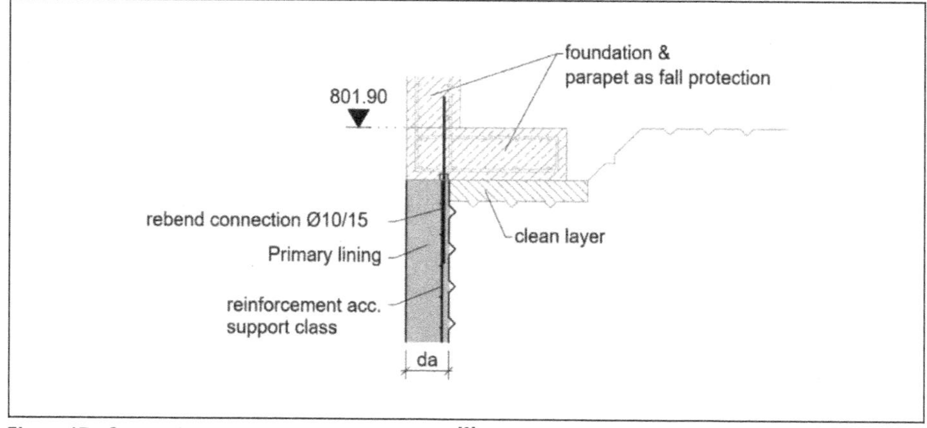


Figure 17. Connection primary lining at shaft head [3]

**Figure 18. Sheet pile driving**

**Figure 19. Sheet piles before and after excavation**

Figure 20). The mixture was applied manually (see Figure 21). Water and accelerator were added to the basic damp mixture in the jet. Distribution of accelerator was controlled automatically by means of a distribution unit installed on the surface. In cold periods, hot water was supplied to the jet to heat the concrete mixture.

The problem of dust emission during application of sprayed concrete was solved by a flexible blowing duct, which was pulled nearly to the bottom during the shotcrete application and significantly improved the space ventilation. The crew was additionally equipped with special dust masks when spraying concrete.

## Opening of the Shaft from the Cavern

After finishing the shaft sinking at the end of January 2021, the three pumps in the boreholes where deactivated. The water level in the shaft was allowed to rise to the groundwater table. Before connecting the ventilation cavern to the shaft, it was necessary to pump out the shaft completely. The connection was completed in November 2021 (see Figure 22).

Figure 20. Distribution of concrete mixture to the pump

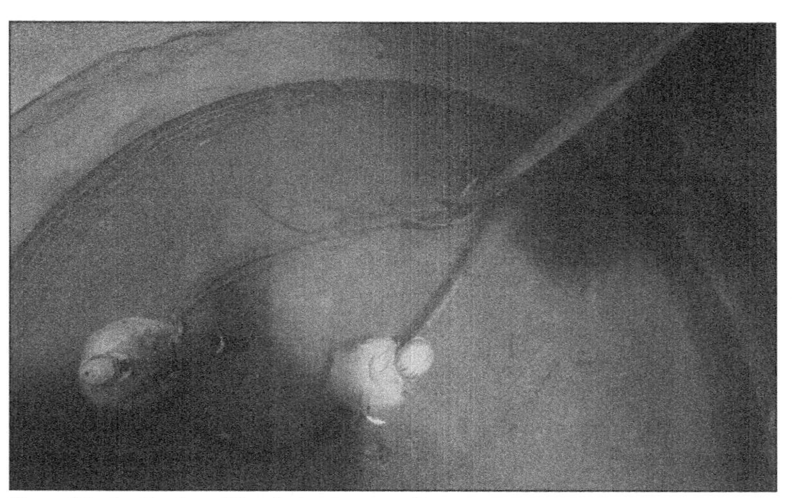

Figure 21. Application of the shotcrete manually

## FINAL LINING IN THE SHAFT

According to the principles of the NATM, which generally state that the excavation method should be readily adaptable to the project's conditions as the excavation proceeds, the primary support shall be selected in accordance with the observed rock mass behavior. Similarly, the secondary and final lining should be amended based on the conditions of the surrounding ground. For the shaft the final lining is designed as a watertight reinforced concrete with a thickness of 400 mm (16 inches). The lining will be concreted in seven pieces of 10.0 m (33 feet) long blocks and one 5.05 m (16.5 feet) long block, which links the guide-wall on the surface. The transition to the rectangular cross-section with the 4.40×4.50 m (14.4×14.8 feet) net internal space follows in the lower part. The execution of the final lining is scheduled for summer 2022 (see Figure 23).

Figure 22. Opening of the shaft from the cavern

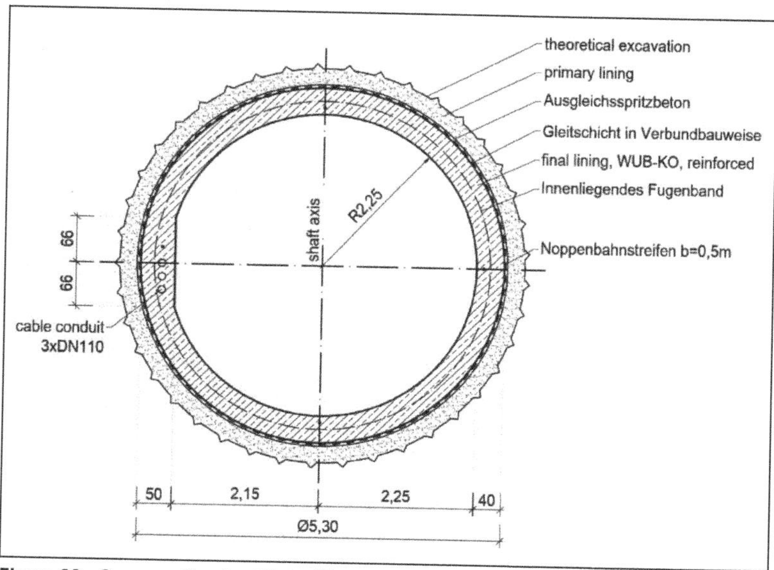


Figure 23. Cross section final lining[3]

## CONCLUSION

The Kramer tunnel project is still in progress, and it remains a key project for the people in Garmisch-Partenkirchen for relief of traffic. The excavation of the shaft ventilation shaft was a challenge in regards to safety measures and logistics.

The application of the modern crane equipped with a camera system allowed for fast and safe transport of material, muck, and persons. From the beginning, it was clear that mucking would be the bottleneck of the operation. The limited space at the shaft bottom allowed only using a small muck box. Despite this limitation, the excavation was successfully carried out quickly and safely. Critical to improving the safety of the persons working at the shaft bottom was the installation of the protective shield (roof) to protect against falling objects during transportation of muck or material.

Another point that improved the safety of the work was the use of the sheet piles in class SCH 6. The encountered rock environment corresponded largely to the prognosis. From the assumed excavation classes only the classes SCH 4 and SCH 6 were used. In the class SCH 4, the maximum length of excavation round achieved amounted to 1.3 m (4.3 feet).

The decision to use a shotcrete mixture applied by hand, which was transported directly from the batching plant, proved to be clearly more economical and logistically more practical.

The realization of the shaft sinking outside of the period of the nesting of protected birds was successfully adhered to thanks largely to the timely and thorough planning of all technical and working procedures, and the selection of an appropriate set of machines and materials to be used. The deployment of highly qualified work crews and technicians in continuous operation was also crucial.

**Figure 24. A view from the shaft bottom during excavation**

## REFERENCES

Patzak J., Schramm T., Josefik D., Kramer Tunnel B23 new Garmisch-Partenkirchen, Magazine of the czech tunnelling association and slovak tunnelling association, Tunel 2/2020.

Staatliches Bauamt Weilheim, Kramertunnel online https://kramer-tunnel.de/.

Technisches Büro Bemo Tunnelling, Rauter G., Steiner S., Ausführungsplanung Schacht 2020.

Bezirksregierung Arnsberg Abteilung Bergbau und Energie in NRW, Technische Anforderungen an Schacht- und Schrägförderanlagen (TAS), Stand: Dezember 2005.

Staatliches Bauamt Weilheim, Ausführungsplanung Zufahrt Abluftbauwerk 2019.

Gradnik R., Zeindl M., Arnold R., Tunnelling the Alps, RETC 2021 Las Vegas.

Kubek J., Josefik D., Kramer Tunnel Ventilation Shaft, Magazine of the czech tunnelling association and slovak tunnelling association, Tunel 2/2021.

# Pipe Canopy Presupport with SEM Tunneling Under Active Roadway and Rail Tracks

**Ryan O'Connell** ▪ Kilduff Underground Engineering
**Todd Kilduff** ▪ Kilduff Underground Engineering

## ABSTRACT

Pipe canopies continue to gain momentum as a reliable methodology for construction of excavations below critical facilities with minimal surface settlement tolerances, such as railroads or roadways. Two tunnel projects designed by Kilduff Underground Engineering in the Salt Lake City, UT area, each at varying stages of completion, had project design criteria for selection of a tunneling methodology to minimize surface settlement. Both projects consist of a pedestrian tunnel, one crossing beneath a major roadway in downtown Salt Lake City and the second beneath UPRR tracks. Both projects were in difficult, coarse alluvium/fan ground conditions with project criteria that resulted in minimal cover above the tunnel. For each project, a pre-support pipe canopy was designed, requiring the placement of steel, canopy pipes driven the extent of the tunnel length. The pipe canopy is installed using a trenchless technology and then excavated with SEM tunneling and shoring provided by traditional support of excavation methods. The pipe canopy pre-support method offers a low-impact method to install an underpass beneath critical structures by adding significant stiffness to the subgrade to mitigate surface settlement during the main tunnel excavation sequence. These case studies seek to communicate the challenges associated with pipe canopy installation in site-specific ground conditions and explains how these challenges were overcome. This paper aims to define the project risks and describes the pipe canopy methodology as a viable solution to mitigate surface settlement in low cover tunneling scenarios.

## INTRODUCTION

Pipe canopies, also known as pipe roofs, are a support technique installed prior to the main excavation sequence of a tunnel. Given pipe canopies are installed prior to the main excavation, they are referred to as a pre-support method. The technique is useful in situations where a large but relatively short tunnel does not warrant the use of a tunnel boring machine to control the face of an excavation. The pipe canopy is a structure composed of multiple pipes installed in the crown or walls of a tunnel alignment. Given the modular nature of the structure, canopies can be flat-backed or arched, and geometries can be non-circular providing partial or full coverage above and around the tunnel alignment.

Pipe canopies are used in one of two principal situations: (1) when minimal disruption of the ground surface is required (i.e., minimal surface settlement) or (2) in poor ground conditions when pre-support can improve control of ground that is exposed during traditional tunneling, such as the Sequential Excavation Method (SEM). Pipe canopies are commonly associated with low-cover tunnels where the potential for settlement is greatest. In these settings, it is common that pipes are larger in diameter (approximately 2-foot or greater) and can be mechanically connected to make an extremely stiff raft of pre-support. Alternatively, pipe canopies can be used in deeper settings or where poor ground, e.g., cohesionless soil, running ground, or fractured rock etc., is more at risk of running than inducing settlement at the ground surface. Permeation grouting through the canopy pipes into the formation can also cement grains in the ground between pipes to reduce ground loss by creating a rigid, grouted pipe canopy, removing the requirement for a mechanical connection of the pipes. Smaller diameter pipes with grout communicated between pipes in non-cobble to boulder ground can also provide the required stiffness to improve the ground's ability to arch between the pipes as a larger excavation is made beneath the canopy. In either setting or arrangement of canopy pipes, an excavation and support sequence follows installation of the pipe canopy, so the pipe canopy is reinforced with a reaction in the vertical axis within the tunnel excavation, reducing the unsupported span of the canopy and minimizing ground movement.

Pipe canopies can be installed by a variety of trenchless methods, including auger boring, micro tunnel boring machine, and down-the-hole hammers. Equipment selection is a function of the ground conditions and project requirements. Typical applications of the pipe canopy concept are to provide low-cover crossings beneath railroads, runways, or roadways commonly to create pedestrian tunnels and utilidors, though the method has been applied to larger-diameter road and metro tunnels. For both case studies described in this paper, the ground conditions, length of tunnel and project constraints dictated the need for a pipe canopy as pre-support.

## CASE STUDY 1: NORTH TEMPLE TUNNEL

KUE was the Engineer of Record for the design package of a 130-linear foot, pedestrian tunnel built to cross North Temple Street in Salt Lake City between the Temple Square and an underground parking structure in the basement of the neighboring conference center. KUE was awarded the design contract in February 2020 and delivered the final design on schedule in July 2020. Tunnel works began in early December 2020 and completed in May 2021. During the project, the Temple Square was in a renovation phase and the area around the Temple had been excavated and shored approximately 40 feet below street level. The tunnel was launched through a soldier pile and lagging wall with a termination at the conference center foundation wall. The tunnel was excavated approximately 19 feet below North Temple Street, 10 feet below an active 66-inch reinforced concrete storm drain, and 4.5 feet below an 18-inch vitrified clay sanitary sewer.

To minimize the impact of tunneling operations on the surrounding utilities and North Temple Street, a grouted pipe canopy was installed as pre-excavation ground support to stabilize the native gravelly soils prior to tunneling. Following pre-support, the tunnel excavation dimensions were approximately 21 feet wide by 21 feet high with internal finished dimensions of 17.5 feet wide by 17.5 feet high. Tunneling was completed by a bench and top heading SEM, supported by a fiber-reinforced shotcrete initial liner and traditionally reinforced final liner with a waterproofing membrane in-between. The pipe canopy pre-support and tunnel were driven at 0% grade and the inclination of the future pedestrian walkway would be achieved with internal structures constructed by others.

### Case Study 1: Geotechnical Conditions

Anticipated geotechnical units in the tunnel alignment were characterized by three geotechnical borings drilled by others from ground surface to 14 feet below invert of the tunnel. Figure 1 shows the interpreted geotechnical strata in longitudinal section view.

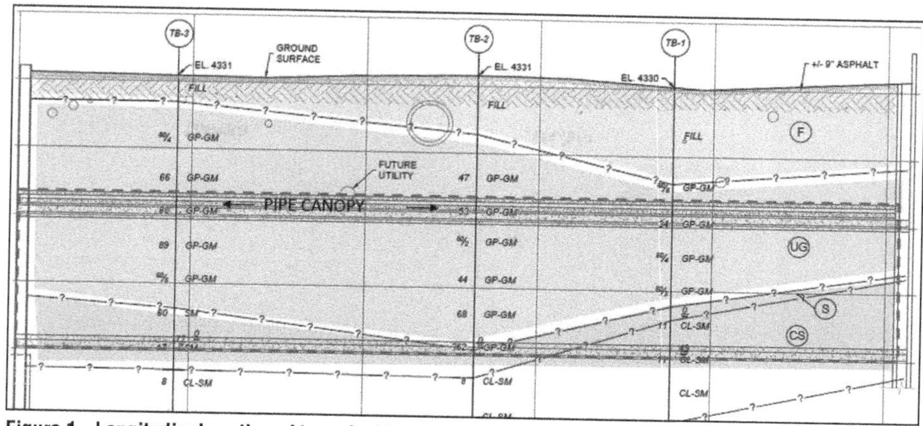


**Figure 1. Longitudinal section of tunnel with a summary of geotechnical conditions. Green/gray superficial zone is roadway-associated fill, blue is the upper gravel unit, and pink is the group of interbedded fines. Tunnel horizon outlined by the red, dashed line**

For this paper, the geotechnical units of concern can be grouped as the dense to very-dense upper gravel and the lower interbedded fines. The upper gravel body was 18 to 31.5 feet thick from ground surface and dominated the top heading of the tunnel and the zone in which the pipe canopy pipes were installed. This unit was a thick body of poorly graded gravels with silt and sand, with blow counts ranging from 34 to refusal during investigation. In the field, this unit had a high percentage of cobbles on the order of 6 to 8 inches in diameter and was consistent in composition throughout the entirety of the tunnel drive.

The percentage of lower interbedded fines, consisting of medium stiff to stiff lean clay and sandy silt, in the bench heading changed significantly over the alignment. An approximately 3 by 3-foot wedge of the silty sand unit was present in the lower right part of the bench heading at the tunnel break-in and grew in size and moisture content with depth into the tunnel unit dominating the bench heading for approximately the last quarter of the drive. A design water table was assumed approximately below excavated invert due to active dewatering of the project site and no consistent groundwater was experienced during tunneling. Localized water which drained from the silty sand rarely ponded to a significant amount and the potential for flowing ground was mitigated with timely application of shotcrete to the exposed face of the bench heading.

## Case Study 1: Initial Support & Excavation Design

The pre-support pipe canopy design was driven by the anticipated ground conditions, tunnel profile geometry, position of the sanitary sewer, and the intent to place the canopy pipes horizontally versus on an inclination or two separate and overlapping canopies installed from either end of the tunnel alignment. The canopy pipes were 0.43-inch-thick, 7 ⅝-inch-diameter steel pipes. The inner row of canopy pipes was concentric to the arches of the tunnel profile geometry. The spacing between the pipes was determined after review of the anticipated ground conditions so that they were close enough to induce interlock of grains and ground arching to minimize potential for ground to run between neighboring pipes. Given the geologic uncertainty and grain size distribution, the pipe spacing was close and micro-fine cement was recommended as the best chance of permeation through the formation. An additional row of

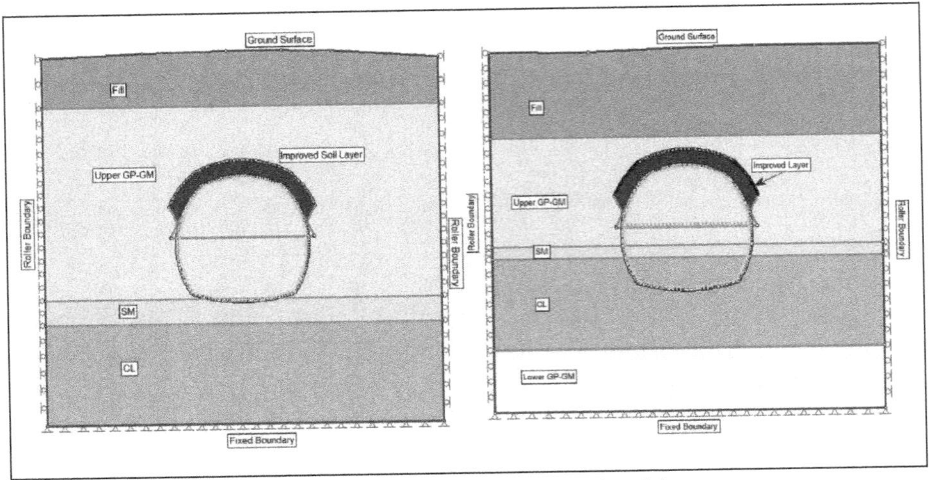


**Figure 2. Model domains in RS2 for the initial liner and pre support models**

canopy pipes was designed with pipes in the outer row placed above the gaps in the inner row of pipes to aid in providing good grout permeation and to minimize potential for ground loss as the tunnel was excavated beneath.

The tunnel initial support system was composed of 6-piece lattice girders on 4-foot center-to-center spacing encapsulated in 10 inches of 5,000 PSI macro-synthetic fiber-reinforced shotcrete. A numerical analysis was conducted to examine the global stability of the tunnel to verify the adequacy of the tunnel initial liner and the pipe canopy, and to predict the surface settlement of the North Temple Street above the tunnel alignment using the two-dimensional finite element modeling software RS2.

Two models were composed to examine the impacts of the tunnel on the varying geotechnical units, shown in Figure 2. Given guidance in literature to avoid discrete modeling of pipes in a 2D plain strain model, the pipe canopy was modeled as an "improved zone" with composite material properties considering the native ground, steel and grout properties. Excavation and support of the tunnel was modeled, including the design groundwater table. No ground relaxation was applied to the Mohr-Coloumb model as it was used to conservatively estimate the initial liner strength. The capacity of the support elements at each stage were examined using combined moment and thrust diagrams which verified the support elements as acceptable as an initial, or temporary, support not subject to seismic loading. Maximum theoretical surface settlement was found to be 0.02 inch.

The final tunnel liner was composed of 9 inches of 5,000 psi shotcrete reinforced with two mats of # 4 steel bars on 8-inch by 8-inch spacing. A beam-spring analysis was completed to simulate the final liner as a stand-alone structure, independent of the pre-support pipe canopy and initial liner. RISA 3D structural software was utilized to simulate the physical response of the shotcrete lining to the simulated loads. The resulting forces in the shotcrete members from the beam-spring model formed the basis for final structural design. The liner was analyzed for combined moment and thrust, longitudinal, shear, and seismic deformation capacities. The 9-inch-thick reinforced shotcrete final lining was determined to have adequate structural capacity for the loading scenarios.

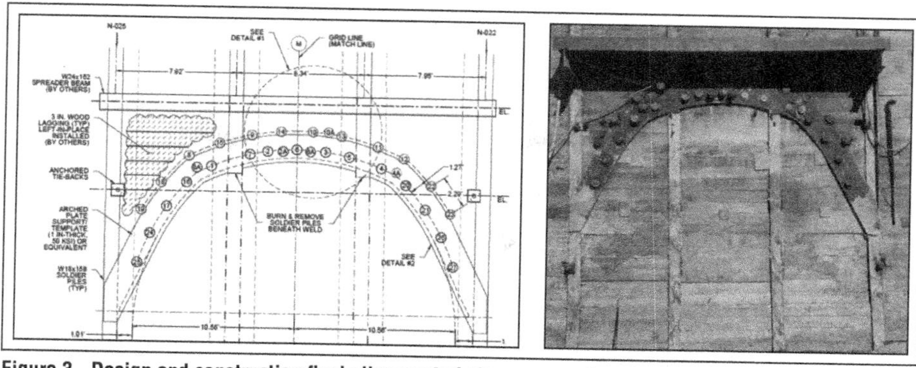


Figure 3. Design and construction final alignment of pipe canopy pipes

## Case Study 1: Construction Phase

The canopy pipes were installed using a modified tie-back drill rig constructed by Drill Tech and a QL60 down the hole hammer by Atlas Copco. Early in the drilling program, survey of the holes indicated significant vertical deviation and the drilling method was re-evaluated after 3 pipes were installed out of tolerance. A Symmetrix drill bit and casing advancement system by Terraroc was substituted with positive results.

Despite the hammer and drill bit's successful progress in the coarse ground, nine of the original 27 planned pipes were not installed to full length. Pipes terminated early were stopped between 44 feet and 94 feet of the 125-foot target depth and were terminated for various reasons, including upward displacement out of tolerance and failed casing connections. Five additional pipes were added to improve coverage of the canopy. Backfill, or permeation, grouting through the canopy pipes was completed in six days throughout the drilling program as the drilling platform level was reduced to install pipes lower down on the tunnel face. Grout was pumped through the perforations in the canopy tubes (which were absent in the initial 13 feet of the casing string past the shoring wall) until communication was observed in the annulus surrounding each pipe.

In total, 27 pipe canopy pipes were planned in two rows reaching approximately between the 10 and 2 o'clock positions of a clock-face. A structural arch template was designed to allow for support of the soldier beams at the entrance to the tunnel, which would be cut during tunneling, and also to provide a guide for the location of each of the canopy pipes. Figure 3 shows the design and final arrangement of pipe canopy pipes prior to tunnel excavation.

The tunnel excavation was performed by the SEM by splitting the tunnel face into a top heading and bench heading with the springline of the tunnel face creating equal separation of the headings. The excavation of the top heading was performed with a modified road header which was able to create the curved geometry of the tunnel profile. The bottom heading was excavated with a standard excavator bucket. By design, each heading was advanced in 4-foot-long rounds and the bench heading trailed the top heading by two rounds to ensure face stability in each heading.

Initial support was provided by lattice girder and fiber-reinforced shotcrete, see Figure 4. Shotcrete was initially placed by remote-controlled nozzle on a robot but this was found to be applying with too little finesse for the cohesionless ground and caused significant rebound and ground loss. Initial lining was applied by hand

**Figure 4. Bench heading at STA 1+14 showing lattice girder in lower bench prior to shotcrete installation and final shotcrete surface post installation in upper bench**

by nozzlemen for better control of application. The lattice girders were used as the guide for the tunnel direction, confirmation of excavation thickness and initial support shotcrete thickness.

As the tunnel progressed, pipe canopy tubes in the crown and left-hand side of the tunnel profile were observed to have deviated into the excavation profile. This was resolved with notches cut into lattice girders and progressed to full removal of sections of canopy tubes, when necessary. As tubes were removed, the ground still benefitted from the permeation grout which cemented the formation together and complete ground collapse was avoided. Installation of initial liner, waterproofing membrane and finale liner is shown in Figure 5.

## Case Study 1: Instrumentation and Monitoring

Monitoring of the project settlement above the tunnel alignment consisted of five 5-point surface settlement arrays, four deep settlement points, and five utility monitoring points measured for movement relative to predetermined early-warning, threshold, and shutdown levels. During excavation and initial support 5-point convergence monitoring arrays around the perimeter of the initial support were measured daily on 12 foot longitudinal spacing to measure ground relaxation into the tunnel.

The largest surface settlement of 0.293 inches was measured in the center of the third array situated in the middle of North Temple Street, approximately 60 feet along the alignment. This surface settlement was below threshold of 0.5 inches but above the theoretical surface settlement of 0.02 inches. The four deep settlement monitoring points installed approximately 10 feet below ground surface measured settlement ranged between 0.197 inches and 0.466 inches which was below the threshold value of 0.75 inches. Convergence monitoring data inside the tunnel showed no significant convergence.

Installation of the final liner was split in two phases to support installation of the waterproofing: following waterproofing of the invert, the final liner in the invert was poured. Using the invert slab as a working surface, the waterproofing was installed in the arches and back of the tunnel which then allowed for the final liner to be prepared and shot into place. The final liner consisted of two mats of #4 rebar and used a

Figure 5. Tunnel entrance following final liner installation and construction completion

combination of 2-inch dobies and anchors placed in the crown and mid-arches to suspend the rebar cages and keep their shape. Anchor points were drilled through the waterproofing membrane and were fabricated with membrane collars which were then heat welded to the membrane in place to ensure integrity of the waterproofing system and assist in final liner construction.

## Case Study 1: Lessons Learned

Early during the pipe canopy installation program, significantly high degrees of torque were required by the drill rig to rotate and advance the casing string. Survey of the holes indicated vertical deviation and the drilling method was re-evaluated and revised to a drill bit and casing advancement system more suitable to the coarse alluvium ground conditions. Despite the improved success, nine of the original 27 planned pipes were not installed to full length and terminated due to upward displacement out of tolerance, downward displacement into the tunnel excavation and failed casing connections. Five additional pipes were added to improve coverage of the canopy. There was one instance of pipe canopy failure when excavation encountered a pipe did not go to depth and a gap between pipes larger than design was present. Ground momentarily ran in the crown of this area and was supported by the miners. The pipe appeared to deviate from a large cobble exposed in the crown.

Issues with pipe canopy installation resulting in significant deviation and ground loss during excavation highlighted the need to evaluate pipe canopy design and installation in similar coarse alluvium environments. A configuration of large-diameter casing pipes, such as 24-inch diameter, for the pipe canopy may have led to less deviation of the casing, as well as the possibility of processing up to large cobbles. Case Study 2 presented below is another example of where coarse alluvium conditions are expected and thus, the canopy design was updated to large-diameter casing pipes to better mitigate potential installation issues and ground loss during excavation.

Options exist for down-the-hole (DTH) hammer drilling, such as the relatively innovative technology pioneered by Geonex, to install pipe casing with minimal deviation. This DTH technology can manage up to 48-inch casing diameter and a project like Case Study 1 would have been suitable for this equipment. In pipe canopy projects where coarse alluvium consisting of cobbles and boulders is a possibility, the correct drilling rig is of the utmost importance to ensure proper pipe installation and integrity of the structure.

## CASE STUDY 2: MAPLETON TUNNEL

KUE is the Engineer of Record for the preliminary concept design package of a proposed 67-linear foot, pedestrian tunnel crossing beneath two active UPRR tracks for a planned trail to connect the Mapleton Lateral Canal Parkway Trail and the Spanish Fork Dripping Rock Trail. KUE was awarded the preliminary design in August of 2021 and delivered on schedule in November 2021, with subsequent revision rounds in March and May of 2022. Project delays have caused award of final design to be revised with final design work slated to be issued in Q2 or Q3 of 2023. The tunnel is proposed to be launched through a sheet pile wall south of the tracks and terminating north between the UPRR tracks and US-6 highway. The tunnel's proposed excavation depth of cover is approximately 12 feet below the tracks that will remain active during construction.

To minimize the impact of tunneling operations on the active UPRR tracks and right-of-way area, a grouted pipe canopy has been proposed as pre-excavation ground support to stabilize the native soils prior to tunneling. Following pre-support, the proposed tunnel excavation dimensions are approximately 16.25 feet wide by 12.25 feet high with internal finished dimensions of 14.25 feet wide by 10.25 feet high. Tunneling is proposed to be completed by short excavation rounds and supporting the exposed ground by a fiber-reinforced shotcrete initial liner and steel sets on 4-foot centers. Finally, a traditional reinforced concrete final liner with a waterproofing drainboard in-between will be placed. The pipe canopy pre-support and tunnel are expected to be driven at 0% grade and any future inclination of the trail would be achieved with internal structures constructed by others.

### Case Study 2: Geotechnical Conditions

Anticipated geotechnical units in the tunnel alignment were characterized by two preliminary geotechnical borings and test pits which were drilled from ground surface to depths of between 11 and 28 feet below invert of the tunnel. Figure 6 shows the assumed geotechnical strata in longitudinal section view.

For this paper, the geotechnical units of concern can be grouped as the upper interbedded fines with gravel unit and the lower gravel unit. The upper fines body is anticipated to be 5 to 18 feet thick and preside in the top heading of the tunnel and the zone in which the pipe canopy pipes are to be installed. This unit is defined as an interbedded body of medium to very stiff sandy silt and clay with gravel, with blow counts ranging from 6 to 18 during investigation. The lower gravel body is anticipated to be 9 to 14 feet thick presiding mainly in the lower bench and invert of the tunnel but interbedded into the top heading. This unit is defined as a thick body of very dense gravel with coarse grained sands, with blow counts ranging from 47 to refusal during investigation. Groundwater is not expected during installation of the pipe canopy or

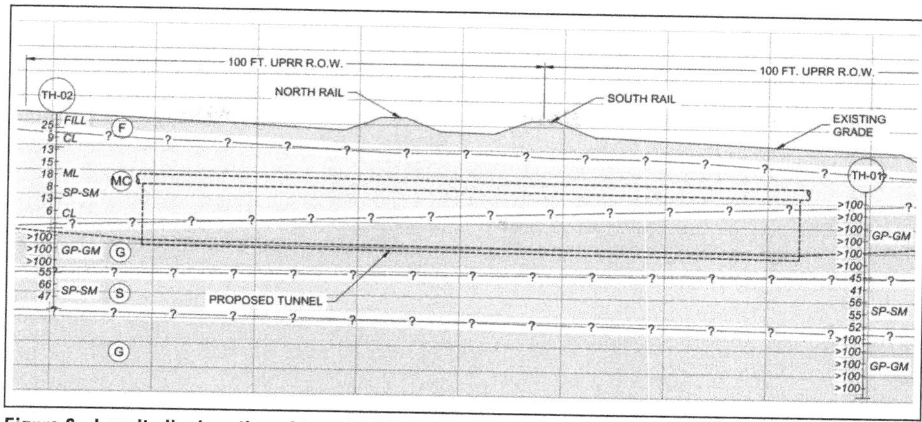


**Figure 6. Longitudinal section of tunnel with a summary of geotechnical conditions. Gray/brown superficial zone is railroad-associated fill, blue is the upper sandy silt with gravel unit, green is the interbedded sand with gravel unit and red is the group of gravels**

excavation of the tunnel, however significant grades at ground surface poses a risk for high surface water flows during storm events.

## Case Study 2: Initial Support & Excavation Design

Anticipated ground conditions and general tunnel profile geometry are similar to case study 1, but case study 2 pre-support pipe canopy design was also driven by the shallow depth from railroad tracks and the intent to place one row of canopy pipes in a box shape from one end of the tunnel alignment. The canopy pipes are anticipated to be 0.5-inch-thick, 24-inch-diameter steel pipes. The pipe design size is in part a consequence of lessons learned in Case Study 1, where the smaller diameter pipes caused deviations during installation. Upsizing the pipes should mitigate multiple installation issues resulting in line and grade deviations. The canopy consists of one row of pipes radially set around the crown and walls of the tunnel. The spacing of the pipes is designed to be close enough so that grout will communicate between each pipe and form the pre-support arch. Extending the row of canopy pipes down each tunnel wall is designed to aid in providing good grout permeation around the perimeter of the excavation and to minimize potential for horizontal ground convergence into the tunnel.

The initial conceptual support design is composed of steel sets on 4-foot center-to-center spacing with 6-inches of 5,000 PSI fiber-reinforced shotcrete sprayed in the bays between steel sets. A 6-inch-thick invert slab will also be poured at the initial support stage to set the column posts laterally. The final/ composite liner will consist of the initial liner with a secondary layer of steel-bar-reinforced concrete and the steel sets to be encased in 5-inches of fiber-reinforced shotcrete. A final reinforced concrete invert slab with a minimum of 6-inches-thick will also be installed. Numerical analyses were conducted to examine the global stability of the tunnel to verify the adequacy of the pipe canopy, tunnel initial liner, tunnel final/ composite liner and to predict the surface settlement of the railroad tracks above the tunnel alignment using the two-dimensional finite element modeling software RS2.

Two models were composed to examine the impacts of the pipe canopy and tunnel liners on the expected surface settlement, shown in Figure 7. Similar to case study 1, the pipe canopy was modeled as an "improved zone" with composite material properties

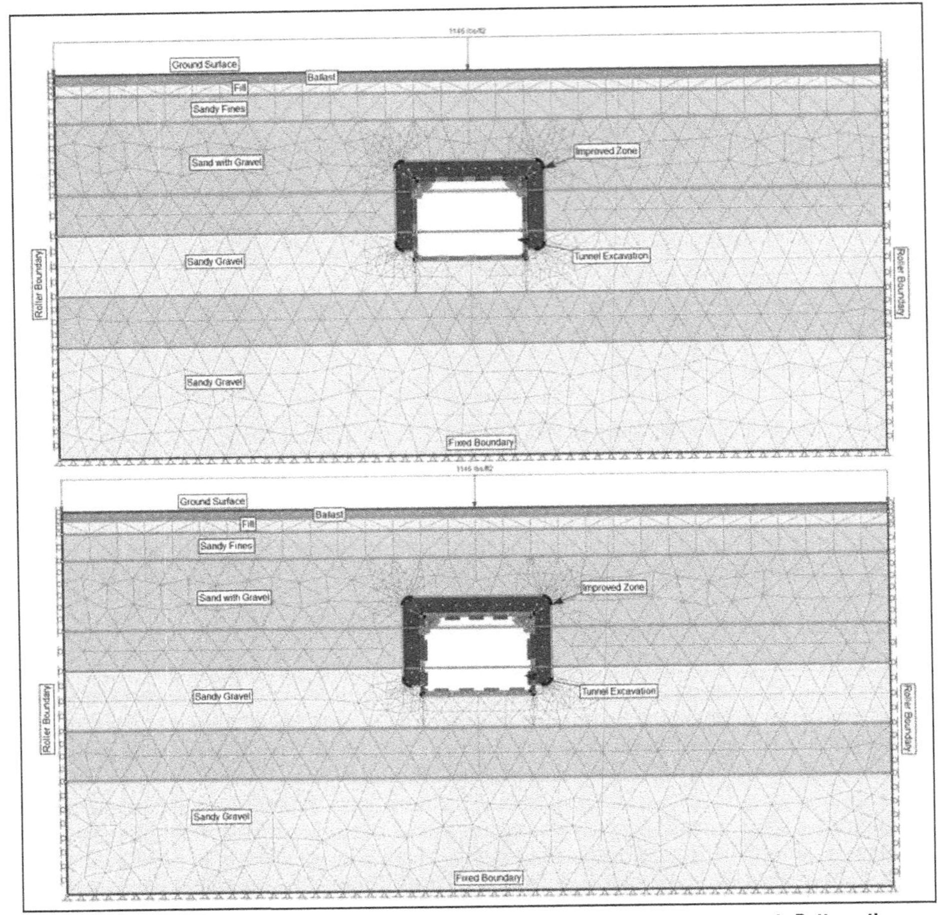


**Figure 7. Model domains in RS2 for, Top: the initial liner with pipe canopy pre support, Bottom: the final/ composite liner with pipe canopy pre support**

considering the native ground, steel and grout properties. Excavation and support of the tunnel was modeled as two separate models (initial and final supports added separately versus as one composite liner). Again, no ground relaxation was applied to the Mohr-Coloumb model as it was used to conservatively estimate the initial liner strength. The capacity of the support elements at each stage were examined using combined moment and thrust diagrams which verified the support elements as acceptable as an initial, or temporary, support not subject to seismic loading. Maximum theoretical surface settlement was found to be 0.095, which is well below UPRR guidelines of 0.25 inch.

## Case Study 2: Construction Phase

At the time of writing, the project is still in conceptual design and has not been adopted by the future general contractor. The canopy pipes are proposed to be installed using a modified tie-back drill rig and a down the hole hammer. Because survey of the case study 1 drill holes indicated significant vertical deviation, drilling methods

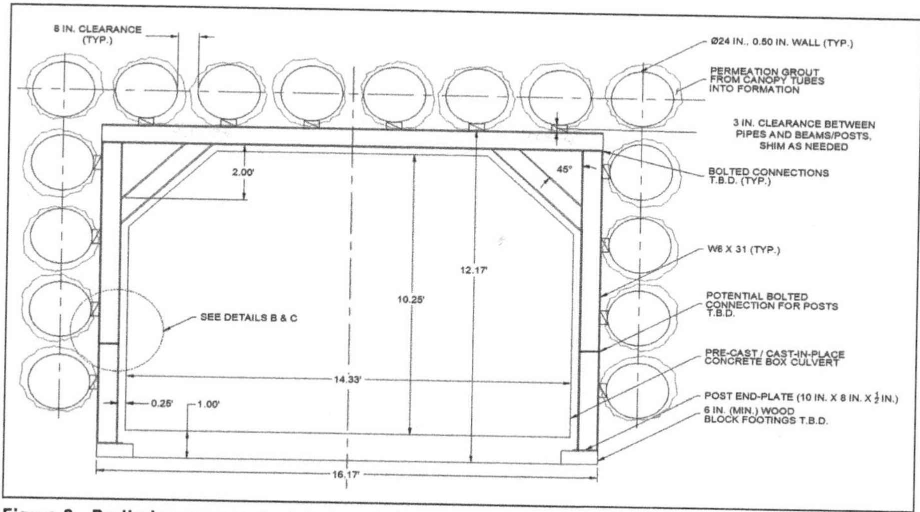

Figure 8. Preliminary conceptual design of pipe canopy pipes

and procedures on case study 2 will be carefully evaluated during the pipe canopy installation phase.

Backfill, or permeation, grouting through the canopy pipes will be completed throughout the drilling program as the drilling platform level is reduced to install pipes lower down around the tunnel face. It is assumed installed pipes will be perforated in place and then grout pumped through the perforations in the canopy tubes until communication is observed in the annulus surrounding each pipe.

At this stage of the conceptual design, 16 pipe canopy pipes are planned with eight pipes along the crown and four pipes down each wall. The principal focus of the pipe canopy layout is to protect the ground above the crown of the tunnel from settling, provide a protective canopy for miners to operate beneath, and stabilize the ground from converging horizontally into the tunnel from the walls. Figure 8 presents the preliminary conceptual design of pipe canopy pipes prior to tunnel excavation.

The tunnel excavation is intended to be performed in 4-foot-long rounds, where initial liner is installed after each tunnel extension to ensure face stability and mitigate ground convergence. Based on preliminary geotechnical data acquired, the heading can be excavated with a standard excavator bucket. Design of heading advanced may be modified based on future geotechnical findings and/or observations during construction. A conceptual rendering of the completed tunnel is shown in Figure 9.

## Case Study 2: Instrumentation and Monitoring

Monitoring of the project settlement above the tunnel alignment is planned to consist of two 7-point rail settlement arrays installed as rail clips or adhesive targets on the railroad tracks, two deep settlement points, and ten surface monitoring points along the braced excavations at tunnel launch and reception. Monitoring points will be measured for movement relative to predetermined early-warning, threshold, and shutdown levels. Agreeable results between numerical analysis and actual recorded settlement at the North Temple Tunnel project suggests similar settlement measurements can be expected at the Mapleton Lateral Canal Parkway Trail tunnel. The theoretical surface

**Figure 9. Conceptual 3D rendering of completed Mapleton Lateral Canal Parkway Trail tunnel**

settlement indicated by the numerical analysis can be realized with proper pipe canopy installation and excavation sequencing.

## CONCLUSION

Pipe canopies are a proven method of initial support installed prior to the main excavation sequence of a tunnel in one of two principal situations: when minimal disruption of the ground surface is required (i.e., minimal surface settlement) or in poor ground conditions when pre-support can improve control of ground that is exposed during traditional tunneling. Typical applications of the pipe canopy concept are to provide low-cover crossings beneath railroads or like structures, commonly to create pedestrian tunnels and utilidors. For both Case Study projects described in this paper, the ground conditions and length of tunnel indicated the need for a pipe canopy as pre-support design.

In Case Study 1, a 130-foot-long tunnel was successfully built using a pre-excavation pipe canopy as a ground support technique prior to the main tunnel construction by the sequential excavation method in Salt Lake City, Utah. The cohesionless, gravel-rich ground created the potential for unstable ground which called for a grouted pipe canopy to be installed to minimize potential for ground loss through the canopy. The pre-support canopy achieved its design purpose to minimize ground settlement and improved standup time in the tunnel face and ribs. With the pre-support in place there were no stoppages due to settlement and the ground was stable between excavation and initial support from the increased stiffness of the ground provided by the grouted canopy. The combination of the pre-, initial, and final supports made for an extremely robust structure.

The preliminary conceptual design in Case Study 2 demonstrates the pipe canopy technology as a viable technology for the construction of excavations below railroads. The combination of pre-support via a pipe canopy, initial and final support installation in the gravel-rich sands and silts results in a robust structure to minimize surface settlement of the railroad tracks. Based on preliminary numerical analysis, the pipe canopy minimizes impact above the tunnel and track settlement is estimated below UPRR settlement guidelines.

PART 

# Shafts, Caverns, and Mining

*Chairs*

**Steve Price**
Walsh Group

**Ehsan Alavi**
Jay Dee Contractors

# Development and Performance of Large Span Caverns at Depth for the LBNF Far Site Project

**S. Pollak** ▪ Arup
**J. Rickard** ▪ Fermi Research Alliance

## ABSTRACT

The Long Baseline Neutrino Facility Far Site in Lead, South Dakota is currently under construction and involves drill and blast excavation of the largest and deepest caverns in North America. This paper will review how the geotechnical risks are being managed through careful consideration of excavation sequence, completion of exploratory pilot tunnels, implementation of a robust instrumentation and monitoring program, and validation against the design. Aside from having a cross section of 5,900 square feet and being situated at a depth of one mile, the caverns also employ a unique permanent rock bolt solution which underwent a significant pre-production pull testing program involving varying installation methods and being subject to blast vibrations. The behavior of the schistose rock mass and performance of the ground support are being continually monitored throughout the construction phase.

## PROJECT OVERVIEW

The Long Baseline Neutrino Facility Far Site (LBNF) project is located deep underground at the Sanford Underground Research Facility (SURF) in Lead, South Dakota. SURF is a dedicated underground scientific facility located at the former site of the Homestake Gold Mine. The new LBNF facility will house the Deep Underground Neutrino Experiment (DUNE), an international flagship experiment which will seek to enhance our understanding of the universe, the matter which comprises it, and ultimately the role that the neutrino plays in all of it. The project is funded by the US Department of Energy and managed by the Fermi Research Alliance. Fermilab is the host laboratory for DUNE, in partnership with a wider, global consortium of agencies and institutions. Four massive, liquid argon filled particle cryostat detectors will be sited nearly one mile underground at SURF and be on the receiving end of one of the world's most intense neutrino beams emitted from the LBNF Near Site facility on the grounds of Fermilab in Batavia, Illinois. Arup is the Engineer of Record for the Excavation contract and Thyssen Mining is the contractor.

This paper will discuss the design and construction observations made during excavation of the three large caverns at SURF which will house the DUNE and its supporting infrastructure.

## CAVERN GEOMETRY

The general arrangement of the LBNF Far Site project is shown in Figure 1 and consists of two main detector caverns housing the four cryostats, separated by a central utility cavern which contains supporting infrastructure and controls for DUNE. Numerous ancillary drifts, tunnels, and chambers surround the cavern complex and are formed by either new excavation or enlargement of the existing Homestake era drifts. The main caverns are mailbox shaped and have dimensions of 65 ft. span × 92 ft. high × 500 ft. long. The overall in situ volume per detector cavern is approximately 100,000 cubic

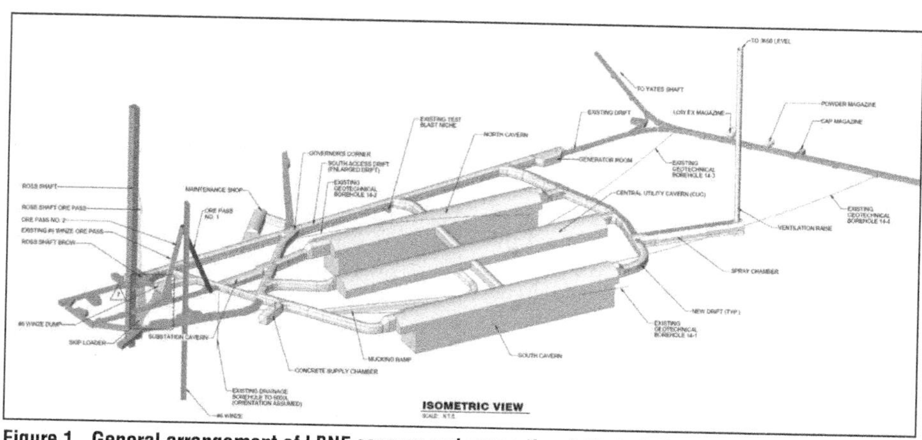

**Figure 1. General arrangement of LBNF caverns and supporting drifts (existing Homestake era mine drifts shown in orange)**

yards. The central utility cavern is of the same span and height as the main detector caverns above springline, but does not contain the deep bench levels below, having an in situ volume of roughly half of the detector caverns. The caverns are constructed at a depth of 4,850 ft. below ground level.

## GEOLOGICAL CONDITIONS

SURF houses numerous neutrino-based physics and other experiments which need to be shielded from cosmic radiation effects. The current lab facility and newly constructed LBNF Far Site sits on the 4850 Level of the former Homestake Gold Mine. The re-purposed mine provides the required infrastructure and experiment conditions necessary to carry out these sensitive observations. The rehabilitation and refurbishment of the waste rock handling system, which was required prior to any excavation associated with the LBNF project, was previously discussed in Pollak et al. (2021).

The host rock formation for the LBNF Far Site is known as the Poorman Formation—a Precambrian age quartz sericite carbonate schist. The genesis of this metamorphic rock originates from a wide variety of protoliths including thinly bedded, carbonate-rich siltstones and claystones, marl, iron formations, and lesser amounts of dolomite. The presence of graphite in much of this formation indicates that the original deposits were rich in organics. A prevalent foliation fabric dips sub-vertically to the NE-E direction and varies from planar to highly contorted. The rock mass is veined with quartz-calcite throughout. During the Tertiary, these rocks were intruded by rhyolite dikes which present a much different rock mass and behavior at depth compared to the schist.

The in situ stress at the site is generally well characterized through anecdotal evidence (borehole breakouts) and previous attempts to measure the tensor using overcoring methods. When resolved to cartesian coordinates, the ratio of vertical to horizontal stress has been assessed to be approximately 1.5, with the horizontal stresses being approximately equivalent. The overburden (vertical) stress at the 4850 Level is taken as 6,200 psi.

The mine is kept dry by a permanent, deep water pumping system which largely removes groundwater concerns from the design, another major benefit of the site. Minor seepage is occasionally present from pockets of perched water which infiltrates down to the 4850 Level from the open pit on surface. These are typically short term flows which dry out as a result of the mine ventilation.

During initial site investigations, which included geological mapping and scanning of existing mine drifts, drilling of four sub-horizontal boreholes (shown by blue traces in Figure 1) into the proposed cavern location, and both in situ and lab tests, it was observed that the rock mass presents an ability to deform over a wide range of scales, from the micro to macro without loss of strength (e.g., fracturing). This favorable characteristic is a result of the high degree of micaceous minerals in its make-up, a defining trait which makes construction of large caverns feasible without risk of bursting or other brittle type failure mechanisms. Conversely, one of the site investigation goals was to identify and avoid, if possible, siting the caverns in a location where rhyolite dikes are present due to their propensity to spall and fracture upon exposure and loss of confinement. Rhyolite is observed to intrude parallel to the foliation trend, exploiting this weaker orientation. Further details of the project site investigation can be found in Hurt et al. (2016).

**Figure 2. Rhyolite/Poorman Formation contact encountered in one of the cavern access drifts**

## Ground Behavior and Calibration

One of the most critical steps in the design process is to identify likely modes of ground behavior upon excavation and the plausible failure mechanisms that result. This stage links the site investigation work to the design analysis and cannot be overlooked. As mentioned earlier, foliation plays a major role in the feasibility of cavern construction, with increased ground support demands (and cost/risk) associated with caverns whose axes runs parallel with the fabric. Fortunately for LBNF, the experiment requirements dictated the alignment of the cavern axes be set in an East-West orientation, perpendicular to foliation trend and resulting in favorable excavation conditions. Along with foliation, the rock mass contains several sets of sub-vertical joints, which combine to form frequent wedges. An observation made in the smaller mine drifts during the site investigation was that the contorted nature of the rock frequently resulted in wedge fall-out, particularly where a fold hinge interacted with foliation to form release planes. Therefore, constraining the size of wedges and blocks exposed in the caverns during excavation was a known design goal through consideration of sequential excavation and timing of ground support installation.

Other modes of behavior considered include stress induced slabbing along foliation, brittle failure of rhyolite, anisotropic deformation, and plastic yielding of the rock mass. An exercise was undertaken during the design stage to "calibrate" the ground model and constitutive model inputs derived from the lab testing program. This involved looking for examples of "failures" and "successes" in the existing mine, such as hourglass shaped pillars, wedge failures, larger rooms which showed no overbreak or loads on existing ground support, and slabbing in tunnels oriented parallel to foliation. The "failures" and "successes" were then numerically backanalyzed with the proposed design inputs. The conclusions which came out of this exercise were that the lab test derived

Figure 3. Slabbing foliation expressed in sidewall of north-south oriented access tunnel

strength envelopes were underrepresenting the in situ rock mass strength. Large zones of "yielding" and deformations indicated by the numerical models did not match with observation. The conclusion was that core extracted under the in situ stress conditions at the 4850 Level underwent some degree of microfracturing upon stress relief, and hence when tested were already in a damaged state. Although the same effect could be said about the boundary of the caverns upon excavation, immediate confinement several feet into the rock provides a much improved rock mass strength. An adjustment to the design parameters was made through an increase in the GSI value. A low confining stress of 150 psi was assumed in derivation of Mohr-Coulomb properties to account for the local deconfinement of the rock mass around the excavation.

On the structural geology side, the size of existing wedges and overbreak in the 4850 Level drifts was observed and mapped. These smaller drifts were ultimately included in 3D numerical models of the caverns and used as a calibration point to ensure blocks generated in the models matched with observed conditions. Once confidence was gained in the overall characterization of the rock mass and behavior modes, the design progressed to excavation sequencing and selection of ground support for the caverns.

## DESIGN OF GROUND SUPPORT AND EXCAVATION SEQUENCE

### Concept

Owing to the favorable conditions at the site with respect to groundwater and rock mass stability, the envisioned cavern support was anticipated to be provided by permanent rock reinforcement and shotcrete. The design life of the support system is required to be 50 years. Although groundwater is not a concern, there are occasional patches of seepage and perched water which filters down to the 4850 Level from the open pit on surface. Therefore, the corrosion risk to permanent ground support needed to be addressed. Another unique requirement of the project is that welded wire fabric must be used above springline in the caverns and be in contact with all installed rock reinforcement in order to satisfy electrical grounding criteria. Shotcrete is used as surficial support only—to provide protection against fall of ground from minor spalls and sloughs which can form in between the rock reinforcement pattern.

The ground conditions and rock mass behavior were characterized as fairly uniform along the length of the caverns. Therefore, a single, prescriptive permanent ground support type was envisioned which addressed all design requirements.

The selection of rock reinforcement type came down to two different systems: a cable bolt or a double corrosion protected rock bolt (CT-Bolt). Double corrosion protected rock bolts provide several advantages over cables including enhanced durability, benchmarks for the required design life, and an immediate supporting "dowel" effect when subjected to joint shear displacements. On the other side, many contractors are more comfortable to install cables, they are preferred for longer applications over bolts, and they provide a greater degree of flexibility and accommodation of larger ground movements, which is deemed a possibility given the cavern size and high in situ stress conditions. The choice of permanent rock reinforcement was made once the anticipated length and compatibility with ground movements could be quantified. A numerical modeling approach was adopted for this purpose.

## Numerical Modeling

The final design of the LBNF cavern ground support was primarily assessed using the 3D distinct element program 3DEC. The software was chosen as it could capture the ground behavior and failure modes previously identified such as kinematic wedge and block formation, as well as block scale strength characteristics of an anisotropic material. The size of the model included all three caverns as well as the center interconnecting tunnel and a parallel, existing drift running alongside the north cavern, which could be used to calibrate the block/wedge size. The constitutive model selected for the rock blocks consisted of a ubiquitous joint model, having strength parameters consistent with those developed from the backanalysis exercise. A discrete fracture network, including foliation parallel discontinuities, was overlain to match the ground model developed from field mapping. The spacing and persistence of the joints in the model were then adjusted and re-generated until a good match was found between the observed wedges and 3DEC generated wedges in the parallel drift tunnel as shown in Figure 4. It was found that assuming fully persistent joints still provided a good match in the calibration to the existing drift. As this represents the most onerous assumption for block size formed in the caverns, up to 30 cubic yards in the sidewall, it was carried forward in the design.

Once calibrated, various ground support types and excavation sequences were assessed. Of note, multiple realizations of the joint network were generated and run in 3DEC to ensure all sensitivities related to the origin of the joint pattern and resulting wedges/blocks were considered. An annulus of reduced stiffness from excavation boundary to 3 ft. away was included in the model to account for blast damage.

There were several objectives to be satisfied by the numerical modeling. These included an assessment of the maximum probable wedge size, determination of the extent of joint opening and plastic yielding due to the U-J model, demonstration of strain compatibility between rock reinforcement and rock mass movements, definition of sequential drift sizes and lead/lag relationships, and development of trigger levels with respect to ground movement. Sensitivity to rhyolite dikes (which implemented

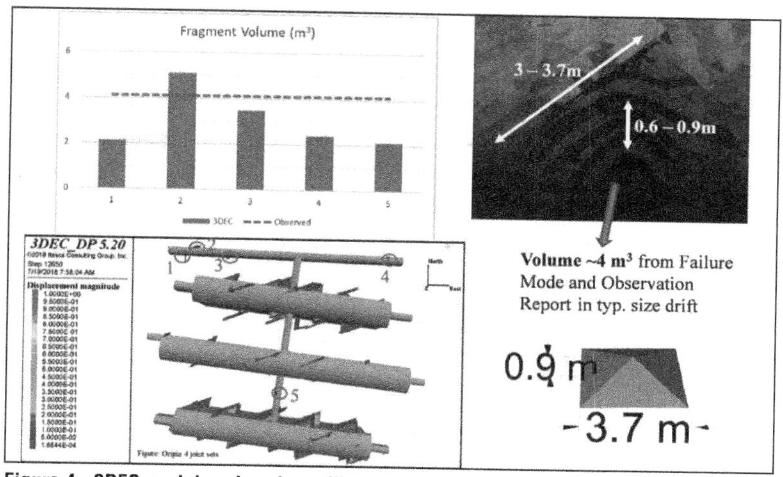


**Figure 4. 3DEC model wedge size calibration exercise based on observed conditions in existing mine**

a cohesion-softening, friction-hardening type strain softening constitutive model in 3DEC) was also considered.

The model results indicated that a 20 ft. long rock reinforcement element was appropriate for the caverns. This length addressed both the plastic yielding zone which was less than 12 ft. thick, as well as the maximum wedge size which was observed to develop in the cavern sidewalls. With a length of 20 ft. defined, both a cable and DCP bolt were still deemed feasible. As stated previously, the DCP bolt offered advantages over the cable bolt, mainly in long term durability and shear performance. The 3DEC model supported this theory, especially in limiting shear displacement of the cavern sidewall wedges. Reduced joint shear displacements help to preserve the overall strength of the rock mass, especially if movement can be kept at or below peak values. Although cables performed well in the top heading, it was decided to adopt the DCP #10 bolt across the entire cavern cross section for uniformity. An initial spacing of 5 ft. × 5 ft. was adopted based on joint spacing and block size distribution. Shotcrete was not included in the model and design based upon analytical approaches for punching shear and WWF embedment. A thickness of 4 in. was prescribed over the entire cavern perimeter.

Once the rock reinforcement type had been selected, the excavation sequence could be finalized.

## Design Excavation Sequence

In cross section, the excavation sequence considers a 30 ft. high top heading split in three equal span partial drifts. The center drift is excavated first, followed by each side drift in a staggered arrangement. The heading height was dictated by the length of rock reinforcement and bolter requirements to install vertically. Three drifts were selected instead of two so as to limit the amount of exposed ground at one time (and the unstable wedge size as a result) as well as keep cycle times below 24 hours. The maximum round length was limited to 15 ft., which was consistent with cycle time requirements and minimizing drill/blast hole deviations in the foliated rock. In the bench, a maximum height of 12 ft. 6 in. is defined for a total of five levels, as this limits

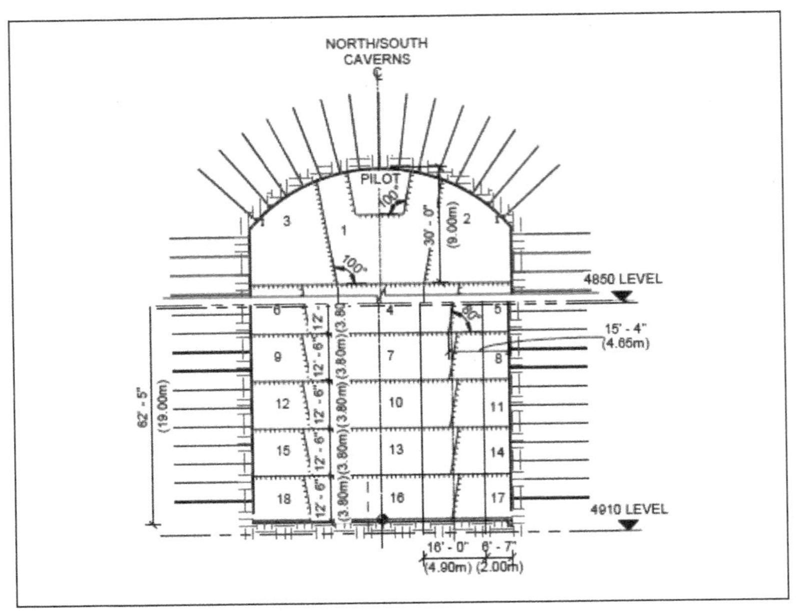


Figure 5. Detector cavern excavation sequence

the exposed ground and maximum unsupported wedge size. A larger center cut of approximately 40 ft. span is allowed for and can be advanced as far as practical by the contractor provided blast PPV and overpressure limits are not exceeded. This leaves in place 13 ft. wide sidewall abutments, which are incrementally blasted in 15 ft. maximum round lengths and supported.

The timing of rock reinforcement installation was also carefully analyzed and defined. A competing philosophy of needing immediate ground support for safety, while not wanting to overstress the system from ground relaxation had to be reconciled. The DCP bolt allows for this as a mechanical end anchor provides immediate support upon install, while waiting to grout the bolt allows for elongation over the full length of the steel in response to relaxation during partial drift excavation. Therefore, the bolt can be installed and point anchored at the face without compromising its long term integrity. The final sequence mandates a minimum 45 ft. stagger distance between active faces in the top heading, and the same distance between the active face and the timing of full column cement grouting of the DCP bolt behind. In practical terms, for any cross section, the center drift bolts have to be grouted before the first side drift is excavated, and likewise for the first side drift bolts prior to excavating the second side drift. The point anchor force was defined to be 20 kip, which was sufficient to support any wedges in the crown for the short term condition. For the bench/sidewall bolts, the delay in grouting is dropped since much of the initial relaxation has occurred from top heading excavation. For the walls, which stand 76 ft. vertically, the immediate support and control of shear displacements along discontinuities is considered critical and require the DCP bolt to be grouted in cycle.

## Design Results

From the staged 3DEC modeling results, the anticipated rock bolt loads were observed to be maximum in the top heading and ranged up to 45 kip, compared against a working load of 70 kip. Sidewall bolt loads were generally 50% of those in the top heading. Bolt strains were noted as less than 0.6%. The maximum anticipated ground movements were observed to be in the cavern sidewalls, amounting to approximately 1 in. (25 mm). Crown movements were anticipated to be less, ranging from 7 mm following the initial top heading center drift, up to 13 mm once the top head full span was developed. With ground movement and support compatibility confirmed, the design was finalized.

### Instrumentation and Monitoring

A robust instrumentation and monitoring program was defined during the design, underpinned by the expected ground movement and behavior modes identified by the design analyses. Aside from typical surface convergence monitoring, deep anchored, multipoint borehole extensometers (MPBXs) are specified for installation in the caverns. A standard array contains five MPBXs in the top heading and between 2–6 horizontal MPBXs in the sidewalls. Each MPBX is made up of five hydraulic bladder filled anchors and a reference head. Depth of anchor points from the cavern surface are at 4 ft., 8 ft., 16 ft., 33 ft., and 66 ft. The MPBXs are mandated to be installed a maximum distance of 45 ft. back from the active heading, so that subsequent partial drift excavations are captured by the instrument.

## RISK MITIGATION

Measures are taken during the design and construction of the caverns to mitigate risk. Each cavern commenced with a small 12 ft. × 12 ft. pilot tunnel which was driven all the way through the full cavern length and thoroughly mapped in order to document joint orientations and confirm the absence of rhyolite. This tunnel also allowed for the installation of the centerline MPBX instrument in the crown since the top of the pilot tunnel aligned with the final cavern top profile at centerline. A geotechnical baseline report (GBR) forms part of the contract documents and baselines items such as rhyolite, slabbing behavior with respect to tunnel orientations, geological overbreak, and areas where round length reduction and/or shotcrete flashcoat is necessary due to minor shears, graphite content, or other geological factors. A robust trigger action and response plan (TARP) is also included in the drawing set which covers potential issues such as blast vibration/overpressure, fall of ground prior to support, and ground movements (post-support). The GBR is linked to the TARP in that a certain number of TARP conditions (threshold, limiting) are baselined for each risk covered. The Engineer maintains a presence on site during excavation to map, carry out design

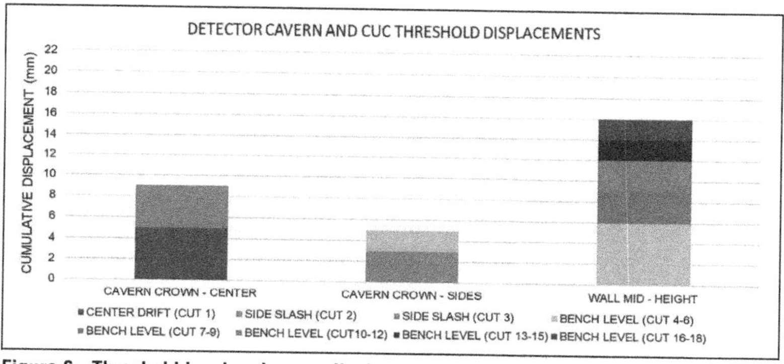


**Figure 6. Threshold level rock mass displacements as determined from staged 3DEC modeling**

**Figure 7. Detector cavern top center heading and pilot tunnel (beyond). A flyrock protected monorail beam is shown running along the crown center**

validation, and most importantly, to respond to situations or issues based on real time data and input which can only be gained by having boots on the ground.

## GROUND SUPPORT PRE-PRODUCTION TESTING

Prior to cavern excavation, a pre-production rock bolt testing campaign was undertaken. This was done to confirm the selected DCP bolt, the FA3 bolt supplied by Jennmar, is compatible with the design requirements and procured bolter (a Sandvik D512i). One unknown which needed to be determined was the torque-tension relationship for the bolt, namely what torque was required to achieve the 20 kip point anchor. Through pull testing, it was found that a torque of 300 ft-lbs was sufficient to mobilize the required point anchor capacity. Another concern is that this anchorage could loosen up when subjected to multiple cycles of blast vibration prior to grouting. Pull testing was carried out on several point anchored bolts installed close to an active drift face. The bolts were pulled in between each of three successive blasts. It was confirmed that at the specified torque, the anchorage did not loosen or slip when subjected to blast vibration. This provided overall confidence that the selected bolt met the project requirements.

## OBSERVATIONS DURING CONSTRUCTION

### Ground Behavior

As anticipated, various ground behaviors have been observed during cavern excavation. The emergence of wedges in the top heading sidewalls has been observed, albeit similarly shaped, smaller scale versions of the maximum sized wedge seen in the 3DEC model. The orientations of joint sets and foliation have been generally consistent with that assumed during design.

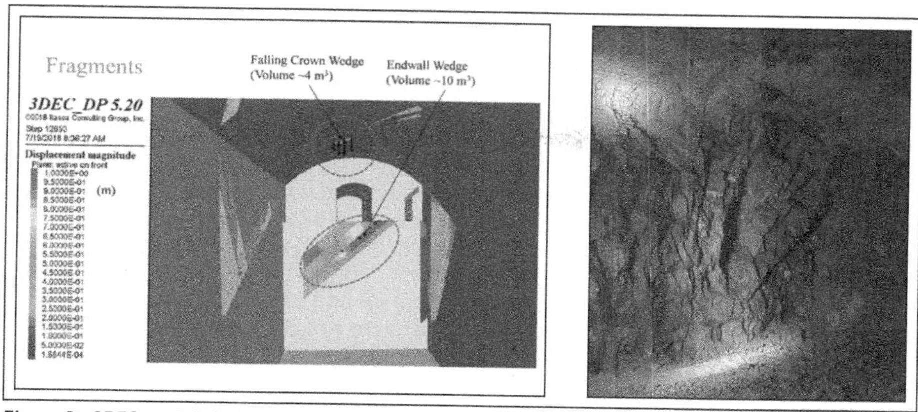


**Figure 8. 3DEC modeled sidewall wedges (left) versus observed top heading sidewall wedges (right) in the cavern**

No rhyolite or other major features were encountered in the cavern pilot tunnel or completed top heading excavations. No spalling or other stress induced mechanisms have been observed around the cavern profile, although some "onion skinning" fractures were seen in the temporary sidewalls of the central top heading drift prior to excavation of side drifts.

In the South Cavern, on the western end, a north-south oriented shear zone extends across the first 75 ft. of the cavern. This shear zone is baselined in the GBR and is visible in other existing drifts on the 4850 Level. The rock mass within the zone is notably more fractured, with an intense, crenulated foliation. Alteration of the mica has occurred where a green, chromium variety (fuchsite) is visible. The zone has been subjected to a shotcrete flashcoat of 2 in. thickness to prevent raveling and intensified monitoring. Otherwise, no change in ground support has been made from the typical pattern.

**Figure 9. Typical shear zone conditions cutting across South Cavern prior to application of shotcrete**

As stated above, rhyolite intrusions were not exposed in the detector caverns, but there was exposure along the left rib of the East Access Drift when it ran sub-parallel with the drift and perpendicular across the spray chamber. During the excavation through the rhyolite, minor brittle failure (no rock bursts) occurred which led to spalling and subsequent overbreak up to 4 ft. When crossing the rhyolite perpendicular in the spray chamber, local spalling occurred in the shoulders of the chamber. When rhyolite was exposed, shotcrete was applied after each round in addition to the bolts and wire mesh to reduce progressive spalling in line with the GBR and TARP.

## Response to Excavation

As the 3DEC models incorporate the MPBX locations, anchor arrangements, and timing of installation, a displacement response of each anchor at each location at each stage can be extracted from the model. This results in a large dataset, but in general the anticipated behavior can be enveloped and used to define threshold and limiting values in the TARP. A typical response of a crown (vertical) MPBX in the cavern is shown in Figure 10. The response shows movement at each stage after the pilot tunnel, namely center cut enlargement, followed by the two side drifts. The measured displacements match well with the 3DEC model, showing approximately 4 mm of displacement after center top heading enlargement, increasing to 8 mm with completion of the full span top heading. Several other observations can be made from the MPBX data. The first two anchors closest to the reference head indicate that the ground is generally moving with the reference head. It could be inferred that this is indicative of the "yielded zone" or perhaps the blast damaged zone or opening of a sub vertical joint (e.g., wedge translation). For purposes of this paper, it will be called the "disturbed" zone. Between the 8 ft. anchor and the 16 ft. anchor, the ground transitions to elastic behavior. This aligns well with the 3DEC model result of a 10–12 ft. plastic zone. It also validates the 20 ft. rock bolt length. Taking the difference in displacement between 3 ft. anchor and 16 ft. anchor, the strain within this region is 6 mm/13,000 mm × 100% = 0.05%, well below the limit of new crack initiation, hence elastic behavior.

The MPBX profile in Figure 11 presents a textbook example of ground movements in relation to staged cavern excavation. Other MPBXs show less movement, on the order of only several mm. The MPBXs installed at the cavern haunches indicate between 2–5 mm of displacement which is in line with the 2–4 mm anticipated. The sensitivity of these instruments is remarkable in that very minor "bumps" in the data (less than 0.1 mm) can be linked to the exact time a blast was initiated underground. Therefore, the data is at least an order of magnitude better compared to manually surveyed surface convergence points.

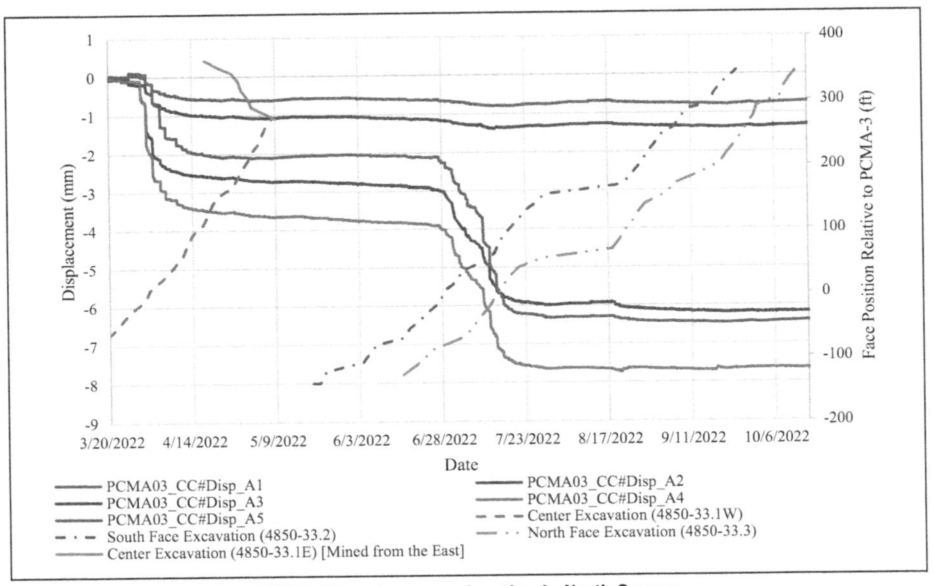


**Figure 10. MPBX Data from PCMA-03 center crown location in North Cavern**

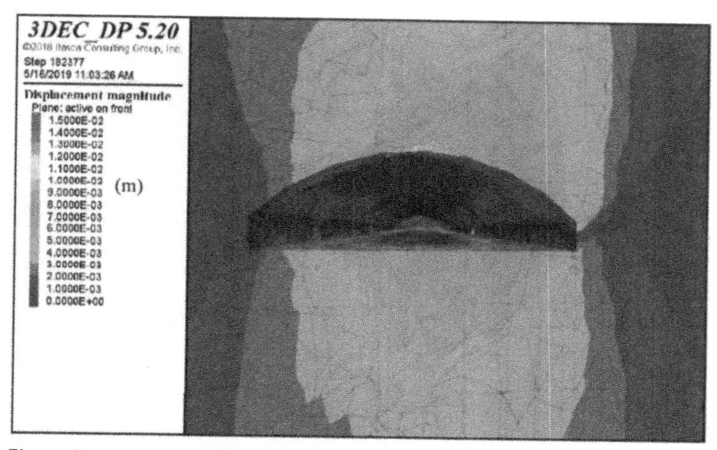


**Figure 11. 3DEC estimated displacements at the completion of the top heading**

## Construction Aspects

The excavation contractor, Thyssen Mining, employed a suite of various sized equipment to excavate the different areas of the project. At the start of the project the only openings were 8 ft. × 8 ft. existing drifts. A small, single boomed Quasar jumbo was used until the existing drifts were enlarged and a workshop excavated to enable larger equipment to be slung underground and rebuilt.

For excavating the detector caverns, the contract required automatic jumbos to drill the blast holes. This specification was included in the contract to reduce operator error in wayward drilling under manual mode which leads to overbreak and associated ground instabilities. Thyssen is using Sandvik 422i jumbos for all the cavern blasthole drilling.

In order to reduce air overpressure on the existing laboratories, trial blasts were undertaken by Arup during the site investigation phase in order to provide a blasting site constant in the Contract Specifications. This enabled Thyssen to calculate the Maximum Instantaneous Chargeweight (MIC) during bid so production rates could be estimated to assist with scheduling and pricing. In addition, blast doors were installed under the advanced works contract, which are closed before each blast to prevent air overpressure from affecting the existing laboratories. To ensure the MIC was not exceeded, Thyssen uses electronic detonators to ensure the precise timing of each blast hole and there prevent overlap of blasthole detonation. During the last 18 months of blasting there have been no exceedance in ground vibration or air overpressure on the existing laboratories.

Ammonium Nitrate Fuel Oil (ANFO) is the bulk explosive used underground which is appropriate as the conditions are dry. All the main production blastholes use ANFO. To reduce blast damage and overbreak around the cavern perimeter, Thyssen uses a cartridged emulsion explosive which is a decoupled explosive charge specifically used in smooth wall blasting. This explosive type has proven excellent for its intended use. During post blast inspections, typically greater than 75% "half barrels" are observed, visual evidence there is little to no blast damage to the cavern profile.

Mucking out of the caverns is undertaken using 4CY muckers which dump onto a grizzly which feeds an ore pass down to the skip loader situated on the 5000L. The rock is loaded into skips and hoisted to the surface. The hoisting, crushing, and conveying system is fully computerized with a hoist operator controlling the skips in conjunction with the rock handling system operator who controls the crusher and conveyor system.

Figure 12. Installation of 20 ft. long FA3 permanent rock bolt in cavern crown using Sandvik 512i

Thyssen's safety protocol states that no person shall work under unsupported ground. Therefore, wire mesh is installed immediately after each round and anchored with 6 ft. long split sets. Not only do the splits provide immediate safety to the excavation, but their use also improves the conformance of the wire mesh to the excavated profile. Following this, a Sandvik 512i is used to install and point anchor the 20 ft. long FA3 permanent rock bolts. The bolter is automatic which ensures the pattern is as per the design, eliminating operator error in missing bolts or incorrect spacing. Even though it is automatic for hole drilling, the bolter still requires manual adjustment to ensure the bolts can be installed and fed into the carousel. With the FA3 bolt having a double corrosion protection plastic sheath, it has been found that this sheath can have problems when being handling in the bolt carousel. In addition, the sheath can get hooked up on the mesh and damaged during installation in the hole. Any damage to the sheath compromises the long term durability and results in abandonment and replacement. After the cement grouting has reached strength, production pull testing is conducted on 2% of all installed permanent bolts. To date, every bolt that has been tested has passed.

Wet shotcrete is applied to the cavern profile to prevent minor rock spalling which could work through the wire mesh openings onto the detectors. The shotcrete is batched underground using a volumetric batch plant. The shotcrete is delivered to site as a dry pre-mixed product delivered in supersacks which are lowered underground in the cage. When required, the pre-mixed shotcrete is deposited into the batch plant hopper and water and admixtures are added. The wet shotcrete mix is then loaded into transmixers which deliver the shotcrete to the spray robot. During the initial phase of drift enlargement, Thyssen used Normet Minimec sprayers. As the caverns have begun to take shape, a larger Getman ProShot has been mobilized for use.

Before ground support installation and after shotcrete application, Thyssen scans the cavern profile to check for compliance, shotcrete thickness, and ensure no underbreak is found. These as-built profile scans, along with rock bolt installation and rock and shotcrete testing forms are submitted as a final as-built package which is used to certify the completion of the works of each individual section.

## CONCLUSION

The overall performance of the rock mass is in line with design expectations through completion of the cavern top headings. This gives a high level of confidence as construction moves into benching, where larger sized wedges and movements are anticipated. Continual evaluation of the ground behavior against the design and adherence to the mandated benching sequence will ensure project goals are met. Lastly, the cooperation and partnering of all key stakeholders has been fundamental to the success enjoyed to date.

## ACKNOWLEDGMENTS

Owner: Fermi Research Alliance, Contractor: Thyssen Mining, Inc., Construction Management Advisor: Kiewit-Alberici JV.

## REFERENCES

Hurt, J., Pollak, S., Havekost, M., Schick, J., and Vardiman, D. 2016. Ground Investigations and Preliminary Design for the LBNF Far Site Conventional Facilities. In *ITA World Tunnel Congress*. San Francisco, CA. SME.

Pollak, S., Seling, B., and Willhite, J. 2021. Pre-Excavation and Early Works Activities for the Long Baseline Neutrino Facility Far Site. In *Rapid Excavation and Tunneling Conference*. Las Vegas. SME.

# Ground Freezing Deep Shaft Excavation Shaft 17B-1 New York City Water Tunnel No. 3 New York, New York

**Andrew Chegwidden** ▪ Keller
**Tara Wilk** ▪ Walsh Construction Company II, LLC

## ABSTRACT

Construction of Shaft 17B-1, a part of The New York City water supply system, required an excavation of approximately 38 m (123 ft) through water-bearing overburden soil and more than 152 m (500 ft) of gneiss bedrock to make a connection to Water Tunnel No. 3. Ground freezing was the specified method to provide temporary earth support and ground water control for the overburden material. The ground freezing process relied on a supplemental geotechnical investigation and comprehensive laboratory testing of the frozen soil to evaluate the potential long-term creep behavior of the soil and the elimination of a temporary lining system. This paper describes that process as well as the drilling and installation methods of the ground freezing system. Additionally, the methodical system of excavation and insulation is discussed that permitted a safe and timely excavation followed by the installation of the permanent concrete liner.

## INTRODUCTION

City Water Tunnel No. 3 (CT3), maintained by the New York City Department of Environmental Protection (DEP), serves as a third means of conveying potable water from Hillview Reservoir to the in-City Distribution Network. The construction of Shaft 17B-1 is a component of Stage 2 of CT3, the Queens-Brooklyn portion. Riser pipes within the shaft will be installed to connect the underlying tunnel and future water distribution chamber at street level.

Shaft 17B-1 required a 17.7 m (58.1 ft) diameter excavation through approximately 38 m (123 ft) of water-bearing overburden soils and more than 152 m (500 ft) of gneiss bedrock. See Figure 1 for the overburden soil profile. Ground freezing was specified as the method of temporary excavation support and groundwater control for the overburden soil. The frozen earth structure was designed and constructed to withstand all lateral earth and hydrostatic pressures and construction surcharge loads without needing a temporary liner.

## SUPPLEMENTAL GEOTECHNICAL INVESTIGATION (SGI)

The specifications required the ground freezing contractor to perform a Supplemental Geotechnical Investigation (SGI) program before the system installation. The goal of the SGI was to classify and confirm the design parameters presented within the geotechnical baseline and data reports. The geology at Shaft 17B-1 can be generalized into five distinct strata presented in Table 1.

The SGI consisted of drilling and sampling one mud rotary boring in the center of the shaft, four sonic core borings, one in the center, and three triangulated around the shaft. Figure 2 shows the drill location of each boring.

**Table 1. Summary of subsurface strata**

| Stratum | Soil Description | Lower Depth | | Lower Depth | |
|---|---|---|---|---|---|
| | | m | ft | m | ft |
| F | Fill | 0.00 | 0.00 | −5.40 | −17.77 |
| MG | Mixed glacial deposits | −5.40 | −17.77 | −18.50 | −60.87 |
| C | Clay | −18.50 | −60.87 | −30.70 | −101.00 |
| DR | Decomposed rock | −30.70 | −101.00 | −37.10 | −122.06 |
| | Gneiss | −37.10 | −122.06 | −100.00 | −328.09 |

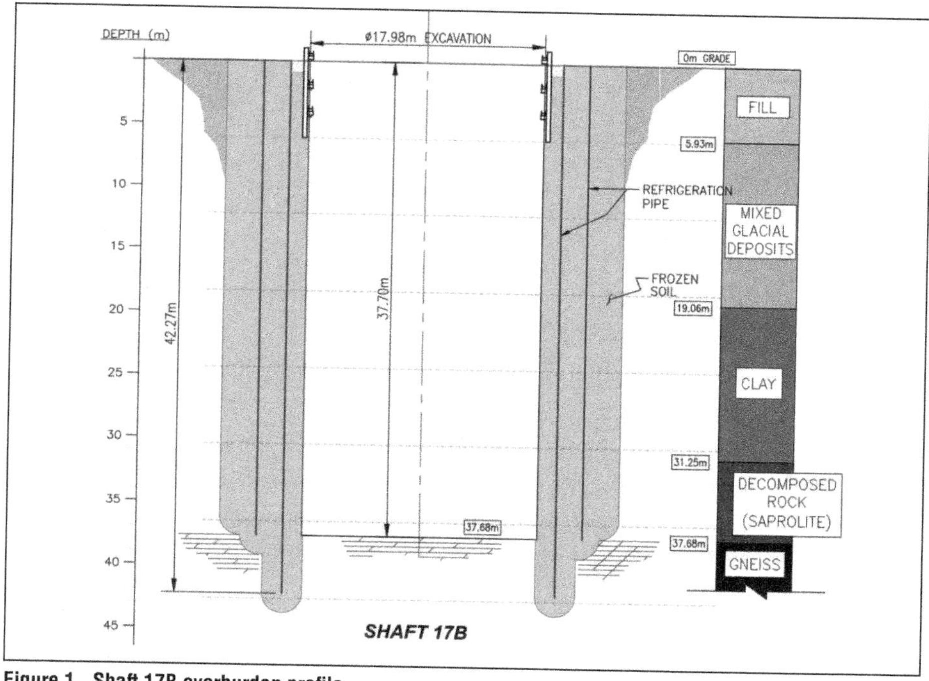


**Figure 1. Shaft 17B overburden profile**

While drilling, the overburden was continuously sampled and visually classified. All soil cores were not preserved; select samples were retained from each stratum and sent to a laboratory for further testing. The samples for frozen soil testing were collected in the mud rotary boring, while those intended for index testing were collected in the sonic core borings. Two samples from each stratum were sent for grain size analysis and Atterberg limits. The sampling was terminated upon encountering the underlying gneiss, and an observation well was installed in each borehole.

The exterior wells were installed with the screen only in the most permeable strata, the glacial deposits. The well in the center of the shaft, PZ-17A, was installed to the full drill depth, screened in all soil strata. Figure 3 shows a section view of the shaft and monitoring wells constructed.

During the mud rotary drilling of the SGI, samples for frozen soil testing were retrieved using a Denison core barrel. The samples were sealed, packaged, and sent to the laboratory to ensure they remained as undisturbed as possible.

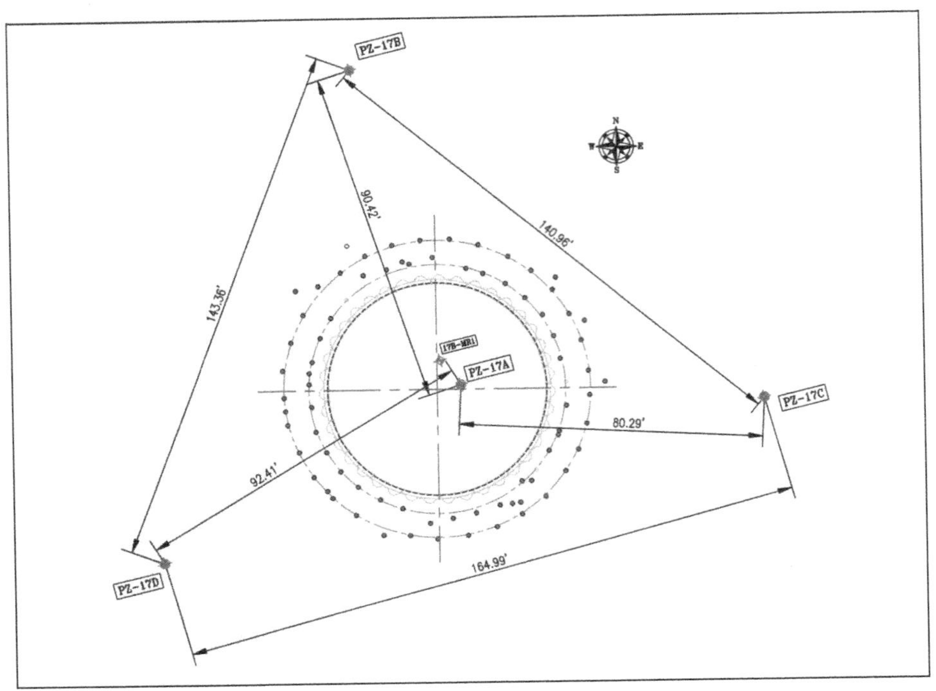


Figure 2. Shaft 17B sampling locations

## FROZEN SOIL TESTING AND DESIGN CONFIRMATION

The project specifications set forth the requirements for the key components of the frozen structure design: groundwater cutoff, structural support, and limitation of time-dependent radial deformation (creep). When ground freezing was previously used for excavation support on Shaft 19B and 20B, excessive creep was observed in the Raritan Clay formation (Sopko et al. 1993; Corwin et al. 1999). With this in mind, a comprehensive frozen soil testing program was deployed during the SGI to ensure the design parameters satisfied each component and mitigated any adverse conditions when incorporated into the Finite Element Method (FEM) analyses.

### Long Term Radial Deformation (Creep)

Long-term radial deformation of the frozen structure was limited to 2% per the specifications. While there are several methods to analyze the creep behavior of frozen soils, the Klein method has proven successful (Klein 1978). The parameters used in this methodology rely on the results of constant stress creep tests and constant rate of strain compression tests.

At the time of the bid, minimal information related to creep properties was available, and the design was completed on a preliminary basis. Plaxis 3D was the software used for the FEM model for structural and creep deformation analysis. Several freeze pipe geometries were considered in the early design stages.

Ultimately, to reduce the amount of creep, a second row of freeze pipes was added to reach the design thickness of the frozen structure, 5 m (16.4 ft). This thickness would

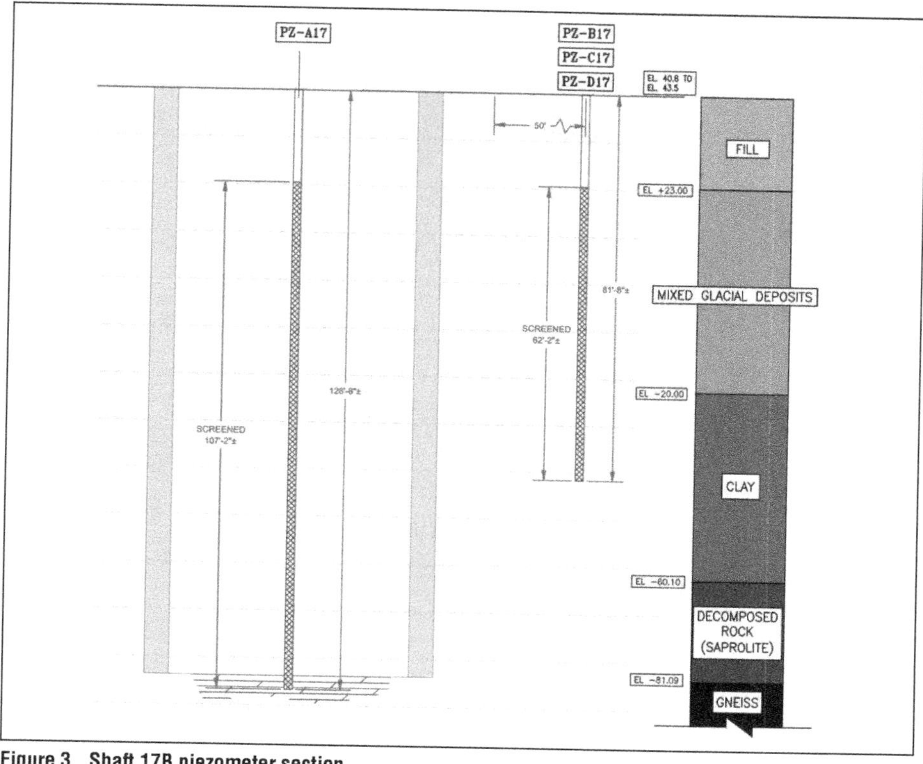


**Figure 3. Shaft 17B piezometer section**

reduce the internal stresses within, minimize the total deformation experienced, and allow the formation of the frozen structure.

## Structural Support

Frozen soil strength is highly dependent on the material properties, the temperature of the sample, and the length of time exposed to an applied load. The strength of a sample decreases with time when a constant load is applied. By performing unfrozen soil triaxial testing and constant rate of strain compression tests previously described the soil strength within the frozen wall can be determined. The soil strength and all lateral earth pressure, hydrostatic, and construction surcharge loads are put into the Plaxis 3D FEM model to confirm that the frozen structure would support the excavation.

The frozen soil testing proved valuable in confirming the parameters used for the FEM analysis.

## Thermal Properties and Groundwater Cutoff

The ability to achieve closure of a frozen structure, when all frozen columns have overlapped to create an impermeable, homogeneous mass, is not dependent on the frozen soil strength properties. Instead, the thermal properties of the soil can be derived from the index testing results described in the SGI (Harlan et al. 1978). In addition to the index testing, thermal conductivity and heat capacity readings were taken onsite on the sonic core samples with a TEMPOS Thermal Properties Analyzer. The onsite data

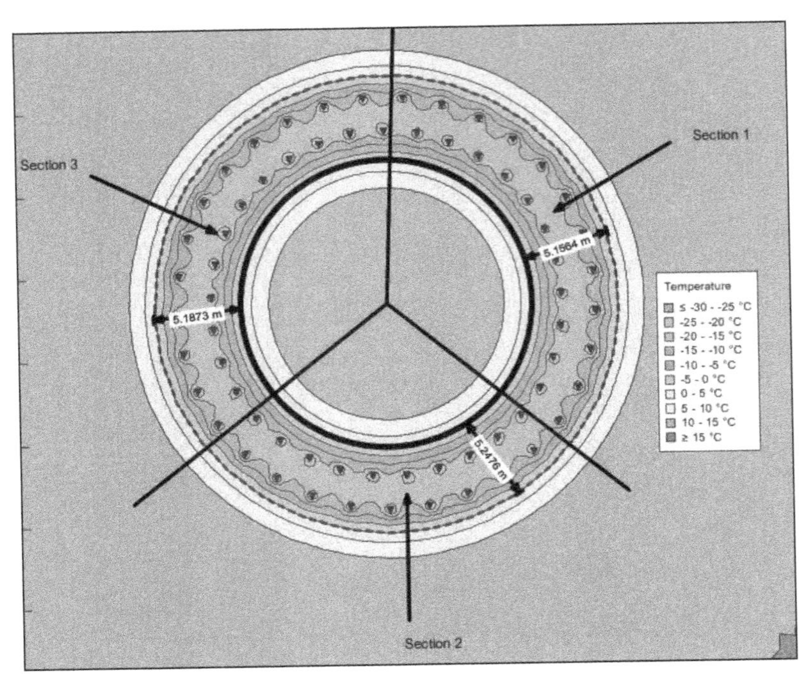


**Figure 4. Shaft 17B day 82 isotherms**

recorded verified the parameters with repeatability across a wide range of the soils compared to the limited samples sent to the laboratory.

Thermal conductivity and heat capacity of each soil stratum, both derived and measured, were input into TEMP/W software. The 2D FEM thermal analysis ultimately confirmed the quantity and spacing of freeze pipes needed to form a continuous frozen structure of adequate thickness to meet the structural requirements within the time allotted in the project schedule. Figure 4 shows the projected isotherms in the clay strata at freeze day 82.

The frozen structure's thermal and structural design was ultimately governed by the clay strata. By nature, most clays are susceptible to creep and tend to have lower thermal conductivity and compressive strength when frozen than coarse grain soils.

## FREEZE PIPE DRILLING

Before the installation of the ground freezing system, interlocking steel sheeting was driven around the shaft circumference from the existing grade to approximately 1 m (3 ft) below the water table. The sheeting provided support to the unsaturated soils above the water table.

The base design for the ground freezing system called for 66 freeze pipes and four temperature monitoring pipes equidistantly spaced around the shaft in two rows. The inner row of freeze pipes was installed to a depth of approximately 42.6 m (140 ft), keying a minimum of 4.5 m (15 ft) into the underlying gneiss, creating a continuous groundwater cutoff. The outer row of freeze pipes was installed to the top of competent rock, a depth of 38.1 m (125 ft), solely to form a thicker frozen structure and mitigate creep within the clay strata. Therefore, the outer row was not required to be installed to the same depth as the inner row. Figure 5 shows the freeze pipe layout for Shaft 17B-1.

The freeze pipes were drilled to the target depth using mud rotary drilling methods. The drilling mud was mixed at the surface, pumped down the drill steel, and out of the bit. As the bit advanced, the cuttings became suspended, returning to the surface and recycled through a mud cleaning system to separate the cuttings.

The greatest challenges to successfully installing the freeze pipes were in the glacial deposit strata, containing several large boulders, gravel zones, and nested cobbles, making it difficult to maintain the stability and verticality of the borehole. At target depth, the drill rods were retracted, and the 100 mm (4 in) nominal refrigeration pipe was installed. Figure 6 is a photo of the boulders removed during the excavation of Shaft 17B-1.

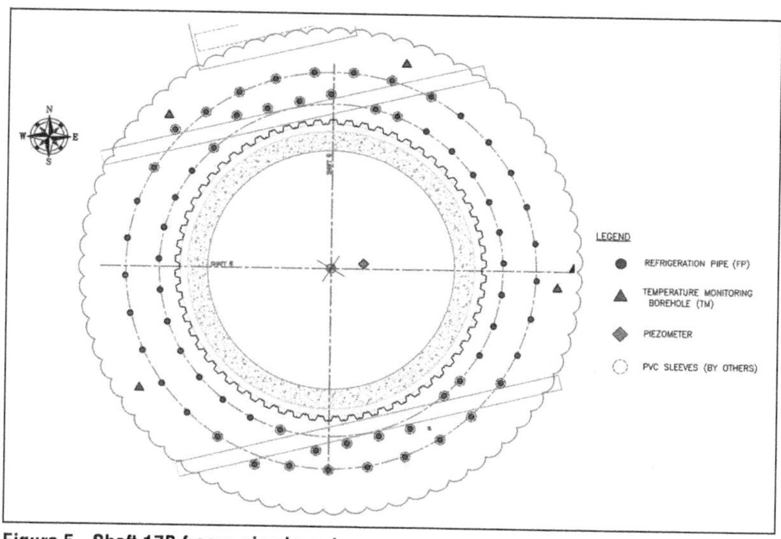


**Figure 5. Shaft 17B freeze pipe layout**

**Figure 6. Boulders removed from the excavation of Shaft 17B**

A robust quality assurance and control (QA/QC) program is equally important to the design as a comprehensive SGI program. Two QA/QC measures were implemented to confirm a successful refrigeration pipe installation; pressure testing and as-built deviation surveys.

Pressure testing confirmed that the welded joints would withstand the system's designed working pressure and prevent the loss of calcium chloride brine in the formation. Each refrigeration pipe was filled with water, and pressure was applied to the water column. If no pressure or water loss had occurred at the end of the specified time, the refrigeration pipe was placed into service.

As-built deviation surveys of the refrigeration pipes were conducted with an inclinometer developed by RST Instruments. A 75 mm (3 in) temporary casing was installed within each refrigeration pipe, and the inclinometer probe was lowered to the tip of each freeze pipe. As it was retracted, incremental readings recorded the pipe position at each corresponding depth relative to the pipe location at the working surface.

In conjunction with the northing and easting location at working surface, the deviation is used to plot the trajectory of each pipe. As-built conditions are then input into the FEM thermal analysis to confirm that the refrigeration pipe spacing provides enough cooling load to support the full growth of the frozen structure. Due to the difficult drilling conditions at Shaft 17B-1, six additional freeze pipes were required to target an area of spacing larger than required by the base design.

## GROUND FREEZING PROCESS & MONITORING

Ground freezing for Shaft 17B-1 was achieved using a secondary cooling system with anhydrous ammonia as the primary refrigerant and calcium chloride brine as the secondary refrigerant. The heat exchanger was located within each of the 430 kW mobile refrigeration plants, containing the ammonia exclusively within the refrigeration plant. Typically supplied at -32 degrees Celsius, brine was circulated to the closed loop system from the refrigeration plants, through the manifold, and to each refrigeration pipe where the heat would be extracted from the ground. Figure 7 shows a process diagram for the brine manifold at Shaft 17B-1.

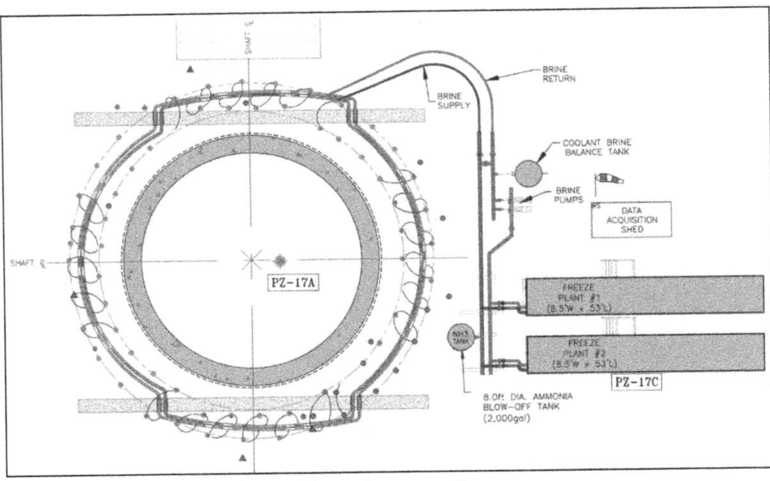


Figure 7. Shaft 17B secondary refrigerant process diagram

A real-time, cloud-based data acquisition system was used to monitor the operating parameters and temperatures of the soil and ground freezing system. The monitoring system allowed engineers and operators to examine the data trends and detect anomalies in real time, regardless of whether they were onsite or remote.

### Ground Temperature Data

Four temperature monitoring pipes were installed through each soil strata to monitor the ground temperatures and freeze growth. Before installing the digital temperature cable (DTC) in each pipe, all sensors were bench tested and calibrated in an ice bath. Baseline temperatures of the ground were recorded before the commissioning of the system.

### System Operation Data

Temperature, flow, and pressure were measured at the refrigeration plant on the distribution manifold supply and return side, ensuring all pipes received a minimum of 56 liters per minute (15 gallons per minute). Each refrigeration pipe was equipped with a DTC sensor to measure the temperature differential from the supply side of the manifold to the return side of the pipe. Refrigeration plant operating parameters confirmed that no alarm conditions existed and that the brine was chilled efficiently.

### Groundwater Level Data

Piezometers installed during the SGI were equipped with a pressure transducer to monitor groundwater elevation trends before and during the ground freezing operation. Due to the challenging site geometry, an independent wireless network was created onsite using a gateway and nodes to collect the data from each pressure transducer.

## FORMATION

Closure of the frozen structure, groundwater cutoff, was achieved on day 56 of freezing, consistent with the FEM thermal analysis completed during the design phase. This was confirmed by the ground temperature data and recorded rise of the center piezometer due to the frozen structure-induced increase in pore water pressure within the shaft. Figure 8 shows the groundwater elevations with the recorded rise in the center piezometer, PZ-17A.

Before excavation, a pump was installed within the center piezometer to confirm no connection between the exterior groundwater and unfrozen pore water within the frozen structure.

Full formation of the frozen structure to the design wall thickness was achieved on day 82 of freezing. As noted, the design wall thickness was governed by the clay stratum, allowing the fill and glacial deposit excavation to commence before the full structural formation at depth.

## SHAFT EXCAVATION

Located in a commercialized area of Queens, New York, the triangular shaped site is only 46 m (150 ft) from an active railroad track. The proposed location of Shaft 17B-1 is toward the street side of the property, only offset from the property line by 3 m (10 ft), requiring the sidewalk be temporarily moved to sink the shaft. Additionally, the shaft was designed with three diameters, reducing a total of 8.8 m (29 ft) from largest diameter to smallest diameter as the excavation progresses to depth. Equipment and construction methods were selected with these challenges in mind.

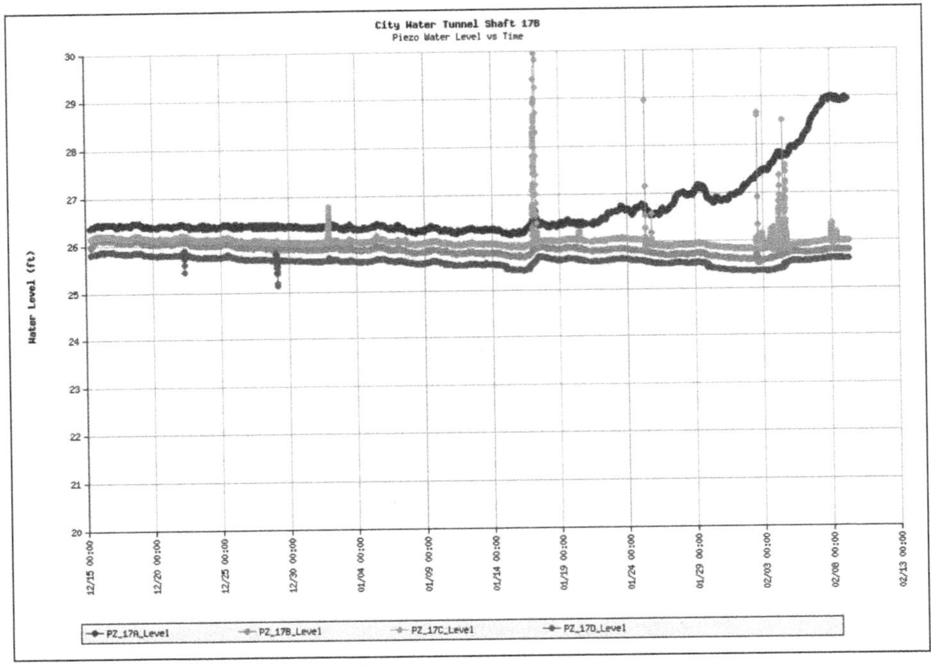


**Figure 8. Shaft 17B groundwater elevation data**

**Figure 9. Shaft 17B project site aerial during construction**

Excavation of Shaft 17B-1 in the overburden section included mucking, hammering the frozen ground at the perimeter, hanging the wire mesh, and applying spray foam. Two excavators were used to sink the shaft, including the CAT 325FL for hammering and the Volvo ECR145EL for excavating excavated and filling the 17-CY muck bucket. The muck bucket was moved in and out of the shaft using a Liebherr HS8200 200-ton Crane, which dumped the muck in the designated area for loading and disposal offsite.

### Shaft Access and Egress

Shaft access and egress were established using a headframe system and were critical to delivering materials and transporting personnel into the shaft. The system consisted of seven winches, two runway rails, two double-main beams spanning the shaft, a headframe platform, and a personnel hoist. The system accommodated all shaft work phases throughout the multiple diameters. The two main beams running across the shaft are installed on the end trucks resting on rails, allowing the main beams to be moved in and out to meet the required diameter configuration. Sheaves are installed between the double-main beams, which can be moved to different locations as needed to accommodate the picking points on the item being hoisted. The winch cables run to the sheaves horizontally from the winches and then vertically down the shaft to hoist the concrete formwork & work decks.

Maintaining safety was expected; therefore, several control measures were added. The personnel hoist cage was designed with spring-loaded locking brakes. In the event of a hoist cable failure, the brakes engage and clamp onto the guide cables, locking the cage in place. Guardrails were installed around the headframe platform, mitigating fall hazards, and around moving parts to mitigate pinch point hazards. Video cameras were installed at the winch locations, allowing the hoist house operator to identify potential cable obstructions before maneuvering the winches. There were also load-sensing pins on the sheaves, alerting the operator if a load was pulled with the winches. Shaft inset covers were on each side of the shaft to prevent anything from falling down the shaft. The covers were supported by a steel sheet pile collar that protrudes 62.5 cm (42 in) above grade around the entire shaft.

### Crane Selection

The Shaft 17B-1 Liebherr HS8200 Crane was a 200-ton with 38.1 m (125 ft) of boom, engineered and fabricated in Germany, and was the first of its kind used in New York City. The crane is a duty cycle, which means that was designed built for heavy duty cycle work. Crane selection considered the heaviest loads, the cable length, and the winch speed required to facilitate mucking and concrete operations extending to a depth of over 213 m (700 ft). Usually, only a tower crane could service an excavation efficiently to this depth, but the HS8200 was engineered for these depths. Due to the proximity of the railway, full-time flagging was required if the crane's boom had a potential to foul or breach the track limits. To avoid the flagging requirement and foul potential, the crane was placed adjacent to the shaft at the furthest distance from the tracks, and the boom length was restricted to 38.1 m (125 ft).

### The Excavated Material

The crane removes the 17-CY muck bucket from the shaft and swings over to the sheeting wall of the muck bin. It engages the puck into the tripping device and tips the bucket over, emptying it into the muck bin. The muck bin sheeting wall was specially designed to contain excavated material, maximizing the use of the small project site. The 17-CY muck bucket has a 2.54 cm (1 in) chain with a puck hanging from the bottom, allowing excavated material to be dumped without personnel in the muck bin area. The trip device eliminates the need for someone to connect rigging to the bottom of the bucket to empty the material, keeping hands away from potential pinch points and muscle strains while lifting heavy rigging for each bucket. The muck bin/muck bucket system also saves significant time.

The excavated material was loaded onto dump trucks using a CAT 950 M Loader and hauled offsite to approved disposal facilities. The trucks drove on the temporary road through the project site. Due to site constraints, the loading area was tight, with a high potential for struck-by incidents. Therefore, the loading area was delineated by signs, and a dedicated spotter was assigned to loading. Furthermore, the loader was equipped with ScanLink, an RFID-based personal detection system that detects people behind mobile equipment. Personnel fixed the RFID stickers on the inside of their hard hats. An alarm would sound, notifying personnel and the operator when an individual was within 7.6 m (25 ft) of the loader (Figure 10).

**Figure 10. Shaft 17B-1 Excavators hammering and loading them muck bucket**

## INSULATION AND PROTECTION OF FROZEN STRUCTURE

The temporary liner was installed between the frozen ground and the waterproofing membrane to provide insulation to the exposed frozen wall maintaining the integrity of the frozen structure and later acting as the substrate for the waterproofing membrane. After considering several options for the liner, wire mesh and the QUIK-SHIELD® 112 XC spray foam was selected.

QUIK-SHIELD® 112 XC is a flexible insulation capable of withstanding the creep of the frozen structure that can maintain its strength and structure while subject to forces of creep. For example, using shotcrete as the temporary liner result in fallout upon applying the hot shotcrete to the cold face of the frozen wall. Even if shotcrete was applied and cured, a creep load could crack or break the shotcrete above those working in the shaft below, creating a potentially catastrophic event. Liner plate as a temporary liner was also considered, but like shotcrete, it is rigid, and once a substantial creep load is applied to the liner plate, bolts and corrugated plates could snap and potentially strike workers below.

Therefore, 1.8 m (6 ft)-high, 1.3 cm (0.5 in) wire mesh with 60d penny nails was secured to the frozen wall face. It was vital to secure the wire mesh as tight as possible to the frozen face. Then QUIK-SHIELD® was applied 112 XC to face of the frozen wall using the Graco E-30 Reactor Spray Foam Mid-production (Figure 11).

## MONITORING FOR CREEP

To fulfill the contract monitoring requirements, a subcontractor was hired to furnish, install, and monitor instrumentation. To monitor creep, five Convergence Monitoring Points (CMPs) were installed at six elevations: +10', –20', –30', –40', –50', and –70' in Shaft 17B-1. The CMPs consisted of a threaded convergence bolt installed into the frozen ground and a high-quality precision optical monitoring target (prisms) with an eye for the tape of extensometer. The CMPs were surveyed weekly to measure potential movement inward that would be caused by creep. Monthly, the extensometer measurements were collected between two CMPs to check for inward movement. CMPs were monitored until the installation of the waterproof membrane. Email notifications

**Figure 11. Shaft 17B-1 temporary liner**

**Figure 12. Shaft 17B-1 CMP were installed through temporary liner**

were sent for yellow level alerts and red level alerts at 1.3 cm (0.5 in) to 2.5 cm (1 in) movement exceedances, respectively.

The monitoring subcontractor furnished and installed four inclinometers outside the shaft's perimeter from surface to top of rock to monitor movement of the ground. The inclinometer would collect and post data automatically hourly to the web-based program, ARGUS Monitoring Software. Email notifications were sent for yellow level alerts and red level alerts at 2.5 cm (1 in) to 5.1 cm (2 in) movement exceedances, respectively. The team did re-baseline the inclinometer alerts level after freeze formation (Figure 12).

## WATERPROOFING INSTALLATION

The subcontracted waterproofing installer furnished and installed the Sikaplan® WP 1100-31 HL2 waterproof membrane and Sika Greensteak 796 water barrier. The membrane is a flexible, homogeneous, and 3.0 mm (0.1 in) thick. The water barrier was installed laterally at every construction joint, typically every 4 m (13 ft) vertically. In the riser valve chamber, access stairwell and elevator shaft, and pipe chamber shaft, the water barrier was installed vertically every 3 m (10 ft). The Klug Construction Systems, LLC Geotextile KCS-TF22 Non-FR is a staple fiber, nonwoven, felt-like material that was installed between the sprayfoam and waterproof membrane as a protection layer for the membrane (Figure 13).

**Figure 13. Shaft 17B-1 waterproofing membrane and waterproofing barrier installed**

## The Surface Preparation

The spray foam was required to meet the smoothness criteria as per Guidelines Tunnel Waterproofing OBV Specification of 4.1.1.2. The qualified supervisor identified and spray-painted areas to shaved or filled with more sprayfoam. To shave, using the ProCUTTER 27.5" Closed Cell Spray Foam Insulation Cutting & Removal Tool 27CCC and the SFS FoamZall FoamSaw—Spray Foam Insulation Trimming Foam Saw w/ 36" Open Cell Blade to remove material. In locations where the foam was damaged and missing or there were small holes, the gaps were filled with foam to provide continuity.

## The Installation

Following excavation to elevation –129.09', the operation transitioned from excavation to waterproofing membrane installation. To reach the entire height of the exposed shaft walls, a consultant was used to design and fabricate a work deck to provide craft access to the shaft walls during multiple phases of work. This minimizes work performed with a man-cage suspended from the crane. The work deck is lowered into the shaft with the crane and connected to the 4-SSK360 winches on the headframe. The work deck is maneuvered vertically using the winches to access different elevations in the shaft. To prevent any fall hazards and the need to tie off, guardrails were included around the entire work deck. However, if a guardrail panel ever needed to be removed and an employee needed to be tied off, the guardrails were also designed as tie-off points. The work deck also supports power distribution for shaft lighting, robodrills, shotcrete pumps, and material storage (Figure 14).

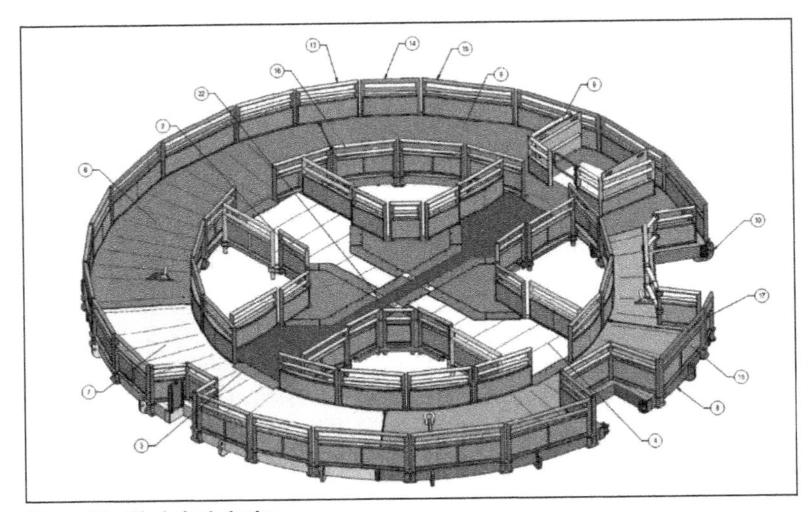

**Figure 14. Work deck design**

**Figure 15. Shaft 17B rebar wall quadrant (left) and formwork (right) being lifted with the HS8200 Crane**

Working off the work deck, tunnel workers anchored geotextile felt with Hilti HPS-1 anchors and an exposed PVC disc. Once secured, the waterproofing membrane was hot air welded to the exposed PVC discs near the top of the roll. After spot welding, welding of all vertical seams continued around the shaft's perimeter concurrently, lowering the work deck as needed.

Since the temporary liner doubled as the substrate for the waterproof membrane, the spray foam had to be thick enough to provide adequate insulation for the frozen structure; however, not too thick that the waterproofing membrane could not be anchored into the frozen soil. The spray foam was applied to limit the void space between itself and the frozen wall. It was assumed that a 4.8 mm (0.2 in.) diameter 10 cm (4 in)-long Hilti HPS-1 anchor with a 10 cm (4 in) diameter PVC disc would be adequate support. Due to overbreak, some larger voids had areas with a thicker layer than 5.1 cm (2 in.) of spray foam applied. To account for these areas, 9.5 cm (0.4 in) diameter 30.5 cm (12 in.)-long galvanized steel spike nails were acquired. A second PVC disc was welded to protect the waterproof membrane from the head of the spike.

## CONSTRUCTION OF FINAL LINER

### The Reinforcement Steel

After the waterproofing, rebar was installed for the shaft. The general contractor determined that 4 m (13 ft)-high placement lift was most efficient based on the total vertical feet in this section. The rebar was cut, bent, and preassembled into rebar wall quadrants which were then lowered into the shaft. In the shaft, the quadrants were placed into position and secured to BA anchors and the previous placement dowels bars. Once secured, walls were spliced together, adding spacers or chairs to ensure correct placement (Figure 15).

### The Formwork

After the rebar, the shaft Everest formwork was installed. The prefabricated 4 m (13 ft)-high steel formwork was lowered into the shaft via crane and then transferred to the headframe winch system. For the first placement in a cycle, the formwork sat atop the muck pile of a level bulkhead. For subsequent placements, the formwork sat on adjustable anchor blocks (shoes), supported by embedded anchors that are tied into the previous concrete placement.

Figure 16. Shaft 17B-1 rebar and formwork

## The Remedial Grout Hoses

The next step was setting the remedial grout hoses so that the hoses would not be fully encapsulated by the concrete. The hoses were heat welded on one side to the waterproofing membrane. The hose was tied to the reinforcing steel to prevent excessive movement during the placement and possible tearing of the waterproofing membrane. The opposite side of the hose was installed in a sealed box to prevent concrete from entering the box.

## The Concrete Placement

Concrete was placed in the form by a 4.6 $m^3$ (6 $yd^3$) concrete bucket, lowered down the shaft, and received by the crew placing the concrete. Due to the vertical rebar spacing and splice length, three vertical splice lengths were replaced with three couplers at eight locations to place concrete at the top of the forms. The concrete consolidation was accomplished with form vibrators and steel-headed flexible shaft vibrators. Once the concrete reached the top of the placement, it was leveled off, and a keyway was cut in with lumber. A circumferential waterstop was welded before the placement and was wet stuck into the top of the finished concrete at the correct elevation.

After the concrete placement and achievement of required strength, the form was removed and moved up with the headframe winches to the next placement to reset the process (Figure 16).

Between each placement and before placing concrete, grout was used as a joint bonding agent to ensure adequate concrete adhesion from one placement to the next. Discharged into a 4.6 $m^3$ (6 $yd^3$) concrete bucket, grout was crane-lowered into the shaft similar to the concrete placement procedure described above.

## CONCLUSION

As a method of ground improvement for deep shaft excavation, ground freezing presents a highly technical, yet effective approach to achieving groundwater cutoff and temporary earth support. With experienced engineers, an effective soil investigation program and comprehensive monitoring system, the formation of the frozen structure at Shaft 17B-1 was achieved.

Excavation of the shaft was completed without a temporary liner. Only a wire mesh and insulation were applied to protect the frozen soil, allowing for the successful and safer construction of the final liner while providing schedule and cost savings. Selection of equipment and means and methods took in consideration of several constraints and challenges.

## REFERENCES

Corwin, A.B., Maishman, D., Schmall, P.C., Lacy, H.S., Ground Freezing for the Construction of Deep Shafts, 1999, *Proceedings Rapid Excavation and Tunneling Conference*, Orlando, FL.

Harlan, R.L. and J.F. Nixon. 1978. Ground thermal regime. Chap. 3 in *Geotechnical Engineering for Cold Regions*, ed. O. B. Andersland and D. M. Anderson. New York: McGraw Hill, pp. 103–63.

Klein, J. and Jessberger, H. L. 1978. Creep stress analysis of frozen soils under multiaxial states of stress. *Proc. 1st ISGF*, Bochum, 1, 217–226, and in *Engng Geol. (1979)*, 13, 353–365.

Sopko., J.A., DelVescovo, A., Ground Freezing as a Principal Remedial Method to Control Ground Water Inflows and Excavation Support for Deep Shaft Construction, 1993, *Proceedings Rapid Excavation and Tunneling Conference*, Boston, MA.

# Large Diameter Shaft Excavation Support Design and Blasting Methods in a Dense Urban Environment for the Pawtucket Tunnel Project

**Andrew R. Klaetsch** ▪ Mueser Rutledge Consulting Engineers PLLC
**Frederic Souche** ▪ Civil & Building North America (CBNA)
**Brian Hann** ▪ Barletta Co
**Nick Goodenow** ▪ Stantec Consulting

## ABSTRACT

This paper describes support of excavation (SOE) design and construction for three large diameter shafts excavated through glacial soil and controlled blasting in complex sedimentary rock formations. SOE in soil consists of unreinforced secant pile rings designed to resist lateral pressure in circumferential compression, eliminating costly steel core beam reinforcing. Controlled blasting at the shafts, adits, and tunnels in densely populated neighborhoods requires managing risks to the public, buildings, and utilities. Controlled blasting approaches limited adverse impacts to receptors nearby while achieving adequate production and fragmentation. A site-specific observational approach was used to adjust blasting parameters, limiting the impact on local stakeholders while maintaining daily progress and managing overbreak.

## INTRODUCTION

The Phase III Pawtucket tunnel project in Pawtucket, RI includes a 30-foot inside finished diameter rock tunnel mined by tunnel boring machine from a 66-foot diameter Launch Shaft located near the Narragansett Bay Commission (NBC) Bucklin Point Wastewater Treatment Facility in East Providence to a 47-foot diameter Receiving Shaft located 2.2 miles to the north along the Blackstone River in Pawtucket. The tunnel will connect by suction header to a new Pump Station constructed in an 84-foot diameter shaft at the tunnel southern terminus. Excavation depth ranges from about 136 feet at the Receiving Shaft to about 152 feet at the Launch Shaft. Over two thirds of the excavation at each shaft is by drill and blast methods in rock. The Launch Shaft and Tunnel Pump Station (TPS) Shaft site, as well as the Receiving Shaft site, border on residential and commercial property in the City of Pawtucket. Shaft sites are shown in Figure 1.

The tunnel, shafts, and ancillary structures comprise Phase III of the NBC combined sewer overflow abatement program, the first phase of which went into construction in 2001. The listed co-authors each participated in the overall collaborative efforts of the Bridging Team, Design-Builder, and Design Team: Stantec provided Program Management services for the NBC Bridging Team; The Phase III design-build contract was awarded to CBNA, a Bouygues-Barletta joint venture; Mueser Rutledge Consulting Engineers PLLC (MRCE) provided support of excavation design services for the large diameter shafts as part of the CBNA design team headed by AECOM. Shaft excavation was completed in 2022, and tunneling operations are underway at the time of this writing.

Figure 1. TPS and launch shaft site (left) and receiving shaft site (right)

## GEOLOGIC CONDITIONS

The Phase III Pawtucket Tunnel alignment and large diameter shaft sites are located within a geologic region known as the Narragansett Basin, which was formed by continental plate collision events over 300 million years ago (Skehan, Rast, and Mosher 1986). Two sedimentary rock units, the Rhode Island Formation and the Wamsutta Formation, are present along the planned tunnel alignment, with the contact between the two formations located within a few hundred feet of the Receiving Shaft site at the north end of the alignment. The Rhode Island Formation is described as predominantly gray sandstone and siltstone, with lesser amounts of shale and conglomerate. The formation is known to contain coal and graphitic shale as minor components of the overall rock mass (Skehan et al 1982). The Wamsutta Formation contains sequences of dense conglomerate, as well as sandstone and shale. Overburden soils are glacial in origin, including till and outwash or glaciolacustrine deposits.

Subsurface conditions at the shaft sites based on the findings of the project subsurface investigation and supplemental investigation by the design team are described in more detail below.

### Tunnel Pump Station Shaft and Launch Shaft Site

Overburden soil encountered at the TPS and Launch Shaft site consists of a surficial layer of fill overlying glacial outwash and glacial till. The overburden profile varies from about 40 feet to about 60 feet in thickness in the vicinity of the shafts. Outwash soils are medium dense to dense granular mixtures of sand, silt, and gravel with minor amounts of clay. Glacial till, where encountered, was generally less than five feet thick. Groundwater levels vary from about 8 feet to 20 feet below existing grade and are influenced by the nearby Seekonk and Blackstone Rivers and by seasonal recharge.

Bedrock of the Rhode Island Formation was encountered underlying glacial soil strata. The uppermost approximately five feet of rock was found to be highly to completely weathered. The remainder of the rock mass within the depths of the planned shafts was described as moderately hard and slightly weathered to fresh. The predominant rock types are sandstone and siltstone. Notably, graphitic shale or coal were not encountered in quantity considered significant to overall rock mass stability.

### Receiving Shaft Site

Overburden soil encountered at the Receiving Shaft site consists of a surficial layer of fill overlying dense glacial till. The overburden profile varies from about 20 feet to about 25 feet in thickness in the vicinity of the Receiving Shaft. Till soil consists of variable mixtures of gravel, sand, and silt with significant presence of cobbles and boulders.

Groundwater levels vary from about 10 feet to 15 feet below existing grade, consistent with the water in the Blackstone River upstream of the nearby Central Falls Dam.

Bedrock of the Wamsutta Formation was encountered underlying the glacial till. The uppermost approximately five feet to ten feet of the rock formation was found to be of poor quality, generally improving with depth. The predominant rock types are sandstone and conglomerate, with lesser amounts of siltstone.

## EXCAVATION SUPPORT DESIGN

### Excavation Support in Soil

Circular secant pile walls were selected as the method of temporary excavation support in soil for the three large diameter shafts, as this method is well suited for the rock depth at each shaft, and due to the need to establish groundwater cutoff. To achieve groundwater cutoff and support lateral soil pressures, secant piles were socketed into unweathered bedrock and designed to have sufficient overlap at depth to form a continuous circular shell, taking into account installation tolerances. Installation of the secant piles was followed by a pre-excavation grouting program to effectively seal the secant-rock interface. The secant wall design included no internal bracing, allowing for rapid shaft excavation upon completion of the grouting program.

#### *Analysis Model and Applied Loads*

The analysis model for the temporary secant walls consisted of a circumferential structure of uniform thickness and diameter. Loads to be resisted by the secant walls during excavation included soil and groundwater pressures, as well as localized construction surcharge pressures near the surface. Soil pressures were approximated as at-rest pressure and groundwater pressure was calculated based on hydrostatic conditions.

The loading conditions resulted in approximately uniform pressure around the secant wall circumference, increasing with depth. A closed-form solution analysis was used to establish initial dimensions and concrete strength, followed by 3D finite element analyses with soil structure interaction considering various construction surcharge cases. It was determined that an unreinforced circular secant wall of sufficient thickness to resist the pressures in circumferential compression was a feasible option, eliminating the need for steel core beams and internal bracing. Localized, near-surface bending moments resulting from non-uniform construction surcharge loads are resisted by a cast-in-place reinforce concrete ring beam at the top of the secant pile walls.

#### *Finite Element Modeling*

Finite element analyses were carried out in Plaxis3D using soil and rock parameters provided in the project GBR-C as well as supplemental borings made by the design team at the shaft locations. The analyses simulated all stages of construction: secant wall construction; shaft excavation; TBM assembly (Launch Shaft); TBM recovery (Receiving Shaft). Each model stage considered unique construction equipment surcharge loads to determine the forces and lateral deflection of the secant pile walls and cap beams.

Construction surcharge considered in the finite element analyses included several crane types and positions relative to the shafts, muck piles, TBM components, and other general surface surcharge loads. Six surcharge cases were analyzed for the TPS shaft, twenty for the Launch Shaft, and eighteen for the Receiving Shaft. A summary of load cases is shown in Table 1. Due to the relatively close proximity of the TPS

**Table 1. Summary of surcharge cases, TPS and launch shafts**

| Shaft | Phase | Crane/ Muck Storage Pile | Surcharge Case |
|---|---|---|---|
| TPS Shaft | Excavation (Phase 1) | MLC300 Crawler Crane without Muck Storage Pile | 1A 1B 1C |
| | | MLC300 Crawler Crane with Muck Storage Pile | 1D 1E 1F |
| Launch Shaft | Excavator (Phase 1) | MLC300 Crawler Crane without Muck Storage Pile | 1A 1B 1C |
| | | MLC300 Crawler Crane with Muck Storage Pile | 1D 1E 1F |
| | | 100 Ton Crawler Crane without Muck Storage Pile | 1G 1H 1I |
| | | 100 Ton Crawler Crane with Muck Storage Pile | 1J 1K 1L |
| | TBM Assembly (Phase 2) | LR 1600 Crawler Crane without Muck Storage Pile | 2A 2B 2C 2D 2C 2D |
| | | MLC300 Crawler Crane without Muck Storage Pile | 2E 2F 2G 2H |

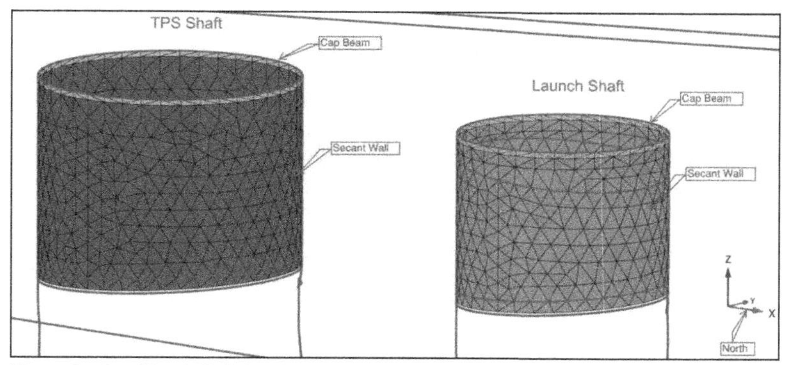


**Figure 2. Combined 3D finite element model showing deformed mesh of TPS and launch shafts**

and Launch Shafts, surcharge cases were first analyzed in separate finite element models to identify governing cases, then a combined model was utilized to simulate the interaction between the two shafts during construction and excavation (Figure 2).

### *Analyses Results*

The TPS Shaft and Launch Shaft designs utilized 1,180 mm diameter secant piles of unreinforced concrete with minimum compressive strengths of 4,400 psi and 2,900 psi, respectively. The Receiving Shaft design consisted of 1,000 mm diameter secant piles with minimum compressive strength of 2,600 psi. Results of the finite element analyses described above indicated the anticipated lateral deflection of secant walls and cap beams at each shaft would be less than 0.2 inches. Results of the individual analyses of the TPS and Launch Shaft were similar to those of the combined models, indicating that excavation at one shaft would have little effect on the adjacent shaft. These results were corroborated during construction by an instrumentation and monitoring program.

## Excavation Support in Rock

The project criteria documents recommended initial rock support in the form of grouted rock dowels and wire mesh reinforced shotcrete. Alternatives were considered during detailed design, including use of fiber reinforcement in lieu of wire mesh, and friction bolts in lieu of grouted dowels. After consideration of alternatives and based on the preference of CBNA, final design was advanced consistent with the initial support system described in the project criteria documents.

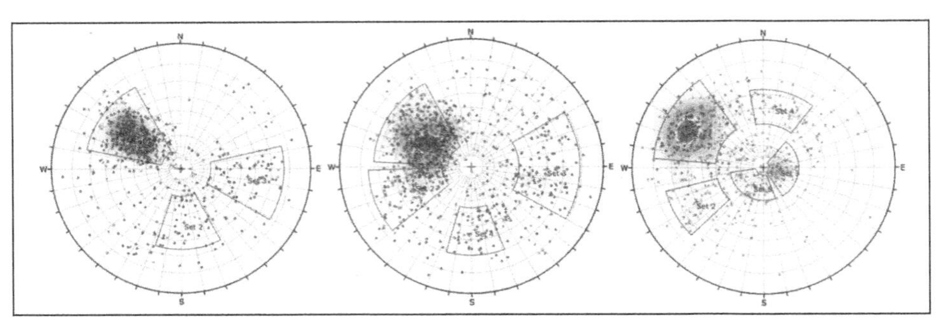


**Figure 3. Lower hemisphere polar plots: TPS shaft (left); launch shaft (center); receiving shaft (right)**

Initial rock support demand in the large diameter shafts was analyzed using the UnWedge software by Rocscience. The analyses were informed by discontinuity data obtained from borehole televiewer logging and strength parameters determined from laboratory testing conducted during the initial project subsurface investigation and the supplemental investigation by the Design Team.

### *Rock Discontinuities*

Rock discontinuity data input to the wedge analyses was obtained by borehole televiewer logging of three borings nearest to each large diameter shaft. The interpreted data provided the dip angle, dip direction, and aperture of each discontinuity detected by the televiewer. Data were summarized in stereonet plots generated using the commercial software program DIPS, as shown in Figure 3. Each point on the stereonet indicates the dip and orientation of an individual fracture, enabling identification of fractures sets with similar dip angle and orientation.

Fracture orientation was found to be fairly consistent across the TPS and Launch Shaft site, as indicated by the similar orientation of identified fracture sets shown in the stereonet plots in Figure 3. A fourth set was added to the analyses of the Launch Shaft, as the orientation of bedding-parallel fractures was found to be spread over a broader range, possibly indicating some localized folding of rock strata. Bedding-parallel fractures at the Receiving Shaft are generally of similar orientation and slightly greater dip than those at the TPS and Launch Shaft. Five fracture sets were considered in the Receiving Shaft analyses, due to the significant presence of near horizontal fracture planes, represented by points plotting near the center of the stereonet.

### *Analyses and Design*

Rock wedge analyses were performed separately for each large diameter shaft based on the fracture orientations and shaft diameters unique to each. Joint friction was set to the lower bound of the range of direct shear test results obtained during the subsurface investigation, and joint cohesion was conservatively assumed to be zero. Constant water pressure was applied along discontinuities consistent with the results of a shaft seepage analysis performed by the Design Team. In each shaft, the analyses demonstrated that adequate support was provided by grouted dowels 10 feet to 12 feet in length at a six foot to seven foot spacing, coupled with five inches of shotcrete.

Initial support design at each shaft was modified slightly to indicate a tighter dowel spacing within the zone of pre-excavation grouting at the base of the secant pile ring, for the purpose of providing additional support where rock may be of relatively poorer

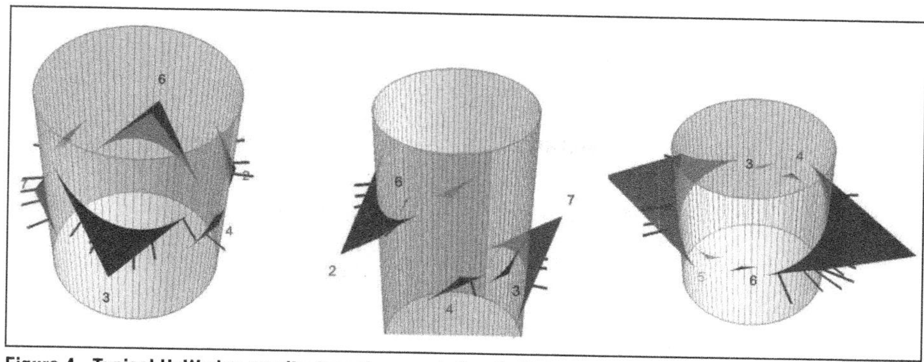


**Figure 4. Typical UnWedge results; tunnel pump station (left); launch shaft (center); receiving shaft (right)**

quality or contain greater fracture frequency. Additional procedural requirements were included in the design, such as line drilling and excavation by mechanical means only near the top of rock to reduce risk of undermining the secant pile SOE system or damaging the grouted cutoff zone. Below this critical zone, the remainder of each shaft was excavated by drill and blast methods. Blast design is discussed in the following sections.

## CONSTRUCTION OF SHAFTS AND ADITS USING CONTROLLED DETONATION

Hard rock excavation by drilling and blasting generates noise and vibrations which can affect nearby residents and industrial activities in the vicinity. Lacking understanding of risks associated with blasting projects may lead to over-conservative design assumptions, resulting in unnecessary costs. Alternatively, underestimating vibration risks can result in unexpected damage to buildings, complaints from the public, unforeseen costs, and excavation delays. By applying observational approach strategy, the cost-effectiveness can be enhanced without generating uncontrollable risks. Blasting at the Pawtucket Tunnel Project requires fundamental understanding of vibration propagation in soil and rock and their interaction with structures. Relatively inexpensive vibration monitoring and data acquisition systems are available today, which provide valuable information about wave propagation in the ground and dynamic interaction of structures and foundations. This fundamental approach was used to successfully implement a blasting program with limited impacts to production, while limiting vibration, noise and flyrock, relevant to nearby residents and structures.

### Blasting Impact Assessments

The contract requires blasting at three separate areas along the tunnel alignment:

- Main Site: launch shaft, pump shaft starter tunnel, tail tunnel, OF218 adit (where TBM is launched)
- Drop shaft 213 and adit tunnel (where MTBM is launched)
- Receiving shaft (where TBM is retrieved)

#### *Main Site; Launch Shaft, Pump Shaft, Starter Tunnel, Tail Tunnel and OF218 Adit*

Figure 6 includes a 500-foot radius indicating nearby properties and structures considered most likely to be affected. The blast design specifically included these areas where commercial and residential structures are located nearby. Designing the blasting process around the main site limited the vibration to the USBM RI 8507.

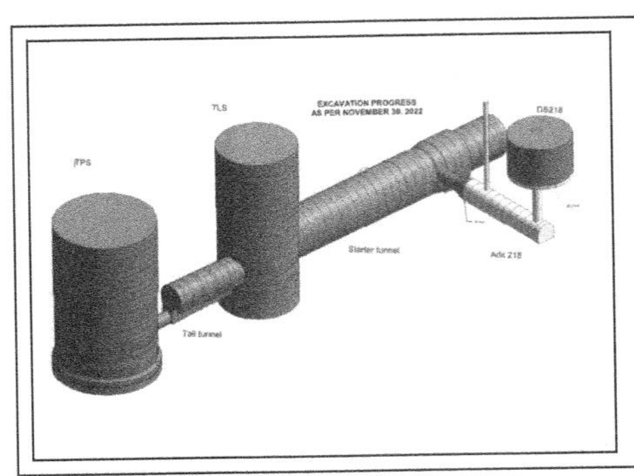

Figure 5. BIM model of main site drill and blast arrangement (left) and installation of TBM thrust frame in blasted top heading

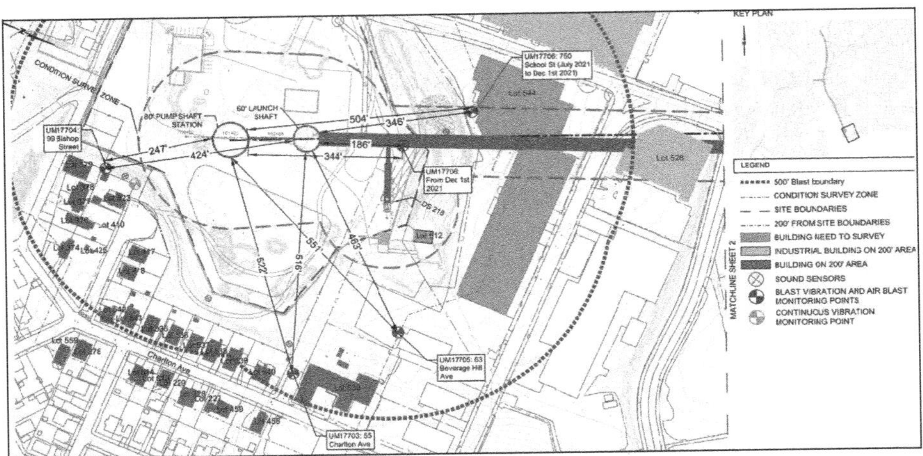


Figure 6. Main site blasting impact assessment & monitoring (air overpressure & vibrations) stations

### *Drop Shaft OF213 Drop Shaft and Adit*

Properties and structures are considered when designing the blasting process at OF213 drop shaft (see Figure 7):

- Lot 616: 71' to drop shaft center
- 12" cast iron gas line: 54' to drop shaft center

The other properties and structures within the 500' blast zone are more distant and therefore less sensitive to vibrations and air overpressure than lot 616 and the gas line (Figure 8).

### *Receiving Shaft*

See Figures 9 and 10.

## Blast Monitoring

Vibrations and air overpressure are monitored by a specific type of seismograph, the Instantel Micromate (Figure 11).

Figure 7. BIM model of drop shaft 213 (left) and blasting process

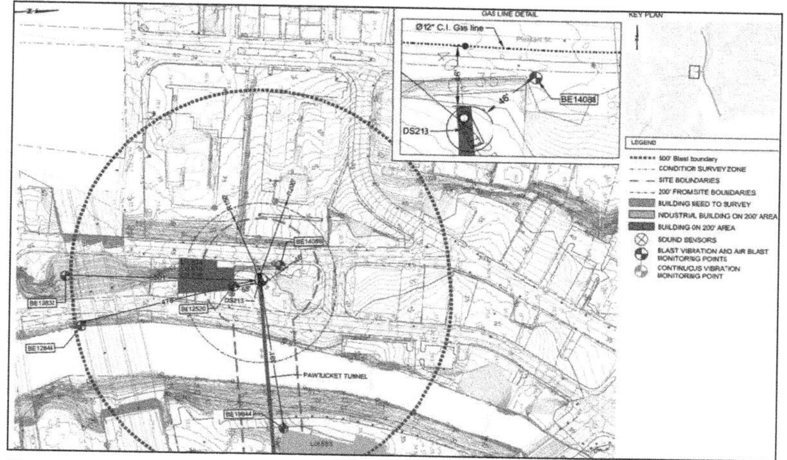


Figure 8. DS213 blasting impact assessment & monitoring (air overpressure & vibrations) stations

Figure 9. BIM model of receiving shaft (left) and partial blast covers installation

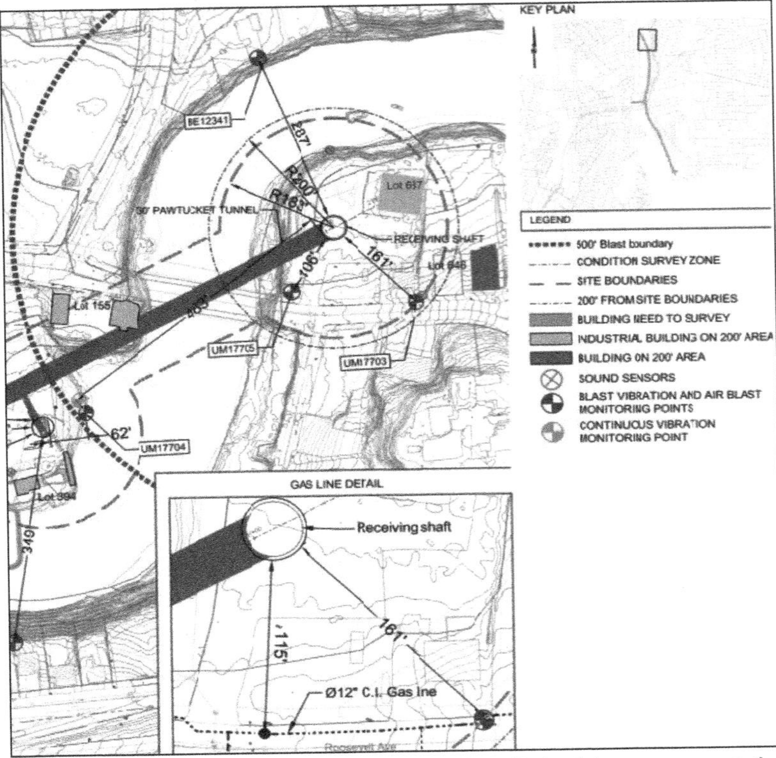


Figure 10. Receiving shaft blasting assessment and monitoring (air overpressure and vibrations) stations

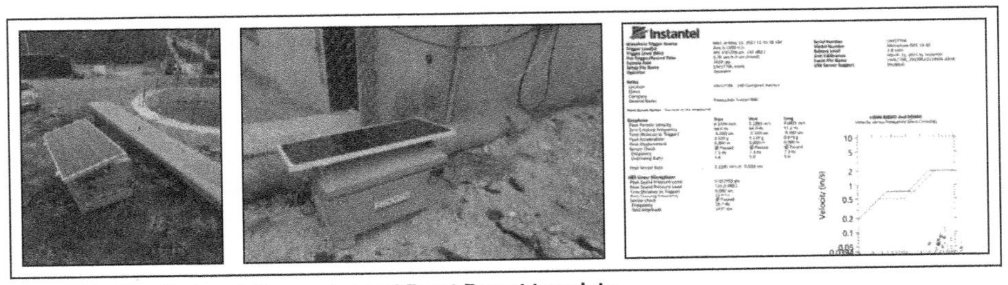


Figure 11. Monitoring stations setup and Event Report template

### *Vibration Monitoring*

Ground and air vibrations are monitored to assess the potential for building damage using Instantel's Micromate and Minimate seismographs with exterior geophones, temporarily placed under or adjacent to a secure enclosure to provide real-time monitoring. The "Auto Call Home" feature of the seismograph is tied into the project specific website that allows for data visualization and real-time notification via text or email to designated personnel when recordings have exceeded the limiting criteria. At main site, total of five seismographs are installed, five on DS213 site and four on Receiving Shaft site.

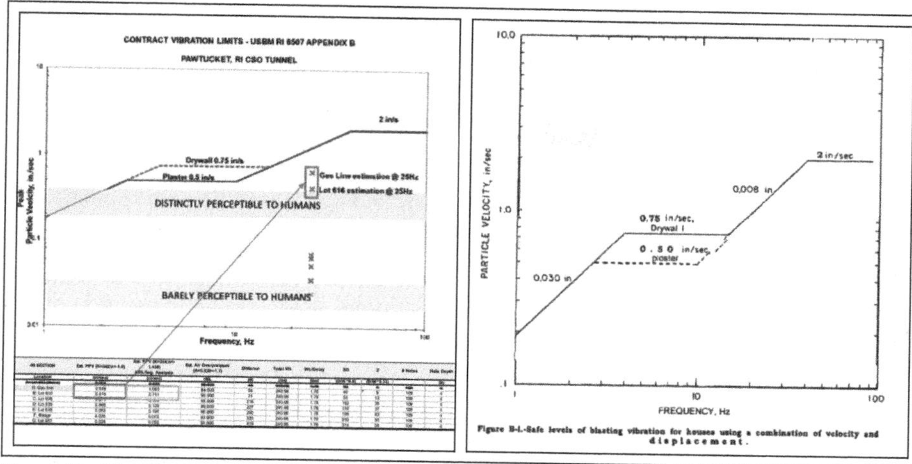


**Figure 12. Vibrations estimations at DS213 (left) and USBM RI 8507 vibration curve**

During each blast round, the seismographs record blast vibration levels, measured as the Peak Particle Velocity (PPV) in inches per second (in/sec), the related vibration frequency in Hertz (Hz) or cycles per second, the Air Overpressure (AOP) in pound per square inch (psi) converted to decibels (dBL) on the Linear Scale, the elastic displacement in inches, and the noise level in dBA on the A Scale. Existing baseline vibration and noise levels are determined at each of the locations where the potential impact on critical structures and the public is most severe. The purpose for the baseline vibration and noise measurements is to provide CBNA Barletta JV with general information about the existing vibration and noise levels from various sources so that appropriate judgments can be made about the potential impacts of blasting and construction on adjacent buildings. An Event Report generated from the seismograph presents actual data, as well as time-history wave forms of the velocity, AOP and noise, and a Compliance Report that shows the actual vibration and frequency data plotted against the required safe limits of the U.S. Bureau of Mines (USBM) Report of Investigation RI 8507, published in 1980. This report recommends a maximum safe level of PPV for residential structures based on the frequency of blast vibrations. Above 40 Hz USBM recommends a maximum PPV of 2.0 inches per second, to minimize the probability of cosmetic damage to interior walls of residential structures. These limits are based on the frequency and the PPV of the blast vibrations and are safe limits for preventing cosmetic damage to the weakest building materials (hairline cracking, or the extension of existing hairline cracks to plaster and sheetrock walls) within residential structures. These criteria are consistent with the blasting restrictions required in the Contract (Figure 12).

### *Air Overpressure Monitoring*

Air blast overpressures, in units of pounds per square inch, (psi) are measured with the seismograph on a linear scale and are typically converted and reported as the sound equivalent in decibels (dBL). The acoustic overpressures (air blast noise) at adjacent structures are not to exceed 0.013 psi, which is equivalent to approximately 133 dBL. Air blasts produce an atmospheric pressure wave transmitted from the blast funneled upward through the shaft and outward into the surrounding area. This pressure wave consists of audible sound that can be heard, and concussion or sub-audible sound (<20 Hz) that cannot be heard. In general, a blast is felt by the pressure more than heard by the noise.

We have used a similar propagation equation PPV = K (D/W1/3)-b for estimating air overpressure with a site K factor of 0.1, and a site attenuation rate of –1.1 (average for confined construction blasting as per ISEE's Blaster's Handbook) in our blast designs. All estimated air overpressures in our designs have been kept below the limit of 133 dBL, or 0.013 psi, which is cited in the specification and is the same limit recommended by the U.S. Bureau of Mines to prevent damage to windows. This would be the first damage if it were to occur. Actual breakage of glass would not be expected until air blast overpressures have reached at least 140 dBL, much higher than the USBM 133 dBL safe limit. Damage to walls of a house would not be expected until air blast overpressures were more than 175 dBL

### *Safe Limits for Ground and Air Response*

Safe limits for ground vibrations for surface structures have been set at the following:

- Ground Vibrations for Surface Structures: USBM RI 8507 Appendix B, 1980
- Ground Vibrations for Engineered Structures and Underground Utilities: 2.0 in/sec
- Air Overpressure: 133 dBL

No blast shall exceed the ground and air vibration limits as per the observational approach adopted by CBNA Barletta JV.

Blast covers are installed prior to each blast, primarily to prevent the ejection of flyrock and to direct air over pressure. The shaft covers consist of rigid steel frame panels each covered with expanded metal and wire mesh with sufficient structural capacity to support persons, anticipated equipment, and materials. The covers also deter vandalism and prevent unauthorized or accidental entry of persons or objects into the shaft excavation. To limit air overpressure, the top of the shaft vents at the shaft collar through an overhang of the blast cover.

After blasting, the shaft cover is removed from the top of the shaft and the ventilation fan exchanges the air inside the shaft. The underground dust limit for silica is PEL < 25 micro grams/cubic meter over 8 hours, thus the need for wet drill. After fumes and dust have cleared, the BIC thoroughly inspects the blast area for indications of misfires, need for scaling, and other hazardous conditions resulting from the effects of the blast. Only after determining the blast area is safe, the "All Clear" signal is sounded (Figure 13).

**Figure 13. Blast cover panels being installed; launch shaft (left) and DS213 (right)**

## Pre and Post Construction Surveys, Notifications, and Communication

In accordance with the Contract, pre-construction condition survey was performed at all buildings, structures and utilities (not controlled by the project) within 80' offset to tunnel C/L, 150' radius from small shafts and 165' radius from large shafts. A pre-construction condition survey plan was provided for each of the five sites that delineates these areas. In addition, CBNA Barletta JV choose to invest in strategic targeted pre-construction survey offerings and or notifications beyond these specified limits.

Pre-construction surveys are documented with recorded or written notations, photographs, videos, or other methods, of the interior and exterior of subject properties. The surveys are conducted in the presence of the owner of the inspected structure or his designated agent. Post-construction surveys will be completed at locations where a damage claim has been reported or where instrumentation data indicates that the limits have been reached or exceeded.

Residents within the survey area are provided with written notification of the commencement of blasting operations with an explanation of blast warning signal sequence code. Neighbors providing contact information (email/phone) who request daily notification are notified of blasting events. Positive public relationship is essential to the successful build of this project (Figure 14).

Drilling and blasting operations coincide with project construction workdays, Monday through Friday. Blast events are scheduled between the hours of 7:00 AM and 5:00 PM. Blasting cannot be conducted at times different from those announced in the blasting schedule except in emergency situations, such as electrical storms or public safety required unscheduled detonation. Blasting is scheduled to minimize disturbance of public, commercial and government operations.

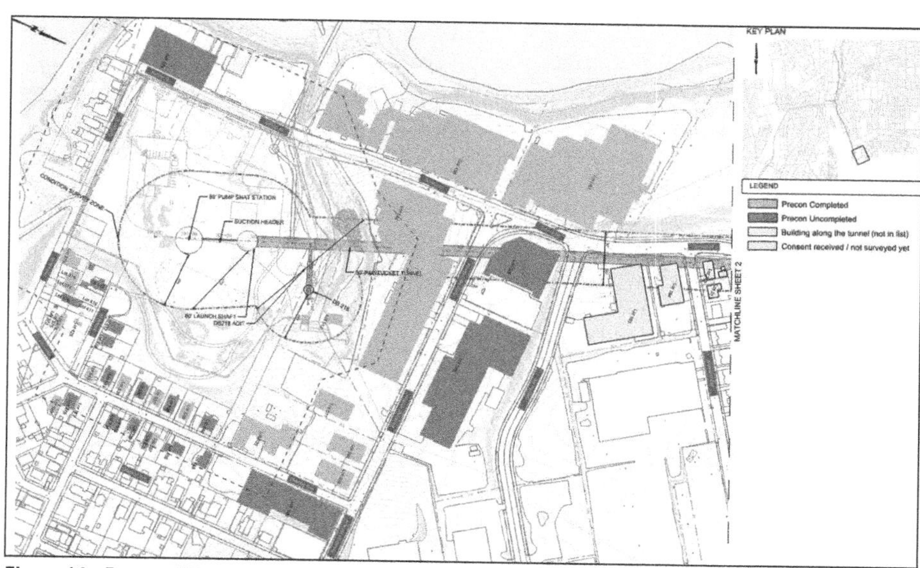


**Figure 14. Pre-condition survey status at main site**

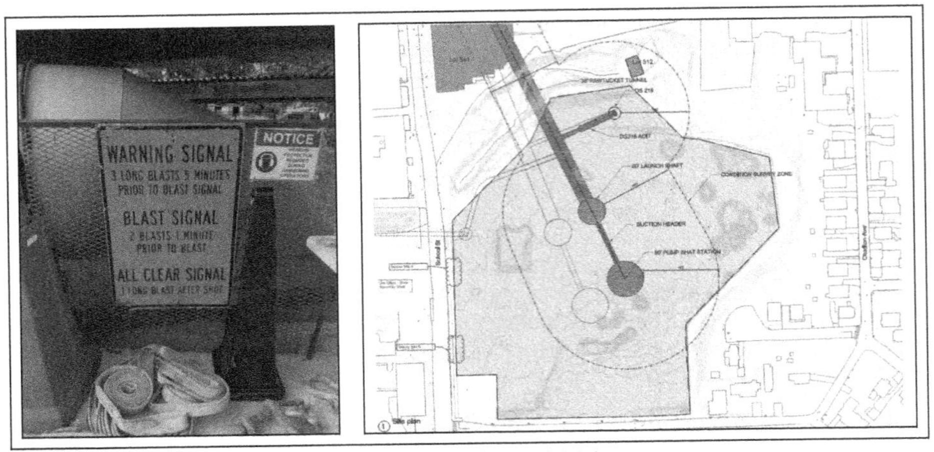


**Figure 15. Warning signal sign (left) and blast security area (right)**

## Blast Area Security, Warning Signs and Signals

Given the proximity of the blasting activities to the surrounding residents, site safety is paramount to the successful completion of the project. We have developed a Site Security Plan for each of the five areas identifying at a minimum, the blast area, equipment requiring removal, blast area access points, sentry locations and designated "safe area(s)." Blast area and blast signal code signs are posted. Areas in which charged holes are awaiting firing are guarded and posted or flagged against unauthorized entry. The secured blast area perimeter may include portions of surface roadways. At the main site, pedestrian and vehicular traffic on School Street was briefly interrupted for the event.

Each blast is preceded by a security check of the controlled area and then a series of warning signals. The project site superintendent, local authority, and Fire Department are in direct contact to ensure the safest possible blast operations. All personnel in the vicinity of the blast area are warned. A sign displaying the warning signal sequence are conspicuously posted at each area. The warning signal is designed to be audible at a distance of 250' from blast site at the surface. No blast is fired until the area has been secured and determined to be safe. The warning signal sequence is as follows:

1. 3 Audible Signal Pulses at 5 Minutes to Blast
2. 2 Audible Signal Pulses at 1 Minute to Blast
3. 1 Audible Signal pulse—All Clear

The blast site is examined by the blaster prior to the all-clear signal to determine that it is safe to resume work (Figure 15).

## Blasting Designs and Observational Approach

### *Main Site—Larger Shafts*

Two test blast designs and one production blast design for the launch shaft (first blasts' location) have been designed. Results of the test blasts are used to evaluate safety relative to vibrations, air overpressure, flyrock and direction of throw. Results also assist with subsequent production blast designs, in determining blast energy efficiency, the effectiveness of fragmentation, and potential for overbreak (Figure 16).

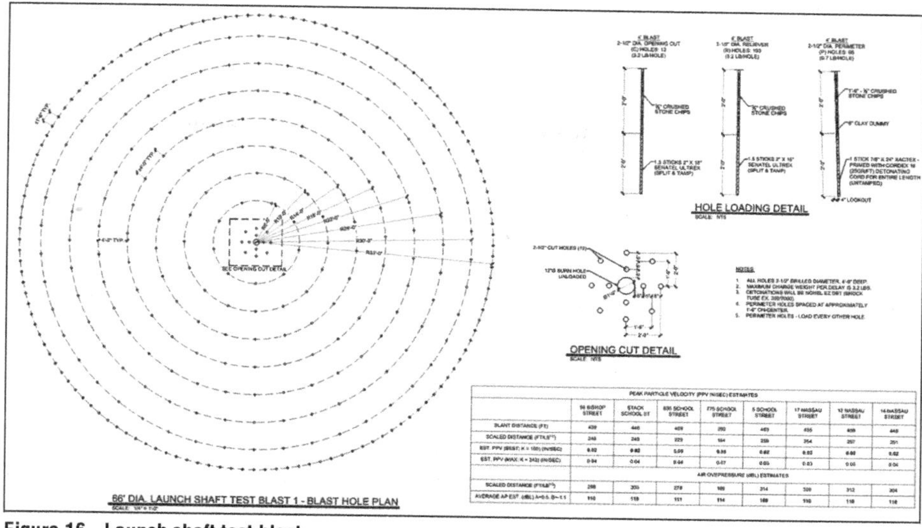


**Figure 16. Launch shaft test blast**

Our blast designs for each of the structures are consistent relative to explosive materials and detonators used, geometry and timing. Blast designs typically include drilling 1-¾" to 2-½" diameter holes to depths of between 4' and 8' for test blasts and 12' for production blasts at a 2' to 4' by 2' to 4' burden and hole-to-hole spacing, respectively. For large shafts at main site, the test blasts are full face but production blasts are limited to half face shots to limit mucking out time and improve overall cycle. Each blast round for the Starter and Tail tunnels are blasted in three shots, Top Heading and two Benches (see Figure 17).

Explosives used are Austin Powder's Emulex 927 (1.5", 2" diameter by 16" long) detonator sensitive and water-resistant packaged emulsion for reliever holes and Austin Powder's Emuline (⅞" by 16") packaged emulsion primed with A-Cord detonator cord for the perimeter holes. The initiation system are Austin Powder's Shock Star LP non-electric 200/5000 dual detonators. These explosive materials have been selected for their very good fume characteristics (Fume Class 1) and their excellent water resistance.

For the larger shaft blast designs, the powder factor (PF) was determined to be between 1.2 and 1.5 lbs./cy for the blasts. The maximum charge weight per 8 msec delay for the Launch shaft is 15.1 lbs. In general, perimeter control has been provided using holes spaced at 18" on-center with either every hole loaded, or every other hole loaded with decoupled explosives. A 12" diameter unloaded hole drilled at the center to full depth will be used for the opening cut generating the second free face at the four shaft locations.

Ground and air vibrations have been estimated for each of the blast designs using the industry standard regression equation as shown in the blast designs. We understand that GBR-C recommends the vibration regression equation with K=393.4 and n=−1.46. However, Industry Standard considers K=160, 242 (maximum) and n=−1.6 as conservative and are proposed to be used. All explosive packaging and waste materials will be either shredded or burned and removed from the site (Figure 18).

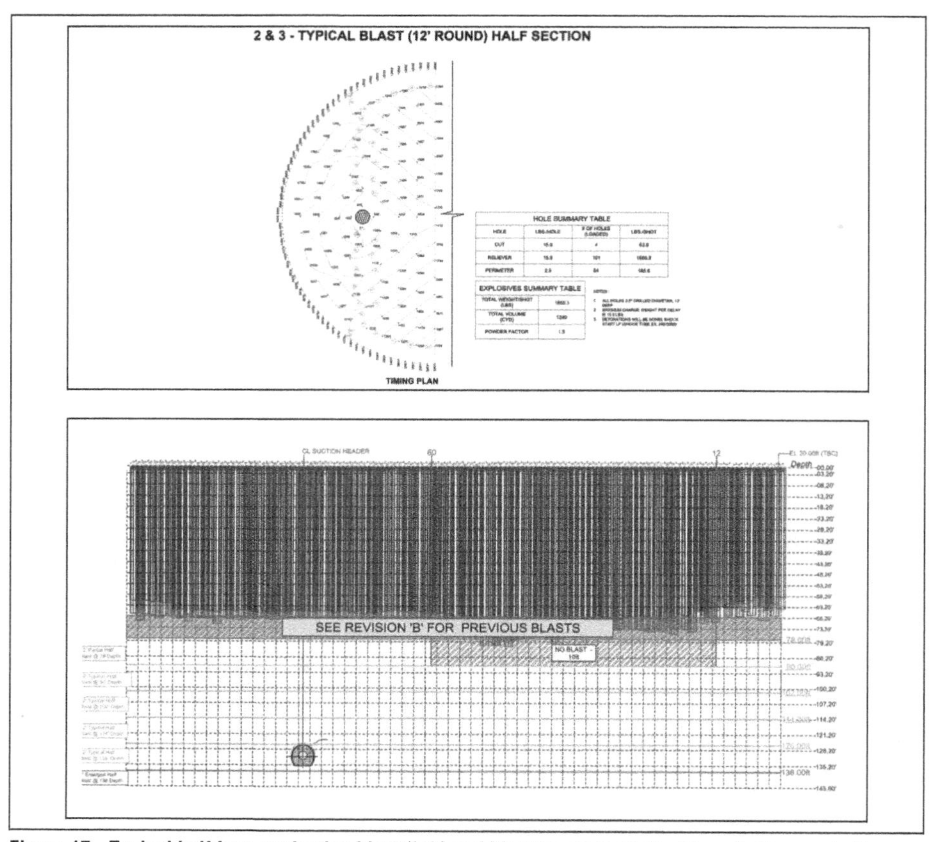


Figure 17. Typical half face production blast (left) and blast levels in shaft with as-built secant piles overlayed (right)

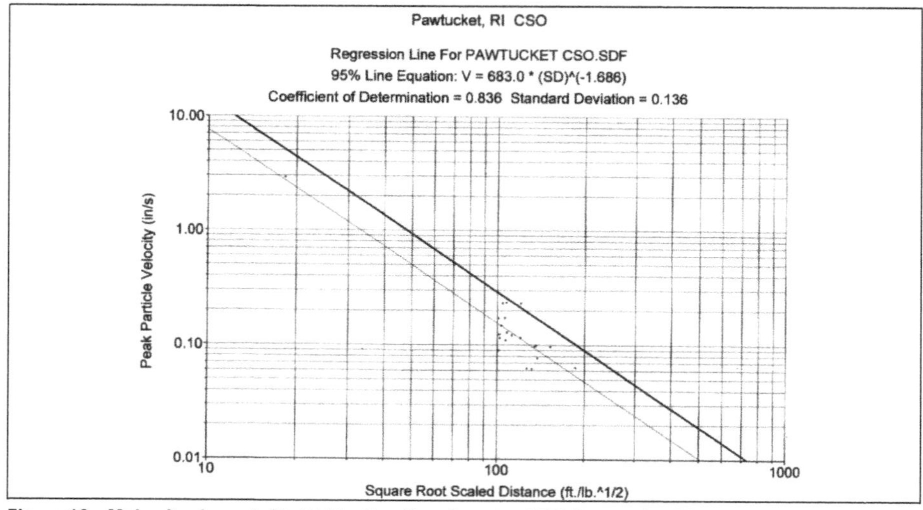


Figure 18. Main site: Launch Shaft Vibration Results using 95% Regression Line

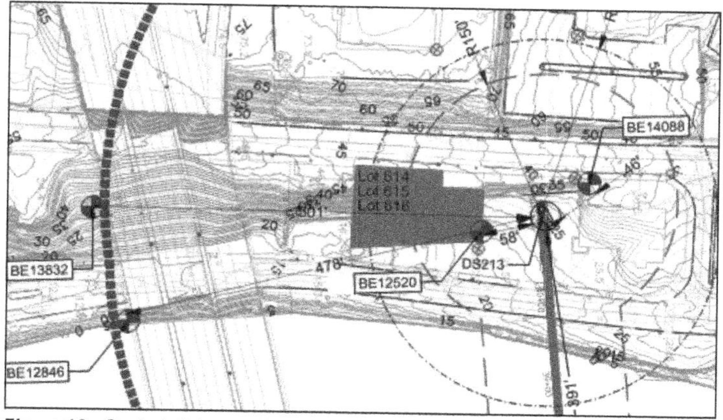


Figure 19. Scaled distance of nearby structures to OF213 drop shaft

### *Off Sites—Smaller Shafts Within Dense Urban Environment*

The most critical site in term of vibrations and air over pressure—DS213—is used to highlight the observational approach:

- Residential unit (lot 616) located at 58' from the blast location.
- 12" cast iron gas line located at 46' from the blast location

Original estimation and records of vibrations and air overpressure for 4 ft deep blasts, reduced and full-face sections were used for the test blasts. We can see in below table that estimated maximum vibration estimated at 0.992 in./sec (K=214 and n=–1.6) and air over pressure 93 dB(L). Recorded vibration was not more than 0.816 in/sec and air over pressure 120 dB(L). Which validates the test blast (Figure 20).

As three test blasts confirmed the original estimations, the decision was made to increase the blast depth to 6 ft for production blasts. Similarly to the test blasts, production blasts maximum vibration was recorded at 1.401 in/sec (at 80 Hz) and air over pressure below the 133 dB(L) limit (Figure 21).

Eventually, for the last deeper production blasts, and as the scaled distance increased significantly, blast depth was increased to 10 ft. production blasts maximum vibration was recorded at 0.633 in/sec (at 80Hz) (Figure 22 and 23).

## SUMMARY AND CONCLUSION

Unreinforced, circular secant pile walls for temporary SOE in large diameter shafts were successfully constructed under the direction of CB3A, performed by Keller NA. The elimination of steel elements within the secant piles proved cost effective, and performed as designed.

Favorable vibration results from blasting related activities were achieved due to a highly involved and knowledgeable construction team. The initial blast design was consistently modified and adjusted between blasts where sensitive receptors were identified. All blasts were measured below the required limits.

The CB3A Design-Build team created an efficient and effective shaft construction and blasting approach to manage identified risks and requirements while maximizing means and methods to best accomplish the work.

| 4ft REDUCED SECTION (Emuline) Blast 01 (June 22nd 2022) | Recorded PPV | Initial Est. PPV (K=214;n=-1.6) | Est. PPV ( K=393.4;n=-1.46) GBR | Est. Air Overpressure (A=0.5;B=-1.1) | Distance | Total Wt. | Wt./Delay | SD | Z | # Holes | Hole Depth |
|---|---|---|---|---|---|---|---|---|---|---|---|
| Location | | (in/sec) | (in/sec) | dBL | (ft) | (lbs) | (lbs) | (D/W^0.5) | (D/W^0.33) | | (ft) |
| BE14088: Gas line | 0.196 | 0.671 | 2.043 | 84.000 | 46 | 24.58 | 1.57 | 37 | 16 | 42 | 4 |
| BE12520: Lot 616 | 0.462 | 0.463 | 1.456 | 93.000 | 58 | 24.58 | 1.57 | 46 | 20 | 42 | 4 |
| BE13832: bridge | No value (Threshold not reached) | 0.015 | 0.063 | 71.000 | 501 | 24.58 | 1.57 | 400 | 431 | 42 | 4 |
| BE12846: Bridge | | 0.016 | 0.067 | 80.000 | 478 | 24.58 | 1.57 | 381 | 411 | 42 | 4 |
| BE19844: Lot 553 | | 0.022 | 0.090 | 77.000 | 391 | 24.58 | 1.57 | 312 | 336 | 42 | 4 |

| 4ft REDUCED SECTION (Emulex 1-1/2) Blast 02 (July 06th 2022) | Recorded PPV | Est. PPV (K=214;n=-1.6) | Est. PPV ( K=393.4;n=-1.46) GBR | Est. Air Overpressure (A=0.5;B=-1.1) | Distance | Total Wt. | Wt./Delay | SD | Z | # Holes | Hole Depth |
|---|---|---|---|---|---|---|---|---|---|---|---|
| Location | | (in/sec) | (in/sec) | dBL | (ft) | (lbs) | (lbs) | (D/W^0.5) | (D/W^0.33) | | (ft) |
| BE14088: Gas line | 0.574 | 0.992 | 2.919 | 84.000 | 46 | 56.32 | 2.56 | 29 | 12 | 42 | 4 |
| BE12520: Lot 616 | 0.816 | 0.685 | 2.081 | 93.000 | 58 | 56.32 | 2.56 | 36 | 15 | 42 | 4 |
| BE13832: bridge | No value (Threshold not reached) | 0.022 | 0.089 | 71.000 | 501 | 56.32 | 2.56 | 313 | 366 | 42 | 4 |
| BE12846: Bridge | | 0.023 | 0.096 | 80.000 | 478 | 56.32 | 2.56 | 299 | 349 | 42 | 4 |
| BE19844: Lot 553 | 0.027 | 0.032 | 0.128 | 77.000 | 391 | 56.32 | 2.56 | 244 | 286 | 42 | 4 |

| 4ft REDUCED SECTION (Emulex 1-1/2) Blast 03 (July 26th 2022) | Recorded PPV | Est. PPV (K=214;n=-1.6) | Est. PPV ( K=393.4;n=-1.46) GBR | Est. Air Overpressure (A=0.5;B=-1.1) | Distance | Total Wt. | Wt./Delay | SD | Z | # Holes | Hole Depth |
|---|---|---|---|---|---|---|---|---|---|---|---|
| Location | | (in/sec) | (in/sec) | dBL | (ft) | (lbs) | (lbs) | (D/W^0.5) | (D/W^0.33) | | (ft) |
| BE14088: Gas line | 0.318 | 0.992 | 2.919 | 84.000 | 46 | 93.11 | 2.56 | 29 | 10 | 56 | 4 |
| BE12520: Lot 616 | 0.466 | 0.685 | 2.081 | 93.000 | 58 | 93.11 | 2.56 | 36 | 13 | 56 | 4 |
| BE13832: bridge | No value (Threshold not reached) | 0.022 | 0.089 | 71.000 | 501 | 93.11 | 2.56 | 313 | 366 | 56 | 4 |
| BE12846: Bridge | | 0.023 | 0.096 | 80.000 | 478 | 93.11 | 2.56 | 299 | 349 | 56 | 4 |
| BE19844: Lot 553 | 0.019 | 0.032 | 0.128 | 77.000 | 391 | 93.11 | 2.56 | 244 | 286 | 56 | 4 |

**Figure 20. DS 213 test blasts (3) results**

| 6ft SECTION (Emulex 1-1/2) Blast 05 (Aug 25th 2022) | Recorded PPV | Est. PPV (K=214;n=-1.6) | Est. PPV (K=393.4;n=-1.46) GBR | Est. Air Overpressure (A=0.5;B=-1.1) | Distance | Total Wt. | Wt./Delay | SD | Z | # Holes | Hole Depth |
|---|---|---|---|---|---|---|---|---|---|---|---|
| Location | | (in/sec) | (in/sec) | dBL | (ft) | (lbs) | (lbs) | (D/W^0.5) | (D/W^0.33) | | (ft) |
| BE14088: Gas line | 0.7 | 1.186 | 3.435 | 84.000 | 46 | 221.18 | 3.2 | 26 | 8 | 89 | 6 |
| BE12520: Lot 616 | 1.401 | 0.933 | 2.449 | 93.000 | 58 | 221.18 | 3.2 | 32 | 10 | 89 | 6 |
| BE13832: bridge | No value (Threshold not reached) | 0.026 | 0.105 | 71.000 | 501 | 221.18 | 3.2 | 280 | 340 | 89 | 6 |
| BE12846: Bridge | | 0.028 | 0.113 | 80.000 | 478 | 221.18 | 3.2 | 267 | 324 | 89 | 6 |
| BE19844: Lot 553 | | 0.039 | 0.151 | 77.000 | 391 | 221.18 | 3.2 | 219 | 265 | 89 | 6 |

| 6ft SECTION (Emulex 1-1/2) Blast 06 (Sept 9th 2022) | Recorded PPV | Est. PPV (K=214;n=-1.6) | Est. PPV (K=393.4;n=-1.46) GBR | Est. Air Overpressure (A=0.5;B=-1.1) | Distance | Total Wt. | Wt./Delay | SD | Z | # Holes | Hole Depth |
|---|---|---|---|---|---|---|---|---|---|---|---|
| Location | | (in/sec) | (in/sec) | dBL | (ft) | (lbs) | (lbs) | (D/W^0.5) | (D/W^0.33) | | (ft) |
| BE14088: Gas line | 0.379 | 1.186 | 3.435 | 84.000 | 46 | 221.18 | 3.2 | 26 | 8 | 89 | 6 |
| BE12520: Lot 616 | 0.556 | 0.933 | 2.449 | 93.000 | 58 | 221.18 | 3.2 | 32 | 10 | 89 | 6 |
| BE13832: bridge | No value (Threshold not reached) | 0.026 | 0.105 | 71.000 | 501 | 221.18 | 3.2 | 280 | 340 | 89 | 6 |
| BE12846: Bridge | | 0.028 | 0.113 | 80.000 | 478 | 221.18 | 3.2 | 267 | 324 | 89 | 6 |
| BE19844: Lot 553 | | 0.039 | 0.151 | 77.000 | 391 | 221.18 | 3.2 | 219 | 265 | 89 | 6 |

| 6ft SECTION (Emulex 1-1/2) Blast 07 (Sept 16th 2022) | Recorded PPV | Est. PPV (K=214;n=-1.6) | Est. PPV (K=393.4;n=-1.46) GBR | Est. Air Overpressure (A=0.5;B=-1.1) | Distance | Total Wt. | Wt./Delay | SD | Z | # Holes | Hole Depth |
|---|---|---|---|---|---|---|---|---|---|---|---|
| Location | | (in/sec) | (in/sec) | dBL | (ft) | (lbs) | (lbs) | (D/W^0.5) | (D/W^0.33) | | (ft) |
| BE14088: Gas line | 0.41 | 1.186 | 3.435 | 84.000 | 46 | 221.18 | 3.2 | 26 | 8 | 89 | 6 |
| BE12520: Lot 616 | 0.778 | 0.933 | 2.449 | 93.000 | 58 | 221.18 | 3.2 | 32 | 10 | 89 | 6 |
| BE13832: bridge | No value (Threshold not reached) | 0.026 | 0.105 | 71.000 | 501 | 221.18 | 3.2 | 280 | 340 | 89 | 6 |
| BE12846: Bridge | | 0.028 | 0.113 | 80.000 | 478 | 221.18 | 3.2 | 267 | 324 | 89 | 6 |
| BE19844: Lot 553 | | 0.039 | 0.151 | 77.000 | 391 | 221.18 | 3.2 | 219 | 265 | 89 | 6 |

| 6ft SECTION (Emulex 1-1/2) Blast 08 (Sept 27th 2022) | Recorded PPV | Est. PPV (K=214;n=-1.6) | Est. PPV (K=393.4;n=-1.46) GBR | Est. Air Overpressure (A=0.5;B=-1.1) | Distance | Total Wt. | Wt./Delay | SD | Z | # Holes | Hole Depth |
|---|---|---|---|---|---|---|---|---|---|---|---|
| Location | | (in/sec) | (in/sec) | dBL | (ft) | (lbs) | (lbs) | (D/W^0.5) | (D/W^0.33) | | (ft) |
| BE14088: Gas line | 0.367 | 1.186 | 3.435 | 84.000 | 46 | 221.18 | 3.2 | 26 | 8 | 89 | 6 |
| BE12520: Lot 616 | 0.608 | 0.933 | 2.449 | 93.000 | 58 | 221.18 | 3.2 | 32 | 10 | 89 | 6 |
| BE13832: bridge | No value (Threshold not reached) | 0.026 | 0.105 | 71.000 | 501 | 221.18 | 3.2 | 280 | 340 | 89 | 6 |
| BE12846: Bridge | | 0.028 | 0.113 | 80.000 | 478 | 221.18 | 3.2 | 267 | 324 | 89 | 6 |
| BE19844: Lot 553 | | 0.039 | 0.151 | 77.000 | 391 | 221.18 | 3.2 | 219 | 265 | 89 | 6 |

| 6ft SECTION (Emulex 1-1/2) Blast 09 (Oct 7th 2022) | Recorded PPV | Est. PPV (K=214;n=-1.6) | Est. PPV (K=393.4;n=-1.46) GBR | Est. Air Overpressure (A=0.5;B=-1.1) | Distance | Total Wt. | Wt./Delay | SD | Z | # Holes | Hole Depth |
|---|---|---|---|---|---|---|---|---|---|---|---|
| Location | | (in/sec) | (in/sec) | dBL | (ft) | (lbs) | (lbs) | (D/W^0.5) | (D/W^0.33) | | (ft) |
| BE14088: Gas line | 0.416 | 1.186 | 3.435 | 84.000 | 46 | 221.18 | 3.2 | 26 | 8 | 89 | 6 |
| BE12520: Lot 616 | 0.57 | 0.933 | 2.449 | 93.000 | 58 | 221.18 | 3.2 | 32 | 10 | 89 | 6 |
| BE13832: bridge | No value (Threshold not reached) | 0.026 | 0.105 | 71.000 | 501 | 221.18 | 3.2 | 280 | 340 | 89 | 6 |
| BE12846: Bridge | | 0.028 | 0.113 | 80.000 | 478 | 221.18 | 3.2 | 267 | 324 | 89 | 6 |
| BE19844: Lot 553 | | 0.039 | 0.151 | 77.000 | 391 | 221.18 | 3.2 | 219 | 265 | 89 | 6 |

**Figure 21. DS 213 6 ft production blasts results**

| 10ft SECTION (Emulex 1-1/2) Blast 10 (Oct 20th 2022) | Recorded PPV | Est. PPV (K=214;n=-1.6) | Est. PPV (K=393.4;n=-1.46) GBR | Est. Air Overpressure (A=0.5;B=-1.1) | Distance | Total Wt. | Wt./Delay | SD | Z | # Holes | Hole Depth |
|---|---|---|---|---|---|---|---|---|---|---|---|
| Location | | (in/sec) | (in/sec) | dBL | (ft) | (lbs) | (lbs) | (D/W^0.5) | (D/W^0.33) | | (ft) |
| BE14088: Gas line | 0.455 | 1.186 | 3.435 | 84.000 | 46 | 221.18 | 3.2 | 26 | 8 | 89 | 10 |
| BE12520: Lot 616 | 0.633 | 0.933 | 2.449 | 93.000 | 58 | 221.18 | 3.2 | 32 | 10 | 89 | 10 |
| BE13832: bridge | No value (Threshold not reached) | 0.026 | 0.105 | 71.000 | 501 | 221.18 | 3.2 | 280 | 340 | 89 | 10 |
| BE12846: Bridge | | 0.028 | 0.113 | 80.000 | 478 | 221.18 | 3.2 | 267 | 324 | 89 | 10 |
| BE19844: Lot 553 | | 0.039 | 0.151 | 77.000 | 391 | 221.18 | 3.2 | 219 | 265 | 89 | 10 |

| 10ft SECTION (Emulex 1-1/2) Blast 10 (Nov 9th 2022) | Recorded PPV | Est. PPV (K=214;n=-1.6) | Est. PPV (K=393.4;n=-1.46) GBR | Est. Air Overpressure (A=0.5;B=-1.1) | Distance | Total Wt. | Wt./Delay | SD | Z | # Holes | Hole Depth |
|---|---|---|---|---|---|---|---|---|---|---|---|
| Location | | (in/sec) | (in/sec) | dBL | (ft) | (lbs) | (lbs) | (D/W^0.5) | (D/W^0.33) | | (ft) |
| BE14088: Gas line | 0.392 | 1.186 | 3.435 | 84.000 | 46 | 221.18 | 3.2 | 26 | 8 | 89 | 10 |
| BE12520: Lot 616 | 0.588 | 0.933 | 2.449 | 93.000 | 58 | 221.18 | 3.2 | 32 | 10 | 89 | 10 |
| BE13832: bridge | No value (Threshold not reached) | 0.026 | 0.105 | 71.000 | 501 | 221.18 | 3.2 | 280 | 340 | 89 | 10 |
| BE12846: Bridge | | 0.028 | 0.113 | 80.000 | 478 | 221.18 | 3.2 | 267 | 324 | 89 | 10 |
| BE19844: Lot 553 | | 0.039 | 0.151 | 77.000 | 391 | 221.18 | 3.2 | 219 | 265 | 89 | 10 |

**Figure 22. DS 213 10 ft production blasts results**

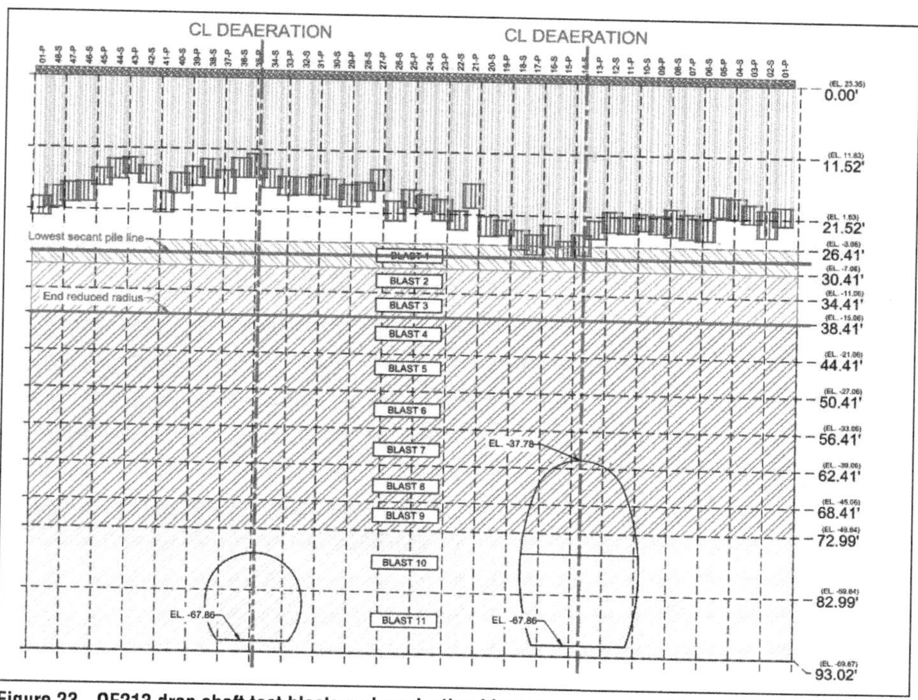


**Figure 23. OF213 drop shaft test blasts and production blasts unwrapped profile**

## REFERENCES

Skehan, J. W., D. P. Murray, J. D. Raben, and H. B. Chase, Jr. 1982. Exploration and exploitation of the Narrangasett coal basin. *Geotechnology in Massachusetts*. Ed. O.C. Farquhar; 381–399. Conference proceedings; March 1980. Amherst: University of Massachusetts Graduate School.

Skehan, J. W., N. Rast, and S. Mosher. 1986. Paleoenvironmental and tectonic controls of sedimentation in coal-forming basins of southeastern New England. In *Paleoenvironmental and tectonic controls in coal-forming basins of the United States*. GSA Special Paper 210. Ed. P. C. Lyons and C. L. Rice; pp. 9–30. Boulder, Colorado: Geological Society of America.

PART 

# Tunnel Rehabilitation

*Chairs*

**Youyou Cao**
STV Inc

**Alston Noronha**
Black and Veatch

# Project Clean Lake's First Large Diameter EPB TBM in Cleveland, Ohio

**Brian Negrea** ▪ McNally Tunneling Corp.
**Lance Jackson** ▪ McNally Tunneling Corp.
**Erica McGlynn** ▪ McNally Tunneling Corp.

## ABSTRACT

There was no need for the people of Cleveland to ask "Where does the raw sewage go? We want to know!" because it was clear. The raw sewage was being discharged into Lake Erie during heavy rainfall events due to the outdated design of the combined sewer system. In efforts to resolve this ongoing issue for the City of Cleveland, the Northeast Ohio Regional Sewer District developed Project Clean Lake in collaboration with the US EPA. Project Clean Lake is a 25-year, $3 billion program consisting of seven large-diameter CSO storage tunnels to be constructed beneath the city's existing infrastructure. The first four tunnels have been completed using traditional hard rock TBM excavation methods. The fifth tunnel, the Shoreline Storage Tunnel Project, presents unique challenges and complexities as it will be the first large-diameter EPB TBM to dig under the City of Cleveland. The project itself consists of one tunnel drive 14,100' (2.7 miles) in length, two pipe jacking tunnels, three large slurry wall shafts, two diversion structures, and three regulator reconstructions. Upon completion of the project, 12 permitted CSO locations along Lake Erie will be taken offline, reducing overflow volumes by approximately 350 million gallons per year. This paper will discuss the current status of construction and the chosen means and methods, including the myriad of nuanced differences between hard rock and EPB TBM tunneling methods. Particular attention will be given to the site preparation, setup, and organization; TBM assembly; and TBM launching practices.

## INTRODUCTION

Project Clean Lake is a $3 billion program by the Northeast Ohio Regional Sewer District (NEORSD) to reduce CSO discharges to waterways in the greater Cleveland area, including Lake Erie. The program consists of seven large-diameter CSO storage tunnels, contracted separately, along with many smaller consolidation and relief sewers that capture and convey overflows to the main storage tunnels.

Prior to the Shoreline Storage Tunnel (SST) project, the large-diameter tunnels for Project Clean Lake had been constructed strictly in rock conditions. The first four tunnels built for the NEORSD, namely the Euclid Creek Tunnel (ECT), Dugway Storage Tunnel (DST), Doan Valley Storage Tunnel (DVT) and Westerly Storage Tunnel (WST), were all mined through shale, a fine-grained sedimentary rock commonly found in northeastern Ohio. In contrast, the SST tunnel profile encompasses a multitude of clays, silts, and tills with varying geological properties depending on the alignment section being considered. Numerous considerations need to be made in order to mitigate risk factors in an effort to minimize the impact of construction to the general public. To achieve this, subsurface investigations have been used to inform the project team on a variety of key decisions that will affect the tunnel drive's level of success. These key decisions include:

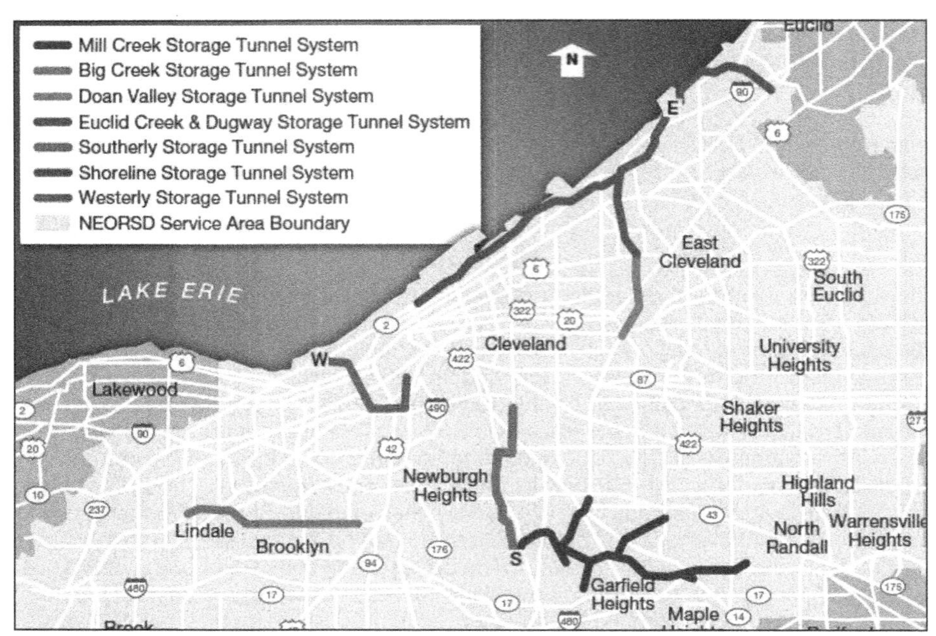


Figure 1. Project Clean Lake overview

- Selection of TBM type
- Range of anticipated excavation pressures and face stability conditions
- Cutterhead tooling selection and associated tool wear and consumption estimates
- Frequency of planned maintenance interventions and locations at which hyperbaric interventions are necessary
- Optimal ground conditioning products and dosages based on soil conditions

McNally/Kiewit SST JV (MKJV), a joint venture of McNally Tunneling Corporation and Kiewit Infrastructure Co., was awarded the SST project for a total contract value of $201,580,000.00. Notice to Proceed (NTP) for the project was issued on July 19, 2021. The SST contract provided flexibility by allowing the Contractor to rely on its experience when determining whether to utilize a Slurry Tunnel Boring Machine or an Earth Pressure Balance machine (EPB TBM). Based on technical assessments performed during the bid phase of the project, MKJV began coordination with Herrenknecht on the design of an EPB machine following the issuance of NTP.

## SST OVERVIEW

The SST project features a single tunnel of 14,100' that can be broken into six reaches, which will be discussed in depth in the following sections.

Aside from the tunnel, the project features three shaft sites (SST-1, SST-2 and SST-3) and three regulator sites (Regulator E-33, Regulator E-34A and Regulator E-36). SST-1 serves as the launch site for the TBM and encompasses one slurry wall shaft, one overflow structure, and a handmine tunnel, which serves as the connection between the SST and DST tunnel systems. SST-2 is an intermediate site and features one slurry wall shaft, a diversion structure, and a gate control vault. SST-3 will serve

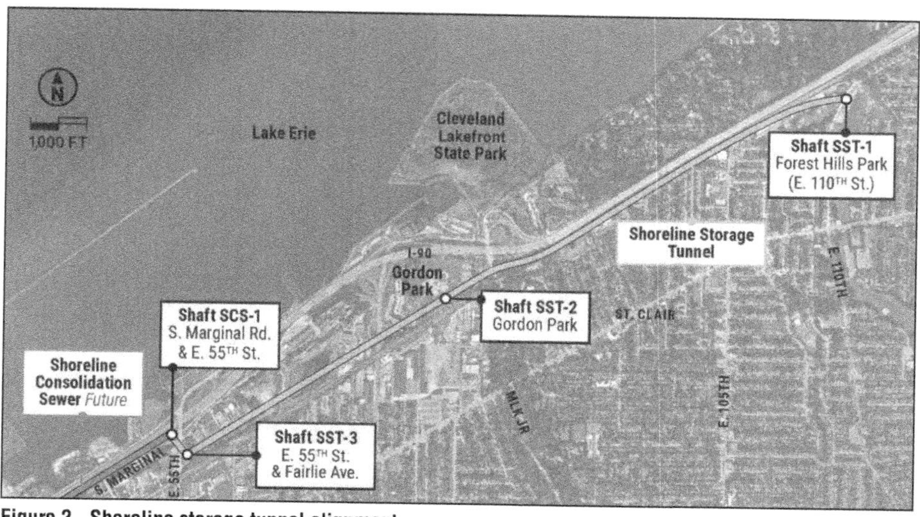


**Figure 2. Shoreline storage tunnel alignment**

as the retrieval site for the SST TBM, it includes one slurry wall shaft, a gate and screening structure, a diversion structure, and a gate control vault. The three regulator sites all feature a modification to the functionality of an existing regulator. Beyond the three regulator sites, additional alterations are to be made to off-site regulators in the project vicinity to adapt the existing system to the new storage tunnel. These modifications vary in complexity and are intended to reduce the amount of combined sewage overflow events to occur once the tunnel system is brought online.

## GROUND CONDITIONS

There were six Engineering Soil Units (ESUs) identified in the GBR. However, only three of those six are expected to be encountered within the SST tunnel profile. The Plastic Clays & Silts (PCS) unit constitutes the majority of the alignment and is described as having slow to no dilatancy, moderate to high toughness, and moderate to high dry strength. The Non-Plastic Silts and Fine Sands (NSF) unit is found predominantly within the first 3,000 of tunnel and is described as having a rapid dilatancy, low to no toughness, and no to low dry strength. The Till and Till-Like Deposits (TLD) unit is found exclusively in the first 1,000' of tunnel, containing both plastic silt and non-plastic to low plasticity silt. It's estimated that 65% of TLD is very dense and the remaining 35% is very stiff.

The project's GBR divides the SST alignment into four reaches, as shown in Figure 3. Further subsurface investigation was performed along the alignment by MKJV's engineering consultant, Maidl Tunnel consultants, and as a result it was determined that the SST alignment could be divided into six reaches. A summary of these reaches can be found in Figure 4. They will be discussed at length in the following sections and can be delineated by three main criteria:

1. The composition of the tunnel face and overburden
2. Feasibility of free-air interventions
3. Presence of sand and silt lenses

| | |
|---|---|
| Reach$_{GBR}$ 1 | SST-1 Shaft (Launch Shaft) to Station (STA) 10+00 |
| Reach$_{GBR}$ 2 | STA 10+00 to STA 30+00 |
| Reach$_{GBR}$ 3 | STA 30+00 to STA 105+00 |
| Reach$_{GBR}$ 4 | STA 105+00 to SST-3 (Reception Shaft) |

Figure 3. SST Reaches defined by the GBR

| Reach No. | Start | End | Length [ft] | Length [m] | Main features |
|---|---|---|---|---|---|
| 1 | 00+00 | 10+00 | 1000 | 305 | Mixed face of NSF with Till and Till-like Deposits (TLD). |
| 2 | 10+00 | 22+00 | 1200 | 366 | Full face of NSF with PCS in crown |
| 3 | 22+00 | 30+00 | 800 | 244 | Mixed face of PCS and NSF |
| 4 | 30+00 | 65+00 | 3500 | 1067 | Full face of PCS with occasional sand lenses in tunnel face |
| 5 | 65+00 | 102+00 | 3700 | 1128 | Full face sandy PCS |
| 6 | 102+00 | 141+00 | 3900 | 1189 | Full face PCS with very soft consistency |

Figure 4. SST Reaches defined by Maidl's subsurface investigation

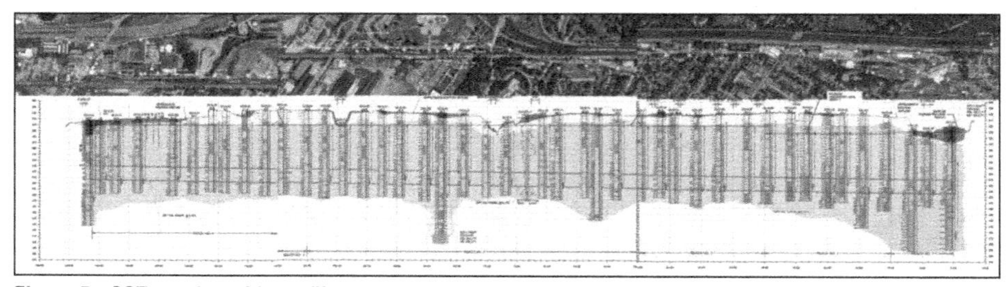

Figure 5. SST stratigraphic profile

## TBM TUNNELING METHODS

All tunneling methods share the same primary objective of excavating routes underground and beneath any existing infrastructure. Each method has its own advantages and disadvantages based on the differing geotechnical properties along a specific alignment. Since the Shoreline Storage Tunnel is the first large-diameter soft ground tunnel of the 25-year Project Clean Lake program, the TBM tunneling method options were limited to Earth Pressure Balance (EPB) TBM operations or Slurry TBM operations. Both methods utilized fully shielded TBMs with active face support systems. These two options were investigated and considered after the initial review of the GBR. The GBR identified that 76% of the tunnel alignment consists of plastic clays and silts (PCS), while the remaining 24% consists of non-plastic sand and gravel (NSG) and till–like deposits (TLD).

### EPB TBM Tunneling Methods

EPB TBM tunneling methods utilize a combination of three different systems to achieve and maintain the required active support pressure within the excavation chamber and at the tunnel face: the foam injection system, the air injection system, and the integrated muck handing system.

The foam injection system introduces soil conditioners into the excavated material, which then mix in the excavation chamber to form earth paste. This earth paste has a highly viscous consistency which contributes to the required supporting pressure being applied to the excavation face. The paste is also used to create a plug in the muck handling system.

In addition to injecting conditioners, the EPB TBM utilizes an air injection system which is used to develop an air bubble within the excavation chamber. This air bubble also assists in increasing and decreasing the required face support pressure.

The muck handling system revolves around the design of the screw conveyor, which is a large, sealed, cylindrical steel tube containing a rotating auger. The screw conveyor inlet usually starts at the invert location of the excavation chamber and proceeds at an incline through the shields to a higher discharge elevation which is typically above spring line. The combination of the aforementioned earth paste plug between the auger flights and the geometry of the screw conveyor creates hydrostatic pressure resistance, which assists in controlling the pressure within the excavation chamber. Increasing and decreasing the material flow through the screw conveyor will directly impact this pressure.

## Slurry TBM Tunneling Methods

Slurry TBM tunneling methods are similar to EPB methods in that they also incorporate a combination of ground conditioning, air injection, and muck handling systems to achieve the required active face support pressures. While both EPB and slurry TBM tunneling methods are fully shielded TBM designs, the technology behind the ground conditioning and muck handling systems for each are completely different.

The ground conditioning system for slurry tunneling methods consists of the transfer of a bentonite/water mixture into the excavation chamber. This mixture can also contain ground conditioning polymers depending on the current soil or rock characteristics. Once transferred, the bentonite water mixture is combined with the excavated spoils, creating the slurry mixture. Next, the slurry mixture is transported via slurry pump and piping to a slurry treatment plant located on surface. At the slurry treatment plant, the suspended solids are separated from the bentonite/water mixture then stockpiled on surface until further disposal. The bentonite that was separated from the mixture is recycled into the ground conditioning bentonite/water mixture for further excavation.

The muck handling system consists of in/out utility pipes which deliver the bentonite/water mixture and extract the slurry from the excavation chamber. The flow is produced, controlled, and maintained by large centrifugal-style pumps. The flow rate of materials entering and exiting the excavation chamber will affect the active face support pressures.

This method also includes an air injection system which assists in controlling the pressure during the exchange of these materials. There is also a buffer wall within the excavation chamber which is designed to assist with the differing hydrostatic pressures during the material exchange.

## Tunneling Methods Comparison

When considering the differences between the EPB and slurry methods described above and comparing them to open-face rock TBM tunneling methods traditionally used in Cleveland, it is obvious that the technology and design behind each method is vastly different. While both EPB and slurry methods require maintaining an active support face pressure, the open-mode rock method typically does not, as the excavated face is supported by the strength of the surrounding rock.

Material conditioning methods for EPB and slurry excavation in soft ground and open-mode hard rock methods also differ greatly. The material conditioning aspects for both EPB and slurry tunneling methods are critical to producing the correct material consistency to maintain the excavation and material handling operations as described previously. Hard rock material conditioning may consist of foam injection mainly to provide lubrication to cutting tools. This material conditioning is also used as a dust reduction agent where needed. While these are important aspects of hard rock tunneling, they are not required for the excavation and material handling operations to progress.

The next operation to be compared is the material handling from the tunnel heading to the surface stockpile. The material handling operations for the EPBM and open-mode hard rock tunneling methods are similar, as both methods have the option of handling the material using either a tunnel/shaft continuous conveyor system or a muck box handling system. In contrast, the slurry TBM method only has one option, which is to transport the bentonite water conditioner and slurry discharge mixture using in/out utility piping.

EPB TBMs are also capable of operating in open mode, which makes this tunneling method more versatile and advantageous when compared to slurry TBMs. Being able to operate in both closed and open modes is advantageous as it allows the TBM to advance through varying types of soil characteristics that may be encountered throughout a tunnel drive. Another factor that should be considered is the process for entering the excavation chamber for inspection and maintenance. Entering the pressurized excavation chamber also requires an air lock and a hyperbaric intervention procedure. This becomes an extensive, time-consuming operation which requires additional resources, site set up, personnel training, and significant preparation work.

## SST Tunneling Method Decision

Among other factors such as cost and availability, choosing a particular tunneling method is dependent on the project's existing ground conditions and geotechnical characteristics. The case was no different for the SST project, which consists of 76% PCS and 24% NSG/TLD along the tunnel alignment. Based on this information, MKJV was leaning toward the EPB TBM tunneling method. The TBM manufacturer, Herrenknecht Tunneling Systems, proposed the EPB TBM during the bid phase, which reinforced MKJV's initial leanings as Herrenknecht is an extremely credible and experienced manufacturer of cutting-edge EPB and slurry TBMs.

As previously mentioned, MKJV also contracted the third-party engineering services of Maidl Tunnelconsultants (MTC) to assist in the investigation of the GBR in efforts to fully understand the characteristics of the soil that is anticipated throughout the tunnel alignment. In addition to developing a report that further analyzes the soil characteristics, MTC also provided their expert insight regarding the mucking and disposal, TBM selection, anticipated TBM performance, ground conditioning, face stability and pressure calculations, and cutting tool life estimations. When evaluating the TBM selection for this project, MTC referenced the German DAUB (German Tunneling Committee) criteria. Tables 1a and 1b demonstrate the results of MTC's report and investigation. MTC also concluded that EPB technology was best suited to handle the differing tunnel sections.

With MTC's conclusion and Herrenknecht's proposal reinforcing MKJV's initial opinion MKJV proceeded with the selection of the EPB TBM tunneling methods. After the TBM selection was finalized, MKJV reviewed the muck handling and segment transportation options. Based on previous experience, means and methods, and a cycle

**Table 1a. Assessment criteria for TBM selection in soft ground**

| Tunneling Section | TS 1 | TS 2 | TS 3 | TS4 | TS 5 | TS 6 |
|---|---|---|---|---|---|---|
| Fines content, % | >40% | >40% | >40% | >40% | >40% | >40% |
| Permeability (max), m/s | $10^{-4}$ | $10^{-4}$ | $10^{-4}$ | $10^{-8}$ | $10^{-8}$ | $10^{-8}$ |
| Consistency (Ic) | > 1.0 | 0.0–1.0 | 0.0–1.0 | 0.8–1.1 | 0.8–1.1 | 0.1–0.5 |
| Relative density | Dense | Dense | Dense | Dense | Dense | Dense |
| Support pressure, bar | 1–4 | 1–4 | 1–4 | 1–4 | 1–4 | 1–4 |
| Swelling potential | None | None | None | None | None | None |
| Abrasivity | High | Medium | Medium | Medium | Low | Low |

**Table 1b. Suitability of TBM types per tunneling section**

| Tunneling Section | | TS 1 | TS 2 | TS 3 | TS 4 | TS 5 | TS 6 |
|---|---|---|---|---|---|---|---|
| Slurry Shield | SLS | o | o | o | o | o | o |
| EPB Shield | EPB | o | + | + | + | + | + |

(+: Main application range; o: Application feasible; application critical)

analysis, MKJV decided to proceed with a continuous conveyor system for handling the muck. For the handling of segments, utilities, consumables, and personnel, MKJV choose to use multi-service vehicles (MSVs). Although MKJV has more experience working with rail-based trains, the means and methods and cost analyses identified MSVs as the best option for this project.

## SITE PREPARATION

The initial site preparation stages of any project are crucial and set the foundation for the work to follow. The primary mining site for this project is the SST-1 site, which consists of 8.4 acres of previously-occupied park green space. The size of this site is larger than generally allotted for these projects. Stripping the topsoil and forming a small stockpile mound around the perimeter of the site was the fist step of the site preparation. This was done in effort to keep the existing topsoil onsite to be reused at the end of the project.

The next stage was to identify the existing underlying utilities and perform a ground pressure surcharge analysis for all the equipment anticipated to be traveling on surface. Once this was complete, MKJV identified which utilities required further protection and installed crane mats, concrete slabs, or barricaded these areas accordingly.

The next step was to determine the temporary utility requirements for construction. This required operational foreshadowing and analysis to determine future requirements, including identifying the existing power, water, and sewer utilities' locations and planning how to run temporary lines throughout the site. Once the utility runs were installed, 12 to 18 inches of #1 and #2 stone topped with crushed road base (ODOT 304) was spread over the site to accommodate the heavy construction traffic.

The next preparation stage was the development of detailed layout plans for each site setup phase: initial layout, TBM assembly, TBM launch, initial mining, reset operations, and production mining. While it may seem obvious to proceed in the chronological order previously stated, it often makes more sense to consider the production mining site setup phase first. This allows for primary consideration of the end result and can help determine which key operations can be installed initially without impacting other preliminary phases. Any key operation that can be installed early and kept

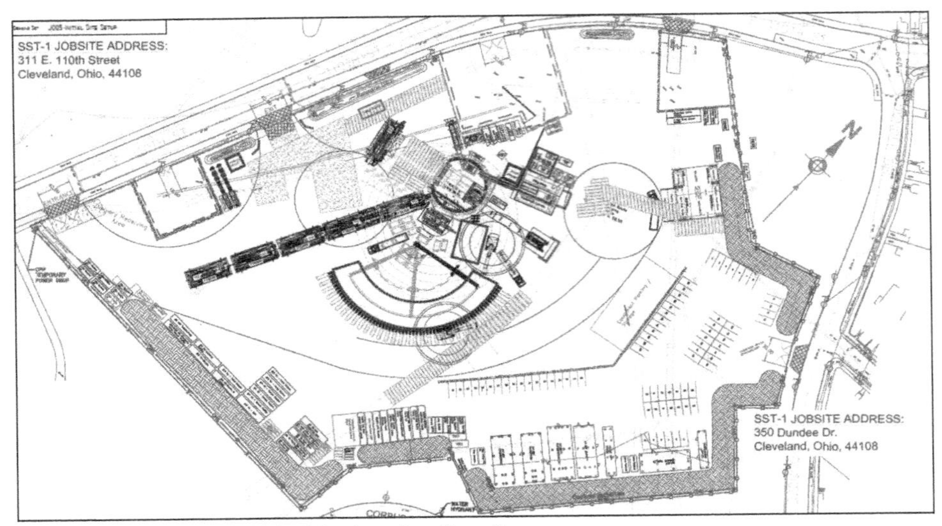


**Figure 6. SST-1 jobsite initial mining layout configuration**

off the critical path will only benefit the schedule in the long run. This also allows more time for testing, commissioning, preparation, or potential unexpected repair (Figure 6).

## Mining Site Setup

The site setup operation can be broken down into four primary focuses: production mining, TBM assembly, initial mining, and the reset operations. When considering the production and initial mining phases, the key systems are muck handling, segment delivery/storage, grouting, air plant, and cooling system. For this particular project and with the SST-1 site being relatively large, it was possible to get all of these key systems installed during the early stages of the site setup. Based on the general principles of the EPB machine, establishing the air plant setup was critical. The air plant, grout plant, and cooling plant were all strategically staged in close proximity of each other in efforts to contain all primary utility supplies in one area. In addition to these key components, the continuous conveyor vertical belt storage unit was also installed close to these systems. Although this storage unit is not required until the production mining phase, early installation did not impact any of the initial mining or TBM assembly phases and therefore was more beneficial for future critical path work during the reset operation.

The production mining surface mucking operation consists of an overland conveyor which transfers the muck to a radial stacking conveyor to discharge into radial containment muck bin. The muck is then loaded into haul trucks using an excavator which travels along the outside perimeter of the radial pit. The muck is hauled off site to a designated disposal site. Based on the initial mining arrangement, a second smaller muck pit was required, as during this phase muck was handled from the shaft bottom using a crane and traditional muck box methods (Figure 7).

The segment delivery for this project consists of receiving rings from the supplier, offloading them using a crane with segment clamps, and storing the complete rings on site. Every three rings for this project are broken up amongst four truck loads, as one complete ring exceeds the allowable shipping weight. This caused additional segment

Figure 7. SST-1 jobsite initial mining photo

handling during the initial stages, as the stacking of complete rings had to be divided into separate phases. The limited ring storage area surrounding the crane during initial mining made this particularly difficult. As segments arrived, they were stacked accordingly in two half stacks, which together formed one complete stack. Stacking the rings strategically is critical for future delivery and staging on the TBM. Therefore, staging the rings accordingly on surface will save time later as the project progresses.

The site setup for the TBM assembly was broken down into two staging areas. The first staging area was designated for the primary TBM components and was located directly above the tunnel alignment surrounding the assembly crane. The secondary staging area was designated for the TBM back gantry system and was staged around the center of the jobsite and adjacent to the previously installed radial muck bin. Keeping these staging areas clear and well-marked is important to avoid traffic jams and delays as the components are being delivered. Both areas will become segment ring storage areas and haul routes during the production mining phase. For this project, it was possible for all the TBM components to be staged on site prior to the initial assembly phases. If space allows, this can be beneficial as the TBM components are sometimes delivered out of order according to the assembly sequence, which could cause unwanted double handling of components and potential schedule impacts if not stored in the proper sequence.

Considerations for the site reset operation were very similar to those of the TBM assembly. There were areas on the jobsite that were dedicated to temporary storage of components, such as the continuous conveyor parts. These temporary storage areas can be later used for storage of segments as the project progresses. There were also several pre-assembly staging areas which allowed for assembly of conveyor components. This drastically reduces the reset operation's overall duration. Having access to these staging areas was a critical aspect of the site setup for this reset operation.

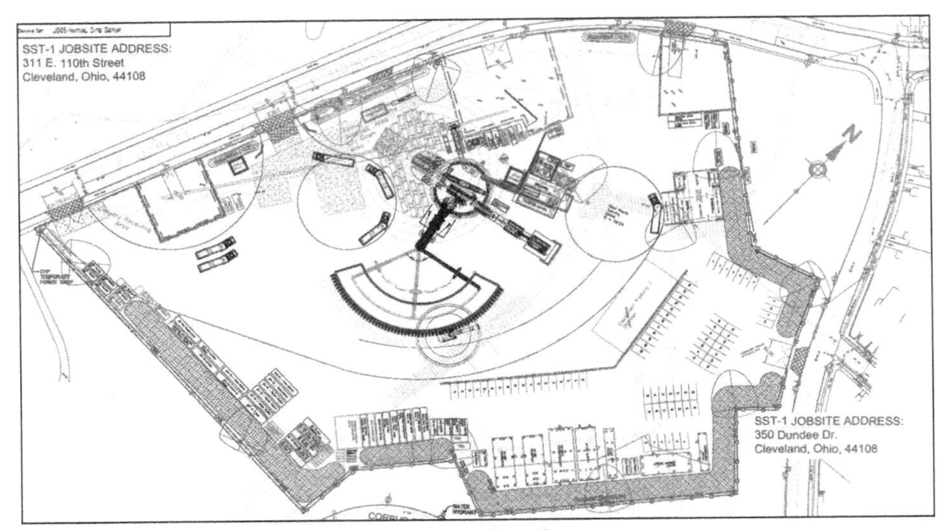


**Figure 8. SST-1 jobsite production mining layout configuration**

## Site Organization

Having an organized job site is one key component to having a successful project. Not only does it save time, but it also provides a safe working environment. For this project, the primary focus was getting all the previously mentioned primary systems and operations in place, then to further focus on secondary areas. These secondary areas consist of the utility extension laydown and storage area, the consumable storage areas, and the equipment and spare parts storage areas. Another key factor of site organization is the designation of haul routes as a tool used to separate the different operations. The utility extension laydown and storage area was intentionally staged further away from the shaft as these are easy materials to move and bring over to the shaft as needed. These utilities also take up a significant area, which isn't possible —or desirable—around the shaft. The consumable storage area consists of several climate-controlled containers to store the oils, greases, and admixtures required for mining. These containers are positioned around the perimeter of the site for the same reason as the utility extension pipes. The TBM spare parts were also stored on-site in a container located near the consumables, ideally organizing all the TBM related materials together. Equipment was staged near the jobsite shop for easy access and maintenance as needed. The haul routes for this project were divided into two primary routes. The first route was designated for the mucking operation with additional service for cement, bentonite, and admixture deliveries. The second haul route was designated for segment deliveries and tunnel staging equipment/materials. Dividing the haul routes in this manner allowed for an organized jobsite without creating or intensifying traffic jams (Figure 8).

# TBM ASSEMBLY

The TBM was delivered to the job site disassembled in 51 primary deliveries. In addition to the primary loads, there were also 20 containers with parts, hardware, hoses, and cable that arrived from the TBM manufacturer. As mentioned previously, the TBM assembly area was divided into two different staging areas. This helped eliminate confusion during the TBM assembly process. The primary TBM bulk items consisting of the TBM cutting wheel, main drive, and shields were delivered in 20 separate shipments. These components were staged strategically around the heavy lift assembly crane, which was positioned directly above the tunnel alignment. The 20 heavy bulk components were then staged sequentially and in order by weight around the crane within the appropriate picking radius. The heavy lift crane used for both the offload and placement of these components was a 660 ton crawler crane. This crane was sized for the three heaviest picks in correlation with their picking radii. Each of the components were lowered into the shaft bottom and assembled on a steel launching cradle, one by one. Because this TBM consisted of a four-section shield design, air winches were used to assist with the alignment of the shield joints during assembly.

Because a starter or tail tunnel was not an economical option in these ground conditions, the entire TBM was assembled within the tight quarters of the 61-ft diameter shaft. In addition to working around the shaft diameter constraints, there was also the previously installed TBM sealing can, which protruded an additional 3 ft into the shaft and added further constraint when working in the shaft. Adding to the complexity of this assembly was the installation of the screw conveyor. This assembly revolved around a specific sequence which needed to be followed precisely to ensure that all the components would fit accordingly. To assist with the sequence, it was decided that the screw conveyor be divided into two sections and installed/connected during different phases of the assembly. MKJV worked closely with Herrenknecht to determine the best sequencing option for this shaft assembly process. Although it took additional planning, temporary supports, and field weldment/installation, it was concluded that separating the screw conveyor in two sections was the best option based on the available working area. This is a hurdle that is not generally encountered during the assembly of a hard rock TBM, where there are no screw conveyors and usually additional options for starter and tail tunnels (Figure 9).

In addition to assembling the main TBM cutter head components underground in the tight shaft quarters, a secondary operation proceeded on surface, assembling the TBM backup gantry system. While there were no significant space constraints to combat during this assembly process, there were a few other hurdles that were successfully overcome. This backup gantry system was offloaded and assembled using

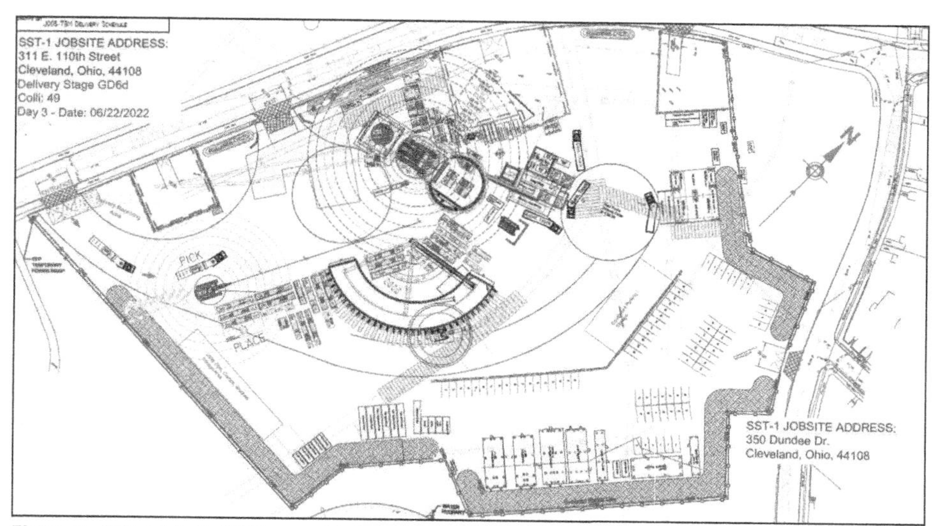


Figure 9. SST-1 jobsite TBM assembly layout configuration

a 130-ton crawler crane with the assistance of an 80-ton crawler crane. Both cranes were owned by MKJV and had been slated for assisting this project from the start.

The plan for launching the TBM was to utilize the completely assembled gantry system on surface, which would be connected to the TBM cutter head in the shaft bottom using an umbilical system. Based on this plan, the gantry system components would need to be assembled and positioned in line on surface as they would be in the tunnel behind the cutter head. This process and the required space were accounted for during the site setup phase of the project. Although this takes up a large area on surface, this was the most cost-effective option and ample space was available on the 8.4-acre site.

Because the TBM has an open gantry that is designed to travel in the tunnel on individual rubber bogies, the assembly of the gantry system on surface was more difficult when compared to that of a closed gantry system. Since the tunnel radius is not formed on surface and forming a temporary radius would be very time consuming and costly, it was decided to assemble the gantry system without the rubber bogies installed. These bogies will be installed during the reset operation when lowering each gantry into the shaft bottom. This worked as an advantage, as it allowed easier access into the backup gantries because they were not as high off the ground surface as they otherwise would have been with the bogies installed. During this stage, the gantries were individually blocked up with dunnage and lined up to gantry five of six total. Gantry six was primarily used for ventilation and utility extension purposes, and therefore it is not required for the launch and initial mining operation. Each gantry was plumbed and connected accordingly to achieve appropriate functionality. The floating walks between gantries were also installed in efforts to make each gantry as accessible as possible.

## TBM LAUNCH

Due to the tight shaft launch, the TBM was assembled with umbilicals connecting the surface gantry system to the completely assembled cutter head staged in the shaft bottom. The umbilicals were staged and supported by a large beam which spanned the shaft on the left side direction of drive of the tunnel alignment. A combination of electric cables and hydraulic hoses were looped seven times in a festoon orientation, which paid out manually as the TBM cutterhead advanced. This umbilical arrangement was designed to allow up to 430 ft of tunnel excavation. This length of excavation would accommodate the entire assembled TBM back up gantry system, with additional room to spare to install the continuous conveyor systems. Handling the umbilicals manually form the supporting beam in the shaft and into the tunnel was much like eating a bowl of spaghetti at times, and this was unfortunately inevitable based on the nature of this operation. Once in the tunnel, the umbilicals were supported using temporary brackets similar to the horizontal tunnel conveyor brackets. These brackets assisted in making this operation manageable. The electric and hydraulic umbilicals were intentionally kept separated in different cable trays in efforts to keep the system organized. Installing and handling the umbilicals was undoubtably one of the most time-consuming parts of launching the EPM TBM from the shaft (Figure 10).

While it is common to launch smaller diameter TBMs from spaces similar in size to the 61-ft diameter SST-1 shaft, it is rare to launch a 26-ft diameter TBM from this size shaft. With the length of the TBM from the cutter head to the end of the tail shield equaling approximately 43 ft and the length from the cutter head to the end of the screw conveyor equaling 58 ft, it is not hard to imagine the limited available working space around the TBM in this shaft. The area to the left of the TBM was completely

**Figure 10. SST-1 shaft umbilical handling**

consumed by the umbilical storage, which served to reduce the size of the shaft even further. Also, at the rear of the TBM between the back shaft wall and the tail shield is the TBM launch frame, which is nestled in around the screw conveyor and only one foot away from touching that rear shaft wall. At the right side of the shaft, you will find the stair tower which is the primary access into/out of the shaft along with a 72-in ventilation bag line running down the shaft next the stair tower. Also on the right side of the shaft are all the utilities, which are secured to the shaft wall. All these different systems are required for launching the TBM, however they also occupy much of the extremely limited shaft real estate that is leftover to function and support the TBM.

The TBM was launched by thrusting off the aforementioned launch frame, which was positioned and secured to the rear shaft wall. This frame was designed to handle the thrust forces delivered by the lower half of the TBM thrust cylinders only. This was designed based on building only half rings through the shaft to allow access to the tunnel from the shaft. Building these half rings consisted of building three segments with the TBM erector, which were then used to thrust the TBM forward. Proceeding this way would allow the required access for equipment into/out of the tunnel. The launch frame was also designed to include several struts and supports which would be installed later as the machine became completely buried. During this stage, a steel transition ring was installed to start building the complete rings through the sealing can and into the mined tunnel. The struts would span the shaft between the transition ring and launch frame, while still maintaining the required access into the tunnel. Based on these launching techniques, the design of the launch frame and transition ring were significantly more intricate when compared to similar hard rock TBM launching methods.

The initial muck handling operation was also impacted by the tight constraints presented by the shaft's limited space. With the TBM being separated from the backup gantry system on surface, the initial muck handling process is completely different from the final configuration. Initial muck handling reverted to traditional, varying-size muck box handling while suspended from a crane. This process started with small muck boxes that would be gradually replaced with larger boxes as the TBM progressed and allowed more space. There were also various combinations of temporary muck chutes involved with this process. These chutes would direct the flow of materials from the screw conveyor to be discharged into whichever box was currently being used. This became a messy situation quickly, as the plug in the screw conveyor was not formed yet and the pressurized materials were being forced through the auger flight and began shooting out the discharge gate as it opened. This process continued until the machine was completely buried and there was enough access to utilize a temporary belt conveyor system, which was suspended off the screw conveyor discharge location. During this stage, the MSV was introduced into the process in a short mode configuration. The MSV would support the muck boxes as they were filled from the temporary belt conveyor configuration. Once the box was full, the MSV would transport the material back to the shaft where the crane would lift, dump, and rest the box on the MSV. This cycle will continue till the completion of the initial mining and the rest of the TBM in is final configuration.

## CONCLUSION

In conclusion, the differences between the various tunneling methods are apparent when compared to one other. Although each method possesses its own advantages and disadvantages, method selection is fully dependent on a spectrum of characteristics that should be analyzed for each project. While the best option for this project was to proceed with the EPB TBM tunneling methods, this process presented numerous obstacles. Through careful and thorough ongoing site management, constant communication with our internal and external experts, our project team has continued to successfully overcome these challenges and turned them into lessons learned. Our team hopes that the presentation of this project is helpful for others as they consider similar projects or potential mining solutions.

# Corrosion and Leakage Remediation for WMATA Yellow Line Steel Tunnels

**James Parkes** ▪ Schnabel Engineering, LLC
**Matthew Goff** ▪ Schnabel Engineering, LLC
**Alan Kolodne** ▪ RK&K, LLP
**Steven Kolarz** ▪ RK&K, LLP
**Tatiana Kotrikova** ▪ Washington Metro Area Transit Authority

## ABSTRACT

The WMATA Yellow Line is the section of steel segmentally lined tunnels that have experienced leakage and significant corrosion in isolated areas. A three part rehabilitation design has been developed for curtain grouting, structural repairs, and corrosion protection to provide continued performance. As WMATA's first Construction-Manager-at-Risk (CMAR) project, the design development benefitted from collaboration with the Contractor through the final design development process. Rehabilitation work started in Fall 2022 and will be completed by late Spring 2023. Details of the project, the remediation design, and the process and benefits of the CMAR approach in the design development are presented.

## INTRODUCTION

The WMATA Yellow Line (L-Line) includes a section of twin transit metro rail tracks that run between Pentagon and L'Enfant Plaza Metro Stations. This section includes a bridge over the Potomac River, then transitions through a cut-and-cover portal to an immersed tube tunnel section under Washington Channel, then connects to a low point pump station and vent shaft structure, then transitions to twin bore tunnel sections that connect to the cut-and-cover cross-over structure and L'Enfant Plaza Station as shown in Figure 1. The twin bore tunnels were constructed in 1975 as described in Kuesel (1976) and Coring et al. (1976).

The twin bore tunnels are lined with steel tunnel segments that have experienced significant corrosion, leakage, and deterioration since construction. In 2000, the steel lined tunnels underwent a major maintenance program that included removal of rust and corrosion products from the liner, replacement of corroded bolts, replacement of lead caulking, and re-application of a two-part corrosion protection paint system. Additional spot repairs have been performed over the years including spot panel repairs and localized injection grouting to seal leaks. Joints between the segments exhibit corrosion and portions of the segments have experienced significant section loss, including some areas where the section loss has been severe enough for outside soils to flow into the tunnel. Examples of steel liner corrosion are shown in Figure 2.

WMATA has recently undertaken a program to perform structural inspection, rehabilitation, and leak mitigation of the steel lined tunnels in order to extend the structure life for at least another 50 years. This rehabilitation program also includes additional upgrades on this section of the Yellow Line including upgrades of the systems cables between L'Enfant Plaza and Pentagon Stations, replacement of bearings on the Potomac River bridge, and rehabilitation of concrete in the portal area between the

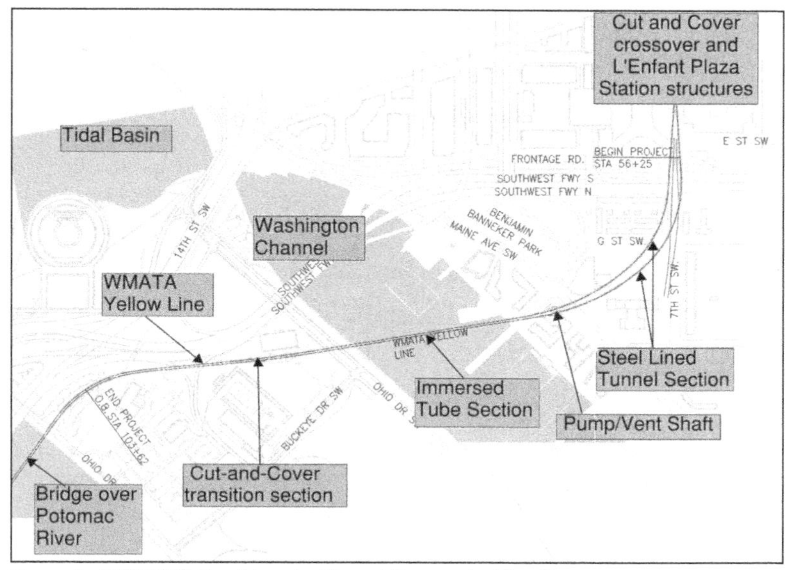


Figure 1. WMATA Yellow Line steel tunnel liner project location

Figure 2. Examples of observed corrosion and soil inflow in the steel lined tunnel

river and Washington Channel. The project is WMATA's first use of the Construction Manager-at-Risk (CMAR) delivery method.

# PROJECT BACKGROUND

## Subsurface Conditions

The Yellow Line Tunnels are located in Coastal Plain Terrace and Cretaceous sediments. Subsurface conditions consist of complex interlayering of sand, clay, silt, and gravel deposits. The principal strata within the tunnel profile consist of Pleistocene Terrace and Cretaceous soils that include stiff clays (CL, CH), medium dense to dense silty or clayey sands (SM, SC), and relatively clean sand and gravel mixtures (SP-SM,

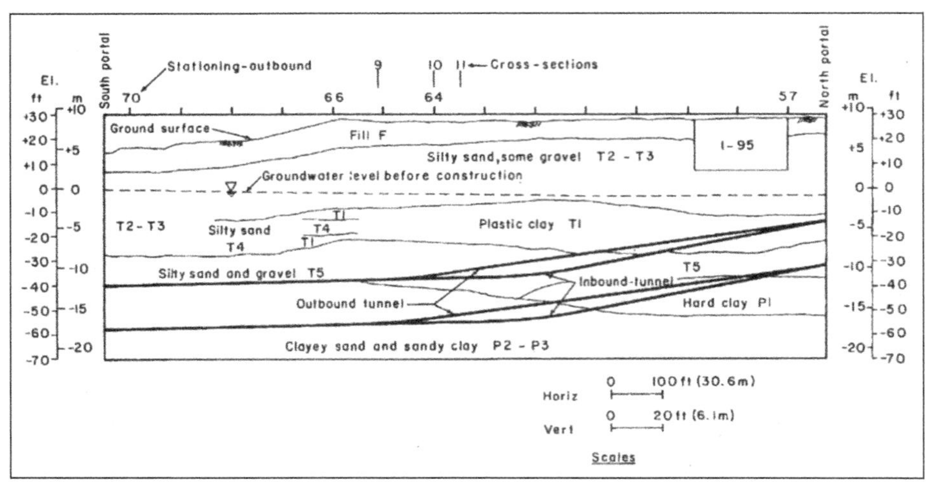


Figure 3. Soil profile along Yellow Line tunnel alignment (Cording et al., 1976)

GP-SM) as shown in Figure 3. Soil permeabilities along the tunnel vary from greater than $10^{-2}$ to less than $10^{-6}$ ft/min with measured groundwater levels over 60 feet above the tunnel invert, indicating the potential for highly pervious soils along the tunnel.

## Steel Tunnel Liner Design and Construction

The circular steel liner of the twin tunnels extend about 1,500 linear feet from about Track 1 (inbound) CM 56+25 to 71+50 and Track 2 (outbound) CM 56+25 to 70+50. The liner consists of prefabricated steel liner plate segments that are bolted together with lead caulking sealant in the joints. The steel segments are comprised of a built-up section of welded steel plates. A 7⁄16" thick steel skin plate forms the semi-circular arch at the segment exterior. Steel plates 1" thick by 6" deep are attached to the edges of the skin plates. On the long edges of the segments, these plates form the circumferential ribs (transverse to the tunnel centerline). On the short edges, these plates form the horizontal flanges or longitudinal joints (parallel to the tunnel centerline). The segments are divided into four bays at 15 degree spacing by three ¾" thick by 5-1⁄16" deep stiffener plates oriented in the longitudinal direction A typical steel liner plate section from the as-built drawings is shown in Figure 4.

The invert of the tunnel includes a reinforced concrete slab that supports the running rails, traction power third rail, and an embedded invert drain. There is also a 24" wide × 27" tall reinforced concrete safety walk. In addition to the concrete slab and safety walk, electrical cables, communication cables, equipment, and fire line systems are attached to the steel tunnel liners. An isometric view of the existing steel tunnel liner is shown in Figure 5.

The tunnels were constructed using open-face articulated tunnel shields with breasting doors and a hydraulic excavator arm. Dewatering was used during construction to lower the groundwater as much as 50 feet in order to maintain face stability. As a result, the tunnels were not subjected to full external water pressure until after the tunneling was completed. Limited permeation grouting was used to stabilize soils in the crown in some areas during mining and where the tunnels passed under the existing 7th street overpass. The grouting was used to provide enough cohesion to allow

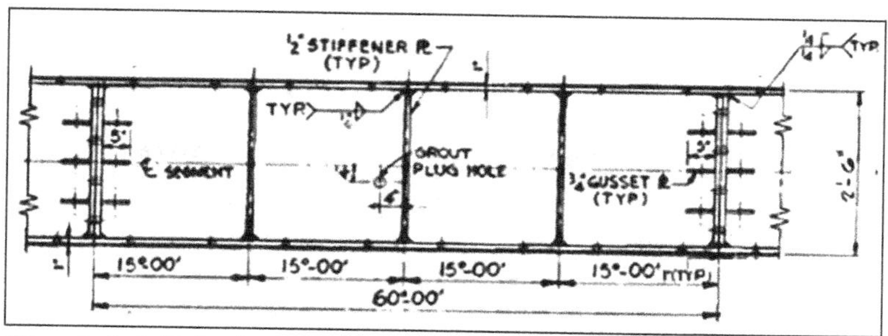


Figure 4. As-built drawing section of Type A steel liner panel

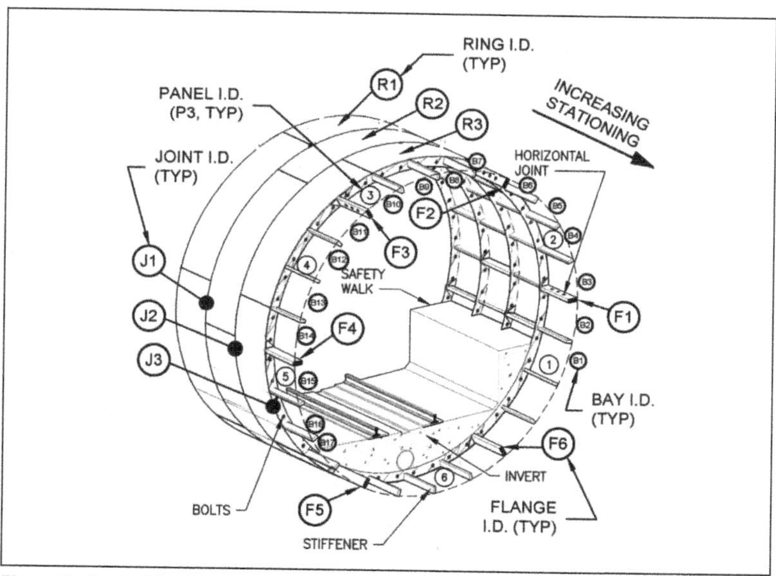


Figure 5. Isometric view of steel lined yellow line tunnel

excavation and installation of the liner; the grouting was not intended to provide long term strength, cohesion, or reduction in permeability.

The annular tail void from the shield around the segmental liner varied from zero at the invert to 6 inches at the crown. This void was cement grouted in two stages – the first was to fill the void between the liner and the tail skin, and the second was after the shield passed to fill the void from the tail skin and any remaining voids. Foam "sausages" were placed within the annular void during first stage grouting to serve as bulkheads to prevent the grout from flowing into the shield (Kuesel, 1976).

## USE OF CMAR METHOD

The rehabilitation of the Yellow Line tunnels is a significant undertaking that requires taking the tunnels out of revenue service for an extended period of time. The tunnel rehabilitation requires removal of the systems cables, which opens the possibility to upgrade or replace elements of those systems. In addition, there are elements of the cut-and-cover concrete portal structure and the Potomac River bridge that require repairs and upgrades. From an operational standpoint, all of this work should be performed during a singular set of revenue service outages in order to make the most efficient use of the outages and minimize the potential for future work or outages on the Yellow Line.

Given the complex and variable nature of the work elements and limited time frame, WMATA chose to explore alternative contracting methods to conventional design-bid-build (DBB). A key aspect for this project is to develop a design that is structurally adequate and cost-effective, with the design period occurring during a time of limited track access (tracks can only be accessed for limited windows during non-revenue night hours). It was desirable to have early contractor involvement during design development in order to incorporate input regarding constructability and cost considerations. This will minimize the potential for delays during construction and allows value engineering to occur during the design development. At the same time, WMATA wanted to retain control of the details of the design in order to maximize consistency with the rest of their system and their operations maintenance programs.

Under the CMAR approach, a general contractor (GC) is engaged based on qualifications during the design, serves as a consultant to provide input during the design, develops a guaranteed maximum price (GMP) for performance of the work, manages or self-performs the construction work, and is responsible as the construction manager (CM) for cost and schedule control in construction (i.e., "at risk"). The pricing for the work, either self-performed or subcontracted, is typically done in an open-book approach so that the owner has confidence that the work is competitively priced. The owner has a separate contract with the design firm and therefore maintains control of the design. An "off ramp" option is typically included for the owner in case they cannot come to terms on the GMP, the owner can choose to exit the CMAR contract and bid the work as a conventional DBB contract. This method may allow for schedule efficiencies because the CMAR procurement is done during the preliminary design, and that the potential for design changes or value engineering in construction (and associated delays) are minimized. The designer can work with the CMAR to develop the design based on their preferred or anticipated means and methods.

WMATA considered the CMAR approach to be the most appropriate for the Yellow Line rehabilitation. It allowed for a qualifications-based selection, which was critical given the complexity and variability of the work elements. It also enabled WMATA to compress the schedule as much as possible, which is critical to minimizing track outages, while maintaining design control. A design-build approach would not have allowed WMATA to retain as much control over the final design. A DB approach would also require a well-defined scope during the early stages of the project; given the limited access to the tunnels and bridge and the potential for unforeseen conditions to be discovered after preliminary design, this limitation created additional project risks. WMATA therefore chose to use the CMAR approach. Kiewit Infrastructure Co. (Kiewit) was selected as the CMAR contractor and was engaged after the 30% design submittal, about one year prior to the scheduled track outage.

## DEVELOPMENT OF REMEDIATION DESIGN

### Preliminary Investigation and Design

The rehabilitation design was led by prime consultant RK&K, LLP with key subconsultants Schnabel Engineering (structural design), Burns Engineering, Inc. (systems design), Alvi Associates, Inc. (portal remediation design), Sheladia Associates, Inc. (estimates, scheduling, and BIM modeling), and Precision Measurements, Inc.

**Figure 6. Grinding of steel liner coating for ultrasonic testing of steel tunnel liner**

(survey). The scope of the tunnel remediation design was to develop a leak remediation and structural rehabilitation design that would extend the life of the tunnels at least another 50 years with at least 25 years before the next major maintenance milestone (such as repainting the tunnel). The design development process involved review of previous inspections, a preliminary inspection and design, followed by final design and construction using the CMAR approach.

The preliminary inspection (as well as previous inspections) included visual inspections and a rating scale combined with the use of non-destructive testing (NDT) consisting of ultrasonic testing (UT) (Figure 6) and caliper measurements to determine the thickness of steel skin plates, ribs, stiffeners, and flanges. The inspection program found that generally the tunnels are more corroded near the interfaces with the shafts, although corrosion is present throughout the lengths of the tunnels. Corrosion is primarily in areas where water infiltration (current or past) is observed through ribs or flanges.

Preliminary analyses, including finite element modeling (FEM), indicated that the skin plates and the ribs are the most critical structural elements in the tunnel; in effect, the skin plates and ribs share external loads from soil and water, with the ribs stiffening the skin plates and preventing them from buckling failure. Inspection findings indicated that the skin plates have experienced the greatest section loss, in some areas as much as 50% or more. Section losses for the ribs were minimal, generally less than 10%. The remediation design therefore needed to provide structural reinforcement or replacement of the corroded skin plates.

A variety of structural remediation options were developed and evaluated using a weighted decision matrix. Options considered varied from isolated structural repairs in corroded areas, partial encapsulation with shotcrete or concrete (to form a composite liner), to full liner replacement or encapsulation. Considerations included initial construction cost, future maintenance costs, service life, constructability, impact on systems and operations, precedence and past performance, and future inspection and maintenance. Risks of flooding and collapse ruled out any options that required removal of the existing liner. Internal clearances effectively ruled out the use of overlays or encapsulation. Considerations for precedence, constructability, and cost narrowed the options down to structural steel repairs for isolated areas. These consist of options to address areas of severe corrosion and include welding new ½" steel skin plates to structurally replace severely compromised skin plates, replacement of compromised stiffeners, replacement of bolts, isolated repairs for panels with corroded grout ports, and replacement of corroded steel seal plates at the interfaces with the concrete cut-and-cover structures.

Leak mitigation options were considered as part of the evaluation of the structural repair options. Leak mitigation options included isolated grout injections at leak locations, installation of membranes in conjunction with structural overlay or replacement systems, and full curtain grouting. Full curtain grouting was selected in conjunction with the structural repair option, with additional considerations for localized injection grouting for isolated leaks following curtain grouting. A performance specification approach was selected for addressing the curtain grouting.

Additional elements of the tunnel remediation design included development of a cathodic protection system that includes installation of external deep grounding beds and reference electrodes connected to the steel liner and removal of the existing coating system and corrosion through blast cleaning and re-coating the liner with a high performance corrosion protection coating system. The coating system is required to be a three part system consisting of (1) a coat of zinc rich primer coating applied to blast cleaned surfaces followed by (2) an intermediate epoxy coat followed by (3) a polyurethane top coat. The original corrosion protection paint system consisted of a factory applied inorganic zinc primer only, and the corrosion protection paint system in 2000 consisted of a two layer system. Three layer systems such as these are considered higher quality and are anticipated to provide better performance compared to two-layer systems or zinc primer alone.

Additional program upgrades include replacement of the systems cables, repairs to concrete at the portal structures, and replacement of bridge bearings and fire standpipes on the bridge structure.

## Final Design Development

The final design proceeded with input from Kiewit on the preliminary design and on-going collaboration between the design team, WMATA and their subject matter experts. Weekly calls were held with all parties to discuss progress and raise coordination or technical issues. Breakout technical sessions were held to address specific concerns regarding detail development or proposed modifications for constructability.

The final design development progressed similar to the preliminary design. Additional field inspections and NDT testing were performed to gather more data, confirm earlier assumptions, and further develop the locations of repairs and overall quantities. Additional and more refined analyses and FEM modeling were performed to confirm previous assumptions and initial simplified models, and additional borings and water readings were performed to verify external loading conditions and test for soil and groundwater characteristics that may impact the design.

Finite element analysis during final design included soil-structure interaction modeling in Plaxis 2D and structural analysis modeling in STAAD. The Plaxis model was used

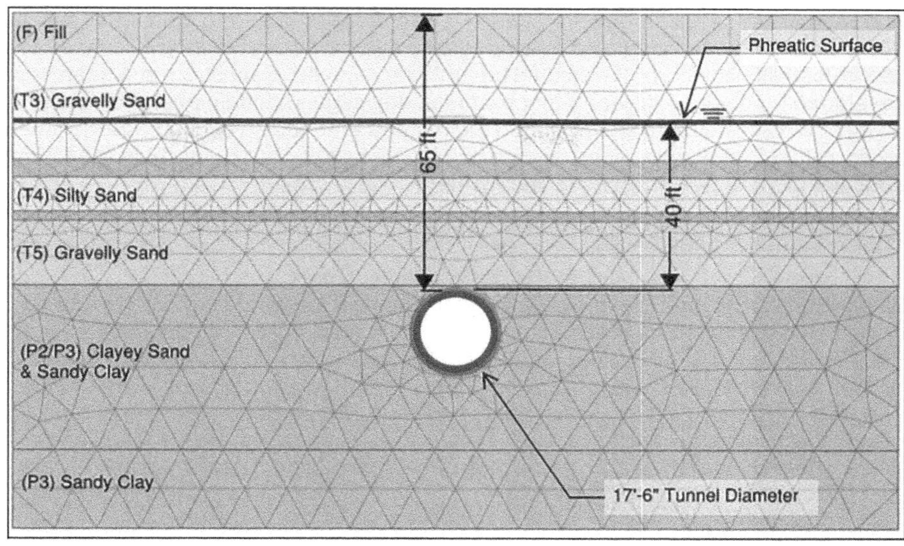


**Figure 7. Plaxis model of tunnel soil-structure interaction**

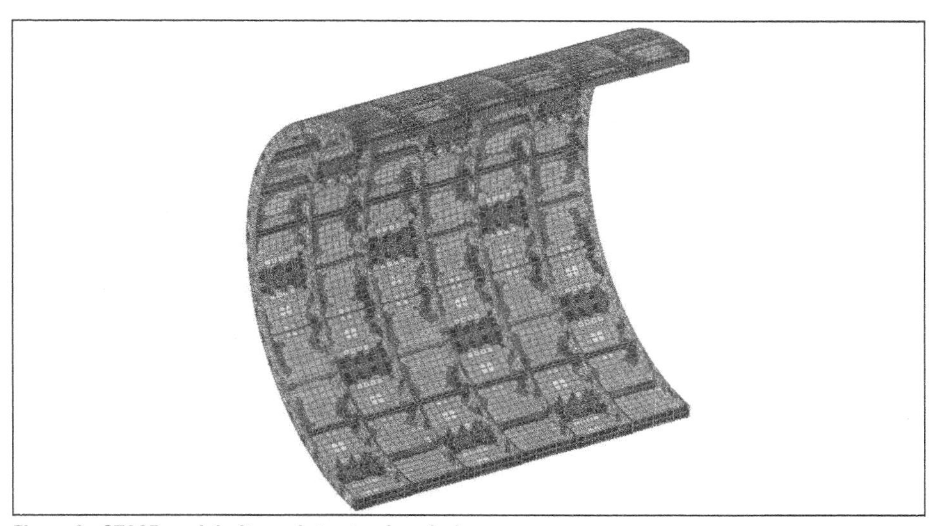

**Figure 8. STAAD model of tunnel structural analysis**

to analyze external design pressures on the steel tunnel liner for the critical design section soil profile along the tunnel alignment. The analysis confirmed that the empirical design equations for external design pressure were conservative for design of the tunnel liner. The calculated external design pressures were applied to the steel tunnel liner using structural analysis software STAAD. The 3D STAAD model was used to analyze critical design stresses in each of the steel liner components (ribs, skin plate, stiffeners, etc.) Graphics from the Plaxis and STAAD models are shown in Figure 7 and Figure 8, respectively.

The regular coordination between WMATA, the design team, and the CMAR resulted in a collaborative working relationship where design issues, technical concerns, or constructability considerations could be further evaluated, discussed, and alternatives developed if needed. Stop-and-plot progress drawing submittals ahead of the formal design milestone submittals facilitated the reviews and helped minimize formal review comments. Final design milestones included 60%, 90%, 100%, and issued for construction (IFC) drawings and specifications.

## Pilot Program and Mock-ups

A key element to the success of the design development was a pilot program undertaken shortly after the 60% design submittal. The pilot program consisted of a field trial of various aspects of the work including trials of curtain grouting, blast cleaning and mechanical removal of existing coatings and corrosion, application of two potential coating systems, bolt removal and replacement, and welding of replacement panels. The field trial was developed collaboratively with all parties to develop objectives, planned activities, staffing, and track outages. A track outage was scheduled for an entire weekend to allow Kiewit full access to the tracks for the pilot program (Figure 9).

This pilot program allowed Kiewit and their subcontractors to develop and test means and methods for performing the work in the tunnels, as well as develop staging, mobilizing, sequencing, and clean-up procedures. The results of the program were available during 90% design development, which allowed updating of the drawings and specifications to incorporate the findings of the program. The results were also informative to the development of the GMP, thus reducing risk of cost issues and unnecessary contingencies. Schedule and logistics planning for the project also benefitted from the use of the pilot program. The CMAR model was a significant benefit to the

Figure 9. Work train for field trial pilot program

project in this regard. Without a contractor on-board at this stage of design, this type of program and the valuable insights from it would not have been possible.

The results of the pilot program included verifying performance of the proposed products and means as well as identifying potential issues and developing modifications to the design, construction methods, or additional mitigation measures to address such issues. For example, one issue that was discovered was that welding of the new skin plates resulted in thermal changes in the tunnel liner that created leakage at the rib and flange joints. The amount of existing lead caulking required for removal and replacement was also reduced based on the field trial results.

The pilot program also allowed testing of the proposed curtain grouting program. Kiewit's subcontractor, Sovereign Hydroseal East, Inc. (Sovereign), proposed a proprietary polymer-based emulsion grout. This grout, combined with pattern injection and monitoring along the tunnel liner, would form an impermeable barrier encapsulating the tunnel. During the grouting program, dye was initially injected through new grout ports drilled in the liner to investigate the flow of water around the tunnel circumferentially as well as longitudinally. The grout product has fluid properties similar to water, so the dye injection program was informative for evaluating how the grout will travel. As the grouting progressed, dye injection verified that the grout was effective because the dye was not observed leaking in areas that were previously grouted. The test grout program allowed for verification and adjustment of grouting procedures (injection hole spacing, injection pressures, etc.). Emergency grouting procedures and injection grouting of leaking joints using drill and tap as well as injectable bolts were also field tested during the pilot program.

Additional mock-up programs were developed based on the results of the pilot program. These included a mock-up of the steel liner segments at Kiewit's facility in Colorado for further testing of weld procedures and detailed measurements of expansion and contraction of the segments during and following welding, and a paint application test program in Baltimore, Maryland to further evaluate application methods, coverage, and post-application defects of coating systems by two manufacturers. The results of these programs allowed for further refinement of the design and means and methods. Procedures to address leaks that may occur during or after welding were developed. Based on the paint test program, the contractor selected their preferred product manufacturer and system for use in construction.

## Additional Design Development Considerations

Additional design phase considerations included development of contingency items and development of final acceptance criteria. Contingency quantities for various structural repairs were necessary because of the limited access to the tunnel during design. Most inspections occurred during nightly non-revenue hours, which are limited to a window of about 5 hours per night. Given requirements for safety coordination and getting personnel on and off the tracks, most inspection work actually occurred in windows of about 2 hours per night in dimly light conditions with system cables and other infrastructure blocking access to portions of the liner. It was therefore necessary to consider contingency quantities for repairs and include such in the documents for development of the GMP.

Final inspection and acceptance criteria were developed based on practices from similar projects as well as considerations for staging of the work. Because some aspects of the work (welding), working conditions, or ambient conditions/temperatures could potentially affect the performance of the leak mitigation program, several inspection walkthroughs were incorporated into the contract requirements.

## CONSTRUCTION

The project is under construction as of the writing of this paper. The Yellow Line was taken out of revenue service for the rehabilitation work beginning in September 2022. The outage includes the tunnel rehabilitation work as well as work on the bridge, portal, and systems. The outage is planned for approximately 8 months, with the majority of the tunnel-related work occurring within the first 5 months. Full time access to the tunnels with adequate lighting and support equipment allowed the CMAR and design personnel the opportunity to perform additional inspection and NDT, review the contingency quantities, and establish the final repair quantities and locations. Additional collaboration during required testing and fit-up operations has allowed the team to develop appropriate field acceptance criteria and make tweaks to cover plate fit up details to minimize the potential for leaks or damage to severely compromised panels.

## SUMMARY AND CONCLUSIONS

WMATA is proactively undertaking a structural rehabilitation and leak remediation program for the steel lined Yellow Line tunnels, in conjunction with systems and bridge repairs and upgrades. This rehabilitation program will extend the service life of the tunnels for another 50 years or more. A program of structural repairs consisting of new liner plate panels, bolt replacements, spot repairs, leak remediation, cathodic protection, and corrosion inhibiting coating system will provide the required structural capacity and protect the liner from future corrosion.

The CMAR procurement method has been key to the successful development and construction of this project. The early involvement of the CMAR contractor created a collaborative work environment for all parties and allowed development of a design focused on constructability while maintaining technical requirements and an aggressive schedule. The pilot program field trial and mock-ups were key to validating and providing information for optimization of the design as well as development of the construction logistics. These aspects of the process also informed the development of the GMP and project schedule and therefore reduce overall risk to the project regarding technical, cost, and schedule considerations.

## ACKNOWLEDGMENTS

The authors would like to acknowledge the contributions of WMATA's technical subject matter engineering (ENGA) staff including Daming Yu, Sam Park, Fred Farhangi, and James Darmody, as well as Kiewit personnel David Neidballa and Claire Ferko for their contributions regarding the construction and CMAR operations.

## REFERENCES

Cording, E.J., Hansmire, W.H., MacPherson, H.H., Lenzini, P.A., and Vonderohe, A.P. (1976). Displacements Around Tunnels in Soil. Washington, D.C.: U.S. Department of Transportation.

Kuesel, T.R. (1976). "Washington Metro's Topless Tunnels." Proc., Rapid Excavation and Tunneling Conference, Las Vegas, NV, 296–310.

# Major Rehabilitation of the Montreal 55 Years Old Lafontaine Immersed Tube Highway Tunnel—Design Considerations

**Jean Habimana** ▪ Hatch
**Bakar Amara** ▪ Hatch
**Laurent Rus** ▪ Singular Structures

## ABSTRACT

The La Fontaine Tunnel in Montreal, Quebec is currently undergoing major structural rehabilitation work and systems upgrade work to comply with current codes, standards and best practices in fire life safety and to extend its lifespan of this 55-year-old immersed tube tunnel for at least another 40 years.

The paper discusses design consideration of ongoing works that include structural analyses to evaluate repair strategies for the post-tensioned reinforced caissons, the scheme to repair two leaky joints between caissons, the design of passive fire protection that involved in situ and real scale laboratory tests, and other tunnel systems upgrades.

## PROJECT BACKGROUND

The La Fontaine road tunnel is an immersed tube tunnel that crosses the St Lawrence River and connects the City of Montréal with Île-Charron and Boucherville along Autoroute 25. Figure 1 shows an aerial view of the tunnel location. The tunnel was constructed between 1963 and 1967 and consists of three main section: Section 1 that is the immersed tube portion located in the center of the tunnel, and Sections 2 and 3 that are the south and north approaches, respectively, that were cast-in-place (Figure 2). Section 1 consists of seven precast tunnel caissons, each approximately 110 m in length, 36.73 m in width, and 7.70 m high, for a total length of 768 m. The installation of precast elements began from the southern riverbank and progressed toward the northern riverbank, with Caisson 1 being the first and Caisson 7 being placed last. Figure 3 shows a construction photo of a caisson being precast in the nearby dry doc. The tunnel has two traffic tubes that are separated by a service corridor. Each tube carries three lanes of traffic in unidirectional traffic flow, one tube for northbound and the other for southbound traffic, respectively. The tunnel is one of the five crossings of the main branch of the Saint Lawrence River in the Montreal urban area and is a major transport link to the nearby Port of Montreal. It is heavily used by heavy goods vehicle that is higher than typically experienced by tunnels as it serves the port. However, transport of dangerous goods cargoes is prohibited within the tunnel.

The Quebec Ministry of Transportation has undertaken major rehabilitation program to do repair and upgrade work to meet current codes and standards and best practices to avoid any major intervention on the tunnel for the next 40 years. This rehabilitation work includes structural systems, ventilation, fire protection, water and drainage, mechanical, pavement structure, electrical and lighting systems, road safety and security for users and workers, communication, traffic signals, management and maintenance of traffic, mitigation measures during construction, as well as public transit systems within the tunnel during and after the rehabilitation work.

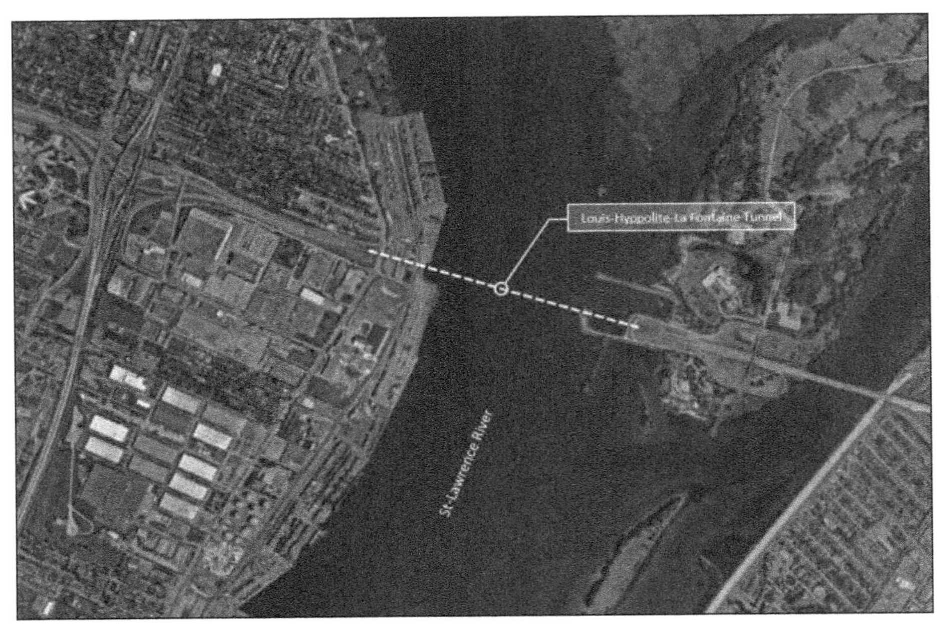


Figure 1. Aerial view of the Lafontaine Tunnel

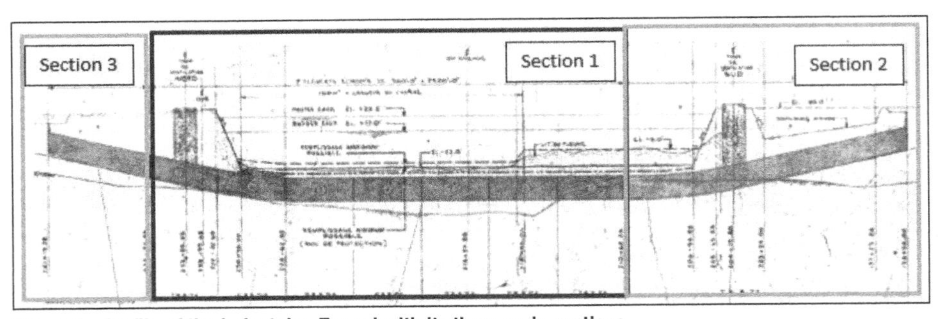


Figure 2. Profile of the Lafontaine Tunnel with its three main sections

The following sections will describe some of the challenges associated in designing the structural repair work. A brief summary is provided for systems upgrades.

## DESIGN OF STRUCTURAL REPAIR

The tunnel precast concrete elements are reinforced and prestressed by posttension both in transverse and longitudinal directions. The main role of the longitudinal post-tension is to hold together precast elements for each caisson, while the role of the transverse post-tension is to balance most of the dead and water weight. There were also temporarily post-tension cables that was used during construction in the mid span of the traffic tube, which was cut after backfilling and re-establishing water level. The longitudinal post-tension cables are fully embedded in the concrete while the transverse ones follow a hyperbolic shape and have limited cover at midspan of the traffic tube as shown on Figure 4. This section describes how damages to the concrete and cables were observed during detailed inspection, summarizes in situ

Figure 3. Construction photo of a caisson element of the Lafontaine Tunnel in a dry doc (provided by MTQ)

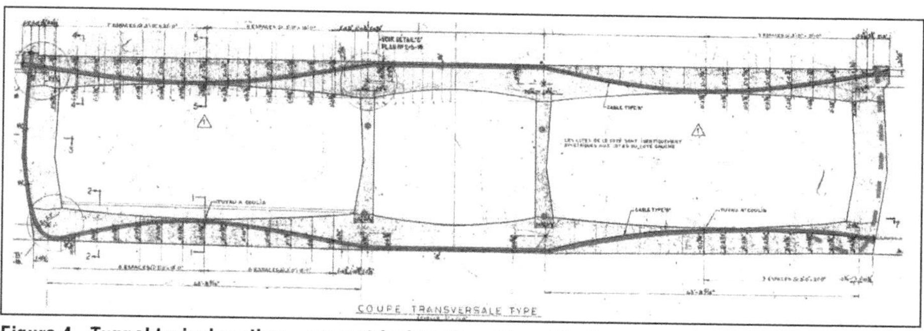


Figure 4. Tunnel typical section—geometrical configuration of the transverse post-tension cables

investigations that were conducted to characterize the level of damage, and gives an overview of numerical modeling that were performed to assess the structural integrity of existing structure during repair work.

In the crown of the tunnel, transverse prestressing cables were placed in bundles of two with each cable having a total of 48 wires inserted in a metal duct. The two bundles of cables are in average spaced 450 mm and they cross the entire width of the tunnel with alternate anchors on east or west side wall, respectively.

## Detailed Inspection and In Situ Tests to Assess Existing Conditions of Transverse Post-Tension Cables

The repair work that is part of the major rehabilitation includes concrete repair of the inside structure to address its degradation over the years. To characterize the level of degradation, a detailed inspection was carried out as part of the final design to assess the existing conditions of the tunnel structure in general and the post-tension cables in particular. During that inspection, significant and repetitive damages were observed on the intrados of the tunnel crown, especially in the mid span area where the post-tension cables have low cover. In general, the damage appeared in the form of concrete spalling, post-tension duct corrosion and localized sectioned prestressing wires. The structural damage was reported qualitatively as per the following damage categories:

- Bursting of concrete and exposed and corroded and including broken condition of the passive reinforcement.
- Bursting of concrete with exposed and corroded post-tension duct.
- Bursting of concrete and damaged post-tension duct with visible and heavily corroded cable wires.
- Bursting of concrete and damaged post-tension duct with broken post-tension cable wires.
- Concrete repair on damaged post-tension cable.
- Previous repair of concrete next to post-tension cables.

To characterize further the level of damage of these cables, an investigation on the state of the post-tensioning cables was carried out with special emphasis on the validation of the effective tension in the post-tensioning cables, the effective stress in the concrete at the crown intrados location, and the condition of the injection grout within the cables. The testing program investigation covered quality of the injection grout, and the residual stress in the post-tension wires and surrounding concrete. It provided enough qualitative and quantitative data on a limited number of investigation points across the different structural elements. To that end, a targeted caisson was selected based on results of the visual inspections. First, a series of inspection observational windows were opened at the midspan of the crown exposing the cables in both "damaged zones" and "healthy zones," to corelate some of the obtained results, and to draw potential trends. Second, a series of in situ tests consisting of a combination of ultrasound computed tomography (also known as MIRA), "crossbow" tests, and slotstress tests were carried out. Their principle is explained below.

The MIRA measurements use ultrasound to measure or detect anomalies within the existing structures. They were used to search for grout injection voids in posttension ducts, combined with observational windows, the latter allowing for the calibration of ultrasound equipment and injection grout quality inspection. They were carried out over a length of approximately 4.8 m and 1.9 m on average for each of the cables investigated, a length located at the midspan of the crown where the post-tensioning cable is at its lowest point of its parabolic configuration.

The "crossbow" test allows to determine the residual tension force, or "effective" force, on a post-tension cable wire. The principle of the test is based on the concept of the crossbow (Figure 5), with two fixed points at the ends, and a central point where a force F, perpendicular to the axis of the wire, is exerted. The greater the force F (a function of the induced displacement "f"), the greater the residual/effective tension in the cable wire.

The average residual tension of the 51 tested wires is 35.1 kN with a standard deviation of 3.6 kN (low), the results of which are within a range between 28.3 kN and 45.6 kN. The average residual/effective stress is 910 MPa, which is compared to the ultimate stress of the wire (fpu = 1620 Mpa) corresponding to 0.562*fpu (range between 0.454*fpu and 0.731*fpu, respectively). The tested wires included both corroded wires (surface corrosion) as well as non-corroded wires, which condition did not influence the obtained results. These values are very well in agreement of those required by the

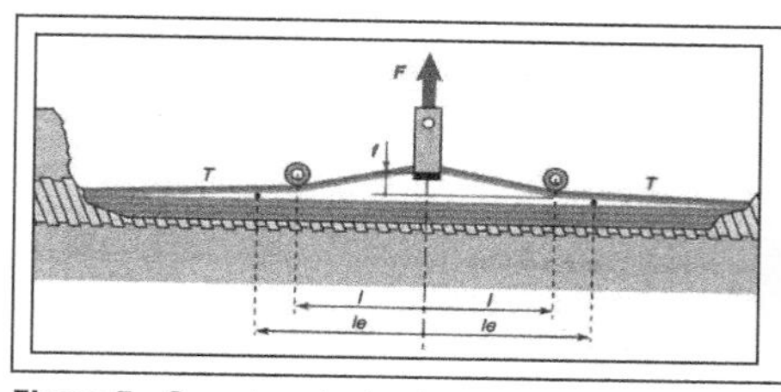

**Figure 5. Crossbow test scheme**

Ministry manual for prestressed concrete elements that mandates a value of 0.55*fpu to account for normal wire relaxation in the structural integrity assessment analyses.

Similarly in situ tests were conducted using a technique known as slotstress that consists of direct on-site measurement of concrete stresses along two points in a concrete section to reproduce the existing section stress profile. The principle of the slotstress technique is like the load cell measurement. The results obtained offer a range of stress values ranging between 0.5 Mpa (1.0 Mpa – 0.5 Mpa) and 3.3 Mpa (2.8 MPa + 0.5 Mpa), that were used to validate numerical modelling results.

### Numerical Modelling to Assess Repair Strategy

The results of the in-situ tests measurements were used to confirm assumptions made during numerical modeling and calibrate key parameters of the models, especially the slotstress results that were used to calibrate design parameters by also incorporating measured water table in the river at the time of the tests.

In addition, assessment of structural integrity was conducted by analyses existing conditions before the demolition and repair work that involve chipping of unsound existing concrete to a certain depth and in some case expose existing rebars to ensure the new repair concrete is fully bonded to existing concrete. Of particular interest was at which extend such demolition could go adjacent to existing post tensioning cables. Several analyses were therefore conducted for each structural caisson element and load conditions under several damage scenarios, some of them beyond what was observed during detailed inspection. The results of these analyses were used to assess the limit of demolition that will not impact structural integrity of the tunnel and if any special pre-support measures, also known as "repair phasing," were required during demolition and curing of repair concrete.

Conducted analyses were carried out using 3D FLAC and SAP200 models that allows to model explicitly incorporate the tunnel geometry of the caissons, the configuration of the post-tensioning cables and soil structure interaction in any loading configuration (Figure 6). Several scenarios representing observed degradation and extrapolated configurations were analyzed, and provisions were included in the design to assure the integrity of the tunnel is maintained throughout the demolition and repair work.

## DESIGN OF TWO LEAKY JOINTS BETWEEN CAISSONS

### Existing Condition and Required Design Criteria for the Repair Strategy

During original construction there was a misalignment due to strong waves in the Saint Lawrence River that resulted in offsets of 35 mm and 15 mm at the joints between caissons 6 and 7 (here after joint 6/7), and joint between caissons 7 and 8 (here after joint 7/8), respectively. The two joints were then filled with concrete but have experience

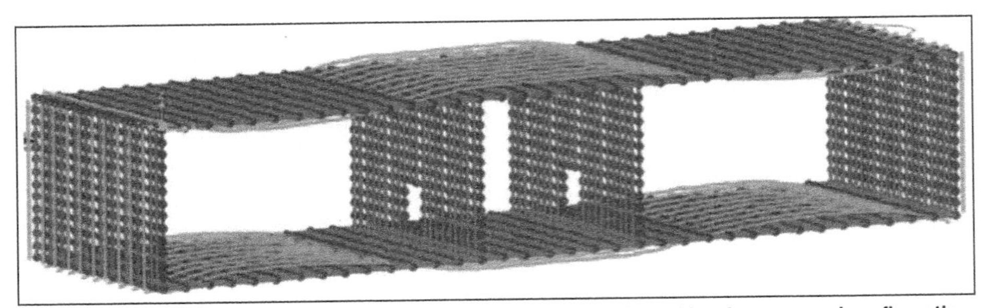

**Figure 6. 3D model that explicitly reflects transverse post-tensioning cables damages and configuration**

water leaks since then. The Ministry has tried several schemes to fix the leak without success and the solution adopted was to manage the water infiltration through a series of channels in transverse direction at the joints location and water is send into the center corridor where a drain takes the water to existing pump station. The drains are equipped with heat trace to make sure no ice is formed during wintertime.

In addition, monitoring of movements along the joints have shown that the joints experience seasonal variation of longitudinal movement with a maximum of lateral movement of 7 mm and 2 mm, for joints 6/7 and 7/8, respectively. The other joints for the rest of the tunnel were monitored and did not experience similar movements.

Given the damage caused by moisture from the drainages and the risk of ice formation during winter season, the Ministry has required a fix to this problem during the ongoing major rehabilitation work. The goal is to find a scheme that will stop the leakage and limit water infiltration. The Ministry has also requested that the joint be able to accommodate 3 times the observed movement, i.e., 21 mm for joint 6/7 and 6 mm for joint 7/8, respectively. Additional requirement was to be able to inspect the proposed repair scheme to ensure it meets the requirements for at least a period of one year.

## Final Design Considerations

The scheme adopted during final design is to use Omega Seal that is traditionally used for new construction of immersed tube tunnel and apply it from the inside of the tunnel along the joints. The joint consists of a rubber membrane that is fabricated by Trellerborg and can withstand high water pressure and accommodate large deformation. Through a collaboration with the Omega seal manufacturer, it was decided to use a 3-piece seal to minimize the welding on site but also accommodate repair sequence. As noted above the tunnel must remain in operation during major rehabilitation work, which implies that the joint cannot be placed at the same time. The three pieces will correspond to major lanes closures and an overlapping length was designed to ensure continuity of membrane.

To finalize sizing and geometry of the seal, a series of in situ measurements were taken to ensure that the seal will cover the leaky area while allowing for tolerances on both side of the joint. These measurements include a laser scan of the joints, coring, ferroscan inside the joint to check the position of rebar and partial demolition of the joint in the corridor portion. A 3D scan of the joints was used to define the exact geometry on site and coordinated with seal manufacturer.

One major constraint was the vicinity of anchors for post-tensioning cables to the joint and any repair option had to avoid any disturbance to these anchors. The selected option has therefore to balance the need to seal off completely the joint while avoiding

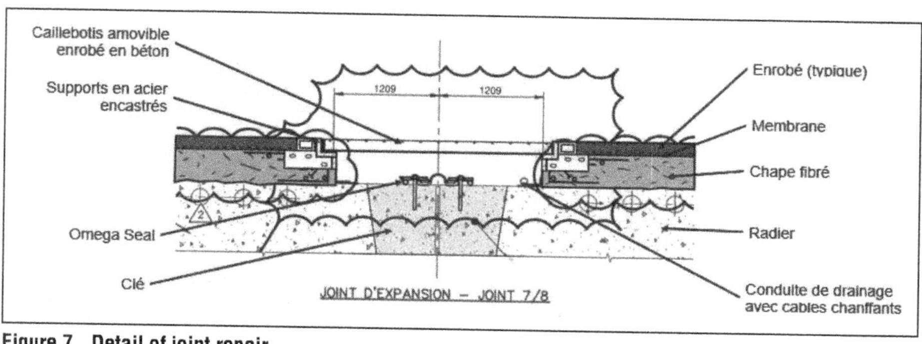


**Figure 7. Detail of joint repair**

the potential negative impact of approaching the anchors. Figure 7 shows a detail of the repair strategy that allows for perioding inspection as required by the contract.

One aspect that had to be analyzed was to ensure the joint has enough shear capacity during local demolition and repair work to resist uplift pressure and shear demand knowing that backfill material at the riverbed has been eroded over the years and it will not be restored during demolition. Thus, a series of bathymetric measurement were conducted at the riverbed to assess the extent of the erosion and the existing backfill thickness that were used in the uplift calculation. The buoyancy assessment calculation was carried using FHWA and ITA guidelines as well as other design manuals. Numerical modelling was conducted to account for reduced capacity of joint during repair work and incorporate repair phases for both the stabilizing screed layer and the reduced backfill material as well as repair phases between the two tubes. It should be noted that the analyses are very conservatives as they ignore the skin friction along the tunnel walls, the live loads as well as equipment.

## DESIGN OF PASSIVE PROTECTION FIRE PROTECTION OF STRUCTURAL

Over the last two decades there have been several serious fires in road tunnels that have caused extensive loss of life and severe collateral loss to the infrastructure that include structural damage and long-term financial effects to the local communities. This has led to a series of updates in codes and regulations requirements, in particular National Fire Protection Association (NFPA) 502, which now includes strict requirements for fire life safety infrastructure and protection measure to the structure of the tunnel. The project used other codes and best practices, in particular the recommendations from the French Ministry of Transportations Center of Tunnels Studies also known by its French acronym as CETU. This section provides details of step taken to design passive protection measures that are being implemented as part of major rehabilitation works.

### In Situ Tests and Assessment of Existing Conditions

During preliminary design stage two fire tests were conducted in the tunnel in summer 2016 to characterize the resistance of the existing concrete to a design fire event that was at that time to be 50 MWA and to predict when spalling will occur after the beginning of the fire in the tunnel. The destructive tests consisted of mobile furnace (Figure 8) that applied the design fire curve to the surface of the concrete with several thermocouples that were installed at different location and depth of the testing area in order to monitor temperature increase within the concrete. The tests showed that spalling will appear within the first 3 minutes when concrete reached a temperature

Figure 8. In situ fire tests conducted in the Lafontaine tunnel

of approximately 400 degrees Celsius. More details on the tests are provided in Habimana et al. 2018.

The results of these tests were used to back calculate the properties of existing concrete such as the effective heat transfer coefficient, the thermal conductivity and the specific heat that are used to conducted thermomechanical analyses to simulate the behavior of the concrete to a design fire event. Finally, these tests were instrumental in building a business case for the need to install passive fire protection measures during final stage of the major rehabilitation wok and they results were used to define performance requirement for the passive fire protection measures.

## Numerical Modeling and Real Scale Laboratory Tests

Since concrete structures show a non-linear behavior under larger strains, a strain-softening constitutive model was chosen to better capture concrete's behavior under fire event. Reinforcements and post-tensioning cables are modeled as cable elements in the FLAC3D commercial software.

In according with project requirement, the Hydrocarbon Modified (HCM) curve was specified as the air temperature in the tunnel during a fire event. For such a fire, the increase in temperature T (°C) as a function of time (t) in minute reads Eq. (1).

$$T = 1{,}280(1 - 0.325^{-0.167t} - 0.675^{e^{-2.5t}}) + 20 \quad \text{(EQ 1)}$$

In addition, real scale laboratory tests were conducted on different panels (Figure 9) that represent the types of material being considered for passive fire protection that include commercially available fire boards from several manufacturer, sacrificial concrete reinforced by polypropylene fibers at various thicknesses and concrete mixes. The results of laboratory tests were used to calibrate the 3D numerical models and assess if the proposed protection measures meet the contract requirements.

Figure 9. Real scale laboratory tests on fire protection panels comparing sample with fiber (noted F) and sample without fiber (noted NF)

## Other Final Design Consideration

There are a variety of other parameters that were considered in comparing different options that were considered for the passive fire protection including resistance to the harsh winter conditions and the de-icing salt environment while maintaining expected esthetics conditions, the interface with equipment and cablings, etc. Other criteria included the interface with repair work in the tunnel crown that minimizing the impact to the existing post tensioning cables as described above.

It was finally decided that a 125 mm thick sacrificial concrete layer reinforced by polypropylene microfibers (2 kg/m$^3$ of concrete) was the option that complies with all requirements and meet considered criteria. On the opposite, non-fibered concrete spalled massively a few minutes after the beginning of the fire trial and was eventually dismissed (Figure 9).

Cast in-situ concrete was retained throughout the tunnel, for the fire passive protection of the walls and the ceiling. In addition, shotcrete was also used, to accommodate the complex geometry of circular ventilation shafts.

# ADDITIONAL UPGRADES

The additional upgrades that are part of the projects are:

- Additional bus lanes at the approaches of the tunnel that extend as far as 35 km in the south of the tunnel to allow bus access during rush hours and minimize impact on commuter during repair work. Additional parking is being provided along the route to incentivize drivers to use public transportation during repair work. These facilities will remain in place once construction is complete.
- Repair of the screed layer in the invert of the tunnel by installing a waterproofing membrane that will prevent future chlorides from deicing salt to penetrate and corrode the invert structure
- Installation of a new ventilation system that will convert the existing semi-transverse ventilation system into a fully longitudinal ventilation system that will improve egress scheme and comply with current codes and best practices. This will among others allow the service corridor to be used both for passenger egress and installation of equipment. No smoke will be permitted to enter the center corridor anymore and special provisions for passengers with reduced mobility will be implemented.

- Tunnel systems upgrades that include a completely new control center with its associated systems that comply with current codes and best practices.
- A new and redundant electrical supply system that will use latest technology to reduce energy consumption while complying the latest requirements and best practices.
- Marine work to restore original design level of backfill and avoid its future erosion by changing backfill characteristics.

## CONCLUSION

The paper discussed three aspects of the major rehabilitation works that are being implemented for the Lafontaine Tunnel in Montreal, i.e., the assessment of structural integrity during concrete repair work accounting for observed damage in post-tensioning cables, the repair of two leaky joints that are a results of a construction defect, and the design of fire protection measures to resist a design fire event in order to comply with latest codes requirements and best standards.

The approached used a combination of cutting-edge testing techniques both in laboratory and on site, numerical modeling, and engineering judgement to select appropriate measures. Several sensitivity analyses were conducted where applicable to account for potential scenarios that can occur during repair work.

Additional upgrades are included in the project to comply with lasts codes and standards as well as best practices for the fire life safety.

## REFERENCES

Habimana, J., Showbary R., Bienefelt J. "In-Situ Fire Test for the Design of Passive Protection for the Lafontaine Tunnel." North American Tunneling Conference, Washington DC, 263–270, 2018.

Federal Highway Administration (FHWA) 2009. *Technical Manual for Design and Construction of Road Tunnels—Civil Elements.*

# A Novel Holistic Approach to Rehabilitation of Underground Structures

**V. Gall** ▪ Gall Zeidler Consultants
**T. Martin** ▪ Gall Zeidler Consultants
**L. Boyd** ▪ Gall Zeidler Consultants

## ABSTRACT

Water infiltration into subgrade infrastructure can cause major impacts on their performance. In addition to damaging the structure, water intrusion leads to deterioration of installations including electrical and mechanical components and in patron discomfort. To remedy these impacts, leak remediation is often carried out to halt water infiltration. Remediation methods include coatings, drainage, injection/stitch grouting, curtain (backside) grouting, and/or internal umbrella systems. Selection of the rehabilitation method depends on the structure's use, owners' priorities, its installations, structural conditions, surrounding ground, and hydrogeologic conditions. Since many factors influence the rehabilitation method chosen, a novel holistic approach is undertaken to understand leakage causes and consequences to develop the most appropriate, efficient, and reliable solution to extend the structure's life. A reconnaissance phase combines geologic, hydrogeologic, and as-built information with detailed digital scans and visual observations to develop a database of existing conditions. This database, called a tunnelband, is used to develop the rehabilitation solution and made part of the contract documents, allowing for an informed bid by specialty contractors. Tunnelband and the preferred rehabilitation system are portrayed in contract documents, which are procured in various contract types depending on the owner's preference and project characteristics. The pool of contractors are required to submit their understanding of this holistic approach by developing and supplying the owner with a detailed workplan. The completed rehabilitation is portrayed in detailed as-built drawings which also provide the owner with an operation and maintenance manual outlining for periodic observation of the structure and checking of its performance. Ultimately, this information is implemented into a "BIM Digital Twin" that is used by the operations and maintenance staff for long-term observations. This proposed novel framework for leak rehabilitation is currently being used successfully in a number of projects throughout the United States.

## INTRODUCTION TO LEAKAGE REMEDIATION

One of the most significant long-term hazards inherent to subgrade infrastructure is water infiltration through and into the structure (Bergeson & Ernst, 2015). If left untreated, water intrusion can create a negative user experience, cause deterioration of critical components within the structure, and can compromise the integrity of the structure itself.

There are a number of different factors that influence water intrusion. Most subgrade infrastructure is wholly or in part below the water table, and over time, exposure to groundwater can lead to the development of pathways for water to flow into the structure in particular in older structures and/or structures where the waterproofing installed as part of the initial construction has become ineffective. This effect can be exacerbated by porous and permeable geology that allows for higher groundwater flow. In concrete-lined structures, a common location for water intrusion to occur is at construction joints; water can also enter through circumferential cracks that developed within the concrete slabs. Masonry-lined tunnels can also exhibit efflorescence and leakage through either cracks in the tunnel structure or in between the masonry units. In metal-lined tunnels, water intrusion may present as corrosion and dampness at joints, and can often be identified as rust or leakage within each segment in more severe cases (McKibbins, Elmer, & Roberts, 2009).

Water infiltration can affect structures of any age, but older infrastructure commonly experiences the most severe effects due to degradation of the lining and waterproofing. Newer structures may also exhibit water intrusion if the waterproofing was not successfully chosen, designed or installed. Given water infiltration can affect subsurface infrastructure regardless of age, type of construction, lining, and ground conditions, a number of remediation methods are employed worldwide to halt water intrusion regardless of the structure or leakage characteristics. Additionally, numerous repair materials exist on the marketplace, with varying characteristics. Several companies have even successfully developed their own proprietary materials for waterproofing, to include in particular grout mixtures.

With so many factors affecting the success of the remediation, it is critical to select a remediation strategy that is appropriate for the structure experiencing water intrusion. The authors propose a novel holistic approach to water intrusion that takes into account not only the existing leaking structure conditions, but fully understands the structural, waterproofing, geology, and hydrogeology of the structure. Any as-built records, from previous rehabilitation efforts and original construction, are also reviewed with the goal of a complete understanding of the water intrusion of the structure. Fully understanding these factors allows for a thorough evaluation of repair methods, and for the most appropriate repair method for both the structure and the owner's operational needs to be chosen.

## REMEDIATION METHODS

Water infiltration is a ubiquitous feature of numerous subsurface structures, and displays many forms depending on varying structural and hydrogeological characteristics. To address water intrusion in these wide range of conditions, a number of repair methods and repair materials have been developed. These include both positive-side (i.e., outside the tunnel structure) and negative-side (i.e., inside the tunnel) methods, and employ different strategies to control, divert, or stop water intrusion.

Several of the predominant methods used in contemporary subgrade water intrusion remediation are presented below. Each of these methods presents a viable strategy for halting water intrusion, and some methods may be used simultaneously.

### Crack Injection

Crack injection, also known as "stitch grouting," is a negative-side remediation method that halts water intrusion through a crack or joint, mostly within concrete structures. Holes are drilled at an angle in an alternating pattern on either side of the joint or crack, and intersect the plane of the deficiency near its midpoint within the slab. Packers are installed, and the crack is flushed with water and then injected with grout such as epoxy, polyurethane, and acrylic resin (Figure 1).

This technique for repairing cracks in concrete can be successful, but there are several limitations of this method. Crack injection is used primarily to repair leaks at discrete cracks and joints, and is not a system-wide water infiltration solution. Additionally,

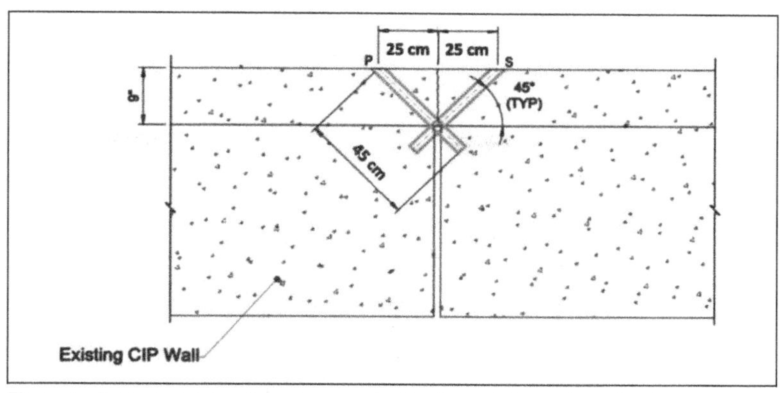


Figure 1. Typical configuration of crack injection holes intersecting a joint

halting water at one crack typically causes the water to migrate to another location nearby where water can find a pathway into the structure. This commonly leads to "chasing" leaks throughout a structure to fully halt water intrusion, and multiple treatments are often required.

## Coatings

Coatings such as crystalline waterproofing are another method for halting leaks through negative-side repair. These coatings can be applied to individual cracks or applied broadly (typically in conjunction with discrete crack repair) to help limit the chasing of leaks. This coating applies a material that reacts with water, which causes a catalytic reaction that produces a non-soluble crystalline. Through the process of diffusion, the crystalline waterproofing chemicals spread throughout the pores and any hairline cracks within in the concrete, allowing this reaction to take place anywhere the crystalline waterproofing material is exposed to water. These insoluble crystals continue to grow until water infiltration is halted.

Figure 2 shows crystalline waterproofing as installed along a crack in concrete. Following application of the repair material along the identified crack, a coating was then applied to the entire structure, which helps to limit new leaks from appearing elsewhere.

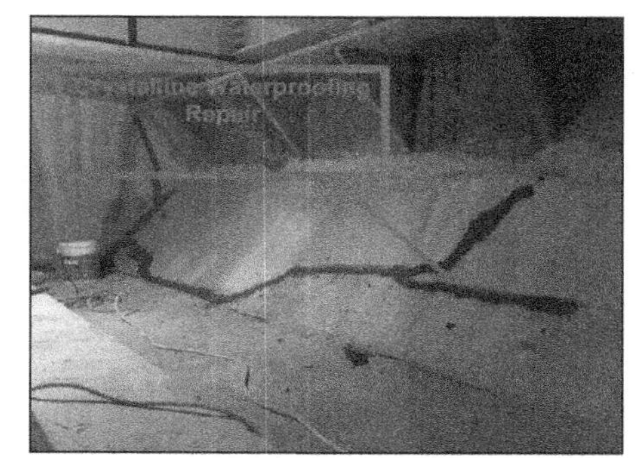

Figure 2. Crystalline waterproofing applied along a crack

While, like crack injection, this method is useful for repairing leaks along discrete cracks, and has the added benefit of the ability to cover a broader area, this is not a system-wide water infiltration solution. Additionally, unlike crack injection, which can be used on cracks with considerable width, the use of coatings such as crystalline waterproofing are generally only recommended for hairline cracks.

## Curtain Grouting

Curtain grouting is a positive-side remediation method which involves treating the outside of the structure's lining and to some extent the surrounding ground to fill any

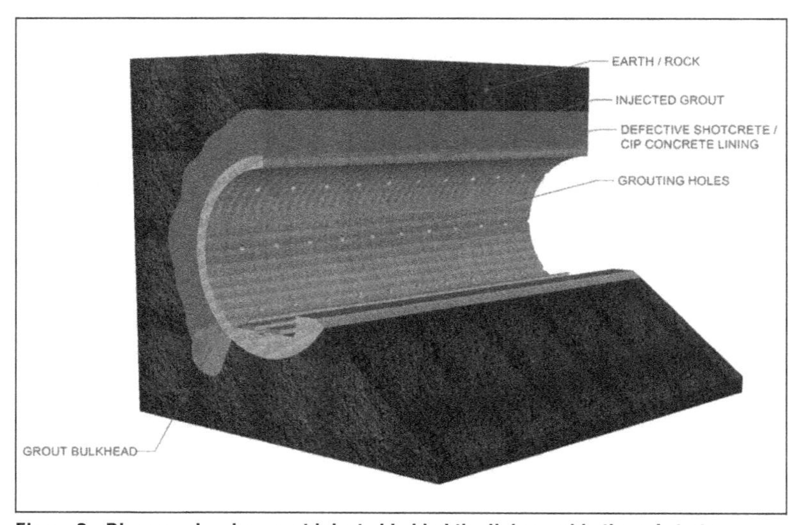


**Figure 3. Diagram showing grout injected behind the lining and in the substrate.**

voids and eliminate the source of water infiltration before it reaches the structure. In this approach, holes are drilled through the lining in a systematic pattern to penetrate the entire thickness of the lining, whether that is steel, cast iron, concrete, shotcrete, brick or another lining type. For steel and cast iron linings existing grout ports may be used for grout injections. These holes reach, at a minimum, the exterior of the structure, and can be drilled into the surrounding substrate as well (Figure 3).

Curtain grouting is a versatile option that can be used in all structure types and practically all ground conditions.

## Lining Replacement and Umbrella Systems

An umbrella system is the most consists of installing a secondary lining following the installation of a waterproofing membrane against the existing liner.

This is typically the most expensive and involved remediation option, and requires the tunnel or structure to be out of service for extended periods of time. Given that, this is an option that should only be considered in cases where the existing lining is unsalvageable and water intrusion is widespread, severe, and systemic. However, installation of waterproofing and drainage systems limits future remediation needs and ensures the longevity of the structure following installation.

## Drainage and Dewatering

Most subsurface structures are constructed with internal drainage systems that divert groundwater through or around the structure. Over time, these drainage systems can become clogged and inoperable, which can lead to additional water intrusion from the water that cannot be adequately diverted. In some scenarios, repairing the drainage system greatly decreases the amount of water intrusion observed.

Dewatering is another method to control the water. Diversions can be constructed that move water through the structure with minimal impact on the structure and its components.

## PROPOSED HOLISTIC APPROACH

Application of the most appropriate water infiltration mitigation strategy is critical to ensure a successful project. With widely varying characteristics that affect water intrusion, and many available remediation strategies available, it is important to carefully craft a leak remediation program that considers these characteristics as well as an understanding of the goals of leak remediation for a particular site. This proposed holistic approach is based on the authors' understanding that there is no universally-appliable leak remediation program, but rather the most appropriate leak remediation strategy for a particular project is derived from a holistic understanding of the structure, the causes of leak remediation, and the owner's goals for the structure.

### Basis of Design

The foundation of the recommended design and project specifications is based on past experience and a thorough review of national and international standards. For example, reference to curtain grouting is made to STUVA (2014). These standards and peer-reviewed papers aid in the development of details included in the project specifications in a grouting program and assist in the understanding of acceptable installation of remediation strategies.

### Inspection and Desktop Study

The most important aspect of leak remediation is gathering all necessary information to make educated recommendations about a project site. Early on in the process, all available data are identified and reviewed to fully understand the structure's layout, the history of its construction, and any factors that could influence the chosen remediation strategy. Some of the most important information to gather at this step is the geology encountered during investigation, the quantity and locations of groundwater inflows, and data from the as-built records to better understand how the structure was built and how the waterproofing, if performed, was applied.

A comprehensive walkthrough and itemization of leaks is performed following this stage, and digital scans are typically performed at this point in the investigation. The digital and visual data are then combined with as-built structural, geological, and hydrological information into a comprehensive database of existing conditions. This database, called a Tunnelband, is used to develop the rehabilitation solution and made part of the procurement documents, allowing for an informed bid by specialty contractors.

During this phase of the work, it is also essential to capture the owner's goals of the project, and understand the purposes of leak remediation. Depending on the structure's uses, the owner's future plans for the structure, and also varying goals in public areas (e.g., station platforms) and private (e.g., ancillary back of house rooms) areas, the recommended remediation strategy may change. For example, an owner of a subsurface mass transit station may want to ensure that no leakage occurs in all areas of the station where the public can access, but may find small amounts of residual leakage acceptable in back rooms, storage facilities, and throughout adjacent tunnels. Understanding the owner's goals, and working with them at every step in the process will aid in ensuring that the leak remediation strategy is consistent with their expectations and will help facilitating a successful project.

### Remediation Selection

Following the gathering of all relevant information, a design phase is undertaken where all of these data are taken together to recommend a comprehensive remediation strategy for the structure. When taking the gathered information into account, there are still several other factors to consider when recommending a path forward.

#### *Structure Types*

While there are some remediation methods that are independent of structure and lining type, many are not. Curtain grouting is a viable option anywhere that a systematic grid for grouting can be developed, but crack injection is generally only appropriate for concrete-lined structures, or in some masonry-lined structures. Replacement of the lining can only be accomplished where sufficient clearance can be maintained after the new lining is installed. Additionally, most coatings can work on all lining types, however special care must be taken when performing this repair method on shotcrete-lined tunnels; due to the rough surface that is generally present, it may need to be smoothened by cementitious coatings prior to application.

#### *Ground Conditions*

Most remediation options are suitable for a wide range of ground conditions, however curtain grouting is much less dependent on the substrate than some of the other remediation methods. Dewatering, for example, is suitable in cases where the substrate is rock; however, this is much more difficult in soft ground.

#### *Impact on Operations During Construction*

For an existing structure, the leak remediation program will impact the typical operations of the structure to a certain extent. If performed within a transit network, it is likely that the work can only be performed during scheduled outages, which may be limited to a few hours at a time at nights and certain times over the weekend. In a road tunnel, an adjacent tunnel may be briefly used for bi-directional traffic, but this is also typically limited to short periods of time. Work must not only be completed during these windows, but at the end of each window, the structure must be clear of any tools and work debris, and be fully operational. For these reasons, lining replacement or installation of an umbrella system can generally not be performed without a prolonged shutdown of the structure, which is not generally possible in many situations. Even the utilization of modern lining overcutting machines behind a protective shield that allow for traffic operation have their limitations and will cause limited shut down periods.

### Procurement

During the design phase, preparing for the procurement and execution phases is essential. Given that any of the more involved remediation methods involve a high level of specialization, a specialized contractor is necessary to assure the success of the remediation project. It is recommended that a prequalification process be undertaken to allow for qualified bidders to be selected prior to the distribution of the Contract Documents.

When prequalifying is not possible, i.e., due to an owner's legal obligations early notice of procurement is crucial for obtaining multiple competitive bids. The nature of these projects typically requires specialty contractors to form partnerships with larger, general contractors. These pursuit partnerships take some time to form, and the earlier notice given to potential bidders, the better.

### During Construction

During construction, it is critical to ensure the presence of a qualified Engineer of Record representing the design intent as well as perform a thorough qualified inspection to verify the remediation was performed in accordance with the Drawing and Specifications (Gall, 2000). The presence of an Engineer of Record on site aids in capturing data regarding the work being performed, and a thorough inspection can demonstrate the work has been acceptably performed, can be used to direct the Contractor to revisit certain areas, and can be used by the owner as part of a repository of data for the work that was performed.

### Leakage Criteria

The definition of acceptable leakage criteria post-construction should be established early in the project life with the owner. Leakage criteria can be uniform throughout the whole structure or varied based on location; an owner may want certain areas to be completely dry. The leakage criteria should be clearly defined in the specifications and drawings. Although a completely dry structure is sometimes preferred, a common baseline for acceptable leakage criteria to be included in the specifications is accepting localized damp spots due to capillary action, but prohibiting dripping or flowing leaks. Another utilized metric is accepting leaks that, when blotting paper is applied, the amount of leakage must be less than a certain diameter on the blotting paper.

### Maintenance after Construction

After construction is completed, the completed rehabilitation is outlined in detailed as-built drawings which also provide the owner with an operation and maintenance manual outlining for periodic observation of the structure and checking of its performance. Ultimately, this information is implemented into a "BIM Digital Twin" that is used by the operations and maintenance staff for long-term observations. While this is important information to have in all cases, this is particularly important when crack repair, dewatering, or drainage is performed. As mentioned previously, crack repair can lead to water finding new pathways through the structure, and continued maintenance should inspect the structure periodically to identify if new leaks are appearing. Additionally, maintenance needs to be performed on a regular basis for drainage lines to keep them in good working order and unclogged.

## CASE HISTORY—WASHINGTON METROPOLITAN AREA TRANSIT AUTHORITY

In 2018, GZ as part of a General Engineering Team under Gannett Fleming, was tasked by Washington Metropolitan Area Transit Authority (WMATA) in the Washington D.C. Metropolitan Area, USA, with performing water intrusion assessments on tunnels in multiple areas of their system. This followed a well-received pilot grouting program which addressed water intrusion in two miles of underground tunnels in Maryland. The goal of the assessments was to optimize future leak remediation work and involved performing inspections, executing a desktop study of all available documents, and preparing a report summarizing the findings. Both visual and digital inspections were carried out, with digital scanning performed, including highly-detailed photogrammetry of the lining. Over 21 miles of tunnels, shafts, and adits were assessed. and recommendations were made to WMATA for locations to prioritize. Several separate leak remediation designs were then developed, and these are in various stages of implementation.

### Implementation of the Proposed Holistic Approach

Prior to gathering information or providing recommendations, GZ met with WMATA to understand their goals of the leak remediation program. Given the size of the WMATA network and the spatial variability of magnitude of water intrusion experienced along the lines, along with the infrastructure present at different locations along the system,

WMATA prioritized certain locations and conditions along the network. Priority was assigned to areas experiencing dripping or flowing leaks and areas where internal infrastructure could eventually be compromised by continued water infiltration. This led to the goal of optimizing the remediation strategy in terms of cost (in both time and money) of remediating the infiltration and the benefits of the leak remediation program.

Once the goals and priorities were well understood and following the gathering of this information, a tunnelband was developed for all of the structures, displaying the geology, hydrogeology, inspection findings, photographs, and construction documents. This wealth of data allowed each section of tunnel to be compared to each other section of tunnel, which were used to both evaluate known problem areas and also identify additional areas that may experience negative impacts from water infiltration in the future. The tunnelband data were then used in conjunction with WMATA's goals and priorities to consider possible remediation methods; after this exercise, unique recommendations were made depending on the structure type, owner priorities, leakage regiment, and access considerations (Figure 4).

### Innovative Inspection Techniques

Given the long distance of the tunnels which were evaluated, conventional inspection techniques would have led to a considerable effort, both in manpower and track time. An estimate predicted over 90 shifts of night inspection would be required, at a minimum. As an alternative, high-resolution scanning was adopted. A hi-rail vehicle was outfitted with equipment to allow for multiple miles of scanning to be completed each night. This shortened the track time required to under three weeks. These scans were then reviewed during normal office hours. This innovation greatly reduced operational impacts on WMATA and field staff required, without sacrificing the accuracy of the inspection.

## BENEFITS OF THE PROPOSED HOLISTIC APPROACH

Typically, a one-size-fits-all leak mitigation solution would be applied that may treat each leak agnostic of leakage characteristics, ground conditions, owner priorities, and the structure of the tunnel. This repair strategy would grout each leak independently without context of the structure's overall leak regiment. Compared to recommendations resulting from the proposed holistic approach, this approach requires greater material and labor costs, may lengthen the duration of the remediation program significantly, and does not significantly improve the results.

In addition to avoiding diminishing returns, this holistic approach can also proactively identify areas that should be addressed sooner to avoid potential negative effects in the future. An understanding of the spatial characteristics of ground conditions can aid in identifying where leakage may be seasonal, or areas where a prolonged multi-year drought can decrease flow from leaks during the duration of the drought. Utilizing a targeted, tailored holistic approach allows for funds, time, and labor to be optimized to meet the Owner goals. Additionally, understanding the structure as a whole, to include geology, hydrogeology, inspection findings, photographs, and construction documents, allows for an optimized and durable remediation solution in line with Owner priorities.

## CONCLUSIONS

Water intrusion into existing underground structures leads to premature deterioration and increased maintenance costs. Often water intrusion is treated with one size fit all remedies, and little care is paid to the cause or characteristics of the water

**Figure 4. (top) Hi-rail vehicle used in scanning; (bottom) Scan from a WMATA tunnel**

infiltration. Commonly, the symptom is treated, rather than the cause of the leakage itself. This can lead to an unsatisfactory remediation effort and requires additional time and money spent to repair the structure to the satisfaction of the owner. Utilizing a holistic approach to tunnel remediation allows the most appropriate methods for the rehabilitation to be implemented, taking all the structural, operational, and stakeholder considerations into account.

## REFERENCES


McKibbins, L., Elmer, R., & Roberts, K. 2009. Tunnels: inspection, assessment, and maintenance. In *CIRIA C671.*

Gall, V., Three pillars for an effective waterproofing system for underground structures. In *Proceedings of the North American Tunneling Conference., Boston, MA, USA, 6–11 June 2000.*

Bergeson, W., & Ernst, E. 2015. Tunnel Operations, Maintenance, Inspection, and Evaluation (TOMIE) Manual. In *United States Department of Transportation Report No. FHWA-HIF-15-005.*

# Tunnel Condition Assessment: State of the Practice

**Saleh Behbahani** ▪ Lyles School of Civil Engineering, Purdue University.
**Tom Iseley** ▪ Division of Construction Engineering and Management, Purdue University.
**Aidin Golrokh** ▪ Advanced Infrastructure Design, Inc.
**Ali Hafiz** ▪ Advanced Infrastructure Design, Inc.
**Kaz Tabrizi** ▪ Advanced Infrastructure Design, Inc.

## ABSTRACT

In the USA many highway tunnels were built during the 1930s through 1960s. This indicates that these tunnels have exceeded their intended design service life. Based on the initial tunnel inventory conducted jointly by the Federal Highway Administration (FHWA) and the Federal Transit Administration (FTA), more than 526 highway tunnels have been identified in the USA. For understanding the condition of the tunnels and risks associated with deteriorating conditions, comprehensive condition assessments are necessary to mitigate risk and remediate works as required. Monitoring the condition and rate of deterioration of tunnels is vital for timely tunnel maintenance and rehabilitation to avoid sudden tunnel collapse. This session will explain the necessary steps for tunnel inspection and review some of the state of the practice information on tunnel inspection techniques.

## INTRODUCTION

A transportation tunnel is "an enclosed roadway for motor vehicle traffic with vehicle access limited to portals, regardless of type of structure or method of construction, that requires, based on the owner's determination, special design considerations to include lighting, ventilation, fire protection systems, and emergency egress capacity" based on the National Tunnel Inspection Standards (NTIS) definition. In 2020, the National Tunnel Inspection Program reported that 526 tunnels exist nationwide. Figure 1 includes the top six states which host the majority of tunnels (more than 300 tunnels) in the US.

It has been reported that the majority of tunnels in the US were built in two different periods; the first period was during the 1930s and 1940s while the second period was during the development of the Interstate Highway System throughout the 1950s and 1960s which shows that many of these tunnels have exceeded their intended design service life.[2] Three major tunnel collapsed disasters due to inadequate tunnel inspection occurred in 1999 (Sunset Tunnel located near Manning (west of Portland), 2006 (the I-90 Central Artery Tunnel in Boston), and 2007 (the I-70 Hanging Lake tunnel in Colorado) respectively. Due to collapsed of these tunnels, major milestones in tunnel inspection occurred in 2006 and 2012, respectively. On July 10, 2006 after ceiling collapse in the Interstate 90 Connector Tunnel in Boston, Massachusetts, the National Transportation Safety Board's Highway Accident Report, NTSB Number HAR-07/02, identified several safety issues including, "Inadequate regulatory requirements for tunnel inspections." The Moving Ahead for Progress in the 21st Century Act (MAP-21) was signed by the President on July 6, 2012 which required the Secretary to establish national standards for tunnel inspections. Therefore, the National Tunnel Inspection Standards (NTIS), the Tunnel Operations Maintenance Inspection and Evaluation (TOMIE) Manual, and the Specifications for National Tunnel Inventory (SNTI) were

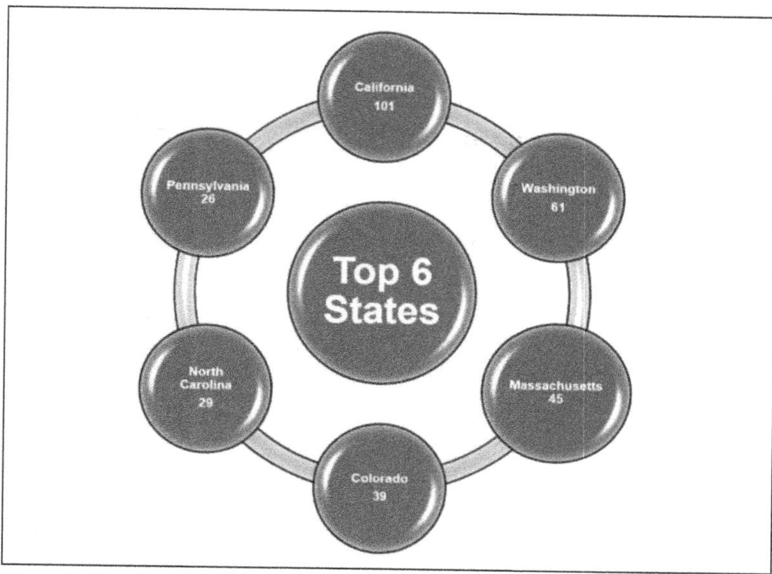


Figure 1. Top six states with majority of tunnels in the US

developed by the Federal Highway Administration (FHWA) to ensure that tunnels continue to provide safe, reliable, and sufficient levels of service.

## BACKGROUND OF TUNNEL INSPECTION

According to the NTIS, five types of highway tunnel inspection have been identified which are initial, routine, damage, in-depth, and special inspections for tunnels. The inspection intervals for in-depth inspections based on the special needs of the tunnel facility is the responsibility of the tunnel inspection organization. While based on the discretion of the tunnel owner, special and damage inspections are performed. Tables 1 and 2 summarize the five types of inspections for highway tunnel and the interval period for initial inspection and routine inspections contained in the NTIS, respectively.

It has been reported that water leakage through the liner is the primary cause of deterioration within an existing tunnel. Water leakage, infiltration, corrosion, spalling, delamination, potential cracking all indicate the water intrusion. Shotcrete, ribbed systems and lagging, segmental linings (precast concrete), and case-in-place concrete are some of the materials which are used for tunnel lining. In soft ground situation, segmental linings are mostly used in conjunction with a tunnel boring machine (TBM).

For tunnel inspection, tunnels must be closed, and this is difficult especially in subway systems. Therefore, the inspection is usually done at night. Depending on the size of tunnel and level of deterioration, field crews might work during several days or weeks at night increasing the cost of inspection. By utilizing the digital tunnel scanning systems such as laser scanning, photogrammetric systems or combination of both systems (hybrid systems) the time which needs to be spent in tunnel can be minimized and the thorough inspection analysis can be done in an office space. Recently, 3D scanning along with high quality image for tunnel condition assessment has gained a lot of attention because it increases the tunnel inspection speed and decreases the

Table 1. Types of highway tunnel inspections[1]

| Inspection Type | Purpose |
|---|---|
| Initial | Establish the inspection file record and the baseline conditions for the tunnel. |
| Routine | Comprehensive observations and measurements performed at regular intervals. |
| Damage | Assess damage from events such as impact, fire, flood, seismic, and blasts. |
| In-Depth | Identify hard-to-detect deficiencies using close up inspection techniques. |
| Special | Monitor defects and deficiencies related to safety or critical findings. |

Table 2. Interval requirements for initial inspection and routine inspections which are contained in the NTIS[1]

| Activity Type | Application | Interval |
|---|---|---|
| Initial Inspection | New tunnel | Prior to opening to traffic to the public. |
| | Existing tunnel | Within 24 months of NTIS effective date. |
| Routine Inspection | Default condition | Every 24 months over lifetime of the tunnel. |
| | Approved written justification | Possibly allow extension up to 48 months. |
| In-depth Inspections | Complex tunnels and for certain structural and functional systems. | Level and frequency to be established by the program manager. |

inspection time on site. As a result, the operation cost will be decreased along with the shutdown times to the regular traffic. This method helps to document and quantify defect in tunnel lining and create a comprehensive data base of tunnel structural conditions for future references.[2–4]

Since manual inspection results depend on human subjectivity in which might lead to inaccuracies and missed detections, automatic detection of tunnel lining remains in high-priority. By the development of deep learning and computer vision techniques, based on convolutional neural networks, new opportunities have been raised for automatic structural health monitoring (SHM). Historically, visual and manual procedures for tunnel inspection have been widely used. Since these procedures are time-consuming and subjective to human error, alternative automated techniques to increase efficiency and reliability in tunnel inspection are needed.

## SCANNING TECHNOLOGY

### 3D Laser Tunnel Scanning Technology (TS3)

The data referenced in this paper are collected by using a 3D laser/Thermal tunnel scanning technology that is developed by SPACETEC Technologies (Figure 2). The device is a three-channel system, simultaneously recording:

1. Reflection intensity of laser on tunnel wall (grayscale, 16-bit resolution, geometrical resolution in circumference: (2*pi*distance)/10000 (~1.5 mm in tunnel with 2.5 m radius); geometrical resolution in tunnel axis @ 300 1/s rotation of laser scanner and 1 m/s travel speed in tunnel direction: 0.3 mm
2. Distance between scanner axis and reflection point resolution
3. Apparent surface temperature of wall at reflection point resolution ~0.1 °C, geometrical resolution ~10 × 10 $mm^2$

One run with the Spacetec scanner (TS3) system produces a 3D profile of the structure (tunnel, retaining wall, etc.), high resolution imagery of the structure, and thermal imaging, on the surface of the structure. In the images produced by the TS3 Scanner, it is possible to identify cracks as small as 0.3 mm in width. One difference to photogrammetry is the absence of focusing depth in the images. Imaging systems with

**Figure 2. TS3 Scanner (left), and TS3 scanner mounted on integrated testing vehicle**

lenses may only focus a certain depth range, whereas the laser always produces a sharp image.

Typically, the scanner is mounted on the rear of a testing vehicle as shown in Figure 2, providing an undisturbed 360° measurement. During field testing, the driving speed of the vehicle was maintained at a constant 3 mph to obtain the best scanning results. The rotating mirror speed of up to 300 Hz is one of the crucial features for measuring speed (Figure 3). This enables fast and non-destructive measurement. Using the combined laser image and thermal image, deficiencies and cracks can be most efficiently identified in the scanned images.

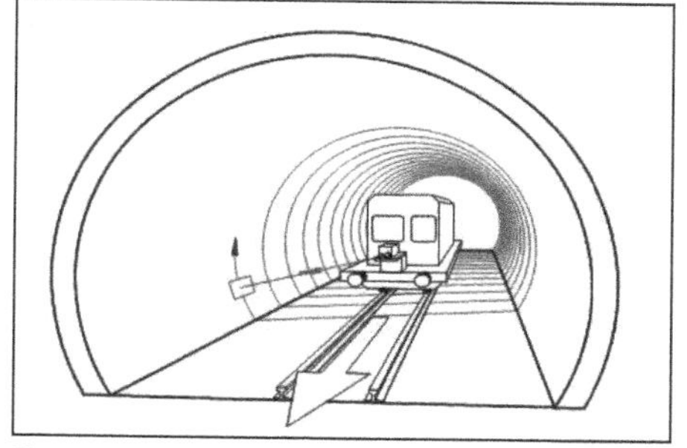

**Figure 3. Scanning principle**

The data visualization is achieved by use of a powerful software package that displays all three channels (visual, thermal and profile) simultaneously, which allows for detailed inspection of the structure of interest (i.e., tunnel, retaining wall, other) on the screen. This helps to analyze and identify deficiencies and compare them on all three channels. Image manipulations such as adjusting the contrast and brightness of the display is also possible with 3D presentations and/or 3D zoom of the image details. Long term monitoring supports the observation of the structure degradation over time with multiple subsequent measurements and recording with an interval of at least one year can be compared.

Visual images are the most frequently used for general documentation of visual distresses. They are precisely referenced along the tunnel length and circumferential position. These images are constructed by the reflectance of the laser light on the tunnel surface and are displayed as gray scale images. They illustrate the visual appearance of the tunnel wall. The images can be used for maintenance and construction activities as the distresses and their position is documented. The locations of concerned areas can also be visually examined.

Thermal imaging results provide a measurement of the surface temperature of the scanned area (i.e., tunnel lining, retaining walls, etc.). Thermal images help identify anomalies that are not seen on visual images. A temperature variation along the surface of the wall difference, which is the result of the heat exchange between the surface of the walls and the air temperature, provides providing clues as to the state of the internal condition of the structure domain. Temperature difference can result from:

- Cooling due to evaporation of water from the surface
- Different emissivity due to material or observation angle changes
- Heat flow changes due to subsurface material variations, e.g., cavities (honeycombs gravel nests below the surface or poor bond between the lining and the rock) or non-homogeneous material composition
- Reaction of the concrete wall material during cooling or heating
- Influence of cold and warm temperatures respectively at the surface

Field verification of areas of anomalies is necessary to confirm and assess the thermal results. In this manner a more precise interpretation of the thermal imaging can be provided. A quasi-stationary heat flow between the air and the backfill of the tunnel wall provides suitable measuring conditions. In the case of unknown heat flow conditions, certain features cannot be clearly identified.

Thermal images typically display areas of temperature differentials (or anomalies) such as:

- Moisture at the surface
- Subsurface moisture
- Delaminations
- Loose or incomplete bonded tiles

Condition monitoring with the Spacetec scanner is based on repeated scans in appropriate intervals. Deterioration processes can be precisely documented, e.g., crack development, spalling or moisture traces. This is a powerful management tool to determine the urgency and frequency of repairs. A typical Unfolded visual image of a 44 m (144 ft.) long section of a retaining wall is shown in Figure 4.

3D-View or 3D-profile can serve many purposes. First and foremost, the existing profile of the structure recorded by TS3 can be used in plans production. Another benefit of this data channel is for closer inspection of the defects as exist on the structure. Since small incremental distances can be measured within the software environment,

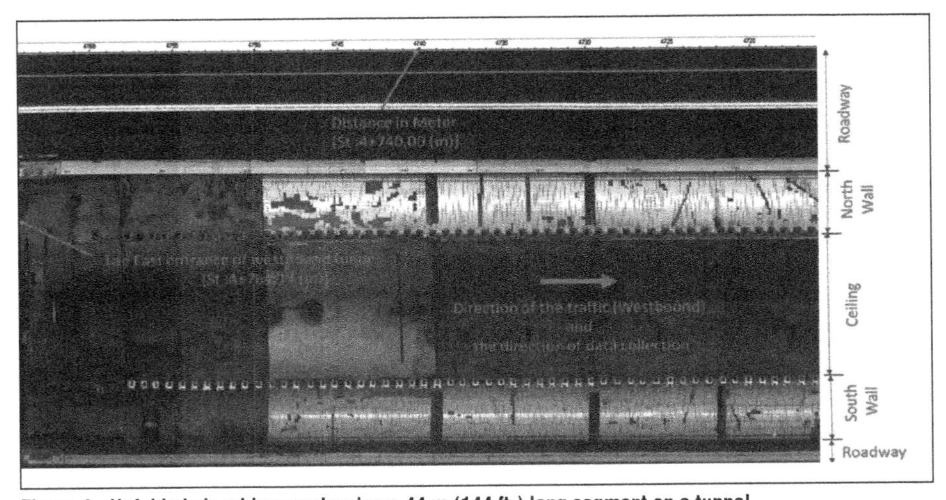


**Figure 4. Unfolded visual image showing a 44 m (144 ft.) long segment on a tunnel**

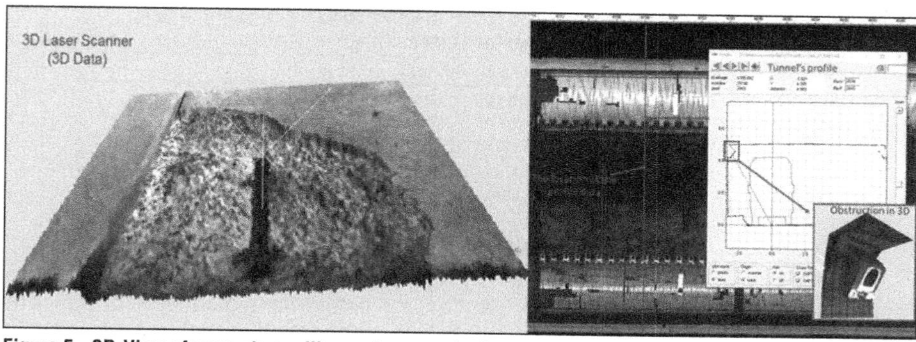


Figure 5. 3D-View of concrete spalling and exposed rebar on retaining wall (left) and 3D sample of obstruction on a tunnel tiled wall (Right)

and because of the high-density scan imagery, small surface defects such as chips, spalling and loose mortar of the concrete wall can be closely examined. Figure 5 illustrates 3D-view samples of areas with concrete spalling identified on the retaining wall as well as obstruction located on tiled wall inside of the tunnel.

## Data Processing and Correction

The recorded data are geometrically corrected, and the 360° display of the domain is projected with a defined scale onto a plane surface for a synchronous display of all three channels. A full dataset consists of a visual, thermal and a 3D-View which is formatted and edited to provide a true to scale display, labeled with distance traveled along the length of each wall. Additionally, each wall is divided into zones that matched the width of each wall panel, as obtained from the available as-built plans of the walls.

Visual images obtained from the laser scanner were used to identify cracks on the walls. These images were also compared to selected images obtained with a traditional digital camera for validation. Similarly, the Visual images from a digital camera and the 3D-View were used to compare surface defects such as concrete spalling.

The thermal data is corrected by the commonly existing air temperature drift along the scanned structure axis. After temperature stabilization, thermal data are displayed with constant air temperature. Therefore, the same phenomena are displayed with the same colors. The data interpretation is based on local temperature differences (indicative of anomalies); thus, an absolute temperature is not needed. Every thermographic surface point corresponds to a color-coded temperature interval with a temperature resolution of 0.1°C (Figure 6) and to one of the 16 colors from black to white. This color palette should give an intuitive physiological indication of cold (black to blue) and warm (red to white) temperatures.

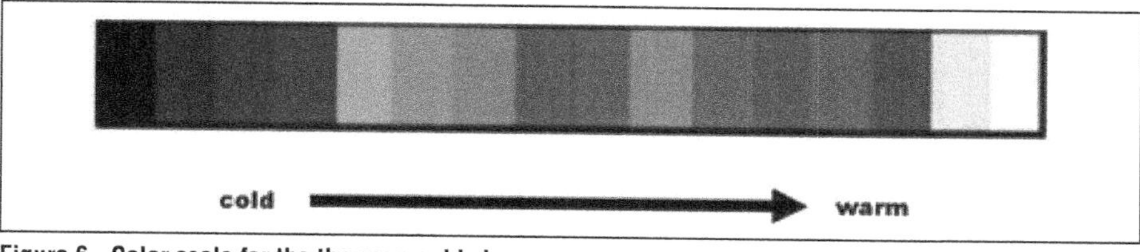


Figure 6. Color scale for the thermographic images

**Table 3. Relationship between surface temperature anomaly and subsurface heat conductivity (when tunnel air is warmer than tunnel wall)**

| Anomaly | Heat Conductivity | Possible Reasons |
|---|---|---|
| Cold | Better | Good thermal contact between rock and wall:<br>▪ Water in wall<br>▪ Higher density of the material |
| Warm | Worse | Bad thermal contact between rock and wall:<br>▪ Loose, less lithified rock<br>▪ Lower density of the material<br>▪ Higher porosity, hollow spaces |

Thermal images reflect the material properties and determine a qualitative correspondence to the nature of the heat source and can be categorized as indicated in Table 3.

The temperature at the tunnel surface reflects the heat conduction between the tunnel air and the materials behind the surface. In general, lateral and thermal resolution of subsurface anomalies decrease with concrete cover. Water intrusions with wet surfaces are always visible in the thermal image with clearly defined patterns in blue. They are caused by the cooling effect due to evaporation and water flowing downwards.

## CASE STUDIES

The following sections showcase three separate projects in which we inspected using TS3, processed, and analyzed data from varying structures. In the first case study the focus of the project was to investigate the amount of water leakage and to map the surface cracks on the inner concrete walls of a tunnel. In the second case study we also succeeded in determining the existing debonding in the body of retaining walls. In the final case study, the tiled walls of a tunnel were scanned. In addition to evaluating the previous anomalies, in this project we also quantified the area of missing tiles. In these case studies, mapped deficiencies were grouped in the following categories:

Main Deficiencies Categories:

- Surface Cracks: Small Crack <⅛", Wide Crack >⅛", and Map Crack
- Thermal anomalies: Cold and Warm
- Leakage: Seepage and Dry Surface
- Spalling and Surface Damage
- Efflorescence

### Case Study #1: Assessment of Concrete Tunnel Wall and Ceiling

This survey was conducted to better quantify the existing damages of a roadway tunnel resulting from possible water infiltration as well as deteriorations. The structure of the tunnel consists of two separate tubes for Northbound and Southbound directions (Northbound and Southbound tubes), each of which having a length of 700 m (or about 2,300 ft). Both tubes are rectangular in shape and accommodate two traffic lanes in each direction. The tubes in each direction have a total width of 11.7 m (or about 38 ft). The tube walls are lined with concrete and the ceiling is covered with precast concrete panels.

Several deficiencies were selected when mapping the recorded data from the laser scanning survey. The National Tunnel Inventory Specifications (NTIS) was followed for mapping the deficiencies. To that end, each tube was divided into four elements: Concrete Slab-on-Grade, Cast-in-Place Concrete (Tunnel Liner-Wall), and Cast-in-Place Concrete (Tunnel Liner-Roof). In addition to main deficiencies categories

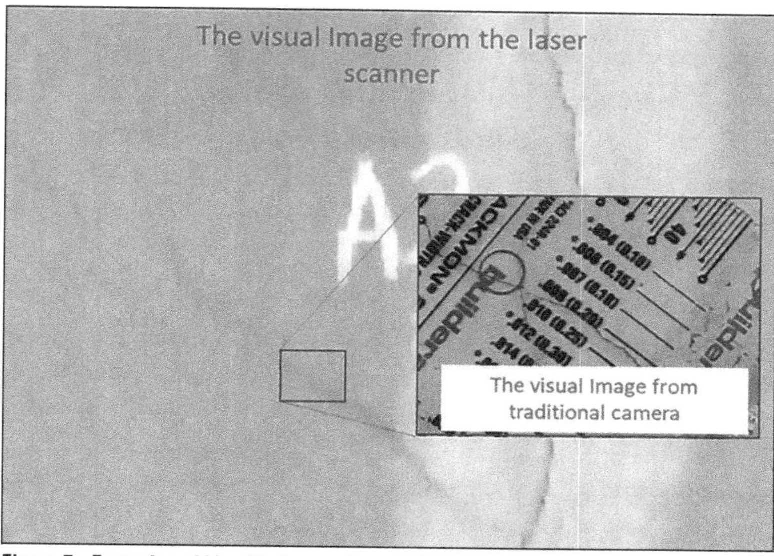


**Figure 7. Examples of identified a small crack (less than 0.3 mm) on tunnel's concrete wall**

that is stated above, the surface cracks were classified into Horizontal Crack, Vertical Crack, Transverse Crack, Longitudinal Crack, Diagonal Crack and Map Cracking or Crack area

A summary of the results of the laser imaging and thermal imaging, obtained for the Northbound and Southbound tubes, were obtained using laser scanner. Moreover, the tunnel has a separate stationing or linear referencing system and the mapping results were summarized for each zone had a length of 100 ft (i.e., one Station) and for each element. As observed, in each image, all distresses and anomalies are superimposed. The anomalies/distresses encountered were mapped based on the criterion set forth by the respective department of transportation.

Figure 7 shows an example of the capability of using the TS3 to identify small cracks. In this example, a crack with width less than 0.3 mm was identified. Figure 8 shows another example that shows how the visual data from the tunnel scanner can be used to identify and classify cracks.

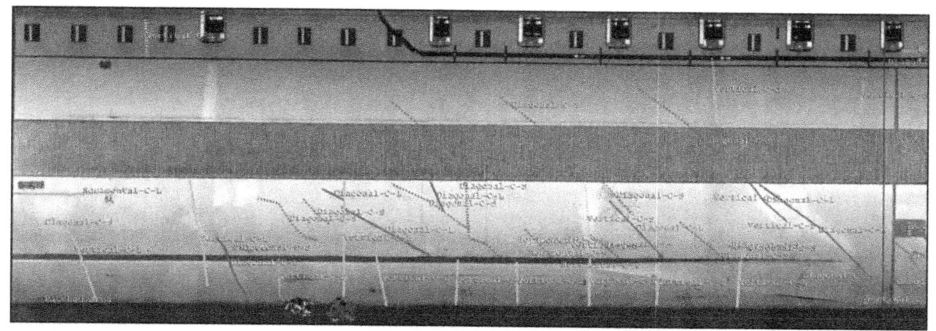

**Figure 8. Examples of identified cracks on tunnel's concrete wall**

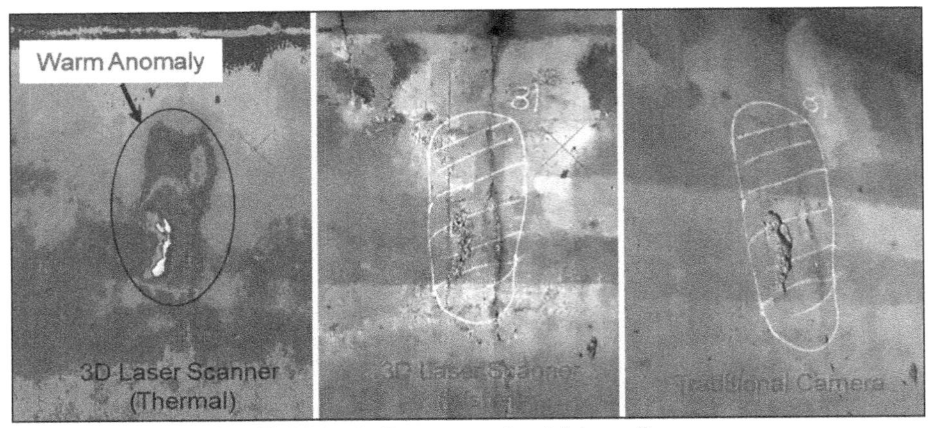


Figure 9. Examples of identified delamination on concrete retaining wall

## Case Study #2: Assessment of Retaining Wall

In this case study, the condition assessment was performed on two concrete retaining walls; Westbound and Eastbound wall. The approximate total length of the Westbound and Eastbound walls are 1,700 ft and 2,900 ft, respectively. The height of the walls varies from 1 ft to 16 ft.

Laser scanning was performed to locate and quantify the existing damage of the retaining walls resulting from water infiltration as well as deterioration. As shown earlier, the main deficiencies categories were used to identify and classified all distresses.

Visual and thermal images were analyzed simultaneously to discern correlations between temperature, related patterns, and visible constructions. The color-coded temperatures and the color resolution are adjusted to the specific temperature anomaly to improve the visibility of the objects.

This case focuses on presenting concrete delamination. Warm areas are always visible in the thermal images as red color. These warm areas could be caused by the delamination in the concrete. Figure 9 shows an example of a warm anomaly and the corresponding mark out (marked by others) on the wall that indicates delamination. This example provides a good verification of the warm anomalies identified with the laser scanner. However, it should be noted that in this case, delamination is shallow (less than 2 in. deep). Identification of deeper delamination (deeper than 2 in.) requires a large temperature gradient between the backfill and the structure surface for an extended period (days).

Finally, statistical summaries indicate the distribution of findings over the length of each retaining wall. These were grouped by data set (i.e., Walls and Zones) and results are presented in feet or square feet depending on the type of anomaly.

## Case Study #3: Assessment of Tiled Tunnel

A roadway tunnel was investigated in this case study. The tunnel is approximately 1 mi. in length and is rectangular in shape except the 47 ft of each end that have an arch shape. The tunnel accommodates two traffic lanes. This tunnel is divided into four elements: North Wall, South Wall, Ceiling and Concrete Arch (West and East concrete Arch). The tube (tunnel) walls are covered with tiles, and the tunnel has a concrete ceiling.

Main Deficiencies Categories were used for mapping the visual and 3D-View data from the laser scanning survey. The deboned tiles and water leakage are main focus to be discussed in this paper. The thermal data was used to identify water leakage which appears in the thermal data as low temperature because of the evaporation of water. Figure 10 shows an example of the appearance of the water leakage in the thermal data.

Warm areas are always visible in the thermal images as red/orange color. These warm areas could be caused by cavities behind the tunnel surface (debonded tiles and delamination in the concrete). Figure 11 shows two examples of a warm anomaly and the corresponding mark out (marked by others) on the wall that indicates debonded tiles.

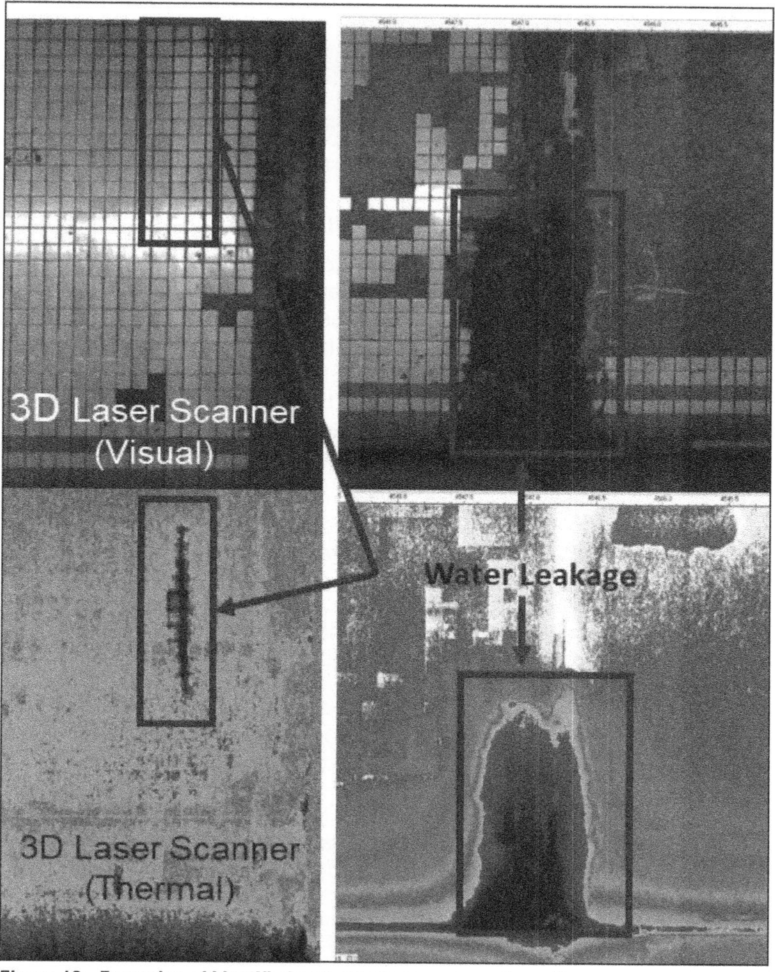


**Figure 10. Examples of identified water leakage on tunnel's wall**

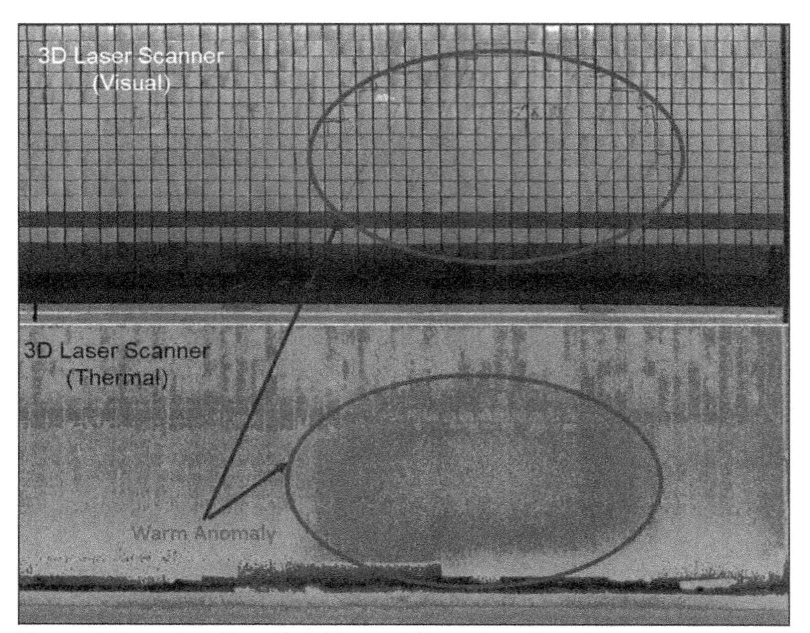


Figure 11. Examples of identified debonded tiles on tunnel's wall

## CONCLUSION

Monitoring the condition and rate of deterioration of tunnels are vital for timely tunnel maintenance and rehabilitation to avoid sudden tunnel collapse. Visual and manual tunnel inspection procedures are a difficult task for tunnel inspectors as it is time-consuming and subjected to human error. Therefore, an alternative is needed to increase efficiency and reliability in tunnel inspection. By utilizing the digital tunnel scanning systems such as laser scanning, the time which needs to be spent in tunnel can be minimized and the thorough inspection analysis can be done in an office space.

## REFERENCES

1. Tunnel Operations, Maintenance, Inspection, and Evaluation (TOMIE) Manual. U.S. Department of Transportation, Federal Highway Administration. Publication No. FHWA-HIF-15-005. July 2015.

2. Steinkühler, J, Mett, M., Dippold, C., and Latimer, K. 360-Degree 3D Digital Assessment of Subsurface Water Infrastructure. North American Tunneling (NAT) Conference, June 19–22, 2022.

3. Kontrus, H., and Mett, M. High-Speed 3D Tunnel Inspection. Rapid Excavation and Tunneling Conference (RETC) 2019.

4. Kontrus, H., Steinkühler, J., and Peal, T. High-Speed 3D Tunnel Inspection in Subway Tunnels: Case Study San Francisco BART. Rapid Excavation and Tunneling Conference (RETC) 2021.

# Index

L



M

## O

## U